ENCYCLOPEDIA OF STATISTICAL SCIENCES

VOLUME 4

**Icing the Tails
to Limit Theorems**

ENCYCLOPEDIA OF STATISTICAL SCIENCES

VOLUME 4

ICING THE TAILS
to LIMIT THEOREMS

A WILEY-INTERSCIENCE PUBLICATION

John Wiley & Sons

NEW YORK · CHICHESTER · BRISBANE · TORONTO · SINGAPORE

Library of Congress Cataloging in Publication Data:
Main entry under title:
Encyclopedia of statistical sciences.

  "A Wiley-Interscience publication."
  Contents: v. 1. A to Circular probable error—
v. 4. Icing the tails to Limit theorems.
  1. Mathematical statistics—Dictionaries.
2. Statistics—Dictionaries.   I. Kotz, Samuel.
II. Johnson, Norman Lloyd.   III. Read, Campbell B.
QA276.14.E5   1982   519.5'03'21   81-10353
ISBN 0-471-05551-4 (v. 4)

Printed in the United States of America

10 9 8 7 6 5 4 3 2 1

CONTRIBUTORS

B. Abraham, *University of Waterloo, Waterloo, Ontario, Canada.* Intervention Model Analysis

B. A. Bailar, *Bureau of the Census, Washington, D.C.* Interpenetrating Subsamples

R. A. Bailey, *The Open University, Milton Keynes, England.* Interaction

E. Balas, *Carnegie-Mellon University, Pittsburgh, Pennsylvania.* Integer Programming

G. A. Barnard, *University of Waterloo, Waterloo, Ontario, Canada.* Likelihood

O. Barndorff-Nielsen, *Aarhus University, Aarhus, Denmark.* Koopman–Darmois–Pitman Family of Distributions

D. J. Bartholomew, *London School of Economics and Political Science, London, England.* Isotonic Inference

A. P. Basu, *University of Missouri, Columbia, Missouri.* Identifiability

T. Berger, *Cornell University, Ithaca, New York.* Information Theory and Coding Theory

H. T. Bhattacharyya, *Southeastern Forest Experiment Station, Research Triangle Park, North Carolina.* Kruskal–Wallis Test

Z. W. Birnbaum, *University of Washington, Seattle, Washington.* Infant Mortality

I. H. Blenkinsop, *Royal Statistical Society, London, England.* Journal of the Royal Statistical Society

H. W. Block, *University of Pittsburgh, Pittsburgh, Pennsylvania.* Inequalities on Distributions: Bivariate and Multivariate

E. Bofinger, *University of New England, New South Wales, Australia.* Least Favorable Configuration

S. A. Book, *California University, Carson, California.* Large Deviations and Applications

K. O. Bowman, *Union Carbide Corporation, Oak Ridge, Tennessee.* Johnson's System of Distribution; Levin's Algorithm

D. R. Brillinger, *University of California, Berkeley, California.* Jump Processes

J. Burbea, *University of Pittsburgh, Pittsburgh, Pennsylvania.* J-Divergences and Related Concepts

P. C. Consul, *University of Calgary, Calgary, Alberta, Canada.* Lagrange and Related Probability Distributions; Lagrange Expansion

P. L. Cornelius, *University of Kentucky, Lexington, Kentucky.* Lattice Designs

R. C. Cornell, *The University of Michigan, Ann Arbor, Michigan.* Kärber Method

P. R. Cox, *The Level House, Mayfield, England.* Life Tables

C. Dagum, *University of Ottawa, Ontario, Canada.* Income Distribution Models; Income Inequality Measures

P. Damiani, *Société de Statistique de Paris, Paris, France.* Journal of the Statistical Society of Paris

J. N. Darroch, *Flinders University, Adelaide, Australia.* Interaction Models

A. P. Dawid, *University College, London, England.* Inference, Statistical; Invariant Prior Distributions

D. O. Dixon, *Southwest Oncology Group, Houston, Texas*. k-Ratio t-Tests, t-Intervals and Point Estimates for Multiple Comparisons

A. Dobson, *University of Newcastle, New South Wales, Australia*. Lexicostatistics

H. E. Doran, *The University of New England, Armisdale, New South Wales, Australia*. Lag Models, Distributed

F. Downton, *The University, Birmingham, England*. Labouchère Systems

D. Dugué, *Sceaux, France*. Lévy, Paul-Pierre

D. B. Duncan, *Johns Hopkins University, Baltimore, Maryland*. k-Ratio t-Tests, t-Intervals and Point Estimates for Multiple Comparisons

P. S. Dwyer,* *East Lansing, Michigan*. Institute of Mathematical Statistics

M. L. Eaton, *University of Minnesota, Minneapolis, Minnesota*. Isotropic Distributions

A. S. C. Ehrenberg, *London Graduate School of Business Studies, London, England*. Lawlike Relationships

C. Eisenhart, *National Bureau of Standards, Washington, D.C.* Law of Error I; Law of Error II; Law of Error III

J. Enger, *The Royal Institute of Technology, Stockholm, Sweden*. Levy Concentration Function

R. A. Evans, *Durham, North Carolina*. *IEEE Transactions on Reliability*

W. B. Fairley, *Analysis and Inference, Inc., Boston, Massachusetts*. Law, Statistics in

L. S. Feldt, *The University of Iowa, Iowa City, Iowa*. Kuder–Richardson Reliability Coefficients 20 and 21

S. E. Fienberg, *Carnegie-Mellon University, Pittsburgh, Pennsylvania*. Iterative Proportional Fitting; *Journal of the American Statistical Association*

J. L. Folks, *Oklahoma State University, Stillwater, Oklahoma*. Inverse Distributions; Inverse Gaussian Distributions

D. A. S. Fraser, *University of Toronto, Toronto, Canada*. Inference, Statistical II

G. H. Freeman, *National Vegetable Research Station. Wellesbourne, Warwick, England*. Incomplete Block Designs

*Deceased

K. S. Fu, *Purdue University, West Lafayette, Indiana*. *IEEE Transactions on Pattern Analysis and Machine Intelligence*

J. D. Gibbons, *The University of Alabama, University, Alabama*. Kolmogorov–Smirnov Symmetry Test

R. A. Gideon, *University of Montana, Missoula, Montana*. Laguerre Series

D. B. Gillings, *University of North Carolina, Chapel Hill, North Carolina*. Inference, Design Based vs. Model Based

N. C. Giri, *University of Montreal, Montreal, Quebec, Canada*. Invariance Concepts in Statistics

R. E. Glaser, *University of California, Livermore, California*. Levene's Robust Test of Homogeneity of Variance

I. Grattan-Guinness, *Hertfordshire, England*. Laplace, Pierre Simon

R. F. Gunst, *Southern Methodist University, Dallas, Texas*. Latent Root Regression

J. Gurland, *University of Wisconsin, Madison, Wisconsin*. Katz System of Distributions

H. L. Harter, *Wright State University, Dayton, Ohio*. Interpolation, Cauchy's Method; Least Squares

J. A. Hausman, *M.I.T., Cambridge, Massachusetts*. Instrumental Variable Estimation

R. G. Heikes, *Georgia Institute of Technology, Atlanta, Georgia*. Inspection Sampling

N. W. Henry, *Virginia Commonwealth University, Richmond, Virginia*. Latent Structure Analysis

C. C. Heyde, *CSIRO, Canberra, Australian Capital Territory, Australia*. Invariance Principles and Functional Limit Theorems; Law of the Iterated Logarithm; Laws of Large Numbers; Limit Theorems, Central

D. V. Hinkley, *Stanford University, Stanford California*. Jackknife Methods

L. K. Holst, *Uppsala University, Uppsala, Sweden*. Limit Theorems

J. R. M. Hosking, *Institute of Hydrology, Wallingford, Oxon, England*. Lagrange Multiplier Test

P. J. Huber, *Harvard University, Cambridge, Massachusetts*. Kinematic Displays

N. T. Jazairi, *York University, Downsview, Ontario, Canada.* Index Numbers; Index of Industrial Production

V. M. Joshi, *The University of Western Ontario, London, Ontario, Canada.* Kingman Inequalities; Likelihood Principle

G. Kallianpur, *University of North Carolina, Chapel Hill, North Carolina.* Indian Statistical Institute

K. Kanneman, *Ottawa, Ontario, Canada.* Intrinsic Rank Test (for *k* Independent Samples)

N. Keyfitz, *Harvard University, Cambridge, Massachusetts.* Keyfitz Method of Life Table Construction; Keyfitz Method of Variance Estimation

G. G. Koch, *University of North Carolina, Chapel Hill, North Carolina.* Inference, Design Based vs. Model Based; Intraclass Correlation Coefficient

H. C. Kraemer, *Stanford University Medical School, Stanford, California.* Kappa Coefficient

P. R. Krishnaiah, *University of Pittsburgh, Pittsburgh, Pennsylvania. Journal of Multivariate Analysis*

A. M. Kshirsagar, *The University of Michigan, Ann Arbor, Michigan.* Lambda Criterion, Wilks's

S. Kullback, *Silver Spring, Maryland.* Kullback Information

J. L. Lebowitz, *Rutgers University, New Brunswick, New Jersey. Journal of Statistical Physics*

R. F. Ling, *Clemson University, Clemson, South Carolina.* Interactive Data Analysis

R. J. Little, *Datametrics Research, Inc., Chevy Chase, Maryland.* Incomplete Data

N. R. Mann, *University of California, Los Angeles, California.* Life Testing

R. S. Mariano, *University of Pennsylvania, Philadelphia, Pennsylvania.* Iterated Maximum Likelihood Estimates

C. Merle, *Association Français de Normalisation, Paris La Défense, France.* International Standardization of Application of Statistics

M. M. Meyer, *Carnegie Mellon University, Pittsburgh, Pennsylvania.* Iterative Proportional Fitting

D. S. Moore, *Purdue University, West Lafayette, Indiana.* Large Sample Theory

R. J. Muirhead, *The University of Michigan, Ann Arbor, Michigan.* Latent Root Distributions

S. Nahmias, *University of Santa Clara, Santa Clara, California.* Inventory Theory

T. V. Narayana, *University of Alberta, Edmonton, Alberta, Canada.* Knock-out Tournaments

J. P. Newhouse, *The Rand Corporation, Santa Monica, California.* Lagged Dependent Variables

J. K. Ord, *The Pennsylvania State University, University Park, Pennsylvania.* Kriging; Laplace Distributions

F. Österreicher, *University of Salzburg, Salzburg, Austria.* Least Favorable Distributions

A. G. Pakes, *The University of Western Australia, Nedlands, Western Australia.* Immigration–Emigration Processes

G. P. Patil, *The Pennsylvania State University, University Park, Pennsylvania.* International Statistical Ecology Program

R. W. Payne, *Rothamsted Experimental Station, Harpenden, Hertfordshire, England.* Identification Keys

S. C. Pearce, *The University, Canterbury, Kent, England.* Kuiper–Corsten Iteration

A. V. Peterson, Jr., *University of Washington, Seattle, Washington.* Kaplan–Meier Estimator

W. R. Pirie, *Virginia Polytechnic Institute, Blacksburg, Virginia.* Jonckheere Tests for Ordered Alternatives

J. Popkin, *Joel Popkin and Company, Washington, D.C.* Labor Statistics

R. F. Potthoff, *Burlington Industries, Inc., Greensboro, North Carolina.* Johnson–Neyman Technique

N. U. Prabhu, *Cornell University, Ithaca, New York.* Integral Equations

D. A. Preece, *Rothamsted Experimental Station, Harpenden, Hertfordshire, England.* Latin Squares

P. Prescott, *The University, Southampton, England*. Influential Observations

D. Raghavarao, *Temple University, Philadelphia, Pennsylvania*. L_2-Designs

R. H. Randles, *University of Florida, Gainesville, Florida*. Klotz Test

M. V. Ratnaparkhi, *Wright State University, Dayton, Ohio*. Inverted Beta Distribution; Inverted Dirichlet Distribution

G. W. Reddien, Jr., *Southern Methodist University, Dallas, Texas*. Interpolation

N. Reid, *University of British Columbia, Vancouver, British Columbia, Canada*. Influence Functions

D. F. Renn, *Journal of the Institute of Actuaries, Government Actuary's Department, London, England*. *Journal of the Institute of Actuaries*

B. H. Renshaw III, *Deputy Director, U.S. Department of Justice, Washington, D.C.* Justice Statistics, Bureau of

S. M. Ross, *University of California, Berkeley, California*. Inspection Paradox

D. B. Rubin, *Datametrics Research, Inc., Chevy Chase, Maryland*. Incomplete Data; Iteratively Reweighted Least Squares

D. Ruppert, *University of North Carolina, Chapel Hill, North Carolina*. Kiefer–Wolfowitz Procedure

A. P. Sage, *University of Virginia, Charlottesville, Virginia*. *IEEE Transactions on Systems, Man and Cybernetics*

A. Sampson, *University of Pittsburgh, Pittsburgh, Pennsylvania*. Inequalities on Distributions; Bivariate and Multivariate

S. C. Saunders, *Washington State University, Pullman, Washington*. Jiřina Sequential Procedures

P. Schmidt, *Michigan State University, East Lansing, Michigan*. Identification Problems

B. M. Schreiber, *Wayne State University, Detroit, Michigan*. L-Class Laws; Lévy–Khinchine Formula

P. K. Sen, *University of North Carolina, Chapel Hill, North Carolina*. *Journal of Statistical Planning and Inference*; Lehmann Tests

E. Seneta, *The University of Sydney, Sydney, New South Wales, Australia*. Liapunov, Alexander Mikhailovich

G. Shafer, *University of Kansas, Lawrence, Kansas*. Lambert, Johann Heinrich

L. R. Shenton, *University of Georgia, Athens, Georgia*. Johnson's System of Distribution; Levin's Algorithm

B. Simon, *California Institute of Technology, Pasadena, California*. Lattice Systems

R. Simon, *National Cancer Institute, Bethesda, Maryland*. Imbalance Functions

G. Simons, *University of North Carolina, Chapel Hill, North Carolina*. Inequalities for Expected Sample Sizes

N. D. Singpurwalla, *George Washington University, Washington, D.C.* Life Testing

P. J. Smith, *University of Maryland, College Park, Maryland*. Inverse Sampling

W. L. Smith, *University of North Carolina, Chapel Hill, North Carolina*. Key Renewal Theorem

R. D. Snee, *E. I. du Pont de Nemours, Wilmington, Delaware*. Industry, Statistics in

M. D. Springer, *University of Arkansas, Fayetteville, Arkansas*. Integral Transforms

D. A. Sprott, *University of Waterloo, Waterloo, Ontario, Canada*. Likelihood

C. Spruill, *Georgia Institute of Technology, Atlanta, Georgia*. Kiefer–Wolfowitz Equivalence Theorem

J. N. Srivastava, *Colorado State University, Boulder, Colorado*. *Journal of Statistical Planning and Inference*

M. A. Stephens, *Simon Fraser University, Burnaby, British Columbia, Canada*. Kolmogorov–Smirnov Statistics; Kolmogorov–Smirnov Type Tests of Fit

F. W. Steutel, *University of Technology, Eindhoven, The Netherlands*. Infinite Divisibility

W. E. Strawderman, *Rutgers University, New Brunswick, New Jersey*. James–Stein Estimators; Likelihood Ratio Tests

A. Stuart, *London School of Economics, London, England*. Kendall's Tau

M. E. Thompson, *University of Waterloo, Waterloo, Ontario, Canada*. Labels

R. C. Tripathi, *The University of Texas, San Antonio, Texas*. Kemp Families of Distributions

C. Villegas, *Simon Fraser University, Burnaby, British Columbia, Canada*. Inner Inference

H. M. Wadsworth, Jr., *Journal of Quality Technology, Georgia Institute of Technology, Atlanta, Georgia*. *Journal of Quality Technology*

G. S. Watson, *Princeton University, Princeton, New Jersey*. Langevin

E. J. Wegman, *Office of Naval Research, Arlington, Virginia*. Kalman Filtering; Kernel Estimators

L. Weiss, *Cornell University, Ithaca, New York*. Inadmissable Decision Rules, Kiefer–Weiss Problem

R. E. Welsch, *M.I.T., Cambridge, Massachusetts*. Influential Data; Leverage

D. W. Wichern, *University of Wisconsin, Madison, Wisconsin*. Lagging Indicators; Leading Indicators (in Economics and Forecasting)

R. A. Wijsman, *University of Illinois, Urbana, Illinois*. Lehmann Alternatives

R. L. Winkler, *Indiana University, Bloomington, Indiana*. Judgments under Uncertainty

S. L. Zabell, *Northwestern University, Evanston, Illinois*. Lexis, Wilhelm

S. Zacks, *Case Western Reserve University, Cleveland, Ohio*. *Journal of Statistical Planning and Inference*

H. J. Zimmerman, *Institut für Wirtschaftswissenschaften, Rhein, Westfalen Technische Hochschule, Aachen, West Germany*. *Journal of Fuzzy Sets and Systems*

ENCYCLOPEDIA OF STATISTICAL SCIENCES

VOLUME 4

**Icing the Tails
to Limit Theorems**

I

ICING THE TAILS

A term used in exploratory data analysis*. It refers to "pulling in" both tails of a distribution of a variable quantity which has extreme values (outliers*) on both ends. Usually, a transformation that is convex on the left and concave on the right is used. Most common are sine functions or odd-numbered roots. For additional information, see ref. 1.

Reference

[1] Hartwig, F. and Dearing, B. E. (1979). *Exploratory Data Analysis*. Sage, Beverly Hills, Calif.

(EXPLORATORY DATA ANALYSIS)

IDEAL POWER FUNCTION

The power function* of a test for which all error probabilities (both of type 1* and type 2*) are zero. This means that the significance level* of the test is zero, and the power with respect to any alternative hypothesis* is 1.

Although such a power function is practically unattainable, it can serve as an "ideal" toward which one strives in constructing test procedures.

IDEMPOTENT MATRIX

A square matrix $\mathbf{A}$ is *idempotent* if $\mathbf{A}^2 = \mathbf{A}$; if the rank of such a matrix $\mathbf{A}$ is r, then $\mathbf{A}$ has r nonzero characteristic roots each equal to 1. The only idempotent matrix of full rank is the identity matrix.

The following properties are useful in the study of quadratic forms* based on normality assumptions. Let $\mathbf{A}$ be $n \times n$ and idempotent: then

1. If $\mathbf{DD'} = \mathbf{I}$ and $\mathbf{D}$ is $n \times n$, then $\mathbf{D'AD}$ is idempotent.
2. If $\mathbf{B}$ is nonsingular $n \times n$, then $\mathbf{BAB}^{-1}$ is idempotent.
3. $\mathbf{I} - \mathbf{A}$ is idempotent.

If $\mathbf{A}$ is also symmetric, the derived matrices in items 1, 2, and 3 are symmetric idempotent. A useful generalization of item 3 is worth stating.

Theorem. If $\mathbf{A}_1, \mathbf{A}_2, \ldots, \mathbf{A}_k$ are $n \times n$ symmetric matrices of ranks $r_1, r_2, \ldots, r_k$, respectively, such that $\sum_{i=1}^{k} \mathbf{A}_i = \mathbf{I}$ and $\sum_{i=1}^{k} r_i = n$, then

1. $\mathbf{A}_i\mathbf{A}_j = \mathbf{0}$, $i \neq j$.
2. $\mathbf{A}_i$ is idempotent; $i = 1, \ldots, k$.

These results lead to characterizations of quadratic forms having chi-square* and Wishart* distributions. In what follows, $\sim$ denotes "is distributed as."

1. Let $\mathbf{x}' = (x_1, \ldots, x_n)$ be a set of independent random variables, where x_i has a normal distribution with mean μ_i and variance σ^2, and $\boldsymbol{\mu}' = (\mu_1, \ldots, \mu_n)$. Then $\mathbf{x}'\mathbf{A}\mathbf{x} \sim \sigma^2\chi_r^2(\boldsymbol{\mu}'\mathbf{A}\boldsymbol{\mu})$, where r = rank of $\mathbf{A}$, $\chi_\nu^2(\lambda)$ denoting noncentral chi-square* with ν degrees of freedom and noncentrality parameter λ, if and only if $\mathbf{A}$ is idempotent.

2. Let $\mathbf{X}' = (\mathbf{x}_1, \ldots, \mathbf{x}_n)$ be a set of independent multivariate normal vectors having common covariance matrix $\boldsymbol{\Sigma}$. Then $\mathbf{X}'\mathbf{A}\mathbf{X}$ has a (noncentral) Wishart distribution with r degrees of freedom and covariance matrix $\boldsymbol{\Sigma}$ if and only if $\mathbf{A}$ is idempotent.

3. Let $\mathbf{x}' = (x_1, \ldots, x_n)$ be distributed as in item 1, and let

$$\mathbf{x}'\mathbf{x} = \mathbf{x}'\mathbf{A}_1\mathbf{x} + \cdots + \mathbf{x}'\mathbf{A}_k\mathbf{x},$$

where r_i = rank of $\mathbf{A}_i$; $i = 1, \ldots, k$; and $\lambda_i = \boldsymbol{\mu}'\mathbf{A}_i\boldsymbol{\mu}$. Then a necessary and sufficient condition that the forms $\mathbf{x}'\mathbf{A}_i\mathbf{x}$ are independent and $\mathbf{x}'\mathbf{A}_i\mathbf{x} \sim \chi_{r_i}^2(\lambda_i)$ for $i = 1, \ldots, k$ is that $\mathbf{A}_1, \ldots, \mathbf{A}_k$ are idempotent; see James [1].

The last result is related to Cochran's theorem, and has a multivariate version in which $\mathbf{x}' \to \mathbf{X}'$ as in item 2, and the $\mathbf{X}'\mathbf{A}_i\mathbf{X}$ have corresponding Wishart distributions.

Reference

[1] James, G. S. (1952). *Proc. Camb. Philos. Soc.*, **48**, 443–446.

Bibliography

See the following works, as well as the reference just given, for more information on the topic of idempotent matrices and their applications in statistics.

Graybill, F. A. (1961). *An Introduction to Linear Statistical Models*. McGraw-Hill, New York, Secs. 4.4–4.6.

Graybill, F. A. (1969). *Introduction to Matrices with Applications in Statistics*. Wadsworth, Belmont, Calif. (See in particular Sec. 12.3.)

Rao, C. R. (1973). *Linear Statistical Inference and Its Applications*, 2nd ed. Wiley, New York, Secs. 3b.4, 8b.2.

Searle, S. R. (1971). *Linear Models*. Wiley, New York, Sec. 2.5.

(CHI-SQUARE DISTRIBUTION
QUADRATIC FORMS
WISHART DISTRIBUTION)

CAMPBELL B. READ

IDENTIFIABILITY

Identifiability problems arise quite naturally when observations, arising in a given situation, could be explained in terms of one of several available models. To illustrate, let us consider the following problem.

Example 1. Let X be normally distributed with mean $E(X) = \mu_1 - \mu_2$. Here $\mu_1 - \mu_2$ could be estimated using X. However, the parameters μ_1 and μ_2 are not uniquely estimable. In fact, one can think of an infinite number of pairs (μ_i, μ_j), $i, j = 1, 2, \ldots$ $(i \neq j)$ such that $\mu_i - \mu_j = \mu_1 - \mu_2$. Here μ_1 and μ_2 can be uniquely estimated only if they are identifiable.

As the example above illustrates, in many statistical problems the underlying statistical distribution is assumed to be completely known except for a set of unknown parameters. Before any inferential procedure can be developed, one needs to assert that the unknown parameters are identifiable. That is, one has to make sure that the given random variable could not have followed a different distribution with a different set of parameters.

A more formal definition of identifiability

could be given by defining *nonidentifiability* first as follows.

Definition. Let U be an observable random variable with distribution function F_θ and let F_θ belong to a family $\mathscr{F} = \{F_\theta : \theta \in \Omega\}$ of distribution functions indexed by a parameter θ. Here θ could be scalar or vector valued. We shall say θ is *nonidentifiable* by U if there is at least one pair (θ, θ'), $\theta \neq \theta'$, where θ and θ' both belong to Ω such that $F_\theta(u) = F_{\theta'}(u)$ for all u. In the contrary case we shall say θ is *identifiable*.

It may happen that θ itself is nonidentifiable, but there exists a nonconstant function $\gamma(\theta)$ which is identifiable. That is, for any θ, θ' belonging to Ω, $F_\theta(u) = F_{\theta'}(u)$ for all u implies that $\gamma(\theta) = \gamma(\theta')$. In this case we say that θ is *partially identifiable*.

In Example 1, $\theta = (\mu_1, \mu_2)$, F_θ is the family of normal distributions and $\gamma(\theta) = \mu_1 - \mu_2$. It may also be possible to introduce additional random variables I, which themselves do not identify θ, so that θ is identifiable by the augmented random variable (U, I). In this case we call the original nonidentifiability problem *rectifiable*. As pointed out later, Example 2 presents an example of a nonidentifiability situation that is rectifiable.

Note that so far we have given a parametric definition of identifiability. But one can also think of a nonparametric definition if, in the definition above, we replace $\{F_\theta\}$ by a class of distribution functions without an indexing parameter. An example illustrating this will be given later in the section "Competing and Complementary Risks."

The problem of identification arises in a number of different fields such as automatic control, biomedical engineering, psychology, system science, etc., where the underlying physical structure could be deterministic. However, here we are restricting our discussion to *statistical identifiability* involving random variables. For a definition in the deterministic case, see Bellman and Åström [8]. Also see Balakrishnan [2], Kalaba and Spingarn [18], Mehra and Lainiotis [21],

Sage and Melsa [24], and the references therein for further examples both in stochastic and deterministic automatic control and system science.

EXAMPLES OF STATISTICAL IDENTIFIABILITY

Statistical identifiability problems arise in a number of different areas whenever the underlying model is not completely specified. Koopmans and Reiersøl [19] give an early survey of the area. Historically, the early problems have been in the areas of linear models*, econometrics*, factor analysis*, and related areas. Thurstone [25, 26] discusses the problems of identifiability in his books on factor analysis. In linear models the problems of identifiability arise under the more familiar terminology of *estimability** as developed by Bose [10].

The identification problem is also met with in the theory of design of experiments*, particularly in the method of confounding* [1]. When confounding is used, the identifiability of certain parameters (second-order interactions, say) is sacrificed to gain certain advantages in the testing of hypotheses and estimation concerning the parameters that remain identifiable (main effects and first-order interactions, say). Reiersøl [23] studied identifiability for related problems for estimating straight lines when both variables are subject to error. Let $Y = \alpha + \beta X$, and we want to estimate α and β but cannot observe Y and X. Instead, we observe $V = Y + \epsilon$ and $U = X + \delta$. Assume that ϵ and δ are normally distributed with the same mean 0 and respective variances σ_ϵ^2 and σ_δ^2, and that X and Y are given constants. It can be shown that one can choose two different sets of constants $(\alpha, \beta, \sigma_\epsilon, \sigma_\delta)$ and $(\alpha', \beta', \sigma_\epsilon', \sigma_\delta')$ such that (U, V) will have the same distribution no matter which set of constants is used. Here the parameters are thus not identifiable by (U, V). Some of Reiersøl's results have been extended to multivariate models by Willassen [29]. For error in variables, see papers by Diestler [11], Diestler and Tintner [12], Geraci [14], and Hsiao [16].

The identification problems arise in many other areas. Some examples are *latent structure** [20], *path analysis** [27], and *stochastic compartmental model** [15]. Van der Genugten [28] has given a survey dealing with identification in *statistical inference*. Puri [22] has given a survey of nonidentifiability of distributions arising out of stochastic modeling, including the models in accident proneness. We describe below two other areas pointing out how the identifiability problem arises.

Structural Equation

A major area where the identification problem has been studied in great detail is identification of parameters in a structural equation which arises often in econometrics. Let

$$A = \begin{bmatrix} B & C \end{bmatrix},$$

$$Z_t = \begin{bmatrix} y_t \\ x_t \end{bmatrix},$$

so that

$$AZ_t = By_t + Cx_t = u_t, \qquad (1)$$

say, is a set of p equations where A is a $p \times (p + k)$ matrix, B is a $p \times p$ matrix of coefficients of current endogenous variables, C a $p \times k$ matrix of coefficients of predetermined variables, Z_t, y_t, x_t, and u_t are column vectors of dimensions $(p + k)$, p, k, and p respectively at time t. Without any loss of generality, assume B to be nonsingular and that $E(u_t) = 0$ and $E(u_t u_t') = \Sigma$, a nonsingular matrix. Also premultiplying (1) by a $p \times p$ nonsingular matrix D, we get a second structural equation

$$(DB)y_t + (DC)x_t = w_t, \qquad (2)$$

where $w_t = Du_t$.

It can be readily seen that the conditional likelihood of y_t for given x_t using model (1) is the same as that obtained from (2), so that the two structures are observationally equivalent and that the underlying model is nonidentifiable. Conditions for identifiability are discussed in standard texts in econometrics such as Johnston [17]. Also see Fisher [13].

To make the problem identifiable additional prior restrictions are imposed in the

matrices B, C, and Σ. Now consider the problem of estimating the parameters of the first equation in (1), out of a system of p equations. If the parameters cannot be estimated, the first equation is called *unidentified* or *underidentified*. If given the prior information there is a unique way of estimating the unknown parameters, the equation is called *just identified*. If the prior information allows the parameters to be estimated in two or more linearly independent ways, the parameters are said to be *overidentified*.

A necessary and sufficient condition for identifiability is given by the *rank condition* that a specified submatrix of A be of rank $p - 1$. Basmann [3, 4, and references cited therein] considers the corresponding *identifiability test*.

Competing and Complementary Risks

The problems of competing and complementary risks arise quite naturally in a number of different contexts. For some of these examples, see Basu [5] and Basu and Ghosh [6, 7]. Let $X_1, X_2, \ldots, X_p$ be p random variables with cumulative distribution functions* (CDF) $F_1, F_2, \ldots, F_p$, respectively. Let $U = \min(X_1, X_2, \ldots, X_p)$ and $V = \max(X_1, X_2, \ldots, X_p)$. Let I be an integer-valued random variable ($I = 1, 2, \ldots, p$). In problems of competing risks* (complementary risks) usually the X_i's are not observable. Instead, we observe either $U(V)$ or the pair (U, I) $\{(V, I)\}$. Here $U(V)$ is called the nonidentified minimum (maximum) and (U, I) (V, I) is called the identified minimum (maximum). In the latter case we not only observe the minimum (maximum) but also know which component of X_1, $X_2, \ldots, X_p$ is the minimum (maximum). The problem here is to uniquely determine the marginal distributions $F_1, F_2, \ldots, F_p$ from that of $U(V)$.

To see how the identifiability problem arises, consider the following example.

Example 2. Let X_i ($i = 1, 2$) be independently distributed random variables with X_i following the negative exponential distribution* with distribution function

$$F_i(x) = 1 - \exp(-\lambda_i x),$$

$$\lambda_i > 0, \quad x > 0 \quad (i = 1, 2).$$

It is well known that $U = \min(X_1, X_2)$ also follows the negative exponential distribution with parameter $\lambda_1 + \lambda_2$. Here λ_1 and λ_2 are not identifiable by U since one can think of an infinite number of pairs of independently and exponentially distributed random variables such that the sum of their parameters is $\lambda_1 + \lambda_2$. Notice that $\theta = (\lambda_1, \lambda_2)$ is partially identifiable since $\gamma(\theta) = \lambda_1 + \lambda_2$ is always identifiable. As remarked before, the identification problem here is rectifiable as explained below.

Basu [5] has given a survey of identifications problems in the parametric case. In the general case Berman [9] has shown that if the X_i's are independently distributed and if we have identified minimum, then the unknown CDFs F_i's can be uniquely determined in terms of the known monotonic functions

$$H_k(x) = P(U \leqslant x, I = k),$$

$$k = 1, 2, \ldots, p,$$

as

$$F_k(x) = 1 - \exp\left\{ -\int_{-\infty}^{x} \left[1 - \sum_{j=1}^{p} H_j(t) \right]^{-1} dH_k(t) \right\},$$

$$k = 1, 2, \ldots, p. \quad (3)$$

Berman's result shows that the nonidentifiability problem in Example 2 is rectifiable when we observe (U, I) rather than U. Basu and Ghosh [6, 7] have considered the cases when the underlying random variables are independent but only U (or V) is observed and when the X_i's are dependent. The following example from Basu and Ghosh [6] shows nonidentifiability in the dependence case.

Example 3. Let X_1 and X_2 be dependent random variables with joint distribution function $F(x_1, x_2)$. Under mild assumptions it can be shown that there exist independent

random variables Y_1 and Y_2 such that $U = \min(X_1, X_2)$ has the same distribution as that of $\min(Y_1, Y_2)$. Thus in the dependence case the nonidentifiability problem cannot be resolved unless one restricts oneself to specific parametric families.

References

[1] Bailey, R. A., Gilchrist, F. H. L., and Patterson, H. D. (1977). *Biometrika*, **64**, 347–354.
[2] Balakrishnan, A. V. (1973). *Stochastic Differential Systems*, Vol. 1. Springer-Verlag, New York. (Discusses system identification at advanced mathematical level.)
[3] Basmann, R. L. (1957). *Econometrica*, **25**, 77–83.
[4] Basmann, R. L. (1960). *J. Amer. Statist. Ass.*, **55**, 650–659.
[5] Basu, A. P. (1981). In *Statistical Distributions in Scientific Work*, Vol. 5, C. Taillie, Patil, and Baldessari, eds. D. Reidel, Dordrecht, Holland, pp. 335–348. (Provides a survey of identification problems in competing risks theory.)
[6] Basu, A. P. and Ghosh, J. K. (1978). *J. Multivariate Anal.*, **8**, 413–429.
[7] Basu, A. P. and Ghosh, J. K. (1980). *Commun. Statist. A*, **9**, 1515–1525.
[8] Bellman, R. and Åström, K. J. (1970). *Math. Biosci.*, **7**, 329–339. (Discusses the concept of structural identifiability.)
[9] Berman, S. M. (1963). *Ann. Math. Statist.*, **34**, 1104–1106.
[10] Bose, R. C. (1949). *Institute of Statistics Mimeo Series 9*, Dept. of Statistics, University of North Carolina, Chapel Hill, N.C.
[11] Deistler, M. (1976). *Int. Econ. Rev.*, **17**, 26–46.
[12] Deistler, M. and Tintner, G. (1981). In *Contributions to Probability*, J. Gani and V. K. Rohatgi, eds. Academic Press, New York, pp. 231–246. (Surveys identifiability problems for errors in the variable model.)
[13] Fisher, F. M. (1966). *The Identification Problem in Econometrics*. McGraw-Hill, New York.
[14] Geraci, V. J. (1976). *J. Econometrics*, **4**, 263–283.
[15] Griffiths, D. (1979). *Technometrics*, **21**, 257–259.
[16] Hsiao, C. (1976). *Int. Econ. Rev.*, **17**, 319–339.
[17] Johnston, J. (1963). *Econometric Methods*. McGraw-Hill, New York. (A good readable survey of identification in structural equations.)
[18] Kalaba, R. and Spingarn, K. (1982). *Control, Identification, and Input Optimization*. Plenum, New York. (An introductory text on identification in control theory.)

[19] Koopmans, T. C. and Reiersøl, O. (1950). *Ann. Math. Statist.*, **21**, 165–181. (A good survey of the field.)

[20] Lazarsfeld, P. F. (1950). *The Interpretation of Some Latent Structures*, Vol. 4: *Studies in Psychology of World War II.*

[21] Mehra, R. K. and Lainiotis, D. G. (1976). *System Identification Advances and Case Studies.* Academic Press, New York. (Provides a survey of system identification.)

[22] Puri, P. S. (1979). In *Optimizing Methods in Statistics*, J. S. Rustagi, ed. Academic Press, New York, pp. 403–417. (A survey of identification in stochastic modeling.)

[23] Reiersøl, O. (1950). *Econometrica*, **18**, 375–398.

[24] Sage, A. P. and Melsa, J. L. (1971). *System Identification.* Academic Press, New York. (An introductory book on system identification.)

[25] Thurstone, L. L. (1935). *The Vectors of Mind.* University of Chicago Press, Chicago. (Considers identification in factor analysis.)

[26] Thurstone, L. L. (1947). *Multiple Factor Analysis.* University of Chicago Press, Chicago. (Considers identification in factor analysis.)

[27] Turner, M. E., Monroe, R. J., and Lucas, H. L. (1961). *Biometrics*, **17**, 120–143.

[28] Van der Genugten, B. B. (1977). *Statist. Neerlandica*, **31**, 69–89. (This paper deals with the concept of identification in inferential statistics.)

[29] Willassen, Y. (1979). *Scand. J. Statist.*, **6**, 89–91.

(COMPETING RISKS
CONFOUNDING
ECONOMETRICS
ERRORS IN VARIABLES
FACTOR ANALYSIS
INFERENCE, STATISTICAL
LATENT STRUCTURE ANALYSIS
PATH ANALYSIS
SURVIVAL ANALYSIS)

A. P. Basu

IDENTIFICATION KEYS

An *identification* (or *diagnostic*) *key* is a device for identifying samples from a set of known *taxa* (e.g., a set of plant or animal species, strains of bacteria, types of machine fault). The main application is in systematic biology [8, 9] and microbiology (e.g., ref. 1), but similar methods have been devised in, for example, computer programming [3, 6, 14] and coding theory. The historical devel-

Table 1   Key to Eleven Common British Trees

1	Texture of bark smooth	2
	Texture of bark rough	4
	Texture of bark corky	Elder
	Texture of bark scored horizontally	Rowan
	Texture of bark scaling	6
2	Leaves not pinnate or lobed	3
	Leaves lobed	Sycamore
	Leaves pinnate	Ash
3	Basic shape of leaf pointed oval	Beech
	Basic shape of leaf heart-shaped	Lime
4	Leaves not pinnate or lobed	5
	Leaves lobed	Oak
	Leaves pinnate	Ash
5	Basic shape of leaf pointed oval	Elm
	Basic shape of leaf broad lanceolate	Sweet chestnut
6	Leaves not pinnate or lobed	Birch
	Leaves lobed	7
7	Position of leaves on stem opposite	Sycamore
	Position of leaves on stem alternate	Plane

opment of keys is discussed by Voss [16], who gives examples dating back to the seventeenth century.

Table 1 shows a key to 11 common British trees, constructed by the computer program Genkey [10]. (This example is derived with minor amendments from that used in the review of Payne and Preece [12].) The key contains seven numbered sets of contrasting statements or *leads*, each set concerned with a particular attribute of the specimen. To use the key, we start with the set numbered 1 and examine the texture of the bark. If the bark is rough, the number 4 on the right directs us to the fourth set of leads, where, for example, "Leaves lobed" would identify the specimen as an oak. Such a sequence of leads is termed a *branch* of the key and the point at which the identification is made is termed an *end point*.

The data from which the key was constructed are shown in Table 2. There is a row for each of the 11 taxa and a column for each attribute. In systematic biology the attributes are termed *characters* and the values that the taxa can take (or *exhibit*) are termed *states*, whereas in machine-fault location, medical diagnosis, and microbiology it is more usual to talk about *tests* and their *results*.

Tests for use in a key may have only a

Table 2   Characteristics of 11 Common British Trees

Taxon	Attribute[a]						
	A	B	C	D	E	F	G
Ash	3	3, 4, 5, 6, 7	—	1	2	.1, 2	1, 2
Beech	1	—	1	2	1	1	1
Birch	1	—	1, 2	2	2	5	1
Elder	3	1, 2, 3, 4	—	1	2	3	2
Elm	1	—	1	2	2	2	2
Lime	1	—	3	2	2	1	2
Oak	2	—	4	2	1	2	1
Plane	2	—	3	2	2	5	1
Rowan	3	5, 6, 7	—	2	2	4	2
Sweet chestnut	1	—	5	2	2	2	1
Sycamore	2	—	3	1	2	1, 5	1

[a] A—Form of leaf: 1, not pinnate or lobed; 2, lobed; 3, pinnate.
B—Number of pairs of leaflets per leaf.
C—Basic shape of leaf: 1, pointed oval; 2, triangular; 3, heart-shaped; 4, oblong; 5, broad lanceolate.
D—Positions of leaves on stem: 1, opposite; 2, alternate.
E—Edges of leaves: 1, not toothed; 2, toothed.
F—Texture of bark: 1, smooth; 2, rough; 3, corky; 4, scored horizontally; 5, scaling.
G—Sexual characteristics of flowers: 1, unisexual; 2, bisexual.

finite number of possible results. In the table these are denoted by positive integers and the corresponding descriptions are listed under CODES. The dashes entered, for example, in column 2 for all taxa except Ash, Elder, and Rowan indicate that the test is *inapplicable*. (It is not appropriate to ask how many leaflets there are per leaf unless the tree has pinnate leaves.) A taxon is said to have a *fixed response* to a test if all members of the taxon always give the same result to the test; for example, specimens of Elder always have pinnate leaves. The response is said to be *variable* if some members of the taxon give one result, others a different one: *completely variable* if all the results can occur (e.g., with Ash both unisexual and bisexual flowers may occur), *partially variable* if there are some results that do not occur (e.g., the leaves of Birch may be pointed oval or triangular but not heart-shaped, oblong, or broad lanceolate). Finally, the response is termed *unknown* if there is no information available about which of the results may occur. Unknown responses are usually treated as though they were completely variable, as no result can be ruled out.

OPTIMAL IDENTIFICATION KEYS

Define α_j to be the prior probability that a specimen to be identified belongs to taxon j, m_i to be the number of possible results obtained from test i, and β_{ijk} to be the probability of obtaining result k to test i, given that the specimen examined belongs to taxon j. (A taxon j with fixed response l to test i will have $\beta_{ijk} = 1$ for $k = l$ and $\beta_{ijk} = 0$ for $k \neq l$.)

The *efficiency* of a key is usually assessed by its expected number of tests per identification:

$$\eta_e = \sum_{l=1}^{d} \left\{ t_l \sum_{j=1}^{N} \alpha_j \prod_{(i,k) \in R_l} \beta_{ijk} \right\},$$

where N = total number of taxa
d = number of end points in the key
t_l = the number of tests to end point l
R_l = set of pairs of indices of the test results observed on the branch to end point l

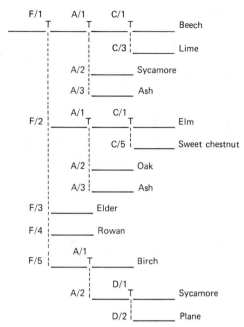

Figure 1 Diagrammatic representation of the key in Table 1. $F/1$, test F result 1 (i.e., texture of bark smooth), etc.

Alternatively, if the tests have different costs, the efficiency would be assessed by the expected cost of identification:

$$\kappa_e = \sum_{l=1}^{d} \left\{ \sum_{(i,k) \in R_l} c_i \right\} \left\{ \sum_{j=1}^{N} \alpha_j \prod_{(i,k) \in R_l} \beta_{ijk} \right\},$$

where c_i is the cost of test i.

A diagrammatic representation of the key in Table 1 is shown in Fig. 1. From this it is easy to see that the first test in the key, test F, divides the taxa into five sets, one set for each result. Set 5, for example, contains the taxa that can give result 5: Birch and Plane (which have fixed response 5) and Sycamore (which is variable, giving either result 5 or result 1). Similarly, tests elsewhere in the key divide sets of taxa into smaller sets.

Picard [13], in the context and terminology of questionnaires, describes the construction of optimal keys and their properties, when there are constraints on the number of different results that tests may have but otherwise tests are available to define

Table 3   Set of Binary Codes

Message	Code
A	0 0
B	0 1
C	1 0
D	1 1 0
E	1 1 1

Figure 2 Binary codes in Table 3, represented as a binary key.

any sets of taxa. The main area where this applies is coding theory (*see* INFORMATION THEORY). The aim here is to efficiently transmit messages, each coded as a sequence of digits. For example, in binary coding, messages are transmitted as sequences of zeros and ones, like those in Table 3; Fig. 2 shows how these can be represented as a binary key. The most important algorithm is that of Huffman [5] which, for tests with equal costs, guarantees construction of the optimum key.

To illustrate the algorithm, consider five messages, A–E, whose probabilities of transmission are

A 0.3   B 0.3   C 0.2
D 0.1   E 0.1

The algorithm first arranges the messages as a list of items in order of probability; then there is a number of steps in each of which the items with smallest probability are combined to form a new item, which is inserted at the appropriate point in the list. For a binary code or key, pairs of items are combined, as shown below. Finally, the codes can be obtained by working right to left and assigning zero to the upper item of a combined pair, one to the lower item. Thus A and B have first digit zero, while C, D, and E have first digit 1. The resulting codes are those in Table 3.

In most biological applications, however, only a limited set of tests is available. The Huffman algorithm cannot then be used as there is no guarantee that tests will be available to make the necessary partitions at each stage. For example, if messages A–E are biological taxa, there may be no single test that separates taxa A and B from taxa C–E. The only known algorithms that then guarantee finding the optimum key involve the enumeration of virtually all possible keys for the taxa and tests under consideration. This is impracticable for even moderately sized problems [4], so approximate algorithms are used. These start by selecting the test that "best" divides the taxa into sets (as described above), then the "best" test for use with each set, and so on. The "best" test is usually taken as that with minimum value of some *selection criterion function*. The two examples below were derived by Payne [11]. Other functions are reviewed by Payne and Preece [12].

The function CM_e is an extension, to tests with unequal costs, of the function (or measure) M_e derived by Brown [2] from Shannon's entropy function [15] (hence the suffix "e"). The second function CM_v also caters for tests with unequal costs; the suffix "v" is to indicate that, for tests with two possible results and equal cost, CM_v is equivalent to the function DV ("dichotomizing value") of Morse [7].

$$(CM_e)_i = c_i + (\bar{c}/\log \bar{m})$$

$$\times \left\{ \sum_{k=1}^{m_i} (p_{ik} + r_i/m_i) \right.$$

$$\times \log(p_{ik} + r_i/m_i) + r_i \log m_i \Big\},$$

$$(CM_v)_i = c_i + (\bar{c}n/2)$$

$$\times \left\{ - \sum_{k=1}^{m_i} (p_{ik} + r_i/m_i) \right.$$

$$\times (1 - r_i - p_{ik}) \Big\},$$

where $\bar{c} =$ average cost of the available tests
$\bar{m} =$ average number of possible results
$n =$ number of taxa in the current set
$p_{ik} =$ proportion of taxa in the current set with fixed response k to test i
$r_i =$ proportion of taxa in the current set that are variable with test i

DIAGNOSTIC TABLES

It has been assumed above that the tests are to be done sequentially; we start with the first test in the key, obtain the result, look at the key to see which test is required next, obtain the result to that test, look at the key again, and so on. If the tests take a long time (as, e.g., in microbiology) it may be preferable to do a set of tests simultaneously. We might still use a key to look up the identity of the specimen (having first determined its results for every test in the key); however, it would be more common to use a *diagnostic table*. This is a table listing the responses of taxa to tests, much like that in Table 2, except that the rows are usually ordered as shown in Table 4.

Table 4   Diagnostic Table for the Trees in Table 2[a]

Tests							
A	D	F	G	B	C	E	Taxon
1	2	1	1	—	1	1	Beech
1	2	1	2	—	3	2	Lime
1	2	2	1	—	5	2	Sweet chestnut
1	2	2	2	—	1	2	Elm
1	2	5	1	—	1,2	2	Birch
2	1	1,5	1	—	3	2	Sycamore
2	2	2	1	—	4	1	Oak
2	2	5	1	—	3	2	Plane
3	1	1,2	1,2	3–7	—	2	Ash
3	1	3	2	1–4	—	2	Elder
3	2	4	2	5–7	—	2	Rowan

[a]The taxa are ordered so that the first rows of the table contain the group of taxa that can give result 1 of the test in column 1 (test A), then the taxa with result 2, and so on; within each group the taxa are ordered according to the test in column 2; then the test in column 3, until each group contains only one taxon. In this table the columns have also been ordered so that the tests in the initial columns, 1–4, comprise an irredundant set—only tests A, D, F, and G are required for identification.

Examination of Table 2 shows that if we do test F we need not do test B, since test B merely distinguishes Elder from Rowan; test B can be deleted without causing any tree to become no longer identifiable. Such a test is termed *redundant* and a set of tests that contains no redundant tests is termed *irredundant*. If any test is deleted from an irredundant set, there will be at least one pair of taxa that are no longer distinguishable. An example is the set of tests (A, D, F, G); if, for example, test A is deleted from the set, Birch cannot be distinguished from Plane, as each has identical results with tests D, F, and G.

It is clear that the test set containing fewest tests, and that with smallest total cost, are both irredundant (since otherwise an improved set could be obtained by deleting one, or more, redundant tests). Payne and Preece [12] review algorithms for finding irredundant test sets, including one that constructs all irredundant sets for the tests and taxa concerned.

References

[1] Barnett, J. A., Payne, R. W., and Yarrow, D. (1979). *A Guide to Identifying and Classifying Yeasts*. Cambridge University Press, Cambridge.

[2] Brown, P. J. (1977). *Biometrika*, **64**, 589–596.

[3] Dixon, P. (1964). *Computers and Aut.*, **13**(Apr.), 14–19.

[4] Garey, M. R. (1972). *SIAM J. Appl. Math.*, **23**, 173–186.

[5] Huffman, D. A. (1952). *Proc. IRE*, **40**, 1098–1101.

[6] Humby, E. (1973). *Programs from Decision Tables*. Computer Monogr. No. 19. Macdonald, London/Elsevier, New York.

[7] Morse, L. E. (1971). *Taxon*, **20**, 269–282.

[8] Morse, L. E. (1975). In *Biological Identification with Computers: Systematics Association Special Vol. No. 7*, R. J. Pankhurst, ed. Academic Press, London, pp. 11–52.

[9] Pankhurst, R. J. (1978). *Biological Identification: The Principles and Practice of Identification Methods in Biology*. Edward Arnold, London.

[10] Payne, R. W. (1978). *Genkey: A Program for Constructing and Printing Identification Keys and Diagnostic Tables*. Rothamsted Experimental Station, Harpenden, England.

[11] Payne, R. W. (1981). *J. Statist. Plann. Infer.*, **5**, 27–36.

[12] Payne, R. W. and Preece, D. A. (1980). *J. R. Statist. Soc. A*, **143**, 253–292.

[13] Picard, C.-F. (1980). *Graphs and Questionnaires*. Mathematical Studies, No. 32. North-Holland, Amsterdam.

[14] Pollack, S. L., Hicks, H. T., and Harrison, W. J. (1971). *Decision Tables: Theory and Practice*. Wiley-Interscience, New York.

[15] Shannon, C. E. (1948). *Bell Syst. Tech. J.*, **27**, 379–423, 623–656.

[16] Voss, E. G. (1952). *J. Sci. Labs Denison Univ.*, **43**, 1–25.

(CLASSIFICATION
DENDRITES
INFORMATION THEORY)

R. W. PAYNE

IDENTIFICATION PROBLEMS

The problem of identification of the parameters of a statistical model is basically the problem of whether or not the values of the parameters are uniquely determined by the likelihood function. Since a set of observations on the random variables in the model can never contain more information than the likelihood function itself, the problem of identification is also basically the question of whether or not the parameters can be estimated, in principle, from observations on the random variables in the model.

Before proceeding to a more formal treatment, a simple example may help to make clear the nature of the problem. Consider the set of simple regression* models

$$y = \beta_0 + \beta_1 x + \epsilon, \qquad (1)$$

$$x = \alpha_0 + \alpha_1 z + u, \qquad (2)$$

each satisfying the usual ideal conditions. Now suppose that x is not observed, so that (1) and (2) cannot be estimated directly. However, we can combine them to obtain

$$y = \gamma_0 + \gamma_1 z + v, \qquad (3)$$

where $\gamma_0 = \beta_0 + \alpha_0 \beta_1$, $\gamma_1 = \alpha_1 \beta_1$, $v = \epsilon + \beta_1 u$. Clearly, γ_0 and γ_1 can be estimated, and just as clearly the γ's do not imply unique values for the α's or β's. The γ's are identified, while the α's and β's are not.

This example is perhaps artificial, but it is also typical in an important way. Given a model, there is generally some basic set of parameters (above, γ's) which determine the distribution of the random variables in the model, and which are identified. The question then becomes whether the parameters of actual interest (above, α's and β's) are uniquely determined, given the basic set of parameters.

DEFINITION OF THE IDENTIFICATION PROBLEM

Following Koopmans and Reiersøl [6], let Y be a vector-valued random variable, with possible values y in some subset of R^n. We define a *model* as a probability distribution $F(y, \theta)$ of known form, with θ specified to belong to some subset Ω of R^m. We define a *structure* to be this model with a given parameter value θ.

Two structures are said to be *observationally equivalent* if they imply the same probability distribution for Y. That is, the structures with parameter values θ_1 and θ_2 are observationally equivalent if $F(y, \theta_1) = F(y, \theta_2)$ for all possible y. Finally, a structure with parameter value θ_0 is *identified* if there is no other θ in Ω which is observationally equivalent.

In some cases, a weaker concept of identifiability is useful. A structure with parameter value θ_0 is said to be locally identified if there exists an open neighborhood of θ_0 which contains no other θ in Ω which is observationally equivalent. This is a useful concept because in models containing nonlinearities, parameters may be locally identified though not (globally) identified.

GENERAL RESULTS

Lack of identification is a reflection of the lack of sufficient information to discriminate between alternative structures. Therefore, it is reasonable that the question of identification can be approached through Fisher's in-

formation* matrix. This connection has been made precisely by Rothenberg [11], who shows, subject to regularity conditions, that θ_0 is locally identified if and only if the information matrix evaluated at θ_0 is nonsingular. The most potentially restrictive of the regularity conditions are that the range of Y must be independent of θ, that the elements of the information matrix must be continuous functions of θ, and that the rank of the information matrix must be constant in some neighborhood of θ_0.

An alternative characterization by Bowden [2] is based on an information theoretic measure. Define the function

$$H(\theta; \theta_0) = E\left\{ \ln\left[\frac{dF(y, \theta)}{dF(y, \theta_0)} \right] \right\}, \quad (4)$$

where the expectation is taken over the distribution $F(y, \theta_0)$. Then, subject to regularity, θ_0 is identified if the only solution in Ω to the equation $H(\theta, \theta_0) = 0$ is $\theta = \theta_0$. Also θ_0 is locally identified if $\theta = \theta_0$ is the only solution in some open neighborhood of θ_0. The regularity conditions required are not very restrictive.

It is clear that unidentified parameters cannot be estimated consistently. In the case that the sample consists of independent, identically distributed (i.i.d.) observations, Le Cam [9] shows, subject to weak regularity conditions, that identification of a parameter implies existence of a consistent estimator of it. This is not necessarily true if the sample is not composed of i.i.d. observations (e.g., the regression case). Gabrielson [4] provides the example of a regression model $y_i = \beta r^i + \epsilon_i$, $i = 1, 2, \ldots$, with $|r| < 1$ (and with the ϵ_i i.i.d.); β is identified but cannot be estimated consistently.

ERRORS IN THE VARIABLES MODEL

A commonly cited example illustrating the problem of identification is the so-called errors in variables model. Consider an exact linear relationship

$$y_i^* = \alpha + \beta x_i^*, \qquad i = 1, 2, \ldots, n, \quad (5)$$

between unobservables y^* and x^*. Suppose

that we observe mismeasured versions of y^* and x^*:

$$y_i = y_i^* + v_i, \qquad (6a)$$

$$x_i = x_i^* + u_i. \qquad (6b)$$

A common set of assumptions is that u_i, v_i, and x_i^* are independent of each other, and are distributed as $N(0, \sigma_u^2)$, $N(0, \sigma_v^2)$, and $N(\mu, \sigma^2)$, respectively. In this case (y_i, x_i) is bivariate normal, and from the five estimable pieces of information—$E(x)$, $E(y)$, var(x), var(y), cov(x, y)—we cannot solve for the six parameters—α, β, μ, σ_u^2, σ_u^2, σ^2. In fact, only $\mu = E(x)$ is identified. A solution is to find more information, often by assuming σ_u^2 / σ_v^2 known. On the other hand, the model is identified without further information if x^* has any distribution other than normal; see Reiersøl [10] or Willassen [13].

IDENTIFICATION OF STRUCTURAL ECONOMETRIC MODELS

Consider as an example the simple model

$$Q = a + bP + \epsilon_1, \qquad (7a)$$

$$Q = \alpha + \beta P + \epsilon_2, \qquad (7b)$$

where (7a) is interpreted as a demand curve, (7b) is interpreted as a supply curve, Q is the quantity sold of the commodity in question, and P is its price. The solutions for Q and P are

$$P = [(\alpha - a) + (\epsilon_2 - \epsilon_1)]/(b - \beta), \qquad (8a)$$

$$Q = [(\alpha b - a\beta) + (b\epsilon_2 - \beta\epsilon_1)]/(b - \beta). \qquad (8b)$$

The variables P and Q are referred to as *endogenous*, being determined by the model. Equations (7a) and (7b) are called *structural equations**, while the solutions (8a) and (8b) are called *reduced-form* equations. (*See* ECONOMETRICS.)

In this case it is clear that the reduced-form parameters are identified, but the structural parameters (a, b, α, β) are not. All we observe is the intersection of the supply and demand curve, as illustrated in Fig. 1; these coordinates (Q, P) are exactly what is given

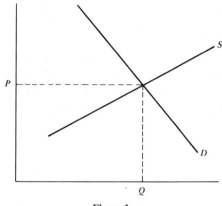

Figure 1

by the reduced-form solution (8). But there exist (in principle) an infinite number of possible demand and supply curves that would generate the point (Q, P). From these two coordinates we cannot calculate the four parameters (two slopes and two intercepts) in which we are interested.

Now consider the amended system in which we assume that the variable W, weather, affects supply (but not demand). We keep the demand curve (7a) but replace (7b) with

$$Q = \alpha + \beta P + \gamma W + \epsilon_2. \qquad (9)$$

The reduced form becomes

$$P = [(\alpha - a) + \gamma W + (\epsilon_2 - \epsilon_1)]/(b - \beta), \qquad (10a)$$

$$Q = [(\alpha b - a\beta) + b\gamma W + (b\epsilon_2 - \beta\epsilon_1)]/(b - \beta). \qquad (10b)$$

Note that weather is treated as *exogenous*, that is, determined outside the system consisting of (7a) and (9).

In this case the parameters (a, b) of the demand curve are identified, although the parameters (α, β, γ) of the supply curve are not. In terms of Fig. 1, the supply curve now shifts, according to the weather. For varying values of W, we therefore generate a number of intersection points, each of which lies on the demand curve. This suffices to determine the demand curve.

As the above should indicate, the question of identification in the present context boils down to the algebraic question of whether it is possible to solve for the structural parameters, given the reduced-form parameters.

Going back to our examples, this algebraic question is easy to resolve. Looking at the reduced form (8), we can see that, in the absence of unusual information on the ϵ's, we have available only the reduced-form parameters

$$(\alpha - a)/(b - \beta) \quad \text{and} \quad (\alpha b - a\beta)/(b - \beta),$$

which are in fact the coordinates of (Q, P) in Fig. 1. From this one cannot solve for a, b, α, or β. However, next turn to the reduced form (10). Here we have available four pieces of information:

$$\Pi_1 = (\alpha - a)/(b - \beta), \qquad (11a)$$

$$\Pi_2 = \gamma/(b - \beta), \qquad (11b)$$

$$\Pi_3 = (\alpha b - a\beta)/(b - \beta), \qquad (11c)$$

$$\Pi_4 = b\gamma/(b - \beta). \qquad (11d)$$

From these we can solve for the parameters (a and b) of the demand equation:

$$b = \Pi_4/\Pi_2, \qquad (12a)$$

$$a = \Pi_3 - \Pi_1\Pi_4/\Pi_2. \qquad (12b)$$

However, it is still impossible to solve for the parameters (α, β, and γ) of the supply equation.

Hopefully, these examples have given some insight into the general nature of the problem of identification of structural parameters. A very complete treatment of the identification of structural parameters can be found in Fisher [3], which also treats identification under other kinds of theoretical restrictions than exclusions of variables from particular equations.

OTHER EXAMPLES OF IDENTIFICATION PROBLEMS

Problems of identification arise in a variety of settings other than those discussed above. Here we mention a few of these.

The problem of identification is encountered in *factor analysis**. Multiplication of the matrix of factor loadings by an arbitrary nonsingular matrix ("rotation") leaves the

likelihood value unchanged. One solution is to impose enough restrictions on the loadings to rule out rotation, as in the models of "confirmatory" factor analysis discussed by Lawley and Maxwell [7, Chap. 7].

There is a literature on the question of whether one can identify the parameters of a pair of distributions, given that only the minimum of the two random variables is observed. A recent contribution, and references to earlier work, can be found in Gilliland and Hannan [5].

Methods for identifying effects and confounding patterns in factorial designs are discussed by Bailey et al. [1].

Some additional examples can be found in the survey paper by Willassen [13].

CONCLUDING REMARKS

If two parameter values imply the same distribution of the data, no observed data can distinguish between them, and there will be no point in attempting to estimate those parameters. Thus consideration of identification is a question that logically precedes estimation.

Although identification is a classical concept, a Bayesian interpretation is also possible. A good survey can be found in Leamer [8, Sec. 5.9]. Loosely speaking, he shows that identification is a necessary and sufficient condition for the likelihood to dominate the prior (and hence for all nondogmatic priors to lead to the same posterior) as sample size increases without bound.

References

[1]   Bailey, R. A., Gilchrist, F. H. L., and Patterson, H. D. (1977). *Biometrika*, **64**, 347–354.

[2]   Bowden, R. (1973). *Econometrica*, **41**, 1069–1074.

[3]   Fisher, F. M. (1966). *The Identification Problem in Econometrics*. McGraw-Hill, New York. (The standard reference for identification in simultaneous econometric models.)

[4]   Gabrielson, A. (1978). *J. Econometrics*, **8**, 261–263.

[5]   Gilliland, D. C. and Hannan, J. (1980). *J. Amer. Statist. Ass.*, **75**, 651–654.

[6] Koopmans, T. C. and Reiersøl, O. (1950). *Ann. Math. Statist.*, **21**, 165–181.

[7] Lawley, D. N. and Maxwell, A. E. (1971). *Factor Analysis as a Statistical Method*, 2nd ed. American Elsevier, New York.

[8] Leamer, E. E. (1978). *Specification Searches*. Wiley, New York.

[9] Le Cam, L. (1956). *Proc. 3rd Berkeley Symp. Math. Statist. Prob.*, Vol. 1. University of California Press, Berkeley, Calif., pp. 128–164.

[10] Reiersøl, O. (1950). *Econometrica*, **18**, 375–389.

[11] Rothenberg, T. J. (1971). *Econometrica*, **39**, 577–592. (Makes the connection between identification and nonsingularity of the information matrix.)

[12] Van der Genugten, B. B. (1977). *Statist. Neerlandica*, **31**, 69–90. (A survey of the problem of identification.)

[13] Willassen, Y. (1979). *Scand. J. Statist.*, **6**, 89–91.

(ECONOMETRICS
EXPLORATORY DATA ANALYSIS
REGRESSION)

PETER SCHMIDT

I-DIVERGENCE   See *J*-DIVERGENCES AND RELATED CONCEPTS

IEEE TRANSACTIONS ON INFORMATION THEORY

This journal, formerly known as the *IRE Transactions on Information Theory*, is published by The Institute of Electrical and Electronics Engineers, Inc., 345 East 47th Street, New York, NY 10017. It is concerned with theoretical and experimental aspects of the transmission, processing, and utilization of information. The boundaries of acceptable subject matter are intentionally not sharply delimited. Rather, it is hoped that as the focus of research activity changes, a flexible policy will permit the journal to follow suit.

The journal publishes reviews of books relevant to its sphere of interest.

Manuscripts submitted for publication are refereed; decision time is usually 1–6 months from date of receipt; it takes about another six months from acceptance to publication.

The contents of the September 1979 issue (Volume IT-25, Number 5) is as follows:

PAPERS

J. I. Capetanakis, "Tree Algorithms for Packet Broadcast Channels," p. 505; I. Rubin, "Access-Control Disciplines for Multi-Access Communication Channels: Reservation and TDMA Schemes," p. 516; J. A. Bucklew and N. C. Gallagher, Jr., "Quantization Schemes for Bivariate Gaussian Random Variables," p. 537; G. Longo and A. Sgarro, "The Source Coding Theorem Revisited: A Combinatorial Approach," p. 544; D. R. Martin, "Universal Source Coding of Finite Alphabet Sources via Composition Classes," p. 549; R. Heim, "On the Algorithmic Foundation of Information Theory," p. 557; P. J. Shlichta, "Higher Dimensional Hadamard Matrices," p. 566; T. M. Cover and A. A. El Gamal, "Capacity Theorems for the Relay Channel," p. 572; T. Hashimoto and S. Arimoto, "Computational Moments for Sequential Decoding of Convolutional Codes," p. 584; J. W. Modestino and A. A. Ningo, "Detection of Weak Signals in Narrowband Non-Gaussian Noise," p. 592; L. P. Devroye and T. J. Wagner, "Distribution-Free Performance Bounds for Potential Function Rules," p. 601.

CORRESPONDENCE

B. E. Hájek, "On the Strong Information Singularity of Certain Stationary Processes," p. 605; O. Moreno, "Symmetries of Binary Goppa Codes," p. 609; T. U. Basar, "Optimum Linear Coding in Continuous-Time Communication Systems with Noisy Side Information at the Decoder," p. 612; D. G. Mead, "The Average Number of Weighings to Locate a Counterfeit Coin," p. 616; T. A. Brown and J. Koplowitz, "The Weighted Nearest Neighbor Rule for Class Dependent Sample Sizes," p. 617; F. M. Davidson and R. T. Carlson, "Point Process Estimators of Gaussian Optical Field Intensities," p. 620; C. W. Helstrom and L. B. Stotts, "Detection of a Coherent Quantum Signal in Thermal Noise," p. 624; J. W. Woods, "Correction to 'Kalman Filtering in Two Dimensions,'" p. 628.

BOOK REVIEWS

B. Prasada, "Digital Image Processing" by W. K. Pratt; "Digital Image Processing" by R. C. Gonzalez and P. Wintz, p. 630.

IEEE TRANSACTIONS ON PATTERN ANALYSIS AND MACHINE INTELLIGENCE

The IEEE Transactions on Pattern Analysis and Machine Intelligence (PAMI) is a new journal published by the IEEE Computer Society, in cooperation with IEEE's Aerospace and Electronic Systems Society, Control Systems Society, Engineering in Medicine and Biology Society, Information Theory Group, Sonics and Ultrasonics Group and Systems, and Man and Cybernetics Society. The journal, starting from 1979, publishes carefully refereed archival papers on all aspects of pattern recognition*, image processing, artificial intelligence, and their applications. Its scope includes:

1. Statistical pattern recognition—feature extraction and selection, classification techniques, statistical estimation and learning, cluster analysis

2. Structural or syntactic pattern recognition—primitive selection, high-dimensional and stochastic grammars for pattern description, linguistic pattern analysis and parsing, grammatical inference

3. Procedures for graph searching and constraint satisfaction

4. Inference systems—theorem proving, production systems, pattern-directed inference systems, planning, problem solving, induction, learning

5. Knowledge representation—assertional and procedural representations, seman-

tic networks, frame systems, knowledge representation languages

6. Analysis of natural languages—morphological analysis, syntax, semantics, programatics, parsing, text understanding

7. Speech and waveform analysis—waveform segmentation, phoneme detection, word recognition, speaker identification, speech understanding

8. Image processing and analysis—representation, coding, enhancement and restoration, segmentation, visual feature detection, shape and texture analysis and machine vision

9. Application of pattern analysis and machine intelligence—speech recognition systems, character and text recognition, biomedical image analysis, remote sensing, robotics, consultation and diagnostic systems, intelligent data base retrieval, other agricultural, business, educational, governmental, industrial, medical, military, archaeological, meteorological, and scientific applications

Four issues were published (quarterly) in 1979. Starting from 1980, the journal has been published bimonthly.

The editorial arrangements of PAMI consist of an Editor in Chief, an Advisory Board, and an Editorial Committee. The Editor in Chief is usually appointed by the IEEE Computer Society on a two-year term. Members of the Advisory Board include representatives from cooperative societies and groups, as well as those invited by the Editor in Chief. The responsibility of the Advisory Board is to form the basic policy of the journal. Members of the Editorial Committee were invited by the Editor in Chief on a two-year term, with the responsibility of jointly executing the basic policy of the journal and actively participating in the paper referee process.

The normal review procedure usually starts with two to three reviewers for each paper. Depending on the requirement of revisions, the normal time for reviewing a paper (including re-reviewing after revisions) is between three and six months. The size of each issue is around 100 printed pages with an average size of a regular paper approximately 12 pages. The journal has a book review section. Regularly published book reviews will appear in 1980. The first Editor in Chief was K. S. Fu, School of Electrical Engineering, Purdue University, West Lafayette, IN 47907. The present Editor in Chief is T. Pavlidis, Bell Laboratories, Murray Hill, NJ 07974.

The table of contents of the January 1980 issue is listed below:

Regular Papers

1. A Convergence Theorem for the Fuzzy Isodata Clustering Algorithm—Jim Bezdek

2. Area Segmentation of Images Using Edge Points—W. A. Perkins

3. Extracting and Labelling Boundary Segments in Natural Scenes—John M. Prager

4. On a Method of Binary-Picture Representation and Its Application to Data Compression—Eiji Kawaguchi

5. A Theoretical Development for the Computer Generation and Display of Piecewise Polynomial Surfaces—Richard Riesenfeld

6. A Real Time Video Tracking System—A. L. Gilbert, M. Giles, G. M. Flachs, R. B. Rogers, and Yee Hsun U

Correspondences

1. On the Sensitivity of the Probability of Error Rule for Feature Selection—Moshe Ben-Bassat

2. Feature Processing by Optimal Factor Analysis Techniques in Statistical Pattern Recognition—G. Della Riccia

3. A Decision Theory Approach to the Approximation of Discrete Probability Densities—D. Kazakos and T. Cotsidas

4. A New System Structure and Classification Rule for Recognition in Partially

Exposed Environments—Belur V. Dasarathy

5. Relaxation Methods in Multispectral Pixel Classification—A. Rosenfeld

6. 3-D Skeletonization Principle and Algorithm—S. Lobreght

7. A Method for Automating the Visual Inspection of Printed Wiring Boards—John F. Jarvis

8. Symbolic Gray Code as a Multi-Key Hashing Function—R. C. T. Lee

K. S. Fu

IEEE TRANSACTIONS ON RELIABILITY

This is the official *Transactions* for the Reliability Society of the Institute of Electrical and Electronics Engineers (IEEE) and has been published since 1952. Up through 1967 it was published aperiodically, with several issues each year. Early editors were Ernie J. Breiding and W. X. Lamb. In 1968, under the editorship of John A. Connor, the *Transactions* was promoted to a quarterly. In 1969 the current editor, Ralph A. Evans, took over the now mature *Transactions*. In 1971 the Electronics Division of the American Society for Quality Control* (ASQC) adopted the *Transactions* as its journal. The *Transactions* has continued to grow in circulation, number of pages, and number of issues. It became a bimonthly (except for February) in 1973, and since then has had five issues per year. It now contains approximately 500 pages each year.

Members of the Reliability Society, IEEE and Electronics Division, ASQC receive the *Transactions* as part of their dues. Individual subscriptions are also available. The editor's address is Ralph A. Evans; 804 Vickers Avenue; Durham, NC 27701. Subscription and membership information is available from IEEE Publishing Services, 345 East 47 Street; New York, NY 10017.

The *Transactions* publishes papers in the general field of product assurance, but has chosen to retain the name "Reliability" rather than to change its name every time some new discipline is added to product assurance or a new buzz word comes along. Papers are welcome in the areas of reliability, maintainability, quality, and effectiveness of processes, hardware, software, human factors, and systems, as well as on related areas such as product liability. Papers ought to provide an insight or tool for a manager, engineer, scientist, or theoretician. The editorial board solicits practical papers, but with no more success than most technical journals. The August issue each year is usually devoted to a special topic such as Maintainability for 1981 and Failure Analysis for 1980.

Because it is one of the very few journals in its field, it publishes a wide spectrum of papers both in mathematical sophistication and variety of topics, from simple tutorials to discussions of statistical distributions and other theory. An interesting, unusual feature of the *Transactions* is its one page of technical editorials in each issue. Reviews of books that are of interest to the readership are also published.

Manuscripts are usually refereed by three people and the editor. Very few papers are rejected. Rather, the editorial policy is to tell the author how the paper must be revised if it is to be accepted. Very often there is a length constraint—it bears a very rough relation to the anticipated utility and interest to the readership. The *Transactions* has a Supplementary Publication Service for authors whose full papers cannot be published. A short paper or extended abstract is put into the *Transactions*, and the full paper is made available from the National Auxiliary Publication Service (NAPS), a separate organization with oversight by the American Society for Information Science (ASIS).

The language of the *Transactions* is English, but the circulation of over 7000 per issue is worldwide. Authors also come from many countries.

Recent issues covered topics such as: Reliability*, Fault Tree and Cut-Set Evaluation Algorithms; Logistics and Support; Maintainability Specification; Inspection and Maintenance Policies; Statistical Estimation; Optimization and Redundancy; Fault-Tolerant Systems; Probability and Mathematics; and Analyses of Special Systems.

RALPH A. EVANS

IEEE TRANSACTIONS ON SYSTEMS, MAN & CYBERNETICS

This journal started publication in 1971. It publishes papers in the general areas of systems science and operations research, systems methodology and design, and systems management. This includes subject fields as mathematical programming, human and behavioral factors, human–machine systems, cybernetics, pattern recognition*, decision analysis, forecasting*, modeling and simulation*, and planning and decision support processes and systems. There are about 980 pages and 12 issues per year; Fig. 1 is a list of contents of the first issue of 1982. Circulation is approximately 6500 copies.

A decision on acceptance or rejection of a paper is usually communicated to the author within four months; actual publication takes about five months, after final revision. (There is a full description of editorial procedures in Vol. 3, July 1973, pp. 305–307 of the journal.)

The current editor of the *Transactions* is Andrew P. Sage, Department of Engineering Science and Systems, School of Engineering and Applied Science, Thornton Hall, University of Virginia, Charlottesville, VA 22901.

A. P. SAGE

IF–THEN TESTS   *See* EDITING STATISTICAL DATA

I, J **DIAGONALS**   *See* CONFOUNDING

ILL-CONDITIONED SYSTEM OF EQUATIONS

A set of linear equations

$$\mathbf{Ax} = \mathbf{b} \tag{1}$$

IEEE TRANSACTIONS ON

SYSTEMS, MAN, AND CYBERNETICS

JANUARY/FEBRUARY 1982   VOLUME SMC-12   NUMBER 1   (ISSN 0018-9472)

A PUBLICATION OF THE IEEE SYSTEMS, MAN, AND CYBERNETICS SOCIETY

EDITORIAL

Cost Effectiveness and the SMC Society in the 1980's . . . . *A. P. Sage*   2

PAPERS

A Systems Engineering Methodology for Structuring and Calibrating Lake Ecosystem Models . . . .
. . . . *G. F. Roberts and F. Di Cesare*   3
An Input/Output Approach to the Structural Analysis of Digraphs . . . . *J. R. Burns and W. H. Winstead*   15
A Time-Varying Approach to the Modeling of Human Control Remnant . . . . *L. D. Metz*   24
A Matrix Operator Approach to Two-Dimensional Signal Processing . . . . *F. N. Bailey*   35
Feature Extraction by System Identification . . . . *B. A. Eisenstein and R. J. Vaccaro*   42

CORRESPONDENCE

Prediction of Air Pollutant Concentrations by Revised GMDH Algorithms in Tokushima Prefecture . . . .
. . . . *T. Yoshimura, R. Kiyozumi, K. Nishino, and T. Soeda*   50
Determination of Model Order for Dynamical System . . . . *R. V. Jategaonkar, J. R. Raol and S. Balakrishna*   56
Near-Nash Feedback Control of a Composite System with a Time-Scale Hierarchy . . . . *Ü. Özgüner*   62
Application of Interaction-Prediction Approach to Load-Frequency Control (LFC) Problem . . . .
. . . . *P. K. Chawdhry and S. I. Ahson*   66
On Complexity and Syntactic Information . . . . *D. E. Boekee, R. A. Kraak, and E. Backer*   71
Multispectral Texture . . . . *A. Rosenfeld, C. Y. Wang, and A. Y. Wu*   79
Building and Road Extraction from Aerial Photographs . . . . *M. Tavakoli and A. Rosenfeld*   84
Image Smooting and Segmentation by Cost Minimization . . . . . . *K. A. Narayanan, D. P. O'Leary, and A. Rosenfeld*   91

BOOK REVIEWS

Modern Power System Analysis, by I. J. Nagrath and D. P. Kothari, *reviewed by R. C. Desai*   96
Simulation of Control Systems, by I. Troch, Ed., *reviewed by S. G. Tzafestas*   96
Methodology in Systems Modelling and Simulation, by B. Zegler, M. Elzas, G. Klir, and T. Oren, Eds., *reviewed by*
    S. G. Tzafestas   96

CONTRIBUTORS   97

IEEE COPYRIGHT FORM   99

Figure 1

is called "ill-conditioned" if small errors (or variations) in the elements of matrix **A** or vector **c** have a large effect on the *exact* solution of the system (1). (For example, in least-squares* equations of regression models the round-off errors result in an ill-conditioned system especially in the case of polynomial regression*.) For additional information, see ref. 1.

Reference

[1] Seber, G. A. F. (1977). *Linear Regression Analysis*. Wiley, New York.

(MATRICES, ILL-CONDITIONED
ROBUST REGRESSION)

ILL-CONDITIONED MATRICES *See*
MATRICES, ILL-CONDITIONED

ILLUSORY ASSOCIATION

Two independent attributes* A and B may both be associated with an attribute C. This may result in association between A and B in the total population, which does not correspond to any real relationship between A and B yielding the so-called illusory association.

(SPURIOUS CORRELATION)

ILLUSORY CORRELATION *See* SPURIOUS CORRELATION

IMBALANCE FUNCTIONS

In an experiment comparing the effects of two treatments it is important that the treatment groups be well balanced with regard to other variables (covariates) that may influence outcome. In a clinical trial*, for example, if most patients receiving treatment A are young and most patients receiving treatment B are old, a superiority in survivals for the first group cannot be interpreted as a difference in treatment efficacy. The random assignment of treatments will tend to ensure

good balance when the number of experimental units is large. Chance imbalances for smaller studies could, however, be sufficiently extreme to preclude adequate treatment comparisons even using a statistical model to account for the covariate effects. This is because model-derived estimators of treatment effects have decreased precision when there are substantial imbalances.

The conventional approach to avoiding chance imbalances for clinical trials is to subdivide the patients into mutually exclusive strata based on the covariates. A randomization* procedure that ensures approximate within stratum balance in the numbers of patients assigned each treatment is employed. Exact balance cannot be assured because there is staggered patient entry and the covariate values of the patients are not known when the study is designed. The stratification* approach is described in more detail by Zelen [13].

When there are many covariates, the stratification approach may not be effective because the number of strata increases multiplicatively with the number of covariates. For example, suppose that the covariates are: sex, age (< 50 versus > 50), tumor size (small, medium, large), tumor histology (three types), ambulatory status of patient (ambulatory versus bedridden), and hospital (five participating hospitals). In this case the number of strata is $2 \times 2 \times 3 \times 3 \times 2 \times 5 = 360$. The stratification approach generally will not ensure good balance for each of six covariates in a study of 150 patients because the number of patients in most of the 360 stratum will be very small. Some covariates may be of limited importance or they may be correlated, but often it is difficult to know this when the study is designed.

An alternative approach to ensuring balance for each covariate was introduced by Harville [5], Taves [10], and Pocock and Simon [8]. The method of Pocock and Simon can be described as follows. Suppose that several patients have already been assigned treatment from the T alternatives ($T \geqslant 2$), and that an assignment must be made for the next patient whose covariate values are known. Let D_{it} denote a measure of imbalance of the treatment groups with regard to covariate i if treatment t is assigned to the new patient. The critical point is that D_{it} is defined ignoring information about covariates other than i. A D_{it} value is calculated for each provisional treatment assignment t and for each covariate $i = 1, \ldots, m$. Let G_t be a function that combines the $D_{it}, \ldots, D_{mt}$ into an overall measure of imbalance if treatment t is assigned to the new patient. Pocock and Simon [8] suggest

$$G_t = \sum_{i=1}^{m} D_{it}. \qquad (1)$$

If some covariates are considered more important than others, then a weighted summation can be used. The G_t values are calculated for each treatment t and ranked in order $G_{(1)} \leqslant G_{(2)} \leqslant \cdots \leqslant G_{(T)}$. The kth ranked treatment is selected for the new patient with probability p_k where the p's are nonnegative biasing constants that sum to unity. For $T = 2$, one might use $p_1 = 2/3$, $p_2 = 1/3$ [2]. Approaches to selecting the p_k parameters are described by Pocock and Simon [8] and by Klotz [6].

As a simple example, suppose that there are two treatments ($T = 2$) and two covariates, sex and age (< 50 versus > 50), indexed $i = 1$ and $i = 2$, respectively. Assume that the distribution of covariates and assigned treatments for the first 27 patients is as follows:

	Treatment 1		Treatment 2	
Age	Male	Female	Male	Female
< 50	3	7	4	6
> 50	2	1	4	0

Let $n_{ijk}(t)$ denote the number of patients with covariate i at level j who received treatment k, in which these quantities are calculated, including the new patient tentatively assigned treatment t. The D_{it} measure must be a function only of the $n_{ijk}(t)$ with i and t held fixed. Pocock and Simon [8] proposed

several imbalance functions, one of which is

$$D_{it} = \text{range}\{ n_{ij_i 1}(t), n_{ij_i 2}(t), \ldots, n_{ij_i T}(t)\}, \qquad (2)$$

where j_i denotes the level of covariate i for the new patient. Range means the largest minus the smallest. For the example above, suppose that the next patient is a male less than 50 years old. There have been five previous males assigned treatment 1 and eight previous males assigned treatment 2. Similarly, 10 previous patients less than 50 years old have been assigned each treatment. For the range, measure $D_{11} = 8 - 6 = 2$, $D_{12} = 9 - 5 = 4$, and $D_{21} = D_{22} = 11 - 10 = 1$. Thus using expression (1), $G_1 = 2 + 1 = 3$, $G_2 = 4 + 1 = 5$, and treatment 1 is the best (lowest ranked) treatment.

The method of Taves [10] is a special case of the general approach described by Pocock and Simon [8]. The range measure (2) is used for D_{it} and the biasing constants are $p_1 = 1$, $p_2 = 0$. This results in a deterministic procedure in which the treatment assigned is always that which most decreases the existing imbalance. Begg and Iglewicz [1] assumed that all covariates are binary and introduced the imbalance function

$$D_{it} = \left[n_{i11}(t) - n_{i12}(t) \right] - \left[n_{i21}(t) - n_{i22}(t) \right].$$

They claim that this measure is more appropriate for approximately minimizing the variance of the estimator of treatment difference derived from a linear model analysis. They also include in their linear model a constant term as a covariate with one level. The method of Harville [5] is more complicated but is similarly based on minimizing the variance of estimated contrasts among the treatments in a linear model. Other imbalance functions have been proposed by Wei [11] and Efron [3].

Freedman and White [4] showed that for some imbalance functions, determination of the best ranked treatment is very simple. For example, using expression (1) and the range function (2) the computations are as follows. For each covariate i assign a score $+1$ to the treatments with the greatest number of previous patients at level j_i, assign score -1 to the treatment with the fewest number of such patients if it is unique and assign score 0 to other treatments. Sum these scores over all covariates and the treatment with the lowest score is the best ranked treatment. White and Freedman [12] also introduced other generalizations of the method of Pocock and Simon [8].

Although Harville [5] considered this "adaptive stratification" approach broadly appropriate to experimentation, most other developments have been focused on applications to clinical trials. These methods are effective for ensuring good balance of treatment groups marginally with regard to many covariates, particularly for small and moderate-size studies. Marginal balance provides credibility for clinicians and also yields improved precision of treatment contrasts when additive models are appropriate. If specific interactions among covariates are known to be important, they can be incorporated in the procedures by redefining variables. The effect of such design methods on the analysis of the experiment, and other references, are given by Simon [9] and by Pocock [7].

References

[1] Begg, C. B. and Iglewicz, B. (1980). *Biometrics*, **36**, 81–90.

[2] Efron, B. (1971). *Biometrika*, **58**, 403–417.

[3] Efron, B. (1980). In *Biostatistics Casebook*, R. G. Miller, B. Efron, B. W. Brown, Jr., and L. E. Moses, eds. Wiley, New York.

[4] Freedman, L. S. and White, S. J. (1976). *Biometrics*, **32**, 691–694.

[5] Harville, D. A. (1974). *Technometrics*, **16**, 589–599.

[6] Klotz, J. H. (1978). *Biometrics*, **34**, 283–287.

[7] Pocock, S. J. (1979). *Biometrics*, **35**, 183–197.

[8] Pocock, S. J. and Simon, R. (1975). *Biometrics*, **31**, 103–115.

[9] Simon, R. (1979). *Biometrics*, **35**, 503–512.

[10] Taves, D. R. (1974). *Clin. Pharmacol. Ther.*, **15**, 443–453.

[11] Wei, L. J. (1978). *J. Amer. Statist. Ass.*, **73**, 559–563.

[12] White, S. J. and Freedman, L. S. (1978). *Brit. J. Cancer*, **37**, 849–857.

[13] Zelen, M. (1974). *J. Chronic Dis.*, **27**, 365–375.

(CLINICAL TRIALS
CONCOMITANT VARIABLES
DESIGN OF EXPERIMENTS
GENERAL BALANCE
GENERAL LINEAR MODEL
INFERENCE, DESIGN BASED
  VS. MODEL BASED
MATCHING
RANDOMIZATION)

RICHARD SIMON

IMMIGRATION–EMIGRATION PROCESSES

Immigration processes originate as models for the development of a population that is augmented by the arrival of individuals who found families that grow independently of each other. The growth of individual families is usually modeled by a branching process*. Thus an immigration-branching process can be regarded as a random superposition of branching processes. Systematic study of such processes began during the 1960s, but the subject has an earlier history of special cases.

An example is the growth in numbers of rare mutant genes or bacteria in an effectively infinite population of nonmutants [7, 8]. Mutation is envisaged as producing a stream of new mutants—the immigrants—which generate growing colonies of similar particles. Interest typically lies in determining the influence of the mutation and growth rates on the distribution of total mutant numbers, and conversely, on using observed mutant numbers to estimate these rates. For such applications it is usually assumed that the colonies grow according to a Galton–Watson process*.

GALTON–WATSON PROCESS WITH IMMIGRATION

Let $\mathcal{Z} = \{Z_t : t = 0, 1, \ldots\}$ be a Galton–Watson process whose offspring probability generating function* is $f(s)$ and set $Z_0 = 1$. For integers $i \geqslant 0$ and $j \geqslant 1$, let $\mathcal{Z}^{(i,j)} =$

$\{Z_t^{(i,j)} : t = 0, 1, \ldots \}$ comprise a collection of independent copies of $\mathscr{Q}$. If $i \geqslant 1$, $Z_t^{(i,j)}$ denotes the size of a family t time units after it was initiated by the jth immigrant in the group of immigrants which arrived at time i. Let $\{N_i : i = 1, 2, \ldots \}$ be a sequence of independent nonnegative integer-valued random variables which are identically distributed with probability generating function* $h(s)$. The size of the group of immigrants arriving at time i is N_i, and to ensure that immigrants can arrive, assume that $h(0) < 1$. Finally, assume that the population size at time $t = 0$ is $k \geqslant 0$.

In accordance with the superposition principle mentioned above, the Galton–Watson process with immigration (GWI) is defined to be the sequence $\mathscr{X} = \{X_t : t = 0, 1, \ldots \}$, where

$$X_t = \sum_{j=1}^{k} Z_t^{(0,j)} + \sum_{i=1}^{t} \sum_{j=1}^{N_i} Z_{t-i}^{(i,j)}. \quad (1)$$

This follows on observing that $\sum_{j=1}^{N_i} Z_{t-i}^{(i,j)}$ is the total number of descendants at time t of the immigrants which arrived at i. The definition ensures that $\mathscr{X}$ is a Markov chain. Its t-step transition probability starting from k has the generating function

$$\Phi_k(s, t) = (f_t(s))^k \prod_{i=0}^{k-1} h(f_i(s)), \quad (2)$$

where $f_0(s) = s$ and $f_i(s) = f(f_{i-1}(s))$.

In 1916, a special case of the GWI was implicitly used by Smoluchowski [9, 29] to describe the fluctuations in numbers of colloidal particles in a small region of a fluid whose molecules subject the particles to independent Brownian motions*. Smoluchowski's theory was subsequently experimentally verified and used to determine Avogadro's constant. The GWI has also been used for modeling the release of quanta of neurotransmitter from an externally stimulated nerve terminal [31]; queue lengths at a computer-controlled intersection [5]; gated queues [17]; an electronic counter used in neurophysiological experiments [16]; plasmid incompatibility [20]; and inflow processes for storage models* [22]. The GWI

occurs as an embedded process* in simple random walk* [6].

If the rate of immigration is not too high, then immigration functions merely as a device for maintaining the population and the asymptotic properties of $\mathscr{X}$ are inherited from those of $\mathscr{Q}$; for which see BRANCHING PROCESS. To describe these it is convenient to distinguish four cases. Let $m = f'(1-) = E(Z_1)$ be the mean number of offspring per individual.

1. **The Subcritical Case, $0 < m < 1$.** If
    $$\mathscr{I} = E(\log^+ N_i) < \infty,$$
    then $\mathscr{X}$ has a limiting distribution whose generating function is $\prod_{t=0}^{\infty} h(f_t(s))$, but if $\mathscr{I} = \infty$, then $X_t \xrightarrow{p} \infty$. This result rests on the observation that $\mathscr{I}$ is finite iff the mean recurrence time of the zero state is finite.

2. **The Critical Case, $m = 1$.** If $E(N_i) < \infty$ and the offspring distribution has a finite variance, then $\{X_t/t\}$ has a limiting gamma distribution. It is possible for $\mathscr{X}$ to have a limiting distribution and this occurs if and only if
    $$\int_0^1 (1 - h(s))/(f(s) - s) \, ds < \infty.$$
    This condition implies that the offspring variance is infinite.

3. **The Supercritical Case, $1 < m < \infty$.** It is known that there is a strictly increasing sequence of constants $\{c_t\}$ for which $\{c_t^{-1} Z_t\}$ has a nondegenerate and nondefective almost sure limit. When $\mathscr{I} < \infty$ the sequence $\{c_t^{-1} X_t\}$ converges almost surely to a positive random variable which has a smooth distribution function. If $\mathscr{I} = \infty$, it is not possible to norm $\mathscr{X}$ to get a nondefective limit which is not degenerate at the origin.

4. **The Infinite Mean Case, $m = \infty$.** There is a function, U, slowly varying at infinity such that $\{e^{-t} U(X(t))\}$ converges almost surely to a nondegenerate and nondefective limit if $E(\log^+ U(N_i)) < \infty$, and to infinity otherwise.

When $m < 1$ and $\mathscr{I} = \infty$ the asymptotic behavior of $\mathscr{X}$ is dominated by the immigration component. In this case the ratio $\lambda(U(X_t))/\lambda(m^{-t}U(X_t))$ converges in law to the standard uniform $[0, 1]$ distribution, for an appropriate choice of $\lambda(\cdot)$ and $U(\cdot)$. This result takes on more transparent forms under additional conditions governing the rate at which $1 - h(1 - e^{-x})$ converges to zero. Analogous results exist for cases 2 to 4 [3, 21].

Let $\lambda = EN_1 < \infty$ and $m < 1$. According to case 1, $\mathscr{X}$ possesses a limiting distribution with mean $\mu = \lambda/(1 - m)$. If $x_t = X_t - \mu$, then $x_{t+1} = mx_t + \delta_t$, where the residual δ_t is uncorrelated with x_t and $E\delta_t = 0$. This relationship is similar to that defining a first-order autoregression and estimators used for these processes can be modified to provide consistent asymptotically normal estimators of λ, m, and μ [9, 26]. Maximum likelihood* estimation can be used for parametrically specified immigration and offspring distributions [4].

OTHER IMMIGRATION–BRANCHING PROCESSES

Continuous-time versions and generalizations of the GWI can be constructed by letting $\mathscr{D}$ be, for example, a Markov branching process* or an age-dependent process of the Bellman–Harris type, or more generally still, of the Crump–Mode–Jagers type—see BRANCHING PROCESSES. Let $\{\tau_i : i = 1, 2, \ldots\}$ be a strictly increasing sequence of positive random variables and for $t \geq 0$, let $U_t = \sup\{i : \tau_i \leq t\}$. The τ_i represent successive arrival times of groups of immigrants— N_i arrive at τ_i—and U_t is the number of immigrant groups arriving during $[0, t]$. The size of the resulting immigration–branching process at time t is given by the following extension of (1):

$$X_t = \sum_{j=1}^{k} Z_t^{(0,j)} + \sum_{i=1}^{U_t} \sum_{j=1}^{N_i} Z_{t-\tau_i}^{(i,j)} \quad (t \geq 0).$$

If $\{\tau_i\}$, $\{N_i\}$, and $\{\mathscr{D}^{(i,j)}\}$ are independent

families and $F(s, t) = E(s^{Z_t})$, then the generating function of X_t is given by the following analog of (2):

$$\Phi_k(s, t) = (F(s, t))^k E\left\{ \prod_{i=1}^{U_t} h(F(s, t - \tau_i)) \right\}.$$

If $\{U_t\}$ is a nonhomogenous Poisson* process with $\Lambda(t) = EU_t$, then

$$\Phi_k(s, t) = (F(s, t))^k \exp\left[-\int_0^t (1 - h(F(s, u)))\Lambda(du) \right].$$

Specializing further, the linear birth–death* process with immigration can be recovered by letting $\{U_t\}$ be a Poisson process and $\mathscr{D}$ be a linear birth–death process.

If $\{U_t\}$ is a Poisson process and $\mathscr{D}$ is a Markov branching process*, the resulting immigration process $\mathscr{X}$ is a Markov chain in continuous time and its asymptotic properties parallel those of the GWI [21]. The asymptotic theory is less complete for more general growth and immigration components. For example, if $\{U_t\}$ is a nonlattice renewal process* with finite expected lifetime and $\mathscr{D}$ is a Bellman–Harris process, then the condition $\int_0^\infty [1 - h(F(0, t))] dt < \infty$ is necessary and sufficient for $\mathscr{X}$ to have a limiting distribution [23, 25]. If $\mathscr{D}$ is a Crump–Mode–Jagers process, then this condition is known only to be sufficient [11].

Extensions of the results under cases 1 and 2 are known for these more general processes [11, 12]. These results have also been extended to multitype and abstract-valued processes [1], to immigration-branching processes with continuous state spaces [13], and to branching random fields* with immigration [10, 27]. Modifications of the simple models have been studied which allow the offspring and immigration distributions to vary with the population size [18]. Many other aspects of these processes have been studied.

EMIGRATION

Emigration in branching process models is usually regarded as synonymous with death and hence its effects can be accounted for by appropriate modification of the offspring distribution. This view implies that the rate of emigration is proportional to the population size. An alternative approach is to regard emigration as "reversed" immigration by considering immigration–branching processes modified to allow N_i to take negative values [15, 19]. Yet another approach [24, 30] to modeling emigration involves a deterministic growth function satisfying a differential equation, for example, exponential growth and logistic* models. This function is subjected to random decrements occurring at random instants at a rate that can depend on the population size. The decrements are interpreted as reductions in population size due to mass emigration—or disasters.

MIGRATION PROCESSES

The term "immigration–emigration" process is occasionally used in place of "migration" process. This is a model for a system of colonies within each of which particles reproduce according to some type of branching process and between which particles migrate at a rate that increases with size of colony of origin [14].

LITERATURE

The literature on immigration–branching processes is widely scattered and there is no single source that can put the reader in touch with even a substantial portion of it. Many of the references cited below contain useful bibliographies.

References

Letters at the end of reference entries denote one of the following categories:

A: monographs
B: applications of the Galton–Watson process with immigration
C: theory of the Galton–Watson process with immigration
D: other immigration–branching processes
E: emigration

[1] Asmussen, S. and Hering, H. (1976). *Math. Scand.*, **34**, 327–342. (D)

[2] Athreya, K. B. and Ney, P. (1972). *Branching Processes*. Springer-Verlag, Berlin. (Contains a short section on immigration and a bibliography which is substantially complete up to 1971.) (A)

[3] Babour, A. D. and Pakes, A. G. (1979). *Adv. Appl. Prob.*, **11**, 63–72. (C)

[4] Bhat, B. R. and Adke, S. R. (1981). *Adv. Appl. Prob.*, **13**, 498–509. (C)

[5] Dunne, M. C. and Potts, R. B. (1967). *Vehicular Traffic Science*, L. C. Edie, R. Herman, and R. Rothery, eds. American Elsevier, New York. (B)

[6] Dwass, M. (1975). *Proc. Amer. Math. Soc.*, **51**, 270–275. (B)

[7] Gladstien, K. and Lange, K. (1978). *Theor. Popul. Biol.*, **14**, 322–328. (B)

[8] Haldane, J. B. S. (1949). *J. R. Statist. Soc. B*, **11**, 1–14. (B)

[9] Heyde, C. C. and Seneta, E. (1972). *J. Appl. Prob.*, **9**, 235–256. (C)

[10] Ivanoff, B. G. (1980). *J. Appl. Prob.*, **17**, 1–15. (D)

[11] Jagers, P. (1975). *Branching Processes with Biological Applications*. Wiley, New York. (Provides a more complete treatment than refs. 2 and 28.) (A)

[12] Kaplan, N. and Pakes, A. G. (1974). *Stoch. Processes Appl.*, **2**, 371–390. (D)

[13] Kawazu, K. and Watanabe, S. (1971). *Theory Prob. Appl.*, **16**, 34–51. (D)

[14] Kelly, F. P. (1979). *Reversibility and Stochastic Networks*. Wiley, New York. (Contains chapters on migration processes.) (A)

[15] Khan, L. V. (1980). *Siberian Math. J.*, **21**, 283–292. (E)

[16] Lampard, D. G. (1968). *J. Appl. Prob.*, **5**, 648–668. (B)

[17] Lebowitz, M. A. (1961). *IBM J. Res. Dev.*, **5**, 204–209. (B)

[18] Levy, J. B. (1979). *Adv. Appl. Prob.*, **11**, 73–92. (D)

[19] Nagaev, S. V. and Khan, L. V. (1980). *Theory Prob. Appl.*, **25**, 514–525. (E)

[20] Novick, R. P. and Hoppensteadt, F. C. (1978). *Plasmid*, **1**, 421–434. (B)

[21] Pakes, A. G. (1979). *Adv. Appl. Prob.*, **11**, 31–62. (Contains many references on discrete time processes.) (C)

[22] Pakes, A. G. (1981). *Stoch. Processes Appl.*, **11**, 57–77. (B)

[23] Pakes, A. G. and Kaplan, N. (1974). *J. Appl. Prob.*, **11**, 652–668. (D)

[24] Pakes, A. G., Trajstman, A. C., and Brockwell, P. J. (1979). *Math. Biosci.*, **45**, 137–157. (E)

[25] Puri, P. S. (1978). *J. Appl. Prob.*, **15**, 726–747. (D)

[26] Quine, M. P. (1976). *Ann. Prob.*, **4**, 319–325. (C)

[27] Radcliffe, J. (1972). *J. Appl. Prob.*, **9**, 13–23. (D)

[28] Sewastjanow, B. A. (1975). *Verzweigungsprozesse.* R. Oldenbourg Verlag, Munich. (Contains a short chapter on the Markov immigration–branching process.) (A)

[29] Smoluchowski, M. V. (1916). *Phys. Zeit.*, **17**, 557, 585. (B)

[30] Tuckwell, H. C. (1980). In *Biological Growth and Spread*, W. Jäger, H. Rost, and F. Tautu, eds. Springer-Verlag, Berlin, pp. 109–118. (E)

[31] Vere–Jones, D. (1966). *Aust. J. Statist.*, **8**, 53–63. (B)

(BIRTH AND DEATH PROCESSES
BRANCHING PROCESSES
GALTON–WATSON PROCESS
MARKOV PROCESSES)

ANTHONY G. PAKES

IMPORTANCE SAMPLING

Generally, sampling is designed to reduce the variance of estimator(s) or parameter(s), for given size of sample. More specifically, it is used, e.g., by Hammersley and Handscomb [1], to denote a form of Monte Carlo* sampling to estimate an integral.

To estimate $\theta = \int_0^1 f(x)\, dx$ one may choose at random values $X_1, \ldots, X_n$ from a standard rectangular distribution* and use the estimator $\hat{\theta} = n^{-1}\{f(X_1) + \cdots + f(X_n)\}$. Alternatively, we may sample from an arbitrarily chosen distribution with density function $g(y)$, $(0 < y < 1)$ and use the estimator $n^{-1}\sum_{j=1}^n \{f(Y_j)/g(Y_j)\} = \tilde{\theta}(g)$. If $g(\cdot)$ and $f(\cdot)$ are functions of similar "shape," so that $g(Y)$ and $f(Y)$ are highly and positively correlated, the variance of their ratio, and so of $\tilde{\theta}(g)$ is reduced.

Although this form of sampling is that most closely associated with the term "importance," other forms of restricted sampling, such as stratified sampling*, can reasonably be regarded as belonging to the same group of techniques.

Reference

[1] Hammersley, J. M. and Handscomb, D. C. (1964). *Monte Carlo Methods*. Wiley, New York.

(MONTE CARLO METHODS
SAMPLING METHODS
STRATIFIED SAMPLING)

IMPROPER DISTRIBUTIONS

The cumulative distribution function* (CDF) of a random variable X is

$$F_X(x) = \Pr\{X \leq x\}.$$

Clearly, $0 \leq F_X(x) \leq 1$ for all values of x, and $F_X(x)$ is a nondecreasing function of x. For *proper* distributions the CDF satisfies the further conditions

$$\lim_{x \to -\infty} F_X(x) = 0; \qquad \lim_{x \to \infty} F_X(x) = 1.$$

If either (or both) of these conditions are not satisfied, the distribution is said to be *improper*. It can be seen, for example, that the reciprocal of any variable with nonzero probability of taking the value zero has an improper distribution. Informally, the variable may be described as "taking infinite value(s) with positive probability."

The term "improper distribution" is also used (with even more relevance) when $F_X(x)$ does not exist, or can take values greater than 1 (or less than zero). The "uniform prior"* used in Bayesian inference* is an example. For such "distributions" it is possible to calculate the *ratios* of probabilities (for the uniform prior

$$\Pr\{a < x < b\}/\Pr\{c < x < d\}$$
$$= (b - a)/(d - c),$$

for example) but the value of "$F_X(x)$" ($\propto \int_{-\infty}^{x} dt$) is formally infinity.

(BAYESIAN INFERENCE
INVARIANT PRIOR DISTRIBUTION)

INADMISSIBLE DECISION RULES

Suppose that a decision is to be chosen, based on an observed random variable X (possibly a vector) whose distribution depends on an unknown parameter θ (possibly a vector). A *decision rule* is a method for basing the decision on X. If δ is a decision rule, let $r(\theta, \delta)$ denote the expected value of the loss when θ is the parameter and the decision rule δ is used. If δ_1, δ_2 are two decision rules, and if $r(\theta, \delta_1) \leqslant r(\theta, \delta_2)$ for all values of θ, with $r(\theta, \delta_1) < r(\theta, \delta_2)$ for at least one value of θ, then we say that δ_1 is a *better decision rule than* δ_2. A decision rule δ is defined to be *inadmissible* if there exists a decision rule that is better than δ.

As an example, consider the following decision problem. We have to decide whether or not to buy a batch of 100 items for $100. The batch contains an unknown number of defectives. Let Y denote the number of defectives in the batch of 100. Each nondefective item can be sold for $2, each defective item is worthless. Before deciding, we observe the number of defectives in a separate batch of two items from the same source as the batch of 100, at a cost of $5 for observing. Let X denote the number of defectives in the batch of two items. Let θ denote the unknown proportion of defectives being turned out by the common source of the batches. Assuming that this source is turning out a large number of items, it is reasonable to assume that X and Y are independent, X with a binomial distribution* with parameters 2, θ, and Y with a binomial distribution with parameters 100, θ. If the decision is to buy the batch of 100, the loss in dollars is $5 + 100 - 2(100 - Y)$. If the decision is not to buy the batch of 100, the loss is 5 dollars. Let δ_1 be the decision rule that acts as fol-

lows: If $X = 0$, buy; if $X = 2$, do not buy; if $X = 1$, toss a fair coin, and buy if the coin comes up heads, do not buy if the coin comes up tails. An easy calculation gives $r(\theta, \delta_1) = 5 + 100(1 - \theta)(2\theta - 1)$. Let δ_2 be the decision rule that acts as follows: If $X = 0$, do not buy; if $X = 2$, buy; if $X = 1$, toss a fair coin, and buy if the coin comes up heads, do not buy if the coin comes up tails. Then $r(\theta, \delta_2) = 5 + 100\theta(2\theta - 1)$. Since $r(\theta, \delta_1) - r(\theta, \delta_2) = -100(2\theta - 1)^2$, it is easily seen that δ_1 is better than δ_2, and thus δ_2 is inadmissible.

In the preceding example, it is clear that δ_2 is a poor decision rule, because as the number of defectives observed increases, we should switch from deciding to buy to deciding not to buy, whereas δ_2 does just the opposite. The following example of a decision rule which looks reasonable but is inadmissible was constructed by Stein [1]. Suppose that $\theta_1, \ldots, \theta_k$ are the unknown means of k normal populations, each population having a common known positive variance. The problem is to estimate $\theta_1, \theta_2, \ldots, \theta_k$. If the estimates are $D_1, D_2, \ldots, D_k$, respectively, the loss is $(D_1 - \theta_1)^2 + (D_2 - \theta_2)^2 + \cdots + (D_k - \theta_k)^2$. Suppose that we take a sample of size n from each population, and let $\overline{X}_1, \overline{X}_2, \ldots, \overline{X}_k$ denote the respective sample means. The decision rule that chooses $D_1 = \overline{X}_1, D_2 = \overline{X}_2, \ldots, D_k = \overline{X}_k$ is inadmissible if $k \geqslant 3$.

Reference

[1] Stein, C. (1956). *Proc. 3rd Berkeley Symp. Math. Statist. Prob.*, Vol. 1. University of California Press, Berkeley, Calif., pp. 197–206.

(ADMISSIBILITY
DECISION THEORY
STEIN EFFECT)

L. WEISS

INCIDENCE MATRIX

In a block design (*see* BLOCKS, RANDOMIZED COMPLETE) let n_{ij} denote the number of observations in the jth block* to which ith treatment* is applied. The matrix $N = (n_{ij})$, $i = 1, \ldots, v$ (v is the number of treatments), $j = 1, \ldots, b$ (b is the number of blocks) is called the *incidence matrix* of the design.

(DESIGN OF EXPERIMENTS)

INCLUSION–EXCLUSION METHOD

This method is used in the proof of theorems that require the probability of events to be determined, when the sample space is finite or countably infinite, and when events A_1, $A_2, \ldots, A_n$ (n possibly infinite) are of interest. The method uses a combinatorial argument, first including all outcomes in succession, then excluding outcomes that have been counted more than once, and repeating the inclusion–exclusion process on any outcomes still not included exactly once, until all relevant outcomes are included correctly.

An example of the method is Feller's proof of the addition law of probability

$$\Pr(A_1 \cup A_2 \cup \cdots \cup A_n)$$

$$= \sum_i \Pr(A_i) - \sum\sum_{i<j} \Pr(A_i \cap A_j)$$

$$+ \sum\sum\sum_{i<j<k} \Pr(A_i \cap A_j \cap A_k) - \cdots$$

$$+ (-1)^{n+1}\Pr(A_1 \cap A_2 \cap \cdots \cap A_n)$$

[1, Sec. IV.1]. Another example [1, Sec. IV.3] is the use of the method in his proof of the law for the probability that exactly m of the events $A_1, A_2, \ldots, A_n$ occur ($1 \leqslant m \leqslant n$). For an extensive discussion of the inclusion-exclusion method, see Takács [2].

References

[1] Feller, W. (1957). *An Introduction to Probability Theory and Its Applications*, 2nd ed., Vol. 1. Wiley, New York, (3rd ed., 1968).

[2] Takács, L. (1967). *J. Amer. Statist. Ass.*, **62**, 102–113.

INCOME DISTRIBUTION MODELS

Research on income distribution (ID) has followed two mainstreams. The first deals with the factor price formation and the corresponding factor shares, i.e., the distribution of income among the factors of production. This approach was initiated by Ricardo [20] in 1817, and further developed by several schools of economic thought, in particular the Marxians, the neoclassicals, and the post-Keynesians.

The second mainstream, which makes an intensive use of probability distribution functions* (PDF), deals with the distribution of income of a population of economic units. This population is defined as a function of socioeconomic or regional characteristics, such as urban and rural families and wage earners of a country. This approach, initiated by Pareto in 1895, analyzes the distribution of the total income of a given population of economic units or income recipients, i.e., the "personal" or size distribution of income, and it is discussed in the following sections. The research on ID deals with the distribution of a *mass of economic units* by size of their income. This approach is integrated with the inquiry on (a) the distribution of its corresponding *mass of income* (*see* LORENZ CURVE), and (b) the degree of income inequality (*see* INCOME INEQUALITY MEASURES).

PARETO'S MODEL

Pareto's research on ID was motivated by his polemic against the French and Italian socialists who were pressing for institutional reforms to reduce the inequality in the distribution of income. Pareto [18, 19] analyzed the characteristics of regularity and permanence in observed income distributions, which indicated that the income elasticity of the survival distribution function (SDF) was constant, that is,

$$- d\left(\log S_X(x)\right)/d \log x = \alpha,$$

$$\alpha > 1, \quad 0 < x_0 \leqslant x, \quad S_X(x_0) = 1, \quad (1)$$

where X is the income variable with range (x_0, ∞). Solving in (1) for $S_X(x)$, Pareto* type I model $S_X(x) = (x/x_0)^{-\alpha}$ is deduced. Pareto type II is obtained after replacing $\log x$ in (1) by $\log(x - \theta)$; thereby, $S_X(x) = ((x - \theta)/(x_0 - \theta))^{-\alpha}$, $\theta < x_0 \leqslant x$. Finally, multiplying the type II by $\exp(-\beta x)$, $\beta > 0$, we obtain Pareto type III. These three types of models are zero modals (strictly decreasing PDF). The truncated characteristic of Pareto's models results from the type of data used, which at that time consisted of the incomes of persons that exceeded a certain limit x_0 fixed by taxation rules.

Pareto type III model has finite rth moment for all r, whereas the types I and II have only for $r < \alpha$. It can be shown that the mathematical expectation of the type I distribution, known as the Pareto law, is $E[X] = \alpha x_0/(\alpha - 1)$, $\alpha > 1$. For $\alpha \leqslant 2$, the variance is infinite and Pareto's law has a heavy right tail (low-order contact). When this is the case, types I and II belong to the Pareto–Lévy class of stable distributions* [15].

Theoretic and empirical research [9, 15] led to the acceptance of the Pareto law as the model of high-income groups. Thereby an important property to be fulfilled by alternative models of ID is their convergence to the Pareto law for high levels of income.

Pareto's contribution stimulated further research in two directions: (a) the specification of new models to fit the *whole* range of income and (b) the measurement of the income inequality within an observed population. Several PDFs were specified as models of ID. In 1898, March proposed the gamma* PDF and fitted it to the distribution of wages in France, Germany, and the United States; in 1925, Amoroso developed the generalized gamma distribution* and applied it to describe the ID in Prussia; in 1974, Salem and Mount fitted the gamma PDF to the income data in the United States. Following Cantelli's research on the conditions to be fulfilled by a PDF in order to be identified as an ID model, Vinci specified the Pearson type V in 1921. Benini, in 1906, proposed a modified Pareto model to account for some observed nonlinearity of the SDF in a double logarithmic scale. However, it was Gibrat's research, published in 1931 [12], that brought to the fore the lognormal distribution* as a widely accepted model of ID for almost 30 years. But its poor goodness of fit, especially in the two tails of the distribution, stimulated the research on alternative models of ID, which are here analyzed through the identification of their corresponding generating systems.

GENERATING SYSTEMS OF ID MODELS

Almost all the models of income distribution, including the three Pareto types, can be deduced from three generating systems. They are (a) the Pearson system, (b) D'Addario's system, and (c) the generalized logistic (or logistic-Burr) system. Champernowne's model, specified in 1952, is the only well-known ID model that does not belong to any of these systems.

Pearson's System*

K. Pearson specified a differential equation from which an important family of PDF are derived. Although the Pearson system was not designed with the specific purpose of generating ID models, some of its PDF were so applied, such as the beta*, the gamma, and the type V. Furthermore, other members were selected as probability generating functions* (PGF) of ID models, namely, (a) the normal distribution chosen by Edgeworth [10], Gibrat [12], Rutherford [21], and Bartels [2]; (b) the Laplace distribution chosen by Fréchet [11]; (c) the gamma distribution selected by D'Addario [6]; and (d) the Student t and the type IV, selected by Kloek and van Dijk [14].

With the exception of Fréchet's transformation, which generates the Pareto types I and II, none of the members of the Pearsonian system converges to the Pareto distribution.

Table 1

$$\text{Members of D'Addario's system} \begin{cases} \lambda = 0 \begin{cases} \begin{aligned} &\beta = 0, \\ &\alpha > 0 \quad \theta = 0 \end{aligned} \begin{cases} p = 1 \to \text{Pareto type I} \\ p = \tfrac{1}{2} \to \text{Log-normal} \end{cases} \\[2ex] \begin{aligned} &\beta = 0, \\ &\alpha > 0 \quad \theta \neq 0 \end{aligned} \begin{cases} p = 1 \to \text{Pareto type II} \\ p = \tfrac{1}{2} \to \text{Displaced lognormal} \end{cases} \\[2ex] \begin{aligned} &\beta = -1, \\ &\alpha \neq 0 \quad \gamma > 0 \end{aligned} \Big\} \to \text{Amoroso} \begin{cases} \alpha = \gamma, & \theta = 0 \to \text{Gamma} \\ \alpha = -\gamma, & \theta = 0 \to \text{Pearson type V} \\ \alpha > 0, & \gamma = 1 \to \text{Weibull} \end{cases} \end{cases} \\ (\lambda = -1, \beta = -1, \alpha = -\gamma, \qquad \text{Davis}) \end{cases}$$

D'Addario's System

Following the idea of transformation function (TF) applied by Edgeworth and Fréchet, D'Addario specified his system by means of the following PGF and TF:

$$g(y) = B(\lambda + \exp y^{1/\gamma})^{-1}$$
$$\theta < x_0 < x < \infty, \quad \gamma > 0 \quad (2)$$
$$y^{\beta} dy/dx = \alpha/(x - \theta)$$
$$\lambda \text{ real}, \beta \text{ real}, \alpha \neq 0, \quad (3)$$

where x is the income variable and $y = \Phi(x)$ is a monotonic function of x. B satisfies the area condition for the corresponding PDF. Table 1 presents the ID models deduced from the system (2)–(3).

The only non-Paretian member of this system that converges to the Pareto law is Davis' model. D'Addario established a formal correspondence between his system and Brillouin's quantum statistic distribution*. The latter reduces to the distribution of Boltzmann's statistic, for $\lambda = 0$ in Table 1, and to the distribution of Bose–Einstein's statistic for $\lambda = -1$ (Davis's ID in D'Addario's system). (*See* FERMI-DIRAC STATISTICS.)

Generalized Logistic-Burr's System

The specification of this system considers the characteristics of regularity and permanence of the income elasticity of observed ID, which is a bounded and decreasing function of the cumulative distribution function (CDF). Hence it takes the form

$$d\log(F(x) - \delta) = \Phi(x)\phi(F)d\log x,$$
$$0 \leqslant x_0 < x, \quad \delta \leqslant F(x_0), \quad (4)$$
$$\Phi(x) > 0, \quad \phi(F) > 0,$$
$$d(\Phi(x)\phi(F))/dx < 0. \quad (5)$$

For each specification of $\Phi(x)$ and $\phi(F)$ satisfying the conditions in (5) a CDF, identified as an ID model, can be deduced. An important class of ID models can be generated from $\phi(F) = 1 - ((F - \delta)/(1 - \delta))^{\beta}$, $\beta > 0$. Table 2 includes well-known members of this system, their authors and years that they were first applied to describe observed ID.

The ID models generated by the three systems can be enlarged by an appropriate use of transformation functions, such as the three types introduced by N. L. Johnson (*see* JOHNSON SYSTEM OF DISTRIBUTIONS) and the Box–Cox transformation*.

PROPERTIES OF ID MODELS

While Edgeworth [10] was the first to propose some desired properties to be fulfilled by ID models, further developments can be found in Fréchet [11], Aitchison and Brown [1], Mandelbrot [15], and Dagum [7]. We single out as the most relevant the following: (a) model foundation, (b) convergence to the Pareto law, and (c) goodness of fit*. Property (b) was already considered above, and property (c) is concerned with the statistical problem of testing the hypothesis that the

Table 2 Income Distribution Models of the Generalized Logistic-Burr System

Income Distribution Model	(β, δ)	$\Phi(x)$	$\phi(F)$	α^{a}	x
Pareto type I (1895)	$(0,0)$	α	$(1-F)/F$	>1	$>x_0>0$
Pareto type II (1896)	$(0,0)$	$\alpha x/(x-\theta)$	$(1-F)/F$	>1	$>x_0>\theta$
Pareto type III (1896)	$(+,0)$	$\beta x + \dfrac{\alpha x}{x-\theta}$	$(1-F)/F$	>0	$>x_0>\theta$
Benini (1906)	$(0,0)$	$2\alpha \log x$	$(1-F)/F$	>0	$>x_0>0$
Weibull (1951)	$(+,0)$	$\beta x(x-\theta)^{\alpha-1}$	$(1-F)/F$	>0	$>\theta$
Fisk (1961)	$(0,0)$	α	$1-F$	>1	>0
Singh–Maddala (1976)	$(+,0)$	α	$(1-(1-F)^{\beta})(1-F)/F$	>0	>0
Dagum (1977), type I	$(+,0)$	α	$1-F^{\beta}$	>1	>0
Dagum (1977), type II[b]	$(+,+)$	α	$1-\left(\dfrac{F-\delta}{1-\delta}\right)^{\beta}$	>1	>0
Dagum (1980), type III[c]	$(+,-)$	α	$1-\left(\dfrac{F-\delta}{1-\delta}\right)^{\beta}$	>1	$>x_0>0$

[a] The condition stated in this column is required for the existence of the mathematical expectation, hence the Lorenz curve.
[b] The positive value of δ is constrained to $0<\delta<1$, and $F(0)=\delta$ is the frequency estimate of the economic units with negative and null income, i.e., $F(0)=P(X \leqslant 0)=\delta$.
[c] In this model, $F(x_0)=0$ and $F(0)=\delta<0$; hence the range of income is the interval (x_0, ∞), where $x_0>0$ is the minimum estimated income, as it is the case when the sample is restricted to the employed economic units, which is the usual practice in developing countries.

observed data were generated by the identified model of ID. We discuss now property (a).

By *model foundation* we understand the extent by which the mathematical form of an ID model is derived from realistic elementary assumptions. Accordingly, the ID models can be grouped into the following three main classes: (a) stochastic, (b) logicoempirical, (c) ad hoc.

An ID model has a *stochastic foundation* when its mathematical form is the outcome of an a priori set of probability assumptions. It has a *logicoempirical foundation* if it is the theoretical counterpart of observed regularities. Finally, we call it an *ad hoc* model when it is proposed for the sole purpose of fitting ID models, providing neither a plausible probability theory basis nor a logicoempirical foundation. Some ID models belong to both classes (a) and (b), such as Pareto types I and II, lognormal, and Rutherford. Cantelli [3] provided a probability theory basis for Pareto types I, II, and III and Champernowne [4] and Mandelbrot [15] developed a Markov chain foundation for Pareto types I and II. Gibrat [12] gave both the

probability theory basis (the law of proportionate effect) and the logicoempirical foundation (the standardized normal variate is observed to behave approximately as a linear function of log income) for the lognormal ID. Rutherford specified a Gram–Charlier type A distribution* of log income as the outcome of a set of assumptions concerning the log income variance of new entrants into the labor force, the life expectancy at entry, and the variance of the shock system.

Models with only a probability theory basis are those specified by Davis [9] and Singh and Maddala [23]. Ord [16] has provided a stochastic justification for Champernowne's and Fisk's models using the forward Kolmogorov equation*.

Models with a logicoempirical foundation are those specified by Pareto (types I and II), Benini, Fisk, and Dagum [7, 8] types I, II, and III. The remainder are *ad hoc* models.

Although in 1930, Pretorius stated that "the superiority of one frequency function over another depends rather on the success with which that function can be applied to graduate data than on the manner in which it originated" [13, p. 150], and Weibull [25] adopted a similar pragmatical view, it can be shown that ID models that have the best goodness of fit have also the property of convergence to the Pareto law and are supported by a logical or a logicoempirical foundation. This is a specific research problem that, unlike the general-purpose objectives in Pretorius's and Weibull's remarks, requires a coherent scientific foundation.

APPLICATION

Since Gibrat's [12] contribution on economic inequality, the lognormal distribution became the most widely applied model to describe the whole range of observed IDs. Pareto type I continued to be considered as the model of the high-income groups [15]. Salem and Mount [22] revived the gamma distribution, Singh and Maddala [23] and Dagum [7, 8] identified new models of ID. To illustrate the application and comparative goodness of fit of alternative models of ID, the four models just mentioned are fitted to the U.S. family ID of 1978 [24], applying the same method of parameter estimation*, i.e., the nonlinear least-squares method*. For this, an algorithm that searches for the minimization of the sum of squares of the deviations of the actual from the fitted ID is used.

Table 3 presents the estimated models.

Dagum type I is a three-parameter model, whose PDF starts from the origin. When the four-parameter model was fitted, the estimate of δ was negative; hence the type III was also identified as a descriptive model and included in Table 3.

Table 4 shows (in percent) the observed and estimated CDF, mean, and median, respectively, the sum of square errors (SSE) for both the PDF and the CDF, the Gini ratio, the Kolmogorov–Smirnov (K-S) statistic, and its critical values at the 5% and 1% significance levels. Dagum type III accepts the goodness of fit at the 5% significance level and type I at the 1% significance level. The other three models do not pass the goodness of fit test either at the 5% or the 1% significance level. Figure 1 presents the observed histogram and its fitting by each one of the specified models.

The F statistic can be used [5] to test the null hypothesis that the reduction in the sum of squared deviations of one fitted model with respect to another is not significant. Accordingly, the reduction of the sum of squares deviations of Dagum type I and III with respect to Singh–Maddala, gamma, and lognormal respectively are significant at the 5% and 1% levels. For instance, between the type III model specified by Dagum and that of Singh–Maddala, we deduce from the row SSE (PDF) in Table 4, and $n = 21$ (number of intervals of income in the observed ID), $F(1, 16) = (6.883 - 1.821)/(1.821/16) = 44.28$, whereas $F_{0.05} = 4.49$.

Table 3   Estimated Models (Income in $10,000)

Log-normal:	$f(x) = \exp[-(\log x - \mu)^2/2\sigma^2]/(\sqrt{2\pi}\,\sigma x)$,
	$\mu = 0.5875$, $\sigma^2 = 0.47718$
Gamma:	$f(x) = \lambda^\alpha x^{\lambda-1}\exp(-\lambda x)/\Gamma(\alpha)$,
	$\alpha = 2.45806$, $\lambda = 1.25028$
Singh–Maddala:	$F(x) = 1 - (1 + ax^b)^{-c}$,
	$a = 0.02511$, $b = 1.6945$, $c = 10.973$
Dagum Type I:	$F(x) = (1 + \lambda x^{-\alpha})^{-\beta}$,
	$\alpha = 3.61826$, $\beta = 0.40269$, $\lambda = 35.1774$
Dagum Type III:	$F(x) = \delta + (1 - \delta)(1 + \lambda x^{-\alpha})^{-\beta}$, $x > x_0$, $x_0 = 0.1348$
	$\alpha = 3.93532$, $\beta = 0.33464$, $\lambda = 60.3955$, $\delta = -0.0184$

Table 4   1978 U.S. Family Income Distribution

Income ($10,000)	Observed CDF (%)	Fitted Cumulative Distribution Function				
		Log-normal	Gamma	Singh and Maddala	Dagum Type I	Dagum Type III
< 0.2	1.78	0.07	0.88	1.79	2.28	1.26
< 0.3	3.25	0.48	2.17	3.51	4.12	3.44
< 0.4	5.56	1.47	4.04	5.65	6.27	5.88
< 0.5	8.16	3.19	6.43	8.13	8.68	8.52
< 0.6	11.12	5.59	9.24	10.89	11.31	11.32
< 0.7	14.21	8.58	12.41	13.89	14.13	14.28
< 0.8	17.54	12.03	15.86	17.07	17.14	17.36
< 0.9	20.78	15.79	19.52	20.39	20.29	20.55
< 1.0	24.00	19.75	23.31	23.82	23.57	23.83
< 1.1	27.52	23.81	27.18	27.32	26.96	27.20
< 1.2	30.77	27.88	31.08	30.85	30.43	30.62
< 1.3	34.21	31.89	34.97	34.40	33.96	34.08
< 1.4	37.56	35.81	38.81	37.92	37.52	37.57
< 1.5	40.70	39.61	42.57	41.40	41.07	41.05
< 1.6	44.41	43.25	46.23	44.82	44.61	44.52
< 1.7	47.85	46.72	49.77	48.16	48.10	47.95
< 1.8	51.22	50.02	53.16	51.41	51.51	51.32
< 2.0	57.60	56.08	59.52	57.57	58.02	57.80
< 2.5	72.12	68.30	72.66	70.77	71.79	71.75
< 5.0	96.40	93.05	97.29	97.17	96.09	96.59
∞	100.00	100.0	100.0	100.0	100.0	100.0
Mean[a]	20,090	25,418	19,660	19,962	20,423	20,110
Median[a]	17,640	17,995	17,069	17,564	17,555	17,621
SSE (CDF)		251.54	40.30	4.06	3.61	1.218
SSE (PDF)		15.39	5.06	6.88	1.66	1.821
Gini ratio		0.375	0.428	0.351	0.369	0.357
K-S[b]		5.626	1.940	1.353	0.875	0.524

[a] In U.S. dollars.
[b] The critical values of the Kolmogorov–Smirnov* statistic, at the 5% and 1% significance levels are (in percent), $D_{0.05} = 0.730$ and $D_{0.01} = 0.876$, for a sample size $n = 34660$.

References

[1] Aitchison, J. and Brown, J. A. C. (1957). *The Lognormal Distribution*. Cambridge University Press, Cambridge.

[2] Bartels, C. P. A. (1977). *Economic Aspects of Regional Welfare*. Martinus Nijhoff, The Hague.

[3] Cantelli, F. P. (1921). *Metron*, **1**(3), 83–91.

[4] Champernowne, D. G. (1953). *Econ. J.*, **53**(2), 318–351.

[5] Cramer, J. S. (1978). *Econometrica*, **46**(2), 459–460.

[6] D'Addario, R. (1949). *G. Econ. Ann. Econ.*, **8**, 91–114.

[7] Dagum, C. (1977). *Écon. Appl.*, **30**(3), 413–436.

[8] Dagum, C. (1980). *Écon. Appl.*, **33**(2), 327–367.

[9] Davis, H. T. (1941). *The Analysis of Economic Time Series*. Principia Press, Bloomington, Ind.

[10] Edgeworth, F. Y. (1898). *J. R. Statist. Soc.*, **61**, 670–700.

[11] Fréchet, M. (1939). *Rev. Inst. Int. Statist.*, **7**, 32–38.

[12] Gibrat, R. (1931). *Les Inégalités Économiques*. Sirey, Paris.

[13] Johnson, N. L. (1949). *Biometrika*, **34**, 149–176.

[14] Kloek, T. and van Dijk, H. K. (1977). *Econ. Appl.*, **30**(3), 439–459.

[15] Mandelbrot, B. (1960). *Int. Econ. Rev.*, **1**, 79–106.

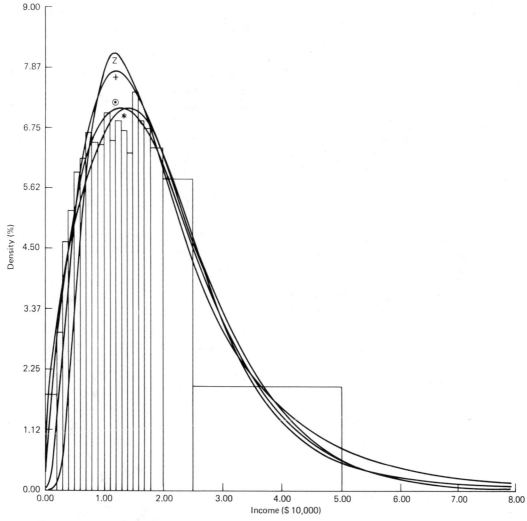

Figure 1 Histogram for four models fitted to the U.S. family ID of 1978. Circle with dot, Singh–Maddala; Z, log-normal; plus, gamma; asterisk, Dagum.

[16] Ord, J. K. (1975). In G. P. Patil, S. Kotz, and J. K. Ord (Eds.) *Statistical Distributions in Scientific Work*, Vol. 2, D. Reidel, Boston.

[17] Pareto, V. (1895). *G. degli Econ.* (*Ser II*), **11**, 59–68.

[18] Pareto, V. (1896). *Écrits sur la courbe de la répartition de la richesse*. Oeuvres complètes de Vilfredo Pareto publiées sous la direction de Giovanni Busino. Librairie Droz, Geneva, 1965.

[19] Pareto, V. (1897). *Cours d'Économie Politique*. New edition by G. H. Bousquet and G. Busino (1964). Librairie Droz, Geneva.

[20] Ricardo, D. (1817). *Principles of Political Economy*. New edition by Piero Sraffa: *Works and Correspondence of David Ricardo*. Cambridge University Press, Cambridge, 1951.

[21] Rutherford, R. S. G. (1955). *Econometrica*, **23**, 277–294.

[22] Salem, A. B. Z. and Mount, J. D. (1974). *Econometrica*, **42**, 1115–1127.

[23] Singh, S. K. and Maddala, G. S. (1976). *Econometrica*, **44**, 963–970.

[24] U. S. Bureau of the Census (1980). *Current Population Reports*, Series P-60, No. 123, Money Income in 1978 of Families and Persons in the United States. U.S. Government Printing Office, Washington, D.C.

[25] Weibull, W. (1951). *J. Appl. Mech.*, **18**, 293–297.

FURTHER READING

See the following works, as well as the references just given, for more information on the topic of income distribution models.

Gini, C. (1912). *Studi Economicogiuridici, Università di Cagliari*, III, 2a. In C. Gini (1955), pp. 211–382.

Gini, C. (1955). *Memorie di Metodologia Statistica*, Vol. 1: *Variabilità e concentrazione*. Edizioni aggiornata a cura di E. Pizetti e. T. Salvemini. Libreria Eredi Virgilio Veschi, Rome.

Johnson, N. L. and Kotz, S. (1970). *Distributions in Statistics: Continuous Univariate Distributions*, 2 vols. Wiley, New York. (An excellent and comprehensive study of probability distribution functions.)

Kaldor, N. (1955–1956). *Rev. Econ. Stud.*, **23**, 83–100. (An excellent survey on the functional distribution of income.)

Lorenz, M. O. (1905). *J. Amer. Statist. Ass.*, **70**, 209–219.

Lydall, H. (1979). *A Theory of Income Distribution*. Oxford University Press, New York.

Acknowledgment

This research was supported by the Social Sciences and Humanities Research Council of Canada under Grant 410-77-0616.

(ECONOMETRICS
PARETO DISTRIBUTIONS)

CAMILO DAGUM

INCOME INEQUALITY MEASURES

Research on income distribution is composed mainly of the following three basic approaches: (a) the distribution of income by size [see INCOME DISTRIBUTION MODELS (IDM)], (b) the distribution of the mass of income (see LORENZ CURVE), and (c) the measurement of the income inequality (II). This study deals with the latter, i.e., income inequality measures (IIMs). They belong to two main classes. One deals with income inequality *within* a population of income receivers, and thus we call it an *intradistribution inequality measure*. The other deals with income inequality *between* populations of income receivers; therefore, we call it an *interdistributions inequality measure*, or economic distance ratio.

INTRADISTRIBUTION INEQUALITY MEASURES (INTRA-DIMs)

Several intra-DIMs have been proposed. They can be analyzed according to (a) the property they fulfill, and (b) the base of their specification. In a seminal monograph, Pareto [33] offered an income inequality interpretation of his parameter $\alpha > 1$, such that when α increases, the income inequality increases. Gini [17] specified the Gini model, which is now known as the Lorenz curve of the Pareto model, and reversed Pareto's interpretation. Later, Gini [18] proposed a general scalar and distribution-free measure of income inequality based on the Gini mean difference*. Recent articles by Samuelson [38] and Chipman [6] revived the issue. Samuelson argues that under certain conditions, Pareto's interpretation can be upheld. For this, Samuelson works with the Pareto–Lévy class of stable distributions* [30, 31], which is a function of four parameters, and not with one member of this class, i.e., the Pareto model. Chipman arrives at an apparent contradictory result because he is ignoring the fact that increasing the value of α while keeping constant the minimum income x^*, the mean income $\mu = E(X) = \alpha x^*/(\alpha - 1)$ decreases.

PROPERTIES OF INTRA-DIMs

Gini's contributions stimulated further research on the distribution of income, wealth, size of firms and other socioeconomic variables. Meanwhile, several other intra-DIMs were proposed. Hence the choice of one of them should be supported by a set of well-accepted properties. Following Dalton's [10] pioneer article we can present the following properties:

1. **Principle of transfers.** Given two income receivers A and B, with incomes $x - d$ and $x(d > 0)$, respectively, the principle of transfers requires that any transfers $0 < h < d$ from B to A should reduce the intra-DIM. For this, the fol-

lowing theorem ensures that property 1 will always be fulfilled [1, 25, 41]:

Theorem 1. An intra-DIM that is the mathematical expectation of a strict convex function of income fulfills the principle of transfers at all levels of income. That is,

$$I = E(V(X)) = \int_0^\infty V(x)\,dF(x), \quad (1)$$

where I stands for an intra-DIM, F for a CDF, and V for a strict convex function of income x.

It follows from Theorem 1 and for all continuously differentiable functions $V(x)$ that the relative sensitivity of the IIM (1) per unit of income transfers from a person with income x to a person with income $x - d$, $d > 0$, is proportional to

$$T(x) = V'(x) - V'(x - d). \quad (2)$$

2. **Principle of proportional addition to incomes (scale independence).** Proportionate addition or subtraction to all incomes should leave the intra-DIM unaffected.

3. **Principle of equal addition to incomes.** Equal additions to all incomes should diminish the intra-DIM and equal subtraction should increase it. This is a corollary to property 2.

4. **Principle of proportional addition to persons.** The intra-DIM should be invariant to proportionate additions to the population of income receivers.

5. **Principle of symmetry.** The intra-DIM should be invariant with respect to any permutation of income among the income receivers.

6. **Principle of normalization.** The range of the intra-DIM should be in the interval $[0,1]$, with zero (one) for perfect equality (inequality).

7. **Principle of operationality.** The intra-DIM should provide a unique, straightforward, and unambiguous estimate of the income inequality by all researchers using the same observed or fitted income distribution, independently of their subjective inequality aversion.

BASES FOR THE SPECIFICATION OF AN INTRA-DIM

The proposed measures of intra-DIM can be grouped in the following two classes: (a)

tion theory foundation. The normative measures are implicitly stated in function of the level of social welfare (SW) such that the intra-DIM is a decreasing function of the level of SW, for a given total of income. However, each positive measure implicitly includes some assumptions about the form of a social welfare function (SWF). They have been discussed, among others, by Dalton [10], Atkinson [1], Sheshinksi [43], Dasgupta et al. [11], Rothschild and Stiglitz [37], Sen [40–42], Kondor [28], Kolm [27], and Kakwani [25].

Table 1 presents a selection of positive and normative intra-DIM; it contains their corresponding mathematical forms and the properties they fulfill. The implied SWFs of this set of intra-DIM, hence their relative sensitivity to transfers at different levels of incomes, can be analyzed by means of formula (2). Table 2 presents the mathematical forms of the IIMs listed in Table 1, corresponding to the Pareto type I model.

Among all the specified intra-DIM, the Gini ratio is the most widely accepted. In 1912 Gini [18] defined it as a function of the Gini mean difference $\Delta = E(|X - Y|)$, and in 1914 [19] he proved that his ratio is equal to twice the area between the identity function $L = F$ and the Lorenz curve,

$$L(x) = \int_0^x y\,dF(y)/E(X). \quad (3)$$

Gini's ratio takes into consideration the income differences between all pairs of income receivers; it has a neat geometrical representation by means of the Lorenz curve, and is sensitive to transfer at all levels of incomes.

The coefficient of variation* fulfills important properties but it has not been widely used as an intra-DIM.

The relative mean deviations* $M(\mu)$ and $M(m)$ (with respect to the mean μ, and the

Table 1

Income Inequality Measure	Mathematical Definition	Property						
		1	2	3	4	5	6	7
Positive measure								
Statistic base								
Coefficient of variation	$CV = \left[\int_0^\infty (x - \mu)^2 dF(x) \right]^{1/2} / \mu$	Yes	Yes	Yes	Yes	Yes	No	Yes
Relative mean deviation (w, r, t, μ)	$M(\mu) = \int_0^\infty (|x - \mu|) dF(x)/2\mu$	No	Yes	Yes	Yes	Yes	Yes	Yes
Relative mean deviation (w.r.t. m)	$M(m) = \int_0^\infty (|x - m|) dF(x)/2\mu$	No	Yes	Yes	Yes	Yes	No	Yes
Variance of the logarithm of income	$\sigma^2 = \int_0^\infty (\log x - \log M_g)^2 dF(x)$	No	Yes	Yes	Yes	Yes	No	Yes
Gini's ratio	$G = E(|X - \mu|)/2\mu = 1 - 2 \int_0^1 L\, dF$	Yes	Yes	Yes	Yes	Yes	Yes	Yes
Information base								
Theil's index	$T = I_0 = \int_0^\infty (x/\mu)\log(x/\mu)\, dF(x)$	Yes	Yes	Yes	Yes	Yes	No	Yes
Hirschman's index	$H = I_1 = \frac{1}{2} \int_0^\infty (x/\mu)(x/\mu - 1) dF(x)$	Yes	Yes	Yes	Yes	Yes	No	Yes
Generalized information index	$I_\beta = \frac{1}{\beta(\beta + 1)} \int_0^\infty (x/\mu)[(x/\mu)^\beta - 1]\, dF(x)$	Yes	Yes	Yes	Yes	Yes	No	No
Normative measure								
Dalton's index	$D = 1 - \int_0^\infty U(x) dF(x)/U(\mu)$	Yes	No	Yes	Yes	Yes	No	No
Atkinson's ratio	$A(\epsilon) = 1 - \left[\int_0^\infty s^{1-\epsilon} dF(x) \right]^{1/(1-\epsilon)} / \mu$	Yes	Yes	Yes	Yes	Yes	Yes	No
Kolm's ratio	$K(\epsilon) = 1 - \log \int_0^\infty \exp[-\epsilon(x - \mu)] dF(x)/\epsilon$	Yes	No	No	Yes	Yes	No	No

median m, respectively) are not sensitive to transfers of incomes on the same side of the mean and median, respectively. It can be proved [34, 35] that

$$M(\mu) = F(\mu) - L(\mu) \quad \text{and}$$

$$M(m) = F(m) - L(m) = \tfrac{1}{2} - L(m), \quad (4)$$

where $M(\mu)$ is the maximum difference between the equidistribution function and the Lorenz curve, hence $M(m) < M(\mu)$.

$M(\mu)$ was first proposed by Bresciani–Turroni [3; 4, p. 429]. More recently, equivalent versions of Bresciani–Turroni relative mean deviation were proposed by Schutz [39], Kuznets [29], and Elteto and Frigyes [12].

The variance of the logarithm of income

acquired relevance with Gibrat's [16] specification of the lognormal model. It has the shortcoming of not being sensitive to transfers among high-income units, i.e., when $x > eM_g$, where M_g is the geometric mean and e is the base of the natural logarithms.

The positive measures based on the theory of information are very attractive for the properties they fulfill. The main objections are: (a) the information theory foundation has no meaning in the field of economic inequality, and (b) perfect equality is achieved when there is maximum entropy*, i.e., maximum disorder, whereas perfect equality can only be thought of as the outcome of maximizing society order, hence SW.

Dalton [10] cogently argued in favor of

Table 2   Income Inequality Measure of the Pareto Type I Model

$[S(x) = 1 - F(x) = (x/x_0)^{-\alpha}, \alpha > 1, 0 < x_0 < x]$

Positive measure
  Statistic base
    Coefficient of variation: $CV = [\alpha(\alpha - 2)]^{-1/2}, \alpha > 2$

Relative mean deviation (w.r.t. μ): $M(\mu) = \alpha^{-\alpha}(\alpha - 1)^{\alpha - 1}$

Relative mean deviation (w.r.t. m): $M(m) = \frac{1}{2}(2^{1/\alpha} - 1)$

Variance of the logarithm of income: $\sigma^2 = 1/\alpha^2$
Gini's ratio: $G = 1/(2\alpha - 1)$

  Information base
    Theil's index: $I_0 = 1/(\alpha - 1) + \log(1 - 1/\alpha)$
Hirschman's index: $I_1 = (CV)^2/2 = [2\alpha(\alpha - 2)]^{-1}, \alpha > 2$
Generalized information
  index $(\alpha > \beta + 1)$: $I_\beta = [(\alpha - 1)^{\beta+1}/\alpha^\beta(\alpha - \beta - 1) - 1]/\beta(\beta + 1)$

Normative measure
  Dalton's ratio: (a) $U(x) = \log x$: $D = [\log\alpha/(\alpha - 1) - 1/\alpha]/\log\mu$
    (b) $U(x) = x^{1-\epsilon}, 0 < \epsilon < 1$: $D_\epsilon = 1 - \alpha^\epsilon(\alpha - 1)^{1-\epsilon}/(\alpha + \epsilon - 1)$
  Atkinson's ratio: (a) $\epsilon = 1$: $A_1 = 1 - M_g/\mu = 1 - (\alpha - 1)e^{1/\alpha}/\alpha$
    (b) $\epsilon > 0, \epsilon \neq 1$: $A_\epsilon = 1 - (\alpha - 1)[\alpha/(\alpha + \epsilon - 1)]^{1/(1-\epsilon)}/\alpha$
  Kolm's ratio: $K_\epsilon = 1 - \log\alpha\epsilon^{\alpha-1}x_0^\alpha e^{\epsilon\mu}[\Gamma(-\alpha) - \Gamma(\epsilon x_0; -\alpha)],$
    $\alpha \neq 2, 3, \ldots$; $\Gamma(\epsilon x_0 : -\alpha)$ is the incomplete gamma function

the specification of an intra-DIM explicitly derived from a SWF. Dalton, Atkinson, and Kolm considered additive, separable (no interpersonal comparison of utilities is permitted) and symmetric SWFs. These extreme constraints outweigh the SW framework of the normative measures. Dalton derived two particular cases; one as a function of the arithmetic and geometric means*, and the other as a function of the arithmetic and harmonic means*. Following Dalton's assumptions, Atkinson [1] specified an intra-DIM which unlike Dalton's is invariant with respect to any positive linear transformation of the individual utility function. His measure is equal to 1 minus the ratio between the mean generating function of order less than 1 ($\epsilon > 0$) and the arithmetic mean* [8]. The parameter ϵ measures the degree of inequality aversion, and is entirely subjective. Similar objections apply to Kolm's measure.

Champernowne [5] compares the performance of several IIMs and investigates their capability to account for the inequality attributable to the lower-, middle-, and upper-income groups, respectively.

ESTIMATION AND HYPOTHESIS TESTING FOR INTRA-DIMS

All intra-DIMs can be estimated either from the observed (distribution-free) or the fitted (parametric) income distribution. Given the size of the sample surveys* on income distributions, large-sample theory is appropriate. Hence the knowledge of the corresponding variance of the intra-DIM estimator is required. Kendall and Stuart [26, Chap. 10] deduce the variance of the coefficient of variation, the Gini mean difference, and the mean deviation, provided that the appropriate moments exists. Following a similar approach, the variance of the remainder measures can be estimated, and the corresponding test of hypothesis performed.

The distribution-free estimates of intra-DIMs are obtained from sample surveys on income distributions published by income classes. As a consequence, they are systematically underestimated. To overcome this source of bias, Gastwirth [13, 15] developed an important approach to estimate lower and upper bounds for the Gini, Theil, and other measures provided that their mathematical specification takes the form

$$I = \int_0^\infty h(x)\,dF(x)/h(\mu) - 1, \qquad (5)$$

where $h(x)$ is a convex function and μ is the mathematical expectation of income. Mehran [32] derived further results for the Gini ratio. It follows from (1) and (5) that

$$V(x) = h(x)/h(\mu) - 1. \qquad (6)$$

It can be verified that all positive measures given in Table 1 fulfill this condition. The normative measures are specified for concave (in general quasi-concave) utility functions $U(x)$. Hence Dalton's intra-DIM also belongs to the class (5), with $V(x) = 1 - U(x)/U(\mu)$.

Gastwirth's bounds also provide a nonstochastic interval estimation of an intra-DIM deduced from a fitted IDM. Therefore, necessary but not sufficient statistical evidence* to entertain the acceptance of the model goodness of fit is that the estimated inequality measure fall inside this interval.

INTERDISTRIBUTIONS INEQUALITY MEASURES (INTER-DIM)

The inter-DIM or economic distance ratio is a scalar representation of the income differential between populations of income receivers. Hence it provides a scalar measure of the relative degree of affluence of one population with respect to another. For this, it complements the information provided by an intra-DIM.

Two main approaches deal with the comparisons of interdistribution inequality between populations with different degrees of economic affluence. One approach performs a disaggregation of the intra-DIM to account for the relative contribution to the overall inequality by each socioeconomic and/or geographic characteristic, and by source of income. This approach follows the methodology of the analysis of variance. Bhattacharya and Mahalanobis [2], Pyatt [36], and Henderson and Rowley [23] deal with the disaggregation of the Gini ratio, whereas Theil [43] deals with his entropy measure.

A second approach specifies a new inter-distribution inequality measure that accounts for the inequality between populations of income receivers. Dagum [9] provided a class D_r (r real) of inter-DIMs and applied the ratios D_0 and D_1 to measure the inequality between regions in Canada and between races in the United States.

The ratio D_r is a normalization of d_r in the interval $[0, 1]$. Given two populations of income receivers with CDF $F_1(x)$ and $F_2(y)$, respectively, where $E(Y) > E(X)$, and introducing the indicator function $I(X, Y)$ such that $I(X, Y) = 0$ if $X > Y$, $I(X, Y) = 1$ if $X < Y$, and $I(X, Y) = \frac{1}{2}$ if $X = Y$ (this case is relevant when dealing with discrete distributions), we define

$$d_r = \left(E\left[I(X, Y)(Y - X)^r \mid E(Y) \right.\right.$$
$$\left.\left. > E(X) \right] \right)^{1/r}, \qquad r \neq 0, \quad (7)$$
$$d_0 = E(I(X, Y) \mid E(Y) > E(X)). \qquad (8)$$

It can be proven that d_r is a monotonic increasing function of r. The decision maker's choice of r will reveal his degree of inter-distributions inequality aversion.

The normalized form of this class of economic distance is

$$D_r = (d_r - d_r^*)/(\Delta - d_r^*),$$
$$r \text{ real.} \quad (9)$$

The symbol d_r^* corresponds to the value of d_r under the *null hypothesis*, i.e., *assuming that the two populations have identical income distributions*; $\Delta = E(|Y - X|)$ is the maximum value that d_r can take, which occurs when the two distributions do not overlap. It can be proven [9] that

$$D_1 = (E(Y) - E(X))/\Delta, \qquad (10)$$

$$D_0 = 2E_2(F_1(Y)) - 1, \qquad (11)$$

where the symbol $E_i(\cdot)$, $i = 1, 2$, stands for the mathematical expectation of the corresponding argument, weighted by the ith PDF.

The measure D_0 can be related to Gini's probability of transvariation [7, 20, 22], Mann–Whitney form of the unpaired two-sample statistic, and Gastwirth's [14] TPROB.

To decide whether the inter-DIM D_r (r real) is significantly different from zero, the null hypothesis $H_0: F_1(y) \equiv F_2(y)$ must be tested. For this, the one-sided two-sample Kolmogorov–Smirnov test* is applied.

References

[1] Atkinson, A. B. (1970). *J. Econ. Theory*, **2**, 244–263.

[2] Bhattacharya, N. and Mahalanobis, B. (1967). *J. Amer. Statist. Ass.*, **62**, 143–161.

[3] Bresciani–Turroni, C. (1910). In *Studi in onore di Biagio Brugi*. Tipografia L. Gaipa, Palermo, Italy, pp. 794–812.

[4] Bresciani–Turroni, C. (1937). *J. R. Statist. Soc. A*, **3**, 421–432.

[5] Champernowne, D. G. (1974). *Econ. J.*, **84**, 787–816.

[6] Chipman, J. S. (1974). *J. Econ. Theory*, **9**, 275–282.

[7] Dagum, C. (1960). *Metron*, **20**, 3–208.

[8] Dagum, C. (1979). *Écon. Appl.*, **32**, 81–93.

[9] Dagum, C. (1980). *Econometrica*, **48**, 1791–1803.

[10] Dalton, H. (1920). *Econ. J.*, **30**, 348–361. (A seminal paper on the properties of IIMs.)

[11] Dasgupta, P., Sen, A. K., and Starrett, D. (1973). *J. Econ. Theory*, **6**, 180–187.

[12] Elteto, O. and Frigyes, E. (1968). *Econometrica*, **36**, 383–396.

[13] Gastwirth, J. L. (1972). *Rev. Econ. Statist.*, **54**, 306–316.

[14] Gastwirth, J. L. (1975). *Amer. Statist.*, **29**, 32–35.

[15] Gastwirth, J. L. (1975). *J. Econometrics*, **3**, 61–70.

[16] Gibrat, R. (1931). *Les Inégalités Économiques*. Sirey, Paris.

[17] Gini, C. (1910). Indici di concentrazione e di dependenza. *Atti della III Riunione della Societa Italiana per il Progresso delle Scienze*. In ref. 21, pp. 3–210.

[18] Gini, C. (1912). Variabilità e mutabilità. *Studi economico-giuridici, Università di Cagliari*, III 2a. In ref. 21, pp. 211–382.

[19] Gini, C. (1914). Sulla misura della concentrazione e della variabilità dei caratteri. *Atti del R. Istituto Veneto di Scienze Lettere ed Arti*. In ref. 21, pp. 411–459.

[20] Gini, C. (1916). Il concetto di transvariazione e le sue prime applicazioni. *G. Econ. Riv. Statist.* In ref. 21, pp. 1–55.

[21] Gini, C. (1955). *Memorie di Metodologia Statistica*, Vol. 1: *Variabilità e Concentrazione*. Edizioni aggiornata a cura di E. Pizetti e T. Salvemini. Libreria Eredi Virgilio Veschi, Rome.

[22] Gini, C. (1960). *Memorie di metodologia statistica*, Vol. 2: *Transvariazione*. Libreria Goliardica, Rome.

[23] Henderson, D. W. and Rowley, J. C. (1978). Decomposition of an Aggregate Measure of Income Distribution. *Discuss. Paper No. 107*, Economic Council of Canada.

[24] Hirschman, A. O. (1945). *National Power and the Structure of Foreign Trade*. University of California Press, Berkeley, Calif. (See also O. C. Herfindahl, Concentration in the Steel Industry. Ph.D. dissertation, Columbia University, 1950.)

[25] Kakwani, N. C. (1980). *Income Inequality and Poverty: Methods of Estimation and Policy Applications*. Oxford University Press, New York.

[26] Kendall, M. G. and Stuart, A. (1958). *Advanced Theory of Statistics*, Vol. 1. Charles Griffin, London.

[27] Kolm, S. C. (1976). *J. Econ. Theory*, **12**, 416–442; *ibid.*, **13**, 82–111.

[28] Kondor, Y. (1975). *Rev. Income Wealth*, **21**, 309–321.

[29] Kuznets, S. (1957). Qualitative Aspects of the Economic Growth of Nations, II: Industrial Distribution of National Product and Labour Force. *Econ. Dev. Cultural Change*, Suppl. to **5**, 1–80.

[30] Lévy, P. (1925). *Calculs des Probabilités*. Gauthier-Villars, Paris.

[31] Mandelbrot, B. (1960). *Int. Econ. Rev.*, **1**, 79–106.

[32] Mehran, F. (1975). *J. Amer. Statist. Ass.*, **70**, 64–66.

[33] Pareto, V. (1896). *Écrits sur la courbe de la répartition de la richesse*. Oeuvres Complètes de Vilfredo Pareto, publièes sous la direction de Giovanni Busino. Librairie Droz, Geneva, 1965.

[34] Pietra, G. (1915). Delle Relazioni tra gli indici di variabilità. *Atti del Reale Istituto Veneto di Scienze, Lettere ed Arti*, **74** 775–804.

[35] Pietra, G. (1948). *Studi di Statistica Metodologica*. A. Giuffré, Milan.

[36] Pyatt, G. (1976). *Econ. J.*, **86**, 243–255. (An excellent discussion of the Gini ratio and its decomposition.)

[37] Rothschild, M. and Stiglitz, J. E. (1973). *J. Econ. Theory*, **6**, 188–204.

[38] Samuelson, P. A. (1965). *Riv. Int. Sci. Econ. Commerc.*, **12**, 246–250.

[39] Schutz, R. R. (1951). *Amer. Econ. Rev.*, **41**, 107–122.

[40] Sen, A. K. (1970). *Collective Choice and Social Welfare*. Holden-Day, San Francisco.

[41] Sen, A. K. (1973). *On Economic Inequality*. Oxford University Press, New York. (An excellent and well-motivated treatment of positive and normative income inequality measures, at an elementary mathematical level.)

[42] Sen, A. K. (1974). *J. Public Econ.*, **4**, 387–403.

[43] Sheshinski, E. (1972). *J. Econ. Theory*, **4**, 98–100.

[44] Theil, H. (1967). *Economics and Information Theory*. North-Holland, Amsterdam. (Chapter IV offers a complete discussion of the entropy index as an IIM as well as its decomposition properties.)

Acknowledgment

This research was supported by the Canada Department of Labour—University Research Program.

(ECONOMETRICS
INCOME DISTRIBUTION MODELS
LOGNORMAL DISTRIBUTION
PARETO DISTRIBUTION)

CAMILO DAGUM

INCOMPLETE BETA FUNCTION (RATIO)

The integral

$$I(p;q;\alpha) = \int_0^\alpha t^{p-1}(1-t)^{q-1}\,dt,$$

$$p > 0; \quad q > 0; \quad 0 \leqslant \alpha \leqslant 1$$

is known as the *incomplete beta function*. The ratio $I(p;q;\alpha)/I(p;q;1)$ is called the incomplete beta function ratio. (The word "ratio" is often omitted.)

Tables of incomplete beta function ratios were compiled by Pearson [6] and extended by Vogler [7]. An efficient computer algorithm for calculating incomplete beta function ratios was developed by Gautschi [2].

Alternative computer algorithms for the incomplete beta integral and the inverse of incomplete beta function ratio using the Newton–Raphson method* were constructed by Majumder and Bhattacharjee [5] and improved by Cran et al. [1].

For more details, see Johnson and Kotz [3,4].

References

[1] Cran, G. W., Martin, K. J., and Thomas, G. E. (1977). *Appl. Statist.*, **26**, 111–114.

[2] Gautschi, W. (1964). *Commun. ACM*, **7**, 143–144.

[3] Johnson, N. L. and Kotz, S. (1969). *Distributions in Statistics: Discrete Distributions*. Wiley, New York, Chap. 1.

[4] Johnson, N. L. and Kotz, S. (1970). *Distributions in Statistics: Continuous Univariate Distributions*, Vol. 2. Wiley, New York, Chap. 24.

[5] Majumder, K. L. and Bhattacharjee, G. P. (1973). *Appl. Statist.*, **22**, 409–414.

[6] Pearson, K., ed. (1934). *Tables of Incomplete Beta Functions*. Cambridge University Press, Cambridge (2nd ed., 1968).

[7] Vogler, L. E. (1964). Percentage Points of the Beta Distribution. Tech. Note 215. National Bureau of Standards, Boulder, Colo.

INCOMPLETE BLOCK DESIGNS

The use of blocks in experimental design is described in many standard statistical textbooks, e.g., Cox [9]. Blocks are used to allow for variability associated with differences between similarly treated units. That is, the experimental units are divided into groups thought to be similar to each other but different from other groups. The experimental units are often called *plots* and the groups of plots *blocks*, this nomenclature deriving from agricultural field trials on blocks of land divided into plots. However, the plots in a block may also be animals from a litter, the litters being blocks, human beings from the same background, samples treated by the same industrial process, or many other things. The unifying feature is that plots within one block share some common property which plots in other blocks may lack. More than one blocking system is possible, an obvious extension being rows and columns, but this article will be restricted to designs with just one blocking system.

If the number of plots available in any block is at least as numerous as the number of experimental treatments, then complete block designs may be used. The simplest of these is the randomized block* design; in the usual version of this, each block contains each treatment exactly once, treatments being allocated at random to plots within each block separately. This is a very common sort of experimental design, indeed probably the commonest of all designs in practice. It has the advantage that the effects of a treatment can be very readily assessed by simply averaging the responses to the treatment over all blocks; formal statistical estimation of variability by means of analysis of variance* is also straightforward. Hence such designs are strongly recommended when they are possible.

Often, however, the variability of the experimental material suggests the use of blocks, but the number of treatments in a trial is more than the number of plots in a block. Examples include the use of twins, human or animal, or the four wheels on a motor vehicle, but there are many situations where this occurs. Some form of *incomplete block design* is then necessary. It is often desirable to group blocks containing less than the full number of treatments into complete replicates, designs with this property being said to be resolvable*. There are separate articles in this encyclopedia on the commoner types of incomplete block design (see the related entries at the end of this article); the present article is intended to give a lead into the others, to suggest which design is likely to be of use in which circumstances, and to describe some of the less common, although still important, types that do not have separate articles.

FACTORIAL DESIGNS

The broad division we shall consider here is into experiments with and without a *factorial* structure. Factorial designs* are used where the experimental treatments themselves form a group that can be ordered by the levels of two or more factors; examples include the

application of different amounts of several fertilizer treatments to plants in a field trial, or the presence or absence of several possible therapeutic agents in a clinical study. If there are many factors, especially if they all have the same number of levels, the block size may be reduced by confounding*. It is almost always important to estimate the *main effect* of any factor, that is, to find the differences between the levels of that factor summed over all levels of all the others: it may be less important to estimate all the *interactions*, that is, the ways in which the combined effects of several factors differ from those expected on the basis of their main effects. It is then possible, and particularly simple when all factors have only two levels, to *confound* the interactions between several factors, or *higher-order* interactions, with blocks. Each block then contains only a subset, usually a simple fraction, of all the treatments, although it is usual to combine the blocks into complete replicates. The basic ideas underlying factorial experimentation were described in considerable detail by Yates [29].

The extreme situation of the use of incomplete blocks in factorial experimentation occurs with fractional replication*, in which only a subset of treatments is used in the whole experiment. Then information on the higher-order interactions is entirely sacrificed in the interests of reducing variability by keeping block size as small as possible. Such designs, especially with all factors at two levels, have been widely used in industrial experimentation, particularly since the comprehensive study of Box and Hunter [3]. The first part of this paper gives an example where difficulties at the filtration stage in a new manufacturing unit were overcome by the use of a 2^7 factorial experiment in eight units, so that only $1/16$ of the possible treatment combinations were used.

GENERAL DESIGNS WITHOUT FACTORIAL STRUCTURE

Even where the treatments are not factorial, they may still be so numerous that it becomes essential to have fewer plots in a block than there are treatments. Extreme cases occur in trials of new chemicals about which little is known, and also at various stages of a plant breeding program: the block size for chemicals may be limited by the number that can be studied in a given work period, while to have all possible plants in a block may result in soil conditions varying unacceptably throughout the block. There are also situations where there is an obvious natural block size, such as the twins or vehicle wheels mentioned earlier, but still many treatments have to be compared. There is not even any necessity to have all blocks the same size, although the formal design and analysis of the trial may be more complicated: an example here would be experimentation using all animals in each litter, despite variations in litter size.

General methods for obtaining the analysis of variance for incomplete block designs were first given by Rao [25] and Tocher [26]. Rao obtained algebraic estimates of treatment effects for a very general class of designs with equal block size and indicated how estimates of variances of these effects might be found, but gave specific formulas only for more restricted designs. By contrast, Tocher derived the variance–covariance matrix* of treatment effects as $\Omega\sigma^2$, where σ^2 is the residual variance and

$$\Omega^{-1} = \mathbf{r}^{\delta} - \mathbf{n}\mathbf{k}^{-\delta}\mathbf{n}' + (1/N)\mathbf{r}\mathbf{r}'.$$

Here $\mathbf{n}$ is the incidence matrix of the design, such that the element in the ith row and jth column represents the number of occurrences of the ith treatment in the jth block, and $\mathbf{n}'$ is its transpose. If $\mathbf{l}$ is a unit vector, $\mathbf{nl} = \mathbf{r}$, the vector of replications, and $\mathbf{n}'\mathbf{l} = \mathbf{k}$, the vector of block sizes: $\mathbf{r}^{\delta}$ is the diagonal matrix with elements those of $\mathbf{r}$, $\mathbf{k}^{-\delta}$ is the diagonal matrix with elements those of $\mathbf{k}^{-1}$, the reciprocal of $\mathbf{k}$, and N is the total number of plots. Then, if $\mathbf{T}, \mathbf{B}$ are vectors of treatment and block totals, respectively, and $\mathbf{Q} = \mathbf{T} - \mathbf{n}\mathbf{k}^{-\delta}\mathbf{B}$, the vector of treatment effects is $\hat{\mathbf{t}}_0 = \Omega\mathbf{Q}$. The residual sum of squares is given by $\mathbf{y}'\mathbf{y} - \mathbf{Q}'\hat{\mathbf{t}}_0 - \mathbf{B}'\mathbf{k}^{-\delta}\mathbf{B}$, where $\mathbf{y}$ is the vector of observed values, and the treatment sum of squares by $\mathbf{Q}'\hat{\mathbf{t}}_0$.

Tocher's approach provides estimates of treatment means and standard errors in any incomplete block design, but there are easier methods available for particular designs with more pattern to them. Nevertheless, methods similar to Tocher's, possibly involving the use of generalized inverses of matrices related to Ω^{-1}, have proved useful in investigating wide classes of experimental design, e.g., the matrix $\mathbf{C} = \mathbf{r} - \mathbf{n}\mathbf{k}^{-\delta}\mathbf{n}'$ [6].

In general, incomplete block designs are such that treatments and blocks are not orthogonal to each other, so that a natural alternative name to incomplete block designs is *nonorthogonal designs*. Pearce [24] classified nonorthogonal designs according to the pattern of Ω^{-1}, showing formally how the analysis of these designs is simplified when Ω^{-1} has particular patterns. Caliński [4] carried this approach further by considering possible patterns for $\mathbf{M}_0 = \mathbf{r}^{-\delta}\mathbf{n}\mathbf{k}^{-\delta}\mathbf{n}' - (1/N)\mathbf{l}\mathbf{r}'$. However, much of the general consideration of suitable matrix properties has led to designs which are the same in practice as those derived from different approaches.

If incomplete block designs have to be used, they should, in some sense, be *optimal*. Optimal designs* are defined according to various criteria, mostly stemming from the work of Kiefer [20]. Some of the most useful special designs meet several of these optimality criteria simultaneously, but there are certain combinations of treatment numbers and block sizes for which there are different optimal designs according to different criteria.

SOME PARTICULAR TYPES OF EQUALLY REPLICATED INCOMPLETE BLOCK DESIGN

We shall now consider some particular types of design: the simplest possible situation for needing an incomplete block design is when all blocks are of equal size and all treatments are equally replicated, i.e., $\mathbf{r} = r\mathbf{l}$, $\mathbf{k} = k\mathbf{l}$, but $t > k$. Even this simplification has led to an enormous literature on types of design, far too big to be comprehensively reviewed here. A natural further requirement is that all treatment comparisons shall have equal *efficiency*, that is, the variances of the comparisons of all pairs of treatments shall be the same; there is equal efficiency if and only if all pairs of treatments concur equally often in a block. With replicates of t treatments on b blocks of k plots each, it is then necessary that $\lambda = r(k-1)/(t-1)$ shall be integral, since λ is the number of concurrences of any pair of treatments within a block. Designs satisfying this condition are called balanced incomplete block designs*, often abbreviated to BIB designs or BIBD. The condition that λ is integral is not sufficient for a BIB design to exist, while a further constraint on the parameters is that $t \leqslant b$; this comes from the fact that $\mathbf{nn}' = (r - \lambda)\mathbf{I} + \lambda\mathbf{ll}'$, where $\mathbf{I}$ is the identity matrix, has rank t and so $\mathbf{n}$ has rank t, while since $\mathbf{n}$ is a $(t \times b)$ matrix its rank $\leqslant b$.

The practical analysis of variance of BIB designs is much simpler than that obtained from Tocher's general methods, since the treatment sum of squares, in general $\mathbf{Q}'\hat{\mathbf{t}}_0$, can here be expressed as a sum of terms which are themselves squares rather than products. It is then possible to partition the treatment sum of squares in just the same way as for a randomized block design, so that a factorial set of treatments can be arranged in a BIB design if required. In BIB designs information can be obtained from differences between blocks as well as within blocks, if it is assumed that block effects are themselves randomly distributed. However, this *recovery of interblock information* [30], although quite workable in theory, is not often of much practical importance.

BIB designs were introduced by Yates [27]: the construction of various possible types of BIB design was thoroughly investigated by Bose [1] and taken further by Hanani [12]. It now seems as if most, possibly all, of the BIB designs suitable for practical experimentation have been discovered. Although BIB designs do not exist for a very wide range of numbers of treatments, block sizes and replications, especially for small r,

they are much used in practice, particularly when the major constraint is a low value of k. BIB designs satisfy all the optimality criteria, so should always be used if they are available. A slight extension of balanced incomplete blocks [24] is to designs with total balance but not necessarily equal block size, in which all diagonal elements of Ω^{-1} have one value and all off-diagonal elements have another value.

Where BIB designs do not exist, there are various possible approaches. An early idea was the use of the pseudo-factorial principle with several replicates of k^2 treatments so arranged that each replicate contains k blocks of k plots each. Such designs, called lattice designs*, were introduced by Yates [28]. A lattice design with $(k+1)$ replicates of k^2 treatments is a BIB design, but it is often desirable in practice to have fewer replicates, although the analysis is then more complicated. Because lattices are only possible when the number of treatments is a square, *rectangular lattices* were introduced by Harshbarger [13] for $k(k+1)$ treatments in blocks of k plots. Both the square and rectangular lattice designs have their uses, but they are restricted in the numbers of possible treatments and block sizes.

Relaxation of the condition of equal efficiency of all treatment comparisons leads to many possible types of design. It may sometimes be important to have some comparisons more efficient than others, while in other trials it may be desirable to have all variances of treatment differences nearly equal. A very general class of designs with more than one level of concurrence of pairs of treatments within a block consists of partially balanced incomplete block designs*, often shortened to PBIB designs or PBIBD, introduced by Bose and Nair [2], which includes lattices as well as many other designs. PBIB designs are such that for any treatment there are exactly n_i treatments having λ_i concurrences with it, these being its ith associates $(i = 1, 2, \ldots)$; further, if two treatments are ith associates, the number of treatments common to the jth associates of one and the kth associates of the other is independent of the two initial treatments. The most important PBIB designs in practice are those with only two associate classes, and these have been classified in various ways; the most numerous are group-divisible designs* in which the treatments are divided into m groups of n treatments each such that all treatments belonging to the same group are first associates and treatments belonging to different groups are second associates. In the 1950s and 1960s many methods of constructing PBIB designs were found, some being very ingenious. A very full review of two-associate PBIB designs was given by Clatworthy [7], listing more than 800 such designs with $r, k \leqslant 10$. Clatworthy pulled together the various methods of construction, and gave methods of analysis for the designs, including recovery of interblock information, but even this large catalog is not complete; for example, Freeman [11] gave 19 new designs, as well as finding alternative derivations of some already known.

A simple way of laying out trials may be to use cyclic designs*, in which the arrangement of treatments of the second and subsequent blocks is a cyclic development of those in the first block. John et al. [18] gave general methods of analysis, including recovery of interblock information, for cyclic designs and listed 460 designs with $r, k \leqslant 10$, 386 of which are neither BIB nor two-associate PBIB designs. They also described their uses in calibration*, i.e., the process of assigning a value for a physical property to an object in terms of values for some reference group. Jarrett and Hall [16] gave a generalization of cyclic incomplete block designs available when $t = mn$.

A quite different approach to the problem of incomplete block designs may be needed in statutory trials of potential new crop varieties. There will usually be not more than three or four replicates of up to 100 varieties, with block size not too important as long as it is less than about 10; however, resolvability is important since field operations will conveniently be conducted on complete replicates. Appropriate designs are known as α-designs [21, 22]. In some α-designs pairs of treatments concur within a block either 0 or

1 times, these being $\alpha(0, 1)$ designs; in others, there are 0, 1, or 2 concurrences, these being $\alpha(0, 1, 2)$ designs. For some numbers of varieties there is no single convenient block size, and it is necessary to have two block sizes differing only by one, but this need not cause difficulties in layout or analysis. An alternative method of constructing resolvable incomplete block designs with zero or one concurrences is sometimes available [20a]; although it cannot be applied for all numbers of replicates, treatments, and block sizes, where it can be used it is easy to implement.

An alternative way of bridging some of the gaps in the PBIB designs is by using *dual* designs, i.e., taking the treatments of a BIB or PBIB design as blocks and the blocks as treatments. Designs for more values of t and relatively small r are then possible, but there is still not a design for every value of t; however, these designs have proved useful for trials where the value of t may be slightly changed [10]. The analysis of α-designs, or of designs that are duals of balanced designs, is more complicated than that of balanced designs, but not unworkably so.

Although they lack full balance, many of the designs just described have good optimality properties. In particular, consider a design with only two concurrences λ_1, λ_2, such that $\lambda_2 = \lambda_1 + 1$, which may or may not be a PBIB design; such a design is called a *regular graph design*, and John and Mitchell [17] (*see also* John and Williams [17a]) proved that some regular graph designs were optimal in one or more of Kiefer's senses, and conjectured that others were. Further, the dual of an optimal design is itself optimal. It thus seems as if, for wide classes of design, there are designs satisfying the optimality criteria exactly, or very nearly.

DESIGNS WITH UNEQUAL REPLICATION OR UNEQUAL BLOCK SIZE

As with complete block designs, most practical incomplete block designs have equal replication of all treatments, but this is not always desirable, especially when many new treatments are to be compared with one standard. *Supplemented* designs may be used in such situations, these being defined as designs in which all treatments are replicated r times except the supplementing one, which has r_0 replicates, and all pairs of treatments concur λ times in blocks, unless one of the pair is the supplementing treatment, when there are λ_0 concurrences [5, 15, 23]. In the commonest supplemented designs in practice the supplementing treatment occurs equally often in each block, but this is not essential. Pearce [24] gave a method for analyzing supplemented designs by partitioning the treatment sum of squares into one component within the main group of treatments and another representing the difference between this group and the supplementing treatment. Incomplete block designs with two groups of treatments having different levels of replication are also possible [8, 25], but are less useful in practice.

In most practical experimental designs all blocks are the same size, although, as stated above, two block sizes differing only by one occur in some α-designs. Also, in an experiment with all blocks initially the same size one or more treatments may be omitted or not recorded. Hedayat and John [14] considered what happens if treatments are omitted from a BIB design and classified such designs as *resistant* or *susceptible* depending on whether the precision of the comparison of the remaining treatments is or is not the same. John [19] showed that if only one treatment is omitted, all remaining pairs of treatments are in fact compared with very nearly the same accuracy.

References

[1] Bose, R. C. (1939). *Ann. Eugen. (Lond.)*, **9**, 353–399. (The first paper on the construction of BIB designs, setting out procedures subsequently followed.)

[2] Bose, R. C. and Nair, K. R. (1939). *Sankhyā*, **4**, 337–372. (The paper that introduced the idea of PBIB designs.)

[3] Box, G. E. P. and Hunter, J. S. (1961). *Technometrics*, **3**, 311–351, 449–458. (Full description of fractional replication.)

[4] Caliński, T. (1971). *Biometrics*, **27**, 275–292.

[5] Caliński, T. and Ceranka, B. (1974). *Biom. Z.*, **16**, 299–305.

[6] Chakrabarti, M. C. (1963). *J. Indian Statist. Ass.*, **1**, 8–23.

[7] Clatworthy, W. H. (1973). *Natl. Bur. Stand. (U.S.), Appl. Math. Ser. 63* (Washington, D.C.). (The definitive listing of PBIB designs with two associate classes.)

[8] Corsten, L. C. A. (1962). *Biometrics*, **18**, 499–519.

[9] Cox, D. R. (1958). *Planning of Experiments*. Wiley, New York. (Still the best general book on experimental design with a minimum of mathematics and a full description of underlying principles.)

[10] Freeman, G. H. (1976). *Biometrics*, **32**, 519–527.

[11] Freeman, G. H. (1976). *Biometrika*, **63**, 555–558.

[12] Hanani, H. (1961). *Ann. Math. Statist.*, **32**, 361–386.

[13] Harshbarger, B. (1947). *Va. Agric. Exp. Stn. Mem. 1*.

[14] Hedayat, A. and John, P. W. M. (1974). *Ann. Statist.*, **2**, 148–158.

[15] Hoblyn, T. N., Pearce, S. C. and Freeman, G. H. (1954). *Biometrics*, **10**, 503–515.

[16] Jarrett, R. G. and Hall, W. B. (1978). *Biometrika*, **65**, 397–401.

[17] John, J. A. and Mitchell, T. J. (1977). *J. R. Statist. Soc. B*, **39**, 39–43.

[17a] John, J. A. and Williams, E. R. (1982). *J. R. Statist. Soc. B*, **44**, 221–225.

[18] John, J. A., Wolock, F. W. and David, H. A. (1972). *Natl. Bur. Stand. (U.S.) Appl. Math. Ser. 62* (Washington, D.C.). (The standard work on cyclic designs.)

[19] John, P. W. M. (1976). *Ann. Statist.*, **4**, 960–962.

[20] Kiefer, J. (1959). *J. R. Statist. Soc. B*, **21**, 272–319. (The paper that introduced the ideas of optimality, although rather mathematical.)

[20a] Khare, M. and Federer, W. T. (1981). *Biom J.*, **23**, 121–132.

[21] Patterson, H. D. and Williams, E. R. (1976). *Biometrika*, **63**, 83–92. (The paper that introduced α-designs.)

[22] Patterson, H. D., Williams, E. R., and Hunter, E. A. (1978). *J. Agric. Sci. Camb.*, **90**, 395–400. (Listing of α-designs.)

[23] Pearce, S. C. (1960). *Biometrika*, **47**, 263–271.

[24] Pearce, S. C. (1963). *J. R. Statist. Soc. A*, **126**, 353–377. (Full description of many types of nonorthogonal design.)

[25] Rao, C. R. (1947). *J. Amer. Statist. Ass.*, **42**, 541–561.

[26] Tocher, K. D. (1952). *J. R. Statist. Soc. B*, **14**, 45–100. (The first attempt to produce general methods of analysis for designs with minimal restrictions.)

[27] Yates, F. (1936). *Ann. Eugen. (Lond.)*, **7**, 121–140. (Introduction of BIB designs.)

[28] Yates, F. (1936). *J. Agric. Sci. Camb.*, **26**, 424–455. (Introduction of lattice designs.)

[29] Yates, F. (1937). *The Design and Analysis of Factorial Experiments*. Commonwealth Agricultural Bureaux, Farnham Royal, Slough, England. (Full account of factorial experimentation.)

[30] Yates, F. (1940). *Ann. Eugen. (Lond.)*, **10**, 317–325.

(ANALYSIS OF COVARIANCE
ANALYSIS OF VARIANCE
BALANCE IN EXPERIMENTAL DESIGN
BALANCED INCOMPLETE BLOCK
  DESIGNS
CYCLIC DESIGNS
DESIGN OF EXPERIMENTS
FRACTIONAL FACTORIAL DESIGNS
GENERAL BALANCE
GROUP-DIVISIBLE DESIGNS
LATTICE DESIGNS
OPTIMALITY
PARTIALLY BALANCED INCOMPLETE
  BLOCK DESIGNS)

G. H. FREEMAN

INCOMPLETE DATA

Incomplete problems occur frequently in statistics. Indeed, one might view inferential statistics in general as a collection of methods for extending inferences from a sample to a population where the nonsampled values are regarded as missing data.

Although some statistical methods for complete data, such as factor analysis*, finite mixture models*, and mixed model analysis of variance*, can be usefully viewed as incomplete data methods [13], we restrict this review to more standard incomplete data problems. For the class of problems reviewed here, we consider "missing data" to be synonymous with "incomplete data." After describing common examples with missing data in the following section, in the section "Methods for Handling Incomplete Data" we describe techniques for handling these problems. In the last section, we discuss the EM algorithm, an ubiquitous algorithm for finding maximum likelihood* (m.l.) estimates from incomplete data. Useful reviews of the analysis of incomplete data are given in Afifi and Elashoff [1], Hartley and Hocking [19], Orchard and Woodbury [34], Dempster et al. [13], and Little [29].

COMMON INCOMPLETE DATA PROBLEMS

We first consider problems where missing values are confined to a single outcome variable y, and interest concerns the distribution of y, perhaps conditional on a set of one or more predictor variables x, that are recorded for all units in the sample. Sometimes we have no information about the missing values of y; at other times we may have partial information, for example, that they lie beyond a known censoring point c. (*See* CENSORED DATA.)

Mechanisms Leading to Missing Values

Any analysis of incomplete data requires certain assumptions about the distribution of the missing values, and in particular how the distributions of the missing and observed values of a variable are related. The work of Rubin [38] distinguishes three cases. If the process leading to missing y values (and, in particular, the probability that a particular value of y is missing) does not depend on the values of x or y, the missing data are called *missing at random* and the observed data are *observed at random*. If the process depends on observed values of x and y but not on missing values of y the missing data are called missing at random, but the observed data are not observed at random. If the process depends on missing values of y, the missing data are not missing at random; in this case, particular care is required in deriving inferences. Rubin [38] formalizes these notions by defining a random variable m that indicates for each unit whether y is observed or missing, and relating these conditions to properties of the conditional distribution of m given x and y. This approach is discussed in the section "The Modeling Approach to Missing Data."

Analysis of Variance

The first incomplete data problem to receive systematic attention in the statistics literature is that of *missing data in designed experiments*; in the context of agriculture trials, this problem is often called the missing plot problem [3, 5]. Designed experiments investigate the dependence of an outcome variable, such as yield of a crop, on a set of factors, such as variety, type of fertilizer, and temperature. Usually, an experimental design is chosen that allows efficient estimation of important effects as well as a simple analysis. The analysis is especially simple when the design matrix is easily inverted, as with complete or fractional replications of factorial designs. The missing data problem arises when at the conclusion of the experiment, the values of the outcome variable are missing for some of the plots, perhaps because no values were possible, as when particular plots were not amenable to seeding, or because values were recorded and then lost. Standard analyses of the resultant incomplete data assume the missing data are missing at random, although in practical situations the plausibility of this assumption needs to be checked. The analysis aims to exploit the "near balance" of the resulting data set to simplify computations. For example, one tactic is to substitute estimates of the missing outcome values and then to carry out the analysis assuming the data to be complete. Questions needing attention then address the choice of appropriate values to substitute and how to modify subsequent analyses to allow for such substitutions. For discussions of this and other approaches, see Healy and Westmacott [21], Wilkinson [47], and Rubin [36, 39].

Censored or Truncated Outcome Variable

We have noted that standard analyses for missing plots assume that the missing data are missing at random, that is, the probability that a value is missing can depend on the values of the factors but not on the missing outcome values. This assumption is violated, for example, when the outcome variable measures time to an event (such as death of an experimental animal, failure of a light bulb), and the times for some units are not recorded because the experiment was terminated before the event had occurred; the resulting data are *censored*. In such cases the analysis must include the information that the units with missing data are censored, since if these units are simply discarded, the resulting estimates can be badly biased.

The analysis of censored samples from the Poisson, binomial, and negative binomial distributions is considered by Hartley [18]. Other distributions, including the normal, log-normal, exponential, gamma, Weibull, extreme value, and logistic, are covered most extensively in the life-testing* literature (for reviews, see Mann et al. [30] and Tsokos and Shimi [46]). Nonparametric estimation of a distribution subject to censoring* is carried out by life table methods, formal properties of which are discussed by Kaplan and Meier [24]. Much of this work can be extended to handle covariate information [2, 11, 16, 26]. The EM algorithm, discussed here in the last section, is a useful computational device for such problems.

A variant of censored values occurs when missing values are known to lie within an interval, as when the data are available in grouped form. The analysis of grouped data* is discussed by Hartley [18], Kulldorff [25], and Blight [7], among others. Another variant of censored data occurs when the number of censored values is unknown. The resulting data are called truncated, since they can be regarded as a sample from a truncated distribution. A considerable literature exists for this form of data [8, 12, 13, 18].

Sample Survey* Data

For the data types discussed in this section, the missing data are not missing at random, but the mechanisms leading to incomplete data are assumed known. For example, the censoring points for censored observations are known. A common and somewhat more intractable problem occurs when the missing data are not missing at random and the mechanism leading to missing data is at best partially known. Incomplete data arising from nonresponse in sample surveys provide an illustration of this kind of problem. For example, nonresponse to a question on household income often depends on the amount of that income, in an unknown way. Restricting the analysis to respondents clearly leads to bias in such situations; given the large samples often available in survey work, this bias is frequently more important than the loss of efficiency of estimation arising from the reduction in sample size.

The effect of survey nonresponse is minimized by (a) designing data collection methods to minimize the level of nonresponse, (b) interviewing a subsample of nonrespondents, and (c) collecting auxiliary information on nonrespondents and employing analytical methods that use this information to reduce nonresponse bias. Models for nonrandomly missing data, as developed by Nelson [32], Heckman [22], and Rubin [40] can also be applied here. Estimates derived from these models, however, are sensitive to aspects of the model that cannot be tested with the available data [17, 19, 41]. A thorough discussion of survey nonresponse* is given in the work of the National Academy of Sciences Panel on Incomplete Data [31].

Multivariate Incomplete Data

The incomplete data structures discussed so far are *univariate*, in the sense that the missing values are confined to a single outcome variable. We now turn to incomplete data structures that are essentially *multivariate* in nature.

Many multivariate statistical analyses, including least-squares* regression*, factor analysis*, and discriminant analysis*, are based on an initial reduction of the data to the sample mean vector and covariance matrix of the variables. The question of how to estimate these moments with missing values in one or more of the variables is, therefore, an important one. Early literature was concerned with small numbers of variables (two or three) and simple patterns of missing data [1, 4]. Subsequently, more extensive data sets with general patterns of missing data were addressed [6, 9, 27, 34, 37, 45].

The reduction to first and second moments is generally not appropriate when the variables are categorical. In this case, the data can be expressed in the form of a multiway contingency table. Most of the work on incomplete contingency tables* has concerned maximum likelihood estimation* assuming a Poisson* or multinomial* distribution for the cell counts. Bivariate categorical data* form a two-way contingency table; if some observations are available on a single variable only, they can be displayed as a supplemental margin. The analysis of data with supplemental margins is discussed by Hocking and Oxspring [23] and Chen and Fienberg [10]. Extensions to log-linear models for higher-way tables with supplemental margins are discussed in Fuchs [15].

Essentially, all the literature on multivariate incomplete data assumes that the missing data are missing at random, and much of it also assumes that the observed data are observed at random. Together these assumptions imply that the process that creates missing data does not depend on any values, missing or observed.

Missing Data in Time Series

Techniques for analyzing time series data often require observations at regular intervals. In practice, many time series are irregularly spaced, by the nature of the collection process or because otherwise regular series become irregular when some values are missing. The problem of maximum likelihood estimation for Gaussian stationary processes with missing data has received some attention [23a, 43a; for other references see the symposium volume in which 43a appears].

State space representations of the likelihood and the EM algorithm are useful computational tools in this work.

METHODS FOR HANDLING INCOMPLETE DATA

A Broad Taxonomy of Methods

Methods for handling incomplete data generally belong to one or more of the following categories:

1. Methods that discard units with data missing in some variables and analyze only the units with complete data (e.g., Nie et al. [33]).

2. **Imputation-Based Procedures.** The missing values are filled in and the resultant completed data are analyzed by standard methods. For valid inferences to result, modifications to the standard analyses are required to allow for the differing status of the real and the imputed values. Commonly used procedures for imputation include *hot deck* imputation (see Ford [14]), where recorded units in the sample are substituted, *mean* imputation, where means from sets of recorded values are substituted and *regression* imputation, where the missing variables for a unit are estimated by predicted values from regression on the known variables for that unit [9]. A variant of imputation methods produces multiple imputations for each missing value and thereby allows simple adjustments to be made to reflect the differing status of real and imputed values [41, 42].

3. **Weighting Procedures.** Randomization inferences from sample survey data without nonresponse are commonly based on *design weights*, which are inversely proportional to the probability of selection. For example, let y_i be the value of the variable y for unit i in the population. Then the population mean is often estimated by

$$\sum \pi_i^{-1} y_i / \sum \pi_i^{-1}, \qquad (1)$$

where the sums are over sampled units, π_i is the probability of selection for unit i, and π_i^{-1} is the design weight for unit i.

Weighting procedures modify the weights to allow for nonresponse. The estimator (1) is replaced by

$$\sum (\pi_i \hat{p}_i)^{-1} y_i / \sum (\pi_i \hat{p}_i)^{-1}, \qquad (2)$$

where the sums are now over sampled units which respond, and $\hat{p}_i$ is an estimate of the probability of response for unit i, usually the proportion of responding units in a subclass of the sample. Weighting is related to mean imputation; for example, if the design weights are constant in subclasses of the sample, then imputing the subclass mean for missing units in each subclass, or weighting responding units by the proportion responding in each subclass, lead to the same estimates of the population mean, although not the same estimates of sampling variance unless adjustments are made to the data with means imputed. A recent discussion of weighting with extensions to two-way classifications is provided by Scheuren [43].

4. **Model-Based Procedures.** A broad class of procedures is generated by defining a model for the incomplete data and basing inferences on the likelihood under that model, with parameters estimated by procedures such as maximum likelihood. Advantages of this approach are: flexibility; the avoidance of adhocery, in that model assumptions underlying the resulting methods can be displayed and evaluated; and the availability of large-sample estimates of variance based on second derivatives of the log-likelihood, which take into account incompleteness in the data. Disadvantages are that computational demands can be large, particularly for complex patterns of missing data, and that little

is known about the small-sample properties of many of the large-sample approximations.

The Modeling Approach to Incomplete Data

Any procedure that attempts to handle incomplete data must, either implicitly or explicitly, model the process that creates missing data. We prefer the explicit approach since assumptions are then clearly stated.

The parametric form of the modeling argument can be expressed as follows [38]. Let y_p denote data that are present and y_m data that are missing. Suppose that $y = (y_p, y_m)$ has a distribution $f(y_p, y_m | \theta)$ indexed by an unknown parameter θ. If the missing data are missing at random, then the likelihood of θ given data y_p is proportional to the density of y_p, obtained by integrating $f(y_p, y_m | \theta)$ over y_m:

$$L(\theta | y_p) \propto \int f(y_p, y_m | \theta) \, dy_m. \quad (3)$$

Likelihood inferences are based on $L(\theta | y_p)$. Occasionally in the literature, the missing values y_m are treated as fixed parameters, rather than integrated out of the distribution $f(y_p, y_m | \theta)$, and joint estimates of θ and y_m are obtained by maximizing $f(y_p, y_m | \theta)$ with respect to θ and y_m (e.g., Press and Scott [35] present a procedure that is essentially equivalent to this). This approach is not recommended since it can produce badly biased estimates which are not even consistent unless the fraction of missing data tends to zero as the sample size increases. Also, the model relating the missing and observed values of y is not fully exploited, and if the amount of missing data is substantial, the treatment of y_m as a set of parameters contradicts the general statistical principle of parsimony [29a].

An important generalization of (3) is to include in the model the distribution of a vector variable m, which indicates whether each value is observed or missing. The full distribution can be specified as

$$f(m, y_p, y_m | \theta, \phi)$$
$$= f(y_p, y_m | \theta) f(m | y_p, y_m, \phi), \quad (4)$$

where θ is the parameter of interest and ϕ relates to the mechanism leading to missing data. This extended formulation is necessary for nonrandomly missing data such as arise in censoring problems.

To illustrate (3) and (4), suppose that the hypothetical complete data $y = (y_1, \ldots, y_n)$ are a random sample of size n from the exponential distribution with mean θ. Then

$$f(y_p, y_m | \theta) = \theta^{-n} \exp(-t_n / \theta).$$

where $t_n = \sum_{i=1}^{n} y_i$ is the total of the n sampled observations. If $r < n$ observations are present and the remaining $n - r$ are missing, then the likelihood ignoring the response mechanism is proportional to the density

$$f(y_p | \theta) = \theta^{-r} \exp(-t_r / \theta), \quad (5)$$

regarded as a function of θ, where t_r is the total of the recorded observations.

Let $m = (m_1, \ldots, m_n)$, where $m_i = 1$ or 0 as y_i is recorded or missing, respectively, $r = \sum m_i$. We consider two models for the distribution of m given y. First, suppose that observations are independently recorded or missing with probability ϕ. Then

$$f(m | y, \phi) = \phi^r (1 - \phi)^{n-r}$$

and

$$f(y_p, m | \theta, \phi)$$
$$= \phi^r (1 - \phi)^{n-r} \theta^{-r} \exp(-t_r / \theta). \quad (6)$$

The likelihoods based on (5) and (6) differ by a factor $\phi^r (1 - \phi)^{n-r}$, which does not depend on θ, provided that θ and ϕ are distinct; i.e., their joint parameter space factorizes into a θ-space and a ϕ-space. Hence we can base inferences on (5), ignoring the response mechanism.

Suppose instead that the sample is censored, in that only values less than a known censoring point c are observed. Then

$$f(m | y, \phi) = \prod_{i=1}^{n} f(m_i | y_i),$$

$$f(m_i | y_i) = \begin{cases} 1 & \text{if } m_i = 1 \text{ and } y_i < c \\ & \text{or } m_i = 0 \text{ and } y_i > c, \\ 0 & \text{otherwise.} \end{cases}$$

The full likelihood is then proportional to

$f(y_p, m|\theta)$

$$= \prod_{i \,:\, m_i=1} f(y_i|\theta)f(m_i|y_i < c) \prod_{i \,:\, m_i=0} \Pr(y_i > c|\theta)$$

$$= \theta^{-r}\exp(-t_r/\theta)\exp[-(n-r)c/\theta]. \quad (7)$$

In this case the response mechanism is not ignorable, and the likelihoods based on (5) and (7) differ. In particular, the maximum likelihood estimate of θ based on (5) is t_r/r, the mean of the recorded observations, which is less than the correct maximum likelihood estimate of θ based on (7), namely $[t_r + (n-r)c]/r$. The latter estimate has the simple interpretation as the total time at risk for the uncensored and censored observations divided by the number of failures (r).

Special Data Patterns: Factoring the Likelihood

For certain special patterns of multivariate missing data, maximum likelihood estimation can be simplified by factoring the joint distribution in a way that simplifies the likelihood. Suppose, for example, that the data have the **monotone** or **nested** pattern in Fig. 1, where y_j represents a set of variables observed for the same set of observations and y_j is more observed than y_{j+1}, $j = 1, \ldots, J - 1$. *The joint distribution of $y_1, \ldots, y_J$ can be factored in the form*

$$f(y_1, \ldots, y_J|\theta) = f_1(y_1|\theta_1)f_2(y_2|y_1, \theta_2) \cdots$$
$$f_J(y_J|y_1, \ldots, y_{J-1}, \theta_J).$$

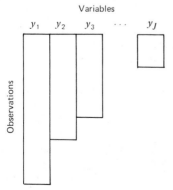

Figure 1 Schematic representation of a monotone (or nested) data pattern.

where f_j denotes the conditional distribution of y_j given $y_1, \ldots, y_{j-1}$, indexed by parameters θ_j. If the parameters $\theta_1, \ldots, \theta_J$ are distinct, then the likelihood of the data factors into distinct complete-data components, leading to simple maximum likelihood estimators for θ [4, 37]. Maximum likelihood estimation with more general patterns of incomplete data can be accomplished by the EM algorithm.

GENERAL DATA PATTERNS: THE EM ALGORITHM

The expectation–maximization (EM) algorithm [13] is an iterative method of maximum likelihood estimation that applies to any pattern of missing data. Let $l(\theta|y_p, y_m)$ denote the log-likelihood of parameters θ based on the hypothetical complete data (y_p, y_m). Let $\theta^{(i)}$ denote an estimate of θ after iteration i of the algorithm. The $(i + 1)$th iteration consists of an E-step and an M-step. The E-step consists of taking the expectation of $l(\theta|y_p, y_m)$ over the conditional distribution of y_m given y_p, evaluated at $\theta = \theta^{(i)}$. That is, the averaged log-likelihood

$$l^*(\theta|y_p, \theta^{(i)})$$
$$= \int l(\theta|y_p, y_m)f(y_m|y_p, \theta^{(i)})dy_m$$

is formed.

The M-step consists in finding $\theta^{(i+1)}$, the value of θ that maximizes l^*. This new estimate, $\theta^{(i+1)}$, then replaces $\theta^{(i)}$ at the next iteration. Each step of EM increases the log-likelihood of θ given y_p, $l(\theta|y_p)$. Under quite general conditions, the algorithm converges to a maximum value of the log-likelihood $l(\theta|y_p)$. In particular, if a unique finite maximum likelihood estimate of θ exists, the algorithm finds it.

An important case occurs when the complete data belong to a regular exponential family*. In this case, the E-step reduces to estimating the sufficient statistics corresponding to the natural parameters of the distribution. The M-step corresponds to maximum likelihood estimation from the hypothetical complete data, with the sufficient statistics replaced by the estimated sufficient statistics from the E-step.

The EM algorithm was first introduced for particular problems (e.g., Hartley [18] for counted data and Blight [7] for grouped or censored data). The regular exponential family case was presented by Sundberg [44]. Orchard and Woodbury [34] discussed the algorithm more generally, using the term "missing information principle" to describe the link with the complete-data log-likelihood. Dempster et al. [13] introduced the term EM, developed convergence properties, and provided a large body of examples. Recent applications include missing data in discriminant analysis [28] and regression with grouped or censored data [20].

The EM algorithm converges reliably, but it has slow convergence properties if the amount of information in the missing data is relatively large. Also, unlike methods like Newton–Raphson that need to calculate and invert an information matrix, EM does not provide asymptotic standard errors for the maximum likelihood estimates as output from the calculations. Its popularity derives from its link with maximum likelihood for complete data and its consequent usually simple computational form. The M-step often corresponds to a standard method of analysis for complete data and thus can be carried out with existing technology. The E-step often corresponds to imputing values for the missing data y_m, or more generally, for the sufficient statistics* that are functions of y_m and y_p, and as such relates maximum likelihood procedures to imputation methods. For example, the EM algorithm for multivariate normal data can be viewed as an iterative version of Buck's [9] method for imputing missing values [6].

Although the EM algorithm is a powerful tool for estimation from incomplete data, many problems remain. For example, nonnormal likelihoods occur more commonly with incomplete data than with complete data, and much remains to be learned about the appropriateness of many incomplete-data methods when applied to real data.

References

[1]   Afifi, A. A. and Elashoff, R. M. (1966). *J. Amer. Statist. Ass.*, **61**, 595–604.

[2]   Aitkin, M. and Clayton, D. (1980). *Appl. Statist.*, **19**, 156–163.

[3]   Anderson, R. L. (1946). *Biometrics*, **2**, 41–47.

[4]   Anderson, T. W. (1957). *J. Amer. Statist. Ass.*, **52**, 200–203.

[5]   Bartlett, M. S. (1937). *J. R. Statist. Soc. B*, **4**, 137–170.

[6]   Beale, E. M. L. and Little, R. J. A. (1975). *J. R. Statist. Soc. B*, **37**, 129–146.

[7]   Blight, B. J. N. (1970). *Biometrika*, **57**, 389–395.

[8]   Blumenthal, S., Dahiya, R. C., and Gross, A. S. (1978). *J. Amer. Statist. Ass.*, **73**, 182–187.

[9]   Buck, S. F. (1960). *J. R. Statist. Soc. B*, **22**, 302–306.

[10] Chen, T. and Fienberg, S. E. (1974). *Biometrics*, **30**, 629–642.

[11] Cox, D. R. (1972). *J. R. Statist. Soc. B*, **34**, 187–220.

[12] Darroch, J. N. and Ratcliff, D. (1972). *Ann. Math. Statist.*, **43**, 1470–1480.

[13] Dempster, A. P., Laird, N. M., and Rubin, D. B. (1977). *J. R. Statist. Soc. B*, **39**(1), 1–38.

[14] Ford, B. N. (1983). In *Incomplete Data and Sample Surveys, Vol. 2, Theory & Bibliographies*, D. B. Rubin, W. G. Madow, and I. Olkin, eds. Academic Press, New York (in press).

[15] Fuchs, C. (1982). *J. Amer. Statist. Ass.*, **77**, 270–278.

[16] Glasser, M. (1967). *J. Amer. Statist. Ass.*, **62**, 561–568.

[17] Greenlees, W. S., Reece, J. S., and Zieschang, K. D. (1982). *J. Amer. Statist. Ass.*, **77**, 251–261.

[18] Hartley, H. O. (1958). *Biometrics*, **14**, 174–194.

[19] Hartley, H. O. and Hocking, R. R. (1971). *Biometrics*, **27**, 783–808.

[20] Hasselblad, V., Stead, A. G., and Galke, W. (1980). *J. Amer. Statist. Ass.*, **75**, 771–779.

[21] Healy, M. and Westmacott, M. (1956). *Appl. Statist.*, **5**, 203–206.

[22] Heckman, J. D. (1976). *Ann. Econ. Social Meas.*, **5**, 475–492.

[23] Hocking, R. R. and Oxspring, H. H. (1974). *Biometrics*, **30**, 469–483.

[23a] Jones, R. H. (1980). *Technometrics*, **27**, 389–395.

[24] Kaplan, E. L. and Meier, P. (1958). *J. R. Statist. Ass.*, **53**, 457–481.

[25] Kulldorff, G. (1961). *Contributions to the Theory*

of Estimation from Grouped and Partially Grouped Samples. Almqvist & Wiksell, Stockholm/Wiley, New York.

[26] Laird, N. and Olivier, D. (1981). J. Amer. Statist. Ass., **76**, 231–240.

[27] Little, R. J. A. (1976). Biometrika, **63**, 593–604.

[28] Little, R. J. A. (1978). J. Amer. Statist. Ass., **73**, 319–322.

[29] Little, R. J. A. (1982). J. Amer. Statist. Ass., **77**, 237–250.

[29a] Little, R. J. A. and Rubin, D. B. (1983). American Statistician, **37**,

[30] Mann, N. R., Schafer, R. E., and Singpurwalla, N. D. (1974). Methods for Statistical Analysis of Reliability and Life Data. Wiley, New York.

[31] National Academy of Sciences (1982). Report of the Panel on Incomplete Data. National Academy of Sciences, Washington, D.C.

[32] Nelson, F. D. (1977). J. Econometrics, **6**, 581–592.

[33] Nie, N. H., Hull, C. H., Jenkins, J. G., Steinbrenner, K., and Bent, D. H. (1975). SPSS, 2nd ed., McGraw-Hill, New York.

[34] Orchard, T. and Woodbury, M. A. (1972). Proc. 6th Berkeley Symp. Math. Statist. Prob., Vol. 1. University of California Press, Berkeley, Calif., pp. 697–715.

[35] Press, S. J. and Scott, A. J. (1976). J. Amer. Statist. Ass., **71**, 366–369.

[36] Rubin, D. B. (1972). Appl. Statist., **21**, 136–141.

[37] Rubin, D. B. (1974). J. Amer. Statist. Ass., **69**, 467–474.

[38] Rubin, D. B. (1976). Biometrika, **63**, 581–592.

[39] Rubin, D. B. (1976). J. R. Statist. Soc. B, **38**, 270–274.

[40] Rubin, D. B. (1977). J. Amer. Statist. Ass., **72**, 538–543.

[41] Rubin, D. B. (1978). In Imputation and Editing of Faulty or Missing Survey Data, U.S. Social Security Administration and Bureau of Census, Washington, D.C., pp. 1–9.

[42] Rubin, D. B. (1980). Handling Nonresponse in Sample Surveys by Multiple Imputations. U.S. Department of Commerce, Bureau of the Census Monograph, Washington, D.C.

[43] Scheuren, F. (1983). In Incomplete Data and Sample Surveys, Vol. 2: Theory and Bibliographies, D. B. Rubin, W. G. Madow, and I. Olkin, eds. Academic Press, New York, (in press).

[43a] Shumway, R. H. (1984). In Proc. Symp. Time Series: Analysis of Irregularly Observed Data, E. Parzen, ed. Lecture Notes in Statistics, Springer-Verlag, New York.

[44] Sundberg, R. (1974). Scand. J. Statist., **1**, 49–58.

[45] Trawinski, I. M. and Bargmann, R. E. (1964). Ann. Math. Statist., **35**, 647–657.

[46] Tsokos, C. P. and Shimi, I. N. (1977). The Theory and Applications of Reliability. Academic Press, New York.

[47] Wilkinson, G. N. (1958). Biometrics, **14**, 360–384.

Acknowledgment

This research was sponsored by the U.S. Army under Contract DAAG29-80-C-0041.

(CENSORING
CONCOMITANT VARIABLES
LIKELIHOOD
LIKELIHOOD RATIO TESTS
SURVEY SAMPLING
SURVIVAL ANALYSIS)

RODERICK J. A. LITTLE
DONALD B. RUBIN

INCOMPLETE GAMMA FUNCTION (RATIO)

Integrals of the form

$$\Gamma_t(\lambda) = \int_0^t e^{-x} x^{\lambda-1} dx,$$

$$\lambda > 0; \quad 0 \leqslant t \leqslant \infty,$$

are known as the incomplete gamma function. The ratio $\Gamma_t(\lambda)/\Gamma_\infty(\lambda)$ is called the incomplete gamma function ratio. (The word "ratio" is often omitted.)

Tables of incomplete gamma function ratios were first compiled by Pearson [4] and extended by Harter [1]. More details are given in Johnson and Kotz [2, 3].

References

[1] Harter, H. L. (1964). New Tables of Incomplete Gamma Function Ratios. Report No. ARL-64-123, Aerospace Research Laboratory, Wright-Patterson Air Force Base, Ohio.

[2] Johnson, N. L. and Kotz, S. (1969). Distributions in Statistics: Discrete Distributions. Wiley, New York, Chap. 1.

[3] Johnson, N. L. and Kotz, S. (1970). Distributions in Statistics: Continuous Univariate Distributions, Vol. 1. Wiley, New York, Chap. 17.

[4] Pearson, K., ed. (1922). *Tables of the Incomplete* Γ-*Function*. H. M. Stationery Office, London, (since 1934: Cambridge University Press, Cambridge).

(CHI-SQUARE DISTRIBUTION
GAMMA DISTRIBUTION)

INCONSISTENT BIAS

The term was introduced by Fisher [2]. This is a bias of an estimator which does not tend to zero as the size of the sample increases. It occurs, in particular, in estimation problems related to stratified sampling*. See, e.g., Deming [1] for more details.

References

[1] Deming, W. E. (1960). *Sample Design in Business Research*. Wiley, New York.

[2] Fisher, R. A. (1922). *Philos. Trans. R. Soc.* **A222**, 309–368.

(ESTIMATION)

INDEPENDENT PROBABILITY *See* CONTINGENCY TABLES

INDEX NUMBERS

The subject of index numbers is one of the oldest in statistics, and one of the most topical in economics. Kendall [16] traces back early works on index numbers to Bishop Fleetwood's *Chronicon Preciosum*, first published in 1707 [7]; Ruggles [25] credits Pigou [23] with being the first to analyze the welfare implications of index numbers; and Maunder [21] lists 2600 entries up to 1968 in his bibliography of index numbers. Because of increased inflationary pressures and renewed interest in the measurement of inflation, the literature on index numbers has been growing exponentially in the past 10 years.

In its broadest sense, an index number is a measure of the magnitude of a variable at one point relative to its value at another point. Suppose that copies of a particular book of identical print and cover sold at a price of \$20 each in 1979, but sell at a price of \$22 in 1980, then the price index in 1980 relative to 1979 is 110% [(22/20) × 100], and thus the price of this book in 1980 is 10% higher than its price in 1979.

The price in 1979 is chosen as a *reference base*, also called *comparison base* or *base year*, and the year 1980 whose price is being compared with the reference base is called the *current year*, where "current" does not necessarily mean the present time. Instead of comparing prices at different dates, we might wish to compare prices in different cities, or prices paid by different groups of households; to a large extent the principle is the same.

THE INDEX NUMBER PROBLEM

In the example above the price of the book is directly observable; consequently, the resultant index number is a pure ratio that does not pose any special theoretical problems. *The index number problem arises when the magnitude of the variable under consideration is nonobservable.* This is the case, for example, when the variable to be compared is the general price level, or its reciprocal, the purchasing power of money. The price of a single or perfectly homogeneous commodity is measurable, but the price level of a complex of heterogeneous commodities is not. The various commodities used in the construction of a price index have no common physical unit, and their prices change in different proportions and different directions. More specifically, the index number problem in economics is how to combine the relative changes in the price or quantity variable of various commodities into a single measure that can be interpreted, for prices, as a measure of the relative change in the general price level, and, for quantities, as a measure of relative welfare. Index number theory is thus part, and as such the oldest part, of the general theory of aggregation*; see Theil [28] and Ijiri [12].

Although index numbers are used in diverse fields, this article will deal only with their use in economics. This is because the main applications of index numbers are in that field, and because almost all the relevant literature concentrates on the design, construction, and use of price and quantity index numbers specifically in economics. Examples of the use of index numbers in other fields such as environmental studies, demography*, technology, transport, and agriculture can be found, respectively, in Doll [3], Jazairi [14], Lave [19], Smith [26], and Kaneda [15].

Price and quantity index numbers are constructs based on both economic and statistical theory; they are designed and constructed by first formulating the concept to be measured on the basis of economic theory, and then by developing an estimator of that concept on the basis of statistical theory. The early history of index numbers focused on price indexes, and distinguished between weighted and unweighted means or, more generally, between the *stochastic* approach and the *aggregative* approach to price indexes. In its original and simplest form, the stochastic approach presupposes systematic and proportional changes in price ratios about their mean. The aggregative approach, in contrast, specifies from the outset the reference aggregate and the target population, e.g., the aggregate of consumer expenditure of the group of wage earners. The aggregative approach is now accepted as the correct approach to the construction of price indexes, although some developed forms of the stochastic approach, especially those based on versions of weighted least squares*, are also found in the current literature on index numbers.

NOTATION

Traditional price and quantity index numbers are limited to the goods and services that are actually purchased in the market. These indexes therefore do not extend to intertemporal comparisons. They also exclude time and leisure, financial assets, goods produced and consumed by the household without entering the market, and environmental variables such as clean air.

We start with a set of commodities $A = \{a_1, a_2, \ldots, a_n\}$ observed in k situations. It is assumed that these commodities are identical in the k situations, and are bought and sold in the same markets. p_{ij} is the price of commodity a_i in situation j, and q_{ij} is the corresponding quantity. The "price" is defined as the actual transaction price per unit, and the "quantity" as the physical quantity. However, in practice, various kinds of "prices" are used, including in particular the average value per unit called the *unit value*. Index numbers based on unit values present a problem in that the unit value of a nonhomogeneous group of commodities could change when the distribution of commodities within the group changes, even if the actual prices of the individual commodities remain unchanged.

The money value of a_i in situation r is $p_{ir}q_{ir}$, and the relative value or value share of a_i is $w_{ir} = p_{ir}q_{ir}/\sum_i p_{ir}q_{ir}$. The ratio of the price of a_i in situation s relative to its price in situation r (p_{is}/p_{ir}) is called *price relative*; similarly, the ratio (q_{is}/q_{ir}) is called *quantity relative*.

The value aggregates of n commodities in k situations are given by

$$\sum_{i=1}^{n} p_{ir}q_{is} \quad \text{for} \quad r, s = 1, 2, \ldots, k.$$

For $r = s$, the value aggregate $\sum_i p_{ir}q_{is}$ is directly observable, and it is called *direct* or *actual value*; for $r \neq s$, $\sum_i p_{ir}q_{is}$ is nonobservable, and called *cross value*.

For example, in the hypothetical data of Table 1, $n = 4$ and $k = 3$; the price relative of milk in 1979 compared with 1978 (p_{21}/p_{20}) is 1.1; the value share of bread in 1980 ($w_{12} = p_{12}q_{12}/\sum_i p_{i2}q_{i2}$) is 0.26; the actual value in 1978 ($\sum_i p_{i0}q_{i0}$) is 753 dollars; and the cross value of the quantities bought

Table 1

Commodity a_i	Unit	Transaction Price per Unit in Year j (dollars)			Quantities Purchased in Year j		
		1978: p_{i0}	1979: p_{i1}	1980: p_{i2}	1978: q_{i0}	1979: q_{i1}	1980: q_{i2}
Bread	Loaf	0.60	0.63	0.65	300	330	360
Milk	Quart	0.50	0.55	0.60	150	150	180
Eggs	Dozen	0.80	0.85	0.90	60	90	120
Beef	Pound	3.00	4.00	5.00	150	120	90

in 1978 valued at the prices of 1980 ($\sum_i p_{i2} q_{i0}$) is 1089 dollars.

P_{rs} and Q_{rs} are the price and quantity index numbers in the current situation s relative to the reference situation r, which is set equal to 100. The choice of the reference base obviously affects the value of the index, and therefore should be exercised with some care. The k situations compared could be time periods and a given group of consumers (or other units) in the same area, geographical areas and a given group of consumers at a given time, or consumer groups in a given area at a given time. In the special, but very important case when $k = 2$, the comparisons between situations r and s are called *binary comparisons*; for $k \geq 3$, they are *multilateral comparisons* produced through a *run* or *series* of index numbers.

STATISTICAL INDEX NUMBERS

The statistical theory of index numbers deals with the functional form, or aggregation function, which is commonly referred to as the choice of formula problem. It also deals with related practical estimation problems such as the classification, collection, and matching of data, and the compilation and dissemination of the indexes. These problems are discussed in ref. 31 and so will be passed over in this article, as will hypotheses testing since it is a rare exercise in the field of index numbers.

Two price index formulas used in the eighteenth century were of the simple forms

$$P_{rs}^* = \sum_i p_{is} / \sum_i p_{ir} \text{ and } P_{rs}^{**} = \frac{1}{n} \sum_i (p_{is}/p_{ir}).$$

The first of these two expressions is the ratio of the means of the prices, which is not invariant under a change in the unit of measurement; the second is the mean of the price ratios, which is still used in practice at least in averaging the price relatives of items within groups of commodities.

Later in the nineteenth century the geometric mean* of price relatives, $P_{rs}^{\dagger}$, where

$$\log P_{rs}^{\dagger} = \frac{1}{n} \sum_i \log(p_{is}/p_{ir}),$$

was proposed as an alternative to P_{rs}^{**}. More important, the idea of weighting appeared implicitly in the writings of Arthur Young [33], and was used explicitly by Joseph Lowe, considered by Sir Maurice Kendall as the father of modern index numbers. In 1822 Lowe [20] used a price index of the form

$$P_{rs}(q_i) = \sum_i p_{is} q_i / \sum_i p_{ir} q_i,$$

where q_i represents the weights applied to the prices. In order to appreciate the significance of Lowe's formula, one must remember that many of the index numbers commonly used are weighted averages of frequency distributions of price relatives and quantity relatives.

Following Lowe in the nineteenth century, Étienne Laspeyres and Hermann Paasche, both of Germany, used formulas similar to Lowe's, except that Laspeyres used $P_{rs}(q_r)$, where the weights are the quantities observed in the reference situation r; and Paasche used $P_{rs}(q_s)$, where the weights are the quantities observed in the current situation s; that is,

$$P_{rs}(q_r) = \sum_i p_{is}p_{ir} \bigg/ \sum_i p_{ir}q_{ir} \quad \text{and}$$

$$P_{rs}(q_s) = \sum_i p_{is}q_{is} \bigg/ \sum_i p_{ir}q_{is}.$$

The Laspeyres and Paasche quantity index numbers, $Q_{rs}(p_r)$ and $Q_{rs}(p_s)$, are analogously defined by

$$Q_{rs}(p_r) = \sum_i q_{is}p_{ir} \bigg/ \sum_i q_{ir}p_{ir} \quad \text{and}$$

$$Q_{rs}(p_s) = \sum_i q_{is}p_{is} \bigg/ \sum_i q_{ir}p_{is}.$$

These formulas, which appear to have been first labelled by Walsh [32] as the Laspeyres and Paasche formulas, are still widely used. They are popular because they afford a very simple but important interpretation. The Laspeyres price index, for example, represents the ratio of the cost of the reference situation "basket" of goods at the prices of the current situation, to its cost at the prices of the reference situation. In Table 1, the Laspeyres price index in 1979 relative to 1978 is 122.51%, and the Paasche price index is 120.13%. The corresponding quantity indexes are 93.63 and 91.81%.

For $k = 2$, the Laspeyres index is base-weighted, and the weighting base is the same as the reference base; the Paasche index is current-weighted. The Paasche price index, $P_{rs}(q_s)$, is the implied price index to match the corresponding Laspeyres quantity index, $Q_{rs}(p_r)$, since $P_{rs}(q_s)$ times $Q_{rs}(p_r)$ is equal to the value or money index $(\sum_i p_{is}q_{is}/\sum_i p_{ir}q_{ir})$. Similarly, the Paasche quantity index, $Q_{rs}(p_s)$, is the implied quantity index to match the Laspeyres price index, $P_{rs}(q_r)$. For $k \geqslant 3$, the run of Laspeyres indexes has fixed weights, which may or may not coincide with the reference base; and the run of Paasche indexes is defined to match the corresponding fixed-weighted Laspeyres indexes.

Many other weighting schemes are found in the literature. A price index formula suggested by Edgeworth, Marshall, and Bowley compares the price level in situation s relative to situation r, using the average quantities of the two situations as weights. This leads to the following formula:

$$P_{rs}(\bar{q}) = \sum_i p_{is}(q_{ir} + q_{is}) \bigg/ \sum_i p_{ir}(q_{ir} + q_{is}).$$

Another alternative is the Walsh index defined as a weighted geometric mean of the price relatives:

$$\prod_{i=1}^{n} (p_{is}/p_{ir})^{w_i},$$

where the weights w_i are defined in terms of the value shares in the k situations.

Test Approach

The aggregation of the price and quantity relatives over the sample of commodities may be required to satisfy certain conditions, such as the retention of particular characteristics of the individual price and quantity relatives under aggregation. Irving Fisher [5] discussed these characteristics in terms of certain "tests," which included in particular the time (or situation) reversal test $(P_{rs}P_{sr} = 1)$, the factor reversal test $(P_{rs}Q_{rs} = \sum_i p_{is}q_{is}/\sum_i p_{ir}q_{ir})$, and the circular test $(P_{rs}P_{st} = P_{rt})$. Fisher himself ultimately decided to abandon the circular test, and Frisch [9], Eichhorn [4], and others found that some of Fisher's tests were inconsistent. However, in his 1922 classic study of index numbers [5], Fisher used the tests criterion to compare the merits of various formulas. He concluded that the geometric mean of the Laspeyres and Paasche formulas, which was already known in the literature, is the "ideal" formula to be used in practice. This has since come to be known as the *Fisher ideal index*, and is still used extensively. (*See* FISHER'S IDEAL INDEX NUMBER.) Applied to Table 1, the Fisher ideal price index in 1979 relative to 1978 is 121.31%, and the corresponding Fisher quantity index is 92.72%.

Chain Index

Direct price comparison of period t relative to period s is of doubtful validity if the current period and the base period are far apart. In general, the more remote the base and current periods, the greater is the change in the commodities and consumer preferences, and the less valid the direct comparison between such two periods. Furthermore, price and quantity indexes for three or more periods are a set of binary comparisons between the current period t and the base period s, without reference to the course of prices and quantities between these two periods. To circumvent these difficulties, the *chain index* defines the price index $\bar{P}_{st}$ in terms of all the consecutive binary comparisons, that is:

$$\log \bar{P}_{st} = \sum_{k=s}^{t-1} \log P_{k,k+1} \quad \text{for} \quad s < t,$$

$$\log \bar{P}_{st} = 0 \quad \text{for} \quad s = t,$$

$$\log \bar{P}_{st} = -\log \bar{P}_{ts} \quad \text{for} \quad s > t.$$

The direct comparison index $P_{k,k+1}$ could, for example, be the Laspeyres base-weighted price index $P_{k,k+1}(q_k)$, hence $\bar{P}_{st}$ is the *chain Laspeyres index*; or it could be the Paasche current-weighted price index $P_{k,k+1}(q_{k+1})$, hence $\bar{P}_{st}$ is the *chain Paasche index*. Every chain index meets the circular test and the time reversal test regardless of the form of the direct comparison index. However, the chain index diverges from the corresponding direct comparison index unless the form of the direct comparison index meets the circular test. In Table 1, for example, the Laspeyres chain price index in 1980 relative to 1978 is 142.56% and the corresponding Laspeyres direct comparison index is 144.62%.

The chain index was originally suggested by Alfred Marshall in 1887, and was later given theoretical justification in terms of the *Divisia index*. In 1925 [2], François Divisia treated, in one of the most remarkable contributions to the pure theory of index numbers, the prices $p_i(t)$ and quantities $q_i(t)$ for all commodities i as continuous functions of time. He defined the continuous Divisia price index $P(t)$ for period t as

$$P(t) = P(0)\exp\left[\int_0^t \sum_i w_i(t) \frac{d \log p_i}{dt} dt\right],$$

and the corresponding Divisia quantity index $Q(t)$ as

$$Q(t) = Q(0)\exp\left[\int_0^t \sum_i w_i(t) \frac{d \log q_i}{dt} dt\right],$$

where $w_i(t)$ is defined as

$$w_i(t) = p_i(t)q_i(t) / \sum_i p_i(t)q_i(t)$$

and $P(0)$ and $Q(0)$ are the base period price and quantity indexes, which may be set equal to unity. (*See* DIVISIA INDICES.) The Divisia index, being line integral, is path dependent; the value of the index at time t relative to time s depends on the course of prices and quantities between these periods. This implies that the Divisia index does not necessarily return to its base period value even if the individual prices and quantities do. Richter [24] and Hulten [11] discuss conditions under which Divisia indexes are path independent.

In practice the chain index of Laspeyres and Paasche forms provides a discrete approximation to the Divisia index. Another form used is the Törnqvist [29] price index $P_{k,k+1}(\bar{w})$, defined by

$$\log P_{k,k+1}(\bar{w})$$
$$= \sum_i \tfrac{1}{2}(w_{ik} + w_{i,k+1})\log(p_{i,k+1}/p_{ik}),$$

where w_{ij} is the value share of the ith commodity in period j. (*See* LOG-CHANGE INDEX NUMBERS.) Errors arising from the discrete approximations to Divisia indexes have recently been analyzed by Trivedi [30], and the practical problems of constructing a chain index are discussed in Forsyth [8].

Best Linear Indexes

An alternative approach to the construction of price and quantity indexes from a complete set of data has been suggested by Theil [27]. Theil starts by assuming, in the spirit of the stochastic approach, that the prices of the n commodities move proportionally, except for random deviations. More generally, for n commodities and t time periods, we form the price matrix P of the order $t \times n$, and the quantity matrix Q, of the same order, and define the value matrix $\mathbf{V} = \mathbf{PQ'}$ of the order $t \times t$. The best linear price and quantity index vectors, $\mathbf{p}$ and $\mathbf{q}$, are obtained by minimizing the sum of squares of the elements of the discrepancy matrix $\mathbf{E} = \mathbf{V} - \mathbf{pq'}$. This minimization procedure establishes $\mathbf{p}$ and $\mathbf{q}$ as the characteristic vectors corresponding to the largest characteristic root of the matrices $\mathbf{VV'}$ and $\mathbf{V'V}$, respectively. For example, in Table 1,

$$\mathbf{p'} = (1 \quad 1.20 \quad 1.40) \quad \text{and}$$

$$\mathbf{q'} = (1 \quad 0.92 \quad 0.85),$$

where unity is the value of the index in the base year 1978. Further developments and applications of the best linear indexes may be found in Kloek and de Wit [17], Banerjee [1], Jazairi [13], and Fisk [6].

Geary [10] suggested another approach to the efficient use of all the relevant price and quantity information. His method has been especially recommended for and applied to international comparisons of real income and purchasing power parities. Thus let π_i be the average "international price" of commodity a_i, and let β_j be the purchasing power of the currency of country j expressed in terms of national currency units per unit of the currency of the base country. For n commodities and k countries, there are $(n + k)$ unknowns defined by the following $(n + k)$ linear homogeneous equations:

$$\pi_i = \sum_j (p_{ij}/\beta_j)\left(q_{ij}/\sum_j q_{ij}\right),$$
$$i = 1, 2, \ldots, n,$$

$$\beta_j = \sum_i \left(p_{ij}q_{ij}/\sum_i \pi_i q_{ij}\right),$$
$$j = 1, 2, \ldots, k,$$

where p_{ij} is the price of commodity i in the national currency units of country j and q_{ij} is the corresponding quantity in physical units. One of these equations is dropped because it is redundant, one of the β_j is set equal to unity as a reference base, and the

Table 2

| Country j | Commodity i | | | |
	p_{1j}	q_{1j}	p_{2j}	q_{2j}
1	10	8	10	2
2	12	4	10	4
3	15	6	25	4
4	20	10	30	8

remaining $(n + k - 1)$ equations for the $(n + k - 1)$ unknowns are solved by means of standard numerical methods for systems of linear homogeneous equations. Table 2 contains data for a hypothetical two-commodity four-country example. The corresponding solutions are

$$\pi_i = (21.638 \quad 27.952) \quad \text{and}$$

$$\beta_j = (0.437 \quad 0.444 \quad 0.786 \quad 1.000),$$

where β_4 has been set equal to unity as the base or "numéraire" country.

The aggregation functions that we have chosen to discuss here are only a few of the many found in the literature. To facilitate cross-reference, the main formulas discussed are summarized in Table 3. Definitions of the various symbols and formulas appearing there are contained in the text.

ECONOMIC INDEX NUMBERS

The object of the economic theory of index numbers is to provide economic interpretation and justification for the statistical index numbers of prices and quantities. In the context of binary comparisons of consumer expenditure over time, the economic theory of index numbers usually refers to the true *cost-of-living index**, and to the true *real-income index*, where "true" means exact. The cost of living index is sometimes referred to as the "constant utility price index." The index of real income is in effect an index of *real consumption*, as it is based on consumer expenditure only. The analysis is valid for a utility-maximizing individual with a fixed preference map.

Table 3   Some Alternative Index Numbers Formulas

Type	Price Index	Quantity Index
Arithmetic mean	$\dfrac{1}{n}\sum_i (p_{is}/p_{ir})$	$\dfrac{1}{n}\sum_i (q_{is}/q_{ir})$
Geometric mean	$\prod_{i=1}^{n} (p_{is}/p_{ir})^{1/n}$	$\prod_{i=1}^{n} (q_{is}/q_{ir})^{1/n}$
Lowe	$\sum_i p_{is}q_i / \sum_i p_{ir}q_i$	$\sum_i p_{is}p_i / \sum_i q_{ir}p_i$
Laspeyres	$\sum_i p_{is}q_{ir} / \sum_i p_{ir}q_{ir}$	$\sum_i q_{is}p_{ir} / \sum_i q_{ir}p_{ir}$
Paasche	$\sum_i p_{is}q_{is} / \sum_i p_{ir}q_{is}$	$\sum_i q_{is}p_{is} / \sum_i q_{ir}p_{is}$
Fisher	$\left[\left(\sum_i p_{is}q_{ir}/\sum_i p_{ir}q_{ir}\right)\left(\sum_i p_{is}q_{is}/\sum_i p_{ir}q_{is}\right)\right]^{1/2}$	$\left[\left(\sum_i q_{is}p_{ir}/\sum_i q_{ir}p_{ir}\right)\left(\sum_i q_{is}p_{is}/\sum_i q_{ir}p_{is}\right)\right]^{1/2}$
Edgeworth	$\sum_i p_{is}(q_{ir}+q_{is}) / \sum_i p_{ir}(q_{ir}+q_{is})$	$\sum_i q_{is}(p_{ir}+p_{is}) / \sum_i q_{ir}(p_{ir}+p_{is})$
Walsh	$\prod_{i=1}^{n}(p_{is}/p_{ir})^{w_i}$	$\prod_{i=1}^{n}(q_{is}/q_{ir})^{w_i}$
Chain	$\bar{P}_{st}=\prod_{k=s}^{t-1}P_{k,k+1}\quad \text{for } s<t$	$\bar{Q}_{st}=\prod_{k=s}^{t-1}Q_{k,k+1}\quad \text{for } s<t$
Divisia	$P(t)=P(0)\exp\left[\int_0^t \sum_i w_i(t)\dfrac{d\log p_i}{dt}dt\right]$	$Q(t)=Q(0)\exp\left[\int_0^t \sum_i w_i(t)\dfrac{d\log q_i}{dt}dt\right]$
Törnqvist	$\prod_{i=1}^{n}[p_{is}/p_{ir}]^{(w_{ir}+w_{is})/2}$	$\prod_{i=1}^{n}[q_{is}/q_{ir}]^{(w_{ir}+w_{is})/2}$
Best linear	$(\mathbf{VV'}-\lambda\mathbf{I})\mathbf{p}=0$	$(\mathbf{V'V}-\lambda\mathbf{I})\mathbf{q}=0$
Geary	$\pi_i=\sum_j (p_{ij}/\beta_j)\left(q_{ij}/\sum_j q_{ij}\right),\quad i=1,2,\ldots,n$	$Q_{kj}=\sum_i p_{ij}q_{ij}/\left\{\beta_j\sum_i p_{ik}q_{ik}\right\},\quad \text{or}$
(Geary-Khamis)	$\beta_j=\left(\sum_i p_{ij}q_{ij}/\sum_i \pi_i q_{ij}\right),\quad j=1,2,\ldots,k$	$\sum_i \pi_i q_{ij}/\sum_i \pi_i q_{ik}$ [†]

[†] Khamis [16a] points out that these two expressions are algebraically identical. Note that π_i and β_j are not price indices in the ordinary sense of the word, and that the price index P_{kj} can, if required, be obtained by dividing β_j by the corresponding official exchange rate.

The cost-of-living index measures the relative minimum costs of the same utility level in two-price situations. Suppose that the current prices refer to year 1 and the base prices refer to year 0; then a binary comparison of the prices in these two years provides two utility levels, u_1 and u_0, and two cost-of-living indexes, $P_{01}(u_0)$ and $P_{01}(u_1)$, defined by

$$P_{01}(u_0)=c(\mathbf{p}_1,u_0)/c(\mathbf{p}_0,u_0),$$
$$P_{01}(u_1)=c(\mathbf{p}_1,u_1)/c(\mathbf{p}_0,u_1),$$

where $c(\mathbf{p}_j,u_r)$ is a cost function increasing in both the price vector $\mathbf{p}_j$ and the scalar measure of welfare u_r, for $j,r=0,1$. The purchase of the base-year vector of goods $\mathbf{q}_0$ at the prices $\mathbf{p}_1$ is not necessarily the cheapest way to reach u_0; thus $\sum_i p_{i1}p_{i0}$ is greater than or equal to $c(\mathbf{p}_1,u_0)$, and hence the base-weighted Laspeyres price index $P_{01}(q_0)$ is greater than or equal to $P_{01}(u_0)$. Similarly, the purchase of the current year vector of the goods $\mathbf{q}_1$ at prices $\mathbf{p}_0$ is not necessarily the cheapest way to reach u_1; thus $\sum_i p_{i0}q_{i0}$ is greater than or equal to $c(\mathbf{p}_0,u_1)$, and hence the current-weighted Paasche price index $P_{01}(\mathbf{q}_1)$ is less than or equal to $P_{01}(u_1)$. The Laspeyres price index $P_{01}(q_0)$ provides an upper bound for the cost-of-living index $P_{01}(u_0)$, and the Paasche price index $P_{01}(q_1)$ provides a lower bound for $P_{01}(u_1)$. Thus a unique cost-of-living index does not exist, except in the special case where u_r can be somehow factored out of $c(\mathbf{p}_j,u_r)$, thereby obtaining a unique cost-of-living index, bounded by $P_{01}(q_0)$ from above and by $P_{01}(q_1)$ from below.

In contrast, the real-income index measures the relative minimum costs of two utility levels at the same price situation, i.e., the ratio of $c(\mathbf{p}_j, u_1)$ to $c(\mathbf{p}_j, u_0)$, which will be denoted by $K_{01}(\mathbf{p}_j)$. This index can be obtained by dividing the value or expenditure index $(\sum_i p_{i1} q_{i1} / \sum_i p_{i0} q_{i0})$ by the true cost-of-living index. The real-income index and the cost-of-living index are thus a matching pair which multiply to the expenditure index. In practice, $P_{01}(u_0)$ is approximated by $P_{01}(q_0)$ and $P_{01}(u_1)$ by $P_{01}(q_1)$. Similarly, $K_{01}(\mathbf{p}_0)$ is approximated by $Q_{01}(\mathbf{p}_0)$ and $K_{01}(\mathbf{p}_1)$ by $Q_{01}(\mathbf{p}_1)$. If the Laspeyres price index is used, the corresponding implied real-income index is the Paasche quantity index; the two multiply to the expenditure index, and since $P_{01}(q_0)$ is an upper bound for $P_{01}(U_0)$, then $Q_{01}(p_1)$ is a lower bound for $K_{01}(\mathbf{p}_1)$. Similarly if the Paasche price index $P_{01}(q_1)$ is used to approximate $P_{01}(U_1)$, the corresponding implied real-income index is the Laspeyres quantity index $Q_{01}(p_0)$, which provides an upper bound for $K_{01}(\mathbf{p}_0)$.

References

[1] Banerjee, K. S. (1963). *Econometrica*, **31**, 712–718. (Intermediate; few references.)

[2] Divisia, F. (1925). *Rev. Écon. Polit.*, **39**(4), 842–861; *ibid.*, (5), 980–1008; *ibid.*, (6), 1121–1151.

[3] Doll, J. P. (1967). *J. Farm Econ.*, **49**, 79–88. (Elementary; references.)

[4] Eichhorn, W. (1976). *Econometrica*, **44**, 247–256. (Intermediate; few references.)

[5] Fisher, I. (1922). *The Making of Index Numbers*, Houghton Mifflin, Boston. (Classic.)

[6] Fisk, P. R. (1977). *J. R. Statist. Soc. A*, **140**, 217–231. (Mathematical; few references.)

[7] Fleetwood, W. (1707). *Chronicon Preciosum*. London. (Of purely historical interest.)

[8] Forsyth, F. G. (1978). *J. R. Statist. Soc. A*, **141**, 348–358. (Elementary; few references.)

[9] Frisch, R. (1930). *J. Amer. Statist. Ass.*, **25**, 397–406. (Intermediate.)

[10] Geary, R. C. (1958). *J. R. Statist. Soc. A*, **121**, 97–99. (Intermediate.)

[11] Hulten, C. R. (1973). *Econometrica*, **41**, 1017–1025. (Mathematical; few references.)

[12] Ijiri, Y. (1971). *J. Amer. Statist. Ass.*, **66**, 766–782. (Intermediate; extensive references.)

[13] Jazairi, N. T. (1971). *Bull. Oxford Inst. Econ. Statist.*, **33**, 181–195. (Intermediate; few references.)

[14] Jazairi, N. T. (1976). *Approaches to the Development of Health Indicators*, Organization for Economic Cooperation and Development, Paris. (Also available in French under the title *Différents Approches pour l'Élaboration d'Indicateurs de Santé*. Elementary; extensive references.)

[15] Kaneda, H. (1967). *J. Farm Econ.*, **49**, 199–212. (Elementary; references.)

[16] Kendall, M. G. (1969). *Int. Statist. Rev.*, **37**, 1–12. (Elementary; references.)

[16a] Khamis, S. H. (1971). *J. R. Statist. Soc. A*, **135**, 96–121.

[17] Kloek, T. and de Witt, G. M. (1961). *Econometrica*, **29**, 602–616. (Intermediate.)

[18] Laspeyres, E. (1864). *Jahrb. Nationaloekon. Statist.* **3**, 81–118. (Of purely historical interest.)

[19] Lave, L. B. (1966). *Technological Change: Its Conception and Measurement*. Prentice-Hall, Englewood Cliffs, N.J. (Intermediate; extensive references.)

[20] Lowe, J. (1822). *The Present State of England in Regard to Agriculture, Trade and Finance*. London. (Of purely historical interest.)

[21] Maunder, W. F., ed. (1970). *Bibliography of Index Numbers*. Athlone Press, London. (Exhaustive from 1707 up to the first half of 1968.)

[22] Paasche, H. (1874). *Jahrb. Nationaloekon. Statist.* **23**, 168–178. (Of purely historical interest.)

[23] Pigou, A. C. (1920). *The Economics of Welfare*. Macmillan, London.

[24] Richter, M. K. (1966). *Econometrica*, **34**, 739–755. (Mathematical; few references.)

[25] Ruggles, R. (1967). In *Ten Economic Studies in the Tradition of Irving Fisher*, W. J. Fellner et al., eds. Wiley, New York, pp. 171–205. (Elementary; useful survey; international comparisons; references.)

[26] Smith, R. T. (1954). *J. Amer. Statist. Ass.*, **49**, 227–239. (Elementary; few references.)

[27] Theil, H. (1960). *Econometrica*, **28**, 464–480. (Intermediate; few references.)

[28] Theil, H. (1962). In *Logic, Methodology and Philosophy of Science*, E. Nagel et al., eds. Stanford University Press, Stanford, Calif., pp. 507–527. (Intermediate; references.)

[29] Törnqvist, L. (1936). *Bank Finland Monthly Bull.*, **10**, 1–8.

[30] Trivedi, P. K. (1981). *Int. Econ. Rev.*, **22**, 71–77. (Intermediate; few references.)

[31] United Nations (1977). Guidelines on Principles of a System of Price and Quantity Statistics. *Sta-*

tist. Papers, Ser. M, No. 59, New York. (Elementary; useful discussion of practical problems.)

[32] Walsh, C. M. (1901). *The Measurement of General Exchange Value.* Macmillan, New York.

[33] Young, A. (1812). *An Enquiry into the Progressive Value of Money in England as Marked by the Price of Agricultural Products.* London. (Of purely historical interest.)

Bibliography

See the following works, as well as the references just given, for more information on the topic of index numbers.

Afriat, S. N. (1977). *The Price Index.* Cambridge University Press, Cambridge. (Intermediate; theory and applications; references.)

Allen, R. G. D. (1975). *Index Numbers in Theory and Practice.* Macmillan, London. (Excellent introduction both for eager beginner and expert; extensive references.)

Banerjee, K. S. (1975). *Cost of Living Index Numbers.* Marcel Dekker, New York. (Elementary; consumer price index; extensive references.)

Deaton, A. (1979). *Rev. Econ. Stud.,* **46**, 391–405. (Intermediate; references.)

Deaton, A. and Muellbauer, J. (1980). *Economics and Consumer Behaviour.* Cambridge University Press, Cambridge. (Intermediate; economic theories of index numbers; extensive references.)

Diewert, W. E. (1976). *J. Econometrics,* **4**, 115–145. (Intermediate; economic theory; mathematical; few references.)

Divisia, F. (1926). *Rev. Écon. Polit.,* **40**(1), 49–81.

Edgeworth, F. Y. (1925). *Econ. J.,* **35**, 379–388.

Fisher, F. M. and Shell, K. (1972). *The Economic Theory of Price Indices.* Academic Press, New York. (Intermediate; economic theory; mathematical; few references.)

Frisch, R. (1936). *Econometrica,* **4**, 1–38. (An exceptionally fine survey; references.)

Griliches, Z., ed. (1961). *Price Indexes and Quality Change.* Harvard University Press, Cambridge, Mass. (Intermediate; collection of important contributions; extensive references.)

Keynes, J. M. (1930). *A Treatise on Money,* Vol. 1. Harcourt Brace, New York.

Konus, A. A. (1939). *Econometrica,* **7**, 10–29; first published in Russian in 1924. (Economic theory.)

Lau, L. (1979). *Rev. Econ. Statist.,* **61**, 73–82. (Intermediate; economic theory; few references.)

Mills, F. C. (1955). *Statistical Methods,* 3rd ed. Holt, Rinehart and Winston, New York, pp. 426–511. (Elementary; references.)

Mudgett, B. D. (1951). *Index Numbers.* Wiley, New York. (Elementary; references.)

Pollak, R. A. (1975). *Int. Econ. Rev.,* **16**, 135–150. (Intermediate; few references.)

Ruist, E. (1968). Index numbers; theoretical aspects. In *International Encyclopedia of the Social Sciences,* Vol. 7. Free Press, New York. Reprinted with a postscript in *International Encyclopedia of Statistics,* Vol. 1, W. H. Kruskal, and J. M. Tanur, eds., Free Press, New York, 1978, pp. 451–456. (Intermediate; few references.)

Samuelson, P. A. and Swamy, S. (1974). *Amer. Econ. Rev.,* **64**, 566–593. (Intermediate; few references.)

Sen, A. K. (1979). *J. Econ. Lit.,* **17**, 1–45. (Elementary; economic theory; critical survey; extensive references.)

Stigler, G. J., Chairman (1961). *The Price Statistics of the Federal Government,* General Series No. 73. National Bureau of Economic Research, New York. (Review and appraisal of U.S. price indexes together with 12 papers on special topics; references.)

Usher, D. (1968). *The Price Mechanism and the Meaning of National Income Statistics.* Oxford University Press, London. (Intermediate; theory and applications; international comparisons.)

Vartia, Y. O. (1976). *Relative Changes and Index Numbers,* Ser. A4, The Research Institute of the Finnish Economy, Helsinki. (Intermediate; extensive references.)

von Hofsten, E. (1952). *Price Indexes and Quality Changes,* George Allen and Unwin, London. (Elementary; references.)

Yule, G. W. and Kendall, M. G. (1968). *An Introduction to the Theory of Statistics,* 14th ed. Hafner, New York, pp. 590–609. (Elementary.)

(CONSUMER PRICE INDEX
DIVISIA INDICES
FISHER'S IDEAL INDEX NUMBER
GROSS NATIONAL PRODUCT DEFLATOR
HEDONIC INDEX NUMBERS
INDEX OF INDUSTRIAL PRODUCTION
WHOLESALE PRICE INDEX)

NURI T. JAZAIRI

INDEX OF DISPERSION

This term is often used for the statistic

$$\sum_{i=1}^{k} \frac{(r_i - n/k)^2}{n/k} = \frac{\sum_{i=1}^{k}(r_i - \bar{r})^2}{\bar{r}}$$

where $r_1, r_2, \ldots, r_k$ are Poisson variables with parameters $\theta_1, \theta_2, \ldots, \theta_k$, $n = \sum_{i=1}^{k} r_i$ and $\bar{r} = n/k = \sum r_i/k$. It is used for testing the hypothesis that all the θ's are equal by comparing its value with the appropriate upper percentage points of a chi-squared* variable with $k - 1$ degrees of freedom. A similar index can be constructed for testing the equality of parameters of k bionomial distributions. For more details see, e.g., Lindley [1].

The index of dispersion is widely used for detection of spatial patterns. If n circular quadrats are independently and randomly placed in a given area and a count made of the number of plants in each, then, under H_0 of randomly distributed plants, X is Poisson and $V(X) = E(X)$. For alternatives involving patches ("clumping") of plants $V(X) > E(X)$ while for regularly spaced plants $V(X) < E(X)$. The power of the index of dispersion test for spatial alternatives to randomness has been studied by several authors. Perry and Mead [2] summarize many of these results and study the power for the case of mosaic alternatives to randomness. *See also* DIVERSITY, MEASURES OF.

References

[1] Lindley, D. V. (1965). *Introduction to Probability and Statistics*, Part 2: *Inference*. Cambridge University Press, Cambridge.

[2] Perry, J. N. and Mead, R. (1979). *Biometrics*, **35**, 613–622.

(BINOMIAL DISTRIBUTION
CHI-SQUARE DISTRIBUTION
DIVERSITY, MEASURES OF
HYPOTHESIS TESTING
POISSON DISTRIBUTION)

INDEX OF FIT

Index of fit is an alternative term for the sample correlation coefficient* used in simple regression analysis*.

INDEX OF INDUSTRIAL PRODUCTION

The index of industrial production is a quantity index number which measures the relative change in the output of the industrial sector of the economy in two situations, usually time periods. The industries covered by the index are typically mining, manufacturing, and gas, electricity, and water. In some countries construction is also included, but industries such as agriculture, transport, finance, and other services are excluded. The index is computed for the total industrial sector, for each industry covered, and for individual parts of these industries, such as machinery and furniture in manufacturing.

Although the index of industrial production is a subindex of the total index of national output, it has historically been distinct because of its special sources and methods, its use as a general indicator of industrial activity and cyclical fluctuations, and its usefulness as a barometer of national power.

THE CONCEPT OF PRODUCTION

The concept of "production" which the index measures refers to the contributions of the industries covered. An industry's contribution to industrial production is its net output or "value added," which is defined as the value of gross output less the value of all inputs of materials and services. The value-added concept is an accounting concept which does not correspond to any observable flow of goods and services, and which could be negative. But the occurrence of negative value added is so rare it is assumed away in practice.

THE VALUE-ADDED INDEX

Suppose that the total number of industries in the economy is $(k + m)$, k being the number of the industries covered by the index, and m the number of those industries excluded from the index; and that each of the k industries of the index produces n_r outputs and uses m_r inputs ($r = 1, 2, \ldots, k$). Then the value-added Laspeyres-type index of industrial production in period t relative to period 0 is given by

$$\frac{\sum_{r=1}^{k}(\sum_{i=1}^{n_r} P_{i0} Q_{it} - \sum_{j=1}^{m_r} p_{j0} q_{jt})}{\sum_{r=1}^{k}(\sum_{i=1}^{n_r} P_{i0} Q_{i0} - \sum_{j=1}^{m_r} p_{j0} q_{j0})}, \quad (1)$$

where capital P and Q denote the price and quantity of gross output, and small p and q denote the price and quantity of input. In this expression the quantities of outputs and inputs of the current period and the base period are valued at their prices in the base period. If, instead, these quantities were valued at their prices in the current period, we obtain a Paasche-type index of value added.

Now suppose that input–output tables are available for the whole economy, and, without loss of generality, assume that each industry is producing only one product, then the Laspeyres-type index of value added in period t relative to period 0 is given by

$$\frac{\mathbf{p}_0'(\mathbf{I} - \mathbf{A}_t)\mathbf{q}_t}{\mathbf{p}_0'(\mathbf{I} - \mathbf{A}_0)\mathbf{q}_0}. \quad (2)$$

In this expression $\mathbf{q}$ is $k \times 1$ vector of the quantities produced by the k industries of the index; $\mathbf{A}$ is $(k + m) \times k$ matrix of input–output coefficients, m being the number of the industries excluded from the index; $\mathbf{I}$ is an identity matrix; and $\mathbf{p}$ is $(k + m) \times 1$ vector of the prices of the final products of all the $(k + m)$ industries in the economy. A Paasche-type index of value added can also be obtained from the input–ouput tables by using the prices of the current period as weights.

LASPEYRES–PAASCHE DIVERGENCE

The divergence between Laspeyres and Paasche indexes is known in the literature of the index of industrial production as the "Gerschenkron effect." Following Gerschenkron [3], this asserts that an index of industrial production using early period prices as weights (typically Laspeyres index) will show a larger rate of increase of output than that using later-period prices (typically Paasche index). If we denote Laspeyres and Paasche quantity indexes by Q_L and Q_P, we

can write

$$\frac{Q_P}{Q_L} = 1 + \left(\frac{Q_P - Q_L}{Q_L}\right). \quad (3)$$

The quantity in parentheses is the divergence between Laspeyres and Paasche indexes. This quantity can be expressed in terms of the weighted coefficient of correlation, ρ, between the price relatives and quantity relatives, and their weighted coefficients of variation, $(\sigma_{\hat{p}}/P_L)$ and $(\sigma_{\hat{q}}/Q_L)$, where P_L is the Laspeyres price index. The weighted variances and covariance of the commodity price and quantity relatives, $\hat{p}$ and $\hat{q}$, are given by

$$\sigma_{\hat{p}}^2 = \sum_i w_{i0}\left(\frac{p_{i1}}{p_{i0}} - P_L\right) \bigg/ \sum_i w_{i0},$$

$$\sigma_{\hat{q}}^2 = \sum_i w_{i0}\left(\frac{q_{i1}}{q_{i0}} - Q_L\right) \bigg/ \sum_i w_{i0},$$

$$\sigma_{\hat{p}\hat{q}} = \sum_i w_{i0}\left(\frac{p_{i1}}{p_{i0}} - P_L\right)\left(\frac{q_{i1}}{q_{i0}} - Q_L\right) \bigg/ \sum_i w_{i0},$$

where $w_{i0} = p_{i0}q_{i0}$, and p_{i0} and q_{i0} are the commodity price and quantity. The weighted correlation is given by

$$\rho = \frac{\sigma_{\hat{p}\hat{q}}}{\sigma_{\hat{p}}\sigma_{\hat{q}}},$$

which upon substitution for $\sigma_{\hat{p}\hat{q}}, \sigma_{\hat{p}}, \sigma_{\hat{q}}$ and rearrangement can be written as

$$\rho = \frac{P_L}{\sigma_{\hat{p}}} \frac{Q_L}{\sigma_{\hat{q}}}\left(\frac{Q_P}{Q_L} - 1\right)$$

or

$$\frac{Q_P}{Q_L} = 1 + \rho \frac{\sigma_{\hat{p}}}{P_L} \frac{\sigma_{\hat{q}}}{Q_L}. \quad (4)$$

Therefore, $Q_P \gtreqless Q_L$ if $\rho \gtreqless 0$. The details of this analysis and an economic interpretation in terms of supply and demand forces can be found in Allen [1] and Ames and Carlson [2].

APPROXIMATE INDEXES OF VALUE ADDED

Direct estimation of a value-added quantity index as, say, in (1) may not be possible in practice because of data constraints. If an approximate Paasche-type price index of value added is available, the numerator of (1) is approximately equal to the current value added deflated by the complementary Paasche-type price index. Instead, an approximate value-added quantity index may be obtained by the method of "double deflation." In this method the first term in the numerator of (1) is approximated by deflating the current value of gross output by an appropriate Paasche price index of gross output, and the second by deflating the current value of input by an appropriate Paasche price index of input.

Alternatively, a Laspeyres-type quantity index of value added can be approximated by extrapolating the base year value added with some proxy indicator of the relative change in net output. In this approach the base year is usually the year for which census value-added data are available. The general form of this approximate index is

$$\frac{\sum_{r=1}^{k} w_{r0} Q_{rt}^*}{\sum_{r=1}^{k} w_{r0}}, \qquad (5)$$

where

$$w_{r0} = \sum_{i=1}^{n_r} P_{i0} Q_{i0} - \sum_{j=1}^{m_r} p_{j0} q_{j0},$$

and Q_{rt}^* is the proxy indicator of the change in the net output of the rth industry in period t relative to period 0. For example, the proxy indicator of the relative change in the net output of the rth industry may be a Laspeyres quantity index of gross output, that is,

$$Q_{rt}^* = \frac{\sum_{i=1}^{n_r} Q_{i1} P_{i0}}{\sum_{i=1}^{n_r} Q_{i0} P_{i0}}, \qquad (6)$$

or a Laspeyres quantity index of the input of materials, that is,

$$Q_{rt}^* = \frac{\sum_{i=1}^{m_r} q_{i1} P_{i0}}{\sum_{i=1}^{m_r} q_{i0} P_{i0}}. \qquad (7)$$

In general, the practical problem usually evolves around the search for suitable proxy indicator. If input–output relationships remain unchanged between the two periods, then a quantity index of gross output, or of input of materials or labor, can be used as a proxy indicator of the change in net output.

COMPILATION PRACTICES AND PRODUCTIVITY INDEXES

Indexes of industrial production, such as those of the United States and the United Kingdom, which are described in great detail in refs. 4 and 5, respectively, are computed monthly from sample data and adjusted periodically on the basis of benchmark industrial census data. The total index is a weighted average of individual indexes of various industries and products. These individual indexes are based on output data and on input data of materials and services. The published indexes are classified into categories and subcategories by industry groupings, such as manufacturing, machinery, and chemicals, and by market or final-use groupings, such as consumer goods, food, and meat products; and adjusted for seasonal variation and trend movement in production levels.

The total and individual indexes of industrial production are used in measuring productivity in the industrial sector. Productivity indexes are obtained as the ratio of net output to one input or to total inputs. For example, a labor productivity index for the manufacturing industry is obtained by dividing an index of labor employed in this industry into the manufacturing production index.

References

[1] Allen, R. G. D. (1963). *Int. Statist. Rev.*, **31**, 281–301.

[2] Ames, E. and Carlson, J. A. (1968). *Oxford Econ. Papers*, **20**, 24–37.

[3] Gerschenkron, A. (1951). *A Dollar Index of Soviet Machinery Output, 1927–28 to 1937*. Rand Corporation, Santa Monica, Calif.

[4] *Industrial Production Measurement in the United States: Concepts, Uses, and Compilation Practices* (1964). Board of Governors of the Federal Reserve System, Washington, D.C.

[5] *The Measurement of Changes in Production* (1976). Central Statistical Office, London. (Describes British practice.)

BIBLIOGRAPHY

Arrow, K. J. (1974). In *Nations and Households in Economic Growth*, P. A. David and M. W. Reder, eds. Academic Press, New York, pp. 3–19. (Few references.)

Christensen, L. R. and Cummings, D. (1981). *Eur. Econ. Rev.*, **16**, 61–94. (International comparison of productivity levels; elementary; references.)

Domar, E. D. (1967). *Quart. J. Econ.*, **81**, 169–187. (Compares various methods; elementary, references.)

Geary, R. C. (1944). *J. R. Statist. Soc. A*, **107**, 251–259. (Concepts and methods; few references.)

Karmel, P. H. and Polasek, M. (1971). *Applied Statistics for Economists*. Pitman, London. (Textbook; contains good discussion of proxy indicators.)

Kendrick, J. W. and Vaccara, N., eds. (1980). *New Developments in Productivity Measurement and Analysis*. University of Chicago Press, Chicago. (Collection of papers on concepts, methods, and applications; elementary; extensive references.)

Maunder, W. F., ed. (1970). *Bibliography of Index Numbers*. Athlone Press, London. (Exhaustive from 1707 up to the first half of 1968.)

Sato, K. (1976). *Rev. Econ. Statist.*, **58**, 434–442. (Concept and measurement of real value added; economic theory; intermediate; few references.)

United Nations (1979). Manual on National Accounts at Constant Prices. *Statist. Papers, Ser. M, No. 64*, New York. (Elementary; few references.)

(CONSUMER PRICE INDEX
GROSS NATIONAL PRODUCT DEFLATOR
INDEX NUMBERS)

NURI T. JAZAIRI

INDIAN STATISTICAL INSTITUTE

HISTORY

Early Years

The Indian Statistical Institute (ISI) was founded in 1931 by Prasanta Chandra Mahalanobis*, who was then (and until his retirement) Professor of Physics at the Presidency College, Calcutta.

During its early years, until about 1947, the Institute derived its income largely from statistical projects. From its inception, most of the statistical research in the Institute originated in direct response to demands from the side of applications. Thus, to give only two examples, the technique of sequential random sampling was conceived for the first time in estimating the acreage and yield of jute in the province of Bengal in 1937. The need for a statistical analysis of certain anthropometric measurements on Anglo-Indians (i.e., persons of mixed British and Indian parentage) eventually led to the formulation of the D^2-statistic, known as the Mahalanobis distance* and later widely used in problems of classification*. Problems of this sort inspired pioneering research in multivariate analysis*, design of experiments*, sample surveys*, and (under the impetus of the first of many visits to the ISI by R. A. Fisher*) statistical inference carried out by the first group of workers who had joined the ISI as faculty or students. This group included S. S. Bose, R. C. Bose, K. R. Nair, S. N. Roy, and, beginning in the early 1940s, C. R. Rao.

1947–1960

With the advent of independence, the ISI entered upon a phase of rapid development which was soon to give it a unique character among institutions of learning in the world. It was as a result of the interaction between two great and remarkable men, Mahalanobis and Nehru, that the scientific and statistical resources of the Institute were placed at the disposal of the country to help in the task of national development. In 1955 the ISI was entrusted by the Prime Minister with initiating studies in economic planning and specifically with the preparation of the draft of the Second Five-Year Plan. Indeed, Nehru's influence on the Institute in the decade of the 1950s was so profound and his encouragement so crucial that he played a very important part in setting the course of the Institute's future development and giving it its present shape.

A division of Economics and Social Sciences was started in Calcutta and in 1954 a Center of the ISI was opened in Delhi to coordinate the Institute's activities with those of the government's Planning Commission. The National Sample Survey (NSS) was established in 1950 for the collection of socioeconomic data, through (continuing) sample surveys covering the entire country and to furnish reliable information on indicators (such as national income) needed by the government for purposes of national planning. The ISI had technical charge of the NSS until the early 1970s when it was taken over by the government. A new division of Statistical Quality Control (SQC) was started in 1953 to provide consultancy services to industry. Eventually, SQC servicing units were set up in 10 major industrial centers.

All these activities resulted in an enormous expansion in the staff of the ISI at the technical as well as nontechnical levels. The original nucleus of the Institute, whose principal concern was research and teaching in statistics, came to be known as the Research and Training Section, which, from the early 1950s onward, was headed by C. R. Rao. The Institute at this time also became known for its liberal academic atmosphere. Because of this, many of the best young scholars in the country were attracted to the ISI even though several among them were interested more in mathematics or economics than in statistics.

THE ISI AS A STATISTICAL UNIVERSITY

This phase of development culminated in an act of parliament, passed in 1959, which declared the ISI to be an "institution of national importance" and empowered it to assume the functions of a university in the field of statistics. At the same time the Institute continued to retain its status as an autonomous institution. Since 1960, the ISI has offered regular courses leading to a three-year Bachelor's degree (B. Stat.) and a two-year Master's degree (M. Stat.) in statistics in addition to awarding the Ph.D. and D.Sc. degrees. Independently of these degree programs, there are numerous diploma courses in SQC, computer science, and other areas.

The teaching curriculum, particularly at the Bachelor's level, emphasizes the applica-

tions of statistics as much as theory and reflects a philosophy of statistics (which Mahalanobis shared with R. A. Fisher and J. B. S. Haldane) that regards statistics (in Fisher's phrase) as a "key technology" of the century.

The Institute also runs the International Statistical Education Center (ISEC), which was founded in 1950 through the joint efforts of the ISI, the International Statistical Institute, UNESCO, and the government of India. About 25 to 30 statistical officers from countries of Asia, Africa, and Latin America come annually for training.

THE ISI AND P. C. MAHALANOBIS

The development of the Indian Statistical Institute, the tradition of free scientific inquiry which it has nurtured over the years, and the view of statistics that underlies its manifold activities—all these are intimately connected with the life and career of its founder. A brief article on P. C. Mahalanobis appears in Volume 5.

THE ISI TODAY: ITS ADMINISTRATIVE SETUP AND ACADEMIC ACTIVITIES

Until his death in 1972, Mahalanobis was secretary and director of the Institute. C. R. Rao held these positions from 1972 until 1976, when a new constitution was adopted under which the two posts were merged into that of director. G. Kallianpur was the first director under the new constitution, from 1976 until 1979. The present holder of that office is B. P. Adhikari.

The affairs of the Institute are managed by a 25-member council presided over by the chairman of the ISI. The director is a member of the council. Besides elected members from within the Institute, the council includes representatives from the government and from other major scientific institutions in the country. P. N. Haksar has been chairman of the ISI since 1973. The general body of the Institute, consisting of all subscribing members, has overall control over the ISI. It is presided over by the president of the ISI, a position held by Subimal Dutt since 1976.

The ISI is now organized into the following divisions:

Theoretical Statistics and Mathematics

Social Sciences and Economics

Physical and Earth Sciences

Biological Sciences

Statistical Quality Control and Operations Research

Library, Documentation, and Information Science

RESEARCH ACTIVITIES

Theoretical Statistics

Statistics and those disciplines of mathematics related to probability theory have been the traditional subjects in which research has been carried out since the early days of the Institute. Research areas in recent years include inference* (parametric and nonparametric), asymptotic theory, sequential analysis*, multivariate analysis, linear models, ergodic theory*, stochastic integrals*, stochastic differential equations*, and topics in mathematical physics.

Mathematics

Set theory, measure theory*, combinatorics* and graph theory*, functional analysis, algebraic topology, and harmonic analysis.

Applied Statistics

Sample surveys, survey methodology, and computer software.

Other major fields in which research has been done since the mid-1950s or the early 1960s are economics and sociology* (perspective planning, economics of agriculture, demography, linguistic studies, psychometry), electronics and geology*, biological sciences (anthropology and human genetics*), and statistical quality control*.

Publications

The Indian Statistical Institute brings out the journal *Sankhyā, the Indian Journal of Statistics*. It was founded in 1933 by P. C. Mahalanobis, who was its editor until his death. The present chief editors are C. R. Rao and G. Kallianpur. The editorial offices of *Sankhyā* are located at the main campus of the Institute in Calcutta. K. B. Goswami is the present editorial secretary. The Institute also publishes, from time to time, a series of monographs in statistics and related fields.

The Indian Statistical Institute today is a many-faceted organization. Besides functioning as a university and as a center for advanced research in statistics, it is:

1. A semiofficial body whose advice on statistical and economic matters is sought by the government

2. A corporation which undertakes large-scale projects

3. A consultative body for both public sector and private sector industry

4. A meeting place for scientists, economists, and other scholars from all parts of the world

The ISI has three main campuses: Calcutta, which is the headquarters; Delhi; and Bangalore (opened in 1978). The addresses of the campuses are as follows:

The Indian Statistical Institute
203, Barrackpore Trunk Road
Calcutta, India 700035

(also the address of *Sankhyā*)

The Indian Statistical Institute
7, S.J.S. Sansanwal Marg,
New Delhi, India 110016

The Indian Statistical Institute
31 Church Street
Bangalore, India 560001

G. Kallianpur

INDIFFERENCE ZONE   *See* LEAST FA-
VORABLE CONFIGURATION

INDIRECT LEAST SQUARES   *See*
ECONOMETRICS

INDIVIDUAL MOVING-AVERAGE INEQUALITY

An inequality on autocorrelations in a moving-average* process (MA process). Consider an MA(q) process

$$z_i = \sum_{j=0}^{q} \theta_j u_{i-j}, \qquad \theta_0 = 1,$$

where u_{i-j} are uncorrelated normally distributed random variables with mean zero and
the same variance σ_u^2 and the θ_j are real
numbers. The autocorrelations of the pro-
cess are given by

$$\rho_k = \sum_{j=0}^{q-k} \theta_j \theta_{j+k} \Big/ \sum_{j=0}^{q} \theta_j^2 \quad (\rho_0 = 1);$$

$$k = 0, 1, \ldots, q.$$

The individual moving-average inequality
asserts that for any MA(q) process

$|\rho_1| \leqslant \cos(\pi/(q+2))$ and

$$|\rho_k| \leqslant \cos\left(\frac{\pi}{[q/k] + 2} \right), \qquad k = 1, \ldots, q.$$

(See, e.g., Granger [4], Anderson [1], Anderson [2], and Davies et al. [3].)

A comprehensive discussion, simple derivations of these inequalities and their extension for the case of a purely seasonal MA(q)
process

$$Z_i = \sum_{j=0}^{Q} \theta_{j,k} u_{i-jk}, \qquad \theta_{0,k} = 1$$

where $Q = [q/k]$, is given in Anderson [2].

References

[1] Anderson, O. D. (1974). *J. Econometrics*, **2**, 189–193.

[2] Anderson, O. D (1978). *Metrika*, **25**, 241–245.

[3] Davies, N., Pate, M. B., and Frost, M. G. (1974). *Biometrika*, **61**, 199–200.

[4] Granger, C. W. J. (1972). Time series modelling and interpretation. Eur. Econometrics Congr., Budapest, 1972.

(AUTOREGRESSIVE-INTEGRATED
  MOVING AVERAGE MODELS
MOVING AVERAGES
TIME SERIES)

INDUCED ORDER STATISTICS   *See*
MULTIVARIATE ORDER STATISTICS

INDUCTIVE LOGIC   *See* LOGIC IN STATIS-
TICAL REASONING

INDUSTRY, STATISTICS IN

Statisticians working in industry have strong
interests in business, science, problem solving, statistics, and computing, and most enjoy working with people. They come from a
wide variety of backgrounds, including all
fields of science, engineering, and mathematics and have an understanding of the logic
of the scientific method. They also find it
helpful to have some background in the
subject-matter area in which they work.

The solution of most problems in industry
involves the collection, analysis, and interpretation of data. Such a situation is a natural one for statisticians, and they get involved in projects at a number of points.
The statistician is often asked to analyze
data that have already been collected; however, the statistician is more effective when
involved at the beginning of a study. Working with a single investigator or a multidisci-
plinary team, the statistician assists with
problem formulation and the development
of an appropriate strategy for collecting the
necessary data. While the data are being
collected, the statistician will monitor the
program to see that the plan is being followed. It is not uncommon for the statistician to be asked to perform interim analyses
or help to revise the program because unexpected difficulties have arisen.

Once the data have been collected, the
investigator(s) and the statistician combine

their skills in the analysis and interpretation of results. At the conclusion of a study the statistician usually prepares a written report which describes the solution process (background of the problem, experimental design, and statistical analysis) and discusses the results. Because of the cooperative nature of problem solving, it is very important that the statistician have good oral and written communication skills.

AREAS OF APPLICATION

Statisticians in industry typically work with people from many different disciplines on a wide variety of problems. It is not unusual for an industrial statistician to be involved in such diverse business areas as research and development, manufacturing, finance, and marketing, all within a short time. The problem areas might include product quality control, process optimization, sales forecasting, business decision theory, consumer preference, product toxicity, mathematical programming*, risk analysis, government product certification testing, market research, inventory control, computing systems, environmental impact, and chemical reaction kinetics. The following examples are typical of many problems on which statisticians in industry are asked to provide guidance.

1. How should the process be operated to produce a good product at minimum cost?
2. Is a new drug safe and effective?
3. Is a particular advertising claim justified?
4. How large an inventory of a particular part is needed?
5. Does the air quality around a manufacturing facility meet environmental standards set by the Environmental Protection Agency?
6. Should a particular product be recalled because of a manufacturing defect?
7. What will next year's sales be?
8. Which of a number of alternative product designs is most appealing to consumers?
9. What is the average life of a light bulb?
10. What sampling plan should be used to ensure that at least 99% of the released product is within product specification?
11. What combinations of ingredients and processing conditions result in a plastic part with maximum strength?
12. How much redundancy is required to ensure that a spaceship guidance system will have a 0.9999 probability of successful operation?

Statisticians must get the investigator in each instance to help them develop an understanding of the technical background of the problem so that they can recommend appropriate statistical methods and communicate effectively.

STATISTICAL METHODS USED

Statisticians use a wide variety of statistical tools because they encounter so many different types of problems in their work. When determining how a process should be operated to produce a good product at minimum cost, the statistician will help the process engineer with the design of experiments* to determine the nature and magnitude of the effects of process operating variables on product properties. Regression analysis* is often used to analyze the data generated by these experiments and to develop prediction equations which are used to find the settings of process operating variables that will produce a product with desired characteristics.

After the process has been operating successfully at the selected set of conditions, a process variation study may be conducted, producing data which will be analyzed by analysis-of-variance* techniques to determine the importance of the various factors that contribute to the random variation in the product properties. This information will be used to develop statistical quality control* procedures to ensure that the process produces the desired product.

In the study of the safety and effectiveness of a new drug, the statistician will help the team of medical researchers determine how many patients should receive the test drug and the placebo and the random order in which the patients should be assigned to the two groups. The statistician will work with computer* scientists to design the data management system that will be used to store and retrieve the data as the study progresses. Data from this kind of a study may be analyzed by a number of statistical methods, including Student's t-test* and a variety of nonparametric* statistical methods. The results of the statistician's work in this instance becomes an integral part of the report which the pharmaceutical firm submits to the Food and Drug Administration to obtain approval to market the new drug.

In both the industrial process and new drug studies the statistician and the investigators make extensive use of graphical* displays to diagnose the characteristics of the data, to determine what statistical analyses should be performed and to aid in their interpretation, and to participate in the presentation of the results to company management and other interested scientists.

Other statistical methods used in industry include qualitative/categorical* data analysis techniques, time-series* analysis, nonlinear estimation*, multivariate* analysis, simulation*, reliability* and life data* analysis, numerical* analysis, and survey sampling*. Examples of the use of statistics in industry can be found in journals such as the *Journal of Quality Technology**, *Technometrics**, and *Applied Statistics**, and in texts on statistical quality control* (e.g., Ott [23]) and the design of experiments (e.g., Box et al. [5]).

HISTORICAL PERSPECTIVE

The history of statistics in industry reveals a steady growth resulting from continued identification of new applications and the development of important innovations in statistical concepts and methods for areas such as quality control, design of experiments, and

data analysis. Shewhart [26] founded the field of statistical quality control* (SQC) in the 1920s. He showed that statistical methods could be used in the study of industrial processes to identify assignable causes for changes in product quality, to develop models for random variation, and to control processes. Shewhart's methods and acceptance sampling inspection procedures, as developed by Dodge and Romig [14] and others (see Dodge [13] and Schilling [25]), were used extensively by industry and the military during World War II and continue to be used by industry throughout the world, most notably by the Japanese. This activity resulted in the founding of the American Society for Quality Control (ASQC) in 1946. ASQC's journals—*Industrial Quality Control*, *Journal of Quality Technology*, and *Quality Progress*—contain many articles on the use of statistics in industry.

Design of experiments began to be used in industry in the 1940s [6, 11] and developed rapidly following the publication of the seminal paper on response surface methodology by Box and Wilson [4] and the books by Bennett and Franklin [1] and Davies [12]. Industry found many application areas for design of experiments, including the evolutionary operation* approach to process improvement [2, 3] and design of mixture experiments* [9, 24, 27, 28].

The advent of the digital computer in the late 1950s and early 1960s made regression analysis a practical data analysis tool. This resulted in its wide use by industry, innovative statistical research by industry [16, 17, 19, 20], and the appearance of several books (e.g., Draper and Smith [15] and Daniel and Wood [10]). Regression analysis is perhaps the most widely used statistical tool in industry.

During the early years of statistics in industry there were few university programs in statistics and few textbooks and journal articles on the subject. The majority of those working in the field learned statistics on the job, having been trained in other disciplines, such as mathematics, chemistry, physics, biology, and engineering. There was a critical need for statisticians in industry to get together to learn from each other's experiences. Central to this activity was the ASQC* Annual Technical Conference, the Fall Technical Conference of the ASQC Chemical Division, and the Gordon Research Conference on Statistics in Chemistry and Chemical Engineering. The meetings of the American Statistical Association* and Biometric Society* also included sessions on statistics in industry.

The ASQC Chemical Division had a very active education program which presented many short courses on the use of statistics in industry, emphasizing basic statistics, statistical quality control, and topics in design of experiments such as factorial and fractional factorial designs, response surface methodology*, and evolutionary operation. These activities by industry and university statisticians working on industrial applications resulted in the founding of *Technometrics* in 1959, a joint effort of the ASQC and the American Statistical Association. The ASQC Chemical Division's short course on applied regression analysis resulted in the first textbook on this subject [15].

Other historical notes are contained in CONSULTING, STATISTICAL.

CAREERS IN INDUSTRY

Statisticians are employed in virtually all industries. Some examples are: petroleum, chemical, electronics, drug, communications, food, steel, automotive, and textiles. The size of these companies ranges from the large corporation which may employ several statisticians to the small company with a single statistician. Many statisticians in industry operate as "entrepreneurs" and run their own statistics consulting business within the company that employs them. This opportunity brings with it the added responsibilities of identifying problems, arranging the necessary funding, and convincing the organization to implement the results [21], [22].

The careers of many statisticians include more than one of the major areas of statisti-

cal activity: teaching, applied research, theoretical research, and applications. Statisticians in industry are no exception. Although the major effect is on problem solving, many statisticians conduct training courses within their companies and do research to develop better statistical methods. It is not uncommon for statisticians in industry to teach in the evening programs of local colleges and universities.

Statisticians in industry are an active group and interact with many people outside their companies. They have a need to keep up with new developments in statistical methodology, which can be accomplished by being active in professional societies and attending and presenting papers at meetings of statistical societies. Also, because of their exposure to a diversity of scientific and business issues, statisticians in industry will sometimes be asked to serve as industry or company representatives. They are often well qualified to serve in this capacity because of their knowledge of the facts and their skills in communicating this knowledge to others.

The rewards of the industrial statistician are many. The pay is comparable with that of other scientists and engineers in industry. Most companies have several different technical ranks and pay levels for statisticians which provide paths of progress and reflect different levels of competence. Job satisfaction is equally important. One may enjoy working on any given problem because he or she enjoys working with the people involved in the study; the statistical methodology being used is of interest; the problem is being closely watched by management; or the scientific area is of major interest. The work is always challenging. Few problems are routine, and almost every one differs from those that appear in textbooks. Thus statisticians in industry must use their knowledge, experience, and imagination to create useful problem solutions within time and funding constraints. Further information on statistics in industry can be found in The Report of the American Statistical Association Committee on Training of Statisticians for Industry

[7, 8] and in the papers of Greenfield [18] and Marquardt [21, 22].

References

[1] Bennett, C. A. and Franklin, N. L. (1954). *Statistical Analysis in Chemistry and the Chemical Industry*. Wiley, New York.

[2] Box, G. E. P. (1957). *Appl. Statist.*, **6**, 3–23.

[3] Box, G. E. P. and Draper, N. R. (1969). *Evolutionary Operation: A Statistical Method for Process Improvement*. Wiley, New York.

[4] Box, G. E. P. and Wilson, K. B. (1951). *J. R. Statist. Soc. B*, **13**, 1–45.

[5] Box, G. E. P., Hunter, W. G., and Hunter, J. S. (1978). *Statistics for Experimenters*. Wiley-Interscience, New York.

[6] Brownlee, K. A. (1949). *Industrial Experimentation*. Chemical Publishing Company, Brooklyn, N.Y.

[7] Committee on Training of Statisticians for Industry (1980). *Amer. Statist.*, **34**, 65–80.

[8] Committee on Training of Statisticians for Industry (1981). *Amer. Statist.*, **35**, 269. (Need for training in statistical quality control is discussed.)

[9] Cornell, J. A. (1981). *Experiments with Mixtures: Designs, Models and the Analysis of Mixture Data*. Wiley-Interscience, New York.

[10] Daniel, C. and Wood, F. S. (1971). *Fitting Equations to Data*. Wiley-Interscience, New York (2nd ed., 1980).

[11] Davies, O. L., ed. (1947). *Statistical Methods in Research and Production*. Oliver & Boyd, Edinburgh (4th ed., O. L. Davies and P. L. Goldsmith, eds., Hafner, New York, 1972).

[12] Davies, O. L., ed. (1954). *The Design and Analysis of Industrial Experiments*. Hafner, New York (2nd ed., Longman, London, 1978).

[13] Dodge, H. F. (1969). *J. Quality Tech.*, **1**, 77–88, 155–162, 225–232; *ibid.*, **2**, 1–8.

[14] Dodge, H. F. and Romig, H. G. (1944). *Sampling Inspection Tables*. Wiley, New York (2nd ed., 1959).

[15] Draper, N. R. and Smith, H., Jr. (1966). *Applied Regression Analysis*. Wiley, New York (2nd ed., 1981).

[16] Efroymson, M. A. (1962). In *Mathematical Methods for Digital Computers*, A. Ralston and H. S. Wilf, eds. Wiley, New York, Chap. 17. (Stepwise regression algorithm is detailed and discussed.)

[17] Gorman, J. W. and Toman, R. J. (1966). *Technometrics*, **8**, 27–51.

[18] Greenfield, A. A. (1979). *Statistician*, **28**, 71–82.

[19] Hoerl, A. E. and Kennard, R. W. (1970). *Technometrics*, **12**, 55–68.

[20] Marquardt, D. W. (1963). *J. Soc. Ind. Appl. Math.*, **2**, 431–441.

[21] Marquardt, D. W. (1979). *Amer. Statist.*, **33**, 102–107.

[22] Marquardt, D. W. (1981). *Amer. Statist.*, **35**, 216–219.

[23] Ott, E. R. (1975). *Process Quality Control: Trouble-shooting and Interpretation of Data*. McGraw-Hill, New York.

[24] Scheffé, H. (1958). *J. R. Statist. Soc. B*, **20**, 344–360.

[25] Schilling, E. G. (1982). *Acceptance Sampling in Quality Control*. Marcel Dekker, New York.

[26] Shewhart, W. A. (1931). *Economic Control of Quality of Manufactured Product*. 50th Anniversary Commemorative Reissue 1981. American Society for Quality Control, Milwaukee, Wis.

[27] Snee, R. D. (1971). *J. Quality Tech.*, **3**, 159–169.

[28] Snee, R. D. (1981). *Technometrics*, **23**, 119–130.

Acknowledgments

The author is pleased to acknowledge the helpful comments and suggestions of W. J. Hill, L. S. Nelson, S. M. Free, Jr., J. W. Gorman, E. G. Schilling, W. S. Cleveland, G. J. Hahn, J. D. Hromi, and the editors, which improved the presentation of this article.

(ACCEPTANCE SAMPLING
CONSULTING, STATISTICAL
DESIGN OF EXPERIMENTS
EDITING STATISTICAL DATA
ENVIRONMENTAL STATISTICS
FDA STATISTICS
INSPECTION SAMPLING
MARKETING STATISTICS
MIXTURE EXPERIMENTS
OPERATIONS RESEARCH, STATISTICS IN
PRINCIPLES OF PROFESSIONAL
    STATISTICAL PRACTICE
QUALITY CONTROL, STATISTICS IN
SAMPLING)

RONALD D. SNEE

INEQUALITIES FOR EXPECTED SAMPLE SIZES

Since the objective of many sequential procedures is to attain efficies lacking in other statistical procedures (sequential or not), much interest centers around the values of expected sample sizes, or average sample numbers (ASNs)* as they are commonly called. Only rarely can the precise value of an ASN be obtained through analytic means, and one must be content with what can be learned through the use of computer simulations, analytic approximations, and inequalities.

Below two types of inequalities are discussed, called here types U and L. Type U are simply those that describe an upper bound to an ASN, and type L, a lower bound; occasionally, they occur simultaneously within the same expression. Type U inequalities usually apply to a *specific* procedure (or class of procedures), while type L usually apply to *all* procedures (or to all procedures within a large class) and thus express fundamental barriers to improvement. Type L inequalities serve the same kind of role that the well-known Cramér–Rao inequality* serves; it provides a "target" toward which designers of statistical procedures can aim. As with the Cramér–Rao inequality, a type L inequality can yield a lower bound that is too small and nowhere near attainable. On the other hand, when one succeeds in producing a sequential procedure whose ASN is close to the lower bound, the procedure is strongly recommended, and so is the lower bound. Also in common with the Cramér–Rao inequality, type L inequalities are usually based, directly or indirectly, on considerations of information.

One simple type L inequality is just a restatement of Wolfowitz's [14] sequential version of the Cramér–Rao inequality:

$$E_\theta N \geqslant \left[I(\theta)\sigma^2 \right]^{-1} \qquad (1)$$

It asserts that an unbiased sequential estimator of a real parameter θ can have as small a variance as σ^2 only if the ASN is as large as that indicated, where N denotes the sample size and $I(\theta)$ denotes the Fisher information* in a single observation. [This reasoning is slightly flawed because the derivation of the Cramér–Rao inequality in a sequential setting is valid only for a large class of unbiased estimators, and violations of (1)

are possible. Nevertheless, it has been shown that (1) must hold for almost every θ. See ref. 11.]

Another simple type L inequality is easily derived (using Jensen's inequality*) from equation (26) of ref. 10: $EN \geqslant a^2\sigma^2/d^2$. It asserts that no statistical procedure (of the kind discussed) can have a smaller expected sample size than that indicated while providing a confidence interval for the normal mean of fixed width $2d$ and with coverage probability $2\Phi(a) - 1$, where Φ denotes the standard normal distribution function. Type U inequalities, appearing in FIXED-WIDTH AND BOUNDED-LENGTH CONFIDENCE INTERVALS, give evidence that the lower bound $a^2\sigma^2/d^2$ is nearly attainable. Numerical evidence for this has been provided as well by Starr [12].

A well-known (and probably the first) type L inequality is due to Wald [13]:

$$E_1 N \geqslant \frac{\alpha_1 \log(\alpha_1/(1 - \alpha_2)) + (1 - \alpha_1) \log((1 - \alpha_1)/\alpha_2)}{\int f_1 [\log(f_1/f_2)] \, d\mu}.$$

$$(2)$$

Here one is concerned with testing whether the common density f (with respect to a σ-finite measure μ) of independent observations equals f_1 or f_2. The inequality asserts that any sequential test with error probabilities as small as α_1 and α_2 requires an ASN $E_1 N$ as large as that indicated when $f = f_1$. A similar bound holds when $f = f_2$. The integral in the denominator of (2) is a Kullback–Leibler information* number. Hoeffding [5] has pointed out that the numerator is also interpretable as information.

The quality of Wald's inequality for $E_1 N$ (and for its analog for $E_2 N$) is attested to by the fact that equality is occasionally attainable in (2) with a sequential probability ratio test* (SPRT). A related fact, which always holds, is that the (simultaneously type U and type L) inequalities

$$E_1 N \geqslant E_1 N^*, \qquad E_2 N \geqslant E_2 N^* \qquad (3)$$

are valid when N^* denotes the sample size of an SPRT with error probabilities α_1 and α_2. (*See* SEQUENTIAL PROBABILITY RATIO TEST.) It follows, of course, that the lower bound $E_1 N^*$ is at least as good (i.e., as large) as the lower bound provided by the right side of (2).

There is a practical need to control the ASN when the model is wrong, i.e., when f neither equals f_1 nor f_2. When $f = f_0$, a third value, the ASN can be quite large, particularly for an SPRT designed to test between f_1 and f_2. Hoeffding [4, 5] has obtained a variety of type L inequalities involving $E_0 N$. The most widely used one appears to be

$$E_0 N \geqslant \zeta^{-2}\left\{\left[(\tau/4)^2 - \zeta \log(\alpha_1 + \alpha_2)\right]^{1/2}\right.$$

$$\left. - \tau/4\right\}^2, \qquad (4)$$

where

$$\zeta = \max(\zeta_1, \zeta_2),$$

$$\zeta_i = \int f_0 \left[\log(f_0/f_i)\right] d\mu, \qquad i = 1, 2,$$

and

$$\tau^2 = \int \left[\log(f_2/f_1) - \zeta_1 + \zeta_2\right]^2 f_0 \, d\mu.$$

Anderson [1] used this inequality to show that his proposed modification of the SPRT performs well in that it produces an ASN for a (troublesome) third density f_0 (within a parameterized family) that is about 3 or 4% larger than the lower bound provided by (4). Since Anderson's formula for $E_0 N$ is not exact, a precise comparison is impossible. Subsequent work by Lorden [8] has yielded similar results with greater accuracy. (See also Hall [3].)

Simons [9] has described a variety of type L inequalities suitable for $k \geqslant 2$ hypotheses, and applied them in the setting of $k = 3$. These represent extensions of Wald's and Hoeffding's bounds for $k = 2$.

There are examples of type U inequalities which could probably just as well be expressed as asymptotic approximations. A particularly nice example of this sort has been given by Lorden [7]:

$$E_\theta N \leqslant n(\theta) + B, \qquad \theta \in (\underline{\theta}, \bar{\theta}). \quad (5)$$

Here, N represents the sample size of a sequential likelihood ratio test for choosing between composite hypotheses, $n(\theta)$ represents the best possible ASN at θ (minimizing at θ), and B is a constant that depends on the error probabilities, but not on θ. (Information concerning the size of B, and how it depends on the error probabilities, can be found in the paper.)

Finally, it should be mentioned that there are risk and Bayes risk analogs of type U and type L equalities (with N replaced by a loss). Type L analogs (for a sequential setting) have been described by Hoeffding [5] and by Ibragimov and Haśminskiĭ [6]; an example of a type U analog can be found in Chernoff's [2, p. 70] work.

References

[1] Anderson, T. W. (1960). *Ann. Math. Statist.*, **31**, 165–197.

[2] Chernoff, H. (1972). *Sequential Analysis and Optimal Design*. Regional Conference Series in Applied Mathematics No. 8. SIAM, Philadelphia.

[3] Hall, J. (1980). *Asymptotic Theory of Statistical Tests and Estimation*. (In honor of Wassily Hoeffding.) Academic Press, New York, pp. 325–350. [Discusses a class of sequential tests which are suggested by Hoeffding's derivation of inequality (4).]

[4] Hoeffding, W. (1953). *Ann. Math. Statist.*, **24**, 127–130.

[5] Hoeffding, W. (1960). *Ann. Math. Statist.*, **31**, 352–368.

[6] Ibragimov, I. A. and Haśminskiĭ, R. (1972). *Selec. Transl. Prob. Statist.*, **13**, 821–824. American Mathematical Society, Providence, R.I.

[7] Lorden, G. (1972). *Ann. Math. Statist.*, **43**, 1412–1427.

[8] Lorden, G. (1976). *Ann. Statist.*, **4**, 281–291.

[9] Simons, G. (1967). *Ann. Math. Statist.*, **38**, 1343–1363.

[10] Simons, G. (1968). *Ann. Math. Statist.*, **39**, 1946–1952.

[11] Simons, G. and Woodroofe, M. (1983). "The Cramér–Rao Inequality Holds Almost Everywhere." *Recent Advances in Statistics: Papers in Honor of Herman Chernoff's Sixtieth Birthday*, M. H. Rizvi, J. S. Rustagi and D. Siegmund, eds. Academic Press, New York.

[12] Starr, N. (1966). *Ann. Math. Statist.*, **37**, 36–50.

[13] Wald, A. (1945). *Ann. Math. Statist.*, **16**, 117–186.

[14] Wolfowitz, J. (1947). *Ann. Math. Statist.*, **18**, 215–230.

(AVERAGE SAMPLE NUMBER (ASN)
FIXED-WIDTH AND BOUNDED LENGTH
    CONFIDENCE INTERVALS
HOEFFDING'S INEQUALITIES
SEQUENTIAL ANALYSIS
SEQUENTIAL PROBABILITY RATIO TEST)

GORDON SIMONS

INEQUALITIES ON DISTRIBUTIONS: BIVARIATE AND MULTIVARIATE

Inequalities have played a central and lasting role in probability theory as well as in mathematics in general. In mathematics, the subject of inequalities has centuries-old roots and many prominent researchers have contributed to its growth. The classic book by Hardy et al. [20], first published in 1934, provides a remarkable compendium of mathematical inequalities. Although the development of probability inequalities is intertwined with the mathematical development, there have been separate major influences as a result of the special needs of probability theory. For instance, one inequality common to both mathematics and probability is the famous Cauchy–Schwarz inequality*, which in its probabilistic form states that $\mathrm{cov}(X, Y) \leqslant (\mathrm{var}\, X)^{1/2}(\mathrm{var}\, Y)^{1/2}$. On the other hand, Chebyshev's* important inequality germinated in the context of probability theory, being developed in order to approximate probabilities of complex events.

The origins of probability inequalities for multivariate distributions are not new (e.g., Boole's inequality and also the Cauchy–Schwarz inequality). However, the dramatic growth of this subject area has taken place in the last half of this century. This growth parallels the major growth of multivariate analysis* itself during this period. Recently, there have been some attempts to impose structure on the area of multivariate probability inequalities. Two notable efforts are the monographs of Marshall and Olkin [29] and Tong [44]. See also the review paper of Eaton [15]. The multivariate inequalities presented here are divided roughly into three groups: inequalities among random variables, stochastic comparisons between random vectors, and moment inequalities. Applications to reliability theory*, simultaneous inference*, and unbiased testing* are discussed.

POSITIVE AND NEGATIVE DEPENDENCE

Various inequalities arise when the entries of a random vector $\mathbf{X} = (X_1, \ldots, X_p)$ are positively dependent. [*See* DEPENDENCE, CONCEPTS OF for a discussion in the bivariate ($p = 2$) case.] For example, there are many circumstances where for all $x_1, \ldots, x_p$,

$$\Pr[X_1 > x_1, \ldots, X_p > x_p]$$
$$\geqslant \prod_{i=1}^{p} \Pr[X_i > x_i]. \qquad (1)$$

This is a type of positive dependence called positive upper orthant dependence (PUOD). A variant, positive lower orthant dependence (PLOD), is obtained by replacing all ">" by "⩽" in (1). In the bivariate case, considered by Lehmann [28], PUOD and PLOD are equivalent; a discussion of the more general case is given by Tong [44]. A different positive dependence concept that implies both PUOD and PLOD is association (*see* DEPENDENCE, CONCEPTS OF for the definition). A simple checkable condition which implies association is called TP$_2$ in pairs. See Block and Ting [9] for a review of this and other positive dependence concepts, their relations, and references.

A concept of negative dependence, negative upper orthant dependent (NUOD), is obtained if "⩾" is replaced by "⩽" in (1). For example, the multinomial* distribution is NUOD. Various other distributions which have the same structure as a multinomial (i.e., essentially $\sum_{i=1}^{p} X_i$ being constant) are also NUOD (see Block et al. [11]). Other concepts of negative dependence are discussed by Karlin and Rinott [21] and Ebrahimi and Ghosh [17].

Concepts derived from (1) can be used to partially order, according to degree of positive dependence, random vectors whose one-dimensional marginal distributions agree. The random vector $\mathbf{X}$ is more PUOD than $\mathbf{Y}$ if

$$\Pr\left[X_1 > t_1, \ldots, X_p > t_p \right]$$
$$\geqslant \Pr\left[Y_1 > t_1, \ldots, Y_p > t_p \right]$$
$$\text{for all } t_1, \ldots, t_p.$$

If "$>$" is replaced by "$\leqslant$" in the above, then $\mathbf{X}$ is said to be more PLOD than $\mathbf{Y}$ (*see* DEPENDENCE, CONCEPTS OF for a discussion in the bivariate case).

When $\mathbf{X} \sim N(\mathbf{O}, \boldsymbol{\Sigma}_X)$, then

1. The density is TP_2 in pairs if and only if $\boldsymbol{\Lambda} = \boldsymbol{\Sigma}_X^{-1}$ satisfies the condition $\lambda_{ij} \leqslant 0$ for all $i \neq j$ [6].
2. $\mathbf{X}$ is associated if and only if $\boldsymbol{\Sigma}_X$ is a nonnegative matrix [32].
3. $\mathbf{X}$ is PUOD and PLOD if and only if $\boldsymbol{\Sigma}_X$ is a nonnegative matrix (this follows from item 2).
4. $\mathbf{X}$ is NUOD if and only if the off-diagonal elements of $\boldsymbol{\Sigma}_X$ are nonpositive [10].

If $\mathbf{Y} \sim N(\mathbf{O}, \boldsymbol{\Sigma}_Y)$, where the diagonal elements of $\boldsymbol{\Sigma}_X$ and $\boldsymbol{\Sigma}_Y$ are the same and $\boldsymbol{\Sigma}_X - \boldsymbol{\Sigma}_Y$ is a nonnegative matrix, then $\mathbf{X}$ is more PUOD [12, Remark 5.1] and also more PLOD than $\mathbf{Y}$. The density of $|X_1|, \ldots, |X_p|$ is TP_2 in pairs if and only if there exists a diagonal matrix $\mathbf{D}$ with elements ± 1, such that the off-diagonal elements of $\mathbf{D}\boldsymbol{\Sigma}_X^{-1}\mathbf{D}$ are all nonpositive [1, 22]. For certain structural conditions on $\boldsymbol{\Sigma}_X$, $|X_1|, \ldots, |X_p|$ are associated [3] and also PUOD [24]; however, $|X_1|, \ldots, |X_p|$ are PLOD [37] for all $\boldsymbol{\Sigma}_X$. Under certain conditions on $\boldsymbol{\Sigma}_X$ and $\boldsymbol{\Sigma}_Y$ it can be shown that $|X_1|, \ldots, |X_p|$ are more PLOD [8] and also more PUOD [40] than $|Y_1|, \ldots, |Y_p|$.

Let $\mathbf{X} \sim N(\mathbf{O}, \boldsymbol{\Sigma}_X)$, $\mathbf{Y} \sim N(\mathbf{O}, \boldsymbol{\Sigma}_Y)$, $s^2 \sim \chi_k^2$, $u^2 \sim \chi_k^2$ and assume $\mathbf{X}$, $\mathbf{Y}$, s^2, and u^2 are all independent. If $\boldsymbol{\Sigma}_X$ is a nonnegative matrix, then the scaled t-vector, $X_1/s, \ldots, X_p/s$ is both PLOD and PUOD. This result follows from Theorem 3.2.1 of Ahmed et al. [2], and the PLOD and PUOD result for $\mathbf{X}$. If

$\boldsymbol{\Sigma}_X - \boldsymbol{\Sigma}_Y$ is a nonnegative matrix and $\boldsymbol{\Sigma}_X$ and $\boldsymbol{\Sigma}_Y$ have the same diagonal elements, then $X_1/s, \ldots, X_p/s$ is both more PLOD and more PUOD than $Y_1/u, \ldots, Y_p/u$ (see Das Gupta et al. [12, Theorem 5.1, Remark 5.1]). Under any conditions that allow $|X_1|, \ldots, |X_p|$ to be associated, $|X_1|/s, \ldots, |X_p|/s$ will be associated (see DEPENDENCE, CONCEPTS OF and Abdel-Hameed and Sampson [1, Lemmas 4.1, 4.2]). Šidák [39] showed that $|X_1|/s, \ldots, |X_p|/s$ is PLOD for arbitrary $\boldsymbol{\Sigma}_X$. The analogous PUOD result has been established for certain special cases of $\boldsymbol{\Sigma}_X$ by, among others Abdel-Hameed and Sampson [1, Theorem 4.2] and Ahmed et al. [2, Sec. 5.6]. If $\boldsymbol{\Sigma}_X - \boldsymbol{\Sigma}_Y$ is a positive semidefinite matrix, then $|X_1|/s, \ldots, |X_p|/s$ is more PLOD than $|Y_1|/u, \ldots, |Y_p|/u$ (this follows from Das Gupta et al. [12, Theorem 3.3]). Many of these results also hold when s^2 and u^2 are arbitrary positive random variables, or when the denominators of the t-vectors are not all the same random variable. These scaled multivariate t-distributions* and their generalizations arise naturally in regression problems*, when the sample regression coefficients are studied.

Let $(X_{i1}, \ldots, X_{ip})'$, $i = 1, \ldots, n$, be independent, identically distributed (i.i.d.) according to $N(\mathbf{O}, \boldsymbol{\Sigma})$, where $\boldsymbol{\Sigma}$ is any covariance matrix such that $|X_{i1}|, \ldots, |X_{ip}|$ are associated. Then $\Sigma X_{i1}^2, \ldots, \Sigma X_{ip}^2$, which can be considered (up to scaling) to be a p-dimensional multivariate χ^2, are associated. (The proof is a direct p-variate extension of Theorem 4.1 of Abdel-Hameed and Sampson [1].) Similar results hold for multivariate F-distributions*.

For distributions with an elliptically symmetric density, Das Gupta et al. [12] give a number of results concerning the random variables being more PLOD and more PUOD. Sampson [35] gives necessary and sufficient conditions in the bivariate elliptically symmetric case for the density to be TP_2. For a random vector $\mathbf{X}$ having a distribution with a covariance scale parameter $\boldsymbol{\Sigma}$, Sampson [34] gives sufficient conditions for the association of $\mathbf{X}$.

BOOLE, BONFERRONI, AND FRÉCHET

For information concerning these specialized multivariate probability inequalities, see their respective entries.

CONVEX SYMMETRIC SET INEQUALITIES

There are a number of probability inequalities involving convex symmetric sets which may be viewed in some sense as generalizations of positive dependence and stochastic ordering. The earliest is given by Anderson [4] (*see* DEPENDENCE, CONCEPTS OF). Khatri [24] showed for a multivariate normal vector $\mathbf{X} \equiv (\mathbf{X}_1 : \mathbf{X}_2)$ with $\text{cov}(\mathbf{X}_1, \mathbf{X}_2)$ having rank one that $\Pr[\mathbf{X}_1 \in C_1, \mathbf{X}_2 \in C_2] \geqslant \Pr[\mathbf{X}_1 \in C_1]$ $\Pr[\mathbf{X}_2 \in C_2]$ for any convex symmetric sets C_1 and C_2. Under certain conditions on the $\text{cov}(\mathbf{X}_1, \mathbf{X}_2)$, Khatri [25] obtained the reverse version of this inequality with " $\geqslant$ " replaced by " $\leqslant$ ", and C_1 and C_2 being complements of convex symmetric sets. Pitt [31] has shown that if $\mathbf{X} = (\mathbf{X}_1, \mathbf{X}_2)'$, where $\mathbf{X} \sim N(\mathbf{O}, \Sigma)$ and C_1, C_2 are convex symmetric sets, then $\Pr[\mathbf{X} \in C_1 \cap C_2] \geqslant \Pr[\mathbf{X}_1 \in C_1] \Pr[\mathbf{X}_2 \in C_2]$. For certain types of convex sets involving quadratic forms, Dykstra [13] has obtained inequalities for the multivariate normal. For instance, if $\Sigma_{X_2} = \mathbf{I}$ and $\text{cov}(\mathbf{X}_1, \mathbf{X}_2)$ arbitrary, then $\Pr(\mathbf{X}_1 \in C_1, \mathbf{X}_2'\mathbf{A}\mathbf{X}_2 \leqslant c_2) \geqslant \Pr(\mathbf{X}_1 \in C_1)\Pr(\mathbf{X}_2'\mathbf{A}\mathbf{X}_2 \leqslant c_2)$ for all convex symmetric sets C_1 and real numbers $c_2 \geqslant 0$, where $\mathbf{A}$ is any matrix satisfying $\mathbf{A}^2 = \mathbf{A}$ (*see* IDEMPOTENT MATRIX).

STOCHASTIC ORDERING

Stochastic ordering is a way of comparing the relative sizes of random variables (vectors). For example, if X and Y are univariate random variables on the same probability space*, one possible definition of the concept of X being less than or equal to Y is to require that $\Pr[X \leqslant Y] = 1$. Because there is a problem when X and Y are not defined on the same space, the usual definition of stochastic ordering is $P\{X > t\} \leqslant P\{Y > t\}$ for all t. This is written $X \overset{\text{st}}{\leqslant} Y$, i.e., X is stochastically less than Y. Moreover, if $X \overset{\text{st}}{\leqslant} Y$, it can be shown that there exist random variables $\tilde{X}$ and $\tilde{Y}$ defined on the same probability space with the same marginal distributions as X and Y, respectively, such that $\Pr[\tilde{X} \leqslant \tilde{Y}] = 1$. It also can be shown that $X \overset{\text{st}}{\leqslant} Y$ if and only if $E(\phi(X)) \leqslant E(\phi(Y))$ for all nondecreasing functions ϕ (see Marshall and Olkin [29, p. 483]).

In the multivariate case, $\mathbf{X}$ is stochastically less than $\mathbf{Y}$, denoted by $\mathbf{X} \overset{\text{st}}{\leqslant} \mathbf{Y}$, if $E(\phi(\mathbf{X})) \leqslant E(\phi(\mathbf{Y}))$ for all nondecreasing functions ϕ, where $\phi : \mathbb{R}^p \to \mathbb{R}^1$. Under the assumption of $\mathbf{X}$ and $\mathbf{Y}$ having identical univariate marginals, this condition implies, but is not equivalent to, $\mathbf{Y}$ being more PUOD than $\mathbf{X}$ [29, p. 486]. (See the next section for an example of multivarate stochastic ordering.) An existence theorem similar to the one in the one-dimensional case holds; i.e., $\mathbf{X} \overset{\text{st}}{\leqslant} \mathbf{Y}$ implies the existence of component-wise ordered random vectors on the same space with the same marginal distributions as $\mathbf{X}$ and $\mathbf{Y}$, respectively. See Arjas and Lehtonen [5] and Marshall and Olkin [29] for the proofs and discussions of these results in both the univariate and multivariate cases.

STOCHASTIC MAJORIZATION AND INEQUALITIES OBTAINED BY MAJORIZATION

One simple way to define a stochastic majorization between random vectors $\mathbf{X}$ and $\mathbf{Y}$ is to require that $\Pr[\mathbf{X} \prec \mathbf{Y}] = 1$, where $\prec$ denotes ordinary majorization*. This definition involves the joint distribution of $\mathbf{X}$ and $\mathbf{Y}$, and hence other definitions are preferable. An alternative definition is to say that $\mathbf{X}$ is stochastically majorized by $\mathbf{Y}$ if $E(\phi(\mathbf{X})) \leqslant E(\phi(\mathbf{Y}))$ for all Schur convex* functions ϕ. See Marshall and Olkin [29, pp. 281–285, 311–317] for other possible definitions of stochastic majorization and their interrelationships. With the use of these concepts, various functions of random vectors corresponding to standard families can be shown to be Schur convex, and useful inequalities can be obtained (see Marshall and Olkin [29, Chap. 11]). Majorization techniques can also be used to show that $E(\phi(\mathbf{X})) \leqslant E(\phi(\mathbf{Y}))$ for other families of functions ϕ (see Marshall and Olkin [29, Chap. 12]). For example, let $Y_1, \ldots, Y_p, Y_1', \ldots, Y_p'$ be $2p$ independent exponential* random variables with means $\lambda_1^{-1}, \ldots, \lambda_p^{-1}, (\lambda_1')^{-1}, \ldots, (\lambda_p')^{-1}$, respectively. Proschan and Sethuraman [33] show that if $\boldsymbol{\lambda} \prec \boldsymbol{\lambda}'$, then $E(\phi(Y_1, \ldots, Y_p)) \leqslant E(\phi(Y_1', \ldots, Y_p'))$ for all nondecreasing ϕ, i.e., $(Y_1, \ldots, Y_p) \overset{\text{st}}{\leqslant} (Y_1', \ldots, Y_p')$. Thus, if $\mathbf{Y}' = (Y_1', \ldots, Y_p')$ comes from a heterogeneous random sample* with means $(\lambda_i')^{-1}$, $i = 1, \ldots, p$, and $\mathbf{Y} = (Y_1, \ldots, Y_p)$ comes from a homogeneous random sample with common mean λ^{-1}, where $\lambda = (\sum_{i=1}^{p} \lambda_i')/p$, then $\mathbf{Y} \overset{\text{st}}{\leqslant} \mathbf{Y}'$, since necessarily $(\lambda, \ldots, \lambda) \prec (\lambda_1', \ldots, \lambda_p')$. This implies that all of the order statistics* of the homogeneous sample are stochastically smaller than the corresponding order statistics of the heterogeneous sample. See MAJORIZATION for other applications.

CHEBYSHEV AND KOLMOGOROV-TYPE INEQUALITIES

A standard univariate version of the Chebyshev inequality is $\Pr[|X - \mu| \leqslant a\sigma] \geqslant 1 - a^{-2}$, where X has mean μ and finite variance σ^2. If $X_1, \ldots, X_p$ are independent with means μ_i and finite variance σ_i^2, $i = 1, \ldots, p$, then

$$\Pr\left[\bigcap_{i=1}^{p} \{|X_i - \mu_i| \leqslant a_i\sigma_i\} \right]$$

$$\geqslant \prod_{i=1}^{p} (1 - a_i^{-2}).$$

If the X_i are dependent, various authors have obtained more general inequalities of which the previous inequality is a special case. One of the first of these was obtained

by Berge [7] in the bivariate case. Let X_1 and X_2 have means μ_1 and μ_2, variances σ_1^2 and σ_2^2, and correlation ρ. Then

$$\Pr\left[|X_1 - \mu_1| \leqslant a\sigma_1, |X_2 - \mu_2| \leqslant a\sigma_2 \right]$$

$$\geqslant 1 - \left[1 + (1 - \rho^2)^{1/2} \right]/a^2.$$

Various multivariate inequalities, including the previous one, can be obtained from the following general result. Let $\mathbf{X} = (X_1, \ldots, X_p)$ have mean vector $\boldsymbol{\mu}$ and covariance matrix $\boldsymbol{\Sigma} = (\rho_{ij}\sigma_i\sigma_j)$. For $a_i > 0$ define the matrix $T = (\tau_{ij})$, where $\tau_{ij} = \rho_{ij}/a_i a_j$. Then

$$\Pr\left[\bigcap_{i=1}^{p} \{|X_i - \mu_i| \leqslant a_i\sigma_i\} \right] \geqslant 1 - \inf_{C \in \mathscr{C}} \mathrm{tr}(CT)$$

where $\mathscr{C} = [\mathbf{C} = (c_{ij}) : \mathbf{C}$ is positive definite and for any $\mathbf{t}$ outside of $\{\mathbf{t} : |t_i| \leqslant 1, i = 1, \ldots, p\}$, $\mathbf{t}'\mathbf{C}\mathbf{t} \geqslant 1]$. Inequalities are then obtained by choosing various $\mathbf{C} \in \mathscr{C}$. As an example, let $\mathbf{C}$ be the identity matrix. Then

$$\Pr\left[\bigcap_{i=1}^{p} \{|X_i - \mu_i| \leqslant a_i\sigma_i\} \right] \geqslant 1 - \sum_{i=1}^{p} a_i^{-2}.$$

For other examples, see Tong [44, pp. 153–154] or Karlin and Studden [23, pp. 517–519]. Both of these references also give bounds for one-sided probabilities; e.g., lower bounds can be obtained on probabilities of the form $\Pr[\bigcap_{i=1}^{p} \{X_i \leqslant \mu_i + a\sigma_i\}]$, where $\mathbf{X} = (X_1, \ldots, X_p)$ has mean vector $\boldsymbol{\mu}$ and variances σ_i^2, $i = 1, \ldots, p$, and for certain $a > 0$. For background and historical references pertaining to the Chebyshev inequality, see Karlin and Studden [23, pp. 467–468]. One of the earliest books to contain the material on multivariate Chebyshev inequalities was Godwin [18].

A Kolmogorov-type inequality* is similar to the above, except that the maximum of partial sums is employed. For example, if $X_1, \ldots, X_n$ are independent and have mean 0 and $S_n = X_1 + \cdots + X_n$ with

$$\sigma(S_n) = \sqrt{\mathrm{var}(S_n)},$$

then $\Pr[\max_{1 \leqslant j \leqslant n} |S_j| (\sigma(S_n))^{-1} \leqslant a] \geqslant 1 -$

a^{-2}. A multivariate version of the Kolmogorov inequality has been obtained by Sen [36]. For multivariate applications of the univariate (independent) result and one-sided analogs, see Tong [44, Sec. 7.3].

MULTIVARIATE MOMENT INEQUALITIES

For the moments and expectations of other functions of multivariate distributions, there are a number of inequalities. The best known states that Σ is the population covariance matrix of any random vector if and only if Σ is nonnegative definite. Moreover, if Σ is positive definite and the sample size large enough, the corresponding sample covariance matrix is positive definite with probability one (see Eaton and Perlman [16]). For suitable bivariate distributions, there exists a canonical expansion (see Lancaster [27]) and a sequence $\{\rho_i\}$ of the canonical correlations. (*See* CANONICAL ANALYSIS.) This sequence $\{\rho_i\}$ can be shown to satisfy certain inequalities, e.g., Griffiths [19] or Thomas and Tyan [43].

Chebyshev has given the following covariance inequality for similarly ordered univariate functions ϕ_1, ϕ_2 of a random vector $\mathbf{X}$ (see Hardy et al. [20, Sec. 2.17] or Tong [44, Lemma 2.2.1]). If ϕ_1, ϕ_2 satisfy the condition that $(\phi_1(\mathbf{x}) - \phi_1(\mathbf{y}))(\phi_2(\mathbf{x}) - \phi_2(\mathbf{y})) \geqslant 0$ for all suitable $\mathbf{x}, \mathbf{y}$ then $\text{cov}(\phi_1(\mathbf{X}), \phi_2(\mathbf{X})) \geqslant 0$. A number of moment inequalities can be obtained from the result that for any nonnegative random vector $\mathbf{X}$, whose distribution is invariant under permutations, it follows that $E \Pi X_i^{\lambda}$ is a Schur-convex function of $(\lambda_1, \ldots, \lambda_p)$ (see Tong [44, Lemma 6.2.4]). For example, if $\mathbf{X} \sim N(\mu \mathbf{e}, \sigma^2 \{(1 - \rho)\mathbf{I} + \rho \mathbf{e} \mathbf{e}'\})$, where $\mathbf{e} = (1, \ldots, 1)'$, then $EX_j^{\Sigma \lambda_i} \geqslant E \Pi_{i=1}^{p} X_i^{\lambda_i} \geqslant E(\Pi_{i=1}^{p} X_i)^{\bar{\lambda}}$ where $\lambda_i \geqslant 0$, $i = 1, \ldots, p$ and $\bar{\lambda} = p^{-1} \Sigma \lambda_i$.

There are several results relating the more PLOD ordering to certain moment inequalities. If (X_1, X_2) is more PLOD than (Y_1, Y_2), then any of the following measures of association—Pearson's correlation*, Kendall's τ*, Spearman's ρ*, or Blomquist's q computed from (X_1, X_2)—are greater than or equal to the corresponding measure based on (Y_1, Y_2) (see Tchen [42]).

Dykstra and Hewett [14] have examined the positive dependence properties of the characteristic roots of the sample covariance matrix (*see* WISHART DISTRIBUTION). If $\mathbf{S}$ is the sample covariance matrix based on a random sample from $N(\mu, \mathbf{I})$, they show that the ordered characteristic roots are associated random variables.

APPLICATIONS

Multivariate probability inequalities are very important for simultaneous confidence* bounds, where lower bounds are sought on probabilities of events such as $\{\bigcap_{i=1}^{p} (|\hat{\theta}_i - \theta_i| \leqslant c_i)\}$, where the estimators $\hat{\theta}_1, \ldots, \hat{\theta}_p$ have some multivariate distribution possibly depending on nuisance parameters*. The basic concept is to bound this probability by probabilities of marginal events, where no parameters are involved. For instance, if $\mathbf{X} \sim N(\mu, \Sigma)$, the fact that $|X_1 - \mu_1|, \ldots, |X_p - \mu_p|$ are PLOD provides conservative simultaneous confidence intervals* for $\mu_1, \ldots, \mu_p$, when X_i, $i = 1, \ldots, p$ are known. General discussions of applications of probability inequalities to simultaneous inference can be found in Miller [30], Krishnaiah [26], and Tong [44]. Also found in the latter are applications of these techniques for establishing unbiasedness* for certain multivariate tests of hypotheses*.

Many of the dependence inequalities are useful in applications to reliability theory*. Consider a nonrepairable binary system consisting of p binary components with lifetimes $T_1, \ldots, T_p$ and having system lifetime T. The system lifetime T is generally a function of the component lifetimes such as $T = \max_{1 \leqslant r \leqslant k} \min_{i \in P_r} T_i$, where the P_r are min path sets (see Barlow and Proschan [6, Chaps. 1, 2, and p. 150]). In general, the T_i are not independent and it is desired to approximate $\Pr[T \in B]$ by the $\Pr[T_i \in B_i]$, $i = 1, \ldots, p$, where B and the B_i's are usually intervals. To do this, various univariate and multivariate inequalities are used. As a simple example, if the T_i are PUOD, then

$$\Pr[\, T > t \,] = Pr\left[\bigcup_{r=1}^{k} \bigcap_{i \in P_r} (T_i > t)\right]$$

$$\geqslant \max_{1 \leqslant r \leqslant k} \Pr\left[\bigcap_{i \in P_r} (T_i > t)\right]$$

$$\geqslant \max_{1 \leqslant r \leqslant k} \prod_{i \in P_r} \Pr[\, T_i > t \,],$$

where the first inequality follows from Boole's inequality and the second follows from PUOD. If the distributions of the T_i are not known but the T_i lie in a class of wearout distributions*, e.g., T_i has increasing failure rate, lower bounds on $\Pr[T > t]$ can be found in terms of the bounds on $\Pr[T_i > t]$. For example, Theorem 6.7 of Barlow and Proschan [6, Chap. 4] can be employed in such an application. Many other applications of this type are contained in Barlow and Proschan [6, Sec. 4.6]. Generalizations of this type of result to multistate systems have been given by Block and Savits [8].

References

[1] Abdel-Hameed, M. and Sampson, A. R. (1978). *Ann. Statist.*, **6**, 1360–1368.

[2] Ahmed, A., Langberg, N., León, R., and Proschan, F. (1979). Partial Ordering of Positive Quadrant Dependence with Applications. Unpublished report.

[3] Ahmed, A., León, R., and Proschan, F. (1981). *Ann. Statist.*, **9**, 168–176.

[4] Anderson, T. W. (1955). *Proc. Amer. Math. Soc.*, **6**, 170–176.

[5] Arjas, E. and Lehtonen, T. (1978). *Math. Operat. Res.*, **3**, 205–223.

[6] Barlow R. E. and Proschan, F. (1975). *Statistical Theory of Reliability and Life Testing*. Holt, Rinehart and Winston, New York. (Many excellent applications of inequalities on distributions to reliability theory are discussed. This book also contains a comprehensive treatment of dependence concepts in the bivariate case.)

[7] Berge, P. O. (1937). *Biometrika*, **29**, 405–406.

[8] Block, H. W. and Savits, T. H. (1982). *J. Appl. Prob.*, **19**, 391–402.

[9] Block, H. W. and Ting, M.-L. (1981). *Commun. Statist. A*, **10**, 742–762.

[10] Block, H. W., Savits, T. H., and Shaked, M. (1981). A Concept of Negative Dependence Using Stochastic Ordering. Unpublished report. (This paper and the following entry provide a comprehensive view on the topic of negative dependence.)

[11] Block, H. W., Savits, T. H., and Shaked, M. (1982). *Ann. Prob.*, **10**, 765–772.

[12] Das Gupta, S., Eaton, M., Olkin, I., Perlman, M., Savage, L., and Sobel, M. (1972). *Proc. 6th Berkeley Symp. Math. Statist. Prob.*, Vol. 2. University of California Press, Berkeley, Calif., pp. 241–265. (One of the first papers extending normal theory inequalities to the broader class of elliptically symmetric distributions.)

[13] Dykstra, R. L. (1980). *J. Amer. Statist. Ass.*, **75**, 646–650.

[14] Dykstra, R. L. and Hewett, J. E. (1978). *Ann. Statist.*, **6**, 235–238.

[15] Eaton, M. L. (1982). *Ann. Statist.*, **10**, 11–43.

[16] Eaton, M. L. and Perlman, M. (1973). *Ann. Statist.*, **1**, 710–717.

[17] Ebrahimi, N. and Ghosh, M. (1981). *Commun. Statist. A*, **10**, 307–337.

[18] Godwin, H. J. (1964). *Inequalities on Distribution Functions*. Hafner, New York. (One of the earlier works on multivariate inequalities.)

[19] Griffiths, R. (1970). *Aust. J. Statist.*, **12**, 162–165.

[20] Hardy, G. H., Littlewood, J. E. and Polya, G. (1952). *Inequalities*. Cambridge University Press, Cambridge. (For a long time this was the fundamental work on inequalities.)

[21] Karlin, S. and Rinott, Y. (1980). *J. Multivariate Anal.*, **10**, 499–516.

[22] Karlin, S. and Rinott, Y. (1981). *Ann. Statist.*, **9**, 1035–1049.

[23] Karlin, S. and Studden, W. J. (1966). *Tchebycheff Systems*, Interscience, New York.

[24] Khatri, C. (1967). *Ann. Math. Statist.*, **38**, 1853–1867.

[25] Khatri, C. (1976). *Gujarat Statist. Rev.*, **3**, 1–12.

[26] Krishnaiah, P. R. (1979). In *Developments in Statistics*, Vol. 2, P. R. Krishnaiah, ed. Academic Press, New York, pp. 157–201.

[27] Lancaster H. (1969). *The Chi-Squared Distribution*. Wiley, New York.

[28] Lehmann, E. L. (1966). *Ann. Math. Statist.*, **43**, 1137–1153.

[29] Marshall, A. W. and Olkin, I. (1979). *Inequalities: Theory of Majorization and Its Applications*. Academic Press, New York. (This is one of the two best references for inequalities on distributions.)

[30] Miller, R. (1981). *Simultaneous Statistical Inference*. Springer-Verlag, New York. (A very complete and readable presentation of simultaneous inference procedures.)

[31] Pitt, L. (1977). *Ann. Prob.*, **5**, 470–474.

[32] Pitt, L. (1982). *Ann. Prob.*, **10**, 496–499.

[33] Proschan, F. and Sethuraman, J. (1976). *J. Multivariate Anal.*, **6**, 608–616.

[34] Sampson, A. R. (1980). *SIAM J. Algebraic Discrete Meth.*, **1**, 284–291.

[35] Sampson, A. R. (1983). *J. Multivariate Anal.*, **13**.

[36] Sen, P. K. (1971). *Ann. Math. Statist.*, **42**, 1132–1134.

[37] Šidák, Z. (1967). *J. Amer. Statist. Ass.*, **62**, 626–633.

[38] Šidák, Z. (1968). *Ann. Math. Statist.*, **39**, 1425–1434.

[39] Šidák, Z. (1971). *Ann. Math. Statist.*, **42**, 169–175.

[40] Šidák, Z. (1975). *Ann. Inst. Statist. Math.*, *Tokyo*, **27**, 181–184.

[41] Slepian, D. (1962). *Bell Syst. Tech. J.*, **41**, 463–501. (One of the first modern papers on multivariate normal inqualities.)

[42] Tchen, A. (1980). *Ann. Prob.*, **8**, 814–827.

[43] Thomas, J. and Tyan, S. (1975). *J. Multivariate Anal.*, **5**, 227–235.

[44] Tong, Y. L. (1980). *Probability Inequalities in Multivariate Distributions*. Academic Press, New York. (This is one of the two best references for inequalities on distributions.)

Acknowledgements

The work of Henry W. Block has been supported by ONR Contract N00014-76-C-0839. The work of Allan R. Sampson is sponsored by the Air Force Office of Scientific Research under Contract F49620-79-C-0161. Reproduction in whole or in part is permitted for any purpose of the U.S. government.

(CHEBYSHEV INEQUALITY
DEPENDENCE, CONCEPTS OF
ELLIPTICALLY SYMMETRIC
    DISTRIBUTIONS
HYPOTHESIS TESTING
KOLMOGOROV INEQUALITY
MAJORIZATION AND SCHUR CONVEXITY
MULTIVARIATE ANALYSIS
MULTIVARIATE NORMAL DISTRIBUTION
SIMULTANEOUS INFERENCE)

HENRY W. BLOCK
ALLAN R. SAMPSON

INEQUALITIES ON DISTRIBUTIONS, UNIVARIATE *See* BERNSTEIN INEQUALITY; BERRY–ESSEEN INEQUALITY; BIRNBAUM–RAYMOND–ZUCKERMAN INEQUALITY; BON-

FERRONI INEQUALITIES; CANTELLI INEQUALITY; CHEBYSHEV'S INEQUALITY; HÁJEK–RÉNYI INEQUALITY; KOLMOGOROV'S INEQUALITY; MEAN-MEDIAN-MODE INEQUALITIES; PROBABILITY INEQUALITY OF RANDOM VARIABLES

INFANT MORTALITY

An infant death is usually defined as the death of a live-born individual occurring at an age of no more than one year. It is of obvious interest to obtain, from whatever data are available, estimates of probabilities of infant deaths for various populations and years, or other periods of time, and to compare these probabilities. The statistics commonly used to estimate such probabilities are known as *infant mortality rates*. Considerable use has been made of infant mortality rates as indicators of the health status or of the quality of health care in nations or socioeconomic strata at given periods of time.

To define the appropriate probabilities and their estimates, let us consider for each individual two quantities: the date of birth T and the life length X. For an individual born at time T, with life length X, the date of death is $T + X$. The pair (T, X) may be interpreted as a two-dimensional random variable, and we admit the possibility that T and X are dependent. For the sake of simplicity, we restrict our discussion to time intervals of one year's length, and consider three consecutive years: the "past," the "present," and the "following" year. Clearly, the infant deaths in the present year may occur to those born in the past year or in the present year.

In Fig. 1, values of T are plotted on the horizontal and values of X on the vertical axis, and the intervals $(-1, 0)$, $(0, 1)$, $(1, 2)$ on the T-axis correspond to the past, the present, and the following year, respectively.

Each of the regions in Fig. 1 denoted by G, H, K, L, M, N, and Q represents an event that can be easily interpreted. For example, G stands for the event "an individual is born in the present year and dies in the present year (hence at age $\leqslant 1$)"; H stands for "an

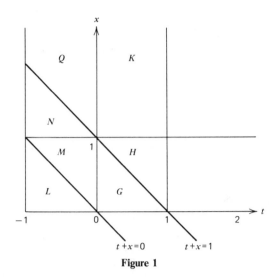

Figure 1

individual is born in the present year and dies in the following year at age $\leqslant 1$"; K stands for "an individual is born in the present year and dies at age > 1"; M stands for "an individual is born in the preceding year and dies in the present year at age $\leqslant 1$."

Two more regions in the (T, X)-plane can be defined as unions of some of the regions in Fig. 1:

B_0 = union of (L, M, N, Q) represents the event "an individual is born in the preceding year,"

B_1 = union of (G, H, K) stands for "an individual is born in the present year."

Let $P\{G\}$ denote the probability of the event described by the region G, $P\{H\}$ the probability of the event H, and so on. A plausible definition of infant death probability in the present year is

p = probability that an individual, born in the present year, will die at age $\leqslant 1$.

One clearly has

$$p = \left[P\{G\} + P\{H\} \right] / P\{B_1\}. \quad (1)$$

Other definitions are possible, but lead to consequences similar to those of (1). We shall therefore limit ourselves to this definition.

To estimate the probability p, one needs data on occurrences of events such as those represented by the regions in Fig. 1. These data are usually obtained from two sources: death certificates, which contain dates of birth and death (hence also life lengths), and birth certificates–both available for the present and the past year and, possibly, earlier years.

If we denote by $\#(R)$ the number of occurrences corresponding to a region R, then death certificates provide the frequencies $\#(G)$, $\#(L)$, $\#(M)$, $\#(N)$, and birth certificates yield $\#(B_0)$, $\#(B_1)$; these data do not enable us to obtain $\#(H)$ or $\#(K)$, since H and K deal with deaths occurring in the following year.

An obvious consistent and unbiased estimate of p is the statistic

$$\hat{p} = \left[\#(G) + \#(H) \right] / \#(B_1). \quad (2)$$

This cannot be computed, however, since we do not know $\#(H)$. In fact, demographic data published by most countries have a further shortcoming: for each year they contain the total number of births and the total number of infant deaths recorded in that year. Thus, for the present year, such data consist of $\#(B_1)$ and of $\#(G) + \#(M)$, without separating the latter into $\#(G)$ and $\#(M)$.

The statistics most frequently used to estimate p are the *infant mortality rate*,

$$r = \frac{\text{number of infant deaths in the present year}}{\text{number of live births in the present year}},$$

and the adjusted infant mortality rate, which we will discuss later.

Using our notations, we can write

$$r = \left[\#(G) + \#(M) \right] / \#(B_1). \quad (3)$$

This statistic differs from the unbiased and consistent estimate (2) by

$$r - \hat{p} = \left[\#(M) - \#(H) \right] / \#(B_1) = b,$$

$$(4)$$

a bias term that is small if and only if $\#(M)$

is close to $\#(H)$, i.e., when the numbers of infants born in one year and dying in the next year at age less than 1 do not change much from one year to the next.

In an attempt to correct for this possible bias, it has been frequent practice to use "separation factors,"

$$f = \#(M)/\left[\#(M) + \#(G)\right]$$
$$1 - f = \#(G)/\left[\#(M) + \#(G)\right] \quad (5)$$

and to compute the *adjusted infant mortality rate*,

$$r^* = \frac{\#(G) + \#(M)}{\#(B_1)}(1 - f)$$
$$+ \frac{\#(G) + \#(M)}{\#(B_0)}f. \quad (6)$$

Since, as mentioned before, most published data do not contain the values $\#(G)$ and $\#(M)$ separately, the factors (5) can usually not be calculated from these data, and are in practice themselves estimated from other demographic studies.

When the estimated values of the separation factors are close to the correct values (5), or when the available data permit computing the expressions (5), the adjusted infant mortality rate (6) reduces to

$$r^* = \frac{\#(G)}{\#(B_1)} + \frac{\#(M)}{\#(B_0)}$$

and the difference between this estimate and (2) is

$$r^* - \hat{p} = \frac{\#(M)}{\#(B_0)} - \frac{\#(H)}{\#(B_1)} = b^*. \quad (7)$$

This bias term is small if and only if the fraction of infants born in one year, who die in the next year at less than one year of age, is approximately the same for the preceding and for the present year.

Expressions (4) and (7) call attention to the fact that r as well as r^* are close to the unbiased and consistent estimate (2) when infant mortality and birthrate are approximately periodic from year to year. There is, however, reason for caution in using either one of these estimates for comparisons or other conclusions when there is some indica-

tion of changes from year to year in mortality or birthrate.

In addition to the theoretical biases due, among others, to changes in birth and death rates, there are further—and often more important—sources of error in evaluating infant mortality. The death and birth records may be incomplete or inaccurate, they may not be easily accessible, and the collecting of data contained in these records may be difficult. Complications of this kind may very differently affect the accuracy of data gathered in different areas or at different times.

Bibliography

Keyfitz, N. (1977). *Introduction to the Mathematics of Population*. Addison-Wesley, Reading, Mass.

Palloui, A. (1979). *Demography*, **16**, 455–473.

Shryock, H. S. and Siegel, J. S. (1971). *The Methods and Materials of Demography*. U.S. Bureau of the Census, Washington, D.C.

(DEMOGRAPHY
FERTILITY
LIFE TABLES
MORTALITY
VITAL STATISTICS)

Z. W. Birnbaum

INFERENCE, DESIGN BASED VS. MODEL BASED

Design-based inference and model-based inference are alternative conceptual frameworks for addressing statistical questions from many types of investigations. These include:

1. Experimental studies of randomly allocated subjects
2. Historical (observational) and follow-up* studies of all subjects in a fortuitous, judgmental, or natural population
3. Sample surveys* of randomly selected subjects

For these situations and others, there is interest in the extent of generality to which conclusions are expressed and the rationale by which they are justified. Some of the underlying inference issues can be clarified by directing attention at the sampling processes for data collection and the assumptions necessary for the data plausibly to represent a defined target population.

A statistical analysis whose only assumptions are random selection of observational units or random allocation of units to experimental conditions may be said to generate *design-based inferences*; i.e., design-based inferences are equivalent to *randomization*-based inferences* as discussed by Kempthorne [20–22], Kish [24, Chap. 14], Lehmann [27, pp. 55–57], and others. Also, such inferences are often said to have *internal validity** (see Campbell and Stanley [3]) when the design is adequate to eliminate alternative explanations for the observed effects other than the one of interest. In this sense, internal validity requires only that the *sampled population** and the *target population** be the same.

Alternatively, if assumptions external to the study design are required to extend inferences to the target population, then statistical analyses based on postulated probability distributional forms (e.g., binomial, normal, Poisson, Weibull, etc.) or other stochastic processes yield *model-based inferences*. These can be viewed as encompassing *Bayesian inferences** and *superpopulation inferences** to the extent that the validity of the claimed generality is *model dependent** via its sensitivity to *model misspecifications**. Also, it is possible to regard Bayesian inferences and superpopulation inferences as providing some unity to the role of design-based and model-based considerations. Thus a focus of distinction here between design-based and model-based inference is the population to which results are generalized rather than the nature of statistical methods. Models can be useful conceptually in either context; also they can shed light on the *robustness** of inferences to their underlying assumptions. The related issue of *external validity* includes substantive justification for the area of application and statistical evaluation of the plausibility of model assumptions. For other pertinent discussion, see Deming [10, Chap. 7], Fisher [12], Godambe and Sprott [15], John-

son and Smith [18], Kempthorne and Folks [23, Chap. 17], Namboodiri [29], and INFERENCE, STATISTICAL and STATISTICS, PHILOSOPHY OF.

The distinctions between design-based inference and model-based inference can be expressed in clearest terms for comparative experimental studies (e.g., *multicenter clinical trials**). Typically, these involve a set of blocks (or sites) which are selected on a judgmental basis. Similarly, the experimental units may be included according to convenience or availability. Thus, these subjects constitute a fixed set of finite local study populations. When they are randomly assigned to two or more treatment groups, corresponding samples are obtained for the potential responses of all subjects under study for each of the respective treatments. By virtue of the research design, randomization model methods (e.g., *Kruskal–Wallis tests**) in CHI-SQUARE TESTS can be used to obtain design-based inferences concerning treatment comparisons without any external assumptions. Illustrative examples are given in CHI-SQUARE TESTS, NUMERICAL EXAMPLES, and LOG-RANK SCORES.

A limitation of design-based inferences for experimental studies is that formal conclusions are restricted to the finite population of subjects that actually received treatment. For agricultural crop studies and laboratory animal studies undertaken at local facilities, such issues merit recognition in a strict sense. However, for *medical clinical trials** undertaken by multiple investigators at geographically diverse locations, it often may be plausible to view the randomized patients as conceptually representative of those with similar characteristics in some large target population of potential patients. In this regard, if sites and subjects had been selected at random from larger eligible sets, then models with *random effects* provide one possible way of addressing both internal and external validity considerations. However, such an approach may be questionable if investigators and/or patients were not chosen by a probability sampling mechanism. In this more common situation, one important consideration for confirming external validity is that *sample coverage* include all relevant subpopulations; another is that treatment differences be homogeneous across subpopulations. More formally, probability statements are usually obtained via assumptions that the data are equivalent to a stratified simple random sample from the partition of this population into homogeneous groups according to an appropriate set of explanatory variables. This stratification is necessary because the patients included in a study may overrepresent certain types and underrepresent others, even though those of each of the respective types might be representative of the corresponding target subpopulations. For categorical (or discrete) response measures, the conceptual sampling process described here implies the product multinomial distribution. As a result, model-based inferences concerning treatment comparisons and their interactions with the explanatory variable stratification can be obtained by using maximum likelihood or related methods as discussed in CHI-SQUARE TEST and CONTINGENCY TABLE. Illustrative examples are given in LOGISTIC REGRESSION. In a similar spirit, *least-squares* methods can be used for model-based inferences when continuous response variables have approximately normal distributions with common variance within the respective strata; and analogous procedures are applicable to other distributional structures (e.g., see Cox [9], McCullagh [28], and Nelder and Wedderburn [30]). The principal advantages of model-based inferences for such situations are their more general scope and the comprehensive information they provide concerning relationships of response measures to treatment and *stratification* variables. Contrarily, their principal limitation is that subjects in a study may not represent any meaningful population beyond themselves. See Fisher [13], Kempthorne [22], Neyman et al. [31], and Simon [37] for further discussion.

For historical (observational) studies, model-based inferences are usually emphasized because the target population is more extensive than the fortuitous, judgmental, or naturally defined group of subjects included.

Also, their designs do not involve either random allocation or random selection, as illustrated by the following examples:

1. A study of driver injury relative to vehicle size, vehicle age, and vehicle model year for all police-reported automobile accidents in North Carolina during 1966 or 1968–1972 (see Koch et al. [26])
2. A nonrandomized prospective study to compare the experience of patients receiving a new treatment with that of a historical control population (see Koch et al. [26])
3. A nonrandomized study to compare nine treatments for mastitis in dairy cows relative to their pretreatment status (*see* CHI-SQUARE TESTS, NUMERICAL EXAMPLES)
4. Market research studies involving *quota sampling** as opposed to random selection (see Kalton [19])

The assumptions by which the subjects are considered representative of the target population and the methods used for analysis are similar to those previously described for experimental studies. Otherwise, design-based inferences are feasible for historical studies through tests of randomization as a hypothesis in its own right, but their use should be undertaken cautiously; specific illustrations are given in CHI-SQUARE TESTS, NUMERICAL EXAMPLES, and Koch et al. [26]. More extensive discussion of various aspects of inference for observational studies appears in Anderson et al. [1], Breslow and Day [2], Cochran [8], Fairley and Mosteller [11], and Kleinbaum et al. [25].

Design-based inferences are often emphasized for sample surveys because the target population is usually the same as that from which subjects have been randomly selected. They are obtained by the analysis of estimates for population averages or ratios and their estimated covariance matrix which are constructed by means of finite population sampling methodology. An illustrative example is given in CHI-SQUARE TESTS, NUMERICAL EXAMPLES.

For sample surveys, the probabilistic interpretation of design-based inferences such as confidence intervals is in reference to repeated selection from the finite population via the given design. In contrast, model-based inferences are obtained from a framework for which the target population is a superpopulation with assumptions characterizing the actual finite population as one realization; and so their probabilistic interpretation is in reference to repetitions of the nature of this postulated sampling process. The latter approach can be useful for situations where the subjects in a sample survey are not necessarily from the target population of interest. For example, Clarke et al. [6] discuss the evaluation of several pretrial release programs for a stratified random sample of 861 defendants in a population of 2578 corresponding to January–March 1973 in Charlotte, North Carolina. Since the entire population here is a historical sample, any sample of it is also necessarily a historical sample. Thus issues of model-based inference as described for historical studies would be applicable. Another type of example involves prediction to a date later than that at which the survey was undertaken; e.g., Cassel et al. [5] studied prediction of the future use of a bridge to be constructed in terms of number of vehicles. Otherwise, it can be noted that statistical methods for design-based inferences often are motivated by a linear model; e.g., a rationale for ratio estimates involves regression through the origin. A more general formulation for which a linear model underlies the estimator and its estimated variance is given in Särndal [35, 36]. Additional discussion concerning aspects of design-based or model-based approaches to sample survey data or their combination is given in Cassel et al. [4], Cochran [7], Fuller [14], Hansen et al. [16], Hartley and Sielken [17], Royall [32], Royall and Cumberland [33], Särndal [34], Smith [38], and LABELS.

The distinction between design-based inference and model-based inference may not be as clear cut as the previous discussion might have suggested. For example, some type of assumption is usually necessary in order to deal with missing data; and stratification undertaken purely for convenient study management purposes (rather than statistical efficiency) is sometimes ignored. Also, a model-based approach may be advantageous for estimation for subgroups with small sample sizes (i.e., *small domain estimation*; see Kalton [19]). For these and other related situations, the issue of concern is the *robustness** of inferences to assumptions.

In summary, design-based inferences involve substantially weaker assumptions than do model-based inferences. For this reason, they can provide an appropriate framework for policy-oriented purposes in an adversarial setting (e.g., legal evidence). A limitation of design-based inferences is that their scope might not be general enough to encompass questions of public or scientific interest for reasons of economy or feasibility. Of course, this should be recognized as inherent to the design itself (or the quality of its implementation) rather than the rationale for inference. In such cases, model-based inferences may provide relevant information given that the necessary assumptions can be justified. It follows that design-based inference and model-based inference need not be seen as competing conceptual frameworks; either they can be interpreted as directed at different target populations and thereby at different statistical questions (e.g., experimental studies), or their synthesis is important to dealing effectively with the target population of interest (e.g., sample surveys).

References

[1] Anderson, S., Auquier, A., Hauck, W. W., Oakes, D., Vandaele, W., and Weisberg, H. I. (1980). *Statistical Methods for Comparative Studies*. Wiley, New York.

[2] Breslow, N. E. and Day, N. E. (1980). *Statistical Methods in Cancer Research*, Vol. 1: *The Analysis of Case Control Studies*. International Agency for Research on Cancer, Lyon.

[3] Campbell, D. T. and Stanley, J. C. (1963). *Handbook on Research on Teaching*, Rand McNally, Chicago, pp. 171–246. (Experimental and quasi-experimental designs for research on teaching.)

[4] Cassel, C. M., Särndal, C. E., and Wretman, J. H. (1977). *Foundations of Inference in Survey Sampling*. Wiley, New York.

[5] Cassel, C. M., Särndal, C. E., and Wretman, J. H. (1979). *Scand. J. Statist.*, **6**, 97–106. (Prediction theory for finite populations when model-based and design-based principles are combined.)

[6] Clarke, S. H., Freeman, J. L., and Koch, G. G. (1976). *J. Legal Stud.*, **5**(2), 341–385. (Bail risk: a multivariate analysis.)

[7] Cochran, W. G. (1946). *Ann. Math. Statist.*, **17**, 164–177. (Relative accuracy of systematic and stratified random samples for a certain class of populations.)

[8] Cochran, W. G. (1972). *Statistical Papers in Honor of George W. Snedecor*, T. A. Bancroft, ed. Iowa State University Press, Ames, Iowa, pp. 77–90.

[9] Cox, D. R. (1972). *J. R. Statist. Soc. B*, **34**, 187–220. [Regression models and life tables (with discussion).]

[10] Deming, W. E. (1950). *Some Theory of Sampling*. Wiley, New York.

[11] Fairley, W. B. and Mosteller, F. (1977). *Statistics and Public Policy*. Addison-Wesley, Reading, Mass.

[12] Fisher, R. A. (1925). *Proc. Camb. Philos. Soc.*, **22**, 700–725. (Theory of statistical estimation.)

[13] Fisher, R. A. (1935). *The Design of Experiments*. Oliver & Boyd, Edinburgh (rev. ed., 1960).

[14] Fuller, W. A. (1975). *Sankhyā C*, **37**, 117–132. (Regression analysis for sample survey.)

[15] Godambe, V. P. and Sprott, D. A. (1971). *Foundations of Statistical Inference*. Holt, Rinehart and Winston, Toronto.

[16] Hansen, M. H., Madow, W. G., and Tepping, B. J. (1978). *Proc. Survey Res. Meth. Sec., Amer. Statist. Ass.*, pp. 82–107. [On inference and estimation from sample surveys (with discussion).]

[17] Hartley, H. O. and Sielken, R. L. (1975). *Biometrics*, **31**, 411–422. (A "super-population viewpoint" for finite population sampling.)

[18] Johnson, N. L. and Smith, H., eds. (1969). *New Developments in Survey Sampling*. Wiley, New York.

[19] Kalton, G. (1983). *Bull. Int. Statist. Inst.* (Models in the practice of survey sampling.)

[20] Kempthorne, O. (1952). *The Design and Analysis of Experiments*. Wiley, New York.

[21] Kempthorne, O. (1955). *J. Amer. Statist. Ass.* **50**, 946–967. (The randomization theory of experimental inference.)

[22] Kempthorne, O. (1979). *Sankhyā B*, **40**, 115–145. (Sampling inference, experimental inference, and observation inference.)

[23] Kempthorne, O. and Folks, L. (1971). *Probability, sity Press, Ames, Iowa.

[24] Kish, L. (1965). *Survey Sampling*. Wiley, New York.

[25] Kleinbaum, D. G., Kupper, L. L., and Morgenstern, H. (1982). *Epidemiologic Research: Principles and Quantitative Methods*. Lifetime Learning Publication, Belmont, Calif.

[26] Koch, G. G., Gillings, D. B., and Stokes, M. E. (1980). *Annu. Rev. Public Health*, **1**, 163–225. (Biostatistical implications of design, sampling, and measurement to health science data analysis.)

[27] Lehmann, E. L. (1975). *Nonparametrics: Statistical Methods Based on Ranks*. Holden-Day, San Francisco.

[28] McCullagh, P. (1980). *J. R. Statist. Soc. B*, **42**, 109–142. (Regression models for ordinal data.)

[29] Namboodiri, N. K. (1978). *Survey Sampling and Measurement*. Academic Press, New York.

[30] Nelder, J. A. and Wedderburn, R. W. M. (1972). *J. R. Statist. Soc. A*, **135**, 370–384. (Generalized linear models.)

[31] Neyman, J., Iwaskiewicz, K., and Kolodziejczyk, S. (1935). *J. R. Statist. Soc.* (Suppl. 1), **2**, 107–154. (Statistical problems in agricultural experimentation.)

[32] Royall, R. M. (1976). *Amer. J. Epidemiol.*, **104**, 463–473. (Current advances in sampling theory: implications for human observational studies.)

[33] Royall, R. M. and Cumberland, W. G. (1981). *J. Amer. Statist. Ass.*, **76**, 66–77. (An empirical study of the ratio estimator and estimators of its variance.)

[34] Särndal, C. E. (1978). *Scand. J. Statist.*, **5**, 27–52. (Design-based and model-based inference in survey sampling.)

[35] Särndal, C. E. (1980). *Biometrika*, **67**, 639–650. (On π-inverse weighting vs. best linear unbiased weighting in probability sampling.)

[36] Särndal, C. E. (1982). *J. Statist. Plann. Infer.*, **7**, 155–170.

[37] Simon, R. (1979). *Biometrics*, **35**, 503–512. (Restricted randomization designs in clinical trials.)

[38] Smith, T. M. F. (1976). *J. R. Statist. Soc. A*, **139**, 183–204. [The foundations of survey sampling: a review (with discussion).]

Acknowledgements

The authors would like to thank Wayne Fuller, Peter Imrey, Graham Kalton, Oscar Kempthorne, Jim Lepkowski, Carl Särndal, and Richard Simon for helpful comments relative to the preparation of this entry. It should be noted that they may not share the views expressed here. This research was partially supported by the U.S. Bureau of the Census through Joint Statistical Agreement JSA-80-19; but this does not imply any endorsement.

GARY G. KOCH
DENNIS B. GILLINGS

INFERENCE, STATISTICAL: I

How far away is the sun? How heavy is this potato? Are there racial differences in IQ scores? What is the relation between the heights of fathers and sons? Does smoking cause cancer? Is this die fair? Has the introduction of a speed limit on this stretch of road made it safer? How potent is this antibiotic? How does the yield of a chemical process depend on temperature and pressure settings? How is wealth distributed in the population?

This is a haphazard assortment of the type of question about the empirical world which the science of statistics can help in answering. First, an experiment or survey is planned and conducted to produce relevant *data*: the potato is weighed several times on a balance; heights of fathers and sons are measured; the diameter of the circle in a bacterial culture cleared by the antibiotic is compared with that of a standard. The choice of what data to collect, and how (i.e., what is to be the *observand*), involves both scientific and statistical considerations, particularly the theory of *experimental design* (*see* DESIGN OF EXPERIMENTS). We take up the story after the experiment or investigation has been selected and conducted. At this stage we ask: What can be learned about the questions of interest? This is the general problem of *scientific inference*.

Scientific inference becomes *statistical inference* when the connection between the unknown "state of nature" and the observand is expressed in probabilistic terms: the measurement error* of a balance may vary between repeated weighings according to a probability distribution of known or unknown form; the values obtained of heights for fathers and sons depend on which random sample from the population happens to be chosen; and the number of road accidents observed depends on uncontrollable chance factors. A *statistical model* describes the way in which the probabilistic structure of the observand is supposed to depend on quantities of interest (and, possibly, on fur-

ther unknown "nuisance" quantities, e.g., an unknown measurement error variance). Such quantities determining, wholly or partly, the *sampling distributions* of the observand are termed *parameters*. A model may be firmly based in a theoretical understanding of the data-generating process, or past experience with similar processes, or the experimental technique employed. Alternatively, it can be ad hoc and tentative, or chosen for ease of interpretation or analysis. Such a model may itself be subject to scrutiny or refinement in the light of experimental data. (According to Stigler [108], this conception of a statistical model as the appropriate setting for statistical inference is due to R. A. Fisher [50]. The section "Postscript" comments on its inadequacies.)

In abstract terms, a statistical model M specifies the observand X, the parameter Θ, and a family $\{P_\theta\}$ of probability distributions for X, P_θ being the supposed distribution of X when $\Theta = \theta$. (These minimal ingredients may be supplemented in some approaches to statistical inference.) Experimentation yields the *observation* $X = x_0$, which is to be used in answering various questions about Θ, on the assumption that M is valid. Alternatively, the validity of M may be assessed, and improvements or changes suggested.

This article concentrates on logical principles of statistical inference, rather than on specific questions and procedures. For these, *see* DECISION THEORY, ESTIMATION, HYPOTHESIS TESTING, INTERVAL ESTIMATION, and similar entries. We shall emphasize the problem of parametric inference, in which M is not challenged, since it is here that our understanding (fragmentary though it is) is deepest. However, this is in no way to minimize the practical importance of model assessment and criticism.

Textbooks that discuss logical aspects of statistical inference are Kempthorne and Folks [73], Barnett [11], and Cox and Hinkley [29]; see also Savage [106], Hacking [62], Godambe and Sprott [61]. The collected works of Neyman and Pearson [90], Neyman [84], and Fisher [15], as well as the book by Fisher [55], are invaluable sources. Fienberg and Hinkley [46] forms a useful guide to Fisher's statistical work.

SCHOOLS OF INFERENCE

The process of arguing from observed data to unobserved parameters, or to underlying laws (models), is one of inductive logic. Since the problem of justifying induction is one of the most controversial in philosophy, it is not surprising that there are various points of view, which group into several major competing schools of thought, on how statistical inference should be performed. These differ in the underlying logical principles they accept, the statements they regard as meaningful, and the questions they are willing to consider.

We describe below the principal current approaches to statistical inference. In doing so, it will be helpful to refer to the following simple example.

Example: Potato-Weighing Experiment. A potato, of unknown weight Θ grams, is to be weighed twice on a balance, producing readings of X_1 and X_2 grams. Thus the observand is $X = (X_1, X_2)$.

Past experience indicates that the balance yields independent unbiased readings, having a normal distribution with standard deviation 10 grams. Hence our model M prescribes that $X_i \sim \mathcal{N}(\Theta, 10^2)$ independently $(i = 1, 2)$.

The experiment is performed and yields observations $(x_1, x_2) = (94, 120)$.

Sampling Theory*

INFERENCE RULES. The sampling theory approach attacks the problem of making appropriate inference from the specific data at hand [i.e., $(X_1, X_2) = (94, 120)$] indirectly. Its principal concern is with *inference rules* or *procedures*. Such a rule attaches an inference statement of some appropriate nature (e.g., a numerical estimate of Θ, or an assertion that Θ lies in some set, or a decision to take some action) to each conceivable value of the ob-

servand X. We may term this statement the *nominal inference*. However when a specific rule is applied, the full inference from the data $(94, 120)$ is generally a combination of the nominal inference which the rule attaches to these data, together with further information, which we shall call *stochastic inference*, about the probabilistic behavior of the rule. This probabilistic behavior also serves as a basis for comparison among rival procedures.

For example, an inference rule for *point estimation** is an *estimator*. For our potato problem, an estimator is a function $\tilde{\theta}$, defined on $\mathbb{R}^2$ with values in $\mathbb{R}^1$. The nominal inference based on data (x_1, x_2) is $\tilde{\theta}(x_1, x_2)$, regarded as an estimate of Θ. Some estimators may appear reasonable, others not. For each value θ of Θ, the distribution of $\tilde{\theta}(X)$ can be found: important features might be the mean and variance of this distribution.

An obvious candidate, for example, is $\tilde{\theta}_0(X) = \bar{X} = \frac{1}{2}(X_1 + X_2)$. This is *unbiased*: $E(\tilde{\theta}_0(X)|\Theta = \theta) \equiv \theta$. Moreover, among all unbiased estimators $\tilde{\theta}$, it minimizes the sampling variance $\text{var}(\tilde{\theta}(X)|\Theta = \theta)$, simultaneously for every value of θ. If these properties are taken as justification for preferring the *rule* $\tilde{\theta}_0$ to all others, the *estimate* of Θ based on the observed data would be 107. This is the nominal inference: it might be qualified by attaching to it, as a stochastic inference, the standard error 7.07 of the estimator $\tilde{\theta}_0(X)$ (in this case independent of the true value of Θ).

In *interval estimation*, the nominal inference is an assertion that Θ belongs to some (data-dependent) set, e.g., the set $\tilde{\theta}(x_1, x_2) = (\bar{x} - 13.86, \bar{x} + 13.86)$. Our data yield the interval $(93.14, 120.86)$ for this rule $\tilde{\theta}$; to this might be attached the stochastic inference that the probability of the interval estimator $\tilde{\theta}(X_1, X_2)$ covering the value θ, when $\Theta = \theta$, is 0.95 (for any θ). This is summarized in the statement that $(93.14, 120.86)$ is a 95% *confidence interval* for Θ. For purposes of comparing rival interval estimators, one might take into account the probability of covering an incorrect value θ' when $\Theta = \theta$, and/or the distribution of the length of the interval. (*See* CONFIDENCE INTERVALS AND REGIONS.)

In *hypothesis testing**, the nominal inference might be to *accept* or to *reject* a null hypothesis H_0. No stochastic inference is normally attached (but see Birnbaum [20]). By contrast, the related *assessment of significance* takes as its nominal inference the value t of a test statistic T, and emphasizes the stochastic inference $\Pr(T \geqslant t \mid H_0)$, the *observed significance level*.

TWO APPROACHES. The early (and still common) practice of sampling-theory inference involved suggesting some reasonable-sounding inference rule, establishing its distributional properties, and applying it to yield the implied nominal and stochastic inference for the data at hand. The fundamental work of Neyman* and Pearson* [86–89] emphasized the importance of choosing among rival procedures by means of well-defined probabilistic criteria, such as the two types of error in testing hypotheses. This approach was extended to general decision problems by Wald* [110], and is well represented by Lehmann [78] and Ferguson [43]. (*See* NEYMAN–PEARSON LEMMA and DECISION THEORY.)

The Neyman–Pearson–Wald (NPW) approach emphasized nominal inference: while stochastic properties are relevant for choosing an inference rule, they are not generally regarded as qualifying its nominal inference for specific data. In contrast, Fisher*, in his many writings, considered stochastic inference as fundamental, and laid special stress on ensuring the relevance of any such inference for the specific data to which it attached. He had little sympathy with probabilistic optimality criteria, and constantly criticized procedures produced by the NPW approach for yielding specific inferences which were, from his viewpoint, ridiculous. An example due to Fieller [44, 45] is a good illustration, although in this case the NPW procedure was, surprisingly, endorsed by Fisher.

Let $X_i \sim \mathcal{N}(\Theta_i, 1)$ independently ($i = 1, 2$). Then a 95% Neyman confidence interval estimator for $\Phi = \Theta_1/\Theta_2$ is $\{\phi : (X_1 - \phi X_2)^2 < 3.84(1 + \phi^2)\}$. For data $(x_1, x_2) = (1.0, 1.5)$, this yields the whole real line. This is clearly unsatisfactory if the "95%" is to be regarded as a stochastic inference, meaningful as some measure of "confidence," for the given data, that the interval contains Φ. But this interpretation is (in theory) strictly to be avoided by NPW practitioners: stochastic inference is not, for them, tailored to specific data, and the "95%" simply describes the overall behavior of the procedure. Neyman [83] even introduced a concept of *inductive behavior** in replacement of *inductive inference* as the philosophical base of statistics, essentially denying the meaningfulness of inference for specific data. This was strongly criticized by Fisher [55].

CONDITIONAL INFERENCE. (The article on this topic by J. Kiefer is particularly relevant here and for the following section and the section "Structural Inference.") A principal concern of Fisher's was that stochastic inferential statements should be derived for the appropriate *frame of reference*. For example, suppose that there was a choice between two balances for weighing the potato, having measurement standard deviations of 10 grams and 50 grams; and that a fair coin was tossed to decide which to use, leading to the choice of the former and the results given. If the whole experiment were to be performed again, the other balance might be chosen. The complete observand is thus $(Y; X_1, X_2)$, where $Y = 10$ or 50, each with probability $\frac{1}{2}$, and, given $Y = y$, $X_i \sim \mathcal{N}(\Theta, y^2)$ independently ($i = 1, 2$). Our data are $(10; 94, 120)$.

The NPW approach would assess the value of an inference rule in terms of its probabilistic properties, as $(Y; X_1, X_2)$ varies according to the foregoing distribution. In particular, even though we have chosen the former balance, we have to consider how we might have reacted had we chosen the latter. The Fisherian view would be that, once the balance has been selected, the appropriate model distributions are those relative to the frame of reference in which Y is fixed at 10: namely, $X_i \sim \mathcal{N}(\Theta, 10^2)$ independently ($i = 1, 2$). Further aspects of the foregoing problem are considered in the section "Conditional and Unconditional Behavior."

In the above, the distribution of Y is known, the same for all values of Θ. Such a statistic is termed *ancillary*. A basic principle of Fisherian stochastic inference is to calculate probabilities conditional on the observed value for a suitable ancillary statistic (but there may be difficulties in choosing between rival ancillaries [12]). The feeling behind this is that data values sharing a common value of an ancillary statistic are more alike in relevant inferential respects; moreover, nothing is lost by ignoring the marginal probabilistic behavior of the ancillary statistic, since it does not distinguish between different values of Θ. (*See* ANCILLARITY.)

This use of an ancillary statistic does not affect nominal inference. Fisher originally introduced it for the case in which nominal inference was given by maximum likelihood estimation*. Conditioning on an ancillary does not change the estimate (nominal inference), but does provide a more relevant assessment of its precision (stochastic inference).

In some problems, particularly those with nuisance parameters, it seems appropriate to condition on a statistic that is not ancillary. For example, one usually conditions on observed X values when studying the regression of Y on X, even when these have arisen from an unknown distribution. Similarly, inference about dependence in a 2×2 contingency table is often conditional on all marginal totals. Barndorff-Nielsen [8] gives a good account of the generalized ancillarity concepts involved in such problems.

RECOGNIZABLE SUBSETS. Another example of the conflict between the NPW and Fisherian outlooks arises when testing the equality of the means of two normal distributions $\mathcal{N}(\mu_1, \sigma_1^2)$ and $\mathcal{N}(\mu_2, \sigma_2^2)$, no relation between σ_1^2 and σ_2^2 being assumed. This is known as *Behrens' problem*. The data may be reduced to the mean and variance estimators: $X_i \sim \mathcal{N}(\mu_i, \sigma_i^2)$, $s_i^2 \sim \sigma_i^2 \chi_{f_i}^2$, all independently. The null hypothesis is $H_0: \mu_1 = \mu_2$, and test procedures leading to one of the decisions "Accept H_0" or "Reject H_0" are

under consideration. (*See* BEHRENS–FISHER PROBLEM.)

A typical NPW requirement for a test is *similarity*: $\Pr(\text{reject } H_0 \mid \mu_1, \mu_2, \sigma_1^2, \sigma_2^2)$ should be a constant preselected value α for all parameter values satisfying H_0. This requirement cannot here be met exactly [82], but Welch [112] produced a rule for which it holds to a very close approximation. This is tabulated in Pearson and Hartley [92, Table 11]. For the case $f_1 = f_2 = 6$, $\alpha = 0.10$, Fisher [56] pointed out that Welch's two-sided test had the property $\Pr(\text{reject } H_0 \mid s_1^2/s_2^2 = 1; \mu_1, \mu_2, \sigma_1^2, \sigma_2^2) \geqslant 0.108$ for all parameter values satisfying H_0. He argued that the nominal rejection probability $\alpha = 0.10$ was therefore inappropriate as a stochastic inference for data such that $s_1^2/s_2^2 = 1$, since the set of such data values could be recognized as yielding rejection probability at least 0.108. He took this as a criticism of Welch's solution (without, however, implying that the conditional rejection probabilities themselves provided a satisfactory stochastic inference). (*See* SIMILAR TESTS.)

Similar considerations apply to confidence intervals. Indeed, the usual interval for a single normal mean, based on the t-distribution, admits a recognizable subset [21, 23]. Fisher's desideratum of the nonexistence of such recognizable subsets has been formalized and studied by Buehler [22], Wallace [111], Pierce [95], and Robinson [101–103]. Very loosely put, the conclusion is that such inconsistencies can only be avoided by procedures that can be given a Bayesian interpretation (see the section "Interval Estimate"), and not always then. To ask this in addition to NPW sampling theory criteria is to impose a restriction so severe that it can be satisfied only in very special cases.

SUFFICIENCY*. Suppose that we were informed of the value of $\bar{X}$ (i.e., 107), but the detailed values $X = (X_1, X_2)$ withheld. Then our residual uncertainty about X is described by the conditional model distributions for X, given $\bar{X} = 107$. In these, $X_1 \sim \mathcal{N}(107, 50)$ (and $X_2 = 214 - X_1$), *irrespective of the value of* Θ. Since this residual distribution does not depend on Θ, it may be claimed that further knowledge of X, once $\overline{X}$ is known, is of no value in making inferences about Θ, and that inferences should thus depend on $\overline{X}$ alone. We say that $\overline{X}$ is *sufficient* for Θ. Sufficiency was introduced by Fisher [48].

From the Fisherian viewpoint, the injunction to base inference on a sufficient statistic is founded on the intuitive requirement that irrelevant information be discarded. However, this same injunction is a direct consequence of NPW optimality criteria, as well as of the likelihood and Bayesian approaches discussed below. The information discarded is *not* irrelevant if the assumed model M is in doubt, and can indeed be used in testing the adequacy of M.

A useful criterion for determining a sufficient statistic is the *Fisher–Neyman factorization theorem*: $T = t(X)$ is sufficient for Θ if the sampling density $f(x|\theta)$ can be written in the form $a(x)b(t(X),\theta)$.

Again, ideas of sufficiency can be generalized to problems with nuisance parameters [8, 30, 32].

Bayesian Inference*

While sampling theory inference uses only those probabilities interpretable as long-term relative frequencies in relevant repetitions of an experiment, the *Bayesian* view is that *any* uncertainty, even about an unknown parameter in a statistical model, or the validity of that model, may be expressed in the language of probability. The implied inference, for any specific data, is thus completely carried by an appropriate distribution for Θ, the *posterior distribution*.

For good accounts of modern Bayesian statistics, see Raiffa and Schlaifer [98], Lindley [80, 81], and De Groot [36].

SUBJECTIVIST BAYESIANISM. This, currently the dominant Bayesian view, interprets probabilities as *degrees of belief*, which may vary from person to person, and which may be assessed operationally by betting behavior. Principal modern sources for this view are

Ramsey [99], Savage [104–106], and de Finetti [34, 35]. Kyburg and Smokler [75] is a valuable collection.

For example, the potato weigher has a good deal of background knowledge about potatoes, and should be able to form a judgment about the weight of this one by looking. Suppose that he or she can describe these prior judgments probabilistically using, say, the normal distribution $\mathcal{N}(80, 20^2)$ for Θ.

His or her *prior density* for Θ is thus

$$\pi(\theta) = \frac{1}{20\sqrt{2\pi}} \exp\left\{ -\frac{1}{800}(\theta - 80)^2 \right\},$$

while the sampling density for $X = (X_1, X_2)$ when $\Theta = \theta$ is

$$f(x_1, x_2 | \theta) = \prod_{i=1}^{2} \frac{1}{10\sqrt{2\pi}}$$
$$\times \exp\left\{ -\frac{1}{200}(x_i - \theta)^2 \right\}.$$

From the combination of the marginal (prior) distribution for Θ and the conditional (sampling) distributions for X given Θ, we can deduce the conditional (so-called *posterior*) distribution for Θ, given X. It has density given by *Bayes' theorem**:

$$\pi(\theta | x_1, x_2) \propto \pi(\theta) f(x_1, x_2 | \theta).$$

Substituting the observed values for (x_1, x_2) yields the relevant posterior distribution for Θ, fully tailored to the data in hand, and adequate to answer any questions about Θ that may be put. For our data, the posterior distribution is $\mathcal{N}(104, (6.67)^2)$.

Subjectivist Bayesian inference is, of course, most open to criticism for its subjectivity. Why should the inference depend on the inferrer? Cannot the data speak for themselves, without prior inputs? It was this concern that, historically, led to the attempt to found inference on the "objective" properties of sampling distributions. To some extent this criticism is met by the *principle of stable estimation* [40], which states, loosely, that as more and more data are gathered, the posterior distribution becomes less and less sensitive to the prior input. For example, if the potato were weighed 100 times, yielding a mean reading 105, the posterior distribution would be close to $\mathcal{N}(105, 1)$ for a very wide range of priors. [The prior $\mathcal{N}(80, 20^2)$ used above yields exact posterior $\mathcal{N}(104.94, (0.999)^2)$.] Where the data are not extensive, the subjectivist might claim that the ideal of "objectivity" is illusory, and that prior opinion should not be disregarded.

LOGICAL BAYESIANISM. An alternative Bayesian view is that a prior distribution can itself be "objective," serving to measure a quasi-logical relationship between the evidence on which it is based (which may be vacuous) and the parameter. Ideas of "logical probability" have been proposed by Keynes [74] and Carnap [24, 25]. However, the actual insertion of values for such probabilities is problematical.

For example, if Θ is an unknown probability about which "nothing is known," an intuitively appealing distribution to represent this ignorance is the uniform distribution on $[0, 1]$. For this seems to treat all values for Θ, about which we are equally uncertain, equally. This argument is known as the "principle of insufficient reason." Although Bayes himself used the uniform prior distribution in his memoir [14], this particular interpretation became current with the work of Laplace* [76]. However, by the early twentieth century, the self-contradictions of this naive approach, and the criticisms of such thinkers as Venn, Boole, and Fisher (see especially Chap. II of Fisher [55]), contributed to the eclipse of Bayesian methods and the development of sampling theory. These objections were effectively disposed of by Jeffreys and his method of constructing "invariant prior distributions"*; more recent workers in the same tradition are Hartigan [63], Novick [91], Zellner [114, 115], Jaynes [66, 67], Villegas [109], and Bernardo [16].

If truly nothing were known about the weight of the potato, or if an "objective" posterior distribution based on experimental data alone were required, we might use the Jeffreys invariant prior density, which for the potato problem is uniform: $\pi(\theta) =$ constant. This cannot be normalized to integrate to 1: it is an *improper* distribution*. Nevertheless, on insertion into Bayes' formula, $\pi(\theta \mid x) \propto \pi(\theta) f(x \mid \theta)$, it yields a proper posterior distribution for Θ, namely (for our data), $\mathcal{N}(107, 50)$.

Bold attempt that it is, Jeffreys' program still seems to contain self-contradictions, for example the *marginalization paradox* of Dawid et al. [33]. (See also Jaynes [68].)

Likelihood Inference* (*See Also* LIKELIHOOD PRINCIPLE)

Let $f(x \mid \theta)$ denote the probability density function of P_θ at x. The *likelihood function* for Θ based on data $X = x_0$ is the function $L(\cdot)$ given by $L(\theta) \propto f(x_0 \mid \theta)$. (Strictly, it is the class of such functions for arbitrary multiplier: only ratios of likelihoods matter.) For the potato problem and its data, we have $L(\theta) \propto \exp\{-\frac{1}{100}(\theta - 107)^2\}$.

If the observand had been the sufficient statistic $\bar{X}$, with distribution $\mathcal{N}(\Theta, 7.07^2)$ and observed value 107, the same likelihood function would result; this follows generally from the Fisher–Neyman criterion. The likelihood function is further unaffected by conditioning on any ancillary statistic.

The likelihood function is all that the Bayesian needs to find the posterior distribution: $\pi(\theta \mid x_0) \propto \pi(\theta) L(\theta)$. (In particular, Bayesian inference, for a fixed prior distribution, is unaffected by reduction to a sufficient statistic or conditioning on an ancillary statistic.) The independent concept of likelihood was introduced, and its importance stressed, by Fisher [47, 49, 50, 55]. Although $L(\cdot)$, if normalized to integrate to 1, can be interpreted as the posterior density for Θ for a uniform prior, Fisher was at pains to avoid such an interpretation: likelihood does *not* have the same logical status as a probability density, and exists independently of any prior distribution. Its direct appeal lies in the idea that a good way to compare values for Θ is by means of the probability they assigned to the outcome that materialized. Likelihood inference usually starts from the premise that only the course of the likelihood function for the observed data is of relevance to inference, thus eschewing any stochastic qualification of its nominal inferences.

Fisher's first application of likelihood ideas was the introduction of *maximum likelihood estimation**: the idea that Θ may be estimated by the value $\hat{\theta}$ which maximizes $L(\theta)$ (107 for our problem). An extension of this interprets $L(\theta_1)/L(\theta_2)$, the *likelihood ratio*, as measuring the *relative support* provided by the data in favor of $\Theta = \theta_1$ as against $\Theta = \theta_2$. The maximum likelihood estimate $\hat{\theta}$ then appears as the best supported value of Θ. [Edwards [39] defines *support* as $\log(L(\theta)/L(\hat{\theta}))$.] The problem of interval estimation may similarly be tackled by fixing c in $(0, 1)$ and quoting the region $\{\theta : L(\theta)/L(\hat{\theta}) \geq c\}$ of values which are relatively well supported.

Likelihood methods are as yet largely unexplored. Complications rapidly arise in multiparameter problems and in the presence of nuisance parameters. Further works on likelihood, and extensions of the idea, are Barnard [5, 6], Barnard et al. [7], Hacking [62], Kalbfleisch and Sprott [72], Kalbfleisch [70], and Edwards [39].

Some likelihood ideas have been absorbed into sampling theory, thereby losing their pure likelihood interpretation. For example, the maximum likelihood estimate can be found for any data outcome, yielding an estimator, the maximum likelihood estimator, whose performance can be assessed in sampling theory terms (it performs very well, in general, for large samples). But such an assessment involves the density $f(x \mid \theta)$ at values for x other than x_0, so that it is not based solely on $L(\cdot)$. Similarly, likelihood interval estimation can be assessed by nonlikelihood methods.

Fiducial Inference*

Fiducial inference is the most problematic of modern theories of inference. Introduced by Fisher [52, 54, 55], mainly by example, its principles have often been obscure and open to various interpretations. The method, when it applies, extracts a probability distribution for Θ on the basis of the data, without having first input any prior distribution. Indeed, this *fiducial distribution* is considered relevant only for the case that Θ is "completely unknown" before the experiment.

In the potato problem, we first restrict attention to the sufficient statistic, $\bar{X}$. We may then note that the distribution of the *pivotal function* $\bar{X} - \Theta$ is $\mathcal{N}(0, 7.07^2)$, whatever Θ may be. The controversial fiducial step is to regard this distribution of $\bar{X} - \Theta$ as still relevant after observing $\bar{X} = 107$. The induced fiducial distribution for Θ is then $\mathcal{N}(107, 7.07^2)$.

Another example is the following. Let R be the sample correlation coefficient

$$\frac{\sum(X_i - \bar{X})(Y_i - \bar{Y})}{\left\{\sum(X_i - \bar{X})^2 \sum(Y_i - \bar{Y})^2\right\}^{1/2}}$$

based on n independent pairs (X_i, Y_i), each having a bivariate normal distribution with population correlation P (capital *rho*!). Let the distribution function of R at r, when $P = \rho$, be $F(r \mid \rho)$. Then, for any given value of P, the distribution of $\Phi = F(R \mid P)$ is uniform on $(0, 1)$. Moreover, if $R = r_0$ is observed, all values for Φ in $(0, 1)$ are still obtainable, as P varies. Then, in the absence of any prior information, the observation $R = r_0$ is regarded, in fiducial logic, as carrying no information about Φ, which is thus still taken to be uniform on $(0, 1)$. The induced fiducial density for P is $-(\partial/\partial\rho)F(r_0 \mid \rho)$.

An attempt to construct a general theory based on Fisher's fiducial ideas is that of Wilkinson [113]. Other relevant references are Lindley [79], Dempster [37], Hacking [62], Pedersen [93], and Dawid and Stone [32a].

Structural Inference*

The theory of structural inference is due to Fraser [57, 59]. Its essential novelty lies in its recognition that an experiment may have structure over and above the family of distributions for X given Θ that it determines, and that such structure should be used, if possible, in inference.

Our model for the potato problem can be strengthened to yield the following *structural model*

$$X_1 = \Theta + E_1$$
$$X_2 = \Theta + E_2$$

with $E_i \sim \mathcal{N}(0, 10^2)$ $(i = 1, 2)$, independently of each other and of the value of Θ.

Here the $\{E_i\}$ are measurement errors, supposed to represent objective characteristics of the measurement process, existing irrespective of what is being measured.

It follow from this structural model that $E_2 - E_1 = X_2 - X_1$. Thus, although E_1 and E_2 are not completely observable, $E_2 - E_1$ is. Having obtained our data (94, 120), we now know that $E_2 - E_1 = 26$. This is the full extent of the logical information in the data about (E_1, E_2), since any values satisfying this relationship are compatible with our data and some value of Θ.

As the distribution of (E_1, E_2) does not involve Θ, $X_2 - X_1 = E_2 - E_1$ is an ancillary statistic, and we take it as appropriate to condition on the observed value of this statistic [53]. Our *reduced* structural model may then be taken as $\bar{X} = \Theta + \bar{E}$, where the relevant distribution of $\bar{E} = \frac{1}{2}(E_1 + E_2)$ is conditional on $E_2 - E_1 = 26$. Further analysis involves the referral of the observed value $\bar{X} = 107$ to this reduced model.

It happens that, with normality, $\bar{E}$ and $E_2 - E_1$ are independent, so that the conditioning described above is irrelevant, we can take $\bar{E} \sim \mathcal{N}(0, 7.07^2)$, and the reduced model just describes the marginal structure of the sufficient statistic $\bar{X}$. This happens in general when a sufficient statistic exists. However, the theory applies just as well if no such sufficient reduction exists, for example if the $\{E_i\}$ have standard Cauchy distributions. In this case the conditioning is necessary, and the relevant reference distribution of $\bar{E}$ will depend on the data (through $X_2 - X_1$).

The reduced model can be used for various sampling theory purposes, such as constructing estimators [96] or tests of signifi-

cance. It may also be used to produce a fiducial distribution, the *structural distribution*, for Θ, using $\bar{E} = \bar{X} - \Theta$ as the pivotal function. With our model assumptions and data, this gives $\Theta \sim \mathcal{N}(107, 7.07^2)$ as in the preceding section. With standard Cauchy errors, Θ would have structural density $g(107 - \theta)$, where g is the density of the average of two standard Cauchy variables, conditional on their difference being 26, i.e.,

$$g(x) \propto \left\{ \left(1 + (x + 13)^2\right)\left(1 + (x - 13)^2\right) \right\}^{-1}.$$

All the foregoing ideas generalize to structural models of the form

$$X = \Theta \circ E,$$

where X and E take values in a space $\mathcal{X}$ on which is given a group G of transformations (the result of applying $g \in G$ to $e \in \mathcal{X}$ being written $g \circ e$), and Θ is known only to take values in G. The structural distribution turns out to be identical with a Bayes posterior with respect to the (usually improper) prior distribution Π over G which corresponds to *right-Haar measure*, i.e., $\Pi(Ag) = \Pi(A)$ for $A \subseteq G$, $g \in G$ [58, 65].

SOME COMPARISONS

Although interpretations differ, the inferences drawn by adherents of different schools can be numerically very similar. In problems with group structure, such as structural models, they often coincide. We illustrate this for the potato problem.

Point Estimate

Consider the estimation of Θ by $\bar{X} = 107$. For a sampling theorist, this is the minimum variance unbiased estimator; for a Bayesian with a uniform prior, it is the mean of the posterior distribution (identical here with the structural distribution). It is also the maximum likelihood estimate.

Interval Estimate

A 95% central confidence interval for Θ is $\bar{X} \pm 1.96 \times 7.07 = (93.15, 120.85)$. The sampling theory interpretation is that an interval constructed by this rule, for all X, can be expected to cover the value of Θ (whatever it may be) with sampling probability 0.95. For the "improper" Bayesian (or structuralist), the posterior probability that the random variable Θ lies between 93.15 and 120.85 is 0.95. The region can also be derived as $\{\theta : L(\theta)/L(\hat{\theta}) \geqslant 0.15\}$: it is "likelihood-based."

Hypothesis Assessment

Consider the hypothesis $H_0 : \Theta \leqslant 100$. The strength of the evidence against H_0 might be measured, for the Bayesian, by the posterior probability that $\Theta > 100$, i.e., 0.84. The sampling theory significance level for testing H_0 against alternative $H_1 : \Theta > 100$ is

$$\Pr(\bar{X} > 107 \mid \Theta = 100) = 0.16 = 1 - 0.84.$$

An interesting parallel in a nonstructural model is for observand X having the Poisson distribution $\mathscr{P}(\Theta)$. The significance level for testing $H_0 : \Theta \leqslant \theta$ against $H_1 : \Theta > \theta$, when $X = x$, may be defined as $p_1 = \Pr(\mathscr{P}(\theta) \geqslant x)$, $p_2 = \Pr(\mathscr{P}(\theta) > x)$, or some compromise, such as $\frac{1}{2}(p_1 + p_2)$. For the Jeffreys prior $\pi(\theta) \propto \theta^{-1/2}$, the posterior probability that $\Theta \leqslant \theta$ always lies between p_1 and p_2. Similar properties hold in some other discrete distributions [1–3].

Asymptotics

Very generally, any suitably regular statistical model can be approximately represented as a normal location model, asymptotically as the number of observands increases [77]. Then the inferences drawn by the various methods will tend to agree, numerically, for very large samples. The maximum likelihood estimator will be asymptotically sufficient for Θ.

For the sample sizes found in practice, it is important to investigate departures from asymptotic limits, for example by considering further terms in asymptotic expansions, or other refinements [10, 41, 51, 100]. In particular, conditioning on asymptotic ancillaries serves to draw sampling theory inferences still closer to Bayesian and likelihood ones [9, 28, 42, 64, 94].

Differences

There are some problems where the different schools must agree to differ. If $X_i \sim \mathscr{N}(\Theta_i, 1)$ independently $(i = 1, \ldots, n)$, a sampling theory unbiased estimator for $\Phi = \sum \Theta_i^2$ is $\sum X_i^2 - n$. The posterior expectation of Φ, for a uniform prior, is $\sum X_i^2 + n$.

Of more practical import is the problem of *sequential analysis**. Let X_i be independently distributed as P_θ when $\Theta = \theta$ $(i = 1, 2, \ldots)$, and suppose that the $\{X_i\}$ are observed one by one. At each stage, a decision is made, based on the data to hand, either to terminate the experiment or to proceed to observe the next X.

Suppose that the final data are $(x_1, x_2, \ldots, x_n)$, and compare (a) the inferences to be drawn from these data in the sequential experiment with (b) those appropriate to the same data for a (conceptual) fixed-size experiment which set out to observe $(X_1, X_2, \ldots, X_n)$. It may be shown that the likelihood function is the same in the two cases. Thus likelihood and subjectivist Bayesian inferences are unaffected by the optional stopping. (The Jeffreys priors, however, which depend on the structure of the experiment, may differ. Also, if the fixed-size experiment constitutes a structural model, this property may be lost in the sequential case.)

There is a drastic effect, however, on sampling theory inference, since the spaces over which the sampling distributions are defined are quite different. Suppose that $X_i \sim \mathscr{N}(\Theta, 1)$, and consider testing $H_0 : \Theta = 0$ against $H_1 : \Theta \neq 0$. For the fixed-sample-size experiment, a test at 5% rejects H_0 if $|\bar{X}| > 1.96n^{-1/2}$. But even if $\Theta = 0$, it is certain (by the law of the iterated logarithm*) that, if sampling continues long enough, this condition will hold for some n. If we perform the sequential experiment which stops as soon as this happens, and then try to make inferences, according to the foregoing rule, as if n had been fixed, we shall always reject H_0, even when it is true. Thus sequential sampling has increased the type I error of this test from 0.05 to 1. From the point of view of Neyman's inductive behavior, it is therefore not possible to regard the inference as unaltered by sequential sampling. Fisher's views on the matter are unrecorded, but in view of his general antipathy to equating the *evidence* against H_0 with the *infrequency* of obtaining such evidence under H_0, he might have been willing to accept the fixed-size test as still appropriate in the sequential case.

Now an alternative description of the stopping rule above is: stop as soon as the posterior probability that $\Theta > 0$, for uniform prior, lies outside $(0.025, 0.975)$. So, even if $\Theta = 0$, the Bayesian is bound to find strong evidence that $\Theta < 0$ or $\Theta > 0$. This paradox has been proposed as an inconsistency in Bayesian inference [4]. A counterargument [26] is that it is irrelevant, since the prior implies that $\Theta \neq 0$ with probability 1. (The paradox fails if there is positive prior probability that $\Theta = 0$.)

TWO SIMPLE HYPOTHESES

Although rarely encountered in practice, this simplest of all statistical models is of fundamental theoretical importance. The parameter Θ takes one of two values, θ_0 or θ_1; the observand X ranges over an arbitrary space $\mathscr{X}$, with distribution P_i when $\Theta = \theta_i$; the density of P_i, with respect to some dominating measure μ on $\mathscr{X}$, is $f_i(\cdot)$. We suppose, for simplicity, that $f_i(x) > 0$ for all x, and define $\lambda(x) = f_1(x)/f_0(x)$. [Note that changing μ to μ^*, say, changes $f_i(x)$ to $f_i^*(x) = f_i(x)g(x)$, where $g = d\mu/d\mu^*$. Thus λ is unaffected.]

Sufficiency

We can write $f_i(x) = a(x)b(\lambda(x), \theta_i)$, where $a(x) = f_0(x)$, $b(\lambda, \theta_0) = 1$, $b(\lambda, \theta_1) = \lambda$. So by the Fisher–Neyman criterion, the *likelihood ratio statistic* $\Lambda = \lambda(X)$ is sufficient for Θ.

Support

Consider the following weak interpretation of the intuitive concept of *support* in terms of likelihood.

WEAK LAW OF SUPPORT. If $\lambda(x_2) > \lambda(x_1)$, we write $x_2 \succ x_1$, and infer that the observation of $X = x_2$ provides more relative support in favor of $\Theta = \theta_1$ (as against $\Theta = \theta_0$) than does the observation of $X = x_1$.

We shall see how this law is acceptable to most schools of inference, each in its own terms.

Sampling Theory Inference

In testing $H_0 : \Theta = \theta_0$ against $H_1 : \Theta = \theta_1$, the Neyman–Pearson lemma* [86] shows that the only admissible tests are *likelihood ratio tests*, which, for some *cutoff* c, accept H_0 if $\lambda(x) < c$, and reject if $\lambda(x) > c$. In particular, if $x_2 \succ x_1$, and H_0 is rejected on data x_1, it is also rejected on data x_2, as the weak law of support would require.

If we consider *significance assessment rules*, which, for some test statistic T, with observed value t, quote $S_T(x) = P_0(T \geqslant t)$ as the observed significance level (small values giving evidence against H_0 and in favor of H_1), then that based on $T = \Lambda$ is optimal (at least if Λ has a continuous distribution): it is *uniformly most sensitive* in that $P_1(S_\Lambda \leqslant \alpha) \geqslant P_1(S_T \leqslant \alpha)$ for all T, and $\alpha \in (0, 1)$ [73]. But if $x_2 \succ x_1$, then $S_\Lambda(x_2) < S_\Lambda(x_1)$, in accordance with the weak law of support.

Bayesian Inference

Let $\omega = \pi(\theta_1)/\pi(\theta_0)$ be the prior odds in favor of $\Theta = \theta_1$ as against $\Theta = \theta_0$, and $\omega(x) = \pi(\theta_1 \mid x)/\pi(\theta_0 \mid x)$ the posterior odds. From Bayes' theorem*, $\omega(x) = \omega.\lambda(x)$. In particular, if $x_2 \succ x_1$, then $\omega(x_2) > \omega(x_1)$, again agreeing with the law.

Likelihood Inference

For data $X = x$, $L(\cdot)$ is a function on $\{\theta_0, \theta_1\}$ with values proportional to $\{f_0(x), f_1(x)\}$. It is thus completely determined (up to proportionality) by $\lambda(x)$. The weak law of support is naturally compatible with the likelihood idea that $\lambda(x)$ [or $\log \lambda(x)$] provides a direct measure of relative support.

Structural Inference

If a structural model $X = \Theta \circ E$ underlies the problem, then the structural distribution will be the same as the Bayes posterior for the prior probabilities $\pi(\theta_0) = \pi(\theta_1) = \frac{1}{2}$. So the structural odds in favor of $\Theta = \theta_1$ are just $\lambda(x)$.

Decision Theory: Double Dichotomy

Suppose that a choice must be made between two actions, a_0 and a_1, a_i being more preferred if $\Theta = \theta_i$. Let l_{ij} denote the loss in deciding a_i when $\Theta = \theta_j$. The risk function of a test ϕ, for the two values θ_0 and θ_1, may be plotted as the point $(r_0, r_1) = (l_{00}, l_{11}) + (b_0 \alpha, b_1 \beta)$, where (α, β) are the type I and type II errors, and $b_0 = l_{10} - l_{00} > 0$, $b_1 = l_{01} - l_{11} > 0$. So comparison of tests, for arbitrary such losses, is on the basis of their error probabilities. The admissible tests are thus just the likelihood ratio tests.

If $\pi_i = \pi(\theta_i)$ $(i = 0, 1)$, for an arbitrary prior distribution, the Bayes criterion, using the $\{\pi_i\}$ as weights, chooses the test ϕ_π that minimizes the *Bayes risk* $\pi_0 r_0 + \pi_1 r_1$, or equivalently $\pi_0 b_0 \alpha + \pi_1 b_1 \beta$. This leads to the test which chooses a_0 or a_1 according as $\lambda(x) <$ or $> (\pi_0 b_0 / \pi_1 b_1)$. In particular, when likelihood ratio tests* are selected by the Bayes criterion (with fixed l's and π's), the relevant *cutoff* does not depend on any features of the sampling distributions—the optimal decision depends only on the observed likelihood function. In contrast, minimax* selection, or the more usual minimization of β for fixed size α, does depend on the sampling distribution. This dependence can lead to paradoxes and anomalies when comparing decisions taken in different experiments [81].

Conditional and Unconditional Behavior [27]

Consider again the random choice between two balances introduced in the section "Conditional Inference," and suppose it is known that Θ must be either $\theta_0 = 100$ or

$\theta_1 = 150$. The likelihood ratio statistic based on the full observand $(Y; X_1, X_2)$ is $\Lambda = \exp\{40(\overline{X} - 125)/Y^2\}$. A likelihood ratio test thus rejects $H_0 : \Theta = 100$ when, for some constant c, $\overline{X} > 125 + cY^2$. Note that the same value of c applies both for $Y = 10$ and for $Y = 50$. The value of c might be chosen (from a NPW perspective) to control, for example at 5%, the overall type I error probability $\alpha = \Pr(\overline{X} > 125 + cY^2 \mid \Theta = 100)$. Here $\alpha = \frac{1}{2}(\alpha_{10} + \alpha_{50})$, where

$$\alpha_y = \Pr(\overline{X} > 125 + cy^2 \mid \Theta = 100, Y = y)$$
$$= \Pr(\mathcal{N}(100, y^2) > 125 + cy^2)$$

is the type I error conditional on $Y = y$ $(y = 10, 50)$. Note that, in this approach, α_{10} and α_{50} are determined by adjusting c to fix their average α, rather than directly.

An alternative approach looks separately at the two conditional problems, in which Y is fixed, without attempting to relate them. Thus α_{10} and α_{50} might independently be controlled (both might be set at 5%, say), and the implied likelihood ratio tests, for the conditional problems, constructed. This would require setting two cutoff constants, k_{10} and k_{50}, and rejecting H_0 if $\bar{x} > k_y$. The value of k_y is determined by the requirement $\Pr(\mathcal{N}(100, y^2) > k_y) = \alpha_y$ $(y = 10, 50)$. This conditional approach appears more in line with Fisherian principles of relevant conditioning.

We note that an unconditional likelihood ratio test arises when k_y has the form $125 + cy^2$, for some value of c, or, equivalently, $k_{50} = 25k_{10} - 3000$. In general, the values of k_{10} and k_{50} selected by the conditional approach will not be related in this way, so that we shall not have a likelihood ratio test overall, and so appear to be ignoring the various arguments in favor of such a test.

Table 1 examines the unconditional and conditional behavior of five possible tests. (Here β denotes a type II error.) Test 1 is constructed to give $\alpha_{10} = \alpha_{50} = 0.05$. However, it is dominated (unconditionally, although not conditionally) by test 2, which is the overall likelihood ratio test at $\alpha = 0.05$. Tests 3 and 4 are the overall likeli-

Table 1

Test	c	k_{10}	k_{50}	$(\alpha_{10}, \beta_{10})$	$(\alpha_{50}, \beta_{50})$	(α, β)
1	—	111.63	158.16	$(0.05, 3 \times 10^{-8})$	$(0.05, 0.59)$	$(0.05, 0.30)$
2	8.135×10^{-3}	125.81	145.34	$(10^{-4}, 3 \times 10^{-4})$	$(0.0999, 0.45)$	$(0.05, 0.23)$
3	-0.134	111.63	-209.2	$(0.05, 3 \times 10^{-8})$	$(1^{-}, 0^{+})$	$(0.525, 2 \times 10^{-8})$
4	0.0133	126.33	158.16	$(10^{-4}, 4 \times 10^{-4})$	$(0.05, 0.59)$	$(0.025, 0.30)$
5	0	125	125	$(2 \times 10^{-4}, 2 \times 10^{-4})$	$(0.24, 0.24)$	$(0.12, 0.12)$

hood ratio tests for which $\alpha_{10} = 0.05$ and $\alpha_{50} = 0.05$, respectively. If one accepts the supremacy of likelihood ratio tests, unconditionally as well as conditionally, the moral appears to be that one should not commit oneself too firmly to a favorite level, such as 0.05, for α, for to do so in some circumstances is to imply that a very different level is appropriate in others.

It may be considered that none of these tests is satisfactory. Test 5, which minimizes $\alpha + \beta$ both conditionally and unconditionally, is perhaps more reasonable. (The equality of α and β is an irrelevant consequence of the symmetry of this specific problem.)

Note that any direct likelihood or Bayesian approach to this problem will fix the likelihood ratio cutoff c, rather than any error probabilities, and so yield the same result whether applied to the conditional or unconditional model. The unconditional NPW argument also seems to point to the value of c as meaningful across different experiments, as does the related analysis of Pitman [97].

In the foregoing framework of testing hypotheses, the distinction between conditional and unconditional testing can be removed by relating k_{10} and k_{50} appropriately. No such escape seems possible when assessing significance. For our data, the observed significance level against H_0 is

$$\Pr(\overline{X} > 107 \mid Y = 10, \Theta = 100) = 0.16.$$

In the overall experiment, this can be obtained as $\Pr(T \geqslant t \mid \Theta = 100)$, with $T = (\overline{X} - 100)/Y$, $t = 0.7$. [Thus $T \sim \mathcal{N}(0, 1)$, both conditionally and unconditionally, under H_0.] But the uniformly most sensitive test statistic overall, for alternative $H_1 : \Theta = 150$, is not T, but $T^* = (\overline{X} - 125)/Y^2$

(effectively the likelihood ratio statistic), with observed value $t^* = -0.18$, conditional observed significance levels 0.16 for $Y = 10$ and 1^- for $Y = 50$, and overall observed significance level 0.58. Of course, T^* depends critically on the choice of H_1, but even allowing H_1 to specify an arbitrary distribution for the readings, on each balance, it is impossible to reproduce the conditional level as an overall level based on a likelihood ratio test. There is thus a direct contradiction between Fisherian principles of relevant conditioning, and NPW criteria of optimality (or Bayesian or likelihood arguments). There is even a clash with the Fisherian desire to use a sufficient statistic, which is T^* in the problem above. The data $(Y; \overline{X}) = (50; -325)$ would yield the same value, -0.18 for T^* as our own, but a very different conditional observed significance level. This suggests that the use of an observed significance level as a measure of evidence against a hypothesis, attractive though it appears, is in conflict with any reasonably general attitudes to inference.

PRINCIPLES OF INFERENCE

One approach to a logic of statistical inference is to set up axioms that appear more or less compelling, and to investigate their consequences. The pioneering work in this field is that of Birnbaum [17–19]; see also Cox and Hinkley [29], Basu [13], Dawid [31], and Godambe [60].

Consider an *inference pattern* I which produces inference $I(\xi, x)$ about Θ from data $X = x$ in experiment ξ. *Conformity principles* are requirements that if (ξ_1, x_1) and (ξ_2, x_2) are suitably related, then $I(\xi_1, x_1)$ and $I(\xi_2,$

x_2) should be identical. We present some of these below; the motivation for the principles has already been touched on.

Sufficiency Principle. For a single experiment ξ, if x_1 and x_2 produce the same value for the (minimal) sufficient statistic, we should make the same inference from both: $I(\xi, x_1) = I(\xi, x_2)$.

Conditionality Principle. Let ξ_1, ξ_2 be experiments, and let ξ be the experiment of tossing a fair coin to choose between ξ_1 and ξ_2, noting the outcome, and performing the chosen experiment. Then require that $I(\xi, (\xi_1, x)) = I(\xi_1, x)$.

Ancillarity Principle. Let S be an ancillary statistic in ξ, and let ξ_s denote the conditional experiment, given $S = s$. Then require that $I(\xi, x) = I(\xi_s, x)$, where $s = S(x)$.

Likelihood Principle*. Let $L_{(\xi, x)}(\cdot)$ denote the likelihood function for Θ obtained for data x in experiment ξ. Then require $I(\xi_1, x_1) = I(\xi_2, x_2)$ whenever (as functions of θ) $L_{(\xi_1, x_1)}(\theta) \propto L_{(\xi_2, x_2)}(\theta)$.

The likelihood principle implies the others above. Conversely, Birnbaum proved that it is implied by the conjunction of the sufficiency and conditionality principles. But although direct likelihood and Bayesian inferences are in accord with the likelihood principle, it is often (as we have seen) impossible to satisfy it by means of sampling theory inferences. This is because such inferences involve features of the sampling distributions in the two experiments ξ_1 and ξ_2, which may be very different. So it seems that statisticians who wish to make inferences in sampling theory terms must reject, or at least modify, one or both of the seemingly harmless principles of sufficiency and conditionality. Durbin [38] and Kalbfleisch [71] attempt such modification. Alternatively, statisticians must restrict their inference patterns, if possible, to ones that are compatible with likelihood ideas: for example, by using a constant cutoff in their likelihood ratio tests,

as in the section "Decision Theory: Double Dichotomy."

POSTSCRIPT

How important are the attitudes, insights, and arguments about inference described here? It is implausible that statisticians will ever come to agree on the fundamental ground rules of statistical inference. Differing schools of thought seem naturally congenial to differing individuals, according to their several views of the nature and purpose of inference, the importance of general applicability and self-consistency, and the demands of applied research. Arguments between rival views often founder on unrecognized differences in interpretation of terms. Fortunately, many practising statisticians, although more or less in agreement with one specific viewpoint, are happy to borrow techniques and insights from other schools when it seems appropriate. Indeed, in the context of applied statistical problem solving, many of the theoretical arguments about the foundations of inference appear sterile and irrelevant.

Practical statisticians must concern themselves with questions such as: What data should be subjected to analysis? What variables should be included? How finely should the observations be stratified? How can causation* and association* be distinguished? Only then are they ready even to think about formulating a statistical model, and in so doing, tackle such puzzles as: How can probability be used to model relationships between variables in a 100% census? Should the model selected take account of my own acts of randomization? What is the meaning of the probability of rain tomorrow? Only when a tentative model has been settled on can a statistician turn to the comparatively less important inferential questions addressed by this article. Although the various schools of inference do supply theory and methods for attacking some of the problems noted above, it is a criticism of the present state of research into the foundations of statistical inference that, by and large, it has been content to take the statistical model as its starting point, so ignoring some of the most important concerns of statistical science. And it is this which, fortunately for this article but unfortunately for statistics, allows "statistical inference" to constitute a small subset of this encyclopedia rather than its totality.

References

[1] Altham, P. M. E. (1969). *J. R. Statist. Soc. B*, **31**, 261–269.

[2] Altham, P. M. E. (1971). *Biometrika*, **58**, 561–576.

[3] Altham, P. M. E. (1971). *Biometrika*, **58**, 679–680.

[4] Armitage, P. (1963). *J. Amer. Statist. Ass.*, **58**, 384–387.

[5] Barnard, G. A. (1949). *J. R. Statist. Soc. B*, **11**, 115–149.

[6] Barnard, G. A. (1966). *Proc. 5th Berkeley Symp. Math. Statist. Prob.*, Vol. 1. University of California Press, Berkeley, Calif., pp. 27–40.

[7] Barnard, G. A., Jenkins, G. M., and Winsten, C. B. (1962). *J. R. Statist. Soc. A*., **125**, 321–372.

[8] Barndorff–Nielsen, O. (1978). *Information and Exponential Families in Statistical Theory*. Wiley, New York. (Advanced. Discusses logical principles of inference in relation to exponential families. Emphasis on generalized sufficiency and ancillarity. No exercises.)

[9] Barndorff–Nielsen, O. (1980). *Biometrika*, **67**, 293–310.

[10] Barndorff–Nielsen, O. and Cox, D. R. (1979). *J. R. Statist. Soc. B*, **41**, 279–312.

[11] Barnett, V. (1973). *Comparative Statistical Inference*. Wiley, New York. (Elementary. Describes and compares sampling theory and Bayesian theories of inference and decision. No exercises.)

[12] Basu, D. (1959). *Sankhyā*, **21**, 247–256.

[13] Basu, D. (1975). *Sankhyā A*, **37**, 1–71.

[14] Bayes, T. (1763). *Philos. Trans.*, **53**, 370–418. (Reprinted in *Biometrika*, **45**, 213–315, 1958.) (The original Bayes' theorem; also an ingenious argument for the "equal distribution of ignorance.")

[15] Bennett, J. H., ed. (1971). *Collected Papers of R. A. Fisher*, Vols. 1–5. University of Adelaide, Adelaide, Australia.

[16] Bernardo, J.-M. (1979). *J. R. Statist. Soc. B*, **41**, 113–147.

[17] Birnbaum, A. (1962). *J. Amer. Statist. Ass.*, **57**, 269–326. (Pathbreaking paper on logical principles of inference.)

[18] Birnbaum, A. (1969). In *Essays in Honor of Ernest Nagel*, S. Morgenbesser, P. Suppes, and M. White, eds. St. Martin's Press, New York.

[19] Birnbaum, A. (1972). *J. Amer. Statist. Ass.*, **67**, 858–861.

[20] Birnbaum, A. (1977). *Synthèse*, **36**, 19–49.

[21] Brown, L. (1967). *Ann. Math. Statist.*, **38**, 1068–1071.

[22] Buehler, R. J. (1959). *Ann. Math. Statist.*, **30**, 845–863.

[23] Buehler, R. J. and Feddersen, A. P. (1963). *Ann. Math. Statist.*, **34**, 1098–1100.

[24] Carnap, R. (1950). *Logical Foundations of Probability*. University of Chicago Press, Chicago.

[25] Carnap, R. (1952). *The Continuum of Inductive Methods*. University of Chicago Press, Chicago.

[26] Cornfield, J. (1970). In *Bayesian Statistics*, D. L. Meyer and R. O. Collier, Jr., eds. F. E. Peacock, Itasca, Ill., pp. 1–28.

[27] Cox, D. R. (1958). *Ann. Math. Statist.*, **29**, 357–372.

[28] Cox, D. R. (1980). *Biometrika*, **67**, 279–286.

[29] Cox, D. R. and Hinkley, D. V. (1974). *Theoretical Statistics*. Chapman & Hall, London. (Advanced. Comprehensive and detailed account of general principles and their specific applications. Many challenging exercises.)

[30] Dawid, A. P. (1975). *J. R. Statist. Soc. B*, **37**, 248–258.

[31] Dawid, A. P. (1977). In *Recent Developments in Statistics*, J. R. Barra, B. van Cutsen, F. Brodeau, and G. Romier, eds. North-Holland, Amsterdam, pp. 245–256.

[32] Dawid, A. P. (1980). In *Bayesian Statistics*, J. M. Bernardo, M. H. De Groot, D. V. Lindley, and A. F. M. Smith, eds. University Press, Valencia, pp. 167–184.

[32a] Dawid, A. P. and Stone, M. (1982). *Ann. Statist.*, **10**, 1054–1074.

[33] Dawid, A. P., Stone, M., and Zidek, J. V. (1973). *J. R. Statist. Soc. B*, **35**, 189–233.

[34] de Finetti, B. (1937). *Ann. Inst. Henri Poincaré*, **7**, 1–68. (Reprinted in English, in Kyburg and Smokler [75]. Contains the famous representation theorem for exchangeable events.)

[35] de Finetti, B. (1975). *Theory of Probability* (English transl.), 2 vols. Wiley, New York. (Thought-provoking synthesis of de Finetti's construction of subjective Bayesian probability and statistics.)

[36] De Groot, M. H. (1970). *Optimal Statistical Decisions*. McGraw-Hill, New York. (Intermediate.

Thorough account of Bayesian inference and decision theory. Exercises.)

[37] Dempster, A. P. (1963). *Ann. Math. Statist.*, **34**, 884–891.

[38] Durbin, J. (1970). *J. Amer. Statist. Ass.*, **65**, 395–398.

[39] Edwards, A. W. F. (1972). *Likelihood*. Cambridge University Press, Cambridge. (Clear account of logic and elementary methods of the likelihood approach.)

[40] Edwards, W., Lindman, H., and Savage, L. J. (1963). *Psychol. Rev.*, **70**, 193–242.

[41] Efron, B. (1975). *Ann. Statist.*, **3**, 1189–1242.

[42] Efron, B. and Hinkley, D. V. (1978). *Biometrika*, **65**, 457–487.

[43] Ferguson, T. S. (1967). *Mathematical Statistics: A Decision Theoretic Approach*. Academic Press, New York. (Advanced. Clear account of NPW decision theory and its relationship with Bayesian inference. Challenging exercises.)

[44] Fieller, E. C. (1940). *J. R. Statist. Soc. Suppl.*, **7**, 1–64.

[45] Fieller, E. C. (1954). *J. R. Statist. Soc. B*, **16**, 175–185.

[46] Fienberg, S. E. and Hinkley, D. V., eds. (1980). *R. A. Fisher: An Appreciation*. Lecture Notes in Statistics No. 1. Springer-Verlag, New York. (Eighteen articles on Fisher and his statistical work, including: "Basic theory of the 1922 mathematical statistics paper" by S. Geisser; "Theory of statistical estimation: the 1925 paper" and "Fisher's development of conditional inference" by D. V. Hinkley; "Fiducial inference" by R. J. Buehler; and "The Behrens–Fisher and Fieller–Creasy problems" by D. L. Wallace.)

[47] Fisher, R. A. (1912). *Messenger Math.* **41**, 155–160. In *Collected Papers* [15], Vol. 1, pp. 53–58. (Introduction to maximum likelihood estimation.)

[48] Fisher, R. A. (1920). *Monthly Notices R. Astron. Soc.*, **80**, 758–770. In *Collected Papers* [15], Vol. 1, pp. 188–201, with author's introductory note [Derives conditional distribution of $\sum |X_i - \overline{X}|$ given $\sum (X_i - \overline{X})^2$ in a normal sample of size 4, and notes its independence of the parameters.]

[49] Fisher, R. A. (1921). *Metron*, **1**, 3–32. In *Collected Papers* [15], Vol. 1, pp. 205–235, with author's introductory note. (Contrasts likelihood function and posterior density.)

[50] Fisher, R. A. (1922). *Philos. Trans. R. Soc. Lond. A*, **222**, 309–368. In *Collected Papers* [15], Vol. 1, pp. 275–335, with author's introductory note. ("The first large-scale attack on the problem of estimation.")

[51] Fisher, R. A. (1925). *Proc. Camb. Philos. Soc.*, **22**, 700–725. In *Collected Papers* [15], Vol. 2, pp.

13–40, with author's introductory note. (Extends and refines Fisher [50].)

[52] Fisher, R. A. (1930). *Proc. Camb. Philos. Soc.*, **26**, 528–535. In *Collected Papers* [15], Vol. 2, pp. 428–436, with author's introductory note. (The fiducial argument, illustrated for the correlation coefficient.)

[53] Fisher, R. A. (1934). *Proc. R. Soc. A*, **144**, 285–307. In *Collected Papers* [15], Vol. 3, pp. 114–137, with author's introductory note. (Sufficiency and ancillarity in relation to exponential families and group-structural models.)

[54] Fisher, R. A. (1935). *Ann. Eugen. (Lond.)*, **6**, 391–398. In *Collected Papers* [15], Vol. 3, pp. 316–324, with author's introductory note. (Pivotal examples of fiducial inference and prediction.)

[55] Fisher, R. A. (1956). *Statistical Methods and Scientific Inference*. Oliver & Boyd, Edinburgh. (The synthesis of Fisher's views on the logic of inference. Typically Fisherian, slippery but stimulating.)

[56] Fisher, R. A. (1956). *J. R. Statist. Soc. B*, **18**, 56–60. In *Collected Papers* [15], Vol. 5, pp. 353–357.

[57] Fraser, D. A. S. (1961). *Biometrika*, **48**, 261–280.

[58] Fraser, D. A. S. (1961). *Ann. Math. Statist.*, **32**, 661–676.

[59] Fraser, D. A. S. (1968). *The Structure of Inference*. Wiley, New York. (Advanced account of structural models and their associated fiducial analysis. Exercises.)

[60] Godambe, V. P. (1979). *J. R. Statist. Soc. B*, **41**, 107–110.

[61] Godambe, V. P. and Sprott, D. A., eds. (1971). *Foundations of Statistical Inference*. Holt, Rinehart and Winston, New York. (Conference proceedings; 28 stimulating papers, with discussion.)

[62] Hacking, I. (1965). *Logic of Statistical Inference*. Cambridge University Press, Cambridge. (Penetrating review of current concepts of inference. Attempts to justify a fiducial argument by means of the "law of likelihood.")

[63] Hartigan, J. (1964). *Ann. Math. Statist.*, **35**, 836–845.

[64] Hinkley, D. V. (1980). *Biometrika*, **67**, 287–292.

[65] Hora, R. B. and Buehler, R. J. (1966). *Ann. Math. Statist.*, **37**, 643–656.

[66] Jaynes, E. T. (1968). *IEEE Trans. Syst. Sci. Cybern.*, **SSC-4**, 227–241.

[67] Jaynes, E. T. (1976). In *Foundations of Probability Theory. Statistical Inference and Statistical Theories of Science*, Vol. 2, W. L. Harper and C. A. Hooker, eds. D. Reidel, Dordrecht, Holland, pp. 175–257.

[68] Jaynes, E. T. (1980). In *Bayesian Analysis in Econometrics and Statistics: Essays in Honor of Harold Jeffreys*, A. Zellner, ed. North-Holland, Amsterdam, pp. 43–87. (With comments by A. P. Dawid, M. Stone, and J. V. Zidek, and Reply.)

[69] Jeffreys, H. (1961). *Theory of Probability*, 3rd ed. Clarendon Press, Oxford (1st ed., 1939). (Highly original logical and mathematical development of a Bayesian methodology based on Jeffreys' philosophy of science.)

[70] Kalbfleisch, J. D. (1971). In *Foundations of Statistical Inference*, V. P. Godambe and D. A. Sprott, eds. Holt, Rinehart and Winston, New York, pp. 378–392.

[71] Kalbfleisch, J. D. (1975). *Biometrika*, **62**, 251–259.

[72] Kalbfleisch, J. D. and Sprott, D. A. (1970). *J. R. Statist. Soc. B*, **32**, 175–208.

[73] Kempthorne, O. and Folks, J. L. (1971). *Probability, Statistics, and Data Analysis*. Iowa State University Press, Ames, Iowa. (Intermediate. Emphasizes logic of statistical methods, with good account of significance assessment. Exercises.)

[74] Keynes, J. M. (1921). *A Treatise on Probability*. Macmillan, London.

[75] Kyburg, H. E., Jr. and Smokler, H. E. (1964). *Studies in Subjective Probability*. Wiley, New York. (Contains source papers by Venn, Borel, Ramsey, de Finetti, Koopman, and Savage.)

[76] Laplace, P. S. de (1820). *Théorie Analytique des Probabilités*, 3rd ed. Courcier, Paris.

[77] Le Cam, L. (1960). *Univ. Calif. Publ. Statist.*, **3**, 37–98.

[78] Lehmann, E. L. (1959). *Testing Statistical Hypotheses*. Wiley, New York. (Advanced. The definitive text of the NPW approach. Exercises.)

[79] Lindley, D. V. (1958). *J. R. Statist. Soc. B*, **20**, 102–107.

[80] Lindley, D. V. (1965). *Introduction to Probability and Statistics from a Bayesian Viewpoint*, Part 2: *Inference*. Cambridge University Press, Cambridge. (Lucid student text on basic Bayesian ideas and methods. Exercises.)

[81] Lindley, D. V. (1971). *Bayesian Statistics: A Review*. SIAM, Philadelphia. (Informative survey.)

[82] Linnik, J. V. (1968). *Statistical Problems with Nuisance Parameters*. Translation of 1966 Russian edition. American Mathematical Society, Providence, R.I. (Advanced monograph on the existence of similar tests and statistics.)

[83] Neyman, J. (1957). *Rev. Int. Statist. Inst.*, **25**, 7–22.

[84] Neyman, J. (1967). *A Selection of Early Statistical Papers of J. Neyman*. Cambridge University Press, Cambridge. (28 papers written before 1946.)

[85] Neyman, J. and Le Cam, L. M., eds. (1965). *Bernoulli, 1713; Bayes, 1763; Laplace, 1813*. Springer-Verlag, Berlin. (Proceedings of International Research Seminar, Berkeley, Calif., 1963.)

[86] Neyman, J. and Pearson, E. S. (1933). *Philos. Trans. R. Soc. A*, **231**, 289–337. In *Joint Statistical Papers* [90], pp. 140–185. (Proves the famous lemma.)

[87] Neyman, J. and Pearson, E. S. (1933). *Proc. Camb. Philos. Soc.*, **24**, 492–510. In *Joint Statistical Papers* [90], pp. 186–202.

[88] Neyman, J. and Pearson, E. S. (1936). *Statist. Res. Mem.*, **1**, 1–37. In *Joint Statistical Papers* [90], pp. 203–239.

[89] Neyman, J. and Pearson, E. S. (1938). *Statist. Res. Mem.*, **2**, 25–57. In *Joint Statistical Papers* [90], pp. 265–299.

[90] Neyman, J. and Pearson, E. S. (1967). *Joint Statistical Papers*. Cambridge University Press, Cambridge.

[91] Novick, M. R. (1969). *J. R. Statist. Soc. B*, **31**, 29–51.

[92] Pearson, E. S. and Hartley, H. O., eds. (1954). *Biometrika Tables for Statisticians*, Vol. 1. Cambridge University Press, Cambridge.

[93] Pedersen, J. G. (1978). *Int. Statist. Rev.*, **46**, 147–170. (Critical survey of Fisher's fiducial writings, and connection with recognizable subsets.)

[94] Peers, H. W. (1978). *Biometrika*, **65**, 489–496.

[95] Pierce, D. A. (1973). *Ann. Statist.*, **1**, 241–250.

[96] Pitman, E. J. G. (1938). *Biometrika*, **30**, 391–421.

[97] Pitman, E. J. G. (1965). In *Bernoulli, 1713; Bayes, 1763; Laplace, 1813*, J. Neyman and L. M. Le Cam, eds. Springer-Verlag, Berlin, pp. 209–216.

[98] Raiffa, H. and Schlaifer, R. (1961). *Applied Statistical Decision Theory*. MIT Press, Cambridge, Mass. (Advanced. Highly original, the bible of Bayesian decision theory. No exercises or index.)

[99] Ramsey, F. P. (1926). In *The Foundation of Mathematics and Other Logical Essays*. Routledge & Kegan Paul, London. (Reprinted in Kyburg and Smokler [75].)

[100] Rao, C. R. (1962). *J. R. Statist. Soc. B*, **24**, 46–72.

[101] Robinson, G. K. (1975). *Biometrika*, **62**, 155–161.

[102] Robinson, G. K. (1979). *Ann. Statist.*, **7**, 742–755.

[103] Robinson, G. K. (1979). *Ann. Statist.*, **7**, 756–771.

[104] Savage, L. J. (1954). *The Foundations of Statistics*. Wiley, New York. (Advanced. Axiomatic

development of a subjectivist Bayesian decision theory.)

[105] Savage, L. J. (1961). The Subjective Basis of Statistical Practice. Duplicated manuscript, University of Michigan, Ann Arbor, Mich.

[106] Savage, L. J. (1962). *The Foundations of Statistical Inference*. Methuen, London. (Report of informal conference, London, 1959. Opened by L. J. Savage: "Subjective probability and statistical practice." Contributions by M. S. Bartlett, G. A. Barnard, D. R. Cox, E. S. Pearson, C. A. B. Smith, and Discussion.)

[107] Savage, L. J. (1976). *Ann. Statist.*, **4**, 441–500.

[108] Stigler, S. M. (1976). *Ann. Statist.*, **4**, 498–500.

[109] Villegas, C. (1971). In *Foundations of Statistical Inference*, V. P. Godambe and D. A. Sprott, eds. Holt, Rinehart and Winston, New York, pp. 409–414.

[110] Wald, A. (1950). *Statistical Decision Functions*. Wiley, New York. (Comprehensive presentation and development of the idea of statistics as the choice of a rule for basing decisions on data. Difficult, no exercises.)

[111] Wallace, D. L. (1959). *Ann. Math. Statist.*, **30**, 864–876.

[112] Welch, B. L. (1947). *Biometrika*, **34**, 28–35.

[113] Wilkinson, G. N. (1977). *J. R. Statist. Soc. B*, **39**, 119–171.

[114] Zellner, A. (1971). *An Introduction to Bayesian Statistics in Econometrics*. Wiley, New York. (Intermediate. Interesting textbook account of Bayesian methods at work. Exercises.)

[115] Zellner, A. (1977). In *New Developments in the Applications of Bayesian Methods*, A. Aykac and C. Brumat, eds. North-Holland, Amsterdam, pp. 211–232.

Acknowledgements

Thanks are due to D. R. Cox, J. M. Dickey, M. Stone, and A. D. McLaren for their helpful comments on an earlier draft of this article.

(ANCILLARITY
BAYESIAN INFERENCE
CONDITIONAL INFERENCE
DECISION THEORY
ESTIMATION
FIDUCIAL INFERENCE
HYPOTHESIS TESTING
INFERENCE, STATISTICAL: II
LIKELIHOOD
LOGIC IN STATISTICAL REASONING
MULTIPLE DECISION THEORY
OPTIMIZATION IN STATISTICS
SIMULTANEOUS INFERENCE
STATISTICAL EVIDENCE
STRUCTURAL INFERENCE
SUFFICIENCY)

A. P. DAWID

INFERENCE, STATISTICAL: II

In its essence statistical inference is the whole of central statistics, omitting perhaps certain branch areas such as *decision theory** and *control theory**. We describe briefly its origins and development and then outline the central areas of the subject.

ORIGINS AND DEVELOPMENT

Statistical inference, viewed as the whole of central statistics, is clearly as old as the theory and methods that have come to constitute present statistics. Its origins tie in with those of *probability theory**, going back more than 200 years and including significant names such as *Bayes**, *de Moivre**, *Gauss**, and *Laplace**. Statistical inference is the theory and methods concerned with the way that background infromation and current data make implications concerning unknowns in a system under investigation. The context can be an experiment in the general scientific sense, an observational investigation, or a historical analysis.

The emergence of an identified area, statistical inference, is relatively recent. Central statistics, of course, has existed for more than 200 years and its rate of development intensified in the 1930s and 1940s largely due to the many contributions of R. A. Fisher*, the clear needs from organized agricultural experimentation, and the diverse pressures from World War II.

The course of development of central statistics, however, changed markedly in the decade following World War II. The publication of *Theory of Games and Economic Behavior*, by J. von Neumann and O. Morgenstern [36] and the statistical research of Abraham Wald* (cumulative summary in

Statistical Decision Functions [37]) cast statistics as a process of *decision making** in the face of uncertainty, as a game between the statistician and nature. *Loss functions** and *risk functions** were viewed progressively in this period as providing *the* answer in statistics. The influence of this casting of statistics was substantial in economic theory where *utility** was the close correspondent of risk, and it pervaded most of statistics except those parts closest to traditional scientific investigations.

The mid-1950s witnessed two publications that opened new alternatives to the preceding decision-oriented development. One of these was *The Foundations of Statistics* by L. J. Savage [33], which built close to decision theory using a personalistic *Bayesian** approach. The other was *Statistical Methods and Scientific Inference* by R. A. Fisher [19], which centrally examined the implications about unknowns that follow from statistical models and data. Each book brought to focus certain directions of development that gained momentum in succeeding years. And each provided a fairly clear alternative to the decision-theoretic approach that had been dominant. These directions concerned with basic and foundational questions involving the "implications" from background information and data came to form the identified area of *statistical inference*.

This identification of the area of statistical inference in some measure leaves specialized theory and techniques to other divisions of statistics—for example, confidence* theory, estimation* theory, and hypothesis testing*. However, to the extent that these are neglected in the global view of statistical inference, so also to that extent is statistical inference delinquent in its basic role of determining the "implications" from the given information in an experiment, an investigation, or an analysis.

The term *implications* has been placed in quotation marks as a reminder that it is not commonly used in the type of context here. One commonly hears of theories of inference and how to infer as if there is some major mystical or nonlogical element in the statistical process. This emphasis the author believes is misleading. The fundamental question focuses on what is implied logically by the available information. Certainly an examination of the subject shows that a variety of *additives* are coupled with the given information but the essential still remains—determining the *implications* from this enlarged given [9].

We now examine basic areas that have come to definition within the framework of statistical inference.

REDUCTION METHODS

In a natural sense at the core of statistical inference are to be found *reduction** methods for simplifying or *reducing* the given information available for analysis. Traditionally, this information has been taken to be a density function model $\{f(y\mid\theta):\theta\in\Omega\}$ and an observed response value. A seemingly obvious but largely neglected related component for inference involves the criteria for the organization and development of statistical models; for some recent discussion, see Fraser [28]. We briefly survey various reduction methods.

Sufficiency*

The most conspicuous reduction method in the statistical literature is that of sufficiency —the use of a *sufficient statistic**; a statistic $t(y)$ is sufficient if the conditional distribution of y given t is independent of the parameter θ. The concept is due to Fisher [16]. With the use of a sufficient statistic is naturally associated a *sufficiency principle**, which prescribes the use of a sufficient statistic on the grounds that a subsequent recording of y itself is equivalent to a value from a θ-free distribution and would thus be uninformative. The principle has had almost total acceptance in the profession and there seems little spirit toward questioning it.

The idea of a best sufficient statistic, one that makes a maximum reduction, has received moderate attention in the literature.

Such a statistic is called *exhaustive* [16] or *minimal sufficient* [32] or *necessary and sufficient* [13]. A direct construction procedure is obtained from notions connected with likelihood and mentioned briefly in the next subsection.

Likelihood*

A second reduction method of long-standing presence in statistics is that of calculating the *likelihood function**. The likelihood function from an observed response is the probability for that observed response as a function of the parameter θ and left indeterminate to a positive multiplicative constant: $L(y \mid \theta) = \{cf(y \mid \theta) : c \in (0, \infty)\}$. The concept of likelihood is due to Fisher [16]. Most statistics textbooks omit the multiplicative constant c, thus running counter to the originator's definition and at the same time not providing a key element needed for certain basic theorems; the importance of including the c and conforming to Fisher's original definition is emphasized in Fraser [25].

As a reduction method we can consider the process of going from a sample space point y to the likelihood function $L(y \mid \cdot)$. We can note that multiple sample points may give rise to the same likelihood function or equivalently a given likelihood function can correspond to a range or contour of sample points, the preimage set. Accordingly, we see that the process gives a reduction on the information available concerning a response value. It was noticed in the profession in the early 1960s and currently is recorded in perhaps just one textbook that this process produced the minimal sufficient statistic; the process is called the *likelihood map** in Fraser [25].

It is of interest that the two fundamental and apparently quite distinct concepts of sufficiency and likelihood both due to Fisher coexisted in their distinct ways for some 30 years in the literature before they were linked in the very fundamental way just mentioned. The linking, however, had been clear to Fisher as a careful rereading of his early papers indicated (see, e.g., Fisher [17]).

Two principles are often associated with likelihood. The *weak likelihood principle** prescribes the reduction from a response value y to the likelihood function $L(y \mid \cdot)$, but with the retention of the statistical model; this is equivalent to the sufficiency principle by virtue of the equivalence of the likelihood map and the minimal sufficient statistic. The *strong likelihood principle** prescribes the reduction to the observed likelihood function with no other information concerning the original response value or the statistical model.

Ancillarity*

A third reduction method involves the use of an *ancillary statistic**. A statistic $a(y)$ is an ancillary if it has a θ-free distribution. The concept is due to Fisher [17]. In an application the reduction occurs by replacing the original distribution for y by the conditional distribution for y given the observed value of $a(y)$. The *ancillarity principle** or *conditionality principle** prescribes this reduction on the grounds that the value for $a(y)$ is obtained from a noninformative θ-free distribution. The principle is usually supported by a range of appealing examples, but we will see in the section "Objective Probability Variable" that these same examples illustrate a more fundamental reduction process.

A fundamental difficulty with ancillarity emerges, however, with the remaining common examples—different ancillaries can occur in the same problem with consequent conflict concerning the application of the principle. As a first example, consider a four-way multinomial with cell probabilities $(2 + \theta)/4$, $(1 - \theta)/4$, $(1 - \theta)/4$, $\theta/4$. Combining cells 1 and 3 and combining cells 2 and 4 gives a first ancillary with respective probabilities $3/4$ and $1/4$ independent of θ; combining cells 1 and 2 and combining cells 3 and 4 gives a second ancillary also with a θ-free distribution (coincidentally the same as the preceding); the two ancillaries give contradictory reductions. As a second example consider the bivariate normal* $(0, 0; 1, 1, \rho)$ for (x, y). The projection to the first coordinate x gives the ancillary x with the normal $(0, 1)$ distribution; the projection to the second coordinate y gives a second ancillary y with the normal $(0, 1)$ distribution (coincidentally the same as the preceding); the two ancillaries gives contradictory reductions.

The direct conclusion from these contradictions is that the ancillarity principle is defective—giving reasonable results for the appealing examples because of some more fundamental characteristic of those examples.

Some Interrelations

Birnbaum [7] examined notions of *statistical evidence** and explored the connection between the sufficiency, the conditionality, and the likelihood (strong) principles. Specifically, he focused on an equivalence of the first two combined and the third of the principles. Some objections to this equivalence were proposed in Fraser [22] on the grounds that various elements of structure beyond density could be violated in a direct use of the principles. Also, the validity of Birnbaum's proof may be in doubt. The provoking aspect of the equivalence is that the first two principles have moderate professional support, whereas the third has virtually no support outside the Bayesian statisticians. The resolution of these difficulties may in fact be simple—the seemingly serious defects in the ancillarity principle may make any consideration of the equivalence irrelevant.

A mixture of sufficiency and ancillarity called sufficiency–ancillarity has been proposed [20] and discussed recently [28]. Various weakenings of this sufficiency–ancillarity have been discussed in depth by Barndorff-Nielson [2] and summarized in Fraser [28].

Invariance*

Invariance appears in statistical inference as a moderately attractive reduction method. The typical applications of invariance are, however, somewhat removed from the standard mathematical–scientific use of invariance to determine symmetries and thus eliminate notational and other arbitrary elements.

In the typical application of a class of transformations, closed under product and inverse and thus a group, is introduced to a statistical problem. The transformations individually may leave unchanged each ingredient of a problem, or key subsets of the problem, or the full set of distributions for the problem. For estimation or for hypothesis testing the *invariance principle** then requires a corresponding property to hold for estimates and for tests. The principle and methods have fairly widespread use but not based on the strong convictions commonly associated with sufficiency and ancillarity.

Separation of Categorical Information

Some recent attention has been given to the separation of categorical 0–1 information and probability or frequency information. By categorical information is meant information that is either yes or no concerning what has been a possibility—a clear delineation of what is known in value as opposed to what is not known in value. The starting point for this type of analysis is the inference base (M, D), consisting of the model M for an investigation and the data D from the investigation.

Separation of categorical information may occur with respect to the parameter space: a sample space value identifies possible parameter values contained in an *options set*, an element of a nontrivial partition of the parameter space. The reduction is obtained by eliminating arbitrary elements from the inference base, leaving the options set as the range for the parameter. For details, see Fraser [28, pp. 49ff.].

An interesting example with much exposure in the literature involves the linear regression model with a singular variance matrix, known or known up to a scale factor σ^2. It can happen from data that certain regression coefficients then become known exactly in value and the remaining coefficients remain subject to the usual tests and confidence procedures. See Feuerverger and Fraser [15]. To treat the full parameter from a statistical viewpoint is artificial and misleading; that is, to talk of estimating something without acknowledging that it is in fact known in value is, to say the least, deceptive.

Objective Probability Variable

In some applications objective variables occur that have a known probability distribution, known in a physical sense and not constructed or contrived so as to have a marginal distribution that omits the parameter dependence that was present in the initiating distribution. An observed value from such a distribution, by requirements of applied probability theory, specifies the conditional model given that observed value. The common appealing examples usually associated with ancillarity provide examples of the present method. For details, see Fraser [28, pp. 54ff.].

A frequent example in the literature concerns the measurement of some physical quantity θ; two measuring instruments are available with quite different precision characteristics and one of them is chosen at random based on equal $(1/2)$ probabilities. The reduction method then gives a model based on the instrument actually used; no ancillarity principle is involved, just a direct use of basic criteria for probability modeling. For some other views of this example, see INFERENCE, STATISTICAL: I.

Objective Function

The recent attention to the analysis of the 0–1 information from an investigation has also focused on the requirements and formalities connected with statistical models for a system and for an investigation. Direct modeling of the basic variation in a system was proposed in Fraser [23; see also ref. 28]; certain formal justifications for this model-

ing process are given in Brenner and Fraser [8] and Brenner et al. [9]. In such modeling the response is a function or presentation of the basic variation, and the model is called a *structural* or *structured* model. Detailed inference procedures follow necessarily with such a model and data; for a survey, see Fraser [28]. In an application involving such a model an observed response by means of the presentation function produces observed values for most "coordinates" of the basic variation. These coordinate values are obtained from the presentation function and data. The resultant statistical model then has two parts—a marginal model for what is observed concerning the basic variation and a conditional model concerning the unobserved component. This typically is a substantial reduction from the initially presented model and various inference procedures follow necessarily and directly; in particular, actual probabilities of a system are available for tests and confidence regions. For examples with discussion, see Fraser [26], and for more general applications, Fraser [28]. Also note that *no* ancillarity principle is used, contrary to common views as, for example, reported in INFERENCE, STATISTICAL: I.

Response Reexpression

The formal ingredients for statistical inference are given by $I = (M, D)$, the model for and data from an investigation. When presented in this form there is a particular choice for the mode of expression for the response variable. The application itself may, however, support other modes of expression for the response; closure under product and inverse then gives a group of such reexpression transformations. The present method of reduction acknowledges the arbitrariness of any particular mode of expression and in effect examines them all—the set of reexpressed inference bases using the reexpression group. This expression-free version of the inference base is then examined to eliminate arbitrary elements. For details, see Fraser [28, pp. 61ff.]. The results correspond to those in the section "Invariance" subject to the relevant choice there of the group and to invariance for each ingredient in the model.

PROBABILITIES FOR UNKNOWNS

A long-standing goal with varying support within the profession has been that of obtaining probability statements concerning unknown parameter values. Various directions are identified within the recent literature.

Objective Priors

Perhaps the most obviously clear case is that in which the unknown parameter value is, in fact, a realization from a separate random system with known (or partially known) probability model. This invites the examination and modeling of the combined system: a *prior distribution** describes the separate random system that produced the unknown parameter value, and the ordinary statistical model then takes the role of the conditional distribution given the realized parameter value. The analysis is routine applied probability theory: the response value is observed; the *conditional distribution* for the parameter given the observed response is then the proper description for the otherwise unknown parameter value; it is called the *posterior distribution** for the parameter. This routine use of probability theory involves calculations that are commonly assembled under the heading *Bayes' theorem**.

For applications that fit the descriptions just given, the support for the preceding analysis is rather strong; in a sense it is just routine analysis of a probability (as opposed to statistical) model. Certain contexts, however, suggest that serious concerns can attach to the procedure; see Fraser [28, pp. 104ff.].

Probabilities in applications can sometimes be estimated from assembled data. In fact, this is perhaps the most obvious source,

the other being symmetry. If a prior distribution is estimated from data, we have a very natural example for the just discussed method of analysis based on an objective prior. This type of application, however, has received some substantial recent attention as an apparent offshoot from *Bayesian inference** and has been given the name *empirical Bayes**. This term, for what in fact is empirical probability modeling, is rather inappropriate.

Personal Priors*

Probabilities for unknown parameter values are the goal of an approach to statistical inference called *Bayesian inference**. In the preceding section, the prior distribution for an unknown parameter value was based on empirical results; in this section the prior distribution is based on personal views of an experimenter or analyst. This *personal prior* distribution can arise in one of several forms:

1. **Flat prior.** A uniform density function with respect to some appropriate or reasonable support measure (on topological groups this can be a right Haar measure or occasionally a left Haar measure).

2. **Locally flat prior.** A density function that is locally uniform over some expected range for the parameter value. The preceding two types of prior are typically chosen on the basis of an argument concerning *insufficient reason** to prefer one parameter value to another.

3. **Personalistic prior.** A prior that represents the differential feelings, one parameter value to another, of the experimenter or analyst. Often such a prior is chosen from a parametric family (*conjugate prior*) that is functionally compatible with likelihood functions that can arise in the application.

The method of combining the prior distribution with the basic statistical model is that discussed in the preceding section; it involves standard probability analysis for the combined model. The difference, however, lies in the support for the prior model, the nominal source of the unknown parameter value. Under the preceding section the support was empirical; under this section the prior enters as an assessment coming from the feelings and judgments of the experimenter or analyst. This raises two questions: whether probabilities are a suitable vehicle for expressing such personal feelings; and given, say, such acceptance of probability as a vehicle, whether the values can be quantified in a meaningful way based on the feelings.

The procedure can, of course, be examined from a purely pragmatic viewpoint—that it produces statistical methods, *a* way of obtaining statistical methods. The merits or demerits of a method would then be examined otherwise. A risk with this approach, however, is that something labeled as a probability may be taken or treated as a probability, when in fact it derives from something as imprecise as *an* indifference feeling or *a* preference of an experimenter or analyst. This raises very serious questions, but the issues will not be addressed further here other than to note that a basic intent of the scientific method has long been to eliminate the personal feelings and judgments of experimenters and investigators. A related question is whether probabilities can always describe an unknown realization, for example, in the context of an *actual* objective random system, where there is incomplete information concerning the realization. Even in a well-defined context of this type, probabilities can be unavailable without special conditions [28, pp. 136ff.]; in other words, information alone can be insufficient to provide probabilities for well-defined random systems.

Internal Randomness

Some other approaches to statistics have sought probabilities for unknown parameter values from the statistical model itself. Historically, the use of the flat prior (item 1) in the preceding section represented an attempt in this direction, as did the original presentation itself of *Thomas Bayes** in 1763 [4]. More recently, however, this flat prior approach would not be viewed as coming from the statistical model alone.

The *fiducial probability** proposed by R. A. Fisher is focused on obtaining probabilities for unknown parameter values from the statistical model itself. A fairly broad survey was given by Fisher [19] in the book *Statistical Methods and Scientific Inference*. The basic procedure involves:

1. A *pivotal function** $P(y, \theta)$, usually required to be dependent on y through an exhaustive/minimal sufficient statistic that has a θ-free distribution.
2. With fixed y the transference to θ of the fixed distribution for the pivotal function giving the *fiducial distribution** for θ.

The early Fisher publications on this led to the sharpening of the methods and definitions for *confidence interval** by Jerzy Neyman*.

Subsequent discussions of the fiducial method seemed to focus on the following questions: the meaning and the interpretation for the fiducial probabilities, as primarily advanced by Fisher himself; nonuniqueness of the pivotal function; inconsistencies under Bayesian-type combination of a first and second system. For certain models with transformation characteristics an objective meaning was found for fiducial probabilities that was acceptable to Fisher (personal correspondence) and for which the nonuniqueness and other difficulties do not arise [21]. However, the prevalent general view in the literature is at least that of an unresolved concept.

The pivotal functions that enter into fiducial probability and into confidence interval derivation also appear in three recent approaches to probabilities for unknown parameter values. Dempster [12] proposed a distribution for the pivotal quantity without a requirement for a distribution for y or θ. Certain aspects of Dempster's approach were developed in Shafer [34]. Beran [5, 6] used a distribution for the pivotal quantity and emphasized the related distribution for y without implications concerning an initial distribution for θ. Barnard [1] has considered pivotal functions with optimum properties relative to a family of distributions for the pivotal function. A pivotal function defined on the full sample space is the mathematical inverse of an expression or presentation function defined on an error, or variation, or pivotal space. Properties and consistency requirements for such presentation functions have been examined extensively [24, 28], especially with respect to the validity of posterior probabilities. Thus the pivotal approach translates as an examination of the *structured* model, a generalization of the *structural* model. Consideration of the latter, however, has generally emphasized the need for objective support for components of the model; see the remainder of this section and the section "Objective Function."

The preceding methods have not focused on the basic or internal randomness or variation of a system; the phrase "internal randomness" was used for the title of this section.

In any statistical application a general and well-founded assumption is that there is a probability model that provides a reasonable approximation for the behavior of the relevant variables under investigation. However, owing to incomplete information, a statistical model is used; the model is a valid or acceptable model if it includes among its possibilities a reasonable approximation to the actual probability model. The statistical model is then relevant or acceptable to the degree that the probability model mentioned is appropriate. A fundamental question then focuses on the degree to which the actual probabilities of the probability model mentioned can be used in the statistical analyses that follow. This was discussed briefly in the section "Objective Function." Our concern in this subsection is with the degree that these probabilities provide probabilities concerning the unknown parameter value.

Probabilities for parameter values from this approach are called structural probabili-

ties and were given extensive coverage in Fraser [23] but largely omitted from Fraser [28] to emphasize various fundamental properties of the variation-based or structural models.

The analysis mentioned in the section "Objective Function" distribution describing the inaccessible coordinates of the basic variation. Together with this is an objective response presentation function $y = \theta v$ that gives the relation $y_0 = \theta v_0$ between the observed response y_0, the unknown parameter value θ, and the inaccessible coordinates v_0 of the basic variation. Structural probabilities are obtained as the implications for θ of the equation $y_0 = \theta v_0$ examined with the probability description for v_0.

Attention in the literature has not approached directly the validity of the preceding procedure, but rather has centered on apparent conflicts with Bayesian and betting assessments; for example, Buehler [10]. However, defects in the betting assessment procedures were indicated in Fraser [27].

OTHER TERMINAL PROCEDURES

A procedure that leads to an acceptable probability distribution for a parameter value can be viewed as a terminal inference procedure, perhaps the ultimate terminal procedure. In this section we mention briefly other terminal procedures, procedures that are typically examined as specialized areas and not directly under a global view of statistical inference.

Confidence intervals and confidence regions* occupy a very central and important place in the statistical literature and form indeed an identified area of study. These are part of statistical inference and deserve greater attention within inference as to their larger implications. Some attention to this is indicated for example by Buehler [10] and Fraser [27] and by the discussion Fraser [28, pp. 91ff.).

Hypothesis testing* theory tends to fall in two distinct directions. One direction points toward decision theory and is concerned largely with the accept–reject Neyman–

Pearson* theory as developed, say, in Lehmann [31]. The other direction focuses on the traditional test of significance in the scientific literature with its fundamental input from Fisher (e.g., refs. 18 and 19). The latter direction is very central in statistical inference but has had little formal consideration under the general heading of statistical inference. In fact, most textbooks in statistics neglect it entirely with exceptions (e.g., Kempthorne and Folks [30] and Fraser [28]).

Estimation theory* has a large and extensive literature and has a central place in statistical inference. It is concerned with obtaining best point estimates or values for the unknown parameter value. The theory generally does not concern itself with associated measures of precision and reliability; such questions once addressed move the theory close to or into confidence theory. Much recent activity in the area of estimation theory has focused on robustness, the sensitivity of the methods to departures from standard model and data assumptions.

ADDENDUM

The degree to which there is a lack of unity in statistics is in some measure indicated by the position of statistical inference as a component area rather than the unifying theory at the top of statistics. The specialized areas of statistics often proceed with their own particular emphases without adequate attention to large implications. The need is strong for integrating overviews and a general examination of methods and implications. The study of inference should rise to this broader challenge.

Some books concerned with statistical inference are Barnett [3], Cox and Hinkley [11], Edwards [14], Fraser [28], Fisher [19], Hacking [29], and Silvey [35].

References

[1]   Barnard, G. A. (1977). *Proc. 41st Sess. ISI*, **47**(1), 543–551. (Discussion, examples.)

[2]   Barndorff-Nielsen, O. (1971). On *Conditional Statistical Inference*. Mimeographed report, Aarhus, Denmark. (Theoretical, detailed, extensive.)

[3]   Barnett, V. (1973). *Comparative Statistical Inference*. Wiley, New York. (Intermediate level, survey.)

[4]   Bayes, T. (1763). *Philos. Trans. R. Soc.*, **53**, 370–395. (Of historical interest.)

[5]   Beran, R. (1971). *Ann. Math. Statist.*, **42**, 157–168. (Theoretical, examples.)

[6]   Beran, R. (1972). *Proc. 6th Berkeley Symp. Math. Statist. Prob.*, Vol. 1. University of California Press, Berkeley, Calif., pp. 1–16. (Theoretical.)

[7]   Birnbaum, A. (1962). *J. Amer. Statist. Ass.*, **57**, 269–306. (Discussion, philosophical.)

[8]   Brenner, D. and Fraser, D. A. S. (1980). *Statist. Hefte*, **21**, 296–304. (Theoretical and technical.)

[9]   Brenner, D., Fraser, D. A. S., and Monette, G. (1981). *Statist. Hefte*, **22**, 231–233. (Discussion.)

[10] Buehler, R. (1973). *Math. Rev. 6514*, **46**, 1120. (Review of Fraser [23].)

[11] Cox, D. R. and Hinkley, D. V. (1974). *Theoretical Statistics*. Chapman & Hall, London. (Broad survey, many examples.)

[12] Dempster, A. (1966). *Ann. Math. Statist.*, **37**, 355–374. (Theoretical.)

[13] Dynkin, E. B. (1950). *Dokl. Akad. Nauk SSSR (N.S.)*, **75**, 161–164. (Advanced level consideration of sufficiency.)

[14] Edwards, A. W. F. (1972). *Likelihood*. Cambridge University Press, Cambridge. (Promotes a pure likelihood approach.)

[15] Feuerverger, A. and Fraser, D. A. S. (1980). *Canad. J. Statist.*, **8**, 41–45. (New light on a specialized regression problem.)

[16] Fisher, R. A. (1922). *Philos. Trans. R. Soc. Lond.*, *A*, **222**, 309–368. (Fundamental paper, advanced level.)

[17] Fisher, R. A. (1925). *Proc. Camb. Philos. Soc.*, **22**, 700–725. (Fundamental paper, advanced level.)

[18] Fisher, R. A. (1948). *Statistical Methods for Research Workers*. Oliver & Boyd, Edinburgh. (Methods book with overtones of sophisticated theory.)

[19] Fisher, R. A. (1956). *Statistical Methods and Scientific Inference*. Oliver & Boyd, Edinburgh. (Loosely knit foray establishing many new directions.)

[20] Fraser, D. A. S. (1956). *Ann. Math. Statist.*, **27**, 838–842. (Theoretical.)

[21] Fraser, D. A. S. (1961). *Biometrika*, **53**, 1–9. (Detailed integration of parts of fiducial theory.)

[22] Fraser, D. A. S. (1963). *J. Amer. Statist. Ass.*, **58**, 641–647. (Discussion in context of examples.)

[23] Fraser, D. A. S. (1968). *The Structure of Inference*. Krieger, Huntington, N.Y. (Theoretical examination of structural models focusing on posterior probabilities.)

[24] Fraser, D. A. S. (1971). *Proceedings of the Symposium on the Foundations of Statistical Inference,*

V. P. Godambe and D. A. Sprott, eds. Holt, Reinhart and Winston, Toronto, pp. 32–55. (Theoretical.)

[25] Fraser, D. A. S. (1976). *Probability and Statistics, Theory and Applications*. DAI (University of Toronto textbook store), Toronto. (Introductory text with current theories and methods.)

[26] Fraser, D. A. S. (1976). *J. Amer. Statist. Ass.*, **71**, 99–113. (Computer program for location-scale analysis with discussion.)

[27] Fraser, D. A. S. (1977). *Ann. Statist.*, **5**, 892–898. (Faults in the betting assessment of probability.)

[28] Fraser, D. A. S. (1979). *Inference and Linear Models*. McGraw-Hill, New York. (Intermediate; requirements for statistical models with detailed consideration of structural models.)

[29] Hacking, I. (1965). *Logic of Statistical Inference*. Cambridge University Press, Cambridge. (A philosophical examination of some parts of inference.)

[30] Kempthorne, O. and Folks, L. (1971). *Probability, Statistics, and Data Analysis*. Iowa State University Press, Ames, Iowa. (An introductory text with current theories and methods.)

[31] Lehmann, E. (1959). *Testing Statistical Hypotheses*. Wiley, New York. (Advanced text.)

[32] Lehmann, E. L. and Scheffé, H. (1950). *Sankhyā*, **10**, 305–340. (Theoretical.)

[33] Savage, L. J. (1954). *The Foundations of Statistics*. Wiley, New York. (Intermediate level, promotes Bayesian approach.)

[34] Shafer, G. (1976). *A Mathematical Theory of Evidence*. Princeton University Press, Princeton, N.J. (Theoretical.)

[35] Silvey, S. D. (1970). *Statistical Inference*. Penguin Books, Middlesex, England. (Survey.)

[36] von Neumann, J. and Morgenstern, O. (1947). *Theory of Games and Economic Behavior*. Princeton University Press, Princeton, N.J. (Advanced, extensive.)

[37] Wald, A. (1950). *Statistical Decision Functions*. Wiley, New York. (Advanced.)

(ANCILLARITY
BAYESIAN INFERENCE
CONDITIONAL INFERENCE
CONFIDENCE INTERVALS AND REGIONS
DECISION THEORY
ESTIMATION
FIDUCIAL INFERENCE
HYPOTHESIS TESTING
INFERENCE, STATISTICAL: I
LIKELIHOOD
LOGIC IN STATISTICAL REASONING
MULTIPLE DECISION THEORY
OPTIMIZATION IN STATISTICS
SIMULTANEOUS INFERENCE
STATISTICAL EVIDENCE
STRUCTURAL INFERENCE
SUFFICIENCY)

D. A. S. FRASER

INFINITE DIVISIBILITY

A random variable (rv) is called infinitely divisible (inf. div.) if for each $n \in \mathbb{N}$ independent, identically distributed (i.i.d.) rvs $X_{n,1}, \ldots, X_{n,n}$ exist such that

$$X \stackrel{d}{=} X_{n,1} + \cdots + X_{n,n}, \qquad (1)$$

where $\stackrel{d}{=}$ denotes equality in distribution. Equivalently, denoting the distribution functions of X and $X_{n,1}$ by F and F_n, one has

$$F = F_n * \cdots * F_n =: F_n^{n*},$$

where $*$ denotes convolution*. In terms of the characteristic functions* (ch.f.'s) φ and φ_n of X and $X_{n,1}$ (and analogously for Laplace–Stieltjes transforms and probability generating functions*) (1) is expressed by

$$\varphi = (\varphi_n)^n. \qquad (2)$$

In fact, φ is the ch.f. of an inf. div. rv if and only if $\varphi \neq 0$, and φ^p is a ch.f. for all $p > 0$ (or all $p = n^{-1}$ with $n \in \mathbb{N}$, or even all $p = 2^{-n}$ with $n \in \mathbb{N}$). Distributions of inf. div. rvs, their distribution functions and densities, their ch.f.'s, and other transforms are also called inf. div.

From (2) it follows that, e.g., the normal distribution, with $\varphi(t) = \exp(-t^2/2)$ is inf. div., and similarly the Poisson distribution with $\varphi(t) = \exp[\mu(e^{it} - 1)]$. By de Finetti's theorem (see ref. 9) every inf. div. distribution is the limit of compound* Poisson distributions.

Infinitely divisible distributions were introduced by de Finetti in 1929, and studied extensively by Lévy [7] in the context of the central limit* problem. The following theorem shows the connection between infinite divisibility and a very general central limit situation (see ref. 8).

Theorem 1.   A rv X is inf. div. if and only if

$$X = d - \lim_{n \to \infty} (X_{n,1} + \cdots + X_{n,k(n)} - A_n),$$

$$(3)$$

where $d - \lim$ denotes convergence in distribution, and where the $X_{n,k}$ are rvs that are independent for fixed n, and "uniformly asymptotically negligible," i.e., $\max P(|X_{n,k}| \geqslant \epsilon) \to 0$ as $n \to \infty$ for every $\epsilon > 0$. Here $k(n) \to \infty$ as $n \to \infty$, and the A_n are norming constants.

More specially, X has a self-decomposable (class L) distribution (see ref. 9) if and only if

$$X = d - \lim_{n \to \infty} \{ (X_1 + \cdots + X_n)/B_n - A_n \},$$

$$(4)$$

where the $X_1, \ldots, X_n$ are independent and X_k / B_n is uniformly asymptotically negligible; X has a stable* distribution (see ref. 2) if and only if the $X_1, \ldots, X_n$ in (4) are i.i.d.

Inf. div. distributions are characterized by their ch.f.'s as follows (see, e.g., ref. 9).

Theorem 2.   A ch.f. φ is inf. div. if and only if

$$\varphi(t) = \exp\left[ict + \int_{-\infty}^{\infty} \left(e^{itx} - 1 - \frac{itx}{1 + x^2} \right) \right.$$

$$\left. \times \frac{1 + x^2}{x^2} \, dK(x) \right], \qquad (i = \sqrt{-1})$$

where c is a real constant, and K, the *spectral measure* function, is bounded and nondecreasing with $K(x) \to 0$ as $x \to -\infty$.

There are several variants (with different spectral measures) of this so-called *canonical representation* by Lévy and Khintchine, especially for distributions on $\mathbb{R}_+$ and on $\mathbb{N}$ (see refs. 2 and 10).

A simple necessary condition for infinite divisibility is: $\varphi(t) \neq 0$ for $t \in \mathbb{R}$; a simple sufficient condition: φ is real, positive, and log-convex* for $t \geqslant 0$.

If φ is an inf. div. ch.f., then $\{\varphi^t; t > 0\}$ is the set (semigroup) of ch.f.'s corresponding to the set $\{S_t; t > 0\}$ of rvs, a stochastic process* with independent stationary incre-

ments* (see ref. 2). The process S_t is called a Lévy process*; it is the natural continuous analog of the partial sum process $S_n = X_1 + \cdots + X_n$, where the X_k are i.i.d.

Although no explicit characterization of infinite divisibility in terms of distribution functions (F) or densities (f) exists, there are several sufficient or necessary conditions in these terms: A distribution on $\mathbb{R}_+$ is inf. div. if $\log f$ is convex, or, more specially, if f is completely monotone*; more complicated conditions on f were given by Bondesson (see ref. 1). A necessary condition is

$$-\log\left[F(-x) + 1 - F(x) \right] = O(x \log x)$$

$$(x \to \infty), \quad (5)$$

with the normal distributions as the only exceptions. From (5) it follows that no bounded rv is inf. div., and that no finite mixture* of normal distributions is inf. div. For inf. div., continuous densities on $\mathbb{R}_+$ (and under certain restrictions for those on $\mathbb{R}$) one has $f(x) \neq 0$, except for an endpoint.

Many of the much used distributions in statistics have been shown to be inf. div., e.g., Student's, Fisher's F, the χ^2, and the log-normal (see refs. 11–13).

Testing for infinite divisibility has been suggested by Katti (see ref. 11), but the main importance of inf. div. distributions for the statistician is in modeling. In certain situations it is known that a rv X is of the form (1), and if one then wishes to use distribution function F as a model for X, one has to know that F is inf. div. (see ref. 11 for examples). Similar considerations led Thorin to his investigations in refs. 12 and 13, where he introduced the class of generalized gamma convolutions, which was studied in detail by Bondesson [1].

Theorems 1 and 2, with appropriate modifications, also hold in $\mathbb{R}_k$ (for Theorem 2 see, e.g., ref. 2); the more special properties do not always carry over easily, and some of the concepts are hard to generalize properly, or uniquely (for some information, see ref. 5 and its references). For the extension of infinite divisibility to distributions on abstract spaces see ref. 4 and its many references. An extensive account of infinitely divisible point processes* (random measures) is given in ref. 6.

Most of the properties mentioned above can be found in one or more of the publications listed below, or in the books and papers referred to therein; refs. 3, 4, 10, and 11 are reviews containing many references.

References

[1]   Bondesson, L. (1979). *Scand. Actuarial J.*, 125–166.

[2]   Feller, W. (1968, 1970). *An Introduction to Probability Theory and Its Applications*, Vol. I (3rd ed.), Vol. II (2nd. ed.), Wiley, New York.

[3]   Fisz, M. (1962). *Ann. Math. Statist.*, **33**, 68–84.

[4]   Heyer, H. (1977). *Probability Measures on Locally Compact Groups*. Ergebnisse der Mathematik und ihrer Grenzgebiete, Vol. 94. Springer-Verlag, New York.

[5]   Horn, R. A. and Steutel, F. W. (1978). *Stoch. Processes Appl.*, **6**, 139–151.

[6]   Kerstan, J., Matthes, K., and Mecke, J. (1978). *Infinitely Divisible Point Processes*. Wiley, New York.

[7]   Lévy, P. (1937). *Théorie de l'Addition des Variables Aléatoires*. Gauthiers-Villars, Paris.

[8]   Loève, M. (1977). *Probability Theory*, 4th ed., Vol. 1. Springer-Verlag, New York.

[9]   Lukacs, E. (1970). *Characteristic Functions*, 2nd ed. Charles Griffin, London.

[10]   Steutel, F. W. (1973). *Stoch. Processes Appl.*, **1**, 125–143.

[11]   Steutel, F. W. (1979). *Scand. J. Statist.*, 57–64.

[12]   Thorin, O. (1977). *Scand. Actuarial J.*, 31–40.

[13]   Thorin, O. (1977). *Scand. Actuarial J.*, 121–148.

(CHARACTERISTIC FUNCTIONS
CONVOLUTION
L CLASS LAWS
LÉVY PROCESS
LIMIT THEOREMS, CENTRAL
POINT PROCESSES
STABLE DISTRIBUTIONS
STOCHASTIC PROCESSES)

F. W. STEUTEL

INFLATED DISTRIBUTIONS

A discrete distribution (F) which is modified by increasing the probability for one value (x_0) of the variable, the remaining probabilities being multiplied by an approximate constant to keep the sum of probabilities equal to 1, is called an "inflated (F) distribution". Denoting $\Pr\{X = x \mid F\}$ by P_x, the modified distribution has

$$\Pr\{X = x_0\} = 1 - \alpha + \alpha P_{x_0} \quad (0 < \alpha < 1),$$

$$\Pr\{X = x\} = \alpha P_x \qquad \text{for all } x \neq x_0.$$

In terms of the crude moments $\{\mu_j'\}$ of (F), the rth crude moment of the inflated F distribution is

$$\tilde{\mu}_r' = (1 - \alpha)x_0^r + \alpha\mu_r'.$$

It follows that if a recurrence relation $g(\mu_1', \mu_2', \dots) = 0$ holds for (F), the moments of the inflated distribution satisfy the recurrence relation

$$g\left(\frac{\tilde{\mu}_1' - (1 - \alpha)x_0}{\alpha}, \frac{\tilde{\mu}_2' - (1 - \alpha)x_0^2}{\alpha}, \dots \right) = 0.$$

$$(1)$$

In particular, if $\sum_j c_j \mu_j' = 0$, then

$$\sum_j c_j \tilde{\mu}_j' = (1 - \alpha)\sum_j c_j x_0^j. \qquad (2)$$

In most applications (F) is a distribution confined to the nonnegative integers, and $x_0 = 0$. The corresponding inflated distribution is sometimes called a distribution "with added zeros". In this case $\tilde{\mu}_r' = \alpha\mu_r'$ and (1) becomes

$$g\left(\alpha^{-1}\tilde{\mu}_1', \alpha^{-1}\tilde{\mu}_2', \dots \right) = 0$$

and (2) becomes $\sum_j c_j \tilde{\mu}_j' = 0$.

If α exceeds 1, we have a *deflated* (F) distribution. Since $\Pr\{X = x_0\} \geqslant 0$, the greatest possible value for α is $(1 - P_{x_0})^{-1}$. If α takes this value, the corresponding probability is zero—x_0 is removed from the set of values of X. If (F) is confined to nonnegative integers and $x_0 = 0$, the resultant distribution is termed "zero-truncated," or more picturesquely, "decapitated"*.

INFLUENCE CURVE   *See* INFLUENCE FUNCTIONS

INFLUENCE FUNCTIONS

The *influence function* or *curve* of an estimator has two main uses in statistics; one is to indicate the sensitivity of the statistic to individual observations, and the other is to compute the asymptotic variance of the statistic, under certain regularity conditions. It was named and studied by Hampel in his thesis and in refs. 5 and 6 in connection with the topic of robust estimation*. The influence function is the kernel of the first derivative of the statistical functional* that defines the estimator, and is defined for each real value x by

$$\mathrm{IC}(x; F, T) = \lim_{t \to 0} t^{-1}\{ T[(1 - t)F + t\delta_x]$$
$$- T(F)\}, \qquad (1)$$

if this limit exists. Here F is a distribution function, usually considered the true underlying distribution function generating the data, δ_x is the distribution function that puts mass 1 at the point x, and T is a functional from the space of distribution functions to the parameter space. In order to discuss the role of the influence function in statistics, it is necessary first to consider statistical functionals and their differentials.

A statistic that is a function of n independent, identically distributed observations from a distribution function F can often be represented as a functional, T, of the empirical cumulative distribution function. For example, the sample mean $n^{-1}\sum X_i$ can be written $T(F_n) = \int x \, dF_n(x)$. In many cases, the parameter being estimated by the statistic is the same functional T of the true underlying cumulative distribution function. This is true in the case of the mean, because $EX = \int x \, dF(x)$. (Estimators that have this property are called Fisher consistent [3, p. 287].) If a statistical functional is differentiable at F, then we can write

$$T(G) = T(F) + dT_F(G - F) + R, \quad (2)$$

where G is some distribution function, dT_F the differential of $T(F)$ and a linear functional operating on $(G - F)$, and R is shorthand for "remainder." There are three main notions of differentiability of functionals; Gâteaux, compact, and Fréchet. They relate to different notions of convergence of R to zero, as G converges to F. Compact differentiability requires that R converge to zero uniformly on compact sets, and Fréchet differentiability requires that R converge to zero uniformly on bounded sets. (Precise definitions are given in Reeds [12, pp. 45–46]; also see Huber [8, Sec. 2.5].) Expansion (2) is called a von Mises expansion, because differentiation of statistical functionals was first proposed by von Mises in 1947 [14]. It is similar to the Taylor series expansion of a function: $f(x) = f(a) + (x - a)f'(a) + o(x - a)$. A thorough study of von Mises expansions is provided in Reeds [12].

The easiest way to compute $dT_F(G - F)$ is to compute the Gâteaux derivative (*see* STATISTICAL FUNCTIONALS)

$$dT_F(G - F) = \lim_{t \to 0} t^{-1}\{ T[F + t(G - F)] - T(F)\}.$$
$$(3)$$

For the asymptotic results described later in this article, it is necessary that T be compactly or Fréchet differentiable, which is a stronger requirement than being Gâteaux differentiable. However, if the compact or Fréchet derivative exists, it equals the Gâteaux derivative, so (3) is still the most convenient computational procedure.

For many statistical functionals (now called von Mises functionals) $dT_F(G - F)$ takes a particularly simple form:

$$dT_F(G - F) = \int \widetilde{\mathrm{IC}}(x; F, T) \, d(G - F)(x).$$
$$(4)$$

Note that $\widetilde{\mathrm{IC}}$ is unique only up to an additive constant, because G and F both have total mass 1. The usual standardization is to modify (4) as follows:

$$dT_F(G - F) = \int \mathrm{IC}(x; F, T) \, dG(x).$$

The kernel $\mathrm{IC}(x; F, T)$ is the influence function of definition (1). It is uniquely defined, and satisfies $\int \mathrm{IC}(x; F, T) \, dF(x) = 0$. If the distribution function G puts all its mass at the point x, then (3) is identical to (1). The distribution function $(1 - t)F + t\delta_x$ is, for small t, almost the distribution function F, but with additional weight at the point x. It is this fact that gives the influence function its name. It measures the effect on the estimator T of a small (infinitesimal) change in the weight that F gives to x, i.e., the influence on T of the point x.

The influence function of the mean is almost trivial to compute: $IC(x; F, T) = x - \int x\, dF(x)$. A more interesting influence function is that of the M-estimator* computed by minimizing $\int \rho(\theta, x)\, dF(x)$. Two classical ρ-functions are $\rho(\theta, x) = (x - \theta)^2$ and $\rho(x) = \log f_\theta(x)$. In general, ρ is a continuously differentiable function and the M-estimator is equivalently defined by $\int \psi(T(F), x)\, dF(x) = 0$, where $\psi(\theta, x) = (\partial/\partial\theta)\rho(\theta, x)$. By differentiating implicitly and rearranging terms, it is not difficult to verify that

$IC(x; F, T)$

$$= -\psi(T(F), x) \Big/ \int \psi'(T(F), x)\, dF(x).$$

(5)

[Here $\psi'(\theta, x)$ means $\partial/\partial\theta\, \psi(\theta, x)$.]

The shape of the influence function provides information about the robustness of the statistic. The sample mean is sensitive to large observations and this is reflected in the fact that the influence function is unbounded. A statistical functional with a bounded influence function is not sensitive to extreme observations, so is robust in this sense. The value $\sup_x |IC(x; F)|$ was called by Hampel the gross error sensitivity. If the influence function is continuous in F, the statistic T is robust to departures from assumptions about the underlying form of F. From (5) we see that the influence function of an M-estimator is a multiple of the function ψ that defines the M-estimator. Thus whatever properties we feel the influence function should have can be built directly into the estimator. This has led to the development of a wide class of redescending M-estimators*. A Monte Carlo study of 68 robust estimators was carried out by Andrews et al. [1], and influence functions for many of these estimators are presented in Chap. 3. Huber [7] provides a readable in-

troduction to influence functions in Chap. 2, and computes influence functions for M-, L-*, and R*-estimators in Chap. 3. A somewhat expanded discussion is provided in Huber [8]. The main work on influence functions and robustness is Hampel [6].

To discuss the application of the influence function to asymptotics, we return to expansion (2), but replace G with F_n. Then

$$T(F_n) = T(F) + \int IC(x; F, T)\, dF_n(x) + R$$

$$= T(F) + n^{-1} \sum IC(X_i; F, T).$$

The second term in this expression is the average of n independent, identically distributed random variables, with mean zero and variance $\sigma^2 = \int IC^2(x; F, T)\, dF(x)$. As long as $\sigma^2 < \infty$ and $\sqrt{n}\, R \to 0$ in probability as $n \to \infty$, the central limit theorem says that $\sqrt{n}\, [T(F_n) - T(F)]$ converges in distribution to a normal random variable with mean 0 and variance σ^2.

If we equip the space of distribution functions with the sup norm, and if T is Fréchet differentiable, then $R = o(\|F_n - F\|_\infty) = o_p(n^{-1/2})$, because $\|F_n - F\|_\infty = O_p(n^{-1/2})$, by the Kolmogorov–Smirnov theorem. For statistical functionals that are not Fréchet differentiable, but only compactly differentiable, the same argument works, but some other theorem than the Kolmogorov–Smirnov theorem is needed to show that $R \to 0$. If T is only Gâteaux differentiable, detailed analysis of R is required, and the influence function proof of asymptotic normality loses its advantage of simplicity. The approach via compact differentiability is studied in Reeds [12]; the argument sketched above can be found in Boos and Serfling [2]. The influence function proof of asymptotic normality* has been rigorously verified for the three main classes of robust estimators; M-estimators, L-estimators, and R-estimators.

The influence function is an infinite sample concept, but a finite sample version of it is related to the jackknife*. This point is discussed in Miller [10, Sec. 4.2; 11]. Another finite-sample approximation to IC is given by Tukey's sensitivity curve [7, p. 11].

Sensitivity curves for many estimators are sketched in Andrews et al. [1], but it seems that in general the jackknife is a more natural approximation to the influence function.

The expansion (2) could be considered a Taylor series expansion of the function $f(t) = T[F + t(G - F)]$, evaluated at $t = 1$, expanded about the point $t = 0$, and there would be no need to consider functional derivatives and von Mises expansions. The advantage of the von Mises approach is that it provides a unified solution to a wide class of problems, and extensions to more general cases than functionals of a distribution function are completely straightforward. For example, Reeds [12, Sec. 1.6] let the functional T depend on some auxiliary parameters, such as the sample size. Recent work on influence functions includes extensions to functionals of more than one distribution function [9, 13] and development of influence functions for testing [4, 9].

References

[1] Andrews, D. F., Bickel, P. J., Hampel, F. R., Huber, P. J., Rogers, W. H., and Tukey, J. W. (1972). *Robust Estimates of Location: Survey and Advances*. Princeton University Press, Princeton, N.J. (Influence functions of many estimates are computed in Chap. 3.)

[2] Boos, D. D. and Serfling, R. J. (1980). A note on differentials and the CLT and LIL for statistical functionals. *Ann. Statist.*, **8**, 618–624.

[3] Cox, D. R. and Hinkley, D. V. (1974). *Theoretical Statistics*. Chapman & Hall, London.

[4] Eplett, W. J. R. (1980). An influence curve for two-sample rank tests. *J. R. Statist. Soc. B*, **42**, 64–70.

[5] Hampel, F. R. (1971). A general qualitative definition of robustness. *Ann. Math. Statist.*, **42**, 1887–1896.

[6] Hampel, F. R. (1974). The influence curve and its role in robust estimation. *J. Amer. Statist. Ass.*, **69**, 383–393. (This is a very readable and informative introduction to influence functions and their applications.)

[7] Huber, P. J. (1977). *Robust Statistical Procedures*. SIAM, Philadelphia. (This monograph is clearly written and contains a wealth of information on various aspects of robustness. The influence curve is discussed in Chap. 2.)

[8] Huber, P. J. (1981). *Robust Statistics*. Wiley, New York. (A major sourcebook for most aspects of robustness. The influence curve is discussed in Chaps. 1 and 3.)

[9] Lambert, D. (1981). Influence functions for testing. *J. Amer. Statist. Ass.*, **76**, 649–657.

[10] Miller, R. G. (1974). The jackknife—a review. *Biometrika*, **61**, 1–15.

[11] Miller, R. G. (1978). The Jackknife: Survey and Applications. *Proc. 23rd Conf. Des. Exp. Army Res. Testing*. ARO Rep. No. 78-2.

[12] Reeds, J. A. (1976). On the Definition of von Mises Functionals. Ph.D. dissertation, Harvard University. (Difficult to obtain, this is a valuable work clarifying von Mises expansions and setting the von Mises approach to asymptotics on a rigorous mathematical footing.)

[13] Reid, N. (1981). Influence functions for censored data. *Ann. Statist.*, **9**, 78–92.

[14] von Mises, R. G. (1947). On the asymptotic distribution of differentiable statistical functions. *Ann. Math. Statist.*, **18**, 309–348. (The first paper on functional expansions in statistics: of historical interest.)

(JACKKNIFE
L-ESTIMATORS
M-ESTIMATORS
REDESCENDING *M*-ESTIMATORS
R-ESTIMATORS
ROBUSTNESS
STATISTICAL FUNCTIONAL)

N. REID

INFLUENTIAL DATA

A subset of data is called *influential* if its deletion would cause major changes in estimates, confidence regions, test and diagnostic statistics, etc. Usually, influential subsets are outside the patterns set by the majority of the data in the context of a model (e.g., linear regression*), likelihood* (Gaussian), and fitting process (maximum likelihood*).

Figure 1 gives some examples of influential subsets of data. The subset AB is influential but its individual elements are not (masking). D and E are influential separately, but the subset DE is not influential. C is not influential for the slope coefficient, but is influential for the intercept coefficient.

Influential subsets may become noninfluential if the model is modified (e.g., by

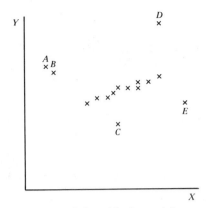

Figure 1   Influential subsets of data.

adding or transforming explanatory variables or using a nonlinear model), if the likelihood is changed (Poisson, binomial, etc.), or if the fitting process is altered (robust or bounded-influence methods). The converse is often also true.

Influential data are generally caused by errors in the data, model failure, or incorrect likelihood assumptions. Legitimate extreme observations may also be influential and it is often important to identify and report this type of data as well.

The detection of influential data is accomplished by arguing that small perturbations of the data (either large changes in a small subset or small changes in a large fraction of the data) should cause small perturbations in the output of an estimation process.

Influence is most often measured by means of an influence function which attempts to measure the effect of adding an observation to a large sample. Let F be the population distribution function and F_n the empirical distribution function. If an observation z is added, the infinitesimal asymptotic influence on the value of an estimate or test statistic $T(F_n)$ is

$$\Omega(z, F, T)$$
$$= \lim_{\alpha \to 0} \frac{T((1-\alpha)F + \alpha\delta_z) - T(F)}{\alpha}, \quad (1)$$

where δ_z denotes the point mass 1 at z. This notion of influence function* was introduced by Hampel [10, 11] and plays two

major roles. It is an essential theoretical tool for the development of robust estimates [12] and it provides a way to help identify influential data.

When the data, model, likelihood, and fitting process are given (e.g., least-squares* linear regression), it is necessary to convert (1) to finite samples. Usually, F is replaced by F_n or $F_{n-1}^{(i)}$, where $F_{n-1}^{(i)}$ is the empirical distribution function of all but the ith observation. In least-squares linear regression, $Y = X\beta + \epsilon$, $T(F_n)$ is b, the usual least-squares estimate, and $T(F_n^{(i)})$ is $b(i)$, the least-squares estimate obtained with the ith observation omitted. If $z = (y, x)$, the use of F_n gives

$$\Omega(z, F_n) = n(X^TX)^{-1}x^T[y - xb] \quad (2)$$

and $F_{n-1}^{(i)}$ gives

$$\Omega(z, F_{n-1}^{(i)}) = (n-1)[X^T(i)X(i)]^{-1}$$
$$\times x^T[y - xb(i)]. \quad (3)$$

When evaluated at $z_i = (y_i, x_i)$, (2) measures the influence of z_i with z_i a part of F_n. In (3), z_i is not a part of the underlying finite sample distribution $F_{n-1}^{(i)}$. This form is particularly useful when z_i is considered to be a "bad" observation and the rest "good."

The definition (1) was for an infinitesimal influence function. If the limit is omitted and α replaced by n^{-1}, another form of finite-sample influence function is obtained by computing

$$n\Big[T\big((1 - 1/n)F_{n-1}^{(i)}$$
$$+ (1/n)\delta_{z_i}\big) - T(F_{n-1}^{(i)})\Big], \quad (4)$$

which for least-squares linear regression is equal to

$$n[b - b(i)] = (X^TX)^{-1}x_i^T(y_i - x_ib(i))$$
$$= \frac{(X^TX)^{-1}x_i^T(y_i - x_ib)}{1 - h_i}, \quad (5)$$

where $h_i = x_i(X^TX)^{-1}x_i^T$. Useful references on finite-sample influence functions are Cook and Weisberg [7], Welsch [21], and Welsch and Samarov [22].

Influence functions are analyzed directly using semigraphic and graphical displays, multivariate methods (clustering, projection pursuit, etc.) or converted to distances with norms proportional to (in the least-squares linear regression case) $s^2(X^T X)^{-1}$, $s^2(i)(X(i)^T X(i))^{-1}$ or a robust covariance matrix. The level of influence deemed to be critical or significant is based largely on heuristics which compare the magnitude of the influence to the stochastic variability of the estimate whose influence function is being examined or to the magnitude of the estimate itself. An overview of these methods is contained in Belsley, et al. [5] and Cook and Weisberg [7]. More formal methods have been proposed by Andrews and Pregibon [1] and Dempster and Gasko–Green [9]. Bayesian approaches to detecting influential subsets of data have been suggested by Johnson and Geisser [13], Box [6], and Bailey and Box [4].

The techniques discussed so far are best suited to measuring the influence of a single isolated observation. Clearly, they are less effective for detecting observations such as A and B in Fig. 1, which may mask each other. Formally, the influence of a subset D with d elements is obtained by adding (with appropriate weights) the influence functions for $z \in D$. However, F_n, $F_{n-d}^{(D)}$, or $F_{n-q}^{(Q)}$ for some fixed subset Q could be used to compute the finite sample influence function in (2), (3), and (5). In general, it is desirable that D be contained in Q so that influence is measured relative to an empirical distribution function which is not based on the (potentially bad) elements of D. To implement these methods is computationally quite expensive. A number of approaches are discussed by Cook and Weisberg [7] and Belsley et al. [5].

An alternative is to attempt to find an estimation procedure (constrained maximum likelihood) that places a bound on the influence of subsets of the data. Such an alternative estimate produces a set of weights on each observation which measures how much downweighting is necessary to bound the overall influence. These weights (when applied to the data) provide the equivalent of a base subset Q (a special case with weights of zero on the observations in Q) which can be used to compute influence functions for each observation (and, by addition of these functions, subsets of observations) in a variety of relatively low cost ways. Bounded-influence procedures for linear regression are discussed by Krasker and Welsch [14] and Samarov and Welsch [19] and diagnostic techniques in Krasker and Welsch [14a].

Finite sample influence function procedures have been adapted to a number of problems. Nonlinear least squares is discussed by Belsley et al. [5] and logistic and generalized linear models* by Pregibon [16, 18]. The influence of subsets of data on transformation diagnostics (Box–Cox procedures, etc.) is analyzed by Pregibon [16] and in papers by Atkinson [2, 3] and Cook and Weisberg [8]. Pregibon [17] has developed techniques to measure influence in censored survival analysis* models. Influential data diagnostics for two-stage least-squares* models have been developed by Kuh and Welsch [15].

Many of the papers that discuss the analysis of influential data also mention computational procedures. Belsley et al. [5], Cook and Weisberg [7], and Velleman and Welsch [20] are especially useful in this regard.

References

[1] Andrews, D. F. and Pregibon, D. (1978). *J. R. Statist. Soc. B*, **40**, 85–93.

[2] Atkinson, A. C. (1981). *Biometrika*, **68**, 13–20.

[3] Atkinson, A. C. (1982). *J. R. Statist. Soc. B*, **44**, 1–36.

[4] Bailey, S. P. and Box, G. E. P. (1980). Modeling the Nature and Frequency of Outliers. *MRC Tech. Summary Rep. No. 2085*, Mathematics Research Center, University of Wisconsin–Madison, Madison, Wis.

[5] Belsley, D. A., Kuh, E., and Welsch, R. E. (1980). *Regression Diagnostics*. Wiley, New York.

[6] Box, G. E. P. (1980). *J. R. Statist. Soc. A*, **143**, 383–430.

[7] Cook, R. D. and Weisberg, S. (1980). *Technometrics*, **22**, 495–508.

[8] Cook, R. D. and Weisberg, S. (1982). *Proc. SAS*

User's Group, SAS Institute, Cary, N.C., pp. 574–579.

[9] Dempster, A. P. and Gasko–Green, M. (1981). *Ann. Statist.*, **9**, 945–959.

[10] Hampel, F. R. (1968). Contributions to the Theory of Robust Estimation. Ph.D. thesis, University of California, Berkeley.

[11] Hampel, F. R. (1974). *J. Amer. Statist. Ass.*, **69**, 383–393.

[12] Huber, P. J. (1981). *Robust Statistics*. Wiley, New York.

[13] Johnson, W. and Geisser, S. (1982). In *Statistics and Probability: Essays in Honor of C. R. Rao*, G. Kallianpur, P. R. Krishnaiah, and J. K. Ghosh, eds. North-Holland, Amsterdam, pp. 343–358.

[14] Krasker, W. S. and Welsch, R. E. (1982). *J. Amer. Statist. Ass.*, **77**, 595–604.

[14a] Krasker, W. S. and Welsch, R. E. (1983). *Proc. Comp. Sci. and Statist.: 14th Symp. Interface*, Springer-Verlag, New York.

[15] Kuh, E. and Welsch, R. E. (1980). *Proc. Workshop on Validation and Assessment Issues of Energy Models*, S. Gass, ed. National Bureau of Standards, Washington, D.C., pp. 445–475.

[16] Pregibon, D. (1979). Data Analytic Methods for Generalized Linear Models. Ph.D. thesis, University of Toronto.

[17] Pregibon, D. (1981). Some Comments on Diagnostics for Non-standard Regression Models. *Tech. Rep. No. 48*, Department of Biostatistics, University of Washington, Seattle, Wash.

[18] Pregibon, D. (1981). *Ann. Statist.*, **9**, 705–724.

[19] Samarov, A. and Welsch, R. E. (1982). In *COMPSTAT 1982: Proceedings in Computational Statistics*, Physica-Verlag, Wien, pp. 412–418.

[20] Velleman, P. F. and Welsch, R. E. (1981). *Amer. Statist.*, **35**, 234–242.

[21] Welsch, R. E. (1982). In *Modern Data Analysis*, R. Launer and A. Siegel, eds. Academic Press, New York, pp. 149–169.

[22] Welsch, R. E. and Kuh, E. (1977). Linear Regression Diagnostics: Tech. Rep. No. 173, NBER Computer Research Center, Massachusetts Institute of Technology, Cambridge, Mass.

Further Reading

See the following works, as well as the references just given, for more information on the topic of influential data.

Barnett, V. and Lewis, T. (1978). *Outliers in Statistical Data*. Wiley, New York.

Cook, R. D. (1977). *Technometrics*, **19**, 15–18.

Cook, R. D. (1979). *J. Amer. Statist. Ass.*, **74**, 169–174.

Cook, R. D. and Weisberg, S. (1982). *Influence and Residuals in Regression*. Chapman & Hall, London.

Cook, R. D. and Weisberg, S. (1982). In *Sociological Methodology 1982*, S. Leinhardt, ed., Chap. 8. Jossey-Bass, San Francisco.

Hawkins, D. M. (1980). *Identification of Outliers*, Chapman and Hall, London.

Hoaglin, D. C. and Welsch, R. E. (1978). *Amer. Statist.*, **32**, 17–22.

Krasker, W. S., Kuh, E., and Welsch, R. E. (1983). In *Handbook of Econometrics*, Z. Griliches and M. D. Intrilligator, eds. North-Holland, Amsterdam, (to appear).

Welsch, R. E. and Peters, S. C. (1978). In *Computer Science and Statistics: Eleventh Annual Symposium on the Interface*. North Carolina State University, Raleigh, N.C., pp. 240–244.

(INFLUENCE FUNCTIONS
INFLUENTIAL OBSERVATIONS
LEVERAGE
OUTLIERS
ROBUST ESTIMATION)

R. E. WELSCH

INFLUENTIAL OBSERVATIONS

Observations are regarded as influential if their omission from the data results in substantial changes to important features of an analysis. Such observations may be outliers* with large residuals* relative to a specified model, or observations that are isolated from the rest of the data in the design space. However, an outlier is not necessarily influential, since it is possible for the fitted model to change very little when an outlier is deleted from the analysis. Identification of influential observations is complicated by the fact that observations may be individually influential or jointly influential with one or more other observations. It is not always the case, though, that jointly influential observations are also individually influential. Potential outliers may be highlighted by an examination of the studentized residuals*, but influential groups of observations, or even individual remote points, may so distort the fitted model that their residuals are relatively small and so do not appear in any way anomalous. The diagrams in Fig. 1 give ex-

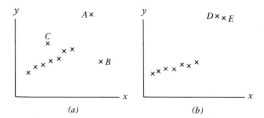

Figure 1 Outliers and influential observations.

amples of different kinds of influential observations relative to a simple straight-line model.

In Fig. 1a, points A, B, and C are outliers but C is not influential. A and B are individually influential but not jointly influential. In Fig. 1b, points D and E are remote points jointly influential but not individually influential.

Although observations may be influential relative to any kind of analysis, interest has centered on the development of procedures to detect influential observations with respect to the general linear model $\mathbf{y} = \mathbf{X\beta} + \boldsymbol{\epsilon}$, where $\mathbf{y}$ is a vector of n observations and $\boldsymbol{\beta}$ is a vector of p unknown parameters. If the least-squares* estimate of $\boldsymbol{\beta}$ using all the data is $\hat{\boldsymbol{\beta}}$ and is $\hat{\boldsymbol{\beta}}_{(ij\dots)}$ when the ith, jth, etc., observations are deleted from the sample, the contribution that these points make to the determination of $\hat{\boldsymbol{\beta}}$ may be judged by comparing $\hat{\boldsymbol{\beta}}$ and $\hat{\boldsymbol{\beta}}_{(ij\dots)}$ in some way. Cook and Weisberg [6] consider a number of procedures based on the empirical influence function* $\hat{\boldsymbol{\beta}}_{(ij\dots)} - \hat{\boldsymbol{\beta}}$. A special case, which assesses the influence of a single observation, is Cook's [4] statistic

$$D_i = (\hat{\boldsymbol{\beta}}_{(i)} - \hat{\boldsymbol{\beta}})'\mathbf{X}'\mathbf{X}(\hat{\boldsymbol{\beta}}_{(i)} - \hat{\boldsymbol{\beta}})/(ps)^2,$$

where s^2 is the residual mean square*. This is a measure of the distance moved by the least-squares estimate when the ith observation is removed from the data. Under the usual assumptions of the linear model, the influence of the ith observation may be assessed by comparing D_i with the percentage points of the F-distribution with p and $n - p$ degrees of freedom. A computationally more convenient form is $D_i = p^{-1}t_i^2 v_i/(1 - v_i)$, where t_i is the ith studentized residual and v_i is the ith diagonal element of $\mathbf{V} =$

$\mathbf{X}(\mathbf{X}'\mathbf{X})^{-1}\mathbf{X}'$. Evidently, t_i^2 is a measure of the degree to which the ith observation may be regarded as an outlier relative to the model. The magnitude of $v_i/(1 - v_i)$, which is equal to the ratio of the variance of the ith predicted value to the variance of the ith residual, indicates the sensitivity of the least-squares analysis to the location of this observation in the design space (see Hoaglin and Welsch [9] for further details). The statistic D_i thus provides a measure of the overall influence of the ith observation.

Cook [5] applied this procedure to the model $y_i = \beta_0 + \beta_1 x_{1i} + \beta_2 x_{2i} + \beta_3 x_{1i}^2 + \epsilon_i$ fitted to a sample of 21 observations discussed in detail by Daniel and Wood [7]. The observations are shown in Fig. 2 plotted in the (x_1, x_2) plane with the y values alongside each point and observations 1, 2, 3, 4, and 21 indicated by the numbers in brackets. Table 1 shows the corresponding values of t_i, v_i, and D_i. It may be seen that observation 21 is the most influential followed by number 4. Observations 1 and 2 have fairly large values of D_i because they are remote in the sample space.

The generalized version of D_i, comparing $\hat{\boldsymbol{\beta}}_{(ij\dots)}$ and $\hat{\boldsymbol{\beta}}$ and assessing the joint influence of a group of observations, may also be expressed as a combination of an outlier measure and a data structure measure. Such decompositions are given by Cook and Weisberg [6] and Draper and John [8]. An alternative general statistic, proposed by

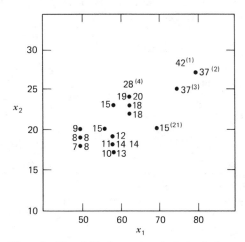

Figure 2 Plot of 21 observations in (x_1, x_2) plane.

Table 1   Values of t_i, v_i, and D_i for a Second-Order Model Fitted to 21 Observations

Observation	y	t_i	v_i	D_i
1	42	0.97	0.409	0.162
2	37	-1.06	0.409	0.193
3	37	1.54	0.176	0.125
4	28	2.27	0.191	0.304
5	18	-0.31	0.103	0.003
6	18	-0.73	0.134	0.021
7	19	-0.84	0.191	0.042
8	20	-0.50	0.191	0.014
9	15	-0.94	0.163	0.043
10	14	0.84	0.139	0.028
11	14	0.84	0.139	0.028
12	13	0.96	0.212	0.062
13	11	-0.17	0.139	0.001
14	12	-0.25	0.092	0.001
15	8	0.17	0.188	0.002
16	7	-0.17	0.188	0.002
17	8	-0.26	0.187	0.004
18	8	-0.26	0.187	0.004
19	9	-0.35	0.212	0.008
20	15	0.68	0.064	0.008
21	15	-2.63	0.288	0.699

Andrews and Pregibon [1] and based on the augmented matrix $\mathbf{X}_1^* = (\mathbf{X} : \mathbf{y})$, measures the proportion of the volume generated by $\mathbf{X}_1^*$ attributable to a given subset of the sample. Draper and John [8] give some numerical comparisons of these test statistics and additional references may be found in Cook and Weisberg [6]. Regression diagnostics including the identification of influential data, are discussed in some detail by Belsley et al. [3]. Graphical diagnostic displays for outlying and influential observations in multiple regression are reviewed by Atkinson [2].

References

[1] Andrews, D. F. and Pregibon, D. (1978). *J. R. Statist. Soc. B*, **40**, 85–93.

[2] Atkinson, A. C. (1982). *J. R. Statist. Soc. B*, **44**, 1–22.

[3] Belsley, D. A., Kuh, E., and Welsch, R. E. (1980). *Regression Diagnostics*. Wiley, New York.

[4] Cook, R. D. (1977). *Technometrics*, **19**, 15–18, 348–350.

[5] Cook, R. D. (1979). *J. Amer. Statist. Ass.*, **74**, 169–174.

[6] Cook, R. D. and Weisberg, S. (1980). *Technometrics*, **22**, 495–508.

[7] Daniel, C. and Wood, F. S. (1971). *Fitting Equations to Data*. Wiley, New York.

[8] Draper, N. R. and John, J. A. (1981). *Technometrics*, **23**, 21–26.

[9] Hoaglin, D. C. and Welsch, R. E. (1978). *Amer. Statist.*, **32**, 17–22.

(INFLUENCE FUNCTIONS
OUTLIERS
RESIDUAL MEAN SQUARE)

P. PRESCOTT

INFORMATION, KULLBACK   *See* KULLBACK INFORMATION

INFORMATION CONTENT

In Bayesian estimation procedures, the information content, IC, of an estimating distribution is equal to the reciprocal of the variance of the distribution. In particular, the information content of the 'diffuse prior' is $1/\infty$ or zero, while the information content of the sampling distribution of the arithmetic mean $\bar{X}$ is $\mathrm{IC}_{\bar{X}} = n/\sigma_x^2$, where σ_x^2 is the variance of the population.

(A PRIORI DISTRIBUTION
BAYESIAN INFERENCE
INFORMATION AND CODING THEORY)

INFORMATION THEORY AND CODING THEORY

Information theory was founded in a celebrated two-part paper by Claude E. Shannon [133]. It is the branch of applied probability theory that treats problems concerning the reproduction at one or more locations of information that was generated elsewhere. Both exact and approximate reproductions are considered. Information theory is applicable to many communication and signal processing problems in engineering and biology. It articulates mathematically with statistical hypothesis testing*, stochastic processes*, ergodic theory, and the algebra of finite (Galois) fields*. Information theory neither subsumes nor is subsumed by the cognate disciplines of information processing and information retrieval that constitute a portion of computer science. It owes debts to early work on statistical physics* by Boltzmann [22], Szilard [150], and von Neumann [159] and on communication theory by Nyquist [112, 113] and Hartley [70], who was perhaps the first to suggest a logarithmic measure of information. Parallel theory developed independently by Norbert Wiener [160, 161] also played a significant part in placing communication theory* on a firm footing rooted in mathematical statistics and led to the development of the closely related discipline of cybernetics.

Information theorists devote their efforts to quantitative examination of the following three questions:

1. What is information?
2. What are the fundamental limitations on the accuracy with which information can be transmitted?
3. What design methodologies and computational algorithms yield practical systems for communicating and storing information that perform close to the aforementioned fundamental limits?

The principal results obtained to date concerning questions 1 and 2 will be surveyed here. The extensive literature addressed to question 3 will not be surveyed here, but some of the major references will be cited.

INFORMATION AND ENTROPY*

Supplying information is equivalent to removing uncertainty. That is,

information supplied = prior uncertainty

− posterior uncertainty.

Indeed, it is senseless to transmit something over a communication link or to retrieve something from a computer memory unless the one for whom it is intended is uncertain about what he or she will receive. Shannon [133] quantified the abstract concept of uncertainty, and thereby that of information, by applying ideas from probability theory. Nonprobabilistic approaches have been proposed subsequently, most notably one based on computational complexity (see, e.g., Kolmogorov [90, 91]), but to date none of these has provided a rich vein for research like that which Shannon's probabilistic approach affords.

Let X be a random variable assuming values x belonging to the set $\mathscr{X}$. For simplicity, consider initially only cases in which $\mathscr{X}$ has finitely many members, say $\mathscr{X} = \{0, 1, \ldots, |\mathscr{X}| - 1\}$. Information theorists refer to $\mathscr{X}$ as an "alphabet" of size $|\mathscr{X}|$, and to each $x \in \mathscr{X}$ as a "letter" of the alphabet. Suppose that you desire to know which value x has been assumed by X. Adopting a probabilistic approach, let us describe your a priori knowledge about x by means of the probability distribution $\{p_X(x), \ x \in \mathscr{X}\}$, where $p_X(x) = \Pr[X = x]$. We shall denote this distribution by $\{p_X(x)\}$ or simply by $\{p\}$ whenever no ambiguity results thereby. Your prior uncertainty is a functional of $\{p\}$, call it $H(\{p\})$.

Most people readily accept the premise that any uncertainty functional H should satisfy axioms A1, A2, and A3 below. Shannon [133] showed that these three axioms imply that H must be of the form

$$H(p) = -K \sum_{x \in \mathscr{X}} p(x) \log p(x),$$

where K is a constant. Choosing K is equivalent to selecting a logarithmic base. It is customary to set $K = 1$ and to use base 2 logarithms, so we shall do so. Then the unit of uncertainty, or equivalently the unit of information, is called a *bit*. Some authors prefer to use base e logarithms, in which case uncertainty and information are measured in nats.

$$H = -\sum_x p(x) \log_2 p(x) \text{ bits}$$

$$= -\sum_x p(x) \ln p(x) \text{ nats.}$$

The three axioms that lead to this formula

for H are:

A1. $H(\{p\})$ is continuous in $\{p\}$.

A2. When all $|\mathscr{X}|$ alternatives are equally likely [$p(x) = 1/|\mathscr{X}|$ for all x], then H increases monotonically with $|\mathscr{X}|$.

A3. If $0 \leqslant \lambda \leqslant 1$ and $\bar{\lambda} = 1 - \lambda$, then
$$H(p(0), \ldots, p(n-1), \lambda p(n), \bar{\lambda} p(n))$$
$$= H(p(0), \ldots, p(n)) + p(n)H(\lambda, \bar{\lambda}).$$

Axiom A3 addresses a situation in which one letter of the alphabet has been partitioned into two subletters and its probability p has been apportioned between them with weights λ and $\bar{\lambda}$. The axiom requires that in such an instance the original uncertainty is augmented by $pH(\lambda, \bar{\lambda})$ because a fraction p of the time it now will be necessary to provide $H(\lambda, \bar{\lambda})$ more bits of information to remove the uncertainty as to which of the two subletters has occurred. The quantity $H(\{p\})$ is called the *entropy* of $\{p\}$. (Since Shannon's communication entropy is minus Boltzmann's statistical mechanics* entropy, some statistical physicists like to refer to Shannon's H as *negentropy*.) Other families of information-theoretic functionals result if Axiom A3 is relaxed in various ways, but they seem to be of little practical significance; for an entry to the literature of such functionals, see Aczél and Daróczy [2].

Now suppose that I observe x (perhaps imperfectly) and then attempt to convey that knowledge to you. Let Y, which assumes values $y \in \mathscr{Y}$, denote the datum that becomes available to you by virtue of my attempt at communication. Thus, for some particular $y \in \mathscr{Y}$, which will be held fixed until further notice, you observe that the event $[Y = y]$ has occurred. Your posterior uncertainty about X must then equal $H(\{P(x|y), \ x \in \mathscr{X}\})$, where $P(x|y) = \Pr[X = x \,|\, Y = y]$. The information about X supplied to you by virtue of your having observed that $[Y = y]$ therefore is

$$I_y = H(\{p(x)\}) - H(\{P(x|y)\})$$
$$= -\sum_x p(x)\log_2 p(x)$$
$$+ \sum_x P(x|y)\log_2 P(x|y).$$

Of particular interest is the average

amount of information I have communicated to you about X. This is calculated by averaging I_y over the distribution $\{q(y), y \in \mathscr{Y}\}$ that governs your observation,

$$I = \sum_{y \in \mathscr{Y}} q(y)I_y$$
$$= -\sum_{xy} q(y)p(x)\log_2 p(x)$$
$$+ \sum_{xy} q(y)P(x|y)\log_2 P(x|y).$$

Note that in the first summation over (x, y) we may replace $q(y)$ by any distribution over $\mathscr{Y}$ without changing the result. In particular, using

$$Q(y|x) \stackrel{\Delta}{=} q(y)P(x|y)/p(x)$$

yields

$$I = \sum_{xy} P(x, y)\log_2\left(\frac{P(x|y)}{p(x)}\right)$$
$$= \sum_{xy} P(x, y)\log_2\left(\frac{P(x, y)}{\pi(x, y)}\right),$$

where $P(x, y) = q(y)P(x|y) = p(x)Q(y|x)$ is the joint distribution $\Pr(X = x, \ Y = y)$ and $\pi(x, y) = p(x)q(y)$ is the product of its marginals. The symmetry of the last form reveals that the amount of information that Y conveys about X equals that which X conveys about Y. Accordingly, I is called the *average mutual information* of the pair of random variables X and Y, often denoted $I(X, Y)$. This is an abuse of notation, of course, since I actually is a functional of the joint distribution of X and Y, not a function of X and Y themselves. It can be proved that $I(X; Y) \geqslant 0$ with equality if and only if X and Y are statistically independent.

We shall find it useful to express I as a functional of $p(x)$ and $Q(y|x)$, namely,

$$I(p, Q) = \sum_{xy} pQ \log\frac{Q}{q},$$

where $q = \sum_x pQ$. The mixed lower- and uppercase notation has been chosen purposely so that no ambiguity results here even though the arguments of the marginal and conditional distributions are omitted. It can be proved that I is concave in p for fixed Q and convex in Q for fixed p.

The operational significance of average mutual information does not derive from the intuitive considerations we have presented thus far. Rather it resides in the central roles that average mutual information plays in the theorems about fundamental limitations on the accuracy of information transmission discussed in the section "Codes and Theorems."

SOURCES AND CHANNELS

Before we can state the basic theorems of information theory, we must define two more concepts:

1. An information source and its rate-distortion function

2. A communication channel and its capacity-fee function

An *information source* generates not just a single random variable of interest but rather an indexed family of them, $\{U_t, t \in T\}$. That is, an information source is a random process. Usually, the parameter set T represents time. If $T = \{0, \pm 1, \pm 2, \ldots\}$ or $\{0, 1, 2, \ldots\}$, we call $\{U_t\}$ a discrete-time source; if $T = (-\infty, \infty)$, $(0, \infty)$, or $[a, b]$, we call $\{U_t\}$ a continuous-time source. Information theory also applies to cases in which the parameter indexes a spatial dimension. Moreover, information theory has been extended, albeit incompletely, to multidimensional parameters (i.e., to cases in which $\{U_t\}$ is a random field). We shall restrict attention here to discrete-time sources.

If for every positive integer n the joint distribution of $U_m, U_{m+1}, \ldots, U_{m+n-1}$ does not depend on m, then $\{U_t\}$ is a *stationary* source. Let $\mathbf{U} = (U_1, \ldots, U_n)$ denote the *block* of letters, or *word*, generated by the source at times 1 through n, and let $p(\mathbf{u}) = \Pr(\mathbf{U} = \mathbf{u})$. Each U_t assumes values in the alphabet $\mathcal{U} = \{0, 1, \ldots, |\mathcal{U}| - 1\}$. Suppose that information about $\mathbf{U}$ is transmitted via a communication system of any kind whatsoever to some destination. Let $\mathbf{V} = (V_1, \ldots, V_n)$ denote the word that is presented to an interested party at this destination as an estimate of $\mathbf{U}$. Assume that each V_t assumes values in the alphabet $\mathcal{V} = \{0, 1, \ldots, |\mathcal{V}| - 1\}$; in most applications $|\mathcal{V}| = |\mathcal{U}|$, but the same theory applies equally well to cases in which $|\mathcal{V}| > |\mathcal{U}|$ or $|\mathcal{V}| < |\mathcal{U}|$. Further assume that the quality of the estimate is assessed by means of a so-called *block distortion measure* $d_n : \mathcal{U}^n \times \mathcal{V}^n \to [0, \infty]$ defined by

$$d_n(\mathbf{u}, \mathbf{v}) = n^{-1} \sum_{b=1}^{n} d(u_t, v_t),$$

where $d : \mathcal{U} \times \mathcal{V} \to [0, \infty]$ is an underlying *single-letter distortion measure*. If d_n is of this form for each n, we say that a *stationary memoryless fidelity criterion* is in force because the penalty $d(u, v)$ assessed for an estimation error of the form $u \to v$ is independent both of time and of the context in which it occurs. We shall consider only stationary, memoryless fidelity criteria here. (The development of theory and especially of practical techniques for dealing with context-dependent fidelity criteria suffers from chronic retardation.)

Let $Q(\mathbf{v} | \mathbf{u})$ denote a conditional probability distribution for $\mathbf{V}$ given $\mathbf{U}$. We write $Q \in \mathcal{Q}_n(D)$ if and only if

$$\sum_{\mathbf{u}, \mathbf{v}} p(\mathbf{u}) Q(\mathbf{v} | \mathbf{u}) d_n(\mathbf{u}, \mathbf{v}) \leqslant D.$$

The *rate-distortion function* of $\{U_t\}$ with respect to d is defined by

$$R(D) = \liminf_{n \to \infty} R_n(D),$$

where

$$R_n(D) = n^{-1} \inf_{Q \in \mathcal{Q}_n(D)} I(p, Q).$$

It can be shown that the inf in the definition of $R_n(D)$ reduces to a minimum and that, if the source is stationary, the lim inf in the definition of $R(D)$ reduces to an ordinary limit.

The significance of the rate-distortion function, as we shall see in the following section, is that $R(D)$ is the minimum rate at which binary digits describing $\{U_t\}$ must be supplied to someone who wishes to produce from these binary digits an estimate of $\{U_t\}$ that has average distortion not exceeding D.

A *communication channel* is a device that transforms one random object, called the channel *input* sequence and henceforth denoted by $\{X_t\}$, into another random object called the channel *output* sequence, henceforth denoted by $\{Y_t\}$. Let ξ index the possible realizations $\{x_t\}$ of the input sequence. From the mathematical perspective a channel is an indexed family $\{Q_\xi(\cdot)\}$ of conditional probability measures governing the realizations of $\{Y_t\}$, where

$$Q_\xi(A) = \Pr\left[\{Y_t\} \in A \mid \{X_t\} = \xi\right].$$

If T denotes the shift operator on the sequence space, then the channel is said to be *stationary* if

$$Q_{T\xi}(TA) = Q_\xi(A) \qquad \text{for all } \xi \text{ and } A.$$

If $Q_\xi(\cdot)$ is a product measure for each ξ, the channel is said to be *memoryless*.

Assume for simplicity that each channel input X_t takes values in a fixed finite set $\mathscr{X} = \{0, 1, \ldots, |\mathscr{X}| - 1\}$ called the *input alphabet*. Further assume that $f: \mathscr{X} \to [0, \infty]$ is a function which specifies the fee $f(x)$ incurred each time letter $x \in \mathscr{X}$ is used as a channel input. Then $f(\cdot)$ extends additively to input words $\mathbf{x} = (x_1, \ldots, x_n)$,

$$f(\mathbf{x}) = n^{-1} \sum_{t=1}^{n} f(x_t).$$

Let $p(\mathbf{x})$ denote a joint distribution for $\mathbf{X} = (X_1, \ldots, X_n)$. We write $p \in \mathscr{P}_n(F)$ if and only if

$$\sum_{\mathbf{x}} p(\mathbf{x})f(\mathbf{x}) \leq F.$$

The capacity-fee function of $\{Q_\xi\}$ with respect to f is defined by

$$C(F) = \limsup_{n \to \infty} C_n(F),$$

where

$$C_n(F) = \sup_{p \in \mathscr{P}_n(F)} I(p, Q)$$

and Q denotes the restriction of $\{Q_\xi\}$ to $1 \leq t \leq n$. The sup in the definition of $C_n(F)$ reduces to a maximum, and, if the channel is stationary, the lim sup in the definition of $C(F)$ reduces to an ordinary limit. We shall see in the following section that $C(F)$ is the maximum rate at which information can be conveyed reliably over the channel with an expected fee per transmitted letter that does not exceed F.

If each channel output Y_t takes values in the finite alphabet $\mathscr{Y} = \{0, 1, \ldots, |\mathscr{Y}| - 1\}$, and the restriction of Q_ξ to $1 \leq t \leq n$ is of the form $\prod_{t=1}^{n} Q(y_t \mid x_t)$, where for each $x \in \mathscr{X} Q(\cdot \mid x)$ is a conditional probability distribution over $\mathscr{Y}$, then Q_ξ is said to be a *discrete memoryless channel* (d.m.c.). Many (but by no means all) channels of practical interest conform well to the d.m.c. model. Of particular interest is the so-called binary symmetric channel with crossover probability p [BSC(p)] defined by $\mathscr{X} = \mathscr{Y} = \{0, 1\}$ and $Q(y \mid x) = (1 - p)\delta_{xy} + p(1 - \delta_{xy})$, which complements either of its inputs with probability p. The capacity of this channel (in the absence of fees, or equivalently in the limit of large F) is readily shown to be $1 - H(p, 1 - p)$, where H was defined in the preceding section.

Arimoto [9] and Blahut [17] independently derived a rapidly converging iterative algorithm for computing $C(F)$ for the general d.m.c. by exploiting the observation that

$$I(p, Q) = \max_{\hat{P}} \sum_{xy} pQ \log_2(\hat{P}/p),$$

where $\hat{P}$ ranges over all conditional distributions for $x \in \mathscr{X}$ given $y \in \mathscr{Y}$. By exploiting the companion formula

$$I(p, Q) = \min_{\hat{q}} \sum_{uv} pQ \log_2(Q/\hat{q}),$$

where $\hat{q}$ ranges over all distributions for $v \in \mathscr{Y}$, Blahut [17] also derived a rapidly converging iterative algorithm for computing $R(D)$ for so-called discrete memoryless sources characterized by $p(\mathbf{u}) = \prod_{t=1}^{n} p(u_t)$.

$C(F)$ is concave in F, with $C(0) = 0$ and $C(F) \to C$ as $F \to \infty$, where the constant C is called the *capacity* of the channel. $R(D)$ is convex in D with $R(D) = 0$ for $D \geq D_{\max} \overset{\Delta}{=} \min_{v \in \mathscr{Y}} \sum_u p(u)d(u, v)$. If there exists for each $u \in \mathscr{U}$ exactly one $v \in \mathscr{V}$ such that $d(u, v) = 0$, then $R(0)$ equals the source *entropy* $H \overset{\Delta}{=} \lim n^{-1}H(p_n)$, where p_n is the distribution governing any n successive letters produced by the stationary source.

Define $D_{\min} = \sum_u p(u)\min_v d(u, v)$.

The theory of $R(D)$ and $C(F)$ can be extended to continuous alphabets. For example, if $\{U_t\}$ is a stationary Gaussian source with spectral density

$$\Phi(f) = \sum_{k=-\infty}^{\infty} \exp(-i2\pi fk)\operatorname{cov}(U_t, U_{t+k}),$$

and $d(u,v) = (u-v)^2$, then $R(D)$ is given parametrically by the equations

$$D(\theta) = \int_{-1/2}^{1/2} df \min[\theta, \Phi(f)]$$

$$R(D(\theta)) = \tfrac{1}{2}\int_{-1/2}^{1/2} df \max[0, \log_2(\Phi(f)/\theta)],$$

where θ traverses the range from 0 to $\sup \Phi(f)$. Similarly, consider a channel the output sequence of which is the sum of the input sequence and an independent, stationary Gaussian noise sequence of spectral density $N(f)$. If $f(x) = x^2$, then $C(F)$ for this channel is given parametrically by

$$F(\theta) = \int_{-1/2}^{1/2} df \max[0, \theta - N(f)]$$

$$C(F(\theta)) = \tfrac{1}{2}\int_{-1/2}^{1/2} df \max[0, \log_2(\theta/N(f))],$$

where θ traverses the range from $\inf N(f)$ to ∞. These results also apply to continuous-time stationary Gaussian sources and channels provided that $\Phi(f) = \int_{-\infty}^{\infty} \exp(-i2\pi f\tau)\operatorname{cov}(U_t, U_{t+\tau})$, $N(f)$ is similarly defined, the integrals in the parametric formulas extend over $(-\infty, \infty)$ rather than $(-\tfrac{1}{2}, \tfrac{1}{2})$, and $R(\cdot)$ and $C(\cdot)$ have units of bits per second rather than bits per letter. For further details, see Shannon [134], Kolmogorov [87], and Gallager [55].

CODES AND THEOREMS

Codes are transformations that recast information sequences in new forms with the intent of enhancing communication system performance. They come in two varieties—source codes and channel codes—each of which can be further subdivided into block codes, variable-length codes, and sliding-block codes. We shall first define block codes for both sources and channels. Then we shall state without proof the block-coding versions of the fundamental theorems on information theory. Extensions to variable-length codes and to sliding-block codes will be sketched subsequently.

As in the preceding section, let $\{U_t\}$ be an information source and let $d : \mathcal{U} \times \mathcal{V} \to [0, \infty]$ be a single-letter distortion measure. A *source code* of *block length* m and *rate* R is a collection of $M = \lfloor 2^{mR} \rfloor$ words from $\mathcal{V}^m$, say $\mathcal{B} = \{\mathbf{v}_1, \mathbf{v}_2, \ldots, \mathbf{v}_M\}$, where $\lfloor \cdot \rfloor$ denotes the integer part operator. (In what follows, $\lfloor \cdot \rfloor$ will be suppressed and should be supplied by the reader whenever a noninteger quantity appears in the role of an integer.) Any mapping ϕ from $\mathcal{U}^m$ into $\mathcal{B}$ is called an *encoding rule*, or *encoder*. The optimum encoder for a given source code $\mathcal{B}$ is the one that sets $\phi(\mathbf{u})$ equal to whichever $\mathbf{v} \in \mathcal{B}$ minimizes $d(\mathbf{u}, \mathbf{v})$. We denote the resulting minimized average distortion by

$$d(\mathcal{B}) = \sum_{\mathbf{u}} p(\mathbf{u}) \min_{\mathbf{v} \in \mathcal{B}} d(\mathbf{u}, \mathbf{v}).$$

Theorem 1A (Source Coding Theorem). Let $\{U_t\}$ be a stationary source and d be a single-letter distortion measure. Let $R(\cdot)$ denote the rate-distortion function of $\{U_t\}$ with respect to the memoryless fidelity criterion induced by d. Given $D \geqslant D_{\min}$ and $\epsilon > 0$, for any $R \geqslant R(D)$ there exists for m sufficiently large a source code $\mathcal{B}$ of block length m and rate less than $R + \epsilon$ for which $d(\mathcal{B}) < D + \epsilon$.

Theorem 1B (Converse Source Coding Theorem). Let $R(\cdot)$ be as in Theorem 1A. Given D, any source code $\mathcal{B}$ whose rate R is less than $R(D)$ must satisfy $d(\mathcal{B}) > D$.

Note that mR binary digits are required to identify a specific word belonging to a source code of block length m and rate R. Therefore, the two parts of Theorem 1 together imply that if block coding is used, then $R(D)$ specifies the minimum number of binary digits per source letter that one must receive in order to be able to estimate $\{U_t\}$ therefrom with fidelity D. Whenever $R(0)$

= H, which we have noted is usually the case, Theorem 1 subsumes the celebrated result that the entropy, H, is the minimum rate at which one must receive binary digits in order to be able to specify the source output therefrom with arbitrarily small error.

Consider a time-discrete channel with input alphabet $\mathscr{X}$, output alphabet $\mathscr{Y}$, and fee schedule $f: \mathscr{X} \to [0, \infty]$. A *channel code* of block length n and rate R is a collection of $N = 2^{nR}$ words from $\mathscr{X}^n$, say $\mathscr{C} = \{\mathbf{x}_1, \ldots, \mathbf{x}_N\}$. Code $\mathscr{C}$ can be used to transmit a message selected randomly from a set of N possible messages by means of n successive uses of the channel. Specifically, if message i is selected, one simply sends in succession the n letters of code word $\mathbf{x}_i$. If we let p_i denote the probability that message i is selected, then the expected fee incurred when code $\mathscr{C}$ is used is

$$f(\mathscr{C}) = \sum_{i=1}^{N} p_i f(\mathbf{x}_i).$$

Any mapping γ from $\mathscr{Y}^n$ into $\mathscr{C}$ is called a *decoding rule*, or decoder. The *probability of error* achieved by decoding rule γ is $P_e = \Pr[\gamma(Y) \neq \mathbf{X}]$. The optimum, or maximum a posteriori, decoding rule is the one that minimizes P_e. Given that $\mathbf{y} \in \mathscr{Y}^n$ has been received, this rule equates $\gamma(\mathbf{y})$ to whichever $\mathbf{x}_i \in \mathscr{C}$ is most likely to have been transmitted. In the event of a tie, $\gamma(\mathbf{y})$ is equated to the lowest-index code word involved in the tie. We shall denote the resulting minimum value of P_e by $P_e(\mathscr{C})$. An explicit formula for it is

$$P_e(\mathscr{C}) = \sum_{i=1}^{N} p_i \sum_{\mathbf{y} \in \Gamma_i} Q_n(\mathbf{y} \mid \mathbf{x}_i),$$

where $\Gamma_i = \{\mathbf{y} : p_i Q_n(\mathbf{y} \mid \mathbf{x}_i) \geqslant p_j Q_n(\mathbf{y} \mid \mathbf{x}_j)$ with strict inequality if $j < i\}$.

Theorem 2A (Channel Coding Theorem). Given a stationary channel with capacity fee function $C(\cdot)$, any $F \geqslant f_{\min} \overset{\Delta}{=} \min_{x \in \mathscr{X}} f(x)$ and any $\epsilon > 0$, there exists for sufficiently large n a source code $\mathscr{C}$ of block length n and rate $R > C(F) - \epsilon$ for which the inequalities $f(\mathscr{C}) < F + \epsilon$ and $P_e(\mathscr{C}) < \epsilon$ both hold.

Theorem 2B (Converse Channel Coding Theorem). Let $C(\cdot)$ and F be as in Theorem 2A. If $R \geqslant C(F)$, then any sequence $\{\mathscr{C}_n\}$ of channel codes of rate R and increasing block length n for which $\lim \sup f(\mathscr{C}_n) < F$ must satisfy $\liminf P_e(\mathscr{C}_n) = 1$.

The two parts of Theorem 2 together imply that if block coding is used, $C(F)$ is the maximum rate (in bits per channel use) at which information can be transmitted reliably over the channel with an expected fee not exceeding F. The number $C \overset{\Delta}{=} \lim_{F \to \infty} C(F)$ is called the capacity of the channel. It represents the maximum rate at which information can be transmitted reliably across the channel when no input letter fees are charged or, equivalently, when infinite resources are available so that transmission fees are not a consideration. (*See* STATISTICAL COMMUNICATION THEORY.)

Theorems 1A and 2A provide asymptotic results about what can be achieved via coding in the limit of large block length. It is natural to enquire into how rapidly the performance of optimum coding algorithms approaches its asymptotic limit as block length increases. Over the last 30 years effort devoted to this question has generated literally hundreds of papers and dissertations about such performance bounds. For entries to this literature, consult Gallager [54], Shannon et al. [138], Slepian [143, 144], and Csiszár and Körner [34]. Combining Theorems 1A and 2A yields the following fundamental result.

Theorem 3A (Information Transmission Theorem). Suppose that we are given a source that produces one letter every τ_s seconds and has rate-distortion function $R(\cdot)$, and a channel that transmits one letter every τ_c seconds and has capacity-fee function $C(\cdot)$. If $C(F)/\tau_c > R(D)/\tau_s$, then it is possible by means of long block codes to transmit over the channel such that an average fee of F or less is incurred and the source data can be estimated from the resulting channel output with an average distortion of D or less.

We also have

Theorem 3B (Converse Information Transmission Theorem). Let $R(\cdot)$, τ_s, $C(\cdot)$, and τ_c be as in Theorem 3A. If $C(F)/\tau_c < R(D)/\tau_s$, it is not possible both to incur an average channel input fee of F or less and to estimate the source data from the resulting channel output with average distortion D or less.

Theorem 3B does not follow from Theorems 1B and 2B because it is not restricted to block codes. It states that no source and channel coding schemes of any kind can achieve fidelity D at cost F if $C(F)/\tau_c < R(D)/\tau_s$. To help develop an appreciation for this distinction, we discuss two popular non-block coding techniques in the next section.

VARIABLE-LENGTH AND SLIDING-BLOCK CODES

Variable-length codes are generalizations of the classical Morse code of telegraphy. In variable-length source coding each source letter $u \in \mathcal{U}$ gets mapped into an element in the set $\mathcal{G}^*$ of finite-length strings from some source coding alphabet $\mathcal{G} = \{0, 1, \ldots, |\mathcal{G}| - 1\}$. Thus a variable-length code is described by specifying a mapping $g: \mathcal{U} \to \mathcal{G}^*$. We shall denote the length of the code string $g(u)$ by $|g(u)|$. A sequence of successive source letters $u_1, u_2, \ldots$ gets mapped into the concatenation of $g(u_1)$, $g(u_2)$, etc. In order for the code to be *uniquely decipherable*, it must be true that, if

$$(u_1, \ldots, u_M) \neq (u'_1, u'_2, \ldots, u'_{M'}),$$

then

$$g(u_1)g(u_2)\cdots g(u_M) \neq g(u'_1)g(u'_2)\cdots g(u'_{M'}).$$

Morse code is not uniquely decipherable when viewed as a sequence of dots and dashes only; it becomes uniquely decipherable when the so-called mark (pause between code strings for successive source letters) is rightfully considered to be a third element of $\mathcal{G}$.

Theorem 4. Given a set $\{n(u), u \in \mathcal{U}\}$ of positive integers indexed by the elements of

$\mathcal{U}$, there exists at least one uniquely decipherable $g: \mathcal{U} \to \mathcal{G}^*$ with $\{|g(u)|, u \in \mathcal{U}\} = \{n(u), u \in \mathcal{U}\}$ if and only if

$$\sum_{u \in \mathcal{U}} |\mathcal{G}|^{-n(u)} \leqslant 1.$$

Moreover, if this inequality is satisfied, at least one of the uniquely decipherable codes with string lengths $\{n(u)\}$ satisfies the *prefix condition* that no $g(u)$ equals the first $|g(u)|$ letters of $g(u')$ for any $u' \neq u$.

The statements comprising Theorem 4 were first proved by Kraft [96] and McMillan [108]; see also Karush [81].

Let $\{p_1(u), u \in \mathcal{U}\}$ denote the marginal distribution that governs each letter U_t produced by the stationary source $\{U_t\}$, and let $H(p_1) = -\sum p_1(u)\log_2 p_1(u)$. Since the elements of $\mathcal{G}^n$ can be labeled by binary strings of length $\lceil n\log_2|\mathcal{G}|\rceil$, a uniquely decipherable variable-length code with average code word length

$$\bar{n} = \sum_u p(u)|g(u)|$$

provides a means for describing $\{U_t\}$ exactly using an average of $\bar{n}\log|\mathcal{G}|$ binary digits per source letter. Huffman [74] has devised an algorithm for producing a uniquely decipherable code that has the smallest $\bar{n}$ and also satisfies the prefix condition. For proofs of the following two basic theorems, see, e.g., Gallager [55].

Theorem 5 (Converse Variable-Length Source Coding Theorem). Let $g: \mathcal{U} \to \mathcal{G}^*$ be a uniquely decipherable variable-length code. Then its average code word length $\bar{n}$ when source letters are distributed according to $\{p_1(u), u \in \mathcal{U}\}$ must satisfy $\bar{n}\log_2|\mathcal{G}| \geqslant H(p_1)$.

Theorem 6 (Variable-Length Source Coding Theorem). Let $\{U_t\}$ be a stationary source with distribution $\{p_m(\mathbf{u}), \mathbf{u} \in \mathcal{U}^m\}$ governing its blocks of m successive letters. Then for every K there exists a uniquely decipherable variable-length code $g: \mathcal{U}^m \to \mathcal{G}^*$ whose average code word length $\bar{n}$ satisfies

$$\bar{n}\log_2|\mathcal{G}| < H(p_m) + 1.$$

Dividing this by m and then letting $m \to \infty$, we see that the average number of code letters per source letter can be made to approach the source entropy H defined at the close of the section "Sources and Channels." For further results about variable-length codes, see Shannon [133], Jelinek and Schneider [79, 80], Gallager [57], Pursley and Davisson [126], and Katona and Nemetz [83].

A *sliding-block coder* of window width $2w + 1$ generates its nth output letter by calculating a function of letters $n - w$ through $n + w$ of the sequence it is encoding or decoding. If this function does not vary with n, then the sliding-block coder has the decided advantage (not shared by block and variable-length codes) that whenever its output is a stationary process so is its input. This partially explains why ergodic theorists consider only sliding-block codes. They allow w to be infinite, but information theorists are interested only in finite w or in asymptotic results as $w \to \infty$. Readers interested in sliding-block codes and sliding-block coding theorems should consult Gray [59], Gray et al. [62], Gray and Ornstein [61], Gray et al. [63], Shields and Neuhoff [140], and Kieffer [86].

The coding results presented above pertain to a particular source and/or a particular channel whose statistical description is completely known. In practice, our statistical knowledge is always incomplete at best. It is an important and perhaps surprising fact that the major results of information theory continue to hold when little more is known than the sizes $|\mathscr{U}|$, $|\mathscr{V}|$, $|\mathscr{X}|$, and $|\mathscr{Y}|$ of the source alphabet, estimation alphabet, channel input alphabet, and channel output alphabet, respectively. Such nonparametric results comprise the theory known as *universal coding*. The objective of universal coding, somewhat oversimplified, is to show that fixed codes of the block, variable-length, or sliding-block variety can be selected knowing only $|\mathscr{U}|$, $|\mathscr{V}|$, $|\mathscr{X}|$, and $|\mathscr{Y}|$ which are such that, as the source and channel are varied over a loosely restricted set, the average fee F and average distortion D that result always satisfy $R(D) \approx C(F)$. For detailed results about universal coding, see Fitingof [52, 53], Kolmogorov [90], Davisson [36], Ziv [172, 173], Pursley and Davisson [126], Ziv and Lempel [174], and Rissanen and Langdon [128].

MULTITERMINAL SOURCES AND CHANNELS

In many situations of interest, information sources correlated with one another are separately observed and encoded by several agents who may or may not be members of the same team. Two examples are reporters for one or more wire services assessing worldwide reaction to some newsworthy event, and meteorologists in a network of weather observatories gathering data from various instruments. How do we generalize the basic concepts of the entropy and the rate-distortion function of a single information source to more complicated situations such as these? It is essential to appreciate that the physical separation of the agents makes such problems fundamentally different from those discussed in the preceding sections. If the multidimensional data all were observed by one agent, we would simply have a classical information source whose alphabet would be the Cartesian product of the alphabets of the component sources.

Let $\{U_{jt}\}$, $1 \leqslant j \leqslant J$, be a collection of sources that are correlated with one another. Assume that there are L agents, and let $a_{jl} = 1$ if agent l can observe $\{U_{jt}\}$ and $a_{jl} = 0$ otherwise. All three cases $L < J$, $L = J$, and $L > J$ can occur. Assume that agent l implements K_l block encodings of all the information available to him, call them

$$\phi_{lk} : \underset{j\,:\,a_{jl}=1}{X} \mathscr{U}_j^n \to \{1, \dots, 2^{nR_{lk}}\},$$

$$1 \leqslant k \leqslant K_l.$$

Let $b_{lkl'} = 1$ if the value assumed by ϕ_{lk} is provided to agent l' and $b_{lkl'} = 0$ otherwise

$(1 \leqslant l, \; l' \leqslant L, \; 1 \leqslant k \leqslant K_l)$. Assume that agent l wishes to obtain an estimate of $\{U_{jt}\}$ that has average distortion not exceeding D_{jl}. Given the arrays $A = (a_{jl})$, $B = (b_{lkl'})$, and $D = (D_{jl})$, the general multiterminal source encoding problem is to determine the lim sup, call it $\mathcal{R}(D)$, of the sequence of sets $\mathcal{R}_n(D)$ defined as follows. $\mathcal{R}_n(D)$ is the set of all arrays $R = (R_{lk}, \; 1 \leqslant l \leqslant L, \; 1 \leqslant k \leqslant K_l)$ for which block length n encoding functions ϕ_{lk} with respective rates R_{lk} exist that permit agent l to estimate source j with average distortion $D_{jl}, \; 1 \leqslant j \leqslant J, \; 1 \leqslant l \leqslant L$. This general problem is exceedingly challenging. Indeed, its solution often depends on whether or not agent l, when calculating ϕ_{lk}, may employ knowledge of the values assumed by other ϕ's either calculated by or communicated to the agent. For some results about general multiterminal source coding problems, see Csiszár and Körner [32], Han and Kobayashi [68], Ahlswede [4, 5], and Shen [139].

Some specific multiterminal source encoding problems have been solved. Probably the most important of these is the celebrated result of Slepian and Wolf [145] and Cover [29]. In this problem agent L wants distortion-free estimates of all $J = L - 1$ jointly ergodic sources $\{U_{jt}\}$ each of which is observed by one and only one agent. That is,

$$a_{jl} = \delta_{jl} \text{ and } K_l = 1 \quad \text{for } 1 \leqslant l \leqslant J = L - 1, \; a_{jL} = 0$$

$$b_{ll'} = \delta_{l'L}, \; D_{jL} = 0 \text{ and } D_{jl} = \infty \quad \text{for } l \neq L.$$

The solution for general L is too cumbersome to report here. For $L = 3$ it assumes the form

$$R = \{(R_{11}, R_{21}): R_{11} + R_{21} \geqslant H_{12},$$

$$R_{11} \geqslant H_{12} - H_2, \, R_{21} \geqslant H_{12} - H_1\},$$

where H_1, H_2, and H_{12} are the entropies of $\{U_{1t}\}$, $\{U_{2t}\}$, and $\{(U_{1t}, U_{2t})\}$, respectively. Other problems in which each D_{jl} equals either 0 or ∞ have been solved by Wyner [166], Ahlswede and Körner [6], Sgarro [132], Körner and Marton [93], Yamamoto [168], and Wyner and Ziv [167]. Certain problems in which some of the D_{jl} satisfy $0 < D_{jl} < \infty$ have been solved or bounded by Wyner and Ziv [167], Berger et al. [14], Witsenhausen and Wyner [162], and Kaspi and Berger [82]. The Wyner–Ziv and Korner–Marton papers are particularly noteworthy for their innovativeness.

Multiterminal channels either accept inputs from more than one agent, or deliver outputs to more than one agent, or both. A conventional telephone line, for example, can support signals in both directions simultaneously but they interfere with one another. Shannon [136] treated such "two-way channels" in what was probably the first paper on multiterminal information theory. His bounds have been sharpened recently by Dueck [44] and by Schalkwijk [131], but the exact capacity region still is not known. [The capacity-fee region $\mathcal{C}(F)$ of a multiterminal channel is a set of channel encoding rate arrays defined in a manner analogous to how the rate-distortion region $\mathcal{R}(D)$ for a multiterminal source was defined above. In the interests of space, its definition will not be detailed here.] Interest in multiterminal information theory was revitalized by Cover [28] in his paper concerning so-called broadcast channels which have a single input terminal and multiple output terminals. The multiple-input single-output channel, or multiaccess channel, first studied by Liao [100] and Ahlswede [3], is encountered in practice in local area networks and in satellite communications. The voluminous literature of multiterminal channel theory was ably summarized by van der Meulen [154]. Noteworthy among the many multiterminal channel papers that have appeared since then are the works of Cover and El Gamal [30], Gelfand and Pinsker [58], Marton [105], El Gamal [45, 46], Dueck [44], Hájek and Pursley [67], and Schalkwijk [131].

Results concerning asymptotic behavior of multiterminal source and channel codes for large block length and/or universal multiterminal coding have been presented by Gallager [56], Viterbi and Omura [157], Koshelev [94], and Zhang [171]. Kieffer [86] has significantly advanced the theory of multiterminal sliding-block codes.

INTERFACE WITH STATISTICS AND MATHEMATICS

For a channel code of rate R and block length n, deciding which one of the 2^{nR} code words was transmitted when $\mathbf{y} = (y_1, \ldots, y_N)$ is received is a problem in multiple hypothesis testing. As remarked in the section "Codes and Theorems," for any $R < C(F)$ the probability that an incorrect decision is made, averaged over the distribution governing which code word was transmitted, can be made to approach 0 as $n \to \infty$ using code words whose expected transmission fee per channel letter does not exceed F. It has been shown that this error probability decays exponentially rapidly with n in the range $0 < R < C(F)$ [49, 54]. Accordingly, attempts have been made to determine the so-called reliability-rate function $E(R, F)$ defined as

$$E(R, F) = \limsup_{n \to \infty} n^{-1} \ln(1/P_e(n, R, F)),$$

where $P_e(n, R, F)$ is the smallest error probability of any channel code of block length n, rate at least R, and expected transmission fee not exceeding F. It turns out that $E(R, F)$ usually is not affected if one imposes the more stringent requirements that $P_e(n, R, F)$ be the maximum rather than the average probability of decoding error over all the transmitted code words and that the code words have per letter transmission fees that are bounded by F uniformly rather than on the average. For most channels there is a critical rate, aptly denoted $R_{\text{crit}}(F)$, such that $E(R, F)$ is known for $R \leqslant R_{\text{crit}}(F)$. For $R < R_{\text{crit}}(F)$, we know upper and lower bounds to $E(R, F)$ that usually are in close percentage agreement, but the exact answer continues to elude all investigators. Generating the exact answer would be tantamount to solving certain asymptotic packing problems of long-standing concerning spheres of large radius in spaces whose dimensionality approaches infinity. (See, among others, Rogers [130], Shannon et al. [138], Gallager [55], Haroutunian [69], Blahut [18, 19], Sloane [146], and Levenshtein [99].) Source coding problems possess similar links with asymptotic problems in covering theory and so-called ϵ-entropy. (See Kolmogorov [90], Vitushkin [158], Posner and McEliece [124], Berger [11], and their reference lists.)

Information measures play an important role in classical binary hypothesis testing problems, too. Consider n i.i.d. observations $Y_1, Y_2, \ldots, Y_r$ each of which is distributed according to $q_0(y)$ if H_0 is true and $q_1(y)$ if H_1 is true. Let $\alpha = \Pr(\text{decide } H_1 \mid H_0 \text{ true})$ and $\beta = \Pr(\text{decide } H_0 \mid H_1 \text{ true})$. If β is bounded away from 1, then [25, 97]

$$n^{-1}\log_2(1/\alpha) \leqslant J(q_1, q_0)$$
$$\stackrel{\Delta}{=} \sum_y q_1(y)\log_2 q_1(y)/q_0(y).$$

Similarly, if α is bounded away from 1 then $n^{-1}\log_2(1/\beta) \leqslant J(q_0, q_1)$. Kullback called J the *discrimination*, but others unfortunately have referred to it by at least a dozen different names. $J(q_0, q_1) \geqslant 0$ with equality if and only if $q_0 \equiv q_1$. J is not symmetric and does not satisfy the triangle inequality, nor does its symmetrized form $J(q_0, q_1) + J(q_1, q_0)$. Note that the mutual information functional I of the section "Information and Entropy" is a special case of J, namely $J(P, \Pi)$ where P is a joint measure on some product space and Π is the product of the marginals of P.

It is possible to drive both α and β to zero exponentially at certain pairs of decay rates. For $0 \leqslant r \leqslant J(q_1, q_0)$, define $e_n(r)$ to be the supremum of $n^{-1}\log_2(1/\beta)$ taken over all decision rules based on n observations that satisfy $n^{-1}\log(1/\alpha) \geqslant r$, and then define $e(r) = \limsup e_n(r)$. Information-theoretic arguments developed independently by Csiszár and Longo [35] and by Blahut [18] show that $e(r)$ is the minimum of $J(q, q_1)$ over all q such that $J(q, q_0) \leqslant r$. Also, $e(r)$ decreases from $(r = 0, e = J(q_0, q_1))$ to $(r = J(q_1, q_0), e = 0)$ in a convex downward fashion.

Since problems concerning statistical estimation of parameters can be formulated, at least loosely, as limits as $M \to \infty$ of appropriate sequences of M-ary hypothesis testing problems, measures of information again make an appearance. In estimation theory, however, the principal role is played by the measure of "information" introduced by Fisher [51] rather than by Shannon theory quantities such as H and I, which were discussed in the section "Information and Entropy," and J, which is under discussion here. For further results linking information theory and statistics, see Kullback [97], Perez [118], Blahut [18], Shore and Johnson [141], and Christensen [26]. For an entry to literature treating the relationship between information theory and the theory of efficient search algorithms, consult Pierce [121], Ahlswede and Wezener [7], Massey [106], and de Faria et al. [37].

Information theory also has had an impact on classical prediction* theory for random sequences and processes as developed either via generalized harmonic analysis [116] or via the Hilbert space projection theorem [31, 43]. According to that theory a stationary process* whose spectral density has finite support is deterministic in the sense that its future can be predicted with zero quadratic mean error based on knowledge of its past. Rate-distortion theory tells us, however, that bits must be provided at a nonzero limiting rate in order to describe a bandlimited Gaussian random process with time-averaged quadratic mean error strictly less than the variance of the process [133]. This apparent contradiction stems from the physically unrealistic assumption in classical prediction theory that the past samples can be observed with unlimited accuracy. It requires an infinite number of bits to describe even a single continuous observation exactly. A physically meaningful theory of prediction must explicitly account for the fact that observations are inherently noisy and in any case cannot be either stored or transmitted with unlimited accuracy. Pinsker [122], Berger [12], and Hájek [66] have developed theories of so-called zero-entropy processes and information-singular processes which remain deterministic even when these practical considerations are taken into account.

Information theory and ergodic theory have interacted to their mutual benefit. Two stationary, finite-alphabet random sequences are isomorphic in the sense of ergodic theory if there is a one-to-one transformation from the space of realizations of one of them (save perhaps for a null set) into that of the other that preserves both measure and the shift operator. The first major contribution to the long-standing problem of developing simple necessary and sufficient conditions for two flows to be isomorphic in the sense of ergodic theory was the discovery by Kolmogorov [88, 89] and Sinai [142] that any two isomorphic random sequences must have the same entropy rate. Ornstein [114] subsequently showed that the entropy invariant also functions as a sufficient condition for isomorphism provided that one restricts attention to the so-called "finitely determined processes" or "B-processes." These processes are the limits, in the sense of Ornstein's [115] $\bar{d}$-metric, of sequences of stationary, finite-order Markov chains. The foundation for this bridge between information theory and ergodic theory was laid by research devoted to various generalizations of the Shannon–McMillan–Breiman theorem which extend the mean and pointwise ergodic theorems so as to encompass the information density functional. (See Perez [117], Moy [111], Dobrushin [40], Pinsker [122], Jacobs [75], Billingsley [16], and Kieffer [85]. Ergodic theory's principal gift to information theory is the theory of sliding-block codes as discussed in the section "Variable Length and Sliding-Block Codes." A collection of papers on ergodic theory and information theory has been edited by Gray and Davisson [60].

The fundamental theorems of information theory detailed in the section "Codes and Theorems" ignited the search for specific classes of source and channel codes that yield performance approximating the theoretical ideal. The theory of error control codes for transmission over noisy, discrete channels has been developed in considerable depth over the past three decades. It now represents probably the major application of the theory of finite, or Galois, fields* to problems of practical interest. It would be inappropriate to attempt to survey here the hundreds, perhaps thousands, of papers written about algebraic coding theory. Instead, we comment briefly on some of the major textbooks on the subject. Berlekamp's [15] celebrated book, which unified the extant literature, also contains significant contributions to the theory of factorization of polynomials over finite fields and applications thereof to fast algorithms for the decoding of cyclic codes. Lin [101] adopts a more straightforward presentation with an engineering slant, as do Peterson and Weldon [120] in their enlarged revision of Peterson's [119] vintage text. MacWilliams and Sloane [104], van Lint [155], and McEliece [107] provide relatively mathematical treatments with heavy emphasis on distance properties of families of codes. Pless [123] emphasizes combinatorial view of algebraic coding, as do Blake and Mullin [21], who also concentrate on links with the theory of block designs. Clark and Cain [27], by stressing applications over theory, have provided a perspective on coding theory that is particularly valuable for the practising digital communications engineer. Blahut's [20] forthcoming text establishes fruitful connections between algebraic coding and digital signal processing in both the time domain and the frequency domain and emphasizes the design of fast decoding algorithms.

The mathematical theory of cryptography and the mathematical theory of communication are, in effect, complementary realizations of the same fundamental insights. Of course, the first to propound this duality was Shannon [135]. More recently, Diffie and Hellman [38, 39] and Merkle [109] introduced the concept of public-key cryptography, and Rivest et al. [129] developed a specific family of algorithms which provides effectively unbreakable, readily implementable public-key cryptographic systems.

Lovasz [103] devised a graph theory* method that resolved a problem of long standing regarding the zero-error capacity of certain channels. This inspired Csiszár and Körner [33] and Ahlswede [4, 5] to develop approaches rooted in graph theory in order to prove various coding theorems of information theory.

Information theory has been applied successfully to the design of systems for the compression of speech data [8, 64] and image data [110, 148]. One can use codes in which the words can be represented on a tree or trellis structure, thereby significantly simplifying the search for code words that satisfactorily match the data.

Attempts have been made to link information theory with physics. Most of the effort has been devoted to the study of quantum communication channels [102], with particular interest in quantum optical communications [72, 73, 169, 170]. A novel aspect here is that the observation process becomes part of the system state and the operators of interest are not self-adjoint. Jarett and Cover [76] have considered relativistic information theory. An early key book by Brillouin [23] explored the scientific implications of information theory.

The natural relationship with statistical mechanics has proved productive, resolution of the Maxwell's demon paradox being probably the most noteworthy accomplishment. (*See* MAXWELL, JAMES CLERK.) The role of maximum entropy in thermodynamics formalism and in spectral estimation has been propounded by Jaynes [77], Tribus [151], and Burg [24]. The thermodynamics of multiphase chemical equilibrium is governed by equations very similar to those encountered in the computation of rate-distortion functions and capacity-fee functions [11, 127]. The role, if any, for information theory in the thermodynamics of open systems and the theory of entropy production propounded by Prigogine and Glansdorff [125] remains to be investigated. It also seems probable that information theory will cross-fertilize with molecular genetics now that it is becoming possible to study the structure and behavior of genetic material in detail.

LITERATURE

Stumpers [149] compiled an information theory bibliography and issued supplements thereto in 1955, 1957, and 1960. Kotz [95],

Wyner [164], and Viterbi [156] summarized major contributions to the field. Dobrushin [42] provided an overview of Soviet accomplishments and generated a compilation of challenging open problems [41] most of which remain unsolved. Multiterminal information theory was summarized by Wyner [165] and more recently by El Gamal and Cover [47] and by Ahlswede [4, 5]. The theories of multiterminal channels and of multiterminal sources have been codified, respectively, by van der Meulen [154] and by Körner [92] and Berger [13]. A collection of key papers on information theory was selected and edited by Slepian [144].

The first book in information theory, by Shannon and Weaver [137], is a compilation of Shannon's seminal papers with a postscript that anticipates applications, a few of which have indeed materialized. The textbook by Feinstein [50] contains the first proof of the exponential decay of error probability as a function of block length for channel codes having rates less than capacity. Khinchin [84], using the mathematical approach of the 1950's Russian school headed by Kolmogorov, introduces categorizations of channels with memory and extends much of the basic theory to them. Kullback [97] stresses properties of information functionals and develops links between information theory and statistical decision theory*. Fano [48] relies heavily on the combinatorial approach he developed in conjunction with Shannon and others in seminars and courses at MIT during the 1950s in order to tighten results about decay rates of source and channel coding error probabilities. Wolfowitz [163] provides constructive proofs rather than existence proofs of the basic coding theorems and treats classes of channels with block memory and/or side information. Ash [10] treats information theory mathematically as a branch of probability theory and therefore as a subbranch of measure theory. Abramson [1], in contrast, affords an easy-going entry to coding and information theory that has been found to be highly appropriate for undergraduate engineering courses. Jelinek [78] provides a thorough treatment of discrete information theory with particular emphasis on buffer-instrumented encoding of sources and convolutional codes with sequential decoding for channels. Gallager [55] presents an eminently readable treatment of the basic material, an elegant proof of the coding theorem for noisy channels, and strong chapters on algebraic codes, continuous channels, and rate-distortion theory; probably more people have been introduced to information theory through Gallager's book than all the others combined. The major role that rate-distortion theory has assumed in information theory was furthered by Berger's [11] textbook devoted exclusively to this subject. Guiasu [65] features connections between information theory, statistical mechanics, algebraic coding, and statistics. McEliece's [107] book, the second half of which is devoted to algebraic coding, was the first to include coding theorems for certain multisources and broadcast channels. Viterbi and Omura [157] extend Gallager's technique for proving coding theorems to multiterminal information theory problems and provide a thorough treatment of maximum likelihood decoding of convolutional codes. Csiszár and Körner [34] develop and unify the entirety of discrete multiterminal information theory within a mathematically rigorous framework.

Two journals devoted to information theory are *IEEE Transactions on Information Theory* and *Problemy Peredachi Informatsii* (Problems of Information Transmission). Two others with high information theory content are *Journal of Combinatorics, Information and Systems Science* and *Information and Control*. A cumulative index of the *IEEE Transactions on Information Theory* was published in September, 1982.

References

[1] Abramson, N. (1968). *Information Theory and Coding*. McGraw-Hill, New York.

[2] Aczél, J. D. and Daróczy, Z. (1975). *On Measures of Information and Their Characterizations*. Academic Press, New York.

[3] Ahlswede, R. (1973). *Proc. 2nd Int. Symp. Inf. Theory*, Tsahkador, Armenia, USSR, Sept. 1971. Hungarian Academy of Sciences, Budapest, Hungary.

[4] Ahlswede, R. (1979). *J. Comb. Inf. Syst. Sci.*, **4**, 76–115.

[5] Ahlswede, R. (1980). *J. Comb. Inf. Syst. Sci.*, **5**, 220–268.

[6] Ahlswede, R. and Körner, J. (1975). *IEEE Trans. Inf. Theory*, **IT-21**, 629–637.

[7] Ahlswede, R. and Wegener, I. (1979). *Suchprobleme*. Teubner, Stuttgart, W. Germany.

[8] Anderson, J. B. and Bodie, J. B. (1975). *IEEE Trans. Inf. Theory*, **IT-21**, 379–387.

[9] Arimoto, S. (1972). *IEEE Trans. Inf. Theory*, **IT-18**, 14–20.

[10] Ash, R. B. (1965). *Information Theory*. Wiley, New York.

[11] Berger, T. (1971). *Rate Distortion Theory: A Mathematical Basis for Data Compression*. Prentice-Hall, Englewood Cliffs, N.J.

[12] Berger, T. (1975). *IEEE Trans. Inf. Theory*, **IT-21**, 502–511.

[13] Berger, T. (1978). *The Information Theory Approach to Communications*, G. Longo, ed. CISM Courses and Lectures No. 229. Springer-Verlag, New York, pp. 171–231.

[14] Berger, T., Housewright, K. B., Omura, J. K., Tung, S.-Y., and Wolfowitz, J. (1979). *IEEE Trans. Inf. Theory*, **IT-25**, 664–666.

[15] Berlekamp, E. R. (1968). *Algebraic Coding Theory*. McGraw-Hill, New York.

[16] Billingsley, P. (1965). *Ergodic Theory and Information*. Wiley, New York.

[17] Blahut, R. E. (1972). *IEEE Trans. Inf. Theory*, **IT-18**, 460–473.

[18] Blahut, R. E. (1974). *IEEE Trans. Inf. Theory*, **IT-20**, 405–417.

[19] Blahut, R. E. (1977). *IEEE Trans. Inf. Theory*, **IT-23**, 656–674.

[20] Blahut, R. E. (1983). *Error Control Codes*. Addison-Wesley, Reading, Mass.

[21] Blake, I. F. and Mullin, R. C. (1975). *The Mathematical Theory of Coding*. Academic Press, New York.

[22] Boltzmann, L. (1896). *Vorlesungen über Gastheorie*. J. A. Barth, Leipzig, Germany.

[23] Brillouin, L. (1956). *Science and Information Theory*. Academic Press, New York.

[24] Burg, J. (1967). *37th Annu. Int. Meet. Soc. Explor. Geophys.* Oklahoma City, Oklahoma.

[25] Chernoff, H. (1952). *Ann. Math. Statist.*, **23**, 493–507.

[26] Christensen, R. (1981). *Entropy Minimax Sourcebook*, Vol. 1: *General Description*. Entropy Limited, Lincoln, Mass.

[27] Clark, G. C., Jr., and Cain, J. B. (1981). *Error-Correction Coding for Digital Communication*. Plenum Press, New York.

[28] Cover, T. M. (1972). *IEEE Trans. Inf. Theory*, **IT-18**, 2–14.

[29] Cover, T. M. (1975). *IEEE Trans. Inf. Theory*, **IT-21**, 226–228.

[30] Cover, T. M. and El Gamal, A. (1979). *IEEE Trans. Inf. Theory*, **IT-25**, 572–584.

[31] Cramér, H. and Leadbetter, M. R. (1967). *Stationary and Related Stochastic Processes: Sample Function Properties and Their Applications*. Wiley, New York.

[32] Csiszár, I. and Körner, J. (1980). *IEEE Trans. Inf. Theory*, **IT-26**, 155–165.

[33] Csiszár, I. and Körner, J. (1981). *IEEE Trans. Inf. Theory*, **IT-27**, 5–12.

[34] Csiszár, I. and Körner, J. (1981). *Information Theory: Coding Theorems for Discrete Memoryless Systems*. Academic Press, New York.

[35] Csiszár, I. and Longo, G. (1971). *Studia Sci. Math. Hung.*, **6**, 181–191.

[36] Davisson, L. D. (1973). *IEEE Trans. Inf. Theory*, **19**, 783–795.

[37] De Faria, J. M., Hartmann, C. R. P., Gerberich, C. L., and Varshney, P. K. (1980). An Information Theoretic Approach to the Construction of Efficient Decision Trees. Unpublished manuscript, School of Computer and Information Science, Syracuse University, New York.

[38] Diffie, W. and Hellman, M. E. (1976). *IEEE Trans. Inf. Theory*, **IT-22**, 644–654.

[39] Diffie, W. and Hellman, M. E. (1979). *Proc. IEEE*, **67**, 397–427.

[40] Dobrushin, R. L. (1959). *Uspekhi Mat. Akad. SSSR*, **14**, 3–104. (Also, *Trans. Amer. Math. Sci. Ser. 2*, **33**, 323–438, 1959.)

[41] Dobrushin, R. L. (1962). *Proc. 4th Berkeley Symp. Math. Statist. Prob.*, Vol. 1. University of California Press, Berkeley, Calif., pp. 211–252.

[42] Dobrushin, R. L. (1972). *IEEE Trans. Inf. Theory*, **IT-18**, 703–724.

[43] Doob, J. L. (1953). *Stochastic Processes*. Wiley, New York.

[44] Dueck, G. (1979). *Inf. Control*, **40**, 258–266.

[45] El Gamal, A. (1978). *IEEE Trans. Inf. Theory*, **IT-24**, 379–381.

[46] El Gamal, A. (1979). *IEEE Trans. Inf. Theory*, **IT-25**, 166–169.

[47] El Gamal, A. and Cover, T. M. (1980). *Proc. IEEE*, **68**, 1466–1483.

[48] Fano, R. M. (1961). *Transmission of Information: A Statistical Theory of Communications*. Wiley, New York.

[49] Feinstein, A. (1954). *IRE Trans. Inf. Theory*, **PGIT-4**, 2–22.

[50] Feinstein, A. (1958). *Foundations of Information Theory*. McGraw-Hill, New York.

[51] Fisher, R. A. (1925). *Proc. Camb. Philos. Soc.*, **22**, 700.

[52] Fitingof, B. M. (1966). *Problemy Per. Inform.*, **2**(2), 3–11 (English transl.: pp. 1–7).

[53] Fitingof, B. M. (1967). *Problemy Per. Inform.*, **3**(3), 28–36 (English transl.: pp. 22–29).

[54] Gallager, R. G. (1965). *IEEE Trans. Inf. Theory*, **IT-11**, 3–18.

[55] Gallager, R. G. (1968). *Information Theory and Reliable Communication*. Wiley, New York.

[56] Gallager, R. G. (1976). *IEEE Int. Symp. Inf. Theory*, Ronneby, Sweden, July 1976.

[57] Gallager, R. G. (1978). *IEEE Trans. Inf. Theory*, **IT-24**, 668–674.

[58] Gelfand, S. I. and Pinsker, M. S. (1978). *Soviet-Czech-Hung. Semin. Inf. Theory*, Tsahkodzor, Armenia, Sept. 1978, and *5th Int. Symp. Inf. Theory*, Tbilisi, July 1979.

[59] Gray, R. M. (1975). *IEEE Trans. Inf. Theory*, **IT-21**, 357–368.

[60] Gray, R. M. and Davisson, L. D., eds. (1977). *Ergodic and Information Theory*. Benchmark Papers in Electrical Engineering and Computer Science, Vol. 19. Dowden, Hutchinson & Ross, Stroudsburg, Pa.

[61] Gray, R. M. and Ornstein, D. S. (1976). *IEEE Trans. Inf. Theory*, **IT-22**, 682–690.

[62] Gray, R. M., Neuhoff, D. L., and Ornstein, D. S. (1975). *Ann. Prob.*, **3**, 478–491.

[63] Gray, R. M., Ornstein, D. S., and Dobrushin, R. L. (1980). *Ann. Prob.*, **8**, 639–674.

[64] Gray, R. M., Gray, A. H., Jr., Rebolledo, G., and Shore, J. E. (1981). *IEEE Trans. Inf. Theory*, **IT-27**, 708–721.

[65] Guiasu, S. (1977). *Information Theory with Applications*. McGraw-Hill, New York.

[66] Hájek, B. E. (1979). *IEEE Trans. Inf. Theory*, **IT-25**, 605–609.

[67] Hájek, B. E. and Pursley, M. B. (1979). *IEEE Trans. Inf. Theory*, **IT-25**, 36–46.

[68] Han, T. S. and Kobayashi, K. (1980). *IEEE Trans. Inf. Theory*, **IT-26**, 277–288.

[69] Haroutunian, E. A. (1968). *Problemy Per. Inform.*, **4**, 37–48.

[70] Hartley, R. V. L. (1928). *Bell Syst. Tech. J.*, **47**, 535.

[71] Heegard, C. (1981). Capacity and Coding for Computer Memory with Defects. Ph.D. thesis, Stanford University, California.

[72] Helstrom, C. W. (1976). *Quantum Detection and Estimation Theory*. Academic Press, New York.

[73] Holevo, A. S. (1973). *J. Multivariate Anal.*, **3**, 337–394.

[74] Huffman, D. A. (1952). *Proc. IRE*, **40**, 1098–1101.

[75] Jacobs, K. (1959). *Math. Ann.*, **137**, 125–135.

[76] Jarett, K. and Cover, T. M. (1981). *IEEE Trans. Inf. Theory*, **IT-27**, 151–160.

[77] Jaynes, E. T. (1957). *Phys. Rev., Pt. I*, **106**, 620–630; *ibid., Pt. II*, **108**, 171–190.

[78] Jelinek, F. (1968). *Probabilistic Information Theory*. McGraw-Hill, New York.

[79] Jelinek, F. and Schneider, K. S. (1972). *IEEE Trans. Inf. Theory*, **IT-18**, 765–774.

[80] Jelinek, F. and Schneider, K. S. (1974). *IEEE Trans. Inf. Theory*, **IT-20**, 750–755.

[81] Karush, J. (1961). *IRE Trans. Inf. Theory*, **IT-7**, 118.

[82] Kaspi, A. and Berger, T. (1982). *IEEE Trans. Inf. Theory*, **IT-28**, 828–841.

[83] Katona, G. O. H. and Nemetz, T. O. H. (1976). *IEEE Trans. Inf. Theory*, **IT-22**, 337–340.

[84] Khinchin, A. I. (1957). *Mathematical Foundations of Information Theory*. Dover, New York.

[85] Kieffer, J. C. (1974). *Pacific J. Math.*, **51**, 203–206.

[86] Kieffer, J. C. (1981). *IEEE Trans. Inf. Theory*, **IT-27**, 565–570.

[87] Kolmogorov, A. N. (1956). *IRE Trans. Inf. Theory*, **IT-2**, 102–108.

[88] Kolmogorov, A. N. (1958). *Dokl. Akad. Nauk*, **119**, 861–864.

[89] Kolmogorov, A. N. (1959). *Dokl. Akad. Nauk*, **124**, 754–755.

[90] Kolmogorov, A. N. (1965). *Problemy Per. Inform.*, **1**, 3–11.

[91] Kolmogorov, A. N. (1968). *IEEE Trans. Inf. Theory*, **IT-14**, 662–664.

[92] Körner, J. (1975). In *Information Theory, New Trends and Open Problems*, G. Longo, ed. CISM Courses and Lectures No. 219. Springer-Verlag, New York.

[93] Körner, J. and Marton, K. (1979). *IEEE Trans. Inf. Theory*, **IT-25**, 60–64.

[94] Koshelev, V. N. (1972). *Problemy Per. Inform.*, **13**(1), 26–32.

[95] Kotz, S. (1966). *J. Appl. Probl.*, **3**, 1–93.

[96] Kraft, L. G. (1949). A Device for Quantizing, Grouping and Coding Amplitude Modulated Pulses. M.S. thesis, Massachusetts Institute of Technology, Cambridge, Mass.

[97] Kullback, S. (1959). *Information Theory and Statistics*. Wiley, New York.

[98] Kuznetsov, A. V., Kasami, T., and Yamamura, S. (1979). *IEEE Trans. Inf. Theory*, **IT-24**, 712–718.

[99] Levenshtein, V. I. (1977). *Problemy Per. Inform.*, **13**(1), 3–18.

[100] Liao, H. (1972). Multiple Access Channels. Ph.D. dissertation, University of Hawaii.

[101] Lin, S. (1970). *An Introduction to Error-Correcting Codes.* Prentice-Hall, Englewood Cliffs, N.J.

[102] Liu, J. W. S. (1970). *IEEE Trans. Inf. Theory,* **IT-16**, 319–329.

[103] Lovasz, L. (1979). *IEEE Trans. Inf. Theory,* **IT-25**, 1–7.

[104] MacWilliams, F. J. and Sloane, N. J. A. (1977). *The Theory of Error-Correcting Codes.* North-Holland, Amsterdam.

[105] Marton, K. (1979). *IEEE Trans. Inf. Theory,* **IT-25**, 306–311.

[106] Massey, J. L. (1976). Topics in Discrete Information Processing. Unpublished manuscript, Dept. of Electrical Engineering, University of Notre Dame, Notre Dame, Ind.

[107] McEliece, R. J. (1977). *The Theory of Information and Coding.* Addison-Wesley, Reading, Mass.

[108] McMillan, B. (1956). *IRE Trans. Inf. Theory,* **IT-2**, 115–116.

[109] Merkle, R. C. (1978). *Commun. ACM,* **21**, 294–299.

[110] Modestino, J. W., Bhaskaran, V., and Anderson, J. B. (1981). *IEEE Trans. Inf. Theory,* **IT-27**, 677–697.

[111] Moy, S.-T. C. (1961). *Pacific J. Math.,* **11**, 706–714, 1459–1465.

[112] Nyquist, H. (1924). *Bell Syst. Tech. J.,* **3**, 324.

[113] Nyquist, H. (1928). *AIEE Trans. Commun. Electron.,* **47**, 617–644.

[114] Ornstein, D. S. (1970). *Adv. Math.,* **4**, 1725–1729.

[115] Ornstein, D. S. (1973). *Ann. Prob.,* **1**, 43–58.

[116] Paley, R. E. A. C. and Wiener, N. (1934). *Amer. Math. Soc. Colloq. Publ.,* **19**.

[117] Perez, A. (1964). In *Transactions of the Third Prague Conference on Information Theory, Statistical Decision Functions, and Random Processes,* J. Kozesnik, ed. Academic Press, New York, pp. 545–574.

[118] Perez, A. (1967). *Proc. Colloq. Inf. Theory,* Debrecen, Hungary, pp. 299–315. (See also *Kybernetika,* **3**, 1–21, 1967.)

[119] Peterson, W. W. (1961). *Error-Correcting Codes.* MIT Press, Cambridge, Mass.

[120] Peterson, W. W. and Weldon, E. J., Jr. (1972). *Error-Correcting Codes.* MIT Press, Cambridge, Mass.

[121] Pierce, J. G. (1979). In *The Maximum Entropy Formalism,* R. D. Levine and M. Tribus, eds. MIT Press, Cambridge, Mass., pp. 339–402.

[122] Pinsker, M. S. (1960). *Information and Information Stability of Random Variables and Processes,* Izdatel'stvo Akademii Nauk SSSR, Moscow. (English transl.: Holden-Day, San Francisco, 1964.)

[123] Pless, V. (1982). *An Introduction to the Theory of Error-Correcting Codes.* Wiley, New York.

[124] Posner, E. C. and McEliece, R. J. (1971). *Ann. Math. Statist.,* **42**, 1706–1716.

[125] Prigogine, I. and Glansdorff, P. (1971). *Thermodynamic Theory of Structure, Stability and Fluctuations.* Wiley-Interscience, New York.

[126] Pursley, M. B. and Davisson, L. D. (1976). *IEEE Trans. Inf. Theory,* **IT-22**, 324–337.

[127] Reiss, H. (1969). *J. Statist. Phys.,* **1**, 107–131.

[128] Rissanen, J. and Langdon, G. G., Jr. (1981). *IEEE Trans. Inf. Theory,* **IT-27**, 12–23.

[129] Rivest, R. L., Shamir, A., and Adleman, L. (1978). *Commun. ACM,* **21**, 120–126.

[130] Rogers, C. A. (1964). *Packing and Covering.* Cambridge University Press, Cambridge, England.

[131] Schalkwijk, J. P. M. (1982). *IEEE Trans. Inf. Theory,* **IT-28**, 107–110.

[132] Sgarro, A. (1977). *IEEE Trans. Inf. Theory,* **IT-23**, 179–182.

[133] Shannon, C. E. (1948). *Bell Syst. Tech. J.,* **27**, 379–423, 623–656.

[134] Shannon, C. E. (1949). *Proc. IRE,* **37**, 10–21.

[135] Shannon, C. E. (1949). *Bell. Syst. Tech. J.,* **28**, 656–715.

[136] Shannon, C. E. (1961). *Proc. 4th Berkeley Symp. Math. Statist. Prob.,* Vol. 1. University of California Press, Berkeley, Calif., pp. 611–644.

[137] Shannon, C. E. and Weaver, W. W. (1949). *The Mathematical Theory of Communication.* University of Illinois Press, Champaign, Ill.

[138] Shannon, C. E., Gallager, R. G., and Berlekamp, E. R. (1967). *Inf. Control,* **10**, 65–103, 522–552.

[139] Shen, S.-Y. (1981). *Chinese Ann. Math.,* **2**, 117–129.

[140] Shields, P. C. and Neuhoff, D. L. (1977). *IEEE Trans. Inf. Theory,* **IT-23**, 211–215.

[141] Shore, J. E. and Johnson, R. W. (1980). *IEEE Trans. Inf. Theory,* **IT-26**, 26–37.

[142] Sinai, J. G. (1962). *Sov. Math. Dokl.,* **3**, 1725–1729.

[143] Slepian, D. (1963). *Bell Syst. Tech. J.,* **42**, 681–707.

[144] Slepian, D., ed. (1974). *Key Papers in the Development of Information Theory.* IEEE, New York.

[145] Slepian, D. and Wolf, J. K. (1973). *IEEE Trans. Inf. Theory,* **19**, 471–480.

[146] Sloane, N. J. A. (1981). *IEEE Trans. Inf. Theory,* **IT-27**, 327–338.

[147] Snyder, D. L. (1975). *Random Point Processes.* Wiley, New York.

[148] Stuller, J. A. and Kurz, B. (1977). *IEEE Trans. Commun.*, **COM-25**, 485–495.

[149] Stumpers, F. L. H. M. (1953). *IRE Trans. Inf. Theory*, **PGIT-2**, Nov. Suppl. 1, **IT-1**, 31–47, Sept. 1955; *ibid.*, Suppl. 2, **IT-3**, 150–166, June 1957; *ibid.*, Suppl. 3, **IT-6**, 25–51, Mar. 1960.

[150] Szilard, L. (1929). *Zeit. Phys.*, **53**, 840–856.

[151] Tribus, M. (1961). *Thermodynamics and Thermostatics*. D. Van Nostrand, Princeton, N.J.

[152] Tsybabov, B. S. (1975). *Problemy Per. Inform.*, **11**(1), 111–113.

[153] Tsybabov, B. S. (1975). *Problemy Per. Inform.*, **11**(3), 21–30.

[154] van der Meulen, E. C. (1977). *IEEE Trans. Inf. Theory*, **IT-23**, 1–37.

[155] van Lint, J. H. (1973). *Coding Theory*. Springer-Verlag, New York.

[156] Viterbi, A. J. (1973). *IEEE Trans. Inf. Theory*, **IT-19**, 257–262.

[157] Viterbi, A. J. and Omura, J. K. (1979). *Digital Communication and Coding*. McGraw-Hill, New York.

[158] Vitushkin, A. G. (1961). *Theory of the Transmission and Processing of Information*. Pergamon Press, Elmsford, N.Y.

[159] von Neumann, J. (1932). *Mathematical Foundations of Quantum Mechanics*. Berlin, Chap. 5.

[160] Wiener, N. (1948). *Cybernetics*. Wiley, New York.

[161] Wiener, N. (1949). *The Interpolation, Extrapolation and Smoothing of Stationary Time Series*. Wiley, New York.

[162] Witsenhausen, H. S. and Wyner, A. D. (1981). *Bell Syst. Tech. J.*, **60**, 2281–2292.

[163] Wolfowitz, J. (1964). *Coding Theorems of Information Theory*, 2nd ed. Springer-Verlag, New York (3rd ed., 1978).

[164] Wyner, A. D. (1969). *SIAM Rev.*, **11**, 317–346.

[165] Wyner, A. D. (1974). *IEEE Trans. Inf. Theory*, **IT-20**, 2–10.

[166] Wyner, A. D. (1975). *IEEE Trans. Inf. Theory*, **IT-21**, 294–300.

[167] Wyner, A. D. and Ziv, J. (1976). *IEEE Trans. Inf. Theory*, **IT-22**, 1–10.

[168] Yamamoto, H. (1981). *IEEE Trans. Inf. Theory*, **IT-27**, 299–308.

[169] Yuen, H. P. and Shapiro, J. H. (1978–1980). *IEEE Trans. Inf. Theory*, Part I, **IT-24**, 657–668; *ibid.*, Part II, **IT-25**, 179–192; *ibid.*, Part III, **IT-26**, 78–92.

[170] Yuen, H. P., Kennedy, R. S., and Lax, M. (1975). *IEEE Trans. Inf. Theory*, **IT-21**, 125–134.

[171] Zhang, Z. (1981). On the Problem of Estimating the Error Bound of Source Coding. M.S. thesis, Nankai University, Tianjin, China.

[172] Ziv, J. (1972). *IEEE Trans. Inf. Theory*, **IT-18**, 384–394.

[173] Ziv, J. (1978). *IEEE Trans. Inf. Theory*, **IT-24**, 405–412.

[174] Ziv, J. and Lempel, A. (1978). *IEEE Trans. Inf. Theory*, **IT-24**, 530–536.

(CLASSIFICATION
COMMUNICATION THEORY,
  STATISTICAL
DISCRIMINANT ANALYSIS
ENTROPY
GAUSSIAN PROCESSES
STOCHASTIC PROCESSES
TIME SERIES)

TOBY BERGER

INITIAL BLOCK   *See* INTRABLOCK SUBGROUP

INNER INFERENCE

Future repetitions used in frequency interpretations* of statistical inferences are not real but hypothetical or simulated, and they should be considered only as a means of learning from the data. In ref. 6 it is suggested that better inferences may be obtained if only future hypothetical samples similar to the data are considered, because in this way the noise may be reduced and we may get a better picture of what the actual sample has to say about the population.

If X is the actual response and Y is a hypothetical future response, the *conditional confidence level* of the actual likelihood set $\mathcal{R}(X)$ is defined to be the conditional probability that the future hypothetical likelihood set $\mathcal{R}(Y)$ will cover θ, the unknown true value of the parameter, given that the future Y is "similar" to the given X, where "similar" means that $\mathcal{R}(Y)$ covers $\hat{\theta}(X)$, the actual maximum likelihood estimate of θ. In symbols, the conditional confidence level of $\mathcal{R}(X)$ is defined to be

$$P_\theta \left\{ \theta \in \mathcal{R}(Y) \,|\, X, \hat{\theta}(X) \in \mathcal{R}(Y) \right\} \quad (1)$$

[6, 7]. This concept is based only on sampling theory and does not assume that any prior for θ is given. In general, the conditional confidence level (1) depends on θ and is unknown. However, for univariate regression problems it follows from the results of refs. 1 and 4 that the conditional confidence level (1) is usually greater than and bounded away from the usual confidence level.

For certain models called *group models*, the usual unconditional confidence level of a likelihood set has a Bayesian interpretation as a posterior probability with respect to a prior, called the *outer prior*, which is invariant under a change of origin in the parameter space (*see* INVARIANT PRIOR DISTRIBUTIONS and ref. 7). In univariate regression models, the outer prior is the conventional prior commonly used to represent ignorance with respect to the unknown parameters. The corresponding outer posterior distributions are formally identical with Fraser's structural distributions [2]. (*See* STRUCTURAL INFERENCE.)

For group models there is another prior, called the *inner prior*, which is invariant under the action of the group. For univariate regression* problems the corresponding inner posterior distributions differ from the usual (outer) posterior distributions only in having more degrees of freedom. Consequently, inner posterior intervals are always shorter than the usual (outer) intervals, sometimes dramatically so [6].

The inner posterior probability of the likelihood set $\mathcal{R}(X)$ has a frequency interpretation as an estimate of the unknown conditional confidence level (1). This estimate is a posterior mean of (1) based on the information $\hat{\theta}(X) \in \mathcal{R}(Y)$ (for a given X) and on the inner prior for θ.

The problem of making inferences concerning the unknown covariance matrix in sampling from a central multivariate normal distribution is a good example for making comparisons between different theories of statistical inference*. This problem can be modeled as a group model using as a group either the group of nonsingular linear transformations [5] or the group of positive lower

triangular transformations [7]. In the second case inner and outer inferences are different, but inner inferences are more appealing because they are equivariant under orthogonal transformations.

It may be conjectured that the future development of inner statistical inference will be closely related to the development of *multigroup models*, that is, models with several groups separately considered as integral parts of the model. There is an illuminating analogy with measure theory, where inner and outer measures coincide over the restricted domain of measurable sets. This analogy suggests that the future task for *logical Bayesian inference** is to develop logical multigroup models that may play the same central role that measurable sets play in measure theory.

As an example, consider the univariate normal model with unknown mean and variance. This classical model can be endowed with the affine group in R^1, thus becoming a unigroup model, or it can be endowed with two groups, the group of translations and the group of multiplications by a positive constant, separately considered as integral parts of the model, thus becoming a bigroup model.

The inner prior for the unigroup model is $d\mu\, d\sigma / \sigma^2$, and this prior generates the *posterior conditionality anomaly* mentioned by Novick [3, p. 30]: namely, the conditional posterior distribution of σ for a given μ is different from the posterior distribution for σ when we assume that μ is known and we use $d\sigma / \sigma$ as the prior for σ. This anomaly disappears if the usual (outer) prior $d\mu\, d\sigma / \sigma$ is used.

A *posterior conditionality principle*, saying in effect that posterior conditionality anomalies should be avoided, is a powerful tool for making inferences for bigroup models. In effect, the above-mentioned univariate normal bigroup model with two parameters, μ and σ, has two *associated submodels*: the models obtained by assuming that μ or σ is the only unknown parameter (because the other is assumed to be known). The inner priors of these submodels, $d\mu$ and $d\sigma / \sigma$, are also outer priors, and to emphasize this fact, they may be called *logical subpriors* of the bigroup model. In general, a bigroup model will be called a *logical bigroup model* if it has logical (both inner and outer) subpriors. According to the posterior conditionality principle, the product of the logical subpriors is the uniquely defined *logical prior* that represents ignorance concerning the two parameters of a logical bigroup model. Therefore, the logical prior for the univariate normal bigroup model with parameters μ and σ is the usual prior $d\mu\, d\sigma / \sigma$.

References

[1] Brown, L. (1967). *Ann. Math. Statist.*, **38**, 1068–1071.

[2] Fraser, D. A. S. (1968). *The Structure of Inference*. Wiley, New York.

[3] Novick, M. R. (1969). *J. R. Statist. Soc. B*, **31**, 29–64.

[4] Olshen, R. A. (1973). *J. Amer. Statist. Ass.*, **68**, 692–698.

[5] Villegas, C. (1971). In *Foundations of Statistical Inference*, V. P. Godambe and D. A. Sprott, eds. Holt, Rinehart and Winston, New York, pp. 409–414. (Discusses inner inferences for the central multivariate normal model.)

[6] Villegas, C. (1977). *J. Amer. Statist. Ass.*, **72**, 453–458. (Discusses inner inferences for the univariate normal model.)

[7] Villegas, C. (1981). *Ann. Statist.*, **9**, 768–776. (Gives the general theory of inner inference for unigroup models.)

(BAYESIAN INFERENCE
INFERENCE, STATISTICAL
INVARIANT PRIOR DISTRIBUTIONS
STRUCTURAL INFERENCE)

C. VILLEGAS

INSPECTION PARADOX

Let $X_1, X_2, \ldots$ be independent random variables having a common distribution function F. If one of the X's is chosen according to some rule, it often turns out that the X selected has a distribution different than F. For instance, if we choose the first X_i whose value is at least as large as A, then the

X selected will have distribution $(F(x) - F(A))/(1 - F(A))$, $A \leqslant x < \infty$. In less obvious situations when the chosen X has a different distribution than F, we say that the *inspection paradox* is at work.

The classical example of the inspection paradox occurs when the X_i, $i \geqslant 1$, are interpreted as the interarrival times of a renewal process*. That is, events are assumed to occur randomly in time and X_i is taken as the time between the $(i-1)$th and ith event. Hence the nth event occurs at time $\sum_{i=1}^{n} X_i$ and $N(t)$, the number of events that occur by time t, is given by

$$N(t) = \max\left\{ n: \sum_{i=1}^{n} X_i \leqslant t \right\}.$$

If we now consider the interarrival interval that contains some fixed time t—that is, $X_{N(t)+1}$—it turns out that

$$P\{X_{N(t)+1} \geqslant x\} \geqslant 1 - F(x).$$

In other words, there is an inspection paradox—namely, that the interval containing the time point t tends to be of larger size than an ordinary interarrival interval. Intuitively, this is so, as larger interarrival intervals have a greater chance of containing the time point t than do smaller ones.

Another version of the inspection paradox occurs if we imagine that busloads of individuals continually arrive at some destination and X_i is the number of the ith bus. If we now choose an individual at random and ask the number of individuals in his or her bus, this number will not have distribution F. The reason is as before: namely, buses that carry more people will be more likely to contain the chosen individual than ones that carry fewer people.

If overlooked, the inspection paradox can lead to serious errors in practical situations. That is, without care, false conclusions can be drawn when what is actually observed is not typical of the population as a whole.

Bibliography

Ross, S. M. (1980). *Introduction to Probability Models*, 2nd ed. Academic Press, New York, pp. 250, 288.

Feller, W. (1971). *An Introduction to Probability Theory and Its Applications*, Vol. 2. Wiley, New York, p. 187.

(ACCEPTANCE SAMPLING
ENGINEERING STATISTICS
INSPECTION SAMPLING
QUALITY CONTROL)

SHELDON M. ROSS

INSPECTION SAMPLING

For processes which result in output that exhibits variation among the various units generated, humankind has long been concerned with the examination of individual units to determine conformance to specifications. In those cases where the purpose of the inspection does not require the examination of every unit of interest, an appropriate sample must be selected. Stigler [26] provides an interesting historical review of the inspection sampling procedures used for monitoring quality of gold and silver coinage at the Royal Mint in London for eight centuries.

The inspection of samples of units for purposes relating to quality evaluation and control of production operations is quite common. Three such purposes (see Juran and Gryna [18, pp. 358–359] for a more complete list) are:

1. To determine if the output from the process has undergone a change from one point in time to another
2. To make a determination concerning a finite population of units (termed a lot) concerning the overall quality of the lot
3. To screen defective items from a sequence or group of production units to improve the resulting quality of the population of interest

Inspection sampling for these purposes, together with appropriate decision rules, constitute what is commonly referred to as *statistical quality control** (SQC).

Inspection sampling also has application for purposes such as signal detection, medical diagnosis*, target search, and tracking. However, in certain formulations of these problems, attention is directed more to the decision rules involved than to the sampling issues.

EARLY DEVELOPMENTS IN SQC

Walter A. Shewhart, Harold F. Dodge, and Harry G. Romig* are generally credited as the original developers of statistical quality control. Their work began in the mid-1920s at the Bell Telephone Laboratories. Hayes and Romig [15, pp. 6–9] note a number of additional individuals who were involved at this time, and credit C. N. Frazee with originating the use of operating characteristic curves with inspection problems in 1916.

Shewhart's efforts dealt with the problem of determining whether variability in output from a particular manufacturing process was due to chance causes or to some definite, assignable sources. These efforts resulted in the systematic way of collecting, organizing, and presenting inspection results referred to as *control charts**.

The work of Dodge and Romig concentrated on inspection situations where decisions concerning the disposition of lots of units are needed. Situations where less than 100% inspection is used and lots are either accepted or rejected on the basis of some criterion is generally referred to as *acceptance sampling**. Dodge and Romig initially considered the case where items inspected are classified as defective or nondefective and rejected lots undergo 100% inspection. The criterion they used for determination of their plans was to minimize inspection required subject to conditions imposed on some aspect of quality of a series of lots subjected to the inspection procedure. The quality aspects specifically considered were probability of accepting lots at the *lot tolerance percent defective* (i.e., the maximum allowable percentage of defectives in a lot) and *average outgoing quality** (AOQ) (the average percent defective in the product after inspection).

The concepts of double sampling* and multiple sampling were also developed at Bell Labs. Under double sampling, after inspection of an initial sample, the decision rules allow for either acceptance or rejection of the lot or for an additional sample to be taken and inspected. An accept or reject decision must be made after this second sample in double sampling whereas in multiple sampling the decision to require additional sampling may be repeated a finite number of additional times, typically five or six. These procedures allow for a decision on the disposition of a lot with smaller total number of items inspected when quality is either very good or very bad.

The results of these efforts were reported in several articles in the *Bell System Technical Journal* which were reprinted and integrated in Dodge and Romig [7]. (*See* DODGE–ROMIG LOT ACCEPTANCE TABLES.)

Although the foregoing acceptance sampling procedures were based on attributes sampling (i.e., inspection that classified items as defective or nondefective), there are advantages of variables-based procedures, in which each quality characteristic of interest on each unit is measured on some interval or ratio scale (e.g., centimeters, ohms, Brinell hardness). Hayes and Romig [15, p. 7] report that although the first published work in this area was the doctoral dissertation of H. G. Romig in 1939, acceptance sampling by variables was applied by Bell Telephone Laboratories as early as 1934. (See also Liberman and Resnikoff [20].) Although variables sampling plans have improved statistical properties, a number of practical problems limit their desirability. In general, the inspection equipment required will be more costly, and inspectors may require greater training. In addition, distributional assumptions are required for each of the quality characteristics of interest.

Abraham Wald's development of the theory of sequential analysis in the early 1940s led to the development of sequential sampling plans, the limiting case of multiple plans. This development is reported in Wald* [29]. Tables for these plans can also be found in Bowker and Goode [3].

Also during this same period of time Dodge [6] describes a procedure for application with attributes inspection where there is a continuous flow of individual items, rather than a sequence of lots, which are available for inspection. These are referred to as continuous sampling plans. Dodge's original plan calls for 100% initial inspection, until a specified consecutive number of nondefective items have been found. When this occurs some fraction of items are inspected until a defective item is inspected, at which time 100% inspection is resumed. A number of variations and alternatives to [6] has been proposed (e.g., Dodge and Torrey [8], Lieberman and Soloman [21], and Wald and Wolfowitz [30]).

SAMPLING SCHEMES

A sampling scheme consists of a set of sampling plans and rules or procedures specifying the way in which these plans are to be applied. Motivated by material requirements of World War II, the U.S. armed forces moved to implement statistically based sampling procedures in their procurement activities. This resulted in a series of tables of plans which included rules for selection of particular sampling plans for different situations, the attempt being to motivate manufacturers to achieve high quality and to reduce inspection when there was evidence of this high quality being achieved.

These tables of plans had sample sizes which were related to lot sizes in what was deemed to be a "reasonable" manner. In addition, the concept of acceptable quality level* (AQL) was introduced to index the plans. The schemes adopted by the American armed forces (British counterparts of these also exist) have evolved to be MIL-STD-105D for acceptance sampling by attributes, MIL-STD-414 for acceptance sampling by variables, and MIL-STD-1235 for continuous sampling plans. (*See* MILITARY STANDARDS, STATISTICS IN.)

PROCESS CONTROL

The Stewart control chart, as well as its early modifications (see Gibra [13]) such as the cumulative sum (cusum) chart* [24], assumed that samples of fixed size would be taken at fixed intervals of time. Although the statistical consequences of changing the size of the sample could readily be ascertained, arguments (see Duncan [10, p. 447]) for samples of four or five generally have prevailed for variables-based Shewhart control charts*. However, for the cusum chart, procedures for design of the sampling and decision parameters are based on average run lengths*. The run lengths considered are the number of samples taken before a shift is indicated when the process is operating correctly and the run length when a shift of a specified size has occurred.

The cusum chart was a departure from the Shewhart-type chart in that it used inspection results from more than the most recent sample. An alternative approach for accomplishing this is incorporated in a model by Girshick and Rubin [14] through a Bayesian approach. This model is also of interest in that it relaxes the Shewhart requirements on sampling intervals and size. By allowing decisions to be made after each item is inspected, and placing no constraints on the minimum intersample interval, the sampling decision becomes the length of the intersample interval rather than choosing a sample size. This leads to optimal process control considerations (see, e.g., Box and Jenkins [4, pp. 486–491] and MacGregor [22]).

ECONOMICALLY BASED SAMPLING

The Girshick and Rubin model also introduced the concept of choosing sampling and decision parameters on economic criteria; however, interest in the development of economic-based models has been centered on traditional control chart procedures. Although fixed sample sizes to be taken at fixed points in time have been shown to be nonoptimal [27], interest in modeling the restricted plans has continued until the present. Montgomery [23] provides a review and literature survey of developments in this area.

Development of models for economically based acceptance sampling plans has proceeded on a parallel basis to that of control charts. Wetherill and Chiu [31] present a comprehensive literature review.

INSPECTION ERRORS

Although it has long been recognized that inspection procedures are not 100% accurate, investigations into the effects of these inaccuracies on the choice and properties of plans did not gain momentum until the late 1960s, although Eagle [11] reports earlier results. Since then numerous investigations on the relationships between different types of inspector error and the statistical properties of several traditional sampling procedures have been reported. In general, it has been found that the statistical properties can be considerably changed in the presence of quite limited inspector errors. In addition, economic models have been developed which incorporate these effects. Dorris and Foote [9] provide a review of work in this area.

SOME ADDITIONAL DEVELOPMENTS

There may be multiple characteristics of interest for each item to be inspected, and nonconformance to specifications on each may have varying degrees of importance. Various classification schemes exist for grouping characteristics in this case (see, e.g., Juran [17, p. 12–20, 12–26]). The classical procedure for handling this in attributes-based inspection is a demerit system. This results in a decision based on a weighted linear combination of inspection results. Case et al. [5] and others have proposed economic models.

Multivariate procedures are necessary when it is impossible to treat the multiple characteristics and independent of each other. Patel [25] has considered the multivariate case for attributes sampling. Multivariate variables sampling, assuming normality, has also received considerable attention (see, e.g., material by Hotelling in Eisenhart et al. [12, pp. 111–184] and by Jackson and Bradley in Krishnaiah [19, pp. 507–518].

Although it is commonly recognized that inspection results may be temporally correlated, limited developments have been reported concerning the classical procedures (which assume independent observations) in the presence of correlation. Berthouex et al. [2] fit an ARMA model to process data before using the residuals on a Shewhart control chart with good results. Johnson and Counts [16] deal with a Shewhart control chart in the presence of a cyclic mechanistic disturbance, Vanlopoulos and Stamboulis [28] consider modifications to the Shewhart control chart limits when data are correlated, and Bagshaw and Johnson [1] use simulation to evaluate the effects of correlation on the average run length of cusum charts.

References

[1] Bagshaw, M. and Johnson, R. A. (1975). *Technometrics*, **17**, 73–80.

[2] Berthouex, P. M., Hunter, W. G., and Pallesen, L. (1978). *J. Quality Tech.*, **10**, 139–149.

[3] Bowker, A. H. and Goode, H. P. (1952). *Sampling Inspection by Variables*. McGraw-Hill, New York.

[4] Box, G. E. P. and Jenkins, G. (1976). *Time Series Analysis*. Holden-Day, San Francisco.

[5] Case, K. E., Schmidt, J. W. and Bennett, G. K. (1975). *AIIE Trans.*, **7**, 363–378.

[6] Dodge, H. F. (1943). *Ann. Math. Statist.*, **14**, 264–279.

[7] Dodge, H. F. and Romig, H. G. (1944). *Sampling Inspection Tables*, Wiley, New York (2nd ed., 1959).

[8] Dodge, H. F. and Torrey, M. N. (1951). *Ind. Quality Control*, **7**, 5–9.

[9] Dorris, A. L. and Foote, B. L. (1978). *AIIE Trans.*, **10**, 184–192.

[10] Duncan, A. J. (1974). *Quality Control and Industrial Statistics*, 4th ed. Richard D. Irwin, Homewood, Ill. (Widely used as a text, especially comprehensive on statistical quality control. Good bibliography.)

[11] Eagle, A. R. (1954). *Ind. Quality Control*, **10**, 10–14.

[12] Eisenhart, C., Hastay, M. W., and Wallis, W. A., eds. (1947). *Techniques of Statistical Analysis*. McGraw-Hill, New York.

[13] Gibra, I. N. (1975). *J. Quality Tech.*, **7**, 183–191. (A survey of control chart developments.)

[14] Girshick, M. A. and Rubin, H. (1952). *Ann. Math. Statist.*, **23**, 114–125.

[15] Hayes, G. E. and Romig, H. C. (1977). *Modern Quality Control*. Bruce, Encino, Calif. (A basic quality control text, includes managerial and nonstatistical topics as well as basic statistical procedures.)

[16] Johnson, E. E. and Counts, R. W. (1979). *J. Quality Tech.*, **11**, 28–35.

[17] Juran, J. M., ed. (1974). *Quality Control Handbook*, 3rd ed. McGraw-Hill, New York.

[18] Juran, J. M. and Gryna, F. M., Jr. (1980). *Quality Planning and Analysis*. McGraw-Hill, New York. (Basic text, especially complete on practical issues of quality control.)

[19] Krishnaiah, P. R., ed. (1966). *Multivariate Analysis*, Vol. 1. Academic Press, New York.

[20] Lieberman, G. J. and Resnikoff, G. J. (1955). *J. Amer. Statist. Ass.*, **50**, 457–516, 1333.

[21] Lieberman, G. J. and Solomon, H. (1955). *Ann. Math. Statist.*, **26**, 686–704.

[22] MacGregor, J. F. (1976). *Technometrics*, **18**, 151–160.

[23] Montgomery, D. C. (1980). *J. Quality Tech.*, **12**, 75–87.

[24] Page, E. S. (1954). *Biometrika*, **41**, 100–115.

[25] Patel, H. J. (1973). *Technometrics*, **15**, 103–112.

[26] Stigler, S. M. (1977). *J. Amer. Statist. Ass.*, **72**, 493–500.

[27] Taylor, H. M., III (1965). *Ann. Math. Statist.*, **36**, 1677–1694.

[28] Vasilopoulos, A. V. and Stanboulis, A. P. (1978). *J. Quality Tech.*, **10**, 20–30.

[29] Wald, A. (1947). *Sequential Analysis*. Wiley, New York.

[30] Wald, A. and Wolfowitz, J. (1945). *Ann. Math. Statist.*, **16**, 30–49.

[31] Wetherill, G. B. and Chiu, W. K. (1975). *Int. Statist. Rev.*, **43**, 191–210.

Further Reading

See the following works, as well as the references just given, for more information on the topic of inspection sampling.

Butterbaugh, G. I. (1946). *A Bibliography of Statistical Quality Control*. University of Washington Press, Seattle.

Butterbaugh, G. I. (1951). *A Bibliography of Statistical Quality Control*, Supplement. University of Washington Press, Seattle.

Jackson, J. E. (1959). *Technometrics*, **1**, 359–377.

Kase, S. and Ohta, H. (1977). *Technometrics*, **19**, 249–257. (Economic-based sampling plans using a GERT model.)

Statistical Research Group, Columbia University (1945). *Sequential Analysis of Statistical Data: Applications*. Columbia University Press, New York.

Taub, T. W. (1976). *J. Quality Tech.*, **8**, 74–80. (A discussion of a particular application of inspection sampling.)

Wetherill, G. B. (1977). *Sampling Inspection and Quality Control*. Chapman & Hall, London. (Historical orientation with emphasis on basic principles and important background theory of statistical quality control.)

(ACCEPTABLE QUALITY LEVEL
ACCEPTANCE SAMPLING
AVERAGE OUTGOING QUALITY
AVERAGE RUN LENGTH
CONTINUOUS SAMPLING PLANS
CONTROL CHARTS
CUSUM CHARTS
DOUBLE SAMPLING
EVOP
MULTIPLE SAMPLING
MULTIVARIATE QUALITY CONTROL
PROCESS CONTROL
QUALITY CONTROL, STATISTICAL
SAMPLING SCHEMES
SEQUENTIAL SAMPLING)

RUSSELL G. HEIKES

INSTITUTE OF MATHEMATICAL STATISTICS

The Institute of Mathematical Statistics was founded in 1935 in response to a growing need in the United States for an organization of people having a special interest in the mathematical aspects of statistics and a conviction that the theoretical aspects of statistics would be greatly advanced by such an organization. Already efforts for meeting the need for the publication of articles with substantial mathematical content had been made by the establishment of the *Annals of Mathematical Statistics* in 1930 by H. C. Carver of the University of Michigan. Dur-

ing the early 1930s he not only edited this journal but personally provided the funds necessary to maintain its solvency. However, an undertaking such as this clearly needed institutional backing, which occurred at the founding of the Institute on September 12, 1935, at Ann Arbor with Carver a strong proponent.

The constitution with by-laws, adopted at the founding meeting, called for the election of officers (president, vice-president, and secretary-treasurer) and for appointed committees on membership and on publications. Not surprisingly, the *Annals of Mathematical Statistics* was designated as the Official Journal of the Institute at this meeting.

Carver was willing to serve the Institute as editor through 1937. The second editor, S. S. Wilks of Princeton University, then served through 1949.

It seems helpful to describe the development of the Institute from its simple beginnings by using the periods 1935–1948, 1949–1971, and 1972–present. Different aspects of growth are then treated in the different periods. They are organization, membership, *Annals* and publications, meetings, committees, and cooperation with other societies. Space does not permit the discussion of each aspect in each period, so selections are made.

Growth changes in organizations were important because they made possible the more efficient attainment of objectives. In the first period, a second vice-presidency was added, a Board of Directors (chiefly officers) functioned between elections, voting was by ballot (which soon became mail ballot), and associate and assistant editors were appointed to the *Annals*.

The number of members increased from 103 in the first published directory to 1101 in 1949. The annual meetings were supplemented with many other meetings, including summer and regional (eastern, central, and west coast) ones. The annual and summer meetings were usually held in conjunction with the national mathematics groups or the national statistics groups. Frequently, meetings were held in conjunction with those of the American Association for the Advancement of Science or some section of it. Extensive cooperation with other societies is shown by the fact that almost every meeting of the Institute was held with some theoretical or applied group(s). A contemplated list of the names of these groups, revealing the variety of interests fostered by the Institute, is too long to be presented here.

The committee system flowered during this period. In 1940 the *Report of the War Preparedness Committee* had national significance and the thoughtful report on *The Teaching of Statistics* was the topic of a meeting, was printed in the *Annals*, received formal Institute approval, and was widely circulated. Although many of the committees were concerned primarily with internal matters, their contributions were also helpful to the Institute in gaining its broader objectives.

By the mid-1940s it became apparent that a new constitution with by-laws was needed. After extensive preparation the new constitution was adopted at the Madison meeting, September 10, 1948, the major change being the replacement of the Board of Directors by an elected Council of at least 12 members (4 each year) with three-year terms and appropriate responsibilities. The president and president-elect, elected by the members, would have one-year terms, while the other officers (secretary, treasurer, and editor), elected by the council, would have three-year terms. The officers serve as an executive committee.

During the second time period, changes in organization were made as needed. The duties of the secretary were separated into those of an executive secretary and those of a program secretary. In addition, a meeting might have its own program chairman, associate secretary (program detail and records), and assistant secretary (physical arrangements for meetings). In 1970 the total membership had increased to 2843 members, including 82 institutional members, 301 student members, 23 life members, 1909 members in the United States and Canada, and 528 elsewhere. All activities of the Institute seemed to be thriving, in 1970, except possibly the *Annals*, which needed help because of its spectacular success.

The important development during the period beginning in 1972 was a new policy on publications, required by the expanding *Annals*. The four issues a year had been increased to six issues a year (reduced again to four per year in 1983), and the editor had been relieved of the managerial duties by the creation of the office of managing editor in the mid-1960s, but much more was needed. The first step, effective in 1972, was the introduction of the *Bulletin*, which incorporated and collected current information, announcements, and records of the type previously published in the *Annals*. The next step was the replacement, beginning with 1973, of the *Annals of Mathematical Statistics* by two journals, the *Annals of Statistics* and the *Annals of Probability*, each with its own editor and staff. This arrangement has been continued.

Other publications of the Institute are:

Directions in Time Series

Current Index to Statistics: Applications, Methods, and Theory (with the American Statistical Association)

Selected Tables in Mathematical Statistics (published by the American Mathematical Society for the Institute)

Statistical Research Monographs (with the University of Chicago)

The Institute has given its mark of approval (Fellow) to some members who are considered to have made appreciable contribution to the general objective of the Institute, specifically "to foster the development and dissemination of the theory and applications of statistics and of probability." A 1978 list of the 445 past and present (deceased and living) Fellows appears in the 1978 Directory of Statisticians published by the American Statistical Association*.

The Institute has issued a 1982 list of members.

There are now nine officers of the Institute. Three of these are presidents (immediate past, present, and future), three are administrative (executive secretary, program secretary, and treasurer), and the remaining three are editors (*Annals of Statistics, Annals of Probability*, and managing editor). The business office staff, together with the officers, including the past president, serve as an executive committee.

The names and addresses of the present nine officers and the names of the members of the Council are listed on the inside back cover of the *Bulletin*. The address of the business office is IMS Business Office, 3401 Investment Blvd., Suite 6, Hayward, CA 94545.

Of course it is recognized that the cumulative use of mathematics in statistics in the last 50 years would have been enormous in any case, but it does seem proper to point out that the Institute of Mathematical Statistics, with its journals which provide outlets and stimulus for papers, with its meetings with discussion providing information and additional stimulation, and with its committees and specialists turning out informative reports and recommendations, has played and promises to continue to play a vital role in the attainment of the general objective.

(*ANNALS OF MATHEMATICAL STATISTICS ANNALS OF PROBABILITY*)

PAUL S. DWYER

INSTITUTE OF STATISTICAL MATHEMATICS

The Institute of Statistical Mathematics is a research organization with the purpose of advancing the study of statistical theory and methodology and of fostering the efficient use of established statistical methods. The organization publishes the *Annals of the Institute of Statistical Mathematics**, *Computer Science Monographs* (in English), and *Proceedings* (in Japanese).

The address of the Institute is:

4-6-7 Minami Azabu
Minato-ku
Tokyo, Japan 106

(*ANNALS OF THE INSTITUTE OF STATISTICAL MATHEMATICS*)

INSTRUMENTAL VARIABLE ESTIMATION

Consider the linear regression specification

$$y = X\beta + \epsilon, \tag{1}$$

where y is an $N \times 1$ vector called the dependent variable, X is an $N \times k$ matrix of regression variables, β is a $k \times 1$ vector of unknown coefficients, and ϵ is an $N \times 1$ vector of stochastic disturbances. Thus we have N observations and k right-hand-side variables. Under the classical Gauss–Markov* assumptions the unknown coefficient vector β is estimated by least squares*. However, the unbiasedness and consistency of least-squares estimation depends on the lack of correlation between X and ϵ because $\beta_{LS} - \beta = (X'X)^{-1}X'\epsilon$. In a wide class of cases we cannot assume that this lack of correlation holds, even in large samples, and instead expect that $X'\epsilon/N$ does not converge in probability to zero as $N \to \infty$, i.e., plim $N^{-1}X'\epsilon \neq 0$ (*see* LAW OF LARGE NUMBERS). Use of least-squares estimation in this situation results in inconsistent estimates. Two types of models in statistics and econometrics often lead to this situation. The first model is the errors-in-variables model. Here, the linear relationship of (1) is assumed to hold for an unobserved variable x_j^* which would be a column vector of the regression matrix. Instead, we observe that $x_j = x_j^* + \eta_j$, where η_j has mean zero and is assumed uncorrelated with x_j^*. If x_j is used in the least-squares regression, it will be correlated with ϵ, which contains η_j by construction. Use of least-squares estimation will result in a downward-biased estimate of β_j as well as bias in all other coefficient estimates. The second model is the simultaneous-equation model of econometrics (*see* ECONOMETRICS). Here one or more columns of X are jointly dependent variables with y, so that they are correlated with ϵ. The stochastic determination of x_j is then

$$x_j = Z\Pi_j + v_j, \tag{2}$$

where Z is an $N \times p$ matrix which is assumed uncorrelated with both ϵ and v_j; Π_j a vector of unknown coefficients; and v_j, the $N \times 1$ vector of stochastic disturbance, is correlated with ϵ. In fact, v_j is a linear combination of stochastic variables including ϵ.

When least-squares estimation of (1) is inappropriate, an alternative estimator is needed. Instrumental variable (IV) estimation provides a consistent estimator for β. For instrumental variable estimation we require a $N \times k$ matrix of instruments W to estimate

$$\beta_{IV} = (W'X)^{-1}W'y. \qquad (3)$$

Usually, only large-sample properties of the IV estimator are readily available, which we now investigate. We take probability limits to find

$$\text{plim}(\beta_{IV} - \beta)$$

$$= \text{plim}(N^{-1}W'X)^{-1}(N^{-1}W'\epsilon). \qquad (4)$$

For β_{IV} to be consistent we therefore require two properties of the instrumental variables:

1. $\text{plim } N(W'X)^{-1}$ exists and is finite, which roughly requires that W be correlated with X.
2. $\text{plim } N^{-1}W'\epsilon = 0$, which implies that the instruments cannot be contemporaneously correlated with the stochastic disturbances.

Instrumental variable estimation was initially proposed by Geary [4] and Reiersøl [12] for the errors-in-variables model. Sargan [13] introduced instrumental variable estimation for simultaneous-equation models.

Where do the instruments arise for estimation? In the errors-in-variables problem the grouping procedures initiated by Wald [14] and investigated by numerous other researchers (see Madansky [11] for references) are instrumental variable estimators. Many other proposed solutions to the errors-in-variables problem also take the instrumental variable estimation form. In the simultaneous-equations model, the Z's of (2) can be used to form the instruments. However, since the number of Z's usually far exceeds

the rank of X, $p > k$, we consider linear combinations of the Z's to form the instruments, $W = ZA$, where A is a $p \times k$ matrix of rank k. A can either be a known matrix or estimated as $\hat{A}$. Any matrix A will lead to consistent estimates as long as $\text{plim } N^{-1}Z'\epsilon = 0$ and A is of full rank, since W will then satisfy the two necessary properties for instruments. (See Hausman [9] for an exposition of the use of IV estimation for simultaneous-equation models.) It turns out to be the case that almost all proposed consistent estimators for the simultaneous-equation model are IV estimators (e.g., Hausman [7]).

The asymptotic distribution of the IV estimator can then be calculated. We normalize (4) to find that

$$\sqrt{N}(\beta_{IV} - \beta)$$

$$= (N^{-1}A'Z'X)^{-1}(N^{-1/2}A'Z'\epsilon). \qquad (5)$$

The first matrix on the right-hand side of (5) has $\text{plim } A'Q$, where $Q = \text{plim } N^{-1}Z'X$, which is assumed to exist and to be finite. We then make sufficient assumptions so that a central limit theorem can be applied to $N^{-1/2}Z'\epsilon$ (see LIMIT THEOREM, CENTRAL). We can then claim that $N^{-1/2}A'Z'\epsilon$ converges in distribution to a normal random vector with distribution $\mathcal{N}(0, \sigma A'MA)$, where σ is the variance of ϵ, which is assumed to be a vector of independent random variables, and $M = \text{plim } N^{-1}Z'Z$, which is also assumed to exist and to be finite. Then using the standard rules on products of random variables where one random variable has a finite probability limit and the other random variable converges in distribution, we find the asymptotic distribution

$$\sqrt{N}(\beta_{IV} - \beta)$$

$$\overset{A}{\sim} \mathcal{N}\left(0, \sigma\left[(A'Q)^{-1}A'MA(Q'A)^{-1}\right]\right).$$

$$(6)$$

If A is replaced by an estimate $\hat{A}$ which has a probability limit equal to A, we obtain identical asymptotic results. Given the formula for the asymptotic covariance matrix for $\hat{\beta}_{IV}$, we would like to find the best choice of A to form the matrix of instruments W.

That is, we want to choose A to minimize, in a matrix sense, the asymptotic covariance. In general, a unique optimum A need not exist, but an optimum choice is $A = (Z'Z)^{-1}Z'X$. For this choice of A we calculate the asymptotic covariance matrix from (6):

$$V(\beta_{IV}) = \sigma[Q'MQ]. \quad (7)$$

We can show that for any other choice of instruments, say $\tilde{W} = Z'E$, which lead to estimates $\tilde{\beta}_{IV}$, that all possible linear combinations $g'\beta$, where g is a $k \times 1$ vector, have at least as small asymptotic variance if A is used:

$$g'[V(\hat{\beta}_{IV}) - V(\tilde{\beta}_{IV})]g \leqslant 0 \quad \text{for all } g. \quad (8)$$

Returning to (1) it is obvious that least squares will be the best IV estimator if the X's satisfy the properties of instrumental variables. Therefore, we might like to propose a test of the hypothesis that plim $N^{-1}X'\epsilon = 0$. A straightforward method to test this hypothesis is to compare the least squares with the IV estimates. Under the null hypothesis least squares is consistent and a test statistic can be calculated [8, 15]:

$$m = (\beta_{LS} - \beta_{IV})'[V(\beta_{IV}) - V(\beta_{LS})]^+ \times (\beta_{LS} - \beta_{IV}), \quad (9)$$

where $[\cdot]^+$ is any generalized inverse*. The statistic m is distributed under the null hypothesis as χ^2 with k degrees of freedom if all the X variables are assumed correlated, or with $l < k$ degrees of freedom if $k - l$ of the columns of X are included among the instrumental variables Z (so that these x_j's are assumed uncorrelated with ϵ).

A second test of interest can be performed when the number of instrumental variables in Z exceed the required number of instruments in W, $p > k$. Then a test of "overidentification" [3, 13] can be made to test the hypothesis plim $N^{-1}Z'\epsilon = 0$. First we form the vector $\hat{q} = Z(Z'Z)^{-1}Z'\hat{\epsilon}$, where $\hat{\epsilon} = y - X\beta_{IV}$. We then form the overidentification test statistic

$$\lambda = \hat{q}'\hat{q}/\hat{\sigma}, \quad (10)$$

where $\hat{\sigma} = \hat{\epsilon}'\hat{\epsilon}/(N - k)$. Under the null hypothesis of no correlation λ is distributed as χ^2 with $p - k$ degrees of freedom. Interpretation of this test statistic requires care since it tests only a restricted subspace and provides only a necessary but not a sufficient test of the hypothesis plim $N^{-1}Z'\epsilon = 0$ (i.e., the test is not consistent for the null hypothesis).

Instrumental variable estimation of simultaneous equation models is widely used in econometrics. It is being used increasingly in "latent" variable models which arise in errors in variables model specifications, (e.g., Jöreskog and Goldberger [10] and Aigner et al. [1]). The IV method of estimation can also be extended to nonlinear models by use of a Gauss–Newton linearization technique [2, 9].

An important example of instrumental variable estimation is in the estimation of the returns to schooling. In (1) the dependent variable is the log of wages while the right-hand-side variables include education, IQ, and sociodemographic variables. The coefficient of the education variable can be interpreted approximately as the rate of return to an additional year of education. An example of IV estimation of the wage equation is contained in Griliches et al. [6]. In Table 1 we give the estimated coefficients and (asymptotic) standard errors of least squares and instrumental variable estimation. The sample consists of 2419 young men from the 1973 wave of the National Longitudinal Survey. The schooling variable was taken as measured with error and the instruments used included all other right-hand-side variables except schooling as well as father's education, mother's education, an index of father's earnings, number of siblings, and an index of culture. We give the

Table 1   Log Wage Equation for 1973

Variable	Least Squares	IV
Years schooling	0.019 (0.004)	0.051 (0.014)
IQ	0.003 (0.001)	0.001 (0.001)
Age	0.042 (0.003)	0.040 (0.003)
Black	0.107 (0.028)	0.115 (0.028)

important right-hand-side variable coefficients. Other right-hand-side variables are: constant, armed forces experience, SMSA, region not south.

Note that the estimated coefficient of schooling has increased by 168% between least-squares and instrumental variable estimation. A Wu–Hausman test of (9) of the consistency of least squares is calculated to be $m = 5.68$, which under the null hypothesis is distributed as χ_1^2. We thus conclude that least-squares estimation leads to a significant downward bias in the estimate of the schooling coefficient. This result has been corroborated when tested over a number of different data sets in a series of published papers in the economics literature.

References

[1] Aigner, D., Hsiao, C., Kapteyn, A., and Wansbeck, T. (1983). In *Handbook of Econometrics*, Z. Griliches and M. Intriligator, eds. North-Holland, Amsterdam.

[2] Amemiya, T. (1983). In *Handbook of Econometrics*, Z. Griliches and M. Intriligator, eds. North-Holland, Amsterdam.

[3] Anderson, T. W. and Rubin, H. (1949). *Ann. Math. Statist.* **20**, 46–63.

[4] Geary, R. C. (1949). *Econometrica*, **17**, 30–59.

[5] Griliches, Z. and Intriligator, M., eds. (1982). *Handbook of Econometrics*. North-Holland, Amsterdam.

[6] Griliches, Z., Hall, B., and Hausman, J. (1978). *Ann. Insée*, 30–31, 137–176.

[7] Hausman, J. (1975). *Econometrica*, **43**, 727–738.

[8] Hausman, J. (1978). *Econometrica*, **46**, 1251–1272.

[9] Hausman, J. (1983). In *Handbook of Econometrics*, Z. Griliches and M. Intriligator, eds. North-Holland, Amsterdam.

[10] Jöreskog, K. and Goldberger, A. (1975). *J. Amer. Statist. Ass.*, **70**, 631–639.

[11] Madansky, A. (1959). *J. Amer. Statist. Ass.*, **54**, 173–206.

[12] Reiersøl, O. (1945). *Ark. Math., Astron. Fys.*, **32**, 1–119.

[13] Sargan, D. (1958). *Econometrica*, **26**, 393–415.

[14] Wald, A. (1940). *Ann. Math. Statist.*, **11**, 284–300.

[15] Wu, D. (1973). *Econometrica*, **41**, 733–750.

(ECONOMETRICS)

JERRY A. HAUSMAN

INSTRUMENTAL VARIABLES

A term used often in econometrics* and social sciences applications (see, e.g., Johnston [2]). Given a linear model

$$\mathbf{Y} = \mathbf{X}\boldsymbol{\beta} + \boldsymbol{\epsilon},$$

where $\mathbf{Y}$ represents the vector of observations and $\mathbf{X}$ is the matrix of observations on dependent (endogenous) and independent (exogenous) variables, respectively, $\boldsymbol{\beta}$ is the vector of unknown coefficients and $\boldsymbol{\epsilon}$ is the vector of error terms. If (unlike in the case of "standard" linear models) at least one of the exogenous variables is correlated with the error term, a set of instrumental variables $\mathbf{Z}$ is introduced that are correlated with $\mathbf{X}$ but not the $\boldsymbol{\epsilon}$.

In this case the instrumental variable estimator $\mathbf{b}^*$ of the vector $\boldsymbol{\beta}$ is given by

$$\mathbf{b}^* = (\mathbf{Z}'\mathbf{X})^{-1}\mathbf{Z}'\mathbf{Y}.$$

If all variables are their own instruments (i.e., are uncorrelated with $\mathbf{U}$) this is simply the ordinary least-squares estimator. For more details on construction of instrumental variables and the properties of $\mathbf{b}^*$ see, e.g., Johnston [2] and Hanushek and Jackson [1].

References

[1] Hanushek, E. A. and Jackson, J. E. (1977). *Statistical Methods for Social Scientists*. Academic Press, New York.

[2] Johnston, J. (1972). *Econometric Methods*, 2nd ed. McGraw-Hill, New York.

(ECONOMETRICS
REGRESSION ANALYSIS
SOCIOLOGY, STATISTICS IN)

INTEGER PROGRAMMING

A linear programming* or nonlinear programming* problem whose variables are constrained to be integer is called a (linear or nonlinear) *integer program*. We consider here only the linear case, although there exist extensions of the techniques to be discussed to nonlinear integer programming.

The integer programming problem can be stated as

$$\min\{cx \mid Ax \geqslant b, x \geqslant 0,$$
$$x_j \text{ integer}, j \in N_1 \subseteq N\}, \quad \text{(P)}$$

where A is a given $m \times n$ matrix, c and b are given vectors of conformable dimensions, $N = \{1, \ldots, n\}$, and x is a variable n-vector. (P) is called a *pure integer program* if $N_1 = N$, a *mixed integer program* if $\emptyset \neq N_1 \neq N$. Integer programming is sometimes called *discrete optimization*.

SCOPE AND APPLICABILITY

Integer programming is the youngest branch of mathematical programming: its development started in the second half of the 1950s. It is the most immediate and frequently needed extension of linear programming. Integrality constraints arise naturally whenever fractional values for the decision variables do not make sense. A case in point is the *fixed-charge problem*, in which a function of the form $\sum_i c_i(x_i)$, with

$$c_i(x_i) = \begin{cases} f_i + c_i x_i & \text{if } x_i > 0 \\ 0 & \text{if } x_i = 0 \end{cases}$$

is to be minimized subject to linear constraints. Such a problem can be restated as an integer program whenever x is bounded and $f_i > 0$, by setting

$$c_i(x_i) = c_i x_i + f_i y_i,$$
$$x_i \leqslant U_i y_i, \quad y_i = 0 \text{ or } 1$$

where U_i is an upper bound on x_i.

By far the most important special case of integer programming is the *0-1 programming problem*, in which the integer-constrained variables are restricted to 0 or 1. This is so because a host of frequently occurring nonlinearities, such as logical alternatives, implications, precedence relations, etc., or combinations thereof, can be formulated via 0-1 variables. For example, a condition such as

$$x > 0 \Rightarrow (f(x) \leqslant a \vee f(x) \geqslant b),$$

where a and b are positive scalars, x is a variable with a known upper bound M, $f(x)$ is a function whose value is bounded from above by $U > 0$ and from below by $L < 0$, and the symbol "$\vee$" means disjunction (logical "or"), can be stated as

$$x \leqslant M(1 - \delta_1),$$
$$f(x) \leqslant a + (U - a)\delta_1 + (U - a)\delta_2,$$
$$f(x) \geqslant b + (L - b)\delta_1 + (L - b)(1 - \delta_2),$$
$$\delta_1, \delta_2 = 0 \text{ or } 1.$$

A linear program with "logical" conditions (conjunctions, disjunctions, and implications involving inequalities) is called a *disjunctive program*, since it is the presence of disjunctions that makes these problems nonconvex. Disjunctive programs can be stated as 0-1 programs and vice-versa, but the disjunctive programming formulation has produced new methods.

Nonconvex optimization problems such as bimatrix games, separable programs involving piecewise linear nonconvex/nonconcave functions, the general (nonconvex) quadratic programming problem, the linear complementarity problem, and many others can be stated as disjunctive or 0-1 programming problems.

A host of interesting combinatorial problems can be formulated as 0-1 programming problems defined on a graph. The joint study of these problems by mathematical programmers and graph theorists has led to the recent development of a burgeoning area of research known as *combinatorial optimization*. Some typical problems studied in this area are edge matching and covering, vertex packing and covering, clique covering, vertex coloring; set packing, partitioning, and covering; Euler tours; and Hamiltonian cycles (the traveling salesman problem). (*See* COMBINATORICS; GRAPH THEORY.)

Applications of integer programming abound in all spheres of decision making. Some typical real-world problem areas where integer programming is particularly useful as a modeling tool, include facility (plant, warehouse, hospital, fire station) location; scheduling (of personnel, production, other activities); routing (of trucks, tankers, airplanes); design of communication (road, pipeline, telephone) networks; capital budgeting; project selection; and analysis of capital development alternatives. As general references on integer programming, see the book by Garfinkel and Nemhauser [13] and the more recent volumes edited by Christofides et al. [8], Hammer et al. [18, 19], and Padberg [33].

RELATION TO STATISTICS

In statistics, integer programming is useful, for instance, in regression analysis*, design of experiments*, stratified sampling*, and cluster analysis.

A common problem in linear *regression analysis** is that of choosing a "best" regression equation. To be specific, let $x_1, \ldots, x_p$ be a set of p independent variables that might affect the value of the (dependent) variable y, and suppose for various reasons that we would like to explain the behavior of y by using at most q of the p independent variables. If for $j = 1, \ldots, p$, $X_{1j}, \ldots, X_{nj}$ are observed values of x_j, and $Y_1, \ldots, Y_n$ are the corresponding values of y, the problem of selecting a "best" subset of size q from among the p variables x_j can be formulated as the mixed integer program

$$\min \sum_{i=1}^{n} |d_i|,$$

$$\sum_{j=1}^{p} X_{ij}\beta_j + d_i = Y_i, \quad i = 1, \ldots, n$$

$$L_j\delta_j \leqslant \beta_j \leqslant U_j\delta_j, \quad j = 1, \ldots, p$$

$$\sum_{j=1}^{p} \delta_j = q,$$

$$\delta_j = 0 \text{ or } 1, \quad j = 1, \ldots, p.$$

Here the β_j are the parameters to be estimated, L_j and U_j are lower and upper bounds on the value of β_j, the d_i represent the deviations of the predicted from the observed values Y_i of y, and the objective is to minimize the mean absolute deviation. The absolute value function is easily turned into a linear function by a transformation well known in the linear programming literature. Having the 0–1 variables δ_j constrained to sum to q has the effect of forcing $p - q$ of them, and thereby $p - q$ of the β_j, to zero. If a least-squares regression is preferred to the objective function above, a similar formulation yields a mixed integer quadratic program. Solving the problem for various values of q one can establish how seriously the restriction on the number of independent variables affects the outcome.

Another area of statistics where integer, and in particular combinatorial programming techniques are useful, is the *design of experiments**. In the construction of balanced incomplete block designs*, in determining the minimal support size for such a design, in the construction of Latin squares* satisfying certain conditions, etc., a number of 0–1 programming problems with 0–1 coefficient matrices of special types have to be solved. An integer programming model widely used in this area as well as in others, is *set covering*. Let $M = \{1, \ldots, m\}$ be a finite set and $\mathcal{F} = \{M_1, \ldots, M_n\}$ a family of subsets of M. The set covering problem is that of finding a minimum cardinality (or minimum-weight) collection of members of $\mathcal{F}$ whose union is M. If the subsets in the collection are also required to form a partition of M, the problem is called *set partitioning*. Let $A = (a_{ij})$ be an $m \times n$ 0–1 matrix whose rows and columns are associated with M and $\mathcal{F}$, respectively, such that $a_{ij} = 1$ if subset M_j contains element i of M, $a_{ij} = 0$ otherwise. The (weighted) set covering problem can then be stated as

$$\min cx,$$
$$Ax \geqslant e,$$
$$x_i = 0 \text{ or } 1, \quad j = 1, \ldots, n,$$

where $e = (1, \ldots, 1)^T$ has m components and c is an n-vector of weights (in the unweighted case $c = (1, \ldots, 1)^T$). The set partitioning problem differs from the above by having $\geqslant$ replaced by $=$. For instance, the problem of finding a pair of orthogonal Latin squares is a specially structured set partitioning problem, in fact a four-index assignment problem.

The area of statistics in which integer and combinatorial programming is most crucial seems to be *cluster analysis** (grouping, typology, numerical taxonomy). Given a set of points in *n*-dimensional space with a distance function defined on it, clustering essentially deals with partitioning the set into subsets (groups, clusters), so as to optimize some function of the distances between the points: minimize the within-groups sums of squared distances, minimize the maximum within-groups distance, maximize the minimum distance between groups, etc. Here the distance may be Euclidean, rectilinear, or whatever. All these problems can be formulated as integer programs, some of which are more tractable than others. There are efficient, "greedy"-type combinatorial algorithms for maximizing the minimum distance within groups which require a number of steps linear in the size of the set to be partitioned. If the objective is to partition the set into no more than two subsets, then a similar linear time algorithm is available for minimizing the maximum within-groups distance. For the remaining cases, the corresponding integer programs cannot be solved in time guaranteed to be polynomial in the size of the set, but can still be solved exactly or approximately for reasonably large sets. One practically significant model for minimizing the within-groups sums of squared distances, with the additional requirement (generalized string property) that every group have a leader such that every point in the group is closer to the leader than to any point outside the group, uses the set-partitioning formulation introduced above, with the following interpretation. The rows of A correspond to points of the set, the columns of A correspond to candidate clusters (groups), and $a_{ij} = 1$ if candidate cluster j contains point i, $a_{ij} = 0$ otherwise. The cost c_j represents the sum of squared distances between the points of the jth candidate cluster. Because of the generalized string property, it can be shown that the number of candidate clusters is $m \times (m - 1)$; hence the problem is of manageable size even for fairly large sets.

For literature on applications of integer programming to statistics, see the book by Arthanari and Dodge [1] and articles by Rao [34] and Mulvey and Crowder [31].

SOLUTION METHODS: OVERVIEW

We denote the optimal objective function value for (P) by $v(P)$, and call it the *value* of (P). We denote by (L) the linear program obtained from (P) by removing the integrality requirements, and call it the *linear programming relaxation* of (P).

Integer programs are notoriously difficult: in the language of computational complexity theory, the general 0–1 programming problem, as well as most of its special cases, is NP-complete. Polynomial time integer programming algorithms do not exist. However, sometimes an integer program can be solved as a linear program; i.e., solving the linear programming relaxation (L) of the integer program (P), one obtains an integer solution. In particular, this is the case when all basic solutions of (L) are integer. For an arbitrary integer vector b, the constraint set $Ax \leqslant b$, $x \geqslant 0$, is known [21] to have only integer basic solutions if and only if the matrix A is totally unimodular (i.e., all nonsingular submatrices of A have a determinant of 1 or -1).

The best known instances of total unimodularity are the vertex-edge incidence matrices of directed graphs and undirected bipartite graphs. As a consequence, shortest-path and network flow problems on arbitrary directed graphs, edge matching (or covering) and vertex packing (or covering) problems on bipartite graphs, as well as other integer programs whose constraint set is defined by the incidence matrix of a directed graph or an undirected bipartite graph, with arbitrary integer right-hand side, are in fact linear programs.

Apart from this important but very special class of problems, the difficulty in solving integer programs lies in the nonconvexity of the feasible set, which makes it impossible to establish global optimality from local conditions. The two principal approaches to solving integer programs try to circumvent this difficulty in two different ways.

The first approach, which in the current state of the art is the standard way of solving integer programs, is *enumerative* (branch and bound*, implicit enumeration). It partitions the feasible set into successively smaller subsets, calculates bounds on the objective function value over each subset, and uses these bounds to discard certain subsets from further consideration. The procedure ends when each subset has either produced a feasible solution, or was shown to contain no better solution than the one already in hand. The best solution found during the procedure is a global optimum. Two early prototypes of this approach are due to Land and Doig [27] and Balas [2].

The second approach, known as the cutting plane method, is a *convexification* procedure: it approximates the convex hull of the set F of feasible integer points by a sequence of inequalities that cut off (hence the term "cutting planes") parts of the linear programming polyhedron, without removing any point of F. When sufficient inequalities have been generated to cut off every fractional point better than the integer optimum, the latter is found as an optimal solution to the linear program (L) amended with the cutting planes. The first finitely convergent procedure of this type is due to Gomory [16].

Depending on the type of techniques used to describe the convex hull of F and generate cutting planes, one can distinguish three main directions in this area. The first one uses algebraic methods, like modular arithmetic and group theory. Its key concept is that of subadditive functions. It is sometimes called the algebraic or group-theoretic approach. The second one uses convexity, polarity, propositional calculus. Its main thrust comes from looking at the 0–1 programming problem as a disjunctive program. It is known as the convex analysis/disjunctive programming approach. Finally, the third direction applies to combinatorial programming problems, and it combines graph theory and matroid theory with mathematical programming. It is sometimes called polyhedral combinatorics.

Besides these two basic approaches to integer programming (enumerative and convexifying), two further procedures need to be mentioned that do not belong to either category, but can rather be viewed as complementary to one or the other. Both procedures essentially *decompose* (P), one of them by partitioning the variables, the other one by partitioning the constraints. The first one, due to Benders [7], gets rid of the continuous variables of a mixed integer program (P) by projecting the feasible set F into the subspace of the integer-constrained variables. The second one, known as Lagrangean relaxation, gets rid of some of the constraints of (P) by assigning multipliers to them and taking them into the objective function.

Each of the approaches outlined here aims at solving (P) exactly. However, since finding an optimal solution tends to be expensive beyond a certain problem size, approximation methods or *heuristics* play an increasingly important role in this area.

Next we briefly review the approaches sketched above, and give some references for each of them.

BRANCH AND BOUND/IMPLICIT ENUMERATION

The following are the basic steps of a typical enumerative algorithm. Start by putting (P) on the list of subproblems, and by setting $\bar{v}(P) = \infty$, where $\bar{v}(P)$ is an upper bound on $v(P)$.

1. Choose, and remove from the list, a subproblem (P_i), according to some criterion specified by the search strategy. If the list is empty, stop: if no solution was found, (P) is infeasible; otherwise, the current best solution is optimal.

2. If (P_i) has constraints involving only 0–1 variables, explore their implications via logical tests to impose as many new constraints of the type $x_i = 0$, or $x_i = 1$ (or

of a more complex type), as possible. If as a result (P$_i$) is shown to be infeasible, discard (P$_i$) and go to 1.

3. Generate a lower bound $\underline{v}(P_i)$ on $v(P_i)$, by solving some relaxation of (P$_i$) (such as the linear programming relaxation, or a Lagrangean relaxation, or either of these two amended with cutting planes). If $\underline{v}(P_i) \geqslant \bar{v}(P)$, discard (P$_i$) and go to 1.

4. Attempt to generate an improved upper bound on $v(P)$ by using some heuristic to find an improved feasible solution. If successful, update $\bar{v}(P)$ and remove from the list all (P$_j$) such that $\underline{v}(P_j) \geqslant \bar{v}(P)$.

5. Split (P$_i$) into two or more subproblems by partitioning its feasible set according to some specified rule. Add the new subproblems to the list and go to 1.

The search strategies that can be used in step 1 range between the two extremes known as "breadth first" (always choose the subproblem with smallest $\underline{v}(P_i)$), and "depth first" (always choose one of the new subproblems just created). The first approach usually generates fewer subproblems, but carries a high cost in terms of storage requirements; therefore, the second one is preferred in most codes. Flexible intermediate rules seem to give the best results.

The branching, or partitioning, rule of step 5, is usually a dichotomy of the form

$$x_k \leqslant \lfloor \bar{x}_k \rfloor \vee x_k \geqslant \lceil \bar{x}_k \rceil,$$

where x_k is some integer-constrained variable whose value $\bar{x}_k$ in the current solution to (P$_i$) is noninteger, while $\lfloor a \rfloor$ and $\lceil a \rceil$ denote the largest integer $\leqslant a$ and the smallest integer $\geqslant a$, respectively. The choice of the variable is important, but no reliable criterion is known for it. "Penalties" and "pseudo-costs" try to assess the change in $v(P_i)$ that will be produced by branching on x_k, with a view of providing a choice that will force the value of at least one of the new subproblems as high as possible.

In problems with some structure, more efficient branching rules are possible. In the

presence of a "multiple-choice" constraint

$$\sum_{j \in Q} x_j = 1, \qquad x_j = 0 \text{ or } 1, \quad j \in Q,$$

for instance, one can branch on the dichotomy

$$x_j = 0, \quad j \in Q_1 \vee x_j = 0, \quad j \in Q \setminus Q_1$$

for some $Q_1 \subset Q$, thus fixing several variables at a time. Other, more sophisticated branching rules have been used for set covering, set partitioning, and traveling salesman problems.

The logical tests of step 2, and/or associated inequalities, whenever applicable, were shown to substantially speed up the procedure. However, by far the most important ingredients of any enumerative procedure are the bounding devices used in steps 3 and 4. The importance of the relaxation used was demonstrated in the case of such special structures as the traveling salesman problem, where the knowledge of deep cutting planes (usually facets of the convex hull of F) has made it possible to replace the common linear programming relaxation (L) by a much "stronger" one, either by amending (L) with cutting planes of the latter type, or by taking those same cutting planes into the objective function in the Lagrangean manner. In either case, the resulting vastly enhanced lower-bounding capability has drastically reduced computing times. Similarly, improvements in the upper-bounding procedure, such as the use of an efficient heuristic to find feasible solutions, were found to affect decisively the performance of branch-and-bound methods. For surveys of this area, see Balas [3], Beale [6], and Spielberg [36].

PARTITIONING THE VARIABLES OR CONSTRAINTS

Benders' partitioning procedure is based on the following result. Consider the problem

$$\min\{cx + dy \mid Bx + Dy = b, x \geqslant 0, y \in Q\}$$

$$(P_1)$$

where B and D are $m \times p$ and $m \times q$ matrices, respectively, c, d, and b are vectors of conformable dimensions, and Q is an arbitrary set (e.g., the set of integer q-vectors) such that for every $y \in Q$, there exists an $x \geq 0$ satisfying $Bx + Dy = b$. Let $U = \{u \mid uB \leq c\}$, and let vert U be the (finite) set of vertices of the polyhedron U. Then (P_1) is equivalent to

$$\min\{w_0 \mid w_0 \geq (d - uD)y + ub,$$

$$u \in \text{vert } U, y \in Q\}, \qquad (P_2)$$

in the sense that if $(\bar{x}, \bar{y})$ solves (P_1), then $\bar{y}$ solves (P_2); and if $\hat{y}$ solves (P_2), there exists an $\hat{x}$ such that $(\hat{x}, \hat{y})$ solves (P_1). Although the inequalities of (P_2) usually outnumber those of (P_1) by far, they can be generated as needed by solving a linear program in the continuous variables x, or its dual (the latter having U as its constraint set). This approach can be useful in particular when B has a structure making it easy to solve the linear programs that provide the constraints of (P_2).

The second type of decomposition procedure, Lagrangean relaxation, partitions the set of constraints $Ax \geq b$ of (P) into $A_1 x \geq b_1$ and $A_2 x \geq b_2$, and formulates the Lagrangean problem

$$L(u) = \min\{(c - uA_2)x + ub_2 \mid A_1 x \geq b_1,$$

$$x \geq 0, x_j \text{ integer},$$

$$j \in N_1 \subseteq N\}.$$

For any u, $L(u)$ is a lower bound on the objective function value of (P). The problem in the variables u of maximizing $L(u)$ subject to $u \geq 0$ is sometimes called the Lagrangean dual of (P). There are several methods for maximizing $L(u)$ as a function of $u \geq 0$, one of them being subgradient optimization. If $\bar{u} \geq 0$ maximizes $L(u)$ and $\bar{x}$ is a minimizing vector in $L(\bar{u})$, then $\bar{x}$ is an optimal solution to (P) if $A_2\bar{x} \geq b_2$ and $\bar{u}(A_2\bar{x} - b_2) = 0$. However, this is usually not the case, since $L(\bar{u})$ and the optimal objective function value of (P) tend to be separated by a so-called duality gap. Nevertheless, since calculating the value of $L(u)$ for

fixed u may be a lot easier than solving (P), this is often a convenient way of generating good lower bounds.

In particular, since $A_2 x \geq b_2$ may consist partly (or wholly) of cutting planes, this is one way of using the latter without vastly increasing the number of inequalities explicitly added to the constraint set. For surveys of these techniques, see Geoffrion [14], Shapiro [35], and Fisher [10].

CUTTING PLANE THEORY

A central problem of integer programming theory is to characterize the *convex hull* of F, the set of integer points satisfying the inequalities of (P). F is called the feasible set, its convex hull (defined as the smallest convex set containing F) is denoted conv F. From a classical result of Weyl [37], it is known that conv F is the intersection of a finite number of linear inequalities. In other words, (P) is equivalent to a linear program. Unfortunately, however, the constraint set of this linear program is in general hard to identify. Only for a small number of highly structured combinatorial optimization problems do we have at this time a linear characterization of conv F, i.e., an explicit representation of conv F by a system of linear inequalities. In the general case, all that we have are some procedures to generate sequences of inequalities that can be shown to converge to such a representation.

One way to solve the general integer program (P) is thus to start by solving (L), the linear programming relaxation of (P), and then to successively amend the constraint set of (L) by additional inequalities (cutting planes), until the whole region between the optimum of (L) and that of (P) is cut off. How much work is involved in this depends on the strength (depth) of the cuts, as well as on the size of the region that is to be cut off, i.e., the size of the gap between $v(L)$ and $v(P)$, the value of L and P. This gap can be very large indeed, as evidenced by a recent result for the class of 0–1 programs called (unweighted) set covering problems (where all entries of A are 0 or 1, and all entries of b and c are 1). For a set covering problem in n variables and an arbitrary number of constraints, the ratio $v(P)/v(L)$ is bounded by $n/4 + 1/2$ for n even, and by $n/4 + 1/2 + 1/(4n)$ for n odd. Furthermore, this is a best possible bound.

As to the strength of various cutting planes, it is useful to address the question from the following angle. Let $F \subset \mathbb{R}^n$, $d \in \mathbb{R}^n$, and $d_0 \in \mathbb{R}$. The set $\{x \in \text{conv } F \mid dx = d_0\}$ is called a *facet* of conv F if $dx \geqslant d_0$ for all $x \in F$ and $dx = d_0$ for n affinely independent points $x \in F$. In the integer programming literature the inequality $dx \geqslant d_0$ defining the facet is also called a facet. Facets are important because among many possible representations of conv F in terms of inequalities, the facets of conv F provide a minimal one. Obviously, they are the strongest possible cutting planes.

Subadditive Cuts

Consider the integer program (P), with $N_1 = N$. Solving the linear programming relaxation (L) of (P) produces a simplex tableau of the form

$$x_i = a_{i0} + \sum_{j \in J} a_{ij}(-x_j), \qquad i \in I, \quad (1)$$

where I and J are the index sets of basic and nonbasic variables, respectively. If a_{i0} is noninteger and we denote $f_{ij} = a_{ij} - \lfloor a_{ij} \rfloor$, $\forall i, j$, one can show that (1), together with the integrality of the variables, implies for every $i \in I$,

$$\sum_{j \in J} f_{ij} x_j \geqslant f_{i0}. \quad (2)$$

The inequality (2) is a cutting plane, since it is satisfied by every integer x that satisfies (1), but is violated for instance by the optimal solution to (L) associated with (1), in which all nonbasic variables are equal to 0. This cut was the basis of Gomory's method of integer forms, the first finitely convergent cutting plane algorithm for pure integer programs. An analogous cut provides a finitely convergent algorithm for mixed integer pro-

grams (with integer-constrained objective function value).

The derivation of the cut (2) is based on simple modular arithmetic. However, the integer program over the polyhedral cone defined by (1), together with the conditions

$$x_j \text{ integer}, j \in I \cup J; \qquad x_j \geqslant 0, j \in J \quad (3)$$

(note that the conditions $x_j \geqslant 0$, $j \in I$ are omitted), is equivalent to an optimization problem over a commutative Abelian group that can be solved as a shortest-path problem [17]. Whenever the vector $\bar{x}$ corresponding to the optimal solution found for the group problem satisfies the conditions $x_j \geqslant 0$, $j \in I$, it is an optimal solution to (P). When this is not the case, $\bar{x}$ provides a lower bound on $v(P)$.

The key concept in Gomory's characterization of the "corner polyhedron," i.e., the convex hull of integer points in the above-mentioned cone, is subadditivity. This has subsequently led to a subadditive characterization of the convex hull of F itself.

A function f defined on a monoid (semigroup) M is subadditive if $f(a + b) \leqslant f(a) + f(b)$ for all $a, b \in M$. Let A be an $m \times n$ matrix with rational entries, let a_j be the jth column of A, and let $X = \{x \mid Ax = b, x \geqslant 0 \text{ integer}\} \neq \emptyset$. Then for any subadditive function f on the monoid $M = \{y \mid y = Ax \text{ for some integer } x \geqslant 0\}$, such that $f(0) = 0$, the inequality

$$\sum_{j=1}^{n} f(a_j) x_j \geqslant f(Ax) \quad (4)$$

is satisfied by every $x \in X$. Conversely, all valid inequalities for X are dominated by an inequality (4) for some subadditive function f on M such that $f(0) = 0$. For literature, see Johnson [24, 25] and Jeroslow [23].

Disjunctive Cuts

A different, geometrically motivated approach derives cutting planes from convexity considerations (intersection or convexity cuts, disjunctive cuts). This approach is directed primarily to the 0–1 programming problem. As mentioned earlier, 0–1 programming is coextensive with disjunctive programming, and the best way of describing the approach is by applying it to the disjunctive program

$$\min\left\{ cx \mid \bigvee_{i \in Q} (A^i x \geqslant b^i, x \geqslant 0) \right\}. \quad \text{(D)}$$

Here Q is an index set, A^i and b^i are $m_i \times n$ and $m_i \times 1$ matrices, and "$\vee$" means that at least one of the systems $A^i x \geqslant b^i$, $x \geqslant 0$, must hold. This is the disjunctive normal form of a constraint set involving logical conditions on inequalities, and any such constraint set can be brought to this form.

The convex hull of a disjunctive set is characterized by the following two results. Let the set be

$$F = \left\{ x \mid \bigvee_{i \in Q} (A^i x \geqslant b^i, x \geqslant 0) \right\},$$

where $A^i, b^i, i \in Q$ are as above, and let Q^* be the set of those $i \in Q$ such that the system $A^i x \geqslant b^i$, $x \geqslant 0$, is consistent. Let $\alpha \in \mathbb{R}^n$ and $\alpha_0 \in \mathbb{R}$. Then the inequality $\alpha x \geqslant \alpha_0$ is satisfied by every $x \in F$ if and only if there exists a set of vectors $\theta^i \in \mathbb{R}^{m_i}$, $\theta^i \geqslant 0$, $i \in Q^*$, such that

$$\alpha \geqslant \theta^i A^i \quad \text{and} \quad \alpha_0 \leqslant \theta^i b^i, \quad i \in Q^*. \quad \text{(5)}$$

Furthermore, if F is full dimensional, Q is finite, and $\alpha_0 \neq 0$, then $\alpha x \geqslant \alpha_0$ is a facet of conv F if and only if $\alpha \neq 0$ is a vertex of the polyhedron

$$F^\# = \{ \alpha \mid \alpha \text{ satisfies (5) for some } \theta^i \geqslant 0, i \in Q^* \}.$$

The first of these results can be used to generate computationally inexpensive cutting planes for a variety of special cases of F, corresponding to logical conditions inherent to the problem at hand, whereas the second result can be used to strengthen any such cut, at an increasing computational cost, up to the point where it becomes a facet of conv F.

Often there is advantage in casting an integer program into the form of a disjunctive program with integrality constraints on some of the variables. For such problems, a procedure called *monoidal cut strengthening* that combines the disjunctive and subadditive approaches can be used to derive a family of cutting planes whose strength versus computational cost ratio compares favorably with cutting planes based on either approach taken separately.

A fundamental question of integer programming theory is whether the convex hull of feasible points can be generated sequentially by imposing the integrality conditions step by step: that is, by first producing all the facets of the convex hull of points satisfying the linear inequalities, plus the integrality condition on, say, x_1; then adding all these facet inequalities to the constraint set and generating the convex hull of points satisfying this amended set of inequalities, plus the integrality condition on x_2; etc. The question also has practical importance, since convex hull calculations for a mixed integer program with a single integer variable are much easier than for one with many integer variables.

To be more specific, suppose that we wish to generate the convex hull of the set

$$X = \{ x \mid Ax \geqslant b, x \geqslant 0, x_j \text{ integer}, j = 1, \ldots, n \}.$$

Let

$$X_0 = \{ x \mid Ax \geqslant b, x \geqslant 0 \}$$

and for $j = 1, \ldots, n$, define recursively

$$X_j = \text{conv}\{ x \in X_{j-1} \mid x_j \text{ integer} \}.$$

Obviously, $X_n \subseteq \text{conv } X$; the question is whether $X_n = \text{conv } X$.

The answer, obtained from disjunctive programming considerations, is that for a general integer program the statement $X_n = \text{conv } X$ is false; but that for a 0–1 program it is true. This is one of the main distinguishing properties of 0–1 programs among integer programs.

For literature, see Balas [4], Glover [15], and Jeroslow [22].

Combinatorial Cuts

Given a graph $C = (V, E)$ with vertex set V and edge set E, a matching in G is a set of pairwise nonadjacent edges of G. If A is the incidence matrix of vertices versus edges of G and a weight w_j is assigned to every edge j, the problem of finding a maximum-weight matching in G is the integer program

$$\max\{wx \mid Ax \leqslant e, x_j = 0 \text{ or } 1, j \in E\},$$

where $e = (1, \ldots, 1)$ has $|V|$ components, and $x_j = 1$ if edge j is in the matching, $x_j = 0$ otherwise. Edmonds [9] has shown that this problem can be restated as a linear program in the same variables, by adding an inequality of the form

$$\sum_{j \in E(S)} x_j \leqslant \tfrac{1}{2}(|S| - 1)$$

for every $S \subseteq V$ such that $|S|$ is odd. Here $E(S)$ is the set of edges with both ends in S.

Unfortunately, the matching polytope is the exception rather than the rule, and for most combinatorial problems such a simple linear characterization of the convex hull of feasible points does not exist. However, certain classes of facets of the convex hull have been identified for several problems.

The vertex packing problem in a graph $G = (V, E)$ with vertex weights c_i, $i \in V$, consists in finding a maximum weight independent (i.e., pairwise nonadjacent) set of vertices. If A is the same incidence matrix as before and T denotes transposition, the vertex packing problem is the integer program

$$\max\{cx \mid A^T x \leqslant e, x_j = 0 \text{ or } 1, j \in V\},$$

where e has $|E|$ components and $x_j = 1$ if vertex j is in the packing, $x_j = 0$ otherwise. Let $I(G)$ denote the packing polytope of G, i.e., the convex hull of incidence vectors of packings in G.

Several classes of facets of $I(G)$ are known. For instance, an inequality of the form

$$\sum_{j \in K} x_j \leqslant 1 \qquad (6)$$

is a facet of $I(G)$ if and only if $K \subseteq V$ is a clique, i.e., a maximal set of pairwise adjacent vertices of G. The class of graphs whose packing polytope $I(G)$ is completely described by this family of inequalities (and the conditions $x_j \geqslant 0, j \in V$) is called *perfect*. A graph is known to be perfect if and only if its complement is perfect. The properties of perfect graphs and their packing polyhedra have been intensely studied dur-

ing the 1960s and 1970s and have, among other things, served as a starting point for a theory of blocking and antiblocking polyhedra developed by Fulkerson [12].

More generally, many classes of facets of $I(G)$ are associated with certain induced subgraphs G' of G. When G' is induced by a clique, the corresponding inequality (6) is, as mentioned above, a facet of $I(G)$. Other induced subgraphs G' yield inequalities that are facets of $I(G')$ rather than $I(G)$, but can be used to obtain corresponding facets for $I(G)$ through a procedure called *lifting*.

Other combinatorial problems for which several classes of facets of the feasible set have been characterized, include the knapsack problem, the traveling salesman problem, etc.

For literature, see the books by Ford and Fulkerson [11] and Lawler [29] and the surveys by Balas and Padberg [5], Hoffman [20], Lovász [30], Klee [26], and Padberg [32].

COMPUTER IMPLEMENTATION

At present all commercially available integer programming codes are of the branch-and-bound type. While they can *sometimes* solve problems with hundreds of integer and thousands of continuous variables, they cannot be *guaranteed* to find optimal solutions in a reasonable amount of time to problems with more than 30–40 variables. On the other hand, they usually find feasible solutions of acceptable quality to much larger problems. These commercial codes, while quite sophisticated in their linear programming subroutines, do not incorporate any of the results obtained in integer programming during the last decade.

A considerable number of specialized branch-and-bound/implicit enumeration algorithms have been implemented by operations research groups in universities or industrial companies. They usually contain other features besides enumeration, such as cutting planes and/or Lagrangean relaxation. Some of these codes can solve general (unstructured) 0–1 programs with up to 80 to 100 integer variables, and structured problems with up to several hundred (assembly line balancing, multiple choice, facility location), a few thousand (sparse set covering or set partitioning, generalized assignment), or several thousand (knapsack, traveling salesman) 0–1 variables.

Cutting plane procedures for *general* pure and mixed integer programs are at present too erratic and slow to compete with enumerative methods. However, for a number of special structures (set covering, traveling salesman problem) where information available about the convex hull of feasible points has made it possible to generate strong inequalities at acceptable computational cost, cutting planes, either by themselves, or in combination with enumerative and/or Lagrangean techniques, have been highly successful.

At the current state of the art, while many real-world problems amenable to an integer programming formulation fit within the stated limits and are solvable in useful time, others substantially exceed those limits. Furthermore, some important and frequently occurring real-world problems, like job shop scheduling and others, lead to integer programming models that are almost always beyond the limits of what is currently solvable. Hence the great importance of approximation methods for such problems.

For literature on computer codes, see Land and Powell [28] and Spielberg [36].

References

[1] Arthanari, T. S. and Dodge, Y. (1981). *Mathematical Programming in Statistics*. Wiley, New York.

[2] Balas, E. (1965). An additive algorithm for solving linear programs with zero–one variables. *Operat. Res.*, **13**, 517–546.

[3] Balas, E. (1975). Bivalent programming by implicit enumeration. In *Encyclopedia of Computer Science and Technology*, Vol. 2, J. Belzer, A. G. Holzman, and A. Kent, eds. Marcel Dekker, New York, pp. 479–494.

[4] Balas, E. (1979). Disjunctive programming. *Ann. Discrete Math.*, **5**, 3–51.

[5] Balas, E. and Padberg, M. (1976). Set partitioning: a survey. *SIAM Rev.*, **18**, 710–760.

[6] Beale, E. M. L. (1979). Branch and bound methods for mathematical programming. *Ann. Discrete Math.*, **5**, 201–220.

[7] Benders, J. F. (1962). Partitioning procedures for solving mixed-variables programming problems. *Numer. Math.*, **4**, 238–252.

[8] Christofides, N., Mingozzi, A., Toth, P., and Sandi, C., eds. (1979). *Combinatorial Optimization*. Wiley, New York.

[9] Edmonds, J. (1965). Maximum matching and a polyhedron with 0–1 vertices. *J. Res. Natl. Bur. Stand.*, **69B**, 125–130.

[10] Fisher, M. L. (1981). The Lagrangean relaxation method for solving integer programming problems. *Manag. Sci.*, **27**, 1–18.

[11] Ford, L. R., Jr. and Fulkerson, D. R. (1962). *Flows in Networks*. Princeton University Press, Princeton, N.J.

[12] Fulkerson, D. R. (1971). Blocking and antiblocking pairs of polyhedra. *Math. Program.*, **1**, 168–194.

[13] Garfinkel, R. S. and Nemhauser, G. L. (1972). *Integer Programming*. Wiley, New York.

[14] Geoffrion, A. (1974). Lagrangean relaxation and its uses in integer programming. *Math. Program. Study 2*, 82–114.

[15] Glover, F. (1975). Polyhedral annexation in mixed integer programming. *Math. Program.*, **9**, 161–188.

[16] Gomory, R. (1958). Outline of an algorithm for integer solutions to linear programs. *Bull. Amer. Math. Soc.*, **64**, 275–278.

[17] Gomory, R. (1969). Some polyhedra related to combinatorial problems. *Linear Algebra Appl.*, **2**, 451–558.

[18] Hammer, P. L., Johnson, E. L., and Korte, B. H., eds. (1979). *Discrete Optimization, Part 1*. Annals of Discrete Mathematics 4. North-Holland, Amsterdam.

[19] Hammer, P. L., Johnson, E. L., and Korte, B. H., eds. (1979). *Discrete Optimization, Part 2*. Annals of Discrete Mathematics 5. North-Holland, Amsterdam.

[20] Hoffman, A. J. (1979). The role of unimodularity in applying linear inequalities to combinatorial theorems. *Ann. Discrete Math.*, **4**, 73–84.

[21] Hoffman, A. J. and Kruskal, J. B. (1958). Integral boundary points of convex polyhedra. In *Linear Inequalities and Related Systems*, H. W. Kuhn and A. W. Tucker, eds. Princeton University Press, Princeton, N.J., pp. 223–246.

[22] Jeroslow, R. G. (1977). Cutting plane theory: disjunctive methods. *Ann. Discrete Math.*, **1**, 293–330.

[23] Jeroslow, R. G. (1979). An introduction to the theory of cutting planes. *Ann. Discrete Math.*, **5**, 71–95.

[24] Johnson, E. L. (1979). On the group problem and a subadditive approach to integer programming. *Ann. Discrete Math.*, **5**, 97–112.

[25] Johnson, E. L. (1980). *Facets, Subadditivity and Duality for Group and Semi-Group Problems.* SIAM, Philadelphia.

[26] Klee, V. (1980). Combinatorial optimization: What is the state of the art? *Math. Operat. Res.*, **5**, 1–26.

[27] Land, A. H. and Doig, A. G. (1960). An automatic method for solving discrete programming problems. *Econometrica*, **28**, 497–520.

[28] Land, A. and Powell, S. (1970). Computer codes for problems of integer programming. *Ann. Discrete Math.*, **5**, 221–269.

[29] Lawler, E. L. (1976). *Combinatorial Optimization: Networks and Matroids.* Holt, Rinehart and Winston, New York.

[30] Lovász, L. (1979). Graph theory and integer programming. *Ann. Discrete Math.*, **4**, 141–159.

[31] Mulvey, J. M. and Crowder, H. P. (1979). Cluster analysis: an application of Lagrangean relaxation. *Manag. Sci.*, **25**, 329–340.

[32] Padberg, M. W. (1979). Covering, packing, and knapsack problems. *Ann. Discrete Math.*, **4**, 265–287.

[33] Padberg, M. W., ed. (1980). *Combinatorial Optimization.* Mathematical Programming Study 12. North-Holland, Amsterdam.

[34] Rao, M. R. (1971). Cluster analysis and mathematical programming. *J. Amer. Statist. Ass.*, **66**, 622–627.

[35] Shapiro, J. F. (1979). A survey of Lagrangean techniques for discrete optimization. *Ann. Discrete Math.*, **5**, 113–138.

[36] Spielberg, K. (1979). Enumerative methods in integer programming. *Ann. Discrete Math.*, **5**, 139–184.

[37] Weyl, H. (1935). Elementäre Theorie der konvexen Polyeder. *Commentarii Math. Helv.*, **7**, 290–306. (English transl.: *Contributions to the Theory of Games*, H. W. Kuhn and A. G. Tucker, eds. Princeton University Press, Princeton, N.J., 1950.)

(BRANCH-AND-BOUND METHOD
COMBINATORICS
DECISION THEORY
DYNAMIC PROGRAMMING
LINEAR PROGRAMMING
MATHEMATICAL PROGRAMMING
NONLINEAR PROGRAMMING
OPTIMIZATION)

EGON BALAS

INTEGRAL EQUATIONS

An integral equation is an equation of the form

$$h(t)f(t) = g(t) + \lambda \int_a^b K(s,t)f(s)\,ds$$
$$(a \leqslant t \leqslant b), \quad (1)$$

where f is the unknown function, while the other functions g, h, K are known, and λ is a nonzero parameter, real or complex. The function K is called the *kernel* of the equation. Regularity conditions such as continuity and integrability are imposed on all these functions. If $g(t) = 0$, then (1) is a homogeneous integral equation. The kernel K is *separable* if

$$K(s,t) = \sum_{i=1}^n a_i(t)b_i(s), \quad (2)$$

where the functions a_i are linearly independent. If $K(s,t) = K_1(t-s)$, where K_1 is a given function of one variable, then (1) is called an integral equation of the *convolution type*.

If $h(t) = 0$, then (1) is called a *Fredholm integral equation* of the first kind, while if $h(t) = 1$, it is a Fredholm equation of the second kind. If the upper limit in (1) is t instead of the constant b, we have a *Volterra integral equation*, of the first or second kind according as $h(t) = 0$ or $h(t) = 1$.

Equation (1) is actually a *linear* equation; a nonlinear equation is of the type

$$f(t) = \int_a^b K(s,t)\big[f(s)\big]^2 ds \qquad (a \leqslant t \leqslant b).$$
$$(3)$$

A *singular* equation is one in which one or both the limits of integration in (1) become infinite, or when the kernel becomes infinite at one or more points within the range of integration.

Integral equations occur in the theory and applications of stochastic processes*. Here the functions f, g, h, K are probability functions [typically cumulative distribution functions* (CDF) or their densities] or else random functions (stochastic processes). In the latter case we view (1) as a *stochastic integral equation*, which holds with probability 1 for sample functions of the unknown stochastic process described in terms of known functions g, h, K. In the same context, together with (1) differential and integrodifferential equations also occur.

The wide variety of integral equations and the full scope of the techniques used in solving them become evident especially in the area of probability models. We demonstrate this by a brief description of the models for population growth, industrial replacement, dams (inventories, storage), and queues.

THE INTEGRAL EQUATION OF RENEWAL THEORY*

Renewal theory had its origin in population analysis (including problems in genetics* and actuarial* problems), industrial replacement, and in the general theory of "self-renewing aggregates." The central topic of discussion in these diverse areas is the integral equation

$$u(t) = g(t) + \int_0^t u(t-s)f(s)\,ds \qquad (t > 0),$$

(4)

where f and g are given nonnegative functions. The main concern is the existence and uniqueness of the solution $u(t)$ of (4), methods for computing $u(t)$ and its behavior as $t \to \infty$. Clearly, (4) is a Volterra equation of the second kind.

The following is a brief description of the problem arising in mathematical biology, as formulated by A. J. Lotka and F. R. Sharpe in 1911. Consider a closed population in which there is no possibility of emigration or immigration. We refer to the female component of the population, and denote by $u(t)$ the rate of (female) births at time t. Assume that the fertility and mortality rates* are constant (time independent). Let $p(x)$ be the proportion of newborn females surviving to age x, and $m(x)$ the reproduction rate of females at age x. Then the average number

of females born during a time interval $(t, t + dt)$ to a female of age t is $f(t)\,dt + o(dt)$, where $f(t) = p(t)m(t)$. We therefore have the equation

$$u(t) = \int_0^\infty u(t-s)f(s)\,ds,$$

which can be written in the form (4) with

$$g(t) = \int_t^\infty u(t-s)f(s)\,ds. \qquad (5)$$

Here $g(t)$ is the rate of birth of females at time t whose mothers were born before time 0, and is assumed to be known. The future growth of the population is described by the birth function $u(t)$ and is determined by the integral equation (4). It should be noted that the integral

$$\int_0^\infty f(s)\,ds \qquad (6)$$

gives the average number of daughters born to a female during her lifetime; for a "healthy" population the integral (6) exceeds unity. (*See also* BIRTH-AND-DEATH PROCESSES.)

In the problem of industrial replacement, items are originally installed at some point of time, and are maintained at a constant level by replacing each item by a new one as soon as it fails. The rate $u(t)$ of replacement at time t is seen to satisfy an equation of the type (4). Here the reproduction (replacement) rate $m(x)$ equals the death (failure) rate, namely, $-p'(x)/p(x)$, so that $f(x) = -p'(x)$. Thus $f(x)$ is the lifetime density of the items and the integral (6) is $\leqslant 1$, strict inequality indicating the possibility of an item living forever. If the lifetimes do not have a density, we replace $f(s)\,ds$ in (4) and (6) by $dF(s)$, where F is the CDF of the lifetimes.

Equation (4) is a special case of the *integral equation of renewal theory*

$$Z(t) = z(t) + \int_0^t Z(t-s)\,dF(s) \qquad (t > 0),$$

(7)

where z is a bounded function vanishing in $(-\infty, 0)$. Its unique solution Z which vanishes in $(-\infty, 0)$ and is bounded over finite intervals is given by

$$Z(t) = \int_0^t z(t-s)\,dU(s), \qquad (8)$$

where $U(t) = \sum_0^\infty F_n(t)$, F_n $(n \geqslant 1)$ being the n-fold convolution of F with itself and $F_0(t) = 0$ for $t < 0$ and $= 1$ for $t \geqslant 0$. Here U is called the renewal function; it satisfies (7) with $z(t) = 1$ for $t \geqslant 0$. Renewal theorems are concerned with the behavior of $U(t)$ and $Z(t)$ as $t \to \infty$. In particular, the elementary renewal theorem states that as $t \to \infty$, $t^{-1}U(t) \to \mu^{-1}$, where μ is the mean lifetime, and the limit is interpreted as zero if $\mu = \infty$. If the lifetimes have a density, then the derivative $U'(t) = u(t)$ exists almost everywhere, and $u(t) \to \mu^{-1}$ as $t \to \infty$.

In (7) the lifetime CDF F is concentrated on $[0, \infty)$; a more general equation with F defined over $(-\infty, \infty)$ occurs in the theory of random walks*.

INTEGRAL EQUATIONS IN DAM MODELS*

We consider the model for a dam of finite capacity c. Let X_{n+1} be the amount of water that has flowed into the dam (the input) during the time interval $(n, n+1]$ $(n \geqslant 0)$. Because of the finite capacity of the dam there is an overflow, and the actual input after the overflow equals $\eta_{n+1} = \min(X_{n+1}, c - Z_n)$, where Z_n is the storage level at time n. Demands for water occur at times $n = 1$, $2, \ldots,$ the amount demanded (the output) at time n being ξ_n. We assume that $\{X_n\}$ and $\{\xi_n\}$ are independent sequences of mutually independent and identically distributed random variables with CDFs $B(x)$ and $A(x)$, respectively. At time $n+1$, an amount of water equal to $f(Z_n + \eta_{n+1}, \xi_{n+1})$ is released from the dam, where the function f is prescribed by the storage policy. After this release, the storage level at time $n+1$ will be $Z_{n+1} = Z_n + \eta_{n+1} - f(Z_n + \eta_{n+1}, \xi_{n+1})$. We shall consider the case where $f(x, y) = \min(x, y)$. We then have

$$Z_{n+1} = \max(0, Z_n + \eta_{n+1} - \xi_{n+1})$$
$$(n \geqslant 0). \quad (9)$$

It is clear from (9) that the sequence of random variables $\{Z_n, n \geqslant 0\}$ forms a time-homogeneous Markov chain* with the state space $[0, c]$. A few easy calculations show that its one-step transition CDF is given by

$$
\begin{aligned}
P(z; x) &= P\{Z_{n+1} \leqslant x \mid Z_n = z\} \\
&= Q(z; x) \\
&\quad + [1 - A(c - x -)][1 - B(c - z)],
\end{aligned}
$$

where

$$Q(z; x) = \int_0^{c-z} dB(v)\,[1 - A(v + z - x -)],$$

the latter being the probability of a transition from the state z to the interval $[0, x]$ with no overflow. Note that $P(z; x) = 1$ for $x \geqslant c$. The stationary CDF of Z_n satisfies the integral equation

$$F(x) = \int_{0-}^c dF(z)\,P(z; x) \qquad (0 \leqslant x \leqslant c) \tag{10}$$

with $F(x) = 0$ for $x < 0$ and $F(x) = 1$ for $x \geqslant c$.

Let us also consider the random variable T, which is the time that the dam with an initial content z takes either to dry up or overflow; thus

$$T = \min\{n \geqslant 1 : Z_{n-1} + X_n > c \text{ or } $$
$$Z_{n-1} + X_n - \xi_n \leqslant 0\} \tag{11}$$

with $Z_0 = z$. The probability that the dam will eventually dry up before overflowing is given by

$$V(z) = P\{T < \infty, Z_{T-1} + X_T - \xi_T \leqslant 0 \mid Z_0 = z\}$$

for $0 < z \leqslant c$. It is easily seen that the function V satisfies the integral equation

$$V(z) = Q(z; 0) + \int_{0+}^c Q(z; dx)\,V(x)$$
$$(0 < z \leqslant c) \tag{12}$$

with $V(0) = 1$ and $V(z) = 0$ for $z > c$.

The integral equations (11) and (12) are of the Fredholm type. In each case the existence and uniqueness of the solution of the type we seek can be established (for example) by functional-analytic techniques. Solutions can be obtained in the case where $\xi_{n+1} = m$ $(< c)$ and X_{n+1} has the gamma density $b(x) = e^{-\mu x}\mu^p x^{p-1}/(p-1)!$ ($\mu > 0$ and p is a positive integer). Here it is seen that $Q(z; dx) = P(z; dx) = b(x - z + m)dx$ for $z - m < x < c - m$, where

$$b(x - z + m) = \sum_{r=0}^{p-1} e^{-\mu(x+m)}\mu^r \frac{(x + m)^r}{r!}$$

$$\times e^{\mu z} \frac{(-\mu z)^{p-1-r}}{(p-1-r)!}. \quad (13)$$

Therefore, the kernels in (10) and (12) are separable. This fact can be used to solve these equations, although the standard procedure is not applicable, since (13) holds only for $x > z - m$.

WIENER–HOPF TECHNIQUE

If $c = \infty$ in the dam model described above, we obtain the equation

$$Z_{n+1} = \max(0, Z_n + X_{n+1} - \xi_{n+1}). \quad (14)$$

The transition CDF becomes $P(z; x) = Q(z; x) = K(x - z)$, where K is the CDF of the random variables $X_{n+1} - \xi_{n+1}$. The integral equation for the stationary CDF of Z_n in this case is given by

$$F(x) = \int_{0-}^{\infty} dF(z) K(x - z) \quad (x \geqslant 0)$$

$$(15)$$

with $F(x) = 0$ for $x < 0$. This integral equation arises in the theory of single-server queueing systems where it is assumed that the successive customers' interarrival times $\{\xi_n\}$ and their service times $\{X_n\}$ are independent sequences of mutually independent and identically distributed random variables. The nth customer's waiting time is then given by Z_n ($n \geqslant 0$). We seek a solution of (15) which is right continuous, monotone nondecreasing, and such that $F(x) = 0$ for $x < 0$ and $F(x) \to 1$ as $x \to \infty$. If $\alpha = E(X_{n+1} - \xi_{n+1})$ is finite, then such a solution exists uniquely if and only if $\alpha < 0$. Let us suppose that this is the case.

Equation (15) is of the convolution type and the Wiener–Hopf technique for solving

it consists of introducing a second function $G(x)$ by setting

$$G(x) = \int_{0-}^{\infty} dF(y) K(x - y) \quad (x \leqslant 0),$$

$$(16)$$

$$G(x) = G(0) \quad (x > 0)$$

and solving for the two unknown functions F and G from (15) and (16). To do this, let $K^*(\omega) = \int_{-\infty}^{\infty} e^{i\omega x} dK(x)$ and similarly for $F^*(\omega)$ and $G^*(\omega)$, with $i = \sqrt{-1}$ and ω real. We note that (15) and (16) define the complete convolution of F and K except for an overlap $F(0)$. Therefore, $F^*(\omega) + G^*(\omega) = F^*(\omega)K^*(\omega) + F(0)$ or

$$F^*(\omega)\left[1 - K^*(\omega)\right] = F(0) - G^*(\omega)$$

$$(\omega \text{ real}). \quad (17)$$

Now from the theory of random walks it is known that there exists a unique factorization

$$1 - K^*(\omega) = D(\omega)\overline{D}(\omega) \quad (\omega \text{ real}), \quad (18)$$

where D and $\overline{D}$ are bounded analytic functions for $\text{Im}(\omega) \geqslant 0$ and $\text{Im}(\omega) \leqslant 0$, respectively, bounded away from zero and such that $D(\omega) \to 1$ as $\text{Im}(\omega) \to \infty$. It follows from (17) and (18) that

$$F^*(\omega) = F(0)D(\omega)^{-1},$$

$$G^*(\omega) = F(0)\left[1 - \overline{D}(\omega)\right]. \quad (19)$$

The functions $D, \overline{D}$ are explicitly known, but have cumbersome expressions. In special cases they can be calculated directly from (18); the analytical technique involved in doing this is fairly deep, but the effort is not without its reward, because it turns out that

$$\lim_{n \to \infty} P\{I_{n+1} < x\} = 1 - G(-x)$$

$$(x > 0) \quad (20)$$

for the random variable

$$I_{n+1} = -\min(0, Z_n + X_{n+1} - \xi_{n+1}). \quad (21)$$

In queueing theory, I_{n+1} is the idle time (if any) that precedes the $(n + 1)$th arrival. This fact provides a rationale for the Wiener–

Hopf technique at least within the context of probability theory.

INTEGRODIFFERENTIAL EQUATIONS

In the continuous-time dam model analogous to the one described earlier, the storage function $Z(t)$ is described by the function

$$Z(t) = Z(0) + X(t) - \int_0^t r[Z(s)]\,ds$$

$$(t > 0), \quad (22)$$

where $X(t)$ is the input during a time interval $(0,t]$, and the release at time t is at a rate $r[Z(t)]$. Here $X(t)$ is a process with stationary independent increments (Lévy process), while $r(x)$ is a continuous nondecreasing function for $x > 0$ and $r(0) = 0$. We shall consider the special case where $r(x) = 1$ for $x > 0$. The process described by (22) is a Markov process* with the state space $[0, \infty)$. It is of the mixed type in which transitions occur continuously and in jumps. Its transition CDF $F(x_0; x, t) = P\{Z(t) \leqslant x \mid Z(0) = x_0\}$ satisfies an integrodifferential equation which is typical of such processes. To see this, suppose, in particular, that $X(t)$ is a compound Poisson process* in which jumps occur at a rate λ and the jump size has CDF $B(x)$. Considering $Z(t)$ over the consecutive intervals $(0,t]$ and $(t, t + dt]$, we obtain the relation

$F(x, t + dt)$

$$= F(x + dt, t)(1 - \lambda\,dt)$$

$$+ \lambda\,dt \int_0^{x + dt} F(x + dt - v, t)\,dB(v)$$

$$+ o(dt),$$

where we have written $F(x, t) \equiv F(x_0; x, t)$ for convenience, for a fixed $x_0 \geqslant 0$. In the limit as $dt \to 0$ this yields the equation

$$\frac{\partial F}{\partial t} = \frac{\partial F}{\partial x} - \lambda F(x, t)$$

$$+ \lambda \int_0^x F(x - v, t)\,dB(v), \quad (23)$$

as expected. This can be solved by the usual transform techniques.

STOCHASTIC INTEGRAL EQUATIONS

In the case of a general input it is more difficult to derive the equation satisfied by $F(x, t)$ because the number of jumps in any finite interval is not finite. However, it is more elegant to consider the relation (22) as a stochastic integral equation for sample functions of the Z process. In the case $r(x) = 1$ for $x > 0$ it is known that it has the unique solution

$$Z(t) = \max\left\{ \sup_{0 \leqslant s \leqslant t} [Y(t) - Y(s-)], Z(0) + Y(t) \right\},$$

$$(24)$$

where $Y(t) = X(t) - t$, so that $Y(t)$ is the net input during $(0, t]$.

LITERATURE

The theory and techniques of linear integral equations are treated by Kanwal [5], who also lists the standard references on the subject. The papers by Feller [2] and Lotka [9, 10] are typical of the historical literature on the integral equation (4) within the context of population analysis and industrial replacement. The integral equation (7) and its generalization to F on $(-\infty, \infty)$ play a central role in Feller's [3, Chaps. VI, XI] treatment of renewal theory and random walks. Equation (10), in the special case where (13) holds, was solved by Prabhu [11]. A discussion of the integral equation (15) and its solutions in special cases will be found in Prabhu [12, pp. 49–59]; details of the Wiener–Hopf technique are given by Feller [3, Chap. XII]. In the more modern treatment the Wiener–Hopf technique is applied directly to investigate (14) and (21); see Prabhu [13, Chap. 1]. Stochastic integral equations such as (22) are investigated by Prabhu [13, Chaps. 3, 4].

FURTHER REMARKS

Integral equations are also encountered in several other contexts in the area of probability and mathematical statistics. Anosov [1], Jadrenko [4], and Krasnitskii [6] treat integral equations that occur in the theory of random fields*. Moments of certain probability distributions can sometimes be obtained by solving integral equations, as shown, for example, by Siegel [14] and Weiner [15]. A stochastic integral equation involving point processes arises in theory of telecommunication traffic—see Le Gall [7, 8].

References

[1] Anosov, D. V. (1978). *Select. Transl. Math. Statist. Prob.*, **14**, 1–6.

[2] Feller, W. (1941). *Ann. Math. Statist.*, **12**, 243–267.

[3] Feller, W. (1971). *An Introduction to Probability Theory and Its Applications*, Vol. 2, 2nd ed. Wiley, New York.

[4] Jadrenko, M. I. (1976). *Theory Prob. Math. Statist.*, **12**, 169–179.

[5] Kanwal, R. P. (1971). *Linear Integral Equations*. Academic Press, New York.

[6] Krasnitskii, S. M. (1973). *Theory Prob. Math. Statist.*, **9**, 115–126.

[7] Le Gall, P. (1974). *Stoch. Processes Appl.*, **2**, 261–280.

[8] Le Gall, P. (1978). *Stoch. Processes Appl.*, **6**, 337–338.

[9] Lotka, A. J. (1939). *Ann. Math. Statist.*, **10**, 1–25.

[10] Lotka, A. J. (1939). *Ann. Math. Statist.*, **10**, 144–161.

[11] Prabhu, N. U. (1958). *Quart. J. Math. (Oxford, 2)*, **9**, 183–188.

[12] Prabhu, N. U. (1965). *Queues and Inventories: A Study of Their Basic Stochastic Processes*. Wiley, New York.

[13] Prabhu, N. U. (1980). *Stochastic Storage Processes*. Springer-Verlag, New York.

[14] Siegel, A. F. (1978). *J. Appl. Prob.*, **15**, 774–789.

[15] Weiner, H. J. (1978). *J. Appl. Prob.*, **15**, 803–814.

(DAM THEORY
DIFFERENCE EQUATIONS
MARKOV PROCESSES
QUEUEING THEORY
RENEWAL THEORY
STOCHASTIC PROCESSES)

N. U. Prabhu

INTEGRAL TRANSFORMS

One of the most powerful tools for deriving the distribution (i.e., the PDF or CDF) of sums, differences, products, quotients, and algebraic functions of continuous random variables (rvs) is the integral transform. Probably the best known of these transforms among statisticians is the characteristic function, which has long been used to derive the distribution of sums and differences of continuous rvs, and which is in fact a *Fourier integral transform*. Its theoretical aspects are discussed in CHARACTERISTIC FUNCTIONS. It is applicable to rvs defined on either a finite, singly infinite, or doubly infinite range. (For a detailed treatment of characteristic functions, see ref. 4.) However, for rvs restricted to either a finite or singly infinite range, the *Laplace integral transform* is usually more easily evaluated than the Fourier, particularly when the transform has poles. When the rv can range over both positive and negative values, the *bilateral Laplace transform* (sometimes referred to as the complex Fourier transform) can be used to advantage over the Fourier transform when the transform has poles.

Equally important—but not as well known—is the fact that the *Mellin integral transform* is a natural tool for deriving the distribution of products, quotients, and rational powers of continuous rvs. As is the case with the Laplace integral transform, the Mellin integral transform applies per se only to rvs that do not take on negative values. One can, however, define a modified Mellin integral transform that is applicable to rvs which can assume both negative and positive values [8].

The aforementioned integral transforms (hereafter referred to simply as transforms) will now be considered separately and briefly. For a more detailed discussion, the reader is referred to the book by Springer [8].

It should perhaps be mentioned at this point that there are additional transforms such as the Z, zeta, Walsh–Hadamard, and

Hankel [8, pp. 31–33]. They apply either to discrete rvs—and hence are not integral transforms—or to continuous rvs which have little—if any—application in theoretical or applied statistics.

The importance of integral transforms stems largely from the fact that if the integral transform of an unknown PDF is known, the function $f(x)$ can be obtained by evaluating the corresponding inversion integral. Actually, there are frequent instances when such is the case; that is, when one knows the integral transform of a PDF $f(x)$ before he or she knows what $f(x)$ is. This situation exists, for example, when one wishes to derive the PDF of a sum, difference, mean, product, rational power, or quotient of specific independent random variables (i.r.v.'s), in which case the transform of the desired PDF is expressible in terms of the known transforms of the PDF's of the specific rvs involved.

THE FOURIER TRANSFORM

If $f(x)$ is a function of a real variable x, its Fourier transform $F_t(f(x))$ is defined as the expected value of e^{itx}: namely,

$$F_t(f(x)) = \int_{-\infty}^{\infty} e^{itx} f(x)\, dx, \qquad (1)$$

where $i = \sqrt{-1}$ and t is a real variable. More specifically, if $f(x)$ is defined and single-valued almost everywhere on the range $-\infty < x < \infty$, and is such that the integral

$$\int_{-\infty}^{\infty} |f(x)| e^{ikx}\, dx$$

converges for some real value k, then $F_t(f(x))$ is the Fourier transform of $f(x)$. It is usually referred to by statisticians as the characteristic function of $f(x)$, and e^{itx} is called the kernel [4]. Conversely, if the Fourier transform is absolutely integrable over the real line $-\infty < t < \infty$, or is analytic (i.e., its derivative exists) in some horizontal strip $-\alpha < it < \beta$ of the complex plane, then $f(x)$ is uniquely determined by the inversion integral (often referred to as the in-

verse Fourier transform)

$$f(x) = \frac{1}{2\pi} \int_{-\infty}^{\infty} e^{-itx} F_t(f(x))\, dt. \qquad (2)$$

The Fourier transform (1) and the inverse Fourier transform (2) constitute a transform pair.

It bears stating that if $f(x)$ is a PDF, its characteristic function always exists and determines the CDF $F_X(x)$ uniquely. In particular,

$$F_X(x) = F(0) + \frac{1}{2\pi} \int_{-\infty}^{\infty} \left(\frac{1 - e^{-itx}}{it} \right) F_t(f(x))\, dt$$

$$(3a)$$

$$F_x(x) = \frac{1}{2} + \frac{1}{2\pi} \int_0^{\infty} \frac{h(t; x)}{it}\, dt \qquad (3b)$$

with $h(t; x) = e^{itx} F_{-t}(f(x)) - e^{-itx} F_t(f(x))$, the form (3a) being particularly convenient for use with nonnegative rvs. The importance of (3a) and (3b) lies in the fact that they enable one to determine the CDF without first determining the PDF.

The following example illustrates the use of the Fourier transform or characteristic function in obtaining the PDF of a sum of i.r.v.'s.

Example 1. Find the PDF $g(w)$ of the sum $W = \sum_{j=1}^{n} X_j$ of n identical normal i.r.v.'s X_j, each having mean 0, variance σ^2, and PDF $f(x) = (\sqrt{2\pi}\,\sigma)^{-1} \exp[-x_j^2/(2\sigma^2)]$.

Since the rvs X_j are independent, we use the well-known fact [8] that the Fourier transform of $g(w)$ is the product of the Fourier transforms of the functions $f(x_j)$; i.e.,

$$F_t(g(w)) = \prod_{j=1}^{n} F_t(f(x_j)). \qquad (4)$$

It follows from (2) that

$$g(w) = \frac{1}{2\pi} \int_{-\infty}^{\infty} e^{-itw} \prod_{j=1}^{n} F_t(f(x_j))\, dt$$

$$= \frac{1}{2\pi} \int_{-\infty}^{\infty} e^{-itw} \left(e^{-t^2\sigma^2/2} \right)^n dt$$

$$= \frac{1}{\sqrt{2\pi}\,\sqrt{n}\,\sigma} e^{-w^2/(2n\sigma^2)}, \qquad -\infty < w < \infty.$$

Thus, as is well known, w is a normal rv with mean 0 and variance $n\sigma^2$.

In the example above, it was not necessary to restrict the means to 0 and to impose the constraint that the variances be identical. These constraints were imposed for the sake of simplicity and brevity. However, if the independence assumption is removed, the problem becomes more difficult. The reason for this is that the joint Fourier transform of $f(x_1, x_2, \ldots, x_n)$ is no longer expressible as the product (4) of the transforms of the normal univariate functions $f(x_j)$, as is illustrated by the following example.

Example 2. Derive the PDF $g(w)$ of the sum $W = X_1 + X_2$ of two bivariate normal rvs with means 0, variances σ_1^2 and σ_2^2, and correlation coefficient ρ.

The joint distribution of X_1 and X_2 is

$f(x_1, x_2)$

$$= \frac{1}{2\pi\sigma_1\sigma_2(1-\rho^2)^{1/2}}$$

$$\times \exp\left[-\frac{1}{2(1-\rho^2)}\left(\frac{x_1^2}{\sigma_1^2} - \frac{2\rho x_1 x_2}{\sigma_1\sigma_2} + \frac{x_2^2}{\sigma_2^2} \right) \right],$$

$$|\rho| < 1, \quad -\infty < x_j < \infty, \quad j = 1, 2.$$

The bivariate Fourier transform or characteristic function, denoted by $F(t_1, t_2)$, is now given by [8]

$$F(t_1, t_2) = \int_{-\infty}^{\infty} \int_{-\infty}^{\infty} \exp[i(x_1 t_1 + x_2 t_2)]$$

$$\times f(x_1, x_2) dx_1 dx_2$$

$$= \exp\left[-\frac{1}{2}(t_1^2\sigma_1^2 + 2\rho\sigma_1\sigma_2 t_1 t_2 + t_2^2\sigma_2^2) \right].$$

To derive $g(w)$, we set $t_1 = t_2 = t$ and evaluate the inversion integral

$$g(w) = \frac{1}{2\pi} \int_{-\infty}^{\infty} e^{-itw} F(t, t) dt$$

$$= \frac{\exp\left[-w^2 / \{2(\sigma_1^2 + 2\rho\sigma_1\sigma_2 + \sigma_2^2)\} \right]}{\sqrt{2\pi}\,(\sigma_1^2 + 2\rho\sigma_1\sigma_2 + \sigma_2^2)^{1/2}}.$$

For a more detailed discussion of sums of dependent rvs, the reader is referred to ref. [8].

THE LAPLACE TRANSFORM

If $f(x)$ is a function of a real variable x and is defined and single-valued almost everywhere for $x \geqslant 0$, and is such that

$$\int_0^{\infty} |f(x)| e^{-kx}$$

converges for some real value k, then $f(x)$ is said to be Laplace transformable, and

$$L_r(f(x)) = \int_0^{\infty} e^{-rx} f(x) dx \qquad (5)$$

is the Laplace transform of $f(x)$, where $r = x + iy$ is a complex variable. Conversely, if the Laplace transform $L_r(f(x))$ is analytic and of the order $O(r^{-k})$ in some half-plane $\operatorname{Re}(r) \geqslant c$, c, k real, $k > 1$, then

$$f(x) = \frac{1}{2\pi i} \int_{c-i\infty}^{c+i\infty} e^{rx} L_r(f(x)) dr. \qquad (6)$$

If $L_r(f(x))$ has poles in the left half plane (LHP) $\operatorname{Re}(r) < c$, and also satisfies the conditions of Jordan's lemma [8], which is almost always the case, (6) may be evaluated by means of the residue theorem [8, pp. 23–26]. Since Jordan's lemma is useful in evaluating inversion integrals involving poles, it will now be stated.

Jordan's Lemma:[1]

(a) *If $f(s) \to 0$ uniformly with respect to $\arg s$ as $|s| \to \infty$ when $\pi/2 \leqslant \arg s \leqslant 3\pi/2$, and if $f(s)$ is analytic when $|s| \to k$ and $\pi/2 \leqslant \arg s \leqslant 3\pi/2$, then*

$$\lim_{a \to \infty} \int_{QKLMP} e^{ms} f(s) ds = 0 \qquad (\text{Fig. } 1a),$$

where k and m are real positive constants.

(b) *If $f(s) \to 0$ uniformly with respect to $\arg s$ as $|s| \to \infty$ when $-\pi/2 \leqslant \arg s \leqslant \pi/2$, and if $f(s)$ is analytic when $|s| \to k$ and $-\pi/2 \leqslant \arg s \leqslant \pi/2$, then*

$$\lim_{a \to \infty} \int_{PTQ} e^{-ms} f(s) ds = 0 \qquad (\text{Fig. } 1b),$$

where k and m are positive real constants.

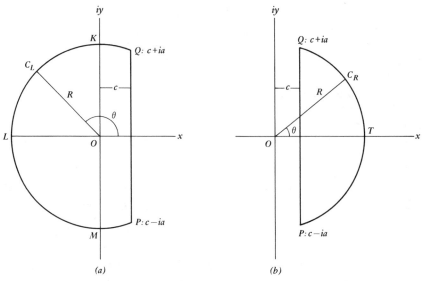

Figure 1 Bromwich contours used in evaluating integrals over the Bromwich path $(c - i\infty, c + i\infty)$.

In our application of this lemma to inversion integrals, $f(s)$ denotes the transform involved, s represents a complex number, and $e^{\pm ms}$ is the kernel. For Laplace inversion integrals, s is replaced with r.

Example 3. Find the PDF $g(w)$ of the sum $W = \sum_{j=1}^{n} X_j$ of n exponential i.r.v.'s each having PDF $f(x_j) = e^{-x_j}$, $x_j \geqslant 0$.

Since the RVs are independent, we have

$$g(w) = \frac{1}{2\pi i} \int_{c-i\infty}^{c+i\infty} e^{rw} \prod_{j=1}^{n} L_r(f(x_j)) \, dr \quad (7a)$$

$$= \frac{1}{2\pi i} \int_{c-i\infty}^{c+i\infty} e^{rw} (1+r)^{-n} \, dr. \quad (7b)$$

Note that the path of integration (called the Bromwich path) can be any vertical line in the complex plane with endpoints $(c, -i\infty)$, $(c, i\infty)$, $c > -1$, since the transform is analytic (and hence has no poles or singularities) in the right half plane (RHP) $\mathrm{Re}(r) > -1$. Also, the conditions of Jordan's lemma are satisfied, since $\pi/2 < \arg r \leqslant 3\pi/2$ and

$$|L_r(g(w))| = |1 + r|^{-n} \to 0 \quad (8)$$

uniformly with respect to $\arg r$ as $R \to \infty$. Hence $g(w) = \sum_j R_j$, where R_j denotes the residue at the jth pole. Since the transform

has only one pole, $g(w) = R_1$, where R_1 denotes the residue at the nth order pole located at $r = -1$. We obtain R_1 by (1) multiplying the integrand in (7b) by $(1 + r)^n$, thereby removing the pole; (2) evaluating the $(n - 1)$st-order derivative of the result at $r = -1$; and (3) dividing the result by $(n - 1)!$. This yields

$$g(w) = \frac{1}{(n-1)!} w^{n-1} e^{-w}, \quad 0 \leqslant w < \infty.$$

THE BILATERAL LAPLACE (COMPLEX FOURIER) TRANSFORM

The bilateral Laplace transform [also called the complex Fourier transform and denoted by $\mathscr{F}_r(f(x))$] can be used to derive the distribution of sums and differences of rvs which may take on both positive and negative values. It has the desirable feature of being expressible in terms of the (unilateral) Laplace transform (5). Specifically, if $f(x)$ is a PDF defined over the range $-\infty < x < \infty$, and we use the notation

$$f(x) = \begin{cases} f^-(x), & -\infty < x < 0 \quad (9a) \\ f^+(x), & 0 \leqslant x < \infty, \quad (9b) \end{cases}$$

then

$$\mathscr{F}_r(f^-(x)) = \int_{-\infty}^0 e^{-rx} f^-(x)\,dx$$

$$= L_{-r}(f^+(x)), \qquad -\infty < x < 0$$

(10a)

$$\mathscr{F}_r(f^+(x)) = \int_0^\infty e^{-rx} f^+(x)\,dx$$

$$= L_r(f^+(x)), \qquad 0 \leqslant x < \infty.$$

(10b)

The corresponding inversion integrals which complete the transform pair are

$$f^-(x) = \frac{1}{2\pi i} \int_{c-i\infty}^{c+i\infty} e^{rx} L_{-r}(f^+(x))\,dr,$$

$$-\infty < x < 0 \quad (11a)$$

$$f^+(x) = \frac{1}{2\pi i} \int_{c-i\infty}^{c+i\infty} e^{rx} L_r(f^+(x))\,dr,$$

$$0 \leqslant x < \infty, \quad (11b)$$

where $(c - i\infty, c + i\infty)$ is a vertical line in the complex plane parallel to the imaginary axis and located in a strip in which both $L_{-r}(f^+(x))$ and $L_r(f^+(x))$ are analytic. The following example is illustrative.

Example 4. Find the PDF $g(w)$ of the sum $W = X_1 + X_2$ of two identical i.r.v.'s having the PDF

$$f(x_i) = \begin{cases} f^-(x_i) = \frac{1}{2} e^{x_i}, \\ \qquad -\infty < x_i < 0, \quad i = 1,2 \\ f^+(x_i) = \frac{1}{2} e^{-x_i}, \\ \qquad 0 < x_i < \infty, \quad i = 1,2. \end{cases}$$

From (10a) and (10b), we have, respectively,

$$\mathscr{F}_r(f^-(x_i)) = \frac{1}{2(1 - r)},$$

$$i = 1,2; \quad -\infty < x_i < 0 \quad (12a)$$

$$\mathscr{F}_r(f^+(x_i)) = \frac{1}{2(1 + r)},$$

$$i = 1,2; \quad 0 \leqslant x < \infty, \quad (12b)$$

so that

$$\mathscr{F}_r(g^-(w)) = \mathscr{F}_r(f^-(x_1))\mathscr{F}_r(f^-(x_2))$$

$$+ \mathscr{F}_r(f^-(x_1))\mathscr{F}_r(f^+(x_2))$$

$$+ \mathscr{F}_r(f^+(x_1))\mathscr{F}_r(f^-(x_2))$$

$$= \frac{1}{4(1 - r)^2} + \frac{1}{2}\left(\frac{1}{1 - r}\right)\frac{1}{1 + r}$$

(13a)

and

$$\mathscr{F}_r(g^+(w)) = \mathscr{F}_r(f^+(x_1))\mathscr{F}_r(f^+(x_2))$$

$$+ \mathscr{F}_r(f^-(x_1))\mathscr{F}_r(f^+(x_2))$$

$$+ \mathscr{F}_r(f^+(x_1))\mathscr{F}_r(f^-(x_2))$$

$$= \frac{1}{4(1 + r)^2} + \frac{1}{2}\left(\frac{1}{1 - r}\right)\frac{1}{1 + r}$$

(13b)

Since the transforms (13a) and (13b) satisfy the conditions of Jordan's lemma, we can use the method of residues, obtaining

$$g^-(w) = \frac{1}{2\pi i} \int_{c-i\infty}^{c+i\infty} e^{rw}\mathscr{F}_r(g^-(w))\,dr,$$

$$-1 < c < 1$$

$$= -\frac{d}{dr}\left[\frac{1}{4} e^{rw}\right]\bigg|_{r=1} + \frac{e^{rw}}{2(1 + r)}\bigg|_{r=1}$$

$$= \frac{e^w}{4}(1 - w), \qquad -\infty < w < 0$$

$$g^+(w) = \frac{1}{2\pi i} \int_{c-i\infty}^{c+i\infty} e^{rw}\mathscr{F}_r(g^+(w))\,dr,$$

$$-1 < c < 1$$

$$= \frac{e^{-w}}{4}(1 + w), \qquad 0 \leqslant w < \infty.$$

THE MELLIN TRANSFORM

Just as the Fourier and Laplace transforms are defined as expected values of e^{itx} and e^{-rx}, respectively, so the Mellin transform $M_s(f(x))$ is defined as the expected value of x^{s-1}, where $s = x + iy$. That is,

$$M_s(f(x)) = \int_0^\infty x^{s-1} f(x)\,dx. \quad (14)$$

More specifically, if $f(x)$ is a real function that is defined and single-valued almost everywhere for $x \geq 0$ and is absolutely integrable over the range $(0, \infty)$, the Mellin transform (14) exists. Conversely, if the Mellin transform exists and is an analytic function of the complex variable s for $c_1 \leq \mathrm{Re}(s) \leq c_2$, where c_1 and c_2 are real, then the inversion integral

$$\frac{1}{2\pi i} \lim_{\beta \to \infty} \int_{w-i\beta}^{w+i\beta} x^{-s} M_s(f(x))\, ds \quad (15)$$

evaluated along any line $c_1 \leq \mathrm{Re}(s) = w \leq c_2$, converges to the function $f(x)$ independently of w. Furthermore, if the Mellin transform has poles and satisfies the conditions of Jordan's lemma, which is almost always the case, then the Mellin transform determines $f(x)$ uniquely.

If $f(x)$ is a PDF, the Mellin transform evaluated at $s = k$ gives the $(k-1)$st moment of $f(x)$ about the origin, $k = 1, 2, 3, \ldots$. In particular $M_s(f(x))|_{s=1} = 1$, which provides a good check on the validity of any derived transform. Note also that the Mellin transform per se is defined only for nonnegative values of the rv. The following example is illustrative.

Example 5. Find the PDF $h(z)$ of the product $Z = \prod_{j=1}^{n} X_j$ of n identical uniform i.r.v.'s X_j having PDF

$$f(x_j) = \begin{cases} 1, & 0 \leq x_j \leq 1, \quad j = 1, 2, \ldots, n \\ 0, & \text{elsewhere.} \end{cases}$$

Since the rvs are independent, we have [8, p. 101]

$$M_s(h(z)) = \prod_{i=1}^{n} M_s(f(x_j)) = \left(\frac{1}{s}\right)^n \quad (16)$$

and

$$h(z) = \frac{1}{2\pi i} \int_{c-i\infty}^{c+i\infty} z^{-s} \left(\frac{1}{s}\right)^n ds, \quad c > 0 \tag{17}$$

$$= \frac{1}{2\pi i} \int_{c-i\infty}^{c+i\infty} e^{(-\ln z)s} \left(\frac{1}{s}\right)^n ds,$$

$$c > 0. \tag{18}$$

Again, since the conditions of Jordan's lemma are satisfied relative to the Bromwich (closed) contour C_L, $c > 0$, the inversion integral is evaluated over C_L by means of residues, yielding $h(z) = h_1(z)$, $0 < z < 1$. If there were also poles in the RHP, integration over C_R would give $h(z) = h_2(z)$, $1 \leq z < \infty$. In some cases, $h_1(z)$ and $h_2(z)$ are identical as, for example, when z is the product of Weibull* i.r.v.'s [8, p. 245], and in other cases they are not, as in the case of the distribution of the quotient of the two uniform i.r.v.'s, to be discussed shortly.

Returning to (17), we have

$$h(z) = h_1(z) = R_1, \quad 0 < z < 1,$$

where R_1 is the nth-order pole at $s = 0$ and there are no other poles. Also,

$$h(z) = h_2(z) = 0, \quad 1 \leq z < \infty,$$

since z is the product of i.r.v.'s restricted to the range $(0, 1)$. It follows that

$$h(z) = \frac{1}{(n-1)!} \frac{d^{n-1}}{ds^{n-1}} \left[z^{-s} \right] \bigg|_{s=0}$$

$$= \left[\ln(1/z) \right]^{n-1} / (n-1)!, \quad 0 < z \leq 1,$$

$$= 0, \quad 1 < z < \infty.$$

Example 6. Derive the PDF $h(y)$ of the quotient $Y = X_1/X_2$ of two uniform i.r.v.'s X_j having PDF

$$f(x_j) = \begin{cases} 1, & 0 \leq x_j \leq 1, \quad j = 1, 2 \\ 0, & \text{elsewhere.} \end{cases}$$

We shall write Y in the form $Y = X_1(1/X_2)$, and denote the PDF of the rv $1/X_2$ by $g(x_2)$. Then, since [8, p. 100] $M_s(g(x_2)) = M_{-s+2}(f(x_2))$, we have

$$h(y) = \frac{1}{2\pi i} \int_{c-i\infty}^{c+i\infty} \frac{y^{-s}}{s(-s+2)}\, ds,$$

$$0 < c < 2.$$

Again, it is easily verified that the conditions of Jordan's lemma are satisfied relative to C_L if $0 \leq y < 1$ and relative to C_R if $1 \leq y < \infty$. Then

$$h(y) = h_1(y), \quad 0 \leq y < 1$$

$$= h_2(y), \quad 1 \leq y < \infty,$$

where

$$h(y_1) = \frac{1}{2\pi i} \int_{c-i\infty}^{c+i\infty} \frac{y^{-s}}{s(-s+2)} ds,$$

$$0 < c < 2, \quad 0 \leqslant y < 1$$

$$= \frac{1}{2}, \quad 0 \leqslant y < 1,$$

$$h_2(y) = \frac{1}{2\pi i} \int_{c-i\infty}^{c+i\infty} \frac{y^{-s}}{s(-s+2)} ds,$$

$$0 < c < 2, \quad 1 \leqslant y < \infty$$

$$= y^{-2}/2, \quad 1 \leqslant y < \infty.$$

The CDF $H(y)$ can be obtained [8, p. 99] directly from its transform without first deriving $h(y)$ by utilizing the fact that

$$M_s(1 - H(y)) = \left[M_{s+1}(h(y)) \right]/s.$$

Thus, for the problem at hand,

$$M_s(1 - H(y)) = \frac{y^{-s}}{s(s+1)(-s+1)},$$

and

$$1 - H_1(y) = \frac{y^{-s}}{(s+1)(-s+1)} \bigg|_{s=0}$$

$$+ \frac{y^{-s}}{s(-s+1)} \bigg|_{s=-1},$$

so that

$$H_1(y) = \frac{y}{2}, \quad 0 \leqslant y < 1.$$

Similarly,

$$1 - H_2(y) = \frac{y^{-s}}{s(s+1)} \bigg|_{s=1}, \quad 1 \leqslant y < \infty,$$

from which

$$H_2(y) = 1 - \frac{1}{2y}, \quad 1 \leqslant y < \infty.$$

Hence,

$$H(y) = \frac{y}{2}, \quad 0 \leqslant y < 1$$

$$= 1 - \frac{1}{2y}, \quad 1 \leqslant y < \infty.$$

Mellin transforms also have application to problems in multivariate analysis involving dependent rvs [8, pp. 151–156]. However, relatively little has been done in this area. A particularly important special case of a Mellin inversion integral is the family of H-functions. (*See* H-FUNCTION DISTRIBUTIONS.) A computer program for evaluating such H-function inversion integrals has been developed and is operational [1].

As has been pointed out, the method of residues is particularly useful in evaluating inversion integrals. Various recursion formulas and algorithms are often involved in evaluating these integrals [8, pp. 109–112]. These formulas frequently involve such things as the computation of the digamma function [usually denoted by $\psi(z)$] and its derivatives, partial fraction expansions, etc. In some instances, numerical evaluation of Fourier transforms and inversion integrals is feasible. There have been several articles in recent journals which expedite the efficient performance of some of these computational aspects. (See, e.g., refs. 2, 5, and 6.)

Finally, it should be pointed out that integral transforms have considerable existing and potential application to the solution of real-world problems, as well as to the solution of various theoretical problems in statistics. Some examples of such problems are given in refs. 3, 7, and 8.

NOTE

1. Reprinted from Springer, *The Algebra of Random Variables*, Wiley (1979), pp. 40–41, with permission of the publishers.

References

[1] Eldred, B. S. (1978). The Application of Integral Transform Theory to the Analysis of Probabilistic Cash Flows. Ph.D. dissertation, University of Texas at Austin. (Includes a very efficient computer program for evaluation of H-function inversion integrals. Program compiled on MNF compiler and run on CDC 6600 with no precision problems. To avoid precision problems on smaller computers, compile program under IBM's Extended H-Compiler.)

[2] Fornberg, B. (1981). A vector implementation of the fast Fourier transform algorithm. *Math. Comp.*, **36**, 189–191. [Vector implementation on the CDC

STAR-100 vector computer of a new FFT (fast Fourier transform) algorithm utilizing trigonometric tables. Execution time three times faster than that of previous FFT algorithms.]

[3] Giffin, W. C. (1975). *Transform Techniques for Probability Modeling*. Academic Press, New York. (Concerned primarily with Laplace, Fourier, and Z transforms and their application to queueing and linear systems analysis. Treats only briefly the evaluation of inversion integrals by residues, relying mostly on the use of partial fraction expansions and tables of transforms.)

[4] Lukacs, E. (1970). *Characteristic Functions*, 2nd ed. Hafner, New York. (Excellent comprehensive treatment of characteristic functions. In the author's words: "In the present monograph we study characteristic functions for their intrinsic mathematical interest and are not concerned with their possible applications. . . . The methods discussed in a considerable part of this monograph should be contained in a mathematical statistician's tool chest.")

[5] Mahoney, J. F. (1981). Partial fraction evaluation by an escalation technique. *Math. Comp.*, **36**, 241–246. (In evaluating inversion integrals, it is occasionally convenient to expand the Mellin transform in partial fractions. An escalation method is here given for performing partial fraction expansions for the case in which the complete list of zeros of the denominator of the proper rational function is known, requiring fewer such arithmetic operations than do other known methods.)

[6] McCullagh, P. (1981). A rapidly convergent series for computing $\psi(z)$ and its derivatives. *Math. Comp.*, **36**, 247–248. [Evaluation of *H*-function inversion integrals involves the use of the Euler psi function $\psi(z)$ and its derivatives $\psi^{(m)}(z)$. The efficiency of this evaluation is considerably enhanced by utilizing the rapidly convergent series presented in this paper.]

[7] Muth, E. J. (1977). *Transform Methods with Applications to Engineering and Operations Research*. Prentice-Hall, Englewood Cliffs, N.J. (Concerned with Laplace and z transforms and their application to certain areas of engineering and operations research.)

[8] Springer, M. D. (1979). *The Algebra of Random Variables*. Wiley, New York. (Presents a self-contained treatment of how to use integral transforms to determine the distribution of sums, differences, products, quotients, and algebraic functions of random variables. Written with the intent of enabling analysts in any field, who understand the basic elements of statistical inference and differential and integral calculus, to learn from one source how to apply integral transform methods to both linear and nonlinear probabilistic models in their fields. Also includes many examples of the application of integral transforms by various authors to distribution problems in statistics.)

(CHARACTERISTIC FUNCTIONS
H-FUNCTION DISTRIBUTIONS
MOMENT GENERATING FUNCTIONS)

MELVIN D. SPRINGER

INTERACTION

In some experiments or processes the yield measured is affected by two or more factors. If the yield is not just the sum of the effects of the separate factors, the factors are said to *interact*. If there are two factors, the difference between the yield and the sum of the effects of the two factors is called the *interaction* of the two factors.

TWO WAY TABLES

Table 1 shows yields (taken from Digby [3]) of four varieties of spring wheat at two different sites. An experimenter interested in the difference between varieties might make a quick inspection of the variety means and conclude that variety C has the highest yield. Similarly, someone interested only in the site differences would conclude that site 2 is best. If now recommendations are made that variety C is best and site 2 is best, a third person, with no access to the data, might infer that the overall highest yield is obtained by growing variety C at site 2. But this conclusion is wrong; Table 1 shows that the best combination is variety C at site 1. It is to guard against such erroneous conclusions that factorial experiments* are performed.

To see what has gone wrong, we examine Table 1 more closely. For variety A, the effect of the site differences is 0.38 tonne/hectare in favor of site 2; for variety B it is only 0.21 tonne/hectare in favor of site 2; while for variety C the site differences produce the opposite effect—0.59 tonne/hectare in favor of site 1. These changes in the site effect with the different varieties are called the *interaction* of sites with varieties. Similarly, the change of the variety differences at the different sites also demonstrates this interaction. Compare these data with

Table 1 Yields of Spring Wheat (tonnes/hectare)

| | Variety | | | | | Adjusted |
	A	B	C	D	Site Mean	Site Mean
Site 1	2.32	2.56	3.72	3.10	2.92	− 0.04
Site 2	2.70	2.77	3.13	3.40	3.00	+ 0.04
Variety mean	2.51	2.66	3.42	3.25	2.96[a]	
Adjusted variety mean	− 0.45	− 0.30	+ 0.46	+ 0.29		

[a]Overall mean.
Source. Digby [3].

Table 2 Fictitious Data Showing No Interaction

| | Variety | | | | | Adjusted |
	A	B	C	D	Site Mean	Site Mean
Site 1	3.05	2.71	3.16	2.92	2.96	− 0.12
Site 2	3.29	2.95	3.40	3.16	3.20	+ 0.12
Variety mean	3.17	2.83	3.28	3.04	3.08[a]	
Adjusted variety mean	+ 0.09	− 0.25	+ 0.20	− 0.04		

[a]Overall mean.

the fictitious data in Table 2. There the site effect is 0.24 in favor of site 2 for *every* variety, so there is no interaction. If the variety and site means are all adjusted by subtracting the overall mean, the yields are all explained by the formula

yield = (overall mean)

+ (adjusted variety mean)

+ (adjusted site mean).

By contrast, if we subtract the overall mean, adjusted variety means, and adjusted site means from the yields in Table 1, we obtain the table of residuals* shown in Table 3. (If we had not introduced rounding errors in calculating the means, the row and column totals would all be zero.) These residuals show what differences there are in the data *after the differences due to varieties and sites have been allowed for.* Thus another way of

viewing the interaction is as the effect that accounts for those differences which are not explained by variety differences or column differences. Since the row and column totals must be zero, the interaction accounts for 3 degrees of freedom* in this example.

More generally, if there are v varieties and s sites, then the $vs − 1$ degrees of freedom are divided as follows:

Varieties	$v − 1$
Sites	$s − 1$
Interaction	$(v − 1)(s − 1)$

Table 3 Residuals from Table 1

| | Variety | | | |
	A	B	C	D
Site 1	− 0.15	− 0.06	+ 0.34	− 0.11
Site 2	+ 0.15	+ 0.07	− 0.33	+ 0.11

The $v - 1$ degrees of freedom for varieties are called the *main effect** of varieties; the $s - 1$ degrees of freedom for sites are called the main effect of sites; the interaction is called the *variety-by-sites interaction.*

In most experiments some allowance must be made for experimental error. If there is a single reading on each variety at each site, the residuals* have the form

interaction + error

so it is impossible to estimate the interaction unless there is a previous estimate of the experimental error. This difficulty may be overcome by taking two or more readings on each variety at each site, because then each interaction component contributes to more than one reading.

The foregoing definitions of main effect and interaction correspond to a linear model* of the form

$$y_{ijk} = \mu + v_i + s_j + w_{ij} + e_{ijk},$$

where

y_{ijk} = yield on the kth plot of

variety i at site j,

μ = overall mean,

v_i = effect of the ith variety

$\left(\text{and } \sum v_i = 0\right)$,

s_j = effect of the jth site

$\left(\text{and } \sum_j s_j = 0\right)$,

w_{ij} = effect of interaction of

variety i with site j

$\left(\text{and } \sum_i w_{ij} = 0 \text{ for all } j,\right.$

$\left. \sum_j w_{ij} = 0 \text{ for all } i\right)$,

e_{ijk} = error on the kth plot of

variety i and site j.

In view of this model, many people talk of interaction as "a departure from additivity."

Statements such as "(the effects of) blocks and treatments are additive" and "there is no block-treatment interaction" are equivalent; they both mean that the linear model has the form

$$y_{ijk} = \mu + t_i + b_j + e_{ijk}$$

where

y_{ijk} = yield on the kth plot of

treatment i in block j,

μ = overall mean,

t_i = effect of the ith treatment,

b_j = effect of the jth block,

e_{ijk} = error on the kth plot of

treatment i in block j.

MEANING OF "INTERACTION"

The definition of "interaction" given above is strictly in terms of marginal totals and residuals, and therefore indirectly in terms of the linear model. This definition does not always agree with the intuitive idea of "in-"teraction," and this can sometimes lead to confusion when consultant statisticians discuss things with their clients. The following examples show some pitfalls.

Example 1. Consider an experiment in which one environment is so severe that nothing grows. It is natural to think that this environment has the "same" effect on every variety, but in terms of residuals the effect is *not* the same. For example, in the fictitious data in Table 4, the environment effect is 3.20 for variety A but only 2.15 for variety

Table 4   Experiment with One Disastrous Environment

	Variety		
	A	B	C
Environment 1	0.00	0.00	0.00
Environment 2	3.20	2.15	2.70

B. These data are *not* fitted by a linear model without an interaction term, so there *is* an environment-by-varieties interaction.

Example 2. It may be that the yield *is* given entirely by the effects of varieties and sites, but in a nonadditive way. For example, the yield on the *i*th variety on the *j*th site may be given by the *product* $v_i s_j$. Because *effects* are defined in terms of an *additive* model, there would be an interaction in this case, even though it might be natural to think there was not.

Example 3. An experiment was performed to test the effect of peppermint conditioning on flies. Successive generations of flies were hatched in a mild peppermint solution and then tested for their affinity to peppermint. Control flies, which had been hatched normally, were similarly tested. The model postulated was that, at the *n*th generation, the proportion of control flies showing an affinity for peppermint was

$$t_n + \text{error}$$

while the proportion of conditioned flies showing an affinity for peppermint was

$$t_n + p_n + \text{error}$$

In this case the main effect of the *n*th generation (i.e., $t_n + \frac{1}{2}p_n$) was of little importance. The main effect of peppermint (i.e., $\sum_n p_n$) *was* important, but more important was the *change of the peppermint effect with time*, i.e., the individual p_n values. Thus the *interaction* was the important effect, so "main effect" does not always mean "important effect."

Example 4. In a similar vein to Example 3, if the *blocks* used for an experiment are not representative of all conditions that might occur in practice, the main effect of a variety is likely to be meaningless. Thus the blocks-by-varieties interaction would contain more important information.

Example 5. Suppose that 12 different varieties are to be tested. It may be convenient to write these in a 3 × 4 array, as in Table 5

Table 5

1	2	3	4
5	6	7	8
9	10	11	12

(which has nothing to do with the layout of the whole experiment), and consider the varieties to be all combinations of three rows with four columns. Then the variety differences are accounted for by the main effect of rows (2 d.f.), the main effect of columns (3 d.f.), and the rows-by-columns interaction (6 d.f.). If the arrangement of varieties in the array was arbitrary, all three effects are of equal status.

Example 6. In the opposite case to Example 5, it may be that the treatments* genuinely consist of all combinations of levels of two treatment factors*. For example, the treatments may be six different fertilizers, consisting of all combinations of three levels of nitrogen with two levels of potash. Now the distinction between the main effect of nitrogen, the main effect of potash, and the nitrogen–potash interaction is probably important. Usually, the interaction is smaller than the main effects. If there is any confounding*, the experimenter must decide whether he or she would rather have more precise information on the main effects (which would probably be the case in an exploratory experiment) or the interaction, which is harder to measure precisely if it is smaller (this might be preferred in a follow-up experiment).

HIGHER-ORDER INTERACTIONS

So far we have concentrated on interactions between two factors: these are called *two-factor interactions* or *second-order interactions*. We can extend the definitions to three or more factors.

Suppose that the treatments are all 12 combinations of three levels of nitrogen with two levels of potash and two levels of phosphate. Then we have

Main effect of nitrogen	2 d.f.
Main effect of potash	1 d.f.
Main effect of phosphate	1 d.f.
Nitrogen–potash interaction	2 d.f.
Nitrogen–phosphate interaction	2 d.f.
Potash–phosphate interaction	1 d.f.

That leaves 2 degrees of freedom unaccounted for, and we define these to be the effect of the *three-factor interaction*. Equivalently, the three-factor interaction is the effect that accounts for the residual differences left when main effects and two-factor interactions have been allowed for. Fourth- and higher-order interactions are defined similarly.

The standard notation for interactions uses letters for each of the factors involved. Thus, if the main effects of nitrogen, potash, and phosphate are denoted by N, K, P, respectively,

NK	Denotes the nitrogen–potash interaction
NP	Denotes the nitrogen–phosphate interaction
K P	Denotes the potash–phosphate interaction
NKP	Denotes the three-factor interaction

DECOMPOSING INTERACTION DEGREES OF FREEDOM

There are two important ways in which the degrees of freedom for an interaction may be decomposed.

Algebraic Decomposition

Suppose that A and B are treatment factors with three levels, denoted by 0, 1, 2. We can define artificial treatment factors $A^i B^j$ ($i, j = 1, 2$) by putting

level of $A^i B^j = i \times$ (level of A)

$+ j \times$ (level of B) modulo 3.

Table 6   Treatment Degrees of Freedom in a 3 $\times$ 3 Experiment

Level of A	0 0 0	1 1 1	2 2 2
Level of B	0 1 2	0 1 2	0 1 2
Level of $A^1 B^1$	0 1 2	1 2 0	2 0 1
Level of $A^2 B^2$	0 2 1	2 1 0	1 0 2
Level of $A^1 B^2$	0 2 1	1 0 2	2 1 0
Level of $A^2 B^1$	0 1 2	2 0 1	1 2 0

It can be seen from Table 6 that the contrasts between the three different levels of $A^1 B^1$ are orthogonal to the main effects of A and of B, so they are part of the AB interaction. The levels of $A^2 B^2$ are twice those of $A^1 B^1$ (modulo 3), so the contrasts for $A^2 B^2$ are the same as those for $A^1 B^1$. Similarly, the contrasts for $A^2 B^1$ are the same as those for $A^1 B^2$: however, these are orthogonal to those for $A^1 B^1$, so we may decompose the interaction AB into two parts, $A^1 B^1$ and $A^1 B^2$, each consisting of two degrees of freedom.

In similar manner, if we have three factors A, B, and C, each with three levels, we may break down the three-factor interaction ABC into four parts, $A^1 B^1 C^1$, $A^1 B^2 C^1$, $A^1 B^1 C^2$, and $A^1 B^2 C^2$, each consisting of 2 degrees of freedom. Again, $A^1 B^2 C^1$ gives the same contrasts as $A^2 B^1 C^2$, for example, but it is conventional to use the notation with an index 1 on the first letter. In fact, it is conventional to omit the index 1 altogether, so the four components of the interaction become ABC, AB^2C, ABC^2, and AB^2C^2. Notice now that ABC denotes *two* different things: the entire three-factor interaction, *and* a 2 d.f. component of that interaction. Unfortunately, the notation is now too well established to change. This decomposition can be extended to interaction of four or more factors.

We can now extend our definition of interaction to allow one or more of the factors to be the artificial factors we have introduced. This is called a *generalized interaction*. It can be shown that the generalized interaction of A and AB^2 is AB (which is the same as $A^2 B^2$); that the generalized interaction of AB^2 and C is AB^2C; that of AB^2 and B is A. In fact, the generalized interaction is obtained by "multiplying" the effects

Table 7   Interaction Components for Three-Level Quantitative Factors

Level of A		0	0	0	1	1	1	2	2	2
Level of B		0	1	2	0	1	2	0	1	2
Main effect of A	linear A	-1	-1	-1	0	0	0	1	1	1
	quadratic A	-1	-1	-1	2	2	2	-1	-1	-1
Main effect of B	linear B	-1	0	1	-1	0	1	-1	0	1
	quadratic B	-1	2	-1	-1	2	-1	-1	2	-1
Inter-action AB	$\text{lin}_A \times \text{lin}_B$	1	0	-1	0	0	0	-1	0	1
	$\text{lin}_A \times \text{quad}_B$	1	-2	1	0	0	0	-1	2	-1
	$\text{quad}_A \times \text{lin}_B$	1	0	-1	-2	0	2	1	0	-1
	$\text{quad}_A \times \text{quad}_B$	1	-2	1	-2	4	-2	1	-2	1

and ignoring A^3, B^3, and C^3. Everything done here for *three* levels may be done for p levels, where p is any prime number.

Although this system looks complicated, it is extremely useful for confounded experiments and fractional factorial* experiments.

Quantitative Factors

If the factors are quantitative, it may be desired to estimate the linear, quadratic, etc., coefficients in the response* equation. For main effects, the appropriate contrasts are provided by the orthogonal polynomials*. This can be extended to interactions by cross-multiplying coefficients, in the manner demonstrated by Table 7. The contrasts so obtained are used to estimate the coefficients of mixed terms such as xy and xy^2 in the response equation.

LITERATURE

Excellent accounts of the meaning of "interaction" may be found in Cox [1, Chap. 6] and Davies [2, Chaps. 4, 7]. These books are both suitable for readers with little mathematical knowledge. They also include discussions of the decomposition of interactions of quantitative factors [1, Chap. 6; 2, Chap. 8].

There is a short account of generalized interactions and the AB^2, etc., decomposition in Chap. 9 of Davies [2]. For the more mathematical reader, Chaps. 16 and 17 of Kempthorne [4] give a detailed explanation of both ways of decomposing interactions.

References

[1] Cox, D. R. (1958). *Planning of Experiments*. Wiley, New York.

[2] Davies, O. L., ed. (1977). *The Design and Analysis of Industrial Experiments*. Oliver & Boyd, Edinburgh.

[3] Digby, P. G. N. (1979). *J. Agric. Sci.*, **93**, 81–86.

[4] Kempthorne, O. (1957). *The Design and Analysis of Experiments*. Wiley, New York.

(ANALYSIS OF VARIANCE
CONFOUNDING
CONTRAST
DEGREES OF FREEDOM
DESIGN OF EXPERIMENTS
FACTORIAL EXPERIMENTS
FRACTIONAL FACTORIAL EXPERIMENTS
GENERAL LINEAR MODEL
MAIN EFFECT
ORTHOGONALITY
RESIDUALS
RESPONSE SURFACE
TREATMENTS)

R. A. BAILEY

INTERACTION MODELS

Interaction models provide simplified structures for the arrays of unknown parameters that arise in *factorial experiments** and in multidimensional *contingency tables**. These two fields of application will be considered side by side, rather more attention being given to contingency tables.

The theory of modeling and analyzing factorial experiments was first developed by Fisher and Yates in the 1930s and the subject now has an enormous literature. Countless books have a chapter or more devoted to it. In a factorial experiment there are, say, d factors $A_1, \ldots, A_d$ and one response Y. If the factors have $r_1, \ldots, r_d$ levels there are $r_1 \times \cdots \times r_d$ different combinations of levels called "cells". As an example, let Y = wheat yield on an experimental plot to which is applied one of two levels of A_1 = nitrogen, one of three levels of A_2 = potash and one of five levels of A_3 = phosphate, giving $2 \times 3 \times 5 = 30$ cells. For each cell we imagine a population of values of Y and let $\eta = E(Y)$ denote the mean, or expected, value in this population. Interest is focused on how η varies from cell to cell. We shall let η denote the array of $r_1 \times \cdots \times r_d$ values of η. In the example the array comprises 30 values and these could be displayed in several ways, as two 3×5 tables or as five 2×3 tables, for example. The $r_1 \times \cdots \times r_d$ parameters in the array η are invariably unknown and the analysis of factorial experiments is concerned with how to make inferences about them. Here we shall only discuss the modeling of the array.

In a "pure-response" d-dimensional contingency table there are d categorical random variables $X_1, \ldots, X_d$ taking $r_1, \ldots, r_d$ values. This time the unknown parameter at each cell is the probability π of that particular combination of response values. For example, a randomly selected married couple is classified according to X_1 = smoking level (zero, low, medium, high) of the wife and X_2 = smoking level of the husband, giving rise to a two-dimensional contingency table

π of $4 \times 4 = 16$ probabilities. The following discussion also applies to d-dimensional contingency tables in which some of the d dimensions correspond to factors and the remainder to responses. In the example above a third dimension could be added by considering the factor A = age group (young, middle, old) of the married couple. We now have a three-dimensional contingency table π of the $3 \times 4 \times 4 = 48$ probabilities π of each combination of smoking levels for each age group. *See* CONTINGENCY TABLES and CATEGORICAL DATA for further information and also the books by Plackett [8], Bishop et al. [2], Fienberg [5], and Upton [9].

In the following two sections interactions are defined in several ways and then, in the sections "Interaction Models" and "Hierarchical Interaction Models As Simple Structures," models are obtained by supposing that some of the interactions are zero. Similarities between the two fields of application are emphasized in the latter section, and in the sections "Interpreting Interactions" and "Decomposable Models" attention is given to dissimilarities. The discussion in these sections is conducted through examples but, in the last section, a framework for discussing the general case is outlined.

INTERACTIONS

Suppose that $d = 3$ and let $\eta_{ijk} = E[Y_{ijk}]$, Y_{ijk} being the response at cell (i, j, k). Let η be the array of $r_1 \times r_2 \times r_3$ values. It can be written as the sum of eight *interaction** arrays

$$\eta = \theta + \theta^1 + \theta^2 + \theta^3$$
$$+ \theta^{23} + \theta^{13} + \theta^{12} + \theta^{123},$$

where $\theta = \theta\mathbf{1}$, $\mathbf{1}$ denoting the array of 1's, θ_{ijk}^1 is constant with respect to (j, k) and equal to θ_i^1, say, $\theta_{ijk}^{23} = \theta_{jk}^{23}$, etc. One way of making the θ arrays uniquely determined from η is to impose the zero-sum conditions exemplified by $\theta_+^1 = 0$, $\theta_{j+}^{23} = \theta_{+k}^{23} = 0$, $\theta_{+jk}^{123} = \theta_{i+k}^{123} = \theta_{ij+}^{123} = 0$, $+$ denoting summation over the missing subscript. The resulting well-known formulas for the θ's are

$$\theta = \eta \ldots, \qquad \theta_i^1 = \eta_{i\cdot\cdot} - \eta \ldots,$$

$$\theta_{jk}^{23} = \eta_{\cdot jk} - \eta_{\cdot j\cdot} - \eta_{\cdot\cdot k} + \eta \ldots,$$

$$\theta_{ijk}^{123} = \eta_{ijk} - \eta_{\cdot jk} - \eta_{i\cdot k} - \eta_{ij\cdot} + \eta_{i\cdot\cdot}$$

$$+ \eta_{\cdot j\cdot} + \eta_{\cdot\cdot k} - \eta \ldots, \text{ etc,}$$

· denoting average over the missing subscript. The interpretation of θ_i^1 as a "main effect" is clear. The interpretation of the higher-order interactions is easily built up recursively. Thus

$$\theta_{ij}^{12} = (\eta_{ij\cdot} - \eta_{\cdot j\cdot}) - (\eta_{i\cdot\cdot} - \eta \ldots),$$

the difference between the effect of $A_1 = i$ within $A_2 = j$ and the main effect* of $A_1 = i$. Similarly,

$$\theta_{ijk}^{123} = (\eta_{ijk} - \eta_{i\cdot k} - \eta_{\cdot jk} + \eta_{\cdot\cdot k})$$

$$- (\eta_{ij\cdot} - \eta_{i\cdot\cdot} - \eta_{\cdot j\cdot} + \eta \ldots).$$

Instead of imposing zero-sum conditions on (1) to get the θ interactions we can impose "zero-reference-cell" conditions and get interactions denoted by ϕ. Thus, with (r_1, r_2, r_3) as the reference cell, we now impose conditions exemplified by $\phi_{r_1}^1 = 0$, $\phi_{jr_3}^{23} = \phi_{r_2 k}^{23} = 0$, $\phi_{r_1 jk}^{123} = \phi_{ir_2 k}^{123} = \phi_{ijr_3}^{123} = 0$. The result is

$$\phi_{ijk} = \phi = \eta_{r_1 r_2 r_3}, \qquad \phi_{ijk}^1 = \phi_i^1 = \eta_{ir_2 r_3} - \eta_{r_1 r_2 r_3},$$

$$\phi_{ijk}^{23} = \phi_{jk}^{23} = \eta_{r_1 jk} - \eta_{r_1 r_2 k} - \eta_{r_1 jr_3} + \eta_{r_1 r_2 r_3}, \text{ etc.}$$

and, starting with the simple contrast ϕ_i^1, an interpretative structure can be built up for the ϕ's as for the θ's.

The θ and ϕ systems of interactions are particular cases of a general system involving arbitrary weights $(w_i^1), (w_j^2), (w_k^3)$ satisfying $w_+^1 = w_+^2 = w_+^3 = 1$. The decomposition of η is now

$$\eta = \psi + \psi^1 + \psi^2 + \psi^3$$

$$+ \psi^{23} + \psi^{13} + \psi^{12} + \psi^{123},$$

where, for instance, $\psi_{ijk}^{12} = \psi_{ij}^{12}$ and $\psi_{i*}^{12} = \psi_{*j}^{12} = 0$, $*$ denoting weighted average. For example, $\psi_{i*}^{12} = \sum_j w_j^2 \psi_{ij}^{12}$. The ψ interactions are related to η by

$$\psi_{ijk} = \psi = \eta_{***}, \qquad \psi_{ijk}^1 = \psi_i^1 = \eta_{i**} - \eta_{***},$$

$$\psi_{ijk}^{23} = \psi_{jk}^{23} = \eta_{*jk} - \eta_{*j*} - \eta_{**k} + \eta_{***}, \text{ etc.}$$

the ψ system reduces to the θ system when $w_i^1 = r_1^{-1}$, etc., and to the ϕ system when $w_i^1 = 0$, $i \ne r_1$, $w_{r_1}^1 = 1$, etc.

The general ψ system is of mathematical rather than interpretative interest.

LOG-LINEAR INTERACTIONS

"Log-linear" interactions for π are simply the linear interactions above applied to $\eta = \log \pi = (\log \pi_{ijk})$. The θ system was introduced by Birch [1] and is widely used. The "main effect" θ_i^1 is given by

$$\theta_i^1 = \frac{1}{r_2 r_3} \sum_{j,k} \log \pi_{ijk} - \frac{1}{r_1 r_2 r_3} \sum_{i,j,k} \log \pi_{ijk}.$$

It is the logarithm of the ratio of two geometric means of the π_{ijk} and thus has very little interpretative value. The ϕ system, introduced by Mantel [7], has more intrinsic interest than the θ system since it involves simple ratios and cross-product ratios (see ODDS RATIO) of probabilities. Thus

$$\phi_i^1 = \log \frac{\pi_{ir_2 r_3}}{\pi_{r_1 r_2 r_3}}, \qquad \phi_{ij}^{12} = \log \frac{\pi_{ijr_3} \pi_{r_1 r_2 r_3}}{\pi_{ir_2 r_3} \pi_{r_1 jr_3}},$$

$$\phi_{ijk}^{123} = \log \frac{\pi_{ijk} \pi_{ir_2 r_3} \pi_{r_1 jr_3} \pi_{r_1 r_2 k}}{\pi_{r_1 r_2 r_3} \pi_{r_1 jk} \pi_{ir_2 k} \pi_{ijr_3}}.$$

As for $\eta = E[Y]$, the general ψ system of interactions is only of mathematical interest.

INTERACTION MODELS

Interaction models for $\eta = E[Y]$ or $\eta = \log \pi$ are obtained simply by imposing the condition that certain of the interaction functions are identically zero.

Continuing to take $d = 3$, consider the two models defined by $\psi^{12} = \psi^{123} = 0$ and by $\psi^1 = \psi^{12} = 0$. The first has the attractive property that it is the same model whichever ψ system is involved. The second lacks this property and is a different model for each different choice of underlying weights (w_i^1), $(w_j^2), (w_k^3)$. What is the cause of the difference?

Define the interaction set of an interaction model to be the set of subsets of $\{1, 2, 3\}$

corresponding to those interaction functions not declared to be zero: $\{\varnothing, 1, 2, 3, 13, 23\}$ and $\{\varnothing, 2, 3, 13, 23, 123\}$ in the two examples, 13 for instance denoting the subset $\{1, 3\}$ and $\varnothing$ the empty set. A *hierarchical* model is one whose interaction set has the property that, if h belongs to it, so do all subsets of h. Thus the first example is hierarchical but not the second. Hierarchical interaction models have the property that they are the same whichever ψ system is used to define them and, for reasons closely related to this, it is they which receive nearly all the attention.

A hierarchical model is clearly determined by the maximal elements of its interaction set and these constitute the *generating class* of the model. Thus the generating class of the model with interaction set $\{\varnothing, 1, 2, 3, 13, 23\}$ is $\{13, 23\}$.

The generating class is the key to a simpler formulation of a hierarchical interaction model. Thus the model with generating class $\{13, 23\}$ can be formulated as

$$\eta = \alpha^{13} + \alpha^{23}$$

and we call it an *additive* model. Here α^{13}, α^{23} are arbitrary arrays subject only to $\alpha_{ijk}^{13} = \alpha_{ik}^{13}$, $\alpha_{ijk}^{23} = \alpha_{jk}^{23}$. This model thus simply says that each η_{ijk} is expressible as the sum of two numbers, one depending only on i, k and the other depending only on j, k. The corresponding *multiplicative* model for π is

$$\pi = \beta^{13} \times \beta^{23},$$

$\times$ denoting element-by-element multiplication.

The connection between the two formulations of hierarchical interaction models will be taken up for general d in the last section.

HIERARCHICAL INTERACTION MODELS AS SIMPLE STRUCTURES

Hierarchical interaction models can be characterized as providing simple structures for $\eta = E[\mathbf{Y}]$ or for π when certain marginal functions are given. Consider $d = 4$. The model with generating class $\{12, 134, 234\}$, namely

$$\eta = \alpha^{12} + \alpha^{134} + \alpha^{234}, \qquad (1)$$

can be arrived at as follows. Suppose that the marginal means $\eta_{ij**}, \eta_{i*kl}, \eta_{*jkl}$ of η_{ijkl}, using arbitrary positive weights $(w_i^1), (w_j^2), (w_k^3), (w_l^4)$, are all given. Then the η which satisfies these marginals and is closest to being uniform, in the sense of minimizing $\sum_{ijkl} w_i^1 w_j^2 w_k^3 w_l^4 (\eta_{ijkl} - \eta_{****})^2$, is unique and satisfies (1).

There is a corresponding characterization of multiplicative (hierarchical log-linear interaction) models. If the marginal totals (π_{ij++}), (π_{i+kl}), (π_{+jkl}) are prescribed and admit a positive array satisfying them, then the π which satisfies them and which is closest to being uniform, in the sense of minimizing negative entropy*

$$\sum_{ijkl} \pi_{ijkl} \log \pi_{ijkl},$$

is unique and satisfies $\pi = \beta^{12} \times \beta^{134} \times \beta^{234}$.

INTERPRETING INTERACTIONS

We have seen in the section "Interaction Models" that hierarchical interaction models can be defined without involving the notion of individual interactions, although this is, of course, not true of nonhierarchical models. This is not to say that interactions are not useful when studying hierarchical models, because they can be used to quantify the difference in fit of two such models to a set of data.

Consider a four-dimensional contingency table for which the observed proportion corresponding to π_{ijkl} is p_{ijkl}. If the multiplicative model $\pi_{ijkl} = \beta_{ij}^{12} \beta_{ikl}^{134} \beta_{jkl}^{234}$ is fitted to $\mathbf{p}$ by maximum likelihood, then the resulting $\hat{\pi}$ satisfies $\hat{\pi}_{ij++} = p_{ij++}$, $\hat{\pi}_{i+kl} = p_{i+kl}$, $\hat{\pi}_{+jkl} = p_{+jkl}$. [This well-known result goes hand and glove with the result of the preceding section that, under the model, π is uniquely determined by $(\pi_{ij++}), (\pi_{i+kl}), (\pi_{+jkl})$.] Suppose that, using a likelihood ratio or other test, $\hat{\pi}$ is judged to adequately fit $\mathbf{p}$ and the question is now asked: Does the simpler model $\pi_{ijkl} = \beta_{ij}^{12} \beta_{ik}^{13} \beta_{il}^{14} \beta_{jkl}^{234}$ also provide an adequate fit to $\mathbf{p}$? Given that π satisfies the first model, it also satisfies the second iff the log-linear interaction array ψ^{134} is $\mathbf{0}$. Thus the answer to the question depends on the magnitude of the interactions $\hat{\psi}_{ijk}^{134}$ derived from $\hat{\pi}$. This is true whichever ψ system is used. In practice, either the θ system or ϕ system is used in computing systems.

There is one important feature of linear interactions which does not carry over to log-linear interactions, and this can be illustrated here. We note that while $\hat{\psi}^{134}$ is a function of the 134 marginal proportions (p_{i+kl}) and the 12 and the 234 marginal proportions, it is not a function of the 134 marginal proportions alone. Moreover, if the difference between the models with generating classes $\{12, 134, 24\}$ and $\{12, 13, 14, 34, 24\}$, which again depends on ψ^{134}, was the matter in question, then the relevant $\hat{\psi}^{134}$ would now be a function of the 134, the 12, and the 24 marginal proportions. By contrast, in a factorial experiment with equal numbers of observations per cell, the estimate of ψ^{134} is the same for any hierarchical model that does not restrict ψ^{134}, and is moreover a function only of the observed 134 marginal means $(y_{i \cdot kl})$.

DECOMPOSABLE MODELS

The property that we have just noted to be lacking in general multiplicative models is present in the subclass of *decomposable* models. These were first explored fully by Haberman [6]. One way of characterizing decomposable models is given in the final section. Meanwhile consider an example in which $d = 5$ and the generating class is $\{125, 134, 145\}$. Simply by being a multiplicative model it causes π_{ijkl} to be some function of $(\pi_{ij++m}), (\pi_{i+kl+}), (\pi_{i++lm})$. However, the special, decomposable, nature of the generating class makes the function particularly simple, namely

$$\pi_{ijkl} = \frac{\pi_{ij++m}\pi_{i+kl+}\pi_{i++lm}}{\pi_{i++l+}\pi_{i+++m}}.$$

Note that the marginal probabilities in the denominator belong to 14 and 15. These two subsets are related to $125, 134, 145$ by $14 = 134 \cap 145$, $15 = 125 \cap 145$. The marginal probability π_{i++++} corresponding to 1

$= 125 \cap 134$ is, in a sense, also present in the denominator but it has canceled with π_{i++++}, arising from $1 = 125 \cap 134 \cap 145$, in the numerator. It is now easy to see that ψ^{134} is a function only of the 134 marginal probabilities (π_{i+kl+}).

The numbers of decomposable models for $d = 2, 3, 4, 5$ are 5, 18, 110, and 1233, corresponding to the numbers of multiplicative models, which are 5, 19, 167, and 7580. For further discussion of the classes of multiplicative and decomposable models, and of the intervening class of *graphical* models, see Darroch et al. [4]. See also Wermuth [10], in whose terminology decomposable models are called multiplicative.

INTERACTIONS AND MODELS FOR GENERAL d

Here we outline a framework in which results about interactions and models can be proved for general values of d. Proofs are given by Darroch and Speed [3].

We continue briefly with $d = 4$ and the additive model for η having generating class $G = \{12, 134, 234\}$, namely,

$$\eta = \alpha^{12} + \alpha^{134} + \alpha^{234}. \quad (2)$$

Let T^{12} denote the linear operator, acting on η, which replaces η_{ijkl} by

$$\eta_{ij**} = \sum_{kl} w_k^3 w_l^4 \eta_{ijkl}.$$

Similarly define T^h for any subset h of $S = \{1, 2, 3, 4\}$, in particular $T^S = I$, the identity operator, and T^ϕ, the operator that replaces η_{ijkl} by η_{****}. These *averaging operators* T^h have the property

$$T^{h_1}T^{h_2} = T^{h_1 \cap h_2}. \quad (3)$$

Now define the operator V^G by

$$I - V^G = (I - T^{12})(I - T^{134})(I - T^{234}). \quad (4)$$

It is easy to prove that η satisfies (2) if and only if

$$(I - V^G)\eta = 0, \quad \text{that is,} \quad \eta = V^G\eta. \quad (5)$$

Because of this it is appropriate to call V^G the *model operator*. On multiplying out the right-hand side of (4) and using (3), it is found that

$$V^G = T^{12} + T^{134} + T^{234}$$
$$- T^1 - T^2 - T^{34} + T^\emptyset,$$

so that (2) is equivalent to

$$\eta_{ijkl} = \eta_{ij**} + \eta_{i*kl} + \eta_{*jkl}$$
$$- \eta_{i***} - \eta_{*j**} - \eta_{**kl} + \eta_{****}.$$

Note that this equivalence is established without any mention of interactions. How do they relate to V^G?

We can now consider general d and for each subset h of $S = \{1, \ldots, d\}$ define the *interaction operator* U^h by

$$U^\emptyset = T^\emptyset, \qquad U^1 = T^1(I - T^{2\cdots d}),$$

$$U^{12} = T^{12}(I - T^{23..d})(I - T^{13..d}),$$

and, for general h, by

$$U^h = T^h \prod_{\sigma \in h} (I - T^{S-\sigma}),$$

where $S - \sigma$ denotes the set S with the element σ removed. If required, the expression for U^h is easily multiplied out giving, for example,

$$U^{12} = T^{12} - T^1 - T^2 + T^\emptyset,$$
$$U^{123} = T^{123} - T^{23} - T^{13} - T^{12}$$
$$+ T^1 + T^2 + T^3 - T^\emptyset.$$

The interaction operators U^h are easy to work with, a fact exemplified by the proof of the basic result

$$\sum_{h \in S} U^h = I. \qquad (6)$$

The left-hand side of (6) is

$$\sum_{h \in S} T^h \prod_{\sigma \in h} (I - T^{S-\sigma})$$

$$= \sum_{h \in S} \prod_{\sigma \in h'} T^{S-\sigma} \prod_{\sigma \in h} (I - T^{S-\sigma})$$

$$= \prod_{\sigma \in S} \left[T^{S-\sigma} + (I - T^{S-\sigma}) \right]$$

$$= \prod_{\sigma \in S} I = I,$$

h' denoting the complement of h in S. Other easily proved properties of the U^h are

$$(U^h)^2 = U^h, \quad U^{h_1} U^{h_2} = 0, \qquad h_1 \neq h_2.$$
$$(7)$$

The connection between V^G and the U^h in the case $G = \{12, 234, 134\}$ is

$$V^G = U^\emptyset + U^1 + U^2 + U^3 + U^4$$
$$+ U^{12} + U^{13} + U^{14}$$
$$+ U^{23} + U^{24} + U^{34} + U^{234} + U^{134}.$$

Thus, because of (5) and (6), the model for η can be expressed as

$$(U^{123} + U^{124} + U^{1234})\eta = 0,$$

and, by virtue of (7), this is equivalent to

$$U^{123}\eta = U^{134}\eta = U^{1234}\eta = 0.$$

In the case of a general G,

$$I - V^G = \prod_{g \in G} (I - T^g),$$

and it is easy to prove that

$$\eta = \sum_{g \in G} \alpha^g \quad \text{if and only if} \quad \eta = V^G \eta.$$

Moreover, if $\bar{G}$ is the *closure* of G, that is,

$$\bar{G} = \{h; h \subset g, g \in G\},$$

then it can be proved that

$$V^G = \sum_{h \in \bar{G}} U^h$$

so that

$$\eta = \sum_{g \in G} \alpha^g \quad \text{if and only if} \quad \eta = \sum_{h \in \bar{G}} U^h \eta$$

$$\text{if and only if} \quad U^h \eta = 0$$

$$\text{for all } h \notin \bar{G}.$$

Finally, we note that G is *decomposable* if, for some order $g_1, \ldots, g_k$ of its elements,

$$(g_1 \cup \cdots \cup g_\alpha) \cap g_{\alpha+1} = g_{\beta_\alpha} \cap g_{\alpha+1},$$

where $\beta_\alpha \in \{1, \ldots, \alpha\}$, for $\alpha = 1, \ldots, k - 1$. Denoting the set $g_{\beta_\alpha} \cap g_{\alpha+1}$ by h_α, it can be shown that the probability $\pi(\mathbf{i})$ at cell $\mathbf{i}$ can be expressed in terms of the marginal probabilities $\pi^{g_\alpha}(\mathbf{i}_{g_\alpha})$, where $\mathbf{i}_h$ is the h subtuple of $\mathbf{i}$, by the formula

$$\pi(\mathbf{i}) = \frac{\prod_{\alpha=1}^{k} \pi^{g_\alpha}(\mathbf{i}_{g_\alpha})}{\prod_{\alpha=1}^{k-1} \pi^{h_\alpha}(\mathbf{i}_{h_\alpha})}.$$

This formula shows that each interaction array ψ^{g_α} depends only on $(\pi^{g_\alpha}(\mathbf{i}_{g_\alpha}))$.

References

[1] Birch, M. W. (1963). Maximum likelihood in three-way contingency tables. *J. Statist. Soc. B*, **25**, 220–233. (Introduced the θ system for contingency tables.)

[2] Bishop, Y. M. M., Fienberg, S. E., and Holland, P. W. (1975). *Discrete Multivariate Analysis: Theory and Practice*. MIT Press, Cambridge, Mass. (A large book which covers much contingency table ground at a leisurely pace. θ system.)

[3] Darroch, J. N. and Speed, T. P. (1984). Additive and multiplicative models and interactions, *Ann. Statist.*, **12** (to appear).

[4] Darroch, J. N., Lauritzen, S. L., and Speed, T. P. (1980). Markov fields and log-linear interaction models for contingency tables. *Ann. Statist.*, **8**, 522–539.

[5] Fienberg, S. E. (1977). *The Analysis of Cross-Classified Data*. MIT Press, Cambridge, Mass. (Lucid, incisive exposition of central contingency table topics. θ system.)

[6] Haberman, S. J. (1974). *The Analysis of Frequency Data*. University of Chicago Press, Chicago. (Source of much fundamental theory of interaction models for contingency tables. Notation and mathematical rigor demanding on the reader. θ system.)

[7] Mantel, N. (1966). Models for complex contingency tables and polychotomous dosage response curves. *Biometrics*, **22**, 83–95. (Introduced the θ system for contingency tables.)

[8] Plackett, R. L. (1974). *Analysis of Categorical Data*. Hafner, New York. (A small book which very concisely covers a large part of contingency table territory. θ system.)

[9] Upton, G. J. G. (1978). *The Analysis of Cross-tabulated Data*, Wiley, Chichester, and Halstead, New York.

[10] Wermouth, N. (1976). Model selection among multiplicative models. *Biometrics*, **32**, 253–263.

(CONTINGENCY TABLES
GLIM
INTERACTION
ITERATIVE PROPORTIONAL FITTING
MULTIDIMENSIONAL CONTINGENCY
   TABLES)

J. N. DARROCH

INTERACTIVE DATA ANALYSIS

The application of statistical methods to the analysis of data has long been recognized as an iterative process in which tentative models and hypotheses, as well as techniques of analysis, often need to be modified in the light of the cumulative evidence obtained. In this broad sense, "interactive data analysis" has much in common with, and can even be said to be synonymous to, "statistical data analysis" or "exploratory data analysis" (*see* EXPLORATORY DATA ANALYSIS). However, because of the essential role "interactive computing" has played toward the effective execution of data analysis, interactive data analysis has, in current usage, acquired the identity of "statistical data analysis in an interactive computing environment." Among the earliest systems developed for interactive data analysis are COMB (Console-Oriented Model Building) and COSMOS (Console-Oriented Statistical Matrix Operator System). Examples illustrating the use of these systems are given by Schatzoff [5].

TWO PRINCIPAL COMPONENTS OF INTERACTIVE DATA ANALYSIS

It can be said that interactive data analysis is comprised of only two essential components, which we now discuss.

The Approach: Flexible and Iterative Probing of Data

Interactive data analysis is the expedient execution of an enlightened data analysis. At the exploratory stage, the analyst does numerical detective work via graphic and semigraphic displays, and often explores a variety of tasks in data editing, smoothing, and reexpression (transformation). At the model building or inferential stage, the analyst acts both as a sponsor and a critic of one or more tentative models. The recurrent themes are: analysis of residuals, analytic and graphic diagnostics, reexpressions, and resistance (robustness) and sensitivity considerations. At the conclusion of an iterative process of probing, confirmatory analysis, if it is done at all, is seldom a major part of the entire analysis. Since the steps involved in interactive data analysis are highly flexible as well as highly individualized (as a function of the problem on hand and the preferred techniques of the analyst) it is not possible to detail the steps of a "typical" analysis by a flowchart. For expository articles elaborating on the philosophy and the general approach to interactive data analysis, see Tukey [7], Ling and Roberts [3], and Box [1].

The Execution: An Interactive Computing Environment

The concept of "interactive computing" has gone through a rapid evolution in the decade of the 1970s. By about 1975, the terms "interactive," "terminal-oriented" or "console-oriented" (or time sharing), and "remote entry of jobs from a terminal" were no longer synonymous. An up-to-date interactive system for data analysis must be capable of supporting a high level of flexible, efficient, human–machine interaction in data analysis. Many such systems have been publicly available only since about 1975. The most widely used systems for interactive data analysis in the early 1980s are self-contained statistical software* packages (e.g., CADA, Consistent System, IDA, GLIM*, MIDAS, MINITAB*, PSTAT, ROSEPACK, and SCSS; see Francis [2]). Other systems are based on the direct use of certain programming languages such as APL, supported by special-purpose routines written in those languages (see, e.g., Schatzoff et al. [6], McNeil [4], and Velleman and Hoaglin [8].

In short, interactive data analysis is a mode of statistical data analysis which makes effective use of the power and flexibility of interactive computing in harmony with the attitude and approach of data analysis.

References

[1] Box, G. E. P. (1976). *J. Amer. Statist. Ass.*, **71**, 791–799. (An expository article on the philosophy of data analysis, with emphasis on "motivated iteration" as the cornerstone for scientific inquiry and the interface between theory and practice.)

[2] Francis, I., ed. (1979). *A Comparative Review of Statistical Software*. International Association for Statistical Computing, Voorsburg, The Netherlands. (An evaluation and presentation of 46 major statistical computing packages. Contains a useful bibliography.)

[3] Ling, R. F. and Roberts, H. V. (1975). *J. Bus.*, **48**, 411–451. [An introductory article on the use of IDA (Interactive Data Analysis) in teaching and research.]

[4] McNeil, D. R. (1977). *Interactive Data Analysis*. Wiley, New York. (Describes and illustrates the use of APL functions and FORTRAN subroutines in an interactive computing environment for exploratory data analysis.)

[5] Schatzoff, M. (1968). *J. Amer. Statist. Ass.*, **63**, 192–208. (A pioneer work on interactive data analysis.)

[6] Schatzoff, M., Bryant, P., and Dempster, A. P. (1975). In *Perspectives in Biometrics*, R. M. Elashoff, ed. Academic Press, New York, pp. 1–28. (Interactive statistical computation with large data structures using APL, with comments on interactive data analysis systems.)

[7] Tukey, J. W. (1962). *Ann. Math. Statist.*, **33**, 1–67. (A classic article on data analysis.)

[8] Velleman, P. F. and Hoaglin, D. C. (1981). *Applications, Basics, and Computing of Exploratory Data Analysis*. Duxbury Press, Boston. (Interactive data analysis, based on Tukey's approach, using APL.)

FURTHER READING

See the following works, as well as the references just given, for more information on the topic of interactive data analysis.

Ling, R. F. (1980). *Commun. ACM*, **23**, 147–154. (An expository article discussing special features in computing packages that facilitate genuine human–machine interaction in interactive data analysis.)

Tukey, J. W. (1977). *Exploratory Data Analysis*. Addison-Wesley, Reading, Mass. (A definitive account of the Tukey approach to exploratory data analysis.)

(EXPLORATORY DATA ANALYSIS
STATISTICAL SOFTWARE)

ROBERT F. LING

INTERBLOCK INFORMATION

As its name leads one to expect, this term usually refers to the analysis of data from experiments in which there is some form of blocking*. The standard linear (parametric) model* for the response (X_{ij}) for the j-th treatment in the i-th block is of form

$$X_{ij} = \beta_i + \tau_j + Z_{ij} \qquad \text{(A)}$$

where the β's and τ's are parameters representing block and treatment effects, respectively, and the Z's are independent random variables identically distributed with expected value zero.

Interblock information (on differences among the treatment effects (τ)) is contained in differences $(X_{ij} - X_{i'j})$ where $i \neq i'$—that is, between responses in *different* blocks. For the analysis of the parametric model (A) above, these quantities cannot be used for inferences on the τ's because of the confounding effect of the block terms:

$$X_{ij} - X_{i'j'} = \beta_i - \beta_{i'} + \tau_j - \tau_{j'} + Z_{ij} - Z_{i'j'}.$$

Although block effects can be removed by using, for example

$$(X_{ij} - X_{i'j'}) - (X_{ij''} - X_{i'j'''})$$

$$= \tau_j - \tau_{j'} - \tau_{j''} + \tau_{j'''} + Z_{ij} - Z_{i'j'} - Z_{ij''} + Z_{i'j'''}$$

this statistic can be expressed in terms of the intrablock differences $(X_{ij} - X_{ij''})$ and $(X_{i'j''} - X_{i'j'''})$, so it does not contribute any extra information.

However, if the β's are replaced by independent identically distributed random variables (also independent of the Z's), interblock differences can be used to contribute additional information on treatment effects. (*See* BLOCKS, BALANCED COMPLETE.)

(COMPONENTS OF VARIANCE)

INTERDECILE RANGE

The interdecile range is computed as the difference between the ninth and first deciles. It comprises the middle 80% of the population (or the sample).

(INTERPERCENTILE DISTANCES
INTERQUARTILE RANGE
QUANTILES)

INTERDEPENDENCE ANALYSIS *See* ELIMINATION OF VARIABLES

INTEREFFECT ORTHOGONALITY

In analyzing factorial designs, the experimenter is interested in drawing conclusions about contrasts* belonging to factorial effects of the treatments.

If the best linear estimators (BLEs) of estimable contrasts belonging to different effects are uncorrelated (mutually orthogonal*), then intereffect orthogonality is valid. In this case the treatment sum of squares (SS) can be split up orthogonally into components due to different effects and then shown in the same ANOVA table*.

Seber [4] derived necessary and sufficient conditions for intereffect orthogonality for completely randomized designs*. Various sufficient conditions under which intereffect orthogonality holds were given by Kurkjian and Zelen [2] and Cotter et al. [1]. A general set of necessary and sufficient conditions for intereffect orthogonality for both connected and disconnected designs in terms of the incidence matrix* of a design were derived by Mukerjee [3].

References

[1] Cotter, S. C., John, J. A., and Smith, T. M. F. (1973). *J. R. Statist. Soc. B*, **35**, 361–367.

[2] Kurkjian, B. and Zelen, M. (1963). *Biometrika*, **50**, 63–73.

[3] Mukerjee, R. (1979). *Bull. Calcutta Statist. Ass.*, **28**, 109–112, 83–108.

[4] Seber, G. A. F. (1964). *Ann. Math. Statist.*, **35**, 705–710.

(FACTORIAL EXPERIMENTS)

INTERFRACTILE DISTANCES *See* INTERPERCENTILE DISTANCES

INTERFRACTILE RANGE *See* INTERQUARTILE RANGE

INTERNAL ADDITIVE CONSISTENCY OF ESTIMATORS

A term used in estimation theory, in particular in connection with sample design estimators. Estimators $\hat{\theta}_1(y)$, $\hat{\theta}_2(y)$, and $\hat{\theta}_3(y)$ are internally additively consistent for θ_1, θ_2, and θ_3 if

$$\hat{\theta}_1(y) + \hat{\theta}_2(y) = \hat{\theta}_3(y)$$

for $\theta_1 + \theta_2 = \theta_3$.

(ESTIMATION, POINT
SURVEY SAMPLING)

INTERNAL LEAST SQUARES

Internal least squares were introduced by Hartley [1] to obtain initial estimates of parameters in a nonlinear (originally exponential) regression model. The essence of the procedure is to relate the dependent variable to its own repeated sums. Estimates of the coefficients of the independent variables* are then obtained by regressing on these sums (using the least-squares* procedure).

Further investigations in this area by Hartley [2], Patterson and Lipton [3], Shah and Khatri [6], and Shah [4] revealed that the estimates obtained using this method are approximately 90% efficient as compared with those obtained using ordinary least-squares procedures. Shah [5] extended this method to the case of unequally spaced data.

References

[1] Hartley, H. O. (1948). *Biometrika*, **35**, 32–45.

[2] Hartley, H. O. (1959). *Biometrika*, **46**, 293–295.

[3] Patterson, H. D. and Lipton, S. (1959). *Biometrika*, **46**, 281–292.

[4] Shah, B. K. (1965). *Technometrics*, **7**, 59–65.

[5] Shah, B. K. (1979). *Biometrics*, **35**, 497–502.

[6] Shah, B. K. and Khatri, C. G. (1963). *J. Indian Statist. Ass.*, **1**, 202–214.

(LEAST SQUARES
NONLINEAR REGRESSION)

INTERNATIONAL STANDARDIZATION: APPLICATION OF STATISTICS

The primary goal of the International Organization for Standardization (ISO) is the removal of technical barriers to trade. ISO thus aims to facilitate commercial exchanges by setting up international agreements in all branches of industry, agriculture, and commerce. Only the technical field of electronics and electrotechnology is not covered by ISO, as this falls under the responsibility of IEC (International Electrotechnical Commission).

Like all Standards, ISO Standards try to find a common ground and to obtain a consensus reconciling the interests of governments, manufacturers, and users. This leads to the publication of International Standards on specifications which define "quality models," Standards of testing methods which supplement them, and finally, Standards describing the tools and methods serving to check the conformity of the elements to the model. It is in this context that statistical methods serving to take account or the variability inherent in any natural product and any manufacturing process occupy a dominant place in international standardization. A very large number of theoretical and technical methods have been developed in recent years, and it is of fundamental importance to guarantee the consistency of this series in order to facilitate its use. The ISO Technical Committee (TC) responsible for examining "applications of statistical methods" is Committee ISO/TC 69.

Nearly 50 countries throughout the world observe the proceedings of this TC, and about 10 of them (European, North American countries, and Japan) participate regularly and actively in the annual meetings of the six most important technical subcommittees which make up the structure of the Committee.

The field of work of each subcommittee (SC) is:

ISO/TC 69/SC1: Terminology and symbols

ISO/TC 69/SC2: Interpretation of statistical data

ISO/TC 69/SC3: Application of statistical methods in standardization—establishment of guidelines for users

ISO/TC 69/SC4: Statistical quality control* (control charts*)

ISO/TC 69/SC5: Acceptance sampling*

ISO/TC 69/SC6: Application of precision data

The major topics dealt with can be divided as follows:

1. Standards dealing with basic techniques:
    a. Essential terminology for clarification of the language
    b. General methodologies such as the problems of estimation and tests, which serve to provide the practitioner with the various procedures of statistical treatment of test results, in the form of a pragmatic approach
2. Standardization dealing with special application techniques:
    a. The evaluation of parameters defining the precision of a testing method and the special applications deriving therefrom
    b. Acceptance and manufacturing inspection which are of interest to all industry sectors

The activities of TC 69 attempt to establish a link between mathematicians and practitioners on the level of statistical techniques.

Following are some details concerning the various activities listed above.

TERMINOLOGY. Work on terminology has dealt essentially with the calculation of probabilities, general statistical terms, and those related to sampling inspection* methods (ISO 3534).

Although this work required many years to accomplish, the responsible Committee is planning to expand the glossary to include control charts, for example, and to revise certain terms such as defects, defective, etc., for which the commercial and indeed legal implications are of fundamental importance owing to the predominant role played at present by economic, safety, and product liability considerations. Our efforts are also devoted to the compilation of a glossary on experimental designs* which, independent of general terms, covers the types of design, factorial experiments*, nested experiments* and analytical methods.

STATISTICAL INTERPRETATION OF DATA. As for the statistical interpretation of data, the published Standards deal with techniques that can be used to estimate the mean or the population variance from samples, and to perform comparative tests relative to the value of these parameters (ISO 2602, ISO 2854, ISO 3207, and ISO 3301).

The analysis of the effectiveness of these comparative tests, which is based on a very comprehensive series of diagrams (ISO 3494), can be used to resolve two routine problems occurring in contractual client/ supplier relationships:

1. Once the sample size to be taken from the batch has been fixed, to define the type II risk associated with the test selected
2. Once the type II risk has been fixed, to determine the sample size to be taken

As these different techniques imply that the distribution of the characteristic analyzed is normal or close to normal*, a choice of graphic and numerical (directional, multidirectional, omnibus) tests serving to check this assumption will be described in ISO/ DIS 5 479.

Standards on estimation and tests on proportions are in preparation.

STATISTICAL ACCEPTANCE INSPECTION TESTS. These are the subject of intense standardization activity. In fact, the standardization of sampling procedures attempts to attain different objectives: a technical objective by the clear definition of basic criteria to be used to check the quality of conformance of a product, the simplification and harmonization of practices, etc.; an economic objective by the search for solutions that achieve a reduction in direct inspection costs. Apart from Standards on attribute* inspection (ISO 2859) and variable inspection (DIS 3951), which are well known because they are a by-product of MIL STD 105D and 414, the Working Group concerned plans to compile a set of rules and sampling plans capable of coping with diversified situations frequently encountered, the use of which can be generalized to include a large number of industry sectors. As ISO 2859 deals essentially with the case of the inspection of continuous series of lots, it proved necessary to deal more specifically with the problems of isolated lots, small lots, costly products, intermittent production, etc.

A draft Standard recently prepared develops a series of plans complementary to those of ISO 2859, indexed by limiting quality (LQ), thus making protection of the client against the acceptance of batches of "poor" quality an essential parameter for the choice of a sampling plan.

The other short-term activities will deal substantially with continuous attribute sampling plans, based on Standard MIL STD 1235A, sampling plans with zero acceptance criteria in connection with problems of product liability and safety, together with sequential sampling* plans.

A guide for the selection of a suitable sampling plan for a given situation will help both client and supplier firms to pinpoint the basic principles of an inspection plan, and to make a choice that meets the real technical, economic, and commercial requirements.

APPLICATIONS OF ACCEPTANCE INSPECTION. As the applications are essentially limited to lots for which a production operation has been terminated, those of in-manufacture

inspection allow the implementation of manufacturing follow-up and the possibility of rational intervention in the production process to maintain quality at the desired level. The work program planned in this area is vast and aims to cover a broad range of techniques, the uses of which may extend beyond the product industries to include, for example, medical, administrative, and commercial sectors.

In this context, the various subjects adopted today are concerned with the following:

1. The preparation of a general document describing the concepts and the philosophy of the different types of control chart*

2. The description of principles, construction methods, and interpretation of the following charts:

   a. Cumulative sum (cusum*) control charts

   b. Control charts of arithmetic means by the technique of quality groups

   c. Control charts of the number of defective items

QUANTITATIVE DEFINITION OF VARIABILITY. A quantitative definition of variability should be associated with every Standard testing method. The determination of the precision* of a testing method in terms of repeatability* and reproducibility* (extreme measurements of variability) forms the subject of Standard ISO 5725, which describes the organization and analysis of interlaboratory tests. This basic work will have sweeping repercussions on any standardization activity concerned with the definition of a reference test.

Since this must be continued by the examination of practical applications of repeatability and reproducibility, international standardization is interested in the following problems:

1. Calibration* and recalibration of equipment, and their statistical implications

nature for the precision of measurement results

2. The acceptability of test results based on samples: namely, the acceptance procedure of one or more results compared with repeatability and reproducibility criteria

3. Principles of determination of specification limits, as the latter must take account of the precision of the method adopted

4. Jointly with the study above, that of the decision procedure on conformity to specifications, especially as a function of the types of requirements, the number of individuals subjected to the test, and the risks associated with the decision

Although it is handled with an undeniable pedagogical bias, the standardization of these various statistical techniques poses some difficulties of interpretation and adaptation with respect to sectorial Standards studies (product Standards), whether national or international, and these difficulties are passed on to the level of practical applications. Hence it appeared essential to prepare a guide for the use of statistical methods in standardization, drafted for the use of nonstatisticians, highlighting the primary areas, such as sampling methods, inspection, measurement errors*, regression*, etc., orienting the use toward appropriate solutions to the problem at hand, and analyzing the traditional types of erroneous interpretation encountered in practice. This study, which meets a need that has been growing in recent years, should help the mission of TC 69: namely, to promote the quality of products and services. It should be noted that standardization is a dynamic and continuous process, and that independent of periodic revisions of Standards which are self-imposed by the ISO Committees in accordance with the development of concepts and techniques, the program is reviewed regularly with new questions considered, while those mentioned above represent the most advanced stage of our investigations.

(ACCEPTANCE SAMPLING
CALIBRATION
CONTROL CHARTS
DESIGN OF EXPERIMENTS
MEASUREMENT ERROR
QUALITY CONTROL, STATISTICAL
REPEATED MEASUREMENTS
SAMPLING)

C. MERLE

INTERNATIONAL STATISTICAL ECOLOGY PROGRAM

The First International Symposium on Statistical Ecology was held in 1969 at Yale University with support from the Ford Foundation and the U.S. Forest Service. The three symposium co-chairmen (G. P. Patil, E. C. Pielou, and W. E. Waters) represented the fields of statistics, theoretical ecology, and applied ecology.

It became clear that a focal forum was necessary to discuss and develop a constructive interface between quantifiable problems in ecology and relevant quantitative methods. As a partial solution to fill this need at professional organizations' level, the director of the symposium (G. P. Patil) made certain recommendations to the presidents of the International Association for Ecology (A. D. Hasler), the International Statistical Institute* (W. G. Cochran), and the International Biometric Society (P. Armitage). The International Association for Ecology (INTECOL) took a timely step in creating a section in its organization, the statistical ecology section. The three societies together set up a liaison committee on statistical ecology. The INTECOL Section and the Liaison Committee together developed the International Statistical Ecology Program. Since its inception in 1970, ISEP (as it has come to be known) has put emphasis on identifying the interdisciplinary needs of statistics and ecology.

The First Advanced Institute on Statistical Ecology in the United States was organized at Pennsylvania State University for six weeks in 1972 with support from the U.S.

National Science Foundation, the U.S. Forest Service, and the Mathematical Social Sciences Board. With support from the UNESCO program of Man and Biosphere, a six-month program was held in Venezuela for participants from Latin America in 1974 under the direction of Jorge Rabinovich. With some initiatives from ISEP, special statistical ecology sessions have been held at the international conferences of the International Statistical Institute and the Biometric Society*.

While plans were being made for the Second International Congress of Ecology, the then Secretary General and current (1982) President of INTECOL (G. A. Knox), and the ISEP Chairman (G. P. Patil) discussed the need for a program in statistical ecology. Out of this emerged the Satellite Program in Statistical Ecology, to further interdisciplinary research and training in statistical ecology and ecological statistics. Recent activities include research workshops and seminars at College Station (Texas), Berkeley, Parma (Italy), and Jerusalem. Funding came from a variety of sources, but especially the NATO Scientific Affairs Division and the U.S. National Marine Fisheries Service. Requests for information on the program's activities, and suggestions, may be addressed to the ISEP Chairman: Professor G. P. Patil, 318 Pond Laboratory, Pennsylvania State University, University Park, Pennsylvania 16802. The present (1982) members of the Statistical Ecology Section are P. Berthet, J. Cancela da Fonseca, L. Orloci, O. Rossi, D. Simberloff, and W. E. Waters. The present members of the Liaison Committee are D. G. Chapman, R. M. Cormack, C. S. Holling, B. Matern, E. C. Pielou, C. R. Rao, D. S. Robson, and W. G. Warren.

A 13-volume series of books summarizes many of the lectures presented at ISEP functions. The first three (eds. G. P. Patil, E. C. Pielou, and W. E. Waters), published by the Pennsylvania State University Press in 1971, treat (1) spatial patterns and statistical distribution; (2) sampling and modeling biological populations and population dynamics; and (3) many-species populations, ecosys-

tems, and system analysis. The remaining volumes, published in 1979 by the International Co-operative Publishing House, deal with (4) statistical distributions in ecological work (eds. J. K. Ord, G. P. Patil, and C. Taillie); (5) sampling biological populations (eds. R. M. Cormack, G. P. Patil, and D. S. Robson); (6) ecological diversity in theory and practice (eds. J. F. Grassle, G. P. Patil, W. K. Smith, and C. Taillie); (7) multivariate methods in ecological work (eds. L. Orloci, C. R. Rao, and W. M. Stiteler); (8) spatial and temporal analysis in ecology (eds. R. M. Cormack and J. K. Ord); (9) systems analysis of ecosystems (eds. G. S. Innis and R. V. O'Neill); (10) compartmental analysis of ecosystem models (eds. J. H. Matis, B. C. Patten, and G. C. White); (11) environmental biomonitoring, assessment, prediction, and management—certain case studies and related quantitative issues (eds. J. Cairns, Jr., G. P. Patil, and W. E. Waters); (12) contemporary quantitative ecology and related ecometrics (eds. G. P. Patil and M. L. Rosenzweig); and (13) quantitative population dynamics (eds. D. G. Chapman and V. F. Gallucci).

For additional information, see ref. 1.

Reference

[1]   Patil, G. P. (1979). *Int. Statist. Rev.*, **47**, 223–228.

(ECOLOGICAL STATISTICS)

G. P. PATIL

INTERNATIONAL STATISTICAL INSTITUTE (ISI)

The International Statistical Institute (ISI) was established in 1885 and is thus one of the oldest international scientific associations functioning in the modern world. The institute is an autonomous society which seeks to develop and improve statistical methods and their applications through the promotion of international activity and co-operation.

The ISI is composed of some 1100 individual elected or ex officio members who are the world's leading statisticians. The membership is drawn from over 120 countries. Membership of the Institute's sections is not restricted to ISI members.

In addition to the activities which are traditionally the concern of learned societies, such as conferences, journals, occasional publications, study committees, etc., the ISI has been actively involved in special programs. Two such, which are making important contributions toward economic and social development are undertaken at present: the World Fertility Survey* and the Statistical Education and Training Program. The ISI operates on an annual budget of approximately $5 million; a large part of the ISI budget is at present earmarked for the World Fertility Survey.

The permanent office is at 428 Prinses Beatrixlaan, 2270 AZ Voorburg (near The Hague). The current Director of the Permanent Office is E. Lunenberg and the current (1979–1981) president of the ISI Bureau is E. Malinvaud.

To promote the ideals and objectives of the Institute, four sections have been established:

1. The International Association for Regional and Urban Statistics (formerly: International Association of Municipal Statisticians, founded in 1957)
2. The Bernoulli Society for Mathematical Statistics and Probability (formerly: International Association for Statistics in Physical Sciences, founded in 1961)
3. The International Association of Survey Statisticians, founded in 1971
4. The International Association for Statistical Computing, founded in 1977

All sections are open to interested persons who wish to participate in their activities and meet the obligations of annual membership dues.

There are four classes of ISI membership, two of which involve election to the Institute.

Honorary members are elected from the existing ranks of ordinary members in recognition that their contributions to statistics merit special honor.

Ordinary members are elected by virtue of their distinguished contributions either to the development or applications of statistical methods or to the administration of statistical services.

The two remaining membership categories are *ex officio* and *corporate* memberships. They have restricted voting rights.

The *general assembly* of the ISI is composed of the individual (honorary, ordinary, and ex officio) members of the Institute and is usually convened biennially. The governance of the Institute is vested in the assembly.

The *bureau* is composed of the Institute's officers (president, president elect, and five vice presidents) who are elected by the general assembly. It is the administrative body of the Institute.

The *permanent office* has a director who is appointed by the Bureau and acts as secretary/treasurer of the ISI. Under the supervision of the Bureau, the director is in charge of the Institute's activities.

The ISI convenes in ordinary or special sessions. As a general rule, the ordinary sessions are held biennially. The first such conference was held in Rome in 1887 and the series has continued except for interruptions during the two world wars. The 44th and 45th sessions are scheduled to take place in Madrid (1983) and The Hague (1985), respectively.

The ISI undertook in 1972 the World Fertility Survey directed by Sir Maurice Kendall in cooperation with the International Union for the Scientific Study of Population, and with the collaboration of the United Nations. Since 1949 the ISI has administered an International Statistical Education Programme, with support from UNESCO and other sources. ISI has produced various teaching aids over the years. These include "A Dictionary of Statistical Terms" by M. G. Kendall and W. R. Buckland, which is now in its fourth edition, and various statistical bibliographies and glossaries. The ISI supports the teaching of statistics by organizing round table meetings, each of which consists of a small, select group of experts in statistical education from various parts of the world.

The ISI has for many years been active in the field of municipal statistics. In cooperation with the International Union of Local Authorities, a number of specialized publications with data on towns and agglomerations with populations of over 100,000 have been produced.

The Institute established a Committee on the Integration of Statistics, which studied gaps existing between the various specialist statistical interest groups and recommended ways and means of promoting closer collaboration, particularly between official and academic statisticians. (The full report of the Committee's deliberations is published in the 1980 volume of the ISI journal, *International Statistical Review**.)

The ISI produces the following periodical publications:

1. **Bulletin of the International Statistical Institute.** The proceedings of the biennial Sessions of the ISI are reproduced in the *Bulletin*, which, consequently, usually appears every two years. The *Bulletin*, of which Volume I was published in 1886, may be ordered from the ISI.

2. **International Statistical Review*.** The *Review* is the main journal of the ISI and, since 1933, it has been published three times a year.

3. **Statistical Theory and Method Abstracts*.** The ISI has established an abstracting service of statistical publications with editors and correspondents all over the world. The abstracts journal has appeared four times per year since 1959.

4. **The ISI Newsletter, *International Statistical Information.*** The circulation is restricted to members of the ISI and of its sections.

5. Two annual directories providing information on statistical societies and agencies are published by the ISI: the *Directory of Statistical Societies* and the *Directory of National Statistical Agencies*.

Additional information concerning the ISI and its activities can be obtained by writing to the permanent office at the address given above.

(BERNOULLI SOCIETY
INTERNATIONAL STATISTICAL REVIEW
STATISTICAL THEORY AND METHOD
  ABSTRACTS
UNITED NATIONS,
  STATISTICAL ASPECTS)

INTERNATIONAL STATISTICAL REVIEW

The *International Statistical Review* is published by the International Statistical Institute* and is, at present (1982), an official journal of the Bernoulli Society for Mathematical Statistics and Probability*, the International Association for Statistical Computing, and the International Association of Survey Statisticians as well as of the International Statistical Institute itself.

The Review publishes (1) original research papers, provided that these are of a wide interest and not too technical; (2) papers providing an integrated critical survey of some field of statistics or probability or giving an exposition of important methodological developments; and (3) papers on the history of statistics and probability. The journal also publishes brief reports on recent activities in the statistical world concerning developments of computer facilities, major survey programs, teaching methods and experiences, etc. All papers are refereed.

The pages of the Review are open to all authors on statistics and probability whether members of the ISI or one of its sections or not, the only limitations on the publishability of a paper being those set by the requirements of quality and the general guidelines described above. The language used must be either English or French.

The Review is published three times a year and the current (1982) subscription price is £16 (U.S. $42). The current (1982) editor is O. E. Barndorff–Nielsen (Aarhus, Denmark) and the publisher is Longman Group Ltd., Longman House, Burnt Mills, Harlow, Essex, England.

The contents of the August 1981 issue (Volume 49, Number 2) is as follows: K. M. Abdelbasit and R. L. Plackett, "Experimental Design for Categorized Data," p. 111; P. Groeneboom and J. Oosterhoff, "Bahadur Efficiency and Small-Sample Efficiency," p. 127; C. C. Heyde, "Invariance Principles in Statistics," p. 143; A. H. Andersen, E. B. Jensen, and G. Schou, "Two-Way Analysis of Variance with Correlated Errors," p. 153; P. J. Bickel, D. M. Chibisov, and W. R. van Zwet, "On Efficiency of First and Second Order," p. 169; H. O. Lancaster, "A Bibliography of Statistical Bibliographies: A Thirteenth List," p. 177; W. T. Federer, "Some Recent Results in Experiment Design with Bibliography—III: Bibliography L–Z," pp. 185–197.

INTERNEIGHBOR INTERVAL (INI)

A measure of spatial dispersion (variability) for two-dimensional distributions uses in geographical applications. It was proposed by Court [1] and is defined as the average distance between neighbors in a given area: For uniform distributions, we have (approximately)

$$\text{INI} = \sqrt{A/N}$$

where N is the number of items (or persons) in an area and A is the area usually expressed in square miles. To express this measure in feet, the value $\sqrt{A/N}$ is multiplied by 5280.

For additional information, see ref. 2.

References

[1] Court, A. (1966). *Yearbook*, *Ass. Pacific Coast Geogr.*, **28**, 180–182.

[2] Norcliffe, G. B. (1977). *Inferential Statistics for Geographers*. Halsted Press, New York.

(GEOGRAPHY, STATISTICS IN SPATIAL DISTRIBUTIONS)

INTERPENETRATING SAMPLES *See* INTERPENETRATING SUBSAMPLES

INTERPENETRATING SUBSAMPLES

Interpenetrated or interpenetrating subsamples are used for three primary purposes:

1. To estimate sampling variances when the sample design is complicated and exact estimators are either not available or are too cumbersome to use
2. To control field work in collecting data for censuses and surveys
3. To measure some of the components of nonsampling variance contributed by enumerators, coders, or data processors

The use of this technique is generally attributed to Mahalanobis*, who instituted interpenetrated subsamples as a characteristic feature of the work in the Indian Statistical Institute*. The technique has been in use in India since 1937. Mahalanobis [9], in his 1946 paper to the Royal Statistical Society*, emphasized the extensive use of the technique for estimating sampling errors for ascertaining gross errors in recording, and for detecting differences among enumerators in collecting data from equivalent subsamples. The examples covered a wide variety of activity, including crop surveys, labor force surveys, food consumption surveys, and public preference surveys.

Deming [3] described the use of *replicated subsamples** as a variance estimation tool. Replicated subsamples are the same as interpenetrated subsamples. Deming recognized the fact that variances estimated from totals of replicated subsamples would contain components of variance arising from nonsampling sources. Hansen et al. [6] described the rationale and method of random groups for estimating variances. The random groups are special cases of interpenetrating subsamples. Cochran [2] describes interpenetrating subsamples as being particularly useful for the study of correlated errors in survey data, and points out that numerous applications of the method, under the name of *replicated sampling**, have been described by Deming [3].

The use of interpenetrated subsamples as a means for estimating variances for complex surveys has a long history. Shah [11] examined several approximations for variance estimation, including the Taylor series expansion, independent replications (interpenetrated subsamples), and the jackknife* method. It has been pointed out by several authors that in the case of a simple random sample of n observations that the precision of the method of random groups is always less than the precision of the usual estimate of variance for simple random sampling, but that the reductions in work and cost may make it desirable to use the random group method. In more complicated sample designs, the method of random groups may yield a good approximation when no exact method of estimating variances is available.

The method of interpenetrated subsamples consists of selecting a sample in the form of two or more samples from the same population, each subsample having full coverage of the entire population. In a general sense the method of interpenetrated subsamples can be illustrated in Fig. 1. There are N units in the population to be sampled and T possible subsamples. If the subsamples are selected with replacement so that they are independent, units may appear in more than one subsample. The sum of the measurements over all units in the jth subsample is denoted by $y_{.j}$. In theory, the case of independent subsamples is more appealing because it leads to unbiased estimators at least for linear estimators. In practice, interpenetrated subsamples are rarely selected independently.

With independent selections, a sample of units is selected from a finite population in accordance with a probability sampling design. The design need not specify that the

Units	Subsamples 1	2	...	j	...	T
1	✓					
2		✓				
3				✓		
.	✓					
.						✓
i				✓ y_{ij}		
.	✓			✓		
.	✓					✓
.						
N						
Total	$y_{.1}$	$y_{.2}$		$y_{.j}$		$y_{.T}$

Figure 1

units within a subsample be selected independently. Thus, the samples of units may be with or without replacement. It may be a multistage design, include stratification*, specify equal or unequal probabilities of selection, and so forth. When the first sample, s_1, is selected, all the units are replaced in the population and a second sample, s_2, is selected in accordance with the same sampling design. This process continues until there are $k \geqslant 2$ samples selected.

There must also be a common measurement process and estimation procedure applied to each of the k samples. If the measurement process or estimation procedure varies, the k samples may not be measuring a common population parameter. However, the measurement process and estimation procedure may be such that independence among the k estimators is lost. For example, if the same interviewer works in more than one subsample, the particular bias of that interviewer may induce a correlation between units in different subsamples. Similarly, if the estimation procedure is such that imputation for missing values is done for the sample as a whole rather than for each subsample separately, the responses from one subsample may be used to replace missing values in another subsample. This procedure may induce a correlation among subsamples.

Consider the case of independent subsamples when a linear estimator such as a mean or total is the measurement of interest. Then, each sample provides an estimate, each with the same expected value, say Y. Let $y_1, y_2, \ldots, y_k$ be the k uncorrelated sample estimators from the k subsamples. Let $\hat{y} = \sum_{j=1}^{k} y_j / k$. Then the $E(\hat{y}) = Y$, and an unbiased estimator of the variance of $\hat{y}$ is

$$\text{var}(\hat{y}) = [(N - n)/N]$$
$$\times \left[\sum_{j=1}^{k} (y_j - \hat{y})^2 / k(k - 1) \right] \quad (1)$$

under the assumption that each y_j has the same variance. Compare this with the standard estimator of the variance of the mean in the case of simple random sampling:

$$\text{var}(\bar{y}) = \frac{N - n}{N} \sum_{i=1}^{n} (y_{ij} - \bar{y})^2 / n(n - 1).$$
$$(2)$$

The estimator in (2) should always be used when there are no special problems in doing so. However, to get a quick estimate of the variance or to save time, money, and labor, there may be an advantage in using (1). Of course, with a more complex estimator, (2) will often not be applicable, but (1) will still provide a useful variance estimate.

In the case of a nonlinear estimator, such as a ratio, the expected value of y_k will not, in general, be equal to the population parameter of interest. Several competing estimators of the parameter may exist. As pointed out by Wolter [12] in the case of a ratio, where the population parameter of interest is $R = Y/X$, one will have $\hat{y}_j$ and $\hat{x}_j (j = 1, \ldots, k)$ as the estimators of Y and X from the k independent subsamples. Let $\hat{r}_j = \hat{y}_j / \hat{x}_j$. Then two estimators of R are

$$\bar{r} = (1/k) \sum_{j=1}^{k} \hat{r}_j$$

and

$$\hat{r} = \left(\sum_{j=1}^{k} \hat{y}_j / k \right) \Big/ \left(\sum_{j=1}^{k} \hat{x}_j / k \right).$$

In general, the two are not the same, but Wolter has shown that the variances of the two estimators are generally of the same order of magnitude. Thus an estimator of

$\text{var}(\bar{r})$ can be used as an estimator of $\text{var}(\hat{r})$. Omitting the finite population correction factors, the two variance estimators are

$$v_1(\bar{r}) = \sum_{j=1}^{k} (\hat{r}_j - \bar{r})^2 / \{k(k - 1)\} \quad \text{and}$$

$$v_2(\hat{r}) = \sum_{j=1}^{k} (\hat{r}_j - \hat{r})^2 / \{k(k - 1)\}.$$

The two are equal when the estimator is linear but for nonlinear estimators $v_1(\bar{r}) \leqslant v_2(\hat{r})$. This is similar to the result for the variances of ratios in stratified sampling*. The variance of the stratum-by-stratum ratios is as small or smaller than the variance of the ratio of averages over all strata combined, when the bias of the ratio estimate is small.

The more usual case with interpenetrating subsamples is that the samples are not independent. Usually, the entire sample is selected at one time, without replacement, and the sample is randomly divided into k groups. Again, estimates are formed for each group and an estimate of the variance similar to that shown in (1) is computed. There is now a covariance among the set of k estimators, so that the expected value

$$E[\text{var}(\hat{y})]$$

$$= E\left[\sum_{j=1}^{k} (y_j - \hat{y})^2 / \{k(k - 1)\} \right]$$

$$= \text{var}(\hat{y})$$

$$- 2 \sum_{j>m}^{k} \sum_{m}^{k} \text{cov}(\hat{y}_j, \hat{y}_m) / \{k(k - 1)\}.$$

For large populations and small sampling fractions, the covariance term will tend to be small and negative.

A question arises about the precision of the interpenetrated subsamples method of estimating variances. Hansen et al. [6] point out that the precision is a function of the size of the subsamples and of the number of subsamples. Shah [11] indicates that the number of subsamples is likely to be small, so that the estimate of variance will have only a limited number of degrees of freedom and hence will tend to be unstable.

Isaki and Pinciaro [8] examined the performance of several variance estimators for a total under a probability proportional to size systematic sampling design with a specific population of mobile-home dealers. The random group estimator was evaluated by means of number of groups and size of group. The investigators found that increasing the number of groups was more important than increasing the size of each group, at least for this specific application.

Mahalanobis [9] used the method of interpenetrating subsamples extensively to control the field work of sample surveys at the Indian Statistical Institute. He pointed out that frequently errors made in the compilation of primary data and in processing were at least as important as sampling error. By comparing the results of the various subsamples, errors in collecting and processing might be detected. An example cited of a labor survey in which five interpenetrated subsamples were selected and analyzed with the finding that high values were the result of one interviewer illustrated the power of the technique as a control on quality.

Recent work in random-digit dialing surveys have made use of the effectiveness of the method of interpenetrated subsampling as a method of controlling quality. O'Neil and Groves [10] described an interpenetrated sample design to study response and completion rates by interviewers when contacting respondents by telephone for surveys using different kinds of interview introductions. Similar kinds of studies have been conducted to identify interviewers with poor response rates.

Interpenetrated subsamples have been used extensively at the U.S. Bureau of the Census* and at Statistics Canada, primarily as a tool for estimating measurement error* contributed by enumerators, other field staff, and processing staff in census and survey statistics. Hansen et al. [7] designed a measurement model which included components of nonsampling error and bias as well as sampling error for means and totals. Interpenetrated subsamples were used in experiments to estimate some of the components of the model. Fellegi [4] extended the

model and reported results from its use in the 1961 Canadian census. Fellegi [5] extended the model again so that estimates of interviewer variance could be based not only on the work of those working in interpenetrated assignments, but on that of other interviewers as well.

At the Bureau of the Census in a 1960 study, a sample of 50 district offices (local offices that are responsible for the taking of a census in a local area) was selected and within each district a sample of enumeration areas was selected. Two interpenetrating subsamples were constituted. The difference between the means of the subsamples was the basis for an estimator of "total variance" including sampling variance, a simple response variance component reflecting basic trial-to-trial variability in response, and a correlated component of response variance reflecting the impact of interviewers and other field staff.

The estimator of total variance of a mean for a size of an enumerator's assignment area was $\mathrm{var}_1(\bar{y}) = \frac{1}{2}(\bar{y}_1 - \bar{y}_2)^2$. An estimator of the sampling variance was based on the average of the estimators of sampling variance within each subsample. This estimator of sampling variance was of the form $\mathrm{var}_2(\bar{y}) = \sum_h^2 \sum_j^n (x_{hj} - \bar{x}_h)^2 / \{2n(n-1)\}$, where x_{hj} denotes the measurement for the jth unit obtained by the hth enumerator. The difference between the estimated sampling variance, $v_2(\bar{y})$, and the estimated total variance, $v_1(\bar{y})$, was an estimator of the correlated component of response variance. In an experiment of this type in the 1950 census it was found that, on the average, the level of the correlated component of response variance was about the same as the sampling variance of a 25% simple random sample of housing units. The 1950 experiment was the basis for the Bureau of the Census moving to reduce the role of enumerators in censuses by encouraging self-enumeration. In 1960, the interpenetrating subsamples experiment showed that the average level of the correlated response variance in 1960 was about one-fourth of what it was in 1950. A further experiment in 1970 showed no further reduction.

The technique was used to study the effect of interviewers in the National Crime Survey, reported by Bailey et al. [1]. In this study, interpenetrated subsamples were designed in each of eight cities and separate estimates of variance were estimated for each city. The size of the estimated correlated component of response variance varied with the city and the kind of victimization item studied, but for many items the interviewer variance was larger than the sampling variance.

The extensive use of random-digit dialing surveys has created a new use for interpenetrated subsamples. A large sample of telephone numbers is divided into subsamples, with only a subsample assigned to each telephone interviewer. Because so many numbers are unproductive, it is not known how many of the numbers will actually be linked to a sample unit. As one subsample is completed, a new one is assigned until the target number of sample units is actually identified. This practice ensures that the sample as implemented is actually a random sample, not just a sample of the units easiest to reach.

The method of interpenetrating subsamples is used extensively for the three basic purposes cited earlier. It is extremely useful in permitting approximate variance estimation, in the estimation of nonsampling variance components, and in identifying possible trouble spots in field data collection or processing.

References

[1] Bailey, L., Moore, T. F., and Bailar, B. A. (1978). *J. Amer. Statist. Ass.*, **73**, 16–23. (An investigation of the effect of interviewers on the variance of victimization data, using interpenetrated subsamples.)

[2] Cochran, W. G. (1978). *Sampling Techniques*. Wiley, New York. (A valuable resource describing the use of interpenetrated subsamples both for estimating variances from complex designs and for nonsampling components.)

[3] Deming, W. E. (1960). *Sample Design in Business Research*. Wiley, New York. (Gives several excellent examples of the use of replicated or interpenetrated subsamples.)

[4] Fellegi, I. P. (1964). *J. Amer. Statist. Ass.*, **59**, 1016–1041. (Gives the extension of the theory of a measurement error model and the use of interpenetrated subsamples in the Canadian census to estimate the parameters of the model.)

[5] Fellegi, I. P. (1974). *J. Amer. Statist. Ass.*, **69**, 496–501. (Extends the measurement error model to include the work of personnel not in interpenetrated subsamples as well as those in interpenetrated subsamples.)

[6] Hansen, M. H., Hurwitz, W. N., and Madow, W. G. (1953). *Sample Survey Methods and Theory*. Wiley, New York. (Gives excellent practical examples of the use of interpenetrated subsamples to estimate variances.)

[7] Hansen, M. H., Hurwitz, W. N., and Bershad, M. A. (1961). *Bull. Int. Statist. Inst.*, **38**, Pt. 2, 359–374. (Presents the basic theory underlying the measurement error model used extensively at the Bureau of the Census and Statistics Canada, emphasizing the use of interpenetrated subsamples.)

[8] Isaki, C. T. and Pinciaro, S. J. (1977). *Proc. Soc. Statist. Sec. Amer. Statist. Ass.*, pp. 308–313. (Shows results of a simulation with systematic sampling with probability proportionate to size and comparisons of several variance estimators, including the random groups or interpenetrated subsamples method.)

[9] Mahalanobis, P. C. (1946). *J. R. Statist. Soc.*, **109**, 326–378. (The groundbreaking work showing uses of the method of interpenetrated subsamples.)

[10] O'Neil, H. J. and Groves, R. M. (1979). *Proc. Sec. Surv. Res. Meth.*, pp. 252–255. (An example of the use of interpenetrated subsamples in telephone interviewing.)

[11] Shah, B. V. (1978). In *Survey Sampling and Measurement*, N. K. Namboodiri, ed. Academic Press, New York, pp. 25–34.

[12] Wolter, K. M. (1980). Introduction to variance estimation. Unpublished notes used in American Statistical Association tutorial on variance estimation, Houston, Tex. (Compares several alternative variance estimators.)

(STRATIFIED SAMPLING SURVEY SAMPLING)

BARBARA A. BAILAR

INTERPERCENTILE DISTANCES

In a frequency distribution let M_x $(0 < x < 1)$ denote a value with a fraction x of the total frequency below it. The difference between M_α and M_β $(\alpha < \beta; 0 < \alpha, \beta < 1)$ is called an interpercentile distance. By giving α and β various values, an infinity of interpercentile distances may be produced. Taking $\alpha = 0.25$, $\beta = 0.75$, we obtain the interquartile distance (the most commonly used measure of dispersion among the family of interpercentile distances).

(INTERDECILE RANGE
INTERQUARTILE RANGE)

INTERPOLATION

Interpolation ist die Kunst zwischen den Zeilen einer Tafel zu lesen   —Thiele

The problem of replacing a real function $f(x)$ by an approximating function $F(\alpha_1, \ldots, \alpha_n; x)$ depending on a fixed finite number n of parameters is fundamental to practical computations and data analysis. Most often one chooses $F(\alpha_1, \ldots, \alpha_n; x)$ to have the form $\sum_{i=1}^{n} \alpha_i \phi_i(x)$, where each $\phi_i(x)$ is a polynomial, because polynomials can be easily evaluated, differentiated, and integrated in a finite number of steps using only the very basic operations of arithmetic. A more recent effective choice for the ϕ_i's is polynomial splines*. However, other choices, such as trigonometric polynomials, rational functions, and other special functions, are possible. The parameters α_i do not have to appear linearly in the definition of F, but they do in the vast majority of cases.

It is generally desired to choose the α_i's so that F is as close as possible to f in some measure of distance, or norm, usually taken to be either $\|f\|_\infty = \max_{a \leqslant x \leqslant b} |f(x)|$ or $\|f\|_p = (\int_a^b |f(x)|^p \, dx)^{1/p}$, where p is typically 1 or 2. Interpolation is a procedure that determines F in a finite and convenient way. Although F may not be the best approximation to f in a selected norm, it is usually satisfactorily close. Let $F(\alpha_1, \ldots, \alpha_n; x)$ lie in a linear space X and let $\lambda_1, \ldots, \lambda_n$ be linear functionals defined on X. Then for a given set of values $y_1, \ldots, y_n$, the general interpolation problem is to find $\alpha_1, \ldots, \alpha_n$ so that

$$\lambda_i(F(\alpha_1, \ldots, \alpha_n; x)) = y_i, \qquad i = 1, \ldots, n. \tag{1}$$

Usually, $\lambda_i f = f(x_i)$ for some real number x_i, although $\lambda_i f = f'(x_i)$ is also a common choice where $f'(x)$ denotes the derivative of f with respect to x.

A basic application is the construction of tables of values of distribution and density functions. The interpolation procedure used determines the spacing and location of the interpolation points. Exponential and sine functions, for example, are used to approximate general distribution functions by Gideon and Gurland [5]. Linear polynomial interpolation and normal probabilities are discussed by Steck [7].

The problem of approximating a function by a polynomial has interested mathematicians throughout the ages. The discovery 90 years ago by Weierstrass that any continuous function can be represented by a polynomial of sufficiently high degree over any finite interval is one of the highlights of the field. Fourier did his work on trigonometric-type approximations about 160 years ago. The subject was not a popular one for researchers until the development of the electronic digital computer provided a need for approximations. Textbooks on the subject changed little from 1920 to 1950. Now, approximation and interpolation theory is a field of great activity. A second impetus to the subject was provided by the introduction of functional analysis methods which gave the subject a unified point of view. This is reflected in the definition of interpolation given above using (1).

Much of numerical analysis* is based on the simple theorem of polynomial interpolation which says essentially that a unique straight line can be passed through two points, a parabola through three, a cubic through four, etc. Indeed, given $n + 1$ distinct points $x_0, x_1, \ldots, x_n$, and $n + 1$ values $y_0, y_1, \ldots, y_n$, there exists a unique polynomial $p_n(x) = a_0 + a_1 x + \cdots + a_n x^n$ of degree less than or equal to n so that

$$p_n(x_i) = y_i, \qquad i = 0, 1, \ldots, n. \tag{2}$$

In (2) the ϕ_i's are $1, x, x^2, \ldots, x^n$ and $\lambda_i f = f(x_i)$. All parameters appearing in (2) can be allowed to be complex numbers. The system (2) is a linear system of equations and can be written in matrix form as

$$
\begin{bmatrix}
1 & x_0 & x_0^2 & \cdots & x_0^n \\
1 & x_1 & x_1^2 & \cdots & x_1^n \\
\vdots & & & & \\
1 & x_n & x_n^2 & \cdots & x_n^n
\end{bmatrix}
\begin{bmatrix}
a_0 \\ a_1 \\ \vdots \\ a_n
\end{bmatrix}
=
\begin{bmatrix}
y_0 \\ y_1 \\ \vdots \\ y_n
\end{bmatrix}.
$$

$$(3)$$

The determinant of the system in (3) is the Vandermonde determinant. The direct solution of (3) is not a computationally effective procedure to produce p_n.

In addition to the interpolation system (2), other important classical examples include simple Hermite interpolation, where p_n is chosen to match the values of both f and f' at selected points, and trigonometric interpolation, where a linear combination of 1, $\cos x, \ldots, \cos nx, \sin x, \sin 2x, \ldots, \sin nx$ is chosen to agree with a function at $2n + 1$ points taken from the interval $-\pi$ to π (both $-\pi$ and π should not be included).

From the standpoint of applications, the important questions are the efficient solution of (2) and the accuracy of the resulting approximation. The polynomial p_n of (2) can actually be written down explicitly using Lagrange polynomials [3].

Evaluation of p_n and its derivatives from the Lagrange form is not particularly efficient, and if n is increased, the new Lagrange polynomials are not related in a simple fashion to the old. The Newton form of the interpolating polynomial is generally considered the most convenient form to use and will be discussed next. However, the Lagrange form can be used for interpolation of a function given in tabular form, and tables [1] can be used to evaluate the functions involved.

In order to give the Newton form for p_n, divided difference formulas need to be introduced. Set

$$
f[x_i, x_k] = \frac{f(x_k) - f(x_i)}{x_k - x_i}, \qquad x_k \neq x_i.
$$

Then general divided differences may be

defined recursively by

$$
f[x_i, x_{i+1}, \ldots, x_{k-1}, x_k]
$$
$$
= \frac{f[x_{i+1}, \ldots, x_{k-1}, x_k] - f[x_i, x_{i+1}, \ldots, x_{k-1}]}{x_k - x_i}.
$$

$$(4)$$

The computation can be arranged as indicated by the following example:

$$
\begin{aligned}
& x_0 : f(x_0) \\
& \qquad\qquad f[x_0, x_1] \\
& x_1 : f(x_1) \qquad\qquad f[x_0, x_1, x_2] \\
& \qquad\qquad f[x_1, x_2] \qquad\qquad f[x_0, x_1, x_2, x_3]. \\
& x_2 : f(x_2) \qquad\qquad f[x_1, x_2, x_3] \\
& \qquad\qquad f[x_2, x_3] \\
& x_3 : f(x_3)
\end{aligned}
$$

The first two columns are the given data points. Then the third, fourth, and fifth columns can be computed in order using (4). The interpolating polynomial $p_n(x)$ can be shown to have the form

$$
\begin{aligned}
p_n(x) = {} & f(x_0) + f[x_0, x_1](x - x_0) \\
& + f[x_0, x_1, x_2](x - x_0)(x - x_1) + \cdots \\
& + f[x_0, x_1, \ldots, x_n](x - x_0)(x - x_1) \\
& \cdots (x - x_{n-1}).
\end{aligned}
$$

$$(5)$$

The formulas (4) and (5) do not assume that the points x_i are ordered, i.e., $x_i < x_{i+1}$. Moreover, the divided difference formulas are symmetric functions of their arguments. Thus from (4) and (5) it follows that an additional interpolating point can be added easily and the degree of the interpolating polynomial increased.

For a function with m continuous derivatives,

$$
f[x_0, x_1, \ldots, x_m] = \frac{f^m(\xi)}{m!}, \qquad (6)
$$

where ξ is in the interval defined by x_0, $x_1, \ldots, x_m$. The error $e(x) = f(x) - p_n(x)$ satisfies

$$
\begin{aligned}
e(x) = {} & f[x_0, x_1, \ldots, x_n, x] \\
& \times (x - x_0)(x - x_1) \cdots (x - x_n),
\end{aligned}
$$

$$(7)$$

and for a smooth function can sometimes be estimated after using (6) to replace the divided difference in (7). Since $e(x)$ looks like a term in (6), the error $f(x) - p_{n-1}(x)$ can be estimated by

$$f[x_0, x_1, \ldots, x_n](x - x_0)(x - x_1) \cdots (x - x_{n-1}).$$

Further details and analyses of the properties of divided difference operators can be found in ref. 3. The formula (6) can be simplified in the case that the points are equally spaced or admit some special ordering. The Newton–Gauss formulas and the Bessel's formula for p_n occur in this way.

High-degree interpolation is of practical interest, since in general a table constructed for high-degree interpolation is often dramatically smaller than one for which linear interpolation is intended to be sufficient. This is particularly important for functions not regularly used or for functions with no known, easily used formula. The size of the table needed can be based on estimates for the error in (7). However, there are several drawbacks to polynomial interpolation. First, it is sensitive to the location of the interpolating points. However, if they are appropriately chosen, the resulting approximation differs very little from the best approximation by polynomials of the same degree. The classical example discovered by Runge in 1901 shows what can happen when the points are equally spaced and their number is fairly large. The function to be interpolated is $f(x) = 1/(1 + 25x^2)$ over the interval $[-1, 1]$. With $n = 10$, eleven interpolating points, and $x_i = -1 + 2i/n$, the error, $e(x)$, near $x = 0.9$ is over 1.5. In fact, p_n defined this way does not converge to f in the norm $\| \cdot \|_\infty$ as n goes to infinity. With equidistant data, the method of least squares with a lower-degree polynomial generally gives a more well behaved curve. If one intends to interpolate using a polynomial of high degree, it is in general best to choose the interpolating points to be the Tchebychev points if possible. These are the zeros of the Tchebychev polynomial of degree $n + 1$ over the interval in question. The Tchebychev polynomial of degree n over the interval $-1 \leqslant x \leqslant 1$ is given by $T_n(x) = \cos(n \cos^{-1} x)$. These polynomials satisfy the recursion $T_{n+1}(x) = 2xT_n(x) - T_{n-1}(x)$, $n \geqslant 1$, with $T(x) = 1$ and $T_1(x) = X$. For the precise location of the zeros of T_n, see ref. 1.

The resulting interpolation is then fairly well conditioned. If there is some doubt, some of the data could be perturbed by say $\pm 10^{-3}$ and the change in the interpolating polynomial observed. Finally, if the degree of the polynomial is large, say $n = 20$, then the form of the representation of the polynomial is important. Evaluation of such a polynomial on a digital computer can be prone to round-off errors. An orthogonal polynomial* expansion is probably called for.

Interpolation by trigonometric polynomials is appropriate if the underlying function f is periodic. These approximations are then valuable in the study of vibrations, sound, light, alternating currents, crystal, and of course time series, which are used in communications theory, control theory, and the study of turbulence. The term "spectral analysis" is often used to describe the development and study of trigonometric polynomial approximation. This approximation can be computed very efficiently using the fast Fourier transform algorithm [2]. (*See* INTEGRAL TRANSFORMS.)

Since the early 1960s, the choice of piecewise polynomial functions for the ϕ_i's has become increasingly popular not only for interpolation, but in the numerical solution of ordinary and partial differential equations. Given a set of points $-\infty < x_0 < x_1 < \cdots < x_n < \infty$, a piecewise polynomial function $p(x)$ is a function that is a polynomial when restricted to any of the subintervals (x_i, x_{i+1}). Thus p may be thought to be double-valued at a mesh point x_i. If p satisfies some overall continuity condition, e.g., if p is continuous or has in addition several continuous derivatives on $(-\infty, \infty)$, then p is called a polynomial spline, or simply a spline*. Accuracy is obtained using piecewise polynomial functions for interpolation by allowing the number of pieces to become large, i.e., n, and having the subinterval lengths $(x_i - x_{i-1})$ simultaneously become small. One popular technique used is simply piecewise interpolation using, for example, the Lagrange formulas where each piece of $p(x)$ has degree 3. By including x_{i-1} and x_i as interpolating points, then only two additional points are needed in each finite subinterval $[x_{i-1}, x_i]$, and the resulting $p(x)$ will be continuous. The other two most popular cases are the Hermite cubics, piecewise polynomial functions of degree 3 with the function having one continuous derivative overall, and the cubic splines, piecewise polynomial functions of degree 3 having two continuous derivatives overall. Cubic splines are perhaps the most popular spline functions. They are easy to calculate and approximate well not only the values of a smooth function, but also its first two derivatives.

Let $h_i = x_i - x_{i-1}$ and $M_i = p''(x_i)$, where p is a cubic spline defined over $x_0 < x_1 < \cdots < x_n$. Then if $p(x_i) = y_i$, it follows that on $[x_i, x_{i+1}]$,

$$p(x) = \frac{(x_{i+1} - x)^3 M_i + (x - x_i)^3 M_{i+1}}{6h_i}$$

$$+ \frac{(x_{i+1} - x)y_i + (x - x_i)y_{i+1}}{h_i}$$

$$- \frac{h_i}{6} \left[(x_{i+1} - x)M_i + (x - x_i)M_{i+1} \right].$$

$$(8)$$

The overall continuity continuities imply that

$$\frac{h_{i-1}}{6} M_{i-1} + \frac{h_i + h_{i-1}}{3} M_i + \frac{h_i}{6} M_{i+1}$$

$$= \frac{y_{i+1} - y_i}{h_i} - \frac{y_i - y_{i-1}}{h_{i-1}} \qquad (9)$$

for $i = 1, \ldots, n-1$. This represents $n-1$ equations in the $n+1$ unknowns $M_0, \ldots, M_n$. There are several end-point conditions currently in use to make the system (9) square. The natural end-point conditions set $M_0 = M_n = 0$. With this choice, the matrix problem (9) is symmetric, positive definite, diagonally dominant, and tridiagonal. Thus (9) can be easily and accurately solved. The M_i's can then be used in (8) to give the cubic spline interpolating a function at the mesh points. Another common choice for the end-point conditions is to set $p'(x_0) = f'(x_0)$ and $p'(x_n) = f'(x_n)$. With these conditions and if the function f being interpolated has four continuous derivatives over

$[x_0, x_n]$, there is a constant $K > 0$ such that

$$\max_{x_0 \leqslant x \leqslant x_n} |f^{(j)}(x) - p^{(j)}(x)| \leqslant K(\max h_i)^{4-j},$$

$$j = 0, 1, 2.$$

Further results on splines in approximation problems, including Fortran programs for their construction, are contained in ref. 4.

Multivariate interpolation is difficult, due in part to the large size of such problems and in part to the fact that there is no Tchebychev set of functions of more than one variable. A set of functions $\{\phi_i(x) \mid i = 1, \ldots, n\}$ is a Tchebychev set on a region if the matrix $(\phi_i(x_j))$, $i, j = 1, 2, \ldots, n$, is nonsingular for every set of n distinct points x_j. If possible, one should use a tensor product of a one-dimensional scheme. This is a limited approach, but it will be applicable and effective, for example, if the region of interest is a rectangle. See refs. 4 and 6.

References

[1] Abramowitz, M. and Stegun, I., eds. (1964). *Handbook of Mathematical Functions*. National Bureau of Standards, Washington, D.C.

[2] Conte, S. and deBoor, C. (1979). *Elementary Numerical Analysis*, 3rd ed. McGraw-Hill, New York.

[3] Davis, P. J. (1963). *Interpolation and Approximation*. Blaisdell, New York.

[4] deBoor, C. (1978). *A Practical Guide to Splines*. Springer-Verlag, New York.

[5] Gideon, R. and Gurland, J. (1971). *J. Amer. Statist. Ass.*, **66**, 577–582.

[6] Rice, J. R. (1969). *The Approximation of Functions*, Vol. 2. Addison-Wesley, Menlo Park, Calif.

[7] Steck, G. P. (1958). *Ann. Math. Statist.*, **29**, 780–800.

(FINITE DIFFERENCES
LEAST SQUARES
SPLINE FUNCTIONS)

G. W. REDDIEN

INTERPOLATION, CAUCHY'S METHOD

As an alternative to the method of least squares*, Cauchy [3] proposed the following method involving successive approximations based on neglecting all but one, two, . . . terms on the right-hand side of the observational equations $y_i = a_i u_i + b_i v_i + c_i w_i + \cdots$, the process continuing until the residuals are comparable with the inevitable errors of observation.

The unknown quantity y, a function of the variable quantity x, being supposed capable of being developed in a converging series

$$au + bv + cw + \cdots \qquad (1)$$

in which u, v, and w represent given functions of the same variable x, if we know n particular values of y corresponding with n particular values $(x_1, x_2, \ldots, x_n)$ of x: if, moreover, we represent by i any one of the integers $1, 2, \ldots, n$ and by $y_i, u_i, v_i, \ldots$ what $y, u, v, \ldots$ become when for x we substitute x_i; then, in order to obtain a sufficient approximation to the general value of y, we shall first determine the coefficient α by means of the formula

$$u = \alpha S u_i \qquad (2)$$

(in which Su_i represents the sum of the absolute values of u_i) and the difference of the first order Δy by means of the formula

$$y = \alpha S y_i + \Delta y. \qquad (3)$$

If the particular values Δy represented by $\Delta y_1, \Delta y_2, \ldots, \Delta y_n$ are comparable with the errors of observation, we may disregard Δy and reduce the approximate value of y to

$$\alpha S y_i.$$

In the contrary case we shall determine β by means of the formulas

$$v = \alpha S v_i + \Delta v, \qquad \Delta v = \beta S' \Delta v_i \qquad (4)$$

($S' \Delta v_i$ being the sum of the absolute values of Δv_i), and the difference of the second order $\Delta^2 y$ by means of the formula

$$\Delta y = \beta S' \Delta y_i + \Delta^2 y, \text{ etc.} \qquad (5)$$

Thus, by supposing the coefficients α, β, . . . determined by the system of equations (2), (4), . . . we shall calculate the orders of differences represented by Δy, $\Delta^2 y$, . . . or, rather, their particular values corresponding with the values $(x_1, x_2, \ldots, x_n)$ of the variable x, until we arrive at a difference the particular values of which are comparable with the unavoidable errors of observation. Then it will be sufficient to represent as zero the value of this difference deduced from the system of equations (3), (5), . . . in order to obtain a sufficient approximation to the general value of y. This general value will be then

$$y = \alpha S y_i \quad \text{or} \quad y = \alpha S y_i + \beta S' \Delta y_i, \ldots$$

according as we shall be able, without appreciable error, to reduce the series (1) to its first term, or its first two terms, ...

Example. Approximate the values of y given in Table 1 (where x is exact, but y may be in error by 0.1) by a polynomial in powers of $x(u = x, v = x^2, \ldots)$. Since the difference between the observed values of y (second column) and those (last column) calculated from a second-degree polynomial ($y = 3x + 2x^2$) does not exceed the error limit, the approximation is satisfactory, and it is not necessary to include a term in x^3.

Bienaymé [2] and Cauchy [4] engaged in a running controversy in the pages of *Comptes Rendus* on the relative merits of least squares

Table 1 Solution by Cauchy's Method of Interpolation

| Data | | | | | | | | | | | |
$u = x$	y	α	$\alpha S y_i$	Δy	$v = x^2$	$\alpha S v_i$	Δv	β	$\beta S' \Delta y_i$	$\Delta^2 y$	y
1	5.1	0.1	9.0	-3.9	1	3	-2	-0.25	-4.0	0.1	5.0
2	13.9	0.2	18.0	-4.1	4	6	-2	-0.25	-4.0	-0.1	14.0
3	27.0	0.3	27.0	0.0	9	9	0	0.	0.0	0.0	27.0
4	44.0	0.4	36.0	8.0	16	12	4	0.5	0.0	0.0	44.0
10	90.0			16.0	30		8				
Su_i	Sy_i			$S'\Delta y_i$	Sv_i		$S'\Delta v_i$				

and Cauchy's method of interpolation. Schott [10] gave a free translation into English of Cauchy's 1837 paper [3] and illustrated the use of Cauchy's method by an example. Bartlett [1] applied Cauchy's method to actual observations in the fields of physics and chemistry. Merriman [9] remarked that Cauchy's method can be used as easily as the method of least squares, although perhaps with less accuracy. In the present century, Cauchy's method has been discussed by several authors, most extensively by Goedseels [5–7] and Linnik [8]. Nevertheless, use of this method has never been widespread and is almost nonexistent today.

References

[1] Bartlett, W. P. G. (1862). *Amer. J. Sci.*, **34**, 27–33.

[2] Bienaymé, I. J. (1853). *C. R. Acad. Sci. Paris*, **37**, 5–13, 309–326.

[3] Cauchy, A. L. (1837). *J. Math. Pures Appl.*, (1), **2**, 193–205. [English transl. of 1835 lithograph, *Philos. Mag.*, (3), **8**, 459–468, 1836.]

[4] Cauchy, A. L. (1853). *C. R. Acad. Sci. Paris*, **36**, 1114–1122; ibid., **37**, 64–69, 100–109, 150–162, 197–206, 264–272, 326–334, 381–385.

[5] Goedseels, E. (1901). *Ann. Soc. Sci. Brux.*, **25**(1), 99–102, 146–149.

[6] Goedseels, E. (1902). *Ann. Soc. Sci. Brux.*, **26**(2), 148–156.

[7] Goedseels, E. (1909). *Théorie des Erreurs d'Observation*, 3rd ed. Charles Peeters, Louvain/Gauthier–Villars, Paris.

[8] Linnik, Yu, V. (1958). *Method of Least Squares and Principles of the Theory of Observations*. Fizmatgiz, Moscow (in Russian). [English transl.: R. C. Elandt (N. L. Johnson ed.), Pergamon Press, New York, 1961.]

[9] Merriman, M. (1877). *Trans. Connecticut Acad. Arts Sci.*, **4**(1), 151–232.

[10] Schott, C. A. (1861). *Report of the U.S. Coast Survey for 1860*, pp. 392–396.

(INTERPOLATION)

H. Leon Harter

INTERQUARTILE RANGE

A common measure of spread* is the interquartile range—the difference ($Q_3 - Q_1$) between the first and third quartiles. It is also called the 50% midrange*.

For the normal distribution $N(\mu, \sigma)$ the (symmetric) interquartile range is between

$$\mu - 0.674\sigma \quad \text{and} \quad \mu + 0.674\sigma.$$

The relation between the standard deviation σ and the interquartile range in the case of the normal distribution is given by

$$\sigma = 0.74(Q_3 - Q_1).$$

Confidence intervals for the interquartile range (under normality assumptions) were constructed by Roy and Cobb [1].

Reference

[1] Roy, S. N. and Cobb, W. (1960). *Ann. Math. Statist.*, **31**, 939–957.

(INTERPERCENTILE DISTANCES
L-ESTIMATORS
QUARTILES
SAMPLE)

INTERVENTION MODEL ANALYSIS

Often, policy changes or decisions in public and private sectors are designed or appear to have been designed to effect some changes in certain response variables occurring in the form of time series*. For example, an advertisement campaign or a price change might affect the monthly sales pattern of a particular item; the creation of the Anti-inflation Board by the Federal Government of Canada in October 1975 was expected to lower the level of inflation in Canada. Such intrusions to a time series are usually referred to as interventions (see ref. 4) and they cannot only change the level of a series abruptly or after a short delay but could also deflect a series going downward, causing it to drift up, or vice versa, or effect some other form of change. Given that such an intervening event has occurred, one may be interested in determining if there is evidence to suggest that a corresponding change has occurred in the time series and if so, in determining the nature and magnitude of this change.

Traditionally a Student's t-test* is used for estimating and testing for a change in the mean levels before and after an intervention. Such a test may not be adequate for the kind of situations cited before where the data are occurring in the form of a time series. This is because (a) in these cases the successive observations are usually serially correlated and often nonstationary, and (b) the form of the change may not be a step as required by the t-test but it could be a gradual increase (decrease), a ramp increase (decrease), or any other form of change. Box and Tiao [4] have provided a strategy for conducting analysis of situations such as those described above and their strategy will be outlined in the next section. Previous related works dealt with only certain special cases, and references to some of these can be found in ref. 5.

TIME SERIES* AND INTERVENTION MODELS

Suppose that the data $\ldots, y_{t-1}, y_t, y_{t+1}, \ldots$ are available as a series obtained at equal time intervals. We then consider the model

$$y_t = f(\beta, I_t) + z_t, \qquad (1)$$

where $f(\beta, I_t)$ is a function of the parameters β and the intervention variables I_t. This function f represents the intervention model in whose absence z_t will be the original time series. z_t can usually be represented by a multiplicative seasonal autoregressive integrated moving-average (ARIMA*) time-series model:

$$\phi(B)\Phi(B^s)(1-B)^d(1-B^s)^D z_t$$
$$= \theta(B)\Theta(B^s)a_t, \qquad (2)$$

where s is the period of seasonality, d and D are nonnegative integers usually taking the values 0, 1, or 2, B is a backward shift operator such that $Bz_t = z_{t-1}$, $B^s z_t = z_{t-s}$; $\phi(B) = 1 - \phi_1 B - \cdots \phi_p B^p$ and $\theta(B) = 1 - \theta_1 B - \cdots - \theta_q B^q$ are polynomials in B; $\Phi(B) = 1 - \Phi_1 B^s - \cdots - \Phi_P B^{Ps}$ and $\Theta(B) = 1 - \Theta_1 B^s - \cdots - \Theta_Q B^{Qs}$ are poly-

nomials in B^s; $\{a_t\}$ for $t = \cdots - 1, 0, 1, \ldots$ is a sequence of independent, identically distributed normal random variables with mean zero and variance σ^2. We also constrain the parameter space such that all the roots of $\phi(u)$, $\Phi(u^s)$, $\theta(u)$, $\Theta(u^s)$ in the complex u-plane lie outside the unit circle. If the data are nonseasonal, the operators $\Phi(B^s)$, $(1-B^s)^D$, and $\Theta(B^s)$ will be absent from (2) and the resulting model is usually referred to as an ARIMA (p, d, q) model. A variety of time series can be modeled by the ARIMA class given in (2). Box and Jenkins [3] suggest a three-stage iterative strategy consisting of specification, estimation, and diagnostic checks for building these models.

Box and Tiao [4] have shown that a variety of interventions can be handled by considering the model

$$f(\beta, I_t) = \frac{\omega(B)}{\delta(B)} I_t(T), \qquad (3)$$

where $\omega(B) - \omega_0 - \omega_1 B - \cdots - \omega_m B^m$, and $\delta(B) = 1 - \delta_1 B - \cdots - \delta_r B^r$ is such that it has roots outside or on the unit circle and β is a vector containing all the ω and δ parameters. The intervention variables $I_t(T)$ in (3) are usually taken as indicator variables indicating the intervention at $t = T$:

$$I_t(T) = S_t(T) = \begin{cases} 0, & t < T; \\ 1, & t \geqslant T \end{cases}$$

$$I_t(T) = P_t(T) = \begin{cases} 0, & t \neq T. \\ 1, & t = T \end{cases}$$

$S_t(T)$ is usually referred to as a step input and this denotes the nonoccurrence and occurrence of interventions, while the pulse input $P_t(T)$ takes the value 1 at the time of intervention and zeros elsewhere. It also follows from the definitions of $S_t(T)$ and $P_t(T)$ that

$$(1-B)S_t(T) = P_t(T). \qquad (4)$$

Model (3) is really a special case of the transferfunction model described in ref. 3 with the input variables being indicator variables. It can also be seen as a distributed lag model discussed extensively in the econometrics* literature with the exogenous variables taken as indicator variables. The trans-

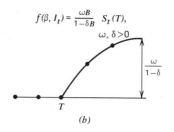

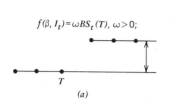

(a)

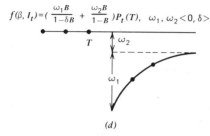

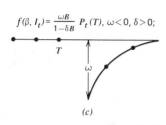

(c)

(d)

Figure 1   Responses to some intervention models.

fer function $\omega(B)/\delta(B)$ is usually estimated from data in the transfer function modeling, whereas in the intervention analysis it is postulated on the basis of the change expected. Some useful intervention models and the corresponding responses are shown in Fig. 1.

Interpretation

For illustration and some interpretation consider the model

$$y_t = \omega S_t(T) + \{(1 - \theta B)/(1 - B)\}a_t \quad (5)$$

in which the intervention is expected to produce a step change of the form given in Fig. 1a. This model can be rewritten as

$$u_t = \omega x_t + a_t, \quad (6)$$

where

$$u_t = \frac{1 - B}{1 - \theta B} y_t$$

$$= y_t - (1 - \theta)$$

$$\times \left(y_{t-1} + \theta y_{t-2} + \theta^2 y_{t-3} + \cdots\right)$$

$$= y_t - \bar{y}_{t-1} \quad \text{(say)},$$

and

$$x_t = \frac{1 - B}{1 - \theta B} S_t(T) = S_t(T) - \bar{S}_{t-1}(T)$$

with $\bar{S}_{t-1}(T)$ defined the same way as $\bar{y}_{t-1}$. Suppose that the $(n + 1)$ time series observations $y_0, y_1, \ldots, y_n$ are available, θ is known, and an intervention occurred at a known time $t = T$. Then one can generate u_t and x_t $(t = 1, 2, \ldots, n)$ which are deviations of y_t and $S_t(T)$ not from the usual arithmetic averages but from certain exponentially weighted averages of previous values. Now model (6) is very much like a regression model through the origin, yielding

$$\hat{\omega} = \sum_{t=1}^{n} x_t u_t \Big/ \sum_{t=1}^{n} x_t^2, \quad V(\hat{\omega}) = \sigma^2 \Big/ \sum_{t=1}^{n} x_t^2. \quad (7)$$

For large n, the finite summations in (7) can be replaced by corresponding infinite summations. After further algebraic simplifications $\hat{\omega}$ reduces to

$$\hat{\omega} = (1 - \theta) \sum_{j=1}^{\infty} \theta^j y_{T+j} - (1 - \theta) \sum_{j=1}^{\infty} \theta^j y_{T-1-j},$$

which is the difference between two exponentially weighted averages, one of observations before time T and the other afterward. It is also interesting to note the similarity of $\hat{\omega}$ with a traditional t statistic in which the difference between two simple averages would be taken.

If $P_t(T)$ is considered instead of $S_t(T)$ in (5), then it can be shown that

$$\hat{\omega} = y_T - \frac{1}{2} \sum_{j=1}^{\infty} (1 - \theta)\theta^{j-1}(y_{T-j} + y_{T+j}),$$

which is a comparison of y_T with an average of two exponentially weighted averages one of the observations before T and the other afterward.

As a final example consider the model

$$y_t = \omega P_t(T) + (1 - \phi B)^{-1} a_t.$$

Proceeding as before it can be shown that

$$\hat{\omega} = y_T - \{\phi/(1 + \phi^2)\}(y_{T-1} + y_{T+1}),$$

which is the difference between y_T and the predictive mean of y_T given ϕ and all other observations. Thus $\hat{\omega}$ would be a reasonable estimate for a sudden change which dissipates completely after the intervention.

Estimation

In the previous illustrations we were considering estimates of ω (intervention parameter) assuming the knowledge of θ or ϕ (time-series model parameters). This implicitly considers the estimation of the time-series model parameters first and then those of the intervention parameters. Often, this is unnecessary, as the parameters of model (1) with f as in (3) and z_t as in (2) can be estimated jointly using likelihood methods. Given a set of $n + d + Ds$ observations $\mathbf{y}$, one can obtain an n-vector $\mathbf{w}$ whose tth element is given by $\mathbf{w}_t = (1 - B)^d(1 - B^s)^D$ $(y_t - f(\boldsymbol{\beta}; I_t))$. The corresponding model for $\mathbf{w}_t$,

$$\mathbf{w}_t = \{\theta(B)\Theta(B^s)/\phi(B)\Phi(B^s)\}a_t,$$

is stationary. Thus the likelihood function* and maximum likelihood* or approximate maximum likelihood estimates can be obtained (see refs. 3 [Chap. 7] and 4). Using the large-sample properties of maximum likelihood estimates, the approximate standard errors of the parameter estimates can be computed and the significance of the parameters checked.

An overall strategy for performing an intervention analysis is outlined in the following steps.

1. Using the preintervention data, build a time-series model for z_t adopting the iterative strategy of specification, estimation, and diagnostic checks (see ref. 3).
2. Given knowledge of the interventions, frame models for change which describe what is expected to occur.
3. Estimate the parameters of the joint model, perform diagnostic checks, modify the structure if necessary, and arrive at a final model.
4. Make appropriate inferences about the parameters.

EXAMPLE: THE CONSUMER PRICE INDEX* (CPI) FOR CANADA

For illustration, we consider the monthly percentage change in the CPI for Canada from February 1967 to December 1977, shown in Fig. 2. The period January 1967–April 1973 is taken as containing no major interventions affecting the series. The identification procedure outlined in ref. 3 leads one to entertain a model of the form

$$(1 - B^{12})z_t = (1 - \theta B)(1 - \Theta B^{12})a_t. \quad (8)$$

In addition, we consider the following two interventions to be important:

1. $I_1(T_1)$—the price boost in the oil in May 1973 by the Organization of Petroleum Exporting Countries (OPEC)
2. $I_2(T_2)$—creation in October 1975 by the Federal Government of Canada of the Anti-inflation Board (AIB) to control inflation

One would expect I_1 to increase the level of the series gradually (see Fig. 1b). However, I_2 might be expected to produce an initial decrease in the CPI immediately following the intervention, followed by a gradual increase in the level with the possibility of a

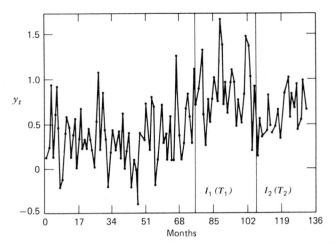

y_t

| | | | | | | | |
|0|17|34|51|68|85|102|119|136|

$I_1(T_1)$ $I_2(T_2)$

Months

Figure 2 Canadian Consumer Price Index monthly percentage change (February 1967–December 1977).

residual loss (see Fig. 1*d*). Thus the model

$$f(\beta; I_1(T_1), I_2(T_2))$$

$$= \frac{\omega_1 B}{1 - \delta_1 B} S_t(T_1)$$

$$+ \left(\frac{\omega_2 B}{1 - \delta_2 B} + \frac{\omega_3 B}{1 - B} \right) P_t(T_2), \quad (9)$$

where

$$S_t(T_1) = \begin{cases} 0, & t < \text{May } 1973 = T_1 \\ 1, & t \geq \text{May } 1973 = T_1 \end{cases}$$

$$P_t(T_2) = \begin{cases} 0, & t \neq \text{November } 1975 = T_2 \\ 1, & t = \text{November } 1975 = T_2, \end{cases}$$

may be entertained for the interventions. Hence for the data for February 1967–December 1977, we entertain the model

$$y_t = f(\beta; I_1(T_1), I_2(T_2))$$

$$+ \left\{ (1 - \theta_1 B)(1 - \Theta B^{12})/(1 - B^{12}) \right\} a_t,$$

$$(10)$$

where f is as given in (9). Estimation of the parameters of this model indicates that $\delta_2 \approx 1$, which implies that the second term in f has the form $\{(\omega_2 + \omega_3)B/(1 - B)\} P_t(T)$ $= \omega_4 B S_t(T)$ [from (4)]. Thus our modified model is

$$y_t = \{\omega_1 B/(1 - \delta_1 B)\} S_t(T_1) + \omega_4 B S_t(T_2)$$

$$+ \{(1 - \theta_1 B)(1 - \Theta B^{12})/(1 - B^{12})\} a_t$$

with $\hat{\omega}_1 = 0.18(0.16)$, $\hat{\delta}_1 = 0.51(0.46)$, $\hat{\omega}_4 = -0.36(0.09)$, $\hat{\theta} = -0.19(0.09)$, $\hat{\Theta} = 0.73$ (0.06), where the quantities in parentheses indicate the standard errors. Analysis of the residuals and their autocorrelations show no obvious model inadequacies. It is clear from the analysis that the OPEC intervention did cause the level to increase gradually and that the AIB had some impact in lowering it.

APPLICATIONS AND EXTENSIONS

The method described here has been applied in various fields, such as business, psychology*, and hydrology* (see, e.g., refs. 5–8). The model (1) can also be extended to allow the inclusion of other exogenous variables and more than one intervention (see refs. 2 and 4). Intervention analysis can also be performed in a multiple time series* where y_t will be a vector series (see ref. 1). Such an analysis can be computationally cumbersome if y_t has more than two components. This is due to the large number of parameters that need to be estimated in a multiple time-series model. Sometimes this difficulty can be partially overcome by adopting the two-stage procedure outlined in ref. 1. In any case, the interpretation of the fitted intervention model is not straightforward in this context.

It should be emphasized that the analysis described assumes that (1) the time-series model and its parameters before and after the intervention are the same, and (2) there are no other events or interventions coinciding with the particular one being considered. It is important to keep these assumptions in mind when drawing conclusions about any intervention.

References

[1] Abraham, B. (1980). *Biometrika*, **67**, 73–80. (Perhaps the only paper in the literature dealing with interventions in a multiple time series.)

[2] Bhattacharyya, M. N. and Layton, A. P. (1979). *J. Amer. Statist. Ass.*, **74**, 596–603. (An Australian case study in intervention analysis: studying the effectiveness of seat belt legislation on the Queensland road toll.)

[3] Box, G. E. P. and Jenkins, G. M. (1976). *Time Series Analysis Forecasting and Control*. Holden-Day, San Francisco. (An excellent reference on time-series analysis.)

[4] Box, G. E. P. and Tiao, G. C. (1975). *J. Amer. Statist. Ass.*, **70**, 70–79. (Perhaps the best comprehensive paper dealing with interventions in a time series.)

[5] Glass, G. V., Wilson, V. L., and Gottman, J. M. (1975). *Design and Analysis of Time Series Experiments*. Colorado Associated University Press, Boulder, Colo. (A good book containing many references to work related to intervention analysis prior to the work of Box and Tiao.)

[6] Hipel, K. W. and McLeod, A. I. (1975). *Water Resour. Res.*, **11**(6), 855–861. (Application of intervention analysis in hydrology.)

[7] Larcker, D. F., Gordon, L. A., and Pinches, G. E. (1980). *J. Financ. Quart. Anal.*, **15**, 267–287. (Application of intervention analysis in business.)

[8] Lettenmaier, D. P. (1980). *Water Resour. Res.*, **16**, 159–171. (Application of intervention analysis in the presence of missing data.)

(AUTOREGRESSIVE-INTEGRATED
    MOVING AVERAGE (ARIMA) MODELS
AUTOREGRESSIVE-MOVING AVERAGE
    (ARMA) MODELS
MULTIPLE TIME SERIES
REGRESSION
STOCHASTIC PROCESSES
TIME SERIES)

BOVAS ABRAHAM

INTERVIEWER VARIANCE *See* VARIANCE, INTERVIEWER

INTRABLOCK INFORMATION

As its name leads one to expect, this term usually refers to the analysis of data for experiments in which there is some form of blocking*. *Intrablock information* is the information contained in differences between responses in the block, as opposed to interblock information*, which is the information contained in differences between responses in different blocks that cannot be expressed as linear functions of intrablock differences.

The standard analysis for a parametric linear model blocked experiment uses only intrablock information for inferences in treatment differences.

(BLOCKS, BALANCED INCOMPLETE
GENERAL LINEAR MODEL
INCOMPLETE BLOCKS)

INTRACLASS CORRELATION COEFFICIENT

Intraclass correlation coefficients are measures of the relative similarity of quantities which share the same observational units of a sampling and/or measurement process. In this regard, two examples of historical interest are the number of ovules in different pods of the same tree for a sample of trees from a particular area [5] and the heights of brothers from the same family for a sample of families [2, 14]. Others currently arise in many areas of application, particularly population genetics* (e.g., animals in the same litter), reliability studies* (e.g., products from the same machine, measurements of a characteristic for the same person), and survey sampling* (e.g., persons in the same neighborhood, persons contacted by the same interviewer).

The most straightforward setting for the formulation of an intraclass correlation coefficient involves a set of D observations y_{ij} on each of N clusters where $i = 1, 2, \ldots, N$ and $j = 1, 2, \ldots, D$; also, the ordering of j within i is arbitrary. For such situations, the early definition of the intraclass correlation coefficient which existed prior to the analysis of variance is

$$\rho_I = \frac{\sum_{i=1}^{N} \sum_{j \neq j'} (y_{ij} - \mu)(y_{ij'} - \mu)}{ND(D-1)v}, \quad (1)$$

where

$$\mu = \left(\sum_{i=1}^{N} \sum_{j=1}^{D} y_{ij} / (ND) \right) \quad \text{and}$$

$$v = \left(\sum_{i=1}^{N} \sum_{j=1}^{D} (y_{ij} - \mu)^2 / (ND) \right)$$

are the finite set mean and variance of all ND observations. The N clusters represent primary sampling units within each of which all observations belong to the same information class since their orderings by j are arbitrary, and this is why the mean μ and variance v are determined from all ND of them. Also, all $ND(D-1)$ within cluster products $\{(y_{ij} - \mu)(y_{ij'} - \mu)\}$ with $j \neq j'$ for $i = 1, 2, \ldots, N$ are included in the numerator of (1), with each reflecting the extent to which the observations in the corresponding pairs deviate from the mean μ in the same direction versus in opposite directions. Thus the expression for ρ_I can be viewed as a product moment correlation coefficient* with respect to the set of all $ND(D-1)$ within cluster pairs $\{(y_{ij}, y_{ij'})\}$, the latter array being known as an *intraclass correlation table*. A specific situation for which ρ_I in (1) is of interest is that of systematic samples* of size D from a finite population of size ND, where the N clusters are the respective possible samples. In this case, ρ_I represents the correlation between pairs of units in the same systematic sample.

In contrast, if the D observations within each cluster belong to different information classes indexed by some criterion $t = 1, 2, \ldots, D$ such as age or time, then the *interclass correlation* between the tth class and the t'th class is the product moment correlation coefficient for the corresponding set of

N pairs $\{(y_{it}, y_{it'})\}$ with $t \neq t'$ fixed and $i = 1, 2, \ldots, N$, i.e.,

$$\rho_{tt'} = \frac{\{\sum_{i=1}^{N} (y_{it} - \mu_t)(y_{it'} - \mu_{t'})\}}{\left[\{\sum_{i=1}^{N} (y_{it} - \mu_t)^2\} \{\sum_{i=1}^{N} (y_{it'} - \mu_{t'})^2\} \right]^{1/2}}, \quad (2)$$

where

$$\mu_t = \left(\sum_{i=1}^{N} y_{it} / N \right) \quad \text{and} \quad \mu_{t'} = \left(\sum_{i=1}^{N} y_{it'} / N \right).$$

From the comparison of (1) and (2), it should be noted that the intraclass correlation ρ_I is expressed in terms of deviations from the mean μ of all observations while an interclass correlation $\rho_{tt'}$ is expressed in terms of deviations from the means μ_t and $\mu_{t'}$ for the tth and t'th classes. Also, the intraclass correlation ρ_I in (1) involves $D(D-1)$ pairs of observations per cluster, whereas an interclass correlation such as $\rho_{tt'}$ in (2) involves only one. Thus when all observations within a cluster belong to the same class, the intraclass correlation provides a more complete measure of their relative similarity than does any interclass correlation based on an arbitrary criterion. For further discussion, see Fisher [2].

Although (1) is of conceptual interest, it is not convenient for computational purposes, particularly when D is large. An important simplification in this regard was provided by Harris [5], who noted that

$$D \sum_{i=1}^{N} (m_{i.} - \mu)^2 = Nv\{1 + (D-1)\rho_I\}, \quad (3)$$

where the $\{m_{i.} = (\sum_{j=1}^{D} y_{ij} / D)\}$ are means of the observations within the respective clusters $i = 1, 2, \ldots, N$. Thus ρ_I could be obtained via

$$\rho_I = (Dv_a - v) / \{(D-1)v\}, \quad (4)$$

where $v_a = \{\sum_{i=1}^{N} (m_{i.} - \mu)^2 / N\}$ is the finite set variance among the cluster means $\{m_{i.}\}$. Since

$$0 \leq v_a \leq v = (v_a + v_w), \quad (5)$$

where $v_w = \{\sum_{i=1}^{N} \sum_{j=1}^{D} [y_{ij} - m_{i.}]^2 / (ND)\}$ is the finite set within cluster variance, it follows that

$$-1/(D-1) \leqslant \rho_I \leqslant 1; \qquad (6)$$

the lower bound occurs when the cluster means $m_{i.}$ are all equal to μ so that v_a in (4) is 0 (e.g., the y_{ij} are within-cluster ranks $1, 2, \ldots, D$); and the upper bound occurs when all observations in each cluster are the same (i.e., $y_{ij} = m_{i.}$) so that $v_w = 0$ in (5) implies that $v = v_a$ in (4). In the former case (which involves negative intraclass correlation), the clusters have maximal internal heterogeneity since $v = v_w$; while in the latter (where $\rho_I = 1$), they have maximal internal homogeneity since $v_w = 0$. Additional insights concerning ρ_I can be gained by noting that if the number of observations per cluster D is potentially infinite (e.g., the number of measurements of the weight of an object), then (4) and (6) simplify to

$$0 \leqslant \rho_I = (v_a/v) = \{1 - (v_w/v)\} \leqslant 1. \quad (7)$$

Thus ρ_I can be interpreted as the proportional extent to which the within-cluster variance v_w is less than the overall variance v; and in this sense, it is a measure of the homogeneity of the observations within such sampling units.

When the y_{ij} are ranks relative to the set of all ND observations (with ties handled via midranks), the Kruskal–Wallis statistic* for the comparison of the cluster means $m_{i.}$ can be written as

$$Q = (ND-1)v_a/v$$
$$= (ND-1)\{1 + (D-1)\rho_I\}/D. \quad (8)$$

For this reason it is sometimes used as a nonparametric measure of intraclass correlation. An alternative measure of intraclass correlation based on ranks has recently been proposed by Rothery [11]. Kendall's τ^* and Spearman's ρ^* are often used as nonparametric measures of interclass correlation.

In many applications, the intraclass correlation ρ_I is estimated from a two-stage random sample* of nd observations y_{kl}, where $k = 1, 2, \ldots, n$ and $l = 1, 2, \ldots, d$. If the n clusters are selected by simple random sampling (without replacement) from the set of N clusters and if the d observations within clusters are also obtained by simple random sampling (without replacement) separately within the n selected clusters, then

$$E\{y_{kl}\} = \mu, \quad \mathrm{var}\{y_{kl}\} = v, \quad \text{and}$$

$$\qquad (9)$$

$$\mathrm{cov}\{y_{kl}, y_{k'l'}\} = \rho_I v \qquad \text{for } k = k', l \neq l'$$
$$= -v_a/(N-1) \quad \text{for } k \neq k', l, l'.$$

Thus ρ_I represents the correlation of randomly selected pairs of observations from within the same cluster. Also, for large N, randomly selected observations from different clusters are essentially uncorrelated.

The sample mean

$$\bar{y} = \left\{ \sum_{k=1}^{n} \sum_{l=1}^{d} y_{kl}/(nd) \right\}$$

is an unbiased estimator of the population mean μ since $E\{\bar{y}\} = \mu$. Its variance has the form

$$\mathrm{var}\{\bar{y}\} = \frac{1}{n}\left(\frac{N-n}{N-1}\right)v_a + \frac{1}{nd}\left(\frac{D-d}{D-1}\right)v_w.$$

$$\qquad (10)$$

Unbiased estimators for the variance components v_a and v_w can be obtained as

$$\hat{v}_a = \left\{\frac{N-1}{Nd}\right\}\left\{s_a^2 - \frac{D-d}{D}s_w^2\right\},$$

$$\hat{v}_w = \{(D-1)s_w^2/D\}, \qquad (11)$$

where $s_a^2 = [d\sum_{k=1}^{n}(\bar{y}_{k.} - \bar{y})^2/(n-1)]$ is the among-clusters mean square with respect to the within-cluster sample means $\{\bar{y}_{k.} = \sum_{l=1}^{d} y_{kl}/d)\}$ and the overall sample mean $\bar{y} = \sum_{k=1}^{n}\sum_{l=1}^{d} y_{kl}/(nd)$, and $s_w^2 = [\sum_{k=1}^{n}\sum_{l=1}^{d}(y_{kl} - \bar{y}_{k.})^2/\{n(d-1)\}]$ is the within-clusters mean square. The estimators $\hat{v}_a$ and $\hat{v}_w$ in (11) can be used to obtain an unbiased estimator for $\mathrm{var}\{\bar{y}\}$ in (10) via

$$\widehat{\mathrm{var}\{\bar{y}\}}$$

$$= \frac{1}{nd}\left\{\left(1 - \frac{n}{N}\right)s_a^2 + \frac{n}{N}\left(1 - \frac{d}{D}\right)s_w^2\right\};$$

$$\qquad (12)$$

and a ratio estimator r_I for ρ_I in (4) can be

similarly constructed as

$$r_I = \frac{\hat{v}_a - \{\hat{v}_w/(D-1)\}}{\hat{v}_a + \hat{v}_w}$$

$$= \frac{\{s_a^2 - s_w^2\} - \{ds_w^2/[(N-1)D]\}}{\{s_a^2 + (d-1)s_w^2\} + \{d(D-1)s_w^2/[(N-1)D]\}}. \tag{13}$$

From (13) it can be noted that the estimator r_I for ρ_I is somewhat more complicated than the sample analog $\tilde{r}_I$ of ρ_I, which has the form

$$\tilde{r}_I = \frac{\sum_{k=1}^{n}\sum_{l\neq l'}(y_{kl}-\bar{y})(y_{kl'}-\bar{y})}{(d-1)\sum_{k=1}^{n}\sum_{l=1}^{d}(y_{kl}-\bar{y})^2}$$

$$= \frac{\{s_a^2 - s_w^2\} - \{s_w^2/(n-1)\}}{\{s_a^2 + (d-1)s_w^2\} + \{(d-1)s_w^2/(n-1)\}}. \tag{14}$$

Thus r_I is not an intraclass correlation coefficient for the sample; and $\tilde{r}_I$ is not the natural ratio estimator for ρ_I. Otherwise, when $d/n \doteq 0$, then $r_I \doteq \tilde{r}_I$.

If N is very large for the population so that $(N-1)/N \doteq 1$ and $n/N \doteq 0$, then (11) and (13) can be simplified to

$$\hat{v}_a = \left\{s_a^2 - \frac{D-d}{D}s_w^2\right\}/d,$$

$$\hat{v}_w = \{(D-1)s_w^2/D\}, \tag{15}$$

$$r_I = (s_a^2 - s_w^2)/\{s_a^2 + (d-1)s_w^2\}. \tag{16}$$

Here, one special case of interest is very large D so that $(D-1)/D \doteq 1$ and $d/D \doteq 0$; and another is small $D = d$. In the former case, (15) can be further simplified to

$$\hat{v}_a = (s_a^2 - s_w^2)/d, \qquad \hat{v}_w = s_w^2. \tag{17}$$

This situation is often called the two-stage nested "random effects" analysis of variance since the observations pertaining to it can be represented as $y_{kl} = \mu + c_k + e_{kl}$; the $\{c_k\}$ and $\{e_{kl}\}$ are mutually uncorrelated random variables with $E\{c_k\} = 0$, $\text{var}\{c_k\} = v_a$, $E\{e_{kl}\} = 0$, and $\text{var}\{e_{kl}\} = v_w$. If the $\{c_k\}$ and the $\{e_{kl}\}$ also have mutually independent normal distributions, then

$$Q_1 = (n-1)s_a^2/(dv_a + v_w)$$

has the $\chi^2(n-1)$ distribution; $Q_2 = n(d-$

$1)s_w^2/v_w$ has the $\chi^2[n(d-1)]$ distribution; and Q_1 and Q_2 are independent. It follows that a $100(1-\alpha)\%$ confidence interval for ρ_I is

$$\frac{(s_a^2/s_w^2) - F_2}{(s_a^2/s_w^2) + (d-1)F_2}$$

$$\leqslant \rho_I \leqslant \frac{(s_a^2/s_w^2) - F_1}{(s_a^2/s_w^2) + (d-1)F_1}, \tag{18}$$

where $F_1 = \mathscr{F}_{\alpha_1}\{(n-1), n(d-1)\}$ and $F_2 = \mathscr{F}_{1-\alpha_2}\{(n-1), n(d-1)\}$ with $\alpha_1 + \alpha_2 = \alpha$. Also, such intervals can be used to test the hypothesis $H_0: \rho_I = 0$ by noting whether they include 0. Further discussion is given in Anderson and Bancroft [1], Kempthorne [6], Scheffé [12], and Searle [13].

For the case where $d = D$ and N is large,

$$\hat{v}_a = s_a^2/D, \qquad \hat{v}_w = (D-1)s_w^2/D, \tag{19}$$

$$r_I = (s_a^2 - s_w^2)/\{s_a^2 + (D-1)s_w^2\}. \tag{20}$$

This situation is usually called a *single-stage cluster sample** since all observations within the selected clusters are included (i.e., there is no subsampling at the second stage). Although estimation for it is straightforward, the formulation of confidence intervals analogous to (18) requires large n for approximate methods to be applicable. Another important consideration is [via (4)] that

$$\text{var}(\bar{y})/\text{var}(\bar{y}_{\text{srs}}) = Dv_a/v$$

$$= 1 + (D-1)\rho_I, \tag{21}$$

where $\text{var}(\bar{y}_{\text{srs}})$ denotes the variance of the mean $\bar{y}_{\text{srs}}$ for a simple random sample of size nD. Since the ratio (21) indicates the consequences of clustering on the estimation of the population mean μ, references in survey sampling such as Kish [8] call it the "design effect"*. Thus, by their relationship in (21), both the "design effect" and the intraclass correlation ρ_I reflect the extent to which within cluster homogeneity implies greater variance for the cluster sample mean $\bar{y}$ than $\bar{y}_{\text{srs}}$. In this sense, Kish [8] interprets ρ_I as a "rate of homogeneity (roh)." Since estimates for $\text{var}\{\bar{y}\}$, $\text{var}\{\bar{y}_{\text{srs}}\}$, and hence the "design

Table 1   Heartrate Determinations at Three Successive Visits for 16 Patients in a Clinical Trial

Patient	Visit 1	Visit 2	Visit 3
1	90	75	70
2	100	95	82
3	60	72	67
4	56	70	75
5	76	75	74
6	80	100	100
7	90	95	88
8	96	100	92
9	85	72	70
10	108	100	102
11	75	85	80
12	74	70	82
13	70	78	72
14	80	78	63
15	68	62	64
16	102	88	100

effect" are often readily available for many sample survey situations, (21) represents a more convenient basis for the estimation of ρ_I for them than (20).

Some aspects of the previous discussion can be illustrated through analysis of the data in Table 1 for heartrate determinations at three successive visits to a medical center of 16 patients in a clinical trial*. Since both N and D for the population of potential determinations can be considered infinite, (16) and (17) are applicable. From $s_a^2 = 448.53$ and $s_w^2 = 50.48$, it follows that

$$\bar{y} = 81.4, \quad \widehat{\text{var}(\bar{y})} = 448.53/48 = 9.34,$$

$$\hat{v}_a = (448.53 - 50.48)/3 = 132.68,$$

$$\hat{v}_w = 50.48,$$

$$r_I = (448.53 - 50.48)/(448.53 + 100.96)$$

$$= 0.724.$$

Also, from (14), $\tilde{r}_I = 0.710$. The use of intraclass correlation is considered more appropriate than interclass correlation for this example since the patients were treated separately at different times rather than simultaneously at the same times; so the orderings of the heartrate determinations by visits are

essentially arbitrary. A 95% confidence interval for ρ_I under the assumption of normal distributions for among-cluster and within-cluster variation is [via (18) with $\alpha_1 = \alpha_2 = 0.025$] the set of values $(0.508 < \rho_I < 0.868)$. Thus the measurement of heartrate can be interpreted as reliable since most of the variance is among patients rather than within patients; the values within patients are relatively homogeneous.

The concepts concerning intraclass correlation which have been included in this entry can be extended in several directions. These can be briefly summarized in two ways.

1. The clusters in a population can have different sizes; and the number of observations per cluster in a sample can vary. The two-stage sampling process can involve stratification and/or unequal probabilities of selection.

2. More than two sources of variation can apply either via multistage cluster samples and/or when data collection involves multiple observers (or interviewers). For these situations, several intraclass correlation coefficients can be formulated on a sampling process or observer basis.

Further discussion of extensions of intraclass correlation in directions such as (1) and (2) are available in many references. Among these, Hansen et al. [3] and Kish [8] are useful for survey sampling applications; Hansen et al. [4] and O'Muircheartaigh [10] for issues pertaining to *interviewer variance**; Kempthorne [7] for genetics applications; and Landis and Koch [9] for applications to *measures of agreement** in *observer reliability studies**.

References

[1]   Anderson, R. L. and Bancroft, T. A. (1952). *Statistical Theory in Research*. McGraw-Hill, New York.

[2]   Fisher, R. A. (1946). *Statistical Methods for Research Workers*. Oliver & Boyd, London.

[3]   Hansen, M. H., Hurwitz, W., and Madow, W.

(1953). *Sample Survey Methods and Theory*, Vols. 1 and 2. Wiley, New York.

[4] Hansen, M. H., Hurwitz, W. N., and Bershad, M. (1959). *Bull. Int. Statist. Inst.*, **38**, 359–374.

[5] Harris, J. A. (1913). *Biometrika*, **9**, 446–472.

[6] Kempthorne, O. (1952). *The Design and Analysis of Experiments*. Wiley, New York.

[7] Kempthorne, O. (1957). *An Introduction to Genetic Statistics*. Wiley, New York.

[8] Kish, L. (1965). *Survey Sampling*. Wiley, New York.

[9] Landis, J. R. and Koch, G. G. (1975). *Statist. Neerlandica*, **29**, 101–123, 151–161.

[10] O'Muircheartaigh, C. A. (1979). In *Analyzing Survey Data: Model Fitting*, Vol. 2, C. A. O'Muircheartaigh and C. Payne, eds. Wiley, New York, Chap. 7.

[11] Rothery, P. (1979). *Biometrika*, **66**, 629–639.

[12] Scheffé, H. (1959). *The Analysis of Variance*. Wiley, New York.

[13] Searle, S. R. (1971). *Linear Models*. Wiley, New York.

[14] Yule, G. U. and Kendall, M. G. (1947). *An Introduction to the Theory of Statistics*. Charles Griffin, London.

Acknowledgments

This research was supported in part by the U.S. Bureau of the Census (JSA-80-19). The author would like to express his appreciation to Julie MacMillan for computing assistance and to Ann Thomas and Lori Turnbull for their typing of the manuscript.

(ANALYSIS OF COVARIANCE
ANALYSIS OF VARIANCE
CORRELATION
DEPENDENCE, MEASURES OF
KENDALL'S TAU
SPEARMAN'S RHO
SURVEY SAMPLING
VARIANCE COMPONENTS)

GARY G. KOCH

INTRINSIC RANK TEST (FOR k INDEPENDENT SAMPLES)

RATIONALE

This inherently nonparametric test may be viewed as a generalization of the k-sample median test to a k-sample, h-quantile test. It may be considered as an alternative to ANOVA* proper, when the normality assumptions for the latter are not even remotely met, or as a substitute for the classical Kruskal–Wallis* (analysis of variance by ranks) test, which relies on squared rank sums, and therefore is highly sensitive to spurious interchanges of ordinal ranks due to measurement uncertainties unrelated to the inference sought.

The inference sought generally is whether or not the null hypothesis (H_0): that the k samples originate from the same parent population, is tenable. In actual practice, however, the notion (H_0) of the k samples "having come from the same parent population" or "being identically distributed" is at best an abstraction. Far more realistic is to speak of (H_0): a specified degree of *commonality* among the sample distributions, the degree being specified by the number of intrinsic rank classes, h, as developed below. With acceptable commonality, under H_0 the k samples are expected to mesh well in the ordinal rank scale and hence capture a share of intrinsic ranks proportional to their sizes. Conversely, under H_0, each intrinsic rank class would capture a share from each sample proportional to its size. The null hypothesis thus corresponds exactly to the conventional expectation of a two-way cross-classification. With an α-significant outcome of the test it is rejected, with the inference that the specified (h) degree of commonality does not obtain at the selected α level.

PROCEDURE

With a common measurement scale of at least ordinal status, the k samples of sizes N_i, $i = 1, 2, \ldots, k$, are combined into one grand sample of size $S = \sum_i N_i$; the original sample identities of the S observations having been retained. The grand sample is then ranked 1 through S on the criterion variable; tied measurements generally pose less of a problem here, as should become evident from what follows.

An intrinsic rank function is then applied to transform the ordinal ranks into intrinsic ranks, one eminently practical closed-form function being

$$j(R) = j(R; S, h) = \left[\!\!\left[\frac{h(R-1)}{S} \right]\!\!\right] + 1, \quad (1)$$

where axb is the integral value of its argument, $1 \leqslant R \leqslant S$ is the ordinal rank, and S and k are as defined above. The integer argument h in formula (1) represents the number of intrinsic rank classes into which the ordinal scale is to be subdivided.

For an intrinsic rank $1 \leqslant j \leqslant h$, and an arbitrary integer $U > S$, let

$$LB(j) = U - \left[\!\!\left[(U-1) - \frac{S(j-1)}{h} \right]\!\!\right],$$

$$(2)$$

$$UB(j) = U - \left[\!\!\left[(U-1) - \frac{Sj}{h} \right]\!\!\right];$$

then $I(j)$: $LB(j) \leqslant R < UB(j)$ defines the jth intrinsic rank class, i.e., the ordinal rank interval mapped into $j(R)$ by formula (1).

The size of the jth intrinsic rank class is then given as

$$M_j = M(j) = UB(j) - LB(j), \quad (3)$$

and depending on whether or not $S \not\equiv 0 \bmod(h)$, the class sizes may differ by at most unity.

An example may clarify this further. Suppose that three samples of sizes $N_1 = 19$, $N_2 = 29$, and $N_3 = 24$ were combined into one grand sample of size $S = 72$, and ranked 1 through 72. If ($h = 5$) intrinsic rank classes are used, then from formulas (2) and (3) we obtain Table 1, where, with (quite arbitrary) $U = 100$, the two interval bounds corresponding to $j = 3$ were computed as

$$LB(3) = 100 - a99 - 72 \cdot (3-1)/5b$$

$$= 100 - a99 - 28.8b$$

$$\dot{=} 100 - a70.2b = 100 - 70 = 30,$$

$$UB(3) = 100 - a99 - 72 \cdot 3/5b$$

$$= 100 - a99 - 43.2b$$

$$= 100 - a55.8b = 100 - 55 = 45,$$

Table 1

j	$LB(j)$	$UB(j)$	$M(j)$
1	1	16	15
2	16	30	14
3	30	45	15
4	45	59	14
5	59	73	14

so that all ordinal ranks satisfying $30 \leqslant R < 45$ are transformed into intrinsic rank 3.

Thus if the second sample has captured ordinal rank 36, then from formula (1),

$$j(36) = a5 \cdot (36 - 1)/72b + 1$$

$$= a2.43b + 1 = 2 + 1 = 3,$$

and so with sample identity 2 and intrinsic rank 3, this particular observation is cross-classified into the (2,3)rd cell of a 3×5 contingency table, incrementing by 1 the cell count a_{23}. Then, by completing this process for all three samples, the 3×5 table Table 2 is completed, with the resulting row sums equal to the sample sizes, N_i, and the column sums, M_j, as computed in Table 1. From what has been explained it follows that the row and column sums of a $k \times h$ contingency table* are determined as soon as the parameters k, N_i, S, and h are defined—an important advantage over polychotomization of the original measurement scale. The number of intrinsic rank classes, h, should be so determined as to permit (under H_0) each sample to contribute at least one observation to each class; and conversely, each class should be able to receive at least one observation from each sample, a necessary condition for which being that

Table 2

		Intrinsic Ranks				
		1	2	3	4	5	
	1	a_{11}	a_{12}	a_{13}	a_{14}	a_{15}	19
Sample	2	a_{21}	a_{22}	a_{23}	a_{24}	a_{25}	29
	3	a_{31}	a_{32}	a_{33}	a_{34}	a_{35}	24
		15	14	15	14	14	72

$h \cdot k \leqslant S$, whence of course $e_{ij} = N_i M_j / S$ $\geqslant 1$ for the conventional cell expectations in the $k \times h$ table.

SIGNIFICANCE EVALUATION

From an observed $k \times h$ table such as the 3×5 table Table 2, and the conventional cell expectations $e_{ij} = N_i M_j / S$, the χ^2-like test statistic is computed:

$$KS = \sum_{i,j} \frac{(a_{ij} - e_{ij})^2}{e_{ij}} = \sum_{i,j} \frac{a_{ij}^2}{e_{ij}} - S, \quad (4)$$

and for well-packed tables, the observed value of KS may be referred to the familiar chi-square distribution with $df = (k - 1)$ $(h - 1)$. For sparser tables the (α, β)-gamma distribution may be used, with the distribution parameters computed from the table parameters [1]; sparse tables may also be exactly evaluated by means of an enumeration algorithm [2].

Tied measurements are assigned consecutive ordinal ranks; problems arise only in the relatively rare event when a string of tied measurements originate from different samples, straddle adjacent rank classes, and cannot be resolved near-proportionally with respective sample sizes, in each intrinsic rank class. The standard remedy is judicious allocation or randomization.

References

[1] Kannemann, K. (1980). *Biom. J.*, **22**, 377–390.
[2] Kannemann, K. (1982). *Biom. J.*, **24**, 679–684.

Bibliography

Gibbons, J. D. (1971). *Nonparametric Statistical Inference*. McGraw-Hill, Toronto. (Kruskal–Wallis test.)

Kannemann, K. (1980). *Biom. J.*, **22**, 229–239. (Detailed computational example.)

Kendall, M. G. and Stuart, A. (1979). *The Advanced Theory of Statistics*, Vol. 2, 3rd. Edn. Charles Griffin, London. (Analysis of variance.)

K. KANNEMANN

INVARIANCE CONCEPTS IN STATISTICS

INTRODUCTION

In statistics the term "invariance" is generally used in its mathematical sense to denote a property that remains unchanged under a particular transformation, and, in practice, many statistical problems possess such a property. For example, the properties of normality and independence remain invariant under an orthogonal transformation of a set of independent normal random variables with the same variance; the problem of testing the hypothesis that the mean of a normal random variable is zero remains invariant under transformations involving a change of scale in which the variables are expressed.

The notion of invariance in statistical decision has an old origin. As discussed in HUNT–STEIN THEOREM, the published work of Hunt and Stein toward the end of World War II has given this notion strong support as to its applicability and meaningfulness in the framework of general classes of statistical decision rules to prove various optimum properties, e.g., minimax*, admissibility*, most stringent*, etc., of many statistical decision rules.

As in other branches of applied sciences, it is a generally accepted principle in statistics that if a problem with a unique solution is invariant under a certain group of transformations, then the solution should also remain invariant under them. The main reason for this intuitive appeal is probably the belief that there should be a unique best way of analyzing a collection of statistical information. However, in cases where the use of an invariant decision rule conflicts violently with the desire to make a correct decision or have a small risk, it must be abandoned. The principle of invariance has been developed primarily in the context of statistical decision problems. We shall treat it first in detail. Other invariant concepts used elsewhere in statistics will be treated later.

BASIC FORMULATION

Let $\mathscr{X} = \{x\}$ be the sample space, $\mathscr{A}$ a σ-algebra of subsets of $\mathscr{X}$ (a collection of subsets of $\mathscr{X}$ containing itself and closed under countable unions and complementation), and $\Omega = \{\theta\}$ be the parametric space. In this notation x or θ or both could be matrices or vectors. We shall denote by P_θ the family of distribution functions on A corresponding to θ in Ω, by D the space of all possible decisions $d(x)$ defined on $\mathscr{X}$ for a decision problem about θ, and by $L(\theta, d(x))$ a real-valued function defined on the Cartesian product space $\Omega \times D \times \mathscr{X}$, the loss function. The principle of invariance involves groups of transformations over the sample space $\mathscr{X}$, the parametric space Ω, and the decision space D (also called action space). Among these the most basic is the group of transformations G on the sample space $\mathscr{X}$. All transformations g in G considered in the context of invariance will be assumed to be:

1. One-to-one from $\mathscr{X}$ onto itself; that is, for every x_1 in $\mathscr{X}$ there exists an x_2 in $\mathscr{X}$ such that $x_2 = g(x_1)$ and $g(x_1) = g(x_2)$ implies $x_1 = x_2$.

2. Bimeasurable, to ensure that whenever X is a random variable with values in $\mathscr{X}$, $g(X)$ is also a random variable with values in $\mathscr{X}$ and for any set $A \in \mathscr{A}$, $g(A)$, $g^{-1}(A)$ ($(g)^{-1}$ is the inverse transformation corresponding to g) both belong to $\mathscr{A}$.

The transformation on the parametric space Ω is the group of induced transformation $\bar{g}$ corresponding to g in G and is defined as follows:

Definition. If the random variable X with values in $\mathscr{X}$ has probability distribution P_θ with θ in Ω, $g(X)$ is also a random variable with values in $\mathscr{X}$ having probability distribution $P_{\theta'}$ with $\theta' = \bar{g}(\theta)$ in Ω. If, in addition, $P_{\theta_1} \neq P_{\theta_2}$ for $\theta_1 \neq \theta_2$, then g determines $\bar{g}$ uniquely. This is also known as the condition of invariance of the family of probabil-

ity distributions $\{P_\theta, \theta \text{ in } \Omega\}$. This can also be expressed as

$$P_\theta\big(X \in g^{-1}(A)\big) = P_{\bar{g}\theta}(X \in A)$$

$$\text{for} \quad A \in \mathscr{A} \quad \text{and} \quad g \in G$$

or for every real-valued integrable function ϕ on $\mathscr{X}$,

$$E_\theta \phi(g(X)) = E_{\bar{g}(\theta)}\phi(X),$$

where E_θ refers to the mathematical expectation when the distribution of the random variable X is given by P_θ.

Example 1. Consider a normal population with unknown mean u and unknown variance σ^2. Let $x_1, \ldots, x_N$ be a sample of size N from the population. The sample space $\mathscr{X}$ is the space of all possible values of $(x_1, \ldots, x_N)$. The parametric space Ω is the space of all values of (u, σ^2). The problem of testing the hypothesis $u = 0$ against the alternatives $u \neq 0$ remains invariant under the group of scale changes which transform each X_i to aX_i, where a is a nonzero real number. The transformation of the parametric space Ω is given by $\bar{g}(u, \sigma^2) = (au, a^2\sigma^2)$. Since for any A in $\mathscr{A}$, $x = (x_1, \ldots, x_N)' \in g^{-1}(A)$ implies that $g(x) \in A$, writing $y_i = ax_i$, $i = 1, \ldots, N$,

$$P\big(g^{-1}(A)\big)$$

$$= \int_{g^{-1}(A)} (2\pi)^{-N/2}(\sigma^2)^{-N/2}$$

$$\times \exp\left\{-\frac{1}{2\sigma^2}\sum_{i=1}^{N}(x_i - u)^2\right\}\prod_{i=1}^{N}dx_i$$

$$= \int_A (2\pi)^{-N/2}(a^2\sigma^2)^{-N/2}$$

$$\times \exp\left\{-\frac{1}{2a^2\sigma^2}\sum_{i=1}^{N}(y_i - au)^2\right\}\prod_{i=1}^{N}dy_i$$

$$= P_{\bar{g}(\theta)}(A).$$

In discussing invariance we are considering groups of transformations only. This is, in no case, a limitation to our discussion. Given any class of transformations leaving the statistical decision problem invariant, it can always be extended to a group, each member of which leaves the problem invariant.

Very often in statistical problems there exists a measure λ on $\mathscr{X}$ such that P_θ is absolutely continuous with respect to λ so that we can write

$$P_\theta(A) = \int_A p_\theta(x)\, d\lambda(x).$$

If it is possible to choose a λ such that it is left invariant with respect to the group G, then the condition of invariance of the family of distributions P_θ can be expressed in terms of the probability density function p_θ with regard to the measure λ as

$$p_{\bar{g}(\theta)}(x) = p_\theta\big(g^{-1}(x)\big).$$

The fact that such a left invariant measure exists in many statistical problems is well known in the literature. The general theory of Haar measures in a large class of topological groups was first given by Haar [6]. In the terminology of Haar $P_\theta(A)$ is called the positive integral of p_θ. The basic result is that for a large class of topological groups there exist a left invariant Haar measure, positive on open sets, finite on compact sets, and unique except for multiplication by a positive constant. Because of the pioneering works of Haar such invariant measures are called invariant Haar measures (*see* HAAR DISTRIBUTIONS).

Example 2. Let G be a subgroup of the permutation group of a finite set $\mathscr{X}$ and for any set A in $\mathscr{A}$, let $\lambda(A) =$ number of points in A. For $g \in G$, gA just permutes the points in A and hence

$\lambda(gA) =$ number of points in the permuted

set $A = \lambda(A)$.

Thus the counting measure λ is an invariant measure under G and is unique up to a positive constant.

Example 3. Let G be E^n, the Euclidean n-space with the usual topology, and the group operation be addition. Then the left and right invariant Haar measure is the ordinary Lebesgue measure $d\lambda(g) = dg$.

INVARIANT DECISION PROBLEM

A decision problem is said to be *invariant* under a group G of transformations on $\mathscr{X}$ onto $\mathscr{X}$ if the family of distributions $\{P_\theta, \theta \in \Omega\}$ is invariant under G and the loss function satisfies $L(\theta, d(x)) = L(\bar{g}(\theta), \tilde{g}d(x))$, where $\tilde{g}$ is the induced transformation on D corresponding to g on $\mathscr{X}$. Note that $\tilde{g}(d(x)) = d(g(x))$. In such cases $\tilde{G} = \{\tilde{g}, g \in G\}$ is also a group of transformations of D onto itself. $\tilde{G}$ is homomorphic to the group G and is also homomorphic to $\bar{G}$.

When a statistical decision problem is invariant under a group of transformations G on $\mathscr{X}$, it is natural to restrict attention to statistical decision rules $\phi(x)$ which are also invariant under G in the sense that $\phi(x) = \phi(g(x))$ for all x in $\mathscr{X}$ and all g in G. (*See* DECISION THEORY.)

MAXIMAL INVARIANT

A function $\phi(x)$ defined on $\mathscr{X}$ is invariant under G if $\phi(x) = \phi(g(x))$ for all x in $\mathscr{X}$ and all g in G. It is a *maximal invariant* under G if it is invariant under G and if $\phi(x) = \phi(y)$, x, y in $\mathscr{X}$ implies that there exists a g in G such that $y = g(x)$.

Let Y be a space. Suppose that $T(x)$ is a measurable mapping from $\mathscr{X}$ to Y. Let h be a one-to-one function from Y to Z. If $\phi(x)$ with values in Y is a maximal invariant on $\mathscr{X}$, $\phi \cdot h$ is a maximal invariant on $\mathscr{X}$ with values in Z. This fact is often utilized in writing the maximal invariant in a simpler form.

Defining, for any x in $\mathscr{X}$, the totality of points $g(x)$ with g in G as the orbit of x, it follows from above that a function $\phi(x)$ is invariant under G if and only if it is constant on each orbit, and it is a maximal invariant under G if it is constant on each orbit and takes different values on different orbits. All maximal invariants are equivalent in the sense that their sets of constancy coincide.

In problems of statistical inference, when we restrict our attention to invariant tests, what is needed is a ratio of probability density functions of the maximal invariant. Stein [14] gave a method of finding this ratio using the invariant Haar measure on the group.

Let C be a group operating on a topological space (sample space) $\mathscr{X}$ and let u be a measure in $\mathscr{X}$ invariant under G. Suppose that there are given two probability densities p_1 and p_2 with respect to u, that is,

$$P_1(A) = \int_A p_1(x) \, du(x),$$

$$P_2(A) = \int_A p_2(x) \, du(x),$$

where P_1 and P_2 vanish simultaneously. Let P_i^* be the distribution of the maximal invariant $T(x)$ on $\mathscr{X}$ with respect to G when $\mathscr{X}$ has distribution P_i. Then under certain conditions Stein's method is given by

$$\frac{dP_2^*}{dP_1^*}(T(x)) = \frac{\int_G p_2(gx) \, d\lambda(g)}{\int_G p_1(gx) \, d\lambda(g)},$$

where λ is a left invariant Haar measure in G.

Since then two different developments of Stein's idea have appeared. Wijsman [17] developed a theory of cross section of orbits and used it as a general tool in obtaining a factorization of the invariant measure needed in Stein's method. Wijsman's idea was further developed by Koehn [8]. They both use differential geometry and Lie group theory. Schwartz [12] gave an alternative proof of Stein's method.

Example 4. In Example 1 the maximal invariant in the sample space and the corresponding maximal invariant in the parametric space are $t^2 = N\bar{x}^2(\sum_{i=1}^N (x_i - \bar{x})^2/(N-1))^{-1}$ with $\bar{x} = (1/N)\sum_{i=1}^N x_i$ and $\delta^2 = Nu^2/\sigma^2$, respectively, and the PDF of t^2 is

$$f_{t^2}(\delta^2) = \frac{\exp(-\tfrac{1}{2}\delta^2)}{(N-1)\Gamma((N-1)/2)}$$

$$\times \sum_{j=0}^\infty \frac{(\tfrac{1}{2}\delta^2)^j (t^2/(N-1))^{j-(1/2)} \Gamma(\tfrac{1}{2}(N+j))}{j!\,\Gamma(j+\tfrac{1}{2})(1 + t^2/(N-1))^{N/2+j}}.$$

It can be shown that $f_{t^2}(\delta^2)/f_{t^2}(0)$ is an increasing function of t^2 so that the usual two-sided t-test* is uniformly most powerful invariant for testing $u = 0$ against $u \neq 0$.

CHARACTERIZATION

In statistical decision problems some simplification is introduced by characterizing the decision rules as functions of a minimal sufficient statistic and thereby reducing the dimension of the sample space to that of the minimal sufficient statistic. However, this characterization does not reduce the dimension of the parametric space. A similar characterization of invariant decision rules in terms of maximal invariants $T(x)$ also exists, namely, a decision rule $\phi(x)$ is invariant under G if and only if there exists a function f such that $\phi(x) = f(T(x))$. In general, f is not a Borel measurable function; however, if the image of the maximal invariant T is Euclidean and T is Borel measurable, then f is Borel measurable.

Let $v(\theta)$ be a maximal invariant under $\overline{G}$ on the parametric space Ω. The distribution of the maximal invariant $T(X)$ where X is a random variable with values in $\mathscr{X}$ depends on Ω only through $v(\theta)$. In other words, the risk function of an invariant decision rule (defined as the mathematical expectation of the loss function) is constant on orbits of Ω with respect to $\overline{G}$. Thus the main advantage of invariant decision rules is to reduce the dimension of the parametric space by reducing the dimension of the sample space to that of the space of maximal invariant.

Although a great deal has been written concerning the theory of tests, decisions, and inference for statistical problems invariant with respect to a certain group of transformations, no great amount of literature exists (see Lehmann and Stein [11] and Wendel [16]) concerning the problem of discerning whether or not a given problem is actually invariant under certain group. Brillinger [3] gave necessary and sufficient conditions that statistical problems must satisfy in order that they be invariant under a fairly large class of transformations groups, Lie transformations groups.

Statistical decisions are made on the basis of sample observations. Sample observations often contain information which is not relevant to the making of the decision. By using the principle of sufficiency* we can discard that part of sample observations which is of no value for any decision problem concerning the parameter. This reduces the dimension of the sample space to that of the space of sufficient statistics, without reducing the dimension of the parametric space. By using the invariance principle, further reduction to the decision space can often be made. The invariance, by reducing the dimension of the sample space to that of the space of maximal invariant, also shrinks the parametric space.

Example 5.   In Example 1 the sufficiency reduces the data to $\bar{x} = (1/N)\sum_i x_i$, $s^2 = (1/N)\sum(x_i - \bar{x})^2$. The joint distribution of $(\bar{X}, S^2)$ depends on (u, σ^2). However, by invariance the data are reduced to t^2 of Example 3 and their distribution depends on a single parameter $\delta^2 = N(u^2/\sigma^2)$.

In view of the fact that both invariance and sufficiency are successful in reducing the dimension of the sample space, one is naturally interested in knowing whether both principles can be used simultaneously and if so, in what order. Under certain conditions this reduction can be carried out by using both principles simultaneously, and the order in which these two principles are used is immaterial in such cases (see Hall et al. [7]). However, one can avoid the task of verifying these conditions by replacing the sample space by the space of sufficient statistics before looking for the group of transformations leaving the statistical problem invariant, and by then looking for the proper group of transformations on the space of sufficient statistics which leaves the problem invariant.

The relative performances of different statistical decision rules are compared by means of their risk functions*. It is therefore of interest to study the implications of the invariance of risk functions of different decision rules rather than the rules themselves. Since the risk function of invariant decision rules depends only on the maximal invariant in the parametric space, any invariant decision rule has an invariant risk function. The converse, that if the risk function of a statistical decision rule is invariant under $\bar{G}$, the rule is invariant under G, does not always hold well. However, if the group G is countable or if there exists a σ-finite measure on G which is right invariant with respect to G and if prior to application of invariance the problem can be reduced to one based on a sufficient statistic on the sample space $\mathscr{X}$ whose distributions constitute a boundedly complete family, the converse holds. A decision rule $\phi(x)$ is said to be "almost invariant" with respect to G if $\phi(x) = \phi(gx)$ for all x in $\mathscr{X} - Ng$, where Ng is a subset of $\mathscr{X}$ depending on g of probability measure 0. If the risk of a statistical decision rule is invariant under $\bar{G}$, the decision rule is almost invariant under G, provided that the family of distributions is boundedly complete.

Apart from the natural justification for the use of invariant decision rules for invariant problems, a powerful support for the principle comes from the famous unpublished Hunt–Stein theorem (*see* HUNT–STEIN THEOREM), which asserts that under certain conditions on the transformation group G acting on the sample space $\mathscr{X}$, there exists an invariant decision rule which is minimax among the class of all rules. In particular, if the transformation group is finite or a compact topological group, the Hunt–Stein theorem holds. Without additional restrictions the theorem does not hold for noncompact groups. The conditions of the Hunt–Stein theorem, whether algebraic or topological, are entirely on the group and are nonstatistical in nature. For admissibility of statistical decision rules through invariance the situation is more complicated. Apart from the finite or the compact groups, statistical structure plays an important role. Various proofs of the famous unpublished result have appeared in the literature. The version of this theorem published by Lehmann [10] is probably close in spirit to the original unpublished one of Hunt and Stein.

In the discussion above we have treated invariance in the framework of statistical decision theory. De Finetti's [4] theory of exchangeability* treats invariance of the distribution of the sample observations under finite permutations. It provides a crucial link between his theory of subjective probability and the frequency concept of probability. Most of the classical methods of statistical analysis take as basic a family of distributions, the true distribution of the sample being an unknown member of the family about which the statistical inference is required. To a subjectivist, however, no probability is unknown. If $x_1, x_2, \ldots$ are the outcomes of a sequence of trials under similar conditions, subjective uncertainty is expressed directly by ascribing to $X_1, X_2, \ldots$ a known joint distribution. If some of the X's are now observed, predictive inference about the others is obtained by conditioning the original distributions on the observations. De Finetti has shown that these two approaches are mathematically equivalent when the subjectivist's joint distribution is invariant under finite permutation.

The weak invariance principle (see Billingsley [2]) is developed to demonstrate the sufficiency of the classical assumptions associated with the weak convergence of stable laws. The weak invariance principle is also called Donsker's theorem [5]. Let $X_1, X_2, \ldots$ be independently and identically distributed random variables with mean 0 and variance σ^2 and let $S_j = \sum_{i=1}^{j} X_i$, $X_n(t) = S_i / \sigma\sqrt{n}$, for $t = i/n$, $i = 1, \ldots, n$. Donsker [5] proved that $\{X_n(t)\}$ converges weakly to Brownian motion*. Lamperti [9] improved it by showing that it holds for a much larger class of functionals under certain assumptions, on the moment of X_i's. A result analogous to Donsker's theorem is proved by Sen [13] for signed-rank statistics. Antille [1] improved Sen's result for signed-rank processes of Wilcoxon* and Van der Waerden type.

Associated with the weak invariance prin-ciple, the concept of strong invariance has been introduced to prove strong convergence results (see Tusnady [15]). Here the term "invariance" is used in the sense that if $X_1, X_2, \ldots$ are independent, identically distributed random variables with $E(X_i) = 0$, var$(X_i) = \sigma^2$, and if h is a continuous function on $[0, 1]$, then the limiting distribution of $h(X_n)$ does not depend on any other property of X_i.

References

[1] Antille, A. (1979). *Zeit. Wahrscheinlichkeitsth. verw. Geb.*, **47**, 315–324.

[2] Billingsley, P. (1968). *Convergence of Probability Measures*. Wiley, New York. (A graduate-level text of mathematical nature.)

[3] Brillinger, D. (1963). *Ann. Math. Statist.*, **34**, 492–500.

[4] De Finetti, B. (1964). *Studies in Subjective Probability*, H. E. Kyburg and H. E. Smoker, eds. Wiley, New York, pp. 93–158.

[5] Donsker, M. (1951). An Invariance Principle for Certain Probability Limit Theorems. *Mem. Amer. Math. Soc.*, **6**.

[6] Haar, A. (1933). *Ann. Math. (2nd Ser.)*, **34**, 147–169.

[7] Hall, W. J., Wijsman, R. A., and Ghosh, J. K. (1965). *Ann. Math. Statist.*, **36**, 575–614.

[8] Koehn, U. (1970). *Ann. Math. Statist.*, **41**, 2045–2056.

[9] Lamperti, J. (1962). *Trans. Amer. Math. Soc.*, **104**, 430–435.

[10] Lehmann, E. (1959). *Testing Statistical Hypotheses*. Wiley, New York. (A graduate-level textbook on statistical testing of hypothesis.)

[11] Lehmann, E. and Stein, C. (1953). *Ann. Math. Statist.*, **24**, 142.

[12] Schwartz, R. (1969). *Properties of Invariant Multivariate Tests*. Ph.D. thesis, Cornell University.

[13] Sen, P. K. (1974). *Ann. Statist.*, **2**, 49–63.

[14] Stein, C. (1956). Some Problems of Multivariate Analysis, Part 1. *Tech. Rep. No. 6*, Dept. of Statistics, Stanford University, Stanford, Calif.

[15] Tusnady, G. (1977). In *Recent Developments in Statistics*, J. R. Barra, F. Brodeau, G. Romier, and B. Van Cutsem, eds. North-Holland, Amsterdam, pp. 289–300.

[16] Wendel, W. G. (1957). *Mich. Math. J.*, **4**, 173–174.

[17] Wijsman, R. A. (1966). *Proc. 5th Berkeley Symp. Math. Stat. Prob.*, Vol. 1. University of California Press, Berkeley, Calif., pp. 389–400.

Further Reading

See the following works, as well as the references just given, for more information on the topic of invariance concepts.

Ferguson, T. S. (1969). *Mathematical Statistics*. Academic Press, New York. [Treats invariance in statistical decision, textbook (graduate level) on decision theoretic mathematical statistics.]

Giri, N. (1975). *Invariance and Minimax Statistical Tests*. Hindustan, Delhi, India. (Treats invariance and minimax multivariate tests.)

Giri, N. (1977). *Multivariate Statistical Inference*. Academic Press, New York. (A graduate-level textbook of multivariate analysis through invariance.)

Giri, N. and Kiefer, J. (1964). *Ann. Math. Statist.*, **35**, 21–35.

Giri, N., Kiefer, J., and Stein, C. (1963). *Ann. Math. Statist.*, **31**, 1524–1535.

(ADMISSIBILITY
ANALYSIS OF VARIANCE
DECISION THEORY
DISCRIMINANT ANALYSIS
EQUIVARIANCE
ESTIMATION, POINT
EXCHANGEABILITY
FACTOR ANALYSIS
HAAR DISTRIBUTIONS
HUNT–STEIN THEOREM
HYPOTHESIS TESTING
MAXIMUM LIKELIHOOD
MINIMAX TESTS
MOST STRINGENT TESTS
MULTIVARIATE ANALYSIS
SUFFICIENCY
VAN DER WAERDEN SIGNED-RANK TEST
WILCOXON SIGNED-RANK TEST)

NARAYAN C. GIRI

INVARIANCE PRINCIPLES AND FUNCTIONAL LIMIT THEOREMS

Invariance principles deal with the convergence of random functions constructed out of a sequence of random variables. The theory has largely grown out of an exploration of the extent to which a random function constructed out of a random walk* process (e.g., by linear interpolation) approximates Brownian motion*. The best results, called strong invariance principles, provide an almost sure (a.s.) bound on the order of magnitude of the error of approximation by some standard random function, while weak invariance principles assert that distributional convergence holds. In either case, a wide variety of limit results involving functionals of the original sequence of random variables can be inferred from corresponding results for the limiting random function.

The concept of the invariance principle evolved from the work of Erdös and Kac [7] on functionals of the sequence $\{S_n = \sum_{i=1}^{n} X_i, n \geqslant 1\}$ of partial sums of independent, and identically distributed (i.i.d.) random variables with zero mean and finite third absolute moment. The central idea of Erdös and Kac was that the asymptotic behavior of functionals such as $\max_{1 \leqslant k \leqslant n} S_k$, $\max_{1 \leqslant k \leqslant n} k^{-1/2} |S_k|$, $\sum_{k=1}^{n} I(S_k \geqslant a)$, and $\sum_{k=1}^{n} \gamma(S_k, S_{k+1})$ [where I is the indicator function and $\gamma(a, b) = 1$ if $ab < 0$ and 0 otherwise] should, as with the classical central limit theorem*, be invariant under change of the distribution of the X_i. They showed that this was the case for the examples above and obtained the asymptotic distributions, under appropriate normalization, by calculation in particular cases where the calculations were straightforward. Unfortunately, their approach required separate treatment for each functional. This shortcoming, however, has been surmounted in more recent work described below on what are now termed invariance principles.

The bases of the weak and strong invariance principles lie largely in the work of Donsker [6] and Skorokhod [22], respectively. The book of Billingsley [2] has had a profound effect in popularizing the weak version, while Breiman [3, Chap. 13] has also been influential. The basic work on the strong version was done by Strassen [23, 24]. A recent survey of the methodology has been provided by Heyde [12].

The prototype for strong invariance principles has been provided by results for the random walk process generated by sums of i.i.d. random variables X_i with zero mean and unit variance. In this case Strassen [23] showed that it is possible to construct a probability space, with processes $S(t)$ and standard Brownian motion $W(t)$ defined on it, such that $S(t)$ is a random polygon whose distribution agrees with $S_n = \sum_{i=1}^n X_i$ for each n and

$$S(t) - W(t) = o(t \log_2 t) \quad \text{a.s.} \quad (1)$$

as $t \to \infty$ where $\log_2 t = \log \log t$. This result contains the familiar strong law of large numbers* and iterated logarithm law* but not the central limit theorem* [which can, however, be readily obtained via a variant of the proof of (1)].

Functional iterated logarithm type results may be obtained from (1) by putting

$$f_n(t) = (2n \log_2 n)^{-1/2} S(nt), \quad 0 \leqslant t \leqslant 1.$$

Then the sequence of functions $\{f_n(t)\}$ is relatively compact in the topology of uniform convergence and has as its set of limit points the set K of all real-valued absolutely continuous functions $h(t)$, $0 \leqslant t \leqslant 1$, with

$$h(0) = 0, \qquad \int_0^1 \left(\dot{h}(t)\right)^2 dt \leqslant 1,$$

the dot denoting derivative. This result is also due to Strassen [23] and can in turn be used to generate many curious special cases.

If $E|X_1|^{2+\delta} < \infty$, some $\delta > 0$, there is a construction giving an error in (1) which is $o(t^{1/2})$ and hence is small enough to permit the central limit theorem to be derived directly in addition to the strong convergence results. For example, Strassen [24] showed that if $E|X_1|^4 < \infty$, the representation could be improved to

$$\limsup t^{-1/4} (\log t)^{-1/2} (\log_2 t)^{-1/4}$$
$$\times |S(t) - W(t)| < K \quad \text{a.s.} \quad (2)$$

for some appropriately chosen K. If only a second moment condition is assumed, then the order of approximation in (1) cannot be improved. However, a minor rescaling does provide a strong invariance principle, due to Major [17], from which all the classical limit results can be obtained.

There has also been a comprehensive investigation of the question of whether (2) is sharp. Improved results under higher moment conditions were obtained by Csörgö and Révész [5] using a so-called quantile transform method*. A recent survey of results in the area has been provided by Major [16].

The results for sums of i.i.d. random variables naturally suggest extensions to martingales* and a variety of weakly dependent processes. General results for martingales which contain the aforementioned ones of Strassen have been provided by Jain et al. [14]. Strong invariance principles for sums of strong mixing, lacunary trigonometric, asymptotic martingale difference sequences, and for certain Markov processes* have been provided by Philipp and Stout [19].

Weak invariance principles deal with weak convergence in an appropriately rich metric space. The ones in most common use are $C([0, 1])$, the space of continuous functions on $[0, 1]$, and $D([0, 1])$, the space of function on $[0, 1]$ which are right continuous with left-hand limits. The general setting is that of a complete separable metric space (X, ρ) (ρ being the metric) with probability measures μ_i, $i = 0, 1, 2, \ldots$, defined on the Borel sets of X. It is said that μ_n *converges weakly* to μ_0 in (X, ρ) if for every bounded continuous function f on X we have $\int f d\mu \to \int f d\mu_0$ as $n \to \infty$. If random elements ξ_n, $n = 0, 1, 2, \ldots$, taking values in X are such that the distribution of ξ_n is μ_n, $n = 0, 1, 2, \ldots$, we write $\xi_n \xrightarrow{d} \xi_0$ if μ_n converges weakly to μ_0.

The most fundamental of the weak invariance principles is Donsker's theorem for sums $S_n = \sum_{i=1}^n X_i$, $n \geqslant 1$, of i.i.d. random variables X_i with $EX_1 = 0$, $EX_1^2 = 1$. This can be framed in $C([0, 1])$ by setting $S_0 = 0$ and

$$S_n(t) = n^{-1/2} \{ S_{[nt]} + (nt - [nt]) X_{[nt]+1} \},$$

$$0 \leqslant t \leqslant 1,$$

where $[x]$ denotes the integer part of x. Then Donsker's theorem asserts that $S_n(t) \xrightarrow{d} W(t)$, standard Brownian motion (e.g., Billingsley [2, p. 68]).

Weak convergence in a suitably rich metric space is of considerably greater use than that in Euclidean space because a wide variety of ordinary convergence in distribution results on the real line can be derived from it with the aid of the continuous mapping theorem. [*Continuous Mapping Theorem* If $\xi_n \overset{d}{\to} \xi_0$ in (X, ρ) and the mapping $h : X \to \mathbb{R}$ (the real line) is continuous (or at least is measurable and $P(\xi_0 \in D_h) = 0$, where D_h is the set of discontinuities of h), then $h(\xi_n) \overset{d}{\to} h(\xi_0)$.] In many applications the limit random element is Brownian motion which has continuous sample paths with probability 1. A general discussion of useful functionals has been provided by Whitt [25] and a survey of applications in applied probability by Iglehart [13]. Rate of convergence results for functional limit theorems are discussed in Hall and Heyde [10, Chap. 4].

The use of invariance principles in statistical contexts as distinct from probabilistic ones is of recent origin. A strong invariance principle has been developed by Csörgö and Révész [5] for the study of the empirical process constructed from the empirical distribution function (*see* EDF STATISTICS). This leads to such practically significant subsidiary results as convergence in distribution for the Kolmogorov–Smirnov and Cramér–von Mises statistics. A comprehensive recent survey of the field of empirical processes has been provided by Gaenssler and Stute [8]. Examples of strong invariance principles in time-series analysis* (estimating the scale parameter in a first-order autoregression), stochastic approximation* (Robbins–Monro procedure), and for record times* are given in Heyde [12].

Weak invariance principles are much more common in the statistical literature and have been established in such contexts as competing risks and survival analysis* [1, 4, 21], change in mean [18], occupancy problems* [20], sequential analysis* [15], likelihood ratios* [11], and U-statistics* (which has a quite extensive literature; see Hall [9] for a recent discussion).

The books referenced are all aimed at a graduate-level audience. A proper treatment of invariance principles has substantial mathematical prerequisites in probability theory*, measure theory*, and topology.

References

[1] Aalen, O. (1978). *Ann. Statist.*, 6, 534–545.

[2] Billingsley, P. (1968). *Convergence of Probability Measures*. Wiley, New York.

[3] Breiman, L. (1968). *Probability*. Addison-Wesley, Reading, Mass.

[4] Breslow, N. and Crowley, J. (1974). *Ann. Statist.*, 2, 437–453.

[5] Csörgö, M. and Révész, P. (1975). *Zeit. Wahrscheinlichkeitsth. verw. Geb.*, 31, 255–259, 261–269.

[6] Donsker, M. (1951). An Invariance Principle for Certain Probability Limit Theorems. *Mem. Amer. Math. Soc.*, 6.

[7] Erdös, P. and Kac, M. (1946). *Bull. Amer. Math. Soc.*, 52, 292–302.

[8] Gaenssler, P. and Stute, W. (1979). *Ann. Prob.*, 7, 193–243.

[9] Hall, P. (1979). *Stoch. Processes Appl.*, 9, 163–174.

[10] Hall, P. and Heyde, C. C. (1980). *Martingale Limit Theory and Its Application*. Academic Press, New York.

[11] Hall, W. J. and Loynes, R. M. (1977). *Ann. Statist.*, 5, 330–341.

[12] Heyde, C. C. (1981). *Int. Statist. Rev.*, 49, 143–152.

[13] Iglehart, D. L. (1974). *Stoch. Processes Appl.*, 2, 211–241.

[14] Jain, N. C., Jogdeo, K., and Stout, W. L. (1975). *Ann. Prob.*, 3, 119–145.

[15] Lai, T. L. (1979). *Ann. Statist.*, 7, 46–59.

[16] Major, P. (1978). *J. Multivariate Anal.*, 8, 487–517.

[17] Major, P. (1979). *Ann. Prob.*, 7, 55–61.

[18] Maronna, R. and Yohai, V. J. (1978). *J. Amer. Statist. Ass.*, 73, 640–645.

[19] Philipp, W. and Stout, W. (1975). Almost Sure Invariance Principles for Partial Sums of Weakly Dependent Random Variables. *Mem. Amer. Math. Soc.*, 2(161).

[20] Sen, P. K. (1979). *Ann. Statist.*, 7, 414–431.

[21] Sen, P. K. (1979). *Ann. Statist.*, 7, 372–380.

[22] Skorokhod, A. V. (1961). *Studies in the Theory of Random Processes*. Kiev University Press, Kiev (in Russian). (English transl.: Addison-Wesley, Reading, Mass., 1965.)

[23] Strassen, V. (1964). *Zeit. Wahrscheinlichkeitsth. verw. Geb.*, **3**, 211–226.

[24] Strassen, V. (1967). *Proc. 5th Berkeley Symp. Math. Statist. Prob.*, Vol. 2. University of California Press, Berkeley, Calif., pp. 315–343.

[25] Whitt, W. (1980). *Math. Operat. Res.*, **5**, 67–85.

(BROWNIAN MOTION
CENTRAL LIMIT THEOREM
COMPETING RISKS
EDF STATISTICS
INVARIANCE CONCEPTS IN STATISTICS
LAW OF ITERATED LOGARITHM
LAWS OF LARGE NUMBERS
LIKELIHOOD RATIO
MARKOV PROCESSES
MARTINGALES
OCCUPANCY PROBLEMS
SEQUENTIAL ANALYSIS
STOCHASTIC APPROXIMATION
TIME SERIES
U-STATISTICS)

C. C. HEYDE

INVARIANT PRIOR DISTRIBUTIONS

The controversy surrounding Bayesian inference*, and its acceptability as a scientific methodology of statistical inference, has centered on its requirement that prior information about statistical parameters be explicitly introduced and described in terms of a probability distribution. (*See* INFERENCE, STATISTICAL for further background on the Bayesian approach.) A common objection is that the seeming arbitrariness and subjectivity of the prior distribution is at variance with the desire that statistical inference be entirely "objective."

The *logical Bayesian* view holds that a prior distribution represents partial logical information about unknown parameters, of the same objective status as a statistical model. In particular, it is supposed that, for any model, there is a specific prior distribution representing "complete ignorance." The program of determining such ignorance priors has been presented most cogently by Jeffreys [18]. (*See* JEFFREYS' NONINFOR-

MATIVE PRIOR.) An important strand in this program is the idea of *invariant prior distributions*.

INVARIANT PRIOR PROBABILITY ASSIGNMENTS

Let P_θ be the distribution of a certain observand X over a space $(\mathscr{X}, \mathscr{A})$, given that a parameter Θ, with possible values in $\tilde{\Theta}$, takes the value θ. Define $\mathscr{P} = \{P_\theta : \theta \in \tilde{\Theta}\}$. We shall assume that $\mathscr{P}$ is dominated by a σ-finite measure μ on $\mathscr{A}$ and write $f(x \mid \theta) = dP_\theta(x)/d\mu$. By the *model* $\mathscr{M} = (X, \Theta, \mathscr{P})$ we shall understand the specification of the variables X and Θ, and distributions $\mathscr{P}$, together, implicitly, with the *parametrization* of $\mathscr{P}$, i.e., the mapping associating the appropriate $P_\theta \in \mathscr{P}$ with the value θ of Θ. We shall term the parametrized family $\mathscr{P}$ the *distribution model* associated with $\mathscr{M}$.

The task set is to associate, with each model $\mathscr{M}$, an ignorance prior distribution $\Pi_{\mathscr{M}}$ for its parameter. The possibility is explicitly allowed that ignorance may not be representable by a "proper" probability distribution, but by a general σ-finite measure giving possibly infinite "probability" to the whole parameter space. For example, ignorance about an unrestricted real parameter might perhaps be represented by Lebesgue measure, the "uniform distribution" on $\mathbb{R}^1$. Bayes's formula, stated as $d\Pi(\theta \mid x) \propto f(x \mid \theta) d\Pi(\theta)$, is formally applicable to such "improper" distributions*, and will often yield proper posterior distributions.

In nineteenth-century applications of Bayes's theorem* it was common to take the uniform distribution, in one or several dimensions, as a suitable representation of ignorance about a parameter with values in a Euclidean space. This practice largely followed Laplace* [19, 20]. However, as pointed out by Fisher* [9, Chap. II], this naive procedure leads to inconsistencies if applied to different parametrizations of the same problem. For example, if Θ is an unknown probability, an alternative parameter is $\Phi = \sin^{-1}\sqrt{\Theta}$. But a uniform distribution for Φ implies a nonuniform density for Θ,

$$\pi(\theta) \propto \theta^{-1/2}(1-\theta)^{-1/2}.$$

Jeffreys attempted to circumvent these difficulties by searching for rules assigning $\Pi_{\mathcal{M}}$ to $\mathcal{M}$ in an invariant way. The main desiderata for such a rule may be set out as follows:

1. **Parameter invariance (PI).** Let $\mathcal{M} = (X, \Theta, \mathcal{P})$ and let $\Phi = \phi(\Theta)$ be a (smooth) invertible function, or *recoding*, of Θ. The model $\mathcal{M}_1 = (X, \Phi, \mathcal{P})$ differs from $\mathcal{M}$ in its parametrization, but describes an equivalent situation. So we should require that $\Pi_{\mathcal{M}_1}(\Phi \in A) = \Pi_{\mathcal{M}}(\Theta \in \phi^{-1}(A))$.

   Assuming Euclidean parameter spaces, and the existence of densities $\pi_{\mathcal{M}}$ and $\pi_{\mathcal{M}_1}$ with respect to Lebesgue measure, this requirement becomes

$$\pi_{\mathcal{M}_1}(\phi) = \pi_{\mathcal{M}}(\theta).|J(\theta)|^{-1},$$

   where $J(\theta)$ is the Jacobian $\det(\partial \phi(\theta)/\partial \theta)$, and $\phi = \phi(\theta)$.

2. **Data invariance (DI).** Now let $Y = y(X)$ be a recoding of X, and let $\mathcal{M}_2 = (Y, \Theta, \mathcal{Q})$ be the induced model for observand Y and parameter Θ. Again the essential situation is unchanged, and we therefore require that $\Pi_{\mathcal{M}_2}(\Theta \in A) = \Pi_{\mathcal{M}}(\Theta \in A)$.

As noted by Dickey [8], these invariance requirements do not relate specifically to ignorance: identical considerations apply for subjective prior distributions representing genuine knowledge. The key additional assumption is:

3. **Context invariance (CI).** If $\mathcal{M} = (X, \Theta, \mathcal{P})$ and $\mathcal{M}' = (X', \Theta', \mathcal{P})$ are two different models having the same distribution model, we should require that $\Pi_{\mathcal{M}}(\Theta \in A) = \Pi_{\mathcal{M}'}(\Theta' \in A)$. In other words, no features of the structure, meaning, or context of a model, other than its distribution model, should be taken into account. This principle thus formalizes ignorance as the irrelevance of context.

When (CI) is assumed, we may write $\Pi_{\mathcal{P}}$

instead of $\Pi_{\mathcal{M}}$. The criteria (PI), (DI), and (CI) together impose strong restrictions on the assignment of prior distributions.

JEFFREYS'S AND HARTIGAN'S RULES

Jeffreys [18, Sec. 3.10] proposed the rule $\pi_{\mathcal{P}}(\theta) = |I(\theta)|^{1/2}$, where $I(\theta)$ is the *Fisher information matrix** of $\mathcal{P}$, with (i, j) entry $E_{\theta}[(\partial l/\partial \theta_i)(\partial l/\partial \theta_j)]$, where $l = l(X, \theta) = \log f(X \mid \theta)$. When it exists, this satisfies conditions (PI), (DI), and (CI). In the one-parameter case, Jeffreys's rule is equivalent to assigning a uniform distribution to that parametrization in which the information is constant, in accordance with a suggestion of Perks [24].

Hartigan [13] considered rules directly associating "inverse" distributions for the parameter with arbitrary data values in a specific model. In these terms, requirement (PI), for example, becomes $\Pi_{\mathcal{M}_1}(\Phi \in A \mid X = x) = \Pi_{\mathcal{M}}(\Theta \in \phi^{-1}(A) \mid X = x)$. When $\Pi_{\mathcal{M}}(\cdot \mid X = x)$ is supposed calculated from a fixed prior $\Pi_{\mathcal{M}}$ using Bayes's theorem, this may be rephrased as *relative parameter invariance* (RPI), requiring that $\Pi_{\mathcal{M}_1}(\Phi \in A) \propto \Pi_{\mathcal{M}}(\Theta \in \phi^{-1}(A))$ [or $\pi_{\mathcal{M}_1}(\phi) \propto \pi_{\mathcal{M}}(\theta) \times |J(\theta)|^{-1}$], where the implicit multiplier, which drops out on forming posteriors, may depend arbitrarily on the models and parametrizations. Similarly, we can introduce (RDI) and (RCI).

Hartigan suggested a rule satisfying (RPI), (RDI), and (RCI) which, for the one-parameter case, yields prior density $\pi_{\mathcal{P}}(\theta)$ with: $(d/d\theta)\log \pi_{\mathcal{P}}(\theta) = E_{\theta}(l_1 l_2)/E_{\theta}(l_2)$, where $l_i = (\partial^i/\partial \theta^i)\log f(X \mid \theta)$ $(i = 1, 2)$. He called this the *asymptotically locally invariant* (ALI) prior density. The rule may be extended to the multiparameter case, yielding simultaneous differential equations that may, however, be insoluble.

Hartigan also introduced several further invariance criteria. These are all satisfied for the Jeffreys and ALI assignment rules. New relatively invariant prior densities may be constructed by the formula $\pi(\theta) \propto \{\pi^J(\theta)\}^{\alpha}\{\pi^H(\theta)\}^{\beta}$, where $\alpha + \beta = 1$, and π^J, π^H are the Jeffreys and ALI densities.

Table 1 gives the Jeffreys and ALI invari-

Table 1 Examples of the Various Invariant Prior Densities

Family of Densities			Type of Invariant Prior Density				
Form	Sample Space	Parameter Space	Relative	Inner	Outer	Jeffreys's	ALI
$(2\pi)^{-1/2}\exp[-\frac{1}{2}(x-\theta)^2]$	$-\infty < x < \infty$	$-\infty < \theta < \infty$	1	1	1	1	1
$\theta^{-1}(2\pi)^{-1/2}\exp(-\frac{1}{2}x^2/\theta^2)$	$-\infty < x < \infty$	$0 < \theta < \infty$	θ^k	θ^{-1}	θ^{-1}	θ^{-1}	θ^{-3}
$\theta_2^{-1}(2\pi)^{-1/2}\exp[-\frac{1}{2}(x-\theta)^2/\theta_2^2]$	$-\infty < x < \infty,$ $0 < \theta_2 < \infty$		θ_2^k	θ_2^{-2}	θ_2^{-1}	θ_2^{-2}	θ_2^{-5}
$|\theta|^{1/2}(2\pi)^{-(1/2)n}\exp(-\frac{1}{2}x^T\theta x)$	$x \in \mathbb{R}^n$	θ positive definite $n \times n$ matrix	$|\theta|^k$	$|\theta|^{-1}$	a	$|\theta|^{-1}$	1
$(2\pi)^{-(1/2)n}\exp[-\frac{1}{2}(x-K\theta)^T(x-K\theta)]$ $K, n\times k$ matrix of rank k	$x \in \mathbb{R}^n$	$\theta \in \mathbb{R}^k$	1	1	1	1	1
$x^{\theta-1}e^{-x}/\Gamma(\theta)$	$0 < x < \infty$	$0 < \theta < \infty$	a	a	a	$[(d^2/d\theta^2)\log\Gamma(\theta)]^{1/2}$	1
$\theta^x e^{-\theta}/x!$	x nonnegative integer	$0 < \theta < \infty$	a	a	a	$\theta^{-1/2}$	θ^{-1}
$\theta_1^{x_1}\theta_2^{x_2}\dots\theta_r^{x_r}n!/(x_1!\dots x_r!)$	x_i integers $x_i > 0,$ $\sum x_i = 1$	$\theta_i > 0, \sum\theta_i = 1$	a	a	a	$(\theta_1\theta_2\dots\theta_r)^{-1/2}$	$(\theta_1\theta_2\dots\theta_r)^-$
$\theta^x(1-\theta)^{n-x}n!/\{x!(n-x)!\}$	x integer, $0 \le x \le n$	$0 < \theta < 1$	a	a	a	$[\theta(1-\theta)]^{-1/2}$	$[\theta(1-\theta)]^{-1}$
$\theta^r(1-\theta)^{x-r}(x-1)!/\{(r-1)!(x-r)!\}$	x integer, $x \ge r$	$0 < \theta < 1$	a	a	a	$(1-\theta)^{-1/2}\theta^{-1}$	$(1-\theta)^{-1}$

[a]Means either that the method is not defined for the family of densities considered, or that it does not determine a prior density for the family.
Source. Hartigan [13], reproduced with permission.

ant priors for some familiar families of distributions. (The columns headed "Relative," "Inner," and "Outer" are explained in the following and in the section "Group Models.")

SELF-CONSISTENCY

Suppose that, for the model $\mathcal{M} = (X, \Theta, \mathcal{P})$, we have a recoding of X, written as $Y = g \circ X$, with the property that, whenever X has a distribution in $\mathcal{P}$, so does Y, and vice versa. We obtain an induced recoding of Θ, $\Phi = \bar{g} \circ \Theta$, such that $X \sim P_\theta$ if and only if $g \circ X \sim P_{\bar{g} \circ \theta}$. Then the model $\mathcal{M}' = (Y, \Phi, \mathcal{P})$ has exactly the same distribution model $\mathcal{P}$ as $\mathcal{M}$. We call g ($\bar{g}$) an *equivariant* recoding of X (Θ), and say that $\mathcal{M}$, or $\mathcal{P}$, is *equivariant* under g and $\bar{g}$. The collection of all equivariant recodings of X (Θ) forms a transformation group $\mathscr{G}$ ($\bar{\mathscr{G}}$).

For $\Phi = \bar{g} \circ \Theta$, $\bar{g} \in \bar{\mathscr{G}}$, it follows from (DI) and (PI) that $\Pi_{\mathcal{M}'}(\Phi \in \bar{g} \circ A) = \Pi_{\mathcal{M}}$ $(\Theta \in A)$. But if criterion (CI) holds, $\Pi_{\mathcal{M}'}$ $(\Phi \in \bar{g} \circ A) = \Pi_{\mathcal{M}}(\Theta \in \bar{g} \circ A)$. So when (PI), (DI), and (CI) all apply, $\Pi_{\mathcal{M}}$ ($= \Pi_{\mathcal{P}}$) must be *invariant* under $\bar{\mathscr{G}}$: that is, $\Pi_{\mathcal{P}}(\bar{g} \circ A) = \Pi_{\mathcal{P}}(A)$ for all $\bar{g} \in \bar{\mathscr{G}}$. Essentially, this argument has been given by Jaynes [17] and Villegas [32].

If $\bar{\mathscr{G}}$ is *transitive* on Θ (so that, for any values θ_1, θ_2 of Θ, there exists $\bar{g} \in \bar{\mathscr{G}}$ with $\theta_2 = \bar{g} \circ \theta_1$), the condition of invariance under $\bar{\mathscr{G}}$ determined $\Pi_{\mathcal{P}}$ uniquely (up to a multiple). This must then agree with Jeffreys's prior, since that certainly satisfies (DI), (PI), and (CI). Frequently, however, $\bar{\mathscr{G}}$ will be small, and there will be numerous invariant distributions. The theory of Brillinger [5] is of relevance to the general characterization of $\bar{\mathscr{G}}$.

If we only insist on the weaker criteria (RPI), (RDI), and (RCI), the self-consistency requirement becomes $\Pi_{\mathcal{P}}(\bar{g} \circ A) = \alpha(\bar{g}) . \Pi_{\mathcal{P}}(A)$ ($g \in \bar{\mathscr{G}}$), for some multiplier α, which must be a homomorphism from $\bar{\mathscr{G}}$ into the multiplicative group of positive reals. Such *relatively invariant* priors will include the ALI prior when it exists.

To clarify ideas, suppose that Θ is the unknown weight, in ounces, of a certain potato, and X is the reading, also in ounces, on a balance used to weigh it. Assume that X is normally distributed about Θ, with unknown standard deviation Φ ounces. Now let $X' = bX$, $\Theta' = b\Theta$, $\Phi' = b\Phi$, where $b = 1/35,840$, be the same quantities measured in tons rather than in ounces. Then the requirement (RCI) demands a proportional formal formula for the prior density of (Θ', Φ') as for that of (Θ, Φ): this is satisfied for the relatively invariant priors $\pi(\theta, \phi) \propto \phi^k$, with invariance if $k = -2$. Clearly, in this context it would *not*, under any reasonable opinion, be irrelevant to the form of the prior distribution whether our measurements were in tons or in ounces. This reflects the fact that we are not entirely ignorant about the weight of potatoes, and demonstrates the strength of the requirement of context invariance.

EXPONENTIAL MODELS

Consider a distribution model with densities constituting a *regular exponential family** of order k:

$$f(x \mid \theta) = \exp\{a(x) + b(\theta) + \phi(\theta)^T t(x)\}$$
$$(\theta \cdot \in \tilde{\Theta}),$$

where

$$\tilde{\Theta} = \left\{ \theta \colon \int \exp\{a(x) + \phi(\theta)^T t(x)\} d\mu(x) < \infty \right\},$$

$\tilde{\Phi} = \phi(\tilde{\Theta})$ is an open convex subset of $\mathbb{R}^k$.

Since the *canonical parameter* $\Phi = \phi(\Theta)$ seems to have a special status, one might assign a suitable prior distribution to Φ, and transfer it to an arbitrary parameter Θ by means of (PI). Specifically, we suppose that a prior distribution over Φ is assigned which depends only on its domain $\tilde{\Phi}$, and on no other feature of the model.

However, if Φ is a canonical parameter, so is any affine transformation of Φ. So the rule above is only self-consistent if the prior assigned is relatively invariant under all affine transformations preserving $\tilde{\Phi}$. The uniform distribution over $\tilde{\Phi}$, which is Hartigan's ALI prior, has this property, and is the only such distribution if $\tilde{\Phi} = \mathbb{R}^k$. For $k = 1$ and $\tilde{\Phi} = (0, \infty)$, $\pi(\phi)$ must have the form ϕ^λ, while for $\tilde{\Phi} = (-1, 1)$, any prior symmetric about 0 is permissible. These results are due to Huzurbazar [16], although his analysis of the case $k > 1$ appears suspect.

An alternative almost identical approach focuses on the *mean-value parameter* $\Psi = \psi(\Theta)$, where $\psi(\theta) = E_\theta(t(X))$, again taking values in a convex subset of $\mathbb{R}^k$, and unique up to an affine transformation.

For the binomial model: $f(x \mid \theta) = \binom{n}{x}\theta^x(1 - \theta)^{n-x}$, $(x = 0, 1, \ldots, n; \ 0 < \theta < 1)$ we have $\Phi = \log\{\Theta/(1 - \Theta)\}$, $\Psi = n\Theta$. The first approach above yields $\pi(\phi) \propto 1$, equivalent to $\pi(\theta) \propto \{\theta(1 - \theta)\}^{-1}$ [compare Jeffreys' prior: $\pi(\theta) \propto \{\theta(1 - \theta)\}^{-1/2}$]. The second approach justifies any prior for Θ symmetric about $\frac{1}{2}$.

In this case a third approach is to assign, say, the discrete uniform distribution directly to the *canonical statistic* X. This arises as the marginal distribution when $\pi(\theta) = 1$. This is essentially the justification of this prior given by Bayes.

For an exponential family any equivariant recoding must induce an affine transformation of both Φ and Ψ, preserving their domains [1], so that the considerations above do not conflict with those of the preceding section. In particular, the uniform distribution for either Φ or Ψ will be relatively invariant under any equivariant recoding.

Huzurbazar [16] extends these ideas to nonregular families, where the range of the distribution varies with the parameter.

GROUP MODELS

In an attempt to weaken (CI), and the strong implications of self-consistency, we might (following Fraser [10]) suppose to be given, along with a model $\mathcal{M} = (X, \Theta, \mathcal{P})$, a subgroup G of the group $\mathcal{G}$ of equivariant recodings of X. Only for $g \in G$ shall we regard the transformed model $\mathcal{M}' = (g \circ X, \bar{g} \circ \Theta, \mathcal{P})$ as equivalent in context to $\mathcal{M}$, and so

require identical, or proportional, prior densities for Θ and $\bar{g} \circ \Theta$. We call $(G, \mathcal{P})$ the *group model* associated with $\mathcal{M}$. An analysis parallel to that of the section "Self-Consistency" now implies only the invariance, or relative invariance, of $\Pi_\mathcal{M}$ under the smaller group $\bar{G} = \{\bar{g} : g \in G\}$.

In typical applications, G and $\bar{G}$ are locally compact topological groups. There then exists a *left Haar* measure μ on the Borel subsets of $\bar{G}$, determined up to a multiple, satisfying $\mu(\bar{g}S) \equiv \mu(S)$ $(\bar{g} \in \bar{G})$, and a *right Haar* measure ν, where $\nu(S) = \mu(S^{-1})$, for which $\nu(S\bar{g}) \equiv \nu(S)$. Moreover, $\nu(\bar{g}S) \equiv \Delta(\bar{g})^{-1}\nu(S)$, where Δ is the *modular function* of $\bar{G}$. For further background, see, e.g., Nachbin [21].

We call the group model *transitive* if $\bar{G}$ is transitive on $\tilde{\Theta}$. Then any measure m on $\bar{G}$ induces a measure Π on $\tilde{\Theta}$ by the rule: $\Pi(A) = m(\{\bar{g} \in \bar{G} : \bar{g} \circ \theta_0 \in A\})$, where $\theta_0 \in \tilde{\Theta}$ is a fixed reference point. If m is left Haar, then Π is invariant under $\bar{G}$, while the condition that Π should not depend on the choice of reference point θ_0 is satisfied when m is right Haar. Villegas [33, 34] terms these induced distributions the *inner* and *outer* priors, respectively. The inner prior, being invariant, agrees with Jeffreys's rule, as is apparent in Table 1. The outer prior is relatively invariant with multiplier $\Delta(\bar{g})^{-1}$.

A quantity $Q \equiv q(X, \Theta)$ is termed *invariant* under G if $q(g \circ x, \bar{g} \circ \theta) \equiv q(x, \theta)$ $(g \in G)$. For a transitive model, the distribution of $Q = q(X, \theta)$ under P_θ is the same for all θ; this remains true conditional on the maximal invariant statistic $A = a(X)$ under G, which is ancillary*. Fraser [10], Stein [25], Hora and Buehler [15], and Bondar [3] show that when the outer prior distribution is used, the posterior distribution of a G-invariant quantity Q is identical with its sampling distribution (conditional on A). For example, for the model $X_i \sim \mathcal{N}(M, \Sigma^2)$ independently $(i = 1, 2, \ldots, n)$, with a typical $g \in G$ operating as $g \circ (X_i) = (a + bX_i)$, $\bar{g} \circ (M, \Sigma) = (a + bM, b\Sigma)$ $(b > 0)$, the outer prior, with $\pi(\mu, \sigma) \propto \sigma^{-1}$, implies that, in the posterior distribution, the t-statistic $n^{1/2}(\bar{x} - M)/s$ has Student's distribution on $(n - 1)$ degrees of freedom. [The inner prior density $\pi(\mu, \sigma) \propto \sigma^{-2}$ yields the "wrong" degrees of freedom n; however, Villegas [33, 34] attempts an objective justification for inference based on such inner priors: *see* INNER INFERENCE.]

The result above implies the identity of Fraser's fiducial* (structural) distribution for Θ with the posterior based on the outer prior distribution. Moreover, equivariant Bayesian confidence intervals* based on this prior will possess the classical confidence property. It also yields best equivariant procedures in decision theory [36].

Such connections with other modes of inference induced Jeffreys and others to abandon the inner prior in favor of the outer prior in group models, as producing more "objective" results. Note, however, that this cannot be done consistently if the strong condition (CI) is assumed, since there exist distribution families equivariant under the action of two different groups, with different modular functions, for which no prior can be outer under both groups. As an example, let $\mathbf{X}$ have the zero-mean p-variate normal distribution, with dispersion matrix Σ. This distribution model is equivariant under the group G of all nonsingular $(p \times p)$ matrices, with $g \circ X = gX$, $g \circ \Sigma = g\Sigma g^T$; and also under its subgroups G_1 (respectively, G_2) of all lower (respectively, upper) triangular matrices with positive diagonal. These yield three different transitive group models, all inducing different outer priors for Σ [23].

SOME DIFFICULTIES

We now present some "paradoxes" relating to the specification of an ignorance prior within a specific model. (Stone and Springer [30] consider the consistency of joint specification of priors in models related to each other.)

Strong Inconsistency

Let $\mathbf{X}_i$ ($i = 1, 2, \ldots, N$) be a random sample from the general p-variate normal distribution $\mathcal{N}(\mathbf{M}, \Lambda)$. This model is equivariant under $\mathbf{X}_i \to \mathbf{AX}_i + \mathbf{b}$, $\mathbf{M} \to \mathbf{AM} + \mathbf{b}$, $\Lambda \to$

$\mathbf{A}\Lambda\mathbf{A}^T$. [$\mathbf{A}$ nonsingular $(p \times p)$, $\mathbf{b} \in \mathbb{R}^p$.] A sufficient statistic is $(\overline{\mathbf{X}}, \mathbf{S})$, where

$$N\overline{\mathbf{X}} = \sum_{i=1}^{N} \mathbf{X}_i,$$

$$(N-1)\mathbf{S} = \sum_{i=1}^{N} \mathbf{X}_i\mathbf{X}_i^T - N\overline{\mathbf{X}}\overline{\mathbf{X}}^T.$$

Relatively invariant priors for this group have density of the form $\pi(\mu, \lambda) \propto |\lambda|^{-(1/2)v}$, being inner for $v = p + 2$ and outer for $v = p + 1$.

Consider $Q_1 = N(\overline{\mathbf{X}} - \mathbf{M})^T\mathbf{S}^{-1}(\overline{\mathbf{X}} - \mathbf{M})$ (Hotelling's T^{2*}) and $Q_2 = N^{1/2}(\overline{X}_1 - M_1)/S_{11}^{1/2}$ (Student's t for a single data component). Then Q_1 is invariant, but Q_2 is not. The sampling distributions for both Q_1 and Q_2 are constant, being respectively $\{p(N-1)/(N-p)\} \cdot F_{p, N-p}$ and t_{N-1}.

It may be shown that the posterior distribution for Q_1 is

$$\{p(N-1)/(N-v+1)\}F_{p, N-v+1},$$

while that for Q_2 is $\{(N-1)/(N-v+1)\}^{1/2} \cdot t_{N-v+1}$ [11]. The choice $v = p + 1$ yields the "correct" distribution for Q_1, as follows from the general results of the section "Group Models"; however, it gives the "wrong" distribution $\{(N-1)/(N-p)\}^{1/2}t_{N-p}$ for Q_2. If we consider the (nonequivariant) interval estimator for μ_1, of the form $(\overline{X}_1 \pm k(S_{11}/N)^{1/2})$, this will have constant sampling coverage probability $\gamma_1 = \Pr(|t_{N-1}| < k)$, and different constant posterior probability

$$\gamma_2 = \Pr\left(|t_{N-p}| < k\{(N-p)/(N-1)\}^{1/2}\right).$$

This conflict between the "objective posterior" and "sampling" interpretations is called "strong inconsistency" by Stone [28].

The choice $v = 2$ eliminates the inconsistency for Q_2, but introduces it for Q_1. No prior can eliminate both inconsistencies simultaneously.

An example with similar behavior is presented by Dempster [7].

Marginalization Paradoxes [6]

Let X_{ij} ($i = 1, 2$; $j = 1, 2, \ldots, N$) be independent, X_{ij} having the normal distribution $\mathcal{N}(M_i, \Sigma^2)$. This model is equivariant under

$(X_{ij}) \rightarrow (a_i + bX_{ij})$, $(M_1, M_2, \Sigma) \rightarrow (a_1 + bM_1, a_2 + bM_2, b\Sigma)$ $(b > 0)$. Relatively invariant prior densities have the form $\pi(\mu_1, \mu_2, \sigma) \propto \sigma^\lambda$, with $\lambda = -3$ giving the inner, and $\lambda = -1$ the outer prior.

Define $\Xi_i = M_i / \Sigma$, $Z_i = \bar{X}_i / S$ $[N\bar{X}_i = \sum_j X_{ij}$, $S^2 = \sum_{i,j}(X_{ij} - \bar{X}_i)^2]$. The marginal posterior distribution for Ξ_1, under a relatively invariant prior, has density at ξ_1 proportional to

$$\int_0^\infty \omega^{2n-4-\lambda} \exp\left[-\tfrac{1}{2}\left\{ \omega^2 + n(z_1\omega - \xi_1)^2 \right\} \right] d\omega, \quad (1)$$

depending only on the value z_1 of Z_1. It seems that it ought, therefore, to be possible to reproduce this marginal posterior if only Z_1 is observed. Now the sampling density of Z_1, which depends only on the value ξ_1 of Ξ_1, is proportional to

$$\int_0^\infty \omega^{2n-2} \exp\left[-\tfrac{1}{2}\left\{ \omega^2 + n(z_1\omega - \xi_1)^2 \right\} \right] d\omega, \quad (2)$$

so that any posterior density for Ξ_1, using Bayes' theorem* with data Z_1 alone, would contain (2) as a factor. However, examining (1), we see that this will not hold, unless $\lambda = -2$—neither inner nor outer.

An almost identical argument applies when (Ξ_1, Ξ_2) are considered jointly, with posterior distribution governed by (Z_1, Z_2); only in this case the choice $\lambda = -3$ is needed to avoid the inconsistency. Thus there can be *no* relatively invariant prior which simultaneously avoids all such paradoxes.

FINITE ADDITIVITY

Many of the problems associated with ignorance priors stem from their impropriety. An alternative approach, still in its infancy, is to insist on propriety, but allow distributions which are only finitely additive. Some relevant theory is given by Heath and Sudderth [14].

We focus on assignments of inverse distributions $\{\Pi_x\}$ for parameter Θ given data $X = x$, which can be regarded as posteriors based on a finitely additive prior. This prior

need not be uniquely determined; moreover, the inverse distributions need not, with finite additivity, be constructed using Bayes' theorem. General coherence* properties for proper priors imply that it is now impossible to have an interval estimator $I(X)$ for Θ for which, simultaneously, $P_\theta(\theta \in I(X)) \leqslant \gamma_1$, all θ, and $\Pi_x(\Theta \in I(x)) \geqslant \gamma_2$, all x, where $\gamma_1 < \gamma_2$. That is, strong inconsistency cannot occur. Similarly, the marginalization paradox is avoided, since the "un-Bayesian" look of the marginal posteriors is deceptive: they *are* true posteriors, based on the appropriate, finitely additive marginal prior distribution [31].

If $(G, \mathscr{P})$ is a transitive group model, the appropriate extension of the theory of the section "Group Models" [based on Hartigan's versions of (PI), (DI), and (CI)] yields the *posterior equivariance* requirement:

$$\Pi_{\mathscr{P}}(\Theta \in \bar{g} \circ A \mid X = g \circ x)$$

$$\equiv \Pi_{\mathscr{P}}(\Theta \in A \mid X = x) \quad (g \in G).$$

This property does hold for formal posteriors based on improper, countably additive, relatively invariant priors. It is, therefore, of interest to enquire when such formal posteriors are true posteriors for some proper, finitely additive prior, since they cannot then be subject to the difficulties of the section "Some Difficulties." This will be so if and only if the formal prior used is outer, and the group $\bar{G}$ has the technical property known as *amenability* [4, 12], implying that it can support a proper right-invariant finitely additive distribution, which then induces the required proper prior. This approach, therefore, gives some justification to the use of the improper outer prior (although no justification for the inner prior), but only for "well-behaved" groups. In particular, the group of the section "Strong Inconsistency" is nonamenable, as must be the case for Q_2 to provide strong inconsistency for the outer prior having $v = p + 1$. For such groups, the posterior equivariance requirement is simply not satisfiable within the framework of finite additivity.

CONCLUSION

Logical Bayesianism is not currently popular. Certainly, the claim that a unique ignorance prior distribution exists for any problem must remain suspect as long as the relevant theory produces a whole range of choices in some problems, and no prior free from all objections in others. Nevertheless, much current Bayesian practice uses, overtly or covertly, priors supposed to represent "vague prior knowledge" with respect to a given model; often an outer prior is supposed. From a subjectivist viewpoint, such formal priors may be regarded as approximations to diffuse but proper real priors, although this argument requires much care and its general validity depends, at least, on the amenability of the underlying group [26, 27, 29]. Alternatively, the need is felt for a "zero" or "reference" prior to which an informative subjective distribution can be compared [2, 22, 35]. The formalization of ignorance thus remains the central object of a continuing quest by the knights of the Bayesian round table: inspiring them to imaginative feats of daring, while remaining, perhaps, forever unattainable.

References

[1] Barndorff–Nielsen, O., Blaesild, P., Jensen, J. L., and Jørgensen, B. (1982). Exponential transformation models. *Proc. R. Soc. Lond. A*, **379**, 41–65.

[2] Bernardo, J.-M. (1979). Reference posterior distributions for Bayesian inference (with Discussion). *J. R. Statist. Soc. B*, **41**, 113–147.

[3] Bondar, J. V. (1972). Structural distributions without exact transitivity. *Ann. Math. Statist.*, **43**, 326–339.

[4] Bondar, J. V. and Milnes, P. (1981). Amenability: a survey for statistical applications of Hunt–Stein and related conditions on groups. *Zeit. Wahrscheinlichkeitsth. verw. Geb.*, **57**, 103–128.

[5] Brillinger, D. R. (1963). Necessary and sufficient conditions for a statistical model to be invariant under a Lie group. *Ann. Math. Statist.*, **34**, 492–500.

[6] Dawid, A. P., Stone, M., and Zidek, J. V. (1973). Marginalization paradoxes in Bayesian and structural inference (with discussion). *J. R. Statist. Soc. B*, **35**, 189–233.

[7] Dempster, A. P. (1963). On a paradox concerning inference about a covariance matrix. *Ann. Math. Statist.*, **34**, 1414–1418.

[8] Dickey, J. M. (1973). Discussion of Dawid, Stone and Zidek (1973). *J. R. Statist. Soc. B*, **35**, 219–221.

[9] Fisher, R. A. (1956). *Statistical Methods and Scientific Inference*. Oliver & Boyd, London.

[10] Fraser, D. A. S. (1961). The fiducial method and invariance. *Biometrika*, **48**, 261–280.

[11] Geisser, S. and Cornfield, J. (1963). Posterior distributions for multivariate normal parameters. *J. R. Statist. Soc. B*, **25**, 368–376.

[12] Greenleaf, P. (1969). *Invariant Means on Topological Groups*. Van Nostrand Reinhold, New York.

[13] Hartigan, J. (1964). Invariant prior distributions. *Ann. Math. Statist.*, **35**, 836–845.

[14] Heath, D. and Sudderth, W. (1978). On finitely additive priors, coherence, and extended admissibility. *Ann. Statist.*, **6**, 333–345.

[15] Hora, R. B. and Buehler, R. J. (1966). Fiducial theory and invariant estimation. *Ann. Math. Statist.*, **37**, 643–656.

[16] Huzurbazar, V. S. (1976). *Sufficient Statistics*. Marcel Dekker, New York.

[17] Jaynes, E. T. (1968). Prior probabilities. *IEEE Trans. Syst. Sci. Cybern.*, **SSC-4**, 227–241.

[18] Jeffreys, H. (1961). *Theory of Probability*, 3rd ed. Clarendon Press, Oxford (1st ed., 1939).

[19] Laplace, P. S. de (1774). Mémoire sur la probabilité des causes par les évènements. *Mém. Acad. R. Sci. Paris (Savants Étrangers)*, **6**, 621–656.

[20] Laplace, P. S. de (1820). *Théorie Analytique des Probabilités*, 3rd ed. Courcier, Paris.

[21] Nachbin, L. (1965). *The Haar Integral*. Van Nostrand, New York.

[22] Novick, M. R. (1969). Multiparameter Bayesian indifference procedures (with Discussion). *J. R. Statist. Soc. B*, **31**, 29–64.

[23] Nussbaum, M. (1976). Structural distributions in the multivariate linear model. *Math. Operat. Statist.*, **7**, 679–683.

[24] Perks, W. (1947). Some observations on inverse probability including a new indifference rule. *J. Inst. Actuaries*, **73**, 285–334.

[25] Stein, C. (1965). Approximation of improper prior measures by prior probability measures. In *Bernoulli, 1713; Bayes, 1763; Laplace, 1813*, J. Neyman and L. M. LeCam, eds. Springer-Verlag, Berlin, pp. 217–240.

[26] Stone, M. (1965). Right Haar measure for convergence in probability to quasiposterior distributions. *Ann. Math. Statist.*, **36**, 440–453.

[27] Stone, M. (1970). Necessary and sufficient condition for convergence in probability to invariant posterior distributions. *Ann. Math. Statist.*, **41**, 1349–1353.

[28] Stone, M. (1976). Strong inconsistency from uniform priors (with Discussion). *J. Amer. Statist. Ass.*, **71**, 114–125.

[29] Stone, M. (1980). Review and analysis of some inconsistencies related to improper priors and finite additivity. In *Logic, Methodology and Philosophy of Science VI*, L. J. Cohen, J. Løs, H. Pfeiffer, and K.-P. Podewski, eds. North-Holland, Amsterdam.

[30] Stone, M. and Springer, B. G. F. (1965). A paradox involving quasi prior distributions. *Biometrika*, **52**, 623–627.

[31] Sudderth, W. D. (1980). Finitely additive priors, coherence, and the marginalization paradox. *J. R. Statist. Soc. B*, **42**, 339–341.

[32] Villegas, C. (1971). On Haar priors. In *Foundations of Statistical Inference*, V. P. Godambe and D. A. Sprott, eds. Holt, Rinehart and Winston, Toronto, pp. 409–414.

[33] Villegas, C. (1977). Inner statistical inference. *J. Amer. Statist. Ass.*, **72**, 453–458.

[34] Villegas, C. (1981). Inner statistical inference II. *Ann. Statist.*, **9**, 768–776.

[35] Zellner, A. (1977). Maximal data information prior distributions. In *New Developments in the Applications of Bayesian Methods*, A. Aykac and C. Brumat, eds. North-Holland, Amsterdam, pp. 211–232.

[36] Zidek, J. V. (1969). A representation of Bayes invariant procedures in terms of Haar measure. *Ann. Inst. Statist. Math.*, **21**, 291–308.

(ANCILLARITY
BAYESIAN INFERENCE
FIDUCIAL PROBABILITY
HAAR DISTRIBUTIONS
SUFFICIENCY)

A. P. Dawid

INVENTORY THEORY

Inventory theory involves a class of mathematical models devoted to the analysis of systems in which stock is maintained to meet an external or an internal demand. The economic motives for maintaining inventories are discussed below.

Economies of Scale

If the marginal cost of producing or ordering units from an outside supplier is a nonin-creasing function of the number of units produced, then it is economical to produce or order in lots and store units for future use. The conventional approach for modeling this phenomenon has been to assume that the cost of ordering y units, say $C(y)$, is of the form

$$C(y) = K\delta(y) + cy,$$

where

$$\delta(y) = \begin{cases} 1 & \text{if } y > 0 \\ 0 & \text{if } y = 0. \end{cases}$$

The fixed cost component, K (often called the setup cost), is incurred if a positive order is placed, independent of the magnitude of the order. The average cost per unit of ordering $y > 0$ units is $[C(y)/y] = (K/y) + c$, which is clearly decreasing in y.

Nonstationarities

If costs of production increase over time, it may be advantageous to retain stocks to avert the higher production costs. Similarly, if the value of the inventory is increasing, stocks are held in order to take advantage of higher future prices. Maintaining inventories may also be advantageous if demands are increasing as a function of time.

Uncertainties

A fundamental motive for carrying inventories is to provide a buffer against uncertainty. The most significant source of uncertainty is future demands. When the demand is random, a consequence of the analysis is that optimal policies retain safety stocks to provide a hedge against the uncertainty. In some systems, uncertainty in supply can be a dominant factor. Another common source of uncertainty is in the time required to replenish stocks.

HISTORICAL BACKGROUND

The simple economic lot size model, which forms the basis for much of the subsequent research in the field, appears to be due to Harris [12], although the term "Wilson lot size formula" is often used because of Wilson's [35] later and more comprehensive analysis. Interest in mathematical inventory models seems to have arisen during World War II, but papers did not start appearing in the open literature until the early 1950s. Dvoretzky et al. [8, 9] discussed the existence of optimal policies and the mathematical structure of inventory models, while Arrow et al. [1] applied renewal theory* to the problem of computing optimal policies. Whitin [34] discussed some of the economic and operational issues of inventory management. He also derived a model which has enjoyed considerable application in practice.

The collection of articles appearing in Arrow et al. [2] provided an important cornerstone to the development of modern inventory theory. The articles treat, in a rigorous and in-depth fashion, the techniques of dynamic programming* and stationary analysis which formed the basis for much of the later research work. The text by Hadley and Whitin [10] provided an excellent reference source for the significant work done up until that time.

Since approximately 1960, well over 1000 papers on inventory control models have appeared in the open literature. These have appeared primarily (although not exclusively) in the operations research journals. The most notable of these are *Management Science, Operations Research*, Naval Research Logistics Quarterly**, and *AIIE Transactions* in the United States, *INFOR* in Canada, *O.R. Quarterly* and *International Journal of Production Research* in Great Britain, *Opsearch* in India, *Cahiers de Recherche Operationelle* in Belgium, and the *Journal of the Japanese OR Society* in Japan.

MODELS WITH KNOWN DEMAND

EOQ Model and Extensions

The classical economic order quantity model may be derived under the following assumptions:

1. Demand is known with certainty and fixed at λ units per unit time.
2. Shortages are not permitted.
3. Costs are levied against:
   a. Ordering at $\$K$ per order
   b. Holding at $\$h$ per unit per unit time

The objective of the analysis is to minimize the average cost per unit time. If batches of size Q are ordered every Q/λ units of time, it follows that the cost rate, $C(Q)$, is given by

$$C(Q) = \frac{K\lambda}{Q} + \frac{hQ}{2}.$$

This follows since one setup is incurred each Q/λ units of time and the average number of units in stock at any point in time is $Q/2$. The minimizing value of Q is

$$Q^* = \sqrt{\frac{2K\lambda}{h}},$$

which is commonly known as the EOQ formula (due to Harris [12]).

There are a number of relatively straightforward extensions of this model which we will only mention. When items are produced internally rather than ordered from an outside supplier, assuming that the rate of production is finite (rather than infinite as we have done above), might be more appropriate. Another common extension is to allow shortages to accrue at a cost per unit backordered per unit time or, if appropriate, at a cost per lost sale. In many circumstances, suppliers might offer quantity discounts for larger orders. The reader interested in these extensions of the simple lot size model should refer to Chap. 2 of Hadley and Whitin [10].

Production Planning Models

When demand is nonstationary, that is, changing over time, the structure of the optimal policy is quite different from the static EOQ model discussed above.

Suppose that known requirements for a

single item in each of N distinct planning periods are $r_1, r_2, \ldots, r_N$. As above, assume that a positive setup cost K is incurred at each placement of an order and a holding cost h is levied against on hand inventory. The holding cost parameter, h, is assumed to be in units of dollars per unit per period, and the holding cost is assumed to be charged against the number of units remaining on hand at the end of each period.

An important result established by Wagner and Whitin [33] is that in optimal policy only orders in periods in which starting inventory is zero. In this way, an optimal policy is completely specified by the periods in which ordering takes place. The optimization can be structured as finding the shortest route through an acyclic network and can be solved by either forward or backward dynamic programming recursions. The extension of this model to allow for backlogging of demand is due to Zangwill [36].

When the order and/or holding costs are convex rather than concave, the property that an optimal policy only orders in periods in which starting stock is zero no longer holds. For this case, Veinott [30] developed a solution procedure that involves starting with a requirements schedule for which an optimal solution is obvious (such as $\mathbf{r} = \mathbf{0}$), and modifying this schedule sequentially by one unit at a time until the actual schedule is achieved. The procedure is based on the fact that the optimal production schedule also changes by one unit at each iteration.

MODELS WITH UNCERTAIN DEMAND

Most of the literature on inventory theory is concerned with systems in which the demand for one or more products is stochastic. Optimal operating policies are fundamentally different from the deterministic solutions discussed above.

An Approximate Continuous Review Model

If inventory levels are reviewed continuously and demands occur on a one-at-a-time basis,

the form of the optimal policy is what is commonly called a lot size-reorder point policy or (Q, r) policy. When the level of on-hand plus on-order stock hits r, an order for Q units is initiated.

There is an approximate (Q, r) model which continues to serve as a basis for many of the operational computer-based inventory systems in use today. We discuss it here more for its practical than theoretical importance.

The original formulation of this model is due to Whitin [34], who assumed that demands were generated by a Poisson process*. Hanssmann [11] discusses the case where demands are normally distributed, while Hadley and Whitin [10] provide the first general and rigorous treatment of this particular model.

We assume that there are costs of holding and setup as above and, in addition, a cost of $\$p$ per unit backordered each cycle. The uncertainty enters the analysis in the form of the demand during procurement lead time which is assumed to be a random variable with PDF $f(x)$.

An approximate expression for the expected cost per unit time as a function of the decision variables (Q, r) is

$$C(Q, r) = \frac{K\lambda}{Q} + h(Q/2 + r - \mu) + \frac{p\lambda}{Q} n(r),$$

where λ is the expected demand rate, μ is the expected demand during lead time, and

$$n(r) = \int_r^\infty (x - r)f(x)dx$$

is the expected number of stockouts per cycle. Since $C(Q, r)$ is convex in the pair of variables (Q, r), the cost minimizing policy satisfies $\partial C/\partial Q = \partial C/\partial r = 0$, which results in the pair of equations

$$Q^* = \sqrt{\frac{2\lambda}{h}\left[K + pn(r^*)\right]},$$

$$1 - F(r^*) = \frac{Q^* h}{p\lambda}.$$

These equations can be solved with reasonable efficiency by an iterative scheme. This model serves as the basis for many of the operational formulas appearing in the literature, including those involving a service constraint rather than a shortage cost (see Nahmias [19]).

Single Period Models with Random Demand

The common "newsboy" model has served as the basis for a large portion of the inventory theory literature on periodic review systems. It is only appropriate when items may be used for only a single period and so has limited applicability in practice.

Suppose that demand in a period for a single item, D, is a random variable with distribution function F and density f. Furthermore, assume that costs are charged against procurement at c per unit, holding at h per unit remaining at the end of the period, and shortage at p per unit of unsatisfied demand. In order to guarantee that it is profitable to hold positive inventory, we assume that $p > c$.

It is easy to show that the expected costs of ordering to y when starting inventory is $x \leqslant y$, say $G(y, x)$, are given by

$$G(y,x) = c(y-x) + h\int_0^y (y-t)f(t)dt$$

$$+ p\int_y^\infty (t-y)f(t)dt$$

$$= G(y) - cx.$$

Since $G(y)$ is convex in y, the minimizing value of y solves

$$G'(y^*) = 0,$$

which is equivalent to $F(y^*) = (p-c)/(p+h)$ when all costs are linear. The optimal policy is to order to y^* if $x < y^*$. This is known as a critical number or base stock policy.

When a positive set up cost for placing an order is included, the order to point is still y^* (which in this case is conventionally called S) but the condition for reordering is changed. Define s to satisfy $G(s) = G(S) + K$, $s < S$. If $x < s$, then $G(x) > G(s) = G(S) + K$, so that it is optimal to order up to S. If $s < x \leqslant S$, then the minimum expected cost that can be achieved is

$G(S) + K > G(x)$, so that it is optimal not to order. Hence the optimal policy is: If $x < s$, order to S. This is the familiar (s, S) policy which is quite similar to its continuous review counterpart, the (Q, r) policy, with $r = s$ and $Q = S - s$.

Dynamic Programming Analysis of Multiperiod Models with Random Demand

Most of the formal inventory theory literature has been concerned with multiperiod stochastic problems which are generally formulated and analyzed by dynamic programming methods. The first dynamic programming formulation of a multiperiod inventory problem is due to Bellman et al. [3]. Extensions and refinements of this model appear in a number of the papers in Arrow et al. [2].

Define $C_n(x)$ as the minimum expected discounted cost when starting stock is x and n periods remain in the planning horizon. Assume initially that there is no setup cost for ordering. Then, in the absence of an order lead time, the functional equations* defining an optimal policy are

$$C_n(x) = \min_{y \geqslant x} \left\{ G(y) - cx + \alpha \int_0^\infty C_{n-1}[s(y,t)]f(t)dt \right\}.$$

The function $s(y, t)$ is known as the transfer function and specifies the amount of inventory on hand one period hence when ordering to y and demand is t. The actual form of $s(y, t)$ depends on the backlogging assumptions. By mathematical induction, one proves that the bracketed term is a convex function of the decision variable y, from which it follows that the optimal policy is a critical number policy in every period. In addition, it can be shown that the critical numbers decrease monotonically in n (i.e., one orders less as the horizon end is approached).

As long as excess demand is backordered, the structure of the optimal policy remains essentially unchanged when a positive lead time for ordering is introduced. In this case one interprets the state variable, x, as the total amount of inventory on hand plus on order. However, when excess demand is lost rather than backordered, the optimal policy is a function of the vector of on-hand and on-order stocks, and it is no longer possible to collapse the state vector to a single scalar without compromising optimality. Nahmias [21] derives a variety of approximations which can be computed as functions of only a single variable.

When a positive setup cost is included in the dynamic problem, proving that an (s, S) policy is optimal in every period is difficult, owing to the fact that the appropriate function to be minimized is no longer convex in the decision variable. Scarf [25] showed, however, that these functions did possess sufficient structure to establish inductively that the (s, S) policy was indeed optimal in every period. The extension of Scarf's results to the infinite horizon problem is due to Iglehart [14].

An interesting point to note is that a positive setup cost for ordering can be considered to be a special case of a concave ordering cost function. Concave functions are of interest since they model economies of scale. Porteus [24] derives the structure of the optimal order policy for a finite horizon problem when the order function is an arbitrary nondecreasing concave function.

Stationary Analysis

Besides dynamic programming, another general technique for analyzing multiperiod inventory problems is via stationary analysis. This approach, which was pioneered by Arrow et al. [1], assumes that one follows a fixed order policy in every period. Utilizing results from renewal theory or Markov chain theory, one then derives an expression for the expected steady-state cost in terms of the policy variables, which can then be minimized with respect to these variables.

Utilizing the fact that when following a stationary (s, S) policy, starting stocks in successive periods forms a Markov process*, Karlin [17] shows that the stationary density

of starting inventory, say $g(x)$, satisfies the pair of equations

$$g(x) = f(S - x) \int_{-\infty}^{s} g(t)dt$$

$$+ \int_{x}^{S} f(t - x)g(t)dt \quad \text{for } s < x < S$$

$$g(x) = f(S - x) \int_{-\infty}^{s} g(t)dt$$

$$+ \int_{s}^{S} f(t - x)g(t)dt \quad \text{for } x < s,$$

where $f(x)$ is the density of one period's demand.

Solving this system of equations requires results from renewal theory. Simple formulas for optimal policies are obtained only when one knows the renewal function explicitly.

Given $g(x)$, one computes an expression for the stationary loss associated with a fixed (s, S) policy by the formula

$$H(s, S) = \int L(x)g(x)dx$$

and

$$L(x) = \begin{cases} K + G(S) - cx & \text{if } x < s \\ G(x) - cx & \text{if } s \leqslant x \leqslant S, \end{cases}$$

where $G(y)$ is defined in the section "Single Period Models with Random Demand." One then finds the pair (s, S) to minimize $H(s, S)$.

As an example of a case where explicit results can be obtained by this technique, suppose that $f(x) = \lambda^{-1}e^{-x/\lambda}$ for $x \geqslant 0$ is the density for one period's demand. Then one obtains the optimal values of (s, S) as

$$S - s = \sqrt{\frac{2K\lambda}{h}} \qquad \text{and}$$

$$e^{-s/\lambda} = \frac{\sqrt{2K\lambda h} + h\lambda}{\lambda(h + p)}.$$

The two striking features of this result—(1) that the order quantity, $S - s$, corresponds exactly to the EOQ formula discussed in the section "EOQ Model and Extensions," and (2) the optimal order quantity is independent of the shortage cost p—hold only when periodic demand is exponential, which rarely occurs in practice.

Multiproduct and Multiechelon Model

In inventory systems in which different products may interact, it is necessary to solve jointly for the order quantities in order not to compromise optimality. A typical way in which different products interact is that they may be either economic substitutes (margarine or butter) or complements (nuts and bolts). Another is that their demands may be correlated.

Although dynamic programming methods can theoretically be utilized for solving multiproduct inventory problems, practically speaking, these methods are not very useful. The problem is that multiple-state dynamic programs are often difficult if not impossible to solve due to the computational effort required.

Veinott [31] developed a method for solving certain kinds of multiproduct inventory problems that does not involve the usual dynamic programming recursion. He has discovered conditions under which an N-period dynamic problem can be decomposed in N one-period problems. When demand is stationary, the solution of an m-product problem may involve only a single search over Euclidean m-space. However, his approach cannot be used when there are setup costs present or certain types of interactions among the products.

Johnson [16] was able to derive the form of the optimal policy for multiproduct systems in which setup costs are present, but the computational scheme suggested appears to be too complex to implement in an actual system.

In many real systems, units are produced at a single plant, then shipped to regional warehouses, which in turn ship items to retail outlets before final demand is met. Clark and Scarf [7] were the first to formulate and solve a mathematical model which specifically considers this phenomenon. In their model, N facilities are arranged in series,

each facility supplies only the next in the series, and external demand occurs at the final installation only. Bessler and Veinott [4] consider a multiechelon system with a fairly general arborescence structure. Recently Schmidt and Nahmias [28] derived the optimal policy for a two-stage assembly system which gives the only optimal policy analysis for a materials requirements planning system under random demand.

A special type of two-echelon model arises in the base depot repair systems observed in managing spare parts, due to the fact that repaired units are recycled back into stock. The prototype analysis of this type of system is due to Sherbrooke [29]. Muckstadt [18] has extended Sherbrooke's METRIC model in order to treat explicitly the relationship between units and their modules and the fact that the failure of a unit may be caused by the failure of a module.

Statistical Issues

The problem of estimating the parameters of the demand distribution is closely connected to the inventory management problem. The usual method of dealing with this issue is to simply decouple the estimation and the inventory control problems and use values of the estimators in formulas for operating policies as if they were known parameters. The mean and MAD (mean absolute deviation) are usually estimated by exponential smoothing* to account for possible changes in the underlying distribution over time (see, e.g., Brown [5]).

Most of the analysis of the joint estimation and inventory management problem has been based on Bayesian analysis. This approach was pioneered by Scarf [25], who assumed that the demand density belongs to the exponential family with a single unknown parameter ω. A prior density on ω is assumed known. Using the fact that if $D_1, D_2, \ldots, D_N$ are N independent observations of demand, then $S = \sum_{i=1}^{N} D_i$ is a sufficient statistic for ω, Scarf proves that the optimal policy is asymptotically "close" to the policy one would obtain by using the maximum likelihood estimator for the unknown parameter in the standard inventory formula.

Iglehart [15] extends Scarf's results to somewhat more general cost functions and allows the demand density to be either the exponential or the range family of densities.

Hayes [13] also discusses the estimation problem but from a slightly different point of view. Let $l(y, \xi)$ be the holding or shortage cost of ordering y when the demand is observed to be ξ. The expected loss is

$$\bar{l}(y) = \int_{\xi} l(y, \xi) \varphi(\xi) d\xi,$$

where $\varphi(\xi)$ is the one period demand density. Since φ is assumed not to be known, there is a policy estimate, say $\hat{y}_d = d(\xi)$ that maps the observed past demands $\xi = (\xi_1, \ldots, \xi_n)$ into an order policy $\hat{y}_d$. The expected total operating cost (ETOC) associated with the policy mapping d may be written

$$\text{ETQC}(d, n) = \int_{\xi} \bar{l}\big[d(\xi) \big] \rho(\xi) d\xi,$$

where $\rho(\xi)$ is the sampling (or posterior) distribution of ξ. Hayes' approach is to find d that minimizes ETOC. Examples assuming ξ is exponential or normal with unknown mean are given.

CONCLUSION

We have attempted in this brief review to outline some of the major developments in inventory theory. In addition to the references given here, the interested reader should be aware of the significant review articles in the field. General reviews on inventory models have been developed by Hanssmann [11], Scarf [27], Veinott [32], and Nahmias [20]. Clark [6] reviews multiechelon inventory models, Nahmias [22] considers reparable item inventory systems, and Nahmias [23] reviews the literature on perishable inventory systems.

References

[1] Arrow, K. J., Harris, T. E., and Marschak, J. (1951). *Econometrica*, **19**, 250–272.

[2] Arrow, K. J., Karlin, S., and Scarf, H., eds. (1958). *Studies in the Mathematical Theory of Inventory and Production*. Stanford University Press, Stanford, Calif.

[3] Bellman, R. E., Glicksberg, I., and Gross, O. (1955). *Manag. Sci.*, **2**, 83–104.

[4] Bessler, S. A. and Veinott, A. F., Jr. (1966). *Naval Res. Logist. Quart.*, **13**, 355–389.

[5] Brown, R. G. (1967). *Decision Rules for Inventory Management*. Dryden Press, Hinsdale, Ill.

[6] Clark, A. (1972). *Naval Res. Logist. Quart.*, **19**, 621–650.

[7] Clark, A. and Scarf, H. (1960). *Manag. Sci.*, **6**, 475–490.

[8] Dvoretzky, A., Kiefer, J., and Wolfowitz, J. (1952). *Econometrica*, **20**, 187–222.

[9] Dvoretzky, A., Kiefer, J., and Wolfowitz, J. (1952). *Econometrica*, **20**, 450–466.

[10] Hadley, G. J. and Whitin, T. M. (1963). *Analysis of Inventory Systems*. Prentice-Hall, Englewood Cliffs, N.J.

[11] Hanssmann, F. (1961). In *Progress in Operations Research*, Vol. 1, R. Sckoff, ed. Wiley, New York, pp. 65–104.

[12] Harris, F. W. (1915). *Operations and Cost.*, Factory Management Series. Shaw, Chicago, Chap. 2.

[13] Hayes, R. H. (1969). *Manag. Sci.*, **15**, 686–701.

[14] Iglehart, D. L. (1963). *Manag. Sci.*, **9**, 259–267.

[15] Iglehart, D. L. (1964). *Manag. Sci.*, **10**, 429–440.

[16] Johnson, E. (1967). *Manag. Sci.*, **13**, 475–491.

[17] Karlin, S. (1958). In *Studies in the Mathematical Theory of Inventory and Production*, K. J. Arrow, S. Karlin, and H. Scarf, eds. Stanford University Press, Stanford, Calif., Chap. 14.

[18] Muckstadt, J. A. (1973). *Manag. Sci.*, **20**, 472–481.

[19] Nahmias, S. (1976). *Naval Res. Logist. Quart.*, **23**, 31–36.

[20] Nahmias, S. (1978). Inventory models. In *The Encyclopedia of Computer Sciences and Technology*, Vol. 9, J. Belzer, A. Holzman, and A. Kent, eds. Marcel Dekker, New York, pp. 447–483.

[21] Nahmias, S. (1979). *Operat. Res.*, **27**, 904–924.

[22] Nahmias, S. (1981). In *Multilevel Production/Inventory Systems: Theory and Practice*, L. Schwarz, ed. TIMS North-Holland, Amsterdam, pp. 253–277.

[23] Nahmias, S. (1982). *Operat. Res.*, **30**, 680–708.

[24] Porteus, E. L. (1971). *Manag. Sci.*, **17**, 411–427.

[25] Scarf, H. E. (1959). *Ann. Math. Statist.*, **30**, 490–508.

[26] Scarf, H. E. (1960). In *Mathematical Methods in the Social Sciences*, K. J. Arrow, S. Karlin, and P. Suppes, eds. Stanford University Press, Stanford, Calif., pp. 196–202.

[27] Scarf, H. E. (1963). In *Multistage Inventory Models and Techniques*, Scarf, Gilford, and Shelly, eds. Stanford University Press, Stanford, Calif., Chap. 7.

[28] Schmidt, C. P. and Nahmias, S. (1981). Optimal Policy for a Two Stage Assembly System under Random Demand. *Working Paper*, University of Santa Clara, Santa Clara, Calif. (to appear in *Operat. Res.*).

[29] Sherbrooke, C. C. (1968). *Operat. Res.*, **16**, 122–141.

[30] Veinott, A. F., Jr. (1964). *Manag. Sci.*, **10**, 441–460.

[31] Veinott, A. F., Jr. (1965). *Manag. Sci.*, **12**, 206–222.

[32] Veinott, A. F., Jr. (1966). *Manag. Sci.*, **12**, 745–777.

[33] Wagner, H. M. and Whitin, T. M. (1958). *Manag. Sci.*, **5**, 89–96.

[34] Whitin, T. M. (1953). *Theory of Inventory Management*. Princeton University Press, Princeton, N.J.

[35] Wilson, R. H. (1934). *Harvard Bus. Rev.*, **13**, 116–128.

[36] Zangwill, W. I. (1966). *Manag. Sci.*, **13**, 105–119.

Bibliography

In this section we provide a brief description of a number of the major texts in the inventory control area. (A number of the references also appear in the preceding section.)

Arrow, K. J., Karlin, S., and Scarf, H., eds. (1958). *Studies in the Mathematical Theory of Inventory and Production*. Stanford University Press, Stanford, Calif. (This book is a collection of original articles primarily by the editors which served as the foundation and stimulus for much of the later work on mathematical inventory theory. There are rigorous proofs and derivations of the structure of optimal policies under a variety of different assumptions. The mathematical techniques of dynamic programming, stationary analysis of Markov processes, renewal theory, control theory, and other methods are exploited in these fundamental papers.)

Arrow, K. J., Karlin, S., and Scarf, H., eds. (1962). *Studies in Applied Probability and Management Science*. Stanford University Press, Stanford, Calif. (This book is similar to the preceding work by Arrow et al. It considers extensions of many of the problems treated in the earlier work.)

Brown, R. G. (1959). *Statistical Forecasting for Inventory Control*. McGraw-Hill, New York. (R. G. Brown has gained recognition as an influential consultant and has been quite successful in discovering those aspects of inventory models which can be applied to real problems. This early text, which was his first, is a brief monograph which focuses primarily on the forecasting problem and discusses the exponential smoothing technique which is generally credited to Brown and which has gained wide acceptance by practitioners.)

Brown, R. G. (1967). *Decision Rules for Inventory Management*. Dryden Press, Hinsdale, Ill. (This text focuses primarily on the inventory rather than the forecasting problem. It is written almost like a novel as we follow the career path of a fictitious MBA student and his introduction of inventory modeling techniques to a large firm. Although the theoretical contribution of Brown's work is minimal, this text should help to provide a better understanding of a variety of important practical issues, such as the interpretation and measurement of alternative service criteria.)

Buchan, J. and Koenigsberg, E. (1963). *Scientific Inventory Management*. Prentice-Hall, Englewood Cliffs, N.J. [This text was written fairly early on by two (then) consultants. It is basically a collection of case studies prepared by the authors and (presumably) by a number of their colleagues.]

Fetter, R. B. and Dalleck, W. C. (1961). *Decision Models for Inventory Management*. Richard D. Irwin, Homewood, Ill. (This is a short and simple text that deals with basic concepts of service level and economic order quantities. There is a good expository discussion of costs and variability and helpful examples.)

Hadley, G. J. and Whitin, T. M. (1963). *Analysis of Inventory Systems*. Prentice-Hall, Englewood Cliffs, N.J. (Hadley and Whitin's book is, in the opinion of this writer, the outstanding overall textbook in inventory theory. All the models treated are rigorously derived and discussed in detail. There is a comprehensive treatment of deterministic EOQ type models, a discussion of the type of continuous review models which form the basis for many actual operating systems, and a brief treatment of dynamic models.)

Hanssmann, F. (1962). *Operations Research in Production and Inventory Control*. Wiley, New York. (Although Hanssmann's book is far less comprehensive than Hadley and Whitin's, he does discuss some useful techniques for implementation.)

Hillier, F. S. and Lieberman, G. J. (1980). *Operations Research*, 3rd ed. Holden-Day, San Francisco. (This general survey book on operations research is one of the outstanding books in the field and contains a concise informative chapter on inventory theory that can serve as a good starting point for the novice.)

Holt, C. C., Modigliani, F., Muth, J. F., and Simon, H. A. (1960). *Planning, Production, Inventories and Work Force*. Prentice-Hall, Englewood Cliffs, N.J. (A good portion of this important early text is based on an actual case study whose analysis forms the basis for the so-called aggregate planning approach to these problems. In addition, the concept of the linear decision rule is introduced.)

Johnson, L. A. and Montgomery, D. C. (1974). *Operations Research, Production Planning, Scheduling and Inventory Control*. Wiley, New York. (The style of presentation of this text is perhaps more akin to a review article with examples rather than a textbook in which models are derived from first principles. However, the book should be useful for a practitioner interested in surveying the area and understanding what would be involved in possibly using any of the algorithms discussed.)

Naddor, E. (1966). *Inventory Systems*. Wiley, New York. [*Inventory Systems* is based on the author's research in the inventory area and deals with a variety of interesting issues (such as the effect of different patterns of demand other than linear during a reorder cycle).]

Peterson, R. and Silver, E. A. (1978). *Decision Systems for Inventory Management and Production Planning*. Wiley/Hamilton, Santa Barbara, Calif. (Peterson and Silver focus on a variety of heuristic models which can presumably be implemented in a practical setting. Techniques for multiproduct systems such as exchange curves are also presented.)

Plossl, G. W. and Wight, O. W. (1967). *Production and Inventory Control*. Prentice-Hall, Englewood Cliffs, N.J. (This text, by two well-known consultants, is geared primarily for the practitioner.)

Scarf, H. E., Gilford, D. M., and Shelly, M. W. (1963). *Multistage Inventory Models and Techniques*. Stanford University Press, Stanford, Calif. (This book is a collection of original research papers on inventory problems which include both fundamental contributions to the theory and an excellent survey article by Scarf.)

Starr, M. K. and Miller, D. W. (1962). *Inventory Control: Theory and Practice*. Prentice-Hall, Englewood Cliffs, N.J. (Although the style of this early text is primarily expository and the results not up to date, the authors do discuss a number of interesting issues not appearing in most other inventory texts such as pricing policies and implementation.)

Wagner, H. M. (1962). *Statistical Management of Inventory Systems*. Wiley, New York. [This early text by Wagner discusses the stationary analysis approach of computing (s, S) policies using Markov chains.]

Wagner, H. M. (1969). *Principles of Operations Research*. Prentice-Hall, Englewood Cliffs, N.J. (Along with Hillier and Lieberman, this represents another widely used operations research survey text. Substantial coverage of inventory models is included with a considerably different emphasis.)

Whitin, T. M. (1953). *Theory of Inventory Management*. Princeton University Press, Princeton, N.J. (Whitin's monograph is of interest primarily for historical reasons and represents a pioneering work predating the later

interest from a variety of different quarters. Of most interest is Whitin's discussion of the inventory management problem in the light of classical economic theory.)

Acknowledgment

The author would like to thank Professor Charles Schmidt for helpful comments.

(DECISION THEORY
DISCOUNTING
DYNAMIC PROGRAMMING
ECONOMETRICS
OPERATIONS RESEARCH
QUALITY CONTROL, STATISTICAL
RENEWAL THEORY
STOCHASTIC PROCESSES)

STEVEN NAHMIAS

INVERSE DISTRIBUTIONS

The expression "inverse distributions" is used in the current literature in three major ways: (1) to denote the distribution of X and Y where they are related through a one-to-one transformation, (2) to describe the distribution of X and Y, where X arises from a simple random sample of size n and Y arises from sampling until some specified phenomenon (such as r successes) occurs, and (3) to describe distributions of X and Y where there is an inverse relationship between certain associated quantities (e.g., the cumulant generating functions).

Aitchison [1] gives several examples under certain one-to-one transformations and advocates the name "inverse distributions." If X and Y are regular n-dimensional random variables where Y is given by the one-to-one transformation $y = g(x)$, then

$$f_Y(y) = f_X\left[g^{-1}(y) \right]|J|$$

and

$$f_X(x) = f_Y\left[g(x) \right]/|J|.$$

Specification of the distribution of either X or Y induces the distribution of the other.

Perhaps the most frequent occurrence of the first type of inverse distribution is when both variables are one-dimensional and $y = g(x) = 1/x$. Usage of the word "inverse" in this sense is not universally accepted; it is more usual, particularly in the literature of Bayesian statistics, to use the expression "inverted." For example, the distribution of $1/X$ where X is gamma distributed is called the inverted gamma* (see Gleser and Healy [3]). The term "inverted beta*" is used in a similar way (see Tiao and Guttman [5]).

One of the best known examples of a pair of inverse distributions is provided by the beta distribution of the first kind and a beta distribution of the second kind. Either distribution can be obtained from the other by a simple one-to-one transformation. It follows, as a special case, that the F distribution* and the beta distribution of the first kind are inverses. If

$$X \sim F(\nu_1, \nu_2)$$

and

$$Y = \frac{\nu_1 X}{\nu_2 + \nu_1 X},$$

then

$$Y \sim B\left(\frac{\nu_1}{2}, \frac{\nu_2}{2}\right).$$

The more interesting examples of inverse distributions in the present sense occur in the multivariate case. If, for example, X_1, $X_2, \ldots, X_n$ are independent random variables with

$$X_i \sim B(p_i, p_{i+1} + p_{i+2} + \cdots + p_{n+1})$$

and we let

$$Y_i = (1 - X_1)(1 - X_2) \cdots (1 - X_{i-1})X_i,$$

then the joint distribution of the Y's is the Dirichlet distribution* $D(p_1, p_2, \ldots, p_n; p_{n+1})$. Since the transformation is one-to-one, the joint distribution of independent beta variables with parameters as given and the Dirichlet transformation may be regarded as inverse distributions (see Tiao and Guttman [5]).

On the other hand, the multivariate inverted Student variable given by Raiffa and Schlaiffer [4] is not a one-to-one function of the multivariate Student variable.

A second major use of the name "inverse distributions" occurs with what is called inverse sampling* (see Tweedie [6]). To understand what is meant by inverse sampling, consider several examples. The binomial distribution* is appropriate when we perform n independent Bernoulli trials* and then count the number of successes. Instead, let us perform independent Bernoulli trials until we observe r successes and then count the number of trials required. This type of sampling is often called inverse binomial sampling. Presumably, this is because the sample size and the random variable have exchanged places; the description has been inverted, so to speak. With inverse binomial sampling, the probability distribution is sometimes called the inverse binomial; more generally, the negative binomial*.

Another major example of inverse distributions from inverse sampling is with the Poisson process*. If we observe the process for a fixed time, T, the number of occurrences is a Poisson random variable with parameter T. If, on the other hand, we observe the process until the first occurrence, the time required has a negative exponential* distribution with parameter λ. Thus the Poisson distribution and the negative exponential distribution are inverses of each other because of the inverse sampling.

Both the inverse binomial and the negative exponential distribution are waiting-time distributions; the first discrete, the second continuous. Another waiting-time distribution that must be discussed under inverse distributions is the inverse Gaussian distribution*. Suppose that $X(t)$ is a Wiener process with positive drift ν, i.e., a process with independent increments with $X(0) = 0$ (see BROWNIAN MOTION) and

$$X(t) \sim N(\nu t, \sigma^2 t).$$

Then the time for $X(t)$ to reach the value a (> 0) for the first time has the density function

$$f_T(t) = \begin{cases} \left(\dfrac{\lambda}{2\pi t^3}\right)^{1/2} \exp\left\{-\dfrac{\lambda(t-\mu)^2}{2\mu^2 t}\right\}, & t > 0 \\ 0, & \text{otherwise,} \end{cases}$$

where

$$\mu = a/\nu \quad \text{and} \quad \lambda = a^2/\sigma^2.$$

For a review article and bibliography on the inverse Gaussian distribution, see Folks and Chhikara [2]. Then with the Wiener process, the normal and inverse Gaussian distributions are inverses because of the inverse sampling. If one samples X at any point in time, a normal distribution is being sampled; on the other hand, if an X value is specified and the time required to attain that value is recorded, the inverse Gaussian distribution is obtained.

The third major use of the name inverse distributions is to describe inverse relationships among moments, cumulants, etc. One of the most interesting usages involves the cumulant generating function*. Let $\phi(t) = \log E[\exp(-tX)]$. Tweedie [6] noted relationships between the pairs of cumulant generating functions for the distributions just mentioned under inverse sampling. The cumulant generating functions for the binomial and the negative binomial are given by

$$\phi_B(t) = n\log(q + pe^{-t}), \quad -\infty < t < \infty,$$

$$\phi_{NB}(t) = -r\log\left(1 - \frac{1 - e^t}{p}\right), \quad t > \log q.$$

Although ϕ_B and ϕ_{NB}^{-1} are not identical and the ranges and domains are not the same, they do have the same functional forms. With $n = r = 1$, however, $\phi_B = \phi_{NB}^{-1}$ and $\phi_B^{-1} = \phi_{NB}$. Actually, the Bernoulli and geometric distributions* are inverses but the practice persists of calling the binomial and negative binomial* inverse distributions.

With the Poisson process, the situation is more striking. The cumulant generating functions for the Poisson and negative exponential are

$$\phi_p(t) = -\lambda(1 - e^{-t}), \quad -\infty < t < \infty$$

$$\phi_{NE}(t) = \log\lambda - \log(\lambda + t), \quad t > -\lambda.$$

It is easily verified that the functional forms of ϕ_p and ϕ_{NE}^{-1} (also ϕ_p^{-1} and ϕ_{NE}) are the same and that they have the same domains and ranges. Thus the Poisson and negative exponential are inverses.

For the normal and inverse Gaussian distributions

$$\phi_N(t) = -\nu t + \sigma^2 t^2/2,$$

$$-\infty < t < \infty$$

$$\phi_{IG}(t) = \frac{\lambda}{\mu}\left[1 - \left(1 + \frac{2\mu^2 t}{\lambda}\right)^{1/2}\right],$$

$$t > -\frac{\lambda}{2\mu^2}.$$

The functional forms of ϕ_N and ϕ_{IG}^{-1} are similar but ϕ_N^{-1} does not exist. If we let $\mu = 1/\nu$ and $\lambda = 1/\sigma^2$ and restrict ϕ_N to $t < \mu/\sigma^2$, then ϕ_N^{-1} exists and $\phi_N = \phi_N^{-1} = \phi_{IG}$. Thus we can say that the normal and inverse Gaussian distributions are inverses.

References

[1] Aitchison, J. (1963). *Biometrika*, **50**, 505–508.

[2] Folks, J. L. and Chhikara, R. S. (1978). *J. R. Statist. Soc. B*, **40**, 263–289.

[3] Gleser, L. J. and Healy, J. D. (1976). *J. Amer. Statist. Ass.*, **71**, 977–981.

[4] Raiffa, H. and Schlaiffer, R. (1961). *Applied Statistical Decision Theory*, Harvard University Press, Boston, Mass.

[5] Tiao, G. C. and Guttman, I. (1965). *J. Amer. Statist. Ass.*, **60**, 793–805.

[6] Tweedie, M. C. K. (1945). *Nature (Lond.)*, **155**, 453.

(INVERSE GAUSSIAN DISTRIBUTION TRANSFORMATIONS)

J. LEROY FOLKS

INVERSE GAUSSIAN DISTRIBUTION

The random variable X has the inverse Gaussian distribution with parameters μ and λ, denoted by $X \sim IG(\mu, \lambda)$, if and only if its density function is given by

$$f_X(x) = \begin{cases} \left(\frac{\lambda}{2\pi x^3}\right)^{1/2} \exp - \left\{\frac{\lambda(x - \mu)^2}{2\mu^2 x}\right\}, & x > 0 \\ 0, & \text{otherwise,} \end{cases}$$

where μ and λ are positive. The unimodal

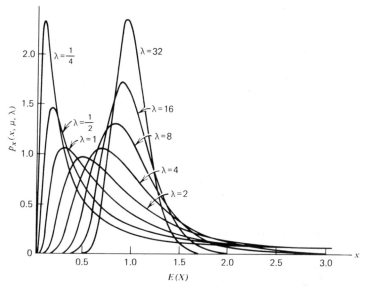

Figure 1 Inverse Gaussian Density Functions $[E(x) = \mu = 1]$. (From ref. 1.)

density function is a member of the exponential family and is skewed to the right. Its shape resembles that of other skewed density functions such as the lognormal*, Weibull*, and gamma*. Johnson and Kotz [4] give the family of curves depicted in Fig. 1 for $\mu = 1$.

All positive and negative moments exist and the mean and variance are given by $E[X] = \mu$ and $\mathrm{var}[X] = \mu^3/\lambda$. The moment generating function is given by

$$m_X(t) = \exp\left\{\frac{\lambda}{\mu}\left[1 - \left(1 - \frac{2\mu^2 t}{\lambda}\right)^{1/2}\right]\right\}.$$

The distribution function can be expressed in terms of the standard normal distribution function, Φ, by

$$F_X(x) = \Phi\left(\sqrt{\frac{\lambda}{x}}\left(-1 + \frac{x}{\mu}\right)\right)$$

$$+ e^{2\lambda/\mu}\Phi\left(\sqrt{\frac{\lambda}{x}}\left(1 + \frac{x}{\mu}\right)\right).$$

This result, first obtained by Schrödinger in 1915, was later obtained independently by Zigangirov in 1962 and Shuster in 1968.

Schrödinger obtained the distribution to the barrier in a Wiener process with positive drift. Let $X(t)$ be a Wiener process with

positive drift. That is, let $X(0) = 0$ and $X(t)$ have independent increments with $X(t) \sim N(\nu t, \sigma^2 t)$, where $\nu > 0$. The time required to reach $x = a(a > 0)$ for the first time is a random variable with the inverse Gaussian distribution where $\mu = a/\nu$ and $\lambda = a^2/\sigma^2$. Schrödinger also obtained maximum likelihood estimates for the parameters.

Wald [9] obtained the distribution as the approximate distribution of the sample size in sequential sampling. For this reason, it is often known as *Wald's distribution*.

Tweedie is responsible for the name inverse Gaussian. He noted the inverse relationship between the cumulant generating function* of the normal and the distribution in question and proposed the name inverse Gaussian. If we let

$$\phi(t) = \log E\left[\exp(-tX)\right],$$

then we find for the normal, $N(\nu, \sigma^2)$, and inverse Gaussian, $IG(\mu, \lambda)$:

$$\phi_N(t) = -\nu t + \sigma^2 t^2/2, \qquad -\infty < t < \infty,$$

$$\phi_{IG}(t) = \frac{\lambda}{\mu}\left[1 - \left(1 + \frac{2\mu^2 t}{\lambda}\right)^{1/2}\right], \qquad t > -\frac{\lambda}{2\mu^2},$$

$$\phi_{IG}^{-1}(t) = -\frac{t}{\mu} + \frac{t^2}{2\lambda}, \qquad t < \frac{\lambda}{\mu}.$$

It is apparent that ϕ_N and ϕ_{IG}^{-1} have the

same functional form, and therefore the name inverse Gaussian was suggested.

There are many striking analogies between the sampling distributions for this distribution and those for the normal; e.g., (1) $\bar{X} \sim \text{IG}(\mu, n\lambda)$, (2) $S = \lambda \sum (1/X - 1/\bar{X}) \sim \chi^2_{n-1}$ independently of $\bar{X}$, (3) minus twice the term in the exponent is distributed as chi-square with n degrees of freedom, and (4) the family of inverse Gaussian density functions is complete. Unlike the normal, standardization is only partially achieved. If we let $Z = \lambda X / \mu^2$, then Z has the inverse Gaussian distribution with parameters ϕ and ϕ^2, where $\phi = \lambda / \mu$. Thus it is possible to use a one-parameter family as the "standardized" family. Wasan and Roy [11] have tabulated the distribution for some values of ϕ. Another property, weakly analogous to those of the normal, is that certain linear combinations of independent IG variables are also inverse Gaussian. If X_i, $i = 1, 2, \ldots, n$, are independent $\text{IG}(\mu_i, \lambda_i)$ then $\sum c_i X_i$ is inverse Gaussian if and only if $\lambda_i / (c_i \mu_i^2)$ is positive and constant, $i = 1, 2, \ldots, n$.

The pair $(\bar{X}, S)$ is complete sufficient for (μ, λ) so that estimators, tests, and decision rules should be functions of $(\bar{X}, S)$. Maximum likelihood* estimators of μ and λ are given by $\hat{\mu} = \bar{X}$ and $\hat{\lambda} = n / S$. Since $\hat{\mu}$ is unbiased, it is minimum variance unbiased. The minimum variance unbiased estimator* of λ is easily found to be

$$\hat{\lambda} = (n - 3) / S.$$

Estimation of other parameters is feasible although the formulas may not be simple. For example, the minimum variance unbiased estimator of σ^2 is rather complex [6].

Chhikara and Folks [1] give a number of useful sampling distributions which are analogous to those from the normal. For example, if we let $V = S / (n - 1)$ and let

$$W = \frac{n^{1/2}(\bar{X} - \mu)}{\mu(\bar{X}V)^{1/2}},$$

then W has the density function

$$h_W(w) = [1 - r(w)] f_{T_{n-1}}(t),$$

where f_T is a Student's t density function and r is an odd function in w. Consequently, the distribution of $|W|$ is the truncated (or folded) Student's t distribution* with $n - 1$ degrees of freedom. Two-sided t tests and two-sided confidence intervals follow immediately for the one-sample and two-sample problems from the results of Chhikara and Folks.

Tweedie [8] gave the analysis of residuals (comparable to an analysis of variance*) for a one way classification*. Shuster and Miura [7] subsequently extended the analysis of residuals to a two-way classification* with interaction*.

The inverse Gaussian distribution has been examined as a model for lifetimes in reliability theory*. Its hazard function has the general nature of increasing to a maximum, then decreasing and approaching a positive, horizontal asymptote. Minimum variance unbiased estimators of the reliability are given by Chhikara and Folks [2].

No satisfactory bivariate inverse Gaussian distribution has as yet been obtained, although Wasan [10] describes a bivariate distribution over the domain $0 < x < y < \infty$.

A generalized inverse Gaussian distribution has been studied intensively by Jørgensen [5]. The three-parameter density function is given by

$$\frac{e^{\lambda/\mu} \mu^{-a} x^{a-1}}{2 K_a(\lambda/\mu)} f_{\text{IG}}(x; \mu, \lambda)$$

where K_a is the modified Bessel function of the third kind with index λ. Maximum likelihood estimates*, sampling distributions, tests, and a study of the hazard function are given.

For additional information, see Folks [3].

References

[1] Chhikara, R. S. and Folks, J. L. (1975). *Commun. Statist.*, **4**, 1081–1091.

[2] Chhikara, R. S. and Folks, J. L. (1977). *Technometrics*, **19**, 461–468.

[3] Folks, J. L. (1978). *J. R. Statist. Soc. B*, **40**, 263–289. (This paper with discussions is the most comprehensive and up-to-date survey of IG distributions to date.)

[4] Johnson, N. L. and Kotz, S. (1970). *Distributions in Statistics: Continuous Univariate Distributions*, Vol. 1. Wiley, New York. (Chapter 15 gives a good survey of results.)

[5] Jørgensen, B. (1980). Statistical Properties of the Generalized Inverse Gaussian Distribution. *Mem. No. 4*, Dept. of Theoretical Statistics, University of Aarhus, Aarhus, Denmark.

[6] Korwar, R. M. (1980). *J. Amer. Statist. Ass.*, **75**, 734–735.

[7] Shuster, J. J. and Miura, C. (1972). *Biometrika*, **59**, 478–481.

[8] Tweedie, M. C. K. (1957). *Ann. Math. Statist.*, **28**, 362–377.

[9] Wald, A. (1944). *Ann. Math. Statist.*, **15**, 283–296.

[10] Wasan, M. T. (1969). First Passage Time Distribution of Brownian Motion with Positive Drift. *Queen's Paper 19*, Queen's University, Canada.

[11] Wasan, M. T. and Roy, L. K. (1969). *Technometrics*, **11**, 591–604.

(BROWNIAN MOTION
INVERSE DISTRIBUTIONS
SEQUENTIAL ANALYSIS
STOCHASTIC PROCESSES
WAITING-TIME DISTRIBUTIONS)

J. Leroy Folks

INVERSE HYPERGEOMETRIC DISTRIBUTION   *See* HYPERGEOMETRIC DISTRIBUTIONS

INVERSE RANKS

This term is used in the sense of ranking the observations from largest to smallest rather than from smallest to largest. This ranking is used, for example, in the case of the Wilcoxon rank-sum test* so that the tabulated lower tail probabilities of the statistic can be used for significance testing.

INVERSE SAMPLING

In inverse sampling, observation proceeds until an event of interest has occurred r times, where r is a preassigned number. Usually, "observation" means conducting trials of a dichotomous experiment one at a time, as in sampling inspection, sample surveys*, etc. However, one can imagine observing a process in which events occur randomly in time, as in radioactive decay.

The most common inverse sampling situation arises in the conduct of independent trials with outcomes "success" and "failure," with common success probability θ. The terms "negative binomial sampling" and "inverse binomial sampling" are often used in this case. Inverse binomial sampling is most commonly used when θ is small, since one can then control the relative error.

Inverse sampling has also been applied to sampling without replacement, to finite population sampling*, to sampling from a Poisson process*, and to the multinomial* case. In addition, comparison of several binomial* populations has been attacked via inverse sampling, with application to clinical trials*.

In all inverse sampling problems the sample size is random, so that inverse sampling may be viewed as a case of sequential analysis*.

HISTORY

Haldane [13, 14] suggested inverse sampling as a method of unbiased estimation of the binomial parameter θ in such a way that the coefficient of variation* of the estimator is independent of θ. He derived the unbiased* estimator of θ and its variance* and showed that the coefficient of variation is nearly independent of θ when θ is small. The term "inverse sampling" comes from Tweedie [27]. He noted that the binomial distribution* of classical sampling and the negative binomial distribution* of inverse sampling satisfy the following relationship between their cumulant* generating functions:

$$K_B(t) = n\log(1 - \theta + \theta e^t) = nH(-t),$$

$$K_{\mathrm{NB}}(t) = -r\log\big(e^{-t}/\theta - (1-\theta)/\theta\big)$$

$$= rH^{-1}(-t),$$

where $H(t) = \log(1 - \theta + \theta e^{-t})$. This inverse functional relationship led Tweedie to call the binomial and negative binomial* an "inverse pair." More generally, if X and Y are random variables such that $\log E[e^{-tX}] = \alpha L(t)$ and $\log E[e^{-tY}] = \beta L^{-1}(t)$, then $E[X] = \alpha \kappa_1$, $\mathrm{var}(X) = \alpha \kappa_2$, $E[Y] = \beta/\kappa_1$, and $\mathrm{var}(Y) = \beta \kappa_2/\kappa_1^3$, where $\kappa_j = (-1)^j L^{(j)}(0)$, $j = 1, 2, \ldots$ The variables X and Y have the same coefficient of variation* if $\beta = \alpha \kappa_1$, where κ_1 and κ_2 are the cumulants derived from L. If the common coefficient of variation is "small," both X/α and β/Y are estimators of κ_1. A pair X, Y with these properties is called a pair of inverse variables. In addition to the binomial–negative binomial pair, Tweedie identified the inverse Gaussian* and inverse Poisson variables.

INVERSE BINOMIAL SAMPLING

If independent trials are conducted until r successes occur, if X is the sample size and if $\Pr[\text{success}] = \theta$, then X follows the negative binomial law [15]:

$$\Pr[X = n] = \binom{n-1}{r-1}\theta^r(1-\theta)^{n-r}$$

with $E[X] = r/\theta$, $\mathrm{var}(X) = r(1-\theta)/\theta^2$. Furthermore, $\hat{\theta} = (r-1)/(X-1)$ is an unbiased estimator of θ if $r > 1$. If $r = 1$, $\hat{\theta} = 1$ if $X = 1$ and $\theta = 0$ if $X > 1$, is unbiased [13, 14]. If $r > 2$,

$$\mathrm{var}(\hat{\theta}) = \frac{\theta^2(1-\theta)}{r}$$

$$+ \theta^2(1-\theta)\sum_{j=1}^{\infty}\binom{r+j}{j+1}^{-1}(1-\theta)$$

$$= (r-1)\theta^r(1-\theta)^{1-r}$$

$$\times \int_0^{1-\theta} t^{r-2}(1-t)^{1-r}\,dt$$

$$= (r-1)\theta^r(1-\theta)^{1-r}$$

$$\times \left[(-1)^{r-1}\log\theta\right.$$

$$\left. + \sum_{i=1}^{r-2}\frac{(-1)^{r-1}}{i}\left(\frac{1-\theta}{\theta}\right)^i\right]$$

[9, 13]. An unbiased estimator of $\mathrm{var}(\hat{\theta})$ is $\hat{\theta}^2(1-\hat{\theta})/(r-1-\hat{\theta}) = \hat{\theta}(1-\hat{\theta})/(X-2)$ [10]. Approximations to $\mathrm{var}(\hat{\theta})$ follow from the simple bounds [17]

$$\frac{\theta^2(1-\theta)}{r} \leqslant \mathrm{var}(\hat{\theta})$$

$$\leqslant \frac{\theta^2(1-\theta)}{r-1-\theta} \leqslant \frac{\theta^2(1-\theta)}{r-2},$$

$$\frac{\theta^2(1-\theta)}{r} \leqslant \mathrm{var}(\hat{\theta})$$

$$\leqslant \frac{\theta^2(1-\theta)}{r} + \frac{2\theta(1-\theta)^2}{r(r+1)}.$$

Sharper but more complicated bounds appear in ref. 20.

In a variety of sequential cases, including inverse sampling, unbiased estimates of $\theta^\alpha(1-\theta)^\beta$ (α, β integers) may be derived by a path-counting (stochastic process*) argument [11]. These estimators are unique in the case of inverse sampling. More generally, a function $h(1-\theta)$ has an unbiased estimator if and only if $h(q)$ has a Taylor series expansion for $0 < q < 1$. The unique minimum variance unbiased estimator* is

$$\frac{(r-1)!}{(X-1)!}\frac{d^{X-r}}{dq^{X-r}}\left(\frac{h(q)}{(1-q)^r}\right)_{q=0}.$$

Unbiased estimators whose variances attain the Cramér–Rao variance bound* exist only for fixed and inverse sampling [9].

The maximum likelihood* estimator of θ is r/X. The maximum likelihood estimator of $g(\theta)$ is $g(r/X)$. As $r \to \infty$, both $\sqrt{r}\{X/(r-\theta)\}$ and $\sqrt{r}(\hat{\theta}-\theta)$ are asymptotically normally distributed with mean 0 and variance $\theta^2(1-\theta)$. The bias and mean squared error of r/X is studied in ref. 3 and it is shown that $\hat{\theta}$ is superior to the maximum likelihood in terms of both variance and mean squared error.

Confidence intervals* and hypothesis testing* for θ can be derived by exploiting the following identity relating binomial and negative binomial probabilities:

$$\sum_{j=r}^{n} \binom{n}{j} \theta^{j}(1 - \theta)^{n-j}$$

$$= \sum_{k=r}^{n} \binom{k-1}{r-1} \theta^{r}(1 - \theta)^{k-r}.$$

This identity has been rederived many times, but seems to have been discovered by Montmort* in 1714 [15, 26].

The most powerful test of $H: \theta \leqslant \theta_0$ rejects for small values of X, according to the Neyman–Pearson lemma*. This is equivalent to rejecting H for large values of $\hat{\theta}$, an intuitively reasonable approach. To find critical values, one must evaluate lower tail probabilities for X. The identity states that lower tail probabilities for X are upper tail probabilities for the binomial distribution, which are well tabulated. Tests of $H: \theta \geqslant \theta_0$ are conducted similarly: one rejects when X is large, and the required upper tail probabilities are computed as lower tail probabilities for the binomial distribution. To test $H: \theta = \theta_0$, one finds lower and upper critical values as above, such that the total tail area is the desired level α.

Exact lower confidence bounds are determined by finding the largest parameter value θ_L such that $H: \theta \geqslant \theta_L$ is accepted for the observed data r, X. This means finding θ_L such that $\sum_{k=X}^{\infty} \binom{k-1}{r-1} \theta_L^{r}(1 - \theta_L)^{k-r} = \alpha$, where $1 - \alpha$ is the desired confidence. This negative binomial upper tail probability is a binomial lower tail probability from the identity above. For computational details, see ref. 10. Results on median unbiased estimation*, approximate confidence limits, and charts for the design of inverse sampling experiments appear in ref. 8.

If one compares fixed sample size and inverse sampling procedures with $n = r/\theta$ (so that expected sample sizes are equal in both cases), inverse sampling is slightly less efficient* for finite sample sizes, but becomes equally efficient as $n, r \rightarrow \infty$. However, inverse sampling has the advantage that the coefficient of variation of $\hat{\theta}$ can be controlled accurately when it is known that

θ is small. Furthermore, when θ is small, inverse sampling guarantees that some successes will be observed, at the cost of very large sample sizes. The accuracy of variance estimates in fixed and inverse sampling is considered in refs. 6 and 23.

INVERSE SAMPLING FROM FINITE POPULATIONS

Suppose that a population contains n members, of which $n\theta$ are "successes." These may be defectives in a lot submitted for inspection, members of a subpopulation of interest in survey sampling, and so on. If n is known and θ is unknown, the inverse sampling procedure is to sample without replacement until r successes are drawn. In this case the trials are dependent and the sample size X has the negative hypergeometric distribution* [15]:

$$\Pr[X = k] = \frac{n\theta - r + 1}{n - k + 1} \binom{n\theta}{r-1}$$

$$\times \binom{n - n\theta}{k - r} \bigg/ \binom{n}{k-1}$$

with $E[X] = (n + 1)r/(n\theta + 1)$ and

$$\text{var}(X) = \frac{(n + 1)(n\theta - r + 1)(n - n\theta)}{(n\theta + 1)^2(n\theta + 2)}.$$

Under inverse sampling, the unbiased estimator of θ is $\hat{\theta} = (r - 1)/(X - 1)$, the maximum likelihood estimator is r/X, and

$$\hat{\theta}(1 - \hat{\theta})/(X - 2)$$

is an unbiased estimator of $\text{var}(\hat{\theta})$, as in inverse binomial sampling. The coefficient of variation of $\hat{\theta}$ is bounded above by

$$\left[r^{-1} - (n\theta + 1)^{-1}\right]\left[(n\theta + 1)/(n\theta + 2)\right] \quad [16].$$

Fixed and inverse sampling procedures are compared in ref. 6.

Inverse sampling has been employed to estimate the size of a finite population via the capture–recapture method*. In this scheme, t members of the population are selected at random, tagged, and released.

Afterward, members are sampled at random, one at a time, until r tagged individuals are caught. Sampling may be performed either with or without replacement. In sampling with replacement, the sample size X has the negative binomial distribution

$$\Pr[X = x] = \binom{x-1}{r-1}\left(\frac{t}{n}\right)^r\left(\frac{n-t}{n}\right)^{x-r}$$

and tX/r is an unbiased estimate of n with variance $n(n-t)/r$. An unbiased estimator of $\text{var}(tX/r)$ is $Xt^2(X-r)/X^2(X+1)$. If one samples without replacement, X has a negative hypergeometric distribution and $X(t+1)/r$ is an unbiased estimator whose variance is approximately n^2/r. One might sample until a fixed number of untagged individuals is found. Although only approximately unbiased estimates of n are possible in this case, the variability of the required sample size is reduced when t is small relative to n. Multistage capture–recapture surveys and other extensions are presented in ref. 7.

Inverse sampling of clusters with probability proportional to size leads to unbiased estimates of a population total [21]. The method is to sample with replacement until r distinct clusters are found. However, the estimator only employs the first $r - 1$ clusters, and thus is inefficient [18]. Related schemes for inverse sampling from finite populations subject to cost constraints have also been proposed [19].

OTHER APPLICATIONS

Inverse sampling schemes for selecting the best binomial population have been developed using both the indifference zone approach [24, 25] and the subset approach [1]. In this setup, if c populations are to be compared, one samples until r successes have been obtained from any one of the populations. An additional complication is introduced, since after completion of a trial, one needs a rule to decide which population should be sampled next. Under inverse sam-

pling, so-called "play the winner"* rules are superior to "vector at a time" rules [2]. For numerical comparisons of inverse sampling versus other sequential schemes, see refs. 4 and 12. The latter monograph also contains references on inverse sampling rules for ranking* by means of tournaments and for selecting* the best Poisson parameter.

Inverse sampling has also been applied to estimation of multinomial* probabilities with k outcomes [28]. Here, one samples until the rth occurrence of outcome k. The frequencies of the first $k - 1$ outcomes, $X_1, \ldots, X_{k-1}$, have the negative multinomial distribution [15]. Inverse sampling is shown to be superior to fixed sampling if one wishes to determine the most probable outcome [5].

A notion of inverse sampling has been proposed for estimating the mean μ of a normal distribution with known variance. In this setup, independent observations X_1, $X_2, \ldots$ are taken until $|\sum_1^N X_i| > k$, a preassigned constant. See ref. 30 for a discussion of various methods of interpreting the estimator $\hat{\mu} = \sum_1^N X_i/N$ and approximations to the distribution of N, in terms of the inverse Gaussian distribution [29].

Inverse sampling of a Poisson process* with intensity λ proceeds as follows. One observes the process until r events have occurred, the observation time T being random. Statistical inference on λ is based on the fact that $2\lambda T$ has the chi-squared distribution* with $2r$ degrees of freedom. For example, T/r is an unbiased estimate of λ^{-1}, the mean time between events [22]. Inverse sampling is applied to ranking and selection* of Poisson processes in ref. 12.

References

[1] Barron, A. M. and Mignogna, E. (1977). *Commun. Statist. A*, **6**, 525–552.

[2] Berry, D. A. and Young, D. H. (1977). *Ann. Statist.*, **5**, 235–236.

[3] Best, D. J. (1974). *Biometrika*, **61**, 385–386.

[4] Bühringer, H., Martin, H., and Schriever, K.-H. (1980). *Nonparametric Sequential Selection Procedures*. Birkhäuser, Boston. (Contains detailed

comparisons of inverse sampling and other sequential schemes for selecting the largest binomial parameter. Extensive references.)

[5] Cacoullos, T. and Sobel, M. (1966). In *Multivariate Analysis*, P. R. Krishnaiah, ed. Academic Press, New York, pp. 423–455.

[6] Chakrabarti, R. P. and Tsai, P. J. (1976). *Proc. Amer. Statist. Ass. (Social Statist. Sec.)*, pp. 236–241.

[7] Chapman, D. G. (1952). *Biometrics*, **8**, 286–306. (Extensive treatment of inverse sampling schemes for estimating the size of a population.)

[8] Crow, E. L. (1975). *Commun. Statist.*, **4**, 397–413.

[9] DeGroot, M. H. (1959). *Ann. Math. Statist.*, **30**, 80–101. (Analyzes sequential procedures for estimating functions of the binomial parameter, emphasizing fixed sample rules, inverse sampling, and truncated inverse sampling.)

[10] Finney, D. J. (1949). *Biometrika*, **36**, 233–235. (Detailed treatment of confidence intervals and hypothesis testing via inverse binomial sampling.)

[11] Girschick, M. A., Mosteller, F., and Savage, L. J. (1946). *Ann. Math. Statist.*, **17**, 282–298. (Outlines path counting/stochastic process interpretation of sequential binomial sampling.)

[12] Gupta, S. S. and Panchapakesan, S. (1979). *Multiple Decision Procedures: Theory and Methodology of Selecting and Ranking Populations*. Wiley, New York. (A thorough survey of its subject, with many applications of inverse sampling.)

[13] Haldane, J. B. S. (1945). *Nature (Lond.)*, **155**, 49–50.

[14] Haldane, J. B. S. (1945). *Biometrika*, **33**, 222–225.

[15] Johnson, N. L. and Kotz, S. (1969). *Distributions in Statistics: Discrete Distributions*. Wiley, New York. (An encyclopedic treatment, with detailed presentation of properties of distributions, applications, and historical notes.)

[16] Knight, W. (1965). *Ann. Math. Statist.*, **36**, 1494–1503. [Treats inverse and sequential sampling for various discrete distributions, including binomial, hypergeometric, and Poisson, via a path counting (stochastic process) approach.]

[17] Mikulski, P. W. and Smith, P. J. (1976). *Biometrika*, **63**, 216–217.

[18] Pathak, P. K. (1964). *Biometrika*, **51**, 185–193.

[19] Pathak, P. K. (1976). *Ann. Statist.*, **4**, 1012–1017.

[20] Sahai, A. (1980). *J. Statist. Plann. Infer.*, **4**, 213–216.

[21] Sampford, M. R. (1962). *Biometrika*, **49**, 27–40.

[22] Sandelius, M. (1950). *Biometrics*, **6**, 291–292.

[23] Scheaffer, R. L. (1974). *Biometrics*, **30**, 187–198.

[24] Sobel, M. and Weiss, G. H. (1971). *J. Amer. Statist. Ass.*, **66**, 545–551.

[25] Sobel, M. and Weiss, G. H. (1972). *Ann. Math. Statist.*, **43**, 1808–1826.

These two papers apply an inverse sampling rule to the problem of selecting the largest binomial parameter, subject to an "ethical constraint."

[26] Todhunter, I. (1869). *A History of the Mathematical Theory of Probability*. Chelsea, New York.

[27] Tweedie, M. C. K. (1945). *Nature (Lond.)*, **155**, 453. (Proposes the term "inverse sampling".)

[28] Tweedie, M. C. K. (1952). *J. R. Statist. Soc. B*, **14**, 238–245.

[29] Tweedie, M. C. K. (1957). *Ann. Math. Statist.*, **28**, 362–377.

[30] Wetherill, G. B. (1975). *Sequential Methods in Statistics*, 2nd ed. Methuen, London.

(BINOMIAL DISTRIBUTION
CAPTURE-RECAPTURE METHODS
EFFICIENCY
GEOMETRIC DISTRIBUTION
HYPOTHESIS TESTING
INVERSE GAUSSIAN DISTRIBUTION
MAXIMUM LIKELIHOOD
NEGATIVE BINOMIAL DISTRIBUTION
POISSON PROCESS
SEQUENTIAL ANALYSIS)

PAUL J. SMITH

INVERSE SINE TRANSFORMATION

See ARC SINE TRANSFORMATION

INVERSE TANH TRANSFORMATION

See FISHER'S *z* TRANSFORMATION

INVERSION FORMULA *See* CHARACTERISTIC FUNCTIONS

INVERTED BETA DISTRIBUTION

An inverted beta distribution (IBD) (also known as a beta distribution of the second kind or a beta-prime distribution), historically, arises as a special case of the Pearsonian system* of frequency curves (in particular, a special case of the Pearson's type VI distribution). Thus an IBD inherits the usefulness of the Pearsonian system in fitting the curves to the empirical data. However, more direct applications of an IBD, in data analysis, are due to its close structural relationship with the beta* and gamma* families of distributions. Six different types of applications of an IBD are:

Type I: Distribution of the Odds Ratio

A random variable, say Y, representing a probability measure (proportion in empirical studies) is often assumed to have a beta distribution. Then, a random variable, $X = Y/(1 - Y)$, (odds ratio*) has an IBD. This relation, in the inverse form, together with the tables of incomplete beta function, is of basic importance in evaluating the probability integrals for the IBD and Snedecor's F-distribution* [11].

Type II: Ratio of Two Independent Gamma Variables

If U and V are independent random variables, each having a gamma distribution with the same scale parameter, then the quotient U/V has an IBD. Since the chi-squared distribution is a special case of the gamma distribution, a similar interrelation holds for the chi-squared variables. Thus, the IBD can be related to an F-distribution (a prominent distribution that arises when sampling from a normal distribution). Further, if X has an IBD with parameters p and q, then $Z = (\log_e(qX/p))/2$ has Fisher's Z distribution. The F and Z distributions are well known for their applications in statistical data analysis. Another logarithmic transformation of an inverted beta variable, $Y = \beta \log_e X + \alpha$, $\beta > 0$, $-\infty < \alpha < \infty$, leads to the compound generalized extreme type I (generalized logistic) distribution [3]. For details concerning these distributions and their structural relationships with the Student's t^*, Cauchy* and other distributions see Johnson and Kotz [6] and Patil et al. [15].

Type III: Mixture of Two Gamma Distributions

Let a random variable X have a gamma distribution with parameters θ(scale) and p(shape). Further, if θ is a random variable having a gamma distribution with parameters β(scale) and q(shape), then the resulting

mixture distribution is the generalized IBD (see the final section) with parameters β, p, and q. Such a mixture distribution is useful as a survival distribution. In particular, for $p = 1$, this distribution reduces to a more familiar survival distribution, namely, the Lomax distribution* [5, 9]. The IBD with parameters β, p, and q also arises as a predictive distribution (a conditional distribution of the future observation (Y), given the observation (x) on a random variable (X) associated with "informative" experiment of interest to a research worker). However, in such cases the parameter β is a linear function of x. For details concerning predictive distributions*, see Aitchison and Dunsmore [1, Chap. 2].

Type IV: IBD as a Prior Distribution

For an IBD with probability density function (PDF) $f_X(x)$, given in the section "Definition and Structure,"

1. $\lim_{x \to \infty} x^{1+\epsilon} f_x(x) < \infty, 0 < \epsilon \leqslant n$.
2. The function $k(x) = x f_X'(x)/f_X(x)$ is nonincreasing in x and further, $\lim_{x \to 0+} k(x) < \infty$.

Because of these properties of an IBD, Ghosh and Parsian [4] have found it useful, as a prior density, in their Bayes minimax* estimation procedure for the multiple Poisson parameters.

Type V: IBD as a Mixing Distribution

A discrete distribution defined by

$$p_X(x) = \int_0^\infty p_{X|\mu}(x) g(\mu) \, d\mu,$$

where

$$p_{X|\mu}(x) = \Gamma\left(\begin{matrix} -k \\ x \end{matrix}\right)\{ \mu/(k + \mu)\}^x$$
$$\times (k/(k + \mu))^k, \qquad x = 0, 1, 2, \ldots$$

and

$$g(\mu) = (\mu/(k + \mu))^p (k/(k + \mu))^q$$
$$\times [\mu B(p, q)]^{-1},$$
$$\mu > 0, \quad (k, p, q) > 0,$$

is considered by Mosimann [13] for building a probability model. Further, he uses this model in the study of the frequently occurring word-frequency curves in quantitative linguistics.

Type VI: Generalizations of IBD

A generalization of an IBD, which is suitable for applications in hydrology* and meteorology*, is considered by Milke and Johnson [12]. The Bradford distribution [8], arising in documentation studies, is a special case of the truncated version of the generalized IBD. Another generalization due to Malik [10], based on the quotient of two independent Stacy's generalized gamma variables [18], is studied by Block and Rao [2]. They consider this generalized IBD for building a bivariate probability model, useful in analysis of the failure-time/warning-time data arising in reliability studies.

DEFINITION AND STRUCTURE

The IBD with parameters p and q is defined by the PDF given by

$$f_X(x) = x^{p-1}(1 + x)^{-(p+q)}/B(p,q),$$
$$x > 0, \quad p > 0, \quad q > 0.$$

Property 1. For $-p < r < q$, the IBD has a finite rth moment given by

$$\mu'_r = B(p + r, q - r)/B(p,q).$$

In particular,

$$E[X] = p/(q - 1), \quad q > 1,$$

$$V(X) = p(p + q - 1)(q - 1)^{-2}(q - 2)^{-1},$$
$$q > 2.$$

Property 2. The modal value of the IBD, for $p \geqslant 1$, is $(p - 1)/(q + 1)$. For $p < 1$, $f_X(x) \to \infty$, as $x \to 0$.

Property 3 [14]. A random variable is said to have a distribution belonging to the log-exponential family if its PDF is of the form

$$f_X(x) = x^\theta a(x)/m(\theta)$$
$$= \exp\{\theta \log x + A(x) - B(\theta)\},$$

where

$$a(x) = \exp\{A(x)\}, \quad m(\theta) = \exp\{B(\theta)\},$$

$$E[\log X] = m'(\theta)/m(\theta) = B'(\theta),$$

prime (') denoting the derivative with respect to θ.

The IBD belongs to the log-exponential family with $\theta = p - 1$. Further, for $\alpha < q$, it is form-invariant under a size bias of order α.

Property 4 [7]. Let X_1 and X_2 be independent random variables each having uniform distribution on $(0, 1)$. Let $Y_1 = a(X_1)^{1/p}$, $Y_2 = b(X_2)^{1/q}$, where a, b, p, and q are positive constants. Then, for every real number k, $0 < k \leqslant \min(a, b)$, the conditional distribution of (Y_1/Y_2), given $Y_1 + Y_2 \leqslant k$, is the IBD with parameters p and q.

This property is useful in generating random numbers from the IBD and F distribution.

Property 5. If X has the IBD with parameters p and q, then $1/X$ has again IBD with parameters q and p.

Property 6. If X has the IBD with parameters p and q, then $F = (v_2/v_1)X$ has the F-distribution with parameters $v_1 = 2p$ and $v_2 = 2q$.

GENERALIZED INVERTED BETA DISTRIBUTION

A random variable Y has the generalized inverted beta distribution (generalized beta distribution of the second kind) if its PDF is

$$f_Y(y) = \frac{(c\beta_2/\beta_1)(\beta_2 y/\beta_1)^{p-1}}{B(p/c, q/c)}$$

$$\times \left[1 + \left(\frac{\beta_2 y}{\beta_1} \right)^c \right]^{-(p+q)/c}$$

$y > 0$; β_1, β_2, p, q, and c are real positive constants.

For details concerning structural and inferential properties of the generalized inverted beta distribution (including its special cases and modified versions) and related applications, see Malik [10], Rao and Garg [17], Block and Rao [2], Milke and Johnson [12], and Tadikamalla [19].

References

[1] Aitchison, J. and Dunsmore, I. R. (1975). *Statistical Prediction Analysis*. Cambridge University Press, Cambridge.

[2] Block, H. W. and Rao, B. R. (1973). *Sankhyā B*, **35**, 79–84.

[3] Dubey, S. D. (1969). *Naval Res. Logist. Quart.*, **16**, 37–40.

[4] Ghosh, M. and Parsian, A. (1981). *J. Multivariate Anal.*, **11**, 280–288.

[5] Harris, C. M. (1968). *Operat. Res.*, **16**, 307–313.

[6] Johnson, N. L. and Kotz, S. (1970). *Distributions in Statistics: Continuous Univariate Distributions*, 2 vols. Wiley, New York.

[7] Kapur, B. D. (1977). Structural Properties of Statistical Distributions Useful in Computer Generation of Random Variables. M.S. thesis, Pennsylvania State University.

[8] Leimkuhler, F. F. (1967). *J. Documentation*, **23**, 197–207.

[9] Lomax, K. S. (1954). *J. Amer. Statist. Ass.*, **49**, 847–852.

[10] Malik, H. J. (1967). *Canad. Math. Bull.*, **10**, 463–465.

[11] Merrington, M. and Thompson, C. M. (1943). *Biometrika*, **33**, 73–88.

[12] Milke, W., Jr. and Johnson, E. S. (1974). *Water Resour. Res.*, **10**, 223–226.

[13] Mosimann, J. E. (1981). Word-frequency curves. Personal communication.

[14] Patil, G. P. and Ord, J. K. (1976). *Sankhyā B*, **38**, 48–61.

[15] Patil, G. P., Boswell, M. T., and Ratnaparkhi, M. V. (1982). *A Modern Dictionary and Classified Bibliography of Statistical Distributions*, Vol. 1: *Univariate Continuous Models*. International Cooperative Publishing House, Fairland, Md. (in press).

[16] Pearson, K. (1895). *Philos. Trans. R. Soc. Lond. A*, **186**, 343–414.

[17] Rao, B. R. and Garg, M. L. (1969). *Canad. Math. Bull.*, **12**, 865–868.

[18] Stacy, E. W. (1962). *Ann. Math. Statist.*, **33**, 1187–1192.

[19] Tadikamalla, P. R. (1980). *Int. Statist. Rev.*, **48**, 337–344.

(BETA DISTRIBUTION
BURR DISTRIBUTIONS
DIRICHLET DISTRIBUTION
INVERSE DISTRIBUTIONS
SURVIVAL ANALYSIS)

M. V. RATNAPARKHI

INVERTED DIRICHLET DISTRIBUTION

An inverted Dirichlet distribution (IDD) is a multivariate generalization of the inverted beta distribution*. The IDD (as a special case) occurs as a joint distribution of the ratios of independent chi-squared variables with common denominator. Thus an IDD is closely related to the multivariate *F*-distribution* [1, pp. 238–243], which plays an important role in analysis of data of various kinds.

An IDD could be derived from independent gamma* variables [6] or from Dirichlet* variables [11]. These results are stated below. For details concerning usefulness of these results, see the section on applications of IDD.

1. Let $Y_0, Y_1, \ldots, Y_s$ be independent random variables, each having a gamma distribution with shape parameters $\alpha_0, \alpha_1, \ldots, \alpha_s$, respectively, and a common scale parameter β. Define $X_i = Y_i / Y_0$, $i = 1, 2, \ldots, s$. Then, the joint distribution of $X_1, X_2, \ldots, X_s$ is inverted Dirichlet.

2. Let $U_1, U_2, \ldots, U_s$ have a Dirichlet distribution [13] with parameters $\alpha_0, \alpha_1, \ldots, \alpha_s$. Define $X_i = U_i / (1 - \sum_1^s U_i)$, $i = 1, 2, \ldots, s$. Then $X_1, X_2, \ldots, X_s$ have an IDD.

DEFINITION AND STRUCTURE

An IDD is defined by the joint probability density function given by

$$f_{X_1,\ldots,X_s}(x_1,\ldots,x_s)$$

$$= C \prod_{i=1}^{s} x_i^{\alpha_i-1} \Big/ \Big(1 + \sum_1^s x_i\Big)^{\alpha}, \quad (1)$$

$0 < x_i < \infty, \quad \alpha_i > 0, \quad i = 0, 1, \ldots, s;$ where

$$C = \Gamma(\alpha)\Big/ \prod_{i=0}^{s} \Gamma(\alpha_i) \quad \text{and} \quad \alpha = \sum_{i=0}^{s} \alpha_i.$$

An IDD is a special case of the Liouville–Dirichlet distributions [2, p. 308].

Property 1.   The joint moment

$$\mu'_{r_1,r_2,\ldots,r_s} = \frac{\Gamma(\alpha_0 - \sum_1^s r_i)}{\Gamma(\alpha_0)} \prod_{i=1}^{s} \left(\frac{\Gamma(\alpha_i + r_i)}{\Gamma(\alpha_i)} \right)$$

is finite for $\alpha_0 > \sum_1^s r_i$. In particular, for $\alpha_0 > 2$,

$$E[X_i] = \alpha_i(\alpha_0 - 1)^{-1},$$

$$\text{var}(X_i) = \frac{\alpha_i(\alpha_0 + \alpha_i - 1)}{(\alpha_0 - 1)(\alpha_0 - 2)}$$

$$\text{cov}(X_i, X_j) = \alpha_i\alpha_j(\alpha_0 - 1)^{-2}(\alpha_0 - 2)^{-1}.$$

Further,

$$E[X_i \mid X_j = x_j] = (1 + x_j)\alpha_i(\alpha_0 + \alpha_j - 1)^{-1},$$

$$\alpha_0 + \alpha_j > 1.$$

Property 2.   Let $X_1, X_2, \ldots, X_s$ have an IDD with PDF given by (1). Then, we have the following:

1. The marginal distribution of $X_1, X_2,$ $\ldots, X_k$, $(k > s)$, is inverted Dirichlet with parameters $\alpha_1, \alpha_2, \ldots, \alpha_k$ and α_0. In particular, the marginal distribution of X_i is inverted beta with parameters α_i and α_0.

2. The random variable $\sum_1^s X_i$ has an inverted beta distribution with parameters $\sum_1^s \alpha_i$ and α_0.

Property 3 [7].   Let $\mathbf{X} = (X_1, X_2, \ldots, X_s)'$ denote a vector of random variables and let α be the parameter vector. Further let $\mathcal{F} = \{\mathbf{X}_{(\alpha)}\}$ denote a family of random vectors indexed by α. If $\mathbf{X}_{(\alpha)}$ has an IDD with parameters $\boldsymbol{\alpha} = (\alpha_1, \alpha_2, \ldots, \alpha_s)'$ and α_0, then, for fixed α_0, $\mathcal{F}$ is a Schur family in $\boldsymbol{\alpha}$.

Property 4.   Let $X_1, X_2, \ldots, X_s$ be independent random variables, X_i having a gamma distribution with parameters θ(scale) and α_i, $i = 1, 2, \ldots, s$. If Θ is a random variable having a gamma distribution with parameters β(scale) $= 1$ and α_0, then the resulting mixture distribution is an IDD.

Property 5.   In property 4, if $\alpha_i = \alpha$, $i = 1, 2, \ldots, s$, then the resulting joint pdf of $(X_1, X_2, \ldots, X_s)$ belongs to a class of multivariate densities known as "positive dependent by mixture." For details concerning this property and related references, see Shaked [10].

APPLICATIONS

Tiao and Guttman [11], as described below, have found an IDD to be useful in Bayesian estimation procedures (pre-posterior analysis) for linear models. (*See* GENERAL LINEAR MODEL.)

Consider a linear model defined by

$$\mathbf{y} = \mathbf{X}\boldsymbol{\beta} + \mathbf{e},$$

where $\mathbf{y}$ is a vector of n observations, $\boldsymbol{\beta}$ a $m \times 1$ parameter vector, $\mathbf{X}$ a $n \times m$ coefficient matrix and $\mathbf{e}$ a $n \times 1$ vector of random errors. Further, assume that $\mathbf{e}$ has a multivariate normal distribution with parameters $\boldsymbol{\mu}$ (mean vector) $= \mathbf{O}$ and $\boldsymbol{\Sigma}$ (covariance matrix) $= \sigma^2\mathbf{I}$, where $\mathbf{I}$ is an $n \times n$ unit matrix.

For the model above, a Bayesian inference-related problem of interest is to find a posterior density function of $(\sigma^2 \mid \mathbf{y})$. A derivation of this density function depends on the knowledge of the joint distribution of the sample quantities **b** and W (estimators of β and residual sum of squares, respectively) having β and σ^2 as parameters. Tiao and Guttman consider a suitable (joint) prior distribution for β and σ^2 and show that the joint distribution of **b** (in canonical form) and W is an IDD. This distribution of **b** and W is further useful in obtaining the distribution of $(\sigma^2 | \mathbf{y})$.

Another application of an IDD, in paleoecological studies [6], arises as follows:

Let $X_1, X_2, \ldots, X_s$ denote the concentrations per unit volume of s kinds of fossil pollen grains found at different depths in sediment. The stochastic variation in X_i, for $i = 1, 2, \ldots, s$, is assumed to be due to depth. Let

$$P_i = X_i / \sum_{i=1}^{s} X_i \quad \text{and} \quad U_i = X_i / X_s,$$

$$\text{for} \quad i = 1, 2, \ldots, s,$$

be the corresponding proportion and ratio variables. The observations on either X_i's, P_i's, or U_i's ($i = 1, 2, \ldots, s$) form a fossil pollen profile. Then the characteristics of this profile are useful in studying the changes that occurred in vegetation and climate in the area of sediment deposition over a period of time.

Now, to obtain a model for the analysis of data collected for studying fossil pollen profile, Mosimann [6] considers the following assumptions:

1. $X_1, X_2, \ldots, X_s$ are positive, nondegenerate, mutually independent random variables.

2. The vector of their proportions, **P**, is independent of $\sum_1^s X_i$.

These assumptions imply that each X_i, $i = 1, 2, \ldots, s$, has a gamma distribution with same scale parameter [3]. Thus the joint distribution of $U_1, U_2, \ldots, U_s$, as mentioned in the introduction, is an inverted Dirichlet, where $u_s \equiv 1$.

However, in practice, rather than having observations on pollen concentrations, it is more natural to have observations on pollen counts. Then, $X_1, X_2, \ldots, X_s$ are discrete random counts which are estimates of the true concentrations. Some models, appropriate for sampling (direct and inverse) from the populations of fossil pollen deposits, have been studied in detail by Mosimann [4, 5]. He also considers the usefulness of the indices P_i / P_s $(= U_i / U_s)$, $i = 1, 2, \ldots, s$, and their covariance structure in the analysis of pertinent data. In particular, he observes that for inverse counts with large sample sizes the $\text{cov}(X_i, X_j)$ tends to $\text{cov}(U_i, U_j)$, where $(U_1, U_2, \ldots, U_s)$ have an IDD.

CONCLUDING REMARKS

An IDD is related to a number of multivariate/matrixvariate distributions. For details, see Johnson and Kotz [1], Patil et al. [8], Roux [9], and Waal [12]. The probability integrals and related approximations of IDD are studied by Tiao and Guttman [11].

References

[1] Johnson, N. L. and Kotz, S. (1972). *Distributions in Statistics: Continuous Multivariate Distributions.* Wiley, New York.

[2] Marshall, A. W. and Olkin, I. (1979). *Inequalities: Theory of Majorization and Its Applications.* Academic Press, New York.

[3] Mosimann, J. E. (1962). *Biometrika*, **49**, 65–85.

[4] Mosimann, J. E. (1963). *Biometrika*, **50**, 47–54.

[5] Mosimann, J. E. (1965). In *Handbook of Paleontological Techniques*, B. Kummel and D. Raup, eds. W. H. Freeman, San Francisco, pp. 636–673.

[6] Mosimann, J. E. (1970). In *Random Counts in Scientific Work*, Vol. 3, G. P. Patil, ed. Pennsylvania State University Press, University Park, Pa., pp. 1–30.

[7] Nevius, S. E., Proschan, F., and Sethuraman, J. (1977). *Ann. Statist.*, **5**, 263–273.

[8] Patil, G. P., Boswell, M. T., Ratnaparkhi, M. V., and Roux, J. J. J. (1982). *A Modern Dictionary and Classified Bibliography of Statistical Distributions*, Vol. 2: *Multivariate Models.* International Co-operative Publishing House, Fairland, Md. (in press).

[9] Roux, J. J. J. (1971). *S. Afr. Statist. J.*, **5**, 27–36.

[10] Shaked, M. (1977). *Ann. Statist.*, **5**, 505–515.

[11] Tiao, G. G. and Guttman, I. (1965). *J. Amer. Statist. Ass.*, **60**, 793–805.

[12] Waal, D. J. de (1970). *Ann. Math. Statist.*, **41**, 1091–1095.

[13] Wilks, S. S. (1962). *Mathematical Statistics*. Wiley, New York.

(BETA DISTRIBUTION
DEPENDENCE, CONCEPTS OF
DIRICHLET DISTRIBUTION
INVERTED BETA DISTRIBUTION)

M. V. Ratnaparkhi

INVERTED WISHART DISTRIBUTION

An inverted Wishart distribution is a continuous distribution of a symmetric random matrix. It is used in Bayesian decision theory and is a natural conjugate for the covariance matrix of a multivariate normal* distribution. *See* WISHART DISTRIBUTION for the explicit form of this distribution. The inverted *complex* Wishart distribution and its applications for constructing spectral estimates of stationary vector processes are discussed by Shaman [2]. For additional information, see LaValle [1].

References

[1] LaValle, I. H. (1970). *An Introduction to Probability, Decision and Inference*. Holt, Rinehart and Winston, New York.

[2] Shaman, P. (1980). *J. Multivariate Anal.*, **10**, 51–59.

(MATRIX-VARIATE DISTRIBUTIONS
WISHART DISTRIBUTION)

INVERTED GAMMA DISTRIBUTION

If a random variable X has a gamma density function

$$f_X(x) = \frac{e^{-\lambda x}(\lambda x)^{\beta - 1}}{\Gamma(\beta)}, \quad x \geqslant 0; \lambda, \beta \geqslant 0$$

the variable $Y = 1/X$ possessing the density

$$f_Y(y) = \frac{e^{-\lambda/y}}{\Gamma(\beta)} \cdot \lambda^\beta \left(\frac{1}{y}\right)^{\beta + 1} \quad y > 0$$

is often referred to as an inverted gamma variable. The expected value and the variance of Y are

$$E(Y) = \frac{\lambda}{\beta - 1} \quad \text{for} \quad \beta > 1$$

and

$$V(y) = \frac{\lambda^2}{(\beta - 1)^2(\beta - 2)} \quad \text{for} \quad \beta > 2$$

respectively. This distribution is encountered in Bayesian reliability applications.

For additional information, see ref. 1.

Reference

[1] Barlow, R. E. and Proschan, F. (1980). (Textbook in preparation; to be published by Wiley, New York.)

(GAMMA DISTRIBUTION
RELIABILITY THEORY)

INVERTIBILITY (IN TIME-SERIES ANALYSIS)

Given an observed stochastic process Y_t functionally related to an unobservable Gaussian noise process ϵ_t, the invertibility of the process is concerned with estimation of the values of ϵ_t that give rise to some observed sequence of Y_t's. More specifically, a model relating the Y_t and ϵ_t processes is invertible if there is an estimation procedure yielding estimates $\hat{\epsilon}_t$ of ϵ_t such that the error $e_t = \hat{\epsilon}_t - \epsilon_t$ tends to zero in a *certain* sense (such as $E(e_t^2) \to 0$ as $t \to \infty$ or e_t tends to zero with probability 1) as the number of Y_t observations increases indefinitely. See Granger and Anderson [2] and Kashyap and Rao [3] for more details.

In linear models the condition of invertibility is used to help to choose a unique model [2].

Babich and Madan [1] define a model to be α-invertible if e_t tends to zero with probability at least α. This generalization allows one to assert invertibility of models not known to be invertible in the ordinary sense.

References

[1] Babich, G. and Madan, D. B. (1980). *Biometrika*, **67**, 704–705.

[2] Granger, C. W. J. and Anderson, A. P. (1978). *Stoch. Processes Appl.*, **8**, 87–92.

[3] Kashyap, R. L. and Rao, A. R. (1976). *Dynamic Stochastic Models from Empirical Data*. Academic Press, New York.

(STOCHASTIC PROCESSES
TIME SERIES)

IRREDUCIBLE MARKOV CHAIN

A Markov chain* is irreducible if every state can be reached from every state. In other words, the chain contains no closed sets except for the set of all states.

(MARKOV PROCESSES)

ISING MODELS   *See* GIBBS DISTRIBUTIONS; LATTICE SYSTEMS

ISOMETRY   *See* ALLOMETRY

ISOTONIC INFERENCE

A great many situations occur in statistical inference where there is prior information of an ordinal kind. For example, in a dosage response experiment the probability of response is usually an increasing function of dose, in life testing*, when failure is due to wear-out, the hazard function increases. In regression* analysis we may know that $E(Y | X = x)$ is a monotonic function of x. This last example includes, as a special case, the one-way analysis of variance*, where x indexes the groups whose means are being compared. In fact, the one-tail test for comparing two means (or other parameters) is the oldest and simplest example of isotonic inference. In all of these examples the statistician has prior knowledge in the form of order restrictions on the parameters or function under investigation. One would expect it to be possible to make better estimates or perform more powerful tests when this information is fully utilized than when it is ignored. The theory of isotonic inference aims to provide means for doing this. Although the practical problems that motivate the study seem to have little in common, they can all be handled within the same theoretical framework.

Two principal approaches to the problem have been tried. One is to introduce parametric families satisfying the required order restrictions thus converting the problem into one to which standard methods apply. For example, in regression analysis we may choose monotonic regression functions such as $a + bx$ or ae^{bx} and estimate them by least squares. With ordered means in the analysis of variance, we can allocate ordered scores to the groups and then use regression methods. This line of approach usually involves stronger assumptions than the prior information warrants. The second approach to isotonic inference aims to make the best possible use of order information but without introducing arbitrary assumptions about the functional forms underlying the ordering.

The systematic study of isotonic inference began in the 1950s, although particular cases such as the one-tail test are much older. Pioneering work on isotonic estimation was done by Ayer et al. [1] and the hypothesis-testing* problem was first treated in Bartholomew [3]. A full account of the subject as it existed in 1971 is contained in Barlow et al. [2], which also includes a full bibliography. Recent work has filled many gaps in the theory and opened up new areas of research, notably in simultaneous inference.

ISOTONIC REGRESSION

The basic mathematical structure underlying the theory can best be expressed in the language of regression. The following brief account aims to convey the basic ideas only. Let $\mathscr{X}$ be a finite set of numbers $(x_1, x_2, \ldots, x_k)$ and let g be a given function on $\mathscr{X}$. A function f on $\mathscr{X}$ is an *isotonic function* if $x, y \in \mathscr{X}$ and $x < y$ imply that $f(x) \leqslant f(y)$. With each point in $\mathscr{X}$ we associate a positive weight w. Then the isotonic regression of g with weights w on $\mathscr{X}$ is denoted by g^* and minimizes

$$\sum_{x \in \mathscr{X}} w(x)\{ g(x) - f(x)\}^2 \qquad (1)$$

in the class of isotonic functions on $\mathscr{X}$.

In ordinary linear regression $f(x) = a + bx$, $g(x)$ is the mean value of the independent variable at x and $w(x)$ is the number of observations on which it is based. With isotonic regression, in the same situation, we allow $f(x)$ to be any member of the class of isotonic rather than linear functions. It turns out that the solution to the problem formulated above has many applications in seemingly unrelated problems.

Algorithms are available for effecting the minimization. The most important case is that of simple order when $x_1 < x_2 < \cdots < x_k$. If $g(x_1) \leqslant g(x_2) \leqslant \cdots g(x_k)$, (1) is obviously minimized by taking $g^*(x_i) = g(x_i)$ for all i. Otherwise, there will be an i for which $g(x_i) > g(x_{i+1})$. This pair is then replaced by their weighted average,

$$\frac{w(x_i)g(x_i) + w(x_{i+1})g(x_{i+1})}{w(x_i) + w(x_{i+1})}. \qquad (2)$$

The procedure is repeated until the resulting averages satisfy the required order restrictions; $g^*(x_i)$ is then equal to the weighted average which includes $g(x_i)$. The following example illustrates the steps when $k = 5$.

x_i	1	2	3	4	5
$g(x_i)$	9	15	12.5	11	16
$w(x_i)$	1	3	2	2	1

First stage $\dfrac{3(15) + 2(12.5)}{5} = 14$

Second stage $\dfrac{5(14) + 2(11)}{7} = 13.1$

$g^*(x_i)$	9	13.1	13.1	13.1	16

At the first stage we average $g(x_2)$ and $g(x_3)$ because $g(x_2) > g(x_3)$ [we could have averaged $g(x_3)$ and $g(x_4)$ but the final result would have been the same]. At the second stage we have four groups with g's and weights as follows:

$$\begin{array}{cccc} 9 & 14 & 11 & 16 \\ 1 & 5 & 2 & 1 \end{array}$$

This is treated exactly as if it were a new problem. There is one order inversion which is removed by averaging 14 and 11. After doing this there are no inversions and the estimation is complete.

Explicit formulas are available for the isotonic regression function in this case, known as the max-min formulas. Let $\mathrm{Av}(s, t)$ denote the weighted average of $g(x_s)$, $g(x_{s+1}), \ldots, g(x_t)$; then

$$g^*(x_i) = \min_{t \geqslant i} \max_{s \leqslant i} \mathrm{Av}(s, t)$$

$$= \max_{s \leqslant i} \min_{t \geqslant s} \mathrm{Av}(s, t)$$

$$= \min_{t \geqslant i} \max_{s \leqslant t} \mathrm{Av}(s, t). \qquad (3)$$

MAXIMUM LIKELIHOOD* ESTIMATION

The isotonic regression problem is of considerable interest in its own right and has direct applications—in multidimensional scaling*, for example. However, its value is increased by the fact that it also provides maximum likelihood estimators for a large class of problems involving ordered parameters. Suppose, for example, that Y_i has a Poisson* distribution with parameter μ_i ($i = 1, 2, \ldots, k$) and it is known that $\mu_1 \leqslant \mu_2 \leqslant \cdots \leqslant \mu_k$. Then the maximum likelihood estimates are those values of μ which maximize the log-likelihood

$$\sum_{i=1}^{k} (y_i \log \mu_i - \mu_i) \qquad (4)$$

subject to $\mu_1 \leqslant \mu_2 \leqslant \cdots \leqslant \mu_k$. It turns out that the required estimates are obtained by solving the isotonic regression problem with $g(x_i) = y_i$. The same kind of thing is true whenever the random variable is a member of an exponential family*. A general method of finding maximum likelihood estimators in such cases involves the following two steps:

1. Find a set of basic estimators (usually the unrestrained maximum likelihood estimators).
2. Find the isotonic regression of these estimators on the set of x's.

In the one-way analysis of variance, for example, the basic estimators would be the sample means and the weights the sample sizes. In a dosage-response experiment, basic estimators would be the proportions responding at a given dose, etc.

When applying the method to the estimation of a monotonic probability density function or hazard function, the maximization takes place in two stages. Consider, for example, the problem of estimating a hazard function $r(u)$ when it is known that this function is nondecreasing. The data are assumed to consist of the order statistics $X_{i:n}$ for a sample of size n. The log-likelihood in this case is

$$\sum_{i=1}^{n} \log r(x_{i:n}) - \sum_{i=1}^{n} \int_{0}^{x_{i:n}} r(u)\,du. \quad (5)$$

The first step is to show that $\hat{r}(u)$ is a step function with jumps at the order statistics. We then only have to estimate $r(u)$ at the points $u = x_{i:n}$, so (5) may be replaced by

$$\sum_{i=1}^{n} \log r_i - \sum_{i=1}^{n-1} (n-i)(x_{i+1:n} - x_{i:n})r_i,$$

$$(6)$$

where $r_i = r(x_{i:n})$. This expression has to be maximized subject to $0 \leqslant r_1 \leqslant r_2 \leqslant \cdots \leqslant r_{n-1} \leqslant \infty$. It turns out (see ref. 2 [pp. 233–234] that this problem can be solved in a manner very similar to that for the Poisson example of (4) starting with the basic estimators

$$\left\{ (n-i)(x_{i+1:n} - x_{i:n}) \right\}^{-1}$$

$$(i = 1, 2, \ldots, n-1).$$

These examples give some hint of the versatility of the approach; a fuller picture will be obtained from ref. 2.

Isotonic estimators have many desirable properties. Usually, they are consistent and have smaller mean square errors than the basic estimators. However, unexpected features arise in the estimation of densities and hazard functions. In one case—that of "star ordered" distributions—the restricted maximum likelihood estimators are not isotonic estimators and neither are they consistent ($g(x)$ is star-shaped on $(0, \infty)$ if $g(\lambda x) \leqslant \lambda g(x)$ for $0 \leqslant \lambda \leqslant 1$). However, the situation can be saved by replacing the maximum likelihood estimator by an isotonic estimator. The question of the consistency of isotonic estimators has been explored further by Hanson et al. [11] and Robertson and Wright [18].

HYPOTHESIS TESTING

The most thoroughly investigated hypothesis testing problem concerns the case of k normal means—the one-way analysis of variance. Suppose that there are k independent means $\bar{y}_1, \bar{y}_2, \ldots, \bar{y}_k$ based on sample sizes $n_1, n_2, \ldots, n_k$ and where the ith mean is $N(\mu_i, \sigma^2/n_i)$ with σ^2 known. The problem is to test the null hypothesis $\mu_1 = \mu_2 = \cdots = \mu_k$ against alternative hypotheses imposing certain order restrictions on the μ's. The most important case, which we shall describe here, is that of simple order when $\mu_1 \leqslant \mu_2 \leqslant \cdots \leqslant \mu_k$. In that case the likelihood ratio test rejects the null hypothesis when

$$\bar{\chi}_k^2 = \sum_{i=1}^{k} n_i(\mu_i^* - \bar{y})^2/\sigma^2 \quad (7)$$

is significantly large. The maximum likelihood estimators μ_i^* are obtained using $\{n_i/\sigma^2\}$ as weights; $\bar{y}$ is the grand mean. The null hypothesis distribution is given by ref. 2 [Theorem 3.1] as

$$\Pr\{\bar{\chi}_k^2 \geqslant C\} = \sum_{l=2}^{k} P(l,k)\Pr\{\chi_{l-1}^2 \geqslant C\}$$

$$(C > 0), \quad (8)$$

where χ_v^2 denotes a chi-squared* random variable with v degrees of freedom. The quantity $P(l,k)$ is the probability that the amalgamation process involved in the estimation leads to exactly l distinct values for the μ^*'s. If the weights are equal, the determination of these probabilities is a combinatorial problem and the answer does not depend on the distribution of the y's. In that case

$$P(l,k) = |S_k^l|/k! \qquad (l = 1, 2, \ldots, k),$$
$$(9)$$

where $|S_k^l|$ is the coefficient of z^l in $z(z+1)$ $(z+2) \cdots (z+k-1)$ (S_k^l is thus the Stirling number of the first kind). A table for $k \leqslant 12$ is in ref. 2 [Table A5]. If the weights are not equal (or, in general, if the order is not simple), the $P(l,k)$'s depend on the form of the distribution of the y's. In the normal case their determination involves integrals of the multivariate normal* distribution. Tables of percentage points of $\bar{\chi}_k^2$ are given in ref. 2 [Tables A2 and A3] for $k \leqslant 12$ when the weights are equal and for $k \leqslant 4$ otherwise. A method of approximation for $k > 4$ has been given by Siskind [19].

When σ^2 is unknown the likelihood ratio statistic is

$$\bar{E}_k^2 = \sum_{i=1}^{k} n_i(\hat{\mu}_i^* - \bar{y})^2 \bigg/ \sum_{i=1}^{k} \sum_{j=1}^{n_i} (y_{ij} - \bar{y})^2$$
$$(10)$$

for which

$$\Pr\{\bar{E}_k^2 \geqslant C\}$$

$$= \sum_{l=2}^{k} P(l,k)\Pr\{\beta_{(1/2)(l-1),(1/2)(N-l)} \geqslant C\}$$
$$(C > 0), \quad (11)$$

where $\beta_{(1/2)(l-1),(1/2)(N-l)}$ is a beta random variable and N is the total sample size. A short table of critical values is given in ref. 2 [Table A4] and a much fuller table in Nelson [16]. Algorithms that can be used for calculating the probability integral are given by Bohrer [4], Bohrer and Chow [5], and Bremner [6].

The power of the isotonic tests has been compared with that of likelihood ratio tests* which ignore the order information and with other tests which use the order information

in different ways [2, Chap. 3; 13, 21]. It appears that it is only possible to improve the isotonic test when further information is available about the spacing of the μ's.

Similar results apply for partial orders. The case where many treatments are compared with a control has been treated in detail by Chase [8].

If the y's are not normal the exact distribution of the likelihood ratio statistic is not known in general. In certain cases, however (see ref. 2 [p. 191, esp. Theorem 4.4]), its distribution may be approximated by the expression given in (8).

Distribution-free versions of the test for equal means have been investigated. Roughly speaking, isotonic tests can be based on ranks in much the same way as the rank equivalents of the usual normal theory tests. Their asymptotic relative efficiency is usually the same as in the case of tests taking no account of order.

Isotonic tests can be developed for the general linear model* where order restrictions exist among main effects* or interactions* (see, e.g., Hirotsu [12]). Robertson and Wegman [17] have extended the theory by treating $H_1: \mu_1 \leqslant \mu_2 \leqslant \cdots \leqslant \mu_k$ as the null hypothesis with alternative $H_2: \mu_1 \neq \mu_2 \neq \cdots \neq \mu_k$. In the normal case with known variances the significance of the likelihood ratio test statistic, T_{12}, can be judged using

$$\Pr\{T_{12} \geqslant C \mid H_1\} \leqslant \sum_{l=1}^{k-1} P(l,k)\Pr\{\chi_{k-l}^2 \geqslant C\}.$$
$$(12)$$

Their results cover the case of unknown, but equal variances and of nonnormal distributions belonging to exponential families. In the latter case (12) serves as an approximation valid as n, the number of members in each group, tends to infinity. An example, based on the Poisson distribution, and tables of significance levels are included.

SIMULTANEOUS INFERENCE*

If the null hypothesis of equality of means (or other parameters) is rejected, the question arises as to whether all or only some of the means differ. The isotonic approaches to this question follow closely the methods used in the absence of order restrictions.

The first approach [15, 20] provides multiple testing procedures for discovering for which pairs of means the hypothesis of equality can be ruled out. Marcus et al. [15] proposed a method based on $\overline{\chi}_k^2$ (or $\overline{E}_k^2$) as follows. If the $\overline{\chi}_k^2$ test applied to all the means rejects the null hypothesis, the most we can assert is that $\mu_1 < \mu_k$. At the second stage, therefore, we consider subsets of the means. For example, if $k = 5$, we can split the sample into two groups $(1, 2)$, $(3, 4, 5)$, say, and test the null hypothesis H_0: $\mu_1 = \mu_2$; $\mu_3 = \mu_4 = \mu_5$ against the alternative H_1: $\mu_1 \leqslant \mu_2$; $\mu_3 \leqslant \mu_4 \leqslant \mu_5$. The likelihood ratio test in this case is the sum of the statistics appropriate for each subgroup. Its distribution can be found by general methods given in [2]. If this hypothesis is rejected, we can infer that either $\mu_1 < \mu_2$ or $\mu_3 < \mu_5$. The authors specify a sequence of such hypotheses to be tested (termed "closed") and a choice of significance levels which ensure a chosen experiment-wise error rate. A table of significance levels is provided for use when the variances are known and the sample sizes equal.

Spjøtvoll [20] approaches the problem in the same spirit as the Newman–Keuls* method by proposing a sequence of tests to be made on pairs of means.

The second approach provides simultaneous confidence intervals for certain kinds of contrasts among the means. Suppose that the basis for ordering the means is the existence of an underlying variable x such that $E(Y \mid x)$ is a nondecreasing function. The values of Y for the ith group correspond to $x = x_i$, where $x_1 \leqslant x_2 \leqslant \cdots \leqslant x_k$. The linear regression coefficient* of $\mu_i = E(Y \mid x = x_i)$ on x_i $(i = 1, 2, \ldots, k)$ is

$$B = \sum_{i=1}^{k} n_i \mu_i (x_i - \overline{x}) \bigg/ \sum_{i=1}^{k} n_i (x_i - \overline{x})^2. \quad (13)$$

Suppose that we wish to find a confidence interval for B which includes the true value of the parameter whatever the x's provided only that they satisfy the order restriction. In

particular, we might wish to choose the x's after we have seen the data so as to make the regression linear. Since the scale of the x's is arbitrary, this problem is essentially the same as finding simultaneous confidence intervals for $\sum_{i=1}^{k} n_i \mu_i c_i$ valid for all c_i's satisfying $\sum_{i=1}^{k} n_i c_i = 0$; $c_1 \leqslant c_2 \leqslant \cdots \leqslant c_k$. Marcus and Peritz [14] give three methods. The first, and simplest, gives one-sided intervals of the form

$$\left[0, \frac{\sum n_i Y_i c_i}{\sum n_i c_i^2} + a_\alpha \right], \quad (14)$$

where a_α is the appropriate percentage point of the $\overline{\chi}_k^2$ distribution. This method makes no use of the ordering of the μ's, so two further, improved methods are given involving the isotonic estimators. They yield intervals of similar form.

An alternative, but very similar, method is given by Williams [22] for use when the sample sizes are equal but σ^2 need not be known. His method does not make use of the ordering of the μ's and it yields intervals of the form

$$\frac{\sum c_i Y_i}{\sum c_i^2} \pm \frac{1}{2} b_{(1/2)\alpha} \frac{\sum |c_i|}{\sum c_i^2}, \quad (15)$$

where $\hat{\sigma}$ is the usual estimator of σ and $b_{(1/2)\alpha}$ is the appropriate percentage point of the studentized distribution of the random variable

$$\max_{1 \leqslant r \leqslant n} \frac{1}{r} \sum_{i=1}^{r} Y_i - \min_{1 \leqslant r \leqslant n} \frac{1}{n - r + 1} \sum_{i=r}^{n} Y_i. \quad (16)$$

A good approximation for the distribution is available in Williams' paper.

Although these results on simultaneous inference apply to normal variables, we anticipate that similar procedures can be devised for other distributions of the exponential family for which the likelihood ratio statistics can be approximated by $\overline{\chi}_k^2$.

References

[1] Ayer, M., Brunk, H. D., Ewing, G. M., Reid, W. T., and Silverman, E. (1955). *Ann. Math. Statist.*, **26**, 641–647. (Treats the maximum likelihood esti-

mation of ordered parameters in bioassay. It contains the pool-adjacent-violators algorithm and the max-min formulas.)

[2] Barlow, R. E., Bartholomew, D. J., Bremner, J. M., and Brunk, H. D. (1972). *Statistical Inference under Order Restrictions*. Wiley, Chichester, England. (A fairly complete account of the subject up to 1971. It gives a full mathematical treatment but contains many examples. Includes tables of significance levels for the various tests, a bibliography, and Complements sections reviewing the literature.)

[3] Bartholomew, D. J. (1959). *Biometrika*, **46**, 36–48. (The first paper to give the likelihood ratio test, $\bar{\chi}^2$, and its null hypothesis distribution.)

[4] Bohrer, R. (1975). *Appl. Statist.*, **24**, 380–384.

[5] Bohrer, R. and Chow, W. (1978). Algorithm AS122, *Appl. Statist.*, **27**, 100–104. (This and the following algorithm by Bremner [6], can be used to find the probability integral of the $\bar{E}^2$ distribution.)

[6] Bremner, J. M. (1978). Algorithm AS123, *Appl. Statist.*, **27**, 104–109.

[7] Brunk, H. D. (1955). *Ann. Math. Statist.*, **26**, 607–616. (The first general treatment of the maximum likelihood estimation of ordered parameters.)

[8] Chase, G. R. (1974). *Biometrika*, **61**, 569–578.

[9] Cran, G. W. (1980). *Appl. Statist.*, **29**, 209–211. (Algorithm for calculating the isotonic regression for simple order.)

[10] Cran, G. W. (1981). *Appl. Statist.*, **30**, 85–91. (Algorithm for calculating the probabilities $P(l, k)$ for the simple order alternative.)

[11] Hanson, D. L., Pledger, G., and Wright, F. T. (1973). *Ann. Statist.*, **1**, 401–421. (A fundamental paper on the properties of estimators.)

[12] Hirotsu, C. (1978). *Biometrika*, **65**, 561–570. (If interaction effects arising in the analysis of variance can be ordered, the problem of testing their significance can be expressed as an isotonic inference problem. This paper develops the theory and gives an example.)

[13] Marcus, R. (1976). *Biometrika*, **63**, 177–183.

[14] Marcus, R. and Peritz, E. (1976). *J. R. Statist. Soc. B*, **38**, 157–165. (A basic paper on the extension of isotonic inference ideas to simultaneous confidence bounds.)

[15] Marcus, R., Peritz, E., and Gabriel, K. R. (1976). *Biometrika*, **63**, 655–660.

[16] Nelson, L. S. (1977). *Biometrika*, **64**, 335–338. (The most extensive table of significance levels for the $\bar{E}^2$-test, superseding Table A4 of ref. 2.)

[17] Robertson, T. and Wegman, E. J. (1978). *Ann. Statist.*, **6**, 485–505.

[18] Robertson, T. and Wright, F. T. (1975). *Ann. Statist.*, **3**, 334–349. (An extension of the basic work in Hansen et al. [11].)

[19] Siskind, V. (1976). *Biometrika*, **63**, 647–654. (Provides approximations to the probability distribution of $\bar{\chi}^2$ for use when the weights are not equal.)

[20] Spjøtvoll, E. (1977). *Biometrika*, **64**, 327–334.

[21] Williams, D. A. (1971). *Biometrics*, **27**, 103–117.

[22] Williams, D. A. (1977). *Biometrika*, **64**, 9–14.

(DISTRIBUTION-FREE METHODS
GENERAL LINEAR MODEL
LIFE TESTING
LIKELIHOOD RATIO TESTS
MAXIMUM LIKELIHOOD ESTIMATION
MULTIDIMENSIONAL SCALING
ORDER STATISTICS
QUADRATIC PROGRAMMING
REGRESSION
RELIABILITY
SIMULTANEOUS INFERENCE)

D. J. Bartholomew

ISOTONIC REGRESSION   *See* ISOTONIC INFERENCE; REGRESSION

ISOTROPIC DISTRIBUTIONS

DEFINITION AND PROPERTIES

An isotropic distribution is a special type of multidimensional distribution which possesses some of the properties of the spherical multivariate normal distribution*. These distributions, often called spherically symmetric distributions, are characterized by their invariance under orthogonal transformation. To define isotropic distributions, suppose that X is a p-dimensional random vector with coordinates $X_1, \ldots, X_p$. Thus X takes values in R^p (p-dimensional Euclidean space).

Definition. The random vector X (and its distribution) is *isotropic* if for all $p \times p$ orthogonal matrices Γ, X and ΓX have the same distribution.

Example 1. If X has a probability density function (PDF) f which can be written as

$$f(x) = h(\|x\|), \qquad x \in R^p, \qquad (1)$$

where $\|x\|$ is the length of the vector x, then X is isotropic. In particular, if $X_1, \ldots, X_p$ are i.i.d. $N(0, \sigma^2)$, then the PDF of X has the form (1) and X has a spherical multivariate normal distribution.

Example 2. Let $\mathscr{S}_{p-1} = \{x \mid \|x\| = 1, \ x \in R^p\}$, so $\mathscr{S}_{p-1}$ is the unit sphere in R^p. For any isotropic random vector X with $\Pr\{X = 0\} = 0$, let $U = X / \|X\|$. Then U takes values in $\mathscr{S}_{p-1}$ and U is isotropic since X is isotropic. Further, all isotropic random vectors with values in $\mathscr{S}_{p-1}$ have the same distribution. This distribution (which is surface Lebesgue measure normalized to be a probability distribution) is often called the *uniform distribution* on $\mathscr{S}_{p-1}$.

Here are some basic properties of isotropic random vectors.

1. If the mean of X exists, then $\mathscr{E}X = 0$. If the variance of X_1 exists, then the covariance matrix of X exists and is $\sigma^2 I_p$, where $\sigma^2 = \mathrm{var}(X_1)$ and I_p is the $p \times p$ identity matrix.

2. For $1 \leq q < p$, the marginal distribution of $X_1, \ldots, X_q$ is isotropic on R^q.

3. For $1 \leq q < p$, the conditional distribution of $X_1, \ldots, X_q$ given $X_{q+1}, \ldots, X_p$ is isotropic.

4. Assume that $P\{X = 0\} = 0$. For $1 \leq q < r < p$, the random variable

$$V = \frac{p - r}{q} \frac{\sum_1^q X_j^2}{\sum_{r+1}^p X_j^2}$$

has an F-distribution* with q and $p - r$ degrees of freedom.

5. If X and $\tilde{X}$ are isotropic and independent, then $X + \tilde{X}$ is isotropic.

6. Let $\eta(u) = \mathscr{E} \exp(iuX_1)$ be the characteristic function of X_1. Then the characteristic function of X is $\eta(\|t\|) = \mathscr{E} \exp(i\sum_1^p t_j X_j)$, $t \in R^p$.

7. For real numbers $a_1, \ldots, a_p$, the random variable $\sum_1^p a_j X_j$ has the same distribution as $(\sum_1^p a_j^2)^{1/2} X_1$.

A discussion of these and other properties of isotropic distributions can be found in Kelker [9], Kariya and Eaton [8], Eaton [3], and Cambanis et al. [3].

A common method of constructing isotropic X's goes as follows. Let U be uniform on $\mathscr{S}_{p-1}$ and let R be a nonnegative random variable independent of U. Then $X = RU$ is isotropic and $\|X\| = R$. In fact, every isotropic distribution can be so represented. This can be used to prove property 4 and to establish other representations involving isotropic distributions [3].

SCALE MIXTURES OF NORMALS*

An interesting class of isotropic distributions can be constructed from the normal distribution as follows. Suppose that Z is a p-dimensional random vector having i.i.d. $N(0, 1)$ coordinates and suppose that the positive random variable R is independent of Z. Then $X = RZ$ is isotropic and X_1 has a density on R^1 given by

$$g(u) = \int_0^\infty r^{-1} \phi(u/r) G(dr), \qquad (2)$$

where G is the distribution function of R and $\phi(u) = (2\pi)^{-1/2} \exp[-\frac{1}{2} u^2]$ is the density of a $N(0, 1)$ distribution. Densities of the form (2) are called *scale mixtures of normals** and G is called the *mixing distribution*. These distributions are discussed in Teichrow [14], Kelker [10], Andrews and Mallows [1], and Efron and Olshen [5]. Andrews and Mallows [1] contains some particularly interesting examples.

If a distribution on R^1 has a density g given by (2), then the random vector $X = RZ$ as defined above has the property that X_1 (and hence each coordinate of X) has g as its marginal density. Conversely, let $\tilde{g}$ be a given density on R^1 and suppose that for each $p = 2, 3, \ldots$, there exists an isotropic p-dimensional random vector X such that X_1 has $\tilde{g}$ as its density. Then $\tilde{g}$ can be written in the form (2). In other words, scale mixtures of normals are the only distributions on R^1 which have the property that for each dimension p, there is an isotropic X with X_1 having the specified distribution on R^1. A number of different proofs of this characterization exist—see Schoenberg [13], Freedman [6], Kingman [11], and Eaton [3].

ROBUSTNESS* CONSIDERATIONS

As mentioned earlier, isotropic distributions possess some of the properties of the spherical normal distribution—in particular, property 4. This suggests that certain properties of statistical procedures known to hold under a normality assumption may continue to hold under the weaker isotropic assumption. An illuminating discussion of such issues related to Student's t-statistic can be found in Efron [4]. The following example is from classical linear regression* theory (see Scheffé [12]).

Consider a standard linear model

$$Y = A\beta + \epsilon \tag{3}$$

where the random observable Y is an n-vector, A is a known $n \times k$ matrix of rank k, β is a $k \times 1$ vector of unknown parameters, and ϵ is a random vector of errors which is assumed to have a mean of zero and a covariance matrix $\sigma^2 I_n$ with σ^2 unknown. The least-squares* estimate for β is $\hat{\beta} = (A'A)^{-1}A'Y$ and the usual estimate of σ^2 is $\hat{\sigma}^2 = (n - k)^{-1}\|Y - A\hat{\beta}\|^2$. Let C be an $s \times k$ matrix of rank s. To test the null hypothesis $H_0: C\beta = 0$, the ratio

$$V = \frac{s^{-1}(C\hat{\beta})'\left[C(A'A)^{-1}C'\right]^{-1}(C\hat{\beta})}{\hat{\sigma}^2}$$

is commonly used. When the error vector ϵ in (3) has a spherical normal distribution and H_0 is true, then V has an F-distribution with s and $n - k$ degrees of freedom. Exactly the same result obtains under the weaker distributional assumption that $P\{\epsilon = 0\} = 0$ and ϵ is isotropic. However, when H_0 is not true, the distribution of V will not be the same under normality as under an arbitrary isotropic distribution. In a Bayesian* context, Hill [7] has argued that

isotropic error distributions, as opposed to normal error distributions, form the basis for the theory of least squares in linear models.

Results of the type just described are called robustness results. A discussion of the case at hand and other robustness questions can be found in Kariya and Eaton [8]. Efron's [4] paper is a more leisurely introduction to this type of problem.

References

[1] Andrews, D. F. and Mallows, C. L. (1974). *J. R. Statist. Soc. B*, **36**, 99–102.

[2] Cambanis, S., Huang, S., and Simons, G. (1981). *J. Multivariate Anal.*, **11**, 368–385.

[3] Eaton, M. L. (1981). *Ann. Statist.*, **9**, 391–400.

[4] Efron, B. (1969). *J. Amer. Statist. Ass.*, **64**, 1278–1302.

[5] Efron, B. and Olshen, R. (1978). *Ann. Statist.*, **6**, 1159–1164.

[6] Freedman, D. (1963). *Ann. Math. Statist.*, **34**, 1194–1226.

[7] Hill, B. M. (1969). *J. R. Statist. Soc. B*, **31**, 89–97.

[8] Kariya, T. and Eaton, M. L. (1977). *Ann. Statist.*, **5**, 206–215.

[9] Kelker, D. (1970). *Sankhyā A*, **32**, 419–430.

[10] Kelker, D. (1971). *Ann. Math. Statist.*, **42**, 802–808.

[11] Kingman, J. F. C. (1972). *Biometrika*, **59**, 492–494.

[12] Scheffé, H. (1957). *The Analysis of Variance*. Wiley, New York.

[13] Schoenberg, I. J. (1938). *Ann. Math.*, **39**, 811–841.

[14] Teichroew, D. (1957). *Ann. Math. Statist.*, **28**, 510–512.

(BAYESIAN INFERENCE
EXCHANGEABILITY
MIXTURE DISTRIBUTIONS
REGRESSION
ROBUSTNESS)

MORRIS L. EATON

ISOTROPY

A term introduced by G. Yule*. A $s \times t$ contingency table* can be subdivided into groups of four adjacent frequency tables

$$(A_m B_n) \qquad (A_{m+1} B_n)$$
$$(A_m B_{n+1}) \qquad (A_{m+1} B_{n+1})$$

called *tetrads* or *fourfold tables*. [There are $(s-1)(t-1)$ tetrads in a $s \times t$ table.] If within each fourfold table the association* is of the same sign, the contingency table is called *isotropic*. The case of independence* is a special case of isotropy.

For additional information, see [1].

Reference

[1] Yule, G. and Kendall, M. G. (1953). *An Introduction to the Theory of Statistics*. Charles Griffin, London.

(ASSOCIATION, MEASURES OF CONTINGENCY TABLES)

ITEM ANALYSIS *See* PSYCHOLOGICAL TESTING THEORY

ITEM EXAMINE SAMPLING *See* MULTIVARIATE MATRIX SAMPLING

ITEM RESPONSE THEORY *See* PSYCHOLOGICAL TESTING THEORY

ITEM SAMPLING *See* PSYCHOLOGICAL TESTING THEORY

ITERATED EXPONENTIAL DISTRIBUTION *See* EXTREME-VALUE DISTRIBUTION

ITERATED MAXIMUM LIKELIHOOD ESTIMATES

In cases where the maximum likelihood* estimator (MLE) cannot be written in closed form, iterative methods (carried through to convergence or terminated after a specified number of steps) have been applied to calculate or approximate the MLE. Suppose that the vector of unknown parameters is θ and the likelihood function is $L(\theta)$. One group of iterative algorithms for maximizing $L(\theta)$ consists of *gradient methods* which start from

an initial point θ_0 and proceed to compute successive approximations to the estimate of θ according to

$$\theta_{p+1} = \theta_p - s_p Q_p \gamma_p,$$

where s_p is a scalar, Q_p is a square matrix which determines the direction of change from θ_p to θ_{p+1}, and γ_p is the gradient or column vector of first partial derivatives of $L(\theta)$ with respect to θ, evaluated at θ_p.

An iterative procedure of this type is the *Newton–Raphson* method, where Q_p is equal to the inverse of the Hessian matrix of L evaluated at θ_p. The Hessian of L, by definition, is the matrix of second-order partial derivatives of L with respect to θ. In this algorithm, the value for θ at the $(p+1)$st stage, with $s_p = 1$, can be interpreted as the solution to a linearization around θ_p of the likelihood equation $(\partial L / \partial \theta = 0)$:

$$0 \approx \partial L(\theta_{p+1}) / \partial \theta$$
$$\approx \partial L(\theta_p) / \partial \theta$$
$$+ \left[\partial^2 L(\theta_p) / \partial \theta \, \partial \theta' \right] (\theta_{p+1} - \theta_p)$$

or

$$0 \approx \gamma_{p+1} \approx \gamma_p + H_p (\theta_{p+1} - \theta_p),$$

where H_p denotes the Hessian of $L(\cdot)$ evaluated at θ_p.

A closely related two-step procedure is the *method of scoring*, where the inverse of the expected value of the Hessian matrix is used as the direction matrix Q_p. In particular, the first step requires the calculation of a consistent estimate $\tilde{\theta}$. This leads to the calculation, in the second step, of the estimate $\hat{\theta}$ as

$$\hat{\theta} = \tilde{\theta} - \left[EH(\tilde{\theta}) \right]^{-1} \left[\gamma(\tilde{\theta}) \right],$$

where $\gamma(\tilde{\theta})$ and $EH(\tilde{\theta})$ are the gradient and the expected value of the Hessian of the likelihood function L, both evaluated at $\tilde{\theta}$. Under regularity conditions (e.g., Rao [30]) this procedure is asymptotically equivalent to MLE. Similarly, the Newton–Raphson method* started with a consistent estimate of θ would lead to estimates, θ_p, at the pth iteration which are asymptotically equivalent to the MLE.

Other iterative algorithms such as steepest ascent*, quadratic hill climbing, Davidon–

Fletcher–Powell, conjugate gradient, the Gauss–Newton, and Marquardt are discussed in numerous references, such as Judge et al. [25] and Goldfeld and Quandt [15]. Berndt et al. [7] also consider the use of the gradient vector of L multiplied by its transpose to form the matrix Q_p.

In cases where the algorithms described above converge, in general, they do not necessarily converge to the global maximum. Also, no general ranking of the algorithms is universally applicable since there is such a wide variety of estimation problems which can be addressed. For a summary of some performance comparisons, see Judge et al. [25], Bard [5, 6], and Goldfeld and Quandt [15].

For some problems, it may be possible to write the likelihood equation ($\partial L / \partial \theta = 0$) in the following form, where $\theta = (\alpha, \beta)$:

$$\alpha = f(\beta) \quad \text{and} \quad \beta = h(\alpha).$$

In these cases the following iterative scheme can be used to calculate the MLE, starting from an initial estimate β_0:

$$\alpha_{p+1} = f(\beta_p) \quad \text{and} \quad \beta_{p+1} = h(\alpha_{p+1}).$$

Convergence properties of such algorithms will have to be investigated on a case-to-case basis. The following are some examples where such algorithms can be utilized for MLE calculation.

1. Given a random sample $(x_1, x_2, \ldots, x_n)$ from the gamma distribution* with density

$$[(x - \gamma)^{\alpha-1} e^{-(x-\gamma)/\beta}] / [\beta^\alpha \Gamma(\alpha)];$$

$$x > \gamma, \alpha > 0, \beta > 0,$$

one obtains the following equations for the maximum likelihood estimates $\hat{\alpha}$, $\hat{\beta}$, $\hat{\gamma}$ (the summations all range over i, from 1 to n):

$$\sum \log(x_i - \hat{\gamma}) - n \log \hat{\beta} - nd \log \Gamma(\hat{\alpha}) / d\alpha$$

$$= 0, \qquad (1)$$

$$\sum (x_i - \hat{\gamma}) / \hat{\beta}^2 - n\hat{\alpha}/\hat{\beta} = 0, \quad (2)$$

$$-(\hat{\alpha} - 1) \sum (x_i - \hat{\gamma})^{-1} + n/\hat{\beta} = 0. \quad (3)$$

One can then solve the foregoing system

iteratively as follows:

a. Given $\hat{\alpha}$ and $\hat{\gamma}$, use (1) to find $\hat{\beta}$.
b. Given $\hat{\alpha}$ and $\hat{\beta}$, use (2) to calculate $\hat{\gamma}$.
c. Given $\hat{\beta}$ and $\hat{\gamma}$, use (3) to calculate $\hat{\alpha}$.

Convergence to MLE follows from the convexity of the likelihood function. For more details, see Johnson and Kotz [21].

2. For the quartic exponential distribution* considered by Matz [28], the likelihood equations for the symmetric case reduce to the form described above. This reduction, however, does not occur for the more general (nonsymmetric) quartic exponential. In this case, Matz [28] proposes a Newton–Raphson method after simplification of the likelihood equations*. In the multivariate exponential distribution* arising from fatal shock models* discussed by Marshall and Olkin [27] and Proschan and Sullo [29], the likelihood equation simplifies to the form

$$\theta = f(\theta). \qquad (4)$$

This immediately suggests the following algorithm (Gauss–Seidel*) of successive approximation

$$\theta_{p+1} = f(\theta_p).$$

See Proschan and Sullo [29] for more details. In the estimation of linkage disequilibrium in randomly mating populations, Hill [20] and Weir and Cockerham [31] point out that the likelihood equation also simplifies to the form (4). In this case, the likelihood equation also reduces to a cubic equation in θ, and Weir and Cockerham [31] suggest solving this cubic equation directly to avoid problems of nonconvergence or convergence to the wrong root.

3. The *Cochrane–Orcutt* procedure developed for the normal linear regression model with first-order autoregressive disturbances iterates back and forth between the regression coefficient estimates on the one hand, and the estimate of the autocorrelation coefficient on the other. This method is asymptotically

equivalent to MLE; it is not exactly MLE because the first observation is not fully utilized. Johnston [22] contains further details concerning this procedure, the exact MLE, and other estimation methods asymptotically equivalent to MLE.

4. Full-information maximum likelihood (FIML) estimation of the *classical normal linear simultaneous equations model*

$$A\mathbf{y}^{(t)} + B\mathbf{x}^{(t)} = u^{(t)}, \quad t = 1, 2, \ldots, T, \quad (5)$$

can also be calculated in an iterative fashion. This model is commonly encountered in *econometrics**. Here $y^{(t)}$ is the $n \times 1$ vector of endogenous or jointly dependent variables, $x^{(t)}$ is the $m \times 1$ vector of exogenous variables, and A and B are, respectively, $n \times n$ and $n \times m$ matrices of structural coefficients. The disturbance terms $u^{(t)}$ are assumed to be independently and identically distributed as multivariate normal with mean zero and covariance matrix Σ. The structural parameters are the unknown elements of A, B, and Σ. Note that identifying restrictions on the system would require in general some of the elements of A and B to be equal to zero. Full-information* maximum likelihood estimation of this model maximizes the likelihood function for $\{y^{(t)}; t = 1, 2, \ldots, T\}$ subject to the identifying restrictions.

Let

$$U' = (u^{(1)}, u^{(2)}, \ldots, u^{(T)}),$$

$$Y' = (y^{(1)}, y^{(2)}, \ldots, y^{(T)}),$$

$$X' = (x^{(1)}, x^{(2)}, \ldots, x^{(T)}),$$

$$y_j = j\text{th column of } Y$$

$$= \text{vector of } T \text{ observations for the}$$

$$jth \text{ endogenous variable},$$

$$u_j = j\text{th column of } U.$$

Introducing the identifying restrictions, we can write the jth equation in (5) as

$$y_j = Y_j\alpha_j + X_j\beta_j u_j = Z_j\delta_j + u_j \,;$$

$$\delta' = (\alpha'_j, \beta'_j), \quad Z_j = (Y_j X_j).$$

Here (y_j, Y_j) and X_j are the submatrices of Y and X corresponding to the endogenous and exogenous variables appearing in the jth equation. We can further write the whole system of (5) in "stacked" form as

$$y = Z\delta + u,$$

where

$$y' = (y'_1, y'_2, \ldots, y'_n),$$

$$u' = (u'_1, u'_2, \ldots, u'_n),$$

$$\delta' = (\delta'_1, \delta'_2, \ldots, \delta'_n),$$

$$Z = \begin{bmatrix} Z_1 & 0 & 0 \\ 0 & Z_2 & 0 \\ \vdots & \vdots & \vdots \\ 0 & 0 & Z_n \end{bmatrix} = \text{diag}(Z_1, Z_2, \ldots, Z_n).$$

An iterative algorithm for FIML is as follows, proceeding from a given estimate $\delta_{(p)}$ at the pth iteration:

$$\sigma_{ij(p)} = [y_i - Z_i\delta_{i(p)}]'[y_j - Z_j\delta_{j(p)}]/T,$$

$$\Sigma_{(p)} = ((\sigma_{ij(p)})),$$

$$\delta_{(p+1)} = \left[\hat{Z}'(\Sigma_{(p)}^{-1} \otimes I)Z\right]^{-1}\left[\hat{Z}'(\Sigma_{(p)}^{-1} \otimes I)y\right],$$

where

$$\hat{Z} = \text{diag}(\hat{Z}_1, \ldots, \hat{Z}_n),$$

$$\hat{Z}_j = (\hat{Y}_j, X_j)$$

and the $\hat{Y}_j$ come from the solution values of the estimated system (5) based on the estimate $\delta_{(p)}$. For a further discussion of this method and others asymptotically equivalent to it, see, e.g., Maddala [26] and Brundy and Jorgenson [9].

Iterative procedures have also been developed for the maximum likelihood estimation of more complicated regression models. For example, such methods are discussed in Box and Jenkins [8], Anderson [4], and Fuller [13] for autoregressive–moving average* processes. Procedures have been suggested by Amemiya [2] and Jorgenson and Laffont [24] for nonlinear simultaneous systems; by Hatanaka [17], Fair [12], and Hendry [19] for simultaneous systems with lagged endogenous variables and/or autocorrelated disturbances; by Zellner [32], Goldberger [14], Chamberlain and Griliches [10], and Jöreskog and Goldberger [23] for simultaneous-equations systems with unobservable components; and by Amemiya [3] and Heckman [18] for simultaneous systems with qualitative or limited dependent variables. Further references are cited in Maddala [26] and Judge et al. [25]. Also, in connection with systems containing unobservable components, *see* LISREL.

In models where the observations can be viewed as incomplete data, the EM (expectation–maximization) algorithm, discussed extensively by Dempster et al. [11], provides a general approach to the iterative computation of maximum likelihood estimates. For more details concerning this algorithm, *see* INCOMPLETE DATA *and* ITERATIVELY REWEIGHTED LEAST SQUARES. The direct application of this procedure varies in difficulty from case to case. The numerous applications discussed and referenced in Dempster et al. [11] include missing value situations; grouped*, censored*, and truncated* data; variance-component* models; factor analysis*; and finite mixture models. Two recent papers containing applications of the EM algorithm are Hartley and Swanson [16] and Aitkin and Wilson [1].

References

[1] Aitkin, M. and Wilson, G. T. (1980). *Technometrics*, **22**, 325–331.

[2] Amemiya, T. (1977). *Econometrica*, **45**, 955–968.

[3] Amemiya, T. (1978). *Econometrica*, **46**, 1192–1206.

[4] Anderson, T. W. (1971). *The Statistical Analysis of Time Series*. Wiley, New York.

[5] Bard, Y. (1970). *SIAM J. Numer. Anal.*, **7**, 157–186.

[6] Bard, Y. (1974). *Nonlinear Parameter Estimation*. Academic Press, New York.

[7] Berndt, E., Hall, B., Hall, R., and Hausman, J. (1974). *Ann. Econ. Social Meas.*, **3**, 653–665.

[8] Box, G. E. P. and Jenkins, G. M. (1970). *Time Series Analysis: Forecasting and Control*, Holden-Day, San Francisco.

[9] Brundy, J. M. and Jorgenson, D. W. (1971). *Rev. Econ. Statist.*, **53**, 207–224.

[10] Chamberlain, G. and Griliches, Z. (1975). *Int. Econ. Rev.*, **16**, 422–449.

[11] Dempster, A. P., Laird, N. M., and Rubin, D. B. (1977). *J. R. Statist. Soc. B*, **39**, 1–38.

[12] Fair, R. (1972). *Rev. Econ. Statist.*, **54**, 444–449.

[13] Fuller, W. A. (1976). *Introduction to Statistical Time Series*. Wiley, New York.

[14] Goldberger, A. S. (1972). *Int. Econ. Rev.*, **13**, 1–15.

[15] Goldfeld, S. and Quandt, R. (1972). *Nonlinear Methods in Econometrics*. North-Holland, Amsterdam.

[16] Hartley, M. and Swanson, E. (1980). Maximum Likelihood Estimation of the Truncated and Censored Normal Regression Models. Unpublished report for the World Bank.

[17] Hatanaka, M. (1976). *J. Econometrics*, **4**, 189–204.

[18] Heckman, J. (1978). *Econometrica*, **46**, 931–960.

[19] Hendry, D. (1971). *Int. Econ. Rev.*, **12**, 257–272; *ibid.*, **15**, 260 (corrections).

[20] Hill, W. G. (1974). *Heredity*, **33**, 229–239.

[21] Johnson, N. L. and Kotz, S. (1970). *Distributions in Statistics: Continuous Univariate Distributions*, Vol. 1. Wiley, New York.

[22] Johnston, J. (1972). *Econometric Methods*, 2nd ed. McGraw-Hill, New York.

[23] Jöreskog, K. G. and Goldberger, A. S. (1975). *J. Amer. Statist. Ass.*, **70**, 631–639.

[24] Jorgenson, D. W. and Laffont, J. J. (1974). *Ann. Econ. Social Meas.*, **3**, 615–640.

[25] Judge, G. G., Griffiths, W. E., Hill, R. C., and Lee, T. S. (1980). *The Theory and Practice of Econometrics*. Wiley, New York.

[26] Maddala, G. S. (1977). *Econometrics*. McGraw-Hill, New York.

[27] Marshall, A. W. and Olkin, I. (1967). *J. Amer. Statist. Ass.*, **62**, 30–44.

[28] Matz, A. W. (1978). *Technometrics*, **20**, 475–484.

[29] Proschan, F. and Sullo, P. (1976). *J. Amer. Statist. Ass.*, **71**, 465–472.

[30] Rao, C. R. (1973). *Linear Statistical Inference and Its Applications*, 2nd ed. Wiley, New York.

[31] Weir, B. S. and Cockerham, C. C. (1979). *Heredity*, **42**, 105–111.

[32] Zellner, A. (1970). *Int. Econ. Rev.*, **11**, 441–454.

Acknowledgment

Partial support is gratefully acknowledged from the following grants to the University of Pennsylvania: National Science Foundation Grant SOC 79-07964 (Department of Economics) and the Oak Ridge National Laboratory Contract 7954 (Wharton Analysis Center for Energy Studies).

(ECONOMETRICS
FULL-INFORMATION ESTIMATORS
GENERAL LINEAR MODEL
MAXIMUM LIKELIHOOD ESTIMATION
REGRESSION ANALYSIS
STRUCTURAL ANALYSIS)

ROBERTO S. MARIANO

ITERATIVE ELLIPSOIDAL TRIMMING (IET)

The IET algorithm is a method for discovering clusters* (or subpopulations) in a given population to study their properties. It was originally devised as an algorithm for robust estimation of covariance matrices [1, 3, 4], and has only recently been utilized by Gillick [2] for discovering clusters.

Given n observations in R^k: $\mathbf{X}_1, \ldots, \mathbf{X}_n$, we start the algorithm with some initial estimates of the mean and covariance $(\tilde{\mu}_0, \tilde{\Sigma}_0)$ of the cluster being sought. (For example, $\tilde{\mu}_0 = \overline{\mathbf{X}}$ and $\hat{\Sigma}_0 = \mathbf{I}_k$, the $k \times k$ identity matrix.) To carry out an iteration one specifies a value of p representing the proportion of the observations to be included in the computation of the next estimate $(\tilde{\mu}, \tilde{\Sigma})$, and then calculates the Mahalanobis distance* D_i^2 of each X_i from the initial estimator $\tilde{\mu}_0$:

$$D_i^2 = (\mathbf{X}_i - \tilde{\mu}_0)' \Sigma_0^{-1} (\mathbf{X}_i - \tilde{\mu}_0).$$

The first iterated estimates are calculated as

$$\tilde{\mu}_1 = [np]^{-1} \sum_{i \in L} \mathbf{X}_i$$

and

$$\hat{\Sigma}_1 = [np]^{-1} \sum_{i \in L} (\mathbf{X}_i - \tilde{\mu}_0)(\mathbf{X}_i - \tilde{\mu}_0)',$$

where $L = \{i : D_i^2 \leq D_{([np])}^2\}$, $D_{(r)}^2$ being the rth-order statistic* of D_i^2 and $[k]$ is the largest integer $\leq k$. The next iteration is performed by again choosing a value of p and treating $(\tilde{\mu}_1, \tilde{\Sigma}_1)$ as the new $(\tilde{\mu}_0, \tilde{\Sigma}_0)$. The value of p may be allowed to change or be kept constant; this decision is often made interactively. Note that IET defines the current cluster rather conservatively as those points within some Mahalanobis distance D^2 of the current $(\tilde{\mu}, \tilde{\Sigma})$. See the references cited below for more details. In Gillick's report [2] a comparison between the IET algorithm and the k-means algorithm* is presented.

For ellipsoidal clusters in high $(k > 2)$-dimensional Euclidean spaces, the IET algorithm seems to yield quite satisfactory results.

References

[1] Devlin, S. J., Gnanadesikan, R., and Kettenring, J. R. (1975). *Biometrika*, **62**, 531–545.

[2] Gillick, L. S. (1980). Iterative Ellipsoidal Trimming. *Tech. Rep. No. 15*, Dept. of Mathematics, Massachusetts Institute of Technology, Cambridge, Mass., Feb. 11.

[3] Gnanadesikan, R. (1977). *Methods for Statistical Data Analysis of Multivariate Observations*. Wiley, New York.

[4] Gnanadesikan, R. and Kettenring, J. R. (1972). *Biometrics*, **28**, 81–124.

(CLASSIFICATION
k-MEANS ALGORITHMS
MAHALANOBIS DISTANCE)

ITERATIVELY REWEIGHTED LEAST SQUARES

Iteratively reweighted least squares (IRLS) refers to an iterative procedure for estimating regression coefficients*: at each iteration, weighted least-squares* computations are performed, where the weights change from iteration to iteration. Although IRLS has been used to estimate coefficients in nonlinear and logistic regressions* (see, e.g., ref. 3a), currently, IRLS tends to be associated with robust regression.

IRLS FOR ROBUST REGRESSION

When using IRLS for robust regression*, the weights are functions of the residuals* from the previous iteration such that points with larger residuals receive relatively less weight than points with smaller residuals. Consequently, unusual points tend to receive less weight than typical points.

IRLS is a popular technique for obtaining estimated regression coefficients that are relatively unaffected by extreme observations. One reason for the popularity of IRLS is that it can be easily implemented using readily available least-squares algorithms. Another reason is that it can be motivated from sound statistical principles (see refs. 4 and 9). A third reason for its popularity is that some experience suggests it is a useful practical tool when applied to real data (see refs. 2 and 7). In order to define precisely IRLS for robust regression, some notation is needed.

WEIGHTED LEAST-SQUARES COMPUTATIONS

Let Y be an $n \times 1$ data matrix of n observations of a dependent variable, let X be the associated $n \times p$ data matrix of n observations of p predictor variables, and let W be an $n \times n$ diagonal matrix of nonnegative weights, which for the moment we assume is fixed. Then the weighted least-squares estimate of the regression coefficient of Y on X is given, as a function of W, by

$$b(W) = (X^TWX)^{-1}(X^TWY), \qquad (1)$$

if (X^TWX) has rank p and is not defined otherwise.

Theoretical justification for the estimator $b(W)$ is straightforward. Suppose that for fixed W, the conditional distribution of Y given X has mean $X\beta$, where β is the $p \times 1$ regression coefficient to be estimated, and variance $\sigma^2 W^{-1}$, where σ^2 is the residual variance, usually also to be estimated. By noting that, for fixed W, $W^{1/2}Y$ has mean $W^{1/2}X\beta$ and variance $\sigma^2 I$, the standard Gauss–Markov* arguments imply that $b(W)$ is the value of β that minimizes the residual sum of squares

$$(Y - X\beta)^TW(Y - X\beta)$$

as well as the minimum variance unbiased estimator of β. If the conditional distribution of Y given X is normal for fixed W, then $b(W)$ is also the maximum likelihood

estimate of β, and the associated maximum likelihood estimate of σ^2 is the weighted sum of squared residuals:

$$
\begin{aligned}
&s(W)^2 \\
&= [Y - Xb(W)]^TW[Y - Xb(W)]/n.
\end{aligned}
\qquad (2)
$$

IRLS is used when the weight matrix is not fixed. Specifically, IRLS applies (1) to obtain $b^{(l+1)}$, the $(l+1)$st iterate of the regression coefficient, from the weight matrix of the previous iteration:

$$b^{(l+1)} = b(W^{(l)}). \qquad (3)$$

To define a specific version of IRLS, we thus need only to define the weight matrix $W^{(l)}$.

THE WEIGHT MATRIX AND ITERATIONS FOR ROBUST REGRESSION

For robust regression, the ith diagonal element in the weight matrix $W^{(l)}$, $W_{ii}^{(l)}$, is a function $w(\cdot)$, of the ith standardized residual obtained by using $b^{(l)}$ to predict Y_i:

$$W_{ii}^{(l)} = w(z_i) = w(-z_i), \qquad (4)$$

where

$$z_i = (Y_i - X_i b^{(l)})/s^{(l)} \qquad (5)$$

and $s^{(l)}$ is the estimate of σ at the lth iteration. A natural form for $s^{(l)}$ based on likelihood criteria is given by (2) with $W^{(l-1)}$ substituted for W, and thus, by (3), with $b^{(l)}$ substituted for $b(W)$:

$$s^{(l)} = s(W^{(l-1)}). \qquad (6)$$

The scalar function $w(\cdot)$ in (4) is a nonnegative and nonincreasing monotone function and thus gives relatively smaller weight to points with larger residuals, e.g., $w(z) = 2/(1 + z^2)$.

With a specified form for $s^{(l)}$ and a specified form for the function $w(\cdot)$, IRLS proceeds by choosing a starting value $W^{(0)}$ e.g., the identity matrix, and then calculating $b^{(1)}$ from (1) and (3), $s^{(1)}$ from such as (2) and (6), and thence $W^{(1)}$ from (4) and (5); from

$W^{(1)}$, the next iterates $b^{(2)}$, $s^{(2)}$, and $W^{(2)}$ are calculated; the procedure can continue indefinitely unless some $s^{(l)} = 0$ or $X^T W^{(l)} X$ has rank less than p. Experience suggests that for many choices of weight functions, the iterations reliably converge.

STATISTICAL JUSTIFICATIONS FOR IRLS

A general statistical justification for IRLS for robust regression arises from the fact that it can be viewed as a process of successive substitution applied to the equations for *M*-estimates* [1, 2, 8–10]. Numerical behavior of IRLS for robust regression is considered in refs. 3, 6, 10, and 11.

A more specialized justification for IRLS, which is consistent with statistical principles of efficient estimation, arises from the fact that some *M*-estimates are maximum likelihood estimates under special distributional forms for the conditional distribution of *Y* given *X*. When *M*-estimates are maximum likelihood estimates, the associated IRLS algorithm is an EM algorithm [4, esp. pp. 19–20], and consequently, general convergence results about EM algorithms apply to IRLS algorithms; important results are that each step of IRLS increases the likelihood and, under weak conditions, IRLS converges to a local maximum of the likelihood function. Details of the relationship between IRLS and EM, including general results on large and small sample rates of convergence, are given in ref. 5.

IRLS/EM FOR THE *t*-DISTRIBUTION

A specific example when IRLS is EM occurs when the specification for the conditional distribution of Y_i given X_i is a scaled *t*-distribution* with *r* degrees of freedom. Then the associated weight function for IRLS is $w(z) = (r + 1)/(r + z^2)$, and the large-sample rate of convergence for IRLS is $3/(r + 3)$. More generally, if $d(z)$ is the probability density function specified for the conditional distribution of Y_i given X_i, then the associated weight function is defined by

$$w(z) = -d'(z)/zd(z) \qquad \text{for} \quad z \neq 0$$

$$= \lim_{z \to 0} -d'(z)/zd(z) \qquad \text{for} \quad z = 0.$$

A small numerical example is given in ref. 5 and summarized here. Ten observations were drawn from a *t*-distribution with 3 degrees of freedom (-0.141, 0.678, -0.036, -0.350, -5.005, 0.886, 0.485, -4.154, 1.415, 1.546). The results of 20 steps of IRLS starting from $W^{(0)} = I$ are given in Table 1. The empirical rate of convergence for both $b^{(l)}$ and $s^{(l)}$ at the 20th iteration is 0.6805, which agrees well with the theoretical small-sample rate of convergence of 0.6806 as calculated in ref. 5; the large-sample rate of convergence is 0.5. Since the rate of convergence of an EM algorithm is proportional to the fraction of information in the observed data (i.e., in *Y* and *X* in the robust regression context) relative to the information in the observed and missing data (i.e., in *Y*, *X*, and *W*), we see that in this example the observed data have relatively more information about β and σ than is typical for sam-

Table 1   Successive Iterations of IRLS

Iteration l	$\beta^{(l)}$	$\sigma^{(l)2}$
1	-0.467496	1.537750
2	0.103069	1.673303
3	0.240781	1.603189
4	0.277822	1.524210
5	0.292411	1.466860
6	0.300280	1.427958
7	0.305188	1.401828
8	0.308413	1.384252
9	0.310571	1.372393
10	0.312027	1.364371
11	0.313012	1.358934
12	0.313680	1.355244
13	0.314133	1.352738
14	0.314442	1.351035
15	0.314651	1.349876
16	0.314794	1.349088
17	0.314890	1.348552
18	0.314956	1.348188
19	0.315001	1.347939
20	0.315032	1.347771

ples from a t-distribution with three degrees of freedom. Further discussion of these points is given in ref. 5.

A MULTIVARIATE EXTENSION

A potentially quite useful and simple generalization of the use of IRLS/EM for the t-distribution has apparently not yet appeared in the literature and illustrates the flexibility of IRLS. Suppose that Y_i is q-variate and X_i is p-variate as before, where β is now $p \times q$, and let the conditional distribution of $Y_i - X_i\beta$ given X_i be a zero-centered multivariate t with r degrees of freedom [3b]. Then the previous notation and equations apply with the following simple modifications: $b(W)$ defined by (1) is now $p \times q$, $s(W)^2$ defined by (2) is now $q \times q$, and the weight function is given by

$$w(z_i) = (r + q)/(r + z_i^2), \qquad (7)$$

where at the lth iteration

$$z_i^2 = \left(Y_i - X_i b^{(l)} \right) \left[s^{(l)2} \right]^{-1} \left(Y_i - X_i b^{(l)} \right)^T. \qquad (8)$$

IRLS begins with a starting value, $W^{(0)}$, e.g., the identity matrix, calculates the $p \times q$ matrix $b^{(1)}$ from (1) and (3), the $q \times q$ matrix $s^{(1)2}$ from (2) and (6), and thence the $n \times n$ diagonal matrix $W^{(1)}$ from (4), (7), and (8); $W^{(1)}$ leads to the next iterates $b^{(2)}$, $s^{(2)2}$, etc.

Under the t-specification, IRLS is EM and so each iteration increases the likelihood of the $p \times q$ location parameter β and the $q \times q$ scale parameter σ^2, and under weak conditions, the iterations will converge to maximum likelihood estimates of β and σ^2. IRLS thus provides a positive-semidefinite estimate of the matrix of partial correlations among the q components of Y_i assuming that the conditional distribution of Y_i given X_i is elliptically symmetric and long tailed (if r is chosen to be small). Some limited experience with real data suggests that this use of IRLS does yield estimates of correlation matrices rather unaffected by extreme observations.

References

[1] Andrews, D. G., Bickel, P. J., Hampel, F. R., Huber, P. J., Rogers, W. H., and Tukey, J. W. (1972). *Robust Estimates of Location: Survey and Advances*. Princeton University Press, Princeton, N.J.

[2] Beaton, A. E. and Tukey, J. W. (1974). *Technometrics*, **16**, 147–185.

[3] Byrd, R. H. and Pyne, D. A. (1979). *Amer. Statist. Ass., Proc. Statist. Comp.*, pp. 68–71.

[3a] Cox, D. R. (1970). *The Analysis of Binary Data*, Methuen, London.

[3b] Dempster, A. P. (1969). *Elements of Continuous Multivariate Analysis*, Addison-Wesley, Reading, MA.

[4] Dempster, A. P., Laird, N. M., and Rubin, D. B. (1977). *J. R. Statist. Soc. B*, **39**, 1–38.

[5] Dempster, A. P., Laird, N. M., and Rubin, D. B. (1980). In *Multivariate Analysis V*, P. R. Krishnaiah, ed. North-Holland, Amsterdam, pp. 35–57.

[6] Dutter, R. (1977). *J. Statist. Comp. Simul.*, **5**, 207–238.

[7] Eddy, W. and Kadane, J. (1982). *J. Amer. Statist. Ass.*, **77**, 262–269.

[8] Holland, P. W. and Welsch, R. E. (1977). *Commun. Statist. A*, **6**, 813–827.

[9] Huber, P. J. (1964). *Ann. Math. Statist.*, **35**, 73–101.

[10] Huber, P. J. (1981). *Robust Statistics*. Wiley, New York.

[11] Klein, R. and Yohai, V. J. (1981). *Commun. Statist. A*, **10**, 2373–2388.

Acknowledgment

Sponsored by the U.S. Army under Contract DAAG29-80-C-0041.

DONALD B. RUBIN

ITERATIVE PROPORTIONAL FITTING

INTRODUCTION AND HISTORICAL REMARKS

The iterative proportional fitting procedure (IPFP) is a commonly used algorithm for maximum likelihood estimation* in log-linear models*. The simplicity of the algorithm and its relation to the theory of log-linear models make it a useful tool, especially for the analysis of cross-classified *categorical data** or *contingency tables**.

To illustrate the algorithm we consider a three-way table of independent Poisson counts, $\mathbf{x} = \{x_{ijk}\}$. Suppose that we wish to fit the log-linear model of no-three-factor interaction* for the mean $\mathbf{m}$, i.e., the model

$$\ln(m_{ijk}) = u + u_{1(i)} + u_{2(j)} + u_{3(k)}$$
$$+ u_{12(ij)} + u_{13(ik)} + u_{23(jk)}. \quad (1)$$

The basic IPFP takes an initial table $\mathbf{m}^{(0)}$ such that $\ln(\mathbf{m}^{(0)})$ satisfies the model (typically we would use $m_{ijk}^{(0)} = 1$ for all i, j and k) and sequentially scales the current fitted table to satisfy the three sets of two-way margins of the observed table, x. The vth iteration consists of three steps which form:

$$m_{ijk}^{(v,1)} = m_{ijk}^{(v-1,3)} \cdot x_{ij+} / m_{ij+}^{(v-1,3)},$$
$$m_{ijk}^{(v,2)} = m_{ijk}^{(v,1)} \cdot x_{i+k} / m_{i+k}^{(v,1)}, \quad (2)$$
$$m_{ijk}^{(v,3)} = m_{ijk}^{(v,2)} \cdot x_{+jk} / m_{+jk}^{(v,2)}.$$

(The first superscript refers to the iteration number, and the second to the step number within iterations. A subscript of $+$ indicates summation over the associated category.) The algorithm continues until the observed and fitted margins are sufficiently close. For a detailed discussion of convergence and some of the other properties of the algorithm, see Bishop et al. [2] or Haberman [16]. A FORTRAN implementation of the algorithm is given in Haberman [14, 15]. (See also the discussion of computer programs for log-linear models in CONTINGENCY TABLES.)

As a computational technique for adjusting tables of counts, the IPFP appears to have been first described by Kruithof [17] (see also Krupp [18]) and then independently formulated by Deming and Stephan [9]. They considered the problem of adjusting (or raking) a table, $\mathbf{n} = \{n_{ijk}\}$, of counts to satisfy some external information about the margins of the table. Deming [8, p. 107] gives an example of a cross-classification, by age and by state, of white persons attending school in New England. The population

cross-classification, $\mathbf{N} = \{N_{ijk}\}$, is unknown but the marginal totals are known. In addition, a sample $\mathbf{n}$ from the population is available. Deming and Stephan's aim was to find an estimate $\mathbf{N}$ which satisfies the marginal constraints and minimizes the χ^2-like distance,

$$\sum (N_{ij} - n_{ij})^2 / n_{ij}. \quad (3)$$

Their erroneous solution (see Stephan [20]) was the IPFP. Although the $\mathbf{N}$ produced by the IPFP need not minimize (3), it does provide an approximate and easily calculated solution.

Over 20 years after the work of Deming and Stephan, Darroch [5] implicitly used a version of the IPFP to find the maximum likelihood estimates in a contingency table but left the details of the general algorithm unclear. Bishop [1] was the first to show how the IPFP could be used to solve the maximum likelihood estimation problem in multidimensional tables. Some further history and other uses of the algorithm, including applications to *doubly stochastic matrices**, are discussed in Fienberg [10].

A COORDINATE-FREE VERSION OF THE IPFP

The basic IPFP is applicable to a class of models much more general than those described solely in terms of margins of a multiway table. Consider an index set $\mathscr{J}$ with J elements and let $\mathbf{x}$ be a table of observed counts which are realizations of independent Poisson* random variables with mean $\mathbf{m}$. Further, let $\mathscr{M}$ be a linear subspace of $\mathbf{R}^J$ with a spanning set $\{\mathbf{f}_k : k = 1, 2, \ldots, K\}$, where each $\mathbf{f}_k$ is a vector of zeros and ones. The calculation of the maximum likelihood estimate $\mathbf{m}$ for the log-linear model*

$$\ln(\mathbf{m}) \in \mathscr{M},$$

begins by taking a starting table $\mathbf{m}^{(0)}$ with $\ln(\mathbf{m}^{(0)}) \in \mathscr{M}$ ($\mathbf{m}^{(0)} = 1$ will always work), and sequentially adjusts the table to satisfy the "margins," i.e., $\langle \mathbf{f}_k, \mathbf{x} \rangle$ for $k = 1, 2, \ldots, K$, the inner products of the data with the spanning vectors. The vth cycle of the procedure takes the current estimate $\mathbf{m}^{(v-1,K)} = \mathbf{m}^{(v,0)}$ and forms

$$\mathbf{m}^{(v,k)} = \mathbf{m}^{(v,k-1)} \frac{\langle \mathbf{f}_k, \mathbf{x} \rangle \mathbf{f}_k}{\langle \mathbf{f}_k, \mathbf{m}^{(v,k-1)} \rangle}$$

$$+ \mathbf{m}^{(v,k-1)} \cdot (\mathbf{1} - \mathbf{f}_k),$$

$$k = 1, 2, \ldots, K \quad (4)$$

(i.e., adjusts the current fitted table so that the margin corresponding to $\mathbf{f}_k$ is correct) to yield $\mathbf{m}^{(v)} = \mathbf{m}^{(v,K)}$. The maximum likelihood estimate is $\lim_v \mathbf{m}^{(v)}$. If one wished to fit the log-affine model

$$\ln(\mathbf{m}) \in \mathbf{t} + \mathscr{M},$$

which is just the translation by $\mathbf{t}$ of the log-linear model $\mathscr{M}$, then using the IPFP with starting values which satisfy this model [e.g., $\mathbf{m}^{(0)} = \exp(\mathbf{t})$] leads to the MLE.

There are many ways to view this basic algorithm and many problems for which the IPFP is of especial use. Although the basic algorithm is limited to linear manifolds, $\mathscr{M}$, with zero–one spanning sets, it is possible to generalize the method to work with any linear manifold. We now look at some topics that relate to the algorithm or its generalizations.

SOME COMPUTATIONAL PROPERTIES

Common alternatives to the IPFP are versions of Newton's method or other algorithms which use information about the second derivatives of the likelihood function and automatically produce an estimate of the variance–covariance matrix of the parameters. While such methods have quadratic convergence properties compared to the linear properties of the IPFP and are often quite efficient (see, e.g., Chambers [3], Haberman [16], or Fienberg et al. [12]), they are of limited use for models of high dimensionality. For example, the model of no-three-factor interaction in a $10 \times 10 \times 10$ table has 271 parameters and this requires $\frac{1}{2} \times 271 \times 272 = 36,856$ numbers to represent the matrix of second derivatives. In contrast, the IPFP requires only about 300 numbers (i.e., the three marginal totals) in addition to the table itself. For many large contingency table problems the IPFP is the most reasonable computational method in use. Of course, for problems with only a small number of parameter Newton's method may be preferable, especially when the model is such that the basic IPFP is not applicable.

It is well known that the IPFP can often be slow to converge. Our experience is that it is generally restrictions on storage rather than computational time which limit an algorithm's usefulness. Thus slow convergence, although disturbing in some contexts, is not necessarily a crucial property.

As we have seen, the basic IPFP is very simple and requires little more than hand calculation. The simplicity of the algorithm allows one to understand and use the mechanics of the calculations to show theoretical results. A good example of this is the theory of decomposable models (models with closed-form estimates) as developed by Bishop et al. [2] or Haberman [16]. For every decomposable model there is an ordering of the margins such that the simple IPFP converges in one iteration.

One of the ideas underlying the IPFP is to sequentially equate a vector of expected values with the sufficient statistics of the model. The IPFP does this one dimension at a time, but there is no reason why several dimensions cannot be simultaneously adjusted. This idea underlies the estimation scheme for partially decomposable graphical models outlined in Darroch et al. [7]. They show that for many models it is possible to fit certain subsets of the marginal totals and to combine the resulting partial estimates using a direct formula.

GENERALIZATIONS OF THE IPFP

A limitation of the basic IPFP is that only certain types of models can be fit. We now consider several methods for extending the

IPFP to cover any log-linear model. For multinomial and Poisson data the problems of maximizing the likelihood function and minimizing the *Kullback–Leibler information* number* can be considered as dual problems which lead to the same estimates (*see* CONTINGENCY TABLES). We now consider generalizations of the IPFP from both these points of view.

Haberman [16] shows that, when viewed from the likelihood perspective, the IPFP is just a version of the cyclic coordinate ascent method of functional maximization. To illustrate Haberman's approach, we choose a fixed set of vectors which span the model space, $\mathcal{M}$, and then we maximize the likelihood along each of these directions in turn. Specifically, we consider a set of vectors $\mathcal{F} = \{\mathbf{f}_k : k = 1, 2, \ldots, K\}$ which span $\mathcal{M}$. If we denote the log-likelihood by $l(\mathbf{m} \mid \mathbf{x})$ and consider an initial estimate $\mathbf{m}^{(0)}$ with $\ln(\mathbf{m}^{(0)})$ in $\mathcal{M}$, then the algorithm proceeds by finding $\mathbf{m}^{(i)}$ such that

$$\ln(\mathbf{m}^{(i)}) = \ln(\mathbf{m}^{(i-1)}) + \alpha_i \mathbf{f}_k ;$$

$$i = k \bmod |K|,$$

where α_i is determined so as to increase the likelihood sufficiently. When $\mathbf{f}_k$ is a vector of zeros and ones

$$\alpha_i = \ln\big(\langle \mathbf{f}_k, \mathbf{x} \rangle / \langle \mathbf{f}_k, \mathbf{m}^{(i-1)} \rangle\big)$$

(i.e., the α_i corresponding to IPFP adjustment maximizes the likelihood in this direction). For arbitrary $\mathbf{f}_k$ there is no direct estimate of α_i and we are left with a one-dimensional maximization problem.

Csiszár [4] considers IPFP as a method for maximizing the Kullback–Leibler information between two probability distributions. When specialized to distributions on finite sets, Csiszár's methods yield a generalized IPFP. The class of algorithms that result from Csiszár's work are dual algorithms to the cyclic ascent methods, except now maximization can be over entire subspaces of $\mathcal{M}$ rather than just vectors. These methods yield powerful theoretical tools and have been instrumental in finding new algorithms which combine some of the advantages of

both Newton's method and the IPFP (see Meyer [19]).

The third generalization of the IPFP we consider is due to Darroch and Ratcliff [6]. This algorithm, known as generalized iterative scaling, was also developed from the information theory perspective, but is not closely related to Csiszár's method. The calculations are similar to those of the basic IPFP; a set of vectors $\mathcal{F}$ which span $\mathcal{M}$ is chosen and the likelihood is increased (but not maximized) in each of these directions in turn. Each iteration can require that the scaling factors be raised to arbitrary powers. These features combine to make the algorithm expensive, as it often takes many iterations to converge and each iteration is complicated.

For some problems it is possible to avoid the complications of the generalized IPFPs by transforming the contingency table into a form where the basic IPFP can be used (see Meyer [19] for details and Fienberg et al. [13] for some examples). This can result in a significant saving in the computational effort and recognition of some of the theoretical advantages (e.g., closed-form estimates) associated with the IPFP. Fienberg and Wasserman [11, Fig. 1] present an example where the convergence rate can be substantially improved by taking advantage of this transformation technique.

References

[1] Bishop, Y. M. M. (1967). Multidimensional Contingency Tables: Cell Estimates. Ph.D. thesis, Harvard University.

[2] Bishop, Y. M. M., Fienberg, S. E., and Holland, P. W. (1975). *Discrete Multivariate Analysis*. MIT Press, Cambridge, Mass.

[3] Chambers, J. M. (1977). *Computational Methods for Data Analysis*. Wiley, New York.

[4] Csiszár, I. (1975). *Ann. Prob.*, **3**, 146–158.

[5] Darroch, J. N. (1962). *J. R. Statist. Soc. B*, **24**, 251–263.

[6] Darroch, J. N. and Ratcliff, D. (1972). *Ann. Math. Statist.*, **43**, 1470–1480.

[7] Darroch, J. N., Lauritzen, S. L., and Speed, T. P. (1980). *Ann. Statist.*, **8**, 522–539.

[8] Deming, W. E. (1943). *Statistical Adjustment of Data*. Wiley, New York.

[9] Deming, W. E. and Stephan, F. F. (1940). *Ann. Math. Statist.*, **11**, 427–444.

[10] Fienberg, S. E. (1970). *Ann. Math. Statist.*, **41**, 907–917.

[11] Fienberg, S. E. and Wasserman, S. S. (1981). *J. Amer. Statist. Ass.*, **76**, 54–57.

[12] Fienberg, S. E., Meyer, M. M., and Stewart, G. W. (1979). Alternative Computational Methods for Estimation in Multinomial Logit Response Models. *Tech. Rep. No. 348*, School of Statistics, University of Minnesota, Minneapolis, Minn.

[13] Fienberg, S. E., Meyer, M. M., and Wassermann, S. S. (1981). In *Looking at Multivariate Data*, V. Barnett, ed. Wiley, Chichester, England, pp. 289–306.

[14] Haberman, S. J. (1972). *Appl. Statist.*, **21**, 218–225.

[15] Haberman, S. J. (1973). *Appl. Statist.*, **22**, 118–126.

[16] Haberman, S. J. (1974). *The Analysis of Frequency Data*. University of Chicago Press, Chicago.

[17] Kruithof, R. (1937). *De Ingenieur*, **52**, E15–E25.

[18] Krupp, R. S. (1979). *Bell Syst. Tech. J.*, **58**(2), 517–538.

[19] Meyer, M. M. (1981). Applications and Generalizations of the Iterative Proportional Fitting Procedure. Ph.D. thesis, School of Statistics, University of Minnesota.

[20] Stephan, F. F. (1942). *Ann. Math. Statist.*, **13**, 166–178.

Acknowledgment

The preparation of this article was partially supported by the Office of Naval Research Contract N00014-80-C-0637 at Carnegie-Mellon University. Reproduction in whole or part is permitted for any purpose of the U.S. government.

(CATEGORICAL DATA
CONTINGENCY TABLES
INFORMATION, KULLBACK
INFORMATION THEORY AND CODING
    THEORY
MULTIDIMENSIONAL CONTINGENCY
    TABLES)

STEPHEN E. FIENBERG
MICHAEL M. MEYER

ITO PROCESSES *See* DIFFUSION PROCESSES

J

JACKKNIFE METHODS

In certain problems of statistical estimation*, theoretical complexity or lack of reliable detailed knowledge of the model may preclude a theoretical approach to calculation of the standard error and the bias* of an estimate. Jackknife methods provide direct numerical approximations of both bias and standard error, and can give reasonably reliable confidence limits. The standard jackknife procedure that we shall describe first turns out to be related to certain theoretical characteristics of estimators, including the influence function* familiar in robust estimation*; this relation leads to useful generalizations and variants of the standard jackknife. The jackknife is a relative of cross-validation* and of various nonparametric methods*.

ONE-SAMPLE JACKKNIFE

To begin with a rather simple problem, suppose that we have a sample of n independent observations $x_1, \ldots, x_n$ from a population that possesses a well-defined characteristic θ, whose unknown value is of interest. We have an estimate t of θ, based on the sample, and

we wish now to determine the bias and standard error of our estimate. In addition, we should like to calculate probable limits for θ, e.g., a 95% confidence interval. However, suppose that we are uncertain as to the distribution of X, or that we are incapable of suitable theoretical approximation of the required moments of T. What can be done? A reasonable starting point in many problems is to assume that the first two moments of T have expansions of the forms

$$E(T) = \theta + \frac{a_1(\theta)}{n} + \frac{a_2(\theta)}{n^2} + \cdots$$

$$\text{var}(T) = \frac{\sigma_1^2(\theta)}{n} + \frac{\sigma_2^2(\theta)}{n^2} + \cdots, \tag{1}$$

and that for large sample size n

$$\{T - E(T)\}/\sqrt{\text{var}(T)} \tag{2}$$

is approximately standard normal. Even without knowledge of the functions $a_1, \ldots, \sigma_1^2, \ldots$ the structure in (1) can be used.

One very elementary approach is to split the data randomly into two equal pieces, supposing for the moment that n is even; calculate the corresponding estimators T_1 and T_2 of θ; and then estimate the leading bias and variance terms of (1) by $-T + \frac{1}{2}(T_1 + T_2)$ and $\frac{1}{4}(T_1 - T_2)^2$, respectively. These estimates are unbiased for $a_1(\theta)/n$ and $\sigma_1^2(\theta)/n$ to first order, meaning that terms in higher powers of n^{-1} are ignored. However, these naive estimates are unreliable.

The standard jackknife methods developed by Quenouille [22] and Tukey [26] provide a sophisticated way of extending the half-sample idea. Define

$$I_j = (n-1)(T - T_{(-j)}), \qquad (3)$$

where $T_{(-j)}$ means T computed from $\{X_i : i \neq j\}$, i.e., the sample with X_j omitted. Then the leading terms for bias and variance in (1) are estimated by $-\bar{I}. = -n^{-1}\sum_{j=1}^{n} I_j$ and $\sum_{j=1}^{n}(I_j - \bar{I}.)^2/\{n(n-1)\}$, respectively. The bias-adjusted form of T is therefore

$$\tilde{T} = T + \bar{I}., \qquad (4)$$

and

$$S = \sqrt{\sum_{j=1}^{n}(I_j - \bar{I}.)^2 / \{n(n-1)\}} \qquad (5)$$

estimates both $SE(T)$ and $SE(\tilde{T})$, ignoring terms beyond first order in (1). The validity of the bias correction is easily checked by using (1) to calculate

$$E(I_j) = (n-1)\left[\left(\frac{1}{n} - \frac{1}{n-1}\right)a_1(\theta) + \cdots\right]$$

$$= -\frac{a_1(\theta)}{n} + O(n^{-2}),$$

where $O(n^{-k})$ means a term essentially bounded by a constant times n^{-k}, so that for large n, $O(n^{-2})$ is negligible relative to $O(n^{-1})$.

The validity of the standard error S is somewhat more subtle. But for a very large class of problems where (1) and (2) do apply, S is valid and one can justify the standard normal approximation for both $(T - \theta)/S$ and $(\tilde{T} - \theta)/S$ for large n. Thus the jackknife method for calculating approx-

imate $100(1 - \alpha)\%$ confidence limits for θ is

$$T \pm k_{1-(1/2)\alpha}S, \qquad (6)$$

or $\tilde{T} \pm k_{1-(1/2)\alpha}S$ if the bias correction is made, where k_ϵ is the upper $100\epsilon\%$ point of the standard normal distribution.

The simple test case $T = \bar{X}$ shows that $I_j = X_j - \bar{X}.$, $\bar{I}. = 0$ (no bias adjustment is needed, of course), and S is the usual sample standard error. Because $T + I_j = X_j$ here, some authors generally refer to

$$P_j = nT - (n-1)T_{(-j)} = T + I_j$$
$$(j = 1, \ldots, n) \quad (7)$$

as *pseudo-values*, and for some purposes the analogy with data values is apt. In general, the P_j are correlated.

Example 1. As an illustration, suppose that ρ is the correlation* for pairs $X = (Y, Z)$. The second and third columns of Table 1 give example data for $n = 15$ law schools,

$Y =$ average score on test A for new admissions

$Z =$ average score on test B for new admissions.

Because (1) and (6) are probably more reliable for $T = \tanh^{-1}r$ than for $r =$ sample correlation itself, we first obtain confidence limits for $\theta = \tanh^{-1}\rho$ and then reexpress these limits on the ρ scale. However, a bias correction for r must be obtained by jackknifing r directly, and *not* by reexpressing the bias-adjusted estimate T, since transformation affects bias. Columns 4 and 5 in Table 1 give values of $r_{(-j)}$ and $I_j^{(r)} = (n-1)(r - r_{(-j)})$; columns 6 and 7 give values of $t_{(-j)} = \tanh^{-1}r_{(-j)}$ and $I_j^{(t)}$. From the calculations at the foot of the table we find that the bias-adjusted correlation estimate is

$$\tilde{r} = r + \bar{I}. = 0.776 + 0.008 = 0.784,$$

and that approximate 95% two-sided confidence limits for θ are

$$1.0352 \pm 1.96 \times 0.438,$$

equivalent to 0.175 and 0.956 on the correlation scale. These limits are somewhat wider than one would obtain under an assumption

Table 1   Jackknife Calculations for Correlation with Law School Data[a]

j	y_j	z_j	$r_{(-j)}$	$I_j^{(r)} = (n-1)(r - r_{(-j)})$	$t_{(-j)} = \tanh^{-1} r_{(-j)}$	$I_j^{(t)} = (n-1)(t - t_{(-j)})$
1	576	3.39	0.892	-1.624	1.432	-5.549
2	635	3.30	0.763	0.182	1.003	0.446
3	558	2.81	0.755	0.294	0.984	0.711
4	578	3.03	0.776	0.	1.035	0.
5	666	3.44	0.731	0.630	0.931	1.461
6	580	3.07	0.780	-0.056	1.045	0.142
7	555	3.00	0.784	-0.112	1.056	-0.286
8	661	3.43	0.736	0.560	0.942	1.310
9	651	3.36	0.751	0.130	0.975	0.840
10	605	3.13	0.776	0.	1.035	0.
11	653	3.12	0.818	-0.588	1.151	-1.617
12	575	2.74	0.785	-0.126	1.058	-0.322
13	545	2.76	0.740	0.504	0.950	1.187
14	572	2.88	0.767	0.126	1.013	0.311
15	594	2.96	0.779	-0.042	1.043	-0.106

$$r = 0.776, \quad T = \tanh^{-1} r = 1.0352$$

Jackknife calculations $\begin{cases} r \text{ scale: } \bar{I}_. = 0.008 \\ T \text{ scale: } \bar{I}_. = -0.098, \; \sum I_j^2 - n\bar{I}_.^2 = 40.2449, \; S = \sqrt{\dfrac{40.2449}{15 \times 14}} = 0.438 \end{cases}$

[a]Table entries have been rounded off.
Source.   Efron [5].

of bivariate normality for X, which assumption would clearly be inappropriate for these data.

The class of problems in which the limits (6) are valid corresponds essentially to those where T is accurately approximable by an average; indeed, it is this fact which leads to (1) and (2). Thus T can be a function of sample moments or of U-statistics* (as long as the function has nonzero derivative), or an estimate of maximum likelihood type [i.e., solution of an equation of the form $\sum a(X_j, T) = 0$ where a is smooth]; but T cannot be a single sample quantile such as a median or extreme value. Miller [16] reviews the validity of the jackknife; Reeds [23] and Miller [18] add useful postscripts.

What makes the jackknife appealing is that the limits (6) are quite robust in the *validity* sense, so that in the correlation example they will be valid for a reasonably large sample from any bivariate distribution.

Of course, this should not be confused with *efficiency robustness* or *resistance* to aberrant values, which the jackknife does not have.

One suggestion for obtaining more reliable confidence limits is to replace the normal percentile k in (6) by the Student-t percentile for $m - 1$ degrees of freedom, where m is the number of distinct I_j values [20, 26]. There is no theoretical support for this. Very often S has a positive bias [5, IV] which tends to correct for variability in S.

With very large samples it may be helpful to make initial (random) groupings of the data, say into g groups each of size $m = n/g$, and then to let X_j represent the jth data group in the earlier discussion. This has the defect of giving an analysis that is not uniquely defined.

For problems where the bias is of special interest, it is useful to know that more refined bias adjustments are possible. Relevant methods are described by Gray et al. [7] and Schucany et al. [24].

RELATION TO THE INFLUENCE FUNCTION

To better understand the jackknife and its generalizations it is useful to look at a series representation for T. Suppose that T is a function of only the sample cumulative distribution function (CDF) $\hat{F}$, which puts probability $1/n$ at each of n distinct observations. That is, $T = t(\hat{F})$, where $\theta = t(F)$ if F is the population CDF of X; e.g., if $T = \bar{X}$, then $t(F) = \int x \, dF(x)$. Then one series of quite general validity is the von Mises expansion

$$T = \theta + n^{-1} \sum_{j=1}^{n} I^{(t)}(X_j ; F)$$

$$+ \tfrac{1}{2} n^{-2} \sum_{j,k=1}^{n} Q^{(t)}(X_j, X_k ; F) + \cdots ,$$

$$(8)$$

where $I^{(t)}(x; F)$ is the influence curve of T.

As the notation suggests, the jackknife component I_j defined in (3) is one approximation to $I^{(t)}(X_j ; F)$; so S^2 in (5) approximates $n^{-1}\mathrm{var}\{I^{(t)}(X; F)\}$, which in turn approximates $\mathrm{var}(T)$ if we ignore "quadratic" and higher-order terms in (8). It is, however, clear that other approximations are possible for $I^{(t)}(x; F)$, for example,

1. $(n + 1)(T_{(+x)} - T)$, where $T_{(+x)}$ is computed from $(X_1, \ldots, X_n, x)$;
2. $I^{(t)}(x; \hat{F})$.

Therefore, other variance approximations are possible, including the empirical delta-method estimate $\sum \{I^{(t)}(X_j ; \hat{F})\}^2 / n^2$. In general, these other approximations underestimate $\mathrm{var}(T)$. One can see also from (8) that the bias adjustment $\bar{I}$. in (4) estimates $-E\{Q^{(t)}(X, X; F)\}/(2n)$, which could therefore be estimated directly. Further details of these ideas, and general discussion, may be found in Mallows [14], Hinkley [9], and Efron [5].

The identification of I_j with the influence curve reinforces the practical value of the I_j as analogs of residuals, helpful in the diag-

nostic mode of analysis. For example, the value $I_1 = -5.549$ in Table 1 would correctly indicate the dominant, possibly unwarranted influence of observation 1 on the correlation analysis—as would be evident from a scatter plot in the example. The relations among I_j, robust estimation, and observation diagnostics are discussed by Devlin et al. [3], Cook [2], and Hinkley and Wang [10].

For a problem where the functions in expansion (8) can be identified, corrections to the normal approximation (6) are, in principle, possible using Edgeworth-type expansions with empirical moments. Little is known about the usefulness of such corrections.

EXTENSIONS TO NONHOMOGENEOUS PROBLEMS

Once the theoretical basis for the one-sample jackknife is understood, it becomes possible to consider variants and extensions in a fairly systematic manner. For example, consider a two-sample statistic T computed from two independent samples $(X_1, \ldots, X_m)$ and $(Y_1, \ldots, Y_n)$ each of independent, identically distributed (i.i.d.) variables. Let $T_{(i,0)}$ be T calculated without X_i, let $T_{(0,j)}$ be T calculated without Y_j, and define

$$I_{i,0} = (m - 1)(T - T_{(i,0)}),$$
$$I_{0,j} = (n - 1)(T - T_{(0,j)}),$$
$$(9)$$

which are analogs of I_j in (3). Then the jackknife estimates of bias and variance of T are, respectively,

$$B = -\bar{I}_{.,0} - \bar{I}_{0,.} = -m^{-1}\sum I_{i,0} - n^{-1}\sum I_{0,j}$$

and

$$S^2 = \frac{1}{m(m-1)} \sum_{i=1}^{m} \left(I_{i,0} - \bar{I}_{.,0}\right)^2$$

$$+ \frac{1}{n(n-1)} \sum_{j=1}^{n} \left(I_{0,j} - \bar{I}_{0,.}\right)^2, \quad (10)$$

these being valid for the general situation in

which

$$E(T) = \theta + a_{10}m^{-1} + a_{01}n^{-1} + \cdots,$$

$$\text{var}(T) = \sigma_{10}^2 m^{-1} + \sigma_{01}^2 n^{-1} + \cdots.$$

Formula (6) again applies for approximate $100(1 - \alpha)\%$ confidence limits. Note the parallel between (9), (10), and the usual analysis for linear statistics $T = \bar{X} \pm \bar{Y}$.

Two points should be made here. First, the variance formula (10) takes no account of similarities that may exist between the distributions of X and Y, such as equal variances. This is another aspect of validity robustness, at the possible expense of inefficiency. Second, it is important to understand the structure of the estimate T: if we had paired samples* it would be necessary to use a one-sample jackknife with (X, Y) as the observation unit. Again, this parallels the analysis of linear statistics.

As a simple example of a two-sample analysis, suppose that one wished to test the hypothesis that rates of city growth were the same in two states A and B from 1970 to 1980, given census* data on 20 representative cities in each state. If the city population measurements are symbolized by

$$X = \left(Z_A^{(1970)}, Z_A^{(1980)} \right)$$

and

$$Y = \left(Z_B^{(1970)}, Z_B^{(1980)} \right)$$

and if growth is measured by

$$\beta_k = E\left(Z_k^{(1980)} \right) / E\left(Z_k^{(1970)} \right)$$

for $k = A, B$, then β_A and β_B are to be compared. For the contrast $\theta = \log(\beta_A / \beta_B)$ with estimate

$$T = T_A - T_B$$

$$= \log\left(\bar{Z}_A^{(1980)} \bar{Z}_B^{(1970)} / \bar{Z}_B^{(1980)} \bar{Z}_A^{(1970)} \right),$$

the jackknife formula (10) for S^2 will be the sum of the one-sample variances for T_A and T_B; see (5). Note that the correlations $\rho(Z_k^{(1970)}, Z_k^{(1980)})$ are automatically taken into account. The hypothesis $\theta = 0$ can now be tested by comparing the Studentized estimate T/S to the standard normal percentiles.

The discussion of the two-sample problem can be extended in a straghtforward way to multisample problems. The earlier remark about possible inefficiency of S^2 would again be relevant, particularly for many small samples.

At first sight, nonhomogeneous problems without replication would seem to be outside the jackknife domain. This is not quite so. For example, consider the regression model

$$y_j = \theta z_j + e_j, \qquad j = 1, \ldots, n,$$

with $(z_1, \ldots, z_n)$ a fixed design. If one artificially treats $X = (Y, Z)$ as a random pair, then the one-sample jackknife is applicable to the symmetric least-squares estimate $T = \sum Z_j Y_j / \sum Z_j^2$. One finds that the bias adjustment $\hat{I}$. is not zero and that the variance estimate in (5) becomes

$$S^2 \doteq n^{-1} \sum \left\{ I^{(t)}(\dot{X}_j; \hat{F}) \right\}^2$$

$$= \sum Z_j^2 (Y_j - TZ_i)^2 / \left(\sum Z_j^2 \right)^2. \quad (11)$$

Of course, if $E(e_j | z_j) = 0$, then T is unbiased, but the jackknife method does not know this. The somewhat curious expression (11) is a natural estimate if nothing is assumed about $\text{var}(e_j | z_j)$, but in most cases it will be somewhat inefficient. For very large samples with smooth designs the jackknife will be quite reliable; but in small samples, especially with skew designs, this will not be so, because the jackknifed design will be too variable. Jackknife methods in regression are discussed by Mallows [14], Miller [17], Hinkley [8], and Fox et al. [6].

OTHER SUBSAMPLING METHODS

There are several other subsampling methods of some importance. Many are described and compared in the excellent study by Efron [5]. Three methods are briefly outlined here.

One very simple extension of the half-sample idea is the following procedure suggested for time series* problems. If $X_1, \ldots, X_n$ is a stationary sequence such that (1) and (2) hold for statistic T, then split the data into g time-ordered groups $(X_1, \ldots, X_m), \ldots, (X_{m(g-1)+1}, \ldots, X_{mg})$, where we suppose, for simplicity, that $n = mg$. Then compute group estimates $T_1^*, \ldots, T_g^*$ corresponding to T. The bias-adjusted estimate will be

$$\tilde{T} = (gT - \bar{T}_\cdot^*)/(g - 1)$$

and the estimated standard error of T or $\tilde{T}$ will be

$$\left\{ \frac{\sum (T_j^* - \bar{T}_\cdot^*)^2}{g(g-1)} \right\}^{1/2}$$

Although this method is clearly not as efficient as the jackknife, it has value in a problem where T is not well defined for a sequence with a missing value.

A more important extension of the half-sample method is the fraction-sampling technique used mostly in sample survey* analysis. Suppose that a sampled population $\mathbb{P}$ is subdivided into H strata $\mathbb{P}_1, \ldots, \mathbb{P}_H$ of sizes $N_1, \ldots, N_H$, and that random samples $(x_{ij}: j = 1, \ldots, n_i, i = 1, \ldots, H)$ are drawn. The statistical estimate T will be a function of the H stratum summaries and prescribed weights; for example, $T = \sum N_i \bar{x}_i / \sum N_i$ estimates the average value of x in $\mathbb{P}$. The fraction-sampling approach is to construct several sets of subsamples, each consisting of fractions of the $n_1, \ldots, n_H$ values, to compute the associated values of T and thence to estimate var(T) from the variation of subsample T's. Thus in the special case $n_i \equiv 2$, half-sampling will be carried out so that each subsample contains one of the two values from each stratum. There are 2^H possible subsamples. If subsample values of T are denoted by $T_i^*, i = 1, \ldots, M$, then the estimated standard error of T will be

$$S^* = \left\{ \sum_{j=1}^{M} (T_j^* - \bar{T}_\cdot^*)^2 / M \right\}^{1/2}. \quad (12)$$

M need not be as large as 2^H. Indeed, Mc-Carthy [15] introduced the method of balanced half-samples, whereby nearly as few as $n = \sum n_i$ subsamples can be taken and yet (12) will still be reasonably accurate pro-

vided that T is not appreciably nonlinear. Further discussion of the subsampling methods in sample survey analysis may be found in Kish and Frankel [12] and Krewski and Rao [13].

The theory and application of jackknife-type methods has been clarified in the systematic study by Efron [4, 5]. Of particular interest is the set of techniques known as "bootstrap* methods." The basic idea is to simulate properties of the statistic T by sampling from an empirical estimate of the underlying probability distribution. To take a simple example, suppose that $X_1, \ldots, X_n$ are of the form $X_j = \mu + \sigma e_j$ and that T is an estimate of θ, which might be μ or σ. If it is reasonable to suppose that the e_j are independent and identically distributed, then their common distribution is estimated by

$$\hat{F}: \text{probability } \frac{1}{n} \text{ at } \hat{e}_j = (x_j - \hat{\mu})/\hat{\sigma},$$

$$j = 1, \ldots, n, \quad (13)$$

for suitably chosen $\hat{\mu}, \hat{\sigma}$. A simulated sample consists of a random sample $e_1^*, \ldots$ from $\hat{F}$ —equivalent to random sampling with replacement from $(\hat{e}_1, \ldots, \hat{e}_n)$—transformed to the x scale by

$$x_j^* = \hat{\mu} + \hat{\sigma} e_j^*, \quad j = 1, \ldots, m, \quad (14)$$

where $m = n$ in the usual application. Each such simulated sample gives a value t^* for the statistic of interest, so that repeated simulations via (14) will yield a frequency distribution (histogram*) of t^* that is the simulated distribution of T. Usually, one would work not with T itself but with a pivotal quantity $Q = (T - \theta)/S$ whose distribution is expected to be stable with respect to θ and other parameters. Then the simulated distribution of Q will be the bootstrap frequency distribution of $Q^* = (T^* - t)/S^*$.

As an illustration, suppose that ρ is a bivariate correlation with $\theta = \tanh^{-1}\rho$ and $T = \tanh^{-1}r$, as in the earlier example. On this scale, $S = 1$ seems reasonable, so $Q = T - \theta$. For the data in Table 1, Fig. 1 gives the bootstrap frequency distribution of $Q^* = T^* - t$ in 1000 samples, obtained by Efron [5]. (In this case x_j^* were sampled

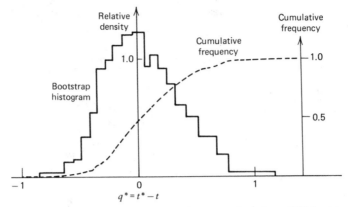

Figure 1 Bootstrap frequency distribution of $T^* - t$ for $T = \tanh^{-1} r$ from 1000 bootstrap simulations of data corresponding to Table 1.

directly from the empirical distribution of x_j, since T is invariant under linear transformation of the data scale.) The graph can be used to read off $100(1 - \alpha)\%$ confidence limits for θ as

$T -$ upper $\frac{1}{2}\alpha$ quantile of $T^* - t$,

$T +$ lower $\frac{1}{2}\alpha$ quantile of $T^* - t$,

which in this numerical example yields 95% limits 0.251 and 0.918 for ρ.

One of the advantages of the bootstrap method is that it avoids restriction of distributional approximations to normal form; some relevant asymptotic theory is described by Bickel and Freedman [1] and Singh [25]. The connection between the bootstrap frequency distribution and the jackknife confidence limit method is, roughly, that the latter employs a normal approximation using the mean and variance of the bootstrap distribution.

Another advantage is that the bootstrap method can incorporate partial model structure at the simulation stage; for example, symmetry of distribution or homogeneity of variation can be incorporated very easily when these are appropriate. The application of bootstrap methods to hypothesis testing* and other problems is illustrated by Hinkley et al. [11].

The most important applications of jackknife and related methods are in complex situations, such as discriminant analysis*, curve fitting*, complex surveys, and so forth. Much of the literature on applications

is included in the excellent bibliography by Parr and Schucany [21]. See also especially Mosteller and Tukey [19, 20] and Efron [5].

References

[1] Bickel, P. J. and Freedman, D. A. (1981). *Ann. Statist.*, **9**, 1196–1217.

[2] Cook, R. D. (1977). *Technometrics*, **19**, 15–18.

[3] Devlin, S. J., Gnanadesikan, R., and Kettenring, J. R. (1975). *Biometrika*, **62**, 531–545.

[4] Efron, B. (1979). *Ann. Statist.*, **7**, 1–26.

[5] Efron, B. (1981). *The Jackknife, the Bootstrap, and Other Resampling Plans.* CBMS Monogr. No. 38. SIAM, Philadelphia.

[6] Fox, T., Hinkley, D. V., and Larntz, K. (1980). *Technometrics*, **22**, 29–33.

[7] Gray, H. L., Watkins, T. A., and Adams, J. E. (1972). *Ann. Math. Statist.*, **43**, 1–30.

[8] Hinkley, D. V. (1977). *Technometrics*, **19**, 285–292.

[9] Hinkley, D. V. (1978). *Biometrika*, **65**, 13–21.

[10] Hinkley, D. V. and Wang, H. L. (1980). *J. R. Statist. Soc. B*, **42**, 347–356.

[11] Hinkley, D. V., Chapman, P., and Runger, G. (1980). Change-Point Models. Unpublished report, University of Minnesota School of Statistics, Minneapolis, Minn.

[12] Kish, L. and Frankel, M. R. (1974). *J. R. Statist. Soc. B*, **36**, 1–37.

[13] Krewski, D. and Rao, J. N. K. (1981). *Ann. Statist.*, **9**, 1010–1019.

[14] Mallows, C. (1975). On some topics in robustness. Unpublished memorandum, Bell Telephone Laboratories, Murray Hill, N.J.

[15] McCarthy, P. J. (1969). *Rev. Inst. Statist. Int.*, **37**, 239–264.

[16] Miller, R. G. (1974). *Biometrika*, **61**, 1–15.

[17] Miller, R. G. (1974). *Ann. Statist.*, **2**, 880–891.

[18] Miller, R. G. (1978). *Proc. 23rd Conf. Des. Exper. Army Res. Dev. Test.*, pp. 371–391.

[19] Mosteller, F. and Tukey, J. W. (1968). In *Handbook of Social Psychology*, G. Lindzey and E. Aronson, eds. Addison-Wesley, Reading, Mass.

[20] Mosteller, F. and Tukey, J. W. (1977). *Data Analysis and Regression.* Addison-Wesley, New York, pp. 133–163.

[21] Parr, W. C. and Schucany, W. R. (1980). *Int. Statist. Rev.*, **48**, 73–78.

[22] Quenouille, M. H. (1956). *Biometrika*, **43**, 353–360.

[23] Reeds, J. A. (1978). *Ann. Statist.*, **6**, 727–739.

[24] Schucany, W. R., Gray, H. L., and Owens, D. B. (1971). *J. Amer. Statist. Ass.*, **72**, 420–423.

[25] Singh, K. (1981). *Ann. Statist.*, **9**, 1187–1195.

[26] Tukey, J. W. (1958). *Ann. Math. Statist.*, **29**, 614 (abstract).

(ADAPTIVE METHODS
BOOTSTRAPPING
INFLUENCE FUNCTION
ROBUSTNESS)

DAVID HINKLEY

JACKSON ESTIMATOR

An estimator of the probability of survival over a specified period (such as a year) based on the observed age distribution of a sample from the population. If there are N_j individuals aged j last birthday (more generally, aged j units plus a fraction of a unit), the estimator is

$$\frac{N_1 + N_2 + \cdots + N_k}{N_0 + N_1 + \cdots + N_{k-1}}$$

where k is the largest k for which N_k exceeds zero. For its validity one needs the assumptions of a stationary population*, constant survival probabilities, and randomness of sample. Although these assumptions are very unlikely to be satisfied, the estimator has been popular, because of its simplicity.

Further Reading

Jackson, C. H. N. (1939). *J. Animal Ecol.*, **8**, 238–246.

(BIRTH-AND-DEATH PROCESSES
CATCH CURVE
CHAPMAN ESTIMATOR
FISHERIES, STATISTICS IN
SURVIVAL ANALYSIS)

JACOBIANS

Given a mapping u continuously differentiable from a domain D in R^n into R^m,

$$\mathbf{u}(\mathbf{x}) = (u_1(x_1, \ldots, x_n), \ldots, u_m(x_1, \ldots, x_n)),$$

$$\mathbf{x} = (x_1, \ldots, x_n),$$

the matrix $\partial(\mathbf{u})/\partial(\mathbf{x})$ with the entries $(\partial u_j / \partial x_k)$, $j = 1, \ldots, m$; $k = 1, \ldots, n$ is called the *Jacobian matrix*. If $n = m$, the determinant of this matrix is called the *Jacobian determinant*, or simply *Jacobian*. Jacobians serve as a basic tool in distribution theory of multidimensional random variables.

A discussion of Jacobians with an emphasis on applications to construction of multivariate distributions is given by Higgins [1, 2]. For additional information, see Johnson and Leone [3].

References

[1] Higgins, J. J. (1975). *Amer. Statist.*, **29**, 43–46.

[2] Higgins, J. J. (1975). *Commun. Statist.*, **4**, 955–966.

[3] Johnson, N. L. and Leone, F. C. (1977). *Statistics and Experimental Design in Engineering and the Physical Sciences*, 2nd ed., Vol. 1. Wiley, New York, p. 165 (Russian edition, 1980).

(GEOMETRY IN STATISTICS
MULTIVARIATE DISTRIBUTIONS)

JACOBI ITERATION *See* GAUSS–SEIDEL ITERATION

JACOBI MATRIX

An infinite matrix with complex-valued entries

$$a_{m,n} = 0 \qquad \text{if } (m - n) \geqslant 2$$

$$a_{n,n+1} = \overline{a_{n+1,n}} \neq 0$$

$$a_{nn} \text{ is real}$$

is called a *Jacobi matrix*. It is closely related to self-adjoint operators* with simple spectra*.

Further Reading

Gantmacher, F. R. (1959). *The Theory of Matrices*, 2 vols. Chelsea, New York.

Stone, M. H. (1932). *Amer. Math. Soc. Colloq. Publ.*

JACOBI POLYNOMIALS

Jacobi polynomials of degree $n, J_n(\alpha, \gamma; x)$ satisfy the differential equation

$$x(1 - x)y'' + \left[\gamma - (\alpha + 1)x\right]y'$$
$$+ n(\alpha + n)y = 0.$$

They are a particular case of the confluent hypergeometric* function

$$J_n(\alpha, \gamma; x) = F(-n, \alpha + n; \gamma; x)$$

and can be explicitly calculated from the expression

$$J_n(\alpha, \gamma; x) = x^{1-\gamma}(1 - x)^{\gamma - a} \frac{\Gamma(\gamma + n)}{\Gamma(\gamma)}$$

$$\times \frac{d^n}{dx^n}\left[x^{\gamma + n - 1}(1 - x)^{\alpha + n - \gamma}\right].$$

They form a family of orthogonal polynomials with the weight function

$$w(x) = \begin{cases} x^{\gamma - 1}(1 - x)^{\alpha - \gamma} & (0 \leqslant x \leqslant 1) \\ 0 & \text{otherwise} \end{cases}$$

where $\text{Re}\,\gamma > 0$ and $\text{Re}(\alpha - \gamma) > -1$.

Chebyshev*, Legendre*, and Gegenbauer polynomials are particular cases of Jacobi polynomials. Jacobi polynomials appear in differential equations of Pearson curves, in curve fitting, and in distribution of the serial correlation coefficient.

Further Reading

Sansone, G. (1959). *Orthogonal Functions*. Interscience, New York.

(CHEBYSHEV-HERMITE POLYNOMIALS
LEGENDRE POLYNOMIALS
PEARSON CURVES
SERIAL CORRELATION COEFFICIENT)

JAMES–STEIN ESTIMATORS

The most widely used estimator of the mean of a normal distribution is the sample mean, $\bar{X}$. This statistic is a maximum likelihood* estimator, is uniformly minimum variance unbiased*, and for a wide variety of loss functions (including squared error) is best invariant*, minimax*, and admissible*. Much the same can be *said* about the use of the vector of sample means, $\bar{\mathbf{X}}$, to estimate the vector of population means when sampling from a multivariate normal population. There is, however, an important exception. If the dimension, p, of the multivariate normal population is at least 3, the vector of sample means may be inadmissible—i.e., there are other estimators whose risk functions are everywhere smaller than the risk of $\bar{\mathbf{X}}$. This remarkable fact was discovered by Charles Stein (see Stein [16]).

In the following we describe the specifics of this result and the developments that followed it. Let X have a (p-dimensional) multivariate normal distribution with mean vector $\boldsymbol{\mu}$ and (known) covariance matrix equal to the identity, I. Let the loss function, $L(\boldsymbol{\mu}, \boldsymbol{\delta})$ be equal to the sum of squared errors—i.e., if $\mathbf{X} = (X_1, \ldots, X_p)'$, $\boldsymbol{\mu} = (\mu_1, \ldots, \mu_p)'$, and $\boldsymbol{\delta} = (\delta_1, \ldots, \delta_p)'$, then

$$L(\boldsymbol{\mu}, \boldsymbol{\delta}) = \sum_{i=1}^{p} (\mu_i - \delta_i)^2 = \|\boldsymbol{\mu} - \boldsymbol{\delta}\|^2.$$

The risk $R(\boldsymbol{\mu}, \boldsymbol{\delta})$ of an estimator $\boldsymbol{\delta}(\mathbf{X})$ is the expected value of the loss, i.e.,

$$R(\boldsymbol{\mu}, \boldsymbol{\delta}) = E_{\boldsymbol{\mu}}(L(\boldsymbol{\mu}, \boldsymbol{\delta}))$$

$$= \left(\frac{1}{\sqrt{2\pi}}\right)^p \int_{R^p} \|\boldsymbol{\delta}(\mathbf{X}) - \boldsymbol{\mu}\|^2$$

$$\times \exp\left(-\tfrac{1}{2}\|\mathbf{X} - \boldsymbol{\mu}\|^2\right) d\mathbf{X}$$

$$= \sum_{i=1}^{p} E_{\boldsymbol{\mu}}(\delta_i(\mathbf{X}) - \mu_i)$$

$$= \sum_{i=1}^{p} \text{var}\,\delta_i(\mathbf{X}) + \sum_{i=1}^{p} \left(\text{bias}(\delta_i(\mathbf{X}))\right)^2,$$

where $\text{bias}(\delta_i(\mathbf{X})) = E_{\boldsymbol{\mu}}(\delta_i(\mathbf{X}) - \mu_i)$.

Remark. Typically, in practice, one would have a sample of observations $\mathbf{X}_1, \ldots, \mathbf{X}_n$, but sufficiency considerations require recording only $\overline{\mathbf{X}}$, which has a multivariate normal distribution with mean μ and covariance matrix $(1/n)\mathbf{I}$. The multiple $1/n$ does not change the problem in an essential way with regard to admissibility, minimaxity, or other decision-theoretic considerations, and hence the setup above covers this apparently more general situation.

Stein [16] studied the class of estimators $\delta^{a,b}$:

$$\delta^{a,b}(\mathbf{X}) = \left(1 - \frac{a}{b + \mathbf{X}'\mathbf{X}}\right)\mathbf{X}$$

and showed that if a is sufficiently small and b is sufficiently large, then $\delta^{a,b}$ has everywhere smaller risk than $\mathbf{X}$. He also indicated that the phenomenon was not restricted to the sum of squared error loss but was true quite generally.

It may seem that the improvement in risk is destined to be negligible since "a small and b large" implies that $\delta^{a,b}$ is little different from $\mathbf{X}$. This turns out not to be the case, however, as was shown in James and Stein [15]. Here the authors were able to develop useful expressions for the risk of estimators of the form

$$\delta^{a,0}(\mathbf{X}) = \left(1 - \frac{a}{\mathbf{X}'\mathbf{X}}\right)\mathbf{X}.$$

They showed that the risk of $\delta^{a,0}(\mathbf{X})$ is strictly less than the risk of $\mathbf{X}$, provided that $0 < a < p - 2$. They also showed that the risk at $\mu = 0$ of $\delta^{p-2,0}(\mathbf{X})$ is equal to 2, a substantial improvement over the constant risk p of the estimator $\mathbf{X}$.

Baranchik [1, 2] studies the class

$$\delta_r(\mathbf{X}) = \left(1 - \frac{r(\mathbf{X}'\mathbf{X})}{\mathbf{X}'\mathbf{X}}\right)\mathbf{X}$$

and showed if $0 \leqslant r(\cdot) \leqslant 2(p - 2)$ and $r(t)$ is nondecreasing in t, then $\delta_r(\mathbf{X})$ is minimax. He also found generalized Bayes minimax estimators of the form $\delta_r(\mathbf{X})$. Strawderman [18] found proper Bayes minimax estimators of the same form for $p \geqslant 5$ and showed that

such do not exist for $p < 5$. (*See* MINIMAX ESTIMATION.)

There are extensions of these results in several directions. Brown [10] showed the inadmissibility of the best invariant estimator of a location vector for $p \geqslant 3$ under quite general conditions on the loss and underlying density. James and Stein [15] gave improved estimators for the case $\mathbf{X} \sim N(\mu, \sigma^2 \mathbf{I})$, σ^2 unknown, when an independent random variable S distributed as σ^2 times a chi-squared variable is available. They also gave results (in the fully invariant case) when the covariance matrix of $\mathbf{X}$ is completely unknown but an independent matrix S is available which distributed as Wishart* $(\mathbf{\Sigma}, m)$. Baranchik's and Strawderman's results also generalize the case where σ^2 is unknown. Explicit improvements for distributions other than the normal were obtained in Strawderman [19], Brandwein and Strawderman [8], and Berger [3], and for losses other than quadratic in Brandwein and Strawderman [9]. (*See* INVARIANCE PRINCIPLES IN STATISTICS.)

Extensions to losses of the form $L(\mu, \delta) = (\delta - \mu)'\mathbf{C}(\delta - \mu)$, where $\mathbf{C}$ is a positive definite matrix, were first obtained by Bhattacharya [6]. Other results for this and related cases were obtained by Bock [7], Brown [11], Berger [4, 5], Efron and Morris [12–14], and others.

A widely useful new technique of proof for such results was developed by Stein [17]. If $\mathbf{X} \sim N(\mu, 1)$, then integration by parts gives $E(\mathbf{X} - \mu)h(\mathbf{X}) = E(h'(\mathbf{X}))$. For example, if $\mathbf{X} \sim N(\mu, I)$,

$$E\left(\frac{\mathbf{X}'(\mathbf{X} - \mu)}{\mathbf{X}'\mathbf{X}}\right) = E\left(\sum_{i=1}^{p} \frac{X_i(X_i - \mu)}{\sum X_j^2}\right)$$

$$= \sum_{i=1}^{p} E\left[\frac{\partial}{\partial X_i}\left[\frac{X_i}{\sum X_j^2}\right]\right]$$

$$= \sum_{i=1}^{p} E\left[\frac{\sum X_j^2 - 2X_i^2}{\left(\sum X_j^2\right)^2}\right]$$

$$= E\left(\frac{p - 2}{\mathbf{X}'\mathbf{X}}\right).$$

Hence

$$E\left[\left\|\left(1 - \frac{a}{\mathbf{X'X}}\right)\mathbf{X} - \boldsymbol{\mu}\right\|^2\right]$$

$$= E\left(\|\mathbf{X} - \boldsymbol{\mu}\|^2 - 2aE\left(\frac{\mathbf{X'(X - \boldsymbol{\mu})}}{\mathbf{X'X}}\right)\right.$$

$$\left. + a^2 E\left(\frac{1}{\mathbf{X'X}}\right)\right.$$

$$= p - \left[2a(p - 2) - a^2\right]E\frac{1}{(\mathbf{X'X})} \ .$$

This will be less than p provided that $0 < a < p - 2$, and hence $\delta^{a,0}(\mathbf{X})$ will have smaller risk than $\mathbf{X}$ for such a's, which is the basic result of James and Stein.

The James–Stein estimator is closely related to a number of other modifications of the estimator $\overline{\mathbf{X}}$—in particular, certain Bayes estimators, empirical Bayes* estimators, and ridge regression* estimators. In fact, the James–Stein estimator can be usefully viewed as an empirical Bayes estimator relative to a multivariate normal prior distribution with mean $\mathbf{0}$ and covariance matrix an unknown multiple of the identity. All of these techniques modify the usual procedures by "shrinking" it toward the origin or some other suitably chosen subspace (such as the mean of all coordinates). For a direct application of the James–Stein estimator, see Efron and Morris [13].

References

[1] Baranchik, A. J. (1964). Multiple Regression and Estimation of the Mean Vector of a Multivariate Normal Distribution. *Tech. Rep. No. 51*, Stanford University, Stanford, Calif.

[2] Baranchik, A. J. (1970). *Ann. Math. Statist.*, **41**, 642–645.

[3] Berger, J. (1975). *Ann. Statist.*, **3**, 1318–1328.

[4] Berger, J. (1976). *Ann. Statist.*, **4**, 223–226.

[5] Berger, J. (1976). *J. Multivariate Anal.*, **6**, 256–264.

[6] Bhattacharya, P. K. (1966). *Ann. Math. Statist.*, **37**, 1818–1825.

[7] Bock, M. E. (1975). *Ann. Math. Statist.*, **3**, 209–218.

[8] Brandwein, A. C. and Strawderman, W. E. (1978). *Ann. Statist.*, **6**, 377–416.

[9] Brandwein, A. C. and Strawderman, W. E. (1980). *Ann. Statist.*, **8**, 279–284.

[10] Brown, L. D. (1966). *Ann. Math. Statist.*, **37**, 1037–1135.

[11] Brown, L. D. (1975). *J. Amer. Statist. Ass.*, **70**, 417–427.

[12] Efron, B. and Morris, C. (1973). *J. Amer. Statist. Ass.*, **68**, 117–130.

[13] Efron, B. and Morris, C. (1973). *J. R. Statist. Soc. B*, **35**, 379–421.

[14] Efron, B. and Morris, C. (1976). *Ann. Statist.*, **4**, 11–21.

[15] James, W. and Stein, C. (1961). *Proc. 4th Berkeley Symp. Math. Statist. Prob.*, Vol. 1. University of California Press, Berkeley, Calif., pp. 361–379.

[16] Stein, C. (1955). *Proc. 3rd Berkeley Symp. Math. Statist. Prob.*, Vol. 1. University of California Press, Berkeley, Calif., pp. 197–206.

[17] Stein, C. (1973). *Proc. Prague Symp. Asympt. Statist.*, pp. 346–381.

[18] Strawderman, W. E. (1971). *Ann. Math. Statist.*, **42**, 385–388.

[19] Strawderman, W. E. (1974). *J. Multivariate Anal.*, **4**, 1–10.

Further Reading

Berger, J. (1979). In *Optimizing Methods in Statistics*, J. S. Rustagi, ed. Academic Press, New York.

Strawderman, W. E. (1972). *Proc. 6th Berkeley Symp. Math. Statist. Prob.*, Vol 1. University of California Press, Berkeley, Calif., pp. 51–55.

Strawderman, W. E. (1973). *Ann. Statist.*, **1**, 1189–1194.

(ADMISSIBILITY
BAYESIAN INFERENCE
EMPIRICAL BAYES THEORY
INVARIANCE PRINCIPLES IN STATISTICS
MEAN SQUARED ERROR
MINIMAX ESTIMATION
REGRESSION ANALYSIS
RIDGE REGRESSION)

WILLIAM E. STRAWDERMAN

J-DIVERGENCES AND RELATED CONCEPTS

Divergences or dissimilarity coefficients serve as measures of discrepancy between distributions or dissimilarity* between populations. A wide class of these measures, commonly called *Jeffreys divergences* [11] or *J-divergences*, is based on entropy* functions and is used in the process of statistical inference. In particular, these measures have been used in a wide variety of studies in anthropology, biology, genetics, communication theory, economics, forecasting, information theory, statistical mechanics, and other fields of research.

BASIC CONCEPTS

One of the most widely used measures of uncertainty* or diversity* of a multinomial* probability distribution, $p \in \mathscr{P}_n$, where

$$\mathscr{P}_n = \left\{ p = (p_1, \ldots, p_n); p_j \geqslant 0, \sum p_j = 1 \right\}$$

is the *Shannon entropy* [18],

$$H_n(p) = -\sum p_i \log p_i.$$

Associated with this entropy is the *Kullback–Leibler information* number

$$I_n(p, q) = \sum p_i \log(p_i/q_i),$$

which serves as a measure of discrepancy or dissimilarity between $p \in \mathscr{P}_n$ and another distribution $q \in \mathscr{P}_n$. Here and in the sequel we employ the convention of $0 \log 0 = 0$ and if one of the q_i is zero, the corresponding p_i is also zero, whence $p_i \log(p_i/q_i) = p_i \log p_i - p_i \log q_i = 0$.

An interpretation for the foregoing two quantities may be given through the language of communication theory*: The quantity $-\log p_i$ is known there as the amount of *self-information* associated with the event E_i whose probability is p_i, and hence $H_n(p)$ is the average amount of information of n events E_i ($i = 1, \ldots, n$). On the other hand, $-\log q_i - (-\log p_i) = \log(p_i/q_i)$ is the *information gain* in predicting the event E_i by the estimation q_i, and hence $I_n(p, q)$ is the average information gain of n events E_i ($i = 1, \ldots, n$). This number, which is also known as the *information divergence* or *I-divergence* of p from q, is asymmetric; i.e., in general, $I_n(p, q) \neq I_n(q, p)$. The symmetric quantity $J_n(p, q) = I_n(p, q) + I_n(q, p)$, namely

$$J_n(p, q) = \sum \{ p_i \log(p_i/q_i) + q_i \log(q_i/p_i) \},$$

is known as the *Jeffreys invariant* or the *J-divergence* between p and q [11–13] (*see also* INFORMATION THEORY).

The concepts discussed above can be extended to the case of continuous probability distributions. Here, however, for simplicity of the exposition, we shall consider only the space of multinomial distributions, observing that the quantities for the continuous case can be obtained by a suitable replacement of the summation by the integral sign.

A measure of entropy is directly conceived as a mapping H of $\mathscr{P}_n$ into $\mathbb{R}_+ \equiv [0, \infty)$, satisfying the following postulates: (1) $H(p) = H(p_1, \ldots, p_n)$ is a continuous function of its n variables and is invariant under their permutations; (2) $H(p, 0) = H(p)$; (3) $H(p) = 0$ if and only if $p \in \mathscr{P}_n$ is degenerate; and (4) $H(p)$, $p \in \mathscr{P}_n$, is maximal if and only if p is the uniform distribution $u^{(n)} \equiv (1/n, 1/n, \ldots, 1/n)$ and its maximum value $H(u^{(n)})$ is increasing with n. (Compare with the axiomatization in INFORMATION THEORY.)

A measure of divergence, on the other hand, is a mapping D of $\mathscr{P}_n \times \mathscr{P}_n$ into $\mathbb{R}_+$, satisfying the following postulates: (1) $D(p, q) = D(p_1, \ldots, p_n; q_1, \ldots, q_n)$ is a continuous function of its $2n$ variables and is invariant under the permutations of the pairs (p_j, q_j), $j = 1, \ldots, n$; (2) $D(p, 0; q, 0) = D(p, q)$; and (3) $D(p, q) = 0$, for $p, q \in \mathscr{P}_n$, if and only if $p = q$.

The measure of entropy is therefore an index of similarity with the uniform distribution, and hence a measure of diversity of uncertainty (the letter H is used to indicate heterogeneity). The measure of divergence, on the other hand, reflects the differences between two distributions.

The Shannon entropy satisfies all the postulates of entropy with the additional *additivity property*; $H_{nm}(p \otimes q) = H_n(p) + H_m(q)$ for $p \in \mathscr{P}_n$ and $q \in \mathscr{P}_m$. Similarly, the *I-divergence* satisfies all three postulates of divergence and the *additivity property*, i.e., $I_{nm}(p \otimes r, q \otimes s) = I_n(p, q) + I_m(r, s)$ for $p, q \in \mathscr{P}_n$ and $r, s \in \mathscr{P}_m$. The same applies for the *J-divergence* $J_n(p, q)$. The Shannon entropy is a concave function on $\mathscr{P}_n$, which meets the intuitive requirement that the average entropy between any $p, q \in \mathscr{P}_n$ is not greater than that between their average, Concavity, therefore, should be an additional desired property of entropy measures. For very analogous reasons, as I_n and J_n are convex on $\mathscr{P}_n \times \mathscr{P}_n$, convexity should be an additional attractive feature of divergences.

PROTOTYPES OF J-DIVERGENCES

A wide variety of divergences have been introduced through the concept of entropy and information; these are so-called *J-divergences*. To obtain a reasonable unified description of these concepts, the following notation will be used: U stands for any interval of $\mathbb{R} \equiv (-\infty, \infty)$ which contains the unit interval $U_0 \equiv [0, 1]$; thus $\mathscr{P}_n = \{x = (x_1, \ldots, x_n) \in U_0^n : \sum x_i = 1\}$. For a C^2-function ϕ on U we consider the functional

$$H_{n,\phi}(x) \equiv -\sum \phi(x_i), \qquad x \in U^n.$$

If ϕ is convex on U_0 and $\phi(0) = \phi(1) = 0$, then $H_{n,\phi}$ defines a genuine measure of entropy $\mathscr{P}_n$, called the ϕ-*entropy*. A basic example of such ϕ is

$$\phi_\alpha(x) \equiv (\alpha - 1)^{-1}(x^\alpha - x), \qquad \alpha > 0,$$

and hence $\phi_1(x) = x \log x$. With the choice of $\phi = \phi_\alpha$, $H_{n,\phi}$ becomes the *Havrda–Charvát entropy of degree* α [8]:

$$H_{n,\alpha}(p) = (1 - \alpha)^{-1}\left(\sum p_i^\alpha - 1\right),$$

$$p \in \mathscr{P}_n.$$

This measure of entropy is nonadditive for $\alpha \neq 1$. On the other hand, the functionally related *Rényi entropy of order* α [17],

$$R_{n,\alpha}(p) \equiv (1 - \alpha)^{-1}\log\left[1 + (1 - \alpha)H_{n,\alpha}(p)\right]$$

is indeed an additive measure of entropy for all $\alpha > 0$. For $\alpha = 1$, these two entropies are identical and they are both equal to the Shannon entropy, i.e., $R_{n,1} \equiv H_{n,1} \equiv H_n$. For $\alpha \neq 1$, on the other hand, it holds that

$$(\alpha - 1)\left[R_{n,\alpha}(p) - H_{n,\alpha}(p)\right] \geq 0,$$

$$p \in \mathscr{P}_n,$$

with equality if and only if p is degenerate. When $\alpha = 2$, $H_{n,2}$ becomes the *Gini–Simpson index of diversity* [7],

$$G_n(p) \equiv H_{n,2}(p) = 1 - \sum p_i^2,$$

which has been used in biological work [16, 20].

Let F be a C^2-function on $U \times U$ and consider

$$D_{n,F}(x, y) \equiv \sum F(x_i, y_i); \qquad x, y \in U^n,$$

which will serve as a prototype for generating divergence measures. Suitable expressions of F in terms of a C^2-function ϕ on U give rise to the following prototypes of J-divergences.

I-Divergences

We assume that U is $\mathbb{R}_+$ and we choose F to be $F(x, y) = y\phi(x/y)$. The resulting $D_{n,F}$ function is denoted by $I_{n,\phi}$. Plainly, $I_{n,\phi}$ is convex on $\mathbb{R}_+^n \times \mathbb{R}_+^n$ if and only if ϕ is convex on $\mathbb{R}_+$. Moreover, if ϕ is convex on $\mathbb{R}_+$ and $\phi(1) = 0$, then $I_{n,\phi}$ gives a genuine *J-divergence* on $\mathscr{P}_n \times \mathscr{P}_n$ which is also convex there. This divergence is also known as the *Csiszár ϕ-divergence* [5], and, in general, is asymmetric. With the choice of $\phi = \phi_\alpha$, $I_{n,\phi}$ becomes the *directed divergence of degree* α [14]:

$$I_{n,\alpha}(p, q) = (\alpha - 1)^{-1}\left(\sum p_i^\alpha q_i^{1-\alpha} - 1\right);$$

$$p, q \in \mathscr{P}_n,$$

which is convex on $\mathscr{P}_n \times \mathscr{P}_n$ but nonadditive for $\alpha \neq 1$. The related *additive divergence of order* α,

$$\hat{I}_{n,\alpha}(p, q) \equiv (\alpha - 1)^{-1}\log[1 + (\alpha - 1)I_{n,\alpha}(p, q)],$$

on the other hand, is indeed an additive measure of divergence for all $\alpha > 0$. When $\alpha = 1$, these two divergences are identical and they are both equal to the Kullback–Leibler information number, i.e., $\hat{I}_{n,1} \equiv I_{n,1} \equiv I_n$. For $\alpha \neq 1$, on the other hand, it holds that

$$(\alpha - 1)\{I_{n,\alpha}(p, q) - \hat{I}_{n,\alpha}(p, q)\} \geq 0;$$

$$p, q \in \mathscr{P}_n,$$

with equality if and only if $p = q$. When $\alpha = \frac{1}{2}$,

$$I_{n,1/2}(p,q) = 2\left[1 - \sum(p_iq_i)^{1/2}\right]$$
$$= \sum(p_i^{1/2} - q_i^{1/2})^2,$$

which is the *Jeffreys invariant* [11] $\mathscr{J}_2(p,q)$, used extensively by Matusita [15] in inference problems. The quantity

$$\rho(p,q) \equiv 1 - 2^{-1}\mathscr{J}_2(p,q) = \sum(p_iq_i)^{1/2};$$
$$p,q \in \mathscr{P}_n$$

is known as the *affinity between p and q* [15]. This can be also interpreted as $\cos\theta$ between the unit vectors $(p_1^{1/2}, \ldots, p_n^{1/2})$ and $(q_1^{1/2}, \ldots, q_n^{1/2})$ in the Euclidean norm, i.e.,

$$\theta = \cos^{-1}\rho(p,q).$$

The quantity $\theta \equiv \theta(p,q)$ defines a proper distance between the distributions p and q, known as the *Hellinger distance** [16] and was proposed by Bhattacharyya [2] as a dissimilarity coefficient between two populations. The *I*-divergences $I_{n,1/2}$ and $\hat{I}_{n,1/2}$ are therefore symmetric and admit the expressions

$$I_{n,1/2}(p,q) = 4\sin^2(\theta/2)$$
$$\hat{I}_{n,1/2}(p,q) = 2\log\sec\theta.$$

J-Divergences

As in the *I*-divergence, U is $\mathbb{R}_+$ but we choose F to be $F(x,y) = y\phi(x/y) + x\phi(y/x)$. The resulting $D_{n,F}$ function is denoted by $J_{n,\phi}$; it may also be obtained via the symmetrization of the *I*-divergence, namely

$$J_{n,\phi}(x,y) = I_{n,\phi}(x,y) + I_{n,\phi}(y,x).$$

Another relationship is

$$J_{n,\phi}(x,y) = I_{n,\Phi}(x,y);$$
$$\Phi(t) \equiv t\phi(t^{-1}) + \phi(t).$$

Moreover, $J_{n,\phi}$ is nonnegative or convex on $\mathbb{R}_+^n \times \mathbb{R}_+^n$ if and only if ϕ is so on $\mathbb{R}_+$. As in the former case, if Φ is convex on $\mathbb{R}_+$ and $\Phi(1) = 0$ (which is always true if ϕ itself has these properties), then $J_{n,\phi}$ gives a *J*-

divergence on $\mathscr{P}_n \times \mathscr{P}_n$ which is also convex there. With the choice of $\phi = \phi_\alpha$, $J_{n,\phi}$ becomes the *J-divergence of degree* α:

$$J_{n,\alpha}(p,q) = (\alpha - 1)^{-1}$$
$$\times\left[\sum(p_i^\alpha q_i^{1-\alpha} + q_i^\alpha p_i^{1-\alpha}) - 2\right],$$

which is always convex on $\mathscr{P}_n \times \mathscr{P}_n$. When $\alpha = 1$, this divergence becomes the previously mentioned Jeffreys invariant, i.e., $J_{n,1} \equiv J_n$. For $\alpha = \frac{1}{2}$ we have $J_{n,1/2} \equiv 2I_{n,1/2}$, which was discussed in the section "*I*-Divergences." Finally, the following inequality [9] between J_n and $J_{n,1/2}$,

$$J_n \geqslant -4\log(1 - \tfrac{1}{4}J_{n,1/2}),$$

is available.

K-Divergences

For any $0 < \lambda < 1$, we choose F as the *Jensen difference*

$$F(x,y) = K_{1,\phi}^{(\lambda)}(x,y) \equiv \lambda\phi(x) + (1-\lambda)\phi(y)$$
$$- \phi[\lambda x + (1-\lambda)y]; \quad x,y \in U$$

and denote by $K_{n,\phi}^{(\lambda)}$ the resulting $D_{n,F}$ function. When $\lambda = \frac{1}{2}$ we simply write $K_{n,\phi}$ for $K_{n,\phi}^{(\lambda)}$. The quantity $K_{n,\phi}^{(\lambda)}$ is nonnegative on $U^n \times U^n$ if and only if ϕ is convex on U, and thus it provides a genuine *J*-divergence on $\mathscr{P}_n \times \mathscr{P}_n$ if ϕ is convex on U_0. For $\pi = (\pi_1, \ldots, \pi_k) \in \mathscr{P}_k$ and k vectors $y_1, \ldots, y_k \in U^n$, $K_{n,\phi}^{(\lambda)}$ admits the self-evident extension in terms of the ϕ-entropy, namely

$$K_{n,\phi}^\pi(y_1, \ldots, y_k) = H_{n,\phi}\left(\sum_{i=1}^k \pi_iy_i\right)$$
$$- \sum_{i=1}^k \pi_iH_{n,\phi}(y_i).$$

A key theorem in this direction is the fact that $K_{n,\phi}^\pi$ is convex on U^{nk} if and only if ϕ and $-(\phi'')^{-1}$ are convex on U [3]. A related but simpler result is that for fixed $y_1, \ldots, y_k \in U^n$, $K_{n,\phi}^\pi(y_1, \ldots, y_k)$ is a concave function with respect to $\pi \in \mathscr{P}_k$ if and only if ϕ is convex on U. In general, $K_{n,\phi}^\pi$, also called the *Jensen difference divergence*, is asymmetric and is symmetric only when π is

$(1/k, \ldots, 1/k)$. With the choice of $\phi = \phi_\alpha$, $K_{n,\phi}^\pi$ becomes the *Jensen difference divergence of degree* α or the *K-divergence of degree* α,

$$K_{n,\alpha}^\pi(y_1, \ldots, y_k) = H_{n,\alpha}\left(\sum_{i=1}^k \pi_i y_i\right)$$

$$- \sum_{i=1}^k \pi_i H_{n,\alpha}(y_i),$$

where $H_{n,\alpha}$ is the previously mentioned Havrda–Charvát entropy of degree α. This is a genuine *J-divergence* on $\mathscr{P}_n^k$, nonnegative on $\mathbb{R}_+^{nk}$ and for a fixed $y_1, \ldots, y_k \in \mathbb{R}_+^n$, $K_{n,\alpha}^\pi(y_1, \ldots, y_k)$ is concave with respect to $\pi \in \mathscr{P}_k$. Moreover, $K_{n,\alpha}^\pi$ is convex on $\mathbb{R}_+^{nk}$ if and only if $\alpha \in [1,2]$ and hence for $n \geqslant 3$, $K^{\pi_{n,\alpha}}$ is convex on $\mathscr{P}_n^k$ if and only if $\alpha \in [1,2]$. On the other hand, $K_{n,\alpha}^\pi$ is convex on $\mathscr{P}_2^k$ if and only if $\alpha \in [1,2] \cup [3, 11/3]$. When $\alpha = 1$, $K_{n,\alpha}^\pi$ is written as K_n^π, i.e.,

$$K_n^\pi(y_1, \ldots, y_k) = H_n\left(\sum_{i=1}^k \pi_i y_i\right)$$

$$- \sum_{i=1}^k \pi_i H_n(y_i),$$

where H_n is the Shannon entropy. When $k = 2$ and $\pi = (\lambda, 1 - \lambda)$, $0 < \lambda < 1$, this divergence is intimately related to the *J*-divergence J_n, i.e.,

$$K_n^{(\lambda)}(x, y) \leqslant \lambda(1 - \lambda) J_n(x, y); \qquad x, y \in \mathbb{R}_+^n$$

and equality holds if and only if $x = y$. The quantity $K_n^\pi(p_1, \ldots, p_k)$ is known as *transinformation* or *mutual information* in information theory, where it is defined as a measure of information on a k-input channel $p_1, \ldots, p_k \in \mathscr{P}_n$ for input distribution $\pi = (\pi_1, \ldots, \pi_k) \in \mathscr{P}_k$ [1]. In biological work [10], $K_n^\pi(p_1, \ldots, p_k)$ is defined to be the *information radius* on the probability distribution $\pi \in \mathscr{P}_k$ associated with $p_1, \ldots, p_k \in \mathscr{P}_n$. It has also a wide range of applications in cluster analysis and analysis of diversity between populations [10, 16, 19]. One may also consider the Jensen difference of $K_{n,\alpha}^\pi$ itself, but the resulting functions are not convex unless $\alpha = 2$. On the other hand,

all higher-order Jensen differences of $K_{n,2}^\pi$ are convex [4]. The Jensen difference $K_{n,2}^\pi(y_1, \ldots, y_k)$ is defined for all $y_i = (y_{i1}, \ldots, y_{in})$ in $\mathbb{R}^n$, $i = 1, \ldots, k$, and admits the expression

$$K_{n,2}^\pi(y_1, \ldots, y_k)$$

$$= \frac{1}{2} \sum_{i,\,j=1}^k \pi_i \pi_j \sum_{m=1}^n (y_{im} - y_{jm})^2.$$

In particular, when $k = 2$ and $\pi = (\lambda, 1 - \lambda)$, $0 < \lambda < 1$,

$$K_{n,2}^{(\lambda)}(x, y) = \lambda(1 - \lambda) \sum_{i=1}^n (x_i - y_i)^2;$$

$$x, y \in R^n,$$

which is the square of the Euclidean distance, modulo the positive factor $\lambda(1 - \lambda)$.

L-Divergences

Again, U is $\mathbb{R}_+$ but we choose F to be

$$F(x, y) = (x - y)(\psi(x) - \psi(y)),$$

$$\psi(x) \equiv \phi(x)/x.$$

The resulting $D_{n,F}$ is denoted by $L_{n,\phi}$. It is nonnegative if and only if ψ is increasing on $\mathbb{R}_+$ and in that case it gives a *J*-divergence on $\mathscr{P}_n \times \mathscr{P}_n$, called the *L-divergence*. This quantity is intimately connected to the *K*-divergence, i.e.,

$$\lambda(1 - \lambda) L_{n,\phi}(x, y)$$

$$= K_{n,\phi}^{(\lambda)}(x, y) - \sum_{i=1}^n [\lambda x_i + (1 - \lambda) y_i] K_{1,\psi}^{(\lambda)}(x_i, y_i).$$

Consequently, one has

$$\lambda(1 - \lambda) L_{n,\phi}(x, y) \geqslant K_{n,\phi}^{(\lambda)}(x, y); \qquad x, y \in \mathbb{R}_+^n,$$

if and only if ψ is concave on $\mathbb{R}_+$, and in that case equality occurs if and only if $x = y$. With the choice of $\phi = \phi_\alpha$, $L_{n,\phi}$ becomes the *L-divergence of degree* α,

$$L_{n,\alpha}(x, y) = (\alpha - 1)^{-1} \sum (x_i - y_i)(x_i^{\alpha-1} - y_i^{\alpha-1}),$$

which is a genuine *J*-divergence for all $\alpha > 0$. It is convex if $\alpha \in [1,2]$. When $\alpha = 1$, $L_{n,\alpha}$ is simply denoted by L_n, in which case

$$L_n(x, y) = \sum (x_i - y_i)(\log x_i - \log y_i),$$

Table 1   *J*-Divergences $D_{n,F}(x, y) = \sum F(x_i, y_i)$, $x, y \in \mathbb{R}^n_+$, **Where** F **Is Related to** ϕ of the ϕ-Entropy $H_{n,\phi}(x) = -\sum \phi(x_i)$, $x \in \mathbb{R}^n_+$

Symbol of $D_{n,F}$	$F(x, y)$	Symbolic Name	Attached Name	Remarks
$I_{n,\phi}$	$y\phi(x/y)$	*I*-divergence	Csiszár's ϕ-divergence	Also called information divergence
$J_{n,\phi}$	$y\phi(x/y) + x\phi(y/x)$	*J*-divergence	Csiszár's symmetric ϕ-divergence	$J_{n,\phi} \equiv I_{n,\Phi}$ with $\Phi(t) \equiv t\phi(t^{-1}) + \phi(t)$
$K^{(\lambda)}_{n,\phi}$	$\lambda\phi(x) + (1-\lambda)\phi(y)$ $- \phi[\lambda x + (1-\lambda)y]$ $(0 < \lambda < 1)$	*K*-divergence	Jensen difference divergence	$K_{n,\phi} \equiv K^{(1/2)}_{n,\phi}$
$L_{n,\phi}$	$(x-y)[\dfrac{\phi(x)}{x} - \dfrac{\phi(y)}{y}]$	*L*-divergence	—	—

Table 2   *J*-Divergences of Degree α, Corresponding to Table 1 with $\phi = \phi_\alpha$, Where $\phi_\alpha(x) = (\alpha - 1)^{-1}(x^\alpha - x)$, $\alpha > 0$

Symbol of $D_{n,F}$	Symbolic Name	Attached Name	$\alpha = 1$	Remarks
$I_{n,\alpha}$	*I*-divergence of degree α	Directed divergence of degree α	$I_n \equiv I_{n,1}$, Kullback–Leibler information number	$I_{n,1/2}$ is Jeffreys' invariant $\mathscr{J}_2$
$J_{n,\alpha}$	*J*-divergence of degree α	Jeffreys' divergence of degree α	$J_n \equiv J_{n,1}$, Jeffreys' invariant	$J_{n,1/2} = 2I_{n,1/2}$
$K^{(\lambda)}_{n,\alpha}$	*K*-divergence of degree α	Jensen difference divergence of degree α	$K^{(\lambda)}_n \equiv K^{(\lambda)}_{n,1}$, Jensen difference divergence, transinformation	$K^{(\lambda)}_{n,2}(x, y) =$ $\lambda(1-\lambda)\sum(x_i - y_i)^2$
$L_{n,\alpha}$	*L*-divergence of degree α	—	$L_n \equiv L_{n,1} \equiv J_n$, Jeffreys' invariant	$L_{n,2}(x, y)$ $= \sum (x_i - y_i)^2$

and hence is identical with the Jeffreys invariant, i.e., $L_n \equiv J_n$. For $\alpha = 2$ we have

$$\lambda(1-\lambda)L_{n,2}(x, y) \equiv K^{(\lambda)}_{n,2}(x, y); \quad x, y \in \mathbb{R}^n,$$

while for $\alpha \neq 2$,

$$(\alpha - 2)\{ K^{(\lambda)}_{n,\alpha}(x, y) - \lambda(1-\lambda)L_{n,\alpha}(x, y)\} \geqslant 0;$$

$$y; y \in \mathbb{R}^n_+,$$

with equality if and only if $x = y$. For convenience, the classification of the various divergences above is provided in tabular form in Tables 1 and 2.

There are several other measures of *J*-divergence that arise from other entropy measures which are also fundamental in statistics and information theory [1, 3, 5, 6, 8, 16]. The convexity properties of these divergences, especially those which are based on the Jensen difference, have been studied by Burbea and Rao [3, 4]. Additional properties of these divergences, as differential geometric properties, etc., may be also found in the Bibliography, where a unified treatment is also given in the Burbea and Rao articles.

References

[1] Aczél, J. and Daróczy, Z. (1975). *On Measures of Information and Their Characterizations.* Academic Press, New York.

[2] Bhattacharyya, A. (1946). A measure of divergence between two multinomial populations. *Sankhyā*, **7**, 401–406.

[3] Burbea, J. and Rao, C. R. (1982). On the convexity of some divergence measures based on entropy functions. *IEEE Trans. Inf. Theory*, **IT-28**, 489–495. (A detailed analysis of the convexity properties of the various generalized divergences, in particular those which are based on Jensen differences.)

[4] Burbea, J. and Rao, C. R. (1982). On the convexity of higher order Jensen differences based on entropy functions. *IEEE Trans. Inf. Theory*, **IT-28**, 961–963.

[5] Csiszár, I. (1972). A class of measures of informativity of observation channels. *Periodica Math. Hung.*, **2**, 191–213.

[6] Gallager, R. G. (1968). *Information Theory and Reliable Communication*. Wiley, New York.

[7] Gini, C. (1912). Variabilità e mutabilità. Studi Economico-Giuridici della Facolta di Giurisprudenza dell Universita di Cagliari, aIII, Parte II.

[8] Havrda, J. and Charvát, F. (1967). Quantification method of classification process: concept of structural α-entropy. *Kybernetika*, **3**, 30–35.

[9] Hoeffding, W. and Wolfowitz, J. (1958). Distinguishability of sets of distributions. *Ann. Math. Statist.*, **29**, 700–718.

[10] Jardine, N. and Sibson, R. (1971). *Mathematical Taxonomy*. Wiley, New York.

[11] Jeffreys, H. (1948). *Theory of Probability*, 2nd ed. Clarendon Press, Oxford.

[12] Kullback, S. (1959). *Information Theory and Statistics*. Wiley, New York.

[13] Kullback, S. and Leibler, R. A. (1951). On information and suffiency. *Ann. Math. Statist.*, **22**, 79–86.

[14] Mathai, A. M. and Rathie, P. N. (1975). *Basic Concepts in Information Theory and Statistics*. Halsted Press, New York.

[15] Matusita, K. (1957). Decision rule based on the distance for the classification problem. *Ann. Inst. Statist. Math. Tokyo*, **8**, 67–77.

[16] Rao, C. R. (1982). Diversity and dissimilarity coefficients: a unified approach. *Theor. Popul. Biol.*, **21**, 24–43.

[17] Rényi, A. (1961). On measures of entropy and information. *Proc. Berkeley Symp. Math. Statist. Prob.*, Vol. 1. University of California Press, Berkeley, Calif., pp. 547–561.

[18] Shannon, C. E. (1948). A mathematical theory of communication. *Bell Syst. Tech. J.*, **27**, 379–423, 623–656.

[19] Sibson, R. (1969). Information radius. *Zeit. Wahrscheinlichkeitsth. verwend. Geb.*, **14**, 149–160.

[20] Simpson, E. H. (1949). Measurement of diversity. *Nature (London)*, **163**, 688.

Bibliography

Aczél, J. and Daróczy, Z. (1975). *On Measures of Information and their Characterizations*. Academic Press, New York. [An axiomatic characterization of "transinformation" or "information radius" and a discussion of its properties (see pp. 196–199).]

Burbea, J. and Rao, C. R. (1982). Entropy differential metric, distance and divergence measures in probability spaces—a unified approach. *J. Multivariate Anal.*, **12**, 575–596. (A study of the local properties of the divergence measures within parametric families of probability distributions from the differential geometric standpoint.)

Campbell, L. L. (1975). *Selecta Statist. Canad.*, **2**, 39–45.

Csiszár, I. (1967). Information-type measures of difference in probability distributions and indirect observations. *Stud. Sci. Math. Hung.*, **2**, 299–318.

Csiszár, I. (1972). A class of measures of informativity of observation channels. *Periodica Math. Hung.*, **2**, 191–213. (A study of the properties of the Csiszár φ-divergence.)

Csiszár, I. (1977). Information measures: a critical survey. *Trans. 7th. Prague Conf. Inf. Theory. Stat. Decision Functions, Random Processes*. D. Reidel, Boston.

Ferreri, C. (1980). *Statistica*, **40**, 155–168.

Gallager, R. G. (1968). *Information Theory and Reliable Communication*. Wiley, New York. (An authoritive text on the subject. See p. 16 for the properties of K_n^π.)

Jeffreys, H. (1948). *Theory of Probability*, 2nd ed. Clarendon Press, Oxford. (A classical monograph. Jeffreys invariants are introduced and studied on pp. 158–167.)

Kullback, S. (1959). *Information Theory and Statistics*. Wiley, New York. [An instructive monograph on the logarithmic measures of information and their application in the process of statistical inference (see pp. 1–66).]

Mathai, A. M. and Rathie, P. N. (1975). *Basic Concepts in Information Theory and Statistics*. Halsted Press, New York. [Contains axiomatic definitions for the basic measures of information and statistics through functional equations (see pp. 35–73).]

Mittal, D. P. (1975). *Metrika*, **22**, 35–46.

(DIVERSITY INDICES
ENTROPY
INFORMATION THEORY AND CODING
  THEORY)

Jacob Burbea

JEFFREYS' DIVERGENCES *See J*-DIVERGENCES AND RELATED CONCEPTS

JEFFREYS' INVARIANT *See* J-DIVERGENCES AND RELATED CONCEPTS

JEFFREYS' PRIOR DISTRIBUTION

Harold Jeffreys [1] suggested using, for prior distribution of a parameter θ, a density function proportional to the square root of the expected value of the square of the log-likelihood function, i.e.,

$$f(\theta) \propto \left\{ E\left[(\partial \log L / \partial \theta)^2 \right] \right\}^{1/2} \quad (1)$$

In the particular case when the only observed value, X, is that of a binomial* variable with parameters n (known) and θ,

$$L = \binom{n}{x} \theta^x (1-\theta)^{n-x}$$

$$(0 \leqslant \theta \leqslant 1; x = 0, 1, \ldots, n).$$

This leads to

$$f(\theta) \propto \left\{ \frac{n}{\theta(1-\theta)} \right\}^{1/2} \quad (0 \leqslant \theta \leqslant 1)$$

i.e., inserting a normalizing factor,

$$f(\theta) = \pi^{-1}\theta^{-1/2}(1-\theta)^{-1/2}$$

$$(0 \leqslant \theta \leqslant 1). \quad (2)$$

This distribution—a standard beta distribution* with parameters $\frac{1}{2}$, $\frac{1}{2}$—is sometimes called "Jeffreys' prior," although the name is more generally applicable to (1).

Perks [2] used the distribution (2) as a "noninformative" prior distribution for a mortality probability. He arrived at this distribution by using the general principle $f(\theta) \propto$ [standard deviation of efficient estimate* of θ]$^{-1}$, which is asymptotically equivalent to (1) under some regularity conditions.

A criticism of this kind of approach to constructing a prior distribution is that the prior depends on the distributions of the variables which are to be observed, so that the alteration in the conditions of observation (design of experiment, measurement instrument, etc.) could change the appropriate prior. (*See* INVARIANT PRIOR DISTRIBUTION for more detail on these controversial topics.)

References

[1] Jeffreys, H. (1946). *Proc. Roy. Soc. Lond. A*, **186**, 453–461.
[2] Perks, W. (1947). *J. Inst. Actuaries*, **73**, 285–334.

(BAYESIAN INFERENCE
FIDUCIAL INFERENCE
INVARIANT PRIOR DISTRIBUTIONS)

JENSEN'S INEQUALITY

This is a generalization of the inequality between arithmetic* and geometric* means. For any convex function ϕ and a random variable* X both defined on a given interval,

$$E\phi(X) \leqslant \phi(EX),$$

where E is the expectation operator (whenever the corresponding expectations exist). In mathematical statistics this inequality is used for proving consistency of maximum likelihood* estimators, among other applications.

Further Reading

Wald, A. (1949). *Ann. Math. Statist.*, **20**, 595–601.
Wolfowitz, J. (1949). *Ann. Math. Statist.*, **20**, 601–602.

JIŘINA SEQUENTIAL PROCEDURES

Let X be a random variable (rv) taking values in a partially ordered sample space $\mathscr{X}, \prec$ for which the equivalence classes have probability zero, i.e.,

$$P[X \prec x, X \succ x] = 0 \quad \text{for all} \quad x \in \mathscr{X}.$$

The order statistics $X_{(1)} \prec X_{(2)} \prec \cdots \prec X_{(n)}$, defined a.s. from any independent, identically distributed (i.i.d.) sample of size n, can be used to form *statistically equivalent blocks* (s.e.b.'s), i.e.,

$$B_j = \{x \in \mathcal{X} : X_{(j-1)} \prec x \prec X_{(j)}\},$$

$$j = 1, \ldots, n+1$$

with the obvious interpretations of $X_{(0)}$ and $X_{(n+1)}$. This nomenclature follows from the fact that any (random) region

$$R = \bigcup_{j \in K} B_j \quad \text{for some } K \subset \{1, 2, \ldots, n+1\}$$

has coverage, (label it $\mu(R) = P[X \in R]$), with a distribution that depends only on the number of elements in K; assume that exactly κ s.e.b.'s are omitted.

The coverage of R has an incomplete beta distribution; i.e., for $\beta \in (0, 1)$,

$$P[\mu(R) > \beta] = \frac{\Gamma(n+1)}{\Gamma(\kappa)\Gamma(n - \kappa + 1)}$$

$$\times \int_\beta^1 u^{n-\kappa}(1-u)^{\kappa - 1} du. \quad (1)$$

Thus R forms a distribution-free tolerance region*, in the sample space $\mathcal{X}$. See ref. 10 and the texts by Wilks [11] and Fraser [3]. Such procedures are attributed to Wilks.

When observations occur sequentially it might be advantageous to use a procedure in which the sample size is not fixed but to have sampling stop whenever the boundary of the tolerance region remains unchanged for a sufficient time. Such sequentially determined tolerance regions are called *Jiřina procedures* after the work of M. Jiřina [4, 5]. His procedure is defined as follows:

Let η, k be preassigned positive integers. Determine one s.e.b. from the first η i.i.d. observations of X; call it R_1. During the jth stage $j = 2, 3, \ldots,$ continue sampling as long as

$$X_{n+i} \in R_{j-1} \quad \text{and} \quad i < k, \quad (2)$$

where n is the number of observations drawn during the preceding $(j-1)$ stages. If (2) holds for $i = k$, stop and set $D = R_{j-1}$. If $X_{n+i} \notin R_{j-1}$ and $i \leqslant k$, determine the new region from the ordered sample of $(n + i)$ observations, call it R_j, by omitting exactly η of the $(n + i + 1)$ s.e.b.'s in such a way that

$R_{j-1} \subset R_j$. We then continue to do our sampling for the $(j + 1)$st stage. This procedure terminates with probability 1; call the region so determined D. The coverage has the distribution

$$P[\mu(D) > \beta] = (1 - \beta)^\eta \exp\left(\eta \sum_{i=1}^k \beta^i / i\right)$$

$$\text{for} \quad 0 < \beta < 1. \quad (3)$$

Various aspects of this procedure were studied by Jiřina and are also discussed in refs. 6 and 7.

In order to compute the assurance with which the coverage of Jiřina's procedure exceeds β, we need tables, or a programmable calculator, to obtain values of

$$\Lambda_k(\beta) = -\sum_1^k \frac{\beta^j}{j} - \ln(1 - \beta)$$

$$= \int_0^\beta \frac{x^k}{1 - x} dx.$$

A short table is given in ref. 7, as are methods to compute the expected sample size of the Jiřina procedure. Methods for the determination of the appropriate sample size to obtain a specified confidence for the coverage were given in refs. 1 and 9 for Wilks' problem. Thus comparisons can be made. If we specify a confidence of $100(1 - \alpha)\%$ that the coverage exceed β, then from (1) and (2), with $\eta = \kappa = 1$ (i.e., an upper or lower confidence bound based on the maximum or minimum observation is sought), we must find the least integers n and k such that $n \geqslant \ln \alpha / \ln \beta$ and $\Lambda_k(\beta) \leqslant \ln(1 - \alpha)$. Let us pick $\alpha = 0.05$, $\beta = 0.09$; then we find a fixed sample size of $n = 29$, and the number of $k = 19$ observations for which the extreme observation remains unchanged.

From results in ref. 7 the expected sample size for the J-procedure with $\eta = 1$ is $\exp\{\sum_1^k 1/j\} = 34.8$. Thus between these alternatives, if the cost of determining the tolerance region depends principally on the number of observations, the J-procedure is expected to be more expensive than Wilks' fixed-sample procedure.

Moreover, if all arrangements of the order observations are equally likely, then the fixed-sample procedure is usually better in the sense of stochastically larger coverage than any sequential procedure with the same expected sample size. However, if the observations determining the boundary must be accepted or rejected as they occur, then only sequential methods can be used. Moreover, when the breaking of records is always kept as a test against trend (see ref. 2), or when there is a known trend in the observations taken serially (see ref. 8), the sequential tolerance bounds may be determined using smaller samples but with the required coverage.

References

[1] Birnbaum, Z. W. and Zuckerman, H. S. (1949). *Ann. Math. Statist.*, **20**, 313–316.

[2] Foster, F. C. and Stuart, A. (1954). *J. R. Statist. Soc. B*, **16**, 1–23.

[3] Fraser, D. A. S. (1957). *Nonparametric Methods in Statistics.* Wiley, New York.

[4] Jiřina, M. (1952). *Czech. Math. J.*, **77**, 211–232; correction, *ibid.*, **78**, 283 (1953), (in Russian).

[5] Jiřina, M. (1961). From *Select. Transl. Math. Statist. Prob.* (*Amer. Math. Soc.*), **1**, 145–156.

[6] Saunders, S. C. (1960). *Ann. Math. Statist.*, **31**, 198–216.

[7] Saunders, S. C. (1963). *Ann. Math. Statist.*, **34**, 847–856.

[8] Saunders, S. C. (1963). *Ann. Math. Statist.*, **34**, 857–865.

[9] Scheffé, H. and Tukey, J. W. (1944). *Ann. Math. Statist.*, **15**, 217.

[10] Wilks, S. S. (1941). *Ann. Math. Statist.*, **12**, 91–96.

[11] Wilks, S. S. (1962). *Mathematical Statistics.* Wiley, New York.

(SEQUENTIAL ANALYSIS)

SAM C. SAUNDERS

JOHNSON–NEYMAN TECHNIQUE

The Johnson–Neyman technique, as it was originally formulated in ref. 3, applies to a situation in which measurements on one Y-variable (dependent variable) and two X-variables (independent variables) are available for the members of two groups. The expected value of the Y-variable is assumed to be a linear function of the X-variables, but not generally the same linear function for both groups. The Y-variable may sometimes be called a criterion or response variable; the X-variables may be called predictor, control, or baseline variables. The Johnson–Neyman technique is used to obtain a point set of values of the X-variables for which one would reject, at a specified level α (such as $\alpha = 0.05$), the null hypothesis that the two groups have the same expected Y-value. This point set, or "region of significance," may be plotted on a graph.

The original formulation of the Johnson–Neyman technique has been generalized so that the number of X-variables may be greater than two, although naturally a point set involving more than two X-variables can no longer be easily plotted. The Johnson–Neyman technique is also applicable when there is just one X-variable, a case that is simple and not uncommon. The detailed formulas that follow will be for a general number of X-variables, to be denoted by r ($r \geqslant 1$).

Suppose that observations are available for n_1 members of the first group and n_2 members of the second group. For the kth member of group j ($k = 1, 2, \ldots, n_j$; $j = 1, 2$), let Y_{jk} denote the observed Y-value and let $\mathbf{X}_{jk}(r \times 1) = (X_{1jk}, X_{2jk}, \ldots, X_{rjk})'$ denote the set of observed values of the r X-variables. With respect to the conditional distribution of Y_{jk} given $\mathbf{X}_{jk}$, it is assumed that the Y_{jk}'s are normally and independently distributed with equal variances, and that

$$E(Y_{jk} \mid \mathbf{X}_{jk}) = \beta_{0j} + \boldsymbol{\beta}_j' \mathbf{X}_{jk},$$

where β_{0j} and $\boldsymbol{\beta}_j(r \times 1) = (\beta_{1j}, \beta_{2j}, \ldots, \beta_{rj})'$ (for $j = 1, 2$) denote unknown regression parameters.

Some additional notation needs to be introduced. For $j = 1, 2$, let $\mathbf{u}_j(n_j \times 1)$ denote a vector consisting of all 1's, and define

$$X_j(r \times n_j) = (\mathbf{X}_{j1}, \mathbf{X}_{j2}, \ldots, \mathbf{X}_{jn_j}),$$

$$Y_j(n_j \times 1) = (Y_{j1}, Y_{j2}, \ldots, Y_{jn_j})',$$

$$\overline{\mathbf{X}}_j(r \times 1) = (1/n_j)\mathbf{X}_j\mathbf{u}_j,$$

$$\overline{Y}_j = (1/n_j)\mathbf{Y}_j'\mathbf{u}_j,$$

$$\mathbf{C}_j(r \times r) = \mathbf{X}_j\mathbf{X}_j' - (1/n_j)(\mathbf{X}_j\mathbf{u}_j)(\mathbf{X}_j\mathbf{u}_j)',$$

and

$$\mathbf{W}_j(r \times 1) = \mathbf{X}_j\mathbf{Y}_j - (1/n_j)(\mathbf{X}_j\mathbf{u}_j)(\mathbf{Y}_j'\mathbf{u}_j).$$

Then the usual estimates of $\boldsymbol{\beta}_j$ and β_{0j} may be written as

$$\mathbf{b}_j(r \times 1) = \mathbf{C}_j^{-1}\mathbf{W}_j$$

and

$$b_{0j} = \overline{Y}_j - \mathbf{b}_j'\overline{\mathbf{X}}_j,$$

respectively. Also, the error mean square may be written as $s_e^2 = S_e^2/f$, where

$$S_e^2 = \sum_{j=1}^{2} \left[\mathbf{Y}_j'\mathbf{Y}_j - (1/n_j)(\mathbf{Y}_j'\mathbf{u}_j)^2 - \mathbf{b}_j'\mathbf{W}_j \right]$$

and

$$f = \sum_{j=1}^{2} (n_j - r - 1).$$

Let $\mathbf{X}(r \times 1) = (X_1, X_2, \ldots, X_r)'$ denote a set of values of the X-variables. Define

$$\Delta(\mathbf{X}) = (\beta_{02} - \beta_{01}) + (\boldsymbol{\beta}_2 - \boldsymbol{\beta}_1)'\mathbf{X},$$

which is the true difference in expected Y-value between the two groups at point $\mathbf{X}$, and

$$D(\mathbf{X}) = (b_{02} - b_{01}) + (\mathbf{b}_2 - \mathbf{b}_1)'\mathbf{X},$$

which is the estimated difference in expected Y-value between the two groups at point $\mathbf{X}$. Also define

$$v(\mathbf{X}) = \sum_{j=1}^{2} \left[(1/n_j) + (\mathbf{X} - \overline{\mathbf{X}}_j)'\mathbf{C}_j^{-1}(\mathbf{X} - \overline{\mathbf{X}}_j) \right].$$

Then the Johnson–Neyman "region of significance," which will be referred to as R, consists of the set of all points $\mathbf{X}$ such that

$$\left[D(\mathbf{X}) \right]^2 - t_{f,1-(1/2)\alpha}^2 v(\mathbf{X})s_e^2 > 0,$$

where $t_{f,1-(1/2)\alpha}$ denotes the $(1 - \frac{1}{2}\alpha)$ fractile

of the t-distribution* with f degrees of freedom. For any point $\mathbf{X}$ in R, one can reject at level α the null hypothesis that $\Delta(\mathbf{X}) = 0$, or, as an alternative interpretation, one can state with at least $100(1 - \alpha)\%$ confidence that $\Delta(\mathbf{X})$ differs from 0 and has the same sign as $D(\mathbf{X})$, inasmuch as the $100(1 - \alpha)\%$ two-sided confidence interval for $\Delta(\mathbf{X})$ does not include the value 0.

Confusion sometimes arises with respect to R. Although one can be at least $100(1 - \alpha)\%$ confident in making a statement about the difference between the two groups for any specified *individual* point $\mathbf{X}$ in R, one cannot be $100(1 - \alpha)\%$ confident in making the statements about the differences simultaneously for *all* points in R.

However, one can define a different point set, to be referred to as R', that *does* allow one to make simultaneous statements (see ref. 5). R' consists of the set of all points $\mathbf{X}$ such that

$$\left[D(\mathbf{X}) \right]^2 - (r + 1)F_{r+1,f,1-\alpha}v(\mathbf{X})s_e^2 > 0,$$

where $F_{r+1,f,1-\alpha}$ denotes the $(1 - \alpha)$ fractile of the F-distribution* with $r + 1$ and f degrees of freedom. With confidence coefficient $\geq 100(1 - \alpha)\%$ one can state simultaneously for *all* points $\mathbf{X}$ in R' that the two groups differ [and that $\Delta(\mathbf{X})$ has the same sign as $D(\mathbf{X})$]. In other words, in the long run not more than $100\alpha\%$ of such regions R' that are calculated will contain any points at all for which $\Delta(\mathbf{X})$ is 0 or has a different sign from $D(\mathbf{X})$.

For a given α-value, R' is smaller than R, and there are points in R that are outside R', but not vice versa. However, R' can be made larger by choosing a somewhat larger α-value for R' than what one would normally choose for R.

Instead of or in addition to using R or R', one can use confidence intervals* that are closely associated with R or R' (see ref. 5). They can provide extra information. Simple confidence intervals are related to R, and simultaneous confidence intervals to R'. Specifically, a $100(1 - \alpha)\%$ simple confidence interval for $\Delta(\mathbf{X})$ for any specified individual point $\mathbf{X}$ (whether inside or outside

R) is given by

$$D(\mathbf{X}) \pm t_{f,1-(1/2)\alpha}\left[v(\mathbf{X})s_e^2\right]^{1/2},$$

and $100(1 - \alpha)\%$ simultaneous confidence intervals for the functions $\Delta(\mathbf{X})$ *for all possible points* $\mathbf{X}$ in the r-dimensional X-space are given by

$$D(\mathbf{X}) \pm \left[(r + 1)F_{r+1,f,1-\alpha}v(\mathbf{X})s_e^2\right]^{1/2}.$$

The point set R or R' can consist of (1) the null set; (2) a single convex set in which either $D(\mathbf{X}) > 0$ throughout or else $D(\mathbf{X}) < 0$ throughout; or (3) two separate convex sets, with $D(\mathbf{X}) > 0$ throughout one of them and $D(\mathbf{X}) < 0$ throughout the other. For $r = 2$, generally set (2) will be the inside of an ellipse and sets (3) will be associated with the two branches of a hyperbola. Sometimes part or all of R or R' will lie well outside the portion of the X-space that is of practical interest; this can limit the usefulness of the results.

It is a common practice to make a preliminary test of the hypothesis $\boldsymbol{\beta}_1 = \boldsymbol{\beta}_2$ before using the Johnson–Neyman technique: If this hypothesis is rejected by the test, then the Johnson–Neyman technique is applied, but if it is not rejected, then analysis of covariance* is applied rather than the Johnson–Neyman technique. Under the assumption that $\boldsymbol{\beta}_1 = \boldsymbol{\beta}_2$, analysis of covariance will provide a confidence interval around ($\beta_{02} - \beta_{01}$), or a test of the hypothesis $\beta_{01} = \beta_{02}$. It is not always recognized, however, that there is nothing incorrect about using R or R' even if $\mathbf{b}_1$ is close to $\mathbf{b}_2$. Although the results from R or R' (or the associated confidence intervals) may often be less definitive than those from analysis of covariance if $\mathbf{b}_1$ does not differ significantly from $\mathbf{b}_2$, this draw-back could sometimes be offset by the fact that the required assumptions for R or R' are not as restrictive as those for analysis of covariance.

A numerical example with $r = 2$ will illustrate the application of the formulas that were given earlier. Suppose that there are $n_1 = 9$ members in the first group and $n_2 = 8$ members in the second group, with observed values of X_{1jk}, X_{2jk}, and Y_{jk} as shown in Table 1. Then

$$\mathbf{C}_1 = \begin{bmatrix} 8522 & 892 \\ 892 & 141 \end{bmatrix} - (1/9)\begin{bmatrix} 268 \\ 27 \end{bmatrix}[268 \quad 27]$$

$$= \begin{bmatrix} 541.556 & 88 \\ 88 & 60 \end{bmatrix},$$

$$\mathbf{C}_2 = \begin{bmatrix} 8361 & 536 \\ 536 & 68 \end{bmatrix} - (1/8)\begin{bmatrix} 255 \\ 18 \end{bmatrix}[255 \quad 18]$$

$$= \begin{bmatrix} 232.875 & -37.75 \\ -37.75 & 27.5 \end{bmatrix},$$

$$\mathbf{W}_1 = \begin{bmatrix} 15{,}404 \\ 1396 \end{bmatrix} - (1/9)\begin{bmatrix} 268 \\ 27 \end{bmatrix}(503) = \begin{bmatrix} 425.778 \\ -113 \end{bmatrix},$$

$$\mathbf{W}_2 = \begin{bmatrix} 12{,}323 \\ 718 \end{bmatrix} - (1/8)\begin{bmatrix} 255 \\ 18 \end{bmatrix}(373) = \begin{bmatrix} 433.625 \\ -121.25 \end{bmatrix},$$

$$\mathbf{b}_1 = \begin{bmatrix} 0.0024243 & -0.0035557 \\ -0.0035557 & 0.0218816 \end{bmatrix}\mathbf{W}_1 = \begin{bmatrix} 1.4340 \\ -3.9865 \end{bmatrix},$$

$$\mathbf{b}_2 = \begin{bmatrix} 0.0055232 & 0.0075818 \\ 0.0075818 & 0.0467714 \end{bmatrix}\mathbf{W}_2 = \begin{bmatrix} 1.4757 \\ -2.3834 \end{bmatrix},$$

$$b_{01} = (503/9) - 1.4340(268/9) - (-3.9865)(27/9)$$
$$= 25.1470,$$

$$b_{02} = (373/8) - 1.4757(255/8) - (-2.3834)(18/8)$$
$$= 4.9497,$$

$$S_e^2 = 29{,}859 - (503^2/9) - 1.4340(425.778)$$
$$- (-3.9865)(-113) + 18{,}677 - (373^2/8)$$
$$- 1.4757(433.625) - (-2.3834)(-121.25)$$
$$= 1042.84,$$

$$f = (9 - 2 - 1) + (8 - 2 - 1) = 11,$$

and

$$s_e^2 = 1042.84/11 = 94.803.$$

Table 1   Data for Numerical Example

	First Group ($j = 1$)									Second Group ($j = 2$)							
k:	1	2	3	4	5	6	7	8	9	1	2	3	4	5	6	7	8
X_{1jk}	31	37	20	19	38	20	39	35	29	31	34	21	30	31	42	33	33
X_{2jk}	2	4	0	4	9	0	2	2	4	1	0	6	1	4	3	1	2
Y_{jk}	49	72	52	42	38	55	86	54	55	45	45	19	49	44	58	66	47

Analysis of covariance might be performed at this point. (If so, one calculates

$$\mathbf{b}(r \times 1) = (\mathbf{C}_1 + \mathbf{C}_2)^{-1}(\mathbf{W}_1 + \mathbf{W}_2)$$

$$= \begin{bmatrix} 1.3331 \\ -3.4427 \end{bmatrix},$$

the estimate of β_1 and β_2 if they are the same as each other, and

$$S_{e*}^2 = \sum_{j=1}^{2} \left[\mathbf{Y}_j'\mathbf{Y}_j - (1/n_j)(\mathbf{Y}_j'\mathbf{u}_j)^2 \right]$$

$$- \mathbf{b}'(\mathbf{W}_1 + \mathbf{W}_2) = 1080.63,$$

the error sum of squares if $\beta_1 = \beta_2$. Then

$$F_{r,f} = \left[(S_{e*}^2 - S_e^2)/r \right]/S_e^2 = 0.199$$

is the F-statistic for testing the hypothesis $\beta_1 = \beta_2$, which cannot be rejected at the 0.05 level since 0.199 is less than $F_{2,11;0.95} = 3.98$. The $100(1 - \alpha)\%$ confidence interval around $(\beta_{02} - \beta_{01})$ under the assumption that $\beta_1 = \beta_2$ is

$$\left(\overline{Y}_2 - \overline{Y}_1 \right) - \mathbf{b}'\left(\overline{\mathbf{X}}_2 - \overline{\mathbf{X}}_1 \right)$$

$$\pm t_{n_1 + n_2 - r - 2, 1 - (1/2)\alpha} \left[(1/n_1) + (1/n_2) \right.$$

$$+ \left. (\overline{\mathbf{X}}_2 - \overline{\mathbf{X}}_1)'(\mathbf{C}_1 + \mathbf{C}_2)^{-1}(\overline{\mathbf{X}}_2 - \overline{\mathbf{X}}_1) \right]^{1/2}$$

$$\times \left[S_{e*}^2/(n_1 + n_2 - r - 2) \right]^{1/2},$$

or -14.642 ± 9.870 if $\alpha = 0.05$.) The point set R for $\alpha = 0.05$ is given by

$$[(4.9497 - 25.1470) + (1.4757 - 1.4340)X_1$$

$$+ (-2.3834 - (-3.9865))X_2]^2$$

$$- 2.2010^2 [(1/9) + 0.0024243(X_1 - 268/9)^2$$

$$+ 2(-0.0035557)(X_1 - 268/9)(X_2 - 27/9)$$

$$+ 0.0218816(X_2 - 27/9)^2$$

$$+ (1/8) + 0.0055232(X_1 - 255/8)^2$$

$$+ 2(0.0075818)(X_1 - 255/8)(X_2 - 18/8)$$

$$+ 0.0467714(X_2 - 18/8)^2](94.803) > 0,$$

or

$$-3.6482X_1^2 - 3.5644X_1X_2 - 28.959X_2^2$$

$$+ 232.20X_1 + 216.92X_2 - 3671.81 > 0,$$

which is the inside of an ellipse. If the point set R' is obtained for, say, $\alpha = 0.10$, the calculations are the same as for R except that $t_{11,0.975}^2 = 2.2010^2 = 4.8443$ is replaced by $3F_{3,11,0.90} = 3 \times 2.6602 = 7.9806$. Then R' is given by

$$-6.0112X_1^2 - 5.9586X_1X_2 - 49.372X_2^2$$

$$+ 383.62X_1 + 399.29X_2 - 6313.13 > 0,$$

which is again the inside of an ellipse, but a smaller ellipse than the one associated with R.

Traditionally, the Johnson–Neyman technique has been used mainly in educational and psychological applications; principal early works are included in the list of references in ref. 2. Typically, the two groups correspond to two competing treatments (instructional methods, for example), although they may also represent something else, such as males and females, or blacks and whites. The members of the two groups may be persons, but sometimes they are classes (in which case it is best for every class to have about the same number of students so that every class average will have about the same variance). Typically, the X-variables will be measures of aptitude or ability, and the Y-variable will be a measure of performance.

In principle, there is no reason why the Johnson–Neyman technique cannot be used in additional fields of application. In fact, there have recently been applications in new fields; e.g., ref. 4 provides an application involving drug evaluation, ref. 7 describes an interesting application in which job dissatisfaction is the Y-variable, and ref. 6 refers to a medical application comparing surgical procedures.

One of the first computer programs for the Johnson–Neyman technique (for $r = 2$) is provided in ref. 2.

A few other points that are related to the Johnson–Neyman technique may be noted. The Johnson–Neyman technique can be extended to handle more than two groups; see ref. 5. In ref. 1, a relatively technical paper, an alternative to the Johnson–Neyman technique is presented. A modification of the Johnson–Neyman technique for the case where the two groups have different variances is considered in ref. 4 (which, incidentally, also presents a good detailed discussion of assumptions made for the Johnson–Neyman technique).

References

[1] Aitkin, M. A. (1973). *Brit. J. Math. Statist. Psychol.*, **26**, 261–269.

[2] Carroll, J. B. and Wilson, G. F. (1970). *Educ. Psychol. Meas.*, **30**, 121–132.

[3] Johnson, P. O. and Neyman, J. (1936). *Statist. Res. Mem.*, **1**, 57–93.

[4] Pigache, R. M., Graham, B. R., and Freedman, L. (1976). *Biol. Psychol.*, **4**, 213–235.

[5] Potthoff, R. F. (1964). *Psychometrika*, **29**, 241–256.

[6] Schafer, W. D. (1981). *Amer. Statist.*, **35**, 179.

[7] Vecchio, R. P. (1980). *Acad. Manag. J.*, **23**, 479–486.

Bibliography

Rogosa, D. (1980). *Psychol. Bull.*, **88**, 307–321. (Comprehensive general article on comparing two regression lines.)

Rogosa, D. (1981). *Educ. Psychol. Meas.*, **41**, 73–84. (Well-presented general article covering the Johnson–Neyman technique.)

(ANALYSIS OF COVARIANCE
CONFIDENCE INTERVALS AND REGIONS
EDUCATIONAL STATISTICS
REGRESSION
SIMULTANEOUS CONFIDENCE
    INTERVALS)

RICHARD F. POTTHOFF

JOHNSON'S SYSTEM OF DISTRIBUTIONS

The Johnson [12] system of distributions is based on a transformed normal variate. If Z is a standardized normal variate (zero mean, $\sigma = 1$), the system is defined by

$$Z = \gamma + \delta f(Y), \qquad (1)$$

where

1. $f(Y) = \ln\{Y + (1 + Y^2)^{1/2}\} = \sinh^{-1}Y$, leading to the S_U system with unbounded range $-\infty < Y < \infty$.

2. $f(Y) = \ln Y$, leading to the lognormal family S_L.

3. $f(Y) = \ln[Y/(1 - Y)]$, leading to the S_B system with bounded range $0 < Y < 1$.

(*Note*: The S refers to system, the subscript to the nature of the range.)

The variate Y is linearly related to a variate X which we wish to approximate in distribution, and

$$Y = (X - \xi)/\lambda, \qquad (2)$$

the parameters being determined from the equivalence of location and scale measures for Y and X; thus from the mean (μ_1') and standard deviation (σ)

$$\mu_1'(Y) = (\mu_1'(X) - \xi)/\lambda, \qquad (3a)$$

$$\sigma(Y) = \sigma(X)/\lambda, \qquad (3b)$$

assuming that the moments of X exist.

The parameters γ, δ may be determined from the equivalence of higher moments (skewness and kurtosis parameters, for example) or the equivalence of percentiles and/or moments.

What is the domain of the system? In terms of skewness ($\sqrt{\beta_1}$) and kurtosis (β_2), the system S_B holds for the region bounded by $\beta_1 = 0$, the "top" line $S_T \equiv \beta_2 - \beta_1 - 1 = 0$, and the lognormal curve (S_L) (see Fig. 1). S_U holds for the corresponding region below S_L. In relation to the Pearson system*, S_B overlaps type I, II, III, and part of type VI; similarly S_U overlaps type IV, V, VII, and part of VI. As for shapes of the density, S_U is unimodal, but S_B may be bimodal (Fig. 2). The skewness and kurtosis uniquely determine a member of the system.

It is clear that other Johnson-type systems can be generated by replacing the normal variate z in (1) by, say, a Laplacian variate [13], logistic variate [4, 30], or other densities, such as the Pearson type IV.

If we have a statistic (such as the sample mean, standard deviation, sample skewness, sample kurtosis, etc.) whose first four moments are known either exactly or approximately, then the Johnson system density with equivalent first four moments may be

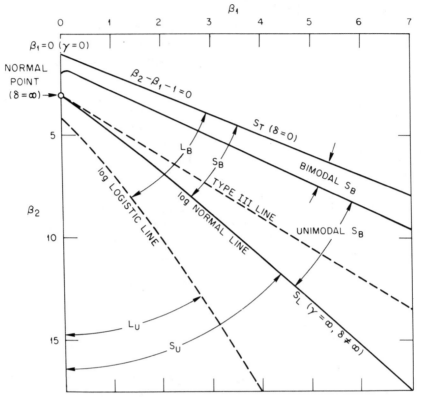

Figure 1 Domains of S_U and S_B (L_U and L_B refer to domains of the system based on a logistic variable).

used as an approximant. Thus

$$x_\alpha = \xi + \lambda \sinh\left[(z_\alpha - \gamma)/\delta\right] \qquad (S_U), \qquad (4a)$$

$$x_\alpha = \xi + \lambda/\left\{1 + \exp\left[(\gamma - z_\alpha)/\delta\right]\right\} \qquad (S_B), \qquad (4b)$$

with z_α a normal deviate at level α, give four-moment approximations to the corresponding x-deviate x_α. It is assumed that the sample is from a specified population. Note that this situation is quite different from the use of the Johnson system on empirical data for which not only moments but percentiles are readily available; the latter are rarely, if ever, known for the case of a theoretical statistic unless we include a precise knowledge of endpoints. Further remarks on this aspect appear in the sequel.

Moments of statistics are rarely known under nonnormality. For example, the low-order moments of the sample skewness and kurtosis are known exactly in normal sampling, but only asymptotically in nonnormal

sampling. However, in the latter case summatory techniques may yield accurate enough results for the application envisaged. The question of the accuracy of the approximants to X is a difficult one. Clearly, there are limitations since only four parameters are involved, whereas distributional characteristics (in general) are infinite. As a useful guide, if there is reason to believe that the statistic studied has a unimodal density free from discontinuities, then the Johnson system approximation evaluated for the probability levels $\alpha = 0.01, 0.05, 0.10, 0.90, 0.95, 0.99$ should show an error of at most 5 to 10%; this has been our experience over several statistics studied over several years. Caution is certainly needed when extreme levels are considered and also if it is suspected that one tail of the density is abrupt.

As with other transformation systems, such as Tukey's transformation of a uniform variate, the Johnson transformations can be used in simulation work when the properties of the distribution of a statistic are consid-

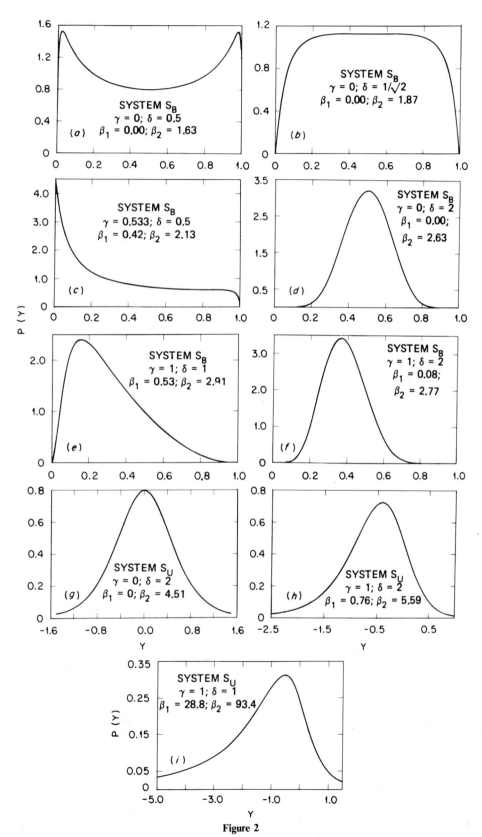

Figure 2

ered under nonnormality. As an illustration, see the study of the power* of tests for departures from normality by Pearson et al. [24].

Finally, the system can be used for the smoothing of empirical data. In his introduction of the system, Johnson [12] studied four large sets of data (1715 cases of cloudiness at Greenwich, a sample of 631,682 concerning the age of mothers at the birth of a child, and extensive data on length and breadth of 9440 beans due to Pretorius), fitted the appropriate Johnson curve by moments, and compared the χ^2 value with those arising from a Pearson curve fit. Moment fitting may be suspect because of the importance it assigns to outliers* and also because of the suspicion that high-order sample moments have large variances and therefore reflect this aspect in the fitting (matching percentiles, with advantageous results vis-à-vis χ^2, have been studied and will be referred to in the sequel). The general problem here, allowing for various sample sizes and appropriate assumptions regarding a null hypothesis, is very far from resolution, and considerable caution is needed. The sample sizes studied by Johnson should be noted carefully.

THE S_U SYSTEM

It has density $p(y)$ given by

$$p(y) = \left[\delta/(2\pi)^{1/2} \right] (1 + y^2)^{-1/2} \exp\left(-\tfrac{1}{2} z^2 \right)$$

$$(-\infty < y < \infty),$$

where $z = \gamma + \delta \sinh^{-1} y$.

Moments

Since $y = \sinh[(z - \gamma)/\delta]$, we have for the rth noncentral moment

$$\mu_r'(Y) = E(Y^r) \qquad (r = 0, 1, \dots)$$

$$= \left[1/(2\pi)^{1/2} \right] \int_{-\infty}^{\infty} (\sinh^r t) \exp\left(-\tfrac{1}{2} z^2 \right) dz$$

$$[t = (z - \gamma)/\delta],$$

and using

$$E[\exp(sZ)] = \exp\left[-s\Omega + s^2/(2\delta^2) \right]$$

$$(\Omega = \gamma/\delta),$$

the first few moments can be found (*see* APPROXIMATIONS TO DISTRIBUTIONS). In particular,

$$\mu_1'(Y) = -\omega^{1/2} \sinh \Omega \qquad \left[\omega = \exp(1/\delta^2) \right] \tag{5a}$$

$$\mu_2(Y) = \tfrac{1}{2}(\omega - 1)(\omega \cosh 2\Omega + 1). \tag{5b}$$

Again for the skewness $\sqrt{\beta_1} = \mu_3/\mu_2^{3/2}$ and kurtosis $\beta_2 = \mu_4/\mu_2^2$, we have

$$\sqrt{\beta_1}(Y) = -\left[\tfrac{1}{2}\omega(\omega - 1) \right]^{1/2} W^{-3/2}$$

$$\times \left[\omega(\omega + 2)\sinh 3\Omega + 3 \sinh \Omega \right] \tag{5c}$$

$$\beta_2(Y) = (a_4 \cosh 4\Omega + a_2 \cosh 2\Omega + a_0)/2W^2, \tag{5d}$$

where

$$W = \omega \cosh 2\Omega + 1$$

$$a_4 = \omega^2(\omega^4 + 2\omega^3 + 3\omega^2 - 3)$$

$$a_2 = 4\omega^2(\omega + 2)$$

$$a_0 = 3(2\omega + 1).$$

Note from (5c) that if $\sqrt{\beta_1} > 0$, then $\Omega < 0$, and $\sqrt{\beta_1} = 0$ has the solution $\Omega = 0$, and $\omega = \omega_0$, where from (5d)

$$\beta_2 = \tfrac{1}{2}(\omega_0^4 + 2\omega_0^2 + 3) \quad \text{or}$$

$$\omega_0 = \sqrt{(2\beta_2 - 2)^{1/2} - 1}. \tag{6}$$

In particular, when in addition $\beta_2 = 3$, we have $\delta_0 = 1$, so that $\delta = \infty$. Further, note that if δ and β_2 are known, then (5d) can be solved for $\cosh 2\Omega$, and so for Ω, which in turn substituted in the skewness equation (5c) leads to a value of $\sqrt{\beta_1}$ corresponding to the couplet (ω, β_2). More specifically, the solution of (5d) for given (ω, β_2) is

$$\tau(\omega, \beta_2) = \omega + 1 + 2\omega \sinh^2 \Omega \tag{7a}$$

$$= -\left[B + \sqrt{B^2 - 4AC} \right]/(2A), \tag{7b}$$

where

$$A = 2(\beta_2 - \omega^4 - 2\omega^3 - 3\omega^2 + 3)$$

$$B = 4(\omega^2 - 1)(\omega^2 + 2\omega + 3)$$

$$C = (\omega^2 - 1)^2(\omega^2 + 2\omega + 3).$$

Moment Solution

Equations (5c) and (5d) cannot be solved explicitly for ω, Ω. An excellent algorithm arises from Johnson's [14] observation, based on contours of δ in the (β_1, β_2) plane, that contours of constant δ are practically linear (Fig. 3) and parallel over a limited region ($\beta_2 < 5$). That being so, he suggested what amounts to the possibility that

$$\psi(\beta_1, \beta_2 ; \omega) = \left[\beta_2 - \tfrac{1}{2}(\omega^4 + 2\omega^2 + 3) \right]/\beta_1 \tag{8}$$

changes little as β_1, β_2 and the corresponding ω vary over a limited domain. Johnson's iterative solution uses this property. Taking $\omega_s > 1$ as an initial value of ω, compute $\tau(\omega_s, \beta_2)$ from (7b). Insert the implied values of ω, Ω into (5c) to evaluate an intermediate value of β_1, say, β_1^*. Then compute an improved ω from

$$\psi(\beta_1, \beta_2 ; \omega) = \psi(\beta_1^*, \beta_2 ; \omega_s) \tag{9}$$

and repeat the process until the desired accuracy is achieved. Fundamentally, Johnson used this approach to construct tables of $-\gamma$ and δ for $\sqrt{\beta_1} = 0.05(0.05)2.0$, and for values of β_2 at intervals of 0.1 (or 0.2) starting at a point in the (β_1, β_2) plane near the S_L curve (see Johnson [14] and the *Biometrika* tables edited by Pearson and Hartley [23]).

The algorithm to determine ω (or δ) is readily programmed on a desk calculator

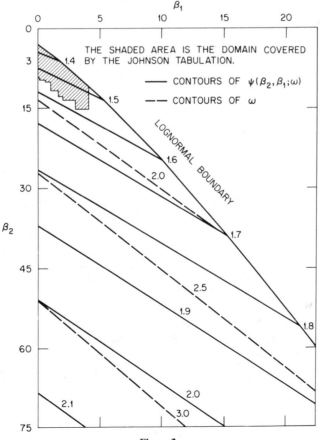

Figure 3

and Ω found as the quadratic root given in (7); keep in mind that $\Omega \lessgtr 0$ according as $\sqrt{\beta_1} \gtrless 0$. A quicker solution uses a rational fraction in β_1 and β_2 to approximate ψ ($= \psi^*$, say) from which there is the approximation

$$\omega = \sqrt{\sqrt{2\beta_2 - 2 - 2\beta_1 \psi^*(\beta_1, \beta_2; \delta)} - 1} \quad (10)$$

with Ω determined from (7) [3]. Note that if β_1 is small and limited accuracy is involved, then the solution for Ω from (7b) may turn out to be incorrectly imaginary. In this case use (5c) iteratively in the form

$$\sinh \Omega_n$$

$$= \frac{-2\beta_1 (\omega \cosh 2\Omega_{n-1} + 1)^3}{[\omega(\omega - 1)]^{1/2} J(\omega, \Omega_{n-1})}$$

$$(n = 1, 2, \ldots)$$

with

$$J(\omega, z) = 2\omega(\omega + 2)\cosh 2z + \omega^2 + 2\omega + 3,$$

$$\sinh \Omega_0 = -\tfrac{1}{3}\left\{2\beta_1 / [\omega(\omega^2 - 1)]\right\}^{1/2}.$$

Further Comments on S_U

Approximations and formulas relating any three of the four parameters γ, δ, β_1, and β_2 are rare. Leslie [19] cast the fundamental equations relating these parameters in the form

$$\beta_1 = \frac{(\omega - 1)m\left[4(\omega + 2)m + 3(\omega + 1)^2\right]^2}{2(2m + \omega + 1)^3} \quad (11)$$

$$\gamma_2 = \beta_2 - 3 = \frac{(\omega - 1)P(m, \omega)}{2(2m + \omega + 1)^2}, \quad (12)$$

where $m = \omega \sinh^2 \Omega$, and $P(\cdot, \cdot)$ is a quadratic in m with polynomials in ω as coefficients. He gave the approximations

$$\omega \sim 1 - \tfrac{1}{3}\beta_1 + \tfrac{1}{4}\gamma_2,$$

$$m \sim \frac{4\beta_1}{9\gamma_2 - 16\beta_1},$$

with quadratic terms.

However, he did not notice an application of Lagrange's expansion* [27] which becomes evident if the parametric form is altered. Briefly, define

$$T = \omega \sinh^2 \Omega / (\omega + 1)^2,$$

$$B_1 = \beta_1 / \left[(\omega^2 - 1)(\omega + 1)^2\right]$$

so that

$$T = 2B_1 \frac{\left[1 + 2(\omega + 1)T\right]^3}{\left[3 + 4(\omega + 2)T\right]^2}, \quad (13)$$

and

$$f(T) = \frac{\gamma_2 - \tfrac{1}{2}(\omega^2 - 1)(\omega^2 + 3)}{(\omega + 1)^2(\omega^2 + 2\omega + 3)}$$

$$= \frac{2(\omega - 1)T\left[1 + T(\omega + 3)\right]}{\left[1 + 2(\omega + 1)T\right]^2}. \quad (14)$$

Then from Lagrange's expansion for the root of an equation, we find that

$$(\omega^2 + 1)^2 = 2\beta_2 - 2 - 6(\omega + 1)^2$$

$$\times (\omega^2 + 2\omega + 3)H(\beta_1, \omega),$$

where

$$H(\beta_1, \omega) = \sum_{n=1}^{\infty} \frac{2^{2n}(3n - 3)!}{3^{3n}n!(2n - 1)!}\left[\frac{\beta_1}{(\omega + 1)^3}\right]^n, \quad (15)$$

which can be used as a check on the solution for ω given β_1, β_2; convergence is quite rapid unless $\beta_1 = (\omega + 1)^3$ approximately, which occurs only on S_L. Moreover, (15) can be used to set up the Maclaurin's series* for $(\omega^2 + 1)^2$ and ω about $\beta_1 = 0$. Note also that Johnson's discovery concerning the near linearity of ω-contours and use of ϕ [expression (8)] becomes less mysterious; for from (15) we can show that

$$\phi(\beta_1, \beta_2; \omega)$$

$$\sim \frac{4}{9}\frac{(f_0^2 + 2f_0 + 3)}{(f_0 + 1)} + \frac{8\beta_1}{243}$$

$$\times \frac{(f_0^2 + 2f_0 + 3)(2f_0^3 + 9f_0^2 + 2f_0 - 3)}{f_0(f_0 + 1)^4(f_0^2 + 1)}$$

and

$$\omega \sim f_0 - \frac{2}{9} \beta_1 \frac{(f_0^2 + 2f_0 + 3)}{f_0(f_0 + 1)(f_0^2 + 1)}$$

$$[f_0 = (\sqrt{4 + 2\gamma_2} - 1)^{1/2}],$$

where f_0 is independent of β_1. In these equations the coefficients of β_1 are numerically less than $5/81$ and $1/3$ for $f_0 \geqslant 1$, so near linearity is assured for small values of β_1.

Quite surprisingly, the Lagrange expansion for T [defined in (14a)] is more complicated than that for the more complicated function $f(T)$. The first few terms defined by

$$T = p_1 z + p_2 z^2 + \cdots, \quad \omega > 1 \atop \{ z = \beta_1 / [(\omega - 1)(\omega + 1)^3] \} \quad (16)$$

are

$$p_1 = 2/3^2,$$

$$p_2 = 2^3(5\omega + 1)/3^5,$$

$$p_3 = 2^7(7\omega^2 + \omega + 1)/3^8,$$

$$p_4 = 2^7(55\omega^3 - 3\omega^2 + 21\omega - 1)/3^{10},$$

$$p_5 = 2^9(1001\omega^4 - 344\omega^3$$
$$+ 750\omega^2 - 152\omega + 41)/3^{12}.$$

Given $(\beta_1, \beta_2; \omega)$ the expansion may be used to check the assessment of Ω, at least for small β_1.

Comments on the S_L Line

The lognormal line, the boundary of the S_U and S_B regions, is given in parametric form as

$$\beta_1 = (\omega - 1)(\omega + 2)^2,$$

$$\beta_2 = \omega^4 + 2\omega^3 + 3\omega^2 - 3. \quad (17)$$

Numerically, we can solve the first equation for ω, given β_1, and then determine β_2. But there is an expression for β_2 in terms of β_1. For

$$\omega = 1 + \frac{\beta_1}{(\omega + 2)^2}$$

so that from Lagrange's theorem, formally,

$$\beta_2 = 3 + \frac{16\beta_1}{3^2} + \frac{13\beta_1^2}{3^5} - \frac{14\beta_1^3}{3^8}$$

$$+ \frac{11\beta_1^4}{3^{10}} - \frac{106\beta_1^5}{3^{14}} + \frac{403\beta_1^6}{3^{17}}$$

$$= 3 + 48\beta_1^* + 18$$

$$\times \sum_{n=2}^{\infty} \frac{(-1)^n(n-1)(9n+8)(3n-5)! \, \beta_1^{*n}}{n!(2n-1)!}$$

$$(\beta_1^* = \beta_1/27), \quad (18)$$

which certainly converges for $\beta_1 \leqslant 4$.

THE S_B SYSTEM

In this case since the range is finite, the system may be fitted either by moments or by utilizing the information given by the endpoints.

S_B Properties

The density is

$$p(y) = \frac{\delta}{\sqrt{2\pi}} \frac{1}{y(1-y)}$$

$$\times \exp\left[-\frac{1}{2}\left(\gamma + \delta \ln \frac{y}{1-y} \right)^2 \right]$$

$$(0 < y < 1). \quad (19)$$

Note that under the mapping $y = 1 - x$, the density changes only in the sign of δ. Moreover, as $y \to 0$ the dominant part of $Y^{-s}p(y)$ is $\exp(-\delta^2 \ln^2 y/2)$ for given s; hence the density has "high contact" at each extremity.

Since in terms of the standard normal z we have $y = 1/(1 + e^{(\gamma-z)/\delta})$, it follows that the median is at $y_m = 1/(1 + e^{\Omega})$. Moreover, as for modality, Johnson [12, pp. 158–159] has shown that the necessary and sufficient conditions for bimodality are

$$\delta < 1/\sqrt{2},$$

$$|\gamma| < \delta^{-1}\sqrt{1 - 2\delta^2} - 2\delta \tanh^{-1}\sqrt{1 - 2\delta^2}.$$

Draper [9] extended the boundary curve of modality started by Johnson [12, p. 157]. It is shown in Fig. 2 and coincides (for the domain shown) approximately with the U-shaped type I curves of the Pearson system; S_B is bimodal "above" the line shown in terms of the parameters (β_1, β_2).

Moment Fitting

For the rth noncentral moment we have

$$\mu'_r(Y) = \frac{1}{\sqrt{2\pi}} \int_{-\infty}^{\infty} e^{-(1/2)z^2}(1 + e^{(\gamma - z)/\delta})^{-r}\, dz$$

$$(r = 0, 1, \ldots), \quad (20)$$

and in general this appears to be a transcendental quantity with some problems of evaluation either by quadrature or other processes (expansions, rational fractions, etc.); to say the least, the solution of the equations for $\sqrt{\beta_1}, \beta_2$ to derive δ, γ is complicated, involving four integrals for the noncentral moments. Few exact solutions are known. However, on the line S_L, it is known [12, p. 175] that for all noncentral moments

$$\mu'_r = \frac{1}{\sqrt{2\pi}} \int_{\gamma}^{\infty} e^{-(1/2)z^2}\, dz,$$

so that from the first three we have, with $\delta = 0$,

$$\gamma = \Phi^{-1}\left[\frac{1}{2} - \frac{1}{2}\sqrt{\frac{\beta_1}{4 + \beta_1}}\right],$$

where $\Phi^{-1}(\cdot)$ is the inverse normal function. This solution is useful for checking purposes.

Further Comments on S_B

In the early development of the subject, Johnson [12] produced a formula for $\mu'_1(Y)$ as the ratio of infinite series involving Jacobi theta functions. He also derived a bivariate recurrence relation for the noncentral moments. Writing $\mu'_r(\gamma, \delta)$ for the rth noncentral moment, the notation showing the dependence on the parameters γ and δ, Johnson [12, expression (59)] proves that

$$\mu'_r(\gamma + \delta^{-1}, \delta) = \left[\exp(-\delta^{-2}/2 - \gamma\delta^{-1})\right]$$
$$\times [\mu'_{r-1}(\gamma, \delta) - \mu'_r(\gamma, \delta)].$$

In this way he was able to set up a tabula-

tion of $\mu'_1, \sqrt{\mu_2}, \beta_1$, and β_2 for $\gamma = 0.0(0.5)$ 2.5, $\delta = 0.5, 1.0, 2.0$.

Fitting S_B

For the user, the steps are:

1. Determine that S_B is the appropriate curve by evaluating, for the given $\sqrt{\beta_1}$, the corresponding β_2 on the log-normal curve; see the section "Further Comments on S_U."

2. Consult the Pearson and Hartley [23] tables which cover values of $\sqrt{\beta_1}, \beta_2$ for $\sqrt{\beta_1} = 0.00(0.05)2.00$, and β_2 at intervals of 0.1 [($\sqrt{\beta_1}, \beta_2$) points in the near vicinity of S_T or S_L are not included; in the first case the distribution may be singular, and in the second case the log-normal curve could be considered]. The tables give $\gamma, \delta, \mu'_1(y)$, and $\sigma(y)$. Interpolation (see ref. 23 [pp. 82–86]) may be necessary.

3. Finally, compute λ and ξ from (3) and set up the relations

$$X = \xi + \lambda/[1 + e^{(\gamma - Z)/\delta}],$$
$$[Z \in N(0, 1)]$$

or

$$Z = \gamma + \delta \ln[(X - \xi)/(\lambda + \xi - X)]$$
$$(\xi < \lambda < \lambda + \xi).$$

(Note that the tabulation in Pearson and Hartley covers, for the most part, the unimodal set of S_B curves. Tabulations involving the bimodal region have been given in a University of North Carolina report by Johnson and Kitchen. For further details of computational approaches, see Draper [9] and Johnson and Kitchen [15].)

Computer algorithms for S_B, S_L, and S_U have been given by Hill et al. [11] using the FORTRAN language. For S_B these authors use a form of Goodwin's [10] approximation to integrals of doubly infinite range with integrands $g(x) = f(x)\exp(-x^2)$, as suggested by Draper [9]. As for precision the authors remark "Single precision arithmetic is generally sufficient, even on machines that use only 32 bits for real number representation." There appears to be no mention of inherent accuracy and the programs, especially for S_B, should be used with caution; for S_U feedback validation is readily available but more complicated for S_B.

Explicit approximations, using polynomial models, have been given by Bowman et al. [5]. Avoiding the narrow regions near S_T and S_L, the approximants, using $\sqrt{\beta_1}$, β_2, and the value of β_2 on S_L for given β_1, give values for γ and δ for the three segments $0 < \beta_1 < 1$, $1 < \beta_1 < 4$, and $4 < \beta_1 < 9$. The mean $\mu_1'(y)$ and variance $\mu_2(y)$ then have to be computed using quadrature. The formulas are suitable for computers or programmable calculators.

End-Point Fitting for Theoretical Distributions

When both endpoints are known the parameters ξ and λ are determined so that we are left with the determination of δ and γ. Thus solutions are needed of

$$\mu_1'(Y) = \{E(X) - \xi\}/\lambda,$$

$$\sigma^2 = \mu_2(Y) = \mu_2(X)/\lambda^2. \quad (21)$$

Johnson and Kitchen [16] have tabulated values of γ, δ for given σ and μ_1', in which μ_1' ranges from 0.01 to 0.50 by intervals of 0.01, and σ lies between 0.05 and 0.49. They remark that when σ is small and δ large, there is the approximation

$$\delta \sim \frac{\mu_1'(1 - \mu_1')}{\sigma} + \frac{1}{4}\sigma\left[\frac{1}{\mu_1'(1 - \mu_1')} - 8\right]$$

$$\gamma \sim \delta \ln\left(\frac{1 - \mu_1'}{\mu_1'}\right) + \frac{1}{\delta}\left(\frac{1}{2} - \mu_1'\right), \quad (22)$$

giving "quite good results."

In illustration they considered the distribution of the correlation coefficient* in sampling from a bivariate normal* population, the range being -1 to 1, their example referring to a case with $\rho = 0.5$ and $n = 3, 5$, 8, and $\rho = 0.8$ with $n = 5$. It turned out that the four-moment fits were better (in the

sense of group expectations) than the endpoint fits, this perhaps being due to the fact that the S_B system has high contact at the extremities for which four moments provide greater flexibility.

End-point fitting has not received much usage, partly because the ranges of many commonly used statistics are not finite.

Empirical Data and Percentile Fitting

It is commonly supposed that four-moment fitting of distributions to samples of various sizes suffers because of the large variances of the third and fourth moments. There may be some truth in this, but it is not well documented. In any event the notion has resulted in investigations of fitting procedures that avoid higher moments or even all moments. The approach has a long history, having been used when the approximating density turned out to be intractable by moments. For example, Kapteyn [18] used percentiles in fitting the transformed normal variate $Z = a(X + b)^k - c$, involving four parameters.

For the Johnson system, it is possible to use four percentiles matched to those of the data. If the method of the preceding section is used, then the end points need precise definition (e.g., the precise extremity of the first interval is certainly required—see Johnson [12], who considers the fitting of S_B to data concerning the degree of cloudiness at Greenwich for the period 1890–1904; the end point could be -0.5 or 0.0 according as the first interval definition is -0.5 to 0.5 or 0.0 to 0.5).

Bukac [6] uses the matching of probability (or frequency) levels at α_2, α_1, $1 - \alpha_1$, and $1 - \alpha_2$, where $0 < \alpha_2 < \alpha_1 < \frac{1}{2}$, and produces a solution equation. When the corresponding normal deviate z_α at level α is such that $z_{\alpha_2}/z_{\alpha_1} = 3$, the solution equation reduces to a quartic. As an example, Bukac considered data on age-specific fertility in Czechoslovakia in 1966.

Slifker and Shapiro [29] produce criteria for discriminating between the systems. If x_1, x_2, x_3, and x_4 are percentiles corresponding to the normal deviates at $3z_0, z_0, -z_0,$ $-3z_0,$ and $m = x_1 - x_2,$ $n = x_3 - x_4,$ $p = x_2 - x_3,$ then the approximate distribution is $S_B, S_L,$ or S_U according as $c < 1,$ $c = 1,$ or $c > 1,$ respectively, where $c = mn/p^2.$

By choosing the normal deviates such that $z_4 - z_3 = z_3 - z_2 = z_2 - z_1$ Mage [20] was able to reduce the solution equation to a quadratic.

ILLUSTRATIONS

Example 1: Triangular Density (Pearson Type I)

Density of X:

$$p(x) = \tfrac{1}{2}x, \qquad 0 < x < 2,$$
$$= 0, \qquad \text{otherwise.}$$

Moments:

$$\mu_1' = \tfrac{4}{3}, \qquad \mu_2 = \tfrac{2}{9},$$
$$\sqrt{\beta_1} = -2\sqrt{2/5}, \qquad \beta_2 = 2.4.$$
$$P_r(X > t) = 1 - t^2/4 \qquad (= \alpha)$$
$$t_\alpha = 2\sqrt{1 - \alpha}.$$

For S_B:

$$\delta = 0.802345$$
$$\gamma = -0.639349$$
$$\mu_1'(Y) = 0.648598$$
$$\mu_2(Y) = 0.511348$$
$$\lambda = \sqrt{\operatorname{var} X / \operatorname{var} Y} = 2.084661,$$
$$\xi = -0.018774$$
$$X = \xi + \lambda/(1 + \exp(\gamma - Z)/\delta).$$

	Probability Levels					
	0.01	0.05	0.10	0.90	0.95	0.99
True	0.200	0.447	0.633	1.897	1.949	1.990
S_B	0.208	0.443	0.627	1.892	1.952	2.015
L_B	0.212	0.435	0.626	1.898	1.963	2.027

Remark. The range of the S_B curves is -0.02 to 2.07, a slight discrepancy. Percentage points agree quite well, although the ordinates at $x = 2$ clearly are discrepant. L_B

is the Johnson-type density based on the logistic distribution [30], implemented on a calculator using [5].

Example 2: First Law of Laplace

Density of X:

$$p(x) = \tfrac{1}{2}\exp -|x|, \qquad -\infty < x < \infty$$

Moments:

$$\mu_1' = 0, \qquad \mu_2 = 2,$$
$$\sqrt{\beta_1} = 0, \qquad \beta_2 = 6.$$
$$P_r(X > t) = \tfrac{1}{2}\int_t^\infty \exp(-y)\,dy \qquad (t > 0)$$
$$t_\alpha = \ln(1/(2\alpha)).$$

For S_U:

$$\delta = 1.610431, \qquad \xi = 0$$
$$\lambda = 1.855133, \qquad \gamma = 0$$
$$X = \lambda \sinh(Z/\delta).$$

	Probability Levels			
	0.75	0.90	0.95	0.99
True	0.693	1.609	2.303	3.912
S_U	0.800	1.636	2.243	3.729
L_U	0.917	1.651	2.245	3.672

Remark. The agreement for the upper levels is acceptable but fades at $\alpha = 0.75,$ as might be suspected because of the cusp at $x = 0.$ L_U is the Johnson-type density based on the logistic.

Example 3. A sample of 15 is drawn from a population with density

$$f(x) = (x/a)^{\rho-1} e^{-(x/a)}/(a\Gamma(\rho)),$$

and $\hat{\rho}, \hat{a}$ are maximum likelihood estimators of ρ and $a,$ when the true values are $\rho = a = 1.$ From tabulations [2] the moments are:

	$\hat{\rho}$	$\hat{a}$
μ_1'	1.2044	1.0613
μ_2	0.2299	0.1631
$\sqrt{\beta_1}$	2.1680	1.0177
β_2	14.4765	4.7287

S_U is the appropriate density for $\hat{\rho}$, whereas for $\hat{a}$ we use S_B since the value of the kurtosis on the lognormal line is 4.8965. In the latter case note that $\mu'_1(y) = 0.077834$, $\sigma^2(y) = 0.00082881$.

S_U for $\hat{\rho}$: $\hat{\rho} = 0.7361$

$$+ 0.3581 \sinh[(Z + 1.3250)/1.4625]$$

S_B for $\hat{a}$: $\hat{a} = -0.0259 + 13.9241$

$$/\{1 + \exp[(6.3843 - Z)/2.5190]\}.$$

Percentage Points

	$\hat{\rho}$		$\hat{a}$	
%	S_U	N^a	S_B	N
1	0.47	0.25	0.40	0.04
5	0.66	0.47	0.53	0.32
10	0.75	0.59	0.61	0.47
90	1.77	1.41	1.60	1.53
95	2.08	1.53	1.82	1.68
99	2.90	1.75	2.29	1.96

[a] N is the basic normal approximation used with, for example, $E\hat{\rho} \sim \rho$, var $\hat{\rho} \sim h(\rho)/n$, $\beta_1 = 0$, $\beta_2 = 3$; $h(\rho)$ is tabulated in Bowman and Shenton [1].

Further illustrations can be found in AP-PROXIMATIONS TO DISTRIBUTIONS and in the following:

1. Johnson [12]
   a. Cloudiness at Greenwich
   b. Age of Australian mothers at birth of child
   c. Length and breadth of bean data due to Pretorius

2. Draper [9]
   a. Bean data in item 1c
   b. Anscombe's approximation to the distribution of "Student's t"

3. Pearson [21, 22]
   a. $\sqrt{b_1}$, and b_2 in normal sampling

4. Johnson [14]
   a. Noncentral t with eight degrees of freedom

5. D'Agostino [7]
   a. The null distribution of what amounts to $\sqrt{b_1}$, with $n > 8$

6. Johnson and Kitchen [15]
   a. Correlation coefficient from the bivariate normal

7. Bukac [6]
   a. Age-specific fertility data

8. Pearson and Hartley [23]
   a. Warp strength of duck cloth, pp. 84–86
   b. Type I density, p. 87

9. D'Agostino and Tietjen [8]
   a. Compares approximations to the null distribution of $\sqrt{b_1}$

10. Shenton and Bowman [25]
    a. $\sqrt{b_1}$ and b_2 in sampling from normal mixtures, and type I densities

11. Shenton and Bowman [26]
    a. Marginal density of $\sqrt{b_1}$ in the joint distribution of $\sqrt{b_1}, b_2$ in general sampling (mainly Pearson type I)

12. Shenton et al. [28]
    a. Quotes S_B applied to the null distribution of $\sqrt{b_1}$ for $n = 4, 6$, and S_U for $n = 8$

13. Mage [20]
    a. Considers the subject in 12a
    b. Hourly average (CO concentration)

14. Slifker and Shapiro [29]
    a. Length of bean data (Pretorius)
    b. Resistances data

15. Shenton and Bowman [27]
    a. Some theoretical cases

GENERAL REMARKS

The Johnson system based on a normal variate provides an approximate normal transformed variate; there are tabulations to aid in solution evaluation for limited values of

$(\sqrt{\beta_1}, \beta_2)$, and quite good approximate solutions, as explicit functions of the skewness and kurtosis which are valid for larger domains than the existing tabulations.

For theoretical statistics, the system provides acceptable percentage points at non-extreme levels ($\alpha \leqslant 0.99$) provided that the distribution approximated is nearly bell shaped; abrupt tails can lead to a loss in approximation, as can multimodality and discontinuities.

For empirical data, moment methods or percentiles can be used, the two approaches subject to the usual problem of the precise response to sampling variations, quite apart from model validity.

References

[1] Bowman, K. O. and Shenton, L. R. (1968). *Properties of Estimators for the Gamma Distribution. Report CTC-1*, Union Carbide Corp., Nuclear Div., Oak Ridge, Tenn.

[2] Bowman, K. O. and Shenton, L. R. (1970). Small Sample Properties of Estimators for the Gamma Distribution. *Report CTC-28*, Union Carbide Corp., Nuclear Div., Oak Ridge, Tenn.

[3] Bowman, K. O. and Shenton, L. R. (1980). *Commun. Statist. B*, **9**, 127–132.

[4] Bowman, K. O. and Shenton, L. R. (1981). In *Statistical Distributions in Scientific Work*, Vol. 5: *Inferential Problems and Properties*, C. Taillie, G. P. Patil, and B. Baldessari, eds. D. Reidel, Dordrecht, Holland, pp. 231–240.

[5] Bowman, K. O., Serbin, C. A., and Shenton, L. R. (1981). *Commun. Statist. B*, **10**, 1–15.

[6] Bukac, J. (1972). *Biometrika*, **59**, 688–690.

[7] D'Agostino, R. B. (1970). *Biometrika*, **57**, 679–681.

[8] D'Agostino, R. B. and Tietjen, G. L. (1973). *Biometrika*, **60**, 169–173.

[9] Draper, J. (1952). *Biometrika*, **39**, 290–301.

[10] Goodwin, E. T. (1949). *Proc. Camb. Philos. Soc.*, **45**, 241–245.

[11] Hill, I. D., Hill, R., and Holder, R. L. (1976). *Appl. Statist.*, **25**, 180–189.

[12] Johnson, N. L. (1949). *Biometrika*, **36**, 149–176.

[13] Johnson, N. L. (1954). *Trab. Estadist.*, **5**, 283–291.

[14] Johnson, N. L. (1965). *Biometrika*, **52**, 547–558.

[15] Johnson, N. L. and Kitchen, J. O. (1971). *Biometrika*, **53**, 223–226.

[16] Johnson, N. L. and Kitchen, J. O. (1971). Tables to Facilitate Fitting S_B Curves, *Mimeo Series No. 683*, Institute of Statistics, University of North Carolina.

[17] Johnson, N. L. and Kitchen, J. O. (1976). *Biometrika*, **58**, 657–668.

[18] Kapteyn, J. C. (1903). *Skew Frequency Curves in Biology and Statistics*. Noordhoff, Groningen.

[19] Leslie, D. C. M. (1959). *Biometrika*, **46**, 229–231.

[20] Mage, D. T. (1980). *Technometrics*, **22**, 247–251.

[21] Pearson, E. S. (1963). *Biometrika*, **50**, 95–112.

[22] Pearson, E. S. (1965). *Biometrika*, **52**, 282–285.

[23] Pearson, E. S. and Hartley, H. O., eds. (1972). *Biometrika Tables for Statisticians*, Vol. 2. Cambridge University Press, Cambridge.

[24] Pearson, E. S., D'Agostino, R. B., and Bowman, K. O. (1977). *Biometrika*, **64**, 231–246.

[25] Shenton, L. R. and Bowman, K. O. (1975). *J. Amer. Statist. Ass.*, **70**, 220–228.

[26] Shenton, L. R. and Bowman, K. O. (1977). *J. Amer. Statist. Ass.*, **72**, 206–211.

[27] Shenton, L. R. and Bowman, K. O. (1980). *J. Statist. Comp. Simul.*, **15**, 89–95.

[28] Shenton, L. R., Bowman, K. O., and Lam, H. K. (1979). *Proc. Statist. Computing Sect., Amer. Statist. Ass.*, pp. 20–29.

[29] Slifker, J. F. and Shapiro, S. S. (1980). *Technometrics*, **22**, 239–246.

[30] Tadikamalla, P. R. and Johnson, N. L. (1982). *Biometrika*, **69**, 461–465.

Acknowledgment

Research sponsored by the Applied Mathematical Sciences Research Program, Office of Energy Research, U.S. Department of Energy under Contract W-7405-eng-26 with the Union Carbide Corporation.

(APPROXIMATION TO DISTRIBUTIONS
CURVE FITTING
FISHER'S z' TRANSFORMATION
FREQUENCY CURVES, SYSTEMS OF
GRAM–CHARLIER SERIES
KURTOSIS
LAGRANGE EXPANSIONS
PEARSON DISTRIBUTIONS
SKEWNESS)

K. O. BOWMAN
L. R. SHENTON

JOINT CONFIDENCE INTERVALS   *See* SIMULTANEOUS INFERENCE

JOINT DISTRIBUTIONS

If $X_1, X_2, \ldots, X_n$ are n random variables defined on the same sample space, then the *joint distribution function* is defined by

$$F(x_1, x_2, \ldots, x_n)$$

$$= \Pr[\, X_1 \leqslant x_1 \text{ and } X_2 \leqslant x_2 \text{ and }$$

$$\cdots \text{ and } X_n \leqslant x_n\,].$$

If F is differentiable with continuous derivatives, then

$$\frac{\partial^n F(x_1, x_2, \ldots, x_n)}{\partial x_1 \partial x_2 \cdots \partial x_n} = f(x_1, x_2, \ldots, x_n)$$

is called the *joint (probability) density function* and one has the relation

$$\int_{a_n}^{b_n} \cdots \int_{a_2}^{b_2} \int_{a_1}^{b_1} f(x_1, x_2, \ldots, x_n)\, dx_1\, dx_2 \ldots dx_n$$

$$= \Pr[a_1 \leqslant X_1 \leqslant b_1, a_2 \leqslant X_2 \leqslant b_2, \ldots,$$

$$a_n \leqslant X_n \leqslant b_n].$$

JOLLY–SEBER ESTIMATOR   *See* CAPTURE–RECAPTURE MODELS

JONCKHEERE TESTS FOR ORDERED ALTERNATIVES

Since the 1950s there has been considerable interest in procedures for ordered location alternatives in multisample data.

Ordered alternatives refer to a generalization of one-sided alternatives in the two-sample problem: If $\{\theta_i : i = 1, \ldots, k\}$ represent location parameters (e.g., medians) for k populations, an ordered alternative is one that specifies a particular ordering of the θ_i, prior to observation of the data. Without loss of generality one may take that a priori ordering to always be $\theta_1 < \theta_2 < \cdots < \theta_k$. (Other orders can be relabeled to conform.)

Two of the earliest works in this area were by Jonckheere, who recognized the similarity to problems of monotone trend, and developed tests for ordered alternatives in the one-way [4] and two-way [5] layouts, based on Kendall's test for rank correlation (*see* RANK CORRELATION).

Terpstra, in a slightly earlier paper [11], had presented a test for the one-way layout that is equivalent to Jonckheere's but in a slightly different form. Presumably due to availability of exact tables, and publication in a more accessible journal, the Jonckheere version has become standard.

THE ONE-WAY LAYOUT (JONCKHEERE–TERPSTRA TEST)

The data consist of a collection $\{X_{ij} : j = 1, \ldots, n_i;\ i = 1, \ldots, k\}$ of independent random variables where X_{ij} has cumulative distribution function (CDF) F_i. The most general form of the ordered alternative problem can be stated as one of stochastic ordering or

$$H_0: \quad F_i(t) = F(t), \quad -\infty < t < \infty; \quad i = 1, 2, \ldots, k, \text{ vs.}$$

$$H_1: \quad F_1(t) \geqslant F_2(t) \geqslant \cdots \geqslant F_k(t), \text{ at least one strict inequality for some } t.$$

If we assume $F_i(t) = F(t - \theta_i)$ for some unknown CDF F, the essential problem becomes one of ordered location shift

$$H_0': \quad \theta_1 = \theta_2 = \cdots = \theta_k \text{ vs.}$$

$$H_1': \quad \theta_1 < \theta_2 < \cdots < \theta_k.$$

It is easy to see that $H_0' = H_0$ and $H_1' \subset H_1$.

Although Jonckheere [4] developed his test as an extension of Kendall's test for rank correlation*, a more convenient, equivalent formulation is now used. Let $M_{i, i'}$ be the Mann–Whitney test statistic (*see* WILCOXON–MANN–WHITNEY RANK SUM TEST) to detect $\theta_i < \theta_{i'}$. Thus

$$M_{i, i'} = \{\text{number of times } X_{i, j} < X_{i', j'},$$

$$j = 1, \ldots, n_i; j' = 1, \ldots, n_{i'}\}$$

$$+ \tfrac{1}{2}\{\text{number of times } X_{i, j} = X_{i', j'},$$

$$j = 1, \ldots, n_i; j' = 1, \ldots, n_{i'}\}.$$

Jonckheere's test rejects $H_0(H_0')$ in favor of $H_1(H_1')$ for large values of

$$J = \sum\sum_{1 \leqslant i < i' \leqslant k} M_{i, i'}$$

$$= M_{1,2} + M_{1,3} + \cdots + M_{1,k} + M_{2,3} +$$

$$\cdots + M_{2,k} + \cdots + M_{k-1,k}.$$

Under the assumptions that F is continuous CDF and that no ties exist in the combined sample, the test is distribution-free and exact tables exist, for example in ref. 3, for $k = 3$, $2 \leq n_i \leq 8$; $k = 4, 5, 6$, $n_1 = n_2 = \cdots = n_k = 2(1)6$.

For larger samples, when F is not continuous, or when ties exist (e.g., due to round-off), an approximate test is used, based on the fact that J is asymptotically normally distributed when $\min_i \{n_i\} \to \infty$. The approximate test rejects H_0 in favor of the a priori ordering when the right-tail α-level critical value of the standard normal is exceeded by

$$Z_J = (J - \mu_J)/\sigma_J .$$

The exact null moments of J are given by

$$\mu_J = \left(N^2 - \sum n_i^2\right)/4$$

and

$$\sigma_J^2 = \left[N^2(2N + 3) - \sum n_i^2(2n_i + 3)\right]/72.$$

The variance formula above must be modified if ties exist among any of the $N = \sum n_i$ observations. If the distinct values of the data are denoted by $\{a_j; j = 1, \ldots, e \leq N\}$ and a_j occurs t_j times, then the exact null variance is

$$\sigma_J^2 = \left[N(N - 1)(2N + 5) - \sum_{i=1}^{k} n_i(n_i - 1)(2n_i + 5)\right.$$

$$\left. - \sum_{j=1}^{e} t_j(t_j - 1)(2t_j + 5)\right]\Big/72$$

$$+ \left[\sum n_i(n_i - 1)(n_i - 2)\right]\left[\sum t_j(t_j - 1)(t_j - 2)\right]$$

$$/[36N(N - 1)(N - 2)]$$

$$+ \left[\sum n_i(n_i - 1)\right] \times \left[\sum t_j(t_j - 1)\right]/[8N(N - 1)].$$

Example 1:

$$X_{1,j} = 5.8, 6.6$$

$$X_{2,j} = 7.9, 6.5, 6.8$$

$$\xi_{3,j} = 8.6, 7.8, 8.2$$

$$M_{1,2} = 5, \quad M_{1,3} = 6, \quad M_{2,3} = 8, \quad J = 19$$

From Table A.8 in ref. 3, $P[J \geq 19] = 0.0143$, indicating good support for the a priori ordering, $\theta_1 < \theta_2 < \theta_3$.

OTHER TESTS AND COMPARISONS

Among other rank tests developed for ordered alternatives in the one-way layout, two are described below, both using the treatment rank sums $R_1, \ldots, R_k$ based on the combined ranking of all N observations. (*See* KRUSKAL–WALLIS TEST.)

If the average ranks by treatment are $R_i. = R_i/n_i$, the Chacko–Shorak procedure [1, 10] pools any adjacent treatments (i and $i + 1$) for which the hypothesized order is reversed (in the sense $R_i. > R_{i+1}.$). If after pooling there are $t \leq k$ remaining groups, with sample sizes n_i^* and rank averages $R_i^*.$, the final test statistic is

$$K = 12 \sum_{i=1}^{t} n_i^* \left[R_i^*. - \tfrac{1}{2}(N + 1)\right]^2 / \{N(N + 1)\}.$$

Existing tables for this test cover only large samples with equal sample sizes (see ref. 7, [p. 236, Table L]).

Jonckheere's approach can also be applied to Spearman's test for rank correlation instead of Kendall's. The most convenient form in this case becomes

$$S = \sum_{i=1}^{k} iR_i .$$

Although small sample tables are not yet available, the standardized version

$$Z_s = (S - \mu_s)/\sigma_s$$

provides a good approximation for a right-tailed test using standard normal tables.

In the expression above the exact null moments of S are

$$\mu_s = (N + 1)\sum_{1}^{k} in_i/4,$$

$$\sigma_s^2 = \left[N(N^2 - 1) - \sum_{1}^{e} t_j(t_j^2 - 1)\right]$$

$$\times \left[N\sum_{1}^{k} i^2 n_i - \left(\sum_{1}^{k} in_i\right)^2\right]\Big/ \{12N(N - 1)\}.$$

The test based on S possesses two important optimality properties when the θ_i are equally spaced and F is a logistic* CDF. In ref. 2, Theorem II.4.6 shows it to be the locally most powerful rank test and Theorem VII.1.3 shows it to be asymptotically most powerful (fixed k, $\min\{n_i\} \to \infty$) among all tests.

In a recent paper Krauth [6] shows that for the same conditions, and $\lim_{N \to \infty} n_i/N = 1/k$, $i = 1, \ldots, k$, the Pitman efficiency of J with respect to S equals 1 (see PITMAN ASYMPTOTIC RELATIVE EFFICIENCY), so that J is also asymptotically optimum in this case.

The ordered shift alternative H_1' is often written as "$\theta_1 \leqslant \theta_2 \leqslant \cdots \leqslant \theta_k$, with at least one inequality strict." Although all of the tests discussed here are consistent for this more general alternative, Potter and Sturm [8] have shown that for very small n_i and some treatments equal, the power of Jonckheere's test J can be bounded substantially below 1. It is conjectured here that the same also holds for test S. Because the Chacko–Shorak test, K, pools treatments that appear to violate the a priori order, it probably avoids the power bounding (see ref. 7 [p. 237] for further discussion and references).

For alternatives H_1' and equally spaced θ_i, the Pitman efficiency of each of the three rank tests with respect to the normal theory likelihood ratio test is $12\sigma_F^2[\int f^2]^2$, where σ_F^2 and f are the variance and density of F. If F is a normal CDF, the efficiency is $3/\pi = 0.955$.

THE TWO-WAY LAYOUT (JONCKHEERE TEST)

In this case $\{X_{ijl} : l = 1, \ldots, n_{ij}; \, j = 1, \ldots, k; \, i = 1, \ldots, b\}$ is a collection of independent random variables and X_{ijl} has CDF F_{ij}. The index j identifies the k treatment levels of interest, and i identifies blocks, which can be considered a nuisance effect.

If $F_{ij}(t) = F_i(t - \theta_j)$, for unknown CDF's $F_1, \ldots, F_b$, then the ordered location shift problem becomes simply H_0' vs. H_1' as defined in the one-way layout.

The test proposed by Jonckheere [5] for this problem can be described as follows: Treat each block separately as a one-way layout, and compute the Jonckheere J-

statistic, yielding $J_1, J_2, \ldots, J_b$. The basic test statistic is simply

$$P = \sum_{i=1}^{b} J_i.$$

Since independence holds among blocks, the null mean and variance of P are easily obtained by summing the corresponding one-way expressions across blocks:

$$\mu_p = \sum_{i=1}^{b} \mu_{J_i}, \qquad \sigma_p^2 = \sum_{i=1}^{b} \sigma_{J_i}^2.$$

An approximate right-tailed normal test is used with $Z_p = (P - \mu_p)/\sigma_p$.

Extensions to the two-way layout are also possible for S (see PAGE'S TEST FOR ORDERED ALTERNATIVES) and for K (see ref. 10). Another important method is that of aligned ranks (see RANK TESTS and ref 9). Page's test with $n_{ij} \equiv 1$ is the best known (see ref. 3 [Chap. 7, Table A.16]).

References

[1] Chacko, V. J. (1963). *Ann. Math. Statist.*, **34**, 945–956. (Introduces the Chacko–Shorack test for equal sample sizes on the one-way layout, and derives Pitman efficiency.)

[2] Hájek, J. and Šidák, Z. (1967). *Theory of Rank Tests*. Academic Press, New York. (An advance mathematical text, developing theoretical optimality properties. Theorems II.4.6 and VII.1.3 apply to Jonckheere's one-way-layout test.)

[3] Hollander, M. and Wolfe, D. A. (1973). *Nonparametric Statistical Methods*. Wiley, New York. [Contains algorithms, examples, and tables for the Jonckheere one-way-layout test (Chap. 6) and the Page two-way-layout test (Chap. 7).]

[4] Jonckheere, A. R. (1954). *Biometrika*, **41**, 133–145. (Presents the Jonckheere test for one-way layouts.)

[5] Jonckheere, A. R. (1954). *Brit. J. Statist. Psychol.*, **7**, 93–100. (Extends the one-way-layout test to two-way layouts with examples.)

[6] Krauth, J. (1979). *Math. Operat. Statist. Ser. Statist.*, **10**, 291–298. (A new formulation of Jonckheere's one-way statistic; efficiencies for both fixed k, $n_i \to \infty$ and bounded n_i, $k \to \infty$.)

[7] Lehmann, E. L. (1975). *Nonparametrics: Statistical Methods Based on Ranks*. Holden-Day, San Francisco. [Examples, tables, comparative discussion, and further references for Jonckheere and Chacko–Shorack tests for one-way layouts (Chap. 5).]

[8] Potter, R. W. and Sturm, G. W. (1981). *Amer. Statist.*, **35**, 249–250. (An example with some treatments equal where the power is bounded below 1.)

[9] Sen, P. K. (1968). *Ann. Math. Statist.*, **39**, 1115–1124. (A general treatment of two-way layouts with $n_{ij} \equiv 1$, using aligned ranks. General and ordered alternatives are examined, and efficiencies derived.)

[10] Shorack, G. R. (1967). *Ann. Math. Statist.*, **38**, 1740–1758. (Extends the test of Chacko [1] to unequal samples and to the two-way layout. Comparisons with other tests, including the normal theory likelihood ratio test.)

[11] Terpstra, T. J. (1952). *Indag. Mat.*, **14**, 327–333. (Introduces a test equivalent to Jonckheere's for the one-way layout.)

W. PIRIE

(DISTRIBUTION-FREE METHODS
PSYCHOLOGY, STATISTICS IN
RANK CORRELATION)

JOURNAL OF FUZZY SETS AND SYSTEMS

The *Journal* was started in 1978 as a quarterly, when fuzzy sets had become a concern and a tool of persons working in a variety of areas: for example, mathematicians in many different fields, such as logic, topology, algebra, etc., were doing research in the area of fuzzy sets. Similarly, the areas of general systems, intelligent systems, artificial intelligence, decision theory* and optimization*, approximate reasoning and multicriteria decision making, sociology*, and other behavioral sciences were applying fuzzy sets. Moreover, the theory of fuzzy sets had already affected a wide variety and disciplines such as control theory, medicine, information processing, genetics*, electrical engineering, operations research*, etc.

Most of the contributions to the theory of fuzzy sets and their applications were distributed over a wide range of scientific journals. Scientists working in this area in different disciplines or on different continents, often did not know of each other and of the contributions. In this situation the *Journal*

was started with the hope to increase considerably the concentrated dissemination of information in this area for the benefit of all scientists and practitioners concerned. The response was so strong that in 1981 the *Journal* had to double its volume and appear bimonthly.

The editors try to balance each single issue as well as possible as to theory and practice with respect to the different disciplines represented in the *Journal*. It is part of the editorial policy to aim at a very fast publication of submitted contributions. Since each submitted paper is refereed by at least two experts and revised by the authors who sometimes use considerable time for the improvement of their paper the time between submission of a paper varies considerably (between 4 and 18 months).

The average number of pages per issue is 110. The number of papers published in 1980 was 18. In 1983 it will be approximately 70. The official language of the *Journal* is English. The subscription price per volume in 1981 is U.S. $73.50 or U.S. $147 per year.

The *Journal* includes different sections:

1. *Regular Papers.* Under this section the *Journal* seeks to publish articles (survey papers, original research, and application papers) of real significance, broad interest, and of high quality. All articles must be readable, well organized, and exhibit good writing style. It is appreciated of a paper satisfying at least one of the following conditions:

   It contains at least one new proof of relevant relationships.

   It contains at least one new algorithm.

   It contains at least one real application of fuzzy sets with numerical results.

   It describes in detail at least one computer program for the implementation of one method or algorithm.

2. *Short Communications.* These are Letters to the Editor, Errata, short but signifi-

cant contributions, etc. This kind of contribution is not put into a queue of papers waiting for publication but is published as soon as it is typeset and proofread.

3. *Book Reviews.* Reviews of new books pertaining to the theory or application of fuzzy sets are invited from specialists in this area and then published immediately.

4. *Bulletin.* Here news about forthcoming events such as conferences, seminars, etc., which are relevant to the area of fuzzy sets are included.

5. *Current Literature.* Under this heading bibliographic data of contributions in the area of fuzzy sets which have not been published in the *Journal of Fuzzy Sets and Systems* are listed for the sake of the subcribers of this journal.

6. *"Who's Who in Fuzzy Sets."* Experience has shown that very often people working in the area of fuzzy sets live quite close to each other without knowing it. To increase communication between these persons, the Editors of the *Journal of Fuzzy Sets and Systems* publish at irregular intervals a list of persons working in this area together with their addresses and special interests.

The address of the editorial office is:

Professor Dr. H.-J. Zimmermann (Principal Editor)
RWTH Aachen
Templergraben 64
5100 Aachen, Federal Republic of Germany

Other Editors are Dr. C. V. Negoita, 695, Park Avenue, New York, (U.S.A.) and Dr. L. A. Zadeh, Computer Science Division, Department of Electrical Engineering and Computer Science, University of California, Berkeley, CA 94729.

The editorial organization consists of the Editorial Board:

M. A. Aizerman, Moscow (USSR)
B. R. Gaines, Toronto (Canada)
J. Goguen, Los Angeles (U.S.)
A. Kaufmann, La Tronche (France)
N. N. Moiseev, Moscow (USSR)
K. Tanaka, Osaka (Japan)

and of the Advisory Editors.

The *Journal* is published by North-Holland Publishing Company, Amsterdam. (*See also* FUZZY SET THEORY.)

H.-J. ZIMMERMANN

JOURNAL OF MULTIVARIATE ANALYSIS

The *Journal of Multivariate Analysis* (*JMA*) was started by P. R. Krishnaiah in 1971 to serve as a central medium for publication of the most important work on a very broad spectrum of topics in the field of multivariate analysis. The main emphasis of the *Journal* is on problems dealing with a finite number of correlated variables. Papers on univariate analysis as well as papers dealing with infinite dimensional cases are also published. The material in the *Journal* is essentially theoretical in nature. Papers dealing with important applications are published if they contain signficant theoretical results.

Editoral decisions on papers are usually made within six months of submission. Many papers were published within 15 months of submission. All papers submitted are refereed and the acceptance rate is less than 50%. Criteria for acceptance of papers include high quality and clarity of presentation. Occasionally, outstanding workers in the field are invited to write expository papers.

JMA is an international journal and the editorial board consists of distinguished workers in the field from Australia, Austria, Canada, France, Hungary, India, Israel, Japan, Romania, the United States, the USSR and West Germany. Papers may be submitted for possible publication either to the editor or to any suitable member of the Editorial Board. The address of the editor is P. R. Krishnaiah, Department of Mathemat-

Table 1   Contents of the June 1981 Issue

L. V. Osipov, On Large Deviations for Sums
of Random Vectors in R_k

Hulmut Strasser, Convergence of Estimates,
Part I

Hulmut Strasser, Convergence of Estimates,
Part II

Peter Hall, Polynomial Expansions of Density
and Distribution Functions of Scale Mixtures

D. F. Nicholls and B. G. Quinn, Multiple Autoregressive Models with Random Coefficients

Ulrich Krengel and Louis Sucheston, Stopping
Rules and Tactics for Processes Index by a Directed Set

R. Kannan, Random Correspondences and
Nonlinear Equations

Eivind Stensholt and Dag Tjøstheim, Factor-
izing Multivariate Time Series Operators

D. R. Jensen and R. V. Foutz, Markov Inequalities on Partially Ordered Spaces

F. Götze, Second-Order Optimality of Randomized Estimation and Test Procedures

Jürg Hüsler, A Note on the Functional Law of
Iterated Logarithm for Maxima of Gaussian Sequences

Malay Ghosh and Ahmad Parsian, Bayes Minimax Estimation of Multiple Poisson Parameters

ics and Statistics, University of Pittsburgh, Pittsburgh, PA 15260. Instructions to authors for preparation of the manuscripts are given at the end of each issue of the *JMA*.

The first volume of the *JMA* consisted of 487 printed pages. At present, one volume per year is published and each volume consists of about 620 printed pages. The issues of the *JMA* are published in March, June, September, and December. All enquiries regarding subscriptions should be addressed to Academic Press, Inc., 111 Fifth Avenue, New York, N.Y. 10003.

The contents of the June 1981 issue are given in Table 1.

P. R. KRISHNAIAH

JOURNAL OF QUALITY TECHNOLOGY, THE

The first issue of the *Journal of Quality Technology* was published by the American Society for Quality Control in January 1969.

Since that time it has been published quarterly in January, April, July, and October of each year. The four issues of each calendar year constitute a volume. Volume 1 (1969) contained 298 pages; since that time volumes have ranged in size from 200 to 256 pages, with Vol. 11 (1979) containing 228 pages. The current (1983) editorial address is: Dr. Peter R. Nelson, Editor, *Journal of Quality Technology*, Searle Research and Development, 4901, Searle Parkway, Skokie, IL 60077.

This journal grew out of an older journal called *Industrial Quality Control*, founded in 1944 by Martin A. Brumbaugh. When the American Society for Quality Control was founded in 1946, it took over publication of *Industrial Quality Control* under the continued editorship of Dr. Brumbaugh, who was succeeded by Mason E. Wescott (1947–1961), Irving W. Burr (1961–1965), and Lloyd S. Nelson (1965–1967). *Industrial Quality Control* was last published in December 1967. In January 1968 the American Society for Quality Control started publication of the monthly journal, *Quality Progress*. This journal publishes news items and feature articles of general interest to members of ASQC as well as other professionals in the field of quality control. The more technical, and particularly the more statistical, articles dealing with the broad areas of quality control and reliability are now published in the *Journal of Quality Technology*.

The founding editor of the *Journal of Quality Technology* was Lloyd S. Nelson. He was followed by H. Alan Lasater (1971–1973), Lawrence D. Romboski (1974–1976), John S. Ramberg (1977–1979), and Harrison M. Wadsworth, Jr. (1980–1982).

The Editorial policy of the *Journal* is to publish manuscripts that are relevant to the technology of quality control or reliability. Since statistics is underlying to this technology, most of the articles are statistical in nature. However, the applications of statistics are stressed rather than statistical theory. Statistical aspects of the technology of quality include acceptance sampling*, control charts*, design of experiments*, regression analysis*, multivariate analyses*, machine capability studies, Bayesian analyses, and many other related topics. The important consideration governing whether a statistical manuscript should be published in the *Journal of Quality Technology* is not the technique used but its applicability for quality and reliability practitioners.

The *Journal* contains, in addition to the feature articles, four departments entitled: Reviews of Standards and Specifications, Computer Programs, Technical Aids, and Book Reviews. Contributions of manuscripts for these departments as well as for feature articles are invited from members and non-members of the American Society for Quality Control. Many authors have been persons from outside the United States, but all articles are published in English. All manuscripts are refereed by one or more referees. Typically, two referees are used.

An Editorial Review Board is appointed by the editor. The members of this board do much of the refereeing, but many other persons, selected by the editor, are also asked to review manuscripts. The contents of a recent issue are given in Table 1.

HARRISON M. WADSWORTH, JR.

Table 1   Contents of Vol. 12, No. 1 (Jan. 1980)

K. E. Case, The p Control Chart under Inspection Error

J. W. Schmidt, G. K. Bennett, and K. E. Case, A Three Action Cost Model for Acceptance Sampling by Variables

H. Ohta, S. Kase, and M. Asao, Evaluation of Inspectors in Sensory Tests—Qualification by Geometric Method and Classification by Bayesian Diagnosis

G. J. Hahn and J. Schmee, Regression Estimates versus Separate Estimation at Individual Test Conditions

D. W. Kroeber, A Graphical Approach to the Design of Sequential Attribute Sampling Plans

A. J. Duncan, A. B. Godfrey, A. B. Mundel, and V. A. Partridge, Single Sampling Plans Indexed by LQLs that Are Compatible with the Structure of MIL-STD-105D

J. H. Sheesley, Comparisons of K Samples Involving Variables or Attributes Data Using Analysis of Means

L. S. Nelson, An Early-Stopping Rule for Variables Sampling

JOURNAL OF STATISTICAL PHYSICS

The *Journal of Statistical Physics*, published by Plenum Press of New York, was started in 1969. The first Editor-in-Chief was Howard Reiss and the editorial board consisted of 30 distinguished scientists drawn from a variety of disciplines, eight of them from outside the United States. In 1976, Joel L. Lebowitz became Editor-in-Chief and the board now consists of 20 scientists, half of them from abroad. The *Journal* was first published as a quarterly, but expanded into a monthly in 1970.

The introduction to the first issue describes the *Journal* as an "international journal concerned with the publication of statistical methods to the solution of problems in the physical, biological, and social sciences."

The areas of interest are given as (1) mathematical and physical foundations of statistical mechanics*; (2) application of statistical mechanics to specific real systems, including both stellar systems and plasmas; (3) noise and fluctuation phenomena; (4) experimental papers dealing with foundations of statistical mechanics (third-law experiments, for example); (5) kinetic theory of transport phenomena; (6) kinetic theory of neutron transport and thermalization; (7) chemical rate theory; (8) biological rate processes; (9) nucleation and the kinetics of phase transformation; (10) application of stochastic methods to chemical, physical, biological, and engineering problems; (11) pattern recognition*; (12) urban problems (traffic control, waste disposal, air pollution, etc.); (13) new mathematical methods in statistics and stochastics; (14) operations research*; and (15) mathematical economics.

The original editorial policy, which put much emphasis on interdisciplinary research as described by categories 11, 12, 14, and 15, was never fully realized. Papers outside the traditional area of statistical mechanics never made up more than a small fraction of the published papers. This led de facto and then officially to a change of policy which emphasizes the physical and mathematical aspects of statistical mechanics. The *Journal* is now recognized as one of the leading journals in the field of statistical mechanics spanning both the rigorous mathematical and the heuristic aspects of the subject.

The average refereeing time is about one month—remarkably short—and the time from acceptance to publication is about seven months. The average length of accepted papers is about 35 pages; an average issue up to 1980 contained about 125 pages, but this has increased to 180 pages from 1981 and 240 pages from 1983. All papers are in English.

JOEL L. LEBOWITZ

JOURNAL OF STATISTICAL PLANNING AND INFERENCE, THE

HISTORICAL BACKGROUND AND ORGANIZATION

The Journal of Statistical Planning and Inference (JSPI) was founded in 1976 by Professor J. N. Srivastava of Colorado State University and is published by North-Holland Publishing Company, Amsterdam. The first issue appeared in February 1977. Six volumes have been published and the seventh volume (1983) is in print. The objective is to publish at least four issues a year whenever sufficient material is available. During the first five years (1976–1980) the Editorial Board consisted of approximately 50 renowned statisticians and combinatorists from various countries, who worked under one Editor-in-Chief, Professor Srivastava. Now, she is joined by two Joint Chief Editors: S. Zacks of the State University of New York at Binghamton and P. K. Sen of the University of North Carolina at Chapel Hill. (During 1981, Dr. A. J. Hoffman of IBM also acted as a Joint Chief Editor.)

AIM AND SCOPE

The aim of the *JSPI* is to provide a medium for the dissemination of information and knowledge in all areas of statistical planning and inference. Statistical planning is a general term for the planning of data collection*, either by experimental designs, sampling surveys*, or through other scientific or technical means. Accordingly, the *Journal* emphasizes all aspects of the traditional design of experiments, sampling from finite populations*, surveillance processes, statistical control processes, sequential analysis*, adaptive processes*, linear models, multivariate analysis*, distribution theory, time-series* analysis, information theory*, and all related parametric and nonparametric statistical inference. This includes, in particular, decision-theoretic and Bayesian approaches as well as the classical ones. The *Journal* also emphasizes areas of combinatorics*, graph theory*, finite groups, coding theory, and finite geometries which are related to the field of statistics. It includes all combinatorial problems of block designs, factorial* designs, design of repeated measurements*, etc. Papers in mathematical programming* and computer simulation* which are related to statistical planning are also considered for inclusion. Innovative papers in applied statistics related to agriculture, biology, industry, social sciences, etc., are much encouraged. However, survey papers, reviews, and short notes are also welcome. Long papers do not have to be condensed beyond the normal regard for brevity and economy of space.

ARTICLES PUBLISHED IN THE FIRST FOUR VOLUMES

Over 100 articles were published in the first four volumes. These articles reflect various areas of interest. Over 40 of the articles are in the area of combinatorics and the classical design of experiments (PBIB, SGDD, Latin squares*, etc.). Eight articles are in the area of optimal experimental designs. The area of sampling surveys is represented by eight papers, which deal mainly in problems of estimation. There are five papers on robustness of designs and seven papers on analysis of contingency tables*, six of which are on the problem of testing in a 2×2 table. The other articles are concerned with problems of linear estimation, projections and generalized inverses*, censoring processes, stochastic approximations*, estimation on graphs, empirical Bayes* procedures, and other subjects. The general picture reveals that almost half of the articles are in the area of design and that the remainder are in various areas of theoretical statistics—parametric and nonparametric.

J. N. Srivastava
S. Zacks
P. K. Sen

JOURNAL OF THE AMERICAN STATISTICAL ASSOCIATION, THE

The *Journal of the American Statistical Association* (*JASA*), with a circulation of almost 18,000, is the most widely read professional statistical periodical in the world. Founded in 1888 under the title *Quarterly Publications of the American Statistical Association, JASA* (as it was officially renamed in 1922), is one of the oldest statistical journals still in existence. At its inception, *JASA* (a quarterly journal) placed primary emphasis on the applications of statistics, and was oriented in large part toward demographic and economics statistics. The table of contents of the major articles that appeared in Vol. 1, reproduced here in Table 1, gives a good indication of the contents of *JASA* in its early years. Such an orientation is not surprising when one realizes the the President of the American Statistical Association (ASA) at the time of the founding of *JASA* was General Francis A. Walker, who was not only simultaneously the (founding) President of

Table 1   Contents of Vol. 1, Nos. 1–8 (1888–1889) (492 pp.)

Issue Numbers	Article Titles
1	Statistics of Water Power Employed in Manufacturing in the United States—George F. Swain
2, 3	Park Areas and Open-Air Spaces in American Cities—E. R. L. Gould
	Key to the Publications of the United States Census, 1790–1887; with Occasional References to Other Statistical Works—Edward Clark
4	Life Insurance in the United States—Walter C. Wright
5	Notes on the Statistical Determination of the Causes of Poverty—Amos G. Warner. Remarks—Samuel W. Dike
	Statistics of Divorce in the United States and Europe—Samuel W. Dike
6	American Railroad Statistics—Arthur T. Hadley
	Statistics of Municipal Finance—Henry B. Gardner
7	Prison Statistics of the United States for 1888—Roland P. Falkner
8	Finance Statistics of the American Commonwealths—Edwin R. A. Seligman
	Divorce in France—Benjamin F. Keller
	Relief of the Poor in Germany—A. G. Warner

the American Economic Association (and President of the Massachusetts Institute of Technology), but also had been the Superintendent of the U.S. Censuses of 1870 and 1880.

Almost all the lead articles in the early issues of the *Journal*, such as those listed in Table 1, were read before the three regular meetings of the Association that took place each year in Boston, and the authors were typically distinguished social scientists (e.g., Hadley was later president of Yale University and Seligman was editer of the original *Encyclopaedia of the Social Sciences*). A notable exception to this practice of printing papers read at meetings occurred in Vol. 2, with the reprinting from the *Transactions of the Academy of Science, St. Louis,* of an article by H. S. Pritchett, "A Formula for Predicting the Population of the United States." This paper was also one of the few to appear before World War I that used algebraic formulas and what we would now call statistical methodology. In it, Pritchett fitted a parabola to U.S. Census counts from 1790 to 1880 using the method of least

squares*, and provided a highly illuminating residual analysis, focusing on the deviation of 1870 from the fitted curve and the effects of the Civil War on population.

While the number of *JASA* papers dealing with mathematical statistical topics continued to grow, especially in the 1920s, *JASA*'s lack of receptivity toward papers with substantial mathematical content ultimately led to creation in 1930 of the *Annals of Mathematical Statistics**, as an ASA journal under the editorship of Professor Harry C. Carver. Although some members of the ASA Board of Directors continued to advocate that the *Annals* be retained as a "section" of *JASA*, those who were opposed to the heavy use of mathematics in statistics won out, and the ASA's sponsorship of the *Annals* ceased with the 1934 volume. (For additional details, see the entry on the *Annals of Statistics*.)

Prior to 1930, *JASA* was the only regular publication of the ASA, and thus it included not only articles on the use of statistics and statistical methodology, but also book reviews, reports on proceedings of ASA meetings, and various news items of interest to the membership. From 1928 to 1935 the complete Proceedings of the Annual Meeting were published as a *JASA* Supplement. The Association resumed the publications of some of the papers presented at the Annual Meeting in the form of separate Proceedings volumes in the 1950s. News items and information on chapter activities appearing in *JASA* until 1934 were shifted in 1935 to the newly created *American Statistical Association Bulletin*, which was replaced by the *American Statistician* in 1947. At about the same time (1945) the ASA, though its Biometrics Section, began the publication of the *Biometrics Bulletin*, later to become *Biometrics**, the official publication of the Biometrics Society*.

During the 1950s and 1960s, there was a gradual shift, both in the ASA and in its journal, *JASA*, toward a more rigorous and mathematical approach to statistical methodology. This shift was accompanied by an expansion of *JASA*, from under 600 pages in

1949 to 1258 pages in 1965 and 1712 pages in 1970, and growing counterpressure from the ASA membership for a renewed emphasis on applications. Beginning with Vol. 65 (1970), *JASA* was formally divided into three sections: (1) Applications, (2) Theory and Methods, and (3) Book Reviews, each section with its own editor appointed by the ASA Board of Directors. A fourth editor whose responsibility was to be coordination and publication of *JASA* has, at least to date, taken on the editorship of the Applications or Theory and Methods Section.

Table 2 contains a complete list of *JASA* Editors from 1888 through 1983. Many of these editors were drawn from areas of statistical application such as demography, economics, and sociology, thus continuing the traditional involvement of both the Association and its *Journal* with social statistics in their broadest sense. A recent issue of *JASA* should be consulted for information on current editorial office addresses.

Table 2   Editors of *JASA* (1888–1983)

Davis R. Dewey		1888–1907	
John Koren		1908–1912, 1918–1919	
William B. Bailey		1913–1917	
William F. Ogburn		1920–1925	
Frank A. Ross		1926–1934, 1941–1945	
Frederick F. Stephan		1935–1940	
William G. Cochran		1946–1950	
W. Allen Wallis		1951–1959	
David L. Wallace (acting editor)		1959	
Clifford Hildreth		1960–1964	
John W. Pratt		1965–1967	
Coordinating		*Applications*	
John W. Pratt	1968–1969	Robert Ferber	1968–1976
Robert Ferber	1970–1976	Stephen E. Fienberg	1977–1980
Stephen E. Fienberg	1977–1979	Donald B. Rubin	1980–1982
Donald B. Rubin	1980–	Joseph B. Kadane	1983–
Theory and Methods		*Book Review*	
John W. Pratt and I. R. Savage	1968–1969	Morris DeGroot	1971–1975
		Stephen E. Fienberg	1976
Bradley Efron	1970–1972	S. James Press	1977–1979
Norman L. Johnson	1973–1975	Herbert T. David	1980–1982
Morris H. DeGroot	1976–1978	Joseph Sedransk	1983–
George T. Duncan (acting)	1979		
Stephen Stigler	1979–1982		
Carl N. Morris	1983–		

Table 3   Partial Contents of Vol. 74 (1979)

Some Problems of Statistics and Everyday Life (Presidential Address)
Invited Papers
Field Experimentation in Weather Modification
Nonparametric Statistical Data Modeling
Methodology, and the Statistician's Responsibility for BOTH Accuracy AND Relevance
Applications
Testing Disease Dependence in Survival Experiments with Serial Sacrifice
Short-Term Forecasting and Seasonal Adjustment
Estimates of Income for Small Places: An Application of James–Stein Procedures to Census Data
The Use of the Jackknife to Estimate Proportions from Toxicological Data in the Presence of Litter Effects
Smooth Pycnophylactic Interpolation for Geographical Regions
Simple Models for the Analysis of Association in Cross-Classifications Having Ordered Categories
The Demand for Urban Rail Transportation
Fair Numbers of Peremptory Challenges in Jury Trials
Distinguishing among Distributions Using Data from Complex Sample Designs
Theory and Methods
A Predictive Approach to Model Selection
Influential Observations in Linear Regression
Fixed-State Assessment of Utility Functions
On Rounding Percentages
Likelihood Function of Stationary Multiple Autoregressive Moving Average Models
Optimal and Adaptive Stopping in the Search for New Species
Balanced Hypotheses and Unbalanced Data
A Structural Probit Model with Latent Variables
Robust Locally Weighted Regression and Smoothing Scatterplots
Normal Bayesian Dialogues

The year 1971 saw a change in size and format of *JASA*, from a 6×9 in. single-column layout to an $8\frac{1}{2} \times 11$ in. double-column layout. The 1971 volume was 940 pages in length, and its size has not increased substantially since then. In 1979, the Theory and Methods Section (84 articles) was just less than double the size of the Applications Section (28 articles), and the Book Review Section carried reviews of over 100 books. Table 3 contains a partial table of contents of the 1979 volume. The topics of the articles spanned the full range of interests of members of the statistical profession, including several articles on statistical programs of government agencies. The same is true of the topics of the book reviews, and the reviews now focus to a large extent on links to and uses of methodology, even when the principal aim of the book is not the exposition of statistical methodology.

Virtually all articles printed in *JASA* are submitted on an unsolicited basis, and are put through a rigorous and confidential refereeing process, typically involving an Associate Editor and two referees. Exceptions to this rule fall into two categories. First, the ASA Presidential Address, delivered at the annual meeting, is traditionally published as the lead article in *JASA* in March of the following year. Second, a limited number of invited papers, intended to review or synthesize developments in a particular area of statistics, are solicited and then subjected to the refereeing process. Some articles published as invited papers are submitted originally on an unsolicited basis. For example, beginning in 1978, two special *JASA* Papers (one from Applications and one from Theory and Methods) have been selected for presentation at the annual meeting from among those submitted to the *Journal* during the previous year, and then published subsequent to reading, with discussion by others.

In recent years, *JASA* has used the following criteria for publication, developed originally by the ASA Committee on Publications and revised somewhat by the Editors:

Applications Section

1. Manuscripts that make a substantive contribution to the particular subject area through the use of an interpretation of sound statistical methods. Of particular interest are papers containing new applications or novel adaptations of statistical methods to the solution of problems in a subject-matter field.

2. Manuscripts that present new statistical data (with information on their reliability and examples of how they are to be used) or a new interpretation of existing statistical data.

3. If related to methodology, manuscripts that illustrate how the methodology is applied in practice or present the results of empirical tests designed to test the applicability and feasibility of the methodology in practical situations.

In all cases, Applications papers must present good statistical practice by current professional standards. Preferably, papers should be of broad interest to those in the particular subject areas.

Theory and Methods Section

1. Original manuscripts on the theory of statistics

2. Original manuscripts on the theory of probability which relate in a reasonably direct manner to existing statistical theory

3. Methodological manuscripts that adapt and extend statistical theory for use in special fields of applications: e.g., economics, medicine, industrial testing, etc.

4. Expository manuscripts offering a unified presentation of some subject not readily available in the existing statistical literature

JASA currently receives approximately 700 new submissions each year. About 20% of these submissions are ultimately published, typically after one or more revisions. Begin-

ning in 1978, the Editors have published annual reports on the processing of manuscripts in the June issue of the *Journal*.

JASA has been indexed five times. The first index published in 1941 covered Vols. 1–34 (1888–1939), and the second published in 1959, covered Vols. 35–50 (1940–1955). Author and book review indexes for Vols. 51–60 (1956–1965), and for Vols. 61–72 (1966–1977) appeared in 1966 and 1977, respectively. Finally, a complete author and permuted title (subject) index for Vols. 51–73 (1956–1978) was published as Part 2 of the June 1979 issue.

Although *The Journal of the American Statistical Association* is an official publication of the American Statistical Association, it draws both its readership and its contributors from all over the world. Authors of papers published in 1979 came from 13 different countries. Thus *JASA* is a journal with an international authorship and reputation.

S. E. FIENBERG

JOURNAL OF THE INSTITUTE OF ACTUARIES

This originated as the *Assurance Magazine*, founded privately by two eminent actuaries, Samuel Brown and Charles Jellicoe, in 1850. (Both the first and second issues—September 1850 and January 1851—begin at page 1.) The Institute of Actuaries* agreed that the *Magazine* could publish the papers and proceedings of the Institute, and from June 1852 the *Magazine* was retitled the *Assurance Magazine and Journal of the Institute of Actuaries*. Volumes 14–24 reversed the order to *Journal of the Institute of Actuaries and Assurance Magazine*, and thereafter the title was shortened to *Journal of the Institute of Actuaries*. Jellicoe edited the *Magazine/Journal* for 18 years (1850–1867) and his successor, Thomas Sprague, for another 15; the latter's work was early enlivened by the use of his own system of phonetic spelling. Legal notes began in 1907 but

Table 1 Contents of Vol. 107 (1980)

Alfred Watson Memorial Lecture: The Social Implications of Advancing Technology.
   By Sir Kenneth Berrill, K.C.B.
   Abstract of the Lecturer's Replies to Questions
Are Objectivism and Subjectivism Compatible Concepts? By P. G. Moore, T.D., Ph.D.,
   F.I.A., F.S.S.
   Abstact of the Discussion on the Preceding
Report of the Maturity Guarantees Working Party
   Editorial Note
   Text of Report
   Appendices
      A Terms of References
      B Bibliography
      C Specifying Univariate Models for the de Zoete Equity Index. By E. J. Godolphin
      D Stock Market Models
      E Simulation Results
      F Notes on Time Series Analysis
   Abstract of the Discussion on the Preceding
The Interaction between Morbidity and Mortality. By A. H. Pollard, M.Sc.,
   M.Sc. (Econ.) Ph.D., F.I.A., F.A.S.S.A.
   Abstract of the Discussion on the Preceding
Compensation for Personal Injury. By G. B. Hey, M.A., F.I.A., F.S.S., P. G. Meins, B.Sc.,
   F.I.A., A.P.M.I., M.B.A., W. I. F. Rowlandson, F.I.A., M.B.C.S., D. E. A. Sanders,
   B.A., F.I.A., and R. C. Wilkinson, B.Sc., F.I.A.
   Abstract of the Discussion on the Preceding
Actuaries and Professional Conduct. By F. B. Corby, M.A., F.I.A.
   Discussion on the Preceding
The Age Pattern of Mortality. By L. Heligman, M.A. and J. H. Pollard, B.Sc., Ph.D.,
   F.I.A., F.S.S., F.A.S.S.A.
Consumer Credit Calculations. By C. D. Daykin, M.A., F.I.A., F.S.S.
Johann Heinrich Lambert (1728–1777). By R. H. Daw, B.Sc., F.I.A.
The Duration of Sickness. By L. G. K. Starke, C.B.E., B.A., F.I.A., F.S.S.
Determination of Interest Rate of Return in Respect of an Arbitrary Cash Flow.
   By G. C. Taylor, B.A., Ph.D. F.I.A., F.I.M.A.
On the Time to Extinction. By S. Haberman, M.A., F.I.A.
Early Uses of Graunt's Life Table. By H. L. Seal, B.Sc., Ph.D., A.I.A., F.F.A., A.S.A.,
   F.C.I.A., M.A.A.A.
Dynamic Response of Insurance Systems with Delayed Profit/Loss Sharing Feedback to
   Isolated Unpredicted Claims. By L. A. Balzer, B.E., B.Sc., Ph.D., A.F.A.I.M.,
   A.F.I.M.A. and S. Benjamin, M.A., F.I.A., F.I.S., A.S.A., F.B.C.S.
Notes on the *Financial Times*—Actuaries Equity Indices in 1979. By J. C. H. Brumwell,
   M.A., F.I.A.
The *F.T*—Actuaries British Government Securities Indices, 1979. By A. D. Wilkie, M.A.,
   F.F.A., F.I.A.
The Recent Trend of Mortality in Great Britain. By C. D. Daykin, M.A., F.I.A., F.S.S.
Notes on the Transactions of the Faculty of Actuaries
Notes on Other Actuarial Journals. By F. W. Eschrich, F.I.A. and R. A. Soward,
   B.A., F.I.A.
Articles, Papers and Publications of Actuarial Interest. By M. D. May, M.A., F.I,A.,
   D. F. Renn, Ph.D., F.I.A. and N. Williams, F.I.A.
Reviews
   UK & European Share Price Behaviour: the Evidence. By P. H. Richards
   Survival Probabilities—The Goal of Risk Theory. By H. L. Seal
   Pension Schemes—A Guide to Principles and Practice. By M. Pilch and V. Wood
   Policymaking for Social Security. By Martha Derthick
   The Analysis of Mortality and Other Acturial Statistics. By B. Benjamin and
      J. H. Pollard
   Pensions: A Practical Guide. By J. S. D. Seres and J. W. Selley
Correspondence
   Corrigenda to *J.I.A*. **106** Part III and **107**, Part II

now appear very infrequently, and, since 1929, reports of annual and special general meetings and the biennial dinner have appeared in a separate *Year Book of the Institute of Actuaries**.

A volume has appeared (in parts) normally one per year, but war and other disruptions have meant that the hundredth volume appeared in 1973, recording the special meetings held to celebrate the 125th anniversary of the foundation of the Institute in 1848. Indexes have been published to Vols. 1–40 (supplanting earlier ones to Vols. 1–10, 1–20, and 21–30), to Vols. 41–55, 56–65, 66–85, and 86–100, the latter two employing an actuarial decimal system of classification.

Today the *Journal* normally appears three times a year, each part containing the papers presented at two of the sessional meetings held a few months previously, together with abstracts of the discussion following their presentation. The first paper of each session, delivered in October, is usually either a presidential address (that for 1980 being devoted to relations with the accountancy profession and in refuting arguments that the code of conduct of the actuarial profession was illegally restrictive in some respects) or a lecture by an eminent nonactuary funded as a memorial to Sir Alfred Watson, the first British Government Actuary of the twentieth century. Other regular items are reports on recent trends in British mortality*, on the financial indexes calculated by the Institute and Faculty of Actuaries and published by the *Financial Times*, notes on other actuarial journals throughout the world (especially the sister journal, the *Transactions of the Faculty of Actuaries in Scotland*) and other publications of actuarial interest, including official ones. Important books are reviewed at length.

Other contributions on actuarial science are welcomed from any source, in the form of notes, articles, or correspondence. Papers are refereed, and there are occasional special issues of the *Journal*, such as that reprinting John Graunt's *Natural and Political Observations ... upon the Bills of Mortality* (Vol. 90, Part 1, No. 384), another on the 125th anniversary meetings of the Institute (Vol.

100, No. 415), and one containing the 1979 Report of the Institute and Faculty Working Party on Maturity Guarantees (Vol. 107, Part II, No. 435; see Table 1).

Issues of the *Journal* vary in length between 100 and 150 pages and are available on microfilm (Vols. 1–86) and microfiche (beginning with Vol. 87). The editorship changes every few years, but the current editor can always be contacted through the Institute of Actuaries, Staple Inn Hall, High Holborn, London WC1V 7QJ, England.

D. F. RENN

JOURNAL OF THE ROYAL STATISTICAL SOCIETY

The *Journal of the Royal Statistical Society* is published by the Royal Statistical Society* and consists of three series, A, B, and C. Series A was established in 1839, Series B in 1934, and Series C (Applied Statistics) in 1954. The current (1983) editors of the Journal are:

Series A: Mr. B. P. Emmett and Mr. D. Newman

Series B: Professor P. Holgate and Professor T. M. F. Smith

Series C: Dr. David A. Williams and Mr. M. Hills

The current address of the editorial office is 25 Enford Street, London W1H 2BH, England. The average time from submission to editors' decision (based on the information received in 1983) is six months.

Series A (General) contains papers read and discussed at Ordinary General Meetings of the Royal Statistical Society (other than those organized by the Research Section). The contributions to the discussions, as well as the papers themselves, are printed, together with considered replies by the authors of the papers to the points raised. Papers and discussions at Special General Meetings are usually included in Series A. It presents the Annual Reports of the Council of the Society, obituaries of distinguished Fellows, book reviews and other items of general interest to Fellows. The remaining pages are devoted to unsolicited papers submitted for publication and not read at meetings of the Society. These are refereed not only for quality but also for suitability, in that they are expected to be potentially interesting to statisticians generally.

Series A is published annually in four parts, each of about 125 pages. An index of contents appears in the fourth part each year. Papers (three copies) should be submitted to the Secretary.

Series B (Methodological) of the *Journal* publishes research work of methodological interest to statisticians and is one of the leading journals in this area. The *Journal* is available to Fellows of the Society free of charge (back numbers may be purchased from William Dawson's of Kent) or may be bought by subscription. It is published annually in three parts and contains around 400 pages. Anyone may submit a paper for consideration for publication; authorship is not restricted to Fellows of the Society. All papers published must satisfy one of several criteria. They may:

1. Describe new methods for data analysis, with theoretical justification, an indication of potentiality and, preferably, an example

2. Compare and evaluate and perhaps review existing techniques

3. Discuss the philosophical basis of statistical theory

4. Build mathematical and stochastic models of phenomena so as to deepen understanding

5. Apply existing methods to illustrate new aspects or limitations

6. Contribute to probability theory *only* if they have a practical aspect

Three types of paper are published in Series B:

1. Read papers, which are the published proceedings of an Ordinary Meeting of the Royal Statistical Society, organized by the Research Section, with discussion; a read paper must satisfy the reviewing committee that it is likely to lead to a good discussion

2. Straightforward papers submitted for publication

3. Short communications related to previous papers or presenting interesting unsolved problems

Papers (four copies) should be submitted to the Secretary; the Editors have an Editorial Panel of Associate Editors to assist in the refereeing process.

Series C (Applied Statistics), aims to publish papers giving a simple presentation of new or recent methodology which will be useful to applied statisticians. Authors are encouraged to include practical examples. The *Journal* also aims to publish interesting practical applications of existing methodology, and reviews or comparisons of existing statistical procedures, providing these highlight novel or little known points of practical use. Short case-study papers that illustrate good statistical practice for the benefit of the novitiate or student in statistics are encouraged. A feature of the *Journal* are the Statistical Algorithms, offprints of which are available from the Society. Reviews of books of particular interest to the practicing statistician are also published. The *Journal* appears three times a year. Papers (three copies) should be submitted to the Secretary.

I. H. BLENKINSOP

JOURNAL OF THE STATISTICAL SOCIETY OF PARIS

The *Journal of the Statistical Society of Paris* is a quarterly international review for French-speaking statisticians. It is published in Paris, with the cooperation of the National Center for Scientific Research and in collaboration with the Statistical Society of France.

THE STATISTICAL SOCIETY OF PARIS AND OF FRANCE

The Statistical Society of Paris was founded on June 5, 1860. The Statistical Society of France was created in 1974; the members and the administrative council are the same for both societies.

THE JOURNAL

The *Journal of the Statistical Society of Paris* has appeared regularly since 1860. Four issues are published each year, at the end of each quarter, representing 300 to 400 pages per year. Special supplementary numbers are occasionally published. 2000 copies of each issue are distributed.

THE BOARD OF EDITORS

The Board of Editors of the *Journal* for 1983 were as follows:

Scientific Director and Editor in Chief: Paul Damiani, Secretary General of the Statistical Society of Paris

Editorial Board: Gerard Calot, Jacques-Michel Durand, Georges Gallais-Hamonno, Henri Guitton, Edmond Malinvaud, Charles Penglaou, Daniel Schwarts

Patrons: J. Bourgeois-Pichat, M. Brichler, M. Bunle, F-L. Closon, P. Delaporte, J. Dubourdieu, D. Dugue, M. Dumas, J. Fourastie, C. Gruson, R. Henon, J. Lamson, E. Morice, A. Sauvy, P. Vendryes, A. Vessereau (all past presidents of the Statistical Society of Paris)

THE JOURNAL FORMAT

The *Journal* contains the texts of communications presented at meetings of the Society, articles, general information, and a bibliography.

There are five or six articles or reports per issue. The maximum length for an article is 20 pages; however, the average length is 10 pages. Time between submission and publication of articles varies from three months to one year depending on the reviewing process.

Articles are published solely in French with résumés in French, English, and German. Articles written in languages other than French can be accepted for publication; they are translated into French either by the author or with the help of the Secretary.

Although theoretical statistics is in the domain of interest of the *Journal*, almost all articles are on applied statistics. For the period 1970–1979, the distribution of types of articles is given in Table 1.

Fields of study covered in the *Journal* during the same period of time are shown in Table 2.

Table 1   Distribution of Articles by Topics

Topics of Articles	Pages per 100
Topics of general interest	33.5
Probability, statistical theory	3.0
Descriptive statistics	18.5
Statistical methods	11.5
Collection and utilization methods	6.0
Applications of statistical methods	27.5

Table 2   Distribution of Articles by Fields of Study

Field	Pages per 100
Teaching	1.5
General problems	10.7
General economics, planning, national accounting	25.9
Price, income	6.1
Investments, finance	16.2
Argiculture, industry, and commerce	10.2
Demography, health	14.2
Humanities, biology, and genetics	4.6
Other	10.6

ADDRESSES AND SUBSCRIPTIONS

For information concerning journal publication, address the Secretariat: Paul Damiani, 18 boulevard A. Pinard, 75675 Paris, Cedex 14, France.

Subscription rates for 1981 are as follows:

France	300F
Foreign countries	330F
Institutional membership	600F
Individual Issues	100F

For subscriptions write to either the Secretary or to the printer: Berger-Levrault, 18 rue des Glacis, 54017 Nancy, Cedex 11, France.

P. DAMIANI

J-SHAPED CURVES

Strictly speaking, this term should be applied to a curve shaped—more or less—like a J, without the horizontal line at the top, as in Fig. 1.

The term is also often used for curves of shape similar to that shown in Fig. 2, which

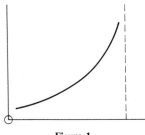

Figure 1

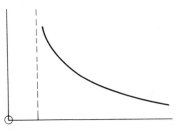

Figure 2

would more appropriately be called reverse J-shaped, or perhaps even L-shaped.

If the curve represents the probability density function* of a distribution, the distribution may also be called J-shaped. Thus the exponential distribution* is (reverse) J-shaped, and the power distribution* with probability density function

$$\alpha x^{\alpha - 1} \qquad (0 < x < 1; \alpha > 1)$$

is J-shaped.

JUDGMENT SAMPLE

A judgment sample is a sample in which the sample units are selected taking into account the personal feeling or opinions of an investigator. Latent prejudices can bias the properties of judgment samples. Moreover, judgment samples cannot usually be analyzed using standard statistical methodology.

(QUOTA SAMPLING
SURVEY SAMPLING)

JUDGMENTS UNDER UNCERTAINTY

Everyone encounters many uncertain situations and makes judgments under uncertainty*. A common example is looking up at the sky and judging how likely it is to rain during the next hour. A driver makes judgments about the quickest route to a particular destination, taking into account distances, possible delays because of traffic, and so on. An investor makes judgments about potential returns from investments in different stocks and bonds, just as a bettor judges how likely each of the horses in a race is to win the race. Physicians and their patients make judgments about the risk involved in various surgical procedures, considering past data about such procedures as well as a patient's age and physical condition. Government officials make judgments about the possible reactions of other countries to decisions involving arms development, tariffs, and many other issues.

Most judgments under uncertainty in everyday life are made in a relatively informal manner. In statistical modeling, however, it is often desirable to formalize certain judgments so that they can be incorporated in the model. The role of judgments under uncertainty in statistical modeling is most obvious in Bayesian inference* and decision theory*, where explicit provisions are made for the inclusion of judgments. Prior probabilities in Bayesian inference may be based on subjective judgments, as may probabilities for uncertain events in decision making under uncertainty. Thus, in the consideration of judgments under uncertainty in statistical modeling, the primary focus is on the expression of judgments in probabilistic form. It should be pointed out, however, that the role of judgment in statistical modeling is much more pervasive than might be suggested by mentioning prior probabilities and decision-making problems. Important modeling issues such as the assumption of normality or the choice of a particular function for a regression* model frequently are resolved on a judgmental basis.

QUANTIFICATION OF JUDGMENTS

Statements such as "it is likely to rain," "the number three horse has the best chance of winning the race," or "a serious accident at this nuclear power plant is highly improbable" illustrate the expression of judgments under uncertainty in a relatively informal manner. Such statements are somewhat ambiguous in that they may be given different interpretations by different people. One person may think that "it is likely to rain" means that rain is almost certain to occur, while another person may think it means that there is only about a 50–50 chance of rain. To avoid such ambiguities, judgments under uncertainty can be quantified. That is, they can be expressed in terms of the formal language of uncertainty: probability. "The probability of rain is 0.80" avoids the ambiguity of "it is likely to rain."

The theory of subjective probability* provides an axiomatic basis for the quantification of judgments [1, 15]. Subjective probabilities assessed in accordance with certain plausible behavioral axioms of coherence* must conform mathematically to a probability measure. In essence, the axioms of coherence are such that it is impossible to set up a series of bets against a person obeying the axioms so that the person is sure to lose regardless of which events actually occur. For example, if someone claims that Pr[rain] = 0.80 and Pr[no rain] = 0.50, this implies that fair odds in favor of rain would be 4 : 1 and fair odds in favor of no rain would be even, or 1 : 1. Betting $4 in favor of rain against $1 (the 4 : 1 odds) and $2.50 in favor of no rain against $2.50 (the even odds) would lead to a loss of $1.50 whether it rains or not. To avoid a sure loss such as this, it is necessary to set Pr[no rain] = 1 − Pr[rain], as prescribed by probability theory.

The axioms of coherence are sufficient conditions for the existence of subjective probabilities. The actual assessment (or elicitation, or encoding) of subjective probabilities is an important but separate issue. Various methods have been proposed and used for the quantification of judgments [3, 17, 18]. The most straightforward approach is simply to ask for the desired probability directly. Some people are not familiar with the notion of probability and feel more comfortable thinking in terms of odds, which can, of course, be converted to probabilities. Devices are also available to enable a person to assess probabilities without having to provide numbers explicitly. For instance, a probability wheel is a disk with two colors (blue and red, say). The proportion of the disk that is blue can be varied from 0 to 1, and the idea is to adjust this proportion until the assessor is indifferent between betting on the event of interest (such as rain) and betting that a pointer mounted on the wheel will come to rest in the blue portion if it is spun. Once this indifference point is reached, the proportion of the wheel that is blue can be read from the reverse side of the wheel. This proportion represents the assessor's probability for the event of interest.

The probability wheel provides a reference event with known probability.

To provide assessors with some incentive to quantify their judgments carefully, bets at the odds implied by the assessed probabilities could be considered. A more frequently used alternative is the notion of a scoring rule which provides the assessor with a score based on the assessed probabilities and on the events actually occurring [11, 16]. For example, if the probability of a single event such as rain is desired, one of the following three scoring rules might be used:

$$S_1 = \begin{cases} 1 - (1 - r)^2 & \text{if rain occurs,} \\ 1 - r^2 & \text{if rain does not occur;} \end{cases}$$

$$S_2 = \begin{cases} \log r & \text{if rain occurs,} \\ \log(1 - r) & \text{if rain does not occur;} \end{cases}$$

$$S_3 = \begin{cases} \dfrac{r}{\left[r^2 + (1 - r)^2\right]^{1/2}} & \text{if rain occurs,} \\[2mm] \dfrac{(1 - r)}{\left[r^2 + (1 - r)^2\right]^{1/2}} & \text{if rain does not occur;} \end{cases}$$

where r represents the assessed probability of rain. These three rules are called quadratic, logarithmic, and spherical scoring rules, respectively. They are all strictly proper scoring rules in the sense that they satisfy the property that an assessor must provide honest assessments in order to maximize the expected score. If a weather forecaster feels that the probability of rain is 0.20, then any assessed probability different from 0.20 will provide a lower expected score.

When the judgments of interest involve a random variable instead of a single event, the uncertainty can be represented in terms of a probability distribution for the random variable. This is often done by assessing various cumulative probabilities and smoothing a curve through them. Certain values of the variable can be selected and cumulative probabilities assessed for these values, or probabilities can be selected and fractiles can be assessed corresponding to the probabilities. An example of the latter approach is the method of successive subdivisions [13]. The real line is first divided into two equally likely regions by assessing the median, each of the two regions is divided

into two equally likely subregions by assessing the first and third quartiles, and so on. One advantage of this approach is that it requires only even-odds judgments, which are easier to think about than judgments involving uneven odds.

EVALUATION OF JUDGMENTS

Subjective probabilities can, and often do, differ from person to person. One physician might judge that the probability of survival for a patient who has just suffered a massive heart attack is 0.40, but a second physician might assess this probability as 0.20. This difference may be due to different past experiences with similar patients, different emphasis on various symptoms and other cues, and other factors. In any event, if the patient survives, the first physician's probability might seem, on an ex post basis, to be "better." If the patient dies, the second physician's assessment might seem "better."

It is difficult to make comparisons on the basis of a single trial. If the two physicians make and record numerous probability assessments over time, it is possible to investigate the calibration* of the probabilities. A calibration curve is a graph of relative frequencies as a function of assessed probabilities. For example, consider all occasions on which a physician has assessed a probability of 0.40. If the event of interest (e.g., survival) has occurred on exactly 40% of these occasions, this represents perfect calibration. Any deviation from 40% indicates a deviation from perfect calibration.

In some psychological studies, subjects have been shown to be poorly calibrated [7]. For example, subjects tend to understate the degree of uncertainty (by assessing probabilities too close to zero or one) in some situations. Biases of this nature have been related to psychological heuristics involving issues such as the availability of information, anchoring on certain values, selective perception, and focusing on "representative" scenarios [4, 6]. However, assessors have been shown to be very well calibrated in other situations. For instance, a large body of evidence indicates that weather forecasters come very close to perfect calibration curves with their forecasts of the probability of rain [12].

Calibration does not by itself indicate "good" judgments on the part of an individual. For example, a weather forecaster could be perfectly calibrated in the long run by just using the past climatological relative frequency of rain as a forecast (assuming no major shifts in weather patterns). If it has rained in July on one-fifth of the days over the past 50 years, then the forecaster would simply use 0.20 as the probability of rain each day in July. This would not be very informative and would not distinguish between days with very different weather patterns. The ideal forecast, of course, would distinguish perfectly, providing a probability of 1 preceding days with rain and a probability of zero preceding days without rain. The state of the art in weather forecasting is such that perfect forecasts like this are not attainable on a regular basis. However, it is desirable for an assessor to provide probabilities close to zero or one while remaining well calibrated. Fortunately, strictly proper scoring rules reflect this desire. In fact, U.S. National Weather Service forecasters are evaluated with a quadratic scoring rule for their forecasts of the probability of rain.

Calibration curves and scores can be used as feedback for probability assessors. Such feedback can help assessors to improve their assessments. In addition, training about potential psychological biases may help assessors to recognize and attempt to avoid the biases. Training and feedback can be quite valuable in improving an individual's judgments and the process of quantifying the judgments.

JUDGMENTS AND STATISTICAL MODELING

The form in which judgments are expressed may depend on the statistical model that is used. For example, in some cases an attempt is made to approximate a person's judgments about a random variable with a member of a particular family of distributions, such as the normal family. The intent in this instance is to make the subsequent analysis more tractable. In Bayesian inference, the analysis is simplified if the prior distribution is a member of the appropriate conjugate family of distributions [2, 14].

In some cases it is very difficult for a person to assess probabilities for the events or variables in a statistical model but much easier to assess probabilities for other events or variables. An assessor may find it hard to assess a prior distribution for the coefficients of a regression model, for instance. An alternative is to ask the assessor for probability distributions for the dependent variable, conditional upon various sets of values of the independent variables. (Such distributions are called predictive distributions.) On the basis of the assessed distributions, inferences can be made about the distribution of the regression coefficients [5]. Here one set of judgments is used to make inferences about another set of judgments that cannot be observed directly. In fact, in the regression example it is possible to go one step further and make inferences about the form of the regression equation. Since the judgments involve observable variables and not model parameters, the process of quantifying judgments is not model specific.

When someone's judgments are quantified, the question of whether they should be taken at face value may be raised. If it is known that a person tends to be poorly calibrated, then it is possible to calibrate the assessed probabilities and to use the calibrated probabilities instead of the assessed probabilities. For instance, if a physician has assessed the probability of survival to be 0.20 on 200 occasions, and 81 of the 200 patients survived, then the next time the physician uses a probability of 0.20, it might seem reasonable to calibrate this probability by shifting it to a value near 0.40. The process of calibration might involve a straightforward use of a calibration curve or it might involve a model of the assessment process itself [8, 9].

In statistical modeling, all relevant information should be considered. Sometimes judgments are available from two or more people about the same event or variable. The probability of a patient surviving might be assessed separately by two or more physicians, for instance. The combining of these separate probabilities to arrive at a single probability (representing, in some sense, the combined judgments of the physicians) is called a *consensus problem* [10, 19]. Various schemes, ranging from simple averages of the probabilities to more sophisticated techniques, have been proposed and studied for the combination of probabilities. An alternative is a behavioral approach involving face-to-face interaction among the assessors.

Judgments are used extensively in statistical modeling, both in an explicit sense (when quantified judgments are used as inputs to a model) and in an implicit sense (in the model-building process itself). A better understanding of the process by which judgments are formed and revised over time could lead to improved methods for the quantification and evaluation of judgments under uncertainty. Since a considerable amount of information is available only in judgmental form, judgments play an important role in problems of inference and decision making under uncertainty.

References

[1]   de Finetti, B. (1974/1975). *Theory of Probability*, 2 vols., A. Machi and A. Smith, trans. Wiley, New York. (An important book on the theory of probability from a subjective viewpoint.)

[2]   DeGroot, M. H. (1970). *Optimal Statistical Decisions*. McGraw-Hill, New York. (A book on statistical decision theory.)

[3]   Hampton, J. M., Moore, P. G., and Thomas, H. (1973). *J. R. Statist. Soc. A*, **136**, 21–42. (A review paper on the quantification and evaluation of judgments.)

[4]   Hogarth, R. M. (1980). *Judgment and Choice*. Wiley, New York. (A book on judgmental inferences and decisions under uncertainty.)

[5]   Kadane, J. B., et al. (1980). *J. Amer. Statist. Ass.*, **75**, 845–854. (A paper on the indirect quantification of judgments.)

[6]   Kahnemann, D., Slovic, P., and Tversky, A., eds. (1981). *Judgment under Uncertainty: Heuristics and Biases*. Cambridge University Press, Cambridge. (A book on psychological biases in judgments under uncertainty.)

[7]   Lichtenstein, S., Fischhoff, B., and Phillips, L. D. (1977). In *Decision Making and Change in Human Affairs*, H. Jungermann and G. de Zeeuw, eds., D. Reidel, Dordrecht, Holland. (Some empirical results concerning the calibration of probability assessors.)

[8]   Lindley, D. V., Tversky, A., and Brown, R. V. (1979). *J. R. Statist. Soc. A*, **142**, 146–180. (A paper on the modeling of the process of quantifying judgments.)

[9]   Morris, P. A. (1974). *Manag. Sci.*, **20**, 1233–1241. (A paper on the modeling of the process of quantifying judgments.)

[10] Morris, P. A. (1977). *Manag. Sci.*, **23**, 679–693. (A paper on the combining of judgments.)

[11] Murphy, A. H. and Winkler, R. L. (1970). *Acta Psychol.*, **34**, 273–286. (A paper on scoring rules in the quantification and evaluation of judgments.)

[12] Murphy, A. H. and Winkler, R. L. (1977). *Appl. Statist.*, **26**, 41–47. (Some empirical results concerning the quantification and evaluation of judgments.)

[13] Raiffa, H. (1968). *Decision Analysis*. Addison-Wesley, Reading, Mass. (A book on modeling decision-making problems under uncertainty.)

[14] Raiffa, H. and Schlaifer, R. (1961). *Applied Statistical Decision Theory*. Harvard Business School, Boston. (A book on statistical decision theory.)

[15] Savage, L. J. (1954). *The Foundations of Statistics*. Wiley, New York. (An axiomatic development of subjective probability.)

[16] Savage, L. J. (1971). *J. Amer. Statist. Ass.*, **66**, 783–801. (A paper on the quantification and evaluation of judgments.)

[17] Spetzler, C. S. and Staël von Holstein, C.-A. S. (1975). *Manag. Sci.*, **22**, 340–358. (A paper on the quantification of judgments.)

[18] Winkler, R. L. (1967). *J. Amer. Statist. Ass.*, **62**, 776–800. (A paper on the quantification of judgments.)

[19] Winkler, R. L. (1968). *Manag. Sci.*, **15**, 61–75. (A paper on the combining of judgments.)

(BAYESIAN INFERENCE
CHANCE
DECISION THEORY
INFERENCE, STATISTICAL
LOGIC IN STATISTICAL REASONING)

ROBERT L. WINKLER

JUMP PROCESSES

Also known as marked point processes and cumulative processes, a jump process is a random process on the real line whose realizations are (right continuous) step functions. An individual realization may be described by a doubly infinite sequence of planar points $\Phi = \{(t_j, u_j)\}_{j=-\infty}^{\infty}$ with $\cdots < t_{-1} < t_0 \leqslant 0 < t_1 < \cdots$ corresponding to the times of events of a point process $N = \{t_j\}_{j=-\infty}^{\infty}$ (see POINT PROCESSES), and with u_j the value of the jump (or mark) at time t_j. It is usual to assume that there are only a finite number of jumps in a finite interval. Smith [8] and Mathes [6] are early references to the concept. Snyder [9] contains a fair amount of material.

Examples of data corresponding to jump processes include sequences of occurrence times of earthquakes with associated Richter magnitudes, times of occurrence of hurricanes with amounts of damage caused, failure times of devices together with repair costs, times of accidents with corresponding insurance awards, and times of arrival of customers at a service facility with waiting times experienced. Specific examples are discussed in Bartlett [1], Boel et al. [2], Snyder [9], and Vere-Jones [11].

The simplest jump process is the (homogeneous) Poisson for which the u_j are identically 1 and the times, $t_{j+1} - t_j$, between successive events are independent, identically distributed exponential variates. Important jump processes include the compound Poisson process and the Markov renewal (or semi-Markov) process. A compound Poisson process* is a jump process with N a homogeneous Poisson process and with the u_j independent of each other and N and having a common distribution. A Markov renewal process is a jump process with N a renewal process (see RENEWAL PROCESSES) and with the distribution of u_{J+1} given $\{(t_j, u_j)\}_{j=-\infty}^{J}$ depending only on u_J.

A jump process is commonly characterized by a pair of random functions $\gamma(t \mid H_t)$, $\rho(u \mid t, H_{t-})$ with $H_t = \{(t_j, u_j); t_j \leqslant t\}$ and

$$\gamma(t \mid H_t)$$
$$= \lim_{h \downarrow 0} \Pr\left[\text{there is a } t_j \text{ in } (t, t+h] \mid H_t\right]/h$$

$$\rho(u \mid t, H_{t-})$$
$$= \lim_{h \downarrow 0} \Pr\left[u_j \text{ in } (u, u+h) \mid t_j = t, H_{t-}\right]/h.$$

The function γ is called the *conditional intensity*, while ρ is called the *transition density*. (This definition assumes that the distribution of u is continuous. The definition in the discrete case is analogous.) The functions γ, ρ provide the evolution of the process. Suppose that one has reached time t; then $\gamma(t \mid H_t)h$ gives the probability that a new point occurs in the next small interval (of length h). Further, $\rho(u \mid t, H_{t-})$ then provides the probability density function of the value of the jump at t given there is a jump at t. These functions are discussed in Boel et al. [2], Rubin [7], and Snyder [9], for example.

A jump process is sometimes described via moment measures $E[\Phi(I_1 \times J_1) \cdots \Phi(I_K \times J_K)]$ with I_k, J_k Borel sets of the real line and $\Phi(I \times J)$ the number of j with $t_j \in I$ and $u_j \in J$. As a jump process may be viewed as a point process in the plane, these moment measures may be viewed as those of a planar point process. Moment measures are discussed in Vere-Jones [11].

A jump process is said to be *stationary* when its probabilistic characteristics are invariant under translations of the time origin. The process $\Phi(t)$ then has stationary increments. If it is further continuous in mean square, then there is a spectral representation

$$\Phi(t) = \int_{-\infty}^{\infty} \left[\exp(i\lambda t) - 1\right]/(i\lambda) \, dZ(\lambda)$$

with Z a complex-valued random function satisfying $\text{cov}(dZ(\lambda), dZ(\mu)) = \delta(\lambda - \mu) dF(\lambda) d\mu$ for a real-valued, nondecreasing function F and with δ the Dirac delta function*. F is called the *spectral measure* of the process Φ. This representation and the corresponding measure are of use in developing sampling properties of statistics of interest, in examining the effects of a variety of operations on the process, and in describing the relationship of a jump process to covarying continuous time series. Spectral considerations are discussed in Bartlett [1] and Brillinger [4].

Two operations important in the theory and practice of jump processes are superposition and deletion. Superposition refers to the simple addition $\Phi(I \times J) = \Phi_1(I \times J) +$

$\Phi_2(I \times J)$ of the counting variates corresponding to two jump processes. The result is itself a jump process. Deletion refers to the random elimination of some of the planar points (t_j, u_j) of a realization of a jump process. The resulting process is again a jump process.

Suppose that the piece $\{(t_j, u_j); \ 0 < t_j < T\}$ of a realization of the process is available for analysis. In the case that expressions are available for the conditional intensity and transition density the likelihood function

$$\exp\left[\int_0^T \log \rho(u_t \mid t, H_{t-}) N(dt) \right.$$
$$\left. + \int_0^T \log \gamma(t \mid H_t) N(dt) - \int_0^T \gamma(t \mid H_t) dt \right]$$

may be used to make inferences concerning the process, for example, to construct maximum likelihood estimates of finite-dimensional parameters. In the case of the homogeneous Poisson, of rate γ, the likelihood function comes down to

$$\gamma^{N(T)} \exp(-\gamma T)$$

and the maximum likelihood estimate of γ is therefore $N(T)/T$. Snyder [9] provides further examples. In the stationary case estimates of the moment measures, and associated densities may be constructed in a direct fashion. Also in the case that the spectral measure F is absolutely continuous, an estimate of its derivative, the power spectrum may be based on the finite Fourier transform values $\sum u_j \exp(-i\lambda t_j)$ with the summation over available values. Brillinger [4] is one reference.

References

[1] Bartlett, M. S. (1967). *Proc. 5th Berkeley Symp. Math. Statist. Prob.*, Vol. 3. University of California Press, Berkeley, Calif., pp. 135–152. [Presents basic motivation and the results of a (cross-) spectral analysis.]

[2] Boel, R., Varaiya, P., and Wong, E. (1975). *SIAM J. Control*, **13**, 999–1061. (Provides the theoretical foundations.)

[3] Brémaud, P. (1981). *Point Processes and Queues.* Springer-Verlag, New York. (Presents the probabilistic foundations.)

[4] Brillinger, D. R. (1972). *Proc. 6th Berkeley Symp. Math. Statist. Prob.*, Vol. 1. University of California Press, Berkeley, Calif., pp. 483–513. (Develops statistical inference for processes with stationary increments, e.g., stationary jump processes.)

[5] Brillinger, D. R. (1982). *Bull. Seismol. Soc. Amer.*, **72**, 1401–1408. (Develops bounds for exceedance probabilities.)

[6] Mathes, K. (1963). *Jahresber. Deutsch. Math. Ver.*, **66**, 66–78. (An early theoretical reference.)

[7] Rubin, I. (1974). *IEEE Trans. Inf. Theory*, **IT-20**, 617–624. (Presents practical descriptions and definitions.)

[8] Smith, W. L. (1955). *Proc. R. Soc. Lond. A*, **232**, 6–31. (An early reference to the concept.)

[9] Snyder, D. L. (1975). *Random Point Processes.* Wiley, New York. (Contains much practical material.)

[10] Vaca, M. V. and Tretter, S. A. (1978). *IEEE Trans. Inf. Theory*, **IT-24**, 289–295. (Obtains recursive equations for minimum mean-squared error estimates.)

[11] Vere-Jones, D. (1970). *J. R. Statist. Soc. B*, **32**, 1–45. (Concerned with modeling, especially of earthquake sequences.)

(POISSON PROCESSES
RENEWAL THEORY
RISK THEORY
STOCHASTIC PROCESSES)

DAVID R. BRILLINGER

JUSTICE STATISTICS, BUREAU OF

The Bureau of Justice Statistics of the Department of Justice is the national repository for statistical information dealing with crime and the operation of criminal justice systems at all levels of government, the source of financial and technical support to state statistical and operating agencies in all 50 states, and the developer of national information policy on such issues as data privacy, confidentiality and security, interstate exchange of criminal records, and related issues.

Establishment of the Bureau of Justice Statistics (BJS) in December 1979 culminated over half a century of recommendations urging the establishment of an independent and objective national center for criminal justice statistics in order to provide basic information on crime to the president, Congress, the judiciary, state and local governments, the general public, and the media. Creation of the Bureau was intended to ensure (1) collection of adequate statistics on crime and the response to crime from federal, state, and local criminal justice agencies; (2) continuous work to improve the accuracy, completeness, and usefulness of these agencies' statistics; (3) assistance in the development of adequate state and local statistical systems; (4) continued attention to policy implications of criminal justice data collection, utilization, and exchange; and (5) the conduct of surveys, censuses, and special studies in response to immediate policy issues confronting the Department of Justice and Congress. The current (1982) director of the Bureau is Steven R. Schlesinger. The mailing address for inquiries and requests for publications is Bureau of Justice Statistics, U.S. Dept. of Justice Statistics, 633, Indiana Avenue, Washington, D.C. 20531.

A NATIONAL REPOSITORY

Following a decade of operation as the statistical office within the Law Enforcement Assistance Administration (LEAA), BJS has in fact become the national repository of criminal justice information either by initiating new statistical series, by assuming responsibility for ongoing data series from other federal agencies, or by reactivating dormant statistical series. Specifically, the *National Crime Survey* was initiated to provide data on the extent and severity of victimizations of American citizens and their households—second only to the decennial census in terms of the number of persons interviewed, data collection, analysis, and evaluation activities of the Bureau of the Census* for the Survey cost $8.9 million; the *National Prisoner Statistics*, which provide data on conditions in prisons and jails and the characteristics of offenders, was trans-

ferred to BJS from the Bureau of Prisons—annual collection, analysis, and publication costs are $1.4 million; *National Court Statistics*, which provide data on the work load and backlog confronting state court systems and administrators, was reestablished after being discontinued by the Bureau of the Census—the costs of collection, state-by-state compilation, and publication are $475,000; *National Parole and Probation Statistics*, which provide data on the characteristics of persons admitted to and released from parole and probation and subsequent recidivism, was taken over from the National Institute of Mental Health—the statistical functions for these series require $1.5 million annually; *National Expenditure and Employment Statistics*, which provide information on the expenditures, manpower and physical resources, and total operational costs of state and local criminal justice systems was assumed from the Bureau of the Census and expanded by BJS—maintenance of this series costs $780,000 annually; *National Prosecutorial Statistics*, providing data on declinations and dismissals, disposition patterns, and sentencing patterns of selected metropolitan prosecutors was initiated by BJS using data generated from the Prosecutors Management Information System (PROMIS)—the collection and publication of comparable data from 15 to 20 cities costs $275,000; *National Juvenile Justice Statistics*, providing information on juvenile detention facilities and juvenile court processing, was taken over from what was then the Children's Bureau of the Department of Health, Education, and Welfare—continuing support of these series would require $650,000.

STATE STATISTICAL SUPPORT PROGRAMS

In creating the Bureau of Justice Statistics, the Congress directed that BJS "shall give primary emphasis to the problems of State and local justice systems" and "shall utilize to the maximum extent feasible State governmental organizations and facilities responsible for the collection and analysis of criminal justice data and statistics." During the prior decade of operation within LEAA, a major funding and assistance effort—the Comprehensive Data Systems program—was established.

As a result the Bureau now supports a statistical analysis capability in over 40 states which provides statistical information services and policy guidance to the governors, executive branch agencies, legislatures and legislative committees, judiciary, press, and public of these states. In addition, state analysis centers have and will continue to play a vital role in collecting and submitting data to this Bureau for national statistical compilations. The Bureau thus is now supporting cooperative programs to maintain state-level statistical analysis capabilities, to provide states access to federal information useful in addressing state problems, to develop transaction statistics on how state and local systems for the administration of justice are functioning, to permit states to work cooperatively on statistical problems such as correctional population projections, and to develop state probation statistics.

The Bureau also supports the operation of state uniform crime reporting agencies in 44 states in order to facilitate the submission and improve the validity and reliability of arrest and clearance data submitted by local police agencies to the Federal Bureau of Investigation. Cooperation with the FBI in support of the Uniform Crime Reports (UCR) Program has been continuous and will assume a new dimension with a jointly sponsored major evaluation of the UCR series to be undertaken in 1981–1983.

INFORMATION SYSTEMS POLICY

The Bureau of Justice Statistics has continued its role in the formulation of criminal justice information policy initiated when the agency was an element of the Law Enforcement Assistance Administration. With the introduction of automation in record keep-ing, the Bureau has been careful to ensure parallel development of appropriate safeguards for the continued confidentiality of the data in these records. Most important, the Bureau's predecessor agency was a leader in balancing the individual's need for the protection of privacy in his or her personal affairs and the growing needs of the law enforcement community for access to data for crime control purposes. Such efforts resulted in the establishment and promulgation of landmark principles relating to the security and confidentiality of criminal history records. As of this time, such principles have been incorporated by almost all states in state legislation and operational principles. The Bureau monitors the status of such state legislative trends and has provided support to all 50 states in this area. In addition, Bureau attention has been directed to specific legal and policy issues associated with media access to criminal justice data and employer use of information. Documents have been released to assist individual states in the development of policy and procedures on these subjects. There will be a major continuing role for the Bureau in monitoring the effectiveness of existing regulations in privacy and security, interstate exchange of criminal records, and related information policy issues.

IMPORTANCE OF THE BUREAU OF JUSTICE STATISTICS FUNCTIONS

With executive branch, congressional, media, and public concern with crime, its victims, and offenders seemingly at new heights, it is essential to maintain an objective source of information concerning criminal behavior. There must be a means to respond to questions related to the nature and extent of crime, the degree of seriousness and violence, the nature of criminal victimization, the efficient processing of accused persons, the costs of operating systems for the administration of justice, the characteristics of detained, sentenced, and released persons, and the operation of federal, state, and local institutions at all levels of government, and the impact of crime on society.

The statistical programs of the Bureau address these and a multitude of other questions being pressed on elected and appointed officials across the United States. Major activities and reports—the annual data on victimization of individuals and households; special victimization reports dealing with the "Hispanic victim," domestic violence, and the seasonality of certain crimes; the full range of information concerning correctional institutions and their handling of inmates; continued support for state analysis centers and for state uniform crime reporting programs, both of which play a vital role in many state bureaucracies; the evaluation, in cooperation with the FBI, of the Uniform Crime Reports Program—all depend on sustained support for the programs of this Bureau. The total annual costs for collection, analysis, dissemination, and continuous evaluation and redesign associated with these programs—and additional statistical functions such as the national criminal justice data archive—are currently in excess of $20 million.

BENJAMIN H. RENSHAW III

JUST IDENTIFIED EQUATIONS

A term used in econometrics* in connection with the estimation of simultaneous equation systems*. *See* OVERIDENTIFICATION for details.

K

KALMAN FILTERING

A Kalman filter is a recursive, unbiased least-squares* estimator of a Gaussian random signal. It has popular applications in guidance problems associated with the aerospace industry, but has roots that date to the early 1940s. At that time Wiener [20] and Kolmogorov [9] addressed a class of problems associated with the estimation of random signals. The solution to the Wiener–Hopf equation*, an integral equation that resulted from this work, is a weighting function which when convolved with the noise-corrupted linear measurements produces an unbiased minimum variance estimator of the random signal. The Wiener–Hopf equation can be solved explicitly only for several special cases, limiting its practical application. In the 1950s, increased usage of digital computers stimulated the idea of generating recursive least-squares estimators. In 1958, Swerling produced a report, later published in the *Journal of Astronautical Sciences* [17], that presented a recursive filtering procedure similar to that now known as Kalman filtering.

Kalman [7] introduced a novel approach to the problem of Wiener and Kolmogorov for random sequences. Working with Bucy, these results were generalized (see Kalman and Bucy [8]) to random processes. The problem of solving the Wiener–Hopf integral equation is circumvented by the Kalman–Bucy approach. They recognized that digital computers are more effective at solving differential equations numerically. Thus they transformed the integral equation to a differential equation to place the computational burden on the computer. The practicality of the Kalman approach has made it immensely popular for aerospace applications as well as many other settings.

STOCHASTIC DIFFERENTIAL EQUATIONS*

We are interested in properties of stochastic systems whose present state x_t can be modeled by a stochastic differential equation. We assume x_t to be a n-dimensional state vector and consider w_t to be a finite-dimensional disturbance at time t. A general differential equation can thus be written as

$$\frac{dx_t}{dt} = f(x_t, w_t, t), \qquad t \geqslant t_0, \qquad (1)$$

where f is a nonlinear, real n-vector function. Equation (1) is a *stochastic differential equation* and w_t is a *random forcing function*. The initial condition can be a fixed constant on a random variable x_{t_0} with a specified distribution. The probability law of w_t is assumed specified.

An important special case of (1) is the stochastic differential equation with an additive white Gaussian forcing function,

$$\frac{dx_t}{dt} = f(x_t, t) + G(x_t, t)w_t, \qquad t \geqslant t_0,$$

$$(2)$$

where x_t is an *n*-dimensional state vector and w_t satisfies

$$Ew_t = 0 \qquad \text{for all } t$$

$$E(w_t w_t^\dagger) = Q(t)\delta(t - \tau),$$

Q an $n \times n$ symmetric nonnegative definite matrix and $\delta(t - \tau)$ the Dirac delta function*. Here the † indicates transpose.

Equation (2) is known as the *Langevin equation*. Now w_t is white noise and is thus neither mean square Riemann integrable nor integrable with probability 1. Equation (2) is therefore not mathematically meaningful as it stands. However, it may be shown that white Gaussian noise* is the formal derivative of Brownian motion* $\{\beta_t, t \geqslant t_0\}$. Then (2) may be considered formally equivalent to

$$dx_t = f(x_t, t)dt + G(x_t, t)d\beta_t, \quad t \geqslant t_0$$

$$(3)$$

or

$$x_t - x_{t_0} = \int_{t_0}^t f(x_\tau, \tau)d\tau + \int_{t_0}^t G(x_\tau, \tau)d\beta_\tau.$$

$$(4)$$

Thus (2) can be made meaningful in terms of (4). The first integral is an ordinary Riemann integral. The second integral of (4) was defined in a mean square sense by Itô [4]. That integral is called the *Itô stochastic integral* and (3) is the *Itô stochastic differential equation*.

Considering the differentials in (3) as small increments we may write (3) as

$$x_{t+\delta t} - x_t = f(x_t, t)\delta t + G(x_t, t)(\beta_{t+\delta t} - \beta_t).$$

Thus if we are given x_t, we see that $x_{t+\delta t}$ depends only on the Brownian motion increment $\beta_{t+\delta t} - \beta_t$. Since Brownian motion increments are independent and by assumption $\{d\beta_t, t \geqslant t_0\}$ is independent of x_t, we may conclude that given x_t, $x_{t+\delta t}$ is independent of $\{x_\tau, \tau \leqslant t\}$. Thus the process x_t generated by (3) or its equivalent formulations is a *Markov process**. We may fashion a solution to the stochastic differential equation therefore in terms of transition probabilities and the density of x_t.

For the Kalman filter setting we specialize (2) even further to

$$\frac{dx_t}{dt} = F(t)x_t + w_t, \qquad t \geqslant t_0, \qquad (5)$$

where $F(t)$ is an $n \times n$ matrix whose elements are continuous functions of t. The initial state x_{t_0} is a random variable such that

$$E(x_{t_0}) = a_0$$

and

$$E\left(\left[x_{t_0} - a_0\right]\left[x_{t_0} - a_0\right]^\dagger\right) = m_0.$$

Also we assume that x_{t_0} is independent w_t, so that

$$E\left[w_t x_{t_0}^\dagger\right] = 0 \qquad \text{for all } t > t_0.$$

The solution to (5) is

$$x_t = \Phi(t, t_0)x_{t_0} + \int_{t_0}^t \Phi(t, \tau)d\beta_\tau, \qquad (6)$$

where $\Phi(t, \tau)$ is the transition matrix and is the solution to the matrix differential equation

$$d\Phi(t, \tau) = F(t)\Phi(t, \tau),$$

$$\Phi(\tau, \tau) = I \qquad \text{for all } \tau.$$

The transition matrix is nonsingular and satisfies the properties

$$\Phi(t_k, t_j)\Phi(t_j, t_i) = \Phi(t_k, t_i) \qquad \text{for all } t_k, t_j, t_i$$

and

$$\Phi^{-1}(t_k, t_j) = \Phi(t_j, t_k).$$

There are many good discussions of stochastic differential equations, including books by Meditch [14], Kushner [12], Arnold [1], Nevel'son and Haśminskiĭ [15], and Åström [2].

Particularly useful from the linear filtering theory perspective is an excellent treatment by Jazwinski [5].

KALMAN FILTERS

In principle, the solution (6) would be sufficient if the process x_t were known exactly. However, it is frequently the case that only a corrupted version of x_t is available. The measurements z_t are typically modeled as

$$z_t = H(t)x_t + v_t, \qquad (7)$$

where v_t is an m-dimensional, Gaussian white noise process with zero mean and covariance matrix $R(t)$. Equations (5) and (7) taken together for the basis of the Kalman filter problem. Consistent with these equations, the problem is to form an estimate $x_t(\tau)$ of the state x_t that is a linear function of all measured data z_s, $t_0 \leqslant s < \tau$ satisfying the requirements

$$x_t(\tau) \text{ is unbiased,} \qquad \text{i.e., } Ex_t(\tau) = Ex_t \tag{8a}$$

and

$$x_t(\tau) \text{ is best in the sense that}$$

$$E\left(\left[X_t - x_t(\tau)\right]^\dagger \left[x_t - x_t(\tau)\right]\right) \text{ is minimized.} \tag{8b}$$

If $t > \tau$, the problem is referred to as a *prediction problem*; if $t = \tau$, as a *filtering problem*; and if $t < \tau$, a *smoothing problem*. The Kalman–Bucy filter deals with the first two of these three problems.

Derivation of the Kalman filter may be found in several places, including the original paper by Kalman [7], an excellent survey by Kailath [6], or in several of the texts on stochastic differential equations cited above. We outline here several key elements of a heuristic derivation in the case of the filtering problem.

The residual may be expressed as

$$z_t - H(t)x_t(t).$$

This expression is sometimes called the *innovations process* and is an approximation for the measurement error. If transformed by

some unknown matrix, $K(t)$, it is also taken as a modification of (5). That is, the estimate $x_t(t)$ is taken to satisfy

$$\frac{dx_t(t)}{dt} = F(t)x_t(t) + K(t)\left[z_t - H(t)x_t(t)\right], \qquad t > t_0, \qquad (9)$$

where the derivative is taken with respect to the subscript t of $x_t(t)$. The matrix $K(t)$ is called the *gain matrix* or *gain function*.

If we take the initial condition to be unbiased, we may write

$$E\left(x_{t_0}(t_0)\right) = E(x_{t_0}). \tag{10}$$

We may also take expectations in (5), (7), and (9), then combine them to obtain

$$\frac{d}{dt} E\left(x_t - x_t(t)\right)$$

$$= \left(F(t) - K(t)H(t)\right)E\left(x_t - x_t(t)\right),$$

whose solution is

$$E\left(x_t - x_t(t)\right) = M(t)E\left(x_{t_0} - x_{t_0}(t_0)\right), \quad (11)$$

where $M(t)$ is the transition matrix associated with the filter dynamics. Thus (10) and (11) taken together imply the unbiasedness of (8a). Next define

$$\bar{x}_t = x_t - x_t(t). \tag{12}$$

It follows that

$$E\left((x_t - x_t(t))^\dagger(x_t - x_t(t))\right) = \text{trace } E\left(\bar{x}_t\bar{x}_t^\dagger\right) \tag{13}$$

and by definition we let

$$P(t) = E\left(\bar{x}_t\bar{x}_t^\dagger\right).$$

To satisfy (8b), it is desired to choose $K(t)$ to minimize $P(t)$. This may be done by choosing

$$K(t) = P(t)H(t)^\dagger R(t)^{-1} \tag{14}$$

and the matrix $P(t)$ is the solution to

$$\frac{dP(t)}{dt} = F(t)P(t) + P(t)F(t)^\dagger + Q(t)$$

$$- P(t)H(t)^\dagger R(t)^{-1}H(t)P(t). \tag{15}$$

Equation (15) is a matrix Ricatti equation. In summary, then, the unbiased minimum variance estimate (Kalman filter), $x_t(t)$, of (5) and (7) is taken as the solution to

$$\frac{dx_t(t)}{dt} = F(t)x_t(t) + K(t)\big[z_t - H(t)x_t(t)\big],$$

where $x_{t_0}(t_0)$ is chosen so that

$$E\big(x_{t_0}(t_0)\big) = E(x_{t_0}).$$

The optimal gain matrix $K(t)$ is found by

$$K(t) = P(t)H(t)^{\dagger}R(t)^{-1}$$

and the error covariance is the solution to the Ricatti equation (15).

Discrete-time formulations of the Kalman filter problem are frequently discussed. In this setting (5) is replaced by

$$x_k = \phi_k x_{k-1} + w_{k-1}, \qquad k = 1, 2, \ldots,$$
$$(16)$$

while (7) becomes

$$z_k = H_k x_k + v_k, \qquad k = 1, 2, \ldots . \quad (17)$$

Thus z_k is modeled as a noisy filtered version of x_k while the dynamical equation is a nonstationary first-order autoregressive process. The theory in discrete time is analogous to that just sketched for continuous time and for that reason is not included in this brief discussion. Most of the texts referenced above contain discussions in this setting.

Following the papers of Kalman [7] and Kalman and Bucy [8], a large number of papers and reports appeared under the title of linear filtering theory. Many of these are detailed in Kalaith [6]. Much of this sequel work served to apply this theory to practical uses, including satellite orbit determination, submarine and aircraft navigation, and other space flight applications, including the Ranger, Mariner, and Voyager missions and the Apollo missions to the moon. See, for example, Bucy and Joseph [3], Tenney et al. [18], Lindgren and Gong [13], and Titus [19].

In addition to the practical applications of the linear filtering theory, there has been a substantial amount of nonlinear filtering theory developed. Stratonovich [16] was pioneering the nonlinear work in the USSR at the same time the Kalman–Bucy work was done in the West. Stratonovich's work was not immediately known in the West and was in part developed independently by Kushner [10, 11] and Wonham [21]. The text by Jazwinski [5] contains a thorough discussion of both nonlinear and linear filtering and is recommended for further details of Kalman filtering and the related topics of stochastic differential equations and nonlinear filtering.

References

[1] Arnold, L. (1974). *Stochastic Differential Equations: Theory and Applications*. Wiley, New York.

[2] Åström, K. J. (1970). *Introduction to Stochastic Control Theory*. Academic Press, New York.

[3] Bucy, R. S. and Joseph, P. D. (1968). *Filtering for Stochastic Processes with Applications to Guidance*. Interscience, New York. (An exposition of Kalman filters with applications to guidance of aircraft and spacecraft.)

[4] Itô, K. (1944). *Proc. Imp. Acad. Tokyo*, **20**, 519–524. (The major paper on stochastic integrals and a very widely quoted paper.)

[5] Jazwinski, A. H. (1970). *Stochastic Processes and Filtering Theory*. Academic Press, New York. (An excellent introduction to linear and nonlinear filtering theory. An engineering approach pleasantly light on measure theory.)

[6] Kailath, T. (1974). *IEEE Trans. Inf. Theory*, **IT-20**, 146–180.

[7] Kalman, R. E. (1960). *Trans. ASME: J. Basic Eng.*, **82D**, 35–45. (This is the inaugural paper and together with the next entry forms the foundation of Kalman filters.)

[8] Kalman, R. E. and Bucy, R. S. (1961). *Trans. ASME: J. Basic Eng.*, **83D**, 95–108. (This paper deals with the continuous-time process and with the preceding entry forms the foundations of Kalman filtering theory.)

[9] Kolmogorov, A. N. (1941). *Bull. Acad. Sci. USSR, Math. Ser.*, **5**, 3–14.

[10] Kushner, H. J. (1964). *J. Math. Anal. Appl.*, **8**, 332–344.

[11] Kushner, H. J. (1964). *SIAM J. Control*, **2**, 106–119.

[12] Kushner, H. J. (1971). *Introduction to Stochastic Control*. Holt, Rinehart and Winston, New York.

[13] Lindgren, A. G. and Gong, K. F. (1978). *IEEE Trans. Aerosp. Electron. Syst.*, **AES-14**, 564–572.

[14] Meditch, J. S. (1969). *Stochastic Optimal Linear Estimation and Control*. McGraw-Hill, New York.

[15] Nevel'son, M. B. and Haśminskiĭ, R. Z. (1973). *Stochastic Approximation and Recursive Estimation*. American Mathematical Society, Providence, R.I.

[16] Stratonovich, R. L. (1960). *Theory Prob. Appl.*, **5**, 156–178. (This is one of the earliest works on nonlinear filtering theory.)

[17] Swerling, P. (1959). *J. Astronaut. Sci.*, **6**, 46–52.

[18] Tenney, R. R., Hebbert, R. S., and Sandell, N. R., Jr. (1977). *IEEE Trans. Aut. Control.*, **AC-22**, 246–261.

[19] Titus, H., ed. (1977). Advances in Passive Target Tracking. *Rep. No. NPS-62Ys-77071*, Naval Postgraduate School, Monterey, Calif. (Restricted distribution.)

[20] Wiener, N. (1949). *The Extrapolation, Interpolation and Smoothing of Stationary Time Series.* Wiley, New York. (Republished as: *Time Series.* MIT Press, Cambridge, Mass., 1964.)

[21] Wonham, W. M. (1963). *IEEE Int. Conv. Rec.*, **11**, 114–124.

(AUTOREGRESSIVE–MOVING AVERAGE PROCESSES
EXPONENTIAL SMOOTHING
INTEGRAL EQUATIONS
STOCHASTIC CONTROL THEORY
STOCHASTIC DIFFERENTIAL EQUATIONS
STOCHASTIC INTEGRALS
STOCHASTIC PROCESSES)

EDWARD J. WEGMAN

KANNEMANN'S INCIDENCE TEST   *See*
INTRINSIC RANK TEST

KANTOROVICH INEQUALITY

If $\mathbf{A}$ is a positive definite $n \times n$ matrix, with eigenvalues* $\alpha_n \geqslant \alpha_{n-1} \geqslant \cdots \geqslant \alpha_1 > 0$, then [1] for any $1 \times n$ vector $\mathbf{x}$ ($\neq \mathbf{0}$)

$$1 \leqslant (\mathbf{x}'\mathbf{A}\mathbf{x})(\mathbf{x}'\mathbf{A}^{-1}\mathbf{x})/(\mathbf{x}'\mathbf{x})^2$$
$$\leqslant \tfrac{1}{4}(\alpha_1 + \alpha_n)^2/(\alpha_1\alpha_n).$$

A more general form is [3]

$$(\mathbf{x}'\mathbf{A}\mathbf{y})(\mathbf{y}'\mathbf{A}^{-1}\mathbf{x})/\{(\mathbf{x}'\mathbf{x})(\mathbf{y}'\mathbf{y})\}$$
$$\leqslant \tfrac{1}{4}(\alpha_1 + \alpha_n)^2/(\alpha_1\alpha_n).$$

Generalizations and statistical applications are described by Khatri and Rao [2].

References

[1] Kantorovich, L. V. (1948). *Uspekhi Mat. Nauk*, **3**, 89–135.

[2] Khatri, C. G. and Rao, C. R. (1981). *J. Multivariate Anal.*, **11**, 498–505.

[3] Strang, W. G. (1960). *Proc. Amer. Math. Soc.*, **11**, 468.

KAPLAN–MEIER ESTIMATOR

The *Kaplan–Meier* (K–M) estimator, or *product-limit* estimator, of a distribution function or survival function is the censored-data generalization of the empirical distribution function.

CENSORED-DATA PROBLEM

The censored-data* problem arises in many medical, engineering, and other settings, especially follow-up studies*, where the outcome of interest is the *time* to some event, such as cancer recurrence, death of the patient, or machine failure. An example of such a follow-up study in a medical setting would be a clinical trial* investigating the efficacy of a new treatment for lung cancer patients; an example in an engineering setting might be a life-testing* experiment investigating the lifetime distribution of electric motors. In the censored-data problem, the (independent) outcomes $X_i \sim F_i(\cdot)$, $i = 1, 2, \ldots, n$, that are pertinent for inference on the distribution functions $F_i(\cdot)$, $i = 1, 2, \ldots, n$ are, unfortunately, not all fully observed. Some of them are partially observed, or *right-censored**, due to curtailment of the follow-up*. The curtailment may be either of a planned or accidental nature; examples of censoring in the medical setting include loss to follow-up, dropout, and termination (or interim analysis) of the study.

Typically observed in these cases are

$$T_i = \min(X_i, C_i)$$
$$\delta_i = I[T_i = X_i], \tag{1}$$

that is, the smaller of the failure time of interest X_i and a censoring time C_i, and the indicator of whether the observed time T_i is the result of censoring ($T_i = C_i$) or not ($T_i = X_i$). Observations T_i for which $\delta_i = 0$ are called *censored times**, and observations T_i

for which $\delta_i = 1$ are called *uncensored times**, or *failures**. The censoring times may be fixed or random. Although the problem is symmetric in X_i and C_i, the aim of the inference is the distribution of the X_i's; the role of the C_i's is that of interfering with full observation of the X_i's.

In censored-data problems the distribution function $F_i(\cdot)$ are often related by a regression model $F(\cdot \mid z_i)$ (e.g., *see* PROPORTIONAL HAZARDS MODEL), or specified as one of k distribution functions (e.g., *see* CENSORED DATA). This article will discuss only the simplest case, the one-sample problem $F_1(\cdot) = \cdots = F_n(\cdot) = F(\cdot)$. The aim of the inference will be the common distribution function $F(\cdot)$ of the X_i's.

KAPLAN–MEIER ESTIMATOR

In the one-sample problem with censored data (t_i, δ_i), $i = 1, 2, \ldots, n$, the Kaplan–Meier [18] estimator of the (common) *survival function** (or *reliability function** in engineering settings) $S(\cdot) = P(X_i > \cdot)$ is

$$
\hat{S}(t) = \begin{cases} \prod_{i \, : \, t_{i:n} \leqslant t} \left(\dfrac{n-i}{n-i+1} \right)^{\delta_{(i)}} \\ \qquad\qquad \text{for } t \leqslant t_{n:n} \\ 0 \qquad\quad \text{if } \delta_{(n)} = 1 \\ \text{undefined} \quad \text{if } \delta_{(n)} = 0 \\ \qquad\qquad \text{for } t > t_{n:n}, \end{cases}
$$

$$(2)$$

where the $t_{i:n}$, $i = 1, 2, \ldots, n$, denote the observed times t_i arranged in increasing order of magnitude $t_{1:n} \leqslant t_{2:n} \leqslant \cdots \leqslant t_{n:n}$, and where $\delta_{(i)}$ denotes the censoring indicator for $t_{i:n}$. In the case of ties among the $t_{i:n}$, the usual convention is that failures $[\delta_{(i)} = 1]$ precede censorings $[\delta_{(i)} = 0]$.

An alternative [but equal to (2)] expression for the K–M estimator (where it is defined) is particularly useful in the presence of tied failure times:

$$
\hat{S}(t) = \prod_{j \, : \, t_{(j)} \leqslant t} \left(1 - \frac{d_j}{n_j} \right), \qquad (3)
$$

where the $t_{(j)}$ denote the *ordered, distinct* failures $t_{(1)} \leqslant t_{(2)} \leqslant \cdots \leqslant t_{(k)}$, the d_j denote the number of failures at $t_{(j)}$, and the n_j denote the number of items $\#\{i : t_i \geqslant t\}$ still alive just before time t.

Table 1 illustrates the computation of the K–M estimator, and Fig. 1 displays the estimator, for a data set of remission durations of leukemia patients. Here, the time X_i of interest is the time from remission to relapse, the data are (t_i, δ_i), $i = 1, 2, \ldots, 21$, and it is desired to estimate the relapse-free survival function $P(X_i > t) =$ probability that an individual is relapse-free (is still in remission) at time t after remission.

The K–M estimator (2), like the empirical distribution function* estimator, is a step function with jumps at those times t_i that are uncensored. If $\delta_i = 1$ for all i, $i = 1, 2, \ldots, n$ (i.e., no censoring occurs), the K–M estimator reduces to a step function with jumps of height d_j / n at each of the $t_{(j)}$, $j = 1, 2, \ldots, k$, which is the usual empirical distribution function. (*See also* EDF STATISTICS.)

Some authors adopt the convention of defining the K–M estimator to be zero for $t > t_{n:n}$ when $\delta_{(n)} = 0$. Whereas such a convention has advantages of definiteness and simplicity, it is arbitrary; it is usually best in data presentations to specify the undefined character of the K–M estimator in this range rather than specify it to be zero. Of course, if we make the reasonable specification that the estimator retain the properties of a survival function in this range, then it must be nonincreasing, nonnegative, and right-continuous.

Under the nonpredictive-censoring assumption discussed below, the K–M estimator can be motivated in several useful ways. This estimator is:

1. The *"generalized maximum likelihood"* estimator* [18] in the same sense that the empirical distribution function is in the case of uncensored data. (The sense in which the empirical distribution function, and its censored-data generalization the K–M estimator, are "maximum likelihood estimators" among the class

Table 1 Illustration of computation of the K–M estimator for the remission data of Freireich et al. [11] from a clinical trial in acute leukemia.[a]

j	Ordered Distinct Failure Times: $t_{(j)}$	Number of Individuals Alive Just before Time $t_{(j)}$: n_j	Number of Individuals Dying at Time $t_{(j)}$: d_j	Factor Contributed at $t_{(j)}$ to K–M Estimator $(1 - \frac{d_j}{n_j})$	K–M Estimator for $t \in [t_{(j)}, t_{(j+1)})$ $\hat{S}(t)$	Greenwood Variance Estimator $\widehat{var}(\hat{S}(t))$
1	6	21	3	18/21	0.857	0.0058
2	7	17	1	16/17	0.807	0.0076
3	10	15	1	14/15	0.753	0.0093
4	13	12	1	11/12	0.690	0.0114
5	16	11	1	10/11	0.627	0.0130
6	22	7	1	6/7	0.538	0.0164
7	23	6	1	5/6	0.448	0.0181
	35				Undefined after $t = 35$	

[a]The (ordered) remission times in weeks on the 21 chemotherapy patients were 6, 6, 6, 6*, 7, 9*, 10, 10*, 11*, 13, 16, 17*, 19*, 20*, 22, 23, 25*, 32*, 32*, 34*, 35* (* denotes a censored observation).

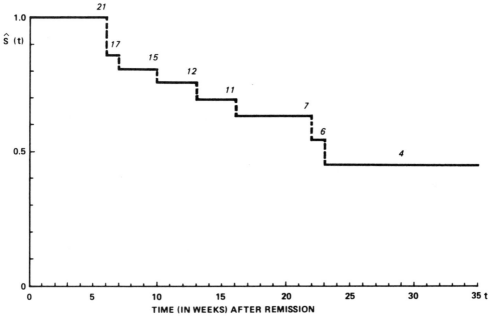

Figure 1 Kaplan–Meier estimator for the acute leukemia remission duration data set of Table 1. (Indicated also are the numbers of individuals alive at various times t.)

of unrestricted distribution functions has been addressed by Kiefer and Wolfowitz [19], Johansen [15], and Scholz [29].)

2. The *limit of life-table** (data grouped in time intervals) *estimators* [4] as the time intervals increase in number and go to zero in length [18]. In fact, the central idea of the K–M estimator as a limit of life-table estimators was present in the early actuarial literature [1a].

3. (Related to item 2) the estimator obtained from a *product of estimators of conditional probabilities* [18].

4. The *"self-consistent" estimator* [5] $\hat{S}(\cdot)$ defined, by analogy with the empirical survival function in the case without censoring, as

$$\hat{S}(t) = \frac{1}{n}\left(\#\{t_i : t_i > t\} + \sum_{t_i \leqslant t} a_i(t)\right),$$

where the fractions

$$a_i(t) = \begin{cases} \hat{S}(t)/\hat{S}(t_i) & \text{if } \delta_i = 0 \\ 0 & \text{if } \delta_i = 1 \end{cases}$$

are estimates of $P(X_i > t \mid T_i = t_i, \delta_i)$.

5. The *redistribute-to-the-right estimator* [5], defined by an algorithm that starts with an empirical distribution that puts mass $1/n$ at each observed time t_i, and then moves the mass of each censored observation by distributing it equally to all observed times to the right of it.

6. A *natural function of two empirical subsurvival functions* [23]. That is, the survivor function $S(\cdot)$ of X can be expressed [under condition (4) below] as a certain function Φ of the subsurvival functions $S_0^*(t) \equiv P(T > t, \ \delta = 0)$ and $S_1^*(t) \equiv P(T > t, \ \delta = 1)$:

$$S(t) = \Phi\big[S_0^*(\cdot), S_1^*(\cdot), t\big].$$

The K–M estimator $\hat{S}(t)$ is just $\Phi[\hat{S}_0^*(\cdot), \hat{S}_1^*(\cdot), t]$, where

$$\hat{S}_0^*(s) \equiv (1/n)\sum_{i=1}^{n} I[t_i > s, \delta = 0]$$

and

$$\hat{S}_1^*(s) \equiv (1/n)\sum_{i=1}^{n} I[t_i > s, \delta = 1]$$

are the empirical subsurvival functions for $S_0^*(\cdot)$ and $S_1^*(\cdot)$, respectively.

APPROPRIATENESS

Of crucial importance to the appropriateness of the K–M estimator (and of most other censored data methods as well) is that for each individual the censoring *must not be predictive of future (unobserved) failure*. Specifically, it must be true for each individual at each time t that

$$\Pr(X \in [t, t + dt) \mid X \geqslant t)$$
$$= \Pr(X \in [t, t + dt) \mid X \geqslant t, C \geqslant t);$$
$$(4)$$

that is, that the instantaneous probability of failure at time t given survival to t is unchanged by the added condition that censoring has not occurred up to time t (e.g., Kalbfleisch and MacKay [16]). As discussed in Chap. 5 of Kalbfleisch and Prentice [17], this condition is equivalent to specifying that for each individual the instantaneous probability of censoring does not depend on the future failure times of this or other individuals.

Unfortunately, the truth of (4) cannot be tested from the censored data (1) alone (Tsiatis [31] and many others). In practice, a judgment about the truth of (4) should be sought based on the best available understanding of the nature of the censoring. For example, end-of-study censoring might typically be expected to meet (4), whereas censoring that is a dropout due to factors related to imminence of failure (e.g., taking as censoring the time of termination of life testing of a machine that shows signs of overheating) would not be expected to meet (4). A judgment on whether certain loss-to-follow-up* circumstances would be expected to satisfy (4) are typically difficult to make, even when the reasons for loss-to-follow-up are known, and thus provide one incentive for strong efforts toward complete follow-up in cohort studies. Inattention to the possibility that the censoring mechanism might be

predictive of failure can be disastrous: the K–M estimator can be grossly in error in the situation where censoring is predictive of failure [22].

PROPERTIES, VARIANCE ESTIMATORS, CONFIDENCE INTERVALS AND CONFIDENCE BANDS

Under random censorship the process $n^{1/2}[\hat{S}(\cdot) - S(\cdot)]$ has the asymptotic distribution of a Gaussian process* (e.g., Breslow and Crowley [2]). Meier [20] discusses corresponding results for the case of fixed censorship. Other aspects of the asymptotic behavior of the K–M estimator have been the subject of numerous recent investigations [1, 8–10, 25, 32, 33].

The asymptotic normality* of $n^{1/2}(\hat{S}(t) - S(t))$ provides a basis for approximating the finite-sample distribution of the K–M estimator $\hat{S}(t)$ by a normal distribution. Alternatively available is a maximum likelihood estimator of this distribution, termed the *bootstrap* distribution [6].

In particular, estimators of the (finite-sample) variance of the K–M estimator $\hat{S}(t)$ at a specified t are readily available. An estimate of the asymptotic variance of $\hat{S}(t)$ provides the well-known Greenwood* [13] estimated variance for $\hat{S}(t)$ (see GREENWOOD'S STATISTIC):

$$\hat{\text{var}}(\hat{S}(t)) = \hat{S}^2(t) \sum_{t_{(j)} \leqslant t} \frac{d_j}{n_j(n_j - d_j)}.$$

Closely related to the Greenwood estimated variance is Efron's [6] bootstrap estimated variance, which is the variance of the K–M estimator's bootstrap distribution. Also, a conservative estimator of the K–M variance is discussed by Peto et al. [24].

Using the Greenwood estimated variance (or one of its alternatives), approximate confidence intervals for $S(t)$ can be obtained, based on the asymptotic normality either of $\hat{S}(t)$ itself [24, 27, 30] or of other functions, such as $\log[-\log \hat{S}(t)]$, that have no range restrictions and/or whose distribution may be more nearly normal.

Simultaneous confidence intervals*, or *confidence bands*, for the survival function $S(\cdot)$, based on the asymptotic equivalence of the K–M process $n^{1/2}(\hat{S}(\cdot) - S(\cdot))$ to Brownian motion processes, have been developed by Gillespie and Fisher [12] and by Hall and Wellner [14].

NONPARAMETRIC QUANTILE ESTIMATION BASED ON THE K–M ESTIMATOR

The entire estimated survival curve $\hat{S}(\cdot)$, together with standard errors or confidence intervals, is usually a good choice for the presentation of survival data with censoring. Nevertheless, summary statistics such as location estimates are sometimes also useful. With censored data the median*, or 0.5 quantile, is a common choice as a location estimator. It is superior to the mean, which is highly sensitive to the right tail of the survival distribution, where estimation tends to be imprecise due to censoring. Other quantiles can be useful in summarizing different aspects of the estimated survival distribution.

In the censored data problem, the maximum likelihood estimator for the pth quantile

$$F^{-1}(p) = S^{-1}(1 - p)$$
$$\equiv \inf\{t : S(t) \leqslant 1 - p\}$$

is conveniently available from the K–M estimator $\hat{S}(\cdot)$:

$$\hat{S}^{-1}(1 - p) \equiv \inf\{t : \hat{S}(t) \leqslant 1 - p\}.$$

The asymptotic distribution of this quantile estimator has been determined by Sander [28] and Reid [26]. However, the asymptotic variance is a function of the failure distribution *density* at the point $S^{-1}(1 - p)$. Because of the difficulty in estimating a density it is difficult to obtain from asymptotic results an approximate estimator for the variance of the quantile estimator. Methods for estimating the finite-sample variance of quantile estimators from censored data include the jackknife* [21] and bootstrap [6] methods.

Approximate confidence limits for a pth quantile $S^{-1}(1 - p)$ based on the asymptotic normality of $\hat{S}(t)$ for a range of t's have been proposed by Brookmeyer and Crowley [3], Emerson [7], and Simon and Lee [30]. Also, Efron [6] has proposed using percentiles of the bootstrap distribution of $\hat{S}^{-1}(1 - p)$ for confidence limits for $S^{-1}(1 - p)$.

References

[1]   Aalen, O. (1978). *Ann. Statist.*, **6**, 534–545.

[1a]   Böhmer, P. E. (1912). *Rapports, Mémoires et Procès-verbaux de Septième Congrès International d'Actuaires*, Amsterdam, Vol. 2, 327–343.

[2]   Breslow, N. E. and Crowley, J. (1974). *Ann. Statist.*, **2**, 437–453.

[3]   Brookmeyer, R. and Crowley, J. (1982). *Biometrics*, **38**, 29–41.

[4]   Cutler, S. J. and Ederer, F. (1958). *J. Chronic Dis.*, **8**, 699–713.

[5]   Efron, B. (1967). *Proc. 5th Berkeley Symp. Math. Statist. Prob.*, Vol. 4. University of California Press, Berkeley, Calif., pp. 831–853.

[6]   Efron, B. (1981). *J. Amer. Statist. Ass.*, **76**, 312–319.

[7]   Emerson, J. (1982). *Biometrics*, **38**, 17–27.

[8]   Földes, A. and Rejtö, L. (1979). Asymptotic Properties of the Nonparametric Survival Curve Estimators under Variable Censoring. Preprint of the Mathematical Institute of the Hungarian Academy of Sciences.

[9]   Földes, A. and Rejtö, L. (1981). *Ann. Statist.*, **9**, 122–129.

[10]   Földes, A., Rejtö, L., and Winter, B. B. (1980). *Periodica Math. Hung.*, **11**, 233–250.

[11]   Freireich, E. O., et al. (1963). *Blood*, **21**, 699–716. (An example of censored data in a medical follow-up setting.)

[12]   Gillespie, M. J. and Fisher, L. (1979). *Ann. Statist.*, **7**, 920–924.

[13]   Greenwood, M. (1926). The natural duration of cancer. *Reports on Public Health and Medical Subjects*. Vol. 33, Her Majesty's Stationary Office, London: 1–26. (Of historical interest, this paper presents the Greenwood estimator of the variance of the K–M estimator.)

[14]   Hall, W. J. and Wellner, J. A. (1980). *Biometrika*, **67**, 133–143.

[15]   Johansen, S. (1978). *Scand. J. Statist.*, **5**, 195–199.

[16]   Kalbfleisch, J. D. and MacKay, R. J. (1979). *Biometrika*, **66**, 87–90.

[17] Kalbfleisch, J. D. and Prentice, R. L. (1980). *The Statistical Analysis of Failure Time Data.* Wiley, New York. (Written for the practicing statistician, this very readable book provides an excellent treatment of the analysis of censored data. Topics include the Kaplan–Meier estimator, the comparison of survival curves, and regression analysis with censored data.)

[18] Kaplan, E. L. and Meier, P. (1958). *J. Amer. Statist. Ass.,* **53**, 457–481. (More than 20 years later, this paper is still an informative and motivated description and discussion of the Kaplan–Meier estimator.)

[19] Kiefer, J. and Wolfowitz, J. (1956). *Ann. Math. Statist.,* **27**, 887–906.

[20] Meier, P. (1975). *Perspectives in Probability and Statistics,* J. Gani, ed. Applied Probability Trust, Sheffield, England.

[21] Miller, R. G. (1974). Jackknifing Censored Data. *Tech. Rep. No. 14,* Dept. of Statistics, Stanford University, Stanford, Calif.

[22] Peterson, A. V., Jr. (1976). *Proc. Natl. Acad. Sci. USA,* **73**, 11–13.

[23] Peterson, A. V., Jr. (1977). *J. Amer. Statist. Ass.,* **72**, 854–858.

[24] Peto, R., et al. (1977). *Brit. J. Cancer,* **35**, 1–39. (This popular paper includes a technical motivation, description, and illustration of the Kaplan–Meier estimator.)

[25] Phadia, E. G. and Van Ryzin, J. (1980). *Ann. Statist.,* **8**, 673–678.

[26] Reid, N. (1979). *Ann. Statist.,* **9**, 78–92.

[27] Rothman, K. J. (1978). *J. Chronic Dis.,* **31**, 557–560.

[28] Sander, J. (1975). The Weak Convergence of Quantiles of the Product-Limit Estimator. *Tech. Rep. No. 5,* Dept. of Statistics, Stanford University, Stanford, Calif.

[29] Scholz, F. W. (1980). *Canad. J. Statist.,* **8**, 193–203.

[30] Simon, R. and Lee, Y. K. (1982). *Cancer Treat. Rep.,* **66**, 67–72.

[31] Tsiatis, A. (1975). *Proc. Natl. Acad. Sci. USA,* **72**, 20–22.

[32] Wellner, J. A. (1982). *Ann. Statist.,* **10**, 595–602.

[33] Winter, B. B., Földes, A., and Rejtö, L. (1978). *Problems Control Inf. Theory,* **7**, 213–225.

Acknowledgment

This work was supported by Grants GM-28314 and CA-15704 from the National Institutes of Health.

(BIOSTATISTICS
CENSORING
CLINICAL TRIALS
FOLLOW UP

LIFE TABLES
RELIABILITY
SURVIVAL ANALYSIS)

Arthur V. Peterson, Jr.

KAPPA COEFFICIENT

The kappa coefficient is a measure of association* used to describe and to test the degree of agreement (reliability* or precision*) in classification. This statistic plays a role for nominal measures analogous to that played by the intraclass correlation coefficient* for interval measures. It serves not only as quantitative documentation of the quality of measure but also is a factor of importance in research design decisions [8].

In its original form [1, 2], each of N sampled items is classified by each of two fixed observers into one of K mutually exclusive categories. More recently, procedures have been extended from two fixed observers to $n \geqslant 2$ fixed observers [6] or to randomly selected observers for each item, possibly unequal in number [9, 10]. Procedures have been extended from K mutually exclusive categories to multiple ordered or unordered choices [9]. As the problem was originally formulated, only null distribution theory (i.e., random choices) was used. Nonnull distribution theory has more recently been emphasized.

Finally, more complex uses for the kappa coefficient have been proposed, including, for example, quantifying agreement when different sets of categories might be used by different observers [7], or when identification of the sources of disagreement is the focus of interest (*see* HIERARCHICAL KAPPA).

THE SIMPLE KAPPA COEFFICIENT

Percentage agreement between subjects, a statistic frequently used to characterize interobserver agreement, is uninterpretable without reference to the number of categories and the frequency of their use. Cohen [1,

2] suggested that one use instead the proportion of interobserver agreement rescaled to correct for chance, the kappa coefficient, defined as follows:

$$\kappa = (p_0 - p_e)/(1 - p_e),$$

$$p_0 = \sum_{ij} p_{ij} w_{ij},$$

$$p_e = \sum_{ij} w_{ij} p_i \cdot p_{\cdot j},$$

where p_{ij} is the proportion of the N items classified into category i by the first observer and into j by the other, with

$$p_i \cdot = \sum_j p_{ij}, \qquad p_{\cdot j} = \sum_i p_{ij}.$$

Here w_{ij} is the strength of agreement between observers for an item classified into category i by the first observer and into category j by the second with $w_{ii} = 1$, and $0 \leqslant w_{ij} \leqslant 1$ otherwise (weighted kappa).

For two fixed observers, when sampling is restricted to samples of N items that would yield the observed marginal frequencies, and if observers' decisions were totally random, then [3] κ is approximately normally distributed with

$$E(\kappa) = 0,$$

$$\mathrm{var}(\kappa) \approx \left(\sum p_{ij} w_{ij}^2 - p_0^2 \right) / N(1 - p_e)^2.$$

In the unweighted form ($w_{ij} = 0$, $i \neq j$):

$$\mathrm{var}(\kappa) = p_0(1 - p_0)/N(1 - p_e)^2.$$

THE GENERAL KAPPA COEFFICIENT

Suppose that N items are sampled and each is evaluated by n_i (not necessarily an equal number) of observers randomly sampled from a pool of observers.

Some measure of agreement is specified for each possible pair of observer choices, with complete agreement yielding a measure of 1 and $-1 \leqslant r \leqslant 1$ otherwise. For example, if choices are restricted to a single category per item, the unweighted form of simple kappa defines $r = 1$ if two observers agree and $r = 0$ otherwise. In the same situation, the weighted form of simple kappa defines $r = 1$ if two observers agree and $r = w_{ij}$ otherwise, where w depends on which two categories are selected. If multiple choices are permitted it has been suggested [5] that r be the ratio of the number of categories both observers mention to the number of categories mentioned by one or the other. If ordered multiple choices are permitted, Kraemer [9] suggests treating each response as a rank ordering (with ties) of the available categories, with r the product-moment correlation coefficient between two rank orders. The appropriate definition of r is dictated by the nature of the specific problem both in terms of what response options are permitted and how disagreements are viewed. There are many candidates.

In any case, let r_i be the average interobserver agreement measure over all pairs of observers evaluating subject i, $i = 1, 2,$ $\ldots, N$, with $r_I = \sum_i r_i / N$. Let r_T be the average interobserver agreement measure over all pairs of observations. The kappa coefficient is defined as

$$\kappa = (r_I - r_T)/(1 - r_T).$$

An $N \to 0$, provided that $n_i/N \to \lambda$, κ is approximately normally distributed with $E(\kappa) = $ population kappa,

$$\mathrm{var}(\kappa) \approx S_r^2 / N(1 - r_T)^2$$

$$\text{where } S_r^2 = \sum_i (r_i - r_I)^2/(N - 1).$$

For small sample sizes, Kraemer [9] suggests use of the jackknife* procedure to obtain confidence intervals* for population kappa.

Illustration

To illustrate the possibilities, Table 1 lists the hypothetical data in which each of five items is classified by each of two observers using three possible categories, A, B, and C. A choice "AB" indicates that A is the primary and B the secondary choice; "A/B" that A and B are considered equally correct choices; "A" that only a primary choice is selected.

Table 1

Item	Choice	Primary Unweighted	Primary Weighted	Choice Unordered	Choice Ordered
1a	A	1.00	1.00	0.50	0.87
1b	AB				
2a	C	0.00	0.00	0.50	0.00
2b	AC				
3a	AB	0.00	0.75	0.33	-0.50
3b	BC				
4a	A	1.00	1.00	1.00	1.00
4b	A				
5a	A/B	—	—	0.50	0.50
		$p_0 = 0.50$	0.69	$r_I = 0.57$	0.37
		$p_e = 0.56$	0.70	$r_T = 0.37$	0.14
		$\kappa = -0.14$	-0.05	$\kappa = 0.31$	0.27

In some situations one would be interested only in the primary choice. If the unweighted form of kappa were used, as indicated above, $\kappa = -0.14$. If it were decided that a disagreement between A and B were not very serious (say, $w_{AB} = 0.75$ as compared to $w_{BC}, w_{AC} = 0$), then $\kappa = -0.05$.

If it were decided that all categories mentioned were of interest, but not the order in which they are mentioned, one might use proportion overlap as r, as suggested by Fleiss. Thus, for example, for items 3a and 3b a total of three categories are mentioned with agreement only on B. Hence $r = 0.33$ for that pair. In this case $\kappa = 0.57$.

Finally, if both categories and the order in which they are mentioned are of interest, one might use the rank-order correlation coefficient. Thus the choice AB imposes a rank order of $(1, 2, 3)$ on the categories (A, B, C); the choice BC, a rank order of $(3, 1, 2)$. The correlation coefficient between these is $r = -0.50$. In this case $\kappa = 0.27$.

References

[1] Cohen, J. (1960). *Educ. Psychol. Meas.*, 37–46.

[2] Cohen, J. (1968). *Psychol. Bull.*, **70**, 213–220.

[3] Everitt, B. S. (1968). *Brit. J. Math. Statist. Psychol*, **21**, 97–103.

[4] Fleiss, J. L. and Cicchetti, D. V. (1978). *Appl. Psychol. Meas.*, **2**, 113–117.

[5] Fleiss, J. L., Spitzer, R. L., Endicott, J., and Cohen, J. (1972). *Arch. Gen. Psychiatry*, **26**, 168–171.

[6] Fleiss, J. L., Nee, J. C. M., and Landis, J. R. (1979). *Psychol. Bull.*, **86**, 974–977.

[7] Hubert, L. (1978). *Psychol. Bull.*, **85**, 183–184.

[8] Kraemer, H. C. (1979). *Psychometrika*, **44**, 461–472.

[9] Kraemer, H. C. (1980). *Biometrics*, **36**, 207–216.

[10] Landis, J. R. and Koch, G. G. (1977). *Biometrics*, **33**, 159–174.

(CATEGORICAL DATA
HIERARCHICAL KAPPA
MEASURES OF AGREEMENT
RANK TESTS)

H. C. KRAEMER

KAPTEYN DISTRIBUTION   *See* LOG-NORMAL DISTRIBUTION

KÄRBER METHOD

The Kärber method is a nonparametric procedure for estimating the median effective dose in a quantal bioassay*. It was described by Kärber [15] but was proposed earlier by Spearman [18], so it is also often referred to as the Spearman–Kärber method. The formula for this estimator, as given in BIOSSAY, STATISTICAL METHODS IN, is

$$\tilde{\mu} = \sum_{i=1}^{k-1} (p_{i+1} - p_i)(x_i + x_{i+1})/2.$$

It is assumed that $p_1 = 0$ and $p_k = 1$. The x_i's are dose measures, usually the logarithm of actual doses, with $x_1 < \cdots < x_k$, where k is the number of doses. The p_i's are observed proportions of responders in an experiment where n_i subjects are tested independently at dose x_i yielding r_i responses, so $p_i = r_i/n_i$, $i = 1, \ldots, k$. Different groups of subjects are tested at different doses, and independence between as well as within groups of subjects is assumed. From the equation for $\tilde{\mu}$ it can be seen that it is a discretized estimator of the mean, μ, of a tolerance distribution*, which is the ED50 ("effective dose, 50%") only if the tolerance distribution is symmetric.

The variance of $\tilde{\mu}$ as given by Cornfield and Mantel [6] is

$$\text{var}(\tilde{\mu}) = \sum_{i=2}^{k-1} (P_i Q_i/n_i)\big[(x_{i+1} - x_{i-1})/2\big]^2,$$

where $P_i = E(p_i)$ and $Q_i = 1 - P_i$. An unbiased estimate of $\text{var}(\tilde{\mu})$ is

$$\text{var}(\tilde{\mu}) = \sum_{i=2}^{k-1} [p_i q_i/(n_i - 1)][(x_{i+1} - x_{i-1})/2]^2$$

provided that $n_i \geqslant 2, i = 1, \ldots, k$.

If the experiment results in a calculated value of $p_1 > 0$ instead of $p_1 = 0$ as assumed, then the practice is often followed of incorporating an x level lower than the smallest x in the experiment into the calculations, for which it is assumed that the corresponding proportion, if observed, would be zero. If $p_k < 1$, the range of the x's is similarly extended upward to a high value of x for which p is set equal to 1. It is also common practice in many quantal bioassay experiments to space the x_i's at equal intervals, d. In this instance the formulas for $\tilde{\mu}$ and $\text{var}(\tilde{\mu})$ simplify to

$$\tilde{\mu} = x_k + (d/2) - d\sum_{i=1}^{k} p_i;$$

$$\text{var}(\tilde{\mu}) = d^2 \sum_{i=2}^{k-1} p_i q_i/(n_i - 1).$$

These formulas were given by Irwin and Cheeseman [13]. Tsutakawa gives an exam-

ple of the calculation of $\tilde{\mu}$ and $\text{var}(\tilde{\mu})$ for equally spaced x_i's in STATISTICAL METHODS IN BIOASSAY.

The properties of $\tilde{\mu}$ have been investigated under the assumptions not only of a common internal d between x_i's but of a common $n_i = n$ for $i = 1, \ldots, k$. Asymptotically, it is also assumed that whenever $k \to \infty$, $d \to 0$. These assumptions are made throughout the remainder of this article.

Variance formulas have been developed and biases evaluated for particular underlying tolerance distributions. Johnson and Brown [14] considered the cumulative tolerance distribution function

$$F(x) = 1 - e^{-\theta e^x}.$$

This is called the one-particle or extreme-value* model. The mean of this distribution, which is the quantity estimated by the Kärber estimator, is $\mu = [-\gamma - \ln \theta]$, where $\gamma = 0.57722$ is Euler's constant*. For this model Cornell [5] has shown that the Kärber estimator can be derived from the estimator proposed by Fisher [9] and tabled by Fisher and Yates [10] by replacing a sum by an integral approximation. The approximation becomes exact as d approaches zero. For sufficiently dense x_i's taken over a wide range, the Kärber estimator $\tilde{\mu}$ and the Fisher estimator are not only equivalent, but they are both unbiased estimators of μ and have the same asymptotic variance, namely,

$$\text{var}(\tilde{\mu}) = d\ln 2/n.$$

Johnson and Brown showed that this is true whether the placement of the x_i's is fixed or random. Moreover, under these circumstances $\tilde{\mu}$ is appropriately normally distributed, so standard normal tables can be used in the calculation of confidence limits on μ. The same results hold when the underlying tolerance distribution is normal except that the equation for $\text{var}(\tilde{\mu})$ becomes

$$\text{var}(\tilde{\mu}) = \sigma d/(\sqrt{\pi}\, n),$$

as given by Gaddum [11] and discussed by Finney [8]. In this expression, σ is the standard deviation of the underlying normal tolerance distribution.

The Spearman–Kärber approach has been extended to estimate the variance σ^2, other moments and also percentiles of a tolerance distribution by Epstein and Churchman [7] and by Chmiel [3]. For constant d between x_i's this estimate as given by Finney [8] is

$$\tilde{\sigma}^2 = d^2 \left[\sum_{i=1}^{k} (2k + 1 - 2i)p_i - \left(\sum_{i=1}^{k} p_i \right)^2 - \tfrac{1}{12} \right].$$

Note that the n_i's do not have to be greater than 1 for this calculation.

The estimator $\tilde{\sigma}$ is calculated from experimental data. For design purposes before experimentation it is necessary to specify σ but not the form of the tolerance distribution. Brown [2] noted that in general the asymptotic variance of $\tilde{\mu}$ exceeds $\sigma d/(2n)$ only slightly, with the amount depending on the form of the tolerance distribution. After multiplication by 2, which allows for a variety of model effects and finite samples, he suggested the approximation var($\tilde{\mu}$) = $\sigma d/n$ for use in experimental design. Alternatively, if the tolerance distribution were assumed to arise from either a one-particle or normal model, the asymptotic variance of $\tilde{\mu}$ for one of those models could be used. Brown advocated as fine a dose mesh as possible with few observations at each dose, that is, small d and n, as opposed to a coarser mesh with more observations per dose.

Properties of $\tilde{\mu}$ have been investigated both as a nonparametric estimate of the mean of an underlying tolerance distribution in the absence of information on that distribution and relative to other procedures based on a known form for the tolerance distribution. Miller [16] shows that $\tilde{\mu}$ is an unbiased estimate of a discretized mean μ_k which is a function of the particular set of x_i's chosen and is not necessarily equal to μ, the true mean of the tolerance distribution, but is the trapezoidal approximation to μ. As $k \to \infty$ and $d \to 0$, $\mu_k \to \mu$ and, in probability, $\tilde{\mu} \to \mu$. For either k or n large, $\tilde{\mu}$ has an asymptotic normal distribution with variance $d^2 \sum_{i=1}^{k} P_i Q_i / n$. Church and Cobb [4] show that $\tilde{\mu}$ is the nonparameter maximum likelihood estimator of μ_k when the restriction $P_1 \leqslant \cdots \leqslant P_k$ is imposed.

The Kärber estimator $\tilde{\mu}$ is also the maximum likelihood* estimator of μ_k when the underlying tolerance distribution is logistic, as shown by Cornfield and Mantel [6]. For this tolerance distribution $\tilde{\mu}$ approaches the maximum likelihood estimate of μ and has an efficiency of one as $k \to \infty$. Brown [1] presented corresponding efficiencies of 0.98 and 0.83 for the normal and one-particle tolerance distributions and zero for the Cauchy distribution*. In general, Miller and Halpern [17] found that estimator $\tilde{\mu}$ performs less well in terms of asymptotic efficiency for contaminated and heavy-tailed distributions than for uncontaminated light-tailed distributions for which its asymptotic efficiency is near 1. Hamilton [12] found that $\tilde{\mu}$ also performs well for small samples for light-tailed symmetric tolerance distributions such as the logistic. He used the mean square error, which equals the variance plus the bias squared, in his comparisons. For heavy-tailed distributions Hamilton recommended a modified estimator which he called the *trimmed* Spearman–Kärber estimator.

References

[1] Brown, B. W., Jr. (1961). *Biometrika*, **48**, 293–302.

[2] Brown, B. W., Jr. (1966). *Biometrics*, **22**, 322–329.

[3] Chmiel, J. J. (1976). *Biometrika*, **63**, 621–626.

[4] Church, J. D. and Cobb, E. B. (1973). *J. Amer. Statist. Ass.*, **68**, 201–202.

[5] Cornell, R. G. (1965). *Biometrics*, **21**, 858–864.

[6] Cornfield, J. and Mantel, N. (1950). *J. Amer. Statist. Ass.*, **45**, 181–210.

[7] Epstein, B. and Churchman, C. W. (1944). *Ann. Math. Statist.*, **15**, 90–96.

[8] Finney, D. J. (1978). *Statistical Methods in Biological Assay*, 3rd ed. Macmillan, New York.

[9] Fisher, R. A. (1921). *Philos. Trans. R. Soc. Lond. A*, **222**, 309–368.

[10] Fisher, R. A. and Yates, F. (1963). *Statistical Tables for Biological, Agricultural and Medical Research*, 6th ed. Oliver & Boyd, Edinburgh.

[11] Gaddum, J. H. (1933). Reports on Biological Standards. III. Methods of Biological Assay Depending on a Quantal Response. *Med. Res. Counc., Spec. Rep. Ser., No. 183.*

[12] Hamilton, M. A. (1979). *J. Amer. Statist. Ass.*, **74**, 344–354.

[13] Irwin, J. O. and Cheeseman, E. M. (1939). *Suppl. J. R. Statist. Soc.*, **6**, 174–185.

[14] Johnson, E. J. and Brown, B. W., Jr. (1961). *Biometrics*, **17**, 79–88.

[15] Kärber, G. (1931). *Arch. Exper. Pathol. Pharmakol.*, **162**, 480–487.

[16] Miller, R. G. (1973). *Biometrika*, **60**, 535–542.

[17] Miller, R. G. and Halpern, J. W. (1979). Robust Estimators for Quantal Bioassay. *Tech. Rep. No. 42*, Division of Biostatistics, Stanford University, Stanford, Calif.

[18] Spearman, C. (1908). *Brit. J. Psychol.*, **2**, 227–242.

(BIOASSAY, STATISTICAL METHODS IN QUANTAL ANALYSIS)

RICHARD G. CORNELL

KARHUNEN–LOÈVE EXPANSION

The Karhunen–Loève (K–L) expansion [5, 6] is used for describing random signals and is concerned with the representation of mn data points, obtained from m experiments each with n observations. The expansion is formed using a set of orthonormal basis functions which can be obtained as a set of eigenvectors of the data covariance matrix. Optimal properties of this expansion are closely related to the properties of least-squares estimation. The truncated series minimizes the mean square error* and the entropy* function. Specifically, the data are given as an $(m \times n)$-dimensional random matrix

$$\mathbf{X} = \begin{bmatrix} x_1(1), & \dots, & x_1(n) \\ \vdots & & \vdots \\ x_m(1), & \dots, & x_m(n) \end{bmatrix}.$$

The K-L expansion is defined by the row representation

$$\mathbf{X} = \mathbf{A}\mathbf{V}^T,$$

where $\mathbf{A}$ is an $m \times r$ random coefficient matrix with $E(\mathbf{A}) = 0$ and the real-valued $n \times r$ matrix $\mathbf{V}$ represents a set of basis functions and contains the orthonormalized n eigenvectors of a positive semidefinite covariance matrix $\mathbf{R}$ defined as

$$\mathbf{R} = E[\mathbf{X}^T \mathbf{P}_1 \mathbf{X}].$$

Here $\mathbf{R}$ is an $n \times m$ matrix and $\mathbf{P}_1$ is an $m \times m$ probability matrix representing the a priori probabilities* associated with the m experiments. The elements p_{ij} of P_1 satisfy $1 \leqslant p_{ii} < 0$, $p_{ij} = 0$, $i \neq j$.

The system "modes" are identified with eigenvalue problems given by $\mathbf{R}\mathbf{V} = \mathbf{V}\Lambda$, where Λ is the diagonal $(n \times n)$ eigenvalue matrix satisfying $\mathbf{A}^T\mathbf{P}_1\mathbf{A} = \Lambda$ and where $\mathbf{R} = E[\mathbf{V}\mathbf{A}^T\mathbf{P}_1\mathbf{A}\mathbf{V}^T]$. Usually, the series is truncated to include only the first k eigenvectors (with eigenvalues ordered in decreasing order of magnitude).

Similarly, the K-L expansion can be defined by a column representation corresponding to the model of m observations resulting from n experiments with a priori probabilities assigned to each column. A double-sided K-L expansion was introduced by Fernando and Nicholson [2]. It takes into account the possibility of correlation between both row and column data associated with either m or n experiments containing either n or m observations, respectively.

For a more detailed discussion and applications in pattern recognition* and prediction*, see, e.g., Fu [3], Andrews [1], Fukunaga [4], and Tou and Gonzalez [7].

References

[1] Andrews, H. C. (1972). *Introduction to Mathematical Methods in Pattern Recognition*. Wiley, New York.

[2] Fernando, K. V. M. and Nicholson, H. (1980). *IEEE Proc.*, **127**, Pt. D, No. 4, 155–160.

[3] Fu, K. S. (1968). *Sequential Methods in Pattern Recognition and Machine Learning*. Academic Press, New York.

[4] Fukunaga, K. (1972). *Introduction to Statistical Pattern Recognition*. Academic Press, New York.

[5] Karhunen, K. (1947). *Ann. Acad. Sci. Fenn. Ser. A*1: Math. Phys., **37**, 1–79.

[6] Loève, M. (1963). *Probability Theory*. D. Van Nostrand, Princeton, N.J.

[7] Tou, J. T. and Gonzalez, R. C. (1974). *Pattern Recognition Principles*. Addison-Wesley, Reading, Mass.

(IMAGE PROCESSING
PATTERN RECOGNITION
TIME-SERIES ANALYSIS)

KARLIN–MCGREGOR THEOREM

The Karlin–McGregor theorem [2] deals with the coincidence probabilities for n particles independently executing a continuous-time Markov process* of a certain type.

A generalization of this theorem to the case when the particles have different stopping times was given by Hwang [1].

References

[1]   Hwang, F. K. (1977). *Ann. Prob.*, **5**, 814–817.
[2]   Karlin, S. and McGregor, J. (1959). *Pacific J. Math.*, **9**, 1141–1164.

(MARKOV PROCESSES)

KATZ SYSTEM OF DISTRIBUTIONS

Let X be a discrete random variable and $P_j = \Pr(X = j)$ the probability that X assumes the value j. In a dissertation, Katz [7] considered a class of distributions of X defined by the relation

$$\frac{P_{j+1}}{P_j} = \frac{Q_1(j)}{Q_2(j)}, \tag{1}$$

where Q_1 and Q_2 are polynomials. In particular, he investigated the system for which

$$\frac{P_{j+1}}{P_j} = \frac{\text{linear function of } j}{\text{quadratic function of } j}. \tag{2}$$

For the case

$$\frac{P_{j+1}}{P_j} = \frac{\alpha + \beta j}{j+1}; \qquad \begin{array}{l} j = 0, 1, 2, \dots \\ \alpha > 0, \quad \beta < 1 \end{array} \tag{3}$$

discussed by Katz [8], he showed that this family comprises the following distributions:

1.   Negative binomial, for $0 < \beta < 1$
2.   Poisson, for $\beta = 0$
3.   Binomial or generalization thereof for $\beta < 0$, according as $-\alpha/\beta$ assumes positive integral or positive nonintegral values

For convenience, this family, generated by (3), will be referred to as a K family.

To gain a proper perspective of the families of distributions described above, it is helpful to consider the family generated by (1) in the framework of Kemp [9] and in terms of the corresponding probability generating function* (PGF) expressed as a hypergeometric function*. On writing

$$\frac{P_{j+1}}{P_j} = \frac{(a_1 + j)(a_2 + j) \cdots (a_p + j)}{(b_1 + j)(b_2 + j) \cdots (b_q + j)} \frac{\lambda}{j+1}, \tag{4}$$

Kemp obtains the corresponding PGF:

$$g(z) = \frac{{}_pF_q(a_1, a_2, \dots, a_p; b_1, b_2, \dots, b_q, \lambda z)}{{}_pF_q(a_1, a_2, \dots, a_p; b_1, b_2, \dots, b_q, \lambda)}, \tag{5}$$

where

$${}_pF_q(a_1, a_2, \dots, a_p; b_1, b_2, \dots, b_q, z)$$
$$= \sum_{j=0}^{\infty} \frac{(a_1)_j (a_2)_j \cdots (a_p)_j}{(b_1)_j (b_2)_j \cdots (b_q)_j} \frac{z^j}{j!}$$

and

$$(a_i)_j = (a_i)(a_i + 1)(a_i + 2) \cdots (a_i + j - 1).$$

The K family is a particular case, with $p = 2$, $q = 1$, $a_1 = \alpha/\beta$, $a_2 = 1$, $b_1 = 1$, $\lambda = \beta$ and with corresponding PGF g_K given by

$$g_K(z) = \frac{{}_2F_1(\alpha/\beta, 1; 1, \beta z)}{{}_2F_1(\alpha/\beta, 1; 1, \beta)}.$$

We also indicate here a few particular cases which bear some relation to the K family and which have appeared in the statistical literature. First, a natural extension of relation (3) is

$$\frac{P_{j+1}}{P_j} = \frac{\alpha + \beta j}{\lambda + j},$$

giving rise to a three-parameter (extended K) family with PGF g_{EK} given by

$$g_{EK}(z) = \frac{{}_2F_1(\alpha/\beta, 1; \lambda, \beta z)}{{}_2F_1(\alpha/\beta, 1; \lambda, \beta)}$$

considered by Tripathi and Gurland [11]. On letting $\beta \to 0$, this PGF becomes

$$\lim_{\beta \to 0} g_{EK}(z) = g_{CB}(z) = \frac{{}_1F_1(1, \lambda, \alpha z)}{{}_1F_1(1, \lambda, \alpha)},$$

the PGF of a family of distributions considered by Crow and Bardwell [1, 2], called hyper-Poisson* and generated by the relation

$$\frac{P_{j+1}}{P_j} = \frac{\alpha}{j + \lambda}.$$

For convenience, we refer to this as the CB family of distributions.

The family generated by (2), and considered by Katz [7], can be regarded as an extension of the CB family by writing

$$\frac{P_{j+1}}{P_j} = \frac{\alpha(j + \gamma)}{(j + \lambda)(j + 1)}$$

$$= \frac{\text{linear function of } j}{\text{quadratic function of } j}$$

and noting that this reduces to the CB family when $\gamma = 1$. This family, designated as E_1CB (extended CB) has been investigated by Tripathi and Gurland [10].

When analyzing data from some discrete distribution and confronted with a wide array of possible distributions it may be required to make a choice from some general family such as, for example, K, EK, or E_1CB. Involved in this inference process is estimation of parameters and tests of pertinent hypotheses. Different types of estimators and tests of hypotheses pertaining to these and related distributions have been investigated by Gurland and Tripathi [4, 5, 11, 12]. References for other related general families of distributions can be found in Johnson and Kotz [6].

References

[1] Bardwell, G. E. and Crow, E. L. (1964). *J. Amer. Statist. Ass.*, **59**, 133–141.

[2] Crow, E. L. and Bardwell, G. E. (1963). *Proc. Int. Symp. Discrete Distrib.*, Pergamon Press, Montreal, pp. 127–140.

[3] Dacey, M. F. (1972). *Sankhyā B*, **34**, 243–250.

[4] Gurland, J. and Tripathi, R. C. (1975). *Statistical Distributions in Scientific Work* (International Conference on Characterizations of Statistical Distributions with Applications), Vol. 1, C. Taillie, G. P. Patil, and B. Baldessari; eds. D. Reidel, Calgary, pp. 59–82.

[5] Gurland, J. and Tripathi, R. C. (1978). *Bull. Soc. Math. Grèce* (*N.S.*), **19**, 217–239.

[6] Johnson, N. L. and Kotz, S. (1969). *Discrete Distributions*. Houghton Mifflin, Boston.

[7] Katz, L. (1945). Characteristics of Frequency Functions Defined by First Order Difference Equations. Dissertation, University of Michigan.

[8] Katz, L. (1963). *Proc. Int. Symp. Discrete Distrib.*, Pergamon Press, Montreal, pp. 175–182.

[9] Kemp, A. W. (1968). *Sankhyā A*, **30**, 401–410.

[10] Tripathi, R. C. and Gurland, J. (1974). Extensions of the Katz Family of Discrete Distributions, Involving Hypergeometric Functions, Tech. Rep. No. 382, Dept. Statistics, University of Wisconsin.

[11] Tripathi, R. C. and Gurland, J. (1977). *J. R. Statist. Soc. B*, **39**, 349–356.

[12] Tripathi, R. C. and Gurland, J. (1979). *Commun. Statist. A*, **8**, 855–869.

JOHN GURLAND

K-DIVERGENCE   *See J*-DIVERGENCES AND RELATED CONCEPTS

KELLEY'S APPROXIMATION

A refinement of an approximation for the upper percentage points of the F-distribution* based on a variant of the Wilson–Hilferty approximation* to the distribution of chi-squared* variables. A computer program for upper percentile points of the F-distribution based on this formula has been published by Jasper [2] and evaluated by Golden et al. [1].

For additional information, see Johnson and Kotz [3] and Kelley [4].

References

[1] Golden, R. R., Weiss, D. J., and Dwass, R. V. (1968). *Educ. Psychol. Meas.*, **26**, 163–165.

[2] Jasper, N. (1965). *Educ. Psychol. Meas.*, **25**, 877–880.

[3] Johnson, N. L. and Kotz, S. (1970). *Distributions in Statistics: Continuous Univariate Distributions*, Vol. 2. Wiley, New York, Chap. 26.

[4] Kelley, T. L. (1948). *The Kelley Statistical Tables*, rev. ed. Harvard University Press, Cambridge, Mass.

(*F*-DISTRIBUTION
WILSON–HILFERTY APPROXIMATION)

KEMP FAMILIES OF DISTRIBUTIONS

The usual urn model form of the hypergeometric distribution* as arising from finite sampling without replacement is

$$P_x = \binom{a}{x}\binom{b}{n-x} \Big/ \binom{a+b}{n}$$

$$
\begin{aligned}
&x = 0, 1, \ldots, n;\\
&0 < n < a + b;\\
&0 < a < a + b;\\
&a, b, n \text{ integers.}
\end{aligned}
\tag{1}
$$

Kemp and Kemp [16] viewed (1) purely as a mathematical frequency function and examined the conditions under which P_x in (1) can represent a probability distribution when a, b, and n are allowed to take real values. On utilizing the concept of extended factorials, they defined

$$\alpha!/(\alpha+\beta)! = (-1)^\beta(-\alpha-\beta-1)!/(-\alpha-1)!$$

$$= (-1)^\beta \Gamma(-\alpha-\beta)/\Gamma(-\alpha),$$

where $\alpha < 0$, and $\beta < 0$ an integer. This allowed a, b, and n in (1) to be real. The distributions thus obtained are called the "generalized hypergeometric series distributions." This name derives from the fact that the corresponding probability generating function* (PGF) can be written as

$$
G(z) = ((a+b-n)!\,b!/\{(a+b)!(b-n)!\})
$$

$$
\times \, _2F_1(-a, -n; b-n+1; z),
\tag{2}
$$

where $_2F_1$ is the hypergeometric function, and the probabilities P_x are the coefficients of z^x in the expansion of (2). The new class contains some distributions for which the P_x's are nonzero for the first R terms but $P_{R+1} = 0$. These are called *terminating* distributions with $P_x > 0$ in the range $0 \leqslant x \leqslant R$ and $P_x = 0$ for $x \geqslant R+1$. It also contains some nonterminating distributions for which $P_x > 0$, $x = 0, 1, 2, \ldots$ The distributions included in this class have been divided into four main types and subtypes as follows ([16]; see also Johnson and Kotz [7]). (In the following J is a nonnegative integer, and where unspecified the range of x is the set of nonnegative integers.)

Type IA(i)	$n - b - 1 < 0$; n integral; $0 \leqslant n - 1 < a$ $(x = 0, 1, \ldots, n)$
Type IA(ii)	$n - b - 1 < 0$; a integral; $0 \leqslant a - 1 < n$ $(x = 0, 1, 2, \ldots, a)$
Type IB	$n - b - 1 < 0$, $J < a < J+1$, $J < n < J+1$
Type IIA	$a < 0 < n$; n integral; $b < 0$; $b \neq -1$ $(x = 0, 1, \ldots, n)$
Type IIB	$a < 0 < a + b + 1$; $J < n < J+1$; $J < n - b - 1 < J+1$
Type IIIA	$n < 0 < a$; $b < n - a$; $b \neq n - a - 1$; a integral $(x = 0, 1, \ldots, a)$
Type IIIB	$n < 0 < a + b + 1$; $J < a < J+1$; $J < n - b - 1 < J+1$
Type IV	$n < 0 < a$; $0 < a + b + 1$

Sarkadi [30] pointed out that the simple transformation $a \to n$, $b \to a + b - n$, $n \to a$ establishes the equivalence of IA(i), IIA, and IIB with IA(ii), IIIA, and IIIB, respectively. He showed that the restriction $b \neq -1$ in IIA (IIIA) was unnecessary. He also brought into the system shifted distributions (i.e., those for which the first k terms have $P_x = 0$).

The usual hypergeometric distribution is a special case of type IA with n, a, and b all integers. The negative hypergeometric distribution* is a special case of type IIA/IIIA with n, a, and b all integers. The Pólya–Eggenberger distribution* also arises as a special case of type IIA/IIIA. For some other distributions and for some limiting cases of the distributions included in this class, the reader is referred to Kemp and Kemp [16].

Kemp [12] utilized the generalized hypergeometric function to further widen the class of discrete distributions. She considered a family of distributions with the PGF

$$G_1(z) = {}_pF_q((a); (b); \lambda z), \tag{3}$$

where $_pF_q((a);(b);z)$ is the generalized hypergeometric function with p numerator, and q denominator parameters $a_1, a_2, \ldots, a_p$, and $b_1, b_2, \ldots, b_q$, respectively, and $C = 1/_pF_q((a);(b);\lambda)$. These are called the *generalized hypergeometric probability distributions* (GHPDs), so named because the probabilities are terms in the expansion of a generalized hypergeometric series. The GHPD includes the distributions considered by Kemp and Kemp [16] discussed above. They also include the power series distributions* defined by Noack [21], which, in turn, include such well-known distributions as binomial*, Poisson*, negative binomial*, and logarithmic distributions*. (For further work on power series distributions, see Khatri [19], Patil [25–27], and Joshi and Patil [8].) The probabilities for (3) satisfy the recurrence relation

$$P_{x+1}/P_x$$
$$= \lambda \prod_{i=1}^{p} (a_i + x) \Big/ \left[(x+1) \prod_{i=1}^{q} (b_i + x) \right].$$
$$(4)$$

Thus the GHPD also includes the distributions considered by Katz [10, 11] and its extensions: the hyper-Poisson* [1], the extended hyper-Poisson*, and the extended Katz [5, 35].

Ord [22] investigated the distributions arising from the difference equations equivalent to (4) with the numerator containing at most two and the denominator containing two factors. He investigated the recurrence relations for the factorial moments* and classified the distributions according to the values of the κ-criterion which depends on the coefficients of skewness and kurtosis (see also Ord [24]).

Kemp and Kemp [14, 17] considered another class of distributions with the PGF:

$$G_2(z) = {}_pF_q((a);(b);\lambda(z-1)). \quad (5)$$

The distributions in this class are called the *generalized hypergeometric factorial-moment distributions* (GHFDs) because their factorial moments are terms in the expansion of

the generalized hypergeometric function. The GHFD contains some well-known terminating distributions, such as negative hypergeometric, Pólya, and some matching distributions*. It also contains nonterminating distributions such as the Poisson, the negative binomial, and many other compound distributions, as will be seen later.

Kemp [13] has pointed out that both GHPD and GHFD may be regarded as particular cases of an even more general family with PGF

$$\frac{_pF_q((a);(b);\lambda z + \mu)}{_pF_q((a);(b);\lambda + \mu)}.$$

This family contains distributions which belong to neither of the two types given above, e.g., Gurland and Tripathi's [5] E$_2$CB distribution with PGF $_1F_1(1;b;\lambda z + \mu)/_1F_1(1;b;\lambda + \mu)$.

The functions $G_1(z)$ and $G_2(z)$ in (3) and (5) are valid PGFs only under certain conditions on the parameters. In the following section, such conditions are examined for the GHPD. In the section "GHF Distributions," similar conditions are examined for the GHFD. Some examples will also be given. Finally, in the section "Some Practical Aspects," some practical aspects of distribution selection and statistical inference are considered.

GHP DISTRIBUTIONS

The function $G_1(z)$ may represent the PGFs of terminating and nonterminating distributions defined on the set of nonnegative integers. It may also be used to obtain shifted and reversed distributions. On the other hand, it may not even be a valid PGF, depending on the values of the parameters. The conditions for $G_1(z)$ to be a valid PGF were examined by Kemp [12] and considered further by Dacey [2]. Kemp also derived the differential equations for various generating functions and used these to obtain the recurrence relations for the associated quantities such as moments and cumulants.

Conditions for Valid PGF

The following theorems are from Kemp [12].

Theorem 1. When $p \leqslant q$, (3) is the PGF of a nonterminating distribution provided that:

1. $\lambda > 0$.
2. $a_i, b_j > 0$ for all i and j, except that one or more pairs (c_k, c_l) of these parameters may take nonintegral negative values such that $[c_k] = [c_l]$, where $[c_k]$ denotes the integral part of c_k.

Some examples of GHPD along with their PGFs are:

> *Poisson*: $C_0 F_0(\lambda z)$
>
> *Hyper-Poisson*: $C_1 F_1(1; b; \lambda z)$
>
> *Displaced Poisson*: $C_1 F_1(1; r + 1; \lambda z)$ [32, 33]
>
> *Extended Hyper-Poisson*: $C_1 F_1(a; b; \lambda z)$ [5]

Theorem 2. When $p = q + 1$, (3) is the PGF of a nonterminating distribution provided that either (a) $0 < \lambda < 1$, and (b) same as prerequisite 2 in Theorem 1; or (a) $\lambda = 1$, (b) $\sum b_j > \sum a_i$, and (c) same as prerequisite 2 in Theorem 1.

Some examples along with their PGFs are:

> *Geometric**: $C_1 F_0(1; qz)$
>
> *Negative binomial*: $C_1 F_0(k; qz)$
>
> *Logarithmic Series*: $\theta z C_2 F_1(1, 1; 2; \theta z)$
>
> *Waring*: $C_2 F_1(1, k; k + \rho + 1; z)$ [6]
>
> *Yule*: $C_2 F_1(1, 1; \rho + 2; z)$ [18]

Some other examples are Kemp and Kemp's [16] type IV, and the factorial distribution [20].

When $p > q + 1$, (3) cannot be the PGF of a nonterminating distribution because the hypergeometric series diverges. The following theorem gives conditions for a terminating distribution.

Theorem 3 [12, Theorem 4]. Terminating distributions can arise for all $p \geqslant 1$, $q \geqslant 0$ provided that at least one of the a_i's is a negative integer and the other parameters (including λ) are suitably chosen.

Some examples are:

> *Binomial*: $C_1 F_0(-n; pz/(p-1))$
>
> *Hypergeometric*: $C_2 F_1(-a, -n; b - n + 1; z)$
>
> *Polya*: $C_2 F_1(-n, M/c; (M - N + cn - c)/c; z)$

Other examples include Kemp and Kemp's [16] type IA and type IIA.

Dacey [2] gave a general method for constructing and identifying the terminating and the nonterminating GHP distributions. The procedure involves adding numerator and/or denominator parameter(s) to the basic distribution $C_0 F_0(\lambda z)$ for nonterminating type, and to $C_1 F_0(-n; -\lambda z)$, $\lambda > 0$ for terminating type. The parameters are added in such a way that the resulting function is a valid PGF (see Dacey [2, Theorems 1 and 2, Properties 4 and 5]). For an extensive list of GHPDs together with their PGFs, see Tables 1 and 2 of Dacey [2].

Methods for constructing *reversed* and *shifted* distributions from GHPDs are given in Theorems 4 and 5, respectively, of Kemp [12]. A reversed distribution has PGF of the form $z^n G(1/z)$, where $G(z)$ is a distribution over the integers $0, 1, \ldots, n$. A shifted distribution has a PGF of the form $z^k G(z)$, where $G(z)$ is a distribution on the integers $0, 1, \ldots$

Truncated forms of GHPD are obtained either by head truncation or by tail truncation, as given in the following theorems.

Theorem 4 [12, Theorem 7]. Truncating the first m frequencies (head truncation) of (3) yields the shifted distribution with PGF

$$z^m C_{p+1} F_{q+1}((a) + m, 1; (b) + m, 1 + m; \lambda z).$$

Theorem 5 [12, Theorem 8]. Truncating all except the first m frequencies (tail truncation) of (3) yields the terminating distribution with PGF

$$C_{p+1}F_{q+1}((a), 1 - m; (b), 1 - m; \lambda z).$$

For example, the head truncation of Poisson distribution gives the PGF $z_1^m F_1(1; 1 + m; \lambda z)$, which is that of a shifted hyper-Poisson distribution. The tail truncation of the negative binomial gives PGF $C_2 F_1(N + 1, - N; - N; z/2)$, which corresponds to the distribution of the number of matches removed in Banach's match-box problem* (see Feller [3, pp. 157, 222]).

Differential Equations Associated with GHPD

Kemp [12] has derived the differential equation for the PGF of GHPD from which the differential equations for other quantities are derived. Let $\theta = z(d/dz)$ be the differential operator. Then the differential equation for the PGF is

$$\theta(\theta + b_1 - 1) \cdots (\theta + b_q - 1)G(z)$$
$$= \lambda z(\theta + a_1) \cdots (\theta + a_p)G(z). \quad (6)$$

From (6), the differential equations for the factorial moment generating function $G(1 + t)$, the moment generating function $G(e^t)$, the cumulant generating function $\ln(G(e^t))$, and the factorial cumulant generating function $\ln G(1 + t)$ can be obtained. These are useful for deriving the recurrence relations for the associated quantities.

As noted earlier, the probabilities satisfy only a first-order recurrence relation. However, the factorial moments satisfy recurrence relations which involve at most $\nu + 1$ lower factorial moments, $\nu = \max(p, q + 1)$. These relations are useful for inference regarding these distributions.

Some Limiting Distributions

Kemp [12] gave the following results, which are useful in obtaining some limiting distributions:

1. $\lim_{d \to \pm \infty} {}_{p+1}F_q((a), c + d; (b); x/d)$
   $= {}_p F_q((a); (b); x)$
2. $\lim_{d \to \pm \infty} {}_p F_{q+1}((a); (b), c + d; dx)$
   $= {}_p F_q((a); (b); x)$
3. $\lim_{d \to \pm \infty} {}_{p+1}F_{q+1}((a), d; (b), c + dk; kx)$
   $= {}_p F_q((a); (b); x)$

Binomial Example:

$$\lim_{n \to \infty, np = \lambda} C_1 F_0(-n, pz/(1 - p))$$
$$= C_0^* F_0(\lambda z), \quad \text{Poisson.}$$

Hypergeometric Example:

$$\lim_{N \to \infty} C_2 F_1(-n, -Np; N - Np + n + 1; z)$$
$$= C_1^* F_0(-n; pz/(1 - p)), \quad \text{binomial.}$$

For more examples, see Kemp [12].

GHF DISTRIBUTIONS

The factorial moment generating function of GHFD is $G_2(1 + t) = {}_2F_1((a); (b); \lambda t)$, and the factorial moments are the coefficients of $t^i/i!$ in the expansion of this series. Although it is possible to give the conditions for $G_1(z)$ to be a valid PGF, it is considerably more difficult to determine when $G_2(z)$ is a valid PGF since this problem involves considerations of convergency of $G_2(z)$ and the conditions for a set of values to be the factorial moments for a valid discrete distribution. (See Potts [29] and Steffensen [34] for some results relating the factorial moments to the probabilities of discrete distributions.) For some distributions such as binomial, Poisson, Pólya, and hypergeometric which are members of the GHP as well as the GHF families, this difficulty does not arise since such conditions are known for the GHP family. However, not all GHFD are GHPD and not all GHPD are GHFD. In fact, a GHFD is a GHPD only if there exists a summation formula for ${}_p F_q((a); (b); -\lambda)$ in terms of products and quotients of gamma functions, which is not always possible (see Slater [31]).

Some Examples

The GHFD contains terminating and nonterminating distributions. Some examples for terminating distributions follow:

Binomial: $_1F_0(-n; p(1-z)), 0 \leqslant p \leqslant 1$
Hypergeometric:
$_2F_1(-n, -Np; -N; 1-z), 0 \leqslant p \leqslant 1$
Negative hypergeometric:
$_2F_1(-n, a; a+b; 1-z), a, b > 0$
Polya:
$_2F_1(-n, M/c; (M+N)/c; 1-z)$
Matching distribution [28]:
$_1F_1(-n; -n; z-1)$

Some examples for nonterminating distributions follow:

Poisson: $_0F_0(\lambda(s-1)), \lambda > 0$
Negative binomial: $_1F_0(k; p(z-1)),$
$k, p > 0$
Poisson Λ Beta: $_1F_1(a; a+b; \lambda(z-1)),$
$a, b, \lambda > 0$
Gurland's [4] *Type H_2 (also see Katti [9]):*
$_2F_1(k, a; a+b; \lambda(z-1)), k, a, b, \lambda > 0$

For some more examples of these distributions, see Kemp and Kemp [14]. Tripathi and Gurland [35] have shown that some GHFD can be interpreted as compound distributions when the basic distributions $_0F_r((b); \lambda(z-1))$ and $_1F_0(-n; \lambda(z-1))$ are compounded with the gamma and/or beta distributions. This compounding operation can be performed in stages to give higher and higher members of the family. Gurland's [4] distributions with PGFs $_1F_1(\alpha; \alpha + \beta; \theta(z-1))$, $_2F_1(k, \alpha; \alpha + \beta; \theta(z-1))$ are obtained in this manner.

Recurrence Relations

The rth descending factorial moment $\mu'_{[r]}$ of the GHFD is

$$\mu'_{[r]} = \Gamma\big[(a+r); (b+r)\big]\lambda^r / \Gamma\big[(a); (b)\big]$$

and satisfies the first-order recurrence relation

$$\mu'_{[r]}/\mu'_{[r-1]} = \lambda \prod_i (a_i + r - 1)/\prod_j (b_j + r - 1).$$

Direct computation of the probabilities may be tedious. However, from the differential equations of the PGF, a useful recurrence relation involving at most $\nu + 1$ probabilities can be obtained, where $\nu = \max(p, q + 1)$. For example, for the Poisson Λ beta distribution the recurrence relation is

$$(x + 2)(x + 1)P_{x+2}$$
$$-(x + 1)(\lambda + x + a + b)P_{x+1}$$
$$-\lambda(a + x)P_x = 0, \quad x \geqslant 0.$$

All the probabilities may be obtained from P_1 and P_0. Differential equations for other generating functions and the corresponding recurrence relations have been considered by Kemp and Kemp [14]. Some examples are also provided.

SOME PRACTICAL ASPECTS

The two families of distributions considered here provide a rich source of flexible models suitable for a wide variety of data. Kemp and Kemp [15] discuss various models such as urn, contagion, stochastic weighting, and STER process models that give rise to these distributions. However, for a given data set, selection of an appropriate model is often a difficult problem. The following discussion may aid in this process.

Over-, Under-, and Equi-Dispersion

A distribution with mean μ and variance σ^2 is over-, equi-, or underdispersed according as $\sigma^2 \gtreqless \mu$. The negative binomial distribution is overdispersed since $\sigma^2 > \mu$. Similarly, the binomial is underdispersed, and the Poisson is equi-dispersed. However, the hyper-Poisson distribution is over-, equi-, and underdispersed according as $b \lesseqgtr 1$. Tripathi and Gurland [37] give conditions for the

GHP and GHF distributions to be over-, equi-, and underdispersed. Since these models are more flexible, they are capable of describing a wide variety of situations.

Graphical Comparison

Ord [23] and Tripathi and Gurland [37] have made graphical comparisons of some members of the GHP family based on the criterion $U_k = kP_k/P_{k-1}$. These comparisons may also be helpful in model selection. For further details, see the papers cited above.

Statistical Inference

Extensive work has been done on fitting the members of the Kemp family involving only one or two parameters, such as binomial, Poisson, negative binomial, hypergeometric, etc. (see Johnson and Kotz [7]). Gurland and Tripathi [5] and Tripathi and Gurland [35] have utilized the recurrence relations for factorial moments and probabilities to fit the extended hyper-Poisson, and the extended Katz distributions by the method of minimum chi-square*. They also investigated the asymptotic relative efficiency* (ARE) of some minimum chi-square estimators which differ from one another in the number of moment and probability relations involved: the estimators that involve a relation based on P_0, the probability of zero count, are highly efficient. They have also utilized such estimators for developing tests of hypotheses for discrete distributions (see Tripathi and Gurland [36]).

Otherwise, little has been done on fitting higher members of the Kemp family; further work on estimation for the family in general is clearly needed.

CONCLUSION

The Kemp family represents a unified approach to a very wide range of discrete distributions. In addition to providing general methods for deriving properties of particular distributions, this approach illumi-

nates the underlying relationships (via their PGFs) between different distributions and assists in practical modeling (in terms of mixing, weighting, etc.).

We understand that work is in progress on extending the approach to bivariate and multivariate situations.

References

[1] Crow, E. L. and Bardwell (1963). In *Classical and Contagious Discrete Distributions*, G. P. Patil, ed. Statistical Publishing Society, Calcutta, pp. 127–140. (An extension of Poisson distribution, called the hyper-Poisson, is considered; it can be over-, under-, and equi-dispersed; easy to read.)

[2] Dacey, M. F. (1972). *Sankhyā B*, **34**, 243–250. (Methods to identify the members of the GHPD are presented; contains a good list of distributions; intermediate level.)

[3] Feller, W. (1957). *An Introduction to Probability Theory and Its Applications.* Wiley, New York. (An excellent title on combinatorial probability and discrete distributions; intermediate level.)

[4] Gurland, J. (1958). *Biometrics*, **14**, 229–249. (An excellent reference on generating contagious distributions by compounding and/or generalizing operations; easy to read.)

[5] Gurland, J. and Tripathi, R. C. (1974). In *A Modern Course on Statistical Distributions in Scientific Work*, Vol. 1, G. P. Patil, S. Kotz, and J. K. Ord, eds. D. Reidel, Dordrecht, Holland, pp. 59–82. (The Katz family and the hyper-Poisson distribution are extended; minimum chi-square estimators are developed for these extensions; easy to read.)

[6] Irwin, J. O. (1963). In *Classical and Contagious Discrete Distributions*, G. P. Patil, ed. Statistical Publishing Society, Calcutta, pp. 159–174.

[7] Johnson, N. L. and Kotz, S. (1969). *Discrete Distributions.* Wiley, New York. (An excellent reference on discrete distributions; a rich source of references.)

[8] Joshi, S. W. and Patil, G. P. (1974). In *A Modern Course on Statistical Distributions in Scientific Work*, Vol. 1, G. P. Patil, S. Kotz, and J. K. Ord, eds. D. Reidel, Dordrecht, Holland.

[9] Katti, S. K. (1966). *Biometrics*, **22**, 44–52. (Interrelations among generalized distributions and their components are discussed; intermediate level.)

[10] Katz, L. (1948). *Ann. Math. Statist.*, **19**, 120.

[11] Katz, L. (1963). *Classical and Contagious Discrete Distributions*, G. P. Patil, ed. Statistical Publishing Society, Calcutta, pp. 175–183. (This is probably the first attempt on extending the binomial, Pois-

son, and negative binomial distributions; easy to read.)

[12] Kemp, A. W. (1968). *Sankhyā A*, **30**, 401–410. (An excellent article which unifies the representation of a wide class of discrete distributions in terms of the generalized hypergeometric functions; a good source of related references and the main source for the present article; intermediate level.)

[13] Kemp, A. W. (1974). *Statist. Rep. Preprints No. 15*, School of Mathematics, University of Bradford, England. (A further unification of a wide class of discrete distributions in terms of the generalized hypergeometric functions; an excellent source of references; easy to read.)

[14] Kemp, A. W. and Kemp, C. D. (1974). *Commun. Statist. A*, **3**(12), 1187–1196. (This article introduces the GHF distributions which unify another class of distributions; contains many known distributions as examples; easy to read.)

[15] Kemp, A. W. and Kemp, C. D. (1975). In *A Modern Course on Statistical Distributions in Scientific Work*, Vol. 1, G. P. Patil, S. Kotz, and J. K. Ord, eds. D. Reidel, Dordrecht, Holland, pp. 31–40. (An excellent discussion of different models that lead to many well-known distributions; a good source of references and examples; easy to read.)

[16] Kemp, C. D. and Kemp, A. W. (1956). *J. R. Statist. Soc. B*, **18**, 202–211. (A first attempt at extending the well-known hypergeometric distribution; intermediate level.)

[17] Kemp, C. D. and Kemp, A. W. (1969). *Bull. Int. Statist. Inst.*, **43**, 336–338.

[18] Kendall, M. G. (1961). *J. R. Statist. Soc. A*, **124**, 1–16.

[19] Khatri, C. G. (1959). *Biometrika*, **46**, 486–490. (This article extends the power series distribution to the multivariate case.)

[20] Marlow, W. H. (1965). *Ann. Math. Statist.*, **36**, 1066–1068.

[21] Noack, A. (1950). *Ann. Math. Statist.*, **21**, 127–132. (The concept of power series distributions originated from this paper; easy to read.)

[22] Ord, J. K. (1967). *Biometrika*, **54**, 649–656. (A system of discrete distributions derived from difference equations is investigated; intermediate level.)

[23] Ord, J. K. (1967). *J. R. Statist. Soc. A*, **130**, 232–238. (A graphical method for comparing a family of discrete distributions is given.)

[24] Ord, J. K. (1972). *Families of Frequency Distributions*. Hafner Press, New York. (An excellent treatment of the systems of discrete distributions derived from Pearsonian difference equation; a

rich source of related references; intermediate level.)

[25] Patil, G. P. (1961). *Sankhyā A*, **23**, 269–280. (On estimation in generalized power series distributions; intermediate level.)

[26] Patil, G. P. (1962). *Ann. Inst. Statist. Math.*, **14**, 179–182.

[27] Patil, G. P. (1963). In *Classical and Contagious Discrete Distributions*, G. P. Patil, ed. Statistical Publishing Society, Calcutta, pp. 183–194.

[28] Patil, G. P. and Joshi, S. W. (1968). *A Dictionary and Bibliography of Discrete Distributions*. Oliver & Boyd, Edinburgh/Hafner, New York. (An excellent source for information on discrete distributions.)

[29] Potts, R. B. (1953). *Aust. J. Phys.*, **6**, 498–499.

[30] Sarkadi, K. (1957). *Magy. Tud. Akad. Mat. Kutatò Int. Közl.*, **2**, 59–69.

[31] Slater, L. J. (1966). *Generalized Hypergeometric Functions*. Cambridge University Press, Cambridge. (A volume on theoretical properties of the generalized hypergeometric function at an advanced level.)

[32] Staff, P. J. (1964). *Aust. J. Statist.*, **6**, 12–20.

[33] Staff, P. J. (1967). *J. Amer. Statist. Ass.*, **62**, 643–654.

[34] Steffensen, J. F. (1923). *Skand. Aktuarietidskr.*, **6**, 73–89.

[35] Tripathi, R. C. and Gurland, J. (1977). *J. R. Statist. Soc. B*, **39**, 349–356. (This article develops minimum chi-square estimators for the extended Katz and the extended hyper-Poisson distributions; easy to read).

[36] Tripathi, R. C. and Gurland, J. (1978). *Bull. Greek Math. Soc.*, **19**, 217–239. (Tests of hypothesis are developed for discrete distributions based on minimum chi-square; easy to read.)

[37] Tripathi, R. C. and Gurland, J. (1979). *Commun. Statist. A*, **8**(9), 855–869. (Over-, under-, and equi-dispersion of GHPD and GHFD is considered. It also deals with how GHFD can be regarded as compound distributions; some graphical methods are also considered; easy to read.)

(CONTINUOUS DISTRIBUTIONS
GENERALIZED HYPERGEOMETRIC
    DISTRIBUTIONS
HYPERGEOMETRIC DISTRIBUTIONS
LAGRANGIAN DISTRIBUTIONS
MIXTURE DISTRIBUTIONS
POWER SERIES DISTRIBUTIONS
URN MODELS)

RAM C. TRIPATHI

KENDALL'S COEFFICIENT OF CONCORDANCE   *See* COEFFICIENT OF CONCORDANCE (*W*)

KENDALL'S TAU

Despite its name, this coefficient of rank correlation* had already been discussed around 1900 by Fechner [4], Lipps [11], and Deuchler [2], and more theoretically in the 1920s by Esscher [3] and Lindeberg [9, 10]; Kruskal [8, Sec. 17] gives an account of its history. Kendall [5] not only rediscovered it independently, but investigated it in the distribution-free* (nonparametric) spirit that informs its use today, and this justifies his eponymy. His monograph [6] contains a full exposition of the theory and a bibliography.

The fundamental notion underlying the use of tau is that of *disarray*: If we observe two variables (x, y) on each member of a sample, it is an elementary notion that if we arrange the x's in increasing order, the extent to which their corresponding y's depart from increasing order indicates the weakness of the correlation between x and y. The simplest indicator of this extent of disarray is the number of interchanges among the y's that will put them in the same (increasing) order as the x's—this is just the number of pairs among the n observations that are in inverse order. We call this number of inversions Q. Since there are $\frac{1}{2}n(n-1)$ distinct pairs in n observations, we have $0 \leqslant Q \leqslant \frac{1}{2}(n-1)$; the lower limit is attained when the y's are already in increasing order so that no interchanging is required, and the upper limit is attained when the y's are in the completely inverse order from the x's, so that every pair has to be interchanged to obtain perfect agreement in order. If we conventionally require a correlation coefficient to be $+1$ when there is perfect positive agreement and -1 when there is perfect negative agreement, we obtain Kendall's tau coefficient

$$\tau = 1 - \frac{4Q}{n(n-1)}.$$

If x and y are independent, Q will be about halfway between its limits and τ therefore will be near zero.

Thus in a sample of $n = 5$ observations, the values

$$y: \quad 14 \quad 5 \quad 8 \quad 11 \quad 7$$
$$x: \quad 19 \quad 41 \quad 12 \quad 26 \quad 17$$

become, on arranging the x's in increasing order,

$$y: \quad 8 \quad 7 \quad 14 \quad 11 \quad 5$$
$$x: \quad 12 \quad 17 \quad 19 \quad 26 \quad 41$$

To put the y-values in increasing order, the value $y = 5$ must be moved four places to the left, making four interchanges, and also the values 7 and 8 must be interchanged and the values 11 and 14 interchanged. Thus $4 + 1 + 1$ interchanges are necessary, and $Q = 6$, so that $\tau = 1 - (4 \cdot 6)/(5 \cdot 4) = -0.2$. It is customary to carry out the operation using the ranks of x and y, which here are

$$y: \quad 3 \quad 2 \quad 5 \quad 4 \quad 1$$
$$x: \quad 1 \quad 2 \quad 3 \quad 4 \quad 5$$

The use of the ranks makes the pattern of disarray clearer, but is not explicitly required —they are, of course, implicit in the ordering in any case.

Another intuitive method of measuring disarray is to find the number of inversions as before, but to weight each inversion by the distance apart of the ranks inverted. In our example, the six inversions of the ranks of y are

$$(3,1), \quad (2,1), \quad (5,1),$$
$$(4,1), \quad (3,2), \quad \text{and} \quad (5,4)$$

and the distances apart of these pairs are 2, 1, 4, 3, 1, and 1, so that the weighted sum of the six inversions is $12 = V$, say. V is always exactly equal to one-half of the sum of squares of the differences between the y-ranks and the corresponding x-ranks, and thus *Spearman's rho** may be defined as

$$\rho = 1 - \frac{12V}{n(n^2 - 1)}$$

and is essentially a weighted form of tau. Despite the weighting difference the two coefficients are almost perfectly correlated, given independence, and then approximately satisfy $3\tau = 2\rho$. The choice between them is one of taste or convenience.

Tau was proposed as a measure of correlation, but it also has uses as a test statistic. When x and y are independently distributed in the population the exact distribution of tau is very simple and easily generated. It has expected value zero, variance equal to $2(2n + 5)/\{9n(n - 1)\}$ and tends to the normal form very quickly, effectively for $n \geqslant 10$. Thus to test independence, we need only see whether τ is outside the limits

$$\pm 1.96\left[2(2n - 5)/9\{n(n - 1)\}\right]^{1/2} \doteqdot \tfrac{4}{3}n^{-1/2}$$

and reject independence at the 5% level if it is, and similarly for other levels. All of these results were obtained by Kendall [5]. More exact tail probabilities are available from ref. 1.

It took only a little longer for it to be realized that tau can also be used as a test against trend for a series of univariate y-observations ordered in time or space—we need only label the time (or space) variable as x, and we return to our previous discussion. More surprisingly, the relative efficiency of tau as a test against trend may be even greater than as a test of bivariate independence—if the underlying distribution in each case is normal, the efficiencies are $(3/\pi)^{1/3} = 98\%$ and $9/\pi^2 = 91\%$, respectively. Such a small shortfall from 100% efficiency is an insurance premium that becomes payable only if the underlying distribution is indeed normal; if it is not, we get a large free increase in the range of validity of our test, since the normal-theory test (based on the regression coefficient and the correlation coefficient respectively) is no longer valid, whereas the test based on tau holds good without any normality assumption.

When there are more than two variables, a *partial tau* may be defined which is analogous to a partial correlation coefficient; Kendall [6] gives the details.

We have implicitly assumed above that x and y may be unambiguously ordered from small to large values. If the variables are continuous, this is almost certainly true apart from round-off errors, but whether for this reason or because the variables are discrete, we have in practice to deal with equal values of x or of y or of both, called *ties*. If there are ties, they are usually dealt with by assigning to each member of the tied group the average rank (called the midrank) that the group would have had if the values of the variables were not exactly equal. If there are ties, whether or not midranks are assigned, tau can no longer attain its limiting values ± 1, and special coefficients (tau$_b$ and tau$_c$) have been proposed for this situation (see Kendall [6]) but perhaps the best form of tau when there are extensive ties is Goodman and Kruskal's coefficient gamma*. The problem of ties becomes particularly acute when we have frequencies arranged in a contingency table with r ordered rows and c ordered columns, for the marginal frequencies then represent the extents of the ties, and it was in this connection that tau$_c$ and gamma were developed. Kendall and Stuart [7, Chap. 33, Secs. 33.36–33.40] discuss the problems and the coefficients in some detail.

References

[1] Best, D. J. and Gipps, P. G. (1974). *Appl. Statist.*, **23**, 98–101.

[2] Deuchler, G. (1914). *Zeit. Pädagog. Psychol. Exper. Pädagog.*, **15**, 114–131, 145–159, 229–242.

[3] Esscher, F. (1924). *Skand. Aktuarietidskr.*, **7**, 201–219.

[4] Fechner, G. T. (1897). *Kollektivmasslehre*. W. Engelmann, Leipzig.

[5] Kendall, M. G. (1938). *Biometrika*, **30**, 81–93.

[6] Kendall, M. G. (1970). *Rank Correlation Methods*, 4th ed. Charles Griffin, London.

[7] Kendall, M. G. and Stuart, A. (1979). *The Advanced Theory of Statistics*. Vol. 2, 4th ed. Charles Griffin, London.

[8] Kruskal, W. H. (1958). *J. Amer. Statist. Ass.*, **53**, 814–861.

[9] Lindeberg, J. W. (1925). *VI Skand. Mathematikerkongr., Copenhagen*, pp. 437–446.

[10] Lindeberg, J. W. (1929). *Nord. Statist. J.*, **1**, 137–141.

[11] Lipps, G. F. (1906). *Die Psychischen Massmethoden*. F. Vieweg und Sohn, Braunschweig, Germany.

(CONCORDANCE
CORRELATION
DISTRIBUTION-FREE METHODS
GOODMAN–KRUSKAL TAU AND GAMMA
NONPARAMETRIC METHODS
RANKING PROCEDURES
SPEARMAN'S RHO)

ALAN STUART

KERNEL ESTIMATORS

Kernel estimators are convolutions of a smooth function with a rough empirical function estimator chosen in such a way as to produce a smooth functional estimator. The underlying idea is to take advantage of the fact that this linear functional transfers continuity properties from the smooth function, the so-called kernel, to the final estimator. Although potentially useful in a variety of settings, kernel methods have been principally exploited in three settings; probability density estimation*, spectral density estimation*, and nonparametric regression*.

In the probability density estimation setting, the estimator usually takes the form

$$\hat{f}_n(x) = \int_{-\infty}^{\infty} K_n(x, y) \, dF_n(y),$$

which may be rewritten

$$\hat{f}_n(x) = \frac{1}{n} \sum_{j=1}^{n} K_n(x, X_j).$$

Here $X_1, \ldots, X_n$ is a random sample, F_n is the empirical distribution function and K_n the kernel. Typically, we choose

$$K_n(x, y) = \frac{1}{h_n} K\left(\frac{x - y}{h_n}\right),$$

where h_n is chosen to approach 0 as n approaches ∞. The sequence h_n is called the *bandwidth* and its choice is one of critical importance to the convergence properties of $\hat{f}_n$. Density estimation is discussed in much more detail in the entry of the same title.

In the time-series* setting, the spectral density is often estimated by

$$f^{(n)}(\omega) = \int_{-\pi}^{\pi} K(\omega - y) I^{(n)}(y) \, dy, \quad (1)$$

where $I^{(n)}$ is the periodogram* based on a time series $X_1, \ldots, X_n$ and, of course, K is the smoothing kernel. In this setting K is typically called a *spectral window*. The periodogram is an inconsistent empirical estimator of the spectral density based on the fact that the spectral density is the Fourier transform of the autocovariance function of a stationary time series. That is, we may write

$$I^{(n)}(\omega) = \frac{1}{2\pi} \sum_{u=-n+1}^{n-1} \gamma^{(n)}(u) e^{-i\omega k},$$

where

$$\gamma^{(n)}(u) = \frac{1}{n} \sum_{j=1}^{n-u} X_{j+u} X_j$$

is the empirical autocovariance function. Because of this Fourier transform duality, (1) may be reexpressed in the lag domain as

$$f^{(n)}(\omega) = \frac{1}{2\pi} \sum_{u=-n+1}^{n-1} \gamma^{(n)}(u) k(u) e^{-i\omega u},$$

$$\qquad (2)$$

where k is called the lag window. The lag window and the spectral window are Fourier transform pairs and in fact kernel spectral estimators are usually computed operationally by some variant of (2). The choice of lag windows was at one time of considerable interest and controversy. A historical perspective can be obtained from Blackman and Tukey [1] and Parzen [5]. More recently, Cogburn and Davis [2] have shown that optimal kernels can be obtained as a function of criterion of optimization. In particular, they determine an optimal kernel as a function of a mixed norm involving fidelity and smoothness criterion and show that the solution yields a smoothing spline.

In the nonparametric regression setting, one is concerned with estimating r in

$$Y = r(X) + \epsilon, \qquad (3)$$

where observations (X_i, Y_i) satisfy (3). If X and Y are random variables, it may be shown that

$$r(x) = E(Y \mid X = x)$$

or, equivalently, that

$$r(x) = \begin{cases} \dfrac{\int_{-\infty}^{\infty} yf(x, y)\, dy}{f(x)}, & f(x) > 0 \\ 0, & f(x) = 0, \end{cases}$$

$f(x)$, $f(x, y)$ being, respectively, the marginal and joint densities. Watson [8] was motivated by this representation together with the then emerging work on kernel density estimators to formulate a nonparametric regression estimator of the form

$$\hat{r}(x) = \frac{\sum_{j=1}^{n} K(x - X_j) Y_j}{\sum_{j=1}^{n} K(x - X_j)}.$$

This is the regression analog of the classic kernel density estimator. Nadaraya [4] discusses both the kernel density estimate and kernel regression. A related estimator was introduced by Priestly and Chao [6]:

$$\hat{r}(x) = \frac{1}{h_n} \sum_{j=1}^{n} Y_j (x_j - x_{j-1}) K\left[\frac{x - x_j}{h_n} \right].$$

This is a nonstochastic x variable case. Finally, we mention the general form by Stone [7]:

$$\hat{r}(x) = \sum_{j=1}^{n} K_{nj}(x) Y_j.$$

A general discussion of nonparametric regression may be found in Johnston [3] or Wegman [9].

References

[1] Blackman, R. B. and Tukey, J. W. (1958). *The Measurement of Power Spectra*. Dover, New York.

[2] Cogburn, R. and Davis, H. T. (1974). *Ann. Statist.*, **2**, 1108–1126.

[3] Johnston, G. J. (1979). *Smooth Nonparametric Regression Analysis*. Ph.D. dissertation, University of North Carolina.

[4] Nadaraya, E. A. (1965). *Theory Prob. Appl.*, **10**, 186–190.

[5] Parzen, E. (1967). *Time Series Analysis Papers*. Holden-Day, San Francisco.

[6] Priestly, M. B. and Chao, M. T. (1972). *J. R. Statist. Soc. B*, **34**, 385–392.

[7] Stone, C. (1977). *Ann. Statist.*, **5**, 595–645.

[8] Watson, G. S. (1964). *Sankhyā A*, **26**, 359–372.

[9] Wegman, E. J. (1980). In *Recent Developments in Statistical Inference and Data Analysis*, K. Matusita, ed. North-Holland, Amsterdam.

(DENSITY ESTIMATION
GRADUATION
INTEGRAL TRANSFORMS
ISOTONIC INFERENCE
REGRESSION, NONPARAMETRIC
SPECTRAL ESTIMATION
SPLINE FUNCTIONS)

EDWARD J. WEGMAN

KESTEN THEOREMS

Kesten theorems [1] deal with accumulation points of $n^{-\alpha} S_n$, where S_n is the sum of n independent, identically distributed random variables (random walk*), and $0 < \alpha < \frac{1}{2}$ or $\alpha = \frac{1}{2}$ or $\alpha = 1$.

In particular, the following theorem, which is closely related to the Kolmogorov zero–one law*, is of special interest being a generalization of Stone's [2] result on the growth of a random walk.

Let $\{X_i\}_{i=1}^{\infty}$ be a sequence of independent, identically distributed random variables with EX_1 well defined (permitting values $+\infty$ or $-\infty$). Let $S_n = \sum_{i=1}^{n} X_i$. If $EX_1^+ = EX_1^- = \infty$, then either

1. $\lim_{n \to \infty} (S_n / n) = +\infty$ almost surely (a.s.), or

2. $\lim_{n \to -\infty} (S_n / n) = -\infty$ a.s., or

3. $\limsup_{n \to \infty} (S_n / n) = +\infty$ and $\liminf_{n \to \infty} (S_n / n) = -\infty$ a.s.

See also Tanny [3] for a simplified proof.

References

[1] Kesten, H. (1970). *Ann. Math. Statist.*, **41**, 1173–1205.

[2] Stone, C. (1969). *Ann. Math. Statist.*, **40**, 2203–2206.

[3] Tanny, D. (1977). *Zeit. Wahrscheinlichkeitsth. verwand. Geb.* **39**, 231–234.

(KOLMOGOROV ZERO–ONE LAW
LAWS OF LARGE NUMBERS
RANDOM WALK)

KEY BLOCK   *See* INTRABLOCK SUBGROUP

KEYFITZ METHOD OF LIFE-TABLE CONSTRUCTION

The study of mortality makes use of calculated survival probabilities and life expectations presented as a life table. The primary data consist of the number of deaths, $_5D_x$, say in one calendar year, and the population alive at midyear $_5P_x$, both for ages x to $x + 4$ at last birthday. Then $_5D_x/_5P_x = {_5}M_x$, the age-specific death rate. The problem of making a life table is to go from this ratio to the probability that a person aged exactly x years will live for a further five years, to exact age $x + 5$. (*See* LIFE TABLES.)

As so stated the problem can have no unique solution. For suppose that the continuous underlying mortality rate is $\mu(a)$, and the number of persons in the exposed population in the small interval a to $a + da$ is $p(a)\,da$; then the quantity $_5M_x$ obtained as datum may be interpreted as

$$_5M_x = \frac{\int_0^5 p(x + t)\mu(x + t)\,dt}{\int_0^5 p(x + t)\,dt}. \qquad (1)$$

This can be considered an equation whose unknown is the mortality function $\mu(x + t)$, with the $p(x + t)$ a nuisance variable of no interest. Any function $\mu(x + t)$, $0 \leqslant t \leqslant 5$, that satisfies (1) provides the probability of survival through the five-year interval as

$$\frac{l_{x+5}}{l_x} = \exp\left[-\int_0^5 \mu(x + t)\,dt\right] = e^{-5\,_5\bar{\mu}_x}, \qquad (2)$$

where $_5\bar{\mu}_x$ is the unweighted average of the $\mu(x + t)$ in the age interval. Thus the problem is to go from a weighted average of the $\mu(x + t)$ contained in (1) to an unweighted average as in (2), in each five-year interval.

Various ways of making a life table that have been used in practice imply various ways of solving (1) for the function $\mu(x + t)$ and then integrating this function to obtain l_{x+5}/l_x. The simplest is to suppose that within the five-year interval $\mu(x + t)$ is constant with respect to t; the constant can only be $_5M_x$ defined above; hence we have

$$\frac{l_{x+5}}{l_x} = \exp(-5\,_5M_x). \qquad (3)$$

To suppose that $p(x + t) = p_x$ is fixed through the interval gives the same result. A third way is to suppose that $p(x + t)$ is proportional to the function sought, $l(x + t)$, and also that this function is a straight line within the interval; on such a pair of assumptions it is easily shown that the answer is

$$\frac{l_{x+5}}{l_x} = \frac{1 - \frac{5}{2}\,_5M_x}{1 + \frac{5}{2}\,_5M_x}. \qquad (4)$$

For the ages, from about 10 to 70, at which the true $l(x + t)$ is concave below, (4) is more precise than (3), which supposes $l(x + t)$ to be concave above.

But we can rescue the exponential (3) by a correction that suitably raises the $_5M_x$. Reed and Merrill [5] suggested adding $_5M_x^2/5$ and tabulated a corrected $_5M_x^*$; their method has probably been used more times than any other way of making a life table. However, it does not allow for the difference between the weighting $p(x + t)$ in the population and $l(x + t)$ in the life table. Greville [2] generalized the Reed–Merrill formula.

If in (1) we take it that both the exposed population and the death rates are straight-line functions of t through the five-year interval, then the corrected $_5M_x^*$ turns out to be

$$_5M_x^* = {_5}M_x + C$$

$$= {_5}M_x + \frac{125}{12\,_np_x}\, p'\left(x + \frac{5}{2}\right)\mu'\left(x + \frac{5}{2}\right), \qquad (5)$$

where the primes signify derivatives, and $p'(x + \frac{5}{2})$ and $\mu'(x + \frac{5}{2})$ may be readily approximated using three consecutive age intervals [4, p. 40]. Notice that this procedure avoids the use of graduation*, which for a long time was regarded as the right way to make a complete life table (i.e., one showing single years of age).

The expression (5) is based on Taylor expansions of $p(x + t)$ and $\mu(x + t)$ taken to the linear terms. Gains in accuracy can be secured by going to higher-order terms, and entering the corresponding higher finite differences in (6). Experimenting by James Frauenthal, who programmed the several ways of making a life table here discussed, showed that the higher-order terms have little effect. The expression for the actual calculation with linear terms only is

$$\frac{l_{x+5}}{l_x} = e^{-5(_5M_x + C)},$$

where

$$C = \frac{(_5P_{x-5} - _5P_{x+5})(_5M_{x+5} - _5M_{x-5})}{48 \, _5P_x}.$$

$$(6)$$

A method that avoids graduation, but assumes that $p(x + t)$ can be fitted within the five-year interval by a stable distribution, provides the pair of equations for each age interval:

$$_5M_x = \frac{e^{-rx}l_x - e^{-r(x+5)}l_{x+5}}{\int_0^5 e^{-r(x+t)}l(x + t)\,dt},$$

$$\frac{_5P_{x+5}}{_5P_{x-5}} = \frac{\int_5^{10} e^{-r(x+t)}l(x + t)\,dt}{\int_{-5}^{0} e^{-r(x+t)}l(x + t)\,dt}.$$

$$(7)$$

The set can be solved for the unknown l_{x+5}/l_x and r by iterating through the ages of the life table [3].

Once the l_{x+5}/l_x are available, by whatever method, they may be cumulatively multiplied, starting with $l_0 = 1$, to give the probability of surviving from birth to age x, and the other columns of the conventional life table follow readily. Generalization to any other interval than five years is straightforward.

The recommended method among those exhibited above is that given by (5) and (6). Most of the others mentioned are extension cases of (5), based on specific assumptions regarding the functions $p(x + t)$ and $\mu(x + t)$.

The approach represented by (5) applies not only when all causes of death are combined, but to individual causes [4, pp. 48–51]. Thus if $l_{x+5}^{(-i)}/l_x^{(-i)}$ is the probability of surviving in the face of all causes but the ith, it can readily be shown that

$$\frac{l_{x+5}^{(-i)}}{l_x^{(-i)}} = \left(\frac{l_{x+5}}{l_x}\right)^R.$$

Chiang [1] developed this expression and took $R = {}_nM_x^{(-i)}/{}_nM_x$. Some gain in accuracy appears to result from

$$R = \frac{_nM_x^{(-i)}}{_nM_x}\left[1 + \frac{1}{24}\left(\frac{_nM_{x+5}^{(-i)} + {}_nM_{x-5}^{(-i)}}{_nM_x^{(-i)}}\right.\right.$$

$$\left.\left. + \frac{_nM_{x+5} + {}_nM_{x-5}}{_nM_x}\right)\right],$$

although the correction is not large. Derivation is by the same Taylor expansion and subsequent approximation to first derivatives that resulted in (6).

References

[1] Chiang, C. L. (1968). *Introduction to Stochastic Processes in Biostatistics*. Wiley, New York.

[2] Greville, T. N. E. (1943). *Rec. Amer. Inst. Actuaries*, **32**, Pt. 1, 29–43.

[3] Keyfitz, N. (1966). *J. Amer. Statist. Ass.*, **61**, 305–311.

[4] Keyfitz, N. (1977). *Applied Mathematical Demography*. Wiley, New York.

[5] Reed, L. J. and Merrell, M. (1939). *Amer. J. Hygiene*, **30**, 33–62.

(BUREAU OF THE CENSUS
GRADUATION
LIFE TABLES
SURVIVAL ANALYSIS)

NATHAN KEYFITZ

KEYFITZ METHOD OF VARIANCE ESTIMATION

A probability sample can perform what to lay persons seems impossible: estimate the error with which its mean approximates to that of the (unknown) population. The theory for doing this provided by Bowley [1] suffices for simple random samples, in which all units in the population have equal probability of being drawn (*see* FINITE POPULATION SAMPLING). In practice we often lack an item by item list of the sampling units, but must choose areas containing several units in a cluster (*see* CLUSTER SAMPLING). Simple random samples of big clusters are subject to large error, and knowing the amount of this error is small compensation for its large size. Hence the many devices to secure efficiency, including stratification, drawing units with unequal probabilities, clustering, often at several nested stages, ratio estimates (*see* STRATIFICATION; MULTISTAGE SAMPLING).

The current population survey (CPS) of the U.S. Bureau of the Census* takes advantage of these and other devices. Its sampling error could indeed be exactly computed on the basis of simple random sampling used for the components of the survey, but the formula for doing so would extend over many hundreds of pages. No one is likely to write out such a formula, and without some other recourse the error would never be calculated, so the chief advantage of probability sampling would be lost. What is needed is a formula that takes account of the myriad specific features of the design, but in some way that avoids explicit recognition of these features in the calculation.

Consider one stratum of the CPS or any other national sample, say the city of Boston. The Boston sample can be thought of as made up of two independent subsamples. Let half the number of unemployed (say) be estimated from the first half-sample as x_{11}, and from the second half-sample as x_{12}. Then the estimate of all the unemployed in Boston is $x_{11} + x_{12}$, and the variance of this is given with one degree of freedom by $(x_{11} - x_{12})^2$, unbiased except for a factor to allow for drawing without replacement, which we will omit here. The detail of the sample design, however complex, can be disregarded as long as we have assurance that each of the two half-samples, multiplied by 2, estimates the stratum independently and without bias [6].

Such a formula has no practical value when applied to a single pair $x_{11} + x_{12}$; we need the variance of $\sum_i^N (x_{i1} + x_{i2})$, the summation over many strata. The variance formulas, like the totals, are additive over the strata, i.e.,

$$\text{var}\left(\sum_i^N (x_{i1} + x_{i2}) \right) = \sum_i^N (x_{i1} - x_{i2})^2.$$

The number of degrees of freedom is equal to the number of strata N, which may be several hundred for a national sample.

In practice, samples are often used to find ratios—the CPS does not actually estimate the number of unemployed, but the ratio of unemployed to people. (The number of people can be precalculated from an earlier census more accurately than it is given by a sample of the size of the CPS.) If the population as estimated for the first half-stratum is p_{11}, and for the second half-stratum is p_{12}, and if RelVar stands for the rel-variance [5], which is the variance divided by the square of the mean, then we have as an approximation to the variance of a ratio:

$$\text{RelVar}\left[\sum_i^N (x_{i1} + x_{i2}) \Big/ \sum_i^N (p_{i1} + p_{i2}) \right]$$

$$\doteq \sum_i^N \left(\frac{x_{i1} - x_{i2}}{\sum(x_{i1} + x_{i2})} - \frac{p_{i1} - p_{i2}}{\sum(p_{i1} + p_{i2})} \right)^2.$$

Expressions of equal simplicity are available to take account of other features of complex samples. For instance, the CPS is multiplied up to the precalculated population of the United States in each separate age–sex group, what may be called poststratification. Sometimes samples are used not to estimate a total or average on one variable, but to find how two variables correlate with one another. It may be useful to know how accurately the variance is estimated. For each of these three problems simple formulas are available [6].

The exposition above has been in terms of strata from each of which two units have been chosen, and the formulas require that the two samples be chosen independently. If in fact the process of stratification has been carried further, so that only one unit is chosen out of each stratum, and we have to pair such units to use the formulas above, then the method gives an overestimate of the variance. In practice this is small, and is the price paid for the (small amount of) accuracy obtained in the last step of stratification. The trade-off is between precision in estimating the mean and that in estimating the variance.

Several other methods are available for

finding the error of a complex sample without analyzing it down to its basic random selection process. The Bureau of the Census for many years divided the entire CPS into two halves at random, rather than stratum by stratum as here. The ingenious jackknife* method of Tukey and Mosteller [2, 9] removes a major component of bias present in all such approaches (*see* BIAS). Subsequent work by Kish [8] extended the procedure to where more than two units are selected from each stratum, and showed that the formulas need not be made much more elaborate. Kish was able to go much further than taking a single difference, as here, and his balanced replications are a flexible way of handling any estimate of variance built up of single degrees of freedom.

A test of the accuracy of the several methods of what may be called practical variance estimation is given by Frankel [4], which also contains a bibliography of abridged methods of variance estimation. These are set in a wider context of Cochran [3] and Kish [7]. The Bureau of the Census has prepared a series of memoranda giving further mathematical properties of the Keyfitz method—for instance, Waksberg [10].

References

[1] Bowley, A. L. (1926). *Bull. Int. Statist. Inst.*, **22**, liv 1, 6–62.

[2] Brillinger, D. R. (1966). *Commentary*, **8**, 74–80.

[3] Cochran, W. G. (1977). *Sampling Techniques*, 3rd ed. Wiley, New York.

[4] Frankel, M. R. (1971). An Empirical Investigation of Some Properties of Multivariate Statistical Estimates from Complex Samples. Doctoral dissertation, Dept. of Sociology, University of Michigan.

[5] Hansen, M. H., Hurwitz, W. N., and Madow, W. G. (1953). *Sample Survey Methods and Theory*, 2 vols. Wiley, New York.

[6] Keyfitz, N. (1957). *J. Amer. Statist. Ass.*, **52**, 503–510.

[7] Kish, L. (1965). *Survey Sampling*. Wiley, New York.

[8] Kish, L. (1970). *J. Amer. Statist. Ass.*, **65**, 1071–1094.

[9] Tukey, J. W. (1958). *Ann. Math. Statist.*, **29**, 614 (abstract).

[10] Waksberg, J. (1966). CPS Keyfitz Variances for Calendar [Year] 1964. Memorandum to W. N. Hurwitz, U.S. Bureau of the Census, Item III-2, Oct. 27. (Ditto.)

(ESTIMATION, POINT
STRATIFICATION
SURVEY SAMPLING
SURVIVAL ANALYSIS)

NATHAN KEYFITZ

KEY RENEWAL THEOREM

Let $X_1, X_2, X_3, \ldots$, be an infinite sequence of independent, identically distributed (i.i.d.) random variables with a common distribution function (df) $F(x) = P\{X_n \leqslant x\}$. Let $S_0 = 0$ and, for $n = 1, 2, \ldots, S_n = X_1 + X_2 + \cdots + X_n$. Assume that

$$\int_{-\infty}^{0} |x| F(dx) < \infty; \qquad (1)$$

then we can always attach a meaning to

$$\mu = \int_{-\infty}^{+\infty} x F(dx). \qquad (2)$$

and shall assume that $0 < \mu \leqslant \infty$. If there is a $\tilde{\omega} > 0$ such that, with probability 1, X_n is always an integral multiple of $\tilde{\omega}$, then we shall say that we have the *periodic* case. Otherwise, we have the *aperiodic* or *continuous* case. In this article we shall assume without further comment that we are dealing with the *continuous* case; it should be noted, however, that there are results for the periodic case which parallel those we describe for the continuous case.

Let $F_n(x)$ be the df of S_n: $F_n(x) = P\{S_n \leqslant x\}$. Then

$$H(x) = \sum_{n=1}^{\infty} \{F_n(x) - F_n(0)\} \qquad (3)$$

is finite for all $-\infty < x < \infty$, and is plainly nondecreasing; it is called the *renewal function*.

If S_n marks instants on the time scale where certain events of interest: $\mathcal{E}$, say, occur, then $H(a) - H(b)$, for any finite $a > b$, gives the expected number of occurrences of

$\mathscr{E}$ in the half-open time interval $(b, a]$. It is helpful, intuitively, to regard $H(dx)$ as the probability that $\mathscr{E}$ will occur in a time increment "dx", although, of course, this is not correct rigorously.

The class $\mathscr{K}$ consists of functions $k(x)$ satisfying the following two properties:

(K1) For any $0 < R < \infty$, $k(x)$ is Riemann integrable over the interval $[-R, +R]$.

(K2) $\sum_{n=-\infty}^{+\infty} \sup_{n < x \leqslant n+1} |k(x)| < \infty$.

The summation in (K2) defines what may be called the Wiener norm $\|k\|_W$ since it was originally introduced by Norbert Wiener in work on Tauberian theorems.

The key renewal theorem then states that for any $k \in \mathscr{K}$, as $x \to \infty$,

$$\int_{-\infty}^{+\infty} k(x-z) H(dz) \to \frac{1}{\mu} \int_{-\infty}^{+\infty} k(z) dz.$$

(4)

If $\mu = \infty$, then the limit on the right-hand side of (4) is to be taken as zero.

The key renewal theorem is useful in studies of a wide range of stochastic processes*, in particular, when examining asymptotic behavior. It was first presented, in a restricted form, in Smith [5]. It was given in the general form just described as Theorem 5 of Smith [6], in which an attempt was made to generalize the key renewal theorem to situations in which the $\{X_n\}$ are not identically distributed.

A proof of (4) can also be found in Feller [3], in which book Feller calls the class $\mathscr{K}$ the class of "directly Riemann-integrable functions" and seems unaware of the, then six-year-old, result of Smith.

By taking $k(x) = 1$ for $0 \leqslant x \leqslant a$, and $k(x) = 0$ otherwise, one can deduce from the key renewal theorem the theorem of Blackwell:

$$H(x + a) - H(x) \to \frac{a}{\mu}, \qquad \text{as } x \to \infty.$$

(5)

This was first given in Blackwell for strictly positive $\{X_n\}$ in 1948 and, for the general case, in 1953 [1, 2].

A proof of something like (4) but with the condition (1) replaced by a weaker condition, whose description would involve too much technicality to be succinctly described here, was given by Feller and Orey [4].

References

[1] Blackwell, D. (1948). *Duke Math. J.*, **15**, 145–150.

[2] Blackwell, D. (1953). *Pacific J. Math.*, **3**, 315–320.

[3] Feller, W. (1966). *An Introduction to Probability Theory and Its Applications*, Vol. 2. Wiley, New York.

[4] Feller, W. and Orey, S. (1961). *J. Math. Mech.*, **10**, 619–624.

[5] Smith, W. L. (1954). *Proc. R. Soc. Edinb. A*, **64**, 9–48.

[6] Smith, W. L. (1960). *Proc. 4th Berkeley Symp. Math. Statist. Prob.*, Vol. 2. University of California Press, Berkeley, Calif., pp. 467–514.

(QUEUEING THEORY
RENEWAL THEORY)

WALTER L. SMITH

KHINCHIN'S INEQUALITY

Let $X_1, X_2, \ldots$ be a sequence of independent identically distributed Bernoulli* random variables with

$$\Pr(X_i = 1) = \Pr(X_i = -1) = \tfrac{1}{2}.$$

Let $\{C_n\}$, $n = 1, 2, \ldots$ be a sequence of numbers. Then for any $0 < p < \infty$ there exist constants A_p and B_p (independent of $\{C_n\}$) such that for any $n \geqslant 1$,

$$A_p \left(\sum_{j=1}^{n} C_j^2 \right) \leqslant \left\| \sum_{j=1}^{n} C_j X_j \right\|_p \leqslant B_p \left(\sum_{j=1}^{n} C_j^2 \right)^{1/2},$$

where $\|X\|_p = (E|X|^p)^{1/p}$.

Khinchin's inequality is a useful tool in the theory of martingales*. (See, e.g., Chow and Teicher [1] for more details.)

Reference

[1] Chow, Y. S. and Teicher, H. (1978). *Probability Theory*. Springer-Verlag, New York.

KIEFER PROCESS

A *Kiefer process* $\{K(y,t); \ 0 \leqslant y \leqslant 1, \ 0 \leqslant t\}$ is a separable *Gaussian process** with $EK(y,t) = 0$, and covariance function

$$EK(y_1, t_1)K(y_2, t_2)$$
$$= \min(t_1, t_2)(\min(y_1, y_2) - y_1 y_2).$$

It can be represented in terms of standard Wiener processes as follows:

$$K(y,t) = W(y,t) - yW(1,t),$$
$$0 \leqslant y \leqslant 1, \quad t \geqslant 0,$$

where $\{W(y,t); \ 0 \leqslant y, \ 0 \leqslant t\}$ is a standard two-time parameter *Wiener process** with $EW(y,t) = 0$ and covariance function

$$EW(y_1, t_1)W(y_2, t_2)$$
$$= \min(y_1, y_2)\min(t_1, t_2).$$

This process plays an important role approximating the *empirical process**

$$\beta_n(x) = n^{1/2}(F_n(x) - F(x))$$

and the sample *quantile process**

$$q_n(y) = n^{1/2}(Q_n(y) - F^{-1}(y))$$

in terms of Gaussian processes (see Csörgö and Yalovsky [1]). ($F_n(x)$ is the *empirical distribution function* (EDF) and $Q_n(y)$ is the least value of x for which $F_n(x) \geqslant y$.)

Reference

[1] Csörgö, M. and Yalovsky, M. (1981). A Numerical Study of the Kiefer and Brownian Bridge Processes with Applications to Tests for Exponentiality. Dept. of Mathematics, Carlton University, Ottawa.

Further Reading

Kiefer, J. (1970). In *Non-parametric Techniques in Statistical Inference*, M. L. Puri, ed. Cambridge University Press, Cambridge.

Kiefer, J. (1970). In *Non-parametric Techniques in Statistical Inference*, M. L. Puri, ed. Cambridge University Press, Cambridge.

Kiefer, J. (1972). *Zeit. Wahrscheinlichkeitsth. verwand. Geb.*, **24**, 1–35.

(BROWNIAN MOTION
EDF STATISTICS

GAUSSIAN PROCESSES
STOCHASTIC PROCESSES
WIENER PROCESSES)

KIEFER–WEISS PROBLEM

The term "Kiefer–Weiss problem" has occasionally been applied to either of two closely related problems in sequential analysis*.

Suppose that $X_1, X_2, \ldots$ are independent and identically distributed random variables, to be observed sequentially. The distribution of X_i depends on an unknown scalar parameter θ. Values θ_0, θ_1 are given, with $\theta_0 < \theta_1$, and we want to test the hypothesis H_0: $\theta \leqslant \theta_0$ against the alternative $H_1 : \theta \geqslant \theta_1$. For the test procedure T, let $P(H_0$ accepted; $\theta; T)$ denote the probability that H_0 will be accepted when T is used and the parameter is θ. Let $E\{N; \theta; T\}$ denote the expected number of X's that T will observe when the parameter is θ. Small positive values α, β are given, and in order to be considered for use, a test procedure T must satisfy the following conditions (called "OC conditions"):

(a) $P(H_0$ accepted; $\theta; T) \geqslant 1 - \alpha$

for all $\theta \leqslant \theta_0$

(b) $P(H_0$ accepted; $\theta; T) \leqslant \beta$

for all $\theta \geqslant \theta_1$.

Lorden [2] calls the problem of constructing a test procedure T^* satisfying the OC conditions, and such that $\max_\theta E\{N; \theta; T^*\} \leqslant \max_\theta E\{N; \theta; T\}$ for every test procedure T satisfying the OC conditions, the "Kiefer–Weiss problem."

If $\bar{\theta}$ is any given value not equal to θ_0 or θ_1, Lorden calls the problem of constructing a test procedure T' satisfying the OC conditions, and such that $E\{N; \bar{\theta}; T'\} \leqslant E\{N; \bar{\theta}; T\}$ for every test procedure T satisfying the OC conditions, the "modified Kiefer–Wiess problem."

The principal use of a solution to the latter problem is in constructing a solution or an approximate solution to the former problem. The former problem is of interest because in sequential analysis we are interested both in controlling the probabilities of making incorrect terminal decisions (accomplished by the OC conditions), and in controlling the expected sample size. Since the value of the parameter is unknown, it seems reasonable to control the maximum expected sample size (the maximum being taken with respect to the unknown parameter). See Kiefer and Weiss [1], Weiss [3], and GENERALIZED SEQUENTIAL PROBABILITY RATIO TESTS for further details.

References

[1] Kiefer, J. and Weiss, L. (1957). *Ann. Math. Statist.*, **28**, 57–75.

[2] Lorden, G. (1976). *Ann. Statist.*, **4**, 281–291.

[3] Weiss, L. (1964). *Ann. Inst. Statist. Math.*, **15**, 177–185.

(SEQUENTIAL ANALYSIS)

L. WEISS

KIEFER–WOLFOWITZ EQUIVALENCE THEOREM

Among others in the theory of *optimal design** of experiments are two criteria of optimality known as G and D optimality. The Kiefer–Wolfowitz theorem, whose proof first appeared in ref. 10 in 1960 (see also refs. 5 and 9), demonstrated that the two criteria lead to the same class of optimal designs in the approximate theory. The same theorem also provided a characterization of these optimal designs in terms of the number of unknown parameters in the regression function. A prototypical application of the latter characterization and interplay between the two types of optimality is exhibited in Fedorov's solution to his Example 1 in Sec. 2.2 of ref. 3. The method of proof of the theorem also suggested certain iterative techniques for the construction of designs whose information matrices converge to the optimal. These were subsequently developed by others.

The statement of the theorem involves a collection $\{f_i\}_{i=1}^k$ of linearly independent real-valued functions defined on a set $\mathscr{X}$. It is assumed that the range of the mapping $\mathbf{f} : \mathscr{X} \to \mathbb{R}^k$ is compact and that $\mathscr{X}$ is equipped with a σ-field which contains all one point subsets of $\mathscr{X}$. Let Ξ denote any collection of probability measures on $\mathscr{X}$ which includes all probability measures on a finite number of points of $\mathscr{X}$. For every $\xi \in \Xi$ the $k \times k$ information matrices $\int \mathbf{f}(x)\mathbf{f}'(x)\,d\xi(x)$ will be denoted by $\mathbf{M}(\xi)$, where $\mathbf{f}'(x) = (f_1(x), \ldots, f_k(x))$. The function $d : \mathscr{X} \times \Xi \to (0, \infty]$ is defined to be $d(x, \xi) = \mathbf{f}'(x)\mathbf{M}^+(\xi)\mathbf{f}(x)$ whenever $\mathbf{f}(x)$ is the range of $\mathbf{M}(\xi)$ and $+\infty$ otherwise. The matrix $\mathbf{M}^+$ denotes the *Moore–Penrose inverse* of $\mathbf{M}$. (*See* GENERALIZED INVERSES.) Writing $|\mathbf{M}|$ for the determinant of $\mathbf{M}$, the Kiefer–Wolfowitz theorem may now be stated.

Theorem 1. The following are equivalent:

(a) $\quad \displaystyle\sup_{x \in \mathscr{X}} d(x, \xi_0) = \inf_{\xi \in \Xi} \sup_{x \in \mathscr{X}} d(x, \xi),$

(b) $\quad |\mathbf{M}(\xi_0)| = \displaystyle\sup_{\xi \in \Xi} |\mathbf{M}(\xi)|,$

(c) $\quad \displaystyle\sup_{x \in \mathscr{X}} d(x, \xi_0) = k.$

Furthermore, the collection of probability measures $\Lambda \subset \Xi$ that satisfy these conditions is nonempty and linear, and $\mathbf{M}(\xi)$ is the same for all $\xi \in \Lambda$.

A subset A of Ξ is said to be *linear* if for all $\alpha \in [0, 1]$ and ξ_1, ξ_2 in A, $\alpha\xi_1 + (1 - \alpha)\xi_2$ is in A whenever it is in Ξ. A fact not given in the statement of the theorem is that there always exist members of Λ whose support contains no more than $k(k + 1)/2$ points (see ref. 3 or 4).

The main statistical importance of the theorem is a consequence of the interpretation of the matrices $\mathbf{M}(\xi)$, the function $d(x, \xi)$, and the functions $f_1, \ldots, f_k$ in the context of a regression* problem. Suppose that for each finite collection of points $\{x_1, \ldots, x_N\} \subset \mathscr{X}$, not necessarily all distinct, an experiment can be performed. The outcome of the experiment is an observation of the random variable $\mathbf{Y} = \mathbf{A}\boldsymbol{\theta} + \boldsymbol{\epsilon}$, where $E(\boldsymbol{\epsilon}) = 0$, $E(\boldsymbol{\epsilon}\boldsymbol{\epsilon}')$ $= \sigma^2\mathbf{I}$, $\boldsymbol{\theta}' = (\theta_1, \ldots, \theta_k)$ is a vector of unknown parameters, and A is the $N \times k$ matrix whose ith row is $\mathbf{f}'(x)$. The variance of the minimum variance linear unbiased estimator of the value $\boldsymbol{\theta}'\mathbf{f}(x)$ of the mean function at x is $(\sigma^2/N)d(x,\xi)$. The measure ξ on $\mathscr{X}$ is determined from the collection $\mathscr{C}$ $= \{x_1, \ldots, x_N\}$ at which observations were taken by equating $N\xi(x_i)$ to the number of times x_i appears in $\mathscr{C}$. The value $+\infty$ has thus been assigned to the variance if $\boldsymbol{\theta}'\mathbf{f}(x)$ is not an estimable function.

A G-optimal design is one that minimizes $\sup_{\mathscr{X}} d(x,\xi)$ for all ξ in Ξ. A D-optimal design maximizes $|\mathbf{M}(\xi)|$ over all ξ in Ξ. The desirability of a design which minimizes the maximum variance seems clear. That a design should be desirable if it maximizes the determinant is less clear. Under normality of $\boldsymbol{\epsilon}$ a design which maximizes $|\mathbf{M}(\xi)|$ has the property that the smallest invariant confidence ellipsoid has minimal volume. For further desirable properties see refs. 3 and 5.

The Kiefer–Wolfowitz theorem demonstrates that the collections of G and D optimal designs coincide. Notice however that certain members of Ξ do not correspond to probability measures which, like the one associated with the experiment above, are called exact designs and assign rational probabilities to each of a finite number of points of support. As was known by Kiefer and Wolfowitz if Ξ in the statement of the theorem is replaced by Ξ_N, the collection of exact designs on N or fewer points, then corresponding statements are not true. Specifically, condition (a)$\Rightarrow$condition (b), condition (a)$\Rightarrow$condition (c), condition (b) $\Rightarrow$condition (a), and condition (b)$\Rightarrow$ condition (c) all fail to hold. An example is provided in Sec. 2.2g of ref. 7. It is always true that conditon (c) implies both conditions (a) and (b). Nevertheless, if a D-optimal design ξ_0 can be found in Ξ supported on m points then a procedure given by Fedorov in ref. 3 enables the construction, starting with ξ_0, and for each $N > m$,

of exact designs $\tilde{\xi}_N \in \Xi_N$ which satisfy both

$$\left(1 - \frac{m}{N}\right)\max_{\Xi_N}|\mathbf{M}(\xi)|^{1/k} \leqslant |\mathbf{M}(\tilde{\xi}_N)|^{1/k}$$

and

$$\max_{\mathscr{X}} d(x,\tilde{\xi}_N) \leqslant \inf_{\Xi_N}\max_{\mathscr{X}} d(x,\xi)\left(1 - \frac{m}{N}\right)^{-1}.$$

In many instances, such as polynomial regression, explicit solutions to D-optimal design problems can be found (see refs. 4 and 13). Otherwise, one of the extant iterative techniques may be employed (see ref. 1, 2, or 3).

Besides the proof of Kiefer and Wolfowitz, there is another using different methods due to Karlin and Studden in ref. 4. Kiefer proved a stronger theorem in ref. 7 relating the deviation from the optimal of the determinant of the information matrix of an arbitrary design ξ to the deviation of $\max_{\mathscr{X}} d(x,\xi)$ from k. Atwood [1] and Wynn [15] subsequently improved these estimates. An elegant geometric characterization of the D optimal designs is due to Sibson and Silvey (see ref. 11 or the discussion of ref. 15). If $\mathscr{R}$ is the convex hull of the two sets $\pm f(\mathscr{X})$ then an ellipsoid $\mathbf{y}'\mathbf{A}\mathbf{y} \leqslant t$ containing $\mathscr{R}$ must satisfy $|\mathbf{A}| \neq 0$. The volume of such an ellipsoid is

$$(\pi t)^{k/2}\left(\Gamma\left(\frac{k+2}{2}\right)|\mathbf{A}|^{1/2}\right)^{-1}$$

and the smallest t for which containment holds is $\max_{\mathscr{X}} \operatorname{tr}(\mathbf{A}\mathbf{M}(x))$. Using Jensen's inequality* it follows that the volume is uniquely minimized by $\mathbf{A} = \mathbf{M}^{-1}(\xi_0)$, where ξ_0 is D-optimal. Thus, as Silvey and Sibson showed, the ellipsoid of minimal volume containing the set $\mathscr{R}$ (see Elfving's theorem in ref. 4) is $\mathbf{y}'\mathbf{M}^{-1}(\xi_0)\mathbf{y} \leqslant k$, where ξ_0 is D-optimal.

Extensions of the Kiefer–Wolfowitz theorem have been made. For the case of estimating $s \leqslant k$ parameters, see ref. 1, 4, 6, or 13. Vector-valued observations are treated in ref. 3 and stochastic process-valued observations in refs. 12 and 14. The theorem has been generalized by Kiefer [8] to Φ optimality.

In addition to the articles cited and their references, the interested reader should consult refs. 2 and 11, which are review articles.

References

[1] Atwood, C. L. (1969). *Ann. Math. Statist.*, **40**, 1570–1602.

[2] Draper, N. R. and St. John, R. C. (1975). *Technometrics*, **17**, 15–23.

[3] Fedorov, V. V. (1972). *Theory of Optimal Experiments*, W. J. Studden and El Klimko, trans./eds. Academic Press, New York.

[4] Karlin, S. and Studden, W. J. (1966). *Ann. Math. Statist.*, **37**, 783–816.

[5] Kiefer, J. (1959). *J. R. Statist. Soc. B*, **21**, 273–319.

[6] Kiefer, J. (1961). *Ann. Math. Statist.*, **32**, 298–325.

[7] Kiefer, J. (1961). *Proc. 4th Berkeley Symp. Math. Statist. Prob.*, Vol. 1. University of California Press, Berkeley, Calif., pp. 381–405.

[8] Kiefer, J. (1974). *Ann. Statist.*, **2**, 849–879.

[9] Kiefer, J. and Wolfowitz, J. (1959). *Ann. Math. Statist.*, **30**, 271–294.

[10] Kiefer, J. and Wolfowitz, J. (1960). *Canad. J. Math.*, **12**, 363–366.

[11] Pazman, A. (1980). *Math. Operat. Statist. Ser. Statist.*, **11**, 415–446.

[12] Spruill, M. C. and Studden, W. J. (1979). *Ann. Statist.*, **7**, 1329–1332.

[13] Studden, W. J. (1980). *Ann. Statist.*, **8**, 1132–1141.

[14] Wahba, G. (1979). Parameter Estimation in Linear Dynamic Systems. *Tech. Rep. No. 547*, Dept. of Statistics, University of Wisconsin, Madison, Wis.

[15] Wynn, H. P. (1972). *J. R. Statist. Soc. B*, **34**, 133–147.

(DESIGN OF EXPERIMENTS
GENERALIZED INVERSES
LINEAR REGRESSION
OPTIMAL DESIGN)

CARL SPRUILL

KIEFER–WOLFOWITZ PROCEDURE

The Kiefer–Wolfowitz (K-W) procedure is a method of locating the point of minimum, or (with the appropriate reformulation) of maximum, of a regression* function, and as such it is a competitor to response surface* methodology. The original work by Kiefer and Wolfowitz [9] was motivated by the Robbins–Monro (R-M) [12] method for locating the root of a regression function (*see* STOCHASTIC APPROXIMATION). The K-W procedure is a stochastic version of iterative gradient methods of minimizing functions.

The basic idea behind the K-W procedure is simple, although the mathematical theory is considerably more complicated. To be concrete, suppose that when the concentration of a hardener has value x during the manufacture of a plastic, then the hardness of the final product is a random variable $y(x)$ the expectation $f(x) = E(y(x))$. Suppose also that f is unimodal* with a maximum at $x = \theta$, and that the maximization of hardness is desired. To locate θ, begin with an initial guess x_1 of θ and try concentrations slightly above and then slightly below x_1 with the resulting hardnesses $y_{1,1}$ and $y_{1,2}$, respectively. If $y_{1,1} > y_{1,2}$, then let x_2 be greater than x_1, while if $y_{1,1} < y_{1,2}$, choose $x_2 < x_1$. If one continues in a suitable fashion, to be described below, then one obtains a sequence x_n converging to θ.

It appears that the K-W procedure has been employed even more rarely than other stochastic approximation procedures, which themselves are not in wide usage. Mead and Pike [11] suggest that stochastic approximation has received less appplication than due because the highly technical nature of the stochastic approximation literature and the lack of finite sample results. Wetherill [17] criticizes the K-W procedure since, unlike the R-M procedure, it is not highly efficient in parametric cases (e.g., where f is quadratic). However, there is no procedure which is known to be efficient in parametric situations and to perform at least as well as the K-W procedure when f is not of parametric form. There are a few instances where the K-W method has been used. Fabian [6] mentions the enthusiasm of friends when they applied stochastic approximation methods to chemical research, but he gives no further account of their experiences. Janáč [7] used a modified K-W process and simulation on an analog computer to design a trailer truck suspension that minimizes driver fatigue.

Perhaps the most widely known alternative to the K-W procedure is the method of steepest ascent* proposed by Box and Wilson [2] in their pioneering paper on response surface* techniques. There has been little in the way of comparison between the two methodologies. Perhaps they should be considered as complementary, rather than competing, since their respective literatures primarily address different sides of the same problem. Response surface methodology is concerned with design questions at each stage, while research on the K-W process focuses on the sequential aspects.

FORMAL DESCRIPTION OF THE PROCEDURE

Let $f(x)$ be a real-valued function of a k-dimensional vector x, and assume that there exists a unique point θ at which f is minimized. The function f is unknown, but for any value of x, one can, for example by performing a suitable experiment, obtain an unbiased estimator* of $f(x)$.

Let D be the gradient of f. Throughout, we assume, as is usually done in the literature, that

$$\inf\{\|D(x)\| : x \notin B\} > 0 \quad \text{and}$$

$$\inf\{f(x) - f(\theta) : x \notin B\} > 0$$

for every open neighborhood B of θ. Then one can locate θ by solving $D(x) = 0$.

The K-W process is a recursively defined sequence of estimators of θ. Let e_i be the ith unit vector $\mathbb{R}^k$, that is, the vector of 0's except for 1 in the ith coordinate. Let a_n and c_n be sequences of positive numbers that converge to 0. Let Y_n be a k-dimensional random vector whose ith coordinate $Y_n^{(i)}$ satisfies

$$E(Y_n^{(i)} | X_1, \ldots, X_n)$$

$$= [f(X_n + c_n e_i) - f(X_n - c_n e_i)]/(2c_n).$$

In practice, one would typically perform $2k$ experiments to obtain unbiased estimates of $f(X_n + c_n e_1), \ldots, F(X_n + c_n e_k), f(X_n - c_n e_1), \ldots, F(X_n - c_n e_k)$ and use these to form Y_n. Then define X_n recursively by

$$X_{n+1} = S_n - a_n Y_n.$$

Kiefer and Wolfowitz [9], who considered $k = 1$ only, showed that if f satisfied certain regularity conditions,

$$\sum_{n=1}^{\infty} a_n = \infty, \qquad \sum_{n=1}^{\infty} a_n c_n < \infty \quad \text{and}$$

$$\sum_{n=1}^{\infty} (a_n/c_n)^2 < \infty,$$

then $X_n \to \theta$ in probability.

Conditions under which $X_n \to \theta$ almost surely were given by Blum [1] and improved results came later from Venter [15]. The asymptotic normality* of $(X_n - \theta - b)$, where b is the asymptotic bias, was studied by Derman [4], Burkholder [3], and Sacks [13]. Schmetterer's [14] and Fabian's [6] review papers give very general results. Fabian shows that if a_n and c_n are properly chosen, if the k-dimensional vector of third-order partial derivatives of f exists and is bounded in a neighborhood of θ, and if the Hessian of f exists and is bounded, then $n^{1/3}(X_n - \theta - b)$ has an asymptotic normal distribution with mean zero.

The rate of convergence, $n^{-1/3}$, is slow compared to the rate, $n^{-1/2}$, for the R-M procedure. Burkholder [3] and Sacks [13] define concepts of local evenness and show that improved rates are possible if f is local even, but this is a highly restrictive condition.

Fabian [5, 6] introduced designs for estimating D which take more observations and are more sophisticated than the $2k$-point design given above. These designs eliminate biases caused by odd-order derivatives. If for an even positive integer s, the $(s + 1)$-order partial derivative of f with respect to x_i exists and is bounded near θ for each $i = 1, \ldots, k$, then under suitable assumptions,

$$n^{s/(2s+2)}(X_n - \theta - b)$$

has an asymptotic normal distribution. Both *b* and the asymptotic variance are complicated but are given in the original paper. Of course, as *s* increases the rate $n^{-s/(2s+2)}$ comes arbitrarily close to the rate $n^{-1/2}$ of the R-M procedure.

Classical K-W-type methods differ in two important ways from modern iterative methods for maximization of deterministic functions. The latter use information about the Hessian, and they proceed in a series of cycles where at the beginning of each cycle a search direction is found. The remainder of the cycle is a search to minimize the function along this line.

Fabian [6] considers replacing the algorithm $X_{n+1} = X_n - a_n Y_n$ by $X_{n+1} = X_n - a_n A_n Y$, where A_n is a sequence of random matrices converging to $H^{-1}(\theta)$, the inverse of the Hessian of *f* at *θ*. This procedure can be considered as a stochastic version of the Newton–Raphson* algorithm. In more than one dimension, premultiplication of Y_n by $H^{-1}(\theta)$, or an estimate of $H^{-1}(\theta)$, results in a change of coordinates which in general is nonorthogonal and which in the deterministic (Newton–Raphson) case improves the speed of convergence, although no results of this kind have been established for the stochastic (Kiefer–Wolfowitz) case.

Kushner and Gavin [10] define a modified K-W process which is composed of a series of cycles where during the *m*th cycle, X_n can move only in a random direction d_m determined at the beginning of the cycle. The unidimensional searches during the cycles are versions of Kesten's [8] accelerated stochastic approximation method. There are no formal results indicating the Kushner and Gavin's method outperforms previous K-W procedures.

References

[1] Blum, J. R. (1954). *Ann. Math. Statist.*, 25, 737–744.
[2] Box, G. E. P. and Wilson, K. B. (1951). *J. R. Statist. Soc. B*, 1–45.
[3] Burkholder, D. L. (1956). *Ann. Math. Statist.*, 27, 1044–1059.
[4] Derman, C. (1956). *Ann. Math. Statist.*, 27, 529–532.
[5] Fabian, V. (1967). *Ann. Math. Statist.*, 38, 191–200.
[6] Fabian, V. (1971). In *Optimizing Methods in Statistics*, J. S. Rustagi, ed. Academic Press, New York, pp. 439–470. (Highly technical but a good review paper.)
[7] Janáč, K. (1971). *Simulation*, 16, 51–58.
[8] Kesten, H. (1958). *Ann. Math. Statist.*, 29, 41–59.
[9] Kiefer, J. and Wolfowitz, J. (1952). *Ann. Math. Statist.*, 23, 462–466.
[10] Kushner, H. J. and Gavin, T. (1973). *Ann. Math. Statist.*, 1, 851–861.
[11] Mead, R. and Pike, D. J. (1975). *Biometrics*, 31, 803–851. (Extensive review. Views stochastic approximation within a larger context.)
[12] Robbins, H. and Monro, S. (1951). *Ann. Math. Statist.*, 22, 400–407.
[13] Sacks, J. (1958). *Ann. Math. Statist.*, 29, 373–405.
[14] Schmetterer, L. (1969). In *Multivariate Analysis*: *Proceedings of the Second International Symposium*, P. R. Krishnaiah, ed. Academic Press, Dayton, Ohio. (Review paper.)
[15] Venter, J. H. (1967). *Ann. Math. Statist.*, 38, 1031–1036.
[16] Wasan, M. T. (1969). *Stochastic Approximation*. Cambridge University Press, Cambridge. (Reviews much of the early work on stochastic approximation, but offers little synthesis or further development.)
[17] Wetherill, G. (1975). *Sequential Methods in Statistics*, 2nd ed. Chapman & Hall, London. (Discusses stochastic approximation and competing sequential procedures.)

(DESIGN OF EXPERIMENTS
OPTIMIZATION
RESPONSE SURFACES
ROBBINS–MONRO PROCEDURES
SEQUENTIAL ANALYSIS
STOCHASTIC APPROXIMATION
STOCHASTIC PROCESSES)

D. RUPPERT

KINEMATIC DISPLAYS

Kinematic displays show computer-generated moving pictures on a TV-like screen. The technique is used most commonly to create spatial effects; the human visual system involuntarily reconstructs the underlying three-dimensional scene if it is shown incrementally rotated two-dimensional views posted on the screen in sufficiently rapid and regular succession.

Kinematic display hardware became available in the late 1960s, and since then has been used widely for the purposes of flight simulation, computer-aided design, and macromolecular structure research. In statistics their use was pioneered by J. W. Tukey, who with M. A. Fisherkeller and J. H. Friedman in 1972–1974 created the experimental PRIM-9 system at the Stanford Linear Accelerator Center [the letters stand for Projection, Rotation, Isolation (separation of clusters), and Masking (see below), and 9 was the highest dimension the system could handle].

The goal behind advanced display techniques is to make fullest possible use of the low-level integrated capabilities of the human visual system: the perception of three-dimensional space, of time, and of the three color dimensions (hue, saturation, and brightness).

For the viewing of three-dimensional scatter plots, kinematic displays are the approach of choice, since with scenes consisting of isolated dots, a rotating or rocking motion gives by far the strongest and most reliable impression of depth, especially when combined with binocular vision (stereo pairs). Other hints for distance from the viewer, or "depth cues," are accommodation, binocular vision, brightness, saturation, perspective, and size of objects.

The principal uses of kinematic displays in statistics are for data analysis, in particular to check multivariate data for unsuspected features; detection of multivariate outliers; visual clustering; interpretation of the output of multidimensional scaling* programs; and carrier and residual analysis in multivariate regression*.

The main advantages of kinematic displays over more traditional graphical methods probably are that structural features of the data can be found much faster, and conversely, that we may convince ourselves much more easily of the opposite, i.e., that no particular structure is present in the data.

Some data sets (like the stack loss data; see Daniel and Wood [1]) hardly would have gained such notoriety in the data analytic literature if kinematic displays had been generally available a few years ago; the peculiarity of certain data points jumps into your eyes if you view this data set on a kinematic display.

An interesting example of a successful use of kinematic display techniques has been published by Reaven and Miller [4], where the data structure—i.e., the difference between healthy individuals, regular diabetics, and "chemical" diabetics—shows up immediately in a three-dimensional scatter plot of certain physiological data: the healthy individuals form the head and the two kinds of diabetics the two ears of a point cloud resembling a flabby-eared rabbit. It would have been much more difficult to infer the structure from merely looking at static two-dimensional plots.

Many of the techniques familiar from stationary two-dimensional plots make good sense also in kinematic displays. For example, lines connecting the points of a scatter plot may be used to indicate a temporal order, or a minimal spanning tree. Different classes of items can be distinguished by markers (symbols of different shapes and colors). For continuous variables, markers of different sizes, shapes, and different color saturation may be used.

In addition to the three variables represented by spatial coordinates shown on the display, some further continuous variables can be encoded kinematically by "masking," i.e., by choosing a hyperplane orthogonal to the displayed dimensions, moving it through the data space at a constant speed, and turning off (or on) points as soon as they are hit by the plane. More exotic possibilities are furnished by "gyring and gimbling," i.e., by letting the points move on small ellipses, with variables encoded into orbital parameters and speed.

However, most people will find it difficult to grasp more than three continuous spatial and one categorical variable—the latter encoded by color (hue)—in a truly simultaneous fashion. All the other proposed methods of encoding additional variables into symbols either seem to require a sequential decoding (looking at one symbol at a time), or they are prone to confounding effects (it is psychologically difficult not to interpret size, saturation, and brightness as depth cues!).

The main *technical requirements* for a kinematic display system are as follows. The picture must be redrawn ("refreshed") at least 30 times per second to avoid irritating flicker. In order to obtain reasonably smooth motion, the picture should be changed ("updated") at least 10 times per second; 4 to 5 updates per second still give a good spatial impression, but if the motion stops for a second or so, the three-dimensional effect disappers and the picture literally falls flat. The human operator of a kinematic display system is integrated into a tight feedback loop; the motion must be under interactive control, so that the operator can stop and explore the vicinity of interesting projections.

The human engineering aspects of the interaction process are crucial. Besides an all-purpose interactive device such as a tablet and pen (i.e., a device that allows the operator to move a tracking cross to "cursor" to any location on the screen in order to pick data points or menu items, and which in addition facilitates x-y-coordinate input), one also needs devices not depending on cursor feedback, such as switches, dials, and ultimately, voice, so that one can interact with the program without taking one's attention from the part of the data cloud under scrutiny. Irregularities in the motion, and especially in the response time, are extremely annoying because they rip open the feedback loop.

The currently available *display hardware* uses either of two techniques. One, the so-called "calligraphic" method, draws line segments by turning on the electron beam and moving it in a continuous motion from the start point to the end point of the segment. The "raster scan" method, on the other hand, sweeps the entire screen surface along horizontal lines, modulating the intensity of the beam, as on an ordinary TV screen.

Kinematic graphics hardware is now (1981) available from many vendors and in a wide price and performance range. At the top end are special-purpose graphics machines such as the Evans & Sutherland Picture System 300 (which can handle pictures consisting of up to 95,000 points and lines). General-purpose microcomputers such as the Apollo can deal with about 1000 points. At the bottom end are home computers such as the Apple II (about 100 points). The display units with the highest resolution are the calligraphic type (typically with 4096 × 4096 addressable points). Raster scan screens have a coarser resolution (rarely better than 1024 × 1024 "pixels" or picture elements in black and white or 512 × 512 in color) because of bandwidth limitations, and oblique lines will show awkward staircase effects.

A statistical data analysis system has lower requirements than, say, molecular modeling, but still, a resolution of at least 512 × 512 pixels, and the ability to display and rotate several hundred data points and lines would seem to be needed. In contrast to most other applications, perspective may be more of a hindrance than a help; also, the elimination of hidden lines and surfaces, so important for applications in architecture or design, seems to be of little use.

Although a demonstration system typically can be rigged up in a few weeks, the *software* effort behind a system designed for doing actual data analysis is very considerable. It is essential that the user have easy access to a general-purpose interactive statistical package: much too often, raw data are not in a form suitable for direct viewing and must be carefully transformed under interactive visual control (e.g., by fitting some equations to the data and inspecting the residuals). Despite the impressive hardware advances of the recent years, it may be difficult to satisfy one's statistical ideas and data analytic wishes within the constraints imposed by the hardware and software environment.

In particular, *hardcopy* is an urgent need posing awkward problems. Static hardcopy is important as a permanent record that can be filed and published, but it falls flat in a literal sense. Videotapes have much poorer resolution than that of a good display unit. Filming, and editing a film, is expensive, the results are not immediately available, and there may be problems with beats between the standard frame frequency 24 per second of the movie camera and the 30 per second of the screen. It is desirable that kinematic data analysis systems offer both a facility for videotaping entire sessions (to allow post-mortems), and a facility for creating a file that allows backtracking to selected branch points and to recreate and modify subsequent parts of the session.

RELATED TOPICS

Projection pursuit (techniques for automated search for interesting projections, indispensable in dimensions higher than three).

LITERATURE

A comprehensive treatment of kinematic computer graphics is still lacking. As an introduction, read Newman and Sproull's *Principles of Interactive Computer Graphics* [3], supplemented by manuals, in particular those for the Evans & Sutherland display systems, and by articles in journals such as *Computer Graphics*. For a general overview of methods for graphic display of high-dimensional data, with extensive references, see Tukey and Tukey [5]. Specific questions of kinematic displays for data analysis are discussed by Donoho et al. [2].

References

[1] Daniel, C. and Wood, F. S. (1980). *Fitting Equations to Data*. Wiley, New York, Chap. 5.

[2] Donoho, D. L., Huber, P. J. and Thoma, H. M. (1981). In *Computer Science and Statistics*: *Proceedings of the 13th Symposium on the Interface*, W. F. Eddy, ed. Springer-Verlag, New York.

[3] Newman, W. M. and Sproull, R. F. (1973). *Principles of Interactive Computer Graphics*. McGraw-Hill, New York.

[4] Reaven, G. M. and Miller, R. G. (1979). *Diabetologia*, **16**, 17–24.

[5] Tukey, P. A. and Tukey, J. W. (1980). In *Proceedings of the Sheffield Conference*, V. Barnett, ed. Wiley, New York.

(CLASSIFICATION
DATA ANALYSIS
DENDRITES
GRAPHICAL REPRESENTATION OF
    STATISTICAL DATA
MULTIDIMENSIONAL SCALING)

Peter J. Huber

KINGMAN INEQUALITIES

The Kingman inequalities are a set of inequalities, two for integer n, which are satisfied by a class of functions known as p-functions. These functions characterize the stochastic process* called "regenerative phenomena."

Definition. A continuous-time stochastic process $\{Z(t), 0 < t < \infty\}$ in which for each t, the random variable $Z(t)$ assumes only the values 0 and 1 is said to be a regenerative phenomenon, if there exists a function $p(t)$, $0 < t < \infty$, such that for every finite sequence $0 = t_0 < t_1 < t_2 < \cdots < t_n$,

$$\Pr\left[Z(t_i) = 1 \text{ for } i = 1, 2, \ldots, n \right]$$

$$= \prod_{r=1}^{n} p(t_r - t_{r-1}). \qquad (1)$$

The p-function, and also the underlying process, are said to be *standard* if $\lim_{t \to 0} p(t) = 1$. Both for theory and applications, it is the subclass of standard p-functions that is important.

The theory of regenerative processes has been developed primarily by J. F. C. Kingman [5]. It has important applications to the theory of the transition functions of a continuous-time Markov chain* [5, pp. 26–27, Chap. VI]. But regenerative processes have also direct practical applications such as in problems relating to queuing theory* and traffic theory [3, pp. 99–101], type II Geiger counters [4], and storage dams [2].

THE INEQUALITIES

Let $S = \{ t_1 < t_2 < t_3 < \cdots < t_n \cdots \}$ be a fixed but arbitrary sequence of epochs. For each integer n, let A_n, B_n be the events defined by $A_n = [Z(t_i) = 0$, for $i = 1, 2, \ldots, n - 1$, $Z(t_n) = 1]$, $B_n = [Z(t_i) = 1$ for some $i, 1 \leqslant i \leqslant n]$. By the disjointness of the A_i, $\Pr(B_n) = \sum_{i=1}^{n} \Pr(A_i)$. Hence

$$\Pr(A_n) \geqslant 0, \tag{2a}$$

$$\sum_{i=1}^{n} \Pr(A_n) \leqslant 1. \tag{2b}$$

The nth order inequalities are obtained simply by substituting in (2a) and (2b), the expression for $\Pr(A_n)$ in terms of the p-function. The latter expression is also easy to derive. From the definition of A_n by taking expectations, we have

$$\Pr(A_n) = E\left\{ Z(t_n) \prod_{i=1}^{n-1} [1 - Z(t_i)] \right\}$$

$$= p(t_n) - \sum_{i=1}^{n-1} p(t_i) p(t_n - t_i)$$

$$+ \sum\sum_{1 \leqslant i < j \leqslant n-1} p(t_i) p(t_j - t_i) p(t_n - t_j) - \cdots$$

$$+ (-1)^{n-1} p(t_1) \prod_{i=1}^{n-1} p(t_{i+1} - t_i) \tag{3}$$

The extreme right-hand side of (3) is obtained by expanding the intermediate right-hand side and substituting for each summand by $E\{\prod_{r=1}^{i} Z(t_r)\} = \prod_{r=1}^{i} p(t_r - t_{r-1})$, which is derived by taking expectations in (1). Following Kingman, denote the right-hand side of (3) by $F(t_1, t_2, \ldots, t_n, p)$. Then the nth-order inequalities are $F(t_1, t_2, \ldots, t_n, p) \geqslant 0$ and $\sum_{i=1}^{n} F(t_1, t_2, \ldots, t_i) \leqslant 1$.

Putting $n = 1$ in (3) and using (2), the first-order inequalities are simply $p(t_1) \geqslant 0$ and $p(t_1) \leqslant 1$ for all $t_1 \geqslant 0$. Taking $n = 2$, the second-order inequalities are

$$p(t_2) - p(t_1) p(t_2 - t_1) \geqslant 0$$

and

$$p(t_1) + p(t_2) - p(t_1) p(t_2 - t_1) \leqslant 1$$

$$\text{for } 0 < t_1 < t_2.$$

But with increasing order the expressions become cumbersome. However, the inequalities are expressible in a simple recursive form due to Griffeath [1, p. 407]. Set $f(t_n) = \Pr(A_n)$, $g(t_n) = 1 - \Pr(B_n)$. Then it can be shown [from (3) or more simply by using the regenerative property*] that the nth-order inequalities assume the form:

$$f(t_n) = p(t_n) - \sum_{i=1}^{n-1} f(t_i) p(t_n - t_i) \geqslant 0,$$

$$g(t_n) = 1 - \sum_{i=1}^{n} p(t_i) \geqslant 0.$$

Note that here $f(t_n)$ and $g(t_n)$ are notational abbreviations, as the probabilities involved depend not only on t_n but also on the preceeding epochs $t_1, t_2, \ldots, t_{n-1}$ in the assumed sequence S.

The Kingman inequalities hold for all p-functions. But because of the continuity condition, they lead to important results for standard p-functions (and hence for the underlying processes). Thus the second-order inequalities imply uniform continuity and strict positiveness of a standard p-function in its domain $(0, \infty)$, [5, p. 32, Theorem 3.2]. (For the behavior of nonstandard p-functions, see ref. 5 [p. 66].) Note also that for a standard p-function, by setting $p(0) = 1$, the inequalities remain valid also for any nondecreasing sequence $0 = t_0 \leqslant t_1 \leqslant t_2 \leqslant \cdots$

References

[1] Griffeath, D. (1973). *Ann. Prob.*, **1**, 406–416.
[2] Kendall, D. G. (1957). *J. R. Statist. Soc. B*, **19**, 207–233.
[3] Kendall, D. G. and Harding, F. F., eds. (1973). *Stochastic Analysis*. Wiley, New York. (Contains contributions at the research level from different authors relating to regenerative and other stochastic processes.)
[4] Kingman, J. F. C. (1970). *Proc. Camb. Philos. Soc.*, **68**, 697–701.
[5] Kingman, J. F. C. (1972). *Regenerative Phenomena*. Wiley, New York. (A definitive work on the subject of p-functions.)

(INFINITE DIVISIBILITY
MARKOV PROCESSES

QUEUEING THEORY
REGENERATIVE PROCESSES
STOCHASTIC PROCESSES)

V. M. JOSHI

KLOTZ TEST

The Klotz test is a distribution-free rank test for the problem of detecting unequal scales between two independent populations when the difference in their locations is known. Assume that $X_1, \ldots, X_m$ is a random sample from a population with distribution function $F(x - \theta_1)$ and $Y_1, \ldots, Y_n$ is an independent random sample from a population with distribution function $F((x - \theta_2)/\eta)$. The distribution function $F(x)$ is assumed to be that of a continuous distribution with $F(0) = \frac{1}{2}$. We wish to test

$$H_0 : \eta = 1 \quad \text{vs.} \quad H_1 : \eta \neq 1.$$

or one-sided versions of the alternative. The population medians, θ_1 and θ_2, are nuisance parameters* and, in order to construct a distribution-free rank test, we assume that $\delta = \theta_2 - \theta_1$ is known. (A variation for δ unknown is discussed later.)

Let R_i denote the rank of X_i among $X_1, \ldots, X_m, Y_1 - \delta, \ldots, Y_n - \delta$, ranking from smallest to largest. Subtracting δ from each of the Y_j's aligns the two samples so that they come from populations with the same locations. If, for instance, $\eta < 1$, the X population is more variable than the Y population and the X_i's will tend to receive the extreme ranks—the smallest and/or largest. If, on the other hand, $\eta > 1$, the Y population is the more variable and the X_i's will tend to receive the middle ranks. The test statistic is

$$K_{m,N} = \sum_{i=1}^{m} \left[\Phi^{-1}(R_i/(N+1)) \right]^2,$$

where $N = m + n$ and $\Phi(\cdot)$ denotes the distribution function of a standard normal distribution*. The terms in the sum are all nonnegative. Extremely small or large ranks contribute large values to the sum, while the middle ranks add only small amounts. Note that the contributions for $R_i = k$ and $R_i = N + 1 - k$ are the same. The test rejects H_0 for

$$H_1 : \eta < 1 \quad \text{if} \quad K_{m,N} > k_{m,N}(1 - \alpha)$$
$$H_1 : \eta > 1 \quad \text{if} \quad K_{m,N} < k_{m;N}(\alpha)$$
$$H_1 : \eta \neq 1 \quad \text{if} \quad K_{m,N} > k_{m,N}(1 - \alpha/2)$$
$$\text{or if} \quad K_{m,N} < k_{m,N}(\alpha/2),$$

where $k_{m,N}(\alpha)$ denotes the (100α)th percentile of the null distribution of $K_{m,N}$. Under H_0, the $\binom{N}{m}$ possible selections of ranks to be associated with the X_i's are all equally likely. This intuitive structure generates a null distribution that is valid for any $F(\cdot)$ as specified, yielding a distribution-free property.

This test was proposed by Klotz [8], who tabulated values for $k_{m,N}(\alpha)$ and developed properties of the procedure. It competes with other rank tests for scale, such as Mood's test, the Ansari–Bradley* test, or Capon's* test, and with parametric tests such as the F-test. The Klotz test is designed to have favorable properties when the underlying population is normal. Because its scores are generated by the inverse of the standard normal distribution function, they are simpler to obtain than ones requiring expected values of normal order statistics* (Capon's test). Klotz tabled the scores for N values up to 20. When both m and n are large the test can be performed by comparing $(K_{m,N} - m\bar{a})/\sigma_N$ to a standard normal distribution, where $a_i = [\Phi^{-1}(i/(N+1))]^2$,

$$\bar{a} = N^{-1} \sum_{i=1}^{N} a_i \quad \text{and}$$

$$\sigma_N^2 = \frac{mn}{N(N-1)} \sum_{i=1}^{N} [a_i - \bar{a}]^2.$$

In many applications the difference in locations is not known. Raghavachari [10] showed that, if the underlying distribution (f) is symmetric and certain other regularity conditions are satisfied, one can estimate δ with a consistent estimator*, such as $\hat{\delta} = \text{median}(Y_j) - \text{median}(X_i)$, and then proceed as indicated above. The test will lose its small-sample distribution-free property, but the large-sample behavior of the test is the same as though δ was known and is good as long as the underlying distributions are symmetric (see Conover et al. [2]).

Example 1. Cumming [3] described an experiment comparing the heart volumes (ml/kg) of male Canadian athletes from different sports who participated in the 1967 Pan American Games. Table 1 shows the data for runners and water polo players. The sample means $\bar{x}_1 = 13.92$ and $\bar{x}_2 = 13.86$ are virtually the same, so we can proceed to test $H_0 : \eta = 1$ vs. $H_1 : \eta \neq 1$ by ranking the X_i's and Y_j's themselves as shown in Table 1. Using Klotz [8], we obtain the $[\Phi^{-1}(\cdot)]^2$ values for the X_i's. Adding them yields $K_{m,N} = 7.609$, which, when compared to Table 4 in Klotz [8], shows that the result is significant at the $\alpha = 0.05$ level, but not at the $\alpha = 0.02$ level.

Pitman asymptotic relative efficiencies of the test were developed in Klotz [8]. The comparison relative to other rank tests is favorable, showing, for instance, superiority to the Ansari–Bradley and Mood tests except when the distribution is very heavy-tailed, like the Cauchy*. Other research showed efficiencies for the exponential distribution* [1] and that the efficiency relative to the F-test can vary between 0 and $+\infty$ with an efficiency value of 1 when the underlying population is normal (see Kotz [8] and Raghavachari [11]). Thus the test is asymptotically efficient for normal populations.

Hwang and Klotz [7] have developed its Bahadur efficiency* relative to the F-test for a normal population finding that the efficiency decreases as η increases away from 1. Small-sample efficiencies are also investigated by Klotz [8]. Extensions of rank procedures for scale alternatives to multisample settings were considered by Puri [9] and a bibliography of tests for scale was given by Duran [4].

References

[1] Basu, A. P. and Woodworth, G. (1967). *Ann. Math. Statist.*, **38**, 274–277. (Correct Klotz's derivation of the efficiency when testing a scale alternative for exponential distributions.)

[2] Conover, W. J., Johnson, M. E., and Johnson, M. M. (1981). *Technometrics*, **23**, 351–361. (Simulation comparison of many sample tests for differences in scales.)

[3] Cumming, G. R. (1975). *J. Sports Med.*, **3**, 18–22. (Source of example data in text.)

[4] Duran, B. S. (1976). *Commun. Statist. A*, **5**, 1287–1312. (Survey of nonparametric tests for scale.)

[5] Gibbons, J. D. (1971). *Nonparametric Statistical Inference*. McGraw-Hill, New York. (Text containing a brief description of the Klotz test.)

[6] Hájek, J. (1969). *Nonparametric Statistics*. Holden Day, San Francisco. (Introductory text containing a description of the Klotz test.)

[7] Hwang, T. Y. and Klotz, J. H. (1975). *Ann. Statist.*, **3**, 947–954. (Compute Bahadur efficiencies of the test when the populations are normal.)

[8] Klotz, J. (1962). *Ann. Math. Statist.*, **33**, 498–512. (Proposes the test, finds its null distribution, as well as large- and small-sample efficiencies.)

[9] Puri, M. L. (1965). *Ann. Inst. Statist. Math. Tokyo*, **17**, 323–330. (Proposes multisample generalizations of two-sample tests for differences in scales.)

[10] Raghavachari, M. (1965). *Ann. Math. Statist.*, **36**, 1236–1242. (Modifies the Klotz test for the unknown locations case, shows asymptotic normality, and that efficiencies are the same as when the medians are known.)

[11] Raghavachari, M. (1965). *Ann. Math. Statist.*, **36**, 1306–1307. (Shows a class of distributions for which the Pitman efficiency of the Klotz test relative to the F-test goes to zero.)

(DISTRIBUTION-FREE METHODS)

RONALD H. RANDLES

Table 1 Heart Volumes (ml/kg)

		Running	Water Polo	
x_i	r_i	$[\Phi^{-1}(r_i(N+1)^{-1})]^2$	y_j	r_j
15.3	12	0.2931	13.5	8
16.1	16	2.4508	15.5	13
11.4	1	2.4508	14.9	11
12.1	3	0.8629	13.4	7
15.7	15	1.4090	12.0	2
12.9	6	0.1424	12.8	5
			14.4	10
			12.3	4
			15.6	14
			14.1	9

k-MEANS ALGORITHMS

The *k*-means algorithms are used for discovering clusters in data sets. There are various versions of *k*-means algorithms, depending on the method in which covariance matrices are estimated and the procedure of reclassifying the means.

Let $X_1, \ldots, X_n$ be n observations in R^k, and $(\tilde{\mu}_1, \tilde{\Sigma}_1), \ldots, (\tilde{\mu}_k, \tilde{\Sigma}_k)$ be the current estimates of the means and covariances of k clusters, respectively. Let D_{ij}^2 be the Mahalanobis distance*

$$D_{ij}^2 = (\mathbf{X}_i - \bar{\mu}_j)'\tilde{\Sigma}_j^{-1}(\mathbf{X}_i - \tilde{\mu}_j).$$

We classify $\mathbf{X}_j$ into the j_0th cluster provided that

$$D_{ij_0}^2 = \min_{1 \leqslant j \leqslant k} D_{ij}^2.$$

If $L_j = \{i : \mathbf{X}_i$ is classified in the jth cluster$\}$, then the subsequent estimates $(\tilde{\mu}_j, \tilde{\Sigma}_j)$ are

$$\tilde{\mu}_j = |L_j|^{-1} \sum_{i \in L_j} \mathbf{X}_j$$

(the notation $|L_j|$ denotes the number of elements in the set L_j), and

$$\tilde{\Sigma}_j = |L_j|^{-1} \sum_{i \in L_j} (\mathbf{X}_j - \tilde{\mu}_j)(\mathbf{X}_j - \tilde{\mu}_j)'.$$

Another version of *k*-means algorithms is to update only the means after reclassifying by Euclidean distance, fixing $\tilde{\Sigma}_j = \mathbf{I}$ (the identity matrix) for all $j = 1, \ldots, k$. Note that *k*-means define the jth cluster as everything closer to $(\tilde{\mu}_j, \tilde{\Sigma}_j)$ than to any of the other cluster centers, and locate the k clusters simultaneously. [Compare this procedure with the iterated ellipsoidal trimming (IET) algorithm*.]

More details are given in Hartigan [3] and Chernoff [1]. In Gillick [2] a detailed comparison between IET algorithms and *k*-means algorithms is presented.

References

[1] Chernoff, H. (1970). *Proc. 6th Berkeley Symp. Math. Statist. Prob.*, Vol. 1. University of California Press, Berkeley, Calif., pp. 621–629.

[2] Gillick, L. S. (1980). Iterative Ellipsoidal Trimming. *Tech. Rep. No. 15*, Dept. of Math., Massachusetts Institute of Technology, Cambridge, Mass.

[3] Hartigan, J. A. (1975). *Clustering Algorithms*. Wiley, New York.

(CLASSIFICATION
ITERATIVE ELLIPSOIDAL TRIMMING
 ALGORITHM)

KNOCK-OUT TOURNAMENTS

Given n objects or players, tournament designs or methods of experimentation involving paired comparisons have been studied by researchers in order to select the best object, or rank the players according to their respective strengths. Selection procedures consist of choosing a subset of random or fixed size containing the best object; the fixed-size case with size 1 corresponds to determining the champion. If π_{ij} is the probability that player i beats player j, a clear-cut ranking of all players is possible when stochastic transitivity (strong stochastic transivity) holds for every triad of different players i, j, k:

$$\pi_{ij} \geqslant \tfrac{1}{2}, \quad \pi_{jk} \geqslant \tfrac{1}{2} \Rightarrow \pi_{ik} \geqslant \tfrac{1}{2}$$

$$[\pi_{ik} \geqslant \max(\pi_{ij}, \pi_{jk})].$$

Assuming transitivity, a round-robin (RR) tournament with n players requires $\binom{n}{2}$ comparisons, while a knock-out (KO) tournament requires only $(n-1)$ comparisons. It is possible to have a tie in RRs, with two or more players achieving the same highest score for general n, whereas the KO always produces a winner. The fundamental problem of tournaments is: Is it possible to improve on RRs by replacing them (with the same number of comparisons) with repeated KOs in which partial balance is somehow introduced? The answer provided by Narayana [8] and Zidek is a qualified, but definite yes, whether small, medium, or asymptotically large numbers of comparisons are made.

It is convenient to divide results on KO tournaments into a classical part, well summarized by David [2], and results derived since then using modern high-speed computers. If players can be ordered according to rank or strength, $1, 2, \ldots, n$, player i always beating player j if $i < j$, then given some permutation of $1, \ldots, n$ we must sort the players in the correct order. To determine the smallest number of comparisons required to sort the worst possible case is obviously a computer problem, and is closely related to the merging algorithm: Two lists of items, a small list of m items and a large list of n items, $m < n$, are supposed sorted and are given; the problem here is to merge the two lists into one sorted list. The Ford–Johnson algorithm (1959, denoted as FJA) for sorting and the Hwang–Lin algorithm (1972, HLA) for merging use binary insertion: i.e., the best method for merging one element into a chain of $2^k - 1$ sorted elements is by repeated comparison with the median element. A maximum of k comparisons are required and this is best possible when merging one element into n, $2^{k-1} \leqslant n \leqslant 2^k - 1$. The reader is referred to two papers by Manacher [3, 4] for a complete bibliography and latest results on FJA and HLA. A brief résumé of some results in Manacher is as follows. The sorting of n items requires at least $E(n) = [\log_2 n!]$ comparisons. No algorithm is known that sorts n items with fewer comparisons than FJA for all n. FJA achieves $E(n)$ for $n \leqslant 11$ and $n = 21, 22$; it is never far away from $E(n)$. If $M(n)$, $(M(m, n))$ is the minimum number of comparisons required to sort n items, (merge two lists), $M(n) \geqslant E(n)$; it is known that $M(12) = E(12) + 1 = 30$. The FJA is not optimal, in that for $n = 189$ (and an infinity of values of $n > 189$) a special method can be devised, involving sorting and merging, which surpasses it. $M(3, 7) = M(3, 8) = 8$; $M(3, 9) = M(3, 10) = 9$; $M(3, 2^d) \leqslant 3d - 1$ for $d > 3$; $M(4, [(5/4)2^d] - 1) \leqslant 4d - 1$ by HLA.

Hartigan (1966) and Maurer [1975] have considered KO models recently; their results can best be described in the context of the Narayana–Zidek *Contributions to the Theory of Tournaments* (Parts I–IV), of which a summary follows. Before reading the technical discussion on KOs that follows, a simple example of tournaments T_1, T_2, T_3, T_4 with only five players may be relevant. In T_1, which leaves the minumum possible byes at every stage, we play two randomly chosen pairs in round 1, leaving one bye. Thus we obtain the first element 2 in the tournament vector $(2, 1, 1)$ of T_1, showing that two pairs play in round 1. After round 1, only the two winners and the bye are left; we play round 2 by choosing a random pair out of these three, again leaving one bye. The second element of $(2, 1, 1)$ represents this playing pair. The final round always has one playing pair only; the finalists here are the winner and bye of round 2. Tournament T_2, when five players are involved, has the vector $(1, 2, 1)$. It represents the quickest reduction to the classical case of 2^t players. With five players only, we thus have to play a single pair in round 1; eliminating the loser, we now have 2^2 players, so that the tournament can proceed as in Wimbledon. In practice, this is the method actually employed. Tournament T_3 is, in every sense, the very opposite of T_1. Here only a single random pair plays in each round, thus leaving a maximum of players in the byes! The vector for T_3, when the total number of players is five, is therefore $(1, 1, 1, 1)$. In every round the loser is eliminated, so that the pair playing the next round is chosen from the byes and the winner *randomly*. Finally, T_4 is like T_3, with the further proviso that the winner of any round meets a bye chosen at random from the byes of the previous round. The winner "persists" in T_4 and John Moon has aptly dubbed this the "king of the castle" tournament, after a well-known nursery game.

Starting with Narayana [5], a theory of tournaments providing comparisons between RRs and KOs has been developed by Narayana, Zidek (NZ), and their students. For latest results and a complete bibliography, we refer to Narayana [6]. Let us consider the simplest KO with $n = 2^t$ players playing the first round, randomly paired off, the winners being randomly paired in subsequent rounds. We assume the one-"outlier" model, i.e., a strong player A with probability $p \geqslant \frac{1}{2}$ of beating $B_1, \ldots, B_{n-1}$, who are of equal strength. Apart from the explicit results provided below, this model has great theoretical interest since it is a least favorable configuration for certain selection problems. If $P_n^{[i]}$ denotes the probability that A meets $B_1, \ldots, B_i$ exactly (i.e., and no others in the tournament), then

$$P_n^{[i]} = p^{i-1}q \Big/ \binom{n-1}{i}, \qquad i = 1, \ldots, t-1$$

$$\tag{1}$$

$$P_n^{[t]} = p^{t-1} \Big/ \binom{n-1}{t}, \qquad \text{and 0 otherwise.}$$

Letting $X_1, \ldots, X_{n-1}$ be the random variables (rvs) which take on the value 1(0) according as A meets $B_1, \ldots, B_{n-1}$ (or not) in the tournament, it is not difficult to see that the X_i's are symmetrically dependent rvs. Clearly,

$$P_n^i = \sum_{v=0}^{n-i-1} P_n^{[i+v]} \binom{n-i-1}{v}, \tag{2}$$

where P_n^i is the probability that A meets B_1 and $B_2 \ldots$ and B_i (and perhaps others). Using (1), in the classical case, it can be shown that

$$\binom{n-1}{i} P_n^i = \left(\frac{p}{q}\right)^{i-1} \frac{p^t}{q}$$

$$\times \left[\frac{1}{p^t} - \binom{t}{0} - \binom{t}{1}\frac{q}{p} - \cdots \right.$$

$$\left. - \binom{t}{i-1}\left(\frac{q}{p}\right)^{i-1} \right]$$

$$(i = 1, \ldots, t). \quad (3)$$

When $p = q = \frac{1}{2}$, this reduces to

$$\binom{n-1}{i} P_n^i$$

$$= \frac{2}{n}\left[n - \binom{t}{0} - \binom{t}{1} - \cdots - \binom{t}{i-1} \right]$$

$$(i = 1, \ldots, t), \quad (4)$$

which for $i = 1$ gives the result dear to graph theorists, namely $P_n^1 = 2/n$; or less trivially, when $p \neq q$, $P_n^1 = (1 - p^t)/q(n-1)$. From (2) inclusion–exclusion* yields

$$Q_n^i = \sum_{v=1}^{n-1} (-1)^{v-1} P_n^v \binom{i}{v}, \tag{5}$$

and also the inverse relations to (2) and (5).

The NZ theory shows that (2) and (5) are valid for at least a general class of random KO tournaments defined as follows. With n players in the same one-outlier model, let us define inductively for every integer $n > 2$, a KO tournament as a vector of positive integers $(m_1, \ldots, m_k)$ satisfying:

$$m_1 + \cdots + m_k = n - 1, \quad m_k = 1;$$

$$2m_1 \leqslant n; \tag{6}$$

$$2m_i \leqslant n - m_i - \cdots - m_{i-1}$$

$$(i \geqslant 2).$$

In the first round of $(m_1, \ldots, m_k)$, $2m_1$ players chosen at random from n, are randomly paired off. Eliminating the m_1 losers, the $(n - 2m_1)$ byes and m_1 winners [i.e., $(n - m_1)$ players in all] play the tournament $(m_2, \ldots, m_k)$. As inclusion–exclusion holds as in (2) and (5), high-speed computers yield accurate results for n large (at least up to 1000). Explicit results or simple recursions can be obtained as in Narayana [6] for the four cases used in practice. Indeed, letting $n = 2^t + k$ $(0 \leqslant k < 2^t)$, consider the tournaments:

T_1: with vector $m_i = [(n + 2^{i-1} - 1)/2^i]$

T_2: with vector $(k, 2^{t-1}, 2^{t-2}, \ldots, 1)$

T_3: with vector $(1, 1, \ldots, 1)$ [that is, $(n - 1)$ 1's]

T_4: same as T_3, but the winner at each encounter plays one of the byes randomly

Not surprisingly, the "king of the castle" tournament T_4, which is *not* a random tournament as in (6), clearly yields symmetrically dependent rvs and hence can be studied using (2) and (5). The general theoretical results of NZ imply that any selection procedure or test used for RRs can also be used for KOs with exchangeable or symmetrically dependent RVs, provided that a win is scored as $+1$ and a loss -1 (not 0, as in David for RRs, since this does not give a sufficient statistic).

In order to get the strong players to play more often and thus obtain better results than the RR in either determining a champion or selection procedures, it is natural to follow a Markov rule in repeated KOs. The winner of the last replication plays one of the players with highest score (assuming that the winner himself has not the unique highest score) to initiate the next "king of the castle" replication; balance can be partially introduced, by making the winner play, in doubtful cases, the player whom he or she played least. The technical details for small, medium, and asymptotic sample sizes are given in Narayana and Hill [8] and show such "partially balanced" KOs always superior to RRs.

A technical point regarding medium sample sizes should serve as a warning to those who glibly produce asymptotic expansions without carefully checking small or medium-size cases. Asymptotically, when the number of replications N is very large, there is no point in introducing partial balance as explained with T_4 in the last paragraph; as $N \to \infty$, randomness automatically takes care of balance. Simulation indicates that it is necessary to introduce balance with KO tournaments, at least partially as indicated, since both small and medium cases do not achieve good results without the balance.

The warning given above applies a fortiori to single KO replications, particularly when we move "far" away from the one-outlier model. Indeed, Chung and Hwang [1] provide results and counterexamples to show that (a) the strongest ranked player may not have the highest probability of winning a tournament if only stochastic transitivity is assumed, and (b) even assuming strong stochastic transitivity, it is difficult to obtain precise quantitative estimates of how well the strongest player will perform in a single KO tournament. Of course, such unrealistic and pathological configurations are not just peculiar to KOs, as David [2, pp. 77–80] has already pointed out similar difficulties for (even repeated) RRs.

The exact reference to papers named here, but not listed in the references, can be found in Manacher [3, 4] or Narayana [6].

References

[1] Chung, F. R. K. and Hwang, F. K. (1978). *J. Amer. Statist. Ass.*, **73**, 593–596.

[2] David, H. A. (1963). *The Method of Paired Comparisons*. Charles Griffin, London.

[3] Manacher, G. K. (1979). *J. ACM*, **26**, 434–440.

[4] Manacher, G. K. (1979). *J. ACM*, **26**, 441–456.

[5] Narayana, T. V. (1968). *C. R. Acad. Sci. Paris*, **267**, 323.

[6] Narayana, T. V. (1979). *Lattice Path Combinatorics with Statistical Applications*. University of Toronto Press, Toronto.

[7] Narayana, T. V. and Agyepong, F. (1980). *Contributions to the Theory of Tournaments IV*. Cahiers BURO, Paris.

[8] Narayana, T. V. and Hill, J. (1974). *Proc. 5th Natl. Math. Conf.*, Shiraz, Iran, pp. 187–221.

(RANKING AND SELECTION
SCORING SYSTEMS IN SPORTS)

T. V. NARAYANA

KNOX'S TESTS

Tests of time and space interaction in epidemiology* proposed by Knox [2–4] based on a (limiting) chi-squared* distribution. A discussion of these tests, together with a description of an amended test criterion, can be found in Abe [1].

References

[1] Abe, O. (1973). *Biometrics*, **29**, 67–77.

[2] Knox, G. (1959). *Brit. J. Prev. Soc. Med.*, **13**, 222–226.

[3] Knox, G. (1963). *Brit. J. Prev. Soc. Med.*, **17**, 121–127.

[4] Knox, G. (1964). *Brit. J. Prev. Soc. Med.*, **18**, 17–24.

(CHI-SQUARED TESTS)

KNUT-VIK DESIGN

A Knut-Vik design of order n is an $n \times n$ array of elements chosen from a set S of n distinct elements (usually referred to as treatments*) such that all the rows, columns, and left and right diagonals are permutations of the set S. These designs are also sometimes referred to as *pandiagonal designs* and are used in design of experiments* applications for eliminating sources of variations along four directions.

A complete enumeration and algebraic description of Knut-Vik designs of order $\leqslant 13$ is given by Atkin et al. [1]. Orthogonal Knut-Vik designs were studied by Hedayat [3], among others. For applications, see basic texts on experimental design such as Federer [2] and Kempthorne [4].

References

[1] Atkin, A. O. L., Hay, L., and Larson, R. G. (1977). *J. Statist. Plann. Infer.*, **1**, 289–297.

[2] Federer, W. T. (1955). *Experimental Design— Theory and Application*. Macmillan, New York.

[3] Hedayat, A. (1977). *J. Comb. Theory A*, **2**, 331–337.

[4] Kempthorne, O. (1952). *Design and Analysis of Experiments*. Wiley, New York.

(DESIGN OF EXPERIMENTS
GRAECO-LATIN SQUARES
LATIN SQUARES)

KNUT-VIK SQUARE *See* DESIGN OF EXPERIMENTS

KOLMOGOROV BACKWARD EQUATION *See* DIFFUSION PROCESSES

KOLMOGOROV CRITERION *See* STRONG LAW OF LARGE NUMBERS

KOLMOGOROV FORWARD EQUATION *See* DIFFUSION PROCESSES

KOLMOGOROV–KHINCHIN THEOREM

The theorem gives a criterion for convergence of a sequence of independent random variables in terms of the behavior of their second moments.

Let $X_1, X_2, \ldots$ be a sequence of independent random variables with $EX_i = 0$, $i = 1$, $2, \ldots$. If $\sum_{j=1}^{\infty} EX_j^2 < \infty$, the series $\sum_{j=1}^{\infty} X_j$ converges with probability 1 (i.e., almost surely*).

Moreover, if X_i, $i = 1, \ldots$, are uniformly bounded, i.e., $p(|X_i| \leqslant c) = 1$, $c < \infty$, $i = 1, \ldots$, then, conversely, the convergence with probability 1 of the series $\sum_{i=1}^{\infty} X_i$ implies that $\sum_{i=1}^{\infty} EX_i^2$ is convergent.

(CONVERGENCE OF SEQUENCES OF
RANDOM VARIABLES)

KOLMOGOROV'S DIFFERENTIAL EQUATION *See* STOCHASTIC PROCESSES

KOLMOGOROV'S INEQUALITY

Let $X_1, \ldots, X_n$ be independent random variables (rvs) with means $\mu_1, \ldots, \mu_n$ and variances $\sigma_1^2, \ldots, \sigma_n^2$, respectively; let $S_1, \ldots, S_n$ be partial sums defined by $S_k = \sum_{i=1}^{k} X_i$; $k = 1, \ldots, n$, and let $S_1, \ldots, S_n$ have means $m_1, \ldots, m_n$ and variances $s_1^2, \ldots, s_n^2$, respectively. Kolmogorov's inequality states that for every $t > 0$,

$$\Pr\{|S_k - m_k| < ts_n \text{ for all } k = 1, \ldots, n\}$$

$$\geqslant 1 - 1/t^2. \tag{1}$$

This result is stronger than Chebyshev's inequality*, which gives the same bound for $\Pr\{|S_n - m_n| < ts_n\}$. Whereas the latter inequality is sufficient to prove the weak law of large numbers for independent, identically distributed (i.i.d.) rvs with finite second moments, inequality (1) is needed to prove the strong law of large numbers for such sequences of rvs ([3, Secs. IX.6, X.5, X.7]; see LAWS OF LARGE NUMBERS).

Feller [2, Sec. VII.8] shows that (1) holds without the condition of independence of $X_1, \ldots, X_n$ if the sequence $\{S_k\}$ of partial sums is a martingale*, that is, if

$$E(S_k \mid S_1, S_2, \ldots, S_{k-1}) = S_{k-1};$$

$$k = 2, \ldots, n;$$

he also gives a version of (1) for submartingales (see also Dunnage [1] and Gilat and Sudderth [4]).

References

[1] Dunnage, J. E. A. (1975). *Quart. J. Math.*, **26**, 361–376. [Analogs of (1) with applications to rates of convergence.]

[2] Feller, W. (1966). *An Introduction to Probability Theory and Its Applications*, Vol. 2, 2nd ed. Wiley, New York.

[3] Feller, W. (1968). *An Introduction to Probability Theory and Its Applications*, Vol. 1, 3rd ed. Wiley, New York.

[4] Gilat, D. and Sudderth, W. D. (1976). *Zeit. Wahrscheinlichkeitsth. verwand. Geb.*, **36**, 67–74. [Generalizations of (1) to martingales.]

(HAJEK–RÉNYI INEQUALITY)

CAMPBELL B. READ

KOLMOGOROV–SMIRNOV STATISTICS

Kolmogorov–Smirnov statistics belong to the wider class of EDF statistics*, so called because they are based on the empirical distribution function of a random sample. The EDF of a random sample $X_1, X_2, \ldots, X_n$ is the function $F_n(x)$ defined by (number of $X_i \leqslant x)/n$; it is the proportion of sample values less than or equal to x, and is an estimate of $F(x)$, the distribution function of x. Let $w_n(x) = \{F_n(x) - F(x)\}$; the Kolmogorov–Smirnov statistics (often called simply Kolmogorov statistics) are defined by

$D^+ = \sup_x \{w_n(x)\}$; $D^- = \sup_x \{-w_n(x)\}$ and $D = \max(D^+, D^-)$; to these we add $V = D^+ + D^-$, a statistic introduced later by Kuiper [8] for observations on a circle. The statistics are sometimes written with a subscript n (e.g., D_n) and are sometimes defined with the factor $\sqrt{n}$ included; also, in the past, K has often been used instead of D. The statistics measure the discrepancy between $F(x)$ and its estimate $F_n(x)$, and so can be used to measure the fit of the sample to a given $F(x)$. Applications to tests of fit are discussed in a separate article (*see* KOLMOGOROV–SMIRNOV-TYPE TESTS OF FIT). Another group of EDF statistics, also used for tests of fit, is the Cramér–von Mises group; two of these are $W^2 = n \int w_n^2(x) \, dF(x)$ and $U^2 = n \int \{w_n(x) - \int w_n(t) \, dF(t)\}^2 \, dF(x)$.

The Kolmogorov–Smirnov statistics have distributions that do not depend on $F(x)$; the same is true for W^2 and U^2. Suppose that, from a random variable X, a new random variable Z is defined by $Z = F(X)$; it is well known that the Z-values are uniformly distributed between 0 and 1, written $U(0, 1)$. The statistics are easily calculated from the values $Z_i = F(X_i)$, $i = 1, \ldots, n$ (*see* KOLMOGOROV–SMIRNOV-TYPE TESTS OF FIT).

DISTRIBUTION THEORY

An immense literature has developed concerning the distribution theory of D^+, D^-, and D, starting with Kolmogorov's paper [7], which introduced D and gave both small-sample and asymptotic formulas. Smirnov [12, 13] gave asymptotic percentage points. The theory is developed by starting with a uniform sample $Z_1, Z_2, \ldots, Z_n$, and using the EDF of the Z_i, $F_n^*(z)$. Many interesting techniques have been used to give distributional results for finite samples, from combinatoric arguments to the use of generating functions* and geometric methods. There are specialized results of unusual elegance, such as the fact that the value of D may occur at a point z uniform between 0 and 1, or the result $P[F_n(z)$ crosses the line $y = cz]$

$= 1/c$ ($c > 1$), a result not dependent on n. Asymptotic theory has also received much attention; the asymptotic distributions take quite simple forms, and the first terms in an expansion in $1/\sqrt{n}$ can be obtained. In earlier work, [e.g., in Kolmogorov [7]] asymptotic results were produced by applying analytic techniques to finite-sample results; many of the proofs have since been simplified using the theory of weak convergence and Doob's observation [2] that $\sqrt{n}\, w_n^*(z)$ tends to a Gaussian process, where $w_n^*(z) = F_n^*(z) - z$. Again unusual results exist, such as the connections between the asymptotic distribution of $\sqrt{n}\, D$ and those of the empirical distribution function (EDF) statistics W^2 and U^2: nD^2/π^2 and U^2 have the same distribution, and $4nD^2/\pi^2$ has the same distribution as the sum of two independent W^2 statistics. Durbin [3] gives a review of distribution theory.

DISTRIBUTION THEORY WHEN $F(x)$ CONTAINS UNKNOWN PARAMETERS

The situation described above, when $F(x)$ is completely specified, and therefore the Z_i are $U(0,1)$, we call case 0. However, $F(x)$ may contain unknown parameters, so that it is defined only to within a family of possible distributions. The parameters can be estimated from the sample, say by maximum likelihood*, and used in $F(x)$ to make the transformation $Z_i = \hat{F}(X_i)$, the symbol ^ indicating the presence of estimates for parameters; the Kolmogorov statistics can then be calculated as before. However, the Z_i are no longer $U(0,1)$, and distribution theory is drastically altered, even asymptotically. Provided that the unknown parameters are location and/or scale parameters, the null distributions of these statistics depend on $F(x)$ and on which parameters are estimated, as well as on n, but do not depend on the true values of the parameters. Exact tables for D^+, D^-, and D for the test for exponentiality, i.e., for $F(x) = 1 - \exp(-x/\theta)$ ($x \geqslant 0$), with θ estimated by $\bar{x}$, have been given by Durbin [4], but in general both exact and

asymptotic theory are difficult, and tables for Kolmogorov statistics, where available, have been given by Monte Carlo* sampling (*see* KOLMOGOROV–SMIRNOV-TYPE TESTS OF FIT). Asymptotic theory, in particular, presents a challenging area, since it is not currently available for the Kolmogorov statistics (it depends on the supremum of a Gaussian process* with mean zero and known covariance function), although asymptotic distributions are known for other EDF statistics*. When unknown parameters are not location or scale, distribution theory will depend on the true values of these parameters. However, another technique is available (the half-sample method; *see* EDF STATISTICS and Durbin [5]), which reduces the distribution theory to the case discussed in the preceding section, where all parameters are known.

KOLMOGOROV–SMIRNOV STATISTICS FOR DISCRETE DISTRIBUTIONS: CASE 0

It is possible to define an analog to the EDF for discrete distributions, based on the histogram of observed data. The possible outcomes are divided into k cells; let p_i be the probability of an observation in cell i, so that, with n observations, $E_i = np_i$ is the expected number in cell i, and let O_i be the observed number. The EDF $F_n(x)$ is defined as (number of observations $\leqslant x)/n$ and a comparison with $F(x) =$ (expected number of observations $\leqslant x)/n$ leads to Kolmogorov–Smirnov statistics

$$S^+ = \max_j \left\{ \sum_{i=1}^{j} (O_i - E_i) \right\},$$

$$S^- = \max_j \left\{ \sum_{i=1}^{j} (E_i - O_i) \right\}$$

and

$$S = \max(S^+, S^-)(1 \leqslant j \leqslant k).$$

The values of S^+, S^-, and S depend on the order of the cells. Pettitt and Stephens [10] have discussed the distributions of S^+, S^-, and S for case 0, i.e., when the p_i are completely specified, and have tabulated the distributions of S for equal p_i; the tables can be used as approximations to the distributions for unequal p_i. The distribution of S is very different from that of its analog nD in case 0, so that tables for nD should not be used with S.

TWO-SAMPLE AND MULTISAMPLE STATISTICS

Analogous to the one-sample statistics, two- or even multisample statistics can be devised to compare several EDFs [6]. Suppose that two samples are given, of sizes n and m, and with EDFs $F_n(x)$ and $G_m(x)$; let $L(x)$ be $F_n(x) - G_m(x)$. Then $D^+ = \sup_x\{L(x)\}$, $D^- = \sup_x\{-L(x)\}$, $D = \max(D^+, D^-)$ and $V = D^+ + D^-$. Again the statistics may be defined with subscripts (e.g., $D_{n,m}$); for more than two samples, there are natural extensions based on the supremum of the difference between all pairs of EDFs. These statistics now lend themselves to tests that two samples come from the same (unknown) distribution, and are nonparametric in nature. The calculations of the statistics depend on the ranks within each sample, compared with the rank in the combined sample, and the statistics have much in common with other rank order statistics. The distribution theory is again largely combinatoric, and there has been an extensive literature, since the early work on two samples by Smirnov [12]; Durbin [3] gives references. Asymptotic distributions of suitably normalized statistics are often the same as the single-sample distributions.

RELATED STATISTICS

An interesting class of statistics is based on looking at the uniformity of Z_i, in case 0, from a different viewpoint, basing tests on the differences between $Z_{(i)}$ and its expected value $i/(n + 1)$. Such statistics are

$C^+ = \max_i\{Z_{(i)} - i/(n + 1)\}$,

$C^- = \max_i\{i/(n + 1) - Z_{(i)}\}$, $(1 \leqslant i \leqslant n)$,

$C = \max(C^+, C^-)$

and $K = C^+ + C^-$. There are obvious parallels with the D-statistics and V, and distribution theory is very similar. The C-statistics appear to have arisen from time-series analysis of the periodogram.

In connection with tests of fit, a number of variations of D have been suggested. In many situations, D has relatively low power, perhaps because it concentrates on one outstanding difference between $F_n(x)$ and $F(x)$, whereas other EDF statistics use the discrepancies all along the line; some authors have proposed statistics that use all values of $F_n(x) - F(x)$, either at the sample points X_i, or at intervals along the line. In general, the theory of such statistics is very complicated. A number of authors (see, e.g., Rényi [11] for references) have discussed statistics derived from $F_n(x)/F(x)$, sometimes conditional on the property $a \leqslant F(x) \leqslant b$, or $a \leqslant F_n(x) \leqslant b$. For these statistics there is much interesting distribution theory, both for finite n and asymptotic. These statistics may be suitable for tests for censored data (*see* KOLMOGOROV–SMIRNOV-TYPE TESTS OF FIT).

When distribution $F(x)$ contains unknown parameters, these may be estimated by other methods than maximum likelihood; provided that the methods are asymptotically efficient, asymptotic theory will be the same as that discussed in the section "Distribution Theory when $F(x)$ Contains Unknown Parameters." Srinivasan [14] has suggested using the Rao–Blackwell theorem* to improve the estimate of $F(x)$, and then to calculate Kolmogorov statistics; however, this leads to very difficult formulas, except for the exponential distribution. An interesting property of Kolmogorov–Smirnov statistics is that they can be used to provide one-sided or two-sided confidence intervals for the parent distribution of a sample (see, e.g., Birnbaum and Tingey [1]). More recently, Littell and Rao [9] have used the statistics to give confidence intervals for unknown parameters, by taking into the intervals those parameter values which give nonsignificant values of D, say, when referred to the appropriate distribution.

References

[1] Birnbaum, Z. W. and Tingey, F. H. (1951). *Ann. Math. Statist.*, 592–596.

[2] Doob, J. L. (1949). *Ann. Math. Statist.*, **20**, 393–403. (The beginning of modern asymptotic theory.)

[3] Durbin, J. (1971). *Distribution Theory for Tests Based on the Sample Distribution Function.* Reg. Conf. Ser. Appl. Math., Vol. 2. SIAM, Philadelphia. (A comprehensive account of work to 1970.)

[4] Durbin, J. (1975). *Biometrika*, **62**, 5–22.

[5] Durbin, J. (1976). Empirical Distributions and Processes. *Lect. Notes Math.*, **566**. Springer-Verlag, Berlin.

[6] Kiefer, J. (1959). *Ann. Math. Statist.*, **30**, 420–447.

[7] Kolmogorov, A. N. (1973). *G. Inst. Ital. Attuari.*, **4**, 83–91 (in Italian). (The beginning of Kolmogorov–Smirnov statistics.)

[8] Kuiper, N. H. (1960). *Ned. Akad. Wet. Proc. A*, **63**, 38–47.

[9] Littell, R. C. and Rao, P. V. (1978). *Technometrics*, **20**, 23–28.

[10] Pettitt, A. N. and Stephens, M. A. (1977). *Technometrics*, **19**, 205–210.

[11] Rényi, A. (1967). *Bull. Intern. Statist. Inst.*, **42**(1), 165–176.

[12] Smirnov, N. V. (1939). *Mat. Sb. (N.S.)*, **6**, 3–26.

[13] Smirnov, N. V. (1948). *Ann. Math. Statist.*, **19**, 279–281.

[14] Srinivasan, R. (1970). *Biometrika*, **57**, 603–611.

(DISTRIBUTION-FREE METHODS
EDF STATISTICS
GOODNESS OF FIT
KOLMOGOROV–SMIRNOV-TYPE TESTS
  OF FIT
ORDER STATISTICS
RANK ORDER STATISTICS)

M. A. STEPHENS

KOLMOGOROV–SMIRNOV SYMMETRY TEST

Kolmogorov–Smirnov symmetry test is the name given in the literature (see, e.g., Gibbons [3]) to a generalization of tests of symmetry proposed by Smirnov [5] and based on a statistic of the Kolmogorov–Smirnov type (*see* KOLMOGOROV–SMIRNOV STATISTICS *and* KOLMOGOROV–SMIRNOV-TYPE TESTS OF FIT). Some particular cases of these generalized tests are discussed in Vol. 1 of this encyclopedia (pp. 340–344) under Butler–Smirnov test* with a comprehensive bibliography. The generalization given here is for a wider range of practical applications.

Suppose that we wish to test the null hypothesis that a random sample is drawn from a continuous population which is symmetric (*see* TESTS OF SYMMETRY) about a specified point, say M. Since a random variable X is symmetric if

$$P(X \leqslant M - x) = P(X \geqslant M + x) \quad \text{for all } x,$$

the corresponding condition for a continuous cumulative distribution function* (CDF) F is

$$F(M - x) = 1 - F(M + x).$$

Therefore, the null hypothesis of symmetry can be written as

$$H_0 : F(M + x) + F(M - x) - 1 = 0. \quad (1)$$

The two-sided alternative is

$$A : F(M + x) + F(M - x) - 1 \neq 0$$

$$\text{for at least one } x.$$

Assume that the sample observations X_1, $X_2, \ldots, X_n$ are measured on at least an ordinal scale. The empirical or sample distribution function, defined as

$$S_n(x) = \text{proportion of sample}$$

$$\text{observations} \leqslant x,$$

can be calculated for all real numbers x and is a point estimate of $F(x)$. Hence a logical test statistic is

$$\sup_x |S_n(M + x) + S_n(M - x) - 1|, \quad (2)$$

where $\sup_x$ means the supremum over x, the largest of the differences in the neighborhood of each observed value of x. The null distribution of (2) is complicated by the fact that the supremum can occur at an x value that is slightly different from any observed value of x. To avoid these complications, the Kolmogorov–Smirnov symmetry test statistic is defined as

$$D_n = \sup_{x \geqslant M} |S_n(M + x)$$
$$+ S_n\big[-(x - M)^-\big] - 1|, \quad (3)$$

where $(x - M)^-$ denotes a value slightly smaller than $x - M$. A large value of D_n would indicate that the empirical distribution function does not meet the condition for symmetry of a cumulative distribution function, and therefore that the CDF which the sample represents is not a symmetric distribution. The appropriate rejection region for H_0 is then large values of D_n, or, if a P-value is to be reported, it should be a right-tail probability. Chatterjee and Sen [2, Table 2] give the exact null distribution of nD_n for $n \leqslant 16$ and expressions for the asymptotic null distribution.

In a practical application, it is not necessary to compute D_n as stated in (3). Instead, the specified value M can be subtracted from each sample observation to form $z_i = x_i - M$ and the test statistic D_n in (3) can be calculated from

$$D_n = \sup_{z \geqslant 0} |S_n(z) + S_n\big[-(z^-)\big] - 1|. \quad (4)$$

The empirical distribution function must be calculated separately only for those $z \geqslant 0$ in the intervals determined by each different observed absolute value of the differences $x_i - M$. The following numerical example for $n = 8$ with $M = 25.0$ should clarify how D_n is calculated.

x:	24.29	24.73	25.12	25.12
	25.21	25.43	25.51	25.79
$z = x - 25.0$:	−0.71	−0.27	0.12	0.12
	0.21	0.43	0.51	0.79

Table 1

Interval Based on Observed z	$nS_n(z)$
$z < -0.71$	0
$-0.71 \leqslant z < -0.27$	1
$-0.27 \leqslant z < 0.12$	2
$0.12 \leqslant z < 0.21$	4
$0.21 \leqslant z < 0.43$	5
$0.43 \leqslant z < 0.51$	6
$0.51 \leqslant z < 0.79$	7
$0.79 \leqslant z$	8

Table 2

Interval Based on Observed $\|z\|$	$nS_n(z)$	$nS_n[-(z^-)]$	$n\|S_n(z) + S_n[-(z^-)] - 1\|$
$z < 0.12$	2	2	4
$0.12 \leqslant z < 0.21$	4	2	2
$0.21 \leqslant z < 0.27$	5	2	1
$0.27 \leqslant z < 0.43$	5	1	2
$0.43 \leqslant z < 0.51$	6	1	1
$0.51 \leqslant z < 0.71$	7	1	0
$0.71 \leqslant z < 0.79$	7	0	1
$0.79 \leqslant z$	8	0	0
$nD_n = 4$,	$P = 0.063$		

This test can be considered a one-sample test for location in the sense that it can be used to test a compound null hypothesis such as:

H_0:   The population CDF is symmetric with median M_0,

because this is equivalent to (1) with M replaced by M_0. In this context, the D_n test is an alternative procedure to the Wilcoxon signed-rank test* (*see* DISTRIBUTION-FREE METHODS). However, if we reject this null hypothesis, we do not know whether we are rejecting the $M = M_0$ part, the symmetry part, or both.

As with other one-sample procedures, the Kolmogorov–Smirnov symmetry test can be used for testing the null hypothesis that for a random sample of paired observations, the differences between pairs come from a continuous population which is symmetric about a specified point M, the median of the population of differences. The differences between pairs are treated as the observations and n is the number of pairs.

Test statistics for one-sided alternatives that specify a direction of asymmetry (positive skewness or negative skewness) can be defined as in (3) but without absolute values. All these tests can be extended to the case of n independent but nonidentically distributed random variables with CDFs $F_1, F_2, \ldots, F_n$ and specified medians $M_1, M_2, \ldots, M_n$, respectively, for the broader null hypothesis

$$H_0 : F_i(M_i + x_i) + F_i(M_i - x_i) - 1 = 0$$

$$\text{for } i = 1, 2, \ldots, n.$$

Koulet and Staudte [4] give lower and upper bounds for power against various alternatives.

References

[1] Butler, C. C. (1969). *Ann. Math. Statist.*, **40**, 2209–2210.

[2] Chatterjee, S. K. and Sen, P. K. (1973). *Ann. Inst. Statist. Math., Tokyo*, **25**, 287–299.

[3] Gibbons, J. D. (1976). *Nonparametric Methods for Quantitative Analysis.* Holt, Rinehart and Winston, New York.

[4] Koul, H. L. and Staudte, R. G. (1976). *Ann. Statist.*, **4**, 924–935.

[5] Smirnov, N. V. (1947). *Akad. Nauk. SSR C. R. (Dokl.) Acad. Sci. URSS*, **56**, 11–14.

(BUTLER–SMIRNOV TEST
KOLMOGOROV–SMIRNOV-TYPE TESTS
   OF FIT
MEDIAN
TESTS OF SYMMETRY)

JEAN DICKINSON GIBBONS

$1 \leqslant i \leqslant n$. The Kolmogorov–Smirnov statistics are

$$D^+ = \max_i \{ i/n - Z_{(i)} \},$$

$$D^- = \max_i \{ Z_{(i)} - (i-1)/n \},$$

$$D = \max(D^+, D^-);$$

to these we add Kuiper's [14] statistic* $V = D^+ + D^-$. Statistic V should be used for tests involving observations on a circle, since its value, in contrast to D, for example, is independent of the origin for x. It may of course also be calculated from observations on a line. The statistics are sometimes defined with a factor $\sqrt{n}$, and are often labeled with a subscript n (e.g., D_n), but these will be omitted. Note the difference in the definitions of D^+ and D^-; it is easy to miscalculate D^-. Examples of the calculations are given in Pearson [21]. A closely related class of statistics, which we shall call the C-class, is defined by

$$C^+ = \max_i \{ i/(n+1) - Z_{(i)} \},$$

$$C^- = \max_i \{ Z_{(i)} - i/(n+1) \},$$

$$C = \max(C^+, C^-), \quad K = C^+ + C^-.$$

(*see* KOLMOGOROV–SMIRNOV STATISTICS).

KOLMOGOROV–SMIRNOV-TYPE TESTS OF FIT

This article is concerned with the practical aspects of tests of fit based on Kolmogorov–Smirnov statistics and statistics similar to these; it is a companion to the entry on Kolmogorov–Smirnov statistics*. The Kolmogorov–Smirnov statistics are derived from the empirical distribution function (EDF) and the entry on EDF statistics* is also relevant. The working definitions of Kolmogorov–Smirnov statistics (often called simply Kolmogorov statistics) in the context of tests of fit, are as follows. Suppose that $F(x)$ is a continuous distribution, to be tested as the parent distribution of a given random sample $X_1, X_2, \ldots, X_n$. Let $X_{(1)}, X_{(2)}, \ldots, X_{(n)}$ be the order statistics and make the substitutions $Z_{(i)} = F(X_{(i)})$,

TESTS OF FIT

The most straightforward problem in goodness-of-fit* testing, if not the most frequently occurring, is when $F(x)$ is completely specified, i.e., contains no unknown parameters to be estimated from the sample. This situation we call case 0. The distribution theory of Kolmogorov statistics and C-statistics for case 0 has been much studied; tables of tail probabilities, and of percentage points for D^+, D^-, and D, were given by Birnbaum and Tingey [4] and Birnbaum [2]. They have been reproduced in many sets of tables (see, e.g., Owen [20]). A condensed table is given in Pearson and Hartley [22, Table 54]. Tables of percentage points for V were given by Stephens [25], tables for C^+, C^-, and C by Durbin [8], and tables for K by Brunk [5] and Stephens [26]. For this case

0, Stephens [28] condensed the tables for most of these statistics by the following device; the calculated statistic (e.g., D) is entered into a formula involving D and n to produce a modified value D^*; D^* is then compared with a set of given percentage points (the asymptotic points of $\sqrt{n}\,D$). Modifications were given for each tail, although in most practical applications only the upper tail will be used; that is, large values of the test statistics will lead to rejection of the tested fit. The modifications and percentage points for D^+, D^-, and D are reproduced in Table 1 in EDF STATISTICS.

Example 1. Suppose that a sample of 20 values is to be tested to be normal with given mean and variance; the 20 Z-values are calculated, and suppose that the formulas given above give $D^+ = 0.286$, $D^- = 0.017$, so that $D = 0.286$. The table gives a modified form, $D^* = 0.286(\sqrt{20} + 0.12 + 0.11/\sqrt{20})$, which equals 1.320; comparison with the points given show that this is significant at about the 6.5% level. Note that if D^+ had been used as test statistic, it would have been significant at about the 3.3% level. This is typical of tail probabilities: the level for a two-tail test using D is about twice that of the larger of the two one-sided statistics D^+ and D^-.

POWER

Statistic D^+ can be used to give a one-sided test, i.e., to test that the distribution is $F_0(x)$ against alternatives $F_A(x)$ for which, ideally, $F_A(x) \geq F_0(x)$ for all x; however, D^+ will give good power if $F_A(x)$ is greater than $F_0(x)$ over most of the range, implying that the true mean is less than that hypothesized. Similarly, D^- will be used to guard against alternatives with a larger mean; the Z_i for these two situations will be closer than expected to 0 or closer to 1, respectively. D gives a test against shifts in mean in either direction. The risk in using a one-sided test is that if the alternative is in the direction opposite to that imagined (e.g., if D^+ is used

in the normal test above when the true mean is greater than that tested), the test statistic will very rarely reject. However, when one-sided tests are in order, D^+ and D^- are quite powerful test statistics, but for two-sided tests, D tends to be less powerful than other EDF statistics of the Cramér–von Mises type (e.g., W^2 and A^2) (*see* EDF STATISTICS). Statistic V, besides its use on the circle, will detect a shift in variance, represented by a cluster of Z_i, or values Z_i in two groups near 0 and 1. The C-statistics have much the same properties as their D counterparts. The D-statistics in case 0 will usually be more powerful than chi-square, when this statistic is used to test a continuous distribution.

USE OF THE LOWER TAIL

The tests in case 0 are essentially tests that the Z_i are uniformly distributed between 0 and 1, written $U(0, 1)$. There are other ways in which data may be transformed to be $U(0, 1)$; for instance, times of successive events, when the intervals are assumed to be exponential, will, if divided by the total time, give a set of ordered uniform points. On some alternatives to the assumptions, the set Z_i might possibly be *superuniform*, i.e., more regular than a uniform sample; then test statistics will be small, and to detect excessive superuniformity the lower tail of, say, D would be used. An interesting example of superuniform observations is provided by the dates of kings and queens of England from 1066 [21]; other examples, particularly relevant to tests for exponentiality and the Poisson process, are given in Seshadri et al. [24].

CONFIDENCE INTERVALS* FOR DISTRIBUTION FUNCTIONS AND FOR PARAMETERS

Suppose, for a sample of size n, that L is found so that $P(D > L) = \alpha$, and boundaries are drawn a vertical distance L above and below the EDF; these provide a confidence band for the true $F(x)$, now assumed unknown, with confidence level $1 - \alpha$. [Strictly, a slight correction in $1 - \alpha$ is needed if values $F(x)$ outside the range $0 \leqslant F(x) \leqslant 1$ are to be omitted.] Statistics D^+ and D^- provide one-sided intervals. A more recent development combines goodness of fit and parameter estimation; parameters are included in a $1 - \alpha$ confidence set if a goodness-of-fit test using these values and an appropriate test statistic (case 0) is nonsignificant at level α. Littell and Rao [17] have examined the use of D in this way.

TESTS OF FIT WITH ESTIMATED PARAMETERS

When $F(x)$ contains unknown location or scale parameters, tests can be made as follows. The parameters are estimated by maximum likelihood; the substitution $Z_{(i)} = \hat{F}(X_{(i)})$, where $\hat{F}(x)$ is $F(x)$ with estimates inserted for the parameters, is now made and Kolmogorov statistics calculated as before.

The theory of this procedure is discussed in KOLMOGOROV–SMIRNOV STATISTICS. For each family tested, tables are needed; these have been provided for D by Lilliefors [15, 16] and Stephens [28] for the normal and exponential distributions (see also Pearson and Hartley [22, Table 54]), by Stephens [29] for the logistic distribution, and by Chandra et al. [6] for the extreme-value and Weibull distributions. The significance points are much smaller than the case 0 values, and use of case 0 tables with estimated parameters, even for large samples, will lead to serious errors in the recorded significance level of a test. The distribution tested will be rejected far less than it should be; this of course may please the tester, but is nevertheless an error. As for case 0, statistics D^+ and D^- are powerful against certain alternatives, although with estimated parameters these are not always easy to classify; D is generally not so powerful as other EDF statistics* but is better than chi-square.

TABLES FOR DISCRETE DISTRIBUTIONS

Kolmogorov statistics can be defined for discrete distributions, by comparing the cumulated histogram* with the expected histogram (*see* KOLMOGOROV–SMIRNOV STATISTICS). Pettitt and Stephens [23] have provided tables for case 0 and for equiprobable cells; they may also be used with unequal probabilities if they are not too different. If a natural ordering of cells exists, the statistics appear to be useful in detecting a trend of values away from the expected—e.g., in detecting a steady decrease in fatalities per year in an operating theater when the null hypothesis is that the death rate remains constant. In this respect the statistics will be superior to the Pearson chi-square test*, which cannot detect trend.

MULTISAMPLE TESTS

Test statistics of Kolmogorov type have been defined for comparing two samples, or, more precisely, for testing that the distribution functions of the two samples are the same (*see* KOLMOGOROV–SMIRNOV STATISTICS). They can be extended for comparisons of k samples. Durbin [9] gives formulas for the D-statistics for the two-sample case and references to tables of probabilities and percentage points; see also Owen [20] and Pearson and Hartley [22, Sec. VIII]. The statistics depend on the ranks of the observations; their power properties compare well with those of the many other statistics dependent on ranks.

TESTS FOR CENSORED DATA*

Many variations of Kolmogorov–Smirnov statistics have been suggested for use with censored data, particularly for tests of case 0, where the parameters of the tested distribution are known. For this case, the substitution $Z_{(i)} = F(X_{(i)})$ may still be made, to give a censored Z-sample which may be tested to be uniform. Tests suitable for *right* or *left censoring** have been given by Barr and Davidson [1], Dufour and Maag [7], Michael and Schucany [18], and Koziol and Byar [13]. Similarly, tests associated with Rényi (*see* KOLMOGOROV–SMIRNOV STATISTICS), involving the ratio $F_n(x)/F(x)$, sometimes over the restricted range $a \leqslant F_n(x) \leqslant b$ or $a \leqslant F(x) \leqslant b$, have been developed. Birnbaum and Lientz [3] discuss some of these tests and give tables; they also show how they may be used to give confidence intervals for a distribution.

Random censoring is an important problem, e.g., with survival data; tests with such data often use the Kaplan–Meier [12] estimate of $F(x)$. Hall and Wellner [11] give a review of earlier work involving Kolmogorov–Smirnov statistics, and show how confidence bounds for the distribution can be found. Tests involving estimated parameters with Kolmogorov–Smirnov statistics have not been much developed for randomly censored data. Koziol and Byar [13] and Fleming et al. [10] discuss tests for two samples.

OTHER TESTS

Various Kolmogorov–Smirnov tests have appeared for special situations; see, e.g., Wood [30] for an application in randomized block* designs. Also, many variations of Kolmogorov–Smirnov statistics have been devised, in addition to those discussed above, often with the motive to increase power against specific alternatives. By and large, however, these have not come into general use.

References

[1] Barr, D. R., and Davidson, T. (1973). *Technometrics*, **15**, 739–757.

[2] Birnbaum, Z. W. (1952). *J. Amer. Statist. Ass.*, **47**, 425–441.

[3] Birnbaum, Z. W. and Lientz, B. P. (1969). *J. Amer. Statist. Ass.*, **64**, 870–877.

[4] Birnbaum, Z. W. and Tingey, F. H. (1951). *Ann. Math. Statist.*, **22**, 592–596.

[5] Brunk, H. D. (1962). *Ann. Math. Statist.*, **33**, 525–532.

[6] Chandra, M., Singpurwalla, D., and Stephens, M. A. (1981). *J. Amer. Statist. Ass.*, **76**, 729–731.

[7] Dufour, R. and Maag, U. R. (1978). *Technometrics*, **20**, 29–32.

[8] Durbin, J. (1969). *Biometrika*, **56**, 1–15.

[9] Durbin, J. (1971). *Distribution Theory for Tests Based on the Sample Distribution Function*. Reg. Conf. Ser. Appl. Math., Vol. 9. SIAM, Philadelphia. (A good review of work to 1970.)

[10] Fleming, T. R., O'Fallon, J. R., O'Brien, P. C., and Harrington, D. P. (1980). *Biometrics*, **36**, 607–625.

[11] Hall, W. J. and Wellner, J. A. (1980). *Biometrika*, **67**, 133–143.

[12] Kaplan, E. L. and Meier, P. (1958). *J. Amer. Statist. Ass.*, **53**, 457–481.

[13] Koziol, J. A. and Byar, D. P. (1975). *Technometrics*, **17**, 507–510.

[14] Kuiper, N. H. (1960). *Ned. Akad. Wet. Proc. A*, **63**, 38–47.

[15] Lilliefors, H. (1967). *J. Amer. Statist. Ass.*, **62**, 399–402.

[16] Lilliefors, H. (1969). *J. Amer. Statist. Ass.*, **64**, 387–389.

[17] Littell, R. C. and Rao, P. V. (1978). *Technometrics*, **20**, 23–27.

[18] Michael, J. R. and Schucany, W. R. (1979). *Technometrics*, **21**, 435–441.

[19] Miller, L. H. (1956). *J. Amer. Statist. Ass.*, **51**, 111–121.

[20] Owen, D. B. (1962). *A Handbook of Statistical Tables*. Addison-Wesley, Reading, Mass.

[21] Pearson, E. S. (1963). *Biometrika*, **50**, 315–325. (An expository paper with good examples.)

[22] Pearson, E. S. and Hartley, H. O. (1972). *Biometrika Tables for Statisticians*, Vol. 2. Cambridge University Press, Cambridge.

[23] Pettitt, A. N. and Stephens, M. A. (1977). *Technometrics*, **19**, 205–210.

[24] Seshadri, V., Csörgö, M., and Stephens, M. A. (1969). *J. R. Statist. Soc. B*, **31**, 499–509.

[25] Stephens, M. A. (1965). *Biometrika*, **52**, 309–321.

[26] Stephens, M. A. (1969). *Ann. Math. Statist.*, **40**, 1833–1837.

[27] Stephens, M. A. (1970). *J. R. Statist. Soc. B*, **32**, 115–122.

[28] Stephens, M. A. (1974). *J. Amer. Statist. Ass.*, **69**, 730–737.

[29] Stephens, M. A. (1979). *Biometrika*, **66**, 591–595.

[30] Wood, C. L. (1978). *Biometrika*, **65**, 673–676.

(EDF STATISTICS
KOLMOGOROV–SMIRNOV STATISTICS
ORDER STATISTICS)

M. A. STEPHENS

KOLMOGOROV'S THREE-SERIES THEOREM

Let $X_1, X_2, \ldots$ be a sequence of independent random variables. In order that the series $\sum_{j=1}^{\infty} X_j$ converge with probability 1, it is necessary and sufficient that for a fixed $c > 0$ the three series

$$\sum EX_i^c, \quad \sum \text{var}(X_i^c), \quad \text{and} \quad \sum \Pr(|X_i| \geqslant c)$$

converge. The notation X^c denotes truncation* at c, i.e.,

$$X^c = \begin{cases} X & \text{if } |X| \leqslant c \\ 0 & \text{if } |X| > c. \end{cases}$$

(CONVERGENCE OF SEQUENCES OF
   RANDOM VARIABLES
TWO-SERIES THEOREM)

KOLMOGOROV'S ZERO–ONE LAW
See ZERO–ONE LAWS

KOOPMAN–DARMOIS–PITMAN FAMILIES

Koopman–Darmois–Pitman families (or Darmois–Koopman families) is an older term for what is now generally known as *exponential families**.

The notion of exponential families first occurred in Fisher [6]. Fisher* argued—in his characteristic, mathematically somewhat imprecise, manner—that families of (one-dimensional) distributions that admit a sufficient reduction have to be exponential, and he also indicated that these families were the only ones supplying uniformly most powerful tests* in the sense of Neyman and Pearson. Many subsequent papers, by other authors, have been concerned with the mathematical questions left open by Fisher in

making the former of the two claims. The first of these papers were by Darmois [4], Koopman [8], and Pitman [11]. For discrete distributions no mathematically and statistically satisfactory formulation has been found, whereas a fairly adequate discussion can be given in the continuous-type case, see Dynkin [5], Brown [3], and Hipp [7] (generalization to higher dimensions is considered in Barndorff-Nielsen and Pedersen [1]). The second of Fisher's claims has been treated fairly recently by Pfanzagl [10]; see also the comments by Neyman and Pearson [9] and Bartlett [2].

References

[1] Barndorff-Nielsen, O. and Pedersen, K. (1968). *Math. Scand.*, **22**, 197–202.

[2] Bartlett, M. S. (1937). *Proc. R. Soc. Lond. A*, **160**, 268–282.

[3] Brown, L. (1964). *Ann. Math. Statist.*, **35**, 1456–1474.

[4] Darmois, G. (1935). *C. R. Acad. Sci. Paris*, **260**, 1265–1266.

[5] Dynkin, E. B. (1951). Necessary and sufficient statistics for a family of probability distributions. (in Russian). (English transl.: *Select. Transl. Math. Statist. Prob.*, **1**, 23–41, (1961).

[6] Fisher, R. A. (1934). *Proc. R. Soc. Lond. A*, **144**, 285–307.

[7] Hipp, C. (1974). *Ann. Statist.*, **2**, 1283–1292.

[8] Koopman, L. H. (1936). *Trans. Amer. Math. Soc.*, **39**, 399–409.

[9] Neyman, J. and Pearson, E. S. (1936). *Statist. Res. Mem.*, **1**, 113–137.

[10] Pfanzagl, J. (1968). *Sankhyā A*, **30**, 147–156.

[11] Pitman, E. J. G. (1936). *Proc. Camb. Philos. Soc.*, **32**, 567–579.

OLE BARNDORFF-NIELSEN

KOOPMAN'S TEST OF SYMMETRY

Given a random sample $(X_1, Y_1), \ldots, (X_n, Y_n)$ from an unknown bivariate distribution function F, to test the hypothesis

$$H_0 : F(x, y) = F(y, x)$$

of bivariate symmetry or, more restrictively, the hypothesis

H_0 : G is symmetric $[1 - G(z) = G(-z)]$

where G is the distribution function $Z = X - Y$ vs. shift alternatives, Koopman [2] proposed a combined test based on a sign test* and a rank sign test*.

For additional information, see Koopman [1].

References

[1] Koopman, P. A. R. (1978). Internal report. Vrije University, Amsterdam. (Contains extensive tables and derivations.)

[2] Koopman, P. A. R. (1979). *Statist. Neerlandica*, **33**, 137–141. (Contains a description of the procedure.)

(RANK SIGN TEST
SIGN TEST
SYMMETRY, TESTS FOR)

K-OUT-OF-*N* SYSTEM   *See* COHERENT STRUCTURE

KOYCK GEOMETRIC LAG   *See* ECONOMETRICS; LAGGED DEPENDENT VARIABLES; LAG MODELS, DISTRIBUTED

k-RATIO *t* TESTS, *t* INTERVALS, AND POINT ESTIMATES FOR MULTIPLE COMPARISONS

The beginning problem in multiple comparisons is the simultaneous testing of all of the $n(n - 1)/2$ differences among n *a priori* undifferentiated treatment means taken a pair at a time. Another common problem is that of simultaneously testing all of the n differences between n treatments and a control. *k*-ratio *t* tests, *t* intervals, and point estimates are a potentially large class of Bayes rules for simultaneous testing problems of this kind and for the corresponding problems of simultaneous interval and point estimation.

Simultaneous Testing: α-Level Approaches

The natural way, at first, to solve any simultaneous testing problem is simply to apply α-level tests to each comparison. This α-*level comparisonwise* approach, however, gives rise to a strong intuitive objection.

Example 1. Suppose that a 0.05-level comparison-wise rule is applied to m differences and only 5% of them prove to be significant. This would be exactly the number of significant differences expected under the joint null hypothesis H_0^m that *all* of the true differences are zero. Such a result would give so much support to H_0^m, especially with m large, that the use of a 0.05-level critical t value for *each* difference would be intuitively much too unconservative.

A first reaction to the *homogeneous outcomes objection* is simply to increase the critical t value to the point at which the probability of wrongly rejecting H_0^m is reduced to the desired α. The Tukey [26] and Bonferroni rule [24] for pairwise differences and the Scheffé [25] rule for all contrasts exemplify this α-*level experimentwise approach.* Krishnaiah [20] gives other examples.

Experimentwise critical values increase dramatically, however, with increasing m (*See* MULTIPLE COMPARISONS). In the known-variance case the Scheffé value for an n-treatment (n Tr) experiment with 100 means is 11.10, more than five times the comparisonwise value, 1.96. Such large values can seem justified in the context of Example 1, where the experiment has encountered treatments which collectively tend to be very homogeneous. Not all outcomes are so homogeneous, however.

Example 2. Suppose that in Example 1 most of the differences have turned out to be *significant* instead of *not* significant. This would give so much evidence against the joint null hypothesis that the critical values of an experimentwise rule would seem far too large, resulting in considerable and unnecessary losses of power. This may be termed the *heterogeneous outcomes objection*.

Various writers have proposed intermediate α-level rules, such as multiple-range* and multiple-*F* tests*, but similar less serious but still strong objections can be raised to each of these. Intuitively, the need is for an adaptive rule: one that can react *a posteriori* to the *collective* nature of the comparisons; one that can vary all the way from being extremely conservative if the overall outcome is very homogeneous to being far more powerful if the outcome is very heterogeneous.

The *k*-ratio approach is a Bayesian one (see the section "Notes on References"), which provides *t* tests and estimates that are adaptive in this way. The adaptivity is due to a dependence on the overall data which takes the form of "shrinkage" in the interval and point estimates, a feature they have more or less in common with the estimation results of Lindley [22, 22a], Lindley and Smith [23]; Efron and Morris [16] and James and Stein [19]; and Box and Tiao [2, Sec. 7.2].

THE *k*-RATIO APPROACH: BASIC CONCEPTS AND ILLUSTRATIONS

The *n*-Treatments Testing Problem

Let $\mathbf{x} = [x_1, \ldots, x_n]$ and s_e^2 denote *n* treatment means and a variance estimate with PDF $f(\mathbf{x} \mid \boldsymbol{\mu}, \sigma_e^2) f(s_e^2 \mid \sigma_e^2, f)$ such that $X_i \sim \text{NID}(\mu_i, \sigma_e^2/r)$ and $S_e^2 \sim \chi_f^2 \sigma_e^2 / f$. Then, given $\mathbf{x}$ and s_e^2, the *n* Tr testing problem, which will be our main example, is usually viewed as that of making two-sided tests of the hypothesis $H_0 : \mu_i = \mu_j$ simultaneously for all $n(n-1)/2$ *combinations* of the means two at a time. However, since there are three decisions $\mu_i < \mu_j$, $\mu_i \approx \mu_j$, $\mu_i > \mu_j$ in each component problem, the *k*-ratio approach defines the joint problem more appropriately [10, Sec. 4.1; 21, Sec. 11] as that of making right-sided tests of the hypotheses $H_0 : \mu_i \leqslant \mu_j$ vs. $H_1 : \mu_i > \mu_j$ for all $n(n-1)$ *permutations* of the means two at a time.

To deal also with the corresponding joint interval estimation problem, the new approach defines the component hypotheses in the extended forms

$$H_0 : \delta \leqslant \delta_0 \quad \text{vs.} \quad H_1 : \delta > \delta_0, \quad (1)$$

where δ is any one of the $n(n-1)$ ordered differences $\mu_i - \mu_j$ and δ_0 is arbitrary.

EXCHANGEABLE* PRIORS.   A basic feature of the new approach lies in the use of exchangeable priors [22]. Thus in the *n* Tr problem with *n a priori undifferentiated* treatments, the $n(n-1)$ differences $\delta = \mu_i - \mu_j$ are all given the *same* prior probability density function (PDF) $p(\delta)$. A good choice for the prior PDF of the "true" means $\boldsymbol{\mu}$ in this problem will often be the model II supernormal population PDF

$$p(\boldsymbol{\mu} \mid \theta, \sigma_\mu^2) : \mu_i \sim \text{NID}(\theta, \sigma_\mu^2). \quad (2)$$

From this the common $p(\delta)$ for the δ's is $\delta \sim N(0, 2\sigma_\mu^2)$.

The Component *n* Tr Test Problem

In solving the basic one-sided problem (1) the *k*-ratio approach uses the linear loss model

$$l(d_0 \mid \delta, \delta_0) = \begin{cases} 0, & \delta \leqslant \delta_0 \\ k_0(\delta - \delta_0), & \delta > \delta_0 \end{cases},$$

$$l(d_1 \mid \delta, \delta_0) = \begin{cases} k_1(\delta_0 - \delta), & \delta \leqslant \delta_0 \\ 0, & \delta > \delta_0 \end{cases}, \quad (3)$$

where d_0 and d_1 are the decisions for H_0 and H_1.

The Known-Variances Component *k*-Ratio *t* Test

Most of the derivation for any *k*-ratio simultaneous rule lies in solving the basic component one-sided test problem. To illustrate, we now obtain the Bayes test for (1) where $\delta = \mu_1 - \mu_2$. Given are the prior PDF (2), the loss model (3), and the variances σ_μ^2 and σ_e^2.

The two-decision theory form for the Bayes critical region for this problem readily reduces [11] to

$$\frac{\int_{\delta_0}^{\infty}(\delta - \delta_0)h(\delta \,|\, d, \sigma_e^2, \Phi)\,d\delta}{\int_{-\infty}^{\delta_0}(\delta_0 - \delta)h(\delta \,|\, d, \sigma_e^2, \Phi)\,d\delta} > k, \quad (4)$$

where h is the posterior density $\delta \sim N(Sd, S\sigma_d^2)$, $d = x_1 - x_2$, $\Phi = \sigma_T^2/\sigma_e^2$ is the "true" F ratio, $\sigma_T^2 = r\sigma_\mu^2 + \sigma_e^2$ is the expected between-treatments mean square, $\sigma_d^2 = 2\sigma_e^2/r$, $S = 1 - 1/\Phi$ and $k = k_1/k_0$ is the loss, or type-1 to type-2 error-seriousness, ratio. S is termed a "shrinkage" factor because of its effect on the posterior mean $\dot{\delta} = Sd$ and variance $\dot{\sigma}_\delta^2 = S\sigma_d^2$ as Φ gets small. (For recent exposition of (4) and related results such as (7) and (11), see Duncan [14].)

Simplifying, (4) reduces to the right-tail test forms

$$(\dot{\delta} - \delta_0)/\dot{\sigma}_\delta > z(k) \quad \text{or} \quad d/\sigma_d > t(k, \Phi, \delta_0),$$
$$(5)$$

where $z(k)$ is the solution for z of $M(z)/M(-z) = k$ in which $M(z) = f(z) + zF(z)$ and $f(\cdot)$ and $F(\cdot)$ are the PDF and CDF of a $N(0,1)$ variable; and the critical t value is

$$t(k, \Phi, \delta_0) = \delta_0/S\sigma_d + z(k)/S^{1/2}. \quad (6)$$

Thus the Bayes rule $(\dot{\delta} - \delta_0)/\dot{\sigma}_\delta > z(k)$ is the same as the corresponding right-tail α-level rule $(d - \delta_0)/\delta_d > z_\alpha$, except for two changes. One, it is based on the posterior PDF for δ rather than the conditional PDF for d, given δ. Two, the critical z value is determined by the choice of a k ratio rather than a conditional tail probability α.

AN EMPIRICAL BAYES* RULE. The known-variances Bayes rule (5) is of little use in practice, of course, because the variances will seldom be known. A natural remedy, at first, is to use in its place the rule $d/s_d > t(k, F, \delta_0)$ obtained by the direct substitution of $s_d^2 = 2s_e^2/r$ for σ_d^2 and $F = s_T^2/s_e^2$ for Φ, where $s_T^2 = r\sum(x_i - \bar{x})^2/(n-1)$. This *empirical Bayes rule*, as we shall see, is adaptive in the desired way and behaves well if the degrees of freedom $q = (n-1)$ and f are large. In small-sample cases, however, it is necessary to note only that the empirical shrinkage factor is $S = 1 - 1/F$ and that F

(unlike Φ) can fall below one to see that this rule can also behave badly.

The *n* Tr *k*-Ratio *t* Test

The k-ratio approach proceeds by the introduction of higher-level priors or *hyperpriors* for the unknown parameters [18a]. Integration with respect to the parameters gives rules which no longer depend on them, which behave well, and which are still Bayes.

To illustrate, we first consider the same problem as in the section "The Known-Variances Component k-Ratio t Test," except with σ_μ^2, σ_e^2, and thus $\sigma_T^2 = r\sigma_\mu^2 + \sigma_e^2$ unknown. The method in this case is to extend the prior $p(\mu\,|\,\theta,\sigma_\mu^2)$ to $p(\mu\,|\,\theta,\sigma_\mu^2)$ $p(\sigma_T^2)p(\sigma_e^2)p(\theta)$, where the new factors are the truncated Jeffreys indifference priors* $p(\sigma_T^2) \propto 1/\sigma_T^2$, $\sigma_T^2 > \sigma_e^2$; $p(\sigma_e^2) \propto 1/\sigma_e^2$, $\sigma_e^2 > 0$; and $p(\theta) \propto 1$. The Bayes critical region is now

$$\frac{\int_{\tau_0}^{\infty}(\tau - \tau_0)h(\tau\,|\,t, F)\,d\tau}{\int_{-\infty}^{\tau_0}(\tau_0 - \tau)h(\tau\,|\,t, F)\,d\tau} > k \quad (7)$$

where

$$h(\tau\,|\,t, F) = \int_0^{\infty}\int_{\sigma_e^2}^{\infty}h(\delta\,|\,d, \sigma_e^2, \Phi)c(\sigma_T^2\,|\,s_T^2, q)$$
$$\times c(\sigma_e^2\,|\,s_e^2, f)\,d\sigma_T^2\,d\sigma_e^2, \quad (8)$$

$$\tau = \delta/s_d, \quad t = d/s_d, \quad c(\sigma^2\,|\,s^2, m)$$

is the conjugate χ^2 PDF, $\sigma^2 \sim ms^2/\chi_m^2$ and $q = n - 1$.

From (8) we see that the noninformative hyperpriors for σ_T^2 and σ_e^2 have served as catalysts, so to speak, in providing two intuitively acceptable conjugate χ^2 priors for σ_T^2 and σ_e^2 based on all the relevant information in $\mathbf{x}$ and s_e^2 through the statistics s_T^2 and s_e^2.

Simplifying, (7) may be written in the right-tail form

$$d/s_d > t(k, F, q, f, \delta_0) \quad (9)$$

where $t(k, F, q, f, \delta_0)$ is the solution for $t = d/s_d$ of the equality LHS = k and LHS is the left-hand side of (7). To illustrate, Table 1 shows the t values for the common case $\delta_0 = 0$, $k = 100$, and a range of F values. (†, no critical t can exist since $t^2 \leqslant qF$.)

Table 1

q, f					F				
	0.8	1.0	1.4	2.0	3.0	6.0	10	25	∞
99, ∞:	7.75	5.81	3.31	2.45	2.11	1.89	1.82	1.76	1.72
6, ∞:	$\dagger$	$\dagger$	2.72	2.42	2.14	1.89	1.82	1.76	1.72
6, 20:	$\dagger$	$\dagger$	2.80	2.59	2.36	2.09	1.99	1.91	1.86

The row 1 values for $q, f = 99, \infty$ are the same as the corresponding empirical Bayes values $t = 1.72/(1 - 1/F)^{1/2}$, except for 2.43, 3.22, ∞, and ∞ for the latter at $F = 2.0, 1.4, 1.0$, and 0.8.

As we shall indicate in the section "On the Choice of k and the Two-Sample t Test," the choice $k = 100$ is equivalent to the choice $\alpha = 0.05$ in an α-level rule. Thus the row 1 values of (10) are comparable with the $n = 100$, $\alpha = 0.05$ comparisonwise value 1.96, and the experimentwise values 4.30, 4.41, and 11.01, of the Tukey, Bonferroni, and Scheffé rules of the section "Simultaneous Testing: α-Level Approaches." The stronger the evidence of overall homogeneity in the form of a low F ratio, the larger and more cautious is the k-ratio critical t value. At sufficiently small values of F the t value gets very large and can thus be even more conservative than those of competing experimentwise rules. On the other hand, the stronger the evidence of overall heterogeneity in the form of a large F ratio, the smaller and less conservative is the t value. For $F > 4.37$ the rule is more powerful than even the α-level comparisonwise rule.

With smaller values of q and f the "sharpness" of the dependence of t on F is "blunted" as may be expected due to the inaccuracies in estimating Φ.

Simultaneous n Tr k-Ratio t Tests

If we now define a joint rule as consisting of the simultaneous application of the k-ratio t test (9) to any two or more of the $n(n-1)$ ordered differences $d_{ij} = x_i - x_j$ in an n Tr problem with $\delta_0 = 0$, we get the comparisonwise rule which has the same initial intuitive

appeal as an α-level comparisonwise rule but which has none of its intuitive homogeneous outcomes objection. The adaptive dependence on F in the component rule takes care of making the joint rule conservative like an experimentwise rule or powerful like an α-level comparisonwise rule as intuition requires.

This joint rule, first presented in part by Duncan [11, 12] and in full for the case $\delta_0 = 0$ by Waller and Duncan [27–29], and then for δ_0 arbitrary in Dixon and Duncan [6], provides a typical example of the simple comparisonwise nature of all rules given by the k-ratio approach. The joint rule is comparisonwise in the sense that the components are the same no matter how many are used in any specific application. The loss structure under which the rule is Bayes, even when used for contrasts [13, Secs. 3, 4.3], will be presented in the section "Loss Additivity."

The k-Ratio as a Critical Posterior Odds

A loss function simpler than (3) would be the zero–one model obtained by replacing $\delta - \delta_0$ and $\delta_0 - \delta$ by 1. The k-ratio approach uses the linear model, however, not only because it seems more appropriate to weight the errors in proportion to their magnitude, but also because the use of linear losses in testing is equivalent to the popular use of squared error losses in estimation (see the following section). If zero–one losses had been used, note that the critical region (7), for example, would have been

$$\Pr[H_1 | t, F, q, f]/\Pr[H_0 | t, F, q, f] > k.$$

For observed values of t and F the LHS is the unweighted posterior odds in favor of H_1. Similarly, the LHS of (7) may be described as the *linearly weighted posterior odds in favor of* H_1.

In significance testing a common practice is to quote P, the smallest value of α at which an observed d would be significant. The weighted posterior odds can be used as a measure of evidence in a similar way. This new statistic, denoted

$$W(t, F \mid q, f, \delta_0)$$

$$= \frac{\int_{\tau_0}^{\infty} (\tau - \tau_0) h(\tau \mid t, F) \, d\tau}{\int_{-\infty}^{\tau_0} (\tau_0 - \tau) h(\tau \mid t, F) \, d\tau} \quad (11)$$

and termed the *posterior odds*, performs better in the context of exchangeable differences, by picking up additional evidence from *all* differences though the *F* ratio.

The *n* Tr *k*-Ratio *t* Interval

To every test there corresponds an interval estimate. The one for $\delta = \mu_1 - \mu_2$ corresponding to the *n* Tr *k*-ratio *t* test is the lower-bounded interval $[\delta_L, \infty)$, where δ_L is the solution for δ_0 of the critical equation $W(t, F \mid q, f, \delta_0) = k$ as used for (9). As such, the interval represents the set of all δ_0 values such that, for any observed *t* and *F*, the hypothesis $H(\delta_0): \delta \leq \delta_0$ is not rejected by the *k*-ratio *t* test.

OPTIMALITY OF THE COMPONENT INTERVAL. Intervals obtained in this way from Bayes tests like (9) are also Bayes in the sense of minimizing Bayes risk, but, surprisingly at first, with respect to *quadratic* rather than linear losses [5]. Specifically, given the same data and densities as for the test (9), δ_L as defined above is the Bayes solution to the lower-bound estimation problem with loss function

$$l(\delta_L, \delta) = \begin{cases} k_1 (\delta_L - \delta)^2, & \delta < \delta_L \\ k_0 (\delta - \delta_L)^2, & \delta \geq \delta_L \end{cases},$$

where k_1 and k_0 are the type 1 and type 2 error coefficients of the testing linear loss function (3). This result readily follows by writing the posterior risk as

$$B(\delta_L) = \int_{-\infty}^{\tau_L} k_1 (\tau_L - \tau)^2 h(\tau \mid t, F) \, d\tau$$

$$+ \int_{\tau_L}^{\infty} k_0 (\tau - \tau_L)^2 h(\tau \mid t, F) \, d\tau, \quad (12)$$

where $\tau_L = \delta_L / s_d$, then minimizing by differentiating with respect to δ_L and equating to zero.

Simultaneous *n* Tr *t* Intervals

If we now define a joint rule as consisting of simultaneously getting the *k*-ratio lower-bounded intervals for any two or more of the set of ordered contrasts in an *n* Tr problem, we have a Bayesian comparisonwise interval estimation rule with intuitively appealing features which correspond exactly to those of the *k*-ratio tests in the section "Simultaneous *n* Tr *k*-Ratio *t* Tests."

A pair of lower-bounded intervals (e.g., for $\delta = \delta_{12}$ and $-\delta = \delta_{21}$) overlap to form a two-sided interval $[\delta_L, \delta_U]$ matching the corresponding two-sided test. Thus the set of $n(n-1)/2$ intervals for the Tukey [26] all-pairwise-differences interval-estimation problem, in the large-sample case for example, have end points $\hat{\delta} \pm \dot{\sigma}_\delta z(k)$, where $\hat{\delta} = S d_{ij}$, $\dot{\sigma}_\delta^2 = S s_d^2$, and $S = 1 - 1/F$, as may be anticipated from the critical region (5). (Dixon and Duncan [6] derived these intervals; Godbold and Duncan [18] developed them further for small-sample application.) We now turn to the loss structure under which these intervals and the corresponding tests are jointly Bayes.

Loss Additivity

The merits of a Bayesian comparison-wise approach can be supported at a more basic level through an *additive losses result* which may be put roughly as follows. Given any joint problem formed by the simultaneous occurrence of several component problems, and given that the loss for each of the joint decisions consists of the *sum* of the losses for its component decisions, then the Bayes rule for the joint problem is provided by the simultaneous application of the Bayes rules for the component problems. More briefly, given that the losses are additive, a joint rule is Bayes provided that the component rules are Bayes. The assumption of additive losses means, for example, that if one joint decision *D* is wrong about twice as many differences as another decision *D'*, then the mistake made by *D* is twice as serious, other things being equal, as the mistake made by D'.

The k-ratio approach is based on the point of view that *additivity of the losses* and *exchangeability** *of the priors* (cf. the section "The *n*-Treatments Testing Problem") are the two main distinguishing features of multiple comparisons* problems.

On the Choice of k and the Two-Sample t Test

Of special interest is the k-ratio two-sample two-tailed t test for the case $q = n - 1 = 1$. In this case $F = t^2$, so the critical value $t(k, F, q, f) = t(k, f)$ varies with only k and f. A surprising coincidence is that at $k = 100$, $t(100, \infty) = 1.96 = t_{\alpha, \infty}$ at $\alpha = 0.05$. The equivalence $t(100, f) \approx t_{0.05, f}$ persists with discrepancies seldom above 0.01 for all values of f. It is for this reason and the near but less close approximations $t(50, f) \approx t_{0.10, f}$ and $t(500, f) \approx t_{0.01, f}$ for all f, that the choices of $k = 50$, 100, and 500 in a k-ratio rule are said to be the equivalents of the choices of $\alpha = 0.10$, 0.05 and 0.01 in a two-sample two-tailed α-level rule.

Thus in any test with $n > 2$, the use of $k = 100$, for example, in calling a difference significant, is requiring the same strength of evidence based on all the data in terms of the posterior odds $W(t, F | q, f, 0)$ as would be required in a two-sample two-tailed t test with $\alpha = 0.05$. Also in getting any interval estimate with $n = 2$ or $n > 2$, the use of $k = 100$, in putting a parameter value outside an interval, is calling for the same strength of evidence in terms of $W(t, F | q, f, \delta_0)$ as would be required in rejecting a parameter value δ_0 in a two-sample two-tailed t test with $\alpha = 0.05$.

Linear Models for Prior Densities

Different problems require different priors leading to different patterns of adaptivity. A simple example consists of the changes needed in going from the *n* Tr problem to the *n*-treatments-versus-control (*n* TvC) problem.

THE *n*TVC PROBLEM. Given a set of means $\mathbf{x} = [x_0, x_1, \ldots, x_n]$ for n treatments and a control (x_0) such that $X_i \sim \mathrm{NID}(\mu_i, \sigma_e^2/r)$, a useful prior density may now be $\mu_0 \sim \mathrm{NID}(\theta_0, \sigma_\mu^2)$ and $\mu_i \sim \mathrm{NID}(\theta_1, \sigma_\mu^2)$, $i = 1, \ldots, n$, where θ_0 and θ_1 come in turn from a higher superpopulation, $\theta_i \sim \mathrm{NID}(\phi, \sigma_\theta^2)$. Using the same linear loss function (3) in testing $\delta \leqslant \delta_0$ vs. $\delta > \delta_0$, where δ is one of the n TvC differences, say $\delta = \mu_1 - \mu_0$, the known-variances Bayes rule is now

$$(\hat{\delta} - \delta_0)/\hat{\sigma}_\delta = (S_0 d_0 + S_1 d_1 - \delta_0)$$
$$/\left(S_0 \sigma_{d_0}^2 + S_1 \sigma_{d_1}^2\right)^{1/2} > z(k), \quad (13)$$

where d_0 and d_1 are the two independent components $d_0 = \bar{x} - x_0$ and $d_1 = x_1 - \bar{x}$ of $d = x_1 - x_0$, S_i is the shrinkage factor $S_i = 1 - 1/\Phi_i$, $i = 0$, 1, Φ_0 and Φ_1 are the true between-groups (TvC) and between-treatments within-group F ratios, and $z(k)$ is the k-ratio critical z value in (5).

The effect of using the critical region (13) for testing $\delta = \mu_1 - \mu_0$ is striking. It can be seen that if the between-treatments F ratio Φ_1 is small, say $\Phi_1 = 1$, then $S_1 = 0$, and the test decision rests completely on the between-groups difference $\bar{x} - x_0$. Individual treatments means x_i can vary about the overall mean $\bar{x}$, but if the collective evidence indicates that their variability is all noise, the decision for each treatment-versus-control difference is determined by $\bar{x} - x_0$ alone. However, as Φ_1 increases toward Φ_0, S_1 increases toward S_0, and the decision rests more and more on the direct difference $x_1 - x_0$. This behavior is in marked contrast to those of the approximate $t_{\alpha/m}$ and exact [15] experimentwise rules for the n TvC problem. Results and tables for k-ratio t tests for this problem are given in Brant and Duncan [4].

Point Estimates

Corresponding to every k-ratio lower bound δ_L we can get a point estimate for δ as the value of δ_L for $k_1 = k_0$ or $k = 1$. From (12) it is clear that this is the Bayes estimate with respect to squared-error loss: the posterior mean. In the known-variances case the estimates are, for example, $\dot{\delta} = Sd$ and $\dot{\delta} = S_0 d_0 + S_1 d_1$, as given for (4) and by (13) for the n Tr and n TvC problems, respectively. For small-sample cases, the point estimate for the n Tr problem is $\dot{\delta} = Sd$ [18], where $S = 1 - R$ and

$$R = \frac{f}{(f-2)F} \left\{ \sum_{r=m}^{n} \binom{n}{r} f^{n-r}(qF)^r \right\}$$

$$\times \left\{ \sum_{r=m+1}^{n} \binom{n}{r} f^{n-r}(qF)^r \right\}^{-1}$$

in which $m = q/2$ and $n = (q + f - 2)/2$, the result being true for even values of q and f. To illustrate, shrinkages obtained for various F values with $q = 2$ and $f = 20$ are as follows:

F:	0	0.1	0.4	1.2	2.0
S:	0.50	0.57	0.54	0.60	0.66
F:	4.0	6.0	10	25	∞
S	0.76	0.82	0.89	0.96	1.0

Concluding Remarks

1. In principle, in the experimentwise approach, the fewer the comparisons to be tested, the more powerful are the tests. Thus to take an extreme example, if the *prior* plan is to test only one contrast $\gamma = \sum c_i \gamma_i$ in an n Tr problem, then, with $\alpha = 0.05$, $n = 100$, and $f = \infty$, the observed contrast $c = \sum c_i x_i$ need exceed only $1.96\sigma_c$ to be significant. If no such plan is made a priori, however, c must exceed $11.10\sigma_c$ to be significant. In contrast, in the k-ratio approach, the corresponding critical value $t(100, F, 99, \infty)$ depends in no way on the number of contrasts being tested. These t values [see (10)] depend only on the overall evidence of heterogeneity through the F ratio and are the same for testing any one or more members from the set of all pairwise differences or the set of all contrasts [13].

2. A need for brevity has not allowed us to emphasize the changes in k-ratio intervals relative to the corresponding α-level intervals, which are even more striking. By and large the new intervals are uni-

formly shorter and considerably more powerful in excluding wrong values of the true contrasts (see Dixon and Duncan [6]).

3. As suggested by the changes required in going from the n Tr to the n TvC problem, the choice of a prior density model is a most important aspect of the k-ratio approach to any new problem, and the variety of such prior models will match that of the data density models. The potential for valuable progress in the use of this approach is considerable and much work remains to be done.

Notes on References

Hitherto, the development of the k-ratio approach has been mainly in terms of the nTr testing problem: The decision-theoretic aspects, including the importance of additive losses, are treated in Duncan [10–12] and Bland and Duncan [1] with origins extending back to Duncan [7–9]. The more Bayesian aspects are begun in the same references starting in 1961 [1, 11, 12] and the full n Tr test development is completed in Waller and Duncan [27–29]. The corresponding minimum Bayes-risk k-ratio n Tr intervals are developed in Dixon and Duncan [6] and Dixon [5]. For a review and a set of course notes see Duncan [13, 14].

References followed by the letter M or T are more methodological or theoretical, respectively.

References

[1] Bland, R. P. and Duncan, D. B. (1964). On a Bayes Rule for Choosing the Largest Mean. *Paper No. 366*, Biostatistics Dept., Johns Hopkins University, Baltimore. (T)

[2] Box, G. E. P. and Tiao, G. C. (1973). *Bayesian Inference in Statistical Analysis*. Addison-Wesley, Reading, Mass.

[3] Brant, L. J. (1978). k-Ratio t Tests: n Treatments versus a Control. Ph.D. thesis, Biostatistics Dept., Johns Hopkins University.

[4] Brant, L. J. and Duncan, D. B. (1982). k-ratio t tests for comparing n treatments with a control. (To be submitted for publication.) (Modification of Brant [3].)

[5] Dixon, D. O. (1976). *J. Amer. Statist. Ass.*, **71**, 406–408. (T)

[6] Dixon, D. O. and Duncan, D. B. (1975). *J. Amer. Statist. Ass.*, **70**, 822–831. (M, T)

[7] Duncan, D. B. (1947). Significance Tests for Differences between Ranked Variates Drawn from Normal Populations. Ph.D. thesis, Iowa State College. (M, T)

[8] Duncan, D. B. (1951). *Va. J. Sci.*, **2**, 171–189. (M)

[9] Duncan, D. B. (1952). *Va. J. Sci.*, **3**, 49–67. (T)

[10] Duncan, D. B. (1955). *Biometrics*, **11**, 1–42. (M)

[11] Duncan, D. B. (1961). *Ann. Math. Statist.*, **32**, 1013–1033. (T)

[12] Duncan, D. B. (1965). *Technometrics*, **7**, 171–222. (M, T)

[13] Duncan, D. B. (1975). *Biometrics*, **31**, 339–359. (M, T)

[14] Duncan, D. B. (1982). Simultaneous Inference for Linear Models. Class notes for Biostatistics 35, Johns Hopkins University, Baltimore. (M, T)

[15] Dunnett, C. W. (1955). *J. Amer. Statist. Ass.*, **50**, 1096–1121.

[15a] Dunnett, C. W. (1964). *Biometrics*, **20**, 482–491.

[16] Efron, B. and Morris, C. (1973). *J. Amer. Statist. Ass.*, **68**, 117–130.

[17] Godbold, J. H. (1976). Small Sample *t* Intervals for Comparisons Suggested by the Data. Ph.D. thesis, Biostatistics Dept., Johns Hopkins University.

[18] Godbold, J. H. and Duncan, D. B. (1982). Small sample *k*-ratio *t* intervals and point estimates. (To be submitted for publication.) (Modification of Godbold [17].)

[18a] Good, I. J. (1981). The robustness of a hierarchical model for multinomials and contingency tables. *Proceedings of a Conference on Scientific Inference, Data Analysis, and Robustness*. Madison. Academic Press, New York.

[19] James, W. and Stein, C. (1961). *Proc. 4th Berkeley Symp. Math. Statist. Prob.*, Vol. 1. University of California Press, Berkeley, Calif., pp. 361–379.

[20] Krishnaiah, P. R. (1979). In *Developments in Statistics*, Vol. 1, P. R. Krishnaiah, ed. Academic Press, New York, pp. 157–201.

[21] Lehmann, E. L. (1950). *Ann. Math. Statist.*, **21**, 1–26.

[22] Lindley, D. V. (1971). In *Foundations of Statistical Inference*, V. P. Godambe and D. A. Sprott, eds. Holt, Rinehart and Winston, Toronto, pp. 435–455.

[22a] Lindley, D. V. (1972). *Bayesian Statistics, a Review*. SIAM, Philadelphia.

[23] Lindley, D. V. and Smith, A. F. M. (1972). *J. R. Statist. Soc. B*, **34**, 1–41.

[24] Miller, R. G. (1966). *Simultaneous Statistical Inference*. McGraw-Hill, New York.

[25] Scheffé, H. (1953). *Biometrika*, **40**, 87–104.

[26] Tukey, J. W. (1951). *Proc. 5th Annu. Conv., Amer. Soc. Quality Control*, pp. 189–197.

[27] Waller, R. A. and Duncan, D. B. (1969). *J. Amer. Statist. Ass.*, **64**, 1484–1503. (M, T)

[28] Waller, R. A. and Duncan, D. B. (1972). *J. Amer. Statist. Ass.*, **67**, 253–255. (M)

[29] Waller, R. A. and Duncan, D. B. (1974). *Ann. Inst. Statist. Math. Tokyo*, **26**, 247–264. (T)

(ANALYSIS OF VARIANCE
BAYESIAN INFERENCE
DECISION THEORY
EMPIRICAL BAYES
EXCHANGEABILITY
JAMES–STEIN ESTIMATOR
MULTIPLE COMPARISONS
SHRINKAGE
SIMULTANEOUS INFERENCE)

DAVID B. DUNCAN
DENNIS O. DIXON

KRAWTCHOUK POLYNOMIALS

These are orthogonal polynomials of the binomial distribution*, related to that distribution similarly as the Charlier polynomials are to the Poisson distribution*. Writing

$$b(x; n, p) = \binom{n}{x} p^x (1-p)^{n-x} \qquad (x = 0, 1, \ldots, n),$$

the *r*th Krawtchouk polynomial is

$$K_r(x; n, p) = \{ p(1-p) \}^r (d^r b / dp^r) b.$$

The orthogonality property is

$$\sum_{x=0}^{n} b(x; n, p) K_r(x; n, p) K_s(x; n, p) = 0$$

if $r \neq s$. Clearly,

$$K_r(x; n, p) = 0 \qquad \text{if} \quad r > n.$$

An explicit formula for the polynomials is

$$K_r(x; n, p) = \sum_{j=0}^{r} (-1)^j \binom{r}{j}$$
$$\times (n - r + 1)^{[j]} p^j x^{(r-j)}$$

[where $a^{[b]} = a(a+1) \cdots (a+b-1)$; $a^{(b)} = a(a-1) \cdots (a-b+1)$], or, equivalently,

$$K_r(x; n, p) = (1 + p\Delta)^{-n+r-1} x^{(r)}.$$

Gonin [1] describes the use of these polynomials (and of polynomials similarly related to negative binomial distributions*) in fitting count data.

For additional information, see Krawtchouk [2].

References

[1] Gonin, H. T. (1961). *Biometrika*, **48**, 115–123.

[2] Krawtchouk, M. (1929). *C. R. Acad. Sci. Paris*, **189**, 620–622.

(APPROXIMATIONS TO DISTRIBUTIONS
ORTHOGONAL POLYNOMIALS)

KRIGING

Kriging is a method of interpolation* for random spatial processes, named after its originator, D. G. Krige. The ideas were developed in the 1950s and 1960s (see Krige [5]) as a solution to the important practical problem of how to estimate the total ore content of a panel (volume of earth and rock that might be mined if sufficiently rich). Subsequently, Matheron [6] and his co-workers developed the theory of regionalized variables to provide a formal framework for inference. For a recent bibliography, see Whitten [10].

Suppose that we wish to estimate the value of Z_V, the average ore content throughout the panel V so that

$$Z_V = \frac{1}{V} \int_V Z(x)\, dx,$$

where $Z(x)$ is the random function describing the ore content at the location with coordinate vector x, and V is used to denote both the panel and its "volume" (in one or more dimensions). It is convenient to retain the mining terminology in describing the method, although it is useful for the analysis of spatial processes in many areas, including geography*, geology*, and meteorology* (see Cliff and Ord [1] and Delhomme [3]).

The data available for estimating Z_V are n drillings or observations, $Z(x_i)$ or simply Z_i, at experimental locations x_i, $i = 1, \ldots, n$. When the mean is constant (μ say), for all locations and the covariance structure is fully specified, the minimum variance linear unbiased estimator for Z_V is

$$\hat{Z}_V = \sum_{i=1}^{n} \lambda_i Z_i,$$

where the $\{\lambda_i\}$ denote a set of weights such that $\sum \lambda_i = 1$. These weights depend on the spatial pattern of the process through the covariances.

When interest focuses on a single point rather than a volume, the estimator is of the same form, although the structure of the λ_i simplifies. The method is then known as punctual (or pointwise) Kriging and provides an exact interpolator at the sample points so that $\hat{Z}_i = Z_i$. Punctual Kriging corresponds to the use of generalized least squares* for this problem as noted, for example, by Journel [4].

A simple example will serve to illustrate these ideas. Suppose that we record the results of two drillings, Z_0 and Z_1, at $x = 0, 1$ on a one-dimensional "seam of ore." If the autocorrelation between Z for two points a distance d apart is γ^d, $0 < \gamma < 1$, it follows that the point estimator at any location x, given Z_0 and Z_1, is of the form

$$\hat{Z}(x) = \lambda_0 Z_0 + \lambda_1 Z_1,$$

where $\lambda_0 = \lambda_0(x)$ and $\lambda_1 = \lambda_1(x)$ depend on x. For example, we obtain the following values:

Location x:	0.5	0.8	1.0	1.5	∞
$\lambda_1(x)$ {when $\gamma = 0.5$	0.50	0.80	1.0	0.85	0.50
{when $\gamma = 0.1$	0.50	0.76	1.0	0.66	0.50

In this case, $\lambda_1(x) = \lambda_0(1 - x)$. When a panel of ore (interval on the line) is to be estimated, $\hat{Z}_V$ has weights $\lambda_i(v)$ such as

Volume v:	[0, 1]	[0, 2]	[1, 2]
$\lambda_1(v)$ {when $\gamma = 0.5$	0.50	0.68	0.86
{when $\gamma = 0.1$	0.50	0.60	0.70

The original problem which led to the development of the technique concerned the (usually upward) bias that resulted when Z_V was estimated using only those drillings that fell within V. From the example, it may be seen that $\hat{Z}_V$ gives appropriate weights to observations outside V, but it does require detailed specification and estimation of the covariance structure.

ESTIMATION

We may estimate the covariances by a grouping method such as

$$c_r = \frac{1}{n_r} \sum \left[Z(x_i) - \hat{\mu} \right]\left[Z(x_j) - \hat{\mu} \right],$$

where $\hat{\mu}$ is the generalized least-squares estimator for μ and the summation is taken over all n_r pairs of locations i and j in distance class r. That is, i and j are a distance d_{ij} apart such that $D_{r-1} < d_{ij} \leqslant D_r$, $D_0 = 0$, and $c_0 = (1/N)\sum[Z(x_i) - \hat{\mu}]^2$. Directional differences (or *anistropy*) can be accommodated by grouping with regard to both direction and distance.

More generally, the spatial process $Z(x)$ will be nonstationary (*see* STATIONARY PROCESSES), and the method must be extended to consider the differences $Z(x) - Z(y)$. The covariation is now described in terms of the *variogram*

$$\gamma(h) = E\left\{ \left[Z(x+h) - Z(x) \right]^2 \right\},$$

which exists even when the process is nonstationary. If $\gamma(0+) > 0$, a *nugget* effect (or local random noise) is said to exist.

Trend terms may be incorporated to describe the nonstationarity and the optimal interpolator (*universal Kriging*) is derived by the same approach as before (see Matheron [6, Chap. 4]).

OTHER APPROACHES

Thus far we have assumed that a model linear in Z is appropriate. However, ore samples and similar data often display considerable skewness; the lognormal distribution* describes the data better, and this choice is well supported theoretically. A model linear in $\ln Z$ would yield biased estimators for the overall mean μ; however, the bias can be removed (see Rendu [7]).

A further approach which does not require explicit distributional assumptions uses Hermite polynomials* and is known as disjunctive Kriging (see Rendu [7, 8]).

Other methods for spatial interpolation include the use of polynomial regression, known in this context as trend surface analysis* (see Watson [9]), and a nonstochastic technique based on Dirichlet cells. Increasingly, practice would seem to favor Kriging, and the computational barriers preventing its application for irregularly spaced data seem to have been overcome (see David, [2]).

References

[1] Cliff, A. D. and Ord, J. K. (1975). *J. R. Statist. Soc. B*, **37**, 297–348. (A general review of spatial models and their application in geography.)

[2] David, M. (1977). *Geostatistical Ore Reserve Estimation*. Elsevier, Amsterdam. (A general review of methodology which includes some computer programs.)

[3] Delhomme, J. P. (1975). In *Display and Analysis of Spatial Data*, J. C. Davis and M. J. McCullagh, eds. Wiley, New York, pp. 96–114. (A useful description of Kriging and its application.)

[4] Journel, A. G. (1977). *J. Int. Ass. Math. Geol.*, **9**, 563–586. (A geometric description of Kriging, linking the method to generalized least squares. Also describes nonlinear methods.)

[5] Krige, D. G. (1962). *J. Inst. Mine Surveyors S. Africa*, **12**, 45–84, 95–136.

[6] Matheron, G. (1971). *The Theory of Regionalized Variables*. Centre de Morphologie Mathématique de Fontainebleau, Paris. (Contains a full account of the general theory, then develops Kriging estimators and their variances.)

[7] Rendu, J.-M. (1979). *J. Int. Ass. Math. Geol.*, **11**, 407–422. (Develops minimum variance unbiased estimators for the mean using the log-normal distribution rather than the normal. Also gives a full discussion of disjunctive Kriging.)

[8] Rendu, J.-M. (1980). *J. Int. Ass. Math. Geol.*, **12**, 305–320. (A continuation of ref. 7.)

[9]   Watson, G. S. (1971). *J. Int. Ass. Math. Geol.*, **3**, 215–226. (An excellent critique of trend surface analysis.)

[10] Whitten, E. H. T. (1981). In *Future Trends in Geomathematics*, R. G. Craig and M. L. Labovitz, eds. Pion, London/Methuen, New York, pp. 48–61. (Includes a bibliography of recent papers; several other papers in this volume relate to this subject area.)

(GEOGRAPHY, STATISTICS IN
GEOSTATISTICS
SPATIAL PROCESSES
STATIONARY PROCESSES
TREND SURFACE ANALYSIS)

J. K. ORD

KRONECKER LEMMA

The Kronecker lemma states:

Let $\{a_n\}$ be a sequence of real numbers such that $\sum a_n$ converges. Then $n^{-1}\sum_{k=1}^{n}ka_k \to 0$ as $n \to \infty$.

It is closely related to the Toeplitz lemma and is useful in proving Kolmogorov's law of large numbers and other limit theorems.

(LAWS OF LARGE NUMBERS
TOEPLITZ LEMMA)

KRONECKER PRODUCT OF DESIGNS

Vartak [2] defines a Kronecker product of designs as follows. If $\mathbf{N}_1$ is the incidence matrix* of a design D_1 and $\mathbf{N}_2$ is the incidence matrix of another design D_2, then the design with the incidence matrix $\mathbf{N}_1 \times \mathbf{N}_2$ where $\times$ denotes the Kronecker product of matrices* is called the Kronecker product of designs D_1 and D_2.

Many details are given in Raghavarao [1]. See also Vartak [3, 4].

References

[1]   Raghavarao, D. (1971). *Constructions and Combinatorial Problems in Design of Experiments*. Wiley, New York.

[2]   Vartak, M. N. (1955). *Ann. Math. Statist.*, **26**, 420–438.

[3]   Vartak, M. N. (1960). *Ann. Math. Statist.*, **321**, 722–778.

[4]   Vartak, M. N. (1963). *J. Indian Statist. Ass.*, **1**, 215–218.

(DESIGN OF EXPERIMENTS
INCOMPLETE BLOCK DESIGNS
KRONECKER PRODUCT OF MATRICES)

KRONECKER PRODUCT OF MATRICES

Let A be a (real- or complex-valued) $m \times n$ matrix (a_{ik}) and B be an $r \times s$ matrix (b_{ji}). The Kronecker product of A and B—usually denoted by $A \otimes B$—is defined as the $mr \times ns$ matrix $C = (c_{\lambda,\mu})$, where $c_{\lambda,\mu} = a_{ik}b_{jl}$, $\lambda = (i, j)$ and $\mu = (k, l)$. It can be expressed as

$$\begin{bmatrix} a_{11}B & a_{12}B & \cdots & a_{1n}B \\ a_{21}B & a_{22}B & \cdots & a_{2n}B \\ \vdots & & & \\ a_{m1}B & a_{m2}B & \cdots & a_{mn}B \end{bmatrix}$$

or

$$\begin{bmatrix} b_{11}A & b_{12}A & \cdots & b_{1s}A \\ b_{21}A & b_{22}A & \cdots & b_{2s}A \\ \vdots & & & \\ b_{r1}A & b_{r2}A & \cdots & b_{rs}A \end{bmatrix}.$$

One of the main properties of the Kronecker product is

$$(A_1 \otimes B_1)(A_2 \otimes B_2) = (A_1A_2) \otimes (B_1B_2),$$

where A_1A_2 and B_1B_2 denote the "usual" products of matrices provided that they are defined.

In statistical theory Kronecker products are used in multivariate analysis and in multivariate distribution theory. For example, the variance–covariance* matrix t-distribution* is given by a Kronecker product of two matrices. The Kronecker product of matrices is also used in experimental design for determination of the existence of certain classes of designs.

(INCIDENCE MATRIX
LINEAR ALGEBRA IN STATISTICS
MATRIX t-DISTRIBUTION
MULTIVARIATE ANALYSIS)

KRONMAL–HARTER ESTIMATES *See*

DENSITY FUNCTION ESTIMATION

KRUSKAL–SHEPARD METHOD *See*

MULTIDIMENSIONAL SCALING

KRUSKAL–WALLIS TEST

It is often of interest to determine if $c(c > 2)$ populations are identical. A nonparametric test based on ranks was suggested by Kruskal and Wallis [16] in addressing this problem. Let the c populations have continuous distribution functions $F_1, F_2, \ldots, F_c$. The null hypothesis for the Kruskal–Wallis test is $H_0: F_i = F_2 = \cdots = F_c$. The alternative hypothesis is generally $H_1: F_i(x) = F(x + \theta_i)$, where F is some continuous distribution function and not all θ_i equal zero. This test may be considered a c-sample extension of the Wilcoxon [26] Mann–Whitney [18] two-sample test for equality of location.

PROCEDURE

Let there be c independent random samples $X_{11}, X_{12}, \ldots, X_{1n_1}; \ldots; X_{c1}, X_{c2}, \ldots, X_{cn_c}$ of sizes $n_1, n_2, \ldots, n_c$ from c populations. Form the combined sample of $N = n_1 + n_2 + \cdots + n_c$ observations, order them, and assign them the ranks $1, 2, \ldots, N$. Let R_{ij} denote the rank of X_{ij} $i = 1, \ldots, c$; $j = 1, \ldots, n_i$. Let $R_i = \sum_{j=1}^{n_i} R_{ij}$ and $\bar{R}_i = R_i/n_i$ denote the sum and mean, respectively, of the ranks in sample i. Observe that the sum of all N ranks is $N(N + 1)/2$, and that under H_0 the expected value of each rank as well as the expected value of the mean rank for each sample is $(N + 1)/2$.

Consider the statistic

$$H = \frac{12}{N(N + 1)} \sum_{i=1}^{c} n_i \left(\bar{R}_i - \frac{N + 1}{2} \right)^2$$

$$= \frac{12}{N(N + 1)} \sum_{i=1}^{c} \frac{R_i^2}{n_i} - 3(N + 1),$$

the second expression being a commonly used computing formula. For fixed sample sizes $n_1, n_2, \ldots, n_c$, the total number of ways of assigning the N ranks to the c samples is $M = N!/\Pi_{i=1}^{c} n_i!$. Under H_0 all outcomes are equally likely with the probability of each outcome being $1/M$. The Kruskal–Wallis test of size α may be defined as: reject H_0 if the number of outcomes resulting in H values greater than the "observed" is less than or equal to m where $m/M = \alpha$.

Extensive tables for the probability levels of H are given in Iman et al. [13]. Tables of critical values of H, generally for $c \leqslant 3$, may be found in Conover [6], Hollander and Wolfe [10], Mosteller and Rourke [20], and Lehmann [17]. When N is large and $n_i/N \to \lambda_i$ where $\lambda_i \neq 0$ Kruskal and Wallis [16] showed that H tends to a chi square distribution with $c - 1$ degrees of freedom. Alternatively the statistic

$$F^* = \frac{\left[\sum_{i=1}^{c} n_i \left(\bar{R}_i - \frac{N + 1}{2} \right)^2 \Big/ (c - 1) \right]}{\left[\sum_{i=1}^{c} \sum_{j=1}^{n_i} \left(R_{ij} - \bar{R}_i \right)^2 \Big/ (N - c) \right]},$$

the usual analysis of variance* F statistic using ranks instead of the original observations X_{ij}, may be computed and its distribution approximated by F with $(c - 1, N - c)$ degrees of freedom. It may be shown that $F^* = (N - c)H/(c - 1)(N - 1 - H)$. Other approximations may be found in Wallace [25] and Iman and Davenport [12].

TIES

Although the underlying distributions F_i are assumed to be continuous, in practice, ties are frequently present. For the Kruskal–

Wallis test the usual procedure of resolving ties is to compute H using the mean rank and replace the test statistic by H',

$$H' = H \bigg/ \left[1 - \sum_{i=1}^{k} (T_i^3 - T_i)/(N^3 - N) \right],$$

where k is the number of tied groups and T_i is the number of observations in tied group i.

PROPERTIES

The Kruskal–Wallis test is consistent. Its asymptotic relative efficiency (ARE) relative to the usual parametric F test is 0.955 when the underlying distribution is normal [1] and is never less than 0.864 [9]. It is greater than 1 for logistic*, exponential*, double exponential*, and other distributions.

Nonparametric competitors to the Kruskal–Wallis test include a c-sample median test due to Brown and Mood [5] (*see* BROWN–MOOD MEDIAN TEST), tests where a generalized scoring function such as normal scores* are used [8], and tests based on "k-plets" suggested by Bhapkar [3] and Bhapkar and Deshpande [4]. For moderate sample sizes simulation studies by Keselman et al. [15] have indicated results generally favorable to the Kruskal–Wallis test when compared with the normal scores test and the classical F test with respect to their sensitivity to heterogeneity of variances.

ORDERED ALTERNATIVES AND MULTIPLE COMPARISONS

If the alternative hypothesis is $F_i(x) = F(x + \theta_i)$, $\theta_1 \leqslant \theta_2 \leqslant \cdots \leqslant \theta_c$, where at least one of the inequalities is strict, the Terpstra–Jonckheere [14, 24] test based on the test statistic J may be used.

$$J = \sum_{i=1}^{c-1} \sum_{i'=1}^{c} U_{ii'},$$

where $U_{ii'}$ is the Mann–Whitney statistic for samples i and i', that is, $U_{ii'} =$

$\sum_{j=1}^{n_i} \sum_{j'=1}^{n_{i'}} (X_{ij}, X_{ij'})$, where $(a, b) = 1$ if $a < b$, $(a, b) = 0$ otherwise. The Terpstra–Jonckheere test is equivalent to a test based on Kendall's correlation coefficient τ where the N observations constitute one sample and the c categories constitute the other sample with all observations tied in each category [2]. For N large the J statistic is approximately normally distributed with mean

$$\left(N^2 - \sum_{i=1}^{c} n_i^2 \right) / 4$$

and variance

$$\left(N^2(2N + 3) - \sum_{i=1}^{c} n_i^2(2n_i + 3) \right) / 72.$$

Tables for the exact probability levels of J may be found in Odeh [21], Hollander and Wolfe [10], and Lehmann [17]. Puri [22] showed the ARE of the J test relative to its corresponding normal theory test are the same as those for the Kruskal–Wallis test. (*See also* JONCKHEERE TEST FOR ORDERED ALTERNATIVES.) Detailed discussions of multiple comparison procedures based on Kruskal–Wallis rankings may be found in Miller [19], Hollander and Wolfe [10], and Gibbons [7].

MULTIVARIATE c-SAMPLE LOCATION TEST

Puri and Sen [23] proposed a general form for a distribution-free* rank permutation test* which, when ranks are used as the scores, may be considered a p-variate Kruskal–Wallis test. The test statistic is

$$L = \sum_{i=1}^{c} n_i \big((T_i - E) V^{-1}(R)(T_i - E)' \big),$$

where R is a $p \times N$ matrix with row k representing a Kruskal–Wallis ranking of variate k, $k = 1, 2, \ldots, p$; T_i and E are $1 \times p$ vectors of mean ranks of sample i and expected mean ranks, respectively, and $V^{-1}(R)$ denotes the inverse of the conditional variance covariance matrix of R. The authors showed the statistic L follows asymptotically a chi-square distribution with $p(c - 1)$ degrees of freedom. A numerical example illustrating the procedure for carrying out the above multivariate test is given in Horrel and Lessig [11].

Example 1.   Let three independent random samples of sizes 3, 4, and 5 drawn from three populations result in the measurements 5.23, 4.41, 4.50; 5.54, 5.17, 5.87, 4.80; 5.15, 4.17, 4.21, 3.93; and 4.45. From the ordered combined sample the three rank sums are found to be 20, 39, and 19. Using the computing formula for H,

$$H = \frac{12}{(12)(13)} \left(\frac{20^2}{3} + \frac{39^2}{4} + \frac{19^2}{5} \right) - 3(13)$$

$$= 6.06.$$

From the Kruskal–Wallis table, $\Pr(H \geqslant 6.06) = 0.037$. The test results in rejecting H_0 for any error rate $\alpha \geqslant 0.037$.

References

[1]   Andrews, F. C. (1954). *Ann. Math. Statist.* **25**, 724–736.

[2]   Barlow, R. E., Bartholomew, D. J., Bremmer, J. M., and Brunck, H. D. (1972). *Statistical Inference Under Order Restrictions.* Wiley, New York. (Intermediate level.)

[3]   Bhapkar, V. P. (1961). *Ann. Math. Statist.*, **32**, 1108–1117.

[4]   Bhapkar, V. P. and Deshpande, J. V. (1968). *Technometrics*, **10**, 578–585.

[5]   Brown, G. W. and Mood, A. M. (1951). *Proc. 2nd Berkeley Symp. Math. Statist. Prob.* University of California Press, Berkeley, Calif., pp. 159–166.

[6]   Conover, W. J. (1971). *Practical Nonparametric Statistics.* Wiley, New York, (Good intermediate-level text.)

[7]   Gibbons, J. D. (1976). *Nonparametric Methods for Quantitative Analysis.* Holt, Rinehart and Winston, New York. (Elementary text.)

[8]   Hájek, J. and Šidák, Z. (1967). *Theory of Rank Tests.* Academic Press, New York. (Advanced text; theoretical.)

[9]   Hodges, J. L., Jr. and Lehmann, E. L. (1956). *Ann. Math. Statist.*, **27**, 324–335.

[10] Hollander, M. and Wolfe, D. A. (1973). *Nonparametric Statistical Methods.* Wiley, New York. (Excellent reference with detailed comments, many tables, and exhaustive bibliography.)

[11] Horrell, J. F. and Lessig, V. P. (1975). *Decis. Sci.*, **6**, 135–141.

[12] Iman, R. L. and Davenport, J. M. (1976). *Commun. Statist.*, **5**, 1335–1348.

[13] Iman, R. L., Quade, D., and Alexander, D. A. (1975). In *Selected Tables in Mathematical Statistics*, Vol. 3, H. L. Harter and D. B. Owen, eds., Markham, Chicago, pp. 329–384. [Extensive tables, $n_i = n_2 = n_3 \leqslant 8$; $c = 3, \max(n_i) \leqslant 6$; $c = 4$, $\max(n_i) \leqslant 4$; $c = 5, \max(n_i) \leqslant 3$.]

[14] Jonckheere, A. R. (1954). *Biometrika*, **41**, 133–145.

[15] Keselman, H. J., Rogan, J. C., and Feir–Walsh, B. J. (1977). *Brit. J. Math. Statist. Psychol.*, **30**, 213–227.

[16] Kruskal, W. H. and Wallis, W. A. (1952). *J. Amer. Statist. Ass.*, **47**, 583–621.

[17] Lehmann, E. L. (1975). *Nonparametrics: Statistical Methods Based on Ranks.* Holden-Day, San Francisco. (Intermediate level; large section on K–W and related tests.)

[18] Mann, H. B. and Whitney, D. R. (1947). *Ann. Math. Statist.* **18**, 50–60.

[19] Miller, R. G., Jr. (1966). *Simultaneous Statistical Inference.* McGraw-Hill, New York. (Intermediate level; clear and systematic description of multiple comparison procedures.)

[20] Mosteller, F. and Rourke, R. E. K. (1973). *Sturdy Statistics.* Addison-Wesley, Reading, Mass. (Elementary text; contains tables up to $c = 4$, $N \leqslant 9$.)

[21] Odeh, R. E. (1971). *Technometrics*, **13**, 912–918.

[22] Puri, M. L. (1965). *Commun. Pure Appl. Math*, **18**, 51–63.

[23] Puri, M. L. and Sen, P. K. (1971). *Nonparametric Methods in Multivariate Analysis*, Wiley, New York. (Advanced text; very theoretical.)

[24] Terpstra, T. J. (1952). *Indagationes Math.*, **14**, 327–333.

[25] Wallace, D. L. (1959). *J. Amer. Statist. Ass.*, **54**, 225–230.

[26] Wilcoxon, F. (1945). *Biometrics*, **1**, 80–83.

(BROWN–MOOD MEDIAN TEST
DISTRIBUTION-FREE METHODS
F TEST
JONCKHEERE TEST FOR NORMAL
   ALTERNATIVES
KENDALL'S TAU
MANN–WHITNEY–WILCOXON TEST
NORMAL SCORES TEST
ONE-WAY CLASSIFICATION
RANK TESTS)

HELEN BHATTACHARYYA

k STATISTICS   *See* FISHER k STATISTICS

KUDER–RICHARDSON RELIABILITY COEFFICIENTS 20 AND 21

Every measurement of an individual's ability, achievement, personality traits, or attitudes only approximates his or her true status, T. All observed measurements, X, must be presumed to include a component of measurement error, E. This condition of fallibility is symbolically stated as

$$X = T + E.$$

For conceptual purposes, T of person i is perceived as the expected value of X over an infinite number of assessments of person i. The error component is assumed to be linearly independent of T within any very large population, that is, $\rho(T, E) = 0$. Moreover, if all persons were measured twice, $\rho(E_1, E_2)$ would also equal zero.

The *reliability coefficient* of a test for a specified population is defined as the ratio $\text{var}(T)/\text{var}(X)$. While the values of X are known for a sample of persons, the values of T are not. Thus $\text{var}(T)$ must be approximated indirectly. Kuder–Richardson formulas 20 and 21 represent methods of approximating this ratio through the use of indirect estimates of $\text{var}(T)$.

In the derivation of these formulas it is presumed that the testing instrument consists of k independently scored exercises. The total score on the test equals the sum of the scores for the separate exercises. Each of these exercises is assumed to measure the same basic ability as every other. But each of the individual exercise scores, like the total score, is presumed to represent the sum of a true and an error component. Symbolically, the total test score (X) and individual exercise scores (X_j) may be represented as

$$X = T + E$$
$$X_1 = T_1 + E_1$$
$$\vdots$$
$$X_k = T_k + E_k,$$

where $X = \sum_{j=1}^{k} X_j$, $T = \sum_{j=1}^{k} T_j$, and $E = \sum_{j=1}^{k} E_j$.

To assume that all exercises measure the same basic ability is to assume that in the examinee population $\rho(T_j, T_{j'}) = 1.0$. The assumptions about E also apply to the E_j. Thus $\rho(T_j, E_j) = 0$ and $\rho(E_j, E_{j'}) = 0$ when $j \neq j'$. If the further assumption is made that $\text{var}(T_j) = \text{var}(T_{j'})$ for all values of j and j', we may deduce the following relationships:

$$\text{var}(T) = \sum_{j}^{k} \text{var}(T_j) + \sum_{j \neq j'}^{k} \sum^{k} \text{cov}(T_j, T_{j'})$$

$$= k^2 \text{var}(T_j);$$

$$\sum_{j \neq j'} \sum \text{cov}(X_j, X_{j'})$$

$$= \sum_{j \neq j'} \sum \left[\text{cov}(T_j, T_{j'}) + \text{cov}(T_j, E_{j'}) \right.$$

$$\left. + \text{cov}(E_j, T_{j'}) + \text{cov}(E_j, E_{j'}) \right]$$

$$= k(k - 1)\text{var}(T_j).$$

Thus the reliability of the total test score equals

$$\frac{\text{var}(T)}{\text{var}(X)} = \frac{k^2 \text{var}(T_j)}{\text{var}(X)}$$

$$= \frac{k^2 \sum\sum_{j \neq j'} \text{cov}(X_j, X_{j'})}{k(k - 1)\text{var}(X)}$$

$$= \left[\frac{k}{k - 1} \right] \left[\frac{\sum\sum_{j \neq j'} \text{cov}(X_j, X_{j'})}{\text{var}(X)} \right].$$

$$(1)$$

Since

$$\text{var}(X) = \sum_{j} \text{var}(X_j) + \sum_{j \neq j'} \sum \text{cov}(X_j, X_{j'}),$$

the sum of all interexercise covariances may be alternatively obtained by subtracting the sum of exercise score variances from total test score variance. In many tests the exercises are scored as 0 for an incorrect or omitted answer and 1 for a correct answer. The variance of scores on exercise j may then be expressed as $\phi_j(1 - \phi_j)$, where ϕ_j is the *proportion* of correct answers to exercise j. The reliability of the total test, ρ, then becomes

$$\rho = \left[\frac{k}{k-1}\right]\left[\frac{\text{var}(X) - \sum_j \phi_j(1 - \phi_j)}{\text{var}(X)}\right]. \tag{2}$$

This formula, derived somewhat differently, was first published by G. F. Kuder and M. W. Richardson [3]. In that publication it appeared as formula 20 and has since been referred to as KR20. In any given application, sample estimates of $\text{var}(X)$ and ϕ_j are used.

A simple approximation of formula (2) can be obtained by using the average value of the ϕ_j, which equals the mean total test score (μ_X) divided by k, as an estimate of every ϕ_j. The resultant expression is Kuder and Richardson's formula 21:

$$\text{KR21} = \frac{k}{k-1}\left[\frac{\text{var}(X) - \mu_X(k - \mu_X)k^{-1}}{\text{var}(X)}\right]. \tag{3}$$

Only the sample estimates of the mean and variance of total test scores are necessary for this approximation, and hence it is easily evaluated. It always underestimates KR20, however, as Tucker [4] first showed. The difference may be expressed

$$\text{KR20} - \text{KR21} = \left[\frac{k^2}{k-1}\right]\left[\frac{\text{var}(\phi)}{\text{var}(X)}\right],$$

where $\text{var}(\phi)$ is the variance of the k values of ϕ_j. Only in the unlikely circumstance that all exercises are equally difficult will this difference equal zero. In practice, it is generally less than 0.05, although large differences are possible.

KR20 and KR21 yield spuriously high estimates when many examinees have had insufficient time to finish. The issue of content homogeneity is also important. When the exercises assess heterogeneous abilities, the assumption $\rho(T_j, T_{j'}) = 1.0$, and quite probably the assumption $\text{var}(T_j) = \text{var}(T_{j'})$ will be violated. Cronbach et al. [1] have shown that straightforward application of KR20 can then result in a serious underesti-

mate of total test reliability. The formula can be adapted to such a test, however, if the evaluator can classify exercises into c "clusters" of two or more exercises that are homogeneous with regard to the abilities they measure. In addition to $\text{var}(X)$ and exercise ϕ_j values, the evaluator must obtain for each examinee a score on each cluster. If item cluster scores are designated Y_h ($h = 1, 2, \ldots, c$) and KR20 evaluated for cluster h is represented by ρ_h,

$$\text{KR20} = 1 - \frac{\sum_h^c \text{var}(Y_h)(1 - \rho_h)}{\text{var}(X)}. \tag{4}$$

In 1941, Hoyt [2] showed that KR20 could be evaluated from the mean squares (MS) obtained from an analysis of variance executed on the examinees by exercises score matrix. The Hoyt version of KR20 is

$$\text{KR20} = \frac{\text{MS}_{\text{examinees}} - \text{MS}_{\text{interaction}}}{\text{MS}_{\text{examinees}}}. \tag{5}$$

This relationship to a two-factor model II analysis of variance provides the foundation for a sampling theory for KR20. A confidence interval for the population value of KR20 can be obtained from the relationship

$$\Pr\left[1 - (1 - r)F_{\nu_1, \nu_2, 1-\epsilon}\right.$$
$$\left. > \rho > 1 - (1 - r)F_{\nu_1, \nu_2, \epsilon}\right] = 2\epsilon - 1. \tag{6}$$

In this expression r is the sample coefficient and $F_{\nu_1, \nu_2, \epsilon}$ is the $100\epsilon\%$ point in the central F distribution with $\nu_1 = N - 1$ and $\nu_2 = (N - 1)(k - 1)$ degrees of freedom. The region of retention for a $2(1 - \epsilon)\%$ level test of a specific hypotheses, ζ, about the KR20 is defined by

$$\Pr\left[1 - (1 - \zeta)F_{\nu_1, \nu_2, \epsilon}^{-1} > r\right.$$
$$\left. > 1 - (1 - \zeta)F_{\nu_1, \nu_2, 1-\epsilon}^{-1}\right] = 2\epsilon - 1. \tag{7}$$

Strictly speaking, these results demand that examinee scores on individual exercises satisfy the assumptions of a two-factor random model of analysis of variance. These assumptions cannot be met fully when exercises are scored 0 or 1. Simulation studies suggest, however, that the theory is remarkably robust in the face of this restriction on the score scale and nonnormality.

References

[1] Cronbach, L. J., Schoneman, P., and McKie, D. (1965). *Educ. Psychol. Meas.*, **25**, 291–312.

[2] Hoyt, C. (1941). *Psychometrika*, **6**, 153–160.

[3] Kuder, G. F. and Richardson, M. W. (1937). *Psychometrika*, **2**, 151–160.

[4] Tucker, L. R. (1949). *Psychometrika*, **14**, 117–119.

Bibliography

See the following works, as well as the references just cited, for more information on the topic of Kuder–Richardson reliability coefficients 20 and 21.

Brogden, H. E. (1946). *Educ. Psychol. Meas.*, **6**, 517–520.

Cronbach, L. J. (1951). *Psychometrika*, **16**, 297–334.

Cronbach, L. J. and Azuma, H. (1962). *Educ. Psychol. Meas.*, **22**, 645–665.

Cureton, E. E. (1958). *Educ. Psychol. Meas.*, **18**, 715–738.

Feldt, L. S. (1965). *Psychometrika*, **30**, 357–370.

Feldt, L. S. (1969). *Psychometrika*, **34**, 363–373.

Feldt, L. S. (1980). *Psychometrika*, **45**, 99–105.

Hakstian, A. R. and Whalen, T. E. (1976). *Psychometrika*, **41**, 219–231.

Jackson, R. W. B. and Ferguson, G. A. (1941). Studies on the Reliability of Tests. *Bull. No. 12*, Dept. of Educational Research, University of Toronto, Toronto.

Kristof, W. (1963). *Psychometrika*, **28**, 221–238.

Kristof, W. (1970). *J. Math. Psychol.*, **7**, 371–377.

Kristof, W. (1974). *Psychometrika*, **39**, 23–30.

Lord, F. M. (1955). *Educ. Psychol. Meas.*, **15**, 325–336.

Lord, F. M. and Novick, M. R. (1968). *Statistical Theories of Mental Test Scores*. Addison-Wesley, Reading, Mass.

Lyerly, S. B. (1958). *Psychometrika*, **23**, 267–270.

Nitko, A. J. and Feldt, L. S. (1969). *Amer. Educ. Res. J.*, **6**, 433–437.

Novick, M. R. and Lewis, C. (1967). *Psychometrika*, **32**, 1–13.

Payne, W. H. and Anderson, D. E. (1968). *Educ. Psychol. Meas.*, **28**, 23–39.

Stanely, J. C. (1957). *14th Yearbook of the National Council on Measurement in Education*, pp. 78–92.

Stanely, J. C. (1968). *Proc. 76th Annu. Conv. Amer. Psychol. Ass.*, pp. 185–186.

Stanely, J. C. (1971). In *Educational Measurement*, 2nd ed., R. L. Thorndike, ed. American Council on Education, Washington, D.C., Chap. 13.

(PSYCHOLOGICAL TESTING PSYCHOLOGY, STATISTICS IN)

Leonard S. Feldt

KUIPER–CORSTEN ITERATION

Kuiper–Corsten iteration is useful for calculating the least-squares* estimates of the parameters when two classifications are crossed with no apparent regularity in the design. The problem first arose with nonorthogonal block designs, but is found in many other contexts, for example when different students make different selections of optional examination papers, some of which may be found harder than others. It will be assumed that (1) the parameters of the two classifications are additive, (2) the design is connected (e.g., there is no subset of students who have confined themselves to certain papers that have been avoided by the rest), and (3) the errors are independent and of constant variance.

First, an incidence matrix* is written down with rows for the classification under study (*A* = treatments, students, etc.) and columns for the other (*B* = blocks, papers, etc.). The elements show the frequency with which the two classifications have intersected. Row and column totals therefore show the number of data in each level of a classification. The initial table is completed by the inclusion of data totals and data means. Thus, suppose that an experiment has been designed in three randomized blocks, each with four treatments, but by a mistake two plots had been interchanged and a third one had been lost. The initial table might look as shown in Table 1. Two simple operations are required:

1. *Projection*, in which a vector for *A* is transformed into one for *B*. Thus, if at some stage in the iteration the vector for *A* was $\mathbf{x} = (x_1, x_2, x_3, x_4)'$, the projection would give a vector of $\mathbf{y} = (y_1, y_2, y_3)'$

Table 1

Classification A	Classification B			Number of A-data	A-totals	A-means
	B_1	B_2	B_3			
A_1	2	0	1	3	11.8	3.933
A_2	0	2	1	3	14.6	4.867
A_3	1	1	1	3	12.2	4.067
A_4	1	1	0	2	9.6	4.800
Number of B-data	4	4	3	11		
B-totals	16.6	18.2	13.4		48.2	
B-means	4.150	4.550	4.467			

for B from the formula

$$y_1 = \tfrac{1}{4}(2x_1 + x_3 + x_4)$$

$$y_2 = \tfrac{1}{4}(2x_2 + x_3 + x_4) \quad \text{or} \quad \mathbf{y} = proj(\mathbf{x})$$

$$y_3 = \tfrac{1}{3}(x_1 + x_2 + x_3)$$

The coefficients are the elements of the appropriate column of the table of incidences.

2. *Dual projection*, in which the two classifications exchange roles, i.e.,

$$z_1 = \tfrac{1}{3}(2y_1 + y_3)$$

$$z_2 = \tfrac{1}{3}(2y_2 + y_3)$$

$$z_3 = \tfrac{1}{3}(y_1 + y_2 + y_3) \quad \text{or} \quad \mathbf{z} = proj'(\mathbf{y})$$

$$z_4 = \tfrac{1}{2}(y_1 + y_2)$$

The iteration is started by subtracting the dual projection of the B means from those of A to form a vector, $\mathbf{v}_1$, thus:

$$\mathbf{v}_1 = \begin{bmatrix} 3.9333 - 4.2555 \\ 4.8667 - 4.5223 \\ 4.0667 - 4.3889 \\ 4.8000 - 4.3500 \end{bmatrix}$$

$$= \begin{bmatrix} -0.322 \\ +0.344 \\ -0.322 \\ +0.450 \end{bmatrix}.$$

It proceeds thus:

$$\mathbf{u}_j = proj(\mathbf{v}_j), \qquad \mathbf{v}_{j+1} = proj'(\mathbf{u}_j).$$

Then $\sum_{j=1}^{\infty} \mathbf{v}_j$ gives the vector of estimates of treatment parameters. Similarly, $\sum_{j=1}^{\infty} \mathbf{u}_j$ subtracted from the B means gives those for blocks. The iteration usually converges

speedily. It may be noted that for all $\mathbf{v}_j$, the result of multiplying each element by the appropriate number of data and adding should give zero [e.g., $3(-0.322) + 3(0.344) + 3(-0.322) + 2(0.450) = 0$]. To avoid the accumulation of rounding errors it is advisable in each cycle to adjust the origin of the elements of $\mathbf{v}_j$ to make that relationship hold.

Given both sets of parameters it is an easy matter to subtract the relevant parametric values, one for A and one for B, from each datum to find residuals and hence the error sum of squared deviations and hence the residual mean square, $\hat{\sigma}^2$. The covariance matrix for the treatment parameters $\mathbf{\Upsilon}\hat{\sigma}^2$ is readily found from renewed use of the iteration. For the first treatment, which has three data out of a total of eleven, $\mathbf{v}_1$ is set equal to $(\tfrac{1}{3} - \tfrac{1}{11}, -\tfrac{1}{11}, -\tfrac{1}{11}, -\tfrac{1}{11})'$. Iterating as before, $\sum_j \mathbf{v}_j$ gives the first column of $\mathbf{\Upsilon}$. Similarly, setting $\mathbf{v}_1$ equal to $(-\tfrac{1}{11}, -\tfrac{1}{11}, -\tfrac{1}{11}, \tfrac{1}{2} - \tfrac{1}{11})'$ gives the fourth, and so on. In the example the vectors of estimated parameters, one for treatments and another for blocks, and the covariance matrix, $\mathbf{\Upsilon}$, are derived respectively as:

$$\begin{bmatrix} -0.50 \\ +0.50 \\ -0.33 \\ +0.49 \end{bmatrix}, \quad \begin{bmatrix} 4.36 \\ 4.26 \\ 4.58 \end{bmatrix},$$

$$\begin{bmatrix} 0.328 & -0.172 & -0.089 & -0.099 \\ -0.172 & 0.328 & -0.089 & -0.099 \\ -0.089 & -0.089 & 0.244 & -0.099 \\ -0.099 & -0.099 & -0.099 & -0.446 \end{bmatrix}.$$

Some would find it easier if all the treatment parameters were increased by the general mean (4.38) and all block parameters decreased by the same amount. The matrix Υ gives correct variances for treatment contrasts; e.g., the variance of the difference between parameters for the first and last treatments is

$$\left[0.328 + 0.446 - 2(-0.099) \right] \hat{\sigma}^2 = 0.972 \hat{\sigma}^2.$$

It should not be used for linear combinations that are not contrasts.

The iteration was first suggested by N. H. Kuiper (*Statistica*, **6**, 149–194, 1952) and extended to row-and-column designs by L. C. A. Corsten (*Meded. Landbouwhogesch.*, *Wageningen*, **58**, 1–92, 1958) and to a wide range of designs by B. A. Worthington (*Biometrika*, **62**, 113–119, 1975). Its use in finding the covariance matrix was suggested by T. Caliński (*Biometrics*, **27**, 275–292, 1971). The algebraic justification was examined by S. C. Pearce, T. Caliński, and T. F. de C. Marshall (*Biometrika*, **61**, 449–460, 1974). Its use with examination results was suggested by P. N. Murgatroyd (*Int. J. Math. Educ. Sci. Techol.*, **6**, 435–444, 1975).

(ANALYSIS OF VARIANCE
BLOCKS, COMPLETE RANDOMIZED
DESIGN OF EXPERIMENTS
GENERALIZED INVERSE
GENERAL LINEAR MODEL
ROW-AND-COLUMN DESIGNS)

S. C. PEARCE

KUIPER'S STATISTIC *See* KOLMO-GOROV–SMIRNOV STATISTICS

KULLBACK INFORMATION

Information in a technically defined sense was first introduced in statistics by R. A. Fisher in 1925 in his work on the theory of estimation. (*See* FISHER INFORMATION.) Shannon and Wiener independently published in 1948 works describing logarithmic measures of information for use in communication theory [4, 6]. These stimulated a tremendous amount of study in engineering circles on the subject of information theory. Information theory is a branch of the mathematical theory of probability and mathematical statistics. As such, it can be and is applied in a wide variety of fields. In spirit and concepts, information theory has its mathematical roots in the concept of disorder or entropy in thermodynamics and statistical mechanics*. (See Kullback [3, pp. 1–3] for many reading references.) Although Wald* did not explicitly mention information in his treatment of sequential analysis*, it should be noted that his work must be considered a major contribution to the statistical applications of information theory.

Whenever we make statistical observations, or design and conduct statistical experiments, we seek information. How much can we infer from a particular set of statistical observations or experiment about the sampled population? It has been shown by Shore and Johnson [5] that the use of any separator other than (1) or (2) below (which they refer to as cross-entropy) for inductive inference when new information is in the form of expected values leads to a violation of one or more reasonable consistency axioms.

Statistical information theory, as a branch of mathematical statistics, is developing into an extensive body of knowledge in its own right, distinct from communication theory*. According to Bartlett [1], this fact "has not, I think, been recognized sufficiently in some of the recent conferences on information theory to which mathematical statisticians *per se* have not always been invited." Because of the growing interest in and broad applicability of information theoretic ideas in diverse fields, papers on the theory and/or applications of information theory have appeared in a broad spectrum of international journals. This has impeded the development of a uniform and consistent terminology.

The information measure and its properties defined and discussed herein, may be found in the technical literature under various names: Kullback–Leibler information, directed divergence, discrimination information, Rényi's information gain, expected weight of evidence, entropy distance, entropy*, cross-entropy. The term "discrimination information" is used herein. Consider a space Ω of points $\omega \in \Omega$. Suppose that the hypotheses H_1 and H_2 imply the probability distributions p and π over Ω, respectively. The mean information for discrimination in favor of H_1 against H_2 when H_1 is true is defined by

$$I(p : \pi) = \sum_{\Omega} p(\omega)\ln(p(\omega)/\pi(\omega)) \quad (1)$$

when the space Ω is discrete and by

$$I(p : \pi) = \int_{\Omega} p(\omega)\ln(p(\omega)/\pi(\omega))\, d\omega \quad (2)$$

when the space Ω is continuous. Natural logarithms are used in the discrimination information because of statistical considerations and properties of the measures. Note that the discrimination information is the expected value of the logarithm of a likelihood ratio, i.e.,

$$I(p : \pi) = E(\ln(p(\omega)/\pi(\omega))\,|\,H_1). \quad (3)$$

The discrimination information measures the divergence between two distributions; the greater the measure, the greater the discrepancy between the distributions. The definition of discrimination information has been extended to general probability measure spaces and sub-sigma algebras thereof.

Some of the properties of the discrimination information measure follow.

1. $I(p : \pi) \geqslant 0$, with equality if and only if $p(\omega) = \pi(\omega)$. This follows from the convexity of $t \ln t$ and the use of Jensen's inequality*.

2. If $I(p : \pi) < \infty$, then $p(\omega)$ is zero whenever $\pi(\omega)$ is zero. $0 \ln 0$ is defined as 0.

3. Discrimination information is additive. Suppose that $X(\omega)$ and $Y(\omega)$ are statistics, and we write $I(p : \pi; X)$ for the discrimination information for the distribution of $X(\omega)$ under $p(\omega)$, $\pi(\omega)$, etc. It

may be shown that

$$I(p : \pi; X, Y)$$
$$= I(p : \pi; X) + I(p : \pi; Y\,|\,X)$$
$$= I(p : \pi; Y) + I(p : \pi; X\,|\,Y),$$

where $I(p : \pi; Y\,|\,X) = E_x I(p : \pi; Y\,|\,X = x)$ and similarly, $I(p : \pi; X\,|\,Y) = E_y I(p : \pi; X\,|\,Y = y)$ and for X and Y independent

$$I(p : \pi; X, Y) = I(p : \pi; X) + I(p : \pi; Y).$$

4. The equality $I(p : \pi) = I(p : \pi; X)$ is necessary and sufficient for $X(\omega)$ to be a sufficient statistic. The likelihood ratio $X(\omega) = p(\omega)/\pi(\omega)$ is a sufficient statistic.

5. If $Y = T(X)$ is a one-to-one transformation, then $I(p : \pi; X) \geqslant I(p : \pi; Y)$ with equality if and only if Y is sufficient for X. If $Y = T(X)$ is a nonsingular transformation, then $I(p : \pi; X) = I(p : \pi; Y)$.

6. The grouping of observations generally causes a loss of information unless the conditional probability of the observations given the grouping is the same under both hypotheses. If $Y = T(X)$, then $I(p : \pi; X) \geqslant I(p : \pi; Y)$ with equality if and only if the conditional probability (density) of X given $T(X) = Y$ is the same under both hypotheses. If there is equality, then $Y = T(X)$ is called a sufficient statistic for discrimination. There can be no gain of information by statistical processing of data. A transformation that considers only a marginal distribution in a multivariate situation (ignores some of the variates) is one that will generally result in a loss of information.

7. Suppose that the space Ω is partitioned into the disjoint sets E_1 and E_2; i.e., $\Omega = E_1 + E_2$ with Ω the sample space of n independent observations. Assume a test procedure such that if the sample point $\omega \in E_1$, one accepts the hypothesis H_1 (reject H_2), and if the sample point

$\omega \in E_2$, one accepts the hypothesis H_2 (reject H_1). We take H_2 as the null hypothesis. E_1 is called the critical region. The probability of incorrectly accepting H_1, the type I error, is $\alpha = \Pr(\omega \in E_1 \mid H_2)$, and the probability of incorrectly accepting H_2, the type II error, is $\beta = \Pr(\omega \in E_2 \mid H_1)$. Let O_n denote a sample of n independent observations and O_1 a single observation; then

$$I(p : \pi; O_n) = nI(p : \pi; O_1)$$
$$\geqslant \beta \ln(\beta/(1 - \alpha))$$
$$+ (1 - \beta)\ln((1 - \beta)/\alpha)$$
$$= 2.649995 \quad \text{for} \quad \alpha = \beta = 0.05,$$

for example.

To illustrate the foregoing statements, consider the following examples.

Example 1. Let H_i imply the binomial distributions $B(p_i, q_i, n)$, $p_i + q_i = 1$, $i = 1, 2$, so that $p(\omega)$ and $\pi(\omega)$ are, respectively, $C_\omega^n p_i^\omega q_i^{n-\omega}$, $\omega = 0, 1, \ldots, n$, and $\Omega = \{0, 1, \ldots, n\}$; then

$$I(p : \pi) = n(p_1\ln(p_1/p_2) + q_1\ln(q_1/q_2)). \tag{4}$$

The discrimination information for n independent binomial observations is n times the discrimination information per binomial observation. The number of "successes" is a sufficient statistic. See Kullback [3, Table II].

Example 2. Let p and π, respectively, be Poisson distributions with parameters m_i, $i = 1, 2$; then

$$I(p : \pi) = m_1\ln(m_1/m_2) + m_2 - m_1. \tag{5}$$

The discrimination (5) for the Poisson distributions is the limit of (4) for the binomial distributions as $n \to \infty$, $p_i \to 0$, $np_i = m_i$, $i = 1, 2$.

Example 3. Let $1 + \omega$ be the number of independent trials needed to get a success when the probability of a success is constant for each trial. If p and π are, respectively,

$p_i q_i^\omega$, $\omega = 0, 1, 2, \ldots$, and $q_i = 1 - p_i$, $i = 1, 2$, then

$$I(p : \pi) = E(1 + \omega \mid H_1)$$
$$\times (p_1\ln(p_1/p_2) + q_1\ln(q_1/q_2)). \tag{6}$$

That is, the discrimination information is the product of the expected number of trials and the discrimination information per trial.

Example 4. Let H_1 and H_2 specify, respectively, the probabilities of two c-valued populations (c categories or classes) H_i: $p_{i1}, p_{i2}, \ldots, p_{ic}$, $p_{i1} + p_{i2} + \cdots + p_{ic} = 1$, $i = 1, 2$; then

$$I(p : \pi) = p_{11}\ln(p_{11}/p_{21}) + \cdots$$
$$+ p_{1c}\ln(p_{1c}/p_{2c}). \tag{7}$$

If we group two categories, the first and second, for example, then

$$I_g(p : \pi) = (p_{11} + p_{12})\ln\left(\frac{(p_{11} + p_{12})}{(p_{21} + p_{22})}\right)$$
$$+ \sum_{j=3}^{c} p_{1j}\ln(p_{1j}/p_{2j}). \tag{8}$$

It follows that

$$I(p : \pi) - I_g(p : \pi)$$
$$= p_{11}\ln\frac{p_{11}/(p_{11} + p_{12})}{p_{21}/(p_{21} + p_{22})}$$
$$+ p_{12}\ln\frac{p_{12}/(p_{11} + p_{12})}{p_{22}/(p_{21} + p_{22})} \geqslant 0, \tag{9}$$

with equality if and only if

$$p_{11}/(p_{11} + p_{12}) = p_{21}/(p_{21} + p_{22}),$$
$$p_{12}/(p_{11} + p_{12}) = p_{22}/(p_{21} + p_{22}).$$

That is, the grouping is sufficient if and only if $p_{11}/p_{12} = p_{21}/p_{22}$.

The discrimination information for a sample O_n of n independent observations (multinomial distributions) is

$$I(p : \pi; O_n) = n\sum_{j=1}^{c} p_{1j}\ln(p_{1j}/p_{2j}). \tag{10}$$

Example 5. Let H_1 and H_2 imply, respectively, the k-variate normal populations

$N(\boldsymbol{\mu}_i, \boldsymbol{\Sigma}_i)$, $\boldsymbol{\mu}_i' = (\mu_{i1}, \mu_{i2}, \ldots, \mu_{ik})$, $\boldsymbol{\Sigma}_i = (\sigma_{irs})$, $i = 1, 2$; $r, s = 1, 2, \ldots, k$; then in matrix notation

$$I(p : \pi) = \tfrac{1}{2}\ln(\det \boldsymbol{\Sigma}_2 / \det \boldsymbol{\Sigma}_1)$$

$$+ \tfrac{1}{2} \operatorname{tr} \boldsymbol{\Sigma}_1(\boldsymbol{\Sigma}_2^{-1} - \boldsymbol{\Sigma}_1^{-1})$$

$$+ \tfrac{1}{2} \operatorname{tr} \boldsymbol{\Sigma}_2^{-1}(\boldsymbol{\mu}_1 - \boldsymbol{\mu}_2)(\boldsymbol{\mu}_1 - \boldsymbol{\mu}_2)'. \quad (11)$$

When $\boldsymbol{\Sigma}_1 = \boldsymbol{\Sigma}_2 = \boldsymbol{\Sigma}$, (11) becomes

$$I(p : \pi) = \tfrac{1}{2} \operatorname{tr} \boldsymbol{\Sigma}^{-1} \boldsymbol{\delta}\boldsymbol{\delta}' = \tfrac{1}{2} \boldsymbol{\delta}' \boldsymbol{\Sigma}^{-1} \boldsymbol{\delta}. \quad (12)$$

Note that $k\boldsymbol{\delta}'\boldsymbol{\Sigma}^{-1}\boldsymbol{\delta}$ is Mahalanobis's generalized distance* where $\boldsymbol{\delta} = \boldsymbol{\mu}_1 - \boldsymbol{\mu}_2$. When $\boldsymbol{\delta} = \boldsymbol{\mu}_1 - \boldsymbol{\mu}_2 = \boldsymbol{0}$, (11) becomes

$$I(p : \pi) = \tfrac{1}{2}\ln(\det \boldsymbol{\Sigma}_2 / \det \boldsymbol{\Sigma}_1)$$

$$- k/2 + \tfrac{1}{2} \operatorname{tr} \boldsymbol{\Sigma}_1 \boldsymbol{\Sigma}_2^{-1}. \quad (13)$$

For single-variate normal populations (11) is

$$I(p : \pi) = \tfrac{1}{2}\ln(\sigma_2^2/\sigma_1^2) - \tfrac{1}{2}$$

$$+ \tfrac{1}{2}\sigma_1^2/\sigma_2^2 + (\mu_1 - \mu_2)^2/2\sigma_2^2 \quad (14)$$

and for $\sigma_1^2 = \sigma_2^2 = \sigma^2$, (14) becomes

$$I(p : \pi) = (\mu_1 - \mu_2)^2/2\sigma^2. \quad (15)$$

The discrimination information for a sample of n independent observations from the single-variate normal populations is n times $I(p : \pi)$ in (14). It may be shown that

$$nI(p : \pi) = I(p : \pi; \bar{x}, s^2)$$

$$= I(p : \pi; \bar{x}) + I(p : \pi; s^2), \quad (16)$$

where $\bar{x}$ is the sample average, s^2 is the sample unbiased estimate of the variance, and

$$I(p : \pi; \bar{x}) = \tfrac{1}{2}\ln(\sigma_2^2/\sigma_1^2) - \tfrac{1}{2} + \tfrac{1}{2}\sigma_1^2/\sigma_2^2$$

$$+ n(\mu_1 - \mu_2)^2/2\sigma_2^2, \quad (17)$$

$$I(p : \pi; s^2) = (n - 1)\left(\tfrac{1}{2}\ln(\sigma_2^2/\sigma_1^2)\right.$$

$$\left. - \tfrac{1}{2} + \tfrac{1}{2}\sigma_1^2/\sigma_2^2\right). \quad (18)$$

For $\sigma_1^2 = \sigma_2^2 = \sigma^2$ we have

$$I(p : \pi; \bar{x}) = n(\mu_1 - \mu_2)^2/2\sigma^2. \quad (19)$$

The principle of *minimum discrimination information estimation* provides a useful basis for the analysis of contingency tables* or,

more generally count data, and leads naturally to log-linear models. For a more detailed exposition and applications, see Gokhale and Kullback [2]. Consider the discrimination information defined in (1) with $\sum_\Omega p(\omega) = \sum_\Omega \pi(\omega) = 1$. For notational convenience we shall take the number of points (cells) in Ω as n. Suppose that the distribution $\pi(\omega)$ in (1) is some fixed distribution and $p(\omega)$ is a member of the family $\mathbb{P}$ of distributions which satisfy the linearly independent constraints

$$\sum_\Omega c_i(\omega)p(\omega) = \theta_i, \quad i = 0, 1, \ldots, r. \quad (20)$$

The rank of the $(r + 1) \times n$ model matrix $\mathbf{C} = (c_i(\omega))$, $\omega = 1, 2, \ldots, n$ is $r + 1 \leqslant n$. To satisfy the natural constraint $\sum_\Omega p(\omega) = 1$, we take $c_0(\omega) = 1$ for all ω and $\theta_0 = 1$. The minimum discrimination information estimate $p^*(\omega)$ is that member of the family of distributions $\mathbb{P}$ which minimizes $I(p : \pi)$ subject to the constraints (20). The minimizing distribution has the representation

$$p^*(\omega) = \exp(\tau_0 + \tau_1 c_1(\omega) + \cdots$$

$$+ \tau_r c_r(\omega))\pi(\omega) \quad (21)$$

or

$$\ln(p^*(\omega)/\pi(\omega))$$

$$= \tau_0 + \tau_1 c_1(\omega) + \cdots + \tau_r c_r(\omega),$$

$$\omega \in \Omega, \quad (22)$$

where the τ's are to be determined so that

$$\sum_\Omega c_i(\omega)p^*(\omega) = \theta_i, \quad i = 0, 1, \ldots, r. \quad (23)$$

The fact that $I(p : \pi)$ is a convex function ensures a unique minimum. The *log-linear representation* is given in (22). If we write $\tau_0 = -\ln M(\tau_1, \ldots, \tau_r)$ where

$$M(\tau_1, \ldots, \tau_r)$$

$$= \sum_\Omega \exp(\tau_1 c_1(\omega) + \cdots + \tau_r c_r(\omega))\pi(\omega),$$

$$(24)$$

then (23) may also be expressed as

$$\theta_i = (\partial/\partial\tau_i)\ln M(\tau_1, \tau_2, \ldots, \tau_r),$$

$$i = 1, \ldots, r. \quad (25)$$

Computer programs are available to implement the determination of $p^*(\omega)$ using (23) or (25). It may be shown that if $p(\omega)$ is any member of the family $\mathbb{P}$ of distributions, then

$$I(p : \pi) = I(p^* : \pi) + I(p : p^*). \quad (26)$$

The Pythagorean-type property (26) plays an important role in the analysis of information.

References

[1] Bartlett, M. S. (1975). *Probability, Statistics and Time: A Collection of Essays*. Chapman & Hall, London.

[2] Gokhale, D. V. and Kullback, S. (1978). *The Information in Contingency Tables*. Marcel Dekker, New York. (Presentation at an intermediate level emphasizing methodology in the analysis of count data; contains many practical examples. Extensive bibliography.)

[3] Kullback, S. (1959). *Information Theory and Statistics*. Wiley, New York (Dover, New York, 1968; Peter Smith Publisher, Magnolia, Mass. 1978.) (First five chapters contain a measure-theoretic presentation of theory. Chapters 6–13 consider applications at an intermediate level. Contains many examples, problems, an extensive bibliography, tables, and a glossary.)

[4] Shannon, C. E. (1948). *Bell Syst. Tech. J.*, **27**, 379–423, 623–656.

[5] Shore, J. E. and Johnson, R. W. (1980). *IEEE Trans. Inf. Theory*, **IT-26**(1), 26–37.

[6] Wiener, N. (1948). *Cybernetics*. Wiley, New York. (Not primarily statistical.)

Bibliography

See the following works, as well as the references just given, for more information on the topic of Kullback information.

Akaike, H. (1977). In *Applications of Statistics*, P. R. Krishnaiah, ed. North-Holland, Amsterdam, pp. 27–41.

Campbell, L. L. (1970). *Ann. Math. Statist.*, **41**, 1011–1015.

Johnson, R. W. (1979). *IEEE Trans. Inf. Theory*, **IT-25**(6), 709–716.

Kullback, S. (1967). *J. Amer. Statist. Ass.*, **62**, 685–686.

Kullback, S. and Leibler, R. A. (1951). *Ann. Math. Statist.*, **22**, 79–86.

Kullback, S. and Thall, P. F. (1977). *J. Comb. Inf. Syst. Sci.*, **2**(2/3), 97–103.

Osteyee, D. B. and Good, I. J. (1974). *Information, Weight of Evidence, the Singularity between Probability Measures and Signal Detection*. Springer-Verlag, New York. (Advanced level. References.)

Rényi, A. (1961). *Proc. 4th Berkeley Symp. Math. Statist. Prob.*, Vol. 1. University of California Press, Berkeley, Calif., pp. 547.

Savage, L. J. (1954). *The Foundations of Statistics*. Wiley, New York. (A theoretical discussion at the beginning graduate level. Bibliography.)

Wilks, S. S. (1962). *Mathematical Statistics*. Wiley, New York. (An extensive exposition at the beginning graduate level. Examples, problems, and bibliography.)

Zacks, S. (1971). *The Theory of Statistical Inference*. Wiley, New York. (Theoretical presentation at an advanced level. Examples, problems, and bibliography.)

(BINOMIAL DISTRIBUTION
CONTINGENCY TABLES
COUNT DATA
ENTROPY
EXPONENTIAL FAMILIES
FISHER INFORMATION
GEOMETRY IN STATISTICS: CONVEXITY
INFERENCE, STATISTICAL—I, II
LIKELIHOOD
NORMAL DISTRIBUTION
POISSON DISTRIBUTION
SEQUENTIAL ANALYSIS
SUFFICIENCY)

S. KULLBACK

KULLBACK–LIEBLER INFORMATION

See KULLBACK INFORMATION

KURTOSIS

Kurtosis is a measure of a type of departure from normality*. (Another measure of departure from normality is skewness*.) The population kurtosis is given by the fourth moment over the square of the variance ($b_2 = m_4 / m_2^2$, where m_2 and m_4 are the second and fourth sample moments around the arithmetic mean, respectively). This ratio equals 3 for any normal distribution. Ratios greater than 3, such as occur for t-distributions*, are indicators of an excess of values in the neighborhood of the mean* as well as far away from it (with a depletion of the "flanks" of the curve representing the distribution). b_2 cannot be less than 1.

Ratios smaller than 3 correspond to curves with a flatter top than the normal (e.g., rectangular*). The sample estimator of kurtosis, g_2, is given by

$$g_2 = \left(m_4 / m_2^2 \right) - 3.$$

Often "peaked" (as compared with normal) distributions have positive kurtosis, and flat-topped-ones have negative kurtosis. For normal populations the sampling distribution of g_2 is skewed and approaches normality very slowly.

Tables of percentage points for sample sizes 50(25)200(50)1000(200)2000 are given in K. Pearson's *Tables for Statisticians and Biometricians* [3]; those for sample sizes 200(50)1000(200)2000(500)5000 are given in Pearson and Hartley's *Biometrika Tables*, Vol. I [4], and are reproduced in various texts on statistics (e.g., Snedecor and Cochran [5]). D'Agostino and Tietjen [1] present percentiles of b_2 for sample sizes less than 50. Geary [2] first proposed a test for kurtosis (*see* DEPARTURES FROM NORMALITY, TESTS FOR for details and for other tests).

Kurtosis is useful for studying the distribution of sample variance when the original population is nonnormal. Some details are given in Snedecor and Cochran [5].

References

[1] D'Agostino, R. B. and Tietjen, G. L. (1971). *Biometrika*, **58**, 669–672.

[2] Geary, R. C. (1936). *Biometrika*, **28**, 295–307.

[3] Pearson, K. (1914). *Tables for Statisticians and Biometricians*. Cambridge University Press, Cambridge.

[4] Pearson, E. S. and Hartley, H. O. (1954). *Biometrika Tables for Statisticians*, Vol. 1. Cambridge University Press, Cambridge.

[5] Snedecor, G. W. and Cochran, W. (1967). *Statistical Methods*, 6th ed. Iowa University Press, Ames, Iowa.

(DEPARTURES FROM NORMALITY FREQUENCY CURVES, SYSTEMS OF SKEWNESS)

L

LABELS

It may be argued that the availability of *labels* is what sets inference in finite population sampling* apart from most of the rest of statistical theory. Their presence greatly enlarges the classes of designs and estimators available to the statistician, and in order to choose among them, or indeed to make inferences at all, one is forced to come to terms with possible relationships of the labels with the characteristic of interest.

Specifically, a finite population of size N consists of units (e.g., people or households) on a list or map, and these units may be *labeled* by name or location, or more simply by successive integers $1, \ldots, N$. In the simplest and most widely discussed problem of inference, each unit i has an associated variate value y_i; the problem is to infer about some function of the population vector $\mathbf{y} = (y_1, \ldots, y_N)$, such as the population mean* $\mu_y = (y_1 + \cdots + y_N)/N$, having somehow drawn a sequence $s^* = (i_1, i_2, \ldots, i_n)$ from $\{1, \ldots, N\}$ and having observed the corresponding y_{i_j} [10]. The full data from the sampling experiment consist of the pair sequence

$$X_{s*} = ((i_1, y_{i_1}), (i_2, y_{i_2}), \ldots, (i_n, y_{i_n})),$$

each sampled y-value being paired with its label.

Because of the availability of labels, the class of possible randomized sampling designs* is extremely wide, since the units of the population* may be assigned arbitrary selection probabilities at each successive draw, and in particular repetitions of units may be allowed or not as desired. However, if we regard the randomized design as providing a family of distributions for X_{s*} indexed by the parameter $\mathbf{y}$, it can be shown [1] that for any inference about $\mathbf{y}$, the reduced data

$$X_s = \{(i, y_i) : i \in s\}$$

are sufficient*, where s is the set of distinct members of the sequence s^*; this is consistent with the intuitive notion that repetitions of pairs in X_{s*} should provide no new information.

It may be noted that without further structure, such as a prior distribution on $\mathbf{y}$ which allows the parameter to be reduced to the population order statistic* $(y_{(1)}, y_{(2)}, \ldots, y_{(N)})$ [17], we cannot say that the sample order statistic $X_s^0 = \{y_i : i \in s\}$ (listed in order with repetitions included) is sufficient in any satisfactory sense; it is easy to show that because of the presence of labels, the conditional distribution of the data X_s, given X_s^0, is heavily dependent on the parameter $\mathbf{y}$. In fact, even with a simple random sampling design it is not true that for every unbiased estimator of μ_y there is a better one which is symmetric (i.e., based on X_s^0). For example, the availability of labels allows unbiased asymmetric linear estimators of the form

$$\sum_{i \in s} b(i,s)y_i,$$

which for some $\mathbf{y}$ have smaller variance than the sample mean, the best unbiased estimator among functions of X_s^0 [3, 14, 15]. For example, let $N = 3$ and $\mathbf{y} = (a, b, 0)$, and consider the estimator given by $\bar{y}_s - y_1/3$ if $s = \{1,2\}$, $\bar{y}_s + y_1/3$ if $s = \{1,3\}$, and $\bar{y}_s$ if $s = \{2,3\}$, where $\bar{y}_s$ is the sample mean. Under simple random sampling without replacement, two draws, this estimator is unbiased, and the difference between its variance and that of $\bar{y}_s$ is $a(2a - 3b)/27$, which is negative if $3b > 2a > 0$.

In contrast, a typical problem from the main body of statistical theory involves the observation of a sequence $x_1, \ldots, x_n$ of values of independent and identically distributed random variables with cumulative distribution function (CDF) F_θ. Such a sequence is often thought of as generated by independent draws from a hypothetical population of x-values with distribution F_θ. With this interpretation we have, as in the finite sampling case, a sequence of observations drawn "at random" from a population, but there is no analog of the *label* present. The sampling design is necessarily simple random sampling with replacement, which requires no knowledge of the labels to implement. It can easily be shown that the order statistic, namely the set of n x-values observed without regard to the order in which they were drawn, is sufficient for inferences about the parameter $\boldsymbol{\theta}$; i.e., given the n observed x-values, the ordered sequences $x_1, \ldots, x_n$ may be taken to be distributed independently of $\boldsymbol{\theta}$, all sequences consistent with the order statistic being equally likely. Thus, in this case, if the class of distributions

indexed by $\boldsymbol{\theta}$ is wide enough for the order statistic to be complete, it may be shown without further qualification that the optimal unbiased estimator of the distribution mean is the sample mean [9].

More fundamentally, it may be argued that the presence of labels affects not only the problem of choice of estimation strategy, but also the logic of sampling inference itself, as may be illustrated by the following example. Consider an urn containing N balls, M of which are white and $N - M$ of which are black. A fixed number n are drawn at random without replacement. The number m of these which are white has a hypergeometric distribution*, which provides a likelihood function* and confidence intervals* for M, and this would certainly be used for inference about M if the balls were indistinguishable apart from colour. Suppose, however, that the balls are also labeled $1, \ldots, N$. Then the full data must include the labels of the balls drawn, and to reduce the data to the order statistic m requires a conscious decision to assume in effect that the labels are uninformative about M. (One way of formalizing this is to assume a superpopulation model* in which the colors have been assigned randomly to the labeled balls.) If one is not willing to make this assumption or some alternative assumption about the relationship of label and color, only the trivial inference that M lies between M and $N - (n - m)$ is possible. Thus we have the paradox, arising often in reasoning based on probabilities, of a statistician with more information being apparently able to say less about the parametric function of interest—unless he is able to justify an explicit assumption about the relationship of the additional variate (the *label*) with the characteristic y. See also URN MODELS.

It seems to the author that the root of the historical controversy surrounding the subject of labels has lain partly in the reluctance of many statisticians to accept this paradox, and partly in a certain lack of agreement as to what constitutes a justification (in the example above) of the use of the hypergeometric distribution of m in estimating M

(see, e.g., Godambe [5–7] and Hartley and Rao [11–13]). The position expressed by Godambe [7] can be summarized as follows. Anything one can do by "ignoring" labels one can also do by assuming a suitable superpopulation. Furthermore, the assumption of superpopulations allows one to deal with many practical situations in which simply ignoring labels leads nowhere. The concept of labels "carrying no information" cannot be defined satisfactorily (see the remarks on sufficiency above), and hence ignoring labels brings about logical as well as practical difficulties.

An interesting example shows that absurd inferences can result from ignoring labels in the sample if one has and attempts to use the information that (1) the y_i take values in a specified set $\{t_1, \ldots, t_k\}$ and (2) y_i is a nondecreasing function of i [4]. For instance, if $k = 2$, $N = 20$, and $X_s = \{(2, t_1), (3, t_2)\}$, it is known with certainty that $\mathbf{y} = (t_1, t_1, t_2, \ldots, t_2)$; if, however, the labels are discarded, so that all that is used is the fact that both t_1 and t_2 have been observed, then (assuming a simple random sample) the relative likelihood for the true $\mathbf{y}$ is only 0.36.

HISTORICAL NOTES

In essence the formalization of the sampling problem used above was introduced in ref. 3. In ref. 10 it was shown that any probability sampling design could be implemented as a sequence of draws for which probabilities were successively specified, while the sufficiency of X_s for inference about $\mathbf{y}$ had been noted in ref. 1.

The best known theorem in ref. 3 states that for most sampling designs there is no homogeneous linear unbiased estimator which has minimum variance for all $\mathbf{y}$. The restriction of homogeneous linearity was later removed in ref. 8; some further proofs of these results have been suggested (e.g., in refs. 2 and 14). In refs. 3 and 8, an alternative criterion based on superpopulation expectations of the sampling variance was introduced to prove optimality theorems for

certain estimators, including the sample mean.

In the case where the y_i are known to take values in a finite set, such as for example $\{0, 1\}$, maximum likelihood estimation for functions of $\mathbf{y}$ based on the distribution of X_s^0 was discussed in refs. 11, 12, 16, and 18. (Reference 16 also contains a completeness theorem for X_s^0 under fixed-size sampling designs.) Since such an analysis assumes that it is possible (or perhaps unavoidable) to discard the sample labels, these papers (particularly ref. 11) touched off the controversy referred to above and carried on principally in refs. 5–7 and 13.

References

[1] Basu, D. (1958). *Sankhyā*, **20**, 287–294.

[2] Basu, D. (1971). In *Foundations of Statistical Inference*, V. P. Godambe and D. A. Sprott, eds. Holt, Rinehart and Winston, Toronto, pp. 203–242.

[3] Godambe, V. P. (1955). *J. R. Statist. Soc. B*, **17**, 269–278.

[4] Godambe, V. P. (1968). *J. R. Statist. Soc. B*, **30**, 243.

[5] Godambe, V. P. (1969). In *New Developments in Survey Sampling*, N. L. Johnson and H. Smith, eds. Wiley-Interscience, New York, pp. 27–58.

[6] Godambe, V. P. (1970). *Amer. Statist.*, **24**, 33–38.

[7] Godambe, V. P. (1975). *Sankhyā C*, **37**, 53–76.

[8] Godambe, V. P. and Joshi, V. M. (1965). *Ann. Math. Statist.*, **26**, 1707–1722.

[9] Halmos, P. R. (1946). *Ann. Math. Statist.*, **17**, 34–42.

[10] Hanumantha Rao, T. V. (1962). *Sankhyā*, **24**, 327–330.

[11] Hartley, H. O. and Rao, J. N. K. (1968). *Biometrika*, **55**, 547–557.

[12] Hartley, H. O. and Rao, J. N. K. (1969). In *New Developments in Survey Sampling*, N. L. Johnson and H. Smith, eds. Wiley-Interscience, New York, pp. 147–169.

[13] Hartley, H. O. and Rao, J. N. K. (1971). *Amer. Statist.*, **25**, 21–27.

[14] Lanke, J. (1973). *Metrika*, **20**, 196–202.

[15] Roy, J. and Chakravarti, I. M. (1960). *Ann. Math. Statist.*, **31**, 392–398.

[16] Royall, R. (1968). *J. Amer. Statist. Ass.*, **63**, 1269–1279.

[17] Sugden, R. A. (1979). *J. R. Statist. Soc. B*, **41**, 269–273.

[18] Wilks, S. S. (1960). *Bull. Int. Statist. Inst.*, **37**, 241–248.

(FINITE POPULATIONS, SAMPLING FROM HYPERGEOMETRIC DISTRIBUTIONS INFERENCE, STATISTICAL, I, II ORDER STATISTICS SAMPLING SCHEMES SUFFICIENCY SURVEY SAMPLING URN MODELS)

M. E. Thompson

LABOR STATISTICS

The term "labor statistics" is used loosely to describe a broad spectrum of statistics that relate to the labor forces throughout the world and the workers that comprise them. Such statistics are of paramount importance in economic and social analysis because of the significance of workers in the process of production and because of the predominance of workers in the fabric of society in most countries.

Labor statistics generally fall into several categories. The first can be termed "labor force and employment status." These data are usually collected from households by means of sample surveys.

In the United States the sample survey is called the Current Population Survey (*see* BUREAU OF LABOR STATISTICS). In some countries such data are collected from data on registrants at employment offices. In this category of data are statistics about the demographic characteristics of the population, the labor force and those employed, as well as those not employed either because they are not in the labor force or because they are in the labor force but unemployed. Such data may also include work experience history, occupation, the characteristics of multiple job holders, and material and other family characteristics that are relevant attributes of those in the labor force and those employed and unemployed. More recently, information compiled about school enrollment and educational attainment has been expanded greatly because the concept of human capital has become more widely recognized, and analysts are attempting to measure such capital.

In recent years the household surveys used to collect these data in the United States and throughout the world have been expanded. In the United States the survey also includes questions that determine the characteristics of those unemployed. These include the duration of unemployment, whether the unemployed person is a new entrant or reentrant into the labor force, or left employment voluntarily or because of economic or other reasons. Data on income are also collected regularly in the U.S. household survey; *see also* SURVEYS, HOUSEHOLD.

Another large body of data relates to those employed. It is usually collected from the employer and is available in detail by industry and by geographical area such as region and state. Also included in statistics about those employed are hours worked and earnings. Sometimes available are data on labor turnover and quit rates in various industries. In connection with earnings, there is an additional body of data, frequently collected as special supplements to ongoing surveys, that consists of compensation studies. These supplemental studies cover such topics as (1) comparisons of earnings by industry, area, and occupation used to study earnings differentials; (2) differentials paid for working different shifts; (3) fringe benefits; (4) union versus nonunion hourly rates; and (5) indexes of straight-time hourly earnings.

To get a more accurate picture of the purchasing power of employee compensation, it is necessary to get some idea of the rate of price inflation. For this purpose, most countries compile and publish a consumer price index* (CPI). Such indexes are designed to measure changes in the cost of a fixed basket of goods and services of the composition typically purchased by workers.

The fixed basket of goods and services is usually obtained by another type of household survey called a family expenditure survey. It is, perhaps, the most complex economic survey conducted because its purpose is to obtain the amount each household in the sample spends on a detailed listing of goods and services. In the United States, some of the information is collected by interview, and some by a diary left with the household. The data, when tabulated, are suitable for weighting the relatives of the prices of the various items priced for the sample.

The family expenditure survey and other data provide a basis for selecting the items priced for an index. A sample of outlets in which the prices are collected is drawn from a number of sources.

Of course, consumer price indexes are not the only overall measures of inflation. Nor are they the only price measures relevant for analyzing all aspects of inflation as it impacts on labor markets. For example, the price index for gross domestic product is a measure of the prices businesses receive for the sale of their output to final users, and therefore a measure of their ability to compensate labor.

Consumer price indexes are usually produced in those government agencies that are responsible for labor statistics. This is distinct from the common practice of calculating other price indexes, such as the wholesale or producer price indexes* and price indexes for the gross domestic product or national output, in other agencies of the government, particularly those that are more commerce oriented.

Another measure of the ability to compensate labor is a set of statistics known as "productivity indexes." Productivity indexes are usually calculated at the national level as well as for particular industrial sectors. Productivity is measured by a ratio. The numerator of the ratio is output, which is usually measured in either number of units (tons, etc.) or by dividing the dollar value of output by a price index. The denominator of the productivity ratio is the input used to produce the output, usually hours worked by labor. Sometimes the productivity measure is broadened to include measures of the input of all factors of production, not only labor but also capital, materials, etc.

Another area of labor statistics has to do with unions and industrial relations. Included in these data are statistics on union membership and its distribution by region and by industry. Data are available on industrial disputes (their number, workers involved, time lost), cases subjected to mediation and similar subjects.

A rapidly growing field of labor statistics is that of occupational injuries and illnesses. Such data are usually available by occupation and by industry.

The agency in the United States with primary responsibility for concepts, definitions, sample design, collection, compilation, and publication of labor statistics is the Bureau of Labor Statistics (BLS) of the U.S. Department of Labor. Bureaus of labor statistics first emerged in the United States as state agencies. Massachusetts established one in 1869. By 1884 fourteen states had such bureaus. In that year, Congress passed an act to establish a bureau of labor in the Department of Interior. Four years later that law was amended to establish the Bureau as an independent Department of Labor without cabinet status; it became part of the cabinet Department of Commerce and Labor created in 1903. In 1913 the Department was split and a new cabinet-level Department of Labor was established. At that time the bureau of labor became the Bureau of Labor Statistics in the new department. It has remained there ever since.

The concepts and definitions of the various series of labor statistics are formulated by BLS in most cases, sometimes with the advice of other governmental and nongovernmental committees. Two committees established and regularly consulted by BLS are the Business and the Labor Research Advisory Committees. Federal statistical oversight agencies, in the past the Office of Management and Budget (OMB), have established continuing or special committees on various topics. An example is the special committee set up by OMB in 1971 to critique the BLS's decision to treat the cost of antipollution devices on automobiles as a price increase in the CPI.

The adoption of concepts and definitions at the BLS or most other statistical agencies is usually the task of a subject matter specialist. For example, the input of economists is necessary to decide whether the labor measure in a productivity ratio or index should be a simple count of hours worked or a weighted figure that takes account of the effect of the increasing educational achievement of the work force. In other words, is an hour of work today the equivalent of or more than the hour worked 50 years ago? Economists address such issues in the context of economic theory and the intended use of the statistical measure. Use of the statistical series is a major consideration in selecting the appropriate economic concept for measurement.

From time to time, committees of nongovernment economists and statisticians have been appointed to review entire programs in the labor statistics area. Five such studies over the past 20 years are notable in their scope. In 1960, the Price Statistics Review Committee of the National Bureau of Economic Research (NBER) rendered a report on the price statistics published by U.S. government agencies, primarily the BLS. The report was requested by the Office of Statistical Standards of the Bureau of the Budget (now OMB). It is entitled *The Price Statistics of the Federal Government* [5]. The U.S. Council on Wage and Price Stability requested that the NBER do a related study in 1977; it is entitled *The Wholesale Price Index: Review and Evaluation* [4]. A third report, *Measurement and Interpretation of Productivity* [2], was published in 1979. It was undertaken by a panel appointed by the National Research Council at the suggestion of the National Center for Productivity and Quality of Working Life, a U.S. agency now abolished. *Counting the Labor Force* [1], a report on employment and unemployment

statistics, was published in 1979 by the National Commission on Employment and Unemployment Statistics established by the Congress. A report of similar scope was published in 1962, by a presidentially appointed committee; it is entitled *Measuring Employment and Unemployment* [3].

Once the concept has been selected, statisticians usually undertake the implementation of the statistical tasks involved in measurement. These include survey design, sample size, and calculation of sampling variance, and the monitoring of data collection to evaluate response rates and determine if bias is being introduced. When the data are compiled, the statistician attaches statistical attributes of the data so that they may be included by the subject matter specialist in describing the behavior of the series.

In labor statistics, conceptual error is frequently as important as statistical error in evaluating a measure. A perfectly designed and collected sample of consumer price statistics will nonetheless be defective if the wrong concept is used in the housing cost component of the index or if the adjustment of prices for quality change is done inappropriately.

The BLS uses certain government wide classifications for its data, such as the U.S. Standard Industrial Classification (SIC) for industry classification, the Dictionary of Occupational Titles for occupations, and the Standard Metropolitan Statistical Areas system for classifying areas.

Most of the information collected by the BLS is reported voluntarily; the BLS preserves the confidentiality of the data it reports. In some instances, the BLS collects data for other government agencies. In other instances, the U.S. Census Bureau collects data for the BLS.

The data of the BLS are published in various press releases and reports covering specific series and general publications, both monthly and annually. Most statistics that are part of regular programs are announced to the public in the form of a press release. Such releases are by subject—CPI, employment and unemployment, productivity, etc. Each contains newly published data for the latest time period as well as revisions to data released over the past year or so. The BLS publishes an annual *Handbook of Labor Statistics* covering most BLS series, with data for some as far back as 1947. Not all the detail is published in the *Handbook*. Such detail is covered in reports and detailed bulletins. Many of the series published annually in the *Handbook* are provided once a month in monthly detail in the *Monthly Labor Review*, which also contains analytical studies.

There is considerable interest in making comparisons of various characteristics of workers among countries. The International Labor Organization (ILO) of the United Nations has been coordinator for that purpose. It publishes technical guides describing each UN member country's labor statistics, and it establishes guidelines for statistical offices to use when calculating labor statistics. Similar efforts have been undertaken by the Statistical Office of the European Community and the Organization for Economic Cooperation and Development (OECD) within the geographical areas with which each is concerned.

A number of statistical series are subject to comparison. An important one is the unemployment rate. Countries use different definitions in calculating the unemployment rate. It is important to know the various definitions in order to make accurate comparisons of labor market conditions.

The age at which persons are assumed to enter the labor force is one source of the difference. The conditions that determine whether a person is unemployed or not in the labor force is another; this depends in some countries on whether the individual actively sought work. Coverage varies especially with respect to institutional population, military, students, and unpaid family workers.

Consumer price indexes also vary by country in concept, scope, and other aspects of construction. In most countries the fixed market basket concept applies but some countries follow the concept of a cost-of-living index more closely than others. In those countries the prices of tangible assets, such as houses, purchased by consumers are calculated on the basis of their rental equivalent or cost of using them as distinct from the total price of buying them. In some countries the scope of the CPI is confined to a certain family of workers—urban wage earners and clerical workers to use examples pertaining to the United States. In some countries the CPI applies only to the capital or other large cities, in others to the country as a whole. There is some variation in the formulas countries use to calculate their indexes and in the way they treat seasonal items such as fruits and vegetables that are not available throughout the year. Some countries conduct and introduce new weights from family expenditure surveys more frequently than others.

The foregoing are important examples of the types of variations that need to be considered in comparing statistics across countries. The International Labor Organization publishes a two-volume *Technical Guide* which is useful in many such comparisons. The first volume is devoted to consumer price indexes. It provides information by country on the title of the index, the index base (i.e., 1967 = 100), the computational formula (i.e., Laspeyres), the source of and data for the expenditure weights, the source of price data and the number of items priced, and the publications containing the price data. The second volume contains similar information for statistics on employment, unemployment, hours of work, and wages.

The ILO also publishes annually the *Year Book of Labor Statistics*. It provides data for the 10 most recent years for 180 countries and territories. The text is in English, Spanish, and French. The information is based mainly on data sent to ILO by national statistical offices. The data are classified according to the International Standard Industrial Classification of All Economic Activities (ISIC) or the International Standard

Classification of Occupation (ISCO), whichever is relevant. The topics for which data are available are (1) the total and economically active population by industry, status, sex, and occupation; (2) employment by industry; (3) unemployment by industry and occupation; (4) hours of work by nonfarm industry; (5) wages for nonagricultural and manufacturing employees; (6) consumer price indexes; (7) industrial accidents; and (8) industrial disputes.

Other relevant ILO publications are *An Integrated System of Wage Statistics: A Manual of Methods* (1979), *Measuring Labour Productivity* (1969), and *Household Income and Expenditure Statistics*, No. 3, 1968–1976, which provides such data on a basis as comparable as possible among countries.

The ILO conducts an international conference of labor statisticians. It has met 12 times between 1923 and 1973 and will meet in 1981. A summary volume contains its recommendations on (1) classification systems (i.e., ISIC); (2) labor force, employment, unemployment, and underemployment data; (3) wages, hours of work, labor costs, and employee income data; (4) consumer prices and price indexes; (5) family living studies; (6) international comparisons of real wages (purchasing power of wages); (7) social security; (8) employee injuries; (9) industrial disputes; and (10) collective bargaining.

Clearly, the ILO plays a key role internationally in labor statistics. The ILO was created by the Treaty of Versailles in 1919, together with the League of Nations. After the League disbanded, the ILO continued in existence, becoming in 1946 the first specialized agency of the United Nations. The ILO received the Nobel Peace Prize in 1969. The bulk of its program consists of the setting of international labor standards and the provision of technical assistance intended to promote national economic and social development and provide employment. The headquarters of the ILO are in Geneva, Switzerland. A branch office is located in Washington, D.C.

References

[1] National Commission on Employment and Unemployment Statistics (1979). *Counting the Labor Force*. U.S. Government Printing Office, Washington, D.C. (Stock No. 052-003-00695-1 or 2).

[2] National Research Council, Committee on National Statistics (1979). *Measurement and Interpretation of Productivity*. National Academy of Sciences, Washington, D.C.

[3] President's Committee to Appraise Employment and Unemployment Statistics (1962). *Measuring Employment and Unemployment*. U.S. Government Printing Office, Washington, D.C.

[4] President's Council on Wage and Price Stability (1977). *The Wholesale Price Index: Review and Evaluation*. U.S. Government Printing Office, Washington, D.C. (Stock No. 052-003-00387-2).

[5] Price Statistics Review Committee of the National Bureau of Economic Research (1961). *The Price Statistics of the Federal Government*. NBER GS No. 73. National Bureau of Economic Research, New York.

(BUREAU OF LABOR STATISTICS
CONSUMER PRICE INDEX
FEDERAL STATISTICS
INDEX NUMBERS
PRODUCER PRICE INDEX
SURVEY SAMPLING
SURVEYS, HOUSEHOLD
WHOLESALE PRICE INDEX)

JOEL POPKIN

LABOUCHÈRE SYSTEMS

The betting system for the even chances (red/black, odd/even, high/low) in roulette, named after the English member of parliament and journalist Henry Du Pre Labouchère (1831–1912), is designed to provide the player with a sequence of bets at the end of which he will have won a fixed amount. In this it is similar in intention to the martingale* (or doubling-up) system, but gives a slower growth in bet size and longer sequences of bets. It suffers from the same flaw as the martingale, in that sooner or later it may require a bet which exceeds either the house limit or the gambler's capital. The gambler then suffers a loss outweighing earlier gains. The system is sometimes called the cross-out, cancellation, or top-and-bottom system, or in one form the split martingale.

The *standard Labouchère system* involves writing down a line of k integers. The size of the first bet (in suitable units) is the sum of the two extreme integers in the line. If the bet wins, these two numbers are deleted; if it loses, the number corresponding to the amount lost is added at the right of the line. The next bet is then the new sum of the extremes, and this process continues until all the numbers, both the original ones and those added, have been deleted. Thus a win results in the deletion of two numbers, while a loss adds only one; analogy with a random walk on the positive integers in which there is unit displacement to the right in the event of a loss and a displacement of two units to the left in the event of a win shows that the process, if unbounded above, is certain to terminate with all numbers deleted unless the probability of a loss exceeds $\frac{2}{3}$. Any completed sequence of bets results in a win equal to the total of the k initial numbers. The system is commonly used with the four numbers 1, 2, 3, and 4 as the starting values; Labouchère himself used the five numbers 3, 4, 5, 6, and 7 and attributed the system to Condorcet (Thorold [2, pp. 60–61] quotes from an article by Labouchère in the newspaper *Truth* for February 15, 1877).

The way in which sequences develop is illustrated by an eight-bet sequence (four wins and four losses) in Table 1. In the split martingale system the initial bet would have been one unit. If it had won, the bet would be repeated; if it had lost, the Labouchère numbers would become 1, 2, 3, 4, and 1 and the standard Labouchère system would then operate.

The *reverse Labouchère system* adds to the line of numbers after a win and deletes the two extreme numbers after a loss, on the fallacious argument that if the Labouchère system gives a large loss to counterbalance a series of small wins, its mirror image will give an occasional larger win to counterbalance a series of small losses. Although this argument ignores the effect of the zero (or zeros) on a roulette wheel, the reverse Labouchère system has the advantage that it can be operated with relatively small capital since large bets are made only when wins from the bank have provided the stake money.

A *parallel Labouchère system* involves operating separate Labouchère processes (either both standard or both reverse) on a pair of opposing even chances (e.g., red and black). These two processes are not independent and no analysis of their joint behavior is known. Although no system can affect the bank's advantage from the zero on a true wheel, a parallel Labouchère system will reduce that advantage if there is a bias on the wheel; it does not require knowledge of the direction of the bias.

Table 1   Typical Labouchère Betting Sequence

Bet Number	Amount Bet	Won or Lost	Total Won or Lost	Labouchère Numbers
0	—	—	—	1 2 3 4
1	5	L	− 5	1 2 3 4 5
2	6	W	+ 1	~~1~~ 2 3 4 ~~5~~
3	6	W	+ 7	~~1~~ ~~2~~ 3 ~~4~~ ~~5~~
4	3	L	+ 4	~~1~~ ~~2~~ 3 ~~4~~ ~~5~~ 3
5	6	L	− 2	~~1~~ ~~2~~ 3 ~~4~~ ~~5~~ 3 6
6	9	L	− 11	~~1~~ ~~2~~ 3 ~~4~~ ~~5~~ 3 6 9
7	12	W	+ 1	~~1~~ ~~2~~ ~~3~~ ~~4~~ ~~5~~ 3 6 ~~9~~
8	9	W	+ 10	~~1~~ ~~2~~ ~~3~~ ~~4~~ ~~5~~ ~~3~~ ~~6~~ ~~9~~

Table 2   Moments of Lengths of Four-Number Labouchère Sequences

	Mean	Variance
Standard (double-zero wheel)	10.28	131.2
Standard (single-zero wheel)	9.42	101.4
Standard or reverse (fair wheel)	8.65	78.9
Reverse (single-zero wheel)	8.00	62.5
Reverse (double-zero wheel)	7.47	50.9

There has been little mathematical analysis of Labouchère systems and few results in closed form are known. Downton [1] used the difference equations for the number of sequences of n games containing m losses which do not contain a completed sequence to compute the probability distributions of the length of completed standard and reverse Labouchère sequences starting with four numbers for true roulette wheels both with a single and with a double zero. These distributions, which have very long tails, do not provide information about the way in which bet sizes build up, which remains an unsolved problem.

Explicit expressions may be obtained for the moments of the distributions using the random walk analogy. For example, if E_k is the expected number of bets in a Labouchère sequence starting with k numbers, and if p and $1 - p$ are the probabilities of winning and losing a bet, respectively, then

$$E_k = 1 + pE_{k-2} + (1 - p)E_{k+1}, \qquad k \geqslant 1,$$

with $E_k = 0$, $k \leqslant 0$. These equations give

$$E_k = \frac{1 + (1 + \alpha)k - (-\alpha)^{-k}}{(3p - 1)(1 + \alpha)},$$

where $\alpha = \sqrt{1/p - 3/4} + 1/2$. Higher moments may similarly be expressed as functions of powers of α. The means and variances of the length of standard and reverse Labouchère bet sequences starting with four numbers for true wheels with a single and double zero, and for a fair wheel with no zero are given in Table 2. In practice these means and variances would be reduced by the truncation of the sequences because of stake limits.

References

[1] Downton, F. (1980). *J. R. Statist. Soc. A*, **143**, 363–366.

[2] Thorold, A. L. (1913). *The Life of Henry Labouchère*. Constable, London. (A comprehensive account of Labouchère's eventful life, probably of more interest to historians than to practicing statisticians.)

(BLACKJACK
GAMBLING, STATISTICS IN
MARTINGALES
RANDOM WALK
ROULETTE
RUNS)

F. DOWNTON

LADDER INDEX

Ladder index is a term used in the theory of random walks*. Let $X_1, X_2, \ldots, X_n$ be an independent sequence of random variables with a common distribution. Define

$$S_0 = 0 \quad \text{and} \quad S_n = X_1 + \cdots + X_n.$$

To visualize the S_n at successive stages, we introduce the integers n such that

$$\max_{0 \leqslant k < n} S_k < S_k.$$

These form the ladder index. They are the positions at which a new high record value of S_n is established.

For more details see, e.g., Billingsley [1].

Reference

[1] Billingsley, P. (1979). *Probability and Measure*. Wiley, New York.

(RANDOM WALKS
RECORDS)

LAGGED DEPENDENT VARIABLES

One uses a lagged dependent variable as an explanatory variable in several situations. The most common is the Koyck [17] transformation, which converts an infinitely long distributed lag with geometrically declining weights into a specification with a lagged dependent variable on the right-hand side of the equation. The simplest case is as follows:

Given that

$$Y_t = \alpha + \beta_0 X_t + \lambda\beta_0 X_{t-1} + \lambda^2\beta_0 X_{t-2} + \cdots + \epsilon_t, \qquad (1)$$

where the ϵ_t's are independent, identically distributed random variables with mean 0 and variance σ^2, and are independent of X, and where $0 \leqslant \lambda < 1$, it follows that

$$\lambda Y_{t-1} = \alpha\lambda + \lambda\beta_0 X_{t-1} + \lambda^2\beta_0 X_{t-2} + \cdots + \lambda\epsilon_{t-1}. \qquad (2)$$

Subtracting (2) from (1) yields

$$Y_t = \alpha(1-\lambda) + \beta_0 X_t + \lambda Y_{t-1} + \epsilon_t - \lambda\epsilon_{t-1}. \qquad (3)$$

Notice that if ϵ_t in (1) is not autocorrelated, the transformation induces autocorrelation. Any autocorrelation in (3) poses an estimation problem, as discussed below. Additional explanatory variables and different lags for different variables can be included in the equation, but estimation should then be done under nonlinear constraints (see Theil [26, Chap. 6]). The deterministic portion of (3) is further elaborated in Jorgenson [14]. Alternative specifications of the lag, such as polynomial lags, can avoid the estimation problems associated with lagged dependent variables, but pose others (*see* LAG MODELS, DISTRIBUTED). The analyst contemplating the use of a model such as (1) should also consider the use of frequency-domain methods, such as cross-spectral analysis. A comprehensive discussion of issues pertaining to distributed lags is found in Dhrymes [2]; see also Griliches [9] and Nerlove [20] for surveys of these issues.

Sims [23] emphasizes the role of dynamic optimization in the type of time-series models that generate lagged dependent variables. For example, a traditional model using a lagged dependent variable is the adaptive expectations model (see Kmenta [16] for a derivation), but this model does not imply optimization of forecasts. For a further discussion of how such models may be inappropriate, especially when used to predict policy consequences, see Sargent [22].

If the error term in the equation to be estimated is not autocorrelated [e.g., if in (3) $\lambda = \rho$, a first-order autocorrelation coefficient], we can enjoy the usual asymptotic properties of least-squares regression, although small sample properties of the ordinary least squares (OLS) estimator may not be very desirable [19, Chaps. 14, 15]. If, however, the error term in the transformed model is autocorrelated, ordinary least squares will be inconsistent. The inconsistency occurs because the value of the lagged dependent variable is not independent of the error term. Consider, for example, (3). From (2) it is clear that the covariance of Y_{t-1} and $\epsilon_t - \lambda\epsilon_{t-1}$ equals $-\lambda\sigma^2$ and not zero, as would be required for OLS to be consistent.

The usual Durbin–Watson test* for detecting autocorrelation cannot be used in cases with a lagged dependent variable, because it is biased toward 2. Durbin has developed two alternative tests for first-order autocorrelation, the h test and the m test [4]. Of these, the m test is preferable because the h test in small samples appears to be biased when the null hypothesis is true and because in certain cases the h test is not computationally feasible [15, 24]. The m test is as follows. If one rejects the null hypothesis that β_1 equals zero in (4), then one concludes that there is autocorrelation present:

$$z_t = \beta_0 + \beta_1 z_{t-1}$$
$$+ \beta_2(\text{vector of predetermined variables})$$
$$+ v_t, \qquad (4)$$

where z_t is the estimated residual from estimating an equation such as (3) and v_t is a standard error term [Equation (3) has only X_t as a predetermined variable; less simple formulations may have additional predetermined variables.] For testing higher-order autocorrelation or moving-average error term processes, a Lagrange multiplier test* described in Godfrey [6, 7] is recommended.

If first-order autocorrelation is present, two estimation methods are recommended; both are consistent and asymptotically efficient. A small-sample comparison has not been made. The first is the method of maximum likelihood [18, 28]. In small samples one must exert care in the treatment of the term involving unobserved values that are prior to the sample period; *see* LAG MODELS, DISTRIBUTED. The second uses instrumental variables* and modifies Hatanaka's [11] procedure by using a different estimator for ρ (asymptotic properties are unaffected). It proceeds as follows: X_{t-1} is used as an instrument for Y_{t-1} in an equation such as (3). The estimated residuals u_t from this equation are consistent, and are used to estimate the autocorrelation coefficient ρ. There is no consensual estimator for ρ; Wallis [27] recommends

$$\hat{\rho} = \frac{\sum_{t=2}^{T} u_t u_{t-1}/(T-1)}{\sum_{t=1}^{T} u_t^2/T} + \frac{3}{T}, \quad (5)$$

where $3/T$ corrects for small-sample bias. Alternatively, Park and Mitchell [21] recommend:

$$\hat{\rho} = \frac{\sum_{t=2}^{T} u_t u_{t-1}}{\sum_{t=2}^{T-1} u_t^2}. \quad (6)$$

The Park and Mitchell estimator minimizes the sum-of-squares conditional on $\hat{\beta}$. No small-sample comparison of these two methods has been undertaken, although each was the best of several methods tried in the two studies. The variables in an equation such as (3) are then transformed by differencing using $\hat{\rho}$, u_{t-1} is included, and least squares is used to estimate:

$$Y_t - \hat{\rho} Y_{t-1} = (1 - \hat{\rho})\alpha + \beta_0(X_t - \hat{\rho} X_{t-1})$$
$$+ \lambda(Y_{t-1} - \hat{\rho} Y_{t-2})$$
$$+ \hat{\rho}_{t-1} u_{t-1}. \quad (7)$$

Hatanaka [11] gives the asymptotic covariance matrix of this estimator, which is simpler than that of another two-step estimator proposed by Wallis [27] and recommended in some econometrics* textbooks. Wallis' estimator is also not asymptotically efficient. For cases of autocorrelation higher than first order and generalization to seemingly unrelated regression*, see Spencer [25]; for the effects of errors in the variables, see Grether and Maddala [8].

All the above has dealt with single equations. For treating lagged endogenous variables in simultaneous equation systems* see Hatanaka [12], Dhrymes and Taylor [3], Hendry and Srba [13], Godfrey [5], Guilkey [10], and Basmann, et al. [1].

References

[1] Basmann, R. L., Richardson, D. H., and Rohr, R. J. (1974). *Econometrica*, **42**, 717–730.

[2] Dhrymes, P. J. (1971). *Distributed Lags: Problems of Estimation and Formulation*. Holden-Day, San Francisco.

[3] Dhrymes, P. J. and Taylor, J. B. (1976). *Int. Econ. Rev.*, **17**, 362–376.

[4] Durbin, J. (1970). *Econometrica*, **38**, 410–421.

[5] Godfrey, L. G. (1976). *Econometrica*, **44**, 1077–1084.

[6] Godfrey, L. G. (1978). *Econometrica*, **46**, 1293–1301.

[7] Godfrey, L. G. (1978). *Econometrica*, **46**, 1303–1310.

[8] Grether, D. M. and Maddala, G. S. (1973). *Econometrica*, **41**, 255–262.

[9] Griliches, Z. (1967). *Econometrica*, **35**, 16–49.

[10] Guilkey, D. K. (1975). *Econometrica*, **43**, 711–717.

[11] Hatanaka, M. (1974). *J. Econometrics*, **2**, 199–220.

[12] Hatanaka, M. (1976). *J. Econometrics*, **4**, 189–204.

[13] Hendry, D. F. and Srba, F. (1977). *Econometrica*, **45**, 969–990.

[14] Jorgenson, D. (1966). *Econometrica*, **34**, 135–149.

[15] Kenkel, J. L. (1974). *Econometrica*, **42**, 763–769.

[16] Kmenta, J. (1971). *Elements of Econometrics*. Macmillan, New York.

[17] Koyck, L. M. (1954). *Distributed Lags and Investment Analysis*. North-Holland, Amsterdam.

[18] Maddala, G. S. (1971) *Econometrica*, **39**, 23–24.

[19] Malinvaud, E. (1966). *Statistical Methods of Econometrics*. North-Holland, Amsterdam.

[20] Nerlove, M. (1972). *Econometrica*, **40**, 221–251.

[21] Park, R. E. and Mitchell, B. M. (1980). *J. Econometrics*, **13**, 185–201.

[22] Sargent, T. J. (1981). *J. Polit. Econ.*, **89**, 213–248.

[23] Sims, C. A. (1974). In *Frontiers of Quantitative Economics*, Vol. 2, M. D. Intriligator and D. A. Kendrick, eds. North-Holland, Amsterdam.

[24] Spencer, B. G. (1975). *J. Econometrics*, **3**, 239–254.

[25] Spencer, D. E. (1979). *J. Econometrics*, **10**, 227–241.

[26] Theil, H. (1971). *Principles of Econometrics*. Wiley, New York.

[27] Wallis, K. F. (1967). *Rev. Econ. Statist.*, **49**, 555–567.

[28] Zellner, A. and Geisel, M. S. (1970). *Econometrica*, **38**, 865–888.

(AUTOREGRESSIVE-INTEGRATED
  MOVING AVERAGE (ARIMA) MODELS
AUTOREGRESSIVE-MOVING AVERAGE
  (ARMA) MODELS
ECONOMETRICS
LAG MODELS, DISTRIBUTED
TIME SERIES)

Joseph P. Newhouse

LAGGING INDICATORS

Lagging indicators are part of a collection of economic time series* designed to provide information about broad swings in measures of aggregate economic activity known as business cycles. (The remaining members of the collection are characterized as leading or coincident indicators; *see* LEADING INDICATORS.) Specifically, lagging indicators are those economic time series whose changes in direction generally *trail* turns in gross national product (GNP: the total market value of all goods and services produced by the economy) or other measures of economic output. Thus lagging indicators are used primarily to *confirm* recent turning points in business cycles.

The first identification of indicators occurred in 1937–1938 at the National Bureau of Economic Research (NBER), a private Cambridge, Massachusetts, firm specializing in business cycle research. NBER initially selected 21 economic time series which seemed to be reliable indicators of economic trends and characterized them as leading, coincident, and lagging indicators of economic conditions (*see* LEADING INDICATORS for a further historical perspective and selection criteria). The most recent (1977) revision of NBER's indicator list gives 111 indicators which can be cross-classified according to timing, economic process, and performance at business cycle peaks and troughs (see ref. 2). There are presently 18 lagging indicators of business cycle peaks and 40 lagging indicators of business cycle troughs. Some lagging indicators, such as the average prime interest rate charged by banks and the average duration of unemployment, are reliable lagging indicators of both (NBER designated) business cycle peaks and troughs.

The current and historical values of all U.S. indicators are published by the U.S. Department of Commerce in *Business Conditions Digest* (see ref. 2).

Groups of indicators are combined to form a broad view *index* of cyclical movement. Composite indices tend to be more reliable and smoother indicators of economic trends than their component series.

In 1981 the index of lagging indicators had six components. The lagging indicators included in the index were: average duration of unemployment (weeks); manufacturing and trade inventories, total (billion $1972); labor cost per unit of output, manufacturing (index: 1967 = 100); average prime interest rate charged by banks (percent); commercial and industrial loans outstanding (million dollars); and the ratio of consumer installment credit to personal income (percent).

Component lagging series are often adjusted before they are included in the index. Moreover, the index of lagging indicators is adjusted to make it compatible with the indices of leading and coincident indicators. (See ref. 2 for a description of the adjustment processes.)

How well do lagging indicators perform? There is very little published empirical evidence documenting the performance of lagging indicators. An analysis covering the 1974–1975 U.S. recession (see ref. 1) showed that the index of lagging indicators began its downturn about the same time the rest of the economy started its upturn. Also, there was no clear upturn in the index of lagging indicators until long after the recession was officially over. In this case the large lags in the turns of the composite index made its usefulness doubtful.

References

[1] Granger, C. W. J. (1980). *Forecasting in Business and Economics*. Academic Press, New York. (Elementary; Chap. 7 contains material on indicators.)
[2] U.S. Department of Commerce (1981). *Business Conditions Digest*. Washington, D.C. (Contains descriptions of indicators and data for many economic time series, including the NBER business cycle indicators.)

Further Reading

Moore, G. H., ed. (1961). *Business Cycle Indicators*, Vol. 1. Princeton University Press, Princeton, N.J. (Contains early contributions to the analysis of business conditions.)

(ECONOMETRICS
INDEX NUMBERS
LABOR STATISTICS
TIME SERIES)

DEAN WICHERN

LAG MODELS, DISTRIBUTED

GENERALITIES

We define a distributed lag model by

$$y_t = \alpha + \sum_{s=0}^{\infty} \beta_s x_{t-s} + u_t, \qquad (1)$$

where y_t is some dependent variable, x_t an independent (exogenous) variable, and u_t a disturbance satisfying

$$E(u_t) = 0, \qquad E(u_t u_s) = \sigma^2 \delta_{ts},$$

$$E(u_t x_{t-s}) = 0. \qquad (2)$$

In addition, it will be assumed that

$$\sum_{s=0}^{\infty} |\beta_s| < \infty. \qquad (3)$$

Models of this kind have had a long history in empirical economics (dating back to Fisher [8]) as a means of modeling inertia induced by psychological factors (e.g., expectations and habit persistence) or technical and institutional rigidities. The reader is referred to Griliches' [2] survey for references to the history of the model.

As an aid to the interpretation of (1), let us suppose that $x_t = 0$ ($t < 0$) and that at $t = 0$ the level of x_t is raised to $\bar{x}$ and sustained at this level. Then, ignoring the disturbance u_t, we have

$$y_0 = \alpha + \beta_0 \bar{x}$$

$$y_1 = \alpha + (\beta_0 + \beta_1)\bar{x}$$

$$y_2 = \alpha + (\beta_0 + \beta_1 + \beta_2)\bar{x},$$

and so on. Thus the complete impact on y_t of a change in x_t is not felt immediately but is *distributed* over time. In fact, it takes an *infinite* period of adjustment before y_t is once again in equilibrium at the new level $\bar{y}$, given by

$$\bar{y} = \alpha + \left(\sum_{s=0}^{\infty} \beta_s \right) \bar{x}. \qquad (4)$$

We note that the assumption (3) guarantees that the new equilibrium exists.

Again, if instead of sustained change in x_t there is a unit impulse in x_t at $t = 0$, it is easily seen that the sequence of lag weights $\beta_0, \beta_1, \beta_2, \ldots$ represents the response of $y_t - \alpha$ over time. If it is believed that this distributed response is induced by frictions of some kind, it may be plausible to assume that the sequence $\{\beta_s\}$ is *smooth*. This assumption, either explicitly or implicitly, underlies many of the attempts to estimate the parameters of (1). A thorough discussion of its validity may be found in Sims [5].

It is convenient at this point to introduce an alternative notation to (1). We define the "lag operator" L by

$$L^r x_t = x_{t-r}, \qquad (5)$$

with $L^0 x_t \equiv I x_t = x_t$, where I is the identity operator. Equation (1) may now be written in the form

$$y_t = \alpha + \left(\sum_{s=0}^{\infty} \beta_s L^s \right) x_t + u_t$$

$$= \alpha + \Phi(L) x_t + u_t, \qquad (6)$$

where

$$\Phi(L) = \beta_0 I + \beta_1 L + \beta_2 L^2 + \cdots, \qquad (7)$$

and is an infinite polynomial in L. The algebra of lag operators is well known and for most purposes L may simply be treated as an ordinary algebraic symbol (see, e.g., Dhrymes [1, Chap. 2]).

Clearly, the model (1) involves an infinite-dimensional parameter space, and cannot therefore be estimated unless restrictions are applied to the parameters β_s. These restrictions can be considered as falling into four groups:

Finite Lag Schemes

It is sometimes assumed that the effect on y_t of a change in x_t works itself out in a *finite* number of periods n, where n is known. Usefulness of this model is somewhat limited as n is rarely, if ever, known and at most only an upper bound for n is available. In addition, if no other restrictions on the β_s are imposed, there will almost certainly be a problem of multicollinearity* (see the section "Estimation of Finite Distributed Lag Schemes"). However, if one is interested mainly in long-run effects, it is the *sum* of the β's which is needed [see (4)] and this may be reasonably precisely estimated, despite the multicollinearity.

Finite Parameter Infinite Lag Schemes

It may be assumed that the infinite sequence $\{\beta_s\}$ is generated by a mechanism involving only a small number of parameters. By far the best-known example is the geometric lag

scheme suggested by Koyck [23]. Under this assumption $\beta_s = \gamma \beta^s$, where $0 < \beta < 1$. Equation (1) now takes the form

$$y_t = \alpha + \gamma \left(\sum_{s=0}^{\infty} \beta^s L^s \right) x_t + u_t$$

$$= \alpha + \left[\gamma I / (I - \beta L) \right] x_t + u_t.$$

Only three parameters α, β, and γ have to be estimated.

Rational Lag Models

It can be shown that the arbitrary, infinite lag polynomial $\Phi(L)$ may be approximated by a rational function of finite polynomials $A(L)/B(L)$. Thus Jorgenson [22] in a well-known investment study approximated (1) by

$$y_t = \alpha + \left[A(L)/B(L) \right] x_t + u_t, \qquad (8)$$

where the degrees of $A(L)$ and $B(L)$ are arbitrarily assigned and small. We note that the Koyck lag scheme mentioned above gives rise to a rational lag model with $A(L) = \gamma I$ and $B(L) = (I - \beta L)$. The transfer function models of Box and Jenkins [15] also take (8) as their starting point; *see* BOX–JENKINS MODEL.

The Jorgenson method has the disadvantage that there is nothing in underlying theory which would provide a guide to the degrees of the polynomials $A(L)$ and $B(L)$. Experience, however, suggests that polynomials of degree no greater than 2 will almost always be adequate (see, e.g., Maddala and Rao [25]). On the other hand, a plus for this approach is that the roots of the polynomial $B(L)$ provide information on two important aspects of distributed lag model behavior— the speed of adjustment in y_t to a change in x_t, and the nature of the time pattern of adjustment (monotonic or cyclical).

If we multiply (8) by $B(L)$, we obtain an alternative form, known as the "dynamic form" or "autoregressive form" and given by

$$B(L) y_t = A(L) x_t + B(L) u_t. \qquad (9)$$

Expressed in this form, the model is characterized by lagged endogenous variables and autocorrelated disturbances.

Models Involving Unobservable Variables

As an advance on simply postulating an equation of the form (1), attempts have been made to model explicitly the inertia factors underlying the distributed lag schemes and the restrictions on the lag coefficients result from a definite theory. Familiar examples are the hypotheses known as adaptive expectations, partial adjustment, permanent income, and dynamic demand. We will consider very briefly the adaptive expectations hypothesis model as a simple example of how such hypotheses give rise to restrictions on the lag coefficients.

Suppose that the supply of a commodity q_t depends linearly on the expected (unobservable) price p_t^*. That is,

$$q_t = \alpha_0 + \alpha_1 p_t^* + u_t, \qquad (10)$$

where u_t is a disturbance whose properties can be unspecified.

The supplier is assumed to adapt his expectations according to the mechanism

$$p_t^* - p_{t-1}^* = (1 - \beta)[p_t - p_{t-1}^*]$$
$$(0 < \beta < 1).$$

That is, in lag operator notation,

$$[I - \beta L]p_t^* = (1 - \beta)p_t. \qquad (11)$$

The price expectation p_t^* can now be eliminated from (10) to give

$$q_t = \alpha_0 + [\alpha_1(1 - \beta)I/(I - \beta L)]p_t + u_t,$$

and it is clear that the distributed lag in p_t is of the geometric form $\beta_s = \alpha_1(1 - \beta)\beta^s$. We note from (11) that under the adaptive expectation hypothesis

$$p_t^* = (1 - \beta)\sum_{s=0}^{\infty} \beta^s p_{t-s},$$

implying that current expected price is a weighted average of *past* prices alone. Reference is made to Nelson [35] for a comment on this.

ESTIMATION OF FINITE-DISTRIBUTED LAG SCHEMES

The model (1) is restricted to

$$y_t = \sum_{s=0}^{n} \beta_s x_{t-s} + u_t, \qquad (12)$$

where n is known. If the x_{t-s} are fixed in repeated samples, then ordinary least squares (OLS) will produce estimates which are best linear unbiased. On the other hand, if the x_{t-s} are stochastic, then (under fairly general conditions on the x_{t-s}) OLS will produce unbiased and consistent estimates. If, in addition, u_t is assumed to be normally distributed, the OLS estimator is also asymptotically efficient. However, the special nature of the regressor variables, being successive observations on the *same* variable x_t, virtually guarantees multicollinearity, and hence imprecise estimates of the β_s. Thus additional assumptions on the nature of the lag distribution are often applied.

By far the most popular technique, proposed originally by Almon [6], is based on the assumption that the lag weights β_s lie on a polynomial of degree $k < n + 1$. That is,

$$\beta_s = \sum_{j=0}^{k} \gamma_j s^j. \qquad (13)$$

Substituting (13) into (12), we have

$$y_t = \sum_{j=0}^{k} \gamma_j z_{tj} + u_t, \qquad (14)$$

where

$$z_{tj} = \sum_{s=0}^{n} s^j x_{t-s}. \qquad (15)$$

There are now $k + 1$ parameters instead of $n + 1$, implying that a further $n - k$ restrictions have been imposed on the β_s.

If (13) is correct, then OLS applied to (14) will produce unbiased, efficient estimates of the γ_j, and hence the β_s. The problem here is that usually neither n nor k is known. Many authors (see Dhrymes [1], Schmidt and Waud [12], and Trivedi and Pagan [14]) have pointed out that arbitrary assumptions about these parameters can have drastic effects on the nature of the implied lag distribution.

Methods for removing the arbitrariness of this choice are thus of crucial importance. Amemiya and Morimune [7] consider optimizing the choice of k by consideration of a loss function, while Godfrey and Poskitt [9] and Trivedi and Pagan [14] propose sequences of nested hypothesis tests.

A further problem (see Schmidt and Waud [12]) is that there is no way within the Almon framework to test for no distributed lag (i.e., $n = 0$). Schmidt [27] has therefore suggested a modification to (13) defined by

$$\beta_s = \alpha^s \sum_{j=0}^{k} \gamma_j s^j \qquad (0 \leqslant \alpha < 1). \quad (16)$$

This scheme is similar to Almon's for small s, and approaches the Koyck distribution as $s \to \infty$. It includes the case of no lag ($\alpha = 0$), and has the interesting property (noted earlier in connection with rational lags) that it can approximate any arbitrary infinite-distributed lag (see Schmidt and Mann [30]).

The restrictive nature of the polynomial assumption has led to considerable research directed toward finding more flexible parameterizations, which nonetheless preserve the spirit of Almon's approach. We will briefly mention two of these, in addition to the Schmidt modification.

First, Shiller [13], recognizing the polynomial assumption as imposing $n - k$ *exact* linear restrictions on the differences of the β_s, proposed that *stochastic* restrictions be applied.

Second, Poirier [11] has suggested the use of splines* as a means of optimally achieving smoothness, and has presented empirical evidence to support his contention that spline lags are more stable than Almon lags.

ESTIMATION OF FINITE PARAMETER, INFINITE LAG MODELS

The infinite lag model is estimated either by applying a finite parameter scheme, or using the rational lag model, perhaps in the auto-regressive form. Apart from the Koyck lag, finite parameter models include the Pascal [25] and the gamma lags [28].

We will continue to use the adaptive expectations model to illustrate general principles, and will assume that the disturbance is generated by a first-order autoregressive process. The model under consideration is thus

$$y_t = \alpha \big[I/(I - \beta L) \big] x_t + \big[I/(I - \rho L) \big] \epsilon_t, \quad (17)$$

and estimation will be considered under four headings. In this treatment we will generally follow Dhrymes [1].

Maximum Likelihood* Methods

If we set $x_t^* = [I/(I - \beta L)]x_t$, then x_t^* is generated by a first-order difference equation* which can be solved to give

$$x_t^* = \beta^t x_0^* + \sum_{s=0}^{t-1} \beta^s x_{t-s}, \quad (18)$$

where $x_0^* = \sum_{s=0}^{\infty} \beta^s x_{-s}$. We note that x_0^*, often called the "truncation remainder," is a summary of all the preperiod values of x_t and is not observable. The splitting of the lag distribution in (18) into terms involving preperiod (unobservable) and sample period (observable) values of x_t is fundamental to all maximum likelihood methods, whether applied to finite parameter or rational lag schemes.

The formulation and maximization of the likelihood function are straightforward and can be found in Dhrymes [1]. Three essential features emerge:

1. Analytic maximization is conditional on the parameters β and ρ. The global maximum is thus obtained by a *search procedure* over the (β, ρ) parameter space.

2. The truncation remainder x_0^* is treated as a parameter and estimated.

3. Maximum likelihood estimation provides consistent, efficient estimates. (The parameter x_0^* is an exception. As it contains only preperiod observations, it can never be estimated consistently.)

Two main variants of the procedure discussed above have been proposed. First, the preperiod values of x_t have been estimated, using some assumption about the nature of the process generating x_t, and second, the parameter x_0^* has been ignored. From (18), the parameter x_0^* is associated multiplicatively with β^t, and hence its effect will diminish as $t \to \infty$. Thus, whichever of the three alternatives above is used, the *asymptotic* properties are unchanged. However, evidence from Maddala and Rao [25] and Schmidt [29] shows that the small-sample properties can be substantially affected by the treatment of x_0^*. These authors suggest that the truncation remainder should not simply be ignored.

An interesting synthesis of maximum likelihood methods can be found in Nicholls et al. [4] or Pagan and Byron [26].

Instrumental Variable* Estimation

Multiplying (17) through by the lag operator $(I - \beta L)$ we obtain

$$y_t = \alpha x_t + \beta y_{t-1} + w_t, \qquad (19)$$

where $w_t = [(I - \beta L)/(I - \rho L)]\epsilon_t$. Application of OLS to (19) will of course yield inconsistent estimators of α and β. Liviatan [24] proposed instrumental variable regression, using x_{t-1} as an instrument for y_{t-1}. This estimator is consistent, but as it takes no account of the covariance structure of w_t, cannot be efficient. The Liviatan estimator has achieved a significant place in distributed lag estimation, as it provides an easy method for obtaining consistent estimates of α and β, which may then be used in two-step procedures (see below).

Two-Step Methods

In order to avoid the search procedure necessary for maximum likelihood, methods have been developed which involve only OLS or, at most, GLS estimation. A consistent procedure (e.g., Liviatan's) is applied as a first step and residuals are generated which are used to estimate the structure of the disturbances. Feasible Aitken estimation then follows.

In (17), the parameter β appears nonlinearly. A linearized version can be conveniently obtained in two ways, leading to different disturbance processes.

First, using the difference equation for the unobservable variable x_t^*

$$y_t = \alpha x_t + \alpha \beta x_{t-1}^* + u_t, \qquad (20)$$

where $(I - \rho L)u_t = \epsilon_t$. Given an initial consistent estimate of β, the variable x_{t-1}^* may be predicted, as suggested by (18), by

$$\tilde{x}_{t-1}^* = \sum_{s=0}^{t-2} \tilde{\beta}^s x_{t-s-1}. \qquad (21)$$

Residuals $\tilde{u}_t = y_t - \tilde{\alpha}x_t - \tilde{\alpha}\tilde{\beta}\tilde{x}_{t-1}^*$ are used to obtain a consistent estimate of ρ, and feasible Aitken estimation is then applied to (20), with $\tilde{x}_{t-1}^*$ replacing x_{t-1}^*. Gupta [19] takes advantage of the autoregressive nature of u_t to avoid Aitken estimation. He writes the model as

$$y_t = \alpha x_t + \alpha \beta x_{t-1}^* + \rho u_{t-1} + \epsilon_t$$

and including $\tilde{u}_{t-1}$ among the regressors of the second step applies OLS.

Second, Dhrymes [16] uses the form (19) with y_{t-1}, on the right-hand side, and the autoregressive–moving average* disturbance w_t. Proceding in exactly the same way as before to obtain consistent estimates of β and ρ, he estimates the structure of w_t.

One-Step Gauss–Newton Estimators

The likelihood function of the general dynamic model

$$A(L)y_t = B(L)x_t + C(L)\epsilon_t \qquad (22)$$

can be concentrated so that maximum likelihood estimates can be obtained by minimizing the sum of squares $S(\theta) = \epsilon'(\theta)\epsilon(\theta)$ (see, e.g., Nicholls, et al. [4] or Godfrey [17]), where θ is used to denote the vector of parameters in (22). Suppose that a consistent estimator $\tilde{\theta}$ is available, and used to generate the vector of residuals $\tilde{\epsilon}(\tilde{\theta})$. Hartley and Booker [20] have shown that a one-step

Gauss–Newton iteration from $\tilde{\theta}$, defined by

$$\hat{\theta} = \tilde{\theta} - \left[\frac{\partial \tilde{\epsilon}(\tilde{\theta})'}{\partial \theta} \frac{\partial \tilde{\epsilon}(\tilde{\theta})}{\partial \theta} \right]^{-1} \frac{\partial \tilde{\epsilon}(\tilde{\theta})'}{\partial \theta} \tilde{\epsilon}(\tilde{\theta}),$$

(23)

where $\partial \tilde{\epsilon}(\tilde{\theta})/\partial \theta$ is the matrix obtained by differentiating the vector $\epsilon(\theta)$ with respect to θ evaluated at $\tilde{\theta}$, produces estimates $\hat{\theta}$ which are asymptotically efficient; *see* NEWTON IT-ERATION EXTENSIONS. This forms the basis for Pagan's [10] suggestion for the construction of *efficient* two-step estimators. Noting the form of the right-hand side of (23), Pagan suggests regressing $- \tilde{\epsilon}(\tilde{\theta})$ on $\partial \tilde{\epsilon}(\tilde{\theta})/\partial \theta$ using OLS to obtain an estimate $\hat{\hat{\theta}}$. An efficient estimator, $\hat{\theta}$, say, is defined as $\hat{\theta} = \tilde{\theta} + \hat{\hat{\theta}}$. Hatanaka's [21] efficient two-step estimator of the dynamic demand model belongs to this class (see Pagan and Byron [26]).

HYPOTHESIS TESTING

Our treatment of infinite lag models has concentrated exclusively on estimation, and this merely reflects the imbalance in the methodology currently available for distributed lag analysis.

Hypothesis tests relating to particular parameters or groups of parameters, such as tests concerning short- or long-run responses or speed of adjustment, may be routinely constructed using the known asymptotic distributions of the various estimators.

It is in the area of model specification, when lagged endogenous variables are present, that developments have been somewhat slow. The reader is referred to Sims' [5] review for a discussion of the issues involved. Recent contributions by Godfrey [17, 18] provide specification tests which are likely to be more appropriate than the well-known Durbin h-test.

AN EXAMPLE

As a simple example, the permanent income hypothesis was applied to data from the Australian economy. Annual observations on private consumption expenditure (C) and gross national expenditure (Y) (both series deflated) for the years 1948–1949 to 1979–1980 were obtained from Australian national accounts.

The permanent income hypothesis states that consumption expenditure is determined by "permanent" income (say Y_t^*) rather than actual income. We will assume then that

$$C_t = \alpha + \beta Y_t^* + u_t.$$

(24)

Furthermore, it is assumed that Y_t^* is adjusted by the adaptive expectations mechanism. That is,

$$Y_t^* - Y_{t-1}^* = \gamma \left[Y_{t-1} - Y_{t-1}^* \right]$$

$$(0 < \gamma < 1). \quad (25)$$

Thus we are assuming that permanent income is adjusted by a fixed proportion γ of the amount by which permanent and actual income differed in the last period.

Substituting (25) into (24), current consumption may be expressed as a distributed lag in actual income as follows:

$$C_t = \alpha + \beta \gamma \left[I/(I - (1 - \gamma)L) \right] Y_{t-1} + u_t.$$

(26)

That is,

$$C_t = \alpha + \beta \gamma \sum_{s=0}^{\infty} (1 - \gamma)^s Y_{t-s-1} + u_t. \quad (27)$$

Expressing $\sum_{s=0}^{\infty} (1 - \gamma)^s Y_{t-s-1}$ as the sum of terms involving only sample period observations on Y_t and those involving preperiod observations, we have that

$$C_t = \alpha + \beta^* X_{1t}(\gamma) + \beta^{**} X_{2t}(\gamma) + u_t,$$

(28)

where

$$X_{1t}(\gamma) = 0 \qquad\qquad \text{if } t = 1$$

$$= \sum_{s=0}^{t-2} (1 - \gamma)^s Y_{t-s-1} \quad (t > 1)$$

$$X_{2t}(\gamma) = (1 - \gamma)^{t-1},$$

$$\beta^* = \beta \gamma,$$

and

$$\beta^{**} = \beta^* \sum_{s=0}^{\infty} (1 - \gamma)^s Y_{-s}.$$

Equation (28) is the estimating equation. Maximum likelihood estimates are obtained by constructing $X_{1t}(\gamma)$ and $X_{2t}(\gamma)$ *conditional on* γ, applying OLS to (28), and searching over γ for the value that produces the minimum residual sum of squares (RSS). The following results were obtained:

γ	RSS
0.2	4.0860×10^6
0.4	2.6378×10^6
0.6	2.9725×10^6
0.8	3.9376×10^6

After taking finer divisions in the value of γ, we finally obtained $\hat{\gamma} = 0.436$, and at this value the estimated (28) is

$$\hat{C}_t = 848.85 + 0.2856\, X_{1t} + 6644.7\, X_{2t};$$
$$\quad\;\; (163.54) \quad\; (0.0031) \qquad\; (327.2)$$

$$R^2 = 0.997, \tag{29}$$

where the numbers in parentheses are standard errors and R^2 is the coefficient of determination. The estimated permanent income model is therefore

$$C_t = 848.85 + 0.6550\, Y_t^*, \tag{30}$$

$$Y_t^* - Y_{t-1}^* = 0.436\big[Y_{t-1} - Y_{t-1}^* \big]. \tag{31}$$

The dynamic (or autoregressive) form is obtained by multiplying (26) by the operator $I - (I - \gamma)L$ and is

$$C_t = 370.1 + 0.564\,C_{t-1} + 0.2856\, Y_{t-1} + v_t,$$
$$\tag{32}$$

where v_t is a disturbance with a first-order moving-average structure.

The implications of the estimated model are the following:

1. If the value of income is changed, there will be no immediate response in consumption. Consumption starts to respond after one period has elapsed.

2. If a unit sustained increase in income occurs, after all adjustment has taken place, consumption will have increased by 0.655 unit.

3. The speed of adjustment of consumption to changes in income is determined by γ.

If Y_t increases by Δy at $t = 0$, it is easily deduced from (27) that C_t increases by $\beta[1 - (1 - \gamma)^t]\Delta y$ after t periods. If we denote by h_t the proportion of total adjustment which has taken place after t time periods, then

$$h_t \equiv \frac{(\Delta C)_t}{(\Delta C)_\infty} = \frac{\beta\big[1 - (1 - \gamma)^t\big]\Delta y}{\beta\,\Delta y}$$
$$= \big[1 - (1 - \gamma)^t\big].$$

Thus

$$t = \ln(1 - h_t)/\ln(1 - \gamma). \tag{33}$$

Setting $h_t = 0.8$, 0.9, and 0.99 (with $\gamma = 0.436$) we deduce from (33) that C_t will adjust 80% after approximately three periods, 90% after approximately four periods and 99% after approximately eight periods.

Further Reading

There are a number of important topics in distributed lag analysis which are outside the scope of a relatively short introductory article. We mention below some of these, together with a selection of references.

1. **Bayesian Methods.** In this approach uncertainty about the unknown parameters β_s of the distributed lag is expressed in terms of subjective probability density functions. Prior knowledge of the β's is incorporated into the analysis via assumed prior distributions, and then sample information is combined with these prior distributions, using Bayes' theorem, to produce posterior distributions for the β_s [34, 36].

2. **Distributed Lag Models Based on Optimizing Assumptions and/or Rational Expectations.** We have seen that attempts have been made to model the underlying causes of distributed lags (e.g., adap-

tive expectations), but as Griliches [2] has remarked that "the theoretical rationalizations offered are often only skin deep." A sounder theoretical base has been sought by postulating behavior based on optimization of some objective function (see, e.g., Nerlove [3]). A major problem here appears to be that the restrictions on the lag coefficients become exceedingly complicated. Another approach, sometimes used in conjunction with the optimization assumption, is to use the "rational expectations hypothesis," in which it is assumed that expectations are formed by taking into account *all* relevant economic interrelationships, not just past values of the variable of interest (usually price). Expectations are formed within the context of an explicit structural model, and the distributed lags arise through an assumption on the behavior of the *exogenous* variables (see Nelson [35], Hansen and Sargent [33], and Wallis [39]).

3. Frequency-Domain (Spectral) Estimation [31, Chap. 4; 32, Chap. 7].

4. Causality and Exogeneity [38].

5. Time Aggregation [37].

References

Book

[1] Dhrymes, P. J. (1971). *Distributed Lags: Problems of Estimation and Formulation*. Holden-Day, San Francisco. (This book has a very thorough treatment of maximum likelihood methods of estimation, and the derivation of the asymptotic properties of the various estimators. Because of its great detail, it is not recommended for a first approach.)

Review Articles

[2] Griliches, Z. (1967). *Econometrica*, **35**, 16–49. (Concentrates mainly on distinguishing between different lag models.)

[3] Nerlove, M. (1972). *Econometrica*, **40**, 221–252. (Nerlove's emphasis is on the development of adequate theories of dynamic economic behavior leading to lag models.)

[4] Nicholls, D. F., Pagan, A. R., and Terrell, R. D. (1975). *Int. Econ. Rev.*, **16**, 113–134. (This survey is particularly helpful for an understanding of the maximum likelihood estimators used for rational distributed lag models.)

[5] Sims, C. (1974). In *Frontiers of Quantitative Economics*, Vol. 2, M. D. Intrilligator and D. A. Kendrick, eds. North-Holland, Amsterdam. (A discussion containing many useful insights into the problem areas of distributed lags.)

Finite Lag Models

[6] Almon, S. (1965). *Econometrica*, **33**, 178–196.

[7] Amemiya, T. and Morimune, K. (1974). *Rev. Econ. Statist.*, **56**, 378–386.

[8] Fisher, I. (1937). *Bull. Inst. Int. Statist.*, **29**, 323–328.

[9] Godfrey, L. G. and Poskitt, D. S., (1975). *Amer. Statist. Ass.*, **70**, 105–108.

[10] Pagan, A. R. (1978). *J. Econometrics*, **8**, 247–254.

[11] Poirier, D. J. (1976). *The Econometrics of Structural Change*. North-Holland, Amsterdam.

[12] Schmidt, P. and Waud, R. N. (1973). *J. Amer. Statist. Ass.*, **68**, 11–19.

[13] Shiller, R. J. (1973). *Econometrica*, **41**, 775–788.

[14] Trivedi, P. K. and Pagan, A. R. (1976). Polynomial Distributed Lags: A Unified Treatment. *Australian National University Working Paper No. 34*.

Infinite Lag Models

[15] Box, G. E. P. and Jenkins, G. M. (1970). *Time Series Analysis: Forecasting and Control*. Holden-Day, San Francisco.

[16] Dhrymes, P. J. (1969). *Int. Econ. Rev.*, **10**, 47–67.

[17] Godfrey, L. G. (1978). *Econometrica*, **46**, 1293–1301.

[18] Godfrey, L. G. (1978). *Econometrica*, **46**, 1303–1310.

[19] Gupta, Y. P. (1969). *Int. Econ. Rev.*, **10**, 112–113.

[20] Hartley, H. O. and Booker, A. (1965). *Ann. Math. Statist.*, **36**, 638–650.

[21] Hatanaka, M. (1974). *J. Econometrics*, **2**, 199–220.

[22] Jorgenson, D. W. (1966). *Econometrica*, **34**, 135–149.

[23] Koyck, L. M. (1954). *Distributed Lags and Investment Analysis*. North-Holland, Amsterdam.

[24] Liviatan, N. (1963). *Int. Econ. Rev.*, **4**, 44–52.

[25] Maddala, G. S. and Rao, A. S. (1971). *Rev. Econ. Statist.*, **52**, 80–88.

[26] Pagan, A. R. and Byron, R. P. (1978). In *Stability and Inflation*, R. Bergstrom, ed. Wiley, New York.

[27] Schmidt, P. (1974). *J. Amer. Statist. Ass.*, **69**, 678–681.

[28] Schmidt, P. (1974). *Int. Econ. Rev.*, **15**, 246–250.

[29] Schmidt, P. (1975). *Rev. Econ. Statist.*, **57**, 387–389.

[30] Schmidt, P. and Mann, W. R., (1977). *J. Amer. Statist. Ass.*, **72**, 442–443.

Miscellaneous

[31] Fishman, G. S. (1969). *Spectral Methods in Econometrics*, Harvard University Press, Cambridge, Mass.

[32] Hannan, E. J. (1970). *Multiple Time Series*. Wiley, New York.

[33] Hansen, L. P. and Sargent, T. J. (1980). *J. Econ. Dynamics Control*, **2**, 7–46.

[34] Leamer, E. E. (1972). *Econometrica*, **40**, 1059–1081.

[35] Nelson, C. R. (1975). *Int. Econ. Rev.*, **16**, 555–561.

[36] Richard, J. F. (1977). In *New Developments in the Application of Bayesian Methods*, K. Aykac, and C. Brumat, eds.

[37] Sims, C. A. (1971). *Econometrica*, **39**, 545–564.

[38] Sims, C. A. (1972). *Amer. Econ. Rev.*, **62**, 540–552.

[39] Wallis, K. F. (1980). *Econometrica*, **48**, 49–73.

Acknowledgment

This work was carried out while the author was a Fellow of the Alexander von Humboldt Foundation at the University of Bonn.

(AUTOREGRESSIVE-INTEGRATED
    MOVING AVERAGE (ARIMA) MODELS
AUTOREGRESSIVE MOVING AVERAGE
    (ARMA) MODELS
ECONOMETRICS
MULTIPLE REGRESSION
SERIAL CORRELATION
SPECTRAL ANALYSIS)

HOWARD E. DORAN

LAGRANGE AND RELATED PROBABILITY DISTRIBUTIONS

The wide class of discrete Lagrangian probability distributions, defined by Consul and Shenton [5–7], consists of many important families such as the generalized Poisson distribution [3], generalized negative binomial distribution* [10], generalized logarithmic series distribution* [11], modified power series distribution* [8], and the generalized power series distribution* [16].

If $g(s)$ is an analytic function of s such that

$$g(0) > 0, \qquad g(1) = 1$$

and

$$\left(\frac{\partial}{\partial s} \right)^{x-1} (g(s))^x \big|_{s=0} \geqslant 0, \qquad x \geqslant 2, \quad (1)$$

then the transformation $s = u\, g(s)$ defines, for the numerically smallest nonzero root, a probability generating function* (PGF) $s = \phi(u)$ whose expansion in powers of u is given by

$$s = \phi(u) = \sum_{x=1}^{\infty} \frac{u^x}{x!} \left(\frac{\partial}{\partial s} \right)^{x-1} (g(s))^x \big|_{s=0}$$

(2)

[*see* LAGRANGE EXPANSIONS, formula (1), where $f(s) = s$].

The formula above is called the basic Lagrangian PGF and the discrete probability distribution represented by it, i.e.,

$$\Pr(X = x) = \frac{1}{x!} \left(\frac{\partial}{\partial s} \right)^{x-1} (g(s))^x \big|_{s=0},$$

$$x \in N \quad (3)$$

is the basic Lagrangian probability distribution (basic LPD) defined on N, a subset of the set of positive integers.

It can be easily seen that numerous values of $g(s)$ satisfying the conditions (1) give particular families of the basic LPD. Table 1 gives some examples.

It has been shown (see Consul and Shenton [7]) that all basic LPDs are closed under convolution and that the cumulants D_k, $k = 1, 2, 3, 4$ of the basic LPD are

$$D_1 = \{1 - g'(s)\}^{-1} \big|_{s=1}$$

$$D_2 = G_2 D_1^3, \qquad D_3 = G_3 D_1^4 + 3 G_2 D_1^5 \quad (4)$$

$$D_4 = G_4 D_1^5 + 10 G_3 G_2 D_1^6 + 15 G_2^3 D_1^7,$$

where G_2, G_3, and G_4 are the cumulants given by the probability generating function $g(s)$.

Table 1

Number	$g(s)$	Probability Distribution
1	$\theta + (1 - \theta)s, 0 < \theta < 1$	Geometric distribution
2	$\theta\{1 - (1 - \theta)s\}^{-1}, 0 < \theta < 1$	Haight distribution
3	$\theta + (1 - \theta)s^2, 0 < \theta < 1$	Simple asymmetric random walk
4	$\{\theta + (1 - \theta)s\}^m, 0 < \theta < 1, m > 1$	Consul distribution
5	$\theta^{m(s-1)}, m > 0$	Borel distribution

Consul and Shenton [7] have given a formula for higher-order cumulants of the basic LPD and have also shown that all basic LPDs are the probability distributions of the busy periods of a single server when the queue is initiated by a single customer and is served on the basis of first come, first served.

THE GENERAL LPD

If $f(s)$ is another analytic function such that $0 \leqslant f(0) < 1, f(1) = 1$, and

$$\left. \frac{\partial^{x-1}}{\partial s^{x-1}} \left\{ (g(s))^x \frac{\partial f(s)}{\partial s} \right\} \right|_{s=0} \geqslant 0, \qquad x \geqslant 1,$$

$$(5)$$

then the general Lagrangian probability distribution is given by [see LAGRANGE EXPANSIONS, formula (1)]

$$\Pr(X = x) = \begin{cases} f(0), & x = 0 \\ \left. \dfrac{1}{x!} \dfrac{\partial^{x-1}}{\partial s^{x-1}} \left\{ (g(s))^x \dfrac{\partial f(s)}{\partial s} \right\} \right|_{s=0}, \\ & x \in N \end{cases}$$

where N is a subset of the set of positive integers.

The PGF of the general LPD is a function of u given by $f(s)$, where $s = u \cdot g(s)$.

The following are some important results for the general LPDs:

1. The general LPDs given by $f(s) = s^n$ are the n-fold convolutions of the basic LPD.

2. The general LPD given by $f(s) = g(s)$ is the basic LPD displaced by one step to the left.

3. The general LPD is obtained by randomizing the index parameter n in the probability distribution obtained from $f(s) = s^n$ according to any other probability distribution.

4. For a given transformation $s = ug(s)$ all LPDs are closed under convolution.

5. If F_1, F_2, F_3, F_4 are the cumulants of the PD given by the PGF $f(s)$, then the cumulants L_k, $k = 1, 2, 3, 4$, of the general LPD are given by

$$L_1 = F_1 D_1$$
$$L_2 = F_1 D_2 + F_2 D_1^2$$
$$L_3 = F_1 D_3 + 3F_2 D_1 D_2 + F_3 D_1^3 \qquad (6)$$
$$L_4 = F_1 D_4 + 3F_2 D_2^2 + 4F_2 D_1 D_3$$
$$\qquad + 6F_3 D_1^2 D_2 + F_4 D_1^4,$$

where D_k, $k = 1, 2, 3, 4$, are the cumulants of the basic LPD.

By giving specific values to $g(s)$ and $f(s)$ we can write the three most interesting families of the general LPD.

1. $g(s) = e^{\alpha(s-1)}$, $f(s) = e^{\theta\alpha(s-1)}$, $0 \leqslant \alpha < 1, \theta > 0$ give the family of generalized Poisson distributions [3, 4] and is discussed under the title of Lagrangian Poisson distribution (LPD) in the next section.

2.

$$g(s) = (1 - \theta + \theta s)^m, \quad \text{and}$$
$$f(s) = (1 - \theta + \theta s)^n \quad \text{or}$$
$$g(s) = (1 - \theta)^{m-1}(1 - \theta s)^{1-m}, \quad \text{and}$$
$$f(s) = (1 - \theta)^n(1 - \theta t)^{-n},$$

where $0 < \theta < 1$, $n \geqslant 1$, $0 \leqslant m\theta < 1$,

$m = 0$, or $m \geqslant 1$ provide the family of the generalized negative binomial distribution which is discussed in a subsequent section as the Lagrangian binomial distribution (LBD).

3. $g(s) = (1 - \theta)^{m-1}(1 - \theta s)^{1-m}$, $f(s) = \log(1 - \theta s)/\log(1 - \theta)$, $0 < \theta < 1$, $m \geqslant 1$, $0 < m\theta < 1$ gives the family of generalized logarithmic series distribution.

It can be easily seen that numerous such families can be obtained by assigning different values to $g(s)$ and $f(s)$ satisfying the conditions $g(0) > 0$, $g(1)$, $0 \leqslant f(0) < 1$, $f(1) = 1$, and (5).

The bivariate and multivariate forms of Lagrangian distributions can be generated by the bivariate and multivariate forms of Lagrangian expansions [see formulas (2), (3), and (4) in that article] and have been given by Shenton and Consul [17] and Consul and Shenton [6] and are partially discussed in the final section.

THE LAGRANGIAN POISSON DISTRIBUTION (LPD)

A number of its properties were discussed under the title of a generalization of the Poisson distribution by Consul and Jain [3], although one of the conditions was incorrect. The LPD is given by

$$\Pr(X = x) = \frac{(1 + \alpha x)^{x-1}}{x!} \frac{(\theta e^{-\alpha\theta})^{x}}{e^{\theta}}, \quad (7)$$

for $x = 0, 1, 2, 3, \ldots$, $\theta > 0$, $0 \leqslant \alpha\theta < 1$.

Janardan et al. [12] have given some interesting applications of this distribution. Consul and Jain [3, 4] have given many other interesting properties.

The first four central moments become

$$\mu_1'(X) = \theta(1 - \alpha\theta)^{-1},$$
$$\mu_2(X) = \theta(1 - \alpha\theta)^{-3},$$
$$\mu_3(X) = \theta(1 + 2\alpha\theta)(1 - \alpha\theta)^{-5}, \quad (8)$$
$$\mu_4(X) = 3\theta^2(1 - \alpha\theta)^{-6}$$
$$+ \theta(1 + 8\alpha\theta + 6\alpha^2\theta^2)(1 - \alpha\theta)^{-7}.$$

Also some of the negative moments given by Kumar and Consul [14] about $-k$ $(k > 0)$ are

$$E\left[(X + k)^{-1}\right]$$

$$= \begin{cases} \dfrac{1}{k} - \dfrac{\alpha\theta}{k + 1} & \text{for } \alpha k = 1; \\[3mm] -\dfrac{\alpha}{1 - \alpha k} + e^{-\theta(1 - \alpha k)} \\[3mm] \times \displaystyle\sum_{i=0}^{\infty} \dfrac{(1 - \alpha k)^{i-1}\theta^{i}}{i!\,(k + i)} & \alpha k \neq 1; \end{cases}$$

$$(9)$$

$$E\left[(X + k)^{-2}\right] = \frac{\alpha\theta}{k + 1}\left[\frac{\alpha\theta}{k + 2} - \frac{1}{k + 1}\right]$$
$$+ \frac{1}{k} E\left[(X + k)^{-1}\right]$$
$$\text{for } \alpha k = 1.$$

If $X_1, X_2, X_3, \ldots, X_n$ is a random sample of size n taken from the LPD, the maximum likelihood estimator of θ is given [9] by

$$\hat{\theta} = \frac{\overline{X}}{1 + \alpha\overline{X}}, \quad (10)$$

which is a biased estimator. The bias and variance of $\hat{\theta}$ [14] are

$$b(\hat{\theta}) = -\frac{\alpha\theta}{n + \alpha},$$
$$\quad (11)$$
$$V(\hat{\theta}) = \frac{n^2\theta}{(n + \alpha)^2}\left[\frac{1}{n} - \frac{\alpha\theta}{n + 2\alpha}\right].$$

If y is the sum of the sample values, the minimum variance unbiased estimators $\{\{l(\theta)\}\}$ of some functions $l(\theta)$ of the parameter θ are given in [15] as

$$\{\{\theta^{m}\}\} = \frac{y!}{(y - m)!} \frac{(n + \alpha m)}{n(n + \alpha y)^{m}},$$

$$y \geqslant m, \quad \text{and zero otherwise}$$

$$\{\{\Pr(X = m)\}\} = \binom{y}{m}\frac{n - 1}{n}$$

$$\times \frac{(1 + m\alpha)^{m-1}\{n - 1 + (y - m)\alpha\}^{y-m-1}}{(n + y\alpha)^{y-1}}$$

for $y \geqslant m$, and zero otherwise.

The following characterization theorems have also been proved:

1. If the sum X of two independent random variables (rvs) X_1 and X_2 follows the LPD with parameters M and λ, then X_1 and X_2 must each have LPDs defined on nonnegative integers.

2. If a nonnegative integral LPD variate Z is subdivided into two components X and Y such that the conditional distribution $\Pr(X = k, Y = n - k \mid Z = n)$ is quasi-binomial with parameters (n, p, θ) then the random variables X and Y are independent and have LPDs.

THE LAGRANGIAN BINOMIAL DISTRIBUTION (LBD)

This was discussed by Jain and Consul [10] as a generalized negative binomial distribution with many applications. However, one of the conditions in that paper is incorrect. The LBD is given by

$$\Pr(X = x) = \frac{m\Gamma(m + \beta x)}{x!\,\Gamma(m + \beta x - x + 1)}$$

$$\times \frac{\left\{\theta(1 - \theta)^{\beta - 1}\right\}^x}{(1 - \theta)^{-m}} \quad (12)$$

for $x = 0, 1, 2, \ldots$, and $0 < \theta < 1$, $\beta = 0$ or $\beta \geq 1$, $0 \leq \theta\beta < 1$, and $m > 0$.

The first four central moments become

$$\mu_1'(X) = m\theta(1 - \beta\theta)^{-1}$$

$$\mu_2(X) = m\theta(1 - \theta)(1 - \beta\theta)^{-3}$$

$$\mu_3(X) = m\theta(1 - \theta)\{1 - 2\theta + \beta\theta(2 - \theta)\}$$

$$\times (1 - \beta\theta)^{-5} \quad (13)$$

$$\mu_4(X) = 3m^2\theta^2(1 - \theta)^2(1 - \beta\theta)^{-6} + m\theta(1 - \theta)$$

$$\times \{1 - 6\theta + 6\theta^2 + 2\beta\theta(4 - 9\theta + 4\theta^2)$$

$$+ \beta^2\theta^2(6 - 6\theta + \theta^2)\}(1 - \beta\theta)^{-7}.$$

The first and second negative moments of the decapitated LBD (truncated on the left

at $x = 0$) are given by [14] as

$$E[X^{-1}] = \frac{(1 - \theta)^m}{1 - (1 - \theta)^m}$$

$$\times \left[\sum_{i=1}^{m} \binom{m}{i}\left(\frac{\theta}{1 - \theta}\right)^i \frac{1}{i}\right.$$

$$- \frac{\beta}{m}\frac{1 - (1 - \theta)^m}{(1 - \theta)^m} - \beta\log(1 - \theta)\right]$$

$$(14)$$

and

$$E[X^{-2}] = \frac{(1 - \theta)^m}{1 - (1 - \theta)^m}\int_0^\theta \frac{1 - \beta\theta}{\theta(1 - \theta)}$$

$$\times \left[\sum_{i=1}^{m} \binom{m}{i}\left(\frac{\theta}{1 - \theta}\right)^i \frac{1}{i}\right.$$

$$- \frac{\beta}{m}\left\{(1 - \theta)^{-m} - 1\right\}$$

$$- \beta\log(1 - \theta)\right]d\theta. \quad (15)$$

Also, the first negative moment about the point $-k(k > 0)$ for the LBD becomes

$$E\left[(X + k)^{-1}\right] = \frac{1}{(1 - \theta)^l}\sum_{i=0}^{\infty}\binom{-l + i}{i}$$

$$\times \left\{\frac{1}{k + i} - \frac{\beta\theta}{k + i - 1}\right\}\theta^i$$

$$(16)$$

where $\beta k - k - m = l < 0$. However, when $l \neq -1, -2, -3, \ldots, -k$ and k is a positive integer,

$$E[(X + k)^{-1}] = \frac{\beta}{\beta k - m} + \frac{m}{\beta k - m}$$

$$\times \left[\sum_{i=1}^{k}\frac{(1 - \theta)^{-1}}{\theta^i(\beta k - k - m)_{(i)}}\right.$$

$$\times \frac{(k - 1)!}{(k - i)!} - \frac{(k - 1)!}{(\beta k - k - m)_{(k)}}$$

$$\times \frac{(1 - \theta)^{k + m - \beta k}}{\theta^k}\right], \quad (17)$$

where $l_{(i)} = l(l + 1)(l + 2)\cdots(l + i - 1)$.

If $X_1, X_2, \ldots, X_n$ is a random sample of size n taken from the LBD and $Y = \sum_{i=1}^{n}X_i$, then the maximum likelihood estimator of θ

is given [9] by

$$\hat{\theta} = \frac{1}{\beta} - \frac{mn}{\beta^2}\left[\frac{1}{Y + mn/\beta}\right]. \quad (18)$$

The actual bias in $\hat{\theta}$ is given by [14]

$$b(\hat{\theta}) = 1 - \theta + k\sum_{i=1}^{k-1}\frac{(-1)^i}{k-i}\left(\frac{1-\theta}{\theta}\right)^k$$

$$+ (-1)^k k \log(1-\theta)\left(\frac{1-\theta}{\theta}\right)^k, \quad (19)$$

which is bounded by

$$-\frac{\theta}{N+2} - \frac{\theta(1-\theta)^N}{(N+1)(N+2)}$$

and

$$-\frac{\theta(1-\theta)^N}{N+1} - \frac{N\theta^2(1-\theta)^N}{N+2},$$

where $N = mn/\beta$.

The minimum variance unbiased estimators $\{\{l(\theta)\}\}$ for functions $l(\theta)$ are given by [15]

$$\{\{\theta^k\}\} = \frac{y!\,(mn + k\beta - k)\Gamma(y\beta + mn - k)}{(y-k)!\,mn\Gamma(y\beta + mn)},$$

$$y \geqslant k,$$

and zero elsewhere

for $k = 1, 2, 3, \ldots;$ (20)

and

$$\{\{\Pr(X=k)\}\}$$

$$= \begin{cases} \dfrac{y!}{mn(mn + \beta y - 1)_{(y-1)}}, & y = k \\[2ex] \dfrac{n-1}{n}\dfrac{\begin{array}{c}y_{(k)}\{m(n-1)\\ +(y-k)-1\}_{(y-k-1)}\end{array}}{(mn + \beta y - 1)_{(y-1)}}, & y > k \\[2ex] 0, & y < k, \end{cases}$$

(21)

where $k = 0, 1, 2, \ldots$, and $m \geqslant k$ if $\beta = 0$.

A number of characterization theorems have also been proved for the LBD by Consul [1]. Some of them are as follows:

1. If a nonnegative integer-valued rv X is reduced by the quasi-hypergeometric damage process to an integer valued rv Y, then the condition

   $\Pr(Y = r) = \Pr(Y = r\,|\,\text{undamaged})$

implies that X must be a Lagrangian binomial variate.

2. Let X and Y be two nonnegative integer-valued rvs whose sum $X + Y = Z$ has an LBD. If the conditional distribution of X, given $Z = n$, is a quasi-hypergeometric distribution with parameters n, d, $a - d$, and θ for all values of Z, then X and Y are both independent Lagrangian binomial variates.

3. If X and Y are independent nonnegative integer-valued rvs and their sum Z has an LBD, then each of X and Y must be distributed as LBD.

4. Let $X_1, X_2, \ldots, X_N$ be a random sample from a population having the first four moments. Also, let $\Lambda = X_1 + X_2 + \cdots + X_N$ and a statistic T be defined in terms of eight subscripts $g, h, \ldots, n$ by

$$T = 7!\sum X_g \cdots X_n(8 - X_n) - 240(N - 7)$$

$$\times \sum X_g \cdots X_l X_m^2(9 - 5X_l + 6X_m)$$

$$+ 6(N - 6)(N - 7)\sum X_g \cdots X_k X_l$$

$$\times (24X_k - 45X_j X_k + 20X_k X_l - 10X_l^2)$$

$$- (N - 5)(N - 6)(N - 7)\sum X_g X_h X_i X_j^2 X_k^2$$

$$\times (12X_i - 72X_h X_i + 8X_i X_k$$

$$+ 6X_j X_k - 3X_k^2), \quad (22)$$

where the summations go over all subscripts $g, h, i, \ldots, n$, which are all different and vary from 1 to N. Consul and Gupta [2] have shown that the population is LBD, if and only if the statistic T has zero regression on the statistic Λ.

THE LAGRANGIAN LOGARITHMIC DISTRIBUTION (LLD)

The LLD was given by Jain and Gupta [11] as a generalized logarithmic series distribution in the form

$$P(X = x) = \frac{\Gamma(\beta x)}{x!\,\Gamma(\beta x - x + 1)}\frac{\{\theta(1-\theta)^{\beta-1}\}^x}{-\log(1-\theta)},$$

for $x = 1, 2, 3, \ldots$ and $0 < \theta < 1$, $\beta \geqslant 1$, $0 < \theta\beta < 1$.

The mean and the variance of the distribution are

$$\mu_1' = -\theta / \{(1 - m\theta)\log(1 - \theta)\},$$

$$\sigma^2 = -\frac{\theta}{(1 - m\theta)^2 \log(1 - \theta)}$$

$$\times \left| \frac{1 - \theta}{1 - m\theta} + \frac{\theta}{\log(1 - \theta)} \right|. \quad (23)$$

The probability generating function of the LLD is given by

$$f(u) = \log(1 - \theta s)/\log(1 - \theta),$$

where $s = u(1 - \theta)^{m-1}(1 - \theta s)^{1-m}$.

Kumar [13] has given the MVU estimator of θ^m, $m = 1, 2, 3, \ldots$, as

$$\{\{\theta^m\}\} = \frac{\sum_{i=1}^{y} L(i, n, \beta) k(y - i)}{L(y, n, \beta)},$$

$$y = n, n + 1, \ldots, \quad (24)$$

where y is the sum of n sample values, $k(i) = m(i\beta - m - 1)_{(i-m)}/\{i(i - m)!\}$ for $i \geqslant m$ and zero otherwise,

$$(a)_{(j)} = a(a - 1) \cdots (a - j + 1),$$

and

$$L(y, n, \beta) = \frac{n!}{y!} \left| \sum_{t=n-1}^{y} \frac{|S_t^{n-1}|(-1)^{y-t-1}}{t!} \right.$$

$$\times \binom{-\beta y + y - 1}{y - n - 1}, \quad (25)$$

where S_t^k are Stirling numbers* of the first kind.

BIVARIATE LAGRANGIAN DISTRIBUTIONS

Let $h(t_1, t_2)$ and $k(t_1, t_2)$ be any two bivariate PGFs defined on nonnegative integers such that $h(0, 0)$ and $k(0, 0)$ are nonzero. In view of the bivariate Lagrange expansion [see formulas (2) and (3) in LAGRANGE EXPANSIONS] any other bivariate PGF $f(t_1, t_2)$, under the transformations

$$t_1 = u \cdot h(t_1, t_2), \qquad t_2 = v \cdot k(t_1, t_2)$$

becomes the PGF in u and v of a bivariate

Lagrangian probability distribution which is given by $P_{0,0} = f(0, 0)$, and for $r + s > 0$, $r \geqslant 0$, $s \geqslant 0$,

$$P_{r,s} = \frac{1}{r! \, s!} \partial_{t_1}^{r-1} \partial_{t_2}^{s-1}$$

$$\times \left\{ h^r k^s \frac{\partial^2 f}{\partial t_1 \partial t_2} + h^r \frac{\partial k^s}{\partial t_1} \frac{\partial f}{\partial t_2} \right.$$

$$\left. + k^s \frac{\partial h^r}{\partial t_2} \frac{\partial f}{\partial t_1} \right\} \bigg|_{t_1 = t_2 = 0}, \quad (26)$$

where h, k, f represent the functions $h(t_1, t_2)$, $k(t_1, t_2)$, and $f(t_1, t_2)$, respectively, and $\partial_{t_1}^{r-1}$ and $\partial_{t_2}^{s-1}$ are abbreviations for

$$\frac{\partial^{r-1}}{\partial t_1^{r-1}} \quad \text{and} \quad \frac{\partial^{s-1}}{\partial t_2^{s-1}}.$$

The bivariate Borel–Tanner distribution, the bivariate Lagrangian Poisson distribution, and the bivariate Lagrangian binomial distribution are easily obtained from the above by choosing particular values of the functions h, k, and f.

If (H_{10}, H_{01}), (K_{10}, K_{01}), and (F_{10}, F_{01}) are the mean vectors of the probability distributions generated by the PGFs $h(t_1, t_2)$, $k(t_1, t_2)$, and $f(t_1, t_2)$, respectively, the mean vector (L_{10}, L_{01}) of the bivariate Lagrangian distribution becomes

$$L_{10} = \frac{F_{10}(1 - K_{01}) + F_{01}K_{10}}{(1 - H_{10})(1 - K_{01}) - H_{01}K_{10}}$$

$$\quad (27)$$

$$L_{01} = \frac{F_{01}(1 - H_{10}) + F_{10}H_{01}}{(1 - H_{10})(1 - K_{01}) - H_{01}K_{10}}.$$

The expressions for the second-order cumulants are somewhat more complicated and are available in Shenton and Consul [17].

The multivariate Lagrangian distributions, their mean vector and variance–covariance matrix, and some applications are available in Consul and Shenton [6].

References

[1] Consul, P. C. (1974). In *Statistical Distributions in Scientific Work*, Vol. 3, C. Taillie, G. P. Patil, and B. Baldessari, eds. D. Reidel, Dordrecht, Holland, pp. 279–290.

[2] Consul, P. C. and Gupta, H. C. (1980). *SIAM J. Appl. Math.*, **39**, 231–237.

[3] Consul, P. C. and Jain, G. C. (1973). *Technometrics*, **15**, 791–799.

[4] Consul, P. C. and Jain, G. C. (1973). *Biom. Zeit.*, **15**, 495–500.

[5] Consul, P. C. and Shenton, L. R. (1971). *SIAM J. Appl. Math.*, **23**, 239–248.

[6] Consul, P. C. and Shenton, L. R. (1973). *Multivariate Statistical Inference*. North-Holland, New York.

[7] Consul, P. C. and Shenton, L. R. (1974). In *Statistical Distributions in Scientific Work*, Vol. 1, C. Taillie, G. P. Patil, and B. Baldessari, eds. D. Reidel, Dordrecht, Holland, pp. 41–57.

[8] Gupta, R. C. (1974). *Sankhyā B*, **36**, 288–298.

[9] Gupta, R. C. (1975). *Commun. Statist.*, **4**, 689–697.

[10] Jain, G. C. and Consul, P. C. (1971). *SIAM J. Appl. Math.*, **21**, 501–513.

[11] Jain, G. C. and Gupta, R. P. (1973). *Trab. Estadist.*, **24**, 99–105.

[12] Janardan, K. G., Kerster, H. W., and Schaeffer, D. J. (1979). *Bioscience*, **29**, 599–602.

[13] Kumar, A. (1979). Some Estimation Problems in Power Series Distributions. Ph.D. thesis, University of Calgary.

[14] Kumar, A. and Consul, P. C. (1979). *Commun. Statist. A*, **8**, 151–166.

[15] Kumar, A. and Consul, P. C. (1980). *Commun. Statist. A*, **9**, 1261–1275.

[16] Patil, G. P. (1962). *Ann. Inst. Statist. Math., Tokyo*, **14**, 179–182.

[17] Shenton, L. R. and Consul, P. C. (1973). *Sankhyā B*, **35**, 229–236.

Bibliography

See the following works, as well as the references just given, for more information on the topic of Lagrange and related probability distributions.

Consul, P. C. (1975). *Commun. Statist.*, **4**, 555–563.

Gupta, R. C. (1977). *Commun. Statist.*, **6**, 977–991.

(GENERATING FUNCTIONS
LAGRANGE EXPANSIONS
LOGARITHMIC SERIES DISTRIBUTIONS
MIXTURES
POWER SERIES DISTRIBUTIONS
QUEUEING THEORY)

P. C. CONSUL

LAGRANGE EXPANSIONS

If $\phi(x)$ is a function that can be expanded in a convergent power series of x and such that $\phi(0) \neq 0$, then the equation $x = y \cdot \phi(x)$ can be easily written in the form

$$y = x/\phi(x) = \sum_{k=0}^{\infty} a_k x^k$$

by the simple operation of division. The inversion of the power series

$$y = \sum_{k=0}^{\infty} a_k x^k \quad \text{to the form} \quad x = \sum_{k=0}^{\infty} b_k y^k$$

was the real problem considered by Lagrange, who, in 1770, provided a more general theorem by which an analytic function $f(x)$ can be expanded in powers of y when $y = x/\phi(x)$ and the coefficient of y^k, for any nonnegative integral value of k, can be specifically determined. The univariate Lagrange's expansion is found in all the classical works on infinite series, differential calculus, and modern analysis like those of Bromwich [1], Goursat [6], and Whittaker and Watson [11].

The Lagrange's expansion is usually defined for complex functions $f(z)$ and $g(z)$ which are analytic on and within a contour C surrounding a point a. If u be another variable such that the inequality

$$|ug(z)| < |z - a|$$

is satisfied at all points z on the perimeter of C, then the equation

$$z = a + ug(z)$$

has one root in the interior of the contour C and any function $f(z)$, which is analytic on and inside the contour C, can be expanded as a power series in u by the formula

$$f(z) = f(a)$$
$$+ \sum_{k=1}^{\infty} \frac{u^k}{k!} \frac{\partial^{k-1}}{\partial z^{k-1}} \left| \{ g(z) \}^k f'(z) \right| \Big|_{z=a},$$

$$(1)$$

where $f'(z)$ is the derivative of $f(z)$ with respect to z and $\phi(a) \neq 0$.

A large number of combinatorial identities have been proved, by considering $a = 0$ in the Lagrange expansion above, by different authors; see Jensen [7], Riordan [9], and Gould [5]. A class of univariate discrete Lagrangian probability distributions was defined by Consul and Shenton [2] (*see* LAGRANGE AND RELATED PROBABILITY DISTRIBUTIONS).

Poincaré [8] gave a bivariate generalization of Lagrange's expansion in a very elegant manner. Although the determination of the coefficients of successive terms is rather complex in the general case the simplest form can be stated as follows.

Let $h(x_1, x_2)$ and $k(x_1, x_2)$ be two bivariate analytic functions of x_1 and x_2 such that $h(0,0)$ and $k(0,0)$ are nonzero and let $f(x_1, x_2)$ be another bivariate analytic function of x_1 and x_2. The transformations

$$x_1 = uh(x_1, x_2), \qquad x_2 = vk(x_1, x_2)$$

define two functions u and v which are bivariate analytic functions of x_1 and x_2 in the neighborhood of the origin. The smallest positive roots of the transformations give, for the function $f(x_1, x_2)$, the following bivariate power series expansion in terms of u and v:

$$f(x_1, x_2) = l(u, v) = \sum_{r=0}^{\infty} \sum_{s=0}^{\infty} u^r v^s K_{r,s}, \quad (2)$$

where $K_{0,0} = f(0,0)$ and for $r \geqslant 0$, $s \geqslant 0$, $r + s > 0$,

$$K_{r,s} = \frac{1}{r!\,s!} \partial_{x_1}^{r-1} \partial_{x_2}^{s-1}$$

$$\times \left(h^r k^s \frac{\partial^2 f}{\partial x_1 \partial x_2} + h^r \frac{\partial k^s}{\partial x_1} \frac{\partial f}{\partial x_2} \right.$$

$$\left. + k^s \frac{\partial h^r}{\partial x_2} \frac{\partial f}{\partial x_1} \right) \Bigg|_{x_1 = x_2 = 0}. \quad (3)$$

The symbols h, k, and f represent the functions $h(x_1, x_2)$, $k(x_1, x_2)$, and $f(x_1, x_2)$, and $\partial_{x_1}^{r-1}$ and $\partial_{x_2}^{s-1}$ are abbreviations for $\partial^{r-1}/\partial x_1^{r-1}$ and $\partial^{s-1}/\partial x_2^{s-1}$, respectively.

Many combinatorial identities have been proved by the above expansion; see Gould

[5]. The bivariate class of discrete Lagrangian probability distributions was considered by Shenton and Consul [10].

A multivariate generalization of Lagrange's expansion was derived by Good [4] and was used by him for the enumeration of trees; however, its form was such that the determination of the coefficients of different terms was not straightforward. Good's form was subsequently modified by Consul and Shenton [3] and was used to obtain multivariate discrete Lagrangian probability distributions.

Let $\mathbf{t}, \mathbf{u}, \mathbf{0}$ represent column vectors so that $\mathbf{t}' = (t_1, t_2, \ldots, t_n)$, $\mathbf{u}' = (u_1, u_2, \ldots, u_n)$, and $\mathbf{0}' = (0, 0, \ldots, 0)$. Also, let $g^{(i)}(\mathbf{t}')$, $i = 1, 2, \ldots, n$, be n multivariate functions which are analytic in the neighborhood of the origin such that $g^{(i)}(\mathbf{0}') \neq 0$. If $t_i = u_i g^{(i)}(\mathbf{t}')$ for $i = 1, 2, 3, \ldots, n$, then any multivariate function $f(\mathbf{t}')$, analytic in the neighborhood of the origin, can be expanded as a power series in u's by the formula

$$f(\mathbf{t}') = \sum \frac{u_1^{m_1} \cdots u_n^{m_n}}{m_1! \cdots m_n!}$$

$$\times \left| D_1^{m_1 - 1} \cdots D_n^{m_n - 1} \| D_v g_{m_v}^{(v)} \mathbf{I} - \mathbf{G} \| f(\mathbf{t}') \right|_{\mathbf{t} = 0},$$

$$(4)$$

where $\sum$ denotes the n-dimensional summation for all m's over nonnegative integers, $\mathbf{I}$ is the $n \times n$ unit matrix,

$$g_{m_v}^{(i)} = \left(g^{(i)}(\mathbf{t}') \right)^{m_v}, \qquad \frac{\partial}{\partial t_j} \left(g^{(i)} \right)^{m_v} = g_{m_v j}^{(i)},$$

$\mathbf{G}$ is the $n \times n$ matrix $(g_{m_v j}^{(i)})$, and $D_i^r = \partial^r/\partial t_i^r$.

References

[1] Bromwich, T. J. I. (1926). *An Introduction to the Theory of Infinite Series*. Macmillan, London.

[2] Consul, P. C. and Shenton, L. R. (1970). *SIAM J. Appl. Math.*, **23**, 239–248.

[3] Consul, P. C. and Shenton, L. R. (1973). In *Multivariate Statistical Inference*, D. G. Kabe and R. P. Gupta, eds. North-Holland, New York.

[4] Good, I. J. (1965). *Proc. Camb. Philos. Soc.*, **61**, 499–517.

[5] Gould, H. W. (1972). *Combinatorial Identities*. Morgantown Printing.

[6] Goursat, H. (1904). *Functions of a Complex Variable*. Ginn, Dover.

[7] Jensen, J. L. W. (1902). *Acta Math.*, **26**, 307–318.

[8] Poincaré, H. (1886). *Acta Math.*, **9**, 321–380.

[9] Riordan, J. (1958). *An Introduction to Combinatorial Analysis*. Wiley, New York.

[10] Shenton, L. R. and Consul, P. C. (1973). *Sankhyā A*, **35**, 229–236.

[11] Whittaker, E. T. and Watson, G. N. (1927). *A Course of Modern Analysis*. Cambridge University Press, Cambridge.

(LAGRANGE AND RELATED PROBABILITY DISTRIBUTIONS)

P. C. CONSUL

LAGRANGE MULTIPLIERS, METHOD OF

A method for finding a conditional relative extremum of a function $f(x_1, \ldots, x_n)$ (a real-valued function defined on a domain G in an n-dimensional Euclidean space R^n) under the conditions that $\phi_1(x_1, \ldots, x_n) = 0, \ldots, \phi_m(x_1, \ldots, x_m) = 0$ $(m < n)$. If the functions $f, \phi_1, \ldots, \phi_m$ are continuously differentiable and the Jacobian* $\partial(\phi_1, \ldots, \phi_m)/\partial(x_{n-m+1}, \ldots, x_m)$ does not vanish in the domain under consideration, then f has a relative extremum at some point $(x_1^0, \ldots, x_n^0)$ given that $\phi_i = 0$ $(i = 1, \ldots, m)$, only if for arbitrary constants $\lambda_1, \ldots, \lambda_n$ (known as *Lagrange multipliers*) the function

$$F(x_1, \ldots, x_n) \equiv f + \lambda_1 \phi_1 + \cdots + \lambda_m \phi_m$$

satisfies $\partial F/\partial x_i = 0$ $(i = 1, \ldots, n)$ and, moreover, $\phi_i = 0$ at $(x_1^0, \ldots, x_n^0)$.

For an application in statistical distribution theory, see, e.g., Chernoff and Lieberman [3]. Applications to maximum likelihood* estimation and hypothesis testing* are given in Aitchison and Silvey [1] and Silvey [4]. For additional information, see Buck [2].

References

[1] Aitchison, J. and Silvey, S. D. (1958). *Ann. Math. Statist.*, **29**, 813–828.

[2] Buck, R. C. (1965). *Advanced Calculus*, 2nd ed. McGraw-Hill, New York.

[3] Chernoff, H. and Lieberman, G. J. (1956). *Ann. Math. Statist.*, **27**, 806–818.

[4] Silvey, S. D. (1959). *Ann. Math. Statist.*, **30**, 389–407.

(LAGRANGE MULTIPLIER TEST)

LAGRANGE MULTIPLIER TEST

Lagrange multipliers are familiar to applied mathematicians from Lagrange's "method of undetermined multipliers*'" for finding extrema of a function subject to constraints. To maximize a (vector) function $f(x)$ subject to the (vector) constraint $g(x) = 0$, form the Lagrangian

$$\mathscr{L}(x, \lambda) = f(x) - \lambda^T g(x),$$

where λ is a vector of Lagrange multipliers, and find the maximum of $\mathscr{L}$ with respect to x and λ by solving the first-order equations

$$0 = \frac{\partial \mathscr{L}}{\partial x} = \frac{\partial f}{\partial x} - \lambda^T \frac{\partial g}{\partial x},$$

$$0 = \frac{\partial \mathscr{L}}{\partial \lambda} = -g(x).$$

If the unconstrained maximum of $\mathscr{L}$ is attained at (x^*, λ^*), then the constrained maximum of f subject to $g = 0$ is attained at $x = x^*$. The elements of the vector λ^* may be interpreted as shadow costs of the constraints; if $\lambda^* = 0$, these costs are zero and $g = 0$ is satisfied at the unconstrained maximum of f.

The natural application of Lagrange's method in statistics is to the maximization of the likelihood* function of a statistical model subject to a constraint on the parameters of the model. This arises in the problem of deciding, on the basis of a number of observations from a statistical model, whether an s-dimensional parameter θ involved in the specification of the model satisfies the r independent constraints $g_i(\theta) = 0$, $i = 1, \ldots, r$. Writing g for the vector of constraints, the problem becomes one of testing the hypotheses

$$H_0 : g(\theta) = 0 \quad \text{vs.} \quad H_1 : g(\theta) \text{ unrestricted.}$$

The maximum-likelihood estimator* of θ on H_0 is then the component $\hat{\theta}$ of the vector $(\hat{\theta}, \hat{\lambda})$ at which the Lagrangian

$$\mathscr{L}(x; \theta, \lambda) = \log L(x; \theta) - \lambda^T g(\theta)$$

attains its maximum. The interpretation of the Lagrange multipliers as costs of the constraint suggests that they may be used to test the hypothesis H_0 vs. H_1, the test being to reject H_0 if $\hat{\lambda}$ is "too large" (in some suitable sense). Such a test has been derived by Aitchison and Silvey [1] and Silvey [10]. Define

$$d = \frac{\partial \log L}{\partial \theta}, \quad S = n^{-1} \frac{\partial^2 \log L}{\partial \theta \, \partial \theta^T},$$

$$H = \frac{\partial g}{\partial \theta^T}$$

and $\hat{d}$, $\hat{S}$, and $\hat{H}$ to be the corresponding quantities evaluated at the constrained maximum likelihood estimator $\hat{\theta}$; then asymptotically on H_0 the statistic

$$\text{LM} = n^{-1} \hat{\lambda}^T \hat{H}^T \hat{S}^{-1} \hat{H} \hat{\lambda}$$

is distributed as χ^2_{s-r}, subject to suitable regularity conditions. The Lagrange multiplier test rejects H_0 if LM exceeds the critical value of its asymptotic χ^2 distribution.

Since the first-order condition $\partial \mathscr{L} / \partial \theta |_{\hat{\theta}} = 0$ for maximization of $\mathscr{L}(x; \theta, \lambda)$ is

$$\hat{d} - \hat{H}\hat{\lambda} = 0,$$

the Lagrange multiplier test statistic may be written as

$$\text{LM} = n^{-1} \hat{d}^T \hat{S}^{-1} \hat{d}.$$

This is known as the efficient-score statistic [9] and is often the most convenient form for the computation of LM.

As an example of a Lagrange multiplier test, consider a sample of size n from a multinomial distribution* with m cells and cell probabilities $\pi_1, \ldots, \pi_m$. If the numbers of observations in each cell are $O_1, \ldots, O_m$,

then the log likelihood is

$$\log L = \sum_{i=1}^m O_i \log \pi_i + f(O_1, \ldots, O_m).$$

Now consider testing the hypothesis H_0 that the true cell probabilities are specified constants, with expected cell counts $E_1, \ldots, E_m$, against the very general alternative H_1 that the cell probabilities are unrestricted except by $\sum \pi_i = 1$. To maintain symmetry among the parameters it is convenient to introduce a Lagrange multiplier μ (not to be confused with the Lagrange multipliers in the eventual test!) and rewrite the log likelihood as

$$\log L = \sum_{i=1}^m O_i \log \pi_i - \mu \left(\sum_{i=1}^m \pi_i - 1 \right)$$
$$+ f(O_1, \ldots, O_m),$$

giving

$$\frac{\partial \log L}{\partial \pi_i} = \frac{O_i}{\pi_i} - \mu.$$

Imposing the condition $\sum \hat{\pi}_i = 1$ on the maximum likelihood estimators $\hat{\pi}_i$ of π_i on H_1 gives $\mu = n$, so the Lagrange multipliers for testing H_0 vs. H_1 are

$$\hat{\lambda}_i = \frac{\partial \log L}{\partial \pi_i} \bigg|_{H_0} = n \left(\frac{O_i}{E_i} - 1 \right),$$

$$i = 1, \ldots, m.$$

Since

$$-n^{-1} \partial^2 \log L / \partial \pi_i \, \partial \pi_j = n^{-1} O_i / \pi_i^2 \xrightarrow{P} \delta_{ij} / \pi_i$$

as $n \to \infty$, the large-sample information matrix* on H_0 is a diagonal matrix with ith diagonal element $\hat{s}_i = E_i / n$, and the Lagrange multiplier test statistic is

$$\text{LM} = n^{-1} \sum_{i=1}^m \hat{\lambda}_i^2 \hat{s}_i = \sum_{i=1}^m (O_i - E_i)^2 / E_i,$$

with an asymptotic χ^2_{m-1} distribution on H_0. This is, of course, Pearson's χ^2 statistic*.

The Lagrange multiplier test is not the only large-sample test of parametric hypotheses; other well-established approaches are the likelihood ratio method*, which tests whether the removal of the constraint results in a significant increase in the likelihood of the model, and Wald's procedure*, which tests whether the unconstrained parameter estimates $\tilde{\theta}$ satisfy the constraint $g(\tilde{\theta}) = 0$. In well-behaved problems all three methods are equivalent; they yield test statistics which have identical asymptotic χ^2 distributions on H_0 and power functions which are identical to order $n^{-1/2}$ [11]. Thus the Lagrange multiplier and Wald tests share the optimality properties—consistency and asymptotic efficiency—of the more familiar likelihood ratio test. In general, there is no uniform ranking of the power properties to order n^{-1} of the three tests [8]. However, for certain linear models, including the linear regression model and autoregressive time-series models, the test statistics LM, LR, and W resulting from the three approaches satisfy the inequality LM $\leq$ LR $\leq$ W, on both the null and alternative hypotheses [2]. In this case the Lagrange multiplier test is the most conservative of the three, being least likely to reject the null hypothesis whether it be true or not.

The asymptotic power properties of the three tests are insufficient to enable any one to be chosen as the best. In practice, therefore, one must choose which test to use according to other criteria, such as small-sample power properties or ease of computation. The advantage of the Lagrange multiplier test is that it does not involve estimation under the alternative model; this can lead to great savings in computation when the alternative model is complicated or nonlinear in structure. For example, Breusch and Pagan [3] have derived a test for heteroscedasticity* in a linear regression* model which requires calculation only of least-squares residuals. More examples are given in Breusch and Pagan [4].

The example given earlier shows that Pearson's χ^2 statistic, originally derived as a pure test of significance* with no alternative hypothesis being specified, is also a Lagrange multiplier test against a particular alternative hypothesis. Three other pure tests of significance which have an interpretation as Lagrange multiplier tests are the portmanteau test* and Quenouille's test* of goodness

of fit* of a time-series model [6], and the Cliff–Ord test* for spatial correlation [5]. Hosking [7] has investigated the relationship between Lagrange multiplier tests and pure significance tests for multivariate time-series models. This relationship between Lagrange multiplier tests and pure tests of significance is not altogether surprising: when the alternative model is very general and wide-ranging it can represent a large variety of deviations from the null model, so the Lagrange multiplier statistic, calculated only from estimates of the null model, might be expected to depend only weakly on the precise form of the alternative model. Thus when the alternative model is very general, the Lagrange multiplier test may be expected to resemble a pure test of significance.

References

[1] Aitchison, J. and Silvey, S. D. (1958). *Ann. Math. Statist.*, **29**, 813–828.

[2] Breusch, T. S. (1979). *Econometrica*, **47**, 203–207.

[3] Breusch, T. S. and Pagan, A. R. (1979). *Econometrica*, **47**, 1287–1294.

[4] Breusch, T. S. and Pagan, A. R. (1980). *Rev. Econ. Stud.*, **47**, 239–253.

[5] Burridge, P. (1980). *J. R. Statist. Soc. B*, **42**, 107–108.

[6] Hosking, J. R. M. (1980). *J. R. Statist. Soc. B*, **42**, 170–181.

[7] Hosking, J. R. M. (1981). *J. R. Statist. Soc. B*, **43**, 219–230.

[8] Peers, H. W. (1971). *Biometrika*, **58**, 577–587.

[9] Rao, C. R. (1948). *Proc. Camb. Philos. Soc.*, **44**, 50–57.

[10] Silvey, S. D. (1959). *Ann. Math. Statist.*, **30**, 389–407.

[11] Wald, A. (1943). *Trans. Amer. Math. Soc.*, **54**, 426–482.

Bibliography

Courant, R. (1936). *Differential and Integral Calculus*, Vol. 2. Blackie, London. (Section III.3 contains a good exposition of Lagrange's method of undetermined multipliers.)

Cox, D. R. and Hinkley, D. V. (1974). *Theoretical Statistics*. Chapman & Hall, London. [Lagrange multiplier tests (under their alias of score tests) are discussed, with examples, in Sec. 9.3.]

(CLIFF–ORD TEST
LAGRANGE MULTIPLIERS, METHOD OF
LIKELIHOOD RATIO TEST
MAXIMUM LIKELIHOOD ESTIMATOR
PEARSON'S χ^2 STATISTIC
PORTMANTEAU TEST
PURE TEST OF SIGNIFICANCE
QUENOUILLE'S TEST
WALD'S PROCEDURE)

J. R. M. HOSKING

LAGRANGE'S INTERPOLATION FORMULA

The polynomial (in x)

$$\sum_{j=1}^{n} \frac{g_j(x)}{g_j(x_j)} y_j,$$

where

$$g_j(x) = \left\{ \prod_{i=1}^{n} (x - x_i) \right\} \Big/ (x - x_j),$$

goes through the points $(x_1, x_1), (x_2, y_2)$, $\ldots, (x_n, y_n)$.

If $y_j = f(x_j)$, this formula can be used to interpolate for values of the function $f(x)$ corresponding to values of x other than $x_1, \ldots, x_n$. When used in this way, it is called *Lagrange's interpolation formula*.

If the x_j's are equally spaced, the formula specializes in various ways, and can be expressed in terms of finite differences.

(EVERETT'S CENTRAL DIFFERENCE
    FORMULA
FINITE DIFFERENCES, CALCULUS OF
INTERPOLATION)

LAGUERRE SERIES

The classical Laguerre polynomials were used in the late 1800s on problems in mathematical physics and in approximating functions*. In particular, they satisfy a differential equation of the Sturm–Liouville type. See Shohat et al. [19] for a bibliography on the early uses of the Laguerre series and books like Sansone [17], Szegö [20],

and Jackson [7] for classical definitions and these first uses of the series. Being interested in approximating functions, Laguerre first wrote on the subject in 1876 [12]; this technique can be seen in Hildebrand [6].

The Laguerre polynomials are orthogonal* over the interval $(0, \infty)$ with a weighting function* of the form

$$w(x) = x^{\alpha} e^{-x}, \qquad \alpha > -1.$$

The rth generalized Laguerre polynomial can be defined by

$$L_r^{(\alpha)}(x) = (x^{-\alpha} e^x / r!) d^r (x^{r+\alpha} e^{-x}) / dx^r$$

$$= \sum_{j=0}^{r} \frac{(-1)^j \Gamma(\alpha + r + 1) x^j}{(r-j)! \, j! \, \Gamma(\alpha + j + 1)}$$

$$\text{for} \quad \alpha > -1.$$

When $\alpha = 0$, $L_r^{(0)}(x)$ is usually denoted by $L_r(x)$ and called a Laguerre polynomial. The orthogonality relationship is given by

$$\int_0^{\infty} e^{-x} x^{\alpha} L_n^{(\alpha)}(x) L_m^{(\alpha)}(x) \, dx$$

$$= \begin{cases} 0, & n \neq m \\ \Gamma(n + \alpha + 1)/n!, & m = n \end{cases}$$

for $m, n = 0, 1, 2, \ldots, \alpha > -1$.

There are many recursive formulas available to generate the Laguerre polynomials which appear in the references already cited. Abramowitz and Stegun [1], in the chapter on orthogonal polynomials, summarize much of the work on orthogonal polynomials, list some coefficients for the Laguerre polynomials, and graph a few of these polynomials.

Laguerre series arise in statistics mainly to represent either the density function* or the cumulative distribution function of random variables involving some type of a linear combination of chi-square variables*. Recent work by Tan and Wong [21, 22] extends some of the results to nonnormal distributions. If, for example, f is a density function of a noncentral chi-squared random variable X,

$$X = \sum_{j=1}^{\nu} (Z_j + \delta_j)^2,$$

where the Z_j are independent standardized

normal variates and the δ_j are constants, then X is usually denoted by $\chi_\nu'^2(\lambda)$ where ν equals the degrees of freedom and $\lambda = \sum_{j=1}^{\nu}\delta_i^2$ is called the noncentrality parameter*. Tiku [23] gives the following Laguerre series to represent the density f of $X/2$,

$$f(x) = \left[\sum_{r=0}^{\infty} a_r L_r^{(\nu/2)}(x)\right] e^{-x}x^{(\nu/2)-1} \Big/ \Gamma(\nu/2),$$

where

$$a_r = \sum_{j=0}^{r}(-1)^j \frac{r!\,\Gamma(\nu/2)E[X^j]}{j!\,(r-j)!\,\Gamma((\nu/2)+j)}.$$

Johnson and Kotz [8] in Vol. 2 of their four-volume series of books on distributions in statistics summarize many of the applications of Laguerre series in representing the distribution of random variables.

In most cases, where the Laguerre series can be used to represent the distribution of a random variable, other series, such as chi-square, power, and the Edgeworth (Hermite), can also be used. The problem is that the ability of any particular series to give an accurate approximation by summing only a moderate number of terms in the series representation of the distribution varies with the parameters of the distribution. Gideon and Gurland [3, 4] have compared the Laguerre series to some of these series in their ability to represent the noncentral chi-square variate and quadratic forms in normal variables. Walker [24] continues work in this area.

Perhaps the greatest use of the Laguerre polynomials comes in representing the distribution of the positive definite quadratic form in the exponent of the exponential function of the density function of the multivariate normal distribution. The distribution of this quadratic form is utilized in the study of the power function of various statistical tests; Patnaik [16] gives some examples. The quadratic form can be reduced through an orthogonal transformation to $Q(Z_1, Z_2, \ldots, Z_n) = \sum_{i=1}^{n}\lambda_i(Z_i + \delta_i)^2$ where the Z_i's are independent standardized normal variables, the λ_i's > 0 are the eigenvalues of the matrix of the quadratic form, and the δ_i's are the noncentrality parameters. The distribu-

tion of the indefinite quadratic form with Laguerre polynomials is more complex; some results can be seen in Gurland [5] and Shah [18].

Many of the approaches in deriving the series involve the characteristic function of the random variable to be represented in the series. The characteristic function is derived, manipulated, and then inverted to obtain the series. Most of these techniques involve the relationship between a gamma* density function and the Laguerre polynomials. For instance, if $g_{\alpha,\lambda}$ is a gamma density,

$$g_{\alpha,\lambda}(x) = \lambda(\lambda x)^\alpha e^{-\lambda x}/\Gamma(\alpha+1), \qquad x \geqslant 0,$$

then

$$d^r(g_{\alpha,\lambda}(x))/dx^r$$
$$= \lambda^{\alpha+r+1}x^\alpha e^{-\lambda x}L_r^{(\alpha)}(\lambda x)/\Gamma(\alpha+r+1)$$

and

$$\int_0^\infty e^{itx}(d^r(g_{\alpha,\lambda}(x))/dx^r)\,dx$$
$$= (-it)^r(1-it/\lambda)^{r-\alpha-1},$$

where $i = \sqrt{-1}$. A general formulation of this technique which includes one-, two-, and three-moment fitting of the random variable being expanded to the first term of the Laguerre series for both the central and noncentral quadratic forms is given in Gideon and Gurland [3]. The Kotz et al. [10, 11] formulation treats the central and noncentral cases separately without using moment fitting. Work is continuing on the use of Laguerre polynomials for quadratic forms in normal variates [9] and in mathematics in general [15].

There are other statistical uses of the Laguerre polynomials. For example, Mills' ratio, the ratio of the tail area of a standard normal variate to its density value, is well approximated by a Laguerre–Gauss quadrature formula. These results by Ray and Pitman appear in Johnson and Kotz [8]. See HOTELLING'S TRACE as an application of a generalized Laguerre expansion for the distribution of Hotelling's trace statistic that arises in the testing of three multivariate hypotheses. Elaboration of this use can be seen in Constantine [2]. When a Poisson process is conditioned on a time parameter λ_t that has a Gaussian distribution, the unconditional counting distribution is based on the Laguerre polynomials and is called the Laguerre distribution. This application appears in Mohanty [14].

References

[1] Abramowitz, M. and Stegun, I. A. (1964). *Handbook of Mathematical Functions with Formulas, Graphs, and Mathematical Tables*. Natl. Bur. Stand. (U.S.) Appl. Math. Ser. 55 (Washington, D.C.).

[2] Constantine, A. G. (1966). *Ann. Math. Statist.*, **37**, 215–225.

[3] Gideon, R. A. and Gurland, J. (1976). *J. Amer. Statist. Ass.*, **71**, 227–232.

[4] Gideon, R. A. and Gurland, J. (1977). *SIAM J. Math. Anal.*, **8**, 100–110.

[5] Gurland, J. (1955). *Ann. Math. Statist.*, **26**, 122–127.

[6] Hildebrand, F. B. (1956). *Introduction to Numerical Analysis*. McGraw-Hill, New York.

[7] Jackson, D. (1941). *Fourier Series and Orthogonal Polynomials*. Carus Math. Monogr. No. 6. The Mathematical Association of America, Washington, D.C.

[8] Johnson, N. L. and Kotz, S. (1970). *Distributions in Statistics: Continuous Univariate Distributions*, Vol. 2. Wiley, New York.

[9] Khatri, C. G. (1977). *S. Afr. Statist. J.*, **11**, 167–179.

[10] Kotz, S., Johnson, N. L., and Boyd, D. W. (1967). *Ann. Math. Statist.*, **38**, 827–837.

[11] Kotz, S., Johnson, N. L., and Boyd, D. W. (1967). *Ann. Math. Statist.*, **38**, 838–848.

[12] Laguerre, E. (1876). *Bull. Soc. Math. Fr.*, **5**, 78–92.

[13] Laguerre, E. (1878). *Bull. Soc. Math. Fr.*, **7**, 72–81.

[14] Mohanty, N. C. (1977). *Proc. 41st Sess. Bull. Inst. Int. Statist.*, **47**, Book 4, 346–349.

[15] Pathak, R. A. (1978). *J. Inst. Math. Appl.*, **21**, 171–180.

[16] Patnaik, P. B. (1949). *Biometrika*, **36**, 202–232.

[17] Sansone, G. (1959). *Orthogonal Functions*. Interscience, New York.

[18] Shah, B. K. (1963). *Ann. Math. Statist.*, **34**, 186–190.

[19] Shohat, J., Hille, E., and Walsh, J. L. (1940). *A Bibliography on Orthogonal Polynomials*. Nat. Res. Counc. Bull. No. 103. National Academy of Science, Washington, D.C.

[20] Szegö, G. (1939). *Orthogonal Polynomials*. Amer. Math. Soc. Colloq. Publ. No. 23, New York.

[21] Tan, W. Y. and Wong, S. P. (1977). *J. Amer. Statist. Ass.*, **72**, 875–880.

[22] Tan, W. Y. and Wong, S. P. (1980). *J. Amer. Statist. Ass.*, **75**, 655–662.

[23] Tiku, M. L. (1965). *Biometrika*, **52**, 415–427.

[24] Walker, J. J. (1979). *J. Amer. Statist. Ass.*, **74**, 389–392.

(GAMMA DISTRIBUTION
HOTELLING'S T^2
MILLS' RATIO
NONCENTRAL CHI-SQUARE
ORTHOGONAL POLYNOMIALS)

RUDY A. GIDEON

LAHIRI'S METHOD OF SELECTION OF SAMPLE UNITS

Lahiri's [1, 2] procedure for the selection of *sample units* with probability proportional to size (pps) (and replacement) consists in choosing a pair of random numbers, the first between 1 and N (where N is the size of the universe) and others between 1 and z_{max}, and z_{max} is the maximum value of z_i (the sizes of the "units" in the universe) obtained on inspection. If for any pair of random numbers chosen, the first number is i and the second number is less than z_i the ith unit is chosen; otherwise, it is rejected (regarded as a dud) and a fresh pair of random numbers chosen. This is continued until the required total number of sample units n has been drawn. (A very large z_{max} may be split up into more than one part and the original unit selected whenever one of the split units is drawn.)

References

[1] Lahiri, D. B. (1951). *Bull. Int. Statist. Inst.*, **33**, Pt. 1, 133–140.

[2] Lahiri, D. B. (1958). *Bull. Int. Statist. Inst.*, **36**, Pt. 3, 144–152.

(SURVEY SAMPLING)

LAMBDA CRITERION, WILKS'S

WILKS'S Λ AND ITS ROLE

Let **A** and **B** be two independent Wishart matrices of order p, with probability density functions (PDFs) $W(A|n - q|\Sigma)$ and $W(B|q|\Sigma)$, respectively, where $W(A|n - q|\Sigma)$ stands for

$$\text{constant } |\mathbf{A}|^{[(n-q)-p-1]/2}\exp\left[-\tfrac{1}{2}\text{tr }\Sigma^{-1}\mathbf{A}\right].$$

$(n - q)$ and q are the degrees of freedom (d.f.) of **A** and **B**, respectively, and Σ is the parameter matrix of the Wishart distributions*. Denoting the determinant of **A** by $|\mathbf{A}|$, we define

$$\Lambda = \frac{|\mathbf{A}|}{|\mathbf{A + B}|},$$

called *Wilks's lambda*. Its distribution depends on only the three parameters n, p, and q; n is the sum of the d.f.'s of **A** and **B**, p is the order of the matrices, and q is the d.f.'s of the matrix **B**. We will denote the PDF of Λ by $\Lambda(n, p, q)$ and

$$\Lambda \sim \Lambda(n, p, q)$$

denotes that Λ has the $\Lambda(n, p, q)$ distribution.

Wilks's Λ criterion plays the same role in multivariate analysis* as the F statistic plays in univariate analysis. In univariate analysis, if we have a general linear model* representing the means of independent normal observations with the same but unknown variance σ^2, many important linear hypotheses about the parameters in the model are tested by dividing the total sum of squares (s.s.) of the observations into two parts, one of which has the $\chi^2\sigma^2$ distribution, irrespective of the values of the parameters, and the other has an independent $\chi^2\sigma^2$ distribution only if the hypothesis is true; otherwise, it has a noncentral $\chi^2\sigma^2$ distribution. The F statistic then compares these two independent $\chi^2\sigma^2$ variables to test the hypothesis. Wilks's Λ extends this idea to multivariate analysis, where we have p multinormal variables with

a variance–covariance matrix Σ and independent observations on these variables with a linear model representing the structure of the mean vectors of these observations. The matrix of the total s.s. and sum of products (s.p.) of these observations is then split into two parts A and B, where A will have the $W(A|n - q|\Sigma)$ distribution irrespective of the values of the parameters in the model and B will have an independent $W(B|q|\Sigma)$ distribution, only if a certain hypothesis about the parameters is true, but otherwise it will have a noncentral Wishart distribution. The hypothesis is then tested by using the Wilks's Λ criterion defined above. The distribution of Λ does not involve the nuisance parameter* Σ, just as F is free of σ^2.

DISTRIBUTION OF WILKS'S Λ

Explicit expressions for the exact PDF of Λ are available in the literature [7, 30, 44, 46], but what is really useful in practice is the Bartlett approximation [4] that

$$W = -m \log_e \Lambda,$$

where

$$m = n - \tfrac{1}{2}(p + q + 1),$$

is asymptotically a χ^2 with pq d.f.'s for large n. The exact percentage points of W are given by Schatzoff [44] and Pillai and Gupta [37] and are also available in *Biometrika Tables for Statisticians*, Vol. 2 [35]. What is tabulated in all these sources is not the $100\alpha\%$ percentage points directly but only a multiplying factor $C_\alpha(p, q, M)$, which depends on α, p, q, and $M = n - p - q + 1$. To obtain the $100\alpha\%$ point of the distribution of W, we have to multiply the $100\alpha\%$ point of the χ^2 distribution* with pq d.f.'s by this multiplying factor $C_\alpha(p, q, M)$. The extra accuracy provided by $C_\alpha(p, q, M)$ is, however, often not necessary if n is sufficiently large. It has been proved (see, e.g., Kshirsagar [27]) that $\Lambda(n, p, q)$ is the same as $\Lambda(n, q, p)$ or that p and q are interchangeable. For the particular cases $p = 1$ or 2 (and due to interchangeability, $q = 1$ or 2), one may use, alternatively, the exact results:

1. For $p = 1$, $\{(1 - \Lambda)/\Lambda\}\{(n - q)/q\}$ has the F distribution* with q and $n - q$ d.f.'s.
2. For $p = 2$, $\{(1 - \Lambda^{1/2})/\Lambda^{1/2}\}\{(n - q - 1)/q\}$ has the F distribution with $2q$ and $2(n - q - 1)$ d.f.'s.

All the results above are derived from the hth moment of Λ, which is obtained from the distributions of A and B (see, e.g., Kshirsagar [27]). Box [6] has given an extended version of Bartlett's approximation stated earlier.

ALTERNATIVE EXPRESSIONS FOR Λ

Many important statistical problems can be viewed as problems of relationship (or lack of relationship) of one p-vector $\mathbf{x}$ of variables with another q-vector $\mathbf{y}$ and Wilks's Λ is useful as a measure of lack of association of $\mathbf{x}$ with $\mathbf{y}$. If $\mathbf{C}_{xx}$ represents the matrix of s.s. and s.p. of observations on $\mathbf{x}$, $\mathbf{C}_{xy}$ represents the matrix of s.p. of observations on $\mathbf{x}$ with those on $\mathbf{y}$, the matrix

$$\mathbf{B} = \mathbf{C}_{xy}\mathbf{C}_{yy}^{-1}\mathbf{C}_{yx}$$

is the "regression s.s. and s.p. matrix" of $\mathbf{x}$ on $\mathbf{y}$ and the matrix

$$\mathbf{A} = \mathbf{C}_{xx} - \mathbf{C}_{xy}\mathbf{C}_{yy}^{-1}\mathbf{C}_{yx} = \mathbf{C}_{xx\cdot y}$$

is called the "error matrix." The d.f.'s of $\mathbf{B}$ are q and of $\mathbf{A}$ are, say, $n - q$. Then under assumption of normality for $\mathbf{x}, \mathbf{y}$, $\mathbf{A}$ will have the $W(\mathbf{A}|n - q|\Sigma)$ distribution but $\mathbf{B}$ will have an independent $W(\mathbf{B}|q|\Sigma)$ distribution only if $\mathbf{x}$ and $\mathbf{y}$ are independent. Wilks's Λ is then also expressible as

$$\frac{|\mathbf{C}_{xx\cdot y}|}{|\mathbf{C}_{xx}|}$$

and interchanging $\mathbf{y}$ and $\mathbf{x}$, it is also

$$\frac{|\mathbf{C}_{yy\cdot x}|}{|\mathbf{C}_{xx}|}.$$

If $p \leqslant q$ and $r_1^2, r_2^2, \ldots, r_p^2$ are the squares of the canonical correlations* between $\mathbf{x}$ and $\mathbf{y}$, Λ is also expressible as

$$\prod_1^b (1 - r_i^2)$$

and yet another expression for Λ is $|\mathbf{L}|$, where the matrix $\mathbf{L}$ is defined by

$$\mathbf{A} = \mathbf{CLC'}, \qquad \mathbf{A} + \mathbf{B} = \mathbf{CC'}.$$

In a sense, L can be described as the ratio of $\mathbf{A}$, the error s.s. and s.p. matrix to $\mathbf{A} + \mathbf{B}$, the total s.s. and s.p. matrix of $\mathbf{x}$ and hence it is a generalization of $1 - r^2$, where r is the correlation coefficient* and R is the multiple correlation coefficient*. It should also be noted that Λ is invariant for linear transformation of $\mathbf{x}$ and $\mathbf{y}$.

USES OF WILKS'S Λ

Wilks's Λ is useful in the following problems.

1. It can be used in a test of equality of the variance–covariance matrices of two independent multivariate normal populations. If samples of sizes n_1 and n_2 are available from these two p-variate normal populations, one can compute $\mathbf{A}$ and $\mathbf{B}$, the matrices of the corrected s.s. and s.p. of the observations for the two samples, and then Λ as defined earlier will have the $\Lambda(n, p, q)$ distribution with

    $$n = n_1 + n_2 - 2, \qquad p, q = n_2 - 1$$

    as the parameters if the hypothesis is true. The hypothesis will be rejected at the $100(1 - \alpha)\%$ point level of significance if W, as defined earlier, exceeds the $100\alpha\%$ point of W.
2. It can also be used in a test of independence of $\mathbf{x}$ and $\mathbf{y}$, when both of them have a multivariate normal distribution or one of them, say $\mathbf{x}$, has a multivariate normal distribution, the other is a fixed vector, and the regression of $\mathbf{x}$ on $\mathbf{y}$ is linear. In this case $\mathbf{C}_{xx}$, $\mathbf{C}_{xy}$, and $\mathbf{C}_{yy}$ are

calculated and the form

$$\Lambda = \frac{|\mathbf{C}_{\mathbf{xx \cdot y}}|}{|\mathbf{C}_{\mathbf{xx}}|}$$

is useful. The parameters n, p, and q for this situation are already given. This is also a test for the hypothesis that the true canonical correlations between $\mathbf{x}$ and $\mathbf{y}$ are all null; and this is also a test for the hypothesis that the matrix of true regression coefficients of $\mathbf{x}$ and $\mathbf{y}$ is null.

3. If a test of significance of only a subset of columns of the matrix of regression coefficients of $\mathbf{x}$ and $\mathbf{y}$ is desired, Wilks's Λ is useful there, too, and this is done by splitting $\mathbf{B} = \mathbf{C}_{\mathbf{xy}}\mathbf{C}_{\mathbf{yy}}^{-1}\mathbf{C}_{\mathbf{yx}}$ suitably (see, e.g., Kshirsagar [27]).

4. If there are $k = q + 1$ independent p-variate normal populations with the same variance–covariance matrix but different mean vectors

$$\boldsymbol{\mu}_1, \boldsymbol{\mu}_2, \ldots, \boldsymbol{\mu}_k,$$

a test of the "multivariate analysis of variance* (MANOVA) hypothesis"

$$\boldsymbol{\mu}_1 = \boldsymbol{\mu}_2 = \cdots = \boldsymbol{\mu}_k$$

is provided by Wilks's Λ, by taking $\mathbf{B}$ as the "between-groups" s.s. and s.p. matrix of $q = k - 1$ d.f.'s and $\mathbf{A}$ as the "within-groups" s.s. and s.p. matrix of d.f.'s $n - q$, where $n = N - 1$ and $N = n_1 + \cdots + n_k$, where n_i is the size of the sample from the ith group or population. This is a typical one-way classification* problem for multivariate analysis and Λ is useful for testing appropriate hypotheses for the general multiway classification for multivariate analysis of variance also. Usefulness of Λ in assessing the degree of separation of the groups is also considered by Gau [13].

5. Wilks's Λ is useful in tests associated with the growth curve model, introduced by Potthoff and Roy [41], namely

$$E(\mathbf{X}) = \mathbf{P}\boldsymbol{\xi}\mathbf{Q},$$

where $\mathbf{X}$ is the $p \times n$ matrix of observations on p multinormal variables, corresponding to m different groups, such that every column of $\mathbf{X}$ has the same

variance–covariance matrix $\boldsymbol{\Sigma}$, which is unknown. $\mathbf{P}$ and $\mathbf{Q}$ are, respectively, $p \times 1$ and $m \times n$ matrices of known elements, representing design matrices within and across individuals [14]. $\boldsymbol{\xi}$ is the matrix of unknown parameters of the growth curves of the m groups. Khatri [17] shows how Wilks's Λ can be used to test the hypothesis

$$\mathbf{C}\boldsymbol{\xi}\mathbf{V} = 0$$

in such a situation. Grizzle and Allen [14] illustrate this method with some interesting practical problems. This test is useful in analyzing longitudinal data, in general.

FACTORS OF WILKS'S Λ AND APPLICATION IN DISCRIMINANT ANALYSIS

Apart from the factorization $\Lambda = \Pi(1 - r_i^2)$ described earlier, there is another, more useful, factorization,

$$\Lambda = t_{11}^2 \cdot t_{22}^2 \cdot \ldots \cdot t_{pp}^2,$$

where t_{ii} are the diagonal elements of a lower triangular matrix $\mathbf{T}$ defined by

$$\mathbf{L} = \mathbf{T}\mathbf{T}', \quad \mathbf{A} = \mathbf{C}\mathbf{L}\mathbf{C}', \quad \mathbf{A} + \mathbf{B} = \mathbf{C}\mathbf{C}',$$

with $\mathbf{C}$ also lower triangular. It is shown that the t_{ii}^2 are independent beta variables [27, 31]. The usefulness of these factors t_{ii}^2, their meaning and significance, and their use in a step-down analysis of the relationship of $\mathbf{x}$ and $\mathbf{y}$ are discussed in Kshirsagar [27]. McHenry [32] and Farmer and Freund [10] have utilized this factorization of Λ for variable selection purposes in multivariate analysis of variance. Rao [42] has utilized this factorization to construct a modified Λ, when observations on one of the p variables are missing.

Bartlett [5] and Williams [48] consider the use of canonical variables for discriminating among several multivariate populations and obtain a test of goodness of fit of a single hypothetical discriminant function using Wilks's Λ. Bartlett has factorized Λ into a direction factor and a "partial" collinearity factor to test the two aspects of the overall goodness-of-fit* hypothesis, namely, whether the hypothetical function agrees with the true one and whether a single function is adequate at all for discrimination. Reference may be made to Kshirsagar [28] for an expository article on this topic and for various related references to work of Kshirsagar [21–27] and Williams [49] where several distributional results, extensions, and generalizations (as, for example, to contingency tables) are considered.

POWER OF WILKS'S Λ AND OTHER MISCELLANEOUS RESULTS

The nonnull distribution of Λ is the distribution when the matrix **A**, defined in the first section, is Wishart as before, but the other matrix **B** has an independent noncentral Wishart distribution. This distribution will obviously be useful in investigating the power of the Wilks's Λ test in the various problems considered. It is, however, too complicated. It is studied by Khatri and Pillai [18], Sugiara and Fujikoshi [45], Pillai et al. [40], and Gupta [15]. Reference may also be made to Pillai and Jayachandran [39], Michail [33], and Ito [16], for comparison of power of Λ with other criteria, in some special cases.

Das Gupta and Perlman [8] have shown that the power of Wilks's Λ strictly increases with p and q in the linear case (i.e., when the noncentrality matrix in the distribution of **B** is of rank 1) and Fujikoshi [12] proves the monotonicity property of the power function.

Wilks's Λ for complex normal variables has been studied by Young [50].

References

[1]   Anderson, T. W. and DasGupta, S. (1964). *Ann. Math. Statist.*, **35**, 206–211.

[2]   Bartlett, M. S. (1934). *Proc. Camb. Philos. Soc.*, **30**, 327–355.

[3]   Bartlett, M. S. (1938). *Proc. Camb. Philos. Soc.*, **34**, 33–48.

[4]   Bartlett, M. S. (1939). *Proc. Camb. Philos. Soc.*, **35**, 180–190.

[5]   Bartlett, M. S. (1951). *Ann. Eugen. (Lond.)*, **16**, 109–127.

[6]   Box, G. E. P. (1949). *Biometrika*, **36**, 317–346.

[7]   Consul, P. C. (1965). *Bull. Acad. R. Belg. (Cl. Sci.)*, **51**, 683–691.

[8]   Das Gupta, S. and Perlman, M. D. (1973). *J. Multivariate Anal.*, **3**, 220–225.

[9]   Das Gupta, S., Anderson, T. W., and Mudholkar, G. S. (1964). *Ann. Math. Statist.*, **35**, 200–220.

[10] Farmer, J. H. and Freund, R. J. (1975). *Commun. Statist.*, **4**, 87–98.

[11] Fujikoshi, Y. (1970). *J. Sci. Hiroshima Univ., Ser. A-1*, **34**, 73–144.

[12] Fujikoshi, Y. (1973). *Ann. Statist.*, **1**, 388–391.

[13] Gau, G. W. (1978). *Decision Sci.*, **9**, 341–345.

[14] Grizzle, J. and Allen, D. M. (1969). *Biometrics*, **25**, 357–381.

[15] Gupta, R. D. (1979). *Statistica*, **39**, 333–342.

[16] Ito, K. (1962). *Biometrika*, **49**, 455–462.

[17] Khatri, C. G. (1966). *Ann. Inst. Statist. Math., Tokyo*, **18**, 75–86.

[18] Khatri, C. G. and Pillai, K. C. S. (1965). *Ann. Math. Statist.*, **36**, 1511–1520.

[19] Kiefer, J. (1966). In *Multivariate Analysis: Proceedings of an International Symposium*, P. R. Krishnaiah, ed. Academic Press, New York.

[20] Krishnaiah, P. R. (1977). Some Recent Developments on Real Multivariate Distributions. Monogr., Dept. Mathematics and Statistics, University of Pittsburgh, Pittsburgh, Pa.

[21] Kshirsagar, A. M. (1961). *Ann. Math. Statist.*, **32**, 104–111.

[22] Kshirsagar, A. M. (1964). *J. Indian Statist. Ass.*, **2**, 1–20.

[23] Kshirsagar, A. M. (1964). *Proc. Camb. Philos. Soc.*, **60**, 217–222.

[24] Kshirsagar, A. M. (1969). *Aust. Math. Soc.*, **10**, 269–273.

[25] Kshirsagar, A. M. (1970). *Ann. Inst. Statist. Math. Tokyo*, **22**, 295–305.

[26] Kshirsagar, A. M. (1970). *Calcutta Statist. Ass. Bull.*, **19**, 123–130.

[27] Kshirsagar, A. M. (1971). *J. R. Statist. Soc., B*, **33**, 111–116.

[28] Kshirsagar, A. M. (1976). In *The Search for Oil: Some Statistical Methods and Techniques*, D. B. Owen, ed. Marcel Dekker, New York, pp. 147–168.

[29] Lawley, D. N. (1959). *Biometrika*, **46**, 59–65.

[30] Mathai, A. M. and Rathie, P. N. (1969). The Exact Distribution of Wilks's Criterion. *Preprint No. 1969-28*, Dept. of Mathematics, Queen's University, Kingston, Ontario.

[31] McHenry, C. (1976). *Commun. Statist.*, **5**, 1047–1053.

[32] McHenry, C. (1978). *Appl. Statist.*, **27**, 291–297.

[33] Michail, M. N. (1965). *Biometrika*, **52**, 149–152.

[34] Narain, R. D. (1950). *Ann. Math. Statist.*, **21**, 293–300.

[35] Pearson, E. S. and Hartley, H. O. (1971). *Biometrika Tables for Statisticians*, Vol. 2. Cambridge University Press, Cambridge.

[36] Pillai, K. C. S. (1977). *Canad. J. Statist.*, **5**, 1–62.

[37] Pillai, K. C. S. and Gupta, H. K. (1969). *Biometrika*, **56**, 109–118.

[38] Pillai, K. C. S. and Jayachandran, K. (1967). *Biometrika*, **54**, 195–203.

[39] Pillai, K. C. S. and Jayachandran, K. (1968). *Biometrika*, **55**, 335–342.

[40] Pillai, K. C. S., Al-Ani, and Jouris, G. M. (1969). *Ann. Inst. Statist. Math. Tokyo*, **21**, 309–320.

[41] Potthoff, R. R. and Roy, S. N. (1964). *Biometrika*, **51**, 313–326.

[42] Rao, C. R. (1956). *J. R. Statist. Soc. B*, **9**, 259–264.

[43] Rencher, A. C. and Larson, S. F. (1980). *Technometrics*, **22**, 349–356.

[44] Schatzoff, M. (1966). *Biometrika*, **53**, 347–358; correction, *ibid*, **54**, 688 (1966).

[45] Sugiura, N. and Fujikoshi, Y. (1969). *Ann. Math. Statist.*, **40**, 942–952.

[46] Wald, A. and Brookner, R. J. (1941). *Ann. Math. Statist.*, **12**, 137–152.

[47] Wilks, S. S. (1932). *Biometrika*, **24**, 471–494.

[48] Williams, E. J. (1951). *Biometrika*, **38**, 17–35.

[49] Williams, E. J. (1952). *Biometrika*, **39**, 274–282.

[50] Young, J. C. (1971). Some Inference Problems Associated with the Complex Multivariate Normal Distribution. *Tech. Rep. No. 102*, Dept. of Statistics, Southern Methodist University, Dallas, Tex.

(MULTIVARIATE ANALYSIS
MULTIVARIATE ANALYSIS OF VARIANCE
WISHART DISTRIBUTION)

ANANT M. KSHIRSAGAR

LAMBDA DISTRIBUTIONS

A family of distributions which contains densities that range from the uniform to very heavy tailed ones, introduced by Tukey [5, 6]. The distribution and the density func-

tions are given implicitly by

$$F_{\lambda^{-1}}(u) = \left[u^{\lambda} + (1 - u)^{\lambda} \right] / \lambda$$

$$f_{\lambda}(F_{\lambda^{-1}}(u)) = \left[u^{\lambda - 1} + (1 - u)^{\lambda - 1} \right]^{-1},$$

$$0 < u < 1, \quad -\infty < \lambda < \infty.$$

For λ positive they have light tails with bounded support. For $\lambda = 0$ the distribution is the logistic, for $\lambda = 0.135$ it is approximately normal, and it is rectangular for $\lambda = 1$ and $\lambda = 2$. For values of λ between 1 and 2 the densities are U shaped, while for $\lambda > 2$ they are peaked (see Joiner and Rosenblatt [2] for more details). The distribution is used in exploratory data analysis* and for construction of score functions for distribution-free tests* (see, e.g., Jones [3]). Generalizations and multivariate extensions of the distribution were studied by Johnson and Kotz [1] and Ramberg et al. [4].

References

[1] Johnson, N. L. and Kotz, S. (1973). *Biometrika*, **59**, 655–661.

[2] Joiner, B. L. and Rosenblatt, J. R. (1971). *J. Amer. Statist. Ass.*, **66**, 394–399.

[3] Jones, D. H. (1979). *J. Amer. Statist. Ass.*, **74**, 822–828.

[4] Ramberg, J. S., Tadikamalla, P. R., Dudewicz, E. J., and Mykyta, E. F. (1979). *Technometrics*, **21**, 201–214.

[5] Tukey, J. W. (1962). *Ann. Math. Statist.*, **33**, 1–67.

[6] Tukey, J. W. (1970). *Exploratory Data Analysis*. Addison-Wesley, Reading, Mass.

(DISTRIBUTION-FREE METHODS
EXPLORATORY DATA ANALYSIS)

LAMBERT, JOHANN HEINRICH

Born: 1728, in Mulhouse, Alsace.

Died: 1777, in Berlin, Germany.

Contributed to: philosophy, photometry, theoretical mathematics, theories of errors and of probability.

Johann Heinrich Lambert, a self-taught and

LAMBERT, JOHANN HEINRICH **467**

remarkably broad scientist, mathematician, and philosopher, was born in Mulhouse, in the Alsace, in 1728. He grew up in impoverished circumstances. At the age of 12 he was forced to leave school in order to help his father, a tailor. But he persisted in his studies, and at the age of 20 he became a tutor to a wealthy Swiss family. During the 10 years he spent with the family he was able to study, travel, and even publish some of his work. In 1765, he finally obtained a permanent position, at the Royal Academy of Berlin. He died in Berlin in 1777, at the age of 49.

Lambert's maternal language was a dialect of German, but it is impossible to assign him a nationality in the modern sense of the term. Alsace is now part of France, but in Lambert's time Mulhouse was a free city associated with Switzerland. In his writing, Lambert used three languages: Latin, French, and German. He was interested in the development of German as a literary and scientific language, and much of his work, including his philosophical treatises, was written in German. But the approximately 50 papers he wrote for the academy at Berlin were in French. In French, his name is Jean-Henri Lambert.

Lambert left his mark on a wide variety of fields, ranging from philosophy and theoretical mathematics to very practical parts of science. He wrote extensively on logic and the philosophy of knowledge. He was the first to demonstrate the irrationality of π and e. He is remembered for his law of cometary orbits, his cosine law in photometry, his map projections, and his hygrometer. He was not the most outstanding scholar of his time in any single field, but he was almost unique among eighteenth-century scholars in the success with which he combined philosophical and broadly practical interests.

Lambert's most substantial contribution to statistics was the work on the theory of errors that is contained in his books *Photometria* (1760) and *Beyträge zum Gebrauche der Mathematik und deren Anwendung* (Vol. 1, 1765). This work was inspired by his own empirical investigations in photometry and geodesy. By the time Lambert began the work, Thomas Simpson had already given a probabilistic argument for the use of averages, and Boscovich and Mayer had already developed algorithms for solving overdetermined sets of equations. Lambert pulled these threads together and created the idea of a general theory of errors. He discussed the problem of determining probability distributions for errors, and stressed the relevance of probability to the general problem of determining unknown constants. His specific proposals for error distributions and methods of estimation were not widely adopted. But his demonstration of the broad relevance of the theory of errors and his formulation of its problems helped set the stage for the more successful work of Laplace* and Gauss*.

In retrospect, the most striking part of Lambert's work on the theory of errors was his formulation, in *Photometria*, of what we now call the method of maximum likelihood* for a location parameter*. Lambert discussed this method only briefly, and he did not use or mention it in his later work. His friend Daniel Bernoulli* published his own account of the method in 1777. The method did not survive as an independent approach to statistical estimation; instead, it was incorporated into Laplace's Bayesian synthesis. (When a uniform prior distribution is adopted for a parameter, the maximum likelihood estimate coincides with the mode of the posterior distribution, which Laplace called the "most probable value" of the parameter.) It was not until the twentieth-century work of R. A. Fisher* that the method of maximum likelihood was reextracted from the Bayesian framework.

Another remarkable anticipation of twentieth-century thought occurs in Lambert's treatment of nonadditive probabilities*. In his philosophical treatise *Neues Organon* published in 1764, Lambert discussed several examples where evidence seems to justify only nonadditive probabilities. One example involves a syllogism in which the major premise is uncertain. If we have only a probability of $\frac{3}{4}$ for the statement that C is an A, then the further information that all A are B justifies a $\frac{3}{4}$ probability that C is B but does not justify any probability at all that C is not B. The numbers $\frac{3}{4}$ and 0 add to less than 1, so this is an example of nonadditive probability. Lambert also corrected and generalized James Bernoulli's rules for combining nonadditive probabilities, rules that could be used, for example, to combine the probabilities provided by the testimony of independent witnesses. As it turns out, Lambert's nonadditive probabilities have the structure of what we now call belief functions, and his rules for combining nonadditive probabilities are special cases of Dempster's rule of combination for belief functions*.

LITERATURE

Lambert's contributions to probability are described in detail in two recent articles in *Archive for History of Exact Sciences*: "J. H. Lambert's Work on Probability," by O. B. Sheynin, Vol. 7 (1971), pp. 244–256, and "Non-additive Probabilities in the Work on Bernoulli and Lambert," by Glenn Shafer, Vol. 19 (1978), pp. 309–370.

Christopher J. Scriba's article on Lambert in *The Dictionary of Scientific Biography*, Vol. 7 (Charles C. Gillispie, ed., Scribner's, New York, 1973), provides an excellent introduction to his life and work and includes a valuable bibliography. Publications too recent to be listed in that bibliography include "Johann Heinrich Lambert, Mathematician and Scientist, 1728–1777," by J. J. Gray and Laura Tilling, in *Historia Mathematica*, Vol. 5 (1978), pp. 13–41; *Le Savant et Philosophe Mulhousien Jean-Henri Lambert (1728–1777)*, by Roger Jaquel, published in 1977 by Editions Ophrys, Paris; "Johann Heinrich Lambert (1728–1777)," by R. H. Daw, in the *Journal of the Institute of Actuaries*, Vol. 107 (1980), pp. 345–363; and two books published in 1980 by Editions Ophrys, Paris: *Colloque International et Interdisciplinaire J. H. Lambert, Mulhouse, 26–30 Septembre 1977*, and *Correspondance entre D. Bernoulli et J.-H. Lambert*.

(BELIEF FUNCTIONS
BERNOULLIS, THE
MAXIMUM LIKELIHOOD
NONADDITIVE PROBABILITY)

GLENN SHAFER

LANGEVIN, PAUL

Born: January, 23, 1872, in Paris, France.

Died: December 19, 1946, in Paris, France.

Contributed to: theory of magnetism, stochastic differential equations.

A brief account of the life and work of this distinguished physicist is given in ref. 2. As a young man he was in close contact with Jean Perrin, Pierre and Marie Curie, J. J. Thompson, E. Rutherford, and others concerned with atomic events which display some randomness*. Perrin showed that cathode rays were a stream of particles (electrons) and used Brownian motion* to show that atoms had a real rather than theoretical existence. Perrin's book [6] was the inspiration for a recent book by Mandelbrot [5] on random phenomena.

In 1905, Langevin [3] used the then new ideas of atomic structure to explain para- and dia-magnetism. For the former he supposed each molecule to have magnetic moment m which, in the absence of an external field, will be oriented at random due to thermal agitation. When a field of strength H is applied, Boltzmann's method gives a probability density for the orientation of the magnetic moments proportional to

$$\exp(m \cdot H / kT),$$

where T is the absolute temperature and k a constant. Writing $m \cdot H = mH \cos \theta$, $\kappa = mH / kT$, he arrived at the density

$$\frac{\kappa}{4\pi \sin \kappa} \exp(\kappa \cos \theta) \qquad (1)$$

much used in the analysis of directional data* in three dimensions. For x and μ unit vectors in a space of q dimensions, the generalization of (1) is

$$(2\pi)^{-q/2}\kappa^{q/2-1}I_{q/2-1}^{-1}(\kappa)\exp(\kappa\mu \cdot x),   \quad (2)$$

the *Langevin distribution*.

In 1908, Langevin [4] described the motion of a particle in a fluid due to Brownian motion* by stochastic differential equations*, a then novel technique. Thus if v_i is the velocity component in the direction of coordinate i, $i = 1, 2, 3$, he set

$$\dot{v}_i = -\alpha v_i + \sigma \xi_i,   \quad (3)$$

where the term $-\alpha v_i$ is the retarding force of Stokes' law and $\sigma\xi$ is the component of a rapidly oscillating random force due to molecular collisions. This latter force is now idealized to be a "white noise." Langevin went on to derive directly this way Einstein's results on diffusion. Langevin's technique of stochastic differential equations has become very important (see, e.g., Arnold [1] and Schuss [7]) and the basic form (3) is known as the *Langevin equation*.

References

[1] Arnold, L. (1974). *Stochastic Differential Equations*. Wiley-Interscience, New York, p. 228.

[2] Gillespie, C. C., ed. (1973). *Dictionary of Scientific Biography*, Vol. 8. Scribner's, New York.

[3] Langevin, P. (1905). *Ann. Chim. Phys.*, **5**, 70–127.

[4] Langevin, P. (1908). *C. R. Acad. Sci. Paris*, **146**, 530–533.

[5] Mandelbrot, B. B. (1977). *Fractals, Form, Chance and Dimension*. W. H. Freeman, San Francisco, p. 365.

[6] Perrin, J. (1923). *Atoms*. Constable, London, p. 231.

[7] Schuss, Z. (1980). *Theory and Applications of Stochastic Differential Equations*. Wiley, New York, p. 321.

[8] Zakusilo, O. K. (1981). Langevin Equations with Poisson Perturbances. *Tech. Rep. No. ONR 81-02*, Statistical Laboratory, University of California, Berkeley, Calif.

G. S. Watson

LANGEVIN DISTRIBUTION   *See* LANGEVIN, PAUL

LANGEVIN EQUATION   *See* DIFFUSION PROCESSES; LANGEVIN, PAUL

LAPLACE, PIERRE SIMON

> **Born:** March 23, 1749, in Beaumont-en-Auge, France.
>
> **Died:** March 5, 1827, in Paris, France.
>
> **Contributed to:** mathematics, theoretical astronomy, probability and statistics.

CAREER ADVANCE

Born into the French bourgeoisie, Laplace died a Marquis, having lived one of the most influential and "successful" careers in science. His scientific life was dominated by three problems, especially connections between them: methods of solution of differential equations, and related techniques; theoretical astronomy; and probability and statistics. All these interests were present in his earliest works, published in the early 1770s, and continued in a stream of papers for the rest of the century. During the 1790s and early 1800s he wrote his major works in astronomy: the *Exposition du Système du Monde* (1st ed. 1796; 6th posthumous ed., 1835), and especially the *Traité de Mécanique Céleste* (Vols. 1–4, 1799–1805).

The Revolution brought Laplace new chances of professional advance, which he grasped readily. A full member of the old Académie des Sciences in 1785, he was active in the scientific class of the Institut de France on its constitution in 1795, and in the restored Académie from 1816. He took the lead in the affairs of the Bureau des Longitudes, laying the emphasis strongly on theoretical astronomy. He exercised similar influence on the École Polytechnique, mainly through its powerful Conseil de Perfectionnement, formed in 1800; for he shifted the thrust of the curriculum toward its theoretical aspects, thus countering Monge's emphasis on practicalities.

When Bonaparte took power in 1799, he made Laplace Minister of the Interior—but he removed him six weeks later for trying to "carry the spirit of the infinitesimal into administration." However, he made Laplace Chancellor of the Senate in 1803, and Count in 1806.

CAREER EMINENCE

During the early 1800s Laplace came to his philosophy of physics, which has since been termed "Laplacian physics." His view was that *all* physical theories should be formulated in terms of binary forces acting between "molecules" of the matter involved. In this spirit he extended the range of his scientific interests, hitherto confined primarily to astronomy and planetary physics, and made useful contributions to capillary theory, sound, optics, and heat diffusion. He and his neighbor at Arcueil, the chemist Berthollet, sponsored a group of young scientists at this time which was called "The Society of Arcueil." Biot and Poisson were among its members.

Laplace was always adept at bending with the political wind. While L. Carnot and Monge lost their seats in the restored Académie des Sciences in 1816, Count Laplace not only kept his but was also elected to the Académie Française that year and even elevated to a Marquisate in 1817. His three connected scientific interests continued to occupy much of his time. His treatise *Théórie Analytique des Probabilités* appeared in editions of 1812, 1814, and 1820 (with supplements in the 1820s), while an *Essai Philosophique sur les Probabilités*, serving as an introduction to the major work, was published in five editions between 1814 and 1825. In the 1820s he also put out a fifth volume of his *Mécanique Céleste*, consisting mostly of reprints or reworkings of earlier papers. He worked right to the end; two substantial papers were published in the year (1827) of his death.

STYLE

The problem of influence on Laplace is particularly hard to solve, partly because of the concurrency of others' work (during the 1770s and 1780s in particular, his work overlapped with that of Lagrange and Legendre in many respects) and especially due to his habit of rarely citing his sources, even his own previous writings. Todhunter has a delicious explanation: "We conclude that he supposed the erudition of his contemporaries would be sufficient to prevent them from ascribing to himself more than was justly due." Laplace often skipped details of passage work, about which his translator Bowditch spoke for all: "Whenever I meet in La Place with the words 'Thus it plainly appears,' I am sure that hours, and perhaps days, of hard study will alone enable me to discover *how* it plainly appears."

CALCULUS

Laplace introduced some new methods of solving differential equations: the methods of cascades and of successive approximations are the most important. By and large, however, he extended or varied techniques developed by others. For example, he was adept at using Lagrange's formal relation

$$u(x + h) - u(x) = \exp[hd/dx - 1]u(x) \tag{1}$$

between differences and differentials to devise means of solving not only differential equations but also difference and mixed equations, and to evolve methods of summing series [1].

ASTRONOMY

Laplace's principal motivation to astronomy was the mathematical analysis of all the known motions, especially the perturbations, of the heavenly bodies. He and Lagrange wished to *prove* the stability of the planetary system, and so refute the catastrophism embodied in the Newtonian tradition. His work also brought him to various aspects of planetary physics: tidal theory, the shape of the earth, planetary rotation, and so on. The connections with probability lay principally in his proposal, rather novel for the time, of using astronomical data to determine constants and numerical values for specific quantities; and in studying "population" problems such as the distribution of comets. (*See* STATISTICS IN ASTRONOMY.) With regard to error analysis, he developed minimax* methods as his preferred linear regression* techniques [2].

EARLY PROBABILITY

One of Laplace's first discoveries in probability itself was Bayes' theorem*, which he published in 1774 [3]. Although Bayes' paper had appeared (posthumously) ten years earlier, it is quite likely that the French were unaware of it. And there is no doubt that Laplace established Bayesian statistics as a body of knowledge in probability theory. His applications of it included standard combinatorial problems (drawing balls from an urn, etc.), and demography* (the ratio of male and female births; and later, population estimation for France). He sometimes used the beta distribution*, and in the process contributed to the theory of the incomplete beta function [4]. Much of his later work switches between Bayesian and non-Bayesian methods, suggesting that for him the distinction between the two did not mark a dichotomy in the way that is often asserted today.

In 1782, Laplace introduced his "generating function"

$$f(t, x) \overset{\text{DF}}{=} \sum_{r=0}^{\infty} y_r(x) t^r \qquad (2)$$

as a technique for solving differential and related equations, and also for use in proba-

bility [5]. In the latter context it stood as a generating function for a discrete random variable, with an integral analog for continuous variables. Transformation of (2) by $t = e^{iu}$ led him later to some consideration of characteristic functions* [6], but he largely missed the bearing of harmonic analysis on such functions for continuous variables, and also the related topic of the Laplace transform. (*See also* INTEGRAL TRANSFORMS.)

LAPLACE TRANSFORM

What, then, did Laplace do with the Laplace transform? The term is something of a misnomer, since he only manipulated the *in*definite integrals

$$\int x^s \phi(x) \, dx \quad \text{and} \quad \int e^{-sx} \phi(x) \, dx \qquad (3)$$

rather than the true Laplace transform. Some connections are there. For example, he began work with (3) in the 1780s in connection with his use of generating functions (2); and his purpose then was linked with asymptotic theory, and included various uses of the error function. For example, in an extensive essay of 1785 on asymptotic theory he obtained expansions of $\int_0^T e^{-t^2} dt$ in powers of T or of T^{-1} according as T was small or large, and obtained definite integrals of related functions, such as [7]

$$\int_{-\infty}^{+\infty} z^{2r} e^{-z^2} \, dz = \frac{(2r)! \sqrt{\pi}}{4^r r!} . \qquad (4)$$

In the 1800s he was aware of Fourier's work; indeed, he eventually encouraged Fourier's study of heat diffusion and trigonometric series and in 1809 used (4) to find a new integral solution to the diffusion equation [8]. This solution in turn led Fourier in 1811 to Fourier transform theory. But the Laplace transform did not come out as correlate theory; in particular, in his *Probabilités* Laplace came to the inverse transform, but, finding it to be an integral along a complex path, did not know how to proceed [9]. Even though Cauchy began to introduce complex variable integrals within a few years, the potential of Laplace transform theory remained hidden for many decades. (*See* INTEGRAL TRANSFORMS.)

LATER STATISTICS

However, Laplace made other advances around 1810. One was his study of a certain urn problem of which the solution led him to the "Hermite polynomials," including the "Gram–Charlier expansion," both misnomers [10]. Inspiration from Fourier may again be evident, as his method imitates Fourier's treatment in 1807 of the (also misnamed) "Bessel function" $J_0(x)$. Another achievement was a form of a central limit theorem*, where again the 1780s studies of the error function played a role (as also, the erudite reader suspects, did some work by Lagrange on determining the mean of a set of observations, Legendre's and Gauss's advocacy of least-squares* regression earlier in the 1800s, and de Moivre*'s previous work). After showing that the sum of two terms symmetrically placed in the binomial expression $(p + q)^{r+s}$ and $2l$ terms apart was given approximately by

$$\sqrt{\frac{2(r + s)}{\pi rs}} \exp\left[-\frac{(r + s)l^2}{2rs} \right], \quad l \text{ large,}$$

$$(5)$$

he used the Euler–MacLaurin* summation formula to prove that "the probability that the difference between the ratio of the number of times that the event a can happen to the total number $[n]$ of attempts and the facility p of its occurrence" was given by

$$\frac{2}{\sqrt{\pi}} \int_0^\alpha e^{-t^2} dt + \frac{\alpha}{l\sqrt{\pi}} e^{-\alpha^2} + O\left(\frac{1}{n}\right), \quad (6)$$

where

$$\alpha^2 = \frac{l^2 n}{2(np + z)(nq - z)},$$

$$p + q = 1, \quad |z| < 1. \quad (7)$$

Thus the probability was $O(1/\sqrt{n})$ [11].

During the last 15 years of his life, Laplace was mostly occupied with problems in Laplacian physics (and astronomy), but he also made further contributions to probability at times, especially in supplements to his treatise. For example, in 1818 he compared absolute deviation with least-squares deviation and in effect carried out an exercise in sufficient statistics (without, however, individuating the concept of sufficiency*) [12]. Again, in his last year of life he handled a multiple regression analysis*, in a paper which exemplified his continually interlocking interests of his whole career, for it was concerned with lunar tides in the atmosphere [13].

PHILOSOPHY OF PROBABILITY

Laplace discussed his philosophy of probability in various places, and took it as the theme of his *Essai*. He saw the parallels between the physical and social sciences, and thus the applicability of probability to each. Comets were distributed; so were births. The planetary system was stable; so were lottery receipts.

Laplace also took probability as the core of his philosophy of science; for he thought of scientific knowledge as only probable, of which the degree of probability would be increased by confirmations. The world was fully determined, but *our* theories could only capture its course partially. "The curve described by a simple molecule of air or any gas is regulated in a manner as certain as the planetary orbits," he wrote, "the only difference between them lies in our ignorance. Probability relates partly to our ignorance, partly to our knowledge" [14].

INFLUENCE

The impact on others of Laplace's contributions to probability and statistics built up more slowly than his mathematical and astronomical achievements. He was working in a much less developed field, and so played pioneer to a greater extent. His "thus-it-plainly-appears" style was especially unfortunate, since the contexts in which the passage work occurred were much less familiar to the reader.

However, Laplace's contributions to probability and statistics were fundamental: not merely some new results but especially a much greater degree of mathematization. He also changed the emphasis of probability from its preoccupation with moral sciences and jurisprudence to include also applications in scientific contexts, whither it had hitherto infrequently strayed. His most important early successors were Quetelet* and Poisson*; after them, both probability and statistics moved to adulthood in the family of sciences, and the heritage from Laplace began to be recognized.

LITERATURE

A seven-volume edition of Laplace's *Oeuvres*, published 1843–1847, contains only the books mentioned in the text. The 14-volume *Oeuvres Complètes* (1878–1912) has another reprinting of these books, with the other seven volumes given over to his papers. The edition is poor even by the standards of works for French mathematicians; notations are sometimes modernized, editorial apparatus and commentary is almost nonexistent, and worst of all, it is *not* complete, lacking his first three papers and his first book [*Théorie du Mouvement et de la Figure Elliptique des Planètes* (1784)]. The only comprehensible edition of *Mécanique Céleste*, Vols. 1–4, is N. Bowditch's English translation *Celestial Mechanics* (4 Vols. 1829–1839, Boston; reprinted in 1966, New York). Most of Laplace's manuscripts were destroyed long ago, although there are a few things here and there; in particular, C. C. Gillispie, 'Mémoires inédits ou anonymes de Laplace . . . ,' *Rev. Hist. Sci.*, 32(1980), 223–280 includes a paper on determining the mean of a set of observations.

The most comprehensive single study of Laplace is C. C. Gillispie (in collaboration with others), "Laplace, Pierre-Simon, Marquis de," in *Dictionary of Scientific Biography*, Vol. 15 (1978, New York), 273–403, including extensive bibliographies of primary and secondary literature. Laplace's work on probability has been studied by various authors: a good start is provided by S. Stigler, "Napoleonic statistics: the work of Laplace," *Biometrika*, 62(1975), 503–517; and O. B. Sheynin, "P. S. Laplace's work on probability," *Arch. Hist. Exact Sci.*, 16(1976), 137–187 and "Laplace's theory of errors," *ibid.*, 17(1977), 1–61.

References

These are taken from *Oeuvres Complètes*, discussed above, in which the books on probability occupy Vol. 7. The volume numbers are to the edition; the dates are of the original publication of the work involved.

[1]   **9**, 315–325 (1780).
[2]   See, e.g., **2**, 246–314 (1799).
[3]   **8**, 27–65 (1774).
[4]   **9**, 422–429 (1781).
[5]   **10**, 549 (1782).
[6]   See, e.g., **12**, 309–319 (1810).
[7]   **10**, 230, 269 (1785); cf. **7**, 104 (1812, 1820).
[8]   **14**, 189–193 (1809).
[9]   **7**, 136–137 (1812, 1820).
[10] **12**, 377–385 (1811).
[11] **7**, 281–287 (1812, 1820).
[12] **7**, 531–580 (1818, 1820).
[13] **13**, 342–353 (1827), and also in a supplement to *Mécanique Céleste* (Vol. 5, pp. 481–505 (1827)).
[14] **7**, viii (1814, 1825).

I. GRATTAN-GUINNESS

LAPLACE DISTRIBUTION

The name of Pierre Laplace* is well known in statistics; his considerable contributions and their impact on the subject are ably chronicled by Stigler [10].

In 1774, Laplace wrote a fundamental paper on symmetric distributions to describe errors of measurement [7]. His first law of error and the one that bears his name was obtained as the distribution whose likelihood is maximized when the location parameter is set equal to the *median* (for an odd number of observations). Laplace's second law of error, corresponding to the sample mean as the maximum likelihood* estimator, is none other than the normal distribution*.

The Laplace distribution has probability density function

$$f_X(x) = \tfrac{1}{2}\phi^{-1}\exp\left[-|x-\theta|/\phi\right], \quad (1)$$

where $\phi > 0$ and $-\infty < x < \infty$. Constants apart, the normal law is obtained if the exponent contains $(x-\theta)^2$ in place of $|x-\theta|$. Thus a straightforward extension of the laws of error* is to use $|x-\theta|^{2/\delta}$ in the exponent, where $\delta > 0$. It is apparent that the two laws correspond to $\delta = 2$ and $\delta = 1$, respectively, and, usually, $\delta \geqslant 1$ in cases of practical interest. This extension was first proposed by Subbotin [11], although minimization of $\sum|x_i-\theta|^{2/\delta}$ has more recently attracted attention as a criterion in robustness* studies. In addition, the Laplace distribution arises as the distribution of the difference between two *exponential* variates, when it is variously known as the *double exponential*, although this term is also used for the extreme value distribution*, the *two-tailed exponential*, or the *bilateral exponential*.

PROPERTIES

It is evident from (1) that the density function is symmetric about θ, whence the mean, median, and mode are all equal to θ. It follows that all odd-order moments about θ are zero and the even-order moments are

$$\mu_{2r} = \phi^{2r}(2r)!, \quad r = 1, 2, \ldots .$$

The variance is $2\phi^2$. The moment ratio

$$\beta_2 = \mu_4/\mu_2^2 = 6$$

reflects the slower rate of decay in the tails of the distribution compared to the normal.

The standard form of the distribution has density

$$f_Z(z) = \tfrac{1}{2}e^{-|z|}, \quad -\infty < z < \infty, \quad (2)$$

where $z = (x-\theta)/\phi$.

The characteristic function for Z is

$$\phi_Z(r) = E(e^{itZ}) = (1+t^2)^{-1}, \quad (3)$$

and it is interesting to note that the Fourier transform pair, (2) and (3), occur in reverse order for the *Cauchy* distribution*.

Further properties and interrelations with other distributions are given in Johnson and Kotz [4, Chap. 23] and Patil et al. [9].

ESTIMATION

We have already remarked that the median is the maximum likelihood estimator for θ (when the sample size, n, is odd). When n is even, equal to $2m$ say, the likelihood function is flat over the interval $[X_{(m)}, X_{(m+1)}]$, where the $X_{(j)}$ are order statistics. By convention, the midpoint of the interval $\tilde{\theta} = \tfrac{1}{2}[X_{(m)} + X_{(m+1)}]$ is usually employed, since $\tilde{\theta}$ is unbiased for θ.

The maximum likelihood estimator ϕ is the median absolute deviation,

$$\hat{\phi} = \frac{1}{n}\sum_{i=1}^{n}|X_i - \theta|. \quad (4)$$

When θ is known, the distribution of $\hat{\phi}$ is chi-squared with $2n$ degrees of freedom so that confidence intervals* for ϕ are readily constructed (see Johnson and Kotz [4, p. 29]). When both parameters are unknown, Kappenman [5] has derived exact confidence intervals by conditioning on the observed values of the $(n-2)$ ancillary statistics. The same approach is used by Kappenman [6] to develop conditional tolerance intervals*.

Approximate sampling distributions are derived by Gallo [1]. The Bayesian predictive distribution for a future observation is given in Ling and Lee [8].

Many studies of robust estimators for the location and scale parameters of symmetric populations have used the Laplace distribution. For recent work on adaptive robust estimators, see Harter et al. [2].

RELATED DISTRIBUTIONS

Just as a family of distributions may be formed around the normal by a suitable class of transformations (e.g., the Johnson system of curves*), a comparable family may be developed for the Laplace distribution [3].

The difference between two independent exponential variates with different scale variates yields an asymmetric Laplace distribution with density function

$$f_X(x) = \begin{cases} \frac{1}{2}\phi_1^{-1}\exp\left[-|x-\theta|/\phi_1\right], & x \geqslant \theta \\ \frac{1}{2}\phi_2^{-1}\exp\left[-|x-\theta|/\phi_2\right], & x < \theta. \end{cases}$$

We note that the density function now has a discontinuity at θ. Other asymmetrical forms and several mixtures based on the Laplace are described by Johnson and Kotz [4, Chap. 23] and Patil et al. [9].

BIVARIATE DISTRIBUTIONS

If X_1, X_2, and X_3 are independent Laplace variates, a bivariate distribution may be formed by setting

$$Y_j = X_j + X_3, \qquad j = 1, 2.$$

However, the properties of the bivariate Laplace do not appear to have been developed in a systematic fashion.

References

[1] Gallo, F. (1979). *Statistica*, **39**, 443–454.

[2] Harter, H. L., Moore, A. H., and Curry, T. F. (1979). *Commun. Statist. A*, **8**, 1473–1492.

[3] Johnson, N. L. (1954). *Trab. Estadíst.*, **5**, 283–291.

[4] Johnson, N. L. and Kotz, S. (1970). *Distributions in Statistics: Continuous Univariate Distributions*, Vol. 2. Wiley, New York.

[5] Kappenman, R. F. (1975). *Technometrics*, **17**, 233–236.

[6] Kappenman, R. F. (1977). *J. Amer. Statist. Ass.*, **72**, 908–909.

[7] Laplace, P. S. (1774). *Mém. Math. Phys.*, **6**, 621–657 (also *Oeuvres complètes*, Vol. 8, Gauthier-Villars, Paris, 1891, pp. 27–65).

[8] Ling, K. D. and Lee, G. C. (1977). *Nanta Math.*, **10**, 13–19.

[9] Patil, G. P., Boswell, M. T., and Ratnaparkhi, M. V. (1982). *A Modern Dictionary and Classified Bibliography of Distributions*, Vol. 1: *Continuous Univariate Models*. International Co-operative Publishing House, Fairland, Md.

[10] Stigler, S. M. (1975). *Biometrika*, **62**, 503–517.

[11] Subbotin, M. T. (1923). *Mat. Sb.*, **31**, 296–301.

(EXPONENTIAL DISTRIBUTION LAWS OF ERROR, I, II, III)

J. K. ORD

LAPLACE–GAUSS DISTRIBUTION *See* NORMAL DISTRIBUTION

LAPLACE'S FIRST LAW *See* LAPLACE DISTRIBUTION

LAPLACE'S LAW OF SUCCESSION

This is sometimes referred to as "Bayes'–Laplace theory." If H is the hypothesis that a population possesses property ϕ and if in a sample of size n (when sampling from an infinite (or large) population), all members are found to possess property ϕ, then the probability that the next member will possess property ϕ is

$$\frac{n+1}{n+2}$$

and that the next m members which possess property ϕ is

$$\frac{n+1}{n+m+1}.$$

(These probabilities are based on uniform assessment of the prior probability.) This "law" has been much discussed and debated in the literature on foundations of statistics and probability, especially in connection with Bayesian inference*.

(AXIOMS OF PROBABILITY
BAYESIAN INFERENCE
LAPLACE, PIERRE SIMON)

LAPLACE'S SECOND LAW   *See* NORMAL
DISTRIBUTION

LAPLACE TRANSFORM   *See* INTEGRAL
TRANSFORMS

LARGE DEVIATIONS AND APPLICATIONS

Consider a sequence of observations X_1, $X_2, \ldots$, which are standardized in such a way that they can be considered to arise from an underlying distribution having mean 0 and variance 1. For each positive integer n, we construct the partial sum $S_n = X_1 + X_2 + \cdots + X_n$ and the sample mean $\bar{X}_n = n^{-1}S_n$. The central limit theorem* asserts that $\bar{P}_n(\lambda) = P\{\bar{X}_n > \lambda/\sqrt{n}\} \to 1 - \Phi(\lambda)$ as $n \to \infty$, where

$$\Phi(\lambda) = (2\pi)^{-1/2} \int_{-\infty}^{\lambda} \exp(-t^2/2)\,dt$$

is the standard normal distribution function for any real number λ. If $\lambda_1, \lambda_2, \ldots$ is a sequence of positive numbers tending to ∞, it follows that $\bar{P}_n(\lambda_n) \to 0$ as $n \to \infty$. In such cases the probability $\bar{P}_n(\lambda_n)$ is called a "large deviation probability" (of the sample mean), and the rate at which it converges to 0 has been the focus of much of the traditional large deviations literature. Although we have defined our large deviation probabilities in terms of the sample mean, the latter can be replaced by almost any statistic of your choosing, and the chances are that there is some published work on the asymptotic behavior of its large deviation probabilities (see, e.g., Rubin and Sethuraman [25], Sievers [29], and Woodroofe [34]).

A large portion of the theory of asymptotic efficiency of tests of statistical hypothe-

ses depends on the calculation of such probabilities, and this dependence has provided the primary impetus toward the extension of large deviations research to wider and wider classes of statistics. If we want to use a statistic $T_n = T(X_1, X_2, \ldots, X_n)$ to conduct a test of a hypothesis H_0 against an alternative H_1, rejecting H_0 if $T_n > c_n$, then as $n \to \infty$ the test ordinarily improves, in the sense that the conditional probability $\alpha_n = P\{T_n > c_n \mid H_0 \text{ is true}\}$, called the significance level of the test, decreases, ideally to 0. The rate at which the large deviation probability α_n tends to 0 is a measure of asymptotic efficiency, i.e., relationship between the sample size n and the type I error probability α_n. Our ability to calculate large deviation probabilities therefore circumscribes our ability to judge asymptotic efficiency of tests in this sense.

The proofs of most strong limit theorems of probability theory require bounds based on large deviation probabilities. As used in the most elementary form of the weak law of large numbers, Chebyshev's inequality* places an upper bound on the probability $P\{\bar{X}_n > \epsilon\}$. Sharper estimates are needed in those limit theorems where "lim inf" and "lim sup" results are proved separately and by different methods. The Borel–Cantelli lemmas* often require that series $\sum_{n=0}^{\infty} P\{S_n > c_n\}$ converge. Such series are sums of sequences of large deviation probabilities, and again our ability to establish their convergence arises from our ability to precisely estimate the probabilities.

FOUNDATIONS

If the underlying distribution is the standard normal and the statistic is the sample mean, the problem of asymptotic behavior of the large deviation probabilities is completely solved. The random variable $\bar{X}_n/\sqrt{n}$ then has itself the standard normal distribution, so that $\bar{P}_n(\lambda_n) = 1 - \Phi(\lambda_n)$. The rate at which the tail of the standard normal tends to 0 as $\lambda_n \to \infty$ is given by Mills' ratio*, first tabulated by Mills [22], which asserts that

$1 - \Phi(\lambda_n) \sim (2\pi\lambda_n^2)^{-1/2} \exp(-\frac{1}{2}\lambda_n^2)$. The symbol "$\sim$" is pronounced "is asymptotic to" and indicates that the ratio of the two sides tends to 1.

The first step beyond the normal distribution was taken by Khintchine [19], who obtained an analog of Mills' ratio for the sample mean of a standardized Bernoulli distribution* valid in the range where $\lambda_n \to \infty$ and $\lambda_n / \sqrt{n} \to 0$. His "new limit theorem" stated that $\bar{P}_n(\lambda_n) \sim (2\pi\lambda_n^2)^{-1/2} \exp[-f_n(\lambda_n)]$, when $\bar{X}_n$ is the sample mean of standardized Bernoulli random variables and $f_n(x)$ is a known function depending only on n and the Bernoulli parameter p. Smirnov [30] sharpened Khintchine's work and, in doing so, created an organizational framework that persists to the present day. The essentials of Smirnov's approach were to obtain a series expansion of the probability $\bar{P}_n(\lambda_n)$ and to classify the number of terms required in the expansion and the magnitude of the remainder term according to the rate at which $\lambda_n \to \infty$. The case $n^{-1/6}\lambda_n \to 0$ in Smirnov's classification guaranteed that

$$\bar{P}_n(\lambda_n) \sim 1 - \Phi(\lambda_n)$$

exactly.

MODERN RESULTS

Modern work on large deviation probabilities began with Cramér [9]. The preeminent position occupied by the Cramér paper in the history of large deviations is due to the fact that Cramér presented a general method for studying large deviation probabilities that is still being used as a research tool today. The method consists of a saddle-point-like approximation leading to the Cramér series, a power series whose coefficients are functions of the semi-invariants of the underlying distribution and which is in many respects a generalization of Khintchine's function $f_n(x)$ and of Smirnov's expansion. Furthermore, Cramér was the first to obtain results for $\lambda_n = c\sqrt{n}$, where c is a positive constant; deviations of this size

have subsequently proved to be the most important in applications.

In the case $\lambda_n = c\sqrt{n}$, Cramér's theorem was restricted to "strongly nonlattice" variables, i.e., those having an absolutely continuous component. This restriction was removed in two stages: first Chernoff [8] showed that $n^{-1}\log \bar{P}_n(c\sqrt{n})$ converges for all distributions having a moment generating function* (MGF) to a constant $\rho(c)$, depending on c and on the underlying distribution. (It turns out that the function ρ, which has come to be known as the "Chernoff function," uniquely determines the underlying distribution; this fact has recently proved to be of great importance in probability theory.) Next, Bahadur and Ranga Rao [3] extended Cramér's original asymptotic representation for $\bar{P}_n(c\sqrt{n})$ to all variables having MGFs. Statulevicius [31] obtained analogs of Cramér's results for a general class of statistics which need not be expressible in terms of sums of observations.

Linnik [21] discovered a far-reaching method of analysis based on characteristic functions instead of MGFs and used it to study the probability $\bar{P}_n(\lambda_n)$ when $\lambda_n / \sqrt{n} \to 0$ and the MGF does not exist. Linnik's work inspired a long list of other papers described most recently by Wolf [33]. Additional Soviet progress in this area is summarized in the expository paper of Nagaev [23], who includes a bibliography of 60 items. Rubin and Sethuraman [25] later showed that $\bar{P}_n(\lambda_n) \sim 1 - \Phi(\lambda_n)$ for $\lambda_n = c\sqrt{\log n}$ whenever the underlying distribution has a moment of order slightly higher than 2. Chapter 14 of the monograph by Ibragimov and Linnik [18] describes one result applicable to the case $\lambda_n / \sqrt{n} \to \infty$.

STATISTICAL APPLICATIONS

Bahadur [2] portrays the role played by large deviation probabilities in the theory of asymptotic efficiency of tests of statistical hypotheses. Details may be found in Bahadur [1]; loosely speaking, the Bahadur "exact slope" of a sequence of test statistics T_n can be considered to be the limit $\lim_{n \to \infty} (-2n^{-1}) \log P\{T_n > c_n \mid H_0$ is true}, where T_n is to be used to test H_0 against H_1, H_0 being rejected if $T_n > c_n$. The exact slope is a measure of the rate at which the significance level α_n tends to 0 as the sample size n increases to ∞. (It is assumed here that $\alpha_n \to 0$ as $n \to \infty$.) Two sequences of statistics for testing the same hypothesis may be compared by computing the ratio of their exact slopes, the ratio being referred to as the "Bahadur asymptotic relative efficiency*" of the first sequence with respect to the second. Exact slopes and Bahadur asymptotic relative efficiencies have been calculated under many different distributional assumptions and in a wide variety of testing situations (e.g., Sievers [29]). For the mathematical statistician, the theory of exact slopes and the theory of large deviation probabilities are one and the same. Bahadur [2] references more than 40 papers published on the subject prior to 1970. Csörgö [10] lists a large number of later papers.

In a paper that generated wide interest, Efron [11] established a criterion called "statistical curvature*," the role of which was to measure how close the usual procedures such as locally most powerful tests* and maximum likelihood* estimators come to being optimal. Such procedures are generally optimal in exponential families* of densities, a class that includes virtually all the common univariate and multivariate densities. Exponential families* have zero curvature, and Efron's work indicated that the farther the curvature of a family is from being zero, the farther from being optimal are those usual procedures. Efron based his analysis largely on the large deviation results of Chernoff [8] and Efron and Truax [12].

Woodroofe [34] studied the large deviation behavior of the likelihood ratio statistic and used his result to investigate type I error probabilities for a sequential likelihood ratio test*. Consider a p-dimensional exponential family whose members are indexed by the parameter $\omega \in \Omega$, and let Ω_q be a q-dimensional submanifold of Ω. We test the hypothesis $H : \omega \in \Omega_q$. Denoting by Λ_n the usual transformed version of the likelihood ratio* statistic, it is shown in the range $\lambda_n \to \infty$ and $\lambda_n / \sqrt{n} \to 0$ that $P\{\Lambda_n \geqslant \lambda_n\}$ $\sim 1 - \Psi(\lambda_n)$, where Ψ is the distribution function of the random variable $\frac{1}{2} W_r^2$, W_r^2 having the chi-square distribution* with $r = p - q$ degrees of freedom. As Woodroofe notes, this large deviation result shows that the previously known and widely used asymptotic approximation for $P\{\Lambda_n \geqslant \lambda_n\}$ is accurate for deviations of order $o(n)$.

PROBABILITY APPLICATIONS

Just as the need to calculate significance levels motivated the study of large deviations on the statistical side, the effort to sharpen the classical strong limit theorems of probability theory spurred a parallel development on the probability side. The first general results on the strong law of large numbers were proved by applying the Borel–Cantelli lemma to a series of the form $\sum_{n=1}^{\infty} \overline{P}_n(\epsilon \sqrt{n})$. Both upper and lower bounds on $\overline{P}_n(\lambda_n)$, where

$$\lambda_n = (1 \pm \delta)\sqrt{2 \log \log n} \,,$$

were needed in Kolmogorov's [20] original law of the iterated logarithm*, for the key step in the proof is again the convergence of a series of large deviation probabilities. Feller [14, 15] published two papers on large deviations, and both of them were expressly tailored to yield extensions of the iterated logarithm law. More recently, V. V. Petrov, who played a dominant role in developing large deviations theory and related limit theorems over the past 25 years, has turned his attention to general analogs of the iterated logarithm law. Chapter 10 of Petrov's [24] monograph presents an account of work done in this area up to 1970.

Bártfai [5] applied Cramér's 1938 large deviation theorem to solve the so-called "stochastic geyser problem." This problem poses the following situation: $Y_1, Y_2, \ldots$ are possibly erroneous observations of a distribution function F in the sense that each

$Y_k = X_k + e_k$, where X_k has distribution F and e_k is an error term of unknown character. The question is: How large an error can we put up with and still be able to determine F from the sequence of observations? Bártfai found that the coefficients of the Cramér series could be used to calculate the MGF of F as long as the accumulated error $E_n = \sum_{k=1}^{n} e_k$ is of order below $\log n$.

In a brilliant paper that remained unnoticed for 15 years, Shepp [28] investigated the limiting behavior of moving averages* $M(K, n) = K(n)^{-1}(S_{n+K(n)} - S_n)$, where K assumed only positive integer values and tends to ∞ in a nondecreasing manner as $n \to \infty$. Noting that

$$M = \limsup_{n \to \infty} M(K, n)$$

is almost surely constant by the zero–one law, Shepp showed that the value of this constant is $\rho^{-1}(r(K))$, where $r(K)$ is the radius of convergence of the series $\sum_{n=1}^{\infty} x^{K(n)}$ and ρ is the Chernoff function. As a special case, it follows that if $K(n)$ is the integer part of $(\log n)/(-\log \rho(\lambda))$ for some constant $\lambda > 0$, then $r(K) = \rho(\lambda)$, so that $M = \lambda$.

Unaware of Shepp's work, Erdös and Rényi [13] derived what they called a "new law of large numbers," again using an argument that can be based on the Chernoff [8] theorem. This result dealt with the behavior of the maximum

$$M_\lambda(N) = \max_{0 \le n \le N - K_\lambda(N)} (S_{n+K_\lambda(N)} - S_n)/K_\lambda(N)$$

of moving averages, where $K_\lambda(N)$ is the integer part of $(\log N)/(-\log \rho(\lambda))$ and ρ is the Chernoff function. The Erdös–Rényi formulation of what is now known as the *Shepp–Erdös–Rényi new law* asserts that, for a wide range of values of $\lambda > 0$, the probability is one that $\lim_{N \to \infty} M_\lambda(N) = \lambda$. The significance of the Shepp–Erdös–Rényi new law, together with a number of generalizations to commonly used test statistics and stochastic processes, is explained in detail by Csörgö [10], who includes a bibliography of over 40 pertinent items. One of the corollaries of the Shepp–Erdös–Rényi law is a simpler proof of the stochastic geyser result under conditions slightly weaker than those of Bártfai. The simpler proof takes advantage of the fact that the Chernoff function uniquely determines the underlying distribution function F.

FURTHER RESULTS

In addition to the classical theorems on sums of independent, identically distributed observations of a one-dimensional random variable, there exists an extensive literature on large deviation probabilities in more general spaces and for special statistics and stochastic processes. A formulation of the large deviation problem for the empirical distribution function in terms of general vector spaces led to an extremely fruitful line of investigation. Influential early papers on these topics are those of Sanov [26], Sethuraman [27], Varadhan [32], and Hoeffding [17]. The current state of the problem, together with large bibliographies of intermediate references may be found in the papers of Groeneboom et al. [16], Bahadur and Zabell [4], and Borovkov and Mogul'skii [6, 7].

Too numerous to mention specifically are the specialized articles on large deviations for order statistics, *U*-statistics*, weighted sums, triangular arrays, random sums, samples from a finite population, renewal processes*, queuing processes*, stationary processes*, Markov chains, variables attracted to a stable law, distributions on a circle, and various special functions of observations. There are also many papers in the Soviet literature on the local behavior of large deviation probabilities.

References

[1] Bahadur, R. R. (1967). *Ann. Math. Statist.*, **38**, 303–324.

[2] Bahadur, R. R. (1971). *Some Limit Theorems in Statistics.* SIAM, Philadelphia.

[3] Bahadur, R. R. and Ranga Rao, R. (1960). *Ann. Math. Statist.*, **31**, 1015–1027.

[4] Bahadur, R. R. and Zabell, S. L. (1979). *Ann. Prob.*, **7**, 587–621.

[5] Bártfai, P. (1966). *Studia Sci. Math. Hung.*, **1**, 161–168.

[6] Borovkov, A. A. and Mogul'skii, A. A. (1978). *Siberian Math. J.*, **19**, 697–709.

[7] Borovkov, A. A. and Mogul'skii, A. A. (1980). *Siberian Math. J.*, **21**, 653–663.

[8] Chernoff, H. (1952). *Ann. Math. Statist.*, **23**, 493–507.

[9] Cramér, H. (1938). *Actualités Scientifiques et Industrielles*, no. 736. Hermann, Paris, pp. 5–23.

[10] Csörgö, S. (1979). *Ann. Statist.*, **7**, 772–787.

[11] Efron, B. (1975). *Ann. Statist.*, **3**, 1189–1242.

[12] Efron, B. and Truax, D. (1968). *Ann. Math. Statist.*, **39**, 1402–1424.

[13] Erdös, P. and Rényi, A. (1970). *J. Anal. Math.*, **23**, 103–111.

[14] Feller, W. (1943). *Trans. Amer. Math. Soc.*, **54**, 361–372.

[15] Feller, W. (1969). *Zeit. Wahrscheinlichkeitsth. verwand. Geb.*, **14**, 1–20.

[16] Groeneboom, P., Oosterhoff, J., and Ruymgaart, F. H. (1979). *Ann. Prob.*, **7**, 553–586.

[17] Hoeffding, W. (1967). *Proc. 5th Berkeley Symp. Math. Statist. Prob.*, Vol. 1. University of California Press, Berkeley, Calif., pp. 203–219.

[18] Ibragimov, I. A. and Linnik, Yu. V. (1971). *Independent and Stationary Sequences of Random Variables*. Walters-Noordhoff, Groningen (esp. Chaps. 6–14).

[19] Khintchine, A. (1929). *Math. Ann.*, **101**, 745–752.

[20] Kolmogorov, A. N. (1929). *Math. Ann.*, **101**, 126–135.

[21] Linnik, Yu. V. (1960). *Proc. 4th Berkeley Symp. Math. Statist. Prob.*, Vol. 2. University of California Press, Berkeley, Calif., pp. 289–306.

[22] Mills, J. P. (1926). *Biometrika*, **18**, 395–400.

[23] Nagaev, S. V. (1979). *Ann. Prob.*, **7**, 745–789.

[24] Petrov, V. V. (1975). *Sums of Independent Random Variables*. Springer-Verlag, New York (esp. Chap. 8).

[25] Rubin, H. and Sethuraman, J. (1965). *Sankhyā A*, **27**, 325–346.

[26] Sanov, I. N. (1957). *Mat. Sb.*, **42**, 11–44. (English transl.: *Select. Transl. Math. Statist. Prob.*, **1**, 213–244.)

[27] Sethuraman, J. (1964). *Ann. Math. Statist.*, **35**, 1304–1316; correction, *ibid.*, **41**, 1376–1380 (1970).

[28] Shepp, L. A. (1964). *Ann. Math. Statist.*, **35**, 424–428.

[29] Sievers, G. L. (1969). *Ann. Math. Statist.*, **40**, 1908–1921.

[30] Smirnov, N. V. (1933). *Mat. Sb.*, **40**, 443–454.

[31] Statulevicius, V. A. (1966). *Zeit. Wahrscheinlichkeitsth. verwand. Geb.*, **6**, 133–144.

[32] Varadhan, S. R. S. (1966). *Commun. Pure Appl. Math.*, **19**, 261–286.

[33] Wolf, W. (1977). *Zeit. Wahrscheinlichkeitsth. verwand. Geb.*, **40**, 239–256.

[34] Woodroofe, M. (1978). *Ann. Statist.*, **6**, 72–84.

(BOREL–CANTELLI LEMMA
LIMIT THEOREMS)

STEPHEN A. BOOK

LARGE-SAMPLE THEORY

Statistical large-sample theory is concerned with the behavior of statistical procedures as the sample size increases without bound. The interest of large-sample theory to users of statistics lies in the relative clarity and simplicity of limiting behavior, and in the use of asymptotic results as approximations for moderate sample sizes. It is common practice to employ the exact distribution of a statistic in small samples but to switch to a large sample approximation as the sample size grows beyond the reach of practical exact calculations. The adequacy of this practice varies with the accuracy of the large-sample approximation, which must be investigated separately for each procedure. For example, a large body of computation and simulation* has shown that the chi-square limiting distribution is a good approximation to the exact null distribution of the Pearson chi-square* statistic even for quite small observed cell frequencies. On the other hand, some adaptive methods*, particularly those that simply insert a general density estimator into statistics whose expressions involve the unknown density function of the population, may require very large sample sizes to approach their limiting behavior.

Two current trends are encouraging a better grounded application of large-sample results in practice. First, routine availability of programmable calculators and computers is encouraging use of exact distributions for sample sizes larger than those in traditional tables, and for procedures not covered by the tables. It is therefore less necessary than in the past to press asymptotic results beyond their reasonable range of applicability. Second, researchers in large-sample theory are making rapid progress in developing asymptotic expansions and other results which provide much better approximations to moderate-sample-size behavior than do such classical methods as the central limit theorem*'s "normal approximation." The convergence of these two trends makes available quite accurate calculations of probabilities and moments when such accuracy is needed. In many cases, however, the first-order approximations given by more classical limit theory are already adequate.

Asymptotic results are often much more general than nonasymptotic results, and in that sense are more applicable. For example, Pearson's chi-square statistic for testing fit to a composite model requires estimation of the parameters of the model from the data. For proper methods of estimation, the limiting null distribution remains chi-square, while the exact distribution is not only intractible but generally varies with the true value of the unknown parameter. Even the question of choosing a "good" estimator has for non-Bayesians only asymptotic answers (such as the maximum likelihood estimator) in any generality. It therefore seems clear that the technical and often apparently unrealistic results of large-sample theory will continue to exercise a heavy influence on statistical practice.

This survey of large-sample theory will divide the field into three aspects: methods from probability theory, large-sample theory of estimators, and large-sample theory of tests. Other statistical procedures in which large-sample theory plays a signficant role include classification*, discriminant analysis*, and ranking and selection*. In addition, some results are available in the setting of general decision theory* (see LeCam [17]).

PROBABILISTIC RESULTS

Statistical large-sample theory relies directly on results from probability theory. Most of these are discussed in individual articles, but an overview here will be helpful. Consider, then, a sequence of statistics T_n with cumulative distribution functions (CDFs) F_n. For convenience, we can think of T_n as a normalized sample sum and of n as the sample size, although similar results apply in much greater generality. The first class of probabilistic results for T_n are *simple limit theorems*, including both pointwise convergence (laws of large numbers*) and central limit theorems*. The latter typically assert that as $n \to \infty$,

$$\sup_x |F_n(x) - \Phi(x)| \to 0, \qquad (1)$$

where Φ is the standard normal CDF. These theorems provide answers to the basic question "What happens in the limit?" for statistical procedures, and are very generally available, at least for independent observations. But simple limit theorems are often not sufficiently accurate for numerical use, and do not distinguish adequately among procedures that are equivalent at this level.

A second level of probabilistic results concers *rates of convergence*, such as the generalization of (1) provided by Berry–Esseen inequalities,

$$\sup_x |F_n(x) - \Phi(x)| = O(n^{-1/2}). \qquad (2)$$

Rates of convergence, although very helpful tools for theorem proving, are of little use in practice, since they provide only general bounds on $F_n - \Phi$ which are too crude for use as approximations in specific cases. Results such as (2) are of statistical interest primarily as essential steps on the way to a third level of refinement, *asymptotic expansions** of F_n such as

$$\sup_x |F_n(x) - \hat{F}_n(x)| = o(n^{-1})$$

$$\hat{F}_n(x) = \Phi(x) + n^{-1/2}A_1(x) + n^{-1}A_2(x).$$

In this typical expansion, the first term is the normal approximation, and A_1 and A_2 correct for skewness and kurtosis. Asymptotic expansions are refined enough to distinguish among procedures equivalent at the level of simple limit theorems, and for numerical work. They are currently undergoing rapid development and increasing application.

In addition to these increasingly precise estimates of $F_n(x) - \Phi(x)$, statisticians have made considerable use of a different refinement of the central limit theorem, the theory of large deviations*. This theory provides limit or order results for the relative error $\{1 - F_n(x)\}/\{1 - \Phi(x)\}$ in tail probabilities as $x_n \to \infty$ at various rates, results which are essential for large-sample comparisons of tests of hypotheses.

Some recent surveys containing references to the probabilistic literature are Bhattacharya [5] on rates of convergence and asymptotic expansions*, and Nagaev [21] on large deviations. While (1) to (3) above were expressed for the common normal case, results are becoming available for statistics, such as empirical distribution function (EDF) statistics and functions of sample covariance matrices, that have nonnormal asymptotic distributions. The annual volumes of the *Current Index to Statistics** record this developing literature under "asymptotic expansions" and "large deviations."

Since the probability literature concentrates on simple statistics such as sample sums, the essential tools of large-sample theory include methods for extending such results as (1) to (3) to more complex statistics. The elementary arithmetic of large-sample theory consists of simple limit theorems, relations among the several modes of convergence, and methods for studying functions of converging sequences of statistics. Taylor series* expansions and other simple analytic tools are applied to deal with functions of statistics. Typical results of elementary large-sample theory include the following:

1. Suppose that X_n converges in law to limiting distribution F, and that Y_n converges in probability to 0. Then $X_n + Y_n$ also converges in law to F. (This is a portion of "Slutsky's theorem*." See Serfling [26, p. 19].)

2. Suppose that real random variables X_n are such that $n^{1/2}(X_n - \mu)$ converges in

law to $N(0, \sigma^2)$, and that f is a function satisfying

$$f(x) = f(\mu) + f'(\mu)(x - \mu) + o(|x - \mu|)$$

as $x \to \mu$. Then $n^{1/2}\{f(X_n) - f(\mu)\}$ converges in law to $N(0, \sigma^2[f'(\mu)]^2)$.

A modern exposition of these elementary methods, with many statistical applications, can be found in Chap. 14 of Bishop et al. [7].

A more recent extension of simple limit theory allows random variables X_n that take values in abstract spaces rather than on the line or in Euclidean space. In particular, a stochastic process can be thought of as a random variable taking values in a function space, and convergence theorems in such settings produce a limit theory for procedures such as EDF statistics* which are functions of a process. Billingsley [6] discusses the case of X_n taking values in a separable metric space.

Just as simple limit theorems are extended and applied by auxiliary techniques, so asymptotic expansions (3) for statistics T_n employed in statistical procedures can be obtained by combining results for sample sums with analytical devices. For parametric procedures, the device has commonly been an expansion of T_n itself into terms each of which is a sample sum, followed by application of a multivariate asymptotic expansion for the joint CDF of these sums. Pfanzagl [23] provides a comprehensive summary of this line of work. Nonparametric procedures have been more often reduced to the sample sum case by conditioning rather than expanding. Albers et al. [1] is an outstanding example of this approach.

The general tools and methods discussed above have produced limit theorems for many special classes of statistics that find application in statistical practice. Such classes include linear combinations of order statistics*, U-statistics* and their generalizations, linear rank statistics (*see* CHERNOFF–SAVAGE THEOREM *and* LINEAR RANK TESTS), Hampel M-estimates*, and EDF statistics*. The limit theory of most of these classes of statistics is systematically surveyed by Serfling [26]. Asymptotic expansions and other refinements are less completely known, particularly for statistics that are not asymptotically normal.

ESTIMATION

The simple limit properties of a sequence of estimators T_n are consistency and the asymptotic distribution of T_n (suitably normalized). The sequence T_n is consistent* for estimating a quantity θ if T_n converges to the true population value of θ as the sample size n increases. In parametric problems, θ is typically a real or vector parameter, and convergence of T_n to θ in probability, in rth mean, and with probability 1 define different types of consistency. In other settings, still other modes of convergence are appropriate, e.g., the uniform convergence of the empiric CDF to the true CDF asserted by the Glivenko–Cantelli theorem*. Inconsistent estimators are unacceptable in most circumstances. Asymptotic distributions are required for error estimates and other inferences. In parametric problems, the most common result is that T_n is asymptotically normal (AN): When θ is the true parameter value, $n^{1/2}(T_n - \theta)$ converges in law to a normal distribution with mean zero. Rates of convergence other than $n^{1/2}$ and nonnormal asymptotic distributions do occur in some circumstances, and convergence in distribution for stochastic processes rather than random variables is important when T_n estimates a function (such as a CDF, a density function, or a survival curve) rather than a parameter. Derivation of simple limit properties for new estimators and under broader conditions (such as censored* or dependent data) continues to occupy much of the literature of large-sample theory.

It often happens that several distinct consistent estimators of the same quantity θ exist in a statistical problem. Comparison of such estimators has proceeded primarily by comparing the concentration of their distributions about the true θ. The limit of the ratio of sample sizes n_1/n_2 required for se-

quences T_{1n} and T_{2n} to achieve equal concentration is the asymptotic relative efficiency* (ARE) of T_{2n} to T_{1n}. For AN sequences, AREs based on several different measures of concentration reduce to comparing the variance (or generalized variance* in dimensions higher than one) of the normal asymptotic distributions. An estimator is therefore said to be *asymptotically efficient* (AE) or *best asymptotically normal* (BAN) if $n^{1/2}(T_n - \theta)$ is AN with minimum possible variance $\sigma^2(\theta)$ for all θ in the parameter space.

Fisher*, with whom many of these ideas originated, hoped to show that for parametric problems the maximum likelihood* estimator (MLE) was consistent, AN and AE. Consistency and asymptotic normality* hold for MLEs in considerable generality, although efforts continue to extend these landmark results (see LeCam [18] for literature). But it turns out that estimators with asymptotically minimum variance do not exist (*see* HODGES SUPEREFFICIENCY), so that the AE part of Fisher's program was initially frustrated. Several major streams of research followed the discovery of superefficiency, however, and now justify the conclusion that Fisher was correct in substance. The first stream seeks to isolate the phenomenon of superefficiency: It occurs only on a set of θ of measure 0 (LeCam), it cannot occur if estimators are required to converge uniformly to their limiting law (Wolfowitz), and so on. Wolfowitz [29] gives an excellent account of these direct efforts to rescue Fisher's program.

A second response to the difficulties posed by superefficiency, due primarily to Hájek and LeCam, has provided a more satisfactory general account of what constitutes asymptotic optimality. This account is in the spirit of Wald's general decision theory. A parametric family of distributions is *locally asymptotically normal* (LAN) if (roughly) the log likelihood ratio has a normal asymptotic distribution with locally constant covariance matrix and a mean that is locally linear in θ. This idea, systematically developed by Le-Cam, is now recognized as the essence of "large-sample regularity." Within the LAN setting, several striking results have been obtained for very general sequences of estimators.

1. There is a lower bound for the local asymptotic minimax* risk [12, 19], which is attained both by MLEs and by Bayes estimators with smooth prior distributions.

2. Hájek [11] showed that the limiting distribution of an arbitrary sequence of estimators is a convolution of the normal limiting law of the MLE, which depends only on the underlying distribution, with another probability law that depends on the choice of estimator.

Results 1 and 2 produce definitive statements of the asymptotic optimality of the MLE in regular cases.

Asymptotic minimax results of the Hajek–LeCam type are perhaps the most significant recent development in large sample theory, for several reasons: They assert asymptotic optimality for general risk functions and over the class of all procedures; the theory can be applied to nonregular (non-LAN) cases, where the MLE may be neither AN nor asymptotically optimal; extensions to robust estimation [3] and nonparametric estimation [20] have been made, and the method is appropriate for general statistical procedures. A systematic exposition by Ibragimov and Haśminskiĭ [16] is now available, and should increase the accessibility of these ideas and results.

In specific settings, other estimators T_n usually share the asymptotic optimality of the MLE $\hat{\theta}_n$. Such estimators are then asymptotically equivalent to the MLE in the sense that $n^{1/2}(T_n - \hat{\theta}_n)$ converges to 0 in probability. Examples include minimum chisquare* estimators in multinomial* problems and Berkson's minimum logit chisquare* estimator in logit regression models. It is therefore natural to consider differences of order n^{-1} and n^{-2} to distinguish among AE estimators. Such considerations of *second-order efficiency** go back to Fisher, and

now have several variations. Asymptotic expansions are required. The general result obtained is that the MLE is superior to its competitors, but there is disagreement over the significance both of the particular results obtained and of second-order efficiency in general. The spirited discussion following Berkson [4] presents the various positions and the literature.

Several additional developments have also contributed to understanding of the asymptotic properties of estimators, and of the large-sample optimality of the MLE. Bahadur has shown that the distance of consistent T_n from θ cannot approach 0 at a rate exceeding a certain exponential rate, which is often attained by the MLE. See Bahadur [2] and Fu [10]. It has already been pointed out that Bayes' estimators share the asymptotic optimality of the MLE. In fact, as the sample size increases, the influence of the specific prior distribution and loss function (provided that these are sufficiently smooth) disappears, and the Bayes estimator and MLE differ at most by order n^{-1} in probability (see Strasser [27]). This result provides a Bayesian justification for the use of the MLE in large samples.

These convincing large-sample endorsements of the MLE do not imply that it is "best" in finite-sample situations, and do not address the question of robustness. The latter issue is important, since the MLE is derived from a specific parametric model in each situation. Beran [3, and references] has shown that adaptive versions of the MLE combine asymptotic optimality with reasonable robustness. Practical experience with these estimators is as yet very limited.

TESTING HYPOTHESES

A sequence of tests based on statistics T_n is consistent against a fixed alternative F if the probability $\beta_n(F)$ of type II error approaches 0 as the sample size n increases and the size α_n remains constant. As in the case of estimation, only consistent procedures are usually considered for use. The simple limit theory of tests focuses on the asymptotic distribution of test statistics T_n under the null hypothesis, since such results give large-sample approximations to the critical points of T_n. A closely related study considers approximate power against alternatives close to the null hypothesis by investigating the asymptotic distribution of T_n under a sequence of alternatives F_n which approach the null hypothesis at a rate such that $\beta_n(F_n)$ has a finite nonzero limit. In parametric problems such as testing $H_0 : \theta = \theta_0$ for a real or vector parameter θ, the appropriate local alternatives are most often of the form $\theta_n = \theta_0 + n^{-1/2}\delta$ for fixed δ. These basic limit results are described in the literature for almost all common test statistics. Some comprehensive references are Hájek and Šidák [13] for rank tests and Neuhaus [22] for tests of fit. In many cases (e.g., most distribution-free tests) the large-sample approximate critical points are in regular use for moderate sample sizes. Broader study of local alternatives has resulted from LeCam's introduction of the concept of contiguity* of probability measures (see Roussas [25] for an exposition).

In parametric problems, likelihood ratio tests* (LRTs) are very commonly used. If L_n is the likelihood ratio, then in general the log-likelihood ratio statistic $-2\log L_n$ has a chi-square limiting distribution under the null hypothesis and a noncentral chi-square* limiting distribution under local alternatives. As in the case of the MLE, other tests are often asymptotically equivalent to the LRT when only simple limit properties are considered. An example is the Pearson chi-square statistic for multinomial problems, which can be expressed as the dominant term in a Taylor series expansion of the log-likelihood ratio* statistic.

Large-sample comparison of tests begins with the consideration of asymptotic relative efficiency (ARE), the limit of the ratio of sample sizes needed for two competing tests to attain the same level of performance against the same alternatives. A number of different AREs have been suggested, based on different assumptions on the behavior of the sequences of sizes α_n, probabilities of type II error β_n, and alternatives F_n. A compact summary and comparison appears in Chap. 10 of Serfling [26]. The two most commonly employed AREs are due to Pitman and Bahadur. It should be noted that each ARE for tests produces a corresponding efficiency measure for confidence intervals based on suitable assumptions about the limiting confidence level and interval length (*see* BAHADUR EFFICIENCY and PITMAN EFFICIENCY).

The Pitman ARE is based on sequences of local alternatives F_n chosen so that α_n and β_n both approach finite nonzero limits. The resulting efficiency does not depend on any particular alternative, and typically is also independent of the limits of α_n and β_n. Computation of Pitman efficiencies requires only simple limit results or such refinements as contiguity* arguments or Berry–Esseen bounds. Rothe [24] presents generalizations and literature. Pitman AREs have played a significant role in assessing the performance of distribution-free alternatives to standard normal-theory tests. For example, the Wilcoxon test for $H_0 : \theta = 0$ against alternatives $\theta > 0$ in a symmetric location parameter family $F(x - \theta)$ has Pitman ARE 0.955 against the sample mean when F is normal. For F Laplace*, the ARE is $\frac{3}{2}$, and Hodges and Lehmann showed that the ARE is at least 0.864 for all continuous F.

Most other measures of ARE for tests, including those of Chernoff, Hodges and Lehmann, and Bahadur, consider performance against fixed alternatives rather than sequences of local alternatives. These AREs require probabilistic results on large deviations for their computation. Bahadur ARE fixes β and the alternative, and studies the rate of convergence to 0 of the size α_n. Its popularity rests in part on an alternative interpretation in terms of the behavior of the actually attained significance level of the test as a random variable. Bahadur [2] gives an extensive exposition. Bahadur ARE, and other AREs for fixed alternatives, distinguish between many test procedures that are equivalent against local alternatives. A striking example is the demonstration by Hoeffding [15] that for several classes of multinomial problems LRTs have strictly better Bahadur performance than the corresponding Pearson chi-square tests. (This is a restatement in current terminology of Hoeffding's results.)

Since the LRT provides most powerful tests of simple hypotheses versus simple alternatives by the Neyman–Pearson lemma*, it is not surprising that LRTs have extensive large-sample optimality properties. Wald [28] already established several such local properties, and Bahadur has shown in some generality that LRTs are asymptotically optimal by his nonlocal criterion. Brown [8] proved in considerable generality that appropriate tests of likelihood ratio type (actually LRTs of possibly larger hypotheses) are at least as good as any given sequence of tests by several nonlocal criteria, including Bahadur's. Thus the large-sample optimality of LRTs for parametric problems is well established.

A finer measure of the relative performance of tests under local alternatives is provided by deficiency, introduced by Hodges and Lehmann [14]. Deficiency is roughly the limit of the difference of sample sizes required to reach the same power against the same alternatives. Tests having the same Pitman ARE (the limit of the *ratio* of sample sizes is 1) can often be distinguished by deficiency. Examples appear in Albers et al. [1]. Evaluation of deficiency requires the use of asymptotic expansions for the powers of the tests.

OTHER TOPICS AND FUTURE PROSPECTS

The emphasis in this survey, as in the statistical literature, has been placed on the study of large-sample properties of statistical procedures that were originally proposed on other grounds. Even likelihood procedures have their origin in the plausibility of maximizing the likelihood function, although their objective justification is almost exclusively asymptotic. There are, however, several classes of statistical procedures whose genesis is specifically in large-sample theory, in the sense that these procedures rely on some convergence property for their conceptual justification. Most adaptive procedures* are of this type, as are empirical Bayes* methods and stochastic approximation* schemes. So too is Wald's method of constructing tests from asymptotically normal estimators, introduced by Wald [28] for MLEs and since widely generalized and much employed. These and other procedures are part of the contribution of large sample theory to statistics.

Much future research in large-sample theory will continue to explore limiting behavior of new or newly popular statistical procedures. Refined use of tools such as asymptotic expansions, combined with exact computation and simulation, will continue to clarify the applicability of large sample approximations in practice. Wider understanding and use of the Hájek–LeCam asymptotic minimax approach offers exciting prospects of a more widely accepted general theory of asymptotic optimality in statistics. It also appears that the search for asymptotic minimax procedures will contribute new statistical methods that are known to be asymptotically optimal and hence are strong candidates for use in practice. Research can be expected to focus on areas where statistics as a whole is evolving rapidly. Such areas at present include robust procedures and computer-based procedures for analysis of large data sets. The generality and clarity afforded by large-sample analysis will continue to illuminate the properties of statistical methods and to influence, but not control, the direction of statistical practice.

References

[1] Albers, W., Bickel, P. J., and Van Zwet, W. R. (1976). *Ann. Statist.*, **4**, 108–156.

[2] Bahadur, R. R. (1971). *Some Limit Theorems in Statistics*. SIAM, Philadelphia. (Exposition and references for Bahadur's influential work on asymptotic comparison of both tests and estimators.)

[3] Beran, R. (1981). *Zeit. Wahrscheinlichkeitsth. verwand. Geb.*, **55**, 91–108. (A fine example of the power of asymptotic minimax ideas.)

[4] Berkson, J. (1980). *Ann. Statist.*, **8**, 457–487. (An attack on uncritical use of the MLE, with comments by leading large-sample theorists.)

[5] Bhattacharya, R. N. (1977). *Ann. Prob.*, **5**, 1–27.

[6] Billingsley, P. (1968). *Convergence of Probability Measures*. Wiley, New York. (Advanced but clear exposition of convergence for processes, stressing EDF and partial sum processes.)

[7] Bishop, Y. M. M., Fienberg, S. E., and Holland, P. (1975). *Discrete Multivariate Analysis*. MIT Press, Cambridge, Mass. (Chapter 14, "Asymptotic Methods," is a 73-page minicourse, independent of the rest of the book except for stress on categorical data in the choice of material.)

[8] Brown, L. D. (1971). *Ann. Math. Statist.*, **42**, 1206–1240.

[9] Chernoff, H. (1956). *Ann. Math. Statist.*, **27**, 1–22. (Old but still useful survey.)

[10] Fu, J. C. (1975). *Ann. Statist.*, **3**, 234–240.

[11] Hájek, J. (1970). *Zeit. Wahrscheinlichkeitsth. verwand. Geb.*, **14**, 323–330.

[12] Hájek, J. (1972). *Proc. 6th Berkeley Symp. Math. Statist. Prob.*, Vol. 1. University of California Press, Berkeley, Calif., pp. 175–194.

[13] Hájek, J. and Šidák, Z. (1967). *Theory of Rank Tests*. Academic Press, New York.

[14] Hodges, J. L., Jr. and Lehmann, E. L. (1970). *Ann. Math. Statist.*, **41**, 783–801.

[15] Hoeffding, W. (1965). *Ann. Math. Statist.*, **36**, 369–408.

[16] Ibragimov, I. A. and Hasminskiĭ, R. Z. (1981). *Statistical Estimation: Asymptotic Theory*, S. Kotz, trans. Springer-Verlag, New York. (Detailed but nonintuitive working out of the ideas of Hájek and LeCam.)

[17] LeCam, L. (1969). *Théorie Asymptotique de la Décision Statistique*. University of Montreal Press, Montreal.

[18] LeCam, L. (1970). *Ann. Math. Statist.*, **41**, 802–828.

[19] LeCam, L. (1972). *Proc. 6th Berkeley Symp. Math. Statist. Prob.*, Vol. 1. University of California Press, Berkeley, Calif., pp. 245–261.

[20] Millar, P. W. (1979). *Zeit. Wahrscheinlichkeitsth. verwand. Geb.*, **48**, 233–252. (Contains the clearest statement and proof of the Hajek–LeCam asymptotic minimax theorem.)

[21] Nagaev, S. V. (1979). *Ann. Prob.*, **7**, 745–789.

[22] Neuhaus, G. (1979). *Math. Operat. Statist.*, **10**, 479–494.

[23] Pfanzagl, J. (1980). In *Developments in Statistics*, Vol. 3, P. R. Krishnaiah, ed. Academic Press, New York, pp. 1–97. (The state of the art in asymptotic expansions and their applications in parametric problems.)

[24] Rothe, G. (1981). *Ann. Statist.*, **9**, 663–669.

[25] Roussas, G. G. (1972). *Contiguity of Probability Measures: Some Applications in Statistics*. Cambridge University Press, Cambridge. (Technical and thorough, but gives no conceptual insight.)

[26] Serfling, R. J. (1980). *Approximation Theorems of Mathematical Statistics*. Wiley, New York. (The best single source for basic results and references. Designed as a graduate textbook, it concentrates on convergence results and omits asymptotic expansions and second-order properties. Well written and comprehensive, more so than the skimpy index suggests.)

[27] Strasser, H. (1977). *Theory Prob. Appl.*, **22**, 349–361.

[28] Wald, A. (1943). *Trans. Amer. Math. Soc.*, **54**, 426–482.

[29] Wolfowitz, J. (1965). *Theory Prob. Appl.*, **10**, 247–260.

(ADAPTIVE METHODS
ASYMPTOTIC EXPANSIONS
ASYMPTOTIC NORMALITY
BAHADUR EFFICIENCY
CHERNOFF–SAVAGE THEOREM
CONSISTENT
CONTIGUITY
CONVERGENCE OF SEQUENCES OF
    RANDOM VARIABLES
CORNISH–FISHER–EDGEWORTH
    EXPANSION
EFFICIENCY, ASYMPTOTIC RELATIVE OF
    ESTIMATORS
EFFICIENCY, SECOND ORDER
GLIVENKO–CANTELLI THEOREM
GRAN–CHARLIER SERIES
HODGES SUPEREFFICIENCY
LARGE DEVIATIONS AND APPLICATIONS
LAWS OF LARGE NUMBERS
LIMIT THEOREM, CENTRAL
LIMIT THEOREMS
OCCUPANCY DISTRIBUTIONS
PITMAN EFFICIENCY
STOCHASTIC APPROXIMATION)

DAVID S. MOORE

LATENT ROOT DISTRIBUTIONS

INTRODUCTION AND HISTORICAL NOTES

If A is a square matrix, the latent roots of A are the roots of the characteristic equation

$$\det(A - \lambda I) = 0.$$

If A is a real symmetric matrix its latent roots are all real; if, in addition, A is positive definite, then its latent roots are all positive. If A is a random matrix (i.e., a matrix whose elements are random variables having a joint distribution), then its latent roots are random variables. Often the requirement that a testing problem in multivariate analysis* be invariant under a natural group of transformations leads to the consideration of functions of these latent roots as test statistics. Consequently, the distributions of the latent roots are an important tool in the derivation of distributions of test statistics.

The statistical theory of latent root distributions based on normal sampling became firmly established in 1939 when Fisher [9], Girshick [11], Hsu [14], Mood [25], and Roy [30], all independently, derived the joint density function of the latent roots of a multivariate beta matrix. (Mood's paper [25] was not published until 1951.) This gave the null distributions of latent roots occurring in multivariate analysis of variance* (MANOVA), canonical* correlation analysis, and in the problem of testing the equality of covariance matrices of two multivariate normal populations. The joint density function of the latent roots of a matrix having the $W_m(n, \lambda I_m)$ distribution (i.e., Wishart* with n degrees of freedom and covariance matrix λI_m) was also obtained by Fisher, Hsu, and Roy. It can be derived by a limiting process from the distribution of the roots of a multivariate beta matrix, as noted by Nanda [28].

Latent root distributions in nonnull situations are considerably more complicated and in general can be expressed in terms of hypergeometric functions with two matrices as arguments; these can be written as infinite series of zonal polynomials* developed by James in a series of papers [17–21] and by Constantine in 1963 [6]. Most of the standard nonnull joint distributions were derived by James [17–20] and Constantine [6] and generalize earlier special cases due to Roy [31], Anderson [1], Bartlett [5], and others.

NULL DISTRIBUTIONS

The latent root distributions of primary importance in multivariate analysis based on normal sampling arise in the following problems:

1. Testing equality of covariance matrices of two m-variate normal distributions. Invariant test statistics here are functions of the latent roots of $S_1(S_1 + S_2)^{-1}$, where S_i has the $W_m(n_i, \Sigma_i)$ distribution ($i = 1, 2$), with S_1 and S_2 being independent. The null hypothesis is $H: \Sigma_1 = \Sigma_2$.

2. Testing the general linear hypothesis in MANOVA. Invariant test statistics are functions of the latent roots of $S_1(S_1 + S_2)^{-1}$, where S_1 is $W_m(n_1, \Sigma, \Omega)$ (i.e., noncentral Wishart with noncentrality matrix Ω), S_2 is $W_m(n_2, \Sigma)$, and S_1 and S_2 are independent. The null hypothesis is $H: \Omega = 0$.

3. Testing independence between two sets of variables. Partitioning the covariance matrix as

$$\Sigma = \begin{bmatrix} \Sigma_{11} & \Sigma_{12} \\ \Sigma_{21} & \Sigma_{22} \end{bmatrix}$$

where Σ_{11} is $p \times p$ and Σ_{22} is $q \times q$ ($p \leqslant q$), the null hypothesis is $H: \Sigma_{12} = 0$. Invariant test statistics based on a sample of size $n + 1$ ($> p + q$) are functions of the sample canonical correlation coefficients whose squares are the latent roots of $S_{11}^{-1}S_{12}S_{22}^{-1}S_{21}$, where S is $W_m(n, \Sigma)$ and is partitioned similarly to Σ.

When the null hypotheses in problems **1** to **3** are true, the joint density functions of the latent roots $l_1, \ldots, l_r$ all have the same form, the result noted by Fisher, Girshick, Hsu, Mood, and Roy in 1939. This can be written as

$$f(l_1, \ldots, l_r; r, \gamma, \delta)$$

$$= K(r, \gamma, \delta) \prod_{i=1}^{r} l_i^{\gamma}(1 - l_i)^{\delta} \prod_{i<j}^{r} (l_i - l_j)$$

$$(1 > l_1 \geqslant \cdots \geqslant l_r > 0), \quad (1)$$

where

$$K(r, \gamma, \delta)$$

$$= \frac{\pi^{(1/2)r^2}\Gamma_r(\gamma + \delta + r + 1)}{\Gamma_r(\frac{1}{2}r)\Gamma_r[\gamma + \frac{1}{2}(r+1)]\Gamma_r[\delta + \frac{1}{2}(r+1)]}$$

with

$$\Gamma_r(a) = \pi^{(1/4)r(r-1)} \prod_{i=1}^{r} \Gamma[a - \frac{1}{2}(i-1)].$$

The parameters r, δ, and γ are different in situations **1** to **3**. In problems **1** and **2** we assume that $n_2 \geqslant m$. If $n_1 \geqslant m$, there are m nonzero roots (with probability 1) and their joint density function is given by (1) with $r = m$, $\gamma = \frac{1}{2}(n_1 - m - 1)$, and $\delta = \frac{1}{2}(n_2 - m - 1)$. If $n_1 < m$, there are n_1 nonzero roots $l_1, \ldots, l_{n_1}$ and their distribution is obtained from the distribution for $n_1 \geqslant m$ via the transformation

$$m \to n_1, \quad n_1 \to m, \quad n_2 \to n_1 + n_2 - m. \quad (2)$$

In **3** the roots are usually denoted by $r_1^2, \ldots, r_p^2$ ($r_1, \ldots, r_p$ are the sample canonical correlation coefficients) and their joint density function is given by (1) with $r = p$, $\gamma = \frac{1}{2}(q - p - 1)$, and $\delta = \frac{1}{2}(n - q - p - 1)$. In addition to **1** to **3**, the following is also of interest:

4. Testing hypotheses about a covariance matrix. Consider testing either of the null hypotheses $H : \Sigma = I_m$ or $H : \Sigma = \lambda I_m$, where λ is unspecified, given a sample covariance matrix S, where nS is $W_m(n, \Sigma)$, with $n \geqslant m$. Invariant test statistics are functions of the latent roots $l_1, \ldots, l_m$ of S. When $\Sigma = \lambda I_m$ their

joint density function is

$$\left(\frac{n}{2\lambda}\right)^{(1/2)mn} \frac{\pi^{(1/2)m^2}}{\Gamma_m(\frac{1}{2}m)\Gamma_m(\frac{1}{2}n)} \exp\left(-\frac{n}{2\lambda}\sum_{i=1}^{m} l_i\right)$$

$$\times \prod_{i=1}^{m} l_i^{(1/2)(n-m-1)} \prod_{i<j}^{m} (l_i - l_j)$$

$$(l_1 \geqslant \cdots \geqslant l_m > 0). \quad (3)$$

A discussion of the various problems described in this section may be found in Anderson [3].

NONNULL DISTRIBUTIONS

When the null hypotheses in problems **1** to **4** are not true, the latent root distributions can be expressed in terms of hypergeometric functions $_pF_q^{(m)}$, with two $m \times m$ matrices R and S as arguments. These are defined as (see James [20])

$$_pF_q^{(m)}(a_1, \ldots, a_p; b_1, \ldots, b_q; R, S)$$

$$= \sum_{k=0}^{\infty} \sum_{\kappa} \frac{(a_1)_\kappa \cdots (a_p)_\kappa}{(b_1)_\kappa \cdots (b_q)_\kappa} \frac{C_\kappa(R)C_\kappa(S)}{k! \, C_\kappa(I_m)}, \quad (4)$$

where $C_\kappa(R)$ is the zonal polynomial* of R corresponding to the partition $\kappa = (k_1, \ldots, k_m)$, $k_1 \geqslant \cdots \geqslant k_m \geqslant 0$, of k, and

$$(a)_\kappa = \prod_{i=1}^{m} \left(a - \frac{1}{2}(i-1)\right)_{k_i},$$

$$(x)_k = x(x+1) \cdots (x + k - 1). \quad (5)$$

For problems **1** to **3** each latent root density function will be written as the null density function multiplied by a "likelihood factor" which depends on population roots and which has the value 1 when the null hypothesis is true.

In problem **1** the joint density function of the latent roots of $S_1(S_1 + S_2)^{-1}$ is, for $n_1 \geqslant m$, $n_2 \geqslant m$,

$$\prod_{i=1}^{m} \alpha_i^{(1/2)n_1}(1 - l_i)^{-(1/2)n}$$

$$\times {}_1F_0^{(m)}\left(\frac{1}{2}n; L(I - L)^{-1}, -A\right)$$

$$\times f(l_1, \ldots, l_m; m, \frac{1}{2}(n_1 - m - 1),$$

$$\frac{1}{2}(n_2 - m - 1))$$

$$(1 > l_1 \geqslant \cdots \geqslant l_m > 0), \quad (6)$$

where $n = n_1 + n_2$, $0 < \alpha_1 \leqslant \alpha_1 \leqslant \cdots \leqslant \alpha_m$ are the latent roots of $(\Sigma_1 \Sigma_2^{-1})^{-1}$, $L = \text{diag}(l_1, \ldots, l_m)$, $A = \text{diag}(\alpha_1, \ldots, \alpha_m)$ and $f(\cdot)$ is the null density function given by (1). The zonal polynomial series for the $_1F_0$ function may not converge for all L; for convergent expressions, see Khatri [23].

In problem **2** the joint density function is, for $n_1 \geqslant m$, $n_2 \geqslant m$,

$$\exp\left(-\tfrac{1}{2}\sum_{i=1}^{m}\omega_i\right) {}_1F_1^{(m)}(\tfrac{1}{2}(n_1+n_2);\tfrac{1}{2}n_1;\tfrac{1}{2}\Omega, L)$$

$$\times f(l_1, \ldots, l_m; m, \tfrac{1}{2}(n_1 - m - 1),$$

$$\tfrac{1}{2}(n_2 - m - 1))$$

$$(1 > l_1 \geqslant \cdots \geqslant l_m > 0), \quad (7)$$

where $\omega_1 \geqslant \cdots \geqslant \omega_m \geqslant 0$ are the latent roots of the noncentrality matrix Ω, $L = \text{diag}(l_1, \ldots, l_m)$ and, without loss of generality, $\Omega = \text{diag}(\omega_1, \ldots, \omega_m)$. For $n_1 < m$ the distribution of the nonzero roots $l_1, \ldots, l_{n_1}$ is obtained from (7) via the transformation (2).

In problem **3** the joint density function of the squares $r_1^2, \ldots, r_p^2$ of the sample canonical correlation coefficients is

$$\prod_{i=1}^{p}(1 - \rho_i^2)^{(1/2)n} {}_2F_1^{(p)}(\tfrac{1}{2}n, \tfrac{1}{2}n; \tfrac{1}{2}q; P^2, R^2)$$

$$\times f(r_1^2, \ldots, r_p^2; p, \tfrac{1}{2}(q - p - 1),$$

$$\tfrac{1}{2}(n - q - p - 1))$$

$$(1 > r_1^2 \geqslant \cdots \geqslant r_p^2 > 0), \quad (8)$$

where $1 > \rho_1 \geqslant \cdots \geqslant \rho_p \geqslant 0$ are the population canonical correlation coefficients (whose squares are the latent roots of $\Sigma_{11}^{-1}\Sigma_{12}\Sigma_{22}^{-1}\Sigma_{21}$), $P = \text{diag}(\rho_1, \ldots, \rho_p)$, $R = \text{diag}(r_1, \ldots, r_p)$.

In problem **4** the joint density function of the latent roots of the sample covariance matrix S is

$$\frac{\pi^{(1/2)m^2}(\tfrac{1}{2}n)^{(1/2)mn}}{\Gamma_m(\tfrac{1}{2}n)\Gamma_m(\tfrac{1}{2}m)} \prod_{i=1}^{m} l_i^{(1/2)(n-m-1)}$$

$$\times \prod_{i<j}^{m}(l_i - l_j)\prod_{i=1}^{m}\alpha_i^{(1/2)n} {}_0F_0^{(m)}(-\tfrac{1}{2}nL, A)$$

$$(l_1 \geqslant \cdots \geqslant l_m > 0), \quad (9)$$

where $0 < \alpha_1 \leqslant \cdots \leqslant \alpha_m$ are the latent roots of Σ^{-1}, $L = \text{diag}(l_1, \ldots, l_m)$, $A = \text{diag}(\alpha_1, \ldots, \alpha_m)$. When $\Sigma = \lambda I_m$ this reduces to (3).

In the nonnull context one further distribution is of interest, arising in MANOVA when the covariance matrix Σ is known. If S has the noncentral $W_m(n, \Sigma, \Omega)$ distribution $(n \geqslant m)$ the joint density function of the latent roots $l_1, \ldots, l_m$ of $\Sigma^{-1}S$ is

$$\frac{\pi^{(1/2)m^2}}{2^{mn/2}\Gamma_m(\tfrac{1}{2}n)\Gamma_m(\tfrac{1}{2}m)}$$

$$\times \exp\left(-\tfrac{1}{2}\sum_{i=1}^{m}l_i\right)\prod_{i=1}^{m}l_i^{(1/2)(n-m-1)}\prod_{i<j}^{m}(l_i - l_j)$$

$$\times \exp\left(-\tfrac{1}{2}\sum_{i=1}^{m}\omega_i\right){}_0F_1^{(m)}(\tfrac{1}{2}n; \tfrac{1}{4}\Omega, L)$$

$$(l_1 \geqslant \cdots \geqslant l_m > 0), \quad (10)$$

where $\omega_1 \geqslant \cdots \geqslant \omega_m \geqslant 0$ are the latent roots of Ω, $L = \text{diag}(l_1 \cdots l_m)$ and, without loss of generality, $\Omega = \text{diag}(\omega_1, \ldots, \omega_m)$.

ASYMPTOTIC RESULTS

In view of the complexity of the nonnull distributions and the computational difficulties involved in evaluating them numerically and hence in studying the effects of population parameters a great deal of emphasis has been placed on the problems involved in deriving asymptotic results for large sample sizes, large population roots, etc. A review of results in this area and an extensive bibliography has been given by Muirhead [26].

Here we give the asymptotic behaviors of the hypergeometric functions* in the nonnull distributions (6) to (10), under certain assumptions concerning the population parameters. These assumptions specify equality of the smallest population roots and as such incorporate null hypotheses of interest in standard multivariate procedures such as discriminant analysis*, canonical correlations* and principal components*. In what follows the notation "$a \sim b$ as $n \to \infty$" means that $a/b \to 1$ as $n \to \infty$.

In (6) we assume that $\alpha_1 < \cdots < \alpha_k$

$< \alpha_{k+1} = \cdots = \alpha_m \, (= \alpha)$; then as $n \to \infty$,

$$_1F_0^{(m)}\left(\tfrac{1}{2}n; L(I - L)^{-1}, -A\right)$$

$$\sim \frac{\Gamma_k(\tfrac{1}{2}m)}{\pi^{(1/2)mk}} \prod_{i=1}^{k} (1 + \alpha_i f_i)^{-(1/2)n}$$

$$\times \prod_{i=k+1}^{m} (1 + \alpha f_i)^{-(1/2)n} \prod_{i=1}^{k} \prod_{\substack{j=1 \\ i<j}}^{m} \left(\frac{2\pi}{nc_{ij}}\right)^{1/2},$$

$$(11)$$

where

$$f_i = l_i/(1 - l_i) \quad \text{and} \quad c_{ij} = \frac{(f_i - f_j)(\alpha_j - \alpha_i)}{(1 + \alpha_i f_i)(1 + \alpha_j f_j)}.$$

In (7), under the assumption that $\Omega = n_2\Delta$, where $\Delta = \text{diag}(\delta_1, \ldots, \delta_m)$ with $\delta_1 > \cdots > \delta_k > \delta_{k+1} = \cdots = \delta_m = 0$, we have as $n_2 \to \infty$ (Glynn [12]),

$$_1F_1^{(m)}\left(\tfrac{1}{2}(n_1 + n_2); \tfrac{1}{2}n_1; \tfrac{1}{2}n_2\Delta, L\right)$$

$$\sim K_n \exp\left\{\tfrac{1}{4}n_2 \sum_{i=1}^{k} (l_i\delta_i)^{1/2}\right.$$

$$\left. \times \left[(l_i\delta_i)^{1/2} + (l_i\delta_i + 4)^{1/2}\right]\right\}$$

$$\times \prod_{i=1}^{k} \left[(l_i\delta_i)^{1/2} + (l_i\delta_i + 4)^{1/2}\right]^{n_2 + (1/2)(n_1 - m + 1)}$$

$$\times (l_i\delta_i + 4)^{-1/4}(l_i\delta_i)^{(1/4)(m - n_1)} \prod_{i=1}^{k} \prod_{j=1}^{m} c_{ij}^{-1/2},$$

$$(12)$$

where

$$K_n = \frac{\Gamma_k(\tfrac{1}{2}n_1)\Gamma_k(\tfrac{1}{2}m)n_2^{(1/2)k[n_2 - m + (1/2) + (1/2)k]}}{\Gamma_k\left[\tfrac{1}{2}(n_1 + n_2)\right]\pi^{(1/4)k(k+1)}}$$

$$\times 2^{(1/2)k[2m - (1/4)k - 3n_2 - n_1 - (3/2)]}e^{-(1/2)n_2k}$$

$$c_{ij} = (\delta_i - \delta_j)(l_i - l_j).$$

In (8), assuming that $1 > \rho_1 > \cdots > \rho_k > \rho_{k+1} = \cdots = \rho_p = 0$, we have, as $n \to \infty$,

$$_2F_1^{(p)}\left(\tfrac{1}{2}n, \tfrac{1}{2}n; \tfrac{1}{2}q; P^2, R^2\right)$$

$$\sim \frac{\Gamma_k(\tfrac{1}{2}q)\Gamma_k(\tfrac{1}{2}p)}{(\tfrac{1}{2}n)^{(1/2)k(p+q-k-1)}\pi^{(1/2)k(k+1)}2^k}$$

$$\times \prod_{i=1}^{k} (1 - r_i\rho_i)^{-n+(1/2)(p+q-1)}(r_i\rho_i)^{(1/4)(p-q)}$$

$$\times \prod_{i=1}^{k} \prod_{j=k+1}^{p} c_{ij}^{-1/2},$$

$$(13)$$

where $c_{ij} = (r_i^2 - r_j^2)(\rho_i^2 - \rho_j^2)$.

In (9), under the assumption that $\alpha_1 < \cdots < \alpha_k < \alpha_{k+1} = \cdots = \alpha_m \, (= \alpha)$, we have as $n \to \infty$,

$$_0F_0^{(m)}\left(-\tfrac{1}{2}nL, A\right)$$

$$\sim \frac{\Gamma_k(\tfrac{1}{2}m)}{\pi^{(1/2)km}} \exp\left[-\tfrac{1}{2}n\left(\sum_{k=1}^{k} \alpha_i l_i + \alpha \sum_{i=k+1}^{m} l_i\right)\right]$$

$$\times \prod_{i=1}^{k} \prod_{j=k+1}^{m} \left(\frac{2\pi}{nc_{ij}}\right)^{1/2},$$

$$(14)$$

where $c_{ij} = (l_i - l_j)(\alpha_j - \alpha_i)$.

Finally, in (10) we assume that $\omega_1 > \cdots > \omega_k > \omega_{k+1} = \cdots = \omega_m = 0$ with $\omega_i \to \infty$ $(i = 1, \ldots, k)$. Then

$$_0F_1^{(m)}\left(\tfrac{1}{2}n; \tfrac{1}{4}\Omega, L\right)$$

$$\sim \frac{2^{(1/2)k(m+n-k-3)}\Gamma_k(\tfrac{1}{2}n)\Gamma_k(\tfrac{1}{2}m)}{\pi^{(1/2)k(k+1)}}$$

$$\times \exp\left[\sum_{i=1}^{k} (\omega_i l_i)^{1/2}\right] \prod_{i=1}^{k} (\omega_i l_i)^{(1/4)(m - n)}$$

$$\times \prod_{i=1}^{k} \prod_{\substack{j=1 \\ i<j}}^{m} c_{ij}^{-1/2},$$

$$(15)$$

where $c_{ij} = (\omega_i - \omega_j)(l_i - l_j)$.

Substitution of the asymptotic behaviors above for the hypergeometric functions in their corresponding distributions yield *asymptotic representations* for the joint density functions of the latent roots. These have been used to study problems of testing equality of subsets of population latent roots; see, for example, James [22], Glynn and Muirhead [13], and Muirhead [26]. These asymptotic representations are, of course, different from asymptotic distributions of suitably standardized roots. For example, consider the latent roots $l_1, \ldots, l_m$ of a sample covariance matrix with joint density function (9) and let $\lambda_1 \geqslant \cdots \geqslant \lambda_m > 0$ be the latent roots of Σ $(\lambda_i = \alpha_i^{-1})$. Putting

$$x_i = \left(\frac{n}{2}\right)^{1/2}\left(\frac{l_i - \lambda_i}{\lambda_i}\right)$$

Girshick [11] observed that if λ_i is a simple root, then x_i is asymptotically independent of x_j for $j \neq i$ and the asymptotic distribution of x_i, as $n \rightarrow \infty$, is $N(0, 1)$. When Σ has multiple latent roots the theory is rather more complicated and was developed by Anderson [4]. Asymptotic distributions of latent roots in the areas of MANOVA and canonical correlations were first derived by Hsu [15, 16] and Anderson [2], with more recent work by Fujikoshi [10]; see Muirhead [26] for extensions and further references.

MARGINAL DISTRIBUTIONS

There are a number of techniques available for studying the distributions of individual latent roots. These procedures are outlined in a review paper by Crowther and Young [8]; see also Pillai [29] for a later review along the same lines. Of all the individual roots the extreme roots appear to be of most importance, particularly in testing problems, since these, or functions of them, arise as test statistics in problems **1** to **4** using the union–intersection principle* for constructing test statistics. For a discussion of this principle in multivariate problems a useful reference is Srivastava and Khatri [32]. Here we give expressions for the marginal distributions of some extreme roots in terms of hypergeometric functions and other series of zonal polynomials. It should be pointed out that these expressions are probably the nicest ones known from an analytical, not computational, point of view.

Consider first the largest and smallest roots l_1 and l_m of a sample covariance matrix. The distribution function of l_1 can be expressed as [6]

$$\Pr[l_1 \leq x]$$

$$= \frac{\Gamma_m\left[\frac{1}{2}(m+1)\right]}{\Gamma_m\left[\frac{1}{2}(n+m+1)\right]} \det\left(\frac{1}{2}nx\Sigma^{-1}\right)^{(1/2)n}$$

$$\times {}_1F_1\left(\frac{1}{2}n; \frac{1}{2}(n+m+1); -\frac{1}{2}nx\Sigma^{-1}\right),$$

$$(16)$$

while if $r = \frac{1}{2}(n - m - 1)$ is a positive inte-

ger,

$$\Pr[l_m > x] = \exp\left(-\frac{1}{2}nx\operatorname{tr}\Sigma^{-1}\right)$$

$$\times \sum_{k=0}^{mr} \sum_{\kappa}{}^* \frac{C_\kappa\left(\frac{1}{2}nx\Sigma^{-1}\right)}{k!},$$

$$(17)$$

where $\sum_\kappa^*$, here and later, denotes summation over those partitions $\kappa = (k_1, \ldots, k_m)$ of k with $k_1 \leq r$. This result is due to Khatri [24].

We next consider the distribution of the largest root l_1 in the MANOVA context (problem **2**), where the joint distribution of all the roots is given by (7). Khatri [24] has shown that when $r = \frac{1}{2}(n_2 - m - 1)$ is an integer,

$$\Pr[l_1 \leq x] = x^{(1/2)mn_1}\exp\left[-\frac{1}{2}(1-x)\operatorname{tr}\Omega\right]$$

$$\times \sum_{k=0}^{mr} \sum_{\kappa}{}^* L_\kappa^\gamma\left(-\frac{1}{2}x\frac{\Omega}{k!}\right)(1-x)^k,$$

$$(18)$$

where $\gamma = \frac{1}{2}(n_1 - m - 1)$ and $L_\kappa^\gamma(R)$ denotes the generalized Laguerre polynomial of the $m \times m$ matrix R (see Constantine [7]).

Finally, we give an expression for the distribution function of the square r_1^2 of the largest sample canonical correlation coefficient where the joint distribution of all coefficients is given by (8). If $r = \frac{1}{2}(n - p - q - 1)$ is an integer, then

$$\Pr[r_1^2 \leq x] = x^{(1/2)pq} \prod_{i=1}^{p} \left(\frac{1 - \rho_i^2}{1 - x\rho_i^2}\right)^{(1/2)n}$$

$$\times \sum_{k=0}^{mr} \sum_{\kappa}{}^* \frac{(1-x)^k}{k!}\left(\frac{1}{2}q\right)_\kappa C_\kappa(I_p)$$

$$\times \sum_{s=0}^{k} \sum_{\sigma} \binom{\kappa}{\sigma} \frac{\left(\frac{1}{2}n\right)_\sigma}{\left(\frac{1}{2}q\right)_\sigma} \frac{C_\sigma(B)}{C_\sigma(I_p)},$$

$$(19)$$

where

$$B = \operatorname{diag}\left[\frac{x\rho_1^2}{1 - x\rho_1^2}, \ldots, \frac{x\rho_p^2}{1 - x\rho_p^2}\right],$$

and $\binom{\kappa}{\sigma}$ is the generalized binomial coefficient defined by

$$\frac{C_\kappa(I + Y)}{C_\kappa(I)} = \sum_{s=0}^{k} \sum_{\sigma} \binom{\kappa}{\sigma} \frac{C_\sigma(Y)}{C_\sigma(I)}. \quad (20)$$

For further expressions for the distributions of extreme latent roots the interested reader is referred to the references in Crowther and Young [8] and Pillai [29].

FURTHER PROBLEMS

Much of the recent work over the last decade has concentrated on the derivation of asymptotic expansions for the joint and marginal distributions of latent roots under various assumptions about the multiplicities of population latent roots. In some instances these have been an important tool used in studying tests of equality of subsets of population latent roots. Such tests are routinely carried out in principle components, multiple discriminant analysis, and canonical correlation analysis. In addition, from the asymptotic distributions it is easy to see how the sample and population roots interact with one another and where the regions of appreciable likelihood are. This is not at all apparent from the exact distributions.

Most of the interesting problems concerning the asymptotic distributions of latent roots have now been studied. Many important problems remain, however, and require thorough investigation. Some of these are now discussed.

1. Much remains to be done in the area of exact calculation of latent root density functions and likelihood functions derived from them. As we have seen in section "Nonnull Distributions," these involve two-matrix hypergeometric functions defined as zonal polynomial series. The numerical evaluation of these series poses severe problems; what is needed are efficient recursive algorithms for zonal polynomials, methods for obtaining more rapidly convergent series, and bounds for remainder terms when a se-

ries is terminated. Very little has been done in this area and without results here there is no way to judge the accuracy of the many asymptotic approximations that have been proposed.

2. The usual approach taken when estimating population latent roots is first to estimate the corresponding matrix and then to find its latent roots. When the matrix is estimated by maximum likelihood the roots, if distinct, are maximum likelihood estimates. It is known that these estimates can be improved on in such aspects as bias reduction and variance stabilization (see, e.g., Muirhead [26]). Decision-theoretic approaches to the estimation of latent roots have not yet been studied. It is not known whether the usual estimates, or simple modifications to them, are admissible, minimax, or Bayes, or whether they can be substantially improved on.

3. The null distributions of statistics used for testing equality of subsets of population latent roots have distributions which depend on nuisance parameters* (unlike the distributions of statistics used for testing equality of all latent roots). No properties of these statistics, beyond their asymptotic distributions, have been studied. It is not even known, for example, whether the tests based on these statistics are unbiased. Here it would be necessary to find workable expressions for the marginal joint distributions of the roots that appear in the test statistics and to study properties of these distributions.

4. Some recent work has concentrated on what happens in principal components and canonical correlations when the distribution being sampled is nonnormal; see Waternaux [33] and Muirhead and Waternaux [27]. The asymptotic distributions of latent roots in these situations and statistics formed from them, turn out to be quite sensitive to the assumption of multivariate normality. What is

needed here is a robust approach to the estimation of latent roots.

References

[1] Anderson, T. W. (1946). *Ann. Math. Statist.*, **17**, 409–431.

[2] Anderson, T. W. (1951). *Proc. 2nd Berkeley Symp. Math. Statist. Prob.* University of California Press, Berkeley, Calif., pp. 103–130.

[3] Anderson, T. W. (1958). *An Introduction to Multivariate Statistical Analysis.* Wiley, New York.

[4] Anderson, T. W. (1963). *Ann. Math. Statist.*, **34**, 122–148.

[5] Bartlett, M. S. (1947). *Ann. Math. Statist.*, **18**, 1–17.

[6] Constantine, A. G. (1963). *Ann. Math. Statist.*, **34**, 1270–1285.

[7] Constantine, A. G. (1966). *Ann. Math. Statist.*, **37**, 215–225.

[8] Crowther, N. A. S. and Young, D. L. (1974). Notes on the Distribution of Characteristic Roots and Functions of Characteristic Roots of Certain Matrices in Multivariate Analysis. *Tech. Rep. No. 92*, Stanford University, Stanford, Calif.

[9] Fisher, R. A. (1939). *Ann. Eugen. (Lond.)*, **9**, 238–249.

[10] Fujikoshi, Y. (1977). *J. Multivariate Anal.*, **7**, 386–396.

[11] Girshick, M. A. (1939). *Ann. Math. Statist.*, **10**, 203–224.

[12] Glynn, W. J. (1977). Asymptotic Distributions of Latent Roots in Canonical Correlation and in Discriminant Analysis with Applications to Testing and Estimation. Ph.D. thesis, Yale University.

[13] Glynn, W. J. and Muirhead, R. J. (1978). *J. Multivariate Anal.*, **8**, 468–478.

[14] Hsu, P. L. (1939). *Ann. Eugen. (Lond.)*, **9**, 250–258.

[15] Hsu, P. L. (1941). *J. Lond. Math. Soc.*, **16**, 183–194.

[16] Hsu, P. L. (1941). *Biometrika*, **32**, 38–45.

[17] James, A. T. (1960). *Ann. Math. Statist.*, **31**, 151–158.

[18] James, A. T. (1961). *Ann. Math.*, **74**, 456–469.

[19] James, A. T. (1961). *Ann. Math. Statist.*, **32**, 874–882.

[20] James, A. T. (1964). *Ann. Math. Statist.*, **35**, 475–501.

[21] James, A. T. (1968). *Ann. Math. Statist.*, **39**, 1711–1719.

[22] James, A. T. (1969). In *Multivariate Analysis: Proceedings of the Second International Symposium*, P. R. Krishnaiah, ed. Academic Press, New York, pp. 205–218.

[23] Khatri, C. G. (1967). *Ann. Math. Statist.*, **38**, 944–948.

[24] Khatri, C. G. (1972). *J. Multivariate Anal.*, **2**, 201–207.

[25] Mood, A. M. (1951). *Ann. Math. Statist.*, **22**, 266–273.

[26] Muirhead, R. J. (1978). *Ann. Statist.*, **6**, 5–33.

[27] Muirhead, R. J. and Waternaux, C. M. (1980). *Biometrika*, **67**, 31–43.

[28] Nanda, D. N. (1948). *Ann. Math. Statist.*, **19**, 340–350.

[29] Pillai, K. C. S. (1976). *Canad. J. Statist.*, **4**, 157–184.

[30] Roy, S. N. (1939). *Sankhyā*, **4**, 381–396.

[31] Roy, S. N. (1942). *Sankhyā*, **6**, 15–34.

[32] Srivastava, M. S. and Khatri, C. G. (1979). *An Introduction to Multivariate Statistics.* North-Holland, New York.

[33] Waternaux, C. M. (1976). *Biometrika*, **63**, 639–645.

Further Reading

Crowther, N. A. S. and Young, D. L. (1974). Notes on the Distributions of Characteristic Roots and Functions of Characteristic Roots of Certain Matrices in Multivariate Analysis of Variance. *Tech. Rep. No. 92*, Stanford University, Stanford, Calif. (Most useful for its discussion of techniques available for studying the distributions of individual latent roots.)

James, A. T. (1964). *Ann. Math. Statist.* **35**, 475–501. (Contains a survey of noncentral distributions of random matrices and latent roots derived from both real and complex normal sampling, as well as a summary of zonal polynomials and hypergeometric functions. A classic paper in the area, very widely referenced.)

Johnson, N. L. and Kotz, S. (1972). *Distributions in Statistics: Continuous Multivariate Distributions.* Wiley, New York. (Chapter 39 of this widely available reference book contains much useful information, particularly about approximations to distributions and tables.)

Krishnaiah, P. R. (1978). *Dev. Statist.*, **1**, 135–169. (Useful for its discussion and references for joint marginal distributions of latent roots and distributions of elementary symmetric functions.)

Muirhead, R. J. (1978). *Ann. Statist.*, **6**, 5–33. (Contains a review of both asymptotic representations for latent root distributions and asymptotic distributions of latent roots. Gives an extensive bibliography in this area.)

Pillai, K. C. S. (1976). *Canad. J. Statist.*, **4**, 157–184.

Pillai, K. C. S. (1977). *Canad. J. Statist.*, **5**, 1–62.

(These papers by Pillai contain an extensive bibliography (409 papers) and are most useful for their discussion of individual latent roots, distributions of test statistics which are functions of latent roots, and references to tables.)

(CANONICAL CORRELATION
COMPONENT ANALYSIS
DISCRIMINANT ANALYSIS
MULTINORMAL DISTRIBUTION
MULTIVARIATE ANALYSIS
WISHART DISTRIBUTION)

ROBB J. MUIRHEAD

LATENT ROOT REGRESSION

Latent root regression is a biased estimation methodology which adjusts the least-squares estimator for "nonpredictive" multicollinearities* which occur among predictor (regressor) variables. A nonpredictive multicollinearity* would occur, for example, if two standardized (see below) predictor variables have a large positive correlation so that $X_1 - X_2 \simeq 0$ but the true theoretical model is only a function of the sum, $X_1 + X_2$, of the two variables; for instance, if the response variable (Y) is adult height and two highly correlated predictor variables are length of tibia (X_1) and length of femur (X_2), both leg measurements, but the prediction of adult height is only a function of the sum of the leg measurements. In this case the multicollinearity, $X_1 - X_2$, would be termed a nonpredictive multicollinearity. Latent root estimators seek to eliminate nonpredictive multicollinearities from the resulting parameter estimates.

Derivations and fundamental properties of latent root estimators are given in refs. 7 and 11, each article providing both an algebraic and a geometric justification for the estimator. Approximate tests for the significance of individual or sets of predictor variables are also derived in these articles. Applications of latent root regression appear in refs. 3, 6, 7, 10, and 12.

The form of the latent root estimator is determined by an identification of multicollinearities among all ($p + 1$) response and predictor variables. Assume a multiple linear regression* model of the form

$$\mathbf{Y} = \beta_0 \mathbf{1} + \mathbf{X}\boldsymbol{\beta} + \boldsymbol{\epsilon},$$

where $\mathbf{Y}$ is an n-dimensional vector of re-

sponse variables, β_0 is an unknown constant, $\mathbf{1}$ is a vector of ones, $\mathbf{X}$ is a full-column-rank matrix of p nonstochastic predictor variables which is standardized ($\mathbf{X}'\mathbf{X}$ is in correlation form), $\boldsymbol{\beta}$ is a p-dimensional vector of unknown regression coefficients, and $\boldsymbol{\epsilon}$ is a vector of unobservable random error terms with $\boldsymbol{\epsilon} \sim N(\mathbf{0}, \sigma^2 \mathbf{I})$. Let $\mathbf{A} = [\mathbf{Z} : \mathbf{X}]$, where $\mathbf{Z} = (\mathbf{Y} - \overline{Y}\mathbf{1})/d_Y$ and $d_Y^2 = \sum(Y_i - \overline{Y})^2$. Then $\mathbf{A}'\mathbf{A}$ is a correlation-form matrix of response and predictor variables. Denote the latent roots of $\mathbf{A}'\mathbf{A}$ by $0 \leqslant \lambda_0 \leqslant \cdots \leqslant \lambda_p$ and the corresponding latent vectors of $\mathbf{A}'\mathbf{A}$ by $\gamma_0, \gamma_1, \ldots, \gamma_p$. Finally, partition γ_j into its first element, γ_{0j}, and the remaining p elements

$$\boldsymbol{\delta}_j' = (\gamma_{1j}, \gamma_{2j}, \ldots, \gamma_{pj}); \quad \text{i.e.,} \quad \gamma_j = (\gamma_{0j}, \boldsymbol{\delta}_j').$$

Examination of the latent roots and latent vectors of $\mathbf{A}'\mathbf{A}$ provides a direct means of assessing the predictive value of multicollinearities among the predictor variables. Since $\lambda_j = \gamma_j' \mathbf{A}'\mathbf{A}\gamma_j$, $\lambda_j \simeq 0 \leftrightarrow \mathbf{A}\gamma_j \simeq 0$; i.e., a small latent root of $\mathbf{A}'\mathbf{A}$ identifies a multicollinearity among the columns of $\mathbf{A}$. If γ_{0j} is also near zero, the multicollinearity involves only the predictor variables since $\lambda_j \simeq 0$ and $|\gamma_{0j}| \simeq 0 \leftrightarrow \mathbf{X}\boldsymbol{\delta}_j \simeq \mathbf{0}$. As illustrated above with the height example, a nonpredictive multicollinearity among the predictor variables is of little or no value in predicting the observed response variables. In refs. 11 and 12, nonpredictive multicollinearities are shown to induce small values of both λ_j and γ_{0j}.

Latent root regression estimates $\boldsymbol{\beta}$ by first identifying nonpredictive multicollinearities (latent vectors $\mathbf{A}'\mathbf{A}$ with $\lambda_j \simeq 0$ and $|\gamma_{0j}| \simeq 0$) and then forming an estimator $\hat{\boldsymbol{\beta}}^* = \sum^* a_j \boldsymbol{\delta}_j$, where the summation ($\sum^*$) extends over all subscripts except those corresponding to the nonpredictive multicollinearities. The a_j are given by

$$a_j = -d_Y \gamma_{0j} \lambda_j^{-1} \left(\sum{}^* \gamma_{0k}^2 \lambda_k^{-1} \right)^{-1},$$

where the summation again extends over all subscripts except those corresponding to nonpredictive multicollinearities. If nonpredictive multicollinearities do not occur in $\mathbf{A}$, the summations above extend over all the latent vectors of $\mathbf{A}'\mathbf{A}$ and $\hat{\boldsymbol{\beta}}^*$ is the ordinary

Table 1

j	λ_j	γ_j'					
5	2.5723	(0.58,	0.41,	0.58,	0.34,	0.05,	0.20)
4	2.0327	(−0.01,	−0.50,	−0.07,	0.34,	0.55,	0.57)
3	0.6888	(0.07,	−0.12,	−0.33,	0.81,	−0.38,	−0.27)
2	0.5332	(0.21,	0.18,	−0.24,	0.06,	0.72,	−0.59)
1	0.1724	(−0.78,	0.37,	0.33,	0.33,	0.19,	−0.03)
0	0.0006	(0.02,	0.63,	−0.62,	−0.00,	0.01,	0.46)

least-squares estimator, $\hat{\beta} = (\mathbf{X'X})^{-1}\mathbf{X'Y}$, of β; otherwise, it is a least-squares estimator in which stochastic restrictions of the form $\delta_j'\hat{\beta}^* = d_Y\gamma_{0j}$ are imposed for nonpredictive multicollinearities.

Exact distributional theory is lacking for the latent root estimator due to the intractable distributions of the latent roots and latent vectors of $\mathbf{A'A}$ and their complicated expression in $\hat{\beta}^*$. Rough cutoffs for λ_j and γ_{0j} were suggested in ref. 11 for identifying nonpredictive multicollinearities in moderate-size data sets (say, $8 \leqslant p \leqslant 15$ and $30 \leqslant n \leqslant 100$): $\lambda_j \leqslant 0.05$ and $|\gamma_{0j}| \leqslant 0.10$. For large sample sizes, asymptotic properties of $\mathbf{A'A}$ suggest refinements in these cutoff values (see ref. 12).

Statistics that are useful for assessing the influence of individual or sets of predictor variables are discussed in refs. 1, 7, and 11. One such statistic is similar to a least-squares t statistic:

$$t_j^* = \hat{\beta}_j^* / \hat{\sigma}\left(g_{jj} - g_{00}^{-1}g_{0j}^2\right)^{1/2},$$

where $g_{uv} = \sum^* \gamma_{uk}\gamma_{vk}\lambda_k^{-1}$ and $\hat{\sigma}$ is the latent root estimator of the error standard deviation. If nonpredictive multicollinearities do not occur in $\mathbf{A}$, this statistic becomes the least-squares t statistic, $t_j = \hat{\beta}_j / \text{s.e.}(\hat{\beta}_j)$, which is appropriate for testing the significance of β_j (see ref. 5 [Sec. 10.2.4]); otherwise, the denominator of t_j^* serves as an approximate standard error of $\hat{\beta}_j^*$ and t_j^* is treated as an approximate t statistic.

To illustrate an application of latent root regression, an analysis of the response and the last five predictor variables in Data Set A.8 of ref. 5 yields the latent roots and latent vectors in Table 1.

A single strong multicollinearity occurs in this data set, as indicated by $\lambda_0 = 0.0006$. Since $\gamma_{00} = 0.02$, this multicollinearity is nonpredictive and the latent root estimate of β is formed from γ_1 through γ_5. The resulting latent root estimates and t^* statistics are shown in Table 2 together with the least-squares estimates and t statistics.

The dramatic differences in the least-squares and latent root estimates and t statistics for the multicollinear predictor variables are common when nonpredictive multicollinearities are eliminated from $\hat{\beta}^*$. It is well known (e.g., ref. 5, [Chap. 9]) that least-squares estimates for coefficients of multicollinear predictor variables tend to be inflated and have signs which are determined more by the multicollinearities than by the correct signs of the true coefficients. Latent root regression eliminates the dominant effects of multicollinearities from the estimates.

The performance of latent root estimators and test statistics relative to the corresponding least-squares estimators and test statistics has been demonstrated in simulation studies (e.g., refs. 4 and 6). These simulations use total squared error, $\sum(\hat{\beta}_j^* - \beta_j)^2$, as a criterion of adequacy. Overall, the latent root

Table 2

Predictor Variable	Least Squares $\hat{\beta}$	t	Latent Root $\hat{\beta}^*$	t^*
X_1	−113.57	−1.77	11.03	6.11
X_2	131.53	2.11	10.79	6.08
X_3	8.62	2.97	9.71	3.25
X_4	2.31	0.71	4.43	1.37
X_5	−90.60	−1.93	0.48	0.19

estimator consistently produces a smaller to- tal squared error than least squares unless (1) the true coefficient vector is parallel (or nearly so) to vectors that define the multi- collinearities in $\mathbf{X}$ and (2) the signal-to-noise ratio, $\beta'\beta/\sigma^2$, is large. These findings are consistent with simulation performance of other biased estimators (e.g., ref. 2). In addi- tion, the statistic t_j^* shows greater sensitivity in ref. 6 to the true value of β_j than the least-squares t statistic when predictor vari- ables are highly multicollinear.

Relationships between the latent root esti- mator and other regression estimators can be established by generalizing the form of $\hat{\beta}^*$. Define a generalized latent root estima- tor as in ref. 9 by letting $\tilde{\beta}^* = \sum a_j \delta_j$ with

$$a_j = -d_Y \gamma_{0j} (\lambda_j + \nu_j)^{-1} \left(\sum \gamma_{0k}^2 (\lambda_k + \nu_k)^{-1} \right)^{-1},$$

$$j = 0, 1, \ldots, p.$$

If $\nu_j = \infty$ for nonpredictive multicollineari- ties and $\nu_j = 0$ otherwise, $\tilde{\beta}^*$ is the usual latent root estimator. If $\nu_j = 0$ for all j, $\tilde{\beta}^*$ is the least-squares estimator; moreover, if $\nu_j = c\lambda_j$ for any constant c and all j the same result holds. If $\nu_j = \theta$, $\theta > 0$, $\tilde{\beta}^*$ is the (simple) ridge regression* estimator (see refs. 8 and 9). In general, if $\tilde{\beta}$ is any regression estimator the generalized latent root estima- tor can be made to be equal to $\tilde{\beta}$ by choos- ing the ν_j so that $a_j = \delta_j' \tilde{\beta}_j - d_Y \gamma_{0j}$. This last result is of interest because of the potential for unifying the study of regression estima- tion but is difficult to exploit for theoretical purposes because of the distributional prob- lems mentioned above.

References

[1] Eplett, W. J. R. (1978). *J. R. Statist. Soc. B*, **40**, 184–185.

[2] Gibbons, D. G. (1981). *J. Amer. Statist. Ass.*, **76**, 131–139.

[3] Gunst, R. F. and Mason, R. L. (1977). *Biometrics*, **33**, 249–260. (Illustrates latent root regression on a medical example in which both singularities and multicollinearities occur in the design matrix, $\mathbf{X}$.)

[4] Gunst, R. F. and Mason, R. L. (1977). *J. Amer. Statist. Ass.*, **72**, 616–628.

[5] Gunst, R. F. and Mason, R. L. (1980). *Regression Analysis and Its Application: A Data-Oriented Ap- proach*. Marcel Dekker, New York. (Chapter 10 discusses fundamental properties of latent root, principal component, and ridge regression estima- tors. Relationships among the three estimators are also noted.)

[6] Gunst, R. F., Webster, J. T., and Mason, R. L. (1976). *Technometrics*, **18**, 75–83.

[7] Hawkins, D. M. (1973). *Appl. Statist.*, **16**, 225– 236. (Derives the latent root estimator by mini- mizing the norm which is vertical to the fitted regression hyperplane.)

[8] Hawkins, D. M. (1975). *Technometrics*, **17**, 477– 480. (Establishes a simple relationship between latent root regression and ordinary ridge regres- sion.)

[9] Hua, T. S. and Gunst, R. F. (1980). *Tech. Rep. No. 143*, Dept. of Statistics, Southern Methodist University, Dallas, Tex. (Explores generalized ver- sions of latent root and ridge estimators and shows how each can be obtained from the other.)

[10] Sharma, S. and James, W. L. (1981). *J. Marketing Res.*, **18**, 154–161.

[11] Webster, J. T., Gunst, R. F., and Mason, R. L. (1974). *Technometrics*, **16**, 513–522.

[12] White, J. W. and Gunst, R. F. (1976). *Technomet- rics*, **21**, 481–488.

(COMPONENT ANALYSIS
ECONOMETRICS
LEAST SQUARES
MULTICOLLINEARITY
MULTIPLE REGRESSION
RIDGE REGRESSION)

RICHARD F. GUNST

LATENT STRUCTURE ANALYSIS

HISTORICAL DEVELOPMENT

The term *latent structure analysis* was coined by Paul F. Lazarsfeld to describe the appli- cation of a class of mathematical models to the study of attitudes. The data for these social–psychological studies are attributes, discrete-valued variables, which are assumed to be indicators of some underlying concept or latent variable. Several models developed by Lazarsfeld during the 1940s appeared for

Table 1   Latent Structure with Two Classes[a]

	Probability of Positive Response within Latent Class:	
	1	2
Do you think you would do more for your country as a soldier or as a worker in a war job? [as a soldier]	0.497	0.175
Do you think the army is giving you a chance to show what you can do? [yes, a very or fairly good chance]	0.844	0.191
How well do you think the army is run? [very well or pretty well]	0.844	0.507
Latent class relative frequency	0.63	0.37

[a] Parameters estimated from data in Lazarsfeld [17]. Positive response to each item is given in parentheses.

the first time in his contribution to *Studies in Social Psychology in World War II* [15, 16]. The significance of this work to the development of the theoretical concept of "attitude" in present-day social psychology has been assessed by Fleming [8]; here the emphasis will be on statistical developments and implications for data analysis.

Lazarsfeld intended latent structure analysis to be a uniquely appropriate methodology for sociological survey analysis. In contrast to factor analysis*, which postulates the existence of factors to explain the correlations among a set of *continuous* variables, latent class models apply explicitly to the discrete variables that are created from responses to questionnaire items.

In the simplest of these models, the observed variables are dichotomies and the latent structure postulates that the population can be divided into two latent classes, an unobserved dichotomy whose existence is only inferred from the existing relationships among the observed items. The parameters of the model are probabilities: the conditional probabilities of making certain responses given membership in each latent class, and the relative sizes of the two

classes. Table 1 illustrates such a latent structure using an example from Lazarsfeld [17]; the latent parameters have been estimated from the responses given by 2660 soldiers using an algorithm that is described in the following section.

An interpretation of this model is that there are two types of soldiers, those with high morale (class 1) and those with low morale (class 2). The tendency to respond positively to each of these three items varies from class to class. The fundamental assumption that serves to define this and all other latent structure models is *local independence*: within a latent class the responses to different items are assumed to be independent. Thus, taking the estimates in Table 1 to be true probabilities, a high-morale soldier would have probability (0.497)(0.844)(0.884) = 0.371 of responding positively to all three items, while a low-morale soldier's probability would be only 0.017.

During the 1950s Lazarsfeld explored the properties of more complex models and attempted to codify the process of model formulation, analysis, parameter estimation, and model testing. (References 17 and 18 are the most important articles from this period.) The extensions include models with more than two classes and with parameters that are constrained in some fashion a priori. They also include models for which the latent variable is continuous rather than discrete. In such models the latent probability of a positive response to an item varies with one's position (x) on a latent continuum, and a parametric form for this *trace-line function* has to be specified as part of the model definition. Local independence remains the key assumption, however; if x is fixed, responses to different items are completely independent. Step functions (the *latent distance* model) and polynomials were among the parametric forms of tracelines studied by Lazarsfeld and his associates. (See Lazarsfeld and Henry [19] for a review of these models.)

The work on the latent continuum models was closely related to that of Lord [20, 21], usually referred to as *test theory*. While Lazarsfeld's latent variables were attitudes, however, Lord's were abilities and traits and his applications were to educational measurement. The conceptual similarities of latent structure analysis, test theory, and classical factor analysis permit them all to be discussed as special cases of a general class of statistical models, the key distinctions being between discrete and continuous variables, as shown in Table 2.

The integration of these models was carried out in an elegant fashion by Anderson [2] in his article "Some scaling models and estimation procedures in the latent class model." A result of this integration is that it is no longer appropriate to speak of latent structure analysis as a distinct body of knowledge. Terms such as "latent trait," "latent variable," and "local independence" can be used without specific reference to Lazarsfeld's work (as in, e.g., the special issue of the *Journal of Educational Measurement*, [14]). The causal model literature, which now spans all the social and behavioral sciences, utilizes discrete and continuous, observed and unobserved variables as the particular application requires. (See the comments of Cooley [6].)

LATENT CLASS MODELS

While latent class models are most closely associated with Lazarsfeld, the computational problems associated with the parameter estimation were a serious obstacle to their application to survey analysis. Even in his own work the model functioned primarily as a metaphor for the data analytic process. Recent developments in computing and in related areas of contingency table analysis have made it much easier to use these models in data analysis, as will be illustrated in this section.

Consider a set of n dichotomous items which serve to define the binary variables X_i, $i = 1, 2, \ldots, n$. If the response to item i is "positive" $X_i = 1$; otherwise, $X_i = 0$. There are 2^n possible response patterns: let Π_s denote the probability of response pattern $s = (X_1, X_2, \ldots, X_n)$ being observed. Suppose that the population consists of m distinct latent classes, with relative frequencies $\nu_1, \nu_2, \ldots, \nu_m$. We can write

$$\Pi_s = \sum_{\alpha = 1}^{m} \Pi_{s\alpha} \nu_a ,$$

where $\Pi_{s\alpha}$ is the probability of response s within class α. The fundamental assumption of local independence specifies that $\Pi_{s\alpha}$ is the product of marginal conditional probabilities: thus, if we let

$$\lambda_{i\alpha} = \Pr(X_i = 1 \mid \text{class } \alpha)$$

Table 2

	Latent Variables	
Observed Variables	Discrete	Continuous
Discrete	Latent class models	Latent trait models
Continuous	Latent profile analysis[a]	Factor analysis models

[a] An interesting variant due to Gibson [9], but one that has not generated applications.

then

$$\Pi_{s\alpha} = \prod_{i=1}^{n} \lambda_{i\alpha}^{X_i} (1 - \lambda_{i\alpha})^{1 - X_i}.$$

Further assumptions are necessary if the latent parameters are to be identifiable (estimable). For example, if there are no constraints on the values of ν_α and $\lambda_{i\alpha}$, and if $m = 2$, then n must be at least 3. Contributions to the identification of latent class models were made by Anderson [1] and Madansky [23], among others.

Several methods of estimating parameters are discussed in Chap. 4 of Lazarsfeld and Henry [19]. Moment estimation and maximum likelihood estimation have been the most widely used. In moment estimation the moments $E(X_i)$, $E(X_iX_j)$, etc., of the observed random variables are expressed as functions of the latent parameters: for example,

$$E(X_iX_j) = \sum_{\alpha=1}^{m} \nu_\alpha \lambda_{i\alpha} \lambda_{j\alpha}, \qquad i \neq j.$$

A subset of these equations is then solved for the parameters. Such a solution, if unique, proves the identifiability of the parameters; when the moments are replaced by their sample estimates (i.e., proportions), consistent estimators of the latent parameters are derived.

Moment methods usually are not efficient and also involve some arbitrary selection of the available data. Maximum likelihood* estimation avoids these statistical drawbacks. McHugh [24] published the Newton–Raphson* algorithm for maximum likelihood estimation assuming simple random sampling but an implementation of this approach on the IBM 7094 in the early 1960s was judged too costly for practical use. The combination of increased computational power and the use of alternate algorithms, however, has changed the situation significantly. (*See also* NEWTON ITERATION EXTENSIONS.)

Goodman [10, 11] has shown that latent class models can be estimated and tested via maximum likelihood using an iterative proportional fitting* algorithm that does not involve any matrix inversion (a major prob-

lem in the earlier Newton–Raphson approach). The algorithm generates the estimates at iteration $t + 1$ from those at iteration t as follows:

$$\nu_\alpha^{(t+1)} = \nu_\alpha^{(t)} \sum_S p_s \Pi_{s\alpha}^{(t)} \Big/ \Pi_s^{(t)}$$

$$\lambda_{i\alpha}^{(t+1)} = \sum_{S_i} p_s \Pi_{s\alpha}^{(t)} \Big/ \Pi_s^{(t)}.$$

Here p_s is the sample proportion showing response pattern s, $\sum_S$ is a sum over all 2^n response patterns, and $\sum_{S_i}$ is the sum over all response patterns for which $X_i = 1$. $\Pi_{S\alpha}^{(t)}$ and $\Pi_s^{(t)}$ are the values of $\Pi_{s\alpha}$ and Π_s computed from the estimates of the λ and ν at the tth iteration. This particular routine can be easily adapted to take into account simple constraints on the parameters, such as the specification that one or more of the ν or λ are fixed a priori or that two or more of them are equal, by resetting the estimates appropriately at the end of each iteration of the equations.

Goodness of fit* of an estimated latent class model can be assessed by chi-square methods. Both the traditional Pearson chi-square* statistic (χ^2) and the likelihood ratio* related statistic are used in the literature (see Fienberg [7, pp. 40–43]. Under the null hypothesis that the right model has been specified, these statistics have asymptotic chi-square distributions with $2^n - k - 1$ degrees of freedom, where k is the number of independent parameters that have been estimated. In the general model with m classes and n items and no special constraints on the parameters, $k = nm + m - 1$.

A simple computer program written by the author in BASIC using this algorithm was applied to the data set in Lazarsfeld [17]. In addition to the three items described in Table 1, numbered 2, 3, and 4, respectively, in the original article, the set contains an additional item:

1. In general, how would you say you feel most of the time? [Usually in good spirits]

The model of two latent classes proved a poor fit to the data, with the Pearson chi-square approximately equal to 41, with 16 –

Table 3   Latent Structure with Three Classes[a]

	Probability of Positive Response within Latent Class:		
	1	2	3
1. In good spirits	0.672	0.157	0.157
2. Do more as a soldier	0.647	0.190	0.190
3. Army gives a chance	0.857	0.857	0.240
4. Army is run well	0.900	0.900	0.522
Latent class relative frequency	0.409	0.177	0.414

[a]Parameters estimated from data in Lazarsfeld [17]. Within-class probabilities constrained as indicated in the text.

$9 - 1 = 6$ degrees of freedom. Lazarsfeld had proposed a three-class model with four constraints:

$$\lambda_{i1} = \lambda_{i2} \quad \text{for} \quad i = 3, 4$$

$$\lambda_{i2} = \lambda_{i3} \quad \text{for} \quad i = 1, 2.$$

This model fits quite well:

$$\chi^2 = 7.75, \quad \text{d.f.} = 16 - 8 - 3 = 5.$$

Tables 3 and 4 give the estimated parameters and the goodness-of-fit analysis.

Goodman's formulation of this algorithm permits the observed variables to have more than two categories, and a computer program implementing the procedure is available [4]. Haberman's [13, Chap. 10] discussion of latent class models within the general context of the log-linear analysis of contingency tables includes a computer program for maximum likelihood estimation using a modified Newton–Raphson algorithm. This method converges more rapidly and also provides estimates of the standard errors of estimated parameters, but is somewhat less convenient to use when dealing with models with constrained parameters. Haberman notes that these estimation procedures must be used with caution, for multiple solutions to the likelihood equations may exist and different initial estimates may lead the programs to converge to different final estimates. It would seem prudent to try different starting estimates and to investigate the stability of a solution whenever applying the latent class model.

Table 4   Goodness-of-Fit Analysis of a Restricted Three-Class Model[a]

Response Pattern, s, for Item:				Observed Frequency, $N p_s$	Expected Frequency, $N \hat{\Pi}_s$[b]
1	2	3	4		
0	0	0	0	286	279.4
1	0	0	0	54	55.4
0	1	0	0	59	68.6
1	1	0	0	25	18.9
0	0	1	0	123	124.3
1	0	1	0	36	43.2
0	1	1	0	56	46.5
1	1	1	0	42	45.4
0	0	0	1	353	355.9
1	0	0	1	98	96.6
0	1	0	1	114	109.8
1	1	0	1	71	76.0
0	0	1	1	439	438.7
1	0	1	1	267	262.5
0	1	1	1	252	258.7
1	1	1	1	385	380.0

[a]Chi-square = 7.75, d.f. = 5.
[b]$\hat{\Pi}_s$ calculated using the estimates of Table 3. Calculations were done using more significant figures than reported here.

LATENT TRAIT MODELS

Suppose that a real number x is associated with each member of a population, and that the value of x is related to the probability that the individual will respond positively to a series of dichotomous items, through a parametric function $p_i(x; \theta)$. When the value of x is not directly observable it is referred to as a latent trait; depending on the context the trait may be an ability (mathematical competence), an attitude (e.g., toward the Army), a personality characteristic (authoritarianism), or some other concept that is theoretically linked with a set of behaviors. The linear and polynomial traceline functions studied by Lazarsfeld have the disadvantage that for sufficiently large or small x values they take on values outside the unit interval. Since $p_i(x; \theta)$ is a probability, the acceptable range of x must be appropriately limited. Parameter estimation is rather difficult in such situations and few successful applications of these models were achieved [19, Chap. 7].

In contrast to this approach, Lord [20] specified that the tracelines have the form of the CDF of a two-parameter normal distribution (see Lord and Novick [22, Chap. 16]). Thus x may vary over the entire real line, since the trace lines are functionally bounded between zero and 1. Subsequently Birnbaum [3] explored the properties of the similar but mathematically more tractable *logistic* trace line

$$p_i(x; a_i, b_i) = 1/\left(1 + \exp\left[a_i(x - b_i)\right]\right).$$

The parameters b_i and a_i are usually interpreted as the difficulty of the item and the discriminating power of the item, respectively. Rasch [25] proposed independently the special case in which all items have the same discriminating power, in effect setting all $a_i = 1$. It should be emphasized that, since "tests" and questionnaires are themselves human constructs, models such as these function as prescriptions for constructing such tests, as well as guides for the analysis of data. Birnbaum [3] provides an excellent explanation of parameter estimation and interpretation for these models. As in the case of latent class models, the development of computer programs for parameter estimation, usually via maximum likelihood, has increased the visibility and the applicability of latent trait models (see ref. 14 for references).

In applications to educational testing, following Lord, the actual x values of the sampled individuals are often considered as explicit parameters of the model, to be estimated from the sample data. This fits quite well with the educational testing context, where the outcome of the testing process is to be a "score" for every individual. In contrast, Lazarsfeld preferred to specify a parametric function for the distribution of x, and to estimate these parameters from the data. For instance, one might assume that the latent trait is normally distributed and thus be more interested in the mean and variance of that distribution than in the individual "abilities" of the persons in the sample. Sanathanan and Blumenthal [26], who refer to these two approaches to modeling as fixed-effects and random-effects, respectively, show how to estimate the mean and variance of the x distribution in the latter case.

CONCLUSION

There are, broadly speaking, two reasons for using latent structure models. On the one hand, it may be important to understand the structure of a fixed set of items: Is it reasonable, for instance, to think of a conceptual variable such as authoritarianism as one-dimensional with a predictable relationship to a specific set of questionnaire items? On the other hand, the goal may be to classify or grade individuals in the population. The sociological applications of latent structure tend to be of the first type; educational applications are more often of the second. Examples of the latter type, to which latent trait models are usually applied, are also likely to contain a much longer list of items: the individual item parameters are only of peripheral interest and highly constrained models (e.g., assuming that *all* items have the same parameters) are common.

There are severe limitations to the application of general latent structure models in exploratory data analysis*. Since maximum likelihood estimation requires examination of all 2^n frequencies generated by n dichotomous items, sample sizes must be relatively large. Furthermore, if models with relatively small numbers of items are to be used as measurement models, a systematic approach to the identification of items as elements of specific domains of content must be taken. While such an agenda was proposed many years ago by Guttman [12], there are no signs of such a trend in the sociological and social psychological literature, despite the increased application of complex statistical methods [5].

References

[1] Anderson, T. W. (1954). *Psychometrika*, **19**, 1–10.

[2] Anderson, T. W. (1959). In *Probability and Statistics*, U. Grenander, ed. Wiley, New York, pp. 9–38.

[3] Birnbaum, A. (1968). In *Statistical Theories of Mental Test Scores*, by F. M. Lord and M. R. Novick. Addison-Wesley, Reading, Mass., Chaps. 17–20.

[4] Clogg, C. C. (1977). Unrestricted and Restricted Maximum Likelihood Latent Structure Analysis: A Manual for Users. *Working Paper 1977-09*, Population Issues Research Office, University Park, Pa. (This program, written in FORTRAN, is easy to use and relatively portable. It applies only to latent class models.)

[5] Clogg, C. C. (1981). In *Factor Analysis and Measurement in Sociological Research*, D. M. Jackson and E. F. Borgatta, eds. Sage, Beverly Hills, Calif.

[6] Cooley, W. W. (1979). In *Advances in Factor Analysis and Structural Equation Models*, K. G. Joreskog and D. Sorbom, eds. Abt, Cambridge, pp. xv–xxviii.

[7] Fienberg, S. E. (1980). *The Analysis of Cross-Classified Categorical Data*, 2nd ed. MIT Press, Cambridge, Mass.

[8] Fleming, D. (1967). *Perspect. Amer. History*, **1**, 287–365.

[9] Gibson, W. A. (1959). *Psychometrika*, **24**, 229–252.

[10] Goodman, L. A. (1974). *Amer. J. Sociol.*, **79**, 1179–1259.

[11] Goodman, L. A. (1974). *Biometrika*, **61**, 215–231.

[12] Guttman, L. (1941). In *The Prediction of Personal Adjustment*, P. Horst, ed. Social Science Research Council, New York, pp. 319–348.

[13] Haberman, S. J. (1979). *Analysis of Qualitative Data*, Vol. 2. Academic Press, New York. (This text, written for statistical specialists, is a thorough description of models and applications. Appendices include computer programs. The chapter on latent structure analysis emphasizes maximum likelihood estimation of parameters of latent class models.)

[14] *Journal of Educational Measurement* (1977). Volume 4, No. 2. (This special issue is devoted to recent developments and applications of latent trait modeling in education research.)

[15] Lazarsfeld, P. F. (1950). In *Measurement and Prediction*, S. Stouffer, ed. Princeton University Press, Princeton, N.J., Chap. 10.

[16] Lazarsfeld, P. F. (1950). In *Measurement and Prediction*, S. Stouffer, ed. Princeton University Press, Princeton, N.J., Chap. 11.

[17] Lazarsfeld, P. F. (1954). In *Mathematical Thinking in the Social Sciences*, P. F. Lazarsfeld, ed. Free Press, New York.

[18] Lazarsfeld, P. F. (1959). In *Psychology: A Study of a Science*, Vol. 3, S. Koch, ed. McGraw-Hill, New York, pp. 476–535.

References 15 to 18 are primarily addressed to sociologists and psychologists with some training in mathematics. There is, therefore, more emphasis on the formulation and interpretation of stochastic models than on sampling and estimation. The outline of the "nine steps of LSA" in the Koch volume, however, would be valuable to the education of any statistical scientist, as it describes the research process from model formulation to testing and back.)

[19] Lazarsfeld, P. F. and Henry, N. W. (1968). *Latent Structure Analysis*. Houghton Mifflin, Boston. (A summary of the previous 20 years' developments, including a complete bibliography. It gives a good idea of Lazarsfeld's orientation wherein the analysis of the implications of a model is often more interesting than its application to data.)

[20] Lord, F. M. (1952). *A Theory of Tests Scores*. Psychometric Monogr. No. 7.

[21] Lord, F. M. (1953). *Educ. Psychol. Meas.*, **13**, 517–549.

[22] Lord, F. M. and Novick, M. R. (1968). *Statistical Theories of Mental Test Scores*. Addison-Wesley, Reading, Mass. (An excellent text that is still current reading for psychometricians. The chapters on latent trait models, including the discussion of the local independence assumption, should be the first source for statisticians interested in learning more about the subject.)

[23] Madansky, A. (1960). *Psychometrika*, **25**, 183–198.

[24] McHugh, R. B. (1956). *Psychometrika*, **21**, 331–347; *ibid.*, **23**, 273–274.

[25] Rasch, G. (1960). *Probabilistic Models for Some Intelligence and Attainment Tests*. Danmarks Paedagogiske Institut, Copenhagen.

[26] Sanathanan, L. and Blumenthal, S. (1978). *J. Amer. Statist. Ass.*, **73**, 794–799.

(COMPONENT ANALYSIS
EDUCATIONAL STATISTICS
FACTOR ANALYSIS
MULTIDIMENSIONAL SCALING
PSYCHOLOGICAL TESTING THEORY
PSYCHOLOGY, STATISTICS IN
SOCIOLOGY, STATISTICS IN)

NEIL W. HENRY

LATIN SQUARES, LATIN CUBES, LATIN RECTANGLES, ETC.

In an $n \times n$ Latin square, each of n symbols appears exactly once in each row and exactly once in each column, e.g.,

$$
\begin{array}{ccccc}
A & B & C & D & E \\
B & A & D & E & C \\
C & E & A & B & D \\
D & C & E & A & B \\
E & D & B & C & A
\end{array}
$$

The symbols (or *elements*) are usually the first n letters of the alphabet, or the integers $0, 1, \ldots, n-1$, or the integers $1, 2, \ldots, n$. The integer n is the *order* of the square. A Latin square of any order greater than or equal to 2 can be obtained most simply by writing the n symbols in their natural order in the first column, and then completing each row cyclically, i.e., with the symbols again in order save that the last symbol is followed, as necessary, by the first; e.g.,

$$
\begin{array}{ccccc}
A & B & C & D & E \\
B & C & D & E & A \\
C & D & E & A & B \\
D & E & A & B & C \\
E & A & B & C & D
\end{array}
$$

However, most Latin squares cannot be derived in this way, and two Latin squares of the same order may differ fundamentally. Thus, although both 5×5 squares above are *reduced* or *standard* (i.e., their symbols are in natural order in both the first row and the

first column), the first contains four *intercalates* (2×2 Latin squares embedded within it, e.g.,

$$
\begin{array}{cc}
A & C \\
C & A
\end{array}
$$

in the first and third rows and columns), whereas the second has no intercalates.

Every Latin square may be regarded as the multiplication table of a quasi-group, some as the multiplication tables of groups*.

The principal reference book on Latin squares is *Latin Squares and their Applications* by Dénes and Keedwell [8]. These authors reported that Latin squares were probably first derived from "problems concerning the movement and disposition of pieces on a chess board. However, the earliest written reference to the use of such squares known to the authors concerned the problem of placing the 16 court cards of a pack of ordinary playing cards in the form of a square so that no row, column, or diagonal should contain more than one card of each suit and one card of each rank. An enumeration by type of the solutions to this problem was published in 1723 [in an edition of *Récréations Mathématiques et Physiques* by Jacques Ozanam (1640–1717)]."

Dudeney [9, Problem 304] recorded the problem under the heading "Bachet's Square": "One of the oldest card puzzles is by Claude Gaspar Bachet de Méziriac (1581–1638), first published in the 1624 [i.e., second] edition of his work [*Problèmes Plaisans et Délectables*]."

The Bachet square apart, the enumeration of Latin squares was first discussed by Euler [10] in a memoir presented to the St. Petersburg Academy of Sciences in 1779.

To facilitate enumeration, various combinatorial concepts must be defined. A *transformation* of a Latin square is any permutation of its rows, or of its columns, or of its symbols, or any combination of such permutations. The rows, columns, and symbols are the three *constraints* of the square. Two squares are said to be *adjugate*, *conjugate*, or occasionally *parastrophic* if a permutation of the constraints of one generates the other (except that *conjugate* has sometimes been restricted to mere transposition of rows and columns); if a square has symbols $1, 2, \ldots, n$, and 2 appears in the cell in row 3 and column 4, then permutation of (say) rows and symbols means that the new square has a 3 in the cell in row 2 and column 4.

An *isomorphism class* of Latin squares is generated from any one of its members by all transformations that involve the same permutation for each constraint. Two squares from the same isomorphism class are *isomorphic*. An *isotopy class* (whose members are *isotopic*) or *transformation set* is defined like an isomorphism class except that there is no restriction on the transformations used. A *main class* or *species* comprises all members of an isotopy class together with all squares adjugate to them.

An *intercalate reversal* involves interchanging the two rows, columns, or symbols of an intercalate [10, Sec. 141]; the Latin square thus obtained may or may not belong to the same main class as the starter square. This last statement is true also for interchanges involving *generalized intercalates*, which are embedded Latin squares larger than 2×2, or embedded rectangles such as

$$
\begin{array}{ccc}
C & D & E \\
D & E & C
\end{array}
$$

whose rows can be interchanged. A *family* of main classes comprises all main classes that can be obtained from one another by one or more successive intercalate reversals. A *domain* of main classes is defined likewise, save

that interchanges involving generalized intercalates are admitted too.

Enumeration details for Latin squares of small order can be summarized as in Table 1. For each order n, the total number of Latin squares is obtained by multiplying the number of reduced squares by $n!(n-1)!$. Tables of Latin squares of small order are given in the statistical tables of Fisher and Yates [13].

Special types of Latin square include diagonal, symmetric, Knut-Vik*, and complete Latin squares. A Latin square whose main diagonal includes every symbol (see the second 5×5 square above) has variously been defined as *left semidiagonal* or simply *diagonal*; correspondingly, the descriptions *diagonal* and *doubly diagonal* have been used for Latin squares with every symbol in each of the two diagonals (see the next 5×5 square below). A *symmetric* Latin square is the same as its transpose; a *totally symmetric* Latin square [2, 3] is invariant under any permutation of its three constraints.

Fisher [12, Sec. 34] used the term *diagonal square* in a different sense, for a Latin square where all the cells on the main diagonal have the *same* symbol, and where each other line parallel to this diagonal has the same symbol throughout. Tedin [30, pp. 191–192] had previously used the epithet *diagonal* for Latin squares where either the main diagonal or the right-to-left diagonal has the foregoing properties.

In the terminology of Fisher [12, Sec. 34], a *Knut-Vik square* is generated cyclically,

Table 1 Enumeration Details for Latin Squares of Small Order

			Number of:			
Order	Domains	Families	Main Classes	Isotopy Classes	Isomorphism Classes	Reduced Squares
2	1	1	1	1	1	1
3	1	1	1	1	1	1
4	1	1	2	2	2	4
5	2	2	2	2	6	56
6	1	4	12	22	109	9,408
7	2	4	147	564	?	16,942,080
8	?	?	?	1,676,257	?	535,281,401,856

the method being "to move each row forward two places instead of one," e.g.,

$$
\begin{array}{ccccc}
A & B & C & D & E \\
D & E & A & B & C \\
B & C & D & E & A \\
E & A & B & C & D \\
C & D & E & A & B
\end{array}
$$

This method produces a Latin square if and only if n is odd, and indeed use of the name *Knut-Vik* seems to have been restricted originally to 5×5 squares such as the above. These are *knight's move* squares in the strict sense that all the cells with the same symbol can be traversed by a series of knight's moves (as on a chessboard) without visiting cells with other symbols. These squares were known in Denmark from about 1872; examples were given by Lindhard [22], who used the term "Springertraek" (knight's move), and by Vik [32], after whom they were named. It seems clear that the name *Knut-Vik square* was never intended to apply to such Latin squares as the following, drawn to the author's attention by P. J. Owens:

$$
\begin{array}{ccccc}
A & B & C & D & E \\
E & D & A & C & B \\
D & C & B & E & A \\
B & E & D & A & C \\
C & A & E & B & D
\end{array}
$$

This square too has the strict knight's-move property, but also has four pairs of identical diagonally adjacent cells (e.g., the third cell in row 1 and the fourth in row 2). Owens has shown that Latin squares with the strict knight's-move property but without any identical diagonally adjacent cells exist only for $n = 5$. He has also shown that, if identical diagonally adjacent cells are allowed, Latin squares with the strict knight's-move property still do not exist except for $n = 5$ and for $n = 4, 8, 12, \ldots$, e.g.,

$$
\begin{array}{cccc}
A & B & C & D \\
C & D & A & B \\
B & A & D & C \\
D & C & B & A
\end{array}
\quad \text{and} \quad
\begin{array}{cccc}
A & B & C & D \\
D & C & B & A \\
B & A & D & C \\
C & D & A & B
\end{array}
$$

The 6×6 and 7×7 "Rösselsprung" (knight's move) Latin squares of Ortlepp [25] have no identical diagonally adjacent cells, but have only a weak knight's move property: All cells with the same symbol can be traversed by knight's moves if rows and columns are both considered to form endless cyclic sequences with the first row (or column) following the last.

Hedayat and Federer [18] used the term *Knut-Vik designs* (not *squares*) for what are sometimes called *totally diagonal Latin squares*, i.e., Latin squares in which each of the n symbols occurs exactly once in each of the n left (broken) diagonals and each of the n right (broken) diagonals. The 5×5 Knut-Vik square displayed above is both a knight's-move square and a totally diagonal square; indeed, for $n = 5$, the set of all totally diagonal Latin squares is also the set of all Latin squares with the strict knight's-move property and with no identical diagonally adjacent cells. An example of a 7×7 totally diagonal Latin square is Ortlepp's 7×7 Rösselsprung square. For $n = 9$ there are no totally diagonal Latin squares, but there are Latin squares with the weak knight's-move property and with no identical diagonally adjacent cells. Some 11×11 totally diagonal Latin squares have the weak knight's-move property, but others do not. Totally diagonal Latin squares exist if and only if the order n is not a multiple of 2 or 3 [16].

To define a *complete* Latin square we first consider the $n - 1$ ordered pairs of consecutive symbols in each of the n columns: If these $n(n - 1)$ ordered pairs are all distinct (with one occurrence of each possible ordered pair of the symbols), then the square is *column complete* or *vertically complete*; the square is complete if it is both *row* (or *horizontally*) *complete* (similarly defined) and column complete, e.g.,

$$
\begin{array}{cccc}
A & B & D & C \\
B & C & A & D \\
D & A & C & B \\
C & D & B & A
\end{array}
$$

Complete $n \times n$ Latin squares are known to exist for all even values of n, for $n = 27$, and for $n = pq$, where p and q are odd primes, p being a divisor of $q - 1$ and having 2 as a primitive root; the smallest such values of n are 21, 39, 55, and 57. Complete Latin squares do not exist for $n = 3, 5$, and 7.

Whether there are any for other odd values of n is unknown, but column-complete Latin squares of orders 9 and 15 have been published [17], and T. W. Tillson has independently obtained such a square of order 9. Freeman [14] enumerated and classified complete Latin squares with small values of n.

Two Latin squares are *mutually orthogonal* if the n^2 ordered pairs of corresponding elements are all distinct (so that each possible ordered pair of the symbols involved occurs exactly once); such squares are often said to constitute a *Graeco-Latin square*.* For values of n that are prime powers, there are *complete sets of mutually (pairwise) orthogonal Latin squares*, the adjective "complete" here connoting a set of $n - 1$ squares [5, 28]. Incomplete sets for other values of n have been obtained by various authors, including Boob and Agrawal [4].

The term *orthogonal symmetric Latin squares* was used by Wallis [33, p. 41] for two symmetric Latin squares with the same symbols and with the same main diagonal on or above which no pair of corresponding symbols (their order being disregarded) occurs more than once. This usage is, however, at variance with the statistical concept of orthogonality, so the alternative term *perpendicular symmetric Latin squares* is perhaps to be preferred (see Lindner and Mendelsohn, [23]).

Addelman [1] extended the concept of Latin square by defining an $n \times n$ *equal frequency square*, $n = tk$, to have each of t symbols occurring exactly k times in each row and column. Similarly, he defined an $n \times n$ *proportional frequency square* with t symbols, $t < n$, to have its symbols occurring in each row and column with frequencies proportional to the symbols' overall frequencies. Hedayat and Seiden [19] used the terms *frequency squares* and *F-squares* synonymously to cover both of Addelman's types of square. *F*-squares orthogonal to Latin squares were considered by Finney [11]. The possibility of using an 8×8 equal frequency square as design for a field experiment with four treatments was suggested by "Student" [29, p. 719].

The different types of *Latin cube*, all natural extensions of the Latin square, are the subjects of terminological confusion. Clarity of exposition is perhaps most easily attained by adopting the basic definitions of Kishen [20, 21].

In each of its three directions, an $n \times n \times n$ cube has n *layers* of size $n \times n$; each layer has n *lines* in each of its two directions. If each of n symbols appears n times in each layer, the arrangement is a *Latin cube of first order*; if all the layers are Latin squares, the first-order cube is *regular* [26]. If each of n^2 symbols occurs once in each layer, the arrangement is a *Latin cube of second order*; this cube is *regular* if, of the n symbols in any one line, none or all occur in each of the $n^2 - 1$ lines parallel to the first one.

A *Latin rectangle* has been defined as a Latin square with some row or rows omitted; an alternative wider definition covers any $r \times s$ array using n symbols ($r < n$, $s \leqslant n$) none of which is repeated in any row or any column. The latter definition includes such arrays as the following ($n = 5$), which cannot be completed as an $n \times n$ Latin square:

$$
\begin{array}{cccc}
A & B & C & D \\
B & A & D & C \\
C & D & E & A
\end{array}
$$

However, any $r \times n$ Latin rectangle ($r < n$) can be completed as an $n \times n$ Latin square [15].

In the design and analysis of comparative experiments, the main role of the Latin square is as a *row-and-column design**, a *row–column design*, or a *design for the two-way elimination of heterogeneity*: the rows and the columns are then used for two mutually orthogonal systems of blocks, and the n symbols are used for the n treatments (or n treatment combinations, if the experiment is factorial). In agricultural experiments, the rows and columns are usually strips of land, with the row strips at right angles to the column strips; the plots are then the intersections of strips in different directions, and the field layout reproduces the standard way of printing a Latin square (except that the plots, and so the experiment as a whole, need not be square). In forestry and orchard

Table 2   Standard Analysis of Variance for a Latin Square

Source of Variation	d.f.	Sum of Squares	Mean Square
Rows	$n-1$	$S_R = \dfrac{1}{n}\sum R_i^2 - \dfrac{G^2}{n^2}$	$S_R/(n-1)$
Columns	$n-1$	$S_C = \dfrac{1}{n}\sum C_j^2 - \dfrac{G^2}{n^2}$	$S_C/(n-1)$
Treatments	$n-1$	$S_T = \dfrac{1}{n}\sum T_k^2 - \dfrac{G^2}{n^2}$	$S_T/(n-1)$
Error	$(n-1)(n-2)$	$S_E = S_{\text{total}} - S_R - S_C - S_T$	$S_E/\{(n-1)(n-2)\}$
Total	n^2-1	$S_{\text{total}} = \sum y_{ijk}^2 - \dfrac{G^2}{n^2}$	—

experiments, each row and column may be a set of adjacent lines of trees. In an industrial experiment whose "plots" are *runs* made on different occasions, the "rows" might be different days, and the "columns" different times of day.

In each practical example quoted so far, two constraints (rows and columns) are used for *block factors* and one (the symbols) for a *treatment factor*. However, a Latin square can also be used as a *main-effect* plan* for two noninteracting sets of n treatments, the plots being grouped in n blocks each of size n. Similarly, a Latin square can provide an unblocked main-effect plan for three treatment factors. (And Cretté de Palluel [7] used a 4×4 Latin square for an experiment on 16 sheep, the factors being breed of sheep, type of food, and date of slaughter.)

Table 2 gives the form of the standard analysis of variance* for a set of yields (or other observations) from a Latin square experiment with a single nonfactorial set of treatments. In the table, y_{ijk} denotes the yield for the plot in row i and column j, this plot having been assigned treatment k; the suffices i, j, and k customarily all take the values 1 to n inclusive. Also, R_i denotes the total of the yields of the n plots in row i; similarly, the quantities C_j and T_k are the column and treatment totals, whereas G is the grand total of all n^2 yields. As in other designs, the *standard error per plot** is the square root of the error mean square M_E. The standard error of a treatment mean* is $\pm\sqrt{M_E/n}$, and the standard error of the

difference (SED) between two treatment means is $\pm\sqrt{2M_E/n}$.

The complete orthogonality of a Latin square makes it an attractive simple experimental design. But, in practice, the restriction that the numbers of rows, columns, and treatments must all be the same can be serious—or indeed impossible to satisfy with the available experimental material and resources. For $n \leqslant 4$, the paucity of degrees of freedom for experimental error* is also important. Furthermore, the standard analysis for a Latin square implies additivity of the effects of the three factors: there are insufficient degrees of freedom for proper identifying of any interactions between these factors, and the presence of such interactions distorts the analysis. If interaction is suspected, confirmation can sometimes be gained by separating one or more degrees of freedom for *nonadditivity** from the error, and comparing the nonadditivity mean square with the new residual mean square. The simplest such procedure involves a form of *Tukey's one degree of freedom for nonadditivity* [31]; more elaborate procedures were given by Milliken and Graybill [24]. Effects of nonadditivities in Latin squares were considered by Wilk and Kempthorne [34], whose work was discussed critically by Cox [6].

Latin squares feature prominently in textbook accounts of the design and analysis of experiments, and are often said to be of particular value in agricultural research. This emphasis on them has led to their uncritical use, especially by agricultural experimenters, who have preferred them to randomized block* designs on grounds that "two blocking systems *must* be better than one"! Such experimenters are liable to overlook the possibility that compact arrays of their plots (perhaps 2×2 or 3×2) would constitute the most appropriate blocks.

Randomization* is required for Latin squares, as for other experimental designs. For a Latin square used as a row-and-column design*, randomization theory shows that a standard validity criterion is satisfied if a random permutation is taken of the rows of any Latin square of the required order, and a random permutation of the columns. However, for example, if the randomization of a 4×4 square were to be restricted to ensure that each of the square's 2×2 corners was a replicate of the treatments, the estimate of experimental error would be vitiated [36].

R. A. Fisher (Sir Ronald Fisher*) introduced a randomized Latin square as the design for an experiment in a forest nursery in 1924. He also supplied a 5×5 Latin square for an experiment laid out in a Welsh forest in 1929: the five "treatments" were five different types of conifer, and the rows were at different altitudes up a hillside running roughly from 380 meters to 550 meters.

Column-complete Latin squares have been used as changeover designs* or crossover designs* for the estimation of "main" and "first residual" effects of the treatments; the rows are then for the different *periods* of the experiment, the columns for the different subjects (e.g., dairy cows in a feeding trial), and the symbols for the treatments (e.g., diets or drugs). For small odd values of n, the column completeness lacking in any single square is attainable with a pair of squares [35], and the pair can be used as a changeover design, e.g.,

Period										
I	A	B	C	D	E	D	E	A	B	C
II	B	C	D	E	A	C	D	E	A	B
III	E	A	B	C	D	E	A	B	C	D
IV	C	D	E	A	B	B	C	D	E	A
V	D	E	A	B	C	A	B	C	D	E

Latin cubes and related arrangements have been proposed as designs for experi-

ments in three dimensions, e.g., in an incubator or refrigerated fruit store, but have been of more statistical interest for their use in the construction of commoner experimental designs.

Latin rectangles are of little statistical interest except for *Youden squares** (which, despite their name, are not square) and for similar partially balanced designs.

Several German agronomic authors have used the name "Lateinisches Rechteck" (Latin rectangle) in a very restricted sense, for what Scott [27, p. 3] and several subsequent authors have called a *semi-Latin square* or "cuadro latino modificado" (modified Latin square). This is an arrangement with r rows, $t = kr$ columns ($k > 1$), and t treatments, the columns being grouped into sets each containing k consecutive columns; each treatment occurs exactly once in each row and exactly once in each set of columns. When used as an experimental design having just one error term, the semi-Latin square is statistically defective, because of the restriction that the grouping of columns imposes on the randomization.

References

[1] Addelman, S. (1967). *J. Amer. Statist. Ass.*, **62**, 226–240.

[2] Bailey, R. A. (1979). *Utilitas Math.*, **15**, 193–216. (Enumerates totally symmetric Latin squares for values of n up to and including 10.)

[3] Bailey, R. A., Preece, D. A., and Zemroch, P. J. (1978). *Utilitas Math.*, **14**, 161–170.

[4] Boob, B. S. and Agrawal, H. L. (1976). *Biometrics*, **32**, 191–193.

[5] Bose, R. C. (1938). *Sankhyā*, **3**, 323–338.

[6] Cox, D. R. (1958). *Biometrika*, **45**, 69–73.

[7] Cretté de Palluel, [F.] (1788). *Mém. d'Agric.*, Trimestre d'Été, pp. 17–23 (with two inset tables). [English transl: *Ann. Agric.*, **14**, 133–139, 1790 (with table facing p. 139).] (Reports the use of a 4×4 Latin square for an experiment on 16 sheep.)

[8] Dénes, J. and Keedwell, A. D. (1974). *Latin Squares and Their Applications*. Akadémiai Kiadó, Budapest/English Universities Press, London/Academic Press, New York. (The principal reference book on Latin squares.)

[9] Dudeney, H. E. (1917). *Amusements in Mathematics*. Nelson, Edinburgh. (Problem No. 304 is "Bachet's square.")

[10] Euler, L. (1782). *Verh. Zeeuwsch Genootsch. Wet. Vlissengen,* **9**, 85–239. (Modern reprint: *Leonhardi Euleri Opera Omnia,* Ser. 1, **7**, 441–457, 1923.) (Discusses the enumeration of Latin squares.)

[11] Finney, D. J. (1945). *Ann. Eugen. (Lond.),* **12**, 213–219.

[12] Fisher, R. A. (1935). *The Design of Experiments.* Oliver & Boyd, Edinburgh (subsequent editions: 1937, 1942, 1947, 1949, 1951, 1960, 1966).

[13] Fisher, R. A. and Yates, F. (1963). *Statistical Tables for Biological, Argicultural and Medical Research,* 6th ed. Oliver & Boyd, Edinburgh (reprinted by Longmans, London). (Includes tables of Latin squares and instructions for randomization.)

[14] Freeman, G. H. (1979). *J. R. Statist. Soc. B,* **41**, 253–262. (Enumerates and classifies complete Latin squares with small values of *n*.)

[15] Hall, M. (1945). *Bull. Amer. Math. Soc.,* **51**, 387–388.

[16] Hedayat, A. (1977). *J. Comb. Theory A,* **22**, 331–337.

[17] Hedayat, A. and Afsarinejad, K. (1978). *Ann. Statist.,* **6**, 619–628.

[18] Hedayat, A. and Federer, W. T. (1975). *Ann. Statist.,* **3**, 445–447.

[19] Hedayat, A. and Seiden, E. (1970). *Ann. Math. Statist.,* **41**, 2035–2044.

[20] Kishen, K. (1942). *Curr. Sci.,* **11**, 98–99.

[21] Kishen, K. (1950). *J. Indian Soc. Agric. Statist.,* **2**, 20–48.

[22] Lindhard, E. (1909). *Tidskr. Landbrugets Planteavl,* **16**, 337–358. (Includes a knight's-move Latin square.)

[23] Lindner, C. C. and Mendelsohn, N. S. (1973). *Aequat. Math.,* **9**, 150–156.

[24] Milliken, G. A. and Graybill, F. A. (1972). *Aust. J. Statist.,* **14**, 129–138.

[25] Ortlepp, H. (1957). *Zeit. Landwirtsch. Vers- Untersuchungswes.,* **3**, 136–150.

[26] Preece, D. A., Pearce, S. C., and Kerr, J. R. (1973). *Biometrika,* **60**, 349–358. (Includes a discussion of Latin cubes.)

[27] Scott, R. A. (1932). *Potato Manurial Trials; Details of Experimental Work, Seasons 1930–32.* Supplement to *Tasmanian J. Agric.,* Aug. 1, 1932. (Gives the first semi-Latin squares.)

[28] Stevens, W. L. (1939). *Ann. Eugen. (Lond.),* **9**, 82–93.

[29] "Student" (1926). *J. Amer. Soc. Agron.,* **18**, 703–719. (Mentions an 8 × 8 equal frequency square for four treatments.)

[30] Tedin, O. (1931). *J. Agric. Sci.,* **21**, 191–208. (Discusses the use of knight's-move Latin squares for agricultural experimentation.)

[31] Tukey, J. W. (1955). *Biometrics,* **11**, 111–113.

[32] Vik, K. (1924). *Meld. Nor. Landbrukshøisk.,* **4**, 129–181. (Gave rise to the name "Knut-Vik" square.)

[33] Wallis, W. D. (1972). In *Combinatorics, Room Squares, Sum-Free Sets, Hadamard Matrices* (by W. D. Wallis, A. P. Street and J. S. Wallis), Lect. Notes Math. 292, A. Dold and B. Eckmann, eds. Springer-Verlag, Berlin, pp. 29–121.

[34] Wilk, M. B. and Kempthorne, O. (1957). *J. Amer. Statist. Ass.,* **52**, 218–236.

[35] Williams, E. J. (1949). *Aust. J. Sci. Res.,* **2**, 149–168.

[36] Yates, F. (1951). *Ann. Inst. H. Poincaré,* **12**, 97–112. (Includes a discussion of randomization of a Latin square.)

(ANALYSIS OF COVARIANCE
ANALYSIS OF VARIANCE
BALANCED INCOMPLETE BLOCKS
BLOCKS, RANDOMIZED
CHANGEOVER DESIGNS
CONFOUNDING
CRISS-CROSS DESIGNS
CROSSOVER DESIGNS
DESIGN OF EXPERIMENTS
FACTORIAL DESIGNS
GRAECO-LATIN SQUARES
LATTICE DESIGNS
MAIN-EFFECTS
RANDOMIZATION
ROW-AND-COLUMN DESIGNS
ROW–COLUMN DESIGNS
SPLIT-PLOT DESIGNS
YOUDEN SQUARES)

D. A. PREECE

LATTICE DESIGNS

Lattice designs comprise a class of incomplete block designs* introduced by Yates [31–34] to increase the precision of treatment comparisons in agricultural crop cultivar trials. The designs are also known as quasi-factorial owing to their construction by analogy to confounding* in factorial experiments*. For example, if k^2 treatments are to be compared, one can arrange them as the points of a two-dimensional lattice (thus the term "lattice design"), and regard the points as representing treatment combinations in a two-factor experiment. Confounding main effects of one pseudo-factor with blocks of k experimental units each in one replicate and similarly confounding main effects of the other pseudo-factor in a second replicate will result in a "simple lattice" design. To illustrate, consider nine treatments arranged in the 3×3 array

$$
\begin{matrix}
1 & 2 & 3 \\
4 & 5 & 6 \\
7 & 8 & 9
\end{matrix}
$$

and superimpose on this a purely fictitious 3×3 factorial arrangement of treatments with rows as levels of pseudo-factor A and columns as pseudo-factor B. A 3×3 simple lattice is obtained by constructing one replicate as a replicate of a split-dot design* with levels of A as main plots (blocks) and similarly constructing a second replicate with levels of B as blocks.

The design is said to be balanced (see BALANCED INCOMPLETE BLOCK DESIGNS) if the number of times a pair of treatments appears together in a block is the same for all such treatment pairs. For the 3×3 case, arranging a third replicate as (1 5 9)(2 6 7) (3 4 8) and a fourth as (1 6 8)(1 4 9) (3 5 7), where the parentheses enclose treatments that are to appear together in a block, gives a 3×3 balanced design with every pair of treatments appearing together in a block once. Each of the third and fourth-replicates is constructed by confounding certain $A \times B$ interaction contrasts with blocks.

An elegant systematic description of the design and analysis of m-dimensional lattices (i.e., k^m treatments) in blocks of k^s experimental units, where k is a prime number or power of a prime, has been developed [21–23]. In general, $n \geq m/s$ replicates are sufficient to provide intrablock information on all treatment effects, provided that the confounding arrangements are judiciously chosen. For the prime-powered case, $k + 1$ distinct confounding arrangements exist for k^2 treatments which, if all used in one experiment, result in a balanced lattice. For any integer k, a triple (three-replicate) lattice exists for k^2 treatments, and, apparently, n replicates of k^n treatments confounded in blocks of size k can be constructed by ar-

ranging each replicate so as to have main effects of one of n pseudo-factors unconfounded, but the remaining $n - 1$ confounded. Basic confounding arrangements may be repeated for additional replication, but use of n distinct confounding arrangements, if they exist, is preferable to repetition. "Rectangular lattices" for $k(k + 1)$ treatments in blocks of size k have also been developed for $n = 2, 3, 4$ or $k + 1$ [15, 17–19], $k + 1$ prime powered if $n > 3$.

Confounding* arrangements for square and triple rectangular lattices and the two-restrictional "lattice squares" (to be discussed later) have been conveniently cataloged [1, 2]. Six confounding arrangements of 81 treatments in blocks of 3 may be found in Federer [11], four arrangements of 32 in blocks of 8 in Federer [12, p. 367], and 10 arrangements of k^5 in blocks of size k^2 for $k = 2$ or 3 in Federer and Robson [14]. A randomized layout is obtained by randomly assigning the treatment designations to the actual treatments, randomly allocating the appropriate sets of treatments to the blocks within each replicate, and randomly allocating the appropriate treatments to the experimental units within each block.

Recent developments allow more choice in the number of treatments that can be accommodated. Raktoe and Federer [26] devised a method of constructing "mixed lattice designs" for $k_1^{n_1} \times k_2^{n_2} \times \cdots \times k_q^{n_q}$ treatments where the k_i are distinct primes [27], but a cataloging of confounding arrangements convenient for the unsophisticated user has not been developed. Williams [29] defined "generalized lattice designs" to include any incomplete block design in which the number of treatments is any multiple of the block size k and the design is resolvable, i.e., the blocks can be arranged into n complete replicates. This definition includes the "α-designs" [25].

Analysis of square lattices with one or more regressor variables confounded with the replicates has been described [3, 5]. Cornelius [4] suggests a factor analytic approach to obtain regressor variables from the data in cases where a qualitative treatment factor is confounded with lattice replicates.

STATISTICAL ANALYSIS

Analysis with recovery of interblock information* is done by first estimating the blocks and intrablock error variance components, σ_b^2 and σ_e^2, and then doing a generalized least-squares (GLS) analysis. (Known values for σ_b^2 and σ_e^2 would be preferred but are ordinarily unavailable. However, Kempthorne [21] concludes that the loss of accuracy resulting from the use of estimates will be negligible if there are at least 10 degrees of freedom for blocks adjusted for treatments and replications. Otherwise, an "intrablock analysis" should be done.) Most of the papers that first introduced a particular design, as well as textbooks (e.g., Cochran and Cox [2], Federer [12], Kempthorne [21], and LeClerg et al. [24]), give worked numerical examples using formulations which are manageable on a desk calculator. Unfortunately, this gives the impression that different designs require different algorithms and, for the cubic and higher-dimensional cases, the computations have probably been sufficiently tedious to discourage use of the designs. A fairly simple general method that works for any one-restrictional lattice or resolvable incomplete block design with or without repetition (possibly unequal) of basic confounding arrangements will be described here. Although tedious on a desk calculator because it involves matrix inversions, it is easily accomplished on a computer.

Table 1 is a hypothetical example of a 3×3 lattice in two replications, which, although it yields too few degrees of freedom

Table 1

		Response,[a]		
Replication	Block	Y		
1	1	16.3(1)	4.2(2)	14.0(3)
	2	8.6(4)	10.3(5)	6.2(6)
	3	9.1(7)	13.5(8)	3.6(9)
2	4	13.5(1)	7.6(4)	8.8(7)
	5	6.8(2)	14.4(5)	14.9(8)
	6	8.5(3)	4.6(6)	2.3(9)

[a]Treatment number in parentheses.

for reliable estimation of σ_b^2 and σ_e^2, will serve the purpose of illustration.

Define the following:

n, t, b, k = number of replications, treatments, total number of blocks in the experiment, and block size, respectively, $t = kb/n$

$\mathbf{Y}$ = the above values of Y written as a 18×1 vector

$\mathbf{T}, \mathbf{B}$ = treatments and blocks design matrices ($T_{ij} = 1$ if the ith observation is taken on the jth treatment, zero otherwise; $B_{ij} = 1$ if the observation occurs in the jth block, zero otherwise

$\bar{\mathbf{Y}} = n^{-1}\mathbf{T}'\mathbf{Y}$ = vector of ordinary treatment means

$\mathbf{D}$ = vector of deviations of the Y's from the ordinary treatment means

$\mathbf{N} = \mathbf{T}'\mathbf{B}$ = incidence matrix[1] ($N_{ij} = 1$ if the ith treatment occurs in the jth block, zero otherwise)

$\mathbf{P} = \mathbf{N}'\mathbf{N}$ (P_{ij} is the number of treatments which are common to the ith and jth blocks)

$\mathbf{U} = k\mathbf{I}_b - n^{-1}\mathbf{P}$

$\mathbf{W}_0$ = any generalized inverse of $\mathbf{U}$ such that $\mathbf{W}_0\mathbf{U}\mathbf{W}_0 = \mathbf{W}_0$, i.e., $\mathbf{W}_0$ must be reflexive

$\mathbf{C}_0 = \mathbf{W}_0\mathbf{B}'\mathbf{D}$

$\text{SSB} = \mathbf{C}_0'\mathbf{B}'\mathbf{D} - \text{SSR}$ = blocks (adjusted) sum of squares, with $b - n$ degrees of freedom, where SSR = replications sum of squares

E_b, E_e = blocks and intrablock error mean squares, with expectations $\sigma_e^2 + n^{-1}(n-1)k\sigma_b^2$ and σ_e^2, respectively

$\theta = \hat{\sigma}_e^2 / \hat{\sigma}_b^2 = k(n-1)E_e/n(E_b - E_e)$

$\mathbf{W} = [\theta\mathbf{I}_b + \mathbf{U}]^{-1}$ = the "weighting matrix"

$\mathbf{C} = \mathbf{W}\mathbf{B}'\mathbf{D}$ = vector of corrections for block differences

$\hat{\boldsymbol{\mu}} = n^{-1}(\mathbf{T}'\mathbf{Y} - \mathbf{NC}) = \bar{\mathbf{Y}} - n^{-1}\mathbf{NC}$ = vector of GLS estimates of the treatment means

$$M_T = [n\hat{\mu}'\hat{\mu} - (\theta + k)^{-1}(\hat{\mu}'NN'\hat{\mu} + \theta G^2/nt)]/(t-1),$$ where G is the grand total of all observations, is the mean square owing to heterogeneity among the $\hat{\mu}_i$'s to be tested against E_e, by an approximate F-test,

$$V = n^{-1}E_e(I_t + n^{-1}NWN')$$
= covariance matrix of $\hat{\mu}$.

In practice, it is unnecessary to form matrices T, B, and N since P can be written down directly, an operation such as $B'D$ simply computes the block totals of the elements of D, and $N'\hat{\mu}$ computes, for each block, the sum of the $\hat{\mu}$'s for those treatments which occur in that block. For the example, with $k = 3$, $b = 6$, $n = 2$, $t = 9$, we have matrices

$$\begin{bmatrix} 3 & 0 & 0 & 1 & 1 & 1 \\ 0 & 3 & 0 & 1 & 1 & 1 \\ 0 & 0 & 3 & 1 & 1 & 1 \\ 1 & 1 & 1 & 3 & 0 & 0 \\ 1 & 1 & 1 & 0 & 3 & 0 \\ 1 & 1 & 1 & 0 & 0 & 3 \end{bmatrix}$$

$$0.5 \begin{bmatrix} 3 & 0 & 0 & -1 & -1 & -1 \\ 0 & 3 & 0 & -1 & -1 & -1 \\ 0 & 0 & 3 & -1 & -1 & -1 \\ -1 & -1 & -1 & 3 & 0 & 0 \\ -1 & -1 & -1 & 0 & 3 & 0 \\ -1 & -1 & -1 & 0 & 0 & 3 \end{bmatrix}$$

$$(2/9) \begin{bmatrix} 5 & 2 & 2 & 3 & 3 & 0 \\ 2 & 5 & 2 & 3 & 3 & 0 \\ 2 & 2 & 5 & 3 & 3 & 0 \\ 3 & 3 & 3 & 6 & 3 & 0 \\ 3 & 3 & 3 & 3 & 6 & 0 \\ 0 & 0 & 0 & 0 & 0 & 0 \end{bmatrix}$$

for P, U, and a suitable choice of W_0, respectively. The ordinary treatment means are $\bar{Y} = [14.90, 5.50, 11.25, 8.10, 12.35, 5.40, 8.95, 14.20, 2.95]'$. Now calculate deviations of the observations from the treatment means and sum these deviations for each block. This gives $B'D = [2.85, -0.75, 0.01, -2.05, 4.05, -4.20]'$ and $C_0 = W_0 B'D = [4.2111, 1.8111, 2.3778, 1.4333, 5.5000, 0.0000]'$. Now SSB is the sum of cross products of elements of C_0 and $B'D$ less SSR, i.e., SSB = 30.2178 - 1.0756 = 29.1422. This allows construction of Table 2, an analysis of variance* table, SSR and SST (owing to replications and treatments, ignoring blocks) being computed as in a randomized complete block design. Intrablock error SS is obtained by subtraction.

If $E_b \leqslant E_e$ the lattice structure may be ignored and the experiment analyzed as a randomized complete block* design. In this example, this does not happen and we have $\theta = 0.25697$. Adding this to the diagonal elements of our previously calculated U and inverting gives

$$W =$$

$$\begin{bmatrix} 1.0792 & 0.5100 & 0.5100 & 0.5974 & 0.5974 & 0.5974 \\ 0.5100 & 1.0792 & 0.5100 & 0.5974 & 0.5974 & 0.5974 \\ 0.5100 & 0.5100 & 1.0792 & 0.5974 & 0.5974 & 0.5974 \\ 0.5974 & 0.5974 & 0.5974 & 1.0792 & 0.5100 & 0.5100 \\ 0.5974 & 0.5974 & 0.5974 & 0.5100 & 1.0792 & 0.5100 \\ 0.5974 & 0.5974 & 0.5974 & 0.5100 & 0.5100 & 1.0792 \end{bmatrix}$$

Hence the block corrections, $C = WB'D = [1.430, -0.619, -0.135, -0.975, 2.497, -2.198]'$. For each treatment, compute the mean of the corrections for the blocks which contain that treatment. This gives $n^{-1}NC = [0.228, 1.964, -0.384, -0.797, 0.939, -1.409, -0.555, 1.181, -1.167]'$. Sub-

Table 2 Analysis of Variance

Source of Variation	d.f.	Sum of Squares	Mean Square
Corrected total	17	314.4978	—
Replications	1	1.0756	1.0756
Treatments (ignoring blocks)	8	280.0178	35.0022
Blocks (adjusted)	4	29.1422	7.2856
Intrablock error	4	4.2622	1.0656

tracting these values from the ordinary treatment means gives the GLS estimates $\hat{\mu} = \overline{Y} - n^{-1}NC = [14.672, 3.536, 11.634, 8.897, 11.411, 6.809, 9.505, 13.019, 4.117]'$.

To obtain M_T and the approximate F-test of heterogeneity of GLS means, first compute $N'\hat{\mu} = [29.842, 27.117, 26.641, 33.074, 27.966, 22.560]'$. The sum of squares of these values is 4720.560, and of the $\hat{\mu}_i$'s, 895.640. The correction term is $G^2/nt = 1553.102$. Thus $M_T = \{(2)(895.640) - (3.25697)^{-1} [4720.560 + (0.25697)(1553.102)]\}/8 = 27.421$. Dividing by $E_e = 1.0656$ gives $F = 25.73$, which, with 8 and 4 d.f., is highly significant.

In constructing the estimated covariance matrix $\mathbf{V}$, note that the (ij)th element of the product $\mathbf{NWN'}$ is just the sum of elements of $\mathbf{W}$ corresponding to certain pairs of blocks, those pairs being all possible combinations of a block which contains the ith treatment paired with a block which contains the jth treatment (pairing a block with itself when $i = j$ or when the ith and jth treatments occur in the same block). Then, division by n^2, addition of $1/n$ if $i = j$, and multiplication by E_e gives the variances and covariances. In the example, the variance of $\hat{\mu}_1$ is $E_e[(W_{11} + W_{14} + W_{41} + W_{44})/n^2 + 1/n] = (1.0656)[(1.0792 + 0.5974 + 0.5974 + 1.0792)/4 + 1/2] = 1.4261$, and the covariance of $\hat{\mu}_1$ and $\hat{\mu}_2$ is $E_e(W_{11} + W_{15} + W_{41} + W_{45})/n^2 = (1.0656)(1.0792 + 0.5974 + 0.5974 + 0.5100)/4 = 0.7417$. A similar computation gives the covariance of $\hat{\mu}_1$ and $\hat{\mu}_5$ as 0.5900. In this example, the variances are all equal and the covariances involve only two distinct values, 0.7417 or 0.5900 for two treatments which do or do not, respectively, occur together in a block. Such variances and covariances may be used for making inference concerning specific comparisons among means. The variance of a difference $\hat{\mu}_i - \hat{\mu}_j$ is $V_{ii} + V_{jj} - 2V_{ij}$, e.g., for $i = 1$, $j = 2$, $1.4261 + 1.4261 - 2(-0.7417) = 1.3688$. Variances computed in this way should be assigned degrees of freedom equal to that for intrablock error in the analysis of variance. Often these variances of a differ-

ence are all very similar and, in that event, all treatment differences may be tested against the average variance of a difference. This is equal to 2(average variance − average covariance), which can be computed, without forming $\mathbf{V}$, as $2n^{-1}E_e[1 + (kb - n)^{-1}(\sum_i\sum_j W_{ij}P_{ij} - b^{-1}nk\sum_i\sum_j W_{ij})]$. In this example $\sum_i\sum_j W_{ij}P_{ij} = 30.1788$, $\sum_i\sum_j W_{ij} = 23.3484$, and the average variance of a difference is 1.5205. Dropping the factor of $2n^{-1}$ ($= 1$ in this example) gives the "effective error variance" of the experiment.

In cases where $t < b$, it may be easier to solve the weighted least-squares equations directly, avoiding the need to compute $\mathbf{W}$. Formulas for $\hat{\mu}$ and $\mathbf{V}$, equivalent to those already given, are $\hat{\mu} = [n\mathbf{I}_t - (\theta + k)^{-1}\mathbf{NN'}]^{-1}[\mathbf{T'} - (\theta + k)^{-1}\mathbf{NB'}]\mathbf{Y}$ and $\mathbf{V} = E_e[n\mathbf{I}_t - (\theta + k)^{-1}\mathbf{NN'}]^{-1}$.

We now briefly describe the intrablock analysis, i.e., analysis without recovery of interblock information. Define $\overline{C}_0$ to be the mean of the elements of C_0 and $\mathbf{W}^*$ a $b \times b$ matrix such that $W_{ij}^* = W_{0ij} - \overline{W}_{0i.} - \overline{W}_{0.j} + \overline{W}_{0..}$ Then, considering block effects fixed, the ordinary "least-squares means" are given by $\tilde{\mu} = \overline{Y} - n^{-1}NC_0 + \overline{C}_0$. In this example $\tilde{\mu} = [14.633, 3.200, 11.700, 9.033, 11.250, 7.050, 9.600, 12.817, 4.317]'$. In this case an exact F-test of heterogeneity of treatment means may be obtained by computing a sum of squares for treatments adjusted for blocks as $SST^* = SST + SSB - SSB^*$, where SSB^* is a sum of squares among blocks within replications ignoring treatments. In the example, $SST^* = 280.018 + 29.142 - 92.849 = 216.311$. Division by 8 d.f. and $E_e = 1.0656$ gives $F = 25.374$. The covariance matrix of $\tilde{\mu}$ is obtained as described for $\mathbf{V}$ with $\mathbf{W}^*$ replacing $\mathbf{W}$. For the example, we obtain 0.7696 for the variances and 0.0592 and -0.1184 for the covariance between two treatments which do and do not, respectively, occur together in a block. If $\mathbf{W}_0$ is chosen as the Moore–Penrose pseudo-inverse of $\mathbf{U}$, then $\mathbf{W}^* = \mathbf{W}_0$.

To this writer's knowledge none of the

widely used statistical software packages contains a program using a general algorithm as described here. Computer programs specifically for the analysis of square or rectangular lattices are readily available. Williams [29] gives a procedure for minicomputers, which avoids the need to do any matrix inversion, but it is based on restrictions which appear to rule out any repetition of confounding arrangements. Williams and Ratcliffe [30], by considering the structure of **P**, obtain formulas for $\hat{\mu}$ ($\hat{\tau}$ in their notation) and **V** which avoid the need for matrix **W**, applicable to square and rectangular lattices with equal, but otherwise arbitrary, repetition of basic confounding arrangements.

Incomplete data or covariates in the analysis, which result in loss of some of the nice properties the analysis otherwise would have, are discussed by Cochran and Cox [2], Cornish [7–9], Cox et al. [10], Federer [12], and Robinson and Watson [28]. Cornelius and Byars [5] and Cornelius et al. [6], however, have analyzed such cases as described by Fuller and Battese [16] for linear models with one-fold nested error structure.

TWO-RESTRICTIONAL LATTICES

"Lattice squares" for k^2 treatments and "lattice rectangles" for ks treatments with both row and column confounding are described and illustrated with numerical examples by Cochran and Cox [2], Kempthorne [21], LeClerg et al. [24], Federer and Raktoe [13], and other references cited by these authors. (The present writer has found the analysis-of-variance table in Federer and Raktoe to disagree with results using a general linear models computer program.) For the prime-powered case, a completely balanced lattice square exists in $k + 1$ replicates and, for odd k, semibalanced (every pseudo-effect and interaction confounded once in either a row or a column) in $(k + 1)/2$ replicates. Kempthorne [21] describes how to devise confounding arrangements for lat-

tice square designs for the nonprime case, the number of possible confounding arrangements depending on how many k-level pseudo-factors can be orthogonally superimposed on a $k \times k$ array. The pseudo-factors are defined by superimposing a Latin square*, or Graeco-Latin square* if it exists, on the $k \times k$ array of treatments. Presumably, work of Raktoe and Federer [26] makes possible the construction of mixed lattice rectangles, but explicit confounding arrangements have not been given.

Provided that each pseudo-effect is unconfounded with either rows or columns in at least one replicate, the computing methods for one-restrictional lattices may be extended to two-restrictional lattices with fairly minor changes. First note that, if each replicate is arranged as k rows and m columns, an analysis regarding either rows or columns (but not both) as blocks may be performed exactly as already given for the one-restrictional case, with k replaced by m when rows are regarded as blocks. Let $\text{SSB}_r, \mathbf{B}_r, \mathbf{N}_r, \mathbf{W}_{0r}$ be $\text{SSB}, \mathbf{B}, \mathbf{N}, \mathbf{W}_0$, respectively, for the one-restrictional case with rows regarded as blocks and analogously define $\text{SSB}_c, \mathbf{B}_c, \mathbf{N}_c, \mathbf{W}_{0c}$ when columns are regarded as blocks. To obtain the two-restrictional analysis, let $\mathbf{B} = [\mathbf{B}_r, \mathbf{B}_c]$, $\mathbf{N} = [\mathbf{N}_r, \mathbf{N}_c]$, $\mathbf{P} = \mathbf{N}'\mathbf{N}$, and $\mathbf{W}_0$ a reflexive generalized inverse of $\mathbf{U} = \mathbf{B}'\mathbf{B} - n^{-1}\mathbf{P}$. $\mathbf{B}'\mathbf{B}$ can be written down directly because $\mathbf{B}_r'\mathbf{B}_r = m\mathbf{I}_{nk}$, $\mathbf{B}_c'\mathbf{B}_c = k\mathbf{I}_{nm}$, and $\mathbf{B}_r'\mathbf{B}_c$ is a matrix whose (ij)th element is 1 if the ith row and jth column of the design belong to the same replicate, zero otherwise. Obtain SS (rows and columns adjusted for treatments), SSB_{rc} (say), as $\mathbf{C}_0\mathbf{B}'\mathbf{D} - \text{SSR}$, where $\mathbf{C}_0 = \mathbf{W}_0\mathbf{B}'\mathbf{D}$. Then $\text{SSB}_{r|c} = \text{SS}$ (rows adjusted for treatments and columns) $= \text{SSB}_{rc} - \text{SSB}_c$, and $\text{SSB}_{c|r} = \text{SSB}_{rc} - \text{SSB}_r$. Division of $\text{SSB}_{r|c}$ and $\text{SSB}_{c|r}$ by their d.f., $n(k - 1)$ and $n(m - 1)$, respectively, gives the rows and columns mean squares, E_r, E_c, with expectations $\sigma^2 + \lambda_r\sigma_r^2, \sigma^2 + \lambda_c\sigma_c^2$, where $\lambda_r = [t(n - 1) - \sum_i \mathbf{U}_{rci}\mathbf{W}_{0c}\mathbf{U}_{rci}']/\{n(k - 1)\}$, $\lambda_c = [t(n - 1) - \sum_j \mathbf{U}_{crj}\mathbf{W}_{0r}\mathbf{U}_{crj}']/\{n(m - 1)\}$, $\mathbf{U}_{rci}$ is the ith row and $\mathbf{U}_{crj}'$ the jth column of $\mathbf{U}_{rc} = \mathbf{B}'_r\mathbf{B}_c - n^{-1}\mathbf{N}'_r\mathbf{N}_c$, which is the upper right $nk \times nm$ submatrix of $\mathbf{U}$. If either $E_r \leqslant E_e$ or $E_c \leqslant E_e$ the blocking by rows or columns, as the case may be, is ignored and a one-restrictional analysis with the other factor as blocks is done. Otherwise, let $\theta_r = \hat{\sigma}_e^2/\hat{\sigma}_r^2 = \lambda_r E_e/(E_r - E_e)$, $\theta_c = \hat{\sigma}_e^2/\hat{\sigma}_c^2 = \lambda_c E_e/(E_c - E_e)$, $\mathbf{H} = \mathrm{diag}[\theta_r\mathbf{I}_{nk}, \theta_c\mathbf{I}_{nm}]$, and $\mathbf{W} = (\mathbf{H} + \mathbf{U})^{-1}$. Now $\mathbf{C}$, $\hat{\boldsymbol{\mu}}$, and $\mathbf{V}$ follow as in the one-restrictional case and

$$M_T = \Big[\, n\hat{\boldsymbol{\mu}}'\hat{\boldsymbol{\mu}} - (\theta_r + m)^{-1}\hat{\boldsymbol{\mu}}'\mathbf{N}_r\mathbf{N}'_r\hat{\boldsymbol{\mu}}$$
$$- (\theta_c + k)^{-1}\hat{\boldsymbol{\mu}}'\mathbf{N}_c\mathbf{N}'_c\hat{\boldsymbol{\mu}} + (t - \theta_r\theta_c)$$
$$\times (\theta_r + m)^{-1}(\theta_c + k)^{-1}G^2/nt\,\Big]/(t-1).$$

The average variance of a difference is $2n^{-1}E_e\{1 + [n(t-1)]^{-1}[\sum_i\sum_j W_{ij}P_{ij} - t^{-1}(m^2\sum_i\sum_j W_{rrij} + 2km\sum_i\sum_j W_{rcij} + k^2\sum_i\sum_j \cdot W_{ccij})]\}$, where W_{rrij}, W_{rcij}, W_{ccij} are obtained from the $nk \times nk$ upper left, $nk \times nm$ upper right, and $nm \times nm$ lower right, respectively, submatrices of $\mathbf{W}$.

A direct solution of the weighted least-squares equations, which may be a more efficient method if $t < n(k + m)$, is given by $\hat{\boldsymbol{\mu}} = [n\mathbf{I}_t - \mathbf{T}'\mathbf{QT}]^{-1}\mathbf{T}'(\mathbf{I}_{nt} - \mathbf{Q})\mathbf{Y}$ and $\mathbf{V} = E_e[n\mathbf{I}_t - \mathbf{T}'\mathbf{QT}]^{-1}$, where $\mathbf{Q} = \mathbf{B}(\mathbf{H} + \mathbf{B}'\mathbf{B})^{-1}\mathbf{B}'$. It can be shown that $\mathbf{T}'\mathbf{QT} = (\theta_r + m)^{-1}\mathbf{N}_r\mathbf{N}'_r + (\theta_c + k)^{-1}\mathbf{N}_c\mathbf{N}'_c - n\phi\mathbf{J}_t$, where $\mathbf{J}_t$ is a $t \times t$ matrix of 1's and

$$\phi = \frac{\theta_r(\theta_c + k) + \theta_c(\theta_r + m)}{(\theta_r + m)(\theta_c + k)\big[(\theta_r + m)(\theta_c + k) - t\big]}$$

Elements of $(\mathbf{I}_{nt} - \mathbf{Q})\mathbf{Y}$ are of the form $[Y - (\theta_r + m)^{-1}$ (row total) $- (\theta_c + k)^{-1}$ (column total) $+ \phi$ (replicate total)].

NOTE

1. Some authors use the term "incidence matrix" for what we here define as "design matrix" (*see*, e.g., GENERAL LINEAR MODEL). However, in the literature relating to lattice and incomplete block designs, the term "incidence matrix" is commonly used as defined here (see, e.g., John [20]).

References

[1] Clem, M. A. and Federer, W. T. (1950). Random Arrangements for Lattice Designs. *Iowa Agric. Exper. Stn. Spec. Rep. No. 5* (Catalogs confounding patterns for two-dimensional lattices and lattice squares. Gives 12 randomized layouts of each confounding pattern.)

[2] Cochran, W. G. and Cox, G. M. (1957). *Experimental Designs*, 2nd ed. Wiley, New York. (Good reference text, with numerical examples, for the nonmathematical reader. Does not use matrix notation. Not particularly oriented toward analysis by modern linear-models computer programs.)

[3] Cornelius, P. L. (1976). *J. Amer. Statist. Ass.*, **71**, 445–450. (Theory paper describing analytical methods for lattice designs with one or more concomitant variables confounded with lattice replicates.)

[4] Cornelius, P. L. (1978). *Crop Sci.*, **18** 627–633. (Describes use of factor-analytic methods to obtain concomitant variables confounded with lattice replicates.)

[5] Cornelius, P. L. and Byars, J. (1976). *Crop Sci.*, **16** 42–49. (A crop cultivar trial with plant density levels confounded with lattice replicates.)

[6] Cornelius, P. L., Taylor, N. L., and Anderson, M. K. (1977). *Crop Sci.*, **17**, 709–713. (Gives results of a lattice experiment with much missing data analyzed as a linear model with nested error structure.)

[7] Cornish, E. A. (1940). *Ann. Eugen. (London.)*, **10**, 137–143. (Estimation of missing values.)

[8] Cornish, E. A. (1940). *Ann. Eugen. (London.)*, **10**, 269–279. (Analysis of covariance without recovery of interblock information.)

[9] Cornish, E. A. (1943). *Aust. Counc. Sci. Ind. Res. Bull. 158* (Recovery of interblock information when data are incomplete.)

[10] Cox, G. M., Eckhardt, R. C., and Cochran, W. G. (1940). *Iowa Agric. Exper. Stn. Res. Bull. XXV* (281), 1–66. (Gives examples of analysis of simple and triple lattices and analysis of covariance in a triple lattice.)

[11] Federer, W. T. (1950). *Cornell Univ. Agric. Exper. Stn. Mem. 299* (k^4 treatments in blocks of size k. Gives numerical example for $k = 3$.)

[12] Federer, W. T. (1955). *Experimental Design*. Macmillan, New York. (Very comprehensive text of 1955 state-of-the-art without matrix notation. Slightly more mathematically sophisticated than Cochran and Cox [12].)

[13] Federer, W. T. and Raktoe, B. L. (1965). *J. Amer. Statist. Ass.*, **60**, 891–904. (Prime-powered lattice rectangles. Analysis-of-variance table in the numerical example apparently contains errors.)

[14] Federer, W. T. and Robson, D. S. (1952). *Cornell Univ. Agric. Exper. Stn. Mem. 309.* (k^5 treatments in blocks of size k^2 with numerical example for $k = 2$.)

[15] Folks, J. L. (1959). *Biometrics*, **15**, 74–86. (Four-replicate rectangular lattices.)

[16] Fuller, W. A. and Battese, G. E. (1973). *J. Amer. Statist. Ass.*, **68**, 626–632. (A method for obtaining generalized least-squares estimates in linear models with nested error structure.)

[17] Harshbarger, B. (1947). *Va. Agric. Exper. Stn. Mem. 1.* (Basic paper on simple rectangular lattices.)

[18] Harshbarger, B. (1949). *Biometrics*, **5**, 1–13. (Triple rectangular lattices.)

[19] Harshbarger, B. (1951). *Va. J. Sci.*, **2**, 13–27. ($k + 1$ replicates of a rectangular lattice with a "near balance" property.)

[20] John, P. W. M. (1971). *Statistical Design and Analysis of Experiments.* Macmillan, New York. (Good reference text for the mathematical reader.)

[21] Kempthorne, O. (1952). *Design and Analysis of Experiments.* Wiley, New York. (Excellent theoretical text. Difficult for the nonmathematical reader.)

[22] Kempthorne, O. and Federer, W. T. (1948). *Biometrics*, **4**, 54–79. (Basic theory paper on prime-powered lattice designs, k^m treatments in blocks of size k.)

[23] Kempthorne, O. and Federer, W. T. (1948). *Biometrics*, **4**, 109–121. (Sequel to the preceding reference, k^m treatments in blocks of size k^s, in lattice squares, and in layouts using combinations of Latin square and split-plot principles.)

[24] LeClerg, E. C., Leonard, W. H., and Clark, A. G. (1962). *Field Plot Technique*, 2nd ed. Burgess, Minneapolis. (Gives numerical examples in cookbook style of desk calculator methods for analysis with recovery of interblock information in simple, triple, and balanced lattices, simple and triple rectangular lattices, balanced and semibalanced lattice squares.)

[25] Patterson, H. D. and Williams, E. R. (1976). *Biometrika*, **63**, 83–92. (Describes algorithm for generating "α-designs.")

[26] Raktoe, B. L. and Federer, W. T. (1972). *Aust. J. Statist.*, **14**, 25–36. (Construction of "mixed" lattice designs. The reader of this and the following reference should have a knowledge of such mathematical concepts as Galois fields, rings, and finite Euclidean and projective geometries.)

[27] Raktoe, B. L., Rayner, A. A., and Chalton, D. O. (1978). *Aust. J. Statist.*, **20**, 209–218. (Sequel to preceding reference.)

[28] Robinson, H. F. and Watson, G. S. (1949). *N. C. Agric. Exper. Stn. Tech. Bull. 88.* (Numerical examples of simple and triple rectangular lattices. Analysis of variance and covariance. Analysis with incomplete data.)

[29] Williams, E. R. (1977). *Aust. J. Statist.*, **19**, 39–42. (Suggests a method of analysis of certain types of lattice designs suitable for minicomputers.)

[30] Williams, E. R. and Ratcliffe, D. (1980). *Biometrika*, **67**, 706–708. (Gives general formulae for analysis of square and rectangular lattices with repeats of the design.)

[31] Yates, F. (1936). *J. Agric. Sci.*, **26**, 424–455. (Designs later to be known as square and cubic lattices first proposed. Analysis without recovery of interblock information presented.)

[32] Yates, F. (1937). *Ann. Eugen. (Lond.)*, **7**, 319–331. (The term "lattice design" suggested. Lattice squares introduced.)

[33] Yates, F. (1939). *Ann. Eugen. (Lond.)*, **9**, 136–156. (Analysis with recovery of interblock information. Worked example of 27 treatments in blocks of size 3.)

[34] Yates, F. (1940). *J. Agric. Sci.*, **30**, 672–687. (Recovery of interblock information in lattice squares.)

Bibliography

Anderson, R. L. and Bancroft, T. A. (1952). *Statistical Theory in Research.* McGraw-Hill, New York. (Part II, approximately two-thirds of the book, deals with analysis of experimental models by least squares. Includes sections on lattice designs.)

Bose, R. C. (1975). In *A Survey of Statistical Design and Linear Models*, J. N. Srivastava, ed. North-Holland, Amsterdam, pp. 31–51. (An elegant presentation of the analysis of incomplete block designs, but contains unfortunately many errors. Difficult for the nonmathematical reader. Anova table needs modification for designs in complete replications.)

Cochran, W. G. (1943). *Iowa Agric. Exper. Stn. Res. Bull. XXVI* (318), 727–748. (Lattice squares with fewer than $(k + 1)/2$ replicates.)

Cornish, E. A. (1941). *J. Aust. Inst. Agric. Sci.*, **7**, 19–26. (Analysis of lattice squares with a missing row, column, or treatment.)

Cornish, E. A. (1944). *Aust. Counc. Sci. Ind. Res. Bull. 175.* (Recovery of interblock information in lattice squares with incomplete data.)

Federer, W. T. (1949). *Biometrics*, **5**, 144–161. (k^3 treatments in blocks of size k. Gives numerical example for $k = 3$ with four replicates.)

Federer, W. T. (1950). *Biometrics*, **6**, 34–58. [Design and analysis of a nonprime (6^2) lattice square in three and six replicates. Gives numerical example.]

Federer, W. T. and Raktoe, B. L. (1966). *J. Amer. Statist. Ass.*, **61**, 821–832. (Gives analysis of lattice squares with arbitrary number of replications and arbitrary repetition of basic confounding patterns.)

Harshbarger, B. and Day, L. L. (1952). *Biometrics*, **8**, 73–84. (Latinized rectangular lattices. Special case where each of $k + 1$ rows extending across all replications contains each treatment once.)

Homeyer, P. G., Clem, M. A., and Federer, W. T. (1947). *Iowa Agric. Exper. Stn. Res. Bull. XXVIII* (347), 25–171. (Desk calculator methods illustrated with numerical examples of square and cubic lattices, semi-balanced and balanced lattice squares. Also describes punched card methods for machines of 1947 vintage.)

John, P. W. M. (1980). *Incomplete Block Designs*. Marcel Dekker, New York. (Construction and analysis, without recovery of interblock information, of incomplete block designs. Difficult for the nonmathematical reader.)

Kempthorne, O. (1947). *J. Agric. Sci.*, **37**, 156–162. (Lattice squares with split plots. Example given is one with split-plot confounding.)

Koch, G. G. (1966). *Univ. N. C. Inst. Statist. Mimeo Ser. No. 464*, Chapel Hill, N.C. [Based on lecture notes of R. C. Bose. A thorough exposition in matrix notation of the analysis of incomplete block designs, emphasizing the relationship between recovery of interblock information and generalized least-squares analysis. A few modifications (the ANOVA table and estimator of σ_b^2, for example) are needed for designs in complete replicates.]

Mazumdar, S. (1967). *Ann. Math. Statist.* **38**, 1293–1295. (Mathematical paper on construction of L-restrictional prime-powered lattice designs, applying theorems in projective geometry and group theory.)

Meier, P. (1954). *J. Amer. Statist. Ass.*, **49**, 786–813. (Unequal repetition of two basic confounding patterns.)

Nair, K. R. (1951). *Biometrics*, **7**, 145–154. (Shows that simple rectangular lattices are partially balanced incomplete block designs, but triple rectangular lattices for $k \geqslant 4$ are not.)

Nair, K. R. (1952). *Biometrics*, **8**, 122–125. (Analysis of simple square and rectangular lattices as partially balanced incomplete block designs.)

Nair, K. R. (1953). *Biometrics*, **9**, 101–106. (Describes construction of n confounding arrangements of $k + 1 \times k$ lattices where $k + 1$ is prime, $2 \leqslant n \leqslant k + 1$.)

Raktoe, B. L. (1967). *Ann. Math. Statist.*, **38**, 1127–1141. (Mathematical paper on construction of balanced L-restrictional prime-powered lattice designs.)

Singh, M. and Dey, A. (1979). *Biometrika*, **66**, 321–326. (A generalization of lattice squares or lattice rectangles in which each of s incomplete blocks has p rows and q columns. Describes analysis without recovery of interblock, interrow or intercolumn information.)

White, D. and Hultquist, R. A. (1965). *Ann. Math. Statist.*, **36**, 1256–1271. (Highly mathematical paper applicable to the construction of confounding plans for mixed lattice designs.)

Williams, E. R. (1977). *Biometrics*, **33**, 410–414. (Gives exact formulae for average variances of a difference in rectangular lattices.)

Zelen, M. (1957). *Biometrics*, **13**, 309–332. (Analysis of covariance for incomplete block designs with or without recovery of interblock information.)

Acknowledgment

This article is published as Journal Article 81-3-18 of the Kentucky Agricultural Experiment Station with the approval of the Director.

(ANALYSIS OF VARIANCE
BALANCED INCOMPLETE BLOCK
  DESIGNS
CONFOUNDING
CHANGEOVER DESIGNS
CROSSOVER DESIGNS
DESIGN OF EXPERIMENTS
FACTORIAL DESIGNS
FACTORIAL EXPERIMENTS
FRACTIONAL FACTORIAL DESIGNS
GENERAL BALANCE
GRAECO-LATIN SQUARES
INCOMPLETE BLOCK DESIGNS
INTERACTIONS
LATIN SQUARES,
  LATIN RECTANGLES, ETC.
LEAST SQUARES
MIXED MODELS
PARTIALLY BALANCED
  INCOMPLETE BLOCK DESIGNS
SPLIT-PLOT DESIGNS
STRIP-PLOT DESIGNS
WEIGHTED LEAST SQUARES
YOUDEN SQUARES)

P. L. CORNELIUS

LATTICE DISTRIBUTION

A discrete distribution is a lattice distribution if its discontinuity points are of the form

$$a + kd, \qquad k = 1, 2, \ldots,$$

where a and d are constants ($d > 0$).

The characteristic function ϕ of a lattice distribution has the property that there exists a $t_0 \neq 0$ such that $|\phi(t_0)| = 1$.

The most common lattice distributions are on the nonnegative *unit lattice* $0, 1, 2, \ldots$.

(CHARACTERISTIC FUNCTIONS)

LATTICE SYSTEMS

Lattice systems are a class of random processes indexed by discrete subgroups of R^ν, such as Z^ν (the lattice of ν-tuples of integers). They have their origin in statistical mechanics* and are of special interest as models of critical phenomena*. They are also of some significance as discrete approximations to Euclidean quantum field theories, (*see* QUANTUM PHYSICS AND FUNCTION INTEGRATION) and also in probability theory: the most elementary multitime Markov chains* are included among the lattice gases.

The simplest and most famous of the lattice systems is the *nearest-neighbor Ising model*. Let ν be an integer and Z^ν the lattice of ν-tuples of integers. For each $\alpha \in Z^\nu$, s_α is a random variable taking the values ± 1. To describe the joint probability distributions, we need some auxiliary functions. Given a finite subset Λ of Z^ν, the symbol

$$\sum_{\langle \alpha\gamma \rangle; \alpha, \gamma \in \Lambda}$$

denotes the sum over all those pairs in Λ with $|\alpha - \gamma| = 1$ (Euclidean distance). The *finite volume Hamiltonian* is the function on $\{-1, 1\}^\Lambda$:

$$H_\Lambda(s_\alpha) = -J \sum_{\langle \alpha\gamma \rangle; \alpha, \gamma \in \Lambda} s_\alpha s_\gamma - h \sum_{\alpha \in \Lambda} s_\alpha,$$

where J and h are parameters. The *finite volume Gibbs measure* is the measure, $d\mu_\Lambda$, on $\{-1, 1\}^\Lambda$ giving weight

$$e^{-\beta H_\Lambda(s_\alpha)} / Z_\Lambda \equiv w$$

to the point $\{s_\alpha\}$. Here Z_Λ is a normalization factor called the *partition function* and is given by

$$Z_\Lambda = \sum_{s_\alpha = \pm 1; \alpha \in \Lambda} e^{-\beta H_\Lambda(s_\alpha)}.$$

β is another parameter (which is redundant since it can be absorbed into J and h). (*See* GIBBS DISTRIBUTION.)

One useful way of thinking of this setup is as the model of magnet: each s_α is viewed as a "spin" pointing up $(+1)$ or down (-1). If $J > 0$, the first term in H_Λ describes a tendency (lower energy and correspondingly higher weight in $d\mu_\Lambda$) for neighboring spins to align parallel and the model is that of a *ferromagnet*. The second term in H_Λ can be viewed as the interaction with an externally applied magnetic field: h is then the product of the magnitude of this field and the magnetic moment. If $\beta = 1/kT$ with k Boltzmann's constant and T the temperature, then $d\mu_\Lambda$ is the measure associated to the *canonical ensemble* according to the rules of statistical mechanics.

Another interpretation is as a "lattice gas." We change variables to $\rho_\alpha = \frac{1}{2}(1 + s_\alpha)$ and interpret $\rho_\alpha = 1$ $(s_\alpha = 1)$ as "occupied" and $\rho_\alpha = 0$ $(s_\alpha = -1)$ as "unoccupied." If $z = \exp(-4\nu J\beta + h\beta)$, then

$$w \equiv \tilde{Z}_\Lambda^{-1} \exp\left(-4\beta J \sum_{\langle \alpha\gamma \rangle; \alpha, \gamma \in \Lambda} \rho_\alpha \rho_\gamma\right) \prod_{\alpha \in \Lambda} z^{\rho_\alpha},$$

where $\tilde{Z}_\Lambda^{-1}$ is a normalization factor. [This is not quite true; $\prod_{\alpha \in \Lambda} z^{\rho_\alpha}$ should really be $\prod_{\alpha \in \Lambda} z_\alpha^{\rho_\alpha}$ with $z_\alpha = \exp(-2n_\alpha J\beta + h\beta)$ and n_α the number of neighbors of α lying in Λ. Thus $z_\alpha = z$ except for boundary α's; the role of such boundary terms is described below.] In this view, $d\mu_\Lambda$ is a weight associated to a *grand canonical ensemble* and z is a *fugacity*. $J > 0$ now corresponds to an attraction between particles in the gas.

Still a third interpretation of the model is as that of an alloy with one of two allowed species at each site.

Given a configuration $t = \{t_\alpha\}_{\alpha \notin \Lambda} \in \{-1, 1\}^{Z^\nu \setminus \Lambda}$, we define the Hamiltonian, partition function, and Gibbs measure with *boundary condition* t_Λ as

$$H_\Lambda(s \mid t) = H_\Lambda(s) - J \sum_{\langle \alpha\gamma \rangle; \alpha \in \Lambda, \gamma \notin \Lambda} s_\alpha t_\gamma$$

$$Z_\Lambda(t) = \sum_{s_\alpha = \pm 1; \alpha \in \Lambda} e^{-\beta H_\Lambda(s \mid t)}$$

$$\int f(s) \, d\mu_\Lambda(s \mid t) = Z_\Lambda(t)^{-1} \sum_{s_\alpha = \pm 1; \alpha \in \Lambda}$$

$$\times f(s) e^{-\beta H_\Lambda(s \mid t)}.$$

Note that all those objects depend only on those t_α with dist$(\alpha, \Lambda) = 1$. The α are often called the *boundary* of Λ and denoted $\partial\Lambda$.

Most of the interesting and subtle phenomena in the model are associated with the passage to an infinite volume limit, say by taking Λ through the sequence of hypercubes $[-n, n]^\nu$. This limit is usually called the *thermodynamic limit*. One important quantity is $\lim_{|\Lambda| \to \infty} |\Lambda|^{-1} \ln Z_\Lambda(t)$ (with $|\Lambda|$ the number of points in Λ), which is known to be independent of the boundary condition, t. We will denote this quantity by $p(\beta, h)$, where we imagine fixing $J = 1$ and make explicit the dependence on β and h. In magnetic language, p is a *free energy* (per unit volume) and in the lattice gas language it is a *pressure*. p is jointly convex in β and βh and so jointly continuous. Even though Z_Λ is manifestly analytic in β and h for each finite Λ, p may *not* be smooth. For example, when $\nu = 2$, $p(\beta, h = 0)$ is explicitly known (the celebrated *Onsager solution*). There is a critical value, β_c, of β so that p is real analytic in β away β_c, but $\partial^2 p / \partial \beta^2$ diverges logarithmically at β_c; this is called a *second-order phase transition* in β. As h is varied, an even more interesting situation results; while an explicit formula is not known, the following can be proven: p is jointly real analytic in the region $h \neq 0$, $\beta > 0$, and C^∞ in the region $\beta < \beta_c$. For $\beta > \beta_c$, p is not C^1 and $\lim_{h \downarrow 0} \partial p / \partial h = -\lim_{h \uparrow 0} \partial p / \partial h \neq 0$ [the first equality comes from $p(\beta, -h) = p(\beta, h)$, which follows from $s_\alpha \to -s_\alpha$ symmetry]. A discontinuous first derivative is called a *first-order phase transition* and is discussed further below. Since $\partial p / \partial h$ is a magnetization (per unit volume) in the magnetic phase, the discontinuity above is the *spontaneous mag-*

netization so typical of ferromagnets. In lattice gas language $\partial p / \partial h$ is a density and the discontinuity is that typical of the change of density in passing from a liquid phase to a gaseous phase.

Of special interest is the behavior of p and its derivatives near β_c. A nonrigorous but extremely stimulating and significant method for studying this behavior in arbitrary dimension ν (and other models) is the *renormalization group method*.

Further insight is obtained by the study of the limits of the Gibbs measures $d\mu_\Lambda$. An *equilibrium state* (for fixed β, j, h) is a measure on $\mathscr{X} \equiv \{-1, 1\}^{Z^\nu}$ obtained by taking arbitrary weak limits of $d\mu_\Lambda(\cdot \mid t)$ and then taking arbitrary weak limits of convex combinations of such limits as the t's are varied but p, J, and h held fixed. This procedure yields exactly the measures that obey the so-called *DLR* (for Dobrushin, Lanford, and Ruelle) *equations*: for each finite Λ, the conditional expectations for functions f of the $\{s_\alpha\}_{\alpha \in \Lambda}$ conditioned on the configuration $\{t_\alpha\}_{\alpha \in \Lambda}$ outside Λ is given by

$$E(f \mid t) = \int f(s) \, d\mu_\Lambda(s \mid t)$$

with $d\mu_\Lambda(\cdot \mid t)$ as given before.

For the nearest-neighbor models we have been discussing, the DLR equations say that the conditional expectation of the interior, $E(f \mid t)$, conditioned on the exterior (t), depends only on the boundary (i.e., $\{t_\alpha\}$ with $\alpha \in \partial\Lambda$), so their states describe certain multidimensional Markov processes.

The group of translations of Z^ν acts in a natural way on $\mathscr{X}$; of particular interest are those equilibrium states which are translation invariant. Extreme points of the (weakly closed, convex) set of translation-invariant equilibrium states are often called *pure phases*.

For the two-dimensional model discussed above, the structure of equilibrium states is well understood. For $h \neq 0$ or $\beta \leqslant \beta_c$, $h = 0$, there is exactly one equilibrium state but for $\beta > \beta_c$, $h = 0$, there is a one-parameter family with two pure phases. All these states are translation invariant. The structure need not be so simple: in three or more dimensions it is known that there are nontranslation-invariant equilibrium states for $h = 0$ and β sufficiently large. And for the three-dimensional model with $h = 0$ and $\sum_{\langle \alpha\gamma \rangle} s_\alpha s_\gamma$ replaced by $\sum_{\langle \alpha\gamma\delta\lambda \rangle} s_\alpha s_\gamma s_\delta s_\lambda$ over all sets of four sites forming a planar square, it is known that there are an *uncountable* infinity of pure phases when β is large!

There is a close connection between multiple phases and first-order phase transitions: indeed, one can construct certain Banach spaces of interactions and for an interaction, Φ, a pressure $p(\Phi)$, convex in Φ, and a notion of equilibrium state for Φ so that there is a unique equilibrium state for Φ_0 if and only if p is (Gateaux) differentiable at Φ_0. (*See* STATISTICAL FUNCTIONALS.) In this theory [1, 2], a major role is played by *entropy** and the *Gibbs variational principle*.

The distinct phases that occur in the two-dimensional model are easy to describe. Let $d\mu_{\Lambda,\pm}$ be the finite volume state with all t_α, $\alpha \notin \Lambda$, equal to ± 1. Then one can show that the $d\mu_{\Lambda,\pm}$ have weak limits $d\mu_\pm$. If $h = 0$ and $\beta > \beta_c$, $d\mu_+ \neq d\mu_-$ and these are the two pure phases. Indeed, $\int s_\alpha \, d\mu_+ = \lim_{h\downarrow 0} dp/dh \neq 0$. Thus even though there are only short-range interactions, the setting of all boundary spins has an effect even in the infinite volume limit: this is called *long-range order*; it is a *cooperative phenomenon*. Notice that even though the basic Hamiltonian with $h = 0$ has the symmetry $s_\alpha \rightarrow -s_\alpha$ (all α), the states $d\mu_\pm$ do not have this symmetry (rather this transformation interchanges $d\mu_+$ and $d\mu_-$). This is called the phenomenon of *spontaneously broken symmetry*. These notions, more properly the notion of spontaneous broken continuous symmetries, play a major role in modern theories of elementary particles.

Having described the nearest-neighbor Ising model, let us briefly describe the form of some of the other popular models.

GENERAL ISING MODELS.   Let $J(A)$ be a translation-invariant function on the finite subsets of Z^ν, let

$$\sigma^A = \prod_{\alpha \in A} \sigma_\alpha$$

and take

$$H_\Lambda = - \sum_{A \subset \Lambda} J(A)\sigma^A.$$

Usually, one requires that

$$\sum_{0 \in A} |J(A)|/|A| < \infty$$

to get sensible infinite volume limits, although this condition eliminates the important *Coulomb lattice gases* where cancellations account for a reasonable thermodynamic limit.

ONE-COMPONENT MODELS.   If one relaxes the condition that S_α take the values ± 1 and allow it to be real-valued, one has a larger class of models. To define the model, one needs not only a Hamiltonian, $H_\Lambda(s_\alpha)$, but also a measure $d\gamma$ on R, called the *a priori measure*, to form the Gibbs state

$$Z_\Lambda^{-1} e^{-\beta H_\Lambda(s)} \prod_{\alpha \in \Lambda} d\gamma(s_\alpha).$$

Technically, the theory is very close in spirit to the ordinary Ising model if $d\gamma$ has compact support; otherwise, one deals with *unbounded spins* and there are many technical problems; for example, there can be "spurious" solutions of the DLR equation which are very singular at spatial infinity. Popular choices include the *spin S Ising* model, where s_α takes values $-2S, -2S + 2, \ldots, 2S - 2, 2S$, and the *lattice φ^4 theory*, where $d\gamma(x) = \exp(-ax^4 - bx^2)dx$. The latter is connected with discrete approximations of quantum field theories.

N-VECTOR MODELS. $\vec{s}_\alpha$ is now a vector-valued random variable, say with values in R^N. Particularly interesting is the case where the a priori measure is the isotropic one on the unit sphere S^{N-1}, often called the N-vector model. The case $N = 3$ with Hamiltonian $H_\Lambda = -\sum_{\alpha\gamma} \vec{s} \cdot \vec{s}_\gamma$ is called the *classical Heisenberg model*. In many ways it is a better model of a magnet than the Ising model. The N-vector models with $N \geqslant 2$ and suitable Hamiltonian have the continuous symmetry group, SO(N), of simultaneous rotations of all spins. One interesting aspect of the theory is that while the discrete $s_\alpha \to -s_\alpha$ symmetry of the Ising model is broken in $\nu \geqslant 2$ dimensions, the continuous symmetry of nearest-neighbor N-vector models is broken only in $\nu \geqslant 3$ dimensions.

SPIN GLASSES. . These are a class of Ising models with Hamiltonian $-\sum J_{\alpha-\gamma} s_\alpha s_\gamma$, but now the J's are also random variables.

SIX- AND EIGHT-VERTEX MODELS. The random variables are now indexed by bonds in the lattice rather than sites; a *bond* is a nearest-neighbor pair. The variable takes two values which inform one in which direction to place an arrow on the bond. These are two-dimensional models. In the six-vertex model, only configurations are allowed with exactly two arrows in and two arrows out at each vertex. In the eight-vertex model, one also allows at each vertex the possibility of all arrows in or all arrows out. With various statistical weightings for given vertices these are models of *ferroelectrics*. One interest of these models is that the pressures have been exactly calculated—first by Lieb for six-vertex models and then by Baxter for eight-vertex models.

LATTICE GAUGE MODELS. These are of extreme interest as discrete versions of the (Euclidean region) non-Abelian gauge theories believed to be fundamental to an understanding of elementary particle interactions. Variables are now indexed by *directed* bonds in Z^ν and take values in some Lie group, G. There is the restriction that if α and $-\alpha$ are the same bond with opposite directions, then $s_{-\alpha} = s_\alpha^{-1}$. Each planar square of bonds is called a *plaquette*, P, and one defines $s^P \equiv s_\alpha s_\beta s_\gamma s_\delta$, where a, β, γ, δ is an ordering of successive sides of P directed so that δ comes out of the site where Δ comes in. If φ is a real character of G, then $\varphi(s^P)$ is independent of which bond among $\alpha, \beta, \gamma, \delta$ is put first. The Hamiltonian is

$$H_\Lambda = c \sum_P \varphi(s^P)$$

the sum being over all plaquettes in Λ. No-

tice that if one assigns a group element h_i to each *site* in the lattice and if we map s_α to $h_i s_\alpha h_j^{-1}$ when α runs from site i to site j, then H is left invariant. This is the group of *gauge transformations* of the model.

QUANTUM MODELS. There is a large class of noncommutative models, where the basic variables s_α are operators in some C^*-algebra rather than functions.

References

[1] Israel, R. (1979). *Convexity in the Theory of Lattice Gases*. Princeton University Press, Princeton, N.J. (A specialized reference on the topic of convexity.)

[2] Ruelle, D. (1969). *Statistical Mechanics: Rigorous Results*. W. A. Benjamin, New York. (The classic work on rigorous results. It is not restricted to lattice systems and it is somewhat out of date as regards these systems.)

Bibliography

See the following works, as well as the references just given, for more information on the topic of lattice systems.

Domb, C. and Green, M. S. (1972). *Phase Transitions and Critical Phenomena*, Vol. 1: *Exact Results*. Academic Press, New York. (This collection of articles contains a wealth of information. Later volumes in the series are also of interest.)

Gruber, C., Hintermann, A., and Merlini, D. (1977). *Group Analysis of Classical Lattice Systems*. Lect. Notes Phys., **60**. Springer-Verlag, Berlin. (A specialized reference on the topic of group analysis.)

Simon, B. (In preparation). *The Statistical Mechanics of Lattice Gases*. Princeton University Press, Princeton, N.J. (This book reviews and synthesizes the literature on general Ising, N-vector, and quantum models.)

Stanley, H. E. (1971). *Introduction to Phase Transitions and Critical Phenomena*. Oxford University Press, New York. (Well-written presentation of much of the basic lore of critical phenomena. It is not mathematically rigorous.)

(QUANTUM PHYSICS AND
    FUNCTION INTEGRATION
STATISTICAL MECHANICS
STATISTICAL PHYSICS)

BARRY SIMON

LATTICE TEST *See* GENERATION OF RANDOM VARIABLES

LAWLEY–HOTELLING TEST *See* HOTELLING T TEST

LAWLIKE RELATIONSHIPS

The lawlike relationships of ordinary science differ categorically from the least-squares regression models of statistics. A lawlike relationship is an empirical generalization which holds to a close degree of approximation for a wide range of different sets of data. In contrast, a least-squares regression equation generally aims to provide a best fit for a single set of data.

PROPERTIES OF LAWLIKE RELATIONSHIPS

Science, technology, and everyday life are full of lawlike relationships, i.e., relationships that are known to hold under a wide range of conditions.

As an example, Boyle's law in physics says that for a given amount of gas, variations in its pressure P and volume V will be such that the product $P \times V$ is approximately constant. The relationship $PV \doteq C$ has been found to hold for different gases, for mixtures of gases (e.g., air); for different amounts of gas; for pressure going up and for pressure going down; for different apparatus and different experimenters; 300 years ago and now; morning, noon, and night; in America, Great Britain, and elsewhere; and so on. Such empirical generalizations usually come about by slowly piecing together the results of almost innumerable experiments and observational studies.

The laws of science are not, however, universal. Typically, $PV \doteq C$ does *not* hold when the temperature changes, when there is a chemical reaction, when there is a leak in the apparatus or physical absorption or condensation of the gas, or when we tried to prove the law at school.

Such laws are in any case deliberate oversimplifications, implying consistent biases. Boyle's law is defined as holding for perfect gases, with perfect gases being defined as substances for which Boyle's law holds.

Correction factors have to be developed, e.g., van der Waals' equation $(P + A/V^2)(V - B) \doteq$ constant for Boyle's law, where A/V^2 is a correction for the mutual attraction of the molecules of the gas, and B a correction term for their volume. This still does not fit perfectly, but the less accurate statement $PV \doteq C$ is used whenever reasonably possible because it is much simpler—one does not have to know the size of the molecules or the values of A and B.

Since the laws of science are not really true, they cannot in themselves be directly causal. They are certainly not causal in terms of the variables appearing in the equation: In Boyle's law, pressure does not cause volume, nor pressure volume. Nor even do *changes* of pressure directly cause *changes* of volume. Instead, if a piston in a suitable apparatus is moved, it does so in such a way that the product PV remains approximately constant. The laws of science are therefore essentially descriptive. But this need not stop us from trying to hypothesize underlying causal mechanisms. (*See* CAUSATION.)

Scientific laws are also not necessarily theory based. Good theory often lags behind. For Boyle's law it took 70 years for it to be generally accepted by the scientific community, and 200 years before a reasonably convincing theoretical explanation first came about (in terms of molecular movements and the kinetic theory of gases).

In summary, the lawlike relationships of science are descriptive generalizations, often at quite a low level. But the variables that do *not* appear in the equation greatly aid our understanding, e.g., that the type of gas, the type of apparatus, etc., do not matter. Such relationships are what we use to predict and control. They are also the building bricks of higher-level theory and explanation.

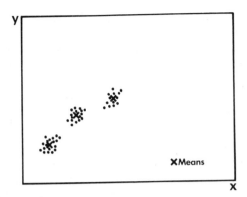

Figure 1 Two or more sets of data.

BETWEEN-GROUP ANALYSIS

To derive a lawlike relationship in two variables x and y, the data must consist of more than one set of readings, i.e., data obtained under different conditions of observation. The experimenter or observer must therefore be able to collect readings in x and y under different conditions, as shown in Fig. 1.

If the means of k different sets of data $(\bar{x}_1, \bar{y}_1), (\bar{x}_2, \bar{y}_2), \ldots, (\bar{x}_k, \bar{y}_k)$, all lie on a straight line (or simple curve), the straight line (or curve) through these means, say $\bar{y} = \alpha + \beta\bar{x}$, is then essentially the line to fit. (A slightly different equation may ultimately be adopted because it is simpler, e.g., that in the inverse-square law of gravity the pull between two bodies varies as $1/d^2$, and not as $1/d^{1.96}$, say).

The coefficients α and β here depend only on the systematic differences between the different pairs of mean values in Fig. 1—the "between-group differences." The individual readings within each set of data—the "within-group" differences—do not directly affect the equation to be fitted. The procedure is therefore referred to as between-group analysis or BGA.

The residual scatter can then be described for the n_1 pairs of individual readings (x, y) in the first set of data, and separately for the n_2 pairs of readings (x, y) in the second set of data, etc., by calculating the mean deviation or the standard deviation of the residuals $(y - \alpha - \beta x)$ in each data set. If the

residual standard deviations are much the same across the different data sets (either in absolute or relative terms), they can be summarized by a single figure, e.g., that $y = \alpha + \beta x$ holds within ± 2 or within $\pm 10\%$.

In practice, the means $(\bar{x}, \bar{y})$ of different sets of data usually do not lie exactly on a straight line (or simple smooth curve), not even for large samples. We may nonetheless fit a straight line (or simple curve) as a deliberately oversimplifying approximation. The problem here is not statistical in the traditional sense because there is no "distribution" of different pairs of means $(\bar{x}, \bar{y})$, nor is it a matter of sampling.

An initial "working solution" can be obtained objectively by picking on the two extreme pairs of means for the slope β, say $(\bar{x}_1, \bar{y}_1)$ as the lowest and $(\bar{x}_k, \bar{y}_k)$ as the highest. This gives an estimate

$$\beta = (\bar{y}_k - \bar{y}_1)/(\bar{x}_k - \bar{x}_1).$$

(If either of the extremes is not unambiguous, one can group the more extreme sets and work out an overall mean for each and fit the slope to these.) The intercept coefficient, α, can be determined by putting the line through the (unweighted) overall means of all the k data sets.

In general, the fit of the resulting line (or curve) to the means $(\bar{x}_1, \bar{y}_1)$ will be approximate rather than exact. There will be some ambiguity over what line precisely to use. This is only because one is deliberately forcing a straight line on to nonlinear data.

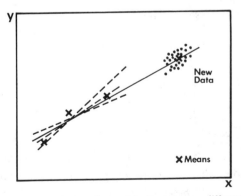

Figure 2 Reducing the ambiguity between different possible lines by more data.

The initial working solution can be adjusted as further data become available, especially data outside the initial range of variation, as illustrated in Fig. 2. Adjustments may also be made so that the equation fits in with other results, so as to provide coherent theory.

Prior Knowledge

One seldom has to derive a lawlike relationship from scratch. In any well-developed area, there generally will be a previous result, e.g., that for children's heights and weights h and w, the means $\bar{h}$ and $\bar{w}$ of different age groups are related by the equation

$$\log \bar{w} = 0.02\bar{h} + 0.76 \pm 0.01,$$

(see, e.g., Ehrenberg [3–5]). One can therefore analyze any new data accordingly: The question is not what line to fit to the new data, but whether or not the earlier relationship holds again, within the average limits of about ± 0.01.

If it does one has a further generalization to the new conditions. The more these conditions are spelt out, the more effective is the further generalization (e.g., that it also holds in Australia for 5-year-old orphans measured at 3 o'clock in the morning).

Nonlinear Relationships

Observable relationships usually do not follow a linear pattern, except possibly as a simplifying approximation over a limited range. Curved relationships are sometimes derived by transforming one or other of the variables as in the relationship $\log w = 0.02h + 0.76$, thus reducing the analysis problem once more to a linear form. For example, Boyle's law can be thought of as linear in the variables P and $1/V$, or in $\log P$ and $\log V$.

In many cases the form of a relationship is suggested by subject-matter theory, often involving advanced mathematics. But at a lower level, we may conjecture that weight increases *proportionately* with height (thus suggesting a $\log w$ transformation) or as the *cube* of height (thus suggesting $\sqrt[3]{w}$). In the early stages of studying a relationship, a suitable mathematical model has usually to be guessed at, often on a trial-and-error basis.

Multivariable Relationships

All relationships involve many variables. With a two-variable relationship like $\log w = 0.02h + 0.76$, other variables, such as age, sex, race, country, time, etc., remain implicit: The equation holds for boys and girls, for Chinese, black, and white children, and so on; i.e., it holds *despite* variation in these other factors.

Sometimes the parameters of a relationship vary symmetrically between different data sets, in a way that is systematically related to another variable. Thus the constant C in Boyle's law $PV \doteqdot C$ is found to vary with temperature T. This leads to the three-variable gas law $PV \doteqdot RT$ or $PV/T \doteqdot R$, where R is again a constant for a given body of gas. R in turn is found to vary for different amounts or kinds of gas in such a way that $PV/T = 1.99M$, where M is the molecular weight of the gas (i.e., approximately 2 for hydrogen, 16 for oxygen, and so on).

Another way in which additional variables can enter an equation is by accounting for some of the observed deviations. An example is van der Waals' equation $(P + A/V^2)(V - B) \doteqdot C$ mentioned earlier.

STATISTICAL METHODS

The derivation of the lawlike relationships of ordinary science differs radically from most traditional *statistical* methods of estimating relationships in the type of question asked, in the type of data analyzed, in the methods of analysis, and in the results.

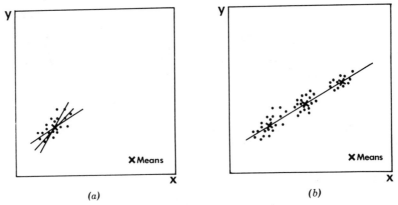

Figure 3   (a) Different possible lines for a given set of data; (b) at most one line for three or more sets of data.

A typical situation considered in statistics is a bivariate normal distribution in x and y. Here many different straight lines could be fitted (e.g., by eye), as shown in Fig. 3a. The question then is to pick one of these equations. This is usually done by using the criterion of least squares. But this approach generally gives at least two regression equations, y on x and x on y. Furthermore, markedly different equations give almost as good a fit as the least-squares regression equation itself [6].

In contrast, when considering a lawlike relationship we must have two or more different sets of data, with differing means. The question then is what straight line (if any) goes through the different pairs of means, as shown in Fig. 3b. The classical problems in regression of deciding between the regressions of y on x and x on y, and between so-called independent and dependent variables, then disappear.

One Variable Controlled

The contrast between lawlike relationships and regression equations is clarified by the case where one variable, x say, is "controlled," e.g., in some experimental setup. There would be n_i readings of y taken at a certain controlled value x_i, with k different controlled values of x_i ($i = 1$ to k).

The data then consist of k different populations (or samples), one for each "con-

trolled" value of x_i. The means of each population are $(\bar{x}_i, \bar{y}_i)$, where $\bar{x}_i \equiv x_i$. If the means $(\bar{x}_i, \bar{y}_i)$, $(\bar{x}_2, \bar{y}_2)$, etc., all lie on a straight line, that is the equation to be fitted. If the means do not lie exactly on a straight line, then with large-sample data there is no exact linear relationship. But one may again decide to simplify by forcing a straight line onto the data, as discussed in the section "Between-Group Analysis" and illustrated in Fig. 4. This case makes it clear that in BGA, the covariation of x and y within each data set cannot directly influence the line to be fitted to the data as a whole.

In the exact linear case, least-squares regression analysis will give the same answer. But this is trivial because there is no problem of choice. Only one line can be fitted,

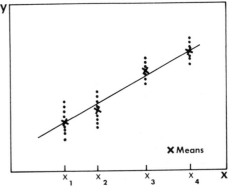

Figure 4   Observations in x and y for controlled values of x.

and no special technique or least-squares principle is needed. If the means are nonlinear, the assumptions of linear least-squares regression analysis do not hold and the analysis would not provide a "best" solution.

Errors in the x-Variable

Usually, there are errors in the controlled x-variable. For example, when a drug is prescribed for different groups of patients at dosages like 5 mg, 12 mg, 27 mg, etc., the actual amounts measured out and/or consumed will vary about these nominally "controlled" values. Or when people's ages are expressed as age last birthday, their true age can be greater by up to 12 months.

When the "true" measurements are known, the data no longer consist of vertical y-arrays at the controlled values of x. Instead, there would be bivariate distributions as in Fig. 5.

With such data, the between-groups type of analysis as discussed in the section "Between-Group Analysis" applies, while standard least-squares regression analysis does not [2].

Functional Relationships

Lawlike relationships resemble the functional relationship of classical statistics. In functional relationships each variable, x and y, is made up of a common "true" part and an independent "error" form, i.e.,

$$x = \text{"true value of } x\text{"} + \text{"error of } x\text{,"}$$

$$y = \alpha + \beta \text{ "true value of } x\text{"} + \text{"error of } y\text{,"}$$

the errors of x and y being assumed independent of each other.

In the statistical literature, functional analysis is generally carried out on a single set of data. But the model is then indeterminate (i.e., there are too many parameters to estimate), unless there is extraneous information.

This extraneous knowledge is sometimes assumed to be prior knowledge of the relative size of the error of measurement of x and y. But there is no reason why the errors in the model above should merely consist of errors of measurement or why these should determine the slope β. But if we consider the functional model in the context of two (or more) different sets of data with means $(\bar{x}_1, \bar{y}_1)$ and $(\bar{x}_2, \bar{y}_2)$, this provides the extraneous information required: the line to be fitted to the first set of data has also to go through $(\bar{x}_2, \bar{y}_2)$.

Wald [7] suggested dividing the given set of data into two subgroups, and Bartlett [1] later suggested forming *three* subgroups. An equation can then be fitted to the different pairs of subgroup means $(\bar{x}_i, \bar{y}_i)$ essentially as in the section "Between-Group Analysis." However, this division into two or three subgroups is arbitrary. The Wald–Bartlett solutions are therefore statistical artifacts and have never been widely used. The contrast is with two or more sets of data collected under different observational conditions, which leads to an empirical generalization.

Analysis of Covariance*

The kind of grouped data relevant to lawlike relationships also arises in classical statistics when considering the analysis of covariance. But the emphasis is usually on comparing the within-groups (or within-cell) regressions with each other, rather than on calculating and interpreting the way the subgroup means $\bar{x}_i$ and $\bar{y}_i$ vary together.

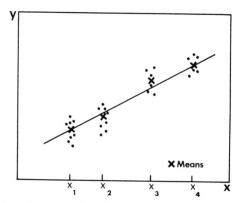

Figure 5 Observations in x and y with errors in x for nominally controlled values of x.

References

[1] Bartlett, M. S. (1949). *Biometrics*, **5**, 207–212.

[2] Berkson, J. (1950). *Biometrics*, **6**, 432–434.

[3] Ehrenberg, A. S. C. (1968). *J. R. Statist. Soc. A*, **131**, 280–302, 315–329.

[4] Ehrenberg, A. S. C. (1975). *Data Reduction*. Wiley, New York (revised reprint, 1978).

[5] Ehrenberg, A. S. C. (1982). *A Primer in Data Reduction*. Wiley, New York.

[6] Ehrenberg, A. S. C. (1982). *J. R. Statist. Soc. A*, **145**, 364–366.

[7] Wald, A. (1940). *Ann. Math. Statist.*, **11**, 284–300.

(ANALYSIS OF VARIANCE
CALIBRATION
STATISTICAL MODELING
STRUCTURAL RELATIONSHIPS)

A. S. C. EHRENBERG

LAW OF DIMINISHING MARGINAL UTILITY   *See* UTILITY

LAW OF SMALL NUMBERS

A term coined by L. von Bortkiewicz* [3]. Given a sequence of independent Bernoulli* trials with probability of success p_k at the kth trial satisfying $kp_k = \lambda$, $k = 1, 2, \ldots$, with a constant λ *independent* of k, then the total number S_n of successes up to the nth trial converges in distribution as $n \to \infty$ to the Poisson distribution* with mean value λ. Additional information can be found in Johnson and Kotz [1] and Maistrov [2].

References

[1] Johnson, N. L. and Kotz, S. (1969). *Distributions in Statistics: Discrete Distributions*. Wiley, New York, Chap. 4, Sec. 2.

[2] Maistrov, L. E. (1974). *Probability Theory: A Historical Sketch*, S. Kotz, trans./ed. Academic Press, New York.

[3] von Bortkiewicz, L. (1898). *Das Gesetz der kleinen Zahlen*. B. G. Teubner, Leipzig.

(BORTKIEWICZ, L. VON
LIMIT THEOREMS
POISSON DISTRIBUTIONS)

LAW OF THE ITERATED LOGARITHM

The *law of the iterated logarithm* (LIL), in its classical form, deals with sums of independent and identically distributed random variables X_i, $i = 1, 2, 3, \ldots$ with $EX_1 = \mu$ and $\operatorname{var} X_1 = E(X_1 - \mu)^2 = \sigma^2 < \infty$. Then, if $S_n = \sum_{i=1}^n X_i$, the classical LIL, which is due to Hartman and Wintner [9], tells us that

$$n^{-1}S_n = \mu + \zeta(n)\left(2\sigma n^{-1}\log\log n\right)^{1/2}$$

where $\zeta(n)$ has its set of almost sure (a.s.) limit points confined to the interval $[-1, 1]$ with $\liminf_{n\to\infty} \zeta(n) = -1$ a.s. and $\limsup_{n\to\infty} \zeta(n) = +1$ a.s. Thus the LIL can be interpreted as a rate of convergence result about the strong law of large numbers* (which asserts that $n^{-1}S_n \to \mu$ a.s. as $n \to \infty$). It provides a detailed assessment of the decreasing fluctuations about μ of almost all sample paths $n^{-1}S_n$, $n = 1, 2, 3, \ldots$. It represents a significant theoretical achievement by virtue of its sharpness, but its practical role is mostly that of a supplement.

The cited version of the LIL above can be restated in a number of equivalent forms. In particular, taking $\mu = 0$ and $\sigma = 1$ without loss of generality and writing $t_n = (2n\log\log n)^{1/2}$,

$$\liminf_{n\to\infty} t_n^{-1}S_n = -1 \text{ a.s.,}$$

$$\limsup_{n\to\infty} t_n^{-1}S_n = +1 \text{ a.s.} \tag{1}$$

and, for $\epsilon > 0$,

$$P\left(S_n > (1 + \epsilon)t_n \text{ i.o.}\right) = 0,$$

$$P\left(S_n > (1 - \epsilon)t_n \text{ i.o.}\right) = 1,$$

$$P\left(S_n < -(1 + \epsilon)t_n \text{ i.o.}\right) = 0,$$

$$P\left(S_n < -(1 - \epsilon)t_n \text{ i.o.}\right) = 1, \tag{2}$$

(Here "i.o." denotes infinitely often.) These last two equivalent forms (1) and (2) suppress the role of the LIL as a rate of convergence result about the strong law of large numbers.

The LIL has its origins in work of Hausdorff [10] and Hardy and Littlewood [8] on the properties of "normal numbers" (i.e., those for which the decimal expansion results in equal limiting relative frequencies for the occurrence of digits). In that context its classical form was discovered by Khintchine [14]. The historical background is discussed in Lamperti [16, Chap. 2, Sec. 11] and Feller [2, Chap. 8].

Perhaps the most profound of the generalizations of the classical LIL are to what are termed functional LILs or invariance principles for the LIL. The first and simplest of these is the following result due to Strassen [22]: Let X_i, $i = 1, 2, 3, \ldots$, be independent and identically distributed random variables with zero mean and unit variance; write $S_n = \sum_{i=1}^{n} X_i$, $n \geq 1$, and let $S(t)$ be the random polygon obtained by linearly interpolating between $S_0 = 0, S_1, S_2 \ldots$. Then, setting

$$f_n(t) = (2n \log \log n)^{1/2} S(nt), \qquad 0 \leq t \leq 1,$$

the sequence of functions $\{ f_n(t) \}$ is relatively compact in the topology of uniform convergence and has as its set of almost sure limit points the set K of all real absolutely continuous functions $h(t)$, $0 \leq t \leq 1$, with

$$h(0) = 0, \qquad \int_0^1 \left(\dot{h}(t) \right)^2 dt \leq 1,$$

the dot denoting a derivative. This result, which can be used to generate many interesting special cases, is comprehensively discussed in Freedman [4, Chap. 1, Sec. 8] and Stout [21, Chap. 5, Sec. 3]. In addition, many functional LILs, including that above, are subsumed by what are called strong invariance principles* which involve the construction of a random function out of the random variables under consideration and its approximation by Brownian motion* and

from which results of strong law of large numbers and central limit type can also be extracted. A survey of results on strong invariance principles appears in Heyde [12].

Many other generalizations of the classical LIL also occur in the literature. For example, results for martingales* which subsume previous results for sums of independent but not identically distributed random variables are given in Stout [21, Chap. 5] and (in functional LIL form) in Hall and Heyde [7, Chap. 4]. LILs for various weakly dependent processes are treated in Philipp and Stout [19] and Hall and Heyde [7, Chap. 5], while the cases of random vectors with values in Euclidean space and Hilbert space are considered in Berning [1] and the Banach valued case in Kuelbs and Philipp [15].

Analogs of the LIL occur in a very wide variety of contexts such as for empirical processes* (e.g., Finkelstein [3], and for a survey including more recent work Gaenssler and Stute [5]), stochastic approximation procedures* (e.g., Heyde [11]), and in renewal theory* (e.g., Iglehart [13]), record value theory* (e.g., Neuts [18] and for more recent work Galambos [6, Chaps. 4, 6]), theory of counters* (e.g., Takács [23]) and adaptive control theory* (e.g., Mandl [17]). Furthermore, ideas related to the LIL have also been used in a variety of statistical contexts. For example, Robbins [20] discusses confidence statements of the form $P(p \in I_n(\epsilon)$ for every $n \geq 1) \geq 1 - \epsilon$, where p is a parameter, $\{ I_n \}$ a sequence of intervals, and $0 < \epsilon < 1$ is arbitrary. Such results can be obtained from "inversion" in statements of the kind $P(S_n \leq c_n$ for some $n \geq 1) \leq \epsilon$, $\{ S_n \}$ denoting successive sample sums.

References

The books by Freedman, Galambos, Hall and Heyde, Lamperti, and Stout are aimed at a graduate-level audience, while that by Feller is accessible at the undergraduate level.

[1] Berning, J. A., Jr. (1979). *Ann. Prob.*, **7**, 980–988.

[2] Feller, W. (1968). *An Introduction to Probability Theory and Its Applications*, Vol. 1, 3rd ed. Wiley, New York.

[3] Finkelstein, H. (1971). *Ann. Math. Statist.*, **42**, 607–615.

[4] Freedman, D. (1971). *Brownian Motion and Diffusion*. Holden-Day, San Francisco.

[5] Gaenssler, P. and Stute, W. (1979). *Ann. Prob.*, **7**, 193–243.

[6] Galambos, J. (1978). *The Asymptotic Theory of Extreme Order Statistics*. Wiley, New York.

[7] Hall, P. and Heyde, C. C. (1980). *Martingale Limit Theory and Its Application*. Academic Press, New York.

[8] Hardy, G. H. and Littlewood, J. E. (1914). *Acta Math.*, **37**, 155–239.

[9] Hartman, P. and Wintner, A. (1941). *Amer. J. Math.*, **63**, 169–176.

[10] Hausdorff, F. (1913). *Grundzüge der Mengenlehre*. Veit, Leipzig.

[11] Heyde, C. C. (1974). *Stoch. Processes Appl.*, **2**, 359–370.

[12] Heyde, C. C. (1981). *Int. Statist. Rev.*, **49**, 143–152.

[13] Iglehart, D. L. (1971). *Zeit. Wahrscheinlichkeitsth. verwand. Geb.*, **17**, 168–180.

[14] Khintchine, A. (1924). *Fund. Math.*, **6**, 9–20.

[15] Kuelbs, J. and Philipp, W. (1980). *Ann. Prob.*, **8**, 1003–1036.

[16] Lamperti, J. (1966). *Probability*. W. A. Benjamin, New York.

[17] Mandl, P. (1977). In *Information Theory, Statistical Decision Functions, Random Processes*, Vol. A, J. Kozesnik, ed. D. Reidel, Dordrecht, Holland, pp. 399–410.

[18] Neuts, M. F. (1967). *J. Appl. Prob.*, **4**, 206–208.

[19] Philipp, W. and Stout, W. F. (1975). Almost Sure Invariance Principles for Partial Sums of Weakly Dependent Random Variables. *Mem. Amer. Math. Soc. No. 161*, iv + 140 pp.

[20] Robbins, H. (1970). *Ann. Math. Statist.*, **41**, 1397–1409.

[21] Stout, W. F. (1974). *Almost Sure Convergence*. Academic Press, New York.

[22] Strassen, V. (1964). *Zeit. Wahrscheinlichkeitsth. verwand. Geb.*, **3**, 211–226.

[23] Takács, L. (1958). *Ann. Math. Statist.*, **29**, 1257–1263.

(ADAPTIVE STATISTICS
CONVERGENCE OF SEQUENCES
 OF RANDOM VARIABLES
INVARIANCE PRINCPLES IN STATISTICS
LIMIT THEOREM, CENTRAL
MARTINGALES
PROCESSES, EMPIRICAL
RECORDS
RENEWAL THEORY
STOCHASTIC APPROXIMATION
STOCHASTIC PROCESSES
STRONG LAW OF LARGE NUMBERS)

C. C. HEYDE

LAWS OF ERROR I: DEVELOPMENT OF THE CONCEPT

INTRODUCTION AND SUMMARY

The *error* of a measurement or observed value Y is, by definition, the difference $Y - \tau$ between the measurement or observed value and the *true value* τ of the quantity concerned. The error of a particular measurement or observed value, y, is, therefore, a fixed number, $y - \tau$. The numerical magnitude and sign of this number are ordinarily unknown and unknowable, because τ, the true value of the quantity concerned, is usually unknown and unknowable.

"Laws of error," i.e., probability distributions assumed to describe the distribution of the errors arising in repeated measurement of a fixed quantity by the same procedure under constant conditions, were introduced in the latter half of the eighteenth century to demonstrate the utility of taking the arithmetic mean* of a number of measurements or observed values of the same quantity as a good choice for the value of the magnitude of this quantity on the basis of the measurements or observations in hand.

The practice of taking the arithmetic mean of two or more measurements or observed values of a single quantity as *the* value of the quantity indicated by these measurements or observations seems to have originated in western Europe more or less simultaneously in several fields—astronomy, metallurgy, navigation—in the latter half of the sixteenth century. Until then the almost universal practice seems to have been to select from among the measurements or observations in hand the one considered to be the "best," e.g., closest to the value adopted by some respected authority, or obtained under optimal measuring or observing conditions, or when the greatest care was exercised, or when the measuring or observing procedure seemed to be behaving especially well. An exception was the brief use, by some eleventh-century Moslem scientists, of the value midway between the extremes, i.e., the midrange*, on the grounds that this is the value from which the largest of the discrepancies of the individual values are minimized, and is in this sense the value with respect to which they are collectively most concordant [1, pp. 89, 202; 9, p. 32].

By the eighteenth century the practice of taking the arithmetic mean "for the truth" had become fairly widespread. Nonetheless, Thomas Simpson (1710–1761) wrote to the president of the Royal Society in March 1755 that some persons of considerable note maintained that one single observation, taken with due care, was as much to be relied on as the mean of a great number. As this appeared to him to be a matter of much importance, he said that he had a strong inclination to ascertain whether by the application of mathematical principles the utility and advantage of the practice might be demonstrated [27]. Measurements, and functions of measurements, such as their arithmetic means, are not amenable to mathematical theory, however, as long as individual measurements are regarded as unique entities, that is, as *fixed numbers* $y_1, y_2, \ldots$. A mathematical theory of measurements, and of functions of measurements, is possible only when particular measurements $y_1, y_2, \ldots$ are regarded as instances characteristic of hypothetical measurements $Y_1, Y_2, \ldots$ that might have been, or might be, yielded by the same measurement process under the same circumstances. Consequently, Simpson hypothesized that the respective chances of the different errors to which any single observa-

tion is subject could be expressed as a discrete probability distribution of error, and introduced the first two discrete "laws of error."

Other discrete laws of error were proposed and studied by Lagrange [10]; continuous laws of error by Simpson [28], Lambert* [12], Laplace* [13, 14], Lagrange* [10], and D. Bernoulli* [2] culminating in the quadratic exponential law of Gauss [6],
$$f_X(x) = (h/\sqrt{\pi})\exp(-h^2x^2),$$
upon which Gauss based his first formulation of the method of least squares, which became almost universally regarded in the nineteenth century as "*the* law of error" (see Fig. 1).

Alternative derivations of Gauss's law were published by Adrian, Laplace, Hagen, Bessel, and Herschel (*see* LAWS OF ERROR II: THE GAUSSIAN DISTRIBUTION). Moderately large scale comparisons of Gauss's law of error with observed distributions of the errors of repeated measurements were published by Bessel, C. S. Peirce, and Airy. The agreement of the observed and theoretical frequencies in these studies were considered to be good, and they were widely interpreted to provide empirical "proof" of the validity of Gauss's law of error as *the* law of error of real-life measurements and observations. Indeed, the French physicist Gabrielle Lippmann (1845–1921) said to the French mathematician Henri Poincaré (1854–1912) one day apropos this law of error: "All the world believes it firmly, because the mathematicians imagine that it is a fact of observation, and the observers that it is a theorem of mathematics" [19, p. v; 20, p. 171].

In due course, however, the work of the Danish astronomer Thorwald Nicolai Thiele (1838–1910), the English biometricians Sir Francis Galton* (1822–1911) and Karl Pearson* (1857–1936), the German economist-demographer Wilhelm Lexis* (1837–1914), the German psychophysicist Gustav Theodore Fechner (1801–1887), and others, shattered belief in the Gaussian or normal distribution as a universal law of errors [*see* LAWS OF ERROR III: LATER (NON-GAUSSIAN) DISTRIBUTIONS]. Today when

LAWS OF ERROR

When	Who	Name	Analytical Expression	Graphical Expression
			Limited Range	
1756 1776	Simpson Lagrange	Discrete Rectangular or Discrete Uniform	$P_r[X=x] = \dfrac{1}{2a+1}$ $x = -a, -a+1, \dots, -1, 0, 1, \dots, a-1, a$	$a=5$
1756 1776	Simpson Lagrange	Discrete Isosceles Triangle	$P_r[X=x] = \dfrac{(a+1)-|x|}{(a+1)^2}$ $x = -a, -a+1, \dots, -1, 0, 1, \dots, a-1, a$	$a=5$
1757 1776	Simpson Lagrange	Continuous Isosceles Triangle	$f_X(x) = \dfrac{1}{a^2}(a-x)$ $-a \le x \le a$	$a=1$
1765 1778	Lambert Daniel Bernoulli	Continuous (Flattened) Semi-circular	$f_X(x) = \dfrac{2}{\pi a^2}\sqrt{a^2-x^2}$ $-a \le x \le a$	$a=1$
1776	Lagrange	Continuous Rectangular or Continuous Uniform	$f_X(x) = \dfrac{1}{a+b}$ $-a \le x \le b$	$a=1$ $b=1.5$
1776 (1781)	Lagrange (Laplace)	Continuous (Asymmetric) Parabolic	$f_X(x) = \dfrac{3}{4a^3}(a^2-x^2)$ $-a \le x \le a$ $\left(f_X(x) = \dfrac{6}{(a+b)^3}(a+x)(b-x), \ -a \le x \le b\right)$	$a=1$

LAWS OF ERROR

When	Who	Name	Analytical Expression	Graphical Expression
Limited Range				
(1781)	Lagrange	Cosine	$f_X(x) = \frac{\pi}{4}\cos\left(\frac{\pi x}{2}\right),\ -1 \le x \le 1$	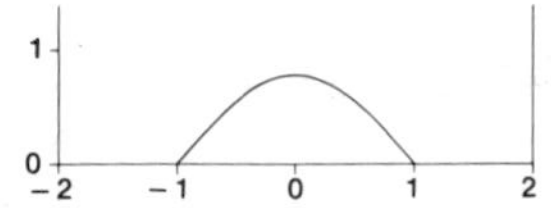
1781	Laplace	Laplace's Limited Range	$f_X(x) = \frac{1}{2a}\log\frac{a}{\lvert x\rvert}$, $-a \le x \le a$	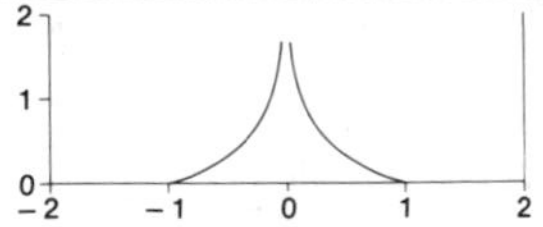
Unlimited Range				
1774	Laplace	Double Exponential (Laplace's First Law of Error)	$f_X(x) = \frac{k}{2}e^{-k\lvert x\rvert},\ -\infty \le x \le \infty$	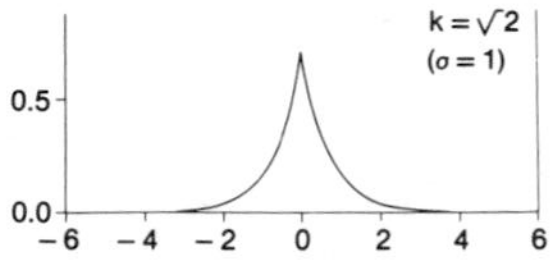 $k=\sqrt{2}$ $(\sigma=1)$
1809	Gauss	Gauss's Law of Error	$f_X(x) = \frac{h}{\sqrt{\pi}}e^{-h^2x^2},\ -\infty \le x \le \infty,\ 0 < h < \infty$	
1810	Laplace	Normal Distribution (Laplace's Second Law of Error)	$f_X(x) = \frac{1}{\sqrt{2\pi\mu_2}}e^{-\frac{x^2}{2\mu_2}},\ -\infty \le x \le \infty,\ 0 < \mu_2(x) < \infty$	$h=\frac{1}{\sqrt{2\mu_2}}=\frac{1}{\sqrt{2}}$ $(\sigma=1)$
1824	Poisson	Cauchy Distribution	$f_X(x) = \frac{1}{\pi\beta\left(1+\frac{x^2}{\beta^2}\right)},\ -\infty \le x \le \infty,\ 0 < \beta < \infty$	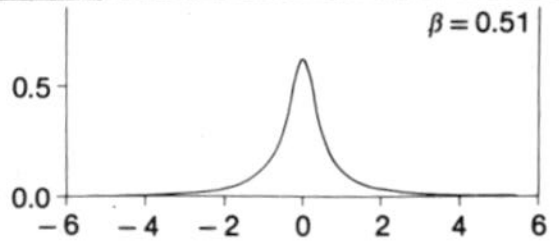 $\beta=0.51$
1853	Cauchy			

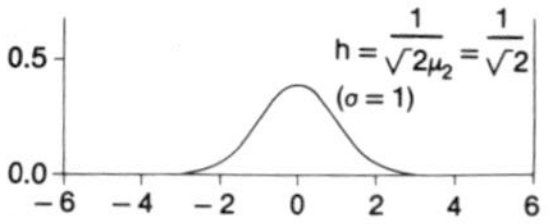

Source: U.S. DEPARTMENT OF COMMERCE
National Bureau of Standards

Figure 1

some particular distribution, Gaussian or otherwise, is adopted as a basis for statistical analysis or inference, it is usually recognized to be only a convenient approximation to whatever the real distribution may be.

SIMPSON

Thomas Simpson, in his path-breaking letter to the president of the Royal Society on the advantage of taking the mean of a number of observations [27], assumed first that the errors to which a measurement of a fixed quantity by a particular measurement process is subject would take the values $-v$, $-v + 1, \ldots, 2, 1, 0, 1, 2, \ldots, v - 1, v$, with equal probabilities, i.e., a *discrete uniform distribution*,

$$\Pr[X = x] = \frac{1}{2v + 1}$$

$$(x = -v, -v + 1, \ldots, -1, 0, 1, \ldots, v),$$

$$(1)$$

where $X = Y - \tau$ is the error of the measurement or observed value Y as a determination of the true value τ. Next, he assumed that the errors would take the values above with probabilities proportional to 1, $2, \ldots, v - 1, v, v + 1, v, \ldots, 2, 1$ respectively, i.e., a *discrete isosceles triangle distribution*,

$$\Pr[X = x] = \frac{v + 1 - |x|}{(v + 1)^2}$$

$$(x = -v, v + 1, \ldots, -1, 0, 1, \ldots, v).$$

$$(2)$$

Utilizing the generating function* techniques that had been employed by Abraham De Moivre* (1667–1754) for the solution of problems relating to tosses of dice and other games of chance, Simpson derived, for each of these distributions, the probability distribution of the *sum* of n independent errors from such a distribution—see Seal [21, pp. 212–213] for explicit expressions—and then from these, the corresponding distributions

of the *arithmetic mean* of n independent errors. Comparing the latter with the corresponding distributions for single errors, he showed by numerical examples that the probability of the arithmetic mean of n, $(n \geqslant 2)$, errors exceeding a particular value is always less than the probability of a single error exceeding the same value, the inequality increasing with increasing n. From these findings he concluded that taking the mean of a number of observations greatly diminishes the chances for all smaller errors and cuts off almost all possibility of any great ones; and the more observations or experiments there are made under the same circumstances, the less will the conclusion be liable to err.

Simpson noted [27, p. 88; 28, p. 67] that the generating function he found for the distribution of the arithmetic mean of n errors distributed independently according to (2) was the same as the generating function of the distribution of $2n$ errors distributed independently according to (1) with double spacing, i.e., according to

$$\Pr[X = x] = \frac{1}{2v + 1}$$

$$(x = -2v, -2(v - 1), \ldots,$$
$$-2, 0, 2, \ldots, 2(v - 1), 2v), \quad (1a)$$

so that (2) is the distribution of the arithmetic mean of two errors distributed independently according to (1a).

At the beginning of a second "Attempt to shew the Advantage arising by Taking the Mean of a Number of Observations . . . " in his last published work [28, pp. 64–75], Simpson states two "suppositions" about errors of measurement or observation that he considers "necessary":

1. That there is nothing in the construction, or position of the instrument whereby the errors are constantly made to tend the same way, but that the respective chances for their happening in excess, and in defect, are either accurately, or nearly, the same.

2. That there are certain assignable limits between which all these errors may be

supposed to fall; which limits depend on the goodness of the instrument and the skill of the observer.

He then repeats his two previous mathematical demonstrations and numerical calculations, and restates his previous concluding remarks. Finally, starting from a discrete isosceles triangle distribution of errors that take the values $-kv, -k(v-1), \ldots, -k, 0, k, \ldots, k(v-1), kv$ with probabilities proportional to $1/k, 2/k, \ldots, v/k, (v+1)/k, v/k, \ldots, 2/k, 1/k$, respectively, and proceeding to the limit as $k \to \infty$ and $v \to 0$ with $kv \to a$, say, he obtains the distribution of the arithmetic mean of n independent errors from a *continuous isosceles triangle distribution*,

$$f(x) = (1/a^2)(a - |x|), \qquad -a \leqslant x \leqslant +a.$$

(3)

(An explicit expression for the distribution of the *sum* of n independent errors so distributed with $a = 1$ is given in Seal [21, p. 213].)

LAGRANGE

Simpson's idea of probability distributions of error was taken up quickly on the Continent. Joseph-Louis, Comte de Lagrange (1736–1813), utilizing the De Moivre–Simpson probability-generating function method of deriving discrete probability distributions, reproduced and elaborated on Simpson's results—without mention of Simpson or De Moivre—in a long memoir "On the utility of the method of taking the mean among the results of several observations" [10], in which he extended the generating-function technique to continuous distributions.

In his opening paragraph, Lagrange states that all observations are, at least in part, inexact due to errors, that consequently it is customary to take the mean of all results "because in this way the different errors are distributed equally among all of the observations," the error of the mean result then becoming the mean of all the errors. He says

that, although the whole world recognizes the utility of this practice to diminish as much as possible the uncertainty that comes from the imperfections of instruments and the inevitable errors of observations, he believes that it will nonetheless be worthwhile to examine and evaluate by mathematical calculation the advantages that one can hope to obtain by this method. This is the object of his memoir.

He begins by supposing that the errors which can slip into each observation are given and that one knows also the number of cases which can give these errors, that is, "la facilité" of each error. (Karl Pearson regarded this to be the first introduction of a term for "frequency" [17, p. 587].) He then stated his assumptions and objectives more precisely: He will suppose that one knows only the limits between which all possible errors are confined and the law of their frequency ("facilité"), which together determine "the law of the frequency of the errors" ("la loi de la facilité des erreurs"); for each of various assumed laws of frequency of errors, he will seek the probability that the error of the mean will be (a) zero, (b) a given value, or (c) comprised between given limits; he will find how to determine *a posteriori* from observations their common law of frequency of errors and the probability that the law thus inferred does not differ from the truth by more than a given amount; and will deduce simple rules for correcting the readings of instruments given the results of repeated checks.

The memoir is divided into discussion of 10 problems numbered I through VIII, X and XI, there being none numbered IX. Isaac Todhunter (1820–1884) restated these problems and Lagrange's solutions primarily as elements in the development of the calculus of probabilities [31, Secs. 556–575], without mentioning the numerical results that throw light on the advantages and shortcomings of "taking the mean" in particular situations, or the role of this memoir in the historical development of probability distributions of error. Karl Pearson provided a much more informative account in his lectures at University College London 1921–1933 [17, pp. 587–612], in which he pointed out and corrected some Lagrange errors overlooked by Todhunter.

Problem I [10, Secs. 1–8] and Problem II [10, Secs. 9–12] relate to the family of simplest possible symmetric probability distributions of errors:

$$\Pr[\,X = x\,] = b, a, b \quad \text{for } x = -1, 0, 1$$

$$(a + 2b = 1) \quad (4)$$

Problem I asks: What is the probability, $\Pr[\overline{X}(n) = 0]$, that the arithmetic mean, $\overline{X}(n)$, of n errors, $X_1, X_2, \ldots, X_n$, independently and identically distributed (i.i.d.) according to (4) will be exactly zero? Two different but equivalent algebraic expressions for this probability are given, derived from different expansions of the probability-generating function. Tables of values of $\Pr[\overline{X}(n) = 0]$ are given for small values of n and (a) equally probable errors ($a = b = \frac{1}{3}$), i.e., the simplest case of Simpson's first hypothesis; (b) a zero error the most probable $[a = 2b = 2(\frac{1}{4})]$, the simplest case of Simpson's second hypothesis; and (c) a zero error the least probable $[b = 2a = 2(\frac{1}{5})]$. These tables, reproduced in Pearson [17, p. 589], show, as Lagrange notes, that, when the errors of observation are i.i.d. according to (4), the probability that the mean error will be zero, i.e., that the arithmetic mean of several observed values will be the correct value, diminishes as the number of observations, n, increases (beyond $n = 2$). The parenthetical qualification is necessary because, as Lagrange also notes, when a zero error is the least probable, case (c) $\Pr[\overline{X}(n) = 0]$ is a maximum for $n = 2$, so that in this case it is most advantageous to take only two observations. Lagrange shows that $\Pr[\overline{X}(n) = 0] \to 0$ as $n \to \infty$, and [10, Sec. 6] that, when a zero error is the most probable ($a = 2b$), $\Pr[X(n) = 0] \simeq 1/\sqrt{\pi n} \to 0$ as $n \to \infty$ (see Pearson [17, p. 591]). Pearson comments [17, p. 589] that, although the probability of obtaining a zero mean error, i.e., a correct result, diminishes as the number of observations, n, increases because the number of

additional possible outcomes increases, yet a correct result, $\overline{X}(n) = 0$, is always of greater probability than any one of the other possible outcomes when errors are i.i.d. according to (4).

In a Scholium [10, Sec. 8] the foregoing findings are interpreted in terms of measured or observed values, Y, that can take only the values $\rho - 1, \rho, \rho + 1$, where $\rho(0 < |\rho| < \infty)$, is the exact ("exacte") value of the quantity measured.

In Problem II an expression is sought for $\Pr[-m/n < \overline{X}(n) < m/n]$ for $m = 0, 1, 2, \ldots, (n - 1)$. A formal algebraic solution is found, and a table of values of these probabilities given (and reproduced in Pearson [17, p. 593]) for the case of equally probable errors ($a = b = \frac{1}{3}$), when $n = 1(1)6$ and $m = 0(1)5$. Lagrange notes that, although $\Pr[-\frac{1}{2} \leqslant \overline{X}(n) \leqslant \frac{1}{2}]$ increases as n increases, this probability is larger for $n = 2$ than for $n = 3$, for $n = 4$ than for $n = 5$, and in general for any even number whatever ($n = 2k$) than for the odd number ($n = 2k + 1$) immediately following, so that it is most advantageous to take the mean of some even number of observations. Neither he nor Pearson notice, however, that this simple pattern and conclusion do not carry over to other cases, e.g., when the limits are $\pm \frac{1}{3}$ or $\pm \frac{2}{3}$. In a Scholium [10, Sec. 12] he discusses very briefly the extension of Problems I and II to the case where the possible errors are still symmetrical ($-1, 0, +1$), but the corresponding probabilities ($c, a,$ and b, respectively) may all be different, that is,

$$\Pr[\,X = x\,] = c, a, b \quad \text{for } x = -1, 0, 1$$

$$(a + b + c = 1) \quad (5)$$

In Problems III [10, Sec. 13] and IV [10, Secs. 14, 15] the probability distribution of errors, X, is taken to be of the form

$$\Pr[\,X = x\,] = b, a, c \quad \text{for } x = -1, 0, r$$

$$(0 < r < 1; a + b + c = 1). \quad (6)$$

Here the possible values of an error are not symmetric except when $r = 1$.

Problem III asks: What is the probability that $\overline{X}(n)$ will be included between given limits? An algorithm is given for evaluating $\Pr[\overline{X}(n) = \mu/n]$ ($\mu = -n, -n+1,$ $\ldots, nr-1, nr$), and a symbolic expression given for $\Pr[-p/n < \overline{X}(n) < q/n](p = 1,$ $2, \ldots; q = 1, 2, 3, \ldots, nr$). Lagrange does not attempt numerical evaluation even for small values of n.

Problem IV has to do with determining the value of μ for which $\Pr[\overline{X}(n) = \mu/n]$ is a maximum. He shows [10, Sec. 14] that if nb, na, and nc are all integers, then the value of $\overline{X}(n)$ of maximum probability corresponds to $\mu = ncr - nb$ and is $cr - b$. But if nb, na, and nc are not whole numbers, then he says that one ought to take the whole numbers β, α, and ν that are closest to them, and consider $(r\nu - \beta)/n$ the value of maximum probability. However, he continues, if one always takes $cr - b$ to be the value of $\overline{X}(n)$ of maximum probability, the difference from $(r\nu - \beta)/n$, if any, will always be very small. (Today, we see no reason to require that nb, etc., be integers, and simply accept $cr - b$ as the value of maximum probability.)

In a corollary [10, Sec. 15], he states that, from the foregoing one can always regard $cr - b$ as the systematic error or bias of the corresponding mean of the observations, $\overline{Y}(n)$, and thus can take this quantity to be the correction to be applied to (i.e., subtracted from) $\overline{Y}(n)$. He comments that when $r = 1$ and $c = b$, as in Problem I, then the correction to the mean result becomes zero; as it will also when $b = rc$. Oddly, he does not note here that $cr - b$ is always the mean or expected value of $\overline{X}(n)$.

In Problem V the possible values of an error, X, and their respective probabilities are taken to be

$$\Pr[X = x] = a, b, c, d, \ldots$$

$$\text{for}\quad x = p, q, r, s, \ldots$$

$$(a + c + d + \cdots = 1). \quad (7)$$

The quantity sought is the correction that one ought to make to the arithmetic mean of many observations. Lagrange says [10, Sec. 16] that it is easy to to show, by a method similar to that of Problem IV, that the mean

error, $\overline{X}(n)$, of maximum probability, will be $ap + bq + cr + ds + \cdots$, and this will be the correction that ought to be made to (i.e., subtracted from) the mean result $\overline{Y}(n)$.

In Corollary I [10, Sec. 17] he states that, if one regards the probabilities $a, b, c, \ldots$ as weights situated at distances $p, q, r, \ldots$ from a fixed point along a straight line, then the distance from the fixed point to the center of gravity of these weights will be the value of the correction that one ought to make. In Corollary II [10, Sec. 18], he states that, if the error of each observation can take all of the values between two given limits, and one knows the curve of frequency of the errors, the ordinates of which represent the frequencies of the errors corresponding to the respective abscissae, then the abscissa corresponding to the center of gravity of the total area under this curve will express the correction to (i.e., to be subtracted from) the mean result, $\overline{Y}(n)$.

Pearson points out [17, p. 598] that Lagrange seems to have "got himself into a quagmire here." In the Corollary to Problem IV, Lagrange evidently considers the mean value of $\overline{X}(n)$, $E[\overline{X}(n)]$, to be acceptable as the correction to be applied to (i.e., subtracted from) the observed mean result, $\overline{Y}(n)$, because $E[\overline{X}(n)]$ happens to be equal —or, from Lagrange's viewpoint, close—to the value of $\overline{X}(n)$ of maximum probability, i.e., its modal value. (Pearson notes that equality of the mean and mode is a property of binomial and multinomial distributions "however skew they may be.") Lagrange, making the transition above to a continuum of errors, now says that the abscissa of the center of gravity of the area under the curve of frequency of the errors, i.e., the mean error, $E[X]$, expresses the correction to be applied to (i.e., subtracted from) an individual observation, Y, or the mean, $\overline{Y}(n)$, of n observations. Pearson considers this to be a change in the correction procedure because $E[X]$ will not generally be equal to, and may differ markedly from, the most probable value (the *mode*) of X. On the other hand, the *central limit theorem**, unknown to Lagrange, tells us that the mean value of the mean error, $E[\overline{X}(n)]$, and the most probable value of $\overline{X}(n)$ will converge as $n \to \infty$, even when the distribution of an individual error, X, is skew. Insofar as the memoir is concerned, however, the distinction is irrelevant, because all of the continuous laws of frequency of errors proposed by Lagrange are symmetrical about $E[X]$, except (9a).

In Problem VI [10, Sec. 19], Lagrange supposes that the readings of an instrument have been checked many (say, n) times, and that the errors $p, q, r, \ldots$ occurred with frequencies $A, B, C, \ldots$, respectively, and asks what ought to be taken for the correction to the instrument. He shows that the values of the underlying probabilities $a, b, c, \ldots$ that maximize the joint probability of occurrence of the observed outcomes are the observed proportions.

$$\hat{a} = A/n, \quad \hat{b} = B/n, \quad \hat{c} = C/n, \quad \ldots$$
$$(A + B + C + \cdots = n),$$

and hence the correction that one ought to apply to (i.e., subtract from) an instrumental reading is $(Ap + Bq + Cr + \cdots)/n$, i.e., the mean of all particular errors that the n verfications gave. In a Corollary [10, Sec. 20], he states if the $\hat{a}, \hat{b}, \hat{c}, \ldots$, are plotted as ordinates on abscissas proportional to p, $q, r, \ldots$ as in Corollary I to Problem V, a *parabolic curve* passed through their extremities, and the center of gravity of the total area under this curve calculated, then the abscissa of this center of gravity will be the required correction. He adds that one sees from this how one can know *a posteriori* the law of frequency of the errors to which an instrument can be subject. In two supplementary remarks [10, Secs. 21, 22], Lagrange aims to find an expression for the probability that the observed proportions $\hat{a}, \hat{b}, \hat{c}, \ldots$ collectively do not differ from the corresponding true proportions $a, b, c, \ldots$ by more than assigned amounts, and, as Karl Pearson has shown [17, pp. 599–603], "came within an ace" of proposing Pearson's χ^2 goodness-of-fit test*.

Problem VII [10, Sec. 26] involves an insignificant generalization of Simpson's first hypothesis:

$$\Pr[X = x] = \frac{1}{\alpha + \beta + 1},$$
$$x = -a, \ldots, -1, 0, 1, \ldots, \beta - 1, \beta$$
$$(0 < \alpha, \beta < \infty) \quad (8)$$

a *discrete uniform distribution* not symmetrical with respect to $x = 0$ unless $\alpha = \beta$. The quantities sought are $\Pr[\overline{X}(n) = \mu/n]$, $(\mu = -n\alpha, -n\alpha + 1, \ldots, n\beta - 1, n\beta)$, and $\Pr[-p/n < \overline{X}(n) < q/n]$ $(p = 1, 2, \ldots, n\alpha;$ $q = 1, 2, 3, \ldots, n\beta)$. Algebraic expressions for each are found. In a Corollary [10, Sec. 27], Lagrange lets α, β, p, and q all tend to infinity in such a way that

$$\beta/\alpha \to \lambda/1, \quad p/\alpha \to r, \quad \text{and} \quad q/\alpha \to s,$$

evaluates the corresponding limiting value of $\Pr[-p/n < \overline{X}(n) < q/n]$, and recognizes that the result obtained, $\Pr[-r/n < X(n) < s/n]$, is the integral from $-r/n$ to s/n of the probability density function, $g_{\overline{X}}(\overline{x})$, of the distribution of the arithmetic mean, $\overline{X}(n)$, of n errors i.i.d. in accordance with the *continuous uniform** or *rectangular distribution*

$$f_X(x) = \frac{1}{\lambda + 1}, \quad -1 \le x \le \lambda, 0 < \lambda < \infty,$$
$$(9)$$

symmetric with respect to $x = (\lambda - 1)/2 = E[X]$.

It is clear that Simpson *could* have derived $g_{\overline{X}}(x)$ for (9) with $\lambda = 1$ from his (1) by the very same limiting process that he used to derive the corresponding result for (3) from his (2). But he *did not*. Therefore, it is customary to ascribe to Lagrange priority for the discovery of the distribution of the arithmetic mean of n random variables i.i.d according to a continuous uniform or rectangular distribution, especially in view of his direct derivation of this result in Example I of Problem X below. It has since been rediscovered again and again [22].

In Problem VIII the assumed probability distribution of error is the *discrete isosceles triangle distribution* (2), Simpson's second hypothesis. Lagrange's analysis [10, Sec. 28] duplicates Simpson's, including transition [10, Sec. 29] to a continuum of errors between -1 and $+1$ having the *continuous isosceles triangle* probability density function (3) with $a = 1$, which he describes in words, and considers to be a hypothesis that conforms more closely to nature than the corresponding limiting distribution of Problem VII.

Problem X [10, Sec. 38] asks: If the error X of an observation can take any value between $-q$ and p with probability given by the continuous probability density function $f_X(x)$, what is the probability that the arithmetic mean error of n independent observations, whose errors are i.i.d. according to $f_X(x)$, will be between $-s/n$ and r/n? To answer this question Lagrange extends, without proof, the definition of a probability-generating function, and its use to derive probability distributions of sums of i.i.d. random variables, from discrete to continuous probability distributions—see Pearson [17, p. 608], Seal [21, p. 214], and Todhunter [31, p. 310]. (A proof is given by Todhunter [31, Sec. 574].)

Lagrange then applies this new mathematical tool to derive the probability distributions of the arithmetic mean, $\overline{X}(n)$, of n errors, $X_1, X_2, \ldots, X_n$, i.i.d. according to four continuous laws of frequency of errors, as follows: The probability distribution of an individual error X is taken in Example I [10, Sec. 29] to be a *continuous uniform* or *rectangular distribution*,

$$f_X(x) = \frac{1}{p + q}$$

$(-q \leqslant x \leqslant p; 0 \leqslant p, q < \infty,$ not both zero),

(9a)

symmetric with respect to $x = (p - q)/2 = E[X]$; in Example II [10, Sec. 40] a *continuous parabolic distribution*

$$f_X(x) = \frac{3}{4p^3}(p^2 - x^2)$$

$(-p \leqslant x \leqslant p, 0 < p < \infty)$ (10)

symmetric with respect to $x = 0 = E[X]$; in

a remark [10, Sec. 41], the *continuous isosceles triangle distribution* (3), with $a = p$, symmetric with respect to $x = 0 = E[X]$; and in Problem XI [10, Sec. 42], a *continuous cosine distribution*,

$$f_X(x) = \tfrac{1}{2}\cos x \qquad (-\pi/2 < x < \pi/2)$$

(11)

symmetric with respect to $x = 0 = E[X]$. Lagrange considers the parabolic distribution of Example II to be "the most simple and the most natural that one can imagine" [10, Sec. 41]. Todhunter misread Example II and says incorrectly [31, Sec 575] that Lagrange found the distribution of $\overline{X}(n)$ for

$$f_X(x) = K\sqrt{c^2 - x^2}, \qquad -c < x < +c,$$

a semicircular law of error.

Lagrange wrote to Leonhard Euler (1707–1783) on February 10, 1975, saying that the foregoing memoir had been sent to press, that he would send Euler a copy, and would appreciate Euler's opinion ("avis") of it [25, p. 46]. Euler evidently considered it to be a work of some importance inasmuch as he presented an *Eclaircissemens . . .* ("Elucidation" or "Explanation") of it to the St. Petersburg Academy on November 27, 1777, which was not published until 11 years later [5]. This *Eclaircissemens* is a quite elementary elucidation and explanation of Lagrange's memoir, evidently written for the benefit of individuals needing help in following the algebraic manipulations involved in Lagrange's application of generating function techniques, and contains nothing new (see Todhunter [31, Sec. 447] and Sheynin [25, p. 47]).

A résumé of Lagrange's memoir was published in the *Encyclopédie Méthodique* [3, pp. 406–409].

LAMBERT

In contrast to Simpson and Lagrange, who initially proposed discrete probability distributions of errors and then introduced continuous distributions of errors more or less as second thoughts, so to speak, Johann Heinrich Lambert (1728–1777) unequivocally regarded measurement errors* to be distributed about zero in accordance with a continuous symmetrical unimodal frequency curve in the sections on errors of measurement in his Latin treatise *Photometria* [11, Secs. 271–306]. These sections are not included in the annotated German translation (Ostwald's *Klassiker der exakten Wissenschaften*, Nos. 31–33, W. Engelmann, Leipzig, 1892) with the comment that they contained nothing of interest to readers, and were brought to light by Oscar Borisovich Sheynin [23], who subsequently provided a fuller English summary [24, Sec. 3.2].

Lambert's discussion starts with consideration of measurements yielded by experiments carried out for the sole purpose of confirming a physical law deduced from basic principles or inferred from other experiments. He describes attributes that such measurements must exhibit with respect to magnitude [11, Sec. 272], configuration and randomness [11, Sec. 274], if they are to be regarded as confirming the law; attributes that should be regarded as casting doubt on the law's validity [11, Sec. 273]; and attributes to be construed as indicating that either the presumed law "deviates slightly from the truth" or the measurements are affected by instrumental defects [11, Sec. 275].

He then turns his attention in Sec. 276 to experimental determination of values of various properties of emitted and reflected light. Since positive and negative deviations are equally *possible*, it follows, he says, that they will also be equally *frequent*, if the experiment is repeated many times [11, Sec. 277]; every time identical deviations occur on both sides, they will cancel [11, Sec. 277]; so that, if the same experiment were to be repeated endlessly, one is fully entitled to conclude that the arithmetic mean of all the measurements ("medium inter omnia") would differ in no way at all from the true value [11, Sec. 279]. However, since no experiment is repeated endlessly, one should inquire what can be inferred from a finite number of experiments because, in this case, it is certainly not at all guaranteed that positive and negative errors of the same size will occur with equal frequency [11, Sec. 280]. Consequently, one must give another method for determining a mean of a finite number of measurements that, with maximum probability, will differ least from the true value [11, Sec. 295].

To this end he assumes [11, Sec. 296] that the distribution of repeated measurements of the same quantity can be represented by a continuous symmetrical unimodal frequency curve as in Fig. 2 (his Fig. 31), with A denoting the origin of the scale of measurement; AC, the true value to be determined by the experiments; CB and CD, the maximum errors on both sides; and ordinates QM, PN, RL, and SK, the "true counts" (i.e., the true relative frequencies) of the errors CP, CQ, CR, and CS, respectively. Why he does not include an explicit ordinate at C, and an expression (CH, say) for the relative frequency of error-free measurements, is puzzling.

Lambert does not employ a functional expression, e.g., $f(x)$ or $\phi(x)$, to characterize the error-frequency distribution in a form amenable to analytic manipulation. Instead, in the succeeding discussion, he regards the error-frequency curve $BNMLKD$ to be of fixed form and translatable to the left and to the right above the horizontal measurement

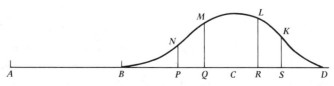

Figure 2   From ref. 11.

axis, with *PN*, *QN*, *RL*, and *SK* always denoting the theoretical or "true" relative frequenices of measurements that are repectively 2*CR* and *CR* units *less*, and *CR* and 2*CR* units *greater* than the true value of the quantity measured, whatever may be the location (always denoted by *C*) of this true value on the measurement axis. Thus, he points out [11, Sec. 297] that if the observed frequences of measured values *P'*, *Q'*, *R'*, and *S'*, say, spaced exactly the same as the ordinates *PN*, *QM*, *RL* and *SK* in the figure, happen to be proportional to the relative frequencies represented by the lengths of these ordinates, respectively, then, by sliding the error-frequency curve left or right until these ordinates stand upon the points *P'*, *Q'*, *R'*, and *S'*, so to speak, the central maximum ordinate will stand on the value of the quantity measured that maximizes the probabilities of the observed frequencies at *P'*, *Q'*, *R'*, and *S'*, and is in this sense the value of the quantity measured of maximum probability given the observed frequencies.

Inasmuch as strict proportionality of observed and "true" relative frequencies at correspondingly spaced values will be "very rare" [11, Sec. 297], he says that a procedure is needed for determining the most probable location of the error-frequency curve and thence of the true value, *C*, when the quantities *AP*, *AQ*, *AR*, and *AS* have been observed *n*, *m*, *l*, and *k* times, respectively [11, Sec. 299]. He notes that a solution is facilitated by the fact that the mutual distances between measurement values corresponding to errors having "true" relative frequencies *PN*, *QM*, *RL*, and *SK* are fixed and known, whatever may be their location on the measurement axis: Once the value on the measurement axis to which any one of these corresponds is known, the values corresponding to the others, and to zero error, i.e., the true value of the quantity measured, are also known, so that the problem may be reduced in this manner to determination of only one unknown [11, Sec. 298]. He then gives [11, Secs. 302–304] a geometric solution in terms of the lengths of subtangents of the curve at abscissae spaced exactly as the

measured values *P'*, *Q'*, *R'*, and *S'*, which is equivalent to finding the value of the location parameter τ that maximizes the product $[f(y_1 - \tau)]^n \cdot [f(y_2 - \tau)]^m \cdot [f(y_3 - \tau)]^l \cdot [f(y_4 - \tau)]^k$, where $f(y - \tau)$ is the probability density function of measured values *Y* when the "true value" is τ, and y_1, y_2, y_3, and y_4 are the values observed with frequencies *n*, *m*, *l*, and *k*, respectively, i.e., what we call the "maximum likelihood" estimate of τ.

To us today it is amazing that Lambert thus introduced the method of maximum likelihood* for finding the location of the center of continuous symmetrical probability distribution of measured values without ever expressing the error-frequency distribution in functional form. Five years later, however, he deduced and applied an explicit expression for the probability distribution of the error of setting the hair or thread line in the eyepiece of an astronomical telescope on the center of the image of a star, from an idealized model of the operation, in his "Anmerkungen und Zusätze zur practischen Geometrie" ("Notes and additions to practical geometry" [12, pp. 1–313]. He says [12, Sec. 428] that the naked eye sees a star as a point, but in the magnification of a telescope its image becomes an area, the different points of which the eye cannot resolve. We can take this area to be circular, he claims, as there are no grounds for assuming that a point has corners [12, Sec. 429]. Consequently, if we let *ADBE* in Fig. 3 (his Fig. 60) be such an area, *AB* its diameter, *C* its center, and let *DE* represent the position of a hairline set upon this area, then [12, Sec. 430] *DF* represents the probability that it will deviate from the center *C* by the

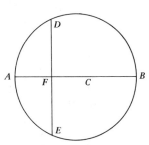

Figure 3   From ref. 12.

amount *FC*. His reasoning is that, inasmuch as the eye cannot distinguish between points within this area, all points within it have an equal probability a priori of being a point through which the line set upon the area passes. The line will pass through and be perpendicular to some diameter. Circular symmetry allows us to rotate the area in any particular instance so that the relevant diameter is horizontal and the line vertical, without loss of generality. The equiprobability assumption implies that the probability that the hairline will be set between *DE* and the parallel line through *C* is equal to the area within the circle between these parallel lines, and hence the probability that it will be *at* a distance *FC* is *DE* = 2*DF*, so that *DF* "represents" the probability.

Lambert does not express this result in functional form at this juncture, but in Sec. 431 he discusses the error of the observed angle between two stars or distant points, whose images he represents [12, Fig. 61] by two circles of equal diameter with centers at *C* and *D* on a horizontal line. He takes [12, Sec. 432] the corresponding errors (with respect to *C* and *D*) of the hairline settings "on" these areas to be i.i.d. as above, the radii of these areas to be unity [12, Sec. 433] and for purposes of calculation indicates the probability of each error as being proportional to an expression of the form $\sqrt{1 - x^2}$, that is, as having a flattened *semicircular* law of error,

$$f_X(x) = (2/\pi r^2)\sqrt{r^2 - x^2} \qquad (-r \leqslant x \leqslant r),$$

$$(12)$$

with $r = 1$.

Taking the joint probability proportional to $\sqrt{1 - x_1^2} \cdot \sqrt{1 - x_2^2}$, where $x_1 = y_1 - C$, $x_1 = y_2 - D$, and y_1 and y_2 denote the observed positions of the left and right hairline settings, respectively, he absorbs [12, Sec. 433] the product $(1 - x_1^2)(1 - x_2^2)$ under a single radical, carries out the multiplication, and expresses the result essentially as a function $u = y_1 - y_2$ and $v = C - D$, develops a series expansion for it, and in subsequent discussion notes [12, Sec. 434, item 7] that

the value of v about which the function is symmetrical is the locus of its maximum ordinate, i.e., is the value of $C - D$ of maximum probability.

The expression "theory of errors" may have been coined by Lambert [24, p. 254]. It appears ("die Theorie der Fehler") without elaboration in the preface to the first volume of his *Beyträge* [12, p. *4], and in Sec. 321 of his "Anmerkungen" [12, p. 226] he states the objectives of this theory to be determination of relationships between errors, of their consequences, of the conditions of measurement, and the accuracy of instruments.

BERNOULLI

Daniel Bernoulli (1700–1782) in a memoir, "The most probable choice between several discrepant observations and the formation therefrom of the most likely induction" [2], written in Latin, considered errors of measurement or observation to be analogous to the horizontal deviations from the vertical line through a set mark of the points of impact of the arrows of an archer aiming his arrows at the set mark with all the care that he can muster. In the case of measurements or observations, the "set mark" is the (unknown) true value of the quantity concerned, which functions as a center of attraction to which the observations are drawn, but whose attraction is opposed by innumerable imperfections and other tiny hidden obstacles that may produce small chance errors in the observations. Some of these will be in the same direction and be cumulative in effect. Others will cancel out. Hence there is a relation between the chances of particular outcomes and the true value that determines the center of attraction; for another position of the true value the chances of the outcomes would be different. Furthermore, there is no doubt that the greatest deviation has limits which are never exceeded, and which are narrowed by the experience and skill of the observer. The probability of a deviation beyond these limits is zero, but inside these limits the probability increases to a maximum at the mark in the center. Hence one should think of a curve—he says "scale" ("scala")—expressing the relation between the distances of the observations from the mark and the corresponding probabilities. To this end imagine a straight line on which various points indicate the results of different observations, and let there be marked thereon some intermediate point to be taken as the true value of quantity concerned. If a perpendicular is erected at each observational point expressing its probability, then the curve drawn through the upper ends of these perpendiculars will be the "scale of probabilities."

If this be accepted, then it can hardly be denied, he says, that the scale of probabilities must have the following characteristics: (a) have two perfectly similar and equal branches, inasmuch as deviations from the true point are "equally easy" ("aequefaciles") in both directions; (b) decrease toward the axis of observations in both directions from the center; (c) have a maximum at the center, where its target will be parallel to the axis of observations; (d) meet the axis of observations at each of the limits of possible deviations, at which extremes all probability vanishes and beyond which no error is possible; and (e) approach the axis of observations steeply toward both limits, its tangents being almost perpendicular to this axis at the extreme points. Bernoulli notes that any ellipse, elongated or compressed, having the line between the error limits as its axis will have the foregoing characteristics and serve quite well; that a semicircle will also do, and lends itself best to numerical calculations. For this and some other less persuasive reasons, he adopts the *semicircular law of error* (12), with $x = y - \tau$, the error of an observed value y of a measurement Y of the *true value* τ, and $\pm r$ the limits that an error will never exceed.

As a rule for determining the radius r he recommended taking the difference between the two extreme observations in hand; adding that it will be sufficient to increase this difference by half to form the diameter of the circle, if several observations have been

made; and that his own practice is to double it for three or four observations, and to increase it by half for more.

In contrast to this somewhat arbitrary and imprecise procedure for determining the maximum error considered possible, he advocated maximization of the product

$$f(x_1)f(x_2) \cdots f(x_n)$$

$$= (2/\pi r^2)^n \prod_{i=1}^{n} \left[r^2 - (y_i - \tau)^2 \right]^{1/2}$$

with respect to τ for the chosen value of r to obtain the value of τ having "the highest probability." Today we recognize this value of "highest probability," $T = T(y_1, y_2, \ldots, y_n)$, to be the maximum likelihood estimate of τ corresponding to the law of error (12). There is absolutely no mention of Lambert's earlier (1765 [12]) introduction of the same law of error (12), nor of Lambert's proposal (1760 [11]) that the "best" mean to take is the function of the observations that maximizes their joint probability, in either Bernoulli's paper or the commentary by Leonhard Euler (1707–1783) on the immediately following pages [4].

For $n = 3$, evaluation of T requires the solution of an equation of the fifth degree consisting of 20 terms; and for $n > 3$, the algebra and arithmetic become unmanageable. However, for $y_1 \leqslant y_2 \leqslant y_3$, Bernoulli showed that T is greater than, equal to, or less than the arithmetic mean of the three values according as the middle value (y_2) is less than, equal to, or greater than the midpoint $\frac{1}{2}(y_1 + y_3)$ between the extremes, respectively. His T thus assigns greater weight to the more distant of the two extreme observations. The actual magnitude of the difference $T - \bar{x}$ depends, however, on the choice of r, the limit an error will never exceed in absolute value, but tends to zero rapidly as $r \rightarrow \infty$, leading Bernoulli to comment wryly that, if one is shocked by these findings and wishes to adhere to the arithmetic mean, all one has to do is to permit the largest conceivable error to be infinitely large.

In an article, "The mean to take between observations" [3] in the *Encyclopédie Mé-*

thodique, Jean Bernoulli—presumably Johann Bernoulli III (1744–1807), nephew of Daniel and executor of the mathematical estate of the Bernoulli family—says that in 1769 he received from Daniel Bernoulli a small Latin manuscript with the same title as the foregoing. Jean Bernoulli's summary indicated that this earlier draft differed considerably from the published version. After mentioning that the ordinary practice is to take the arithmetic mean $\bar{y} = \sum_1^n y_i / n$ of n observations $y_1, y_2, \ldots, y_n$ as the value they indicate for the true value (say, τ) of the quantity concerned, Bernoulli recommends that one take instead the weighted mean $\bar{\bar{y}} = \sum_1^n p_i y_i / \sum_1^n p_i$, with the weight p_i proportional to the probability of occurrence of an error of magnitude $y_i - \tau$ ($i = 1, 2, \ldots, n$), so that observations corresponding to errors of lesser probability are given lesser weight. He recognizes that in order to evaluate the weights p_i it is necessary to know or assume some particular law of error. He says that many considerations indicate selection of a semiellipse or a semicircle, and adopts the latter, whence the weight p_i of the observations y_i is to be taken proportional to $\sqrt{r^2 - (y_i - \tau^2)}$, where r is the magnitude of the (assumed) maximum possible error in either direction. Since the central point, τ, is not known in advance, he recommends using the arithmetic mean $\bar{y}$ as a first approximation, determining the corresponding weights p_i and thence $\bar{\bar{y}}$; repeating the process with $\bar{\bar{y}}$ as starting value, and iterating until the difference between two successive values become negligible. A number of examples of the application of this technique are included. Bernoulli advises against choosing too large a value of r, but does not suggest a rule as in the published version.

LAPLACE

Pierre Simon, Marquis de Laplace (1749–1827), was the first to propose a law of error of unlimited extent in his discussion of a problem "To determine the mean that one ought to take among three given observations of the same phenomenon" [13, Problème III, pp. 634–644; 26, Sec. 2.1]. Referring to Fig. 4 (his Fig. 2), he assumes [13, p. 635] that the point V is the true ("véritable") instant of the phenomenon, and that the probabilities that an observation will be removed from the truth at the distances VP, VP', etc., can be represented by the ordinates of a curve RMM' which decreases following any law and which, on denoting the abscissa VP by x, and the corresponding ordinate PM by y, will be represented by the equation $y = \phi(x)$ [or, $y = f_X(x)$, say]. He says that the (desirable) properties of this curve are:

1. It ought to be divided into two fully equal parts by the ordinate VR, for it is as probable that an observation will differ from the truth to the right as to the left.

2. It ought to have the horizontal axis for asymptote, because the probability that an observation will depart from the truth at an infinite distance is evidently null.

3. The total area under this curve ought to be equal to unity, because it is certain that an observation will fall at one of the points of the horizontal axis.

Adding [13, p. 637] the further requirement that the slope of the curve at a distance x from the truth should be proportional to

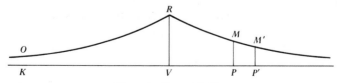

Figure 4   From ref. 13.

the ordinate at that point, that is, $f_X'(x) = \pm \kappa f_X(x)$, he obtains on integration the *double-exponential distribution** of error

$$f_X(x) = (\kappa/2)e^{-\kappa|x|}$$

$$(-\infty < x < \infty, 0 < \kappa < \infty), \quad (13)$$

often called "Laplace's first law of error."

If Y_1, Y_2, Y_3 denote three measurements or observations of a true value, Laplace says [13, p. 636] that the mean ("milieu"), i.e., the function $M(Y_1, Y_2, Y_3)$, of these observations that one ought to take should have one or the other of the following properties: (a) The probabilities of its exceeding or being less than τ should be equal, i.e., $\Pr[M < \tau] = \Pr[M > \tau]$, or (b) the mathematical expectation of its absolute error, $E[|M - \tau|]$, should be a minimum. Laplace shows that these two properties are equivalent, and that $M(Y_1, Y_2, Y_3)$ is the median* of what some today would call the fiducial distribution* of τ given Y_1, Y_2, and Y_3. (For a more recent development along these same lines, see Pitman [18].)

Applying this *minimum mean absolute error estimator* to the case of three observed values y_1, y_2, y_3, when errors, $X_i = Y_i - \tau$ ($i = 1, 2, 3$) are i.i.d. in accordance with his double-exponential law of error, Laplace found that, for $y_1 < y_2 < y_3$, his "mean," M, is greater than, equal to, or less than y_2, the middle value (i.e., the *median*), according as y_2 is less than, equal to, or greater than $\frac{1}{2}(y_1 + y_3)$, the midpoint between the extremes, respectively. M is thus a "corrected median," the correction being in the direction of the more distant of the two extreme observations. Furthermore, $M \to y_2$ as $\kappa \to \infty$ (i.e., very high precision); and $M \to \bar{y}$, the mean of the three values, as $\kappa \to 0$ (i.e., very poor precision).

In Part VIII of a subsequent memoir [14], Laplace developed an inductive procedure, or recursion formula, for determining the probability distribution of the sum of n random variables i.i.d. in accordance with a known continuous law of frequency comprised within given limits. As an application, he found, in Article IX, the distribution of

the sum of n random variables i.i.d. in a *continuous quadratic distribution*, $a + bx + cx^2$, between $-h$ and g, that is, on solving for a, b, and c in terms of $-h$ and g, the completely general *continuous parabolic distribution*

$$f_X(x) = \frac{6}{(g+h)^3}(g-x)(h+x)$$

$$(-h \leqslant x \leqslant g; 0 < h, g < \infty)$$

$$= \frac{6}{(g+h)^3}\left[\left(\frac{g+h}{2}\right)^2 - \left(x - \frac{g-h}{2}\right)^2\right],$$

$$(10a)$$

symmetric with respect to $x = (g - h)/2 = E[X]$. He then discussed the special case in which the parabolic distribution reduces to the *uniform* or *rectangular distribution** (9a), for $q = 0$ and $p = \pi/2$, as a basis for interpreting the arithmetic mean of the angles of inclination of the orbits of comets.

In part X, Laplace applies his recursion method to derive directly the distribution of the sum, and thence of the arithmetic mean, of n errors i.i.d. in accordance with the *continuous isosceles triangular distribution* (3), and acknowledges Lagrange's priority in so doing. Then, following discussions in Parts XI and XII of probability distributions of the skills of players in a game and of the intervals between random points on a line of length a, in which the following distribution is derived, he suggests in Article XIII that when the error X of an individual measurement or observation can take values between $-a$ and $+a$ only, and one is ignorant of the actual law of probability of these errors, one should take

$$f_X(x) = (1/2a)\log(a/|x|),$$

$$-a \leqslant x \leqslant a, \quad 0 \leqslant \alpha \leqslant \infty, \quad (14)$$

as the probability distribution of an error, X, "in research on the mean that one ought to take between the results of many observations." (See also Laplace [15, Book II, Chap. II, final par.].) Later (Part XXXI) he again suggests consideration of this same distribution for individual errors, X_i, with possibly different limits, $\pm a_i$, for each, and its use in the application of his *minimum mean error of estimation* approach, commenting once again (Part XXXII) that "the ordinary rule of arithmetic means" is indicated by this approach, when $a_i \to \infty$ for all i, i.e., all measurements extremely imprecise.

References

[1]   al-Bīrunī. (1025, printed 1962). *Kitāb Tahdīd al-Amākin*, Sultan Fatih No. 3386, P. Boljakoff, ed. Cultural Department of the Arab League, Cairo (Istanbul manuscript). (English transl. by Jamil Ali: *The Determination of the Coordinates of Cities*. American University of Beirut, Beirut, 1967.)

[2]   Bernoulli, D. (1778). *Acta Acad. Sci. Imp. Petropolitanae*, **1**, 3–23 of Memoirs. (English transl.: *Biometrika*, **48**, 3–13, 1961; reprinted in Pearson and Kendall [16, pp. 157–167].)

[3]   Bernoulli, J. (1785). Milieu à prendre entre les observations. *Encyclopédie Méthodique, Mathématiques*, Vol. 2. Panckoucke, Paris/Plomteux, Liège, pp. 404–409. (2nd ed.; *Dictionaire Encyclopédique des Mathématiques*. Panckoucke, Paris, 1789.)

[4]   Euler, L. (1778). *Acta Acad. Sci. Imp. Petropolitanae*, Pars Prior, pp. 24–33 of the Memoirs. [English transl.: by C. G. Allen: *Biometrika*, **48**, 13–18, June 1961; reprinted in Pearson and Kendall [16, pp. 167–172].]

[5]   Euler, L. (1788). *Nova Acta Acad. Sci. Imp. Petropolitanae*, **3**, 289–297 of Memoirs. (Reprinted in *Leonhardi Euleri Opera Omnia*, Ser. 1, Vol. 7. B. G. Teubner, Leipzig, 1923, pp. 425–434.)

[6]   Gauss, C. F. (1809). *Theoria Motus Corporum Coelestium in Sectionibus Conicis Solem Ambientium*. Frid. Perthes et I. H. Besser, Hamburg. (Reprinted in *Carl Friedrich Gauss Werke*, Vol. 7. Perthes, Gotha, 1871, pp. 1–271; reissued, Göttingen, 1906; English transl. by C. H. Davis: *Theory of the Motion of the Heavenly Bodies Moving about the Sun in Conic Sections*. Little, Brown, Boston, 1857; reissued, Dover, New York, 1963.)

[7]   Itard, J. (1973). *Dictionary of Scientific Biography*, Vol. 7. Scribner's, New York.

[8]   Kendall, M. G. and Plackett, R. L., eds. (1977). *Studies in the History of Statistics and Probability*, Vol. 2. Charles Griffin, London/Macmillan, New York.

[9]   Kennedy, E. S. (1973). *A Commentary upon Bīrunī's Kitāb Tahdīd al-Amākin*. American University of Beirut, Beirut.

[10] Lagrange, J. L. (1776). *Miscellanea Taurinensia*, **5** (1770–1773), 167–232 (mathematics portion).

(Written before 1774 [7, p. 568] but not published in printed form until 1776 [30, p. 13]. Reprinted in *Oeuvres de Lagrange*, Vol. 2. Gauthier-Villars, Paris, 1868, pp. 171–234.)

[11] Lambert, J. H. (1760). *Photometria, sive de Mensura et Gradibus Luminis, Colorum et Umbrae* (Photometry, or on measuring and scaling of light, colors, and shade). Detleffsen, Augsburg.

[12] Lambert, J. H. (1765). *Beyträge zum Gebrauche der Mathematik und deren Anwendung*, Part 1. Verlage des Buchladens der Realschule, Berlin.

[13] Laplace, P. S. (1774). *Mém. Math. Phys. Présentés à l'Acad. Royale Sci.*, **6**, 621–657. [Reprinted in *Oeuvres complètes de Laplace*, Vol. 8. Gauthier-Villars, Paris, 1891, pp. 27–65 (Problème III, pp. 41–47).]

[14] Laplace, P. S. (1781). *Mém. Math. Phys. tirées Registres Acad. Royale Sci. Paris*, Année 1778, 227–332. (Reprinted on *Oeuvres complètes de Laplace*, Vol. 9. Gauthier-Villars, Paris, 1893, pp. 383–485.)

[15] Laplace, P. S. (1812). *Théorie Analytique des Probabilités*. Courcier, Paris. 3rd ed., same pagination, with corrections and three Supplements, Courcier, Paris, 1820. 4th ("national") ed., reprint of 3rd ed., with additioinal Supplement (1825) and new pagination (page n of 3rd $\simeq$ page 1.1n of 4th), published as *Oeuvres de Lapace*, Vol. 7. Imprimerie Royale, Paris, 1847; reprinted as *Oeuvres complètes de Laplace*, Vol. 7. Gauthier-Villars, Paris, 1886.]

[16] Pearson, E. S. and Kendall, M. G., eds. (1970). *Studies in the History of Statistics and Probability*. Charles Griffin, London/Hafner, Darien, Conn.

[17] Pearson, K. (1978). *The History of Statistics in the 17th and 18th Centuries*. Charles Griffin, London/Macmillan, New York. (Lectures by Karl Pearson given at University College London during the academic sessions 1921–1933, edited by E. S. Pearson.)

[18] Pitman, E. J. G. (1939). *Biometrika*, **30**, 391–421.

[19] Poincaré, H. (1892). *Thermodynamiques*. Georges Carré, Paris.

[20] Poincaré, H. (1912). *Calcul des Probabilités*, 2nd ed. Gauthier-Villars, Paris. (Revised and augmented by the author.)

[21] Seal, H. L. (1949). *Mitt. Ver. Schweiz. Versich.-Math.*, **49**, 209–228. (Reprinted in Kendall and Plackett [8, Paper 9].)

[22] Seal, H. L. (1950). *J. Inst. Actuaries Students' Soc.*, **10**, 255–258.

[23] Sheynin, O. B. (1966). *Nature (Lond.)*, **211**, 1003–1004.

[24] Sheynin, O. B. (1971). *Arch. History Exact Sci.*, **7**, 244–256.

[25] Sheynin, O. B. (1972). *Arch. History Exact Sci.*, **9**(1), 45–56.

[26] Sheynin, O. B. (1977). *Arch. History Exact Sci.*, **17**(1), 1–61.

[27] Simpson, T. (1756). *Philos. Trans. R. Soc.*, **49**, 82–83.

[28] Simpson, T. (1757). *Miscellaneous Tracts on Some Curious and Very Interesting Subjects in Mechanics, Physical-Astronomy and Speculative Mathematics*. J. Nourse, London.

[29] Stigler, S. M., (1980). *American Contributions to Mathematical Statistics in the Nineteenth Century*, 2 vols. Arno Press, New York.

[30] Taton, R. (1974). *Rev. d'Histoire Sci. Appl.*, **27**, 3–36.

[31] Todhunter, I. (1865). *A History of the Mathematical Theory of Probability from the Time of Pascal to That of Laplace*. Macmillan, London. (The standard reference on the history of probability; not entirely trustworthy. Reprinted by G. E. Stechert, New York, 1931, and Chelsea, New York, 1949, 1965.)

(APPROXIMATIONS TO DISTRIBUTIONS
ARITHMETIC MEAN
LIMIT THEOREMS, CENTRAL
MAXIMUM LIKELIHOOD
NORMAL DISTRIBUTION)

CHURCHILL EISENHART

LAWS OF ERROR II: THE GAUSSIAN DISTRIBUTION

How best to determine objectively the coefficients of a linear relationship $y = a + \beta x$ from observational points (Y_i, x_i) was a matter that received a great deal of attention in the latter half of the eighteenth century. A variety of procedures were proposed and used (see, e.g., Eisenhart [27, pp. 26–27], Sheynin [67, p. 254], Plackett [60, p. 239], and Harter [34, pp. 148–152]). Carl Friedrich Gauss (1777–1855) recognized in 1794 the algebraic and arithmetical advantages of taking for α and β the values a and b that minimize the sum of squared residuals, $\sum_i (Y_i - a - bx_i)^2$, and has said that he used this procedure regularly from 1794 or 1795 [29, art. 186; 60, pp. 240–248]. He said that he attempted to justify his minimum-sum-of-squared-residuals technique by means of the theory of probability in 1797 but soon found

out that determination of the most probable value of an unknown quantity is impossible unless the probability distribution of errors of observation is known explicitly [30, par. 6 (reprinted in *Werke*, Vol. 4, p. 98)]. Therefore, "it seemed to him most natural to proceed the other way round" and seek the probability distribution of errors that "in the simplest case would result in the rule, generally accepted as good, that the arithmetic mean of several values for the same unknown quantity obtained from equally reliable observations shall be considered the most probable value" [30, par. 6].

GAUSS'S DERIVATION

1. To this end, Gauss, as first step, assumed [29, art. 175] that the error, $X = Y - \tau$, of a measurement of direct observation, Y, of a quantity to be distributed in accordance with a single-valued probability density function, $\phi_X(x, \tau) = \phi_X(x)$, that is a symmetric strictly decreasing function of x, analytic in the interval $(-\infty, +\infty)$. He noted [29, art. 175] that, when this is the case, the joint probability (density) function for n independent equally reliable measurements $Y_1, Y_2, \ldots, Y_n$ of the same quantity τ is given by the product

$$\phi(y_1 - \tau) \cdot \phi(y_2 - \tau) \cdots \phi(y_n - \tau) = \Omega,$$

where $\phi(y - \tau) = \phi_X(x)$ for $x = y - \tau$.

2. Next he assumed all of the possible values of τ to be equally probable "previous to the observations" [29, art. 176], and, using the *inverse probability* technique of Thomas Bayes (1702–1761), without reference to Bayes, showed that the probability density function of the a posteriori distribution of τ given the particular observed values $y_1, y_2, \ldots, y_n$ of the measurements or observations $Y_1, Y_2, \ldots, Y_n$ is proportional to the product Ω above (considered as a function of τ for $y_1, y_2, \ldots, y_n$ fixed); and that therefore [art. 177] the most probable value of τ given the y's is that value of τ

which makes Ω a maximum, i.e., the value of τ for which

$$\sum_1^n \frac{d \log \phi(y_i - \tau)}{d\tau} = 0$$

3. He then made his famous pronouncement that one should adopt the traditional arithmetic mean rule as a practical "axiom":

> It has been customary certainly to regard as an axiom the hypothesis that if any quantity has been determined by several direct observations, made under the same circumstances and with equal care, the arithmetical mean of the observed values affords the most probable value, if not rigorously, yet very nearly at least, so that it is always most safe to adhere to it. [29, art. 177].

Since the arithmetic mean of $y_1, y_2, \ldots, y_n$ is the solution of the equation

$$\sum_1^n (y_i - \tau) = 0,$$

if, in keeping with the dictum above, it is also to be the "most probable value," then it must also be a solution of the preceding equation. Pairing and equating corresponding terms of the two summations it is seen that a solution in common requires that

$$\frac{d \log \phi(y_i - \tau)}{d\tau} = k(y_i - \tau)$$

$$(i = 1, 2, \ldots, n),$$

whence

$$\phi(y - \tau) = ce^{-(1/2)k(y-\tau)^2}$$

and k must be positive if the corresponding value of the joint probability density Ω is to be maximum. (By considering the special case of $y_2 = y_3 = \cdots = y_1 - nN$ for n an *arbitrary* positive integer, Gauss showed the foregoing to be the unique solution.)

Setting $k = h^2$ and restoring $x = y - \tau$, the desired law of error is seen to be

$$\phi_X(x) = (h/\sqrt{\pi})e^{-h^2x^2}, \quad -\infty < x < \infty$$

$$(1)$$

in which, Gauss pointed out [29, art. 178], "the constant h can be considered as the measure of precision ('praecisionis') of the observations."

Gauss commented [art. 178] that (1) cannot represent a law of error in full rigor because it assigns probabilities greater than zero to errors outside the range of possible errors, which in practice always has finite limits; that such a feature is unavoidable because one can never assign limits of error with absolute rigor; but this shortcoming is of no importance in the case of (1), because it "decreases so rapidly, when hx has acquired a considerable magnitude, that it can safely be considered as vanishing."

He then achieved his original goal by showing [art. 179] that when errors of observation are independently distributed in accordance with this law (1) and one has a set of observations whose theoretical values are functions of a number of quantities of unknown magnitudes, the most probable values of these quantities are those given by minimizing the sum of the squared differences between the observed and theoretical values, these differences to be multiplied before squaring by the measures of precision h of the corresponding observations, if these are of unequal precision.

The foregoing became known as Gauss's *first* "proof" of the method of least squares* because he published an alternative justification of the method in the 1820s. It was this first justification, however, that struck a responsive chord in physical scientists and Gauss's formulation of the method of least squares became the universally accepted procedure for the analysis of observational and experimental data. Contributing to its acceptance, no doubt, were his recognition of the constant h as a measure of the precision, his provision of a formula for the precision of a linear function of independent observations of equal or unequal precisions, his rule for weighting results of unequal precision so as to obtain the combined result of maximum attainable precision, and his elegant computational procedure for obtaining values of the quantities involved from the data in hand. As a result his probability distribution of error (1) became known as "the law of error."

CRITICISM OF GAUSS'S DERIVATION

Nonetheless, there were some objections to his derivation of it. Hidden in Gauss's assumption 1 is the tacit assumption that the probability of an error X is a function of its magnitude only, i.e., that $\phi_X(x)$ is a function of x only, and not also of τ, the quantity measured, since τ enters only as the location parameter of the probability density function, $\phi_Y(y - \tau)$, of $Y = X + \tau$, a measurement of τ with error X. Julian Lowell Coolidge (1873–1954) considered this tacit assumption "reasonably plausible" inasmuch as we think of random errors as arising from physical causes affecting the operation of measurement or observation, and not dependent on the position of zero on the scale of the measuring or observing instrument [15, p. 114]. Joseph Bertrand (1822–1900) questioned this implicit assumption on the grounds that the probability of an error X being of magnitude x might also depend on the magnitude of the quantity measured, τ [7, p. 564; 8, art. 139]. [If $\phi_X(x)$ does not depend also on τ, this precludes, for instance, the possibility that the standard deviation, σ_Y ($= \sigma_X = 1/h_X\sqrt{2}$), of a measurement Y of τ is a function of τ, e.g., proportional to τ or $\sqrt{\tau}$.] If the probability of an error X being of magnitude x is allowed to be a function also of τ, $\phi_X(x) = g(x, \tau)$, say, then Birger Meidell (1882–1958) and Henri Poincaré (1854–1912) have shown [21, art. 30; 48; 61, arts. 111, 112] that the arithmetic mean, $\bar{y}$, of the observed values, $y_1, y_2, \ldots, y_n$ of n independent measurements $Y_1, Y_2, \ldots, Y_n$ of a quantity τ is the "most probable" value of τ given the observed values for a law of error more general than Gauss's law (1); and in the process demonstrated that Gauss's assumption 2 of a uniform a priori probability density of τ (to which Bertrand had objected also, according

to Poincaré [6, art. 110]) is necessary, if the arithmetic mean Y is to be the "most probable" value of τ a posteriori in their general, or Gauss's special case (see also Coolidge [15, p. 115] and Deltheil [21, p. 60]).

Paolo Pizzetti (1860–1918) commented that, in Gauss's famous statement of the arithmetic mean being the most probable value, his qualifying remark "if not rigorously, yet very nearly at least" ("si non absoluto rigore, tamen proxime saltem") could potentially undermine the favored standing of his law of error (1), because there are many other laws of error, e.g.,

$$f_X(x) = \frac{2\sqrt{h}}{\Gamma(\frac{1}{4})} e^{-h^2 x^4}, \qquad -\infty < x < \infty$$

(2)

that lead to most probable values close to the arithmetic mean and converge rapidly to it as the number of observations, n, increases [20, p. 156; 59, p. 213].

LAPLACE PROVIDES SUPPORT FOR THE LAW

In contrast to these later objections, immediately after Gauss's publication of his law of error (1), Laplace greatly strengthened its acceptability as a practical law of error by his demonstrations that the probability distributions of arithmetic means [42, Part VI; 44, Book II, arts. 18, 19 (English translation of art. 18 in Smith [69])], and of linear functions [43, Part VI; 44, Book II, art. 20] of n independent errors having the same or different precision (or standard deviations) can be approximated (when properly scaled), by Gauss's distribution (1), with the error of the approximation tending to zero as $n \to \infty$, the *central limit theorem**. Hence the method of least squares was seen to yield "most probable values" (under "very general conditions") when the number of independent observations involved is large; and to be supported by the calculus of probabilities without appeal to the Postulate of the Arithmetic Mean, when the number of observations is large. In the Supplement to the 1810 memoir, he showed further that when a value for an unknown quantity has been obtained by taking the arithmetic mean $\overline{Y}$ of n observations whose errors are i.i.d. according to some law of error (having finite standard deviation); by taking the arithmetic mean Y' of n' other observations whose errors are i.i.d. according to a different law of error (with finite SD); by taking the arithmetic mean Y'' of n'' further observations whose errors are i.i.d. according to a third law of error, and so forth, then his method of minimum absolute error of estimation, (Lambert's and) Bernoulli's method of maximizing the probability of obtaining the observations in hand, and Gauss's method of maximum a posteriori probability given the observations in hand, and the method of weighted sum of squared residuals (Gauss's form of the method of least squares) all lead to the same choice of "best" overall mean. And he provided further support for the law of error (1) by showing [44, Book II, art. 23] that, among all the laws of error of the form $f(x) = Ke^{-\psi(x^2)}$, where $\psi(x^2)$ is an arbitrary continuous function of $x^2 = (y - \tau)^2$, the Gaussian distribution (1) is the only one for which the arithmetic mean $\overline{Y}$ of n independent observations is the minimum mean absolute error estimator of τ. Together these findings linked Gauss's distribution to the traditional belief that the arithmetic mean is the "best" estimator without recourse to inverse probability.

ADRAIN'S TWO DERIVATIONS

Robert Adrain (1775–1843), Irish emigrant to America, gave, in a paper [3] completed in 1808 [1, p. 415] but not printed until the autumn of 1809 [36, note 2], two very different derivations of the law of error (1). The first (ref. 3 [pp. 93–95]; reproduced in original notation, but without diagrams, in Abbe [1, pp. 412–414]; and in modern notation, with discussion, in Coolidge [16, pp. 67–68] and Glaisher [31, pp. 76–81]) starts from the assumptions that (a) two different lengths,

say λ_1 and λ_2, are to be measured; (b) the errors X_1 and X_2 made in measuring them will be proportional to their respective magnitudes; but (c) they will be independent, so that their joint probability distribution is the product of their individual probability distributions; and each (d) is assumed to be of the same functional form. Adrain then asks what is the probability of the errors X_1 and X_2 given that $X_1 + X_2 = C$, say. To answer this, he maximizes the probability of the joint occurrence of X_1 and X_2 for $X_1 + X_2 = C$ and arrives at a differential equation for the common functional form of their probability density functions, the "simplest" solution of which is a simple variant of the Gaussian error function (1).

This derivation is open to criticism on two counts. First, the assumption (b) that the errors made in measuring the two lengths will be proportional to the lengths measured "seems very far from being evident, not to say very far from being true, generally" [31, p. 78]. Second, Adrain's differential equation has a more general solution [16, p. 68; 31, p. 81]. "Why the simplest solution should be *the* solution is a mystery" [16, p. 68].

Adrain's second derivation (ref. 3 [pp. 96–98]; reproduced in original notation in Merriman [50, p. 141; 53, p. 164]; and in modern notation, with discussion, in Coolidge [16, pp. 68–69]) starts from the following assumptions:

1. As before, two quantities are measured, the distance and bearing of an object as determined by an observer at a fixed location.

2. Errors in distance and bearing, X_d and X_b, respectively, may be regarded as the rectangular coordinates of a point in a plane, with probability (density) functions, $f_{X_d}(x)$ and $f_{X_b}(x)$, respectively, $-\infty < x < \infty$.

3. Positive and negative errors are equally probable, $f_{X_d}(-x) = f_{X_d}(x)$, $f_{X_b}(-x) = f_{X_b}(x)$.

4. Large errors are less probable than small errors, that is, $f_{X_d}(x)$ and $f_{X_b}(x)$ are decreasing functions of $|x|$.

5. Errors in distance, X_d, and bearing, X_b, are stochastically independent, so that their joint probability (density) function is the product of their respective probability (density) functions, $f_{X_d, X_b}(x_1, x_2) = f_{X_d}(x_1) \cdot f_{X_b}(x_2)$.

He notes that the locus of points for which $f_{X_d, X_b}(x_1, x_2)$ has the same value will be symmetric with respect to both axes, and will intersect any line parallel to either axis in only two points. If now we consider the "simplest possible" case in which

6. X_d and X_b are so scaled that an error of magnitude x has the same probability for each, $f_{X_d}(x) = f_{X_b}(x) = f(x)$, the locus of points for which the values of $f_{X_d, X_d}(x_1, x_2)$ are the same will be a circle, $f_{X_d, X_d}(x_1, x_2) = g(x_1^2 + x_2^2)$.

Hence for a particular circle $x_1^2 + x_2^2 = r_0^2$, say, taking assumption 5 the stochastic independence of X_1 and X_2 into account, one has

$$f(X_1) \cdot f(X_2) = \text{constant}.$$

Expressing the latter in logarithmic form and then taking derivatives of both equations, one obtains

$$2x_1 \, dx_1 + 2x_2 \, dx_2 = 0 \quad \text{or}$$
$$x_1 \, dx_1 = -x_2 \, dx_2,$$

and

$$\frac{d[\log f(x_1)]}{dx_1} \, dx_1 + \frac{d[\log f(x_2)]}{dx_2} \, dx_2 = 0,$$

respectively. Substituting for dx_1, say, in the latter and transposing yields

$$\frac{d[\log f(x_1)]}{dx_1} \frac{1}{x_1} = \frac{d[\log f(x_2)]}{dx_2} \cdot \frac{1}{x_2}$$
$$(x_1 \neq 0, \, x_2 \neq 0)$$

that is,

$$\frac{d[\log f(x)]}{dx} \frac{1}{x} = k$$

or

$$\frac{d[\log f(x)]}{dx} = kx \quad (x \neq 0, \, k \neq 0),$$

where k is some constant. Integrating gives

$$\log f(x) = c + \frac{kx^2}{2}, \quad -\infty < c < \infty,$$

whence

$$f(x) = \exp\left(c + \frac{kx^2}{2}\right) = C \exp\left(\frac{kx^2}{2}\right)$$

In view of assumption 4 that $f(x)$ is a decreasing function of $|x|$, k must be negative, so setting $k = -h^2$, and evaluating C so that $\int_{-\infty}^{+\infty} f(x) \, dx = 1$, the Gaussian error function (1), is obtained.

The weakest link in Adrain's second derivation is assumption 5 of stochastic independence of the errors in direction and bearing, a circumstance that one surely cannot assume to prevail generally in practice.

Immediately following these two derivations of the law of error (1), Adrain proceeded to employ it to deduce the "most probable" solutions to four specific problems [3, pp. 98–109], showing that the described solutions were those given by the technique of minimizing the sum of squared residuals. Inasmuch as

1. We are told [6, p. 277; 16, p. 69] that Adrain had in his library a copy of the book [45] in an Appendix to which Adrien Marie Legendre (1752–1833) introduced the world to the technique of minimum sum of squared residuals solely on the basis of the practical requirements of general applicability, unique arithmetical solutions and ease of computation, without resource to probability or "law of error" considerations, and named it "the method of least squares" ("La Méthode des Moindres Quarrés"), and

2. Adrain's use ot the technique of minimum sum of squared residuals is quite similar to Legendre's,

it seems reasonable to suppose that Adrain's discovery of the law of error (1) was motivated, like Gauss's, by a desire to find a justification for the method of least squares through application of the probability theory of errors. Furthermore, there is no reason to doubt that Adrain devised his derivations without knowledge of Gauss's derivation published in his *Theoria Motus* the preface to which is dated "28 Martii 1809" and copies of which did not reach Paris until May 1809 (Letter from Legendre to Gauss dated 31 May 1809; English translation in Plackett [60, pp. 242–243]).

Owing to the obscurity and scarcity of copies of Adrain's *Analyst*, his derivations remained unknown to developers and students of the theory of errors until unearthed, the first in Abbe [1], the second in Merriman [50, 53], and consequently appear to have had no impact on the development of the theory of errors, which by the 1870s had more or less run its course.

FURTHER "PROOFS" OF GAUSS'S LAW OF ERROR

The usefulness, and popularity of Gauss's development of the method of least squares fostered many additional "proofs" of his law of error (1). They fall into two distinct classes:

1. Those based on somewhat arbitrary but nonetheless quite specific assumptions about the mathematics of errors.

2. Those based on the "hypothesis of elementary errors," i.e., the assumption that the error of any single measurement or observation is the algebraic sum of a very large—essentially, infinite—number of "elementary errors" from different sources, individually negligible in comparison to their sum, and independently distributed in such a manner that a Gaussian distribution of any particular error is assured by the central limit theorem.

"Proofs" Based on Specific Assumptions about the Making and Combination of Errors

HERSCHEL. The English astronomer Sir John F. W. Herschel (1792–1871) gave an informal prose description of a derivation quite similar to Adrain's second derivation in his review [35, p. 17 (or p. 11)] of the French and English editions of Quetelet's *Letters . . . on the Theory of Probabilities . . .* [64]:

> Suppose a ball dropped from a given height, with the intention that it shall fall on a given mark. Fall as it may, its deviation from the mark is *error*, and the probability of that error is the unknown function of its square, i.e. of the sum of the squares of its deviations in any two rectangular directions. Now, the probability of any deviation depending solely on its magnitude, and not on its direction, it follows that the probability of each of these rectangular deviations must be the same function of *its* square. And since the observed oblique deviation is equivalent to the two rectangular ones, supposed concurrent, and which are essentially independent of one another* . . . its probability will be the product of their separate probabilities. Thus the form of our unknown function comes to be determined from this condition, viz., that the product of such functions of two independent elements is equal to the same function of their sum. But . . . this property is the peculiar characteristic of, and belongs only to, the exponential or antilogarithmic function. This, then, is the function of the square of the error, which expresses the probability of committing that error. [35, reprinted in *Essays* (1857), pp. 398–399]

The footnote, added in 1857, reads: "*That is, *the increase or diminution in one of which may take place without increasing or diminishing the other.* On this, the whole force of the proof turns."

The Cambridge mathematician Robert Leslie Ellis (1817–1859) interpreted [28, pp. 325–326] Herschel's prose description to signify that:

1. The probability density function of a deviation r of the point of impact from the mark of aim in any particular direction in the horizontal plane may be expressed by a function $f(r^2) = f(x^2 + y^2)$

for x and y the coordinates of the point of impact with respect to a purely arbitrary pair of orthogonal axes in the plane with origin at the point of aim.

2. The components of r in the x- and y-directions are probability-wise dependent with probability density functions $f(x^2)$ and $f(y^2)$, respectively, "f" being the same function in all three cases so that

3. $$f(x^2) \cdot f(y^2) = f(x^2 + y^2),$$

a functional equation of which the solution was stated, but not explicitly indicated, to be e^{mx^2}, m being a constant.

Ellis pointed out [28, p. 326] that the functional equation above is incorrect; it should read

$$f(x^2) \cdot f(y^2) = f(0) \cdot f(x^2 + y^2),$$

the solution of which is $f(x^2) = Ae^{mx^2}$, A and m both constants. And [pp. 326–327], since we must have $A \int_{-\infty}^{+\infty} f(x^2)\,dx = 1$, m must be negative, $m = -h^2$, say, whence $A = h/\sqrt{\pi}$, so that $f(x^2) = \phi_x(x)$, Gauss's law of error (1).

Ellis criticized Herschel's assumption of the probability-wise independence of the projection of an error r on the x and y axes, saying that "there is no shadow of reason for supposing that the occurrence of a deviation in one direction is independent of that of a deviation in another, whether the two directions are at right angles or not" [28, pp. 325–326]; and adding that "some notion of an analogy with the composition of forces probably prevented [Herschel] from perceiving that, unless it can be shown that a deviation y occurs with the same comparative frequency when x has one value as when it has another, we are not entitled to say that the probability of the concurrence of two deviations x and y is the product of the probabilities of each" [p. 326]. Herschel responded by adding the footnote, when the review was reprinted.

The Adrain–Herschel proof was widely accepted in spite of its questionable basis.

The mathematician and logician George Boole (1815–1864), citing Herschel as its originator, reproduced it [11, p. 628] with Ellis's correction. He restated it as a problem for solution by the "calculus of functions" in his *Calculus of Finite Differences* [12, p. 228; later editions, p. 311], gave the functional equation in the form $f(x) \cdot f(y) = f(0) \cdot f(\sqrt{x^2 + y^2})$, and provided an explicit solution.

Herschel's reasoning was used in three dimensions without attribution, by the physicist James Clerk Maxwell (1831–1879), in his first derivation of the probability distribution of velocities of gas molecules [47, Prop. IV].

Lord Kelvin (Sir William Thomson, 1824–1907) and Peter Guthrie Tait (1831–1901), citing Boole [11], gave a derivation similar to Herschel's in their *Treatise on Natural Philosophy* [72, Vol. 1, art. 391]. They start with dropping a ball; but their derivation involves rotation of the rectangular coordinate system. Their version was picked up from the German translation of their *Treatise* by O. X. Schlömilch (1823–1901), professor in the Polytechnical University, Dresden, who considered it "simple and clear" ("einfache und anschauliche") [65]. Oscar Sheynin tells us [66, Sec. 10] that (a) A. N. Krylov gave [39, Chap. 8] a similar proof, except that he considered a shot fired at a vertical target; (b) Krylov's version is reproduced in Yakovlev [76]; and (c) it has been criticized by Yu. V. Kemnits [37].

W. M. Smart (1889–1975), Regius Professor of Astronomy, University of Glasgow, has provided [68, Sec. 3.06] a new presentation of Herschel's proof that incorporates features of both Ellis's [28] and Boole's [12] versions. He considers the assumption of identical laws of error in perpendicular directions "entirely reasonable for, under the conditions of the experiment, one set of axes is not likely to be more fundamental than any other"; notes that the assumption of probability-wise independence of error components in perpendicular directions is essential to the proof, which would have to be abandoned if it were found to be false; and that: "in the problem as stated, it is clearly necessary to introduce one or more assumptions if a solution is to be achieved."

DONKIN. A very different derivation was given by the Oxford astronomer, W. F. Donkin (1814–1869). He comments [22, p. 159]: "If two observations of the value of an unknown quantity x give respectively $x = a$, $x = b$, and if we have no reason for putting more confidence in one than in the other, then . . . the most probable value must be $(a + b)/2$, since there is no reason for assigning a value nearer to one of the limits a, b than to the other." (A formal proof of this statement [31, p. 92] indicates that it involves a tacit assumption of a symmetric law of error.) "But," Donkin continues, "we cannot, *without making some further assumption*, extend this conclusion to the case of *three* or any greater number of observations." Next, supposing that "the two observations are of the same kind in all respects," he lets $\phi(x - a)\,dx$ denote the [fiducial] probability density function of the unknown value x "arising from the first observation"; $\phi(x - b)\,dx$ the corresponding [fiducial] probability arising from the second; and

$$C\phi(x - a)\phi(x - b)\,dx$$

the joint [fiducial] probability distribution of the unknown value x "arising from both observations combined," with $C = C(a, b)$ "determined by the condition that the integral of the above expression between $\pm \infty$ must $= 1$." "On the other hand, since the most probable value of x arising from the combined observations must be $\frac{1}{2}(a + b)$, it appears a natural and obvious assumption (though I do not pretend that it is *not* an assumption), that *the probability that x is between x and $d + dx$ be expressible in the form*

$$\psi\left(x - \frac{a + b}{2}\right) dx,$$

so that we shall have

$$C\phi(x - a)\phi(x - b) = \psi\left(x - \frac{a + b}{2}\right)."$$

Solving this functional equation [22, p. 160],

he finds

$$\phi(x) = Ce^{Ax + B(x^2/2)};$$

and notes that the condition that $\phi(x)$ be a maximum for $x = 0$ implies that $A = 0$; $\int_{-\infty}^{+\infty}\phi(x)\,dx = 1$, implies that B be negative, $-2h^2$ say, whence $C = h\sqrt{\pi}$ and $\phi(x) = \phi_X(x)$ of (1).

While Donkin's $\phi(x)$ here is of the Gaussian law of error form (1), his $\phi(x - a)\,dx$ is the probability that the unknown value of the quantity concerned lies between x and $x + dx$ given the observed a; and is *not* the probability $\phi(a - x)\,da$ that the value of an observation A will lie between a and $a + da$ given that the unknown value is x.

"Proofs" Based on the Hypothesis of Elementary Errors

As noted by Czuber [20, p. 164], these "proofs" may be divided into three categories according to the assumptions made about the values that the elementary errors may take and their probability distributions:

1. The elementary errors may take specific discrete values only and are i.i.d. according to a particular probability distribution [4, 32, 71].
2. The elementary errors are independently distributed symmetrically about zero according to probability distributions that may be of different form [10].
3. Same as category 2 except that positive and negative errors of equal magnitude are not required to be of equal probability [17, 18, 59].

YOUNG. The hypothesis of elementary errors seems to have been employed first by the English physicist Thomas Young (1773–1829), in his "Remarks on the probabilities of error in physical observations . . . " [77]. In Sec. 1, "On the estimation of the advantage of multiplied observations," Young draws attention to the "apparent constancy" of the frequency of particular outcomes in large aggregates, citing "the occurrence, for example, of almost an equal number of dead letters every year in a general post office" (p. 71). He says that this class of phenomena "may be best explained [as] we shall discover" by demonstration "that the combination of a multitude of independent sources of error, each liable to incessant fluctuation, has a natural tendency, derived from their multiplicity and independence, to diminish the aggregate variation of their joint effect" (p. 71). He calculates some numerical values (pp. 72–74), and then develops (pp. 74–75) formulas for the *mean (absolute) error* and the *probable error* of the arithmetic mean of *n* observations whose errors are i.i.d. in accordance with a binomial distribution with $p = \frac{1}{2}$. He notes that his results "do not materially differ from those of Legendre [sic], Bessel, Gauss, and Laplace," and considers his "mode of investigation . . . to be more simple and intelligible" (p. 79). He does not attempt to find an asymptotic expression for this binomial distribution as $n \to \infty$.

HAGEN. Eighteen years later, the German hydraulic engineer Gotthilf Heinrich Ludwig Hagen (1797–1884) published a derivation of the law of error (1) based on the assumption or hypothesis that "the error of an observation is the algebraic sum of an infinitely large number of elementary errors, all having equal size, and which can just as readily be positive as negative" [32, p. 34; 33, p. 28]. He cited [32, p. 45; 33, p. 28] Young's publication as an instance of earlier use of this hypothesis.

Hagen begins his "proof" by assuming the error X of any single observation to be the algebraic sum of $N = 2n$ elementary errors, $e_1, e_2, \ldots,$ each of which may independently take the values of $-\epsilon$ or $+\epsilon$ with probability $\frac{1}{2}$, so that

$$\Pr[X = 2m\epsilon] = \frac{(2n)!}{(n-m)!(n+m)!} \cdot \frac{1}{2^{2n}}$$

$$(m = 0, \pm 1, \pm 2, \ldots, \pm n),$$

since $(n-m)(-\epsilon) + (n+m)(\epsilon) = 2m\epsilon$. Next he derived an asymptotic expression for the slope of a polygonal segment joining

the tops of two successive ordinates of the binomial distribution above as $N \to \infty$; then obtained by integration a formula akin to (1) for the corresponding continuous probability distribution.

Hagen's proof was made available in English, with some modifications, by Bartholomew Price [63, pp. 376–379], Charles H. Kummell [40, pp. 133–135] and Mansfield Merriman [51, pp. 180–184]. Hagen's proof and these were nonrigorous and involved seemingly arbitrary assumptions, causing Kummell and Merriman to argue over details [41, 52]. W. M. Smart [68, Sec. 2.18] has provided a more rigorous derivation patterned on Hagen's, that should be acceptable today.

TAIT. P. G. Tait published on his own [71] a derivation of the "law of frequency of error" based on the assumption that an error stemming from any source may be considered equivalent to the deviation ($w - np$, say) from the most probable outcome (np) of the observed number (w) of white balls obtained in a large number (n) of random drawings, with replacement, from a bag containing white and black balls in known ratio, $p : q$, $(p + q = 1)$. This analogy seems far fetched and his derivation is simply a redo of Laplace's normal approximation to sums of terms of the binomial expansion of $(p + q)^n$ as $n \to \infty$ [44, Book II, Sec. 16; 73, art. 993].

AIRY. The Astronomer Royal, George Biddell Airy (1801–1892), reproduced [4, Secs. 14–20] Laplace's proof [44, Book II, Sec. 18] of the asymptotic normality of the sum of n errors i.i.d. symmetrically about zero in accordance with Simpson's discrete uniform distribution, and remarked: "The fundamental principle in this investigation is, that an error, as actually occurring in observation, is not of simple origin, but is produced by the algebraical combination of a great many independent causes of error[1], each of which, according to the chance which affects it independently, may produce an error, of either sign and of different magnitude" [4, Sec. 13]. The footnote reads: "[1]This is not the language of Laplace, but it appears to be the understanding on which his investigation is most distinctly applicable to single errors of observation." He does not mention Hagen (or Young).

BESSEL.   The astronomer Friedrich Wilhelm Bessel (1784–1846) published a memoir [10] in which he generalized the Hagen "proof" by assuming that (a) the error of an observation is the sum of a large number of elementary errors, (b) positive and negative elementary errors of the same size have the same probability, (c) the laws of error of the elementary errors may be of different functional form and may range between different finite limits, $\pm a, \pm b, \ldots$, and (d) the mean square errors of the individual elementary errors are of the same order of magnitude, so that no individual error will exceed considerably the others. [Condition (d) is essentially the Lindeberg condition for the central limit theorem.] He remarked [10, Sec. 10]: "Cases in which an error does not result from many independent causes ("Ursachen") working together are probably very rare; even in very simple appearing kinds of observation it is often possible to demonstrate numerous causes of their errors." By way of illustration he then enumerated [cols. 398–400; English translation in Adams [2, pp. 61–63]] 13 potential sources of components of error in the measurement of the distance of a fixed star from the zenith by means of a meridian circle.

CROFTON.   The Irish mathematician, Morgan William Crofton (1826–1915), professor of mathematics at the Royal Military Establishment, Woolwich, states early in the introduction to ref. 17 that his object is to provide a "most general" derivation of "the law of single errors of observations" from the hypothesis of elementary errors, in a manner quite different from Laplace's [44, Book II, Chap. 4] and Poisson's [62, Chap. 4] derivations of what we term the central limit theorem, and to be "most general" in that the laws of error of the respective "elementary errors" (a) need not be all the same, (b) nor

be symmetrical with respect to zero error. Near the end of the introduction he stresses a feature essential for the practical validity of the hypotheses of elementary errors: "It is not enough for our purpose to show, could we do so conclusively, that the error in practice is compounded of a large number of smaller errors; we must also show that they are *independent*, at least for the most part" [17, pp. 178–179]; and follows this comment with an example of error components in astronomical observations which stem from a general rise (or fall) of temperature during the observing period, and therefore are *not* independent.

In this paper he gives two derivations [17, pp. 183–187] in both of which he assumes the (total) error of a single observation to be of the form

$$T = S + \sum_{1}^{n} X_i$$

where S is "any Error of indefinite amplitude," the X_i are "diminutive" errors, each "of indefinitely small importance" compared with S, and distributed independently of each other in accordance with arbitrary laws of error with finite means $\mu_1'(X_i)$ and finite means squares $\mu_2'(X_i)$, whose mean cubes, mean fourth powers, etc., are negligible in comparison with their respective mean squares ($i = 1, 2, \ldots, n$), and S is "the compound finite Error resulting from the combination of all others" and distributed in accordance with a probability density function $f(S)$ having finite mean error and finite mean square error. Starting by considering the addition to S of only one "diminutive" error, X_1, say, he approximates the probability density function $h(s + x_1)$ of "the resulting compound error," $S + X_1$, by a series expansion of $h(s + x)$ in terms of $f(S)$, its first two derivatives, and the first two moments, $\mu_1'(X_1)$ and $\mu_2'(X_1)$, of the added error X_1.

Iterating this process, he shows that as the number n of *added* "minute independent Errors," X_i, is increased indefinitely, the asymptotic normal distribution of their sum dominates the distribution of the total error T, which will tend to be approximately normally distributed with mean equal to the sum of the means of *all* of the added minute errors, and variance equal to the sum of the variances of *all* of the added minute errors. In the first derivation he takes the X_i to be normally distributed, following the suggestion of one of the paper's referees. In the second, he lets the distributional forms of the X_i be arbitrary, but takes S to be normally distributed.

Crofton gave a third derivation in the 9th edition of the *Encyclopaedia Britannica* [18, art. 48]. It begins much as does the second above, except that S is not assumed to be normally distributed. He shows that the asymptotic distribution of the total error T as $n \to \infty$ is a function only of T and the sums of the means and variances of *all* of the added minute errors, M and $\sum^2$, say. M is then altered by an amount δm, leading to a differential equation. All of the minute errors are next increased in the ratio r, this leads to another differential equation. A lot of manipulation and ultimate integration yields the result obtained in the first two instances.

For the asymptotic normal distribution that Crofton reaches in each case to be equivalent to Gauss's law of error (1), it is necessary that the sum of the means of all of the added minute errors be zero. This will be the case if the law of error of each of the added errors is symmetric with respect to zero; or if for laws of error not symmetrical about zero, each of their means is zero; or if not individually zero, their sum is somehow constrained to be zero.

The English economist and statistician Francis Ysidro Edgeworth (1845–1926) employed a somewhat similar argument in his first paper on "the law of error" [23], carried the approximation further in ref. 25, and terming this approach "the method originated by Professor Morgan Crofton" devoted a major section to its exposition and extension in ref. 26 [Sec. II].

PIZZETTI. Czuber [20, pp. 167–168] tells us that generality comparable to Crofton's aim was achieved by Paulo Pizzetti in the first part of the lengthy paper [59] in which he summarizes the work on the theory of errors up to that time, with a bibliography of 503 entries. He made no initial assumptions about the n individual errors $x_1, x_2, \ldots, x_n$ considered beyond that they be distributed independently according to discrete or continuous probability distributions within finite limits (a_i, b_i) $(i = 1, 2, \ldots, n)$ which might, or might not, include or be symmetric with respect to 0. Using a technique equivalent to the method of characteristic functions, Pizzetti derived an expression for the distribution of the sum $x_1 + x_2 + \cdots + x_n$ of n such errors; looked for an asymptotic form of this distribution under the assumptions that the contributions of the individual errors to the variability of the sum be of the same order of magnitude, or, if there were some whose individual variabilities were substantially greater than that of the others, they be finite in number; and thus arrived at the asymptotic normal distribution of the sum of n independent random variables under a version of the Lindberg condition. For Pizzetti's asymptotic normal distribution to be equivalent to Gauss's law of error (1), it is necessary, as in Crofton's cases, for the sum of the means of all of the component errors to be exactly zero.

Of all these "proofs" of the law of error (1), only three—Gauss's, Hagen's, and Herschel's—found their way into expositions of the combination of observations, the method of least squares or the theory of errors. The others passed into oblivion, except in Czuber [19], where all are presented in detail (not always the same as in the original), with mention of some additional secondary sources; and in Czuber [20, Secs. 54, 56–57], where they are discussed in less detail. Some authors prefer the Young–Hagen elementary errors "hypothesis" to Gauss's arithmetic mean "axiom," e.g., Merriman [51, p. 174; 54, arts. 23–36], Brunt [13, art. 8], and Worthing and Geffner [75, Chap. VI, art. 2].

EMPIRICAL SUPPORT

In addition to these efforts to prove mathematically that the Gaussian distribution (1) is *the* law of error governing repeated independent measurements, efforts were made to demonstrate empirically that it is implied by observed distributions of real-life repeated measurement of fixed quantities.

BESSEL. The Prussian astronomer, geodesist, and mathematician Friedrich Wilhelm Bessel (1784–1846) was the first to carry out moderately large scale comparisons of the Gaussian distribution (1) with actual distributions of the errors of repeated measurements. In ref. 9 he analyzed 300 observations of the *declinations* of the six stars that the English astronomer James Bradley (1693–1762) had made at the Greenwich Observatory during the years 1750–1762; 300 of Bradley's observations of the *right ascensions* in time units of four stars; and 470 of Bradley's observations of the *right ascensions* in time units of two stars found from direct comparisons with the sun. Taking as the true values of these quantities the values that he had found from reduction of all of the relevant observations, he determined the deviations from these true values, i.e., the "errors," of each of Bradley's observed values of these quantities, and then presented three tables [9, pp. 19–20] showing the observed and theoretical frequencies of the corresponding absolute errors in successive class intervals according to the Gaussian law of error. In ref. 10 [col. 402] he reproduced these three tables with the observed and theoretical frequencies now in percentage form and gave [col. 403] a new table showing a frequency distribution of the absolute errors of 100 observations of the Pole Star in time units that he had made from 1813 to 1815 with the old transit instrument at the Königsberg Observatory, the value of the standard deviation of this distribution derived from the grouped data, and the corresponding theoretical frequencies.

Bessel's comparisons were not only the first, but the best and most influential. His first, second, and fourth comparisons were reproduced in percentages form in Bertrand [8, art. 148]; all four in Czuber [19, pp. 191–193]; the second in frequency form, in Merriman [54, art. 34]; and the third in

Chauvenet [14], from which it was reproduced in Peirce [58, p. 206]. To the eye, the agreement of the observed and theoretical frequencies in Bessel's four tables is quite good, so that these data were widely interpreted as adding empirical "proof" of the validity of Gauss's law of error (1) as law of error of real-life measurements and observations.

PEIRCE. The American mathematician, physicist, and logician Charles Sanders Peirce (1839–1914) analyzed [58, p. 212] sets of 500 experimental observations made on each of 24 weekdays from July 1 to August 3, 1872, which he presented [pp. 213–224] in 24 separate tables, one for each day with date. Peirce did not carry out numerical comparisons of experience and theory, but relied on a visual appraisal [Plate 27] of (a) numerically smoothed frequency curves of the observations, and (b) "mean curves," one for each day, "drawn by eye so as to eliminate the irregularities entirely."

Many of his smooth "mean" curves "drawn by eye" through the wiggly curves representing the smoothed observational data do resemble somewhat curves of the Gaussian error (or normal) distribution, although some are noticeably asymmetrical. Nonetheless, he seems to have inferred that his results confirmed the practical validity of the Gaussian law of error (1).

Contrary to Peirce's interpretation, his 24 sets of 500 observations are not at all supportive of the hypothesis of an underlying Gaussian law of error. His data were subjected to extensive statistical analysis by the mathematician, physicist, and statistician Edwin Bidwell Wilson (1879–1964) and his coworker Margaret M. Hilferty [74]. They found each of the data sets to be incompatible in one or more respects with the hypothesis of an underlying Gaussian (normal) distribution, and summarized their findings thus: "The upshot of this all is that Peirce had observations which could show as completely as one might desire that the departures of the errors from the normal law was for his series uniformly great" [74, p. 124]. (Some highlights of, and comments on, the

Wilson–Hilferty analysis are given in Mosteller and Tukey [55, Sec. A4; 56, Sec. 1D].

AIRY. Sir George Biddell Airy (1801–1892) included a "Practical Verification of the Theoretical Law for the Frequency of Errors" as an Appendix to the third edition of his *Theory of Errors of Observations* [5, pp. 114–119]. His basic data were 635 observations of the North Polar Distance (NPD) of Polaris made at the Royal Observatory in the years 1869 to 1873. The "errors" of the observations made in particular year were taken to be their deviations from a mean NPD for that year. These within-year deviations for the respective years were all pooled together, and tabulated in a frequency table having class intervals of width 0."05, except toward the extremes, where wider class intervals were adopted.

Sir George then somewhat arbitrarily reduced the frequencies in these wider classes "so as to make them justly comparable with the number of observations in other parts of the series" [5, p. 115], after which it "became evident that there was no marked discordance between the laws of distribution for positive and for negative values: and therefore the corresponding numbers were added together" [p. 116].

Next he smoothed these combined frequencies by the same procedure as Peirce, repeated thrice. This left the first and last frequencies unchanged, the next increased fourfold, and others "on the whole, eight times as large as the original numbers" [p. 116]. He, therefore, increased the end ones to bring them up to par, so to speak, and assembled all in a frequency table, from which he prepared a graph [p. 118], the ordinates of the plotted points being these frequencies, and their abscissas the magnitude of the corresponding absolute "errors." A smooth freehand curve was drawn through the wiggly pattern of the plotted points and extrapolated to zero error, given $y_0 = 124$. Taking, from this data, $\sigma\sqrt{2} = 1.1973$, he plotted the curve

$$y = 124 \exp\left[-x^2/(1.1973)^2 \right],$$

$$0 < x < \infty,$$

on the same graph. It was close to the freehand curve throughout, lying above it for $0.15 \leqslant x \leqslant 1.75$, the difference being greatest (3.97) at $x = 0.90$, but visually is a poor fit to the wiggly data points. Nonetheless, he concluded: "It is evident that the formula represents with all practicable accuracy the observed Frequency of Errors, upon which all the applications of the Theory of Probabilities are founded; and the validity of every investigation in this Treatise is thereby established" [p. 119].

Airy, unlike Peirce, did not report his raw data. His arbitrary adjustments, pooling, and smoothing make it impossible to determine objectively and unequivocally whether his basic observational data are, or are not, in accord with the supposition of an underlying Gaussian law of error. Karl Pearson criticized him vigorously for this in the paper in which he introduced the χ^2 goodness-of-fit test to the world [57, Illus. VI].

These comparisons of observation and measurement data with the Gaussian law of error, and others of smaller scale (e.g., Merriman [54, arts. 19, 33]), were widely interpreted, individually and collectively, to constitute empirical "proof" of the Gaussian law of error (1) as *the* law of error of quality measurements and observations. This was an illusion; and understandable, inasmuch as formal (and statistically sound) tests of goodness of fit, skewness, and kurtosis did not become available until the twentieth century.

References

[1] Abbe, C. (1871). *Amer. J. Sci.*, 3rd Ser., **1**, 411–415. (Reprinted in Stigler [70, Vol. I].)

[2] Adams, W. J. (1974). *The Life and Times of the Central Limit Theorem*. Kaedmon, New York.

[3] Adrain, R. (1809). *The Analyst; or Mathematical Museum*, Vol. 1, 93–109. (Reprinted in Stigler [70, Vol. I]. Microfilm copy included in the American Periodical Series 1800–1850, issued by University Microfilms, Ann Arbor. Date on title page is "1808," but Hogan [36, Note 2] cites evidence to indicate that printed publication probably was delayed until the autum of 1809.)

[4] Airy, G. B. (1861). *On the Algebraical and Numerical Theory of Errors of Observations and the Combination of Observations*. Macmillan, London.

[5] Airy, G. B. (1879). *On the Algebraical and Numerical Theory of Errors of Observations and the Combination of Observations*. 3rd ed. Macmillan, London, pp. 114–119. (Apendix to the third edition of Airy [4].)

[6] Babb, M. J. (1926). *Gen. Mag. Historical Chronicle*, **28**, 272–284. (Reprinted in Stigler [70, Vol. I].)

[7] Bertrand, J. (1888). *Compt. Rend.*, **106**, 563–565.

[8] Bertrand, J. (1889). *Calcul des Probabilités*. Gauthier-Villars, Paris. (2nd ed., "conforme à la première," 1897; reissued, 1907.)

[9] Bessel, F. W. (1818). *Fundamenta astronomiae pro anno MDCCLV deducta ex observationibus viri incomparabilis James Bradley in specula astronomia Grenovicensi per annos 1750–1762 institutis*. F. Nicolov, Regensburg.

[10] Bessel, F. W. (1838). *Astron. Nachr.*, **15**, Cols. 369–404. (Reprinted in *Abhandlungen von Friedrich Wilhelm Bessel*, Vol. 2. W. Engelmann, Leipzig, 1875, pp. 372–391.)

[11] Boole, G. (1857). *Trans. R. Soc. Edinb.*, **21**, 597–652.

[12] Boole, G. (1860). *A Treatise on the Calculus of Finite Differences*. Macmillan, London. (2nd ed., J. F. Moulton, ed. Macmillan, London, 1872; 3rd ed., *idem*, 1880; 3rd ed., with corrections notes on text, hints on exercises, by E. B. Scott, G. E. Stechert, New York, 1932.)

[13] Brunt, D. (1917). *The Combination of Observations*. Cambridge University Press, Cambridge.

[14] Chauvenet, W. (1863). *A Manual of Spherical and Practical Astronomy . . . with an Appendix on the Method of Least Squares*, 2 vols. Lippincott, Philadelphia. Trübner, London. (5th ed., revised and corrected, *idem*, 1891; reissued, paperbound, Dover, New York, 1960.)

[15] Coolidge, J. L. (1925). *An Introduction to Mathematical Probability*. Clarendon Press, Oxford. (Reissued by Dover, New York, 1962.)

[16] Coolidge, J. L. (1926). *Amer. Math. Monthly*, **33**, 61–76. (Reprinted in Stigler [70, Vol. I].)

[17] Crofton, M. W. (1870). *Philos. Trans. R. Soc. Lond.*, **160**, 175–187.

[18] Crofton, M. W. (1885). Probability. *Encyclopaedia Britannica*, 9th ed., Vol. 19. A. & C. Black, Edinburgh.

[19] Czuber, E. (1891). *Theorie der Beobachtungsfehler*. B. G. Teubner, Leipzig.

[20] Czuber, E. (1899). *Jahresber. Dtsch. Math. Ver.*, Vol. 7, No. 2, 1–279.

[21] Deltheil, R. (1930). *Erreurs et Moindres Carrés*. Gauthiers-Villars, Paris.

[22] Donkin, F. W. (1857). *Quart. J. Pure Appl. Math.*, **1**, 152–162.

[23] Edgeworth, F. Y. (1883). *Philos. Mag.*, 5th Ser., **16**(100), 300–309 (Oct. 1883).

[24] Edgeworth, F. Y. (1883). *Philos. Mag.*, 5th Ser., **16**(101), 360–375 (Nov. 1883).

[25] Edgeworth, F. Y. (1896). *Philos. Mag.*, 5th Ser., **41**, 90–99.

[26] Edgeworth, F. Y. (1905). *Trans. Camb. Philos. Soc.*, **20**, 36–66, 113–141. [Continuation of Appendix VII (14 pp.), published privately and circulated with reprints.]

[27] Eisenhart, C. (1964). *J. Wash. Acad. Sci.*, **54**, 24–33. (Reprinted as Paper 4.5 in *Precision Measurement and Calibration: Selected NBS Papers on Statistical Concepts and Procedures*, National Bureau of Standards Special Publication 300, Vol. 1, H. H. Ku, ed. U.S. Government Printing Office, Washington, D.C., Feb. 1969.)

[28] Ellis, R. L. (1850). *Philos. Mag.*, 3rd Ser., **37**, 321–328. (Reprinted in *The Mathematical and Other Writings of Robert Leslie Ellis*. Deighton, Bell, Cambridge, 1863, pp. 53–61.)

[29] Gauss, C. F. (1809). *Theoria Motus Corporum Coelestium in Sectionibus Conicis Solem Ambientium*. Frid. Perthes et I. H. Besser, Hamburg. [Reprinted in *Carl Friedrich Gauss Werke*, Vol. 7. Perthes, Gotha, 1871, pp. 1–271; reissued, Göttingen, 1906; English transl. by C. H. Davis: *Theory of the Motion of the Heavenly Bodies Moving about the Sun in Conic Sections*. Little, Brown, Boston, 1857; reissued, Dover, New York, 1963. Articles 175–179 are reprinted (unnumbered) in the English translation by H. Shapley and H. E. Howarth: *A Source Book in Astronomy*, McGraw-Hill, New York 1929, pp. 188–194; reissued by Harvard University Press, Cambridge, Mass., 1960.]

[30] Gauss, C. F. (1821). *Göttingische gelehrte Anzeigen* (1821), pp. 321–327. (Reprinted in *Carl Friedrich Gauss Werke*, Vol. 4. Göttingen, 1873, pp. 95–100; reissued 1880.)

[31] Glaisher, J. W. L. (1872). *Mem. R. Astron. Soc.*, **39**, 75–124.

[32] Hagen, G. H. L. (1837). *Grundzüge der Wahrscheinlichkeitsrechnung*. Ernst & Korn, Berlin (2nd ed., 1867; 3rd ed., 1882).

[33] Hagen, G. H. L. (1867). 2nd edition of [32].

[34] Harter, H. L. (1974). *Int. Statist. Rev.*, **42**, 147–174.

[35] [Herschel, J. F. W.] (1850). *Edinb. Rev.*, **92**, 1–57 (American two-column edition, pp. 1–30). (Unsigned review, reprinted in Herschel's *Essays from the Edinburgh and Quarterly Reviews*. Longman, Brown, Green, Longmans & Roberts, London, 1857, pp. 365–465.)

[36] Hogan, E. (1977). *Historia Math.*, **4**, 157–172.

[37] Kemnits, Yu. V. (1959). Ob odnom vyvode formuly zakona raspredeleniya veroytanostey sluchaynyx oshibok rezul'tatov izmereniy (On one derivation of the formula for the distribution law

of random error probabilities in measurement results). *Tr. Mosk. in-ta inzhenerov zemleustroystva*, Vyp. 3 (Works of the Moscow Institute of Land Management Engineers, Issue 3).

[38] Kendall, M. and Plackett, R. L., eds. (1977). *Studies in the History of Statistics and Probability*, Vol. 2. Charles Griffin, London/Macmillan, New York.

[39] Krylov, A. N. (1950). *Lektsii o Problizhennykh Vychisleniyakh* (Lectures on approximate computations), 5th ed. Gosudarstv. Izdat. Tekhn.-Teor. Lit., Moscow–Leningrad. [A reissue of the third edition (1935) with only minor changes.]

[40] Kummell, C. H. (1876). *Analyst*, **3**, 133–140, 165–171.

[41] Kummell, C. H. (1877). *J. Franklin Inst.*, 3rd Ser., **74**, 270–274.

[42] Laplace, P. S. (1810). *Mém. Cl. Sci. Math. Phys. Inst. France*, Année 1809, 353–415; Supplement, 559–565. (Reprinted in *Oeuvres complètes de Laplace*, Vol. 12. Gauthier-Villars, Paris, 1898, pp. 301–348 and 349–353.)

[43] Laplace, P. S. (1811). *Mém. Cl. Sci. Math. Phys. Inst. Impérial France*, Année 1810, 279–347. (Reprinted in *Oeuvres complètes de Laplace*, Vol. 12. Gauthier-Villars, Paris, 1898, pp. 357–412.)

[44] Laplace, P. S. (1812). *Théorie Analytique des Probabilités*. Courcier, Paris. [3rd ed., same pagination, with corrections and three Supplements, Courcier, Paris, 1820. 4th ("national") ed., reprint of 3rd ed., with additional Supplement (1825) and new pagination (page n of 3rd $\simeq$ page $1.1n$ of 4th), published as *Oeuvres de Laplace*, Vol. 7. Imprimerie Royale, Paris, 1847; reprinted as *Oeuvres complètes de Laplace* Vol. 7. Gauthier-Villars, Paris, 1886.]

[45] Legendre, A. M. (1805). Sur la méthode des moindres quarrés (On the method of least squares). Appendix (pp. 72–80) to Legendre's *Nouvelles Méthodes pur la Détermination des Orbites des Comètes*. Firmin Didot, Paris. (English transl. of pp. 72–75 by H. A. Ruger and H. M. Walker, in D. E. Smith, [69, pp. 576–579].)

[46] Magie, W. F. (1935). *A Source Book in Physics*. McGraw-Hill, New York. (Reissued, Harvard University Press, Cambridge, Mass., 1963.)

[47] Maxwell, J. C. (1860). *Philos. Mag.*, 4th Ser., **19**, 19–32. (Reprinted in *The Scientific Papers of James Clerk Maxwell*, 2 vols., W. D. Niven, ed. Cambridge University Press, Cambridge, 1890, pp. 377–391; reissued, two volumes in one, Dover, New York, 1952. Proposition IV is reprinted in Magie [46, pp. 259–261].)

[48] Meidell, B. (1908). *Zeit. Math. Phys.*, **56**, 77–85.

[49] Merriman, M. (1877). *Analyst, J. Pure Appl. Math.*, **4**, 33–36. (Reprinted in Stigler [70, Vol. I].)

[50] Merriman, M. (1877). *Analyst, J. Pure Appl. Math.*, **4**, 140–143. (Reprinted, without pagina-

tion, as the last item in Stigler [70, Vol. I]; not listed in Contents.)

[51] Merriman, M. (1877). *J. Franklin Inst.*, 3rd Ser., **74**, 173–187.

[52] Merriman, M. (1877). *J. Franklin Inst.*, 3rd Ser., **74**, 330–334.

[53] Merriman, M. (1877). *Trans. Connecticut Acad. Arts Sci.*, **4**, 151–232. (Index of Authors, pp. 228–230; Index of Subjects, pp. 230–231. Reprinted in Stigler [70, Vol. I].)

[54] Merriman, M. (1884). *A Text-Book on the Method of Least Squares*. Wiley, New York.

[55] Mosteller, F. and Tukey, J. W. (1968). In *Handbook of Social Psychology*, Vol. 2: *Research Methods*, 2nd ed., G. Lindzey and E. Aronson, eds. Addison-Wesley, Reading, Mass., pp. 80–203.

[56] Mosteller, F. and Tukey, J. W. (1977). *Data Analysis and Regression*. Addison-Wesley, Reading, Mass.

[57] Pearson, K. (1900). *Philos. Mag.*, 5th Ser., **50**, 157–175. (Reprinted in *Karl Pearson's Early Statistical Papers*. Cambridge University Press, Cambridge, 1948.)

[58] Peirce, C. S. (1873). On the theory of errors of observation. Appendix No. 21 (pp. 200–224 and plate 27) of *Report of the Superintendent of the U.S. Coast Survey* (for the year ending Nov. 1, 1870). U.S. Government Printing Office, Washington, D.C. (Reprinted in C. S. S. Peirce, *The New Elements of Mathematics*, Vol. 3, C. Eisele, ed. Humanities Press, Atlantic Highlands, N.J., 1976, pt. 1, pp. 639–676; and in Stigler [70, Vol. II].)

[59] Pizzetti, P. (1892). *Atti della Regia Univ. Genovà*, **11**, 113–333.

[60] Plackett, R. L. (1972). *Biometrika*, **59**, 239–251. (Reprinted in Kendall and Plackett [38, Paper 17].)

[61] Poincaré, H. (1912). *Calcul des Probabilités*, 2nd ed. Gauthier-Villars, Paris. (Revised and augmented by the author.)

[62] Poisson, S. D. (1837). *Recherches sur la probabilité des jugements en matière criminelle et en matière civile, précédées des règles générales du calcul des probabilités*. Bachelier, Paris

[63] Price, B. (1865). *A Treatise on Infinitesimal Calculus*. Clarendon Press, Oxford.

[64] Quetelet, A. (1846). *Lettres à S.A.R. le duc régnant de Saxe-Coburg et Goth, sur la théorie des probabilités, appliquée aux sciences morales et politiques*. Hayez, Bruxelles. (English transl. by O. G. Downes.) Charles and Edwin Layton, London, 1849; reissued by Arno Press, New York, 1981.

[65] Schlömilch, O. (1872). *Zeit. Math. Phys.*, **17**, 87–88.

[66] Sheynin, O. B. (1963). *K. Istorii otsenok neposredstvennkh izmerenii i zakona raspredeleniya*

sluchainykh oshibok. Distributed by All-Union Institute of Scientific and Technical Information (VINITI), Moscow. (English transl.: On the History of Treatment of Direct Measurements and Law of Random Error Distribution. TT-67-61690, National Technical Information Service, Springfield, Va.)

[67]   Sheynin, O. B. (1971). *Arch. History Exact Sci.*, **7**, 244–256.

[68]   Smart, W. M. (1958). *Combination of Observations*. Cambridge University Press, Cambridge.

[69]   Smith, D. E. (1929). *A Source Book in Mathematics*. McGraw-Hill, New York. Reprinted in 2 vols. by Dover, New York, 1959.

[70]   Stigler, S. M., ed. (1980). *American Contributions to Mathematical Statistics in the Nineteenth Century*, 2 vols. Arno Press, New York.

[71]   Tait, P. G. (1865). *Trans. R. Soc. Edinb.*, **24**, 139–145.

[72]   Thomson, W. and Tait, P. G. (1867). *Treatise on Natural Philosophy*, Vol. 1. Cambridge University Press, Cambridge (2nd ed., 1888; 3rd, 1896).

[73]   Todhunter, I. (1865). *A History of the Mathematical Theory of Probability from the Time of Pascal to That of Laplace*. Macmillan, London. (The standard reference on the history of probability; not entirely trustworthy. Reprinted by G. E. Stechert, New York, 1931, and Chelsea, New York, 1949, 1965.)

[74]   Wilson, E. B. and Hilferty, M. M. (1929). *Proc. Natl. Acad. Sci. USA*, **15**, 120–125. (Reprinted in Stigler [70, Vol. II].)

[75]   Worthing, A. G. and Geffner, J. (1943). *Treatment of Experimental Data*, Wiley, New York.

[76]   Yakovlev, K. P. (1953). *Matematicheskaya Obrabotka Rezul'tatov Izmereniy* (Mathematical treatment of measurement results), 2nd ed. Gosudarstv. Izdat. Tekhn.-Teor. Lit., Moscow–Leningrad.

[77]   Young, T. (1819). *Philos. Trans. R. Soc. Lond.*, **109**, 70–95.

CHURCHILL EISENHART

LAWS OF ERROR III: LATER (NON-GAUSSIAN) DISTRIBUTIONS

Deep seated belief in the Gaussian law of error [see eq. (1) in LAWS OF ERROR II: THE GAUSSIAN DISTRIBUTION] as the universal law of error of a single measurement or observation led to proposals of mixtures of Gaussian laws of error to explain and cope with sets of measurements of the same quantity that exhibited features inconsistent with the hypothesis of a single underlying Gaussian distribution. The beliefs that arithmetic mean of a set of "equally good" measurements of a fixed quantity is the "best" mean to take, or at least is always "advantageous," and that the "hypothesis of elementary errors" necessarily leads to the Gaussian law of error, were refuted by the invention (1824) of the *Cauchy distribution**.

Whereas the Gaussian law of error or normal distribution to which the distribution of the sum $S(n)$ of n independent random variables with finite means and variances tends in the limit as $n \to \infty$ is symmetric about its mean, the distribution of $S(n)$ may be asymmetric with respect to its mean for all finite values of n, when some or all of the random variables involved are asymmetrically distributed. The derivation (1829) of a penultimate expression for the distribution of $S(n)$ to "a second approximation" that revealed the degree of asymmetry present for large values of n led to series expansions of probability distributions in terms of the normal distribution and its derivatives, and to development of families or systems of frequency curves and probability distributions.

DISCORDANT OBSERVATIONS AND MIXTURES OF GAUSSIAN LAWS OF ERROR

In the latter half of the nineteenth century many scientists held that when the number of large derivations from the mean of a data set exceeds that indicated by the Gaussian law of error, the presence of some faulty or otherwise "abnormal" observations is to be inferred, and these should be identified and excluded. Various procedures for identifying "doubtful observations" that should be rejected were devised. There was much criticism of these rejection rules, and of rejection per se; and rebuttal (see, e.g., Rider [21]).

Other individuals argued that, inasmuch as the object of combining two or more measurements or observed values of a single fixed quantity is to obtain the best possible value of this quantity on the basis of the data in hand, an observation that differs markedly from the rest should not be discarded on this account, but simply assigned lesser weight in computing a weighed mean to be adopted as the "best" value. Clinging to the sanctity of the Gaussian law of error, they proposed that each of the errors $X_i = Y_i - \tau$ of n independent measurements or observed values, Y_i, of a quantity τ be assumed a priori to be distributed in accordance with a Gaussian law of error, but with different precisions, $h_1, h_2, \ldots, h_n$ $(i = 1, 2, \ldots, n)$, that is, that the joint probability distribution of the observations be

$$f_{Y_1, Y_2, \ldots, Y_n}(y_1, y_2, \ldots, y_n)$$

$$= \prod_{i=1}^{n} \frac{h_i}{\sqrt{\pi}} e^{-h_i^2(y_i - \tau)^2}$$

$$(-\infty < y_i < +\infty; 0 < h_i < \infty). \quad (1)$$

Such an assumption may have been the unstated basis of the suggestion of the French mathematician Joseph Diez Gergonne (1771–1859) writing anonymously [1, pp. 197–204] that the simple arithmetic mean, $\bar{Y} = (1/n)\sum_1^n Y_i$, be replaced by a weighted arithmetic mean,

$$\bar{\bar{Y}} = \sum_1^n w_i Y_i \Big/ \sum_1^n w_i,$$

with weights $w_i = 1/(Y_i - \bar{Y})^2$, the precision, h_i $(= 1/\sigma_i\sqrt{2})$, of each observation, Y_i, thus being taken to be inversely proportional to its distance, $|Y_i - \bar{Y}|$, from $\bar{Y}$. (His identity is disclosed by Stigler [24].) The hypothesis that (1) represents a priori the joint probability distribution of the observations is to be found explicitly in Glaisher [9, p. 392], Stone [26, p. 14], and Edgeworth [6, p. 370]—or see Rider [21, pp. 14–17]. James Whitbread Lee Glaisher (1848–1928), English astronomer and mathematician, derives therefrom [9, pp. 393–394] a very strange weighting scheme that on iteration "may never lead to a limit, or that limit may be reached very slowly and tediously" [21, p.

16]. Edward James Stone (1831–1897), English astronomer, and Francis Ysidro Edgeworth (1845–1926), English economist and statistician, independently attempted to maximize (1) simultaneously with respect to τ and h_i, $i = 1, 2, \ldots, n$. The equation for the "most probable" value of τ that each obtained [6, p. 370; 26, p. 15] required iterative solution; and later was found to yield local minima [14, p. 150; 15, p. 6], not the correct maximum likelihood estimate of τ, which is one of the observations and can be unambiguously identified, but which as a "solution" has undesirable properties [14, p. 150; 15, pp. 7–8].

In contrast, the Canadian-born American astronomer, Simon Newcomb (1835–1909), proposed [16, Sec. 5] that a set of n observations be assumed to be i.i.d.

$$f_Y(y) = \frac{1}{\sqrt{\pi}} \prod_{j=1}^{m} p_j h_j e^{-h_j^2(y_i - \tau)^2}$$

$$\left(\sum_{j=1}^{m} p_j = 1\right) \quad (2)$$

p_j being the probability that error of an observation selected at random is distributed according to the Gaussian law with precision constant h_j $(j = 1, 2, \ldots, m)$. Newcomb gave [Sec. 6] a procedure for finding the "best" value of τ, and a modification [Sec. 8] for the case when the observations are numerous. His worked examples show these procedures to be complicated, cumbersome, and tedious—see Rider [21, pp. 18–19] for a concise statement of what is involved.

The Canadian astronomer Robert Meldrum Stewart (1878–1954) simplified matters by restricting the number of different precisions to only two [21, pp. 17–18; 23]; as did also the English geophysicist and astronomer, Harold Jeffreys (1891–), who let one of the components also reflect a small measurement bias by varying about a mean $\delta \neq 0$ [11; 21, p. 19]. (In a later version [12, Sec. 4.41], he allowed the contaminating component to be non-Gaussian.) Such mixtures of Gaussian components in which the one of less precision (smaller h; larger σ) contributes only a small fraction of the total, termed "contaminated normal" distributions, have served as models for recent work on the development of *robust* methods of statistical analysis (see, e.g., Tukey [27] and Andrews et al. [2]).

FAILURE OF THE "ARITHMETIC MEAN RULE" (OR "POSTULATE OF THE ARITHMETIC MEAN") AND OF THE "HYPOTHESIS OF ELEMENTARY ERRORS"

The French mathematical physicist Siméon-Denis Poisson (1781–1840) contrived [18, p. 278] the probability density function

$$f_X(x) = \frac{1}{\pi(1+x^2)}, \qquad -\infty < x < \infty$$

$$(3)$$

today called the *Cauchy distribution* (following its appearance in Cauchy [4, p. 206]), as an example of a law of error (a) for which arithmetic means $\overline{X}(n)$ of n errors thus i.i.d. are distributed exactly according to (3) for $n = 1, 2, 3, \ldots$, and (b) the probability density functions of the corresponding sums $S(n)$

$$f_S(s) = \frac{n}{\pi(n^2+s^2)}$$

do not tend to the Gaussian law of error in the limit as $n \to \infty$. Hence, if $Y_i = \tau + X_i$, $(i = 1, 2, 3, \ldots)$ are measurements of τ with errors X_i, i.i.d. according to (3), then the arithmetic mean Y of two or more such measurements provides a determination of τ no better than that provided by a single measurement chosen at random, contradicting the "arithmetic mean rule" and refuting the postulate of the arithmetic mean. If, on the other hand, $S(n)$ signifies the error of a single measurement or observation equal to the sum of "elementary errors" X_i ($i = 1, 2, \ldots, n$) that are i.i.d. according to (3), the failure of the distribution of $S(n)$ to tend to the Gaussian law of error as $n \to \infty$ implies failure of the "hypothesis of elementary errors."

CORRECTIONS FOR SKEWNESS OR KURTOSIS, AND THE GRAM–CHARLIER AND EDGEWORTH SERIES EXPANSIONS

The objective of Hagen and other supporters of the "hypothesis of elementary errors" as the basis of the Gaussian law of error (see the section "'Proofs Based on the Hypothesis of Elementary Errors" in LAWS OF ERROR II) was to show that the distribution of the deviation $X = X(n) = S(n) - E[S(n)]$ of the sum $S(n)$ of n independently distributed elementary errors from its mean value $E[S(n)]$ tended asymptotically to the symmetric Gaussian law of error $\phi_X(x) = \phi_X(x; h)$ [see eq. (1) in LAWS OF ERROR II] with $h = h(n) = 1/\{\sqrt{2}\,\sigma[S(n)]\}$ as $n \to \infty$, under various assumptions about the probability distributions of the elementary errors—i.e., provide derivations of what we today call the *central limit theorem**. The actual distribution of $X(n)$ may be asymmetric, however, for all finite values of n, when the distributions of some of the elementary errors are asymmetric. S. D. Poisson seems to have been the first to provide an asymmetry "correction" to the Gaussian law of error $\phi_X(x; h)$ for large but finite n by giving [19, Secs. 2–4; 20, Chap. 4] the penultimate form

$$f_X(x; h) = \phi_X(x; h) + a_3\phi_X^{(3)}(x; h) \quad (4)$$

where $\phi_X^{(3)}(x; h)$ is the third derivative of $\phi_X(x; h)$ with respect to x, and a_3 is a negative multiple of h^3 and $\mu_3(S)$, the third centroidal moment of the distribution of $S(n)$, which is equal to the sum of the third centroidal moments of the elementary errors involved. (For i.i.d. elementary errors a_3 will be $O(1/\sqrt{n}\,)$.) Later, Bessel [3, Sec. 9], using the Laplace normal distribution form of the Gaussian law of error, i.e., with $\sigma\,(=1/h\sqrt{2}\,)$ in place of h, gave "correction" terms involving higher-order derivatives and the higher even centroidal moments $\mu_4(S)$ and $\mu_6(S)$ for the case in which the distributions of all the elementary errors are symmetric about zero means, so that $\mu_3(S) = \mu_5(S) = 0$.

From these "corrections" to the Gaussian law of error, the Gram–Charlier type A series* and Edgeworth series* expansions of probability distributions in terms of the normal distribution of its derivatives evolved, as detailed in recent historical accounts [5, 10, 22]. In particular, it seems [10, p. 6] that the work of the "Scandinavian school" (Gram, Charlier, Thiele, etc.) was sparked by discussion of Bessel's approach and results in the textbook on least squares [28] by George Karl Christian Zachariae (1835–1907), director of the Danish Geodetic Institute, whereas Edgeworth repeatedly mentioned Poisson but not Bessel, e.g., Edgeworth [7, pp. 91–92; 8, p. 38].

PEARSON CURVES

In contrast, the *Pearson system of frequency curves** was derived, not by successive modifications of the normal distribution, but rather [17] by deriving a differential equation for the curve through successive ordinates of the *hypergeometric distribution* in much the same manner as Hagen had derived the differential equation of the Gaussian error function from the curve through successive ordinates of a binomial distribution. The normal distribution is consequently a member of the Pearson system.

CONCLUDING REMARKS

Belief in the myth of the Gaussian law of error was slow to die, in spite of the fact that various writers (e.g., Bessel [3, Sec. 2]) derived alternative "laws of error" appropriate to particular measurement circumstances to show that the law could not be universal. Gradually, it became recognized and appreciated that if we adopt a particular probability distribution as a model for data analysis or theoretical developments, it is just a model, a simplification hoped to be an approximation "close enough" for fruitful application. Today we are reluctant to claim that given data are distributed according to

the normal distribution, or any other specific distribution [22, p. 376].

References

[1]   "Abonné" [J. D. Gergonne] (1821). *Ann. Math. Pures Appl.*, **12**, 181–204.

[2]   Andrews, D. F., Bickel, P. J., Hampel, F. R., Huber, P. J., Rogers, W. H., and Tukey, J. W. (1972). *Robust Estimates of Location: Survey and Advances*. Princeton University Press, Princeton, N.J.

[3]   Bessel, F. W. (1838). *Astron. Nachr.*, **15**, Cols. 369–404. (Reprinted in *Abhandlungen von Friedrich Wilhelm Bessel*, Vol. 2. W. Engelmann, Leipzig, 1875, pp. 372–391.)

[4]   Cauchy, A. (1853). *C. R. Acad. Sci. Paris*, **37**, 198–206.

[5]   Cramér, H. (1972). *Biometrika*, **59**, 205–207. (Reprinted in Kendall and Plackett [13].)

[6]   Edgeworth, F. Y. (1883). *Philos. Mag.*, 5th Ser. **16**(101), 360–375 (Nov. 1883).

[7]   Edgeworth, F. Y. (1896). *Philos. Mag.*, 5th Ser., **41**, 90–99.

[8]   Edgeworth, F. Y. (1905). *Trans. Camb. Philos. Soc.*, **20**, 36–66, 113–141. [Continuation of Appendix VII (14 pp.), published privately and circulated with reprints.]

[9]   Glaisher, J. W. L. (1873). *Monthly Notices R. Astron. Soc.*, **33**, 391–402.

[10] Hald, A. (1981). *Int. Statist. Rev.*, **49**, 1–20. (A review in modern notation of the contributions to statistics of T. N. Thiele, and of related work of other members of the Scandinavian School, with a portrait of Thiele.)

[11] Jeffreys, H. (1932). *Proc. R. Soc. Lond. A*, **137**, 78–87.

[12] Jeffreys, H. (1939). *Theory of Probability*. Clarendon Press, Oxford. (3rd ed., 1961.)

[13] Kendall, M. and Plackett, R. L. (1977). *Studies in the History of Statistics and Probability*, Vol. 2. Charles Griffin, London/Macmillan, New York.

[14] Levy, P. S. and Mantel, N. (1974). *J. Statist. Comp. Simul.*, **3**, 147–160.

[15] Mantel, N. (1956). Maximum Likelihood Estimation of the Mean in the Case of Unequal Precisions. M.A. thesis, American University.

[16] Newcomb, S. (1886). *Amer. J. Math.*, **8**, 343–366. (Reprinted in Stigler [25, Vol. II].)

[17] Pearson, K. (1895). *Philos. Trans. R. Soc. Lond. A*, **186**, 343–414. (Reprinted in *Karl Pearson's Early Statistical Papers*. Cambridge University Press, Cambridge, 1948, pp. 41–112.)

[18] Poisson, S. D. (1824). *Connaisance des Tems ...pour l'An 1827.* Bachelier, Paris, pp. 273–302.

[19] Poisson, S. D. (1829). *Connaisance des Tems ...pour l'An 1832.* Bachelier, Paris, Additions 3–22.

[20] Poisson, S. D. (1837). *Recherches sur la probabilité des jugements en matière criminelle et en matière civile, précédées des règles générales du calcul des probabilités.* Bachelier, Paris.

[21] Rider, P. R. (1933). *Criteria for Rejection of Observations.* Wash. Univ. Stud.—N.S., Sci. Tech. No. 8, St. Louis, Mo., Oct. 1933.

[22] Särndal, C. E. (1971). *Biometrika,* **58**, 375–392. (Reprinted in Kendall and Plackett [13].)

[23] Stewart, R. M. (1920). *Popular Astron.,* **28**, 4–6.

[24] Stigler, S. M. (1976). *Historia Math.,* **3**, 71–74.

[25] Stigler, S. M., ed. (1980). *American Contributions to Mathematical Statistics in the Nineteenth Century,* 2 vols. Arno Press, New York.

[26] Stone, E. J. (1873). *Monthly Notices R. Astron. Soc.,* **34**, 9–15.

[27] Tukey, J. W. (1960). In *Contributions to Probability and Statistics: Essays in Honor of Harold Hotelling,* I. Olkin et al., eds. Stanford University Press, Stanford, Calif., Chap. 39.

[28] Zachariae, G. (1871). *De mindste kvadraters Methode* (The method of least squares). Schönemann, Nyborg. (2nd ed., Hansen, Copenhagen, 1887.)

(ARITHMETIC MEAN
CAUCHY DISTRIBUTION
GAUSS, KARL FRIEDRICH
NORMAL DISTRIBUTION)

CHURCHILL EISENHART

LAWS OF LARGE NUMBERS

The *laws of large numbers*, in their classical form, deal with sums of independent and identically distributed random variables X_i, $i = 1, 2, 3 \ldots$ with $E|X_1| < \infty$ and $EX_1 = \mu$. If $S_n = \sum_{i=1}^{n} X_i$, then the classical strong law of large numbers* (SLLN) states that $n^{-1}S_n \to \mu$ almost surely (a.s.) as $n \to \infty$. The weak law of large numbers (WLLN) gives $n^{-1}S_n \to \mu$ in probability as $n \to \infty$. The adjectives "strong" and "weak" refer to the mode of

convergence. A related result of intermediate strength, that of L_p convergence, asserts that $E|n^{-1}S_n - \mu|^p \to 0$ as $n \to \infty$, $0 < p \leqslant 1$. Each of these laws has many generalizations, both in the direction of relaxing the assumption of independence and that of stationarity. Various results and references are given below.

The SLLN embodies the idea of a probability as a strong limit of relative frequencies*; if we take $X_i = I_i(A)$, the indicator function of the set A at the ith trial, then $\mu = P(A)$. On these grounds, the SLLN may be regarded as basic in probability theory. In fact, it constitutes the underpinning of the axiomatic theory as a physically realistic subject as it provides a means of assigning probabilities.

The central limit theorem* and the law of the iterated logarithm* can both be interpreted as results about the rate of convergence in the SLLN. If $\operatorname{var} X_1 = E(X_1 - \mu)^2 = \sigma^2 < \infty$, the former gives convergence in distribution of $n^{1/2}\sigma^{-1}(n^{-1}S_n - \mu)$ to the unit normal law. The latter states that

$$n^{-1}S_n - \mu = \zeta(n)(2\sigma n^{-1} \log \log n)^{1/2},$$

where $\zeta(n)$ has its set of a.s. limit points confined to $[-1, 1]$ with $\limsup_{n \to \infty} \zeta(n) = +1$ a.s. and $\liminf_{n \to \infty} \zeta(n) = -1$ a.s.

The laws of large numbers have a long history. The simplest form of the WLLN, which deals with the case $P(X_i = 1) = p$, $P(X_i = 0) = 1 - p$, $0 < p < 1$, and obtains $n^{-1}S_n \to p$ in probability, was published in Bernoulli [4]. This result opened up, for the first time, the possibility of wide application of probability to statistics through inference in the context of the binomial distribution. However, little further progress was possible until Bayes [2] developed methods of assessing the difference $n^{-1}S_n - p$ via a posterior distribution, and Laplace [13] did the same through confidence intervals based on the central limit theorem*.

The standard approach to SLLNs for independent random variables is based on truncation and the use of the Kolmogorov criterion. This states that if X_i, $i = 1$,

$2, \ldots,$ are independent random variables with zero means and finite variances, then $\sum_{n=1}^{\infty} b_n^{-2} EX_n^2 < \infty$ for some sequence $\{b_n\}$ of positive constants such that $b_n \uparrow \infty$ entails $\lim_{n \to \infty} b_n^{-1} \sum_{i=1}^{n} X_i = 0$ a.s. Indeed, writing $X_i^c = X_i$ if $|X_i| \leqslant b_i$, $X_i^c = 0$ otherwise, and $\bar{X}_i = X_i^c - EX_i^c$, $i = 1, 2, \ldots,$ then $\lim_{n \to \infty} b_n^{-1} \sum_{i=1}^{n} X_i = 0$ a.s. if $\sum_{n=1}^{\infty} P(|X_n| > b_n) < \infty$ and $\sum_{n=1}^{\infty} b_n^{-2} E\bar{X}_n^{-2} < \infty$. For details, the reader should refer, for example, to Loève [14, Secs. 17, 18], or Chow and Teicher [6, Secs. 5.2, 10.1]. In addition, the Kolmogorov criterion continues to hold for the case where $\{\sum_{i=1}^{n} X_i, n \geqslant 1\}$ is a martingale* as a simple consequence of the martingale convergence theorem (e.g., Feller [8, p. 242]). This leads on to a variety of SLLNs for martingales and for various other processes whose increments X_i may be dependent. Similar generalizations are available in the case of the WLLN for which the standard approach is based on truncation and the use of Chebyshev's inequality. Detailed accounts of this subject area may be found in various texts, for example, Révész [15], Stout [16], and Hall and Heyde [10, Chap. 2].

The above-mentioned texts also contain results on convergence of series of the form "$\sum_{n=1}^{\infty} c_n X_n$ converges a.s." These are closely related to SLLNs for, if $c_n \downarrow 0$ as $n \to \infty$, the Kronecker Lemma* (which states that if $\{x_n, n \geqslant 1\}$ is a sequence of real numbers such that $\sum x_n$ converges and $\{b_n\}$ is a monotone sequence of positive constants with $b_n \uparrow \infty$, then $b_n^{-1} \sum_{i=1}^{n} b_i x_i \to 0$ as $n \to \infty$) implies that $c_n \sum_{k=1}^{n} X_k \to 0$ a.s. as $n \to \infty$. The most famous of the series convergence theorems is the three-series theorem* due to Kolmogorov, which deals with the case of independent random variables $\{X_i\}$. The theorem asserts that $\sum_{n=1}^{\infty} X_n$ converges a.s. if and only if for a positive constant c,

$$\sum_{n=1}^{\infty} P(|X_n| > c) < \infty, \qquad (1)$$

$$\sum_{n=1}^{\infty} E(X_n I(|X_n| \leqslant c)) < \infty, \qquad (2)$$

and

$$\sum_{n=1}^{\infty} \left[E(X_n^2 I(|X_n| \leqslant c)) \right.$$
$$\left. - (EX_n I(|X_n| \leqslant c))^2 \right] < \infty, \qquad (3)$$

where I denotes the indicator function.

For a stationary process* $\{X_n, n \geqslant 0\}$ with $E|X_0| < \infty$ and $EX_0 = \mu$, the most important generalization of the classical SLLN is to the so-called ergodic theorem. This asserts that $n^{-1} \sum_{i=1}^{n} X_i \to E(X_0 | \mathscr{I})$ a.s. as $n \to \infty$, where $\mathscr{I}$ denotes the corresponding σ-field of invariant events and $E(X_0 | \mathscr{I})$ is a random variable in general (e.g., Brieman [5, Chap. 6, Sec. 5]). The ergodic theorem, in turn, has a complete extension to (stationary) subadditive processes i.e., processes $\{x_{st}\}$ where $s < t$ with s, t running over the nonnegative integers such that whenever $s < t < u$,

$$x_{su} \leqslant x_{st} + x_{tu}.$$

Details are to be found in Kingman [12] and Hall and Heyde [10, Chap. 7, Sec. 3].

There are also quite a number of papers on laws of large numbers for the case where the random variables assume values in an abstract vector space. Extensions of the standard real line results to Hilbert space are straightforward, but in Banach space there are more difficulties to contend with. Because of the geometric structure of the space, the SLLN fails in some situations. For a survey of results in this area the reader should consult Beck et al. [3], and for a more recent contribution Hoffman-Jørgensen and Pisier [11]. Results on the SLLN for random variables with multidimensional indices are given by Gut [9].

The laws of large numbers play a vital role in asymptotic estimation theory*. An asymptotic condition that is usually regarded as essential is that of consistency [strong consistency], i.e., convergence in probability [almost sure convergence] of the estimator to the true value as the sample size increases to infinity. The LLNs and their generalizations are basic in establishing such consistency in most situations. For example, under mild regularity conditions, strong consistency of the maximum likelihood* estimator in a random sampling situation follows from the classical SLLN (e.g., Cox and Hinkley [7, pp. 288–290]). Corresponding SLLN results for martingales are involved in the general case of stochastic processes (e.g., Basawa and Prakasa Rao [1, Chap. 7] and Hall and Heyde [10, Chap. 6]).

Another fundamental application of the SLLN in estimation theory occurs when a distribution function is estimated. Let X_i, $i = 1, 2, \ldots$, be independent and identically distributed random variables with distribution function F. Then, again using $I(A)$ to denote the indicator function of a set A, the SLLN gives

$$F_n(x) = n^{-1} \sum_{i=1}^{n} I(X_i \leqslant x) \to F(x) \text{ a.s.}$$

as $n \to \infty$, $-\infty < x < \infty$. This result has been strengthened in the Glivenko–Cantelli theorem (e.g., Révész [15, p. 158]) to show that convergence holds uniformly in x. That is, $\sup_x |F_n(x) - F(x)| \to 0$ a.s. as $n \to \infty$.

References

The books by Brieman, Chow and Teicher, Loève, Basawa and Prakasa Rao, Feller, Hall and Heyde, Révész, and Stout are aimed at a graduate-level audience, the first three being largely self-contained. That by Cox and Hinkley is accessible at the advanced undergraduate level.

[1]   Basawa, I. V. and Prakasa Rao, B. L. S. (1980). *Statistical Inference for Stochastic Processes*. Academic Press, London.

[2]   Bayes, T. (1763). *Philos. Trans. R. Soc. Lond.*, **53**, 376–398.

[3]   Beck, A., Giesy, D., and Warren, P. (1975). *Theory Prob. Appl.*, **20**, 127–134.

[4]   Bernoulli, J. (1713). *Ars Conjectandi*. Basileae impensis Thurnisiorum fratrum, Basle.

[5]   Brieman, L. (1968). *Probability*. Addison-Wesley, Reading, Mass.

[6]   Chow, Y. S. and Teicher, H. (1978). *Probability Theory: Independence, Interchangeability, Martingales*. Springer-Verlag, New York.

[7]   Cox, D. R. and Hinkley, D. V. (1974). *Theoretical Statistics*. Chapman & Hall, London.

[8]   Feller, W. (1971). *An Introduction to Probability Theory and Its Applications*, Vol. 2, 2nd ed. Wiley, New York.

[9]   Gut, A. (1978). *Ann. Prob.*, **6**, 469–482.

[10] Hall, P. and Heyde, C. C. (1980). *Martingale Limit Theory and Its Application*. Academic Press, New York.

[11] Hoffman-Jørgensen, J. and Pisier, G. (1976). *Ann. Prob.*, **4**, 587–599.

[12] Kingman, J. F. C. (1968). *J. R. Statist. Soc. B*, **30**, 499–510.

[13] Laplace, P. S. (1812). *Théorie Analytique des Probabilités*. V. Courcier, Paris.

[14] Loève, M. (1977). *Probability Theory I*, 4th ed. Springer-Verlag, New York.

[15] Révész, P. (1968). *The Laws of Large Numbers*. Academic Press, New York.

[16] Stout, W. F. (1974). *Almost Sure Convergence*. Academic Press, New York.

(ASYMPTOTIC NORMALITY
AXIOMS OF PROBABILITY THEORY
BAYESIAN INFERENCE
CONVERGENCE OF SEQUENCES OF
   RANDOM VARIABLES
ESTIMATION
GLIVENKO–CANTELLI THEOREM
LAW OF LARGE NUMBERS
LAW OF THE ITERATED LOGARITHM
LIMIT THEOREMS, CENTRAL
MARTINGALES
MAXIMUM LIKELIHOOD
STATIONARY PROCESSES
SUBADDITIVE PROCESSES)

C. C. HEYDE

LAWS OF MORTALITY   *See* LIFETIME DISTRIBUTIONS; MORTALITY

LAW, STATISTICS IN

Central to many legal trials are the conclusions of factual investigations on which the application of law turns. Is a man the father of a child as alleged? Are salaries of similarly qualified minority and nonminority employees different? Is the product defective within the meaning of the law? Similarly in quasi-judicial administrative hearings of government agencies and commissions complex factual issues are central to the determinations made. Statistics recently has become most prominent in the law through its use in discrimination cases (see Baldus and Cole [1]) and it is also important in areas of business regulatory proceedings, antitrust cases, and environmental regulation, to name several. In other important areas of legal concern, such as legislation and even on occasion in negotiations, factual investigations figure in the application of existing law or the creation of new.

Statistics as a field of science, in contradistinction to statistics as a set of numbers, may be defined as the study of inferential processes used to pass from empirical data to some sort of inference, decision, or conclusion [14]. The field comprises the study of methods of data collection, analysis, and interpretation that are generalizable across disciplines, with special emphasis on inference and decision making in the presence of substantial variability in data and uncertainty in knowledge. The ubiquity of use of statistics in fields of inquiry and, in part for this reason, in court appearances, is reflected in the fact that the classic text on statistical methods by Snedecor and Cochran [20] is the most frequently cited book in the standard index of scientific citations.

Because formal legal proceedings as well as the informal practice of law are often involved with inferring facts from evidence and in reaching decisions in the presence of uncertainty, the potentially close relation between the fields of law and statistics is apparent. Formal legal proceedings may be viewed as the forums where factual investigations relevant to the application of law are "published" and facts are certified as findings of the tribunals. Were all the issues of fact in a dispute agreed upon by the parties, there would in principle be no need of a trial. The judge would acquaint himself or herself with the facts, consult the law, apply it to the agreed fact situation, and reach a verdict. Thus the consideration at trial or in other proceedings of issues of fact is many times at the heart of the formal legal process.

A selection mechanism probably operates to enhance the importance of statistics in formal legal proceedings. Cases reach the trial stage and questions are explored at hearings in part because substantial factual uncertainties or substantial complexities are involved. To draw valid conclusions, the careful application of the appropriate inferential tools is requisite.

Statistics in law has a long history. In 1710, G. W. Leibniz, himself a lawyer, discussed the relation of probabilistic concepts to legal thought in a letter to John Bernoulli

> Leibniz pointed out that lawyers used concepts such as "conjecture," "indication," and "presumption" and that these corresponded to levels of probability which he expressed as "not full," "half full," and "full." Such shadings of meaning led him to conclude that 'Jurists have practiced best of all mankind the art of logic in regard to contingencies, as have mathematicians in regard to necessities.' [5]

In the New World the mathematician Benjamin Peirce's testimony in the Howland will case in 1867 is perhaps the first use of probabilistic inference on record in American courts [17].

There has been a substantial increase in the attention that statisticians and lawyers pay to the uses of statistics and probability in the law. In recent years the natural affinities between their fields have begun to be better appreciated. A useful survey of the uses of statistics in court and legal studies up to 1978 is given by Zeisel [25], and Solomon [21] examines the role of statistics in legal contexts in the federal government of the United States.

The purpose of the present article is to discuss and illustrate the relationship of statistics to aspects of law as a profession and as a discipline. No attempt is made in the brief compass of this article to be exhaustive in considering the scope of statistical applications made in areas of the law. The purpose is to highlight important themes. The following section illustrates the uses of statistics to aid fact-finding in actual investigations conducted in legal contexts. The section "Uses of Statistics to Study Legal Decision Making and Institutions" discusses illustrative examples of the use of statistics in the study of legal decision-making rules and institutions. The section "The Integration of Statistics in Law" discusses four kinds of integration between statistics and law that would help realize the potential contributions of the former to the latter.

USES OF STATISTICS TO AID FACT-FINDING IN LEGAL CONTEXTS

Uses of statistics to resolve factual issues arise in every area of the law. In this section we discuss some specific examples of legal dispute in which statistical concepts and analysis have played prominent roles.

Municipal Services Discrimination

In *Bryan* v. *Koch*,[1] the Mayor of the City of New York, Edward Koch, and others were sued by a resident of Harlem, David Bryan, and others to prevent the planned closing of two city-owned hospitals located in Harlem. Koch and the city were charged with discrimination against minorities (blacks and Hispanics) in the selection of hospitals for closure under a cost-cutting plan of a Mayoral Task Force and the City's Health and Hospitals Corporation, which operates the city hospitals. In testimony for the plaintiffs an expert witness on statistics testified to differences in proportions of minority and nonminority beds closed by the plan. Under a binomial model for selection of beds for closure and a normal theory test of the differences between two proportions, the differences were variously some 10 to 30 standard deviations distant from zero, and he declared that under a null hypothesis of equal probabilities of selecting minority and nonminority beds for closure that the probability the differential observed would have occurred by chance was less than 0.001. The

statistician testifying for the defendants noted that there was no basis in the evidence presented at trial for believing that hospital beds were selected for closure according to the assumptions of a binomial model. In particular, since entire hospitals or portions of them were earmarked for closure, beds were not selected independently but rather in large bunches. The implication of this observation for the statistical analysis and conclusion was substantial. The conclusion of nonparametric tests of differences in proportions of different types of hospitals closed was that these were consistent with race-neutral selection.

The application of a binomial model in this case was obviously erroneous. However, as was pointed out by the judge in the decision, a more subtle question was raised in applying a probability model and an associated test to hospital closure. Unlike the situation with the selection of persons for jury duty, in which a statute mandates random selection for duty in federal courts, there is no presumption that the selection of hospitals for closure would be random. Thus while jurors of different races arguably should turn up in proportions governed by a chance mechanism, there is no such conservative benchmark for hospitals of differing minority composition. Hospital closure is a complex decision. Hospitals, for example, differ widely in operating cost, accreditation status, availability of alternatives, etc. The relevance of a test of significance premised on a model of random selection therefore requires discussion at the very least.

Jury Discrimination

The Supreme Court in *Castaneda* v. *Partida*[2] referred to standard statistical tests in appraising a challenge to nondiscriminatory juror selection with respect to a proportion of Mexican-Americans. The court there referred to the rule of thumb in social science of a "two to three standard deviation" difference as a threshold for statistical proof of a real difference. Numerous lower court opinions have cited the *Castaneda* rule of thumb as a requirement to reject any difference not reaching two or three standard deviations as statistically beyond the pale. However, that rule of thumb is based on a consideration of only one potentially applicable criterion for the importance of a difference, namely the criterion that a particular chance explanation should be ruled out.

Another criterion for statistical proof uses the absolute or percentage size of the difference, such as the "four-fifths rule" adopted in federal regulations on employment discrimination (but see Finkelstein [6]). Even as a criterion for a chance explanation the "2 to 3 standard deviations" rule based on a binomial or normal theory model embodies very strong assumptions about the kind of race-neutral, sex-neutral, or in general group-neutral random selection mechanism that will be called "random." Other group-neutral random selection mechanisms can be defined that nevertheless do not meet the "two to three standard deviation" test as it is customarily defined.

Identification Evidence

There is unfortunately, virtually a tradition of monumental exaggeration of the evidentiary import of the implications of posited probability or statistical models under a null hypothesis. In *People* v. *Collins*[2], a widely cited California case involving an interracial couple charged with robbery on the basis of a witness's description (he had a mustache, she a ponytail and blonde hair, they drove off in a partly yellow automobile, etc.), a mathematics instructor testified at trial to the product rule of probability theory and multiplied together probabilities for each description to conclude that the probability was 1 in 12,000,000 "that any couple possessed the distinctive characteristics of the defendants." The case is discussed in ref. 4. Professor Benjamin Peirce testifying in the Howland will case in 1867, found without qualification that the probability of matching 30 downstrokes in a pair of signatures without forgery was "once in 2,666 millions of millions of millions" and stated:

This number transcends human experience. So vast an improbability is practically an impossibility. Such evanescent shadows of probability cannot belong to actual life. They are unimaginably less than those least things the law cares not for. [17]

Trucking Deregulation

Not all of the uses of statistics in law involve trials or formal proceedings. In the making of new law by statute, agencies and legislatures and the staffs of these bodies continually deal with statistics. An interesting problem in statistical estimation arose in the course of preparation of a bill in the U.S. Congress in 1979 to "deregulate" the trucking industry—more specifically to free in some degree many motor carriers and routes from rate regulation and cargo specification laid down by the Interstate Commerce Commission (ICC).

Staff of the Senate Judiciary Committee with the assistance of staff of the Senate Computer Center and outside advisory economists had prepared an analysis of the degree of "concentration" in the trucking industry between major city pairs (see, for one, Cherry [2]). For example, the "four-firm concentration ratio" for a lane between city pairs is the ratio of the dollar volume of trucking business carried by the top four carriers in the lane in a year to the total dollar volume of trucking business carried in the lane by all carriers. Concentration ratios had been computed for a large number of city-pair lanes from a stratified systematic random sample of way bills. The sample was drawn from data on a computer tape originally obtained from the industry's cooperative Continuing Traffic Survey (CTS), itself a systematic random sample of the population of way bills. In testimony before the Senate Commerce Committee on the deregulation bill in the summer of 1979, Senator Edward Kennedy, then Chairman of the Senate Judiciary Committee, had cited the computed concentration ratios as evidence that the regulated trucking business was highly concentrated and, moreover, most highly concentrated in those areas of the country (the West and Southwest) where ICC regulation could be expected to have the largest anticompetitive concentration-promoting effects.

The basis of Senator Kennedy's testimony was soon disputed by a statistician advising the trucking industry rate bureaus. He contended that the estimates of concentration ratios made by Senate staff were biased, and that any attempt to use the CTS in the estimation of such ratios would involve the user in a statistical "fallacy."

The Senate staff sought advice from a second statistician on the potential bias problem in their estimates. The existence of a bias was confirmed. The sample concentration ratios are upwardly biased estimates of the population values because the sample is used in estimates of the four-firm concentration ratio both to pick the top four firms in a lane and to estimate the four-firm ratio. The source of bias is discussed, for example, by Mosteller and Tukey [18] as the phenomenon that "optimization capitalizes on chance." The dollar volume of the largest dollar volume carrier in the sample is upwardly biased as an estimate of the dollar volume of the largest dollar volume carrier in the population.

The staff was also advised that it was possible to investigate the nature and the size of the bias. Analytical estimates of the size of the bias, although closely related to theory developed for problems of selecting the largest k of n population values (see Gibbons et al. [10]) were not available, but simulation of sample estimates made from the CTS produced estimates of the size of the bias for a range of assumed true populations of dollar volumes by firm. Since the direction of the bias and some information on the relationship of the size of the bias to features of the underlying population were known, adjustments to the sample estimates were possible, reducing the size of the remaining bias.

In the judgment of the economists advising the Senate staff, the sizes of the biases indicated for estimates of concentration ratios would not reverse the qualitative conclu-

sions drawn from the unadjusted estimated ratios. It followed that for their purposes estimated ratios based on the CTS were useful despite the bias problem.

USES OF STATISTICS TO STUDY LEGAL DECISION MAKING AND INSTITUTIONS

A major contribution of statistics to factual inquiries and—in principle though rarely in practice—to decision making is the identification and measurement of sources of error in decision.

Paternity Suits

Paternity suits are an excellent example to consider because the use of blood tests has provided a rare pool of statistical information that permits a quantitative study of decision-making errors. It was Hugo Steinhaus and his student J. Lukasziewicz in Poland in the early 1950s who first studied a sample of paternity cases using blood test results (see refs. 16 and 22). Because the frequency of matches between father and child varies with the proportion of actual fathers in a sample of cases, Steinhaus was able to estimate that 70% of putative fathers in a sample of 1515 cases were actual fathers. Further, using Bayes' theorem, the 70% prior, and likelihoods based on frequencies of matches for different blood tests, he was able to obtain posterior probabilities of fatherhood for each individual case.

Lukasziewicz used these data to estimate that the actual rate of erroneous dismissals (finding against the mother when the man named was the father) was between 0 and 5% of all cases, and the rate of erroneous attributions was between 12 and 17% for a total error of between 12 and 22%. Further, he compared these results to those that would be obtained from certain defined judicial strategies. One strategy would be to dismiss all cases in which the paternity probability was less than a half and find for the mother whenever it was greater than a half. Such a strategy, he was able to calculate for a sample, would result in an estimated rate of erroneous dismissals of 1.4% of all cases and of erroneous attribution of 16.5% for a total error of 17.9%. Such a strategy will minimize total error. An alternative strategy would be to require a higher than average (70%) posterior probability before finding for the mother. Such a strategy has the property of equalizing the sizes of the errors of dismissal and of attribution. He estimated that such error would be 10.9% of all cases under such a strategy, for a total error of 21.8%. The two strategies thus pose as a legal and social policy problem the choice between different distributions of the burdens of error for the parties.

Quantitative analysis of errors in paternity suits has at least two values. First, it measures more precisely than could intuition the import of different blood test results for the likelihood of fatherhood. It is possible that more precise information such as this could help judges to minimize their total errors for any given strategy. Second, it exposes for scrutiny the choice of error distributions that is otherwise made implicitly.

Statistical decision theory permits formal study of burdens of error among the parties. Kaplan [13] has explored how, for a variety of civil and criminal cases, both the probabilities of error and the costs (or disutilities) of errors may separately influence judicial decision making. Of course, legal policy considerations must be subtly interwoven with the use of formal decision theory to achieve a better understanding of its practical value in aiding judicial decision making.

Jury Size and the Rule for Conviction

Scholarly study of jury decision making has spanned two centuries. The literature dates back to the French probabilists, most significantly Poisson [19], whose book presents a substantial applied mathematical analysis of probabilities of conviction and of acquittal for juries of different sizes and employing different rules for the numbers required to convict. In fact, his purpose was to demonstrate the applicability of the emerging theory of probability to real phenomena.

Recent attention in the United States to probability modeling and experimental studies of jury size and conviction rules follows Supreme Court decisions in the early 1970s that upheld six-member in place of twelve-member juries and nonunanimous verdicts in certain cases—the important early decisions being *Williams* v. *Florida*[4] and *Johnson* v. *Louisiana*.[5] The Court did not hear any explicit mathematical treatment of different probabilities of conviction, acquittal, or error but rather relied on intuitive assessments of how the changes would affect the chances of different outcomes. For example, in *Williams* the Court was persuaded that the six-member size was even-handed toward defendants and the prosecution in criminal cases, saying:

> It might be suggested that the 12-man jury gives a defendant a greater advantage since he has more "chances" of finding a juror who will insist on acquittal and thus prevent conviction. But the advantage might just as easily belong to the State, which also needs only one juror out of twelve insisting on guilt to prevent acquittal.

At another point in *Williams* the Court brushed aside the importance of the size of the jury as a factor affecting the reliability of its decision with the following remark: "And, certainly the reliability of the jury as a fact finder hardly seems likely to be a function of its size."

It is possible that probabilistic modeling or experimental studies of jury decision making can contribute to the design of better juries. Since there are now currently very different jury schemes in effect in different states and around the world, such work may eventually have practical value in suggesting changes.

Available models of jury voting and the error characteristics of jury decision are suggestive (see Gelfand and Solomon [9] for a review) but still seem some distance from conclusive evidence for any particular jury size. For example, Gelfand and Solomon propose a mixture of binomial models for reaching probabilities of different numbers of jurors' votes and of guilt. The model uses a unique set of data on the first ballot and final votes of 225 twelve-member juries compiled by Kalven and Zeisel [12] in their landmark study of the American jury. Only 10% of the first ballots were changed and of these 7% were hung juries and only 3% clear reversals. The binomial models rest on probabilities that jurors will vote correctly for conviction or acquittal on the first vote. The comparison between the six- and twelve-member jury, which they find clearly to favor the twelve-member size, may be rather sensitive to the step from the first to the final vote, which is less fully modeled. To challenge the view of the jury as equally rational whether of size six or twelve put forward by the Supreme Court in *Williams* it would seem necessary to parameterize continuously the evidentiary strength of the case and, at least as a benchmark hypothesis, model the jury as reacting rationally to cases according to their strengths. Finally, close study of the techniques that trial lawyers routinely use to influence the compositions of their juries may contribute to realistic models of jury decision making.

Probabilistic model descriptions of the cumulative evaluation of evidence in court linked to meeting judicial standards of proof are discussed by Goldsmith [11], who gives additional references to some of the literature on the topic from lawyers, psychologists, and philosophers. Applications are given by Fairley [3] and Lindley [15].

THE INTEGRATION OF STATISTICS IN LAW

The use of statistics in law requires the integration of a discipline of knowledge with the established institutions and procedures of another field. Four types of integration can be delineated that serve to make this process particularly productive.

Integrating Uses of Statistics with Legal Rules and Procedures

Trials are based on rules of law. Attorneys argue claims that legal conditions have or

have not been violated, and judges rule on these claims. The legal theory of a case may, for instance, connect specific and empirically verifiable circumstances to the violation of a law or to a legally enforceable right.

An example would be a suit for "gross underrepresentation" of a minority group on juries in a locality. This can be established empirically with a substantial statistical disparity in the proportions of the minority and all others on the juries relative to their respective numbers in a population. This would constitute a law violation under a "disparate impact" theory of discrimination. A statistician testifying to the statistical disparities generally does well, however, to steer clear of legal claims that the disparities prove "discrimination" or even "disparate impact." This testimony to the ultimate legal conclusion at issue is on dubious authority and expertise, and furthermore, in an adversary setting it suggests a potential or likelihood of personal bias.

The statistician and substantive expert, who may be one and the same, typically have two major tasks to aid legal proceedings. First, they testify on inferences to be drawn from the evidence to the issues involved in the case. Second, they testify to the strength of the proof for factual conclusions. This expert testimony is circumscribed by the lawyers and the judge according to established rules and procedures.

The legal procedures involved serve a variety of policy objectives including fostering accurate factual conclusions and preserving rights of fairness (due process). These objectives, however, sometimes conflict. To be admitted as evidence in a trial, evidence must first be determined to be relevant to a matter at issue. Evidence that is relevant is further restricted by various exclusionary rules. An important example is the "hearsay" rule, which excludes the admission as proof of accounts of statements made out of court under certain circumstances. In particular, such statements are generally inadmissible when they are offered to establish the truth of the matters contained in them, rather than to prove some other proposition.

The results of sample surveys based on interviews have often been ruled to be inadmissible under the hearsay rule, when they are offered to prove the truthfulness of the respondents' statements. It has been argued, though, that in some circumstances the rule does not apply, so that its application can at times sacrifice valuable information with no counterbalancing gain to the participants or to the judicial process (see Zeisel [25] for further discussion).

Characterization of the strength of evidence is important because different standards of proof apply in different circumstances. The standard that applies in weighing evidence depends on the type of case and on the aspect of the case involved. For example, the standard for proof of guilt in a criminal proceeding is that the allegations have been established "beyond a reasonable doubt." In civil proceedings the standard under which the plaintiff can prevail is stated in milder terms as either proof based on a "preponderance of the evidence" or on "clear and convincing evidence." The criminal standard, although considered to be more stringent than the civil standards, has not been described numerically by legal commentators, but the "preponderance of the evidence" standard has been described by some to mean a greater than 50% probability, using some notion of probability as a degree of belief. A jury in a criminal case cannot use this 50% guideline, and is supposed to acquit a defendant if a "reasonable doubt" remains, even if it believes that the defendant is probably guilty.

Legal standards of proof, however, are not precisely related to explicit theories of inference and strength of evidence, and confusion has arisen in the application of statistical theories of inference to legal evidence. A common form of confusion is that between a conclusion from a test of statistical significance and a conclusion based on the probability that a proposition is true considered as a degree of belief in the proposition. Thus it is not uncommon to read in legal cases that because a difference is not statistically significant, it cannot be present. While the form of the latter statement reflects the customary convention in statistics for reporting on significance tests, it may be mischievously inappropriate in a trial, where the question under the "preponderance of the evidence" test is whether the evidence balances in favor or against a proposition, rather than the degree of certainty. In particular, although they are derived from different approaches to statistical inference and are not directly comparable, a test of significance at the 5% or 1% level suggests to the lay person a higher standard of certainty than does a posterior subjective probability of greater than 50%.

Integrating the Spheres of Legal Judgment and Formal Analysis

The number and variety of cases in which descriptive statistical data, explicit statistical models, and statistical analysis are introduced as evidence have been expanding in recent years. This increase has been attributed to the greater power of data technology, to the expansion of research in all fields, and to increasing methodological sophistication in the use of statistical theory and methods by lawyers and other nonstatisticians. Questions that might previously have been answered by the personal opinion of a lawyer, judge, or expert are now frequently addressed with the aid of a statistical survey, experiment, or analytical investigation. In an adversarial system, the party with the better information, both statistically and otherwise, is likely to be in a superior strategic position.

The province of judgment is not eliminated through the use of statistics, but it is instead merely being more clearly defined when quantified, systematic methods of inference are available. Whether this is beneficial to the legal system is a question that is in some instances debatable. When the underlying attitude in presenting quantitative information is "better precisely wrong than vaguely right," then the phenomenon a colleague (Michael Meyer) refers to as Gresham's law of judgment, "Bad analysis drives out good judgment," can apply. Critical assessment of numerical information and inference is no less necessary than critical assessment of qualitative information and inference.

The search for good methods for combining general knowledge and intuition with immediate observations that avoid problems of subjectivity is much debated within the field of statistics. In fact, this is one of the important foundational problems in the field. Bayesian statistics offers one method for explicitly integrating judgment with the immediate statistical evidence. It has been suggested by Finkelstein and Fairley [7] (see also Fairley [3] that juries would be aided in cases of criminal identification by the guided use of Bayesian statistical analysis of the identity between the accused and the source of identification evidence (such as hair, blood, or fiber sample). This proposal was introduced as a way to help juries to avoid a common fallacy of interpreting the statistical frequency of an identifying characteristic as equivalent to a probability of the guilt or innocence of the accused. It was attacked by Tribe [23] in an article that found it at variance with several legal policy criteria. A debate ensued in which a number of issues involved in the use of mathematics in court were explored (see refs. 8 and 24).

Integrating Statistical and Legal Modes of Thought

The communication of statistical evidence and analysis can be impeded by a lack of knowledge of or experience with statistics by lawyers and judges as well as by poor exposition by statisticians. Less discussed but also prevalent problems are difficulties faced by statisticians who are unfamiliar with legal rules, procedures, and conventions. Unless skillfully guided by lawyer and judge and by his or her own knowledge, the statistician may stray into areas of legal judgment or may be mystified by seemingly arbitrary straitjackets imposed by legal procedures on his or her presentation. To work together effectively, good professional skills are required of both statisticians and lawyers, together with some effort to understand the approaches of both fields.

The difficulties that juries, judges, and lawyers may have with statistical evidence repeat difficulties that others have had. Legal professionals are no more immune to the fallacies and the suspicions of numbers than are many others. Yet numbers, just like words, are subject both to constructive use and to confusion or trickery. The legal system, however, like the scientific world, has a well-institutionalized structure for weeding out error, whether it be a failure of logic or of "numeracy." Statisticians should recognize, moreover, that the difficulties that many find with statistics sometimes derive not from a lack of training but from fundamental unsolved problems either in the field of statistics itself or in the proper translation of theories and methods of statistics to legal contexts.

Integrating Divergent Statistical Analyses

Many judges have experienced great difficulties sorting out, in an adversary environment, different and often incomparable or conflicting statistical formulations of models and choices of data. Based on a survey of the uses of regression models in administrative proceedings, Finkelstein [5] has developed suggested rules of procedure or protocols for judges to use when dealing with regression analyses as evidence. These protocols have obvious generalizations to modeling and estimation in other legal contexts. Three are of particular interest. The first calls for the judge to predesignate, with the advice of the parties (although agreement may be difficult), a minimal data set which clearly warrants use and on which different analyses can be based. The purpose is to provide a base of comparable data that is less complex than the larger sets that parties might eventually wish to use. A second protocol states that a party objecting to a model introduced by another party should demonstrate the numerical significance of the objection whenever possible. Features of model may invite attack that in fact have little effect on the quantity estimated or on the conclusion to be drawn from the application of the model. The third protocol states that a party objecting to a model should produce a superior alternative analysis of the data.

In conclusion, we have seen that statistics as a field and quantative methods and studies generally are used in a variety of actual legal proceedings and in the study of legal decision making methods and institutions. The effective use of statistics in legal contexts requires attention to legal standards of proof, to the relation of statistical evidence to judgments made on other grounds, and to methods or even rules for the presentation of sometimes divergent statistical analyses.

NOTES

1. *Bryan* v. *Koch* 492 F. Supp. 212 (1980).

2. *Castaneda* v. *Partida* 97 S. Ct. 1272, 1977.

3. *People* v. *Collins* 68 Cal. 2d 319, 438 P. 2d 33, 66 Cal. Rptr. 497 (1968).

4. *Williams* v. *Florida*, 399 U.S. 78 (1970).

5. *Johnson* v. *Louisiana*, 406 U.S. 356 (1972).

References

[1] Baldus, D. C. and Cole, J. W. L. (1980). *Statistical Proof of Discrimination*. McGraw-Hill, New York.

[2] Cherry, R. C. (1979). *An Analysis of the 1976 Continuous Traffic Study*. Arthur D. Little, Cambridge, Mass.

[3] Fairley, W. B. (1973). *J. Legal Stud.*, **2**, 493–513.

[4] Fairley, W. B. and Mosteller, F. (1974). *Univ. Chicago Law Rev.*, **41**, 242–253.

[5] Finkelstein, M. O. (1978). *Quantitative Methods in Law*. Free Press, New York.

[6] Finkelstein, M. O. (1980). *Columbia Law Rev.*, **80**, 737–754.

[7] Finkelstein, M. O. and Fairley, W. B. (1970). *Harvard Law Rev.*, **83**, 489–517.

[8] Finkelstein, M. O. and Fairley, W. B. (1971). *Harvard Law Rev.*, **84**, 1801–1809.

[9] Gelfand, A. E. and Solomon, H. (1977). *Jurimetrics*, **17**, 293–313.

[10] Gibbons, J. D., Olkin, I., and Sobel, M. (1977). *Selecting and Ordering Populations: A New Statistical Methodology*. Wiley, New York.

[11] Goldsmith, R. W. (1980). *Acta Psychol.*, **45**, 211–221.

[12] Kalven, H. and Zeisel, H. (1966). *The American Jury*. Little, Brown, Boston.

[13] Kaplan, J. (1968). *Stanford Law Rev.*, **20**, 1065–1092.

[14] Kruskal, W. H. (1978). In *International Encyclopedia of Statistics*, W. Kruskal and J. Tanur, eds., Vol. 2. Free Press, New York, pp. 1071–1093.

[15] Lindley, D. V. (1973). In *Utility, Probability and Human Decision Making*, D. Wendt and C. Vlek, eds. D. Reidel, Dordrecht, Holland, pp. 223–232.

[16] Lukasziewicz, J. (1955). *Zastosow. Mat.*, **2**, 349.

[17] Meier, P. and Zabell, S. (1980). *J. Amer. Statist. Ass.*, **75**, 497–506.

[18] Mosteller, F. and Tukey, J. W. (1977). *Data Analysis and Regression*. Addison-Wesley, Reading, Mass.

[19] Poisson, S. D. (1837). *Recherches sur la probabilité des jugements en criminelle et en matière civile, précédées des règles générales du calcul des probabilités*. Bachelier, Paris.

[20] Snedecor, G. W. and Cochran, W. G. (1980). *Statistical Methods*, 7th ed. Iowa State University Press, Ames, Iowa.

[21] Solomon, H. (1971). *Federal Statistics: Report of the President's Commission*, Vol. 2. U.S. Government Printing Office, Washington, D.C., pp. 497–526.

[22] Steinhaus, H. (1954). *Pr. Wroclaw. Tow. Nauk.*, *Ser. A.*, No. 32, at 5.

[23] Tribe, L. H. (1971). *Harvard Law Rev.*, **84**, 1329–1393.

[24] Tribe, L. H. (1971). *Harvard Law Rev.*, **84**, 1810–1820.

[25] Zeisel, H. (1978). In *International Encyclopedia of Statistics*, W. Kruskal and J. Tanur, eds., Vol. 2. Free Press, New York, pp. 1118–1122.

Acknowledgements

The author gratefully acknowledges comments on an earlier draft of this article by Robert Field, Michael Finkelstein, Richard Goldstein, Peter Kempthorne, and Michael Meyer. These helpful people are not responsible for the views expressed or for any errors.

(BAYESIAN INFERENCE
BIAS
DECISION THEORY
HYPOTHESIS TESTING
INFERENCE, STATISTICAL
PRIOR PROBABILITY
SIGNIFICANCE TESTS
SURVEY SAMPLING)

WILLIAM B. FAIRLEY

LAYARD'S TEST *See* LEVENE'S ROBUST TEST OF HOMOGENEITY OF VARIANCES

L CLASS LAWS

Let $X_1, X_2, \ldots$ be independent random variables (rvs), and consider the normed sums

$$X_n^* = b_n^{-1} \sum_{k=1}^{n} X_k - a_n \qquad (1)$$

where a_n and b_n are constants, $b_n > 0$, and the rvs X_k / b_n, $1 \leqslant k \leqslant n$ are uniformly infinitesimal*. A probability distribution (law) with distribution function (df) F is said to belong to the class L or to be self-decomposable ($F \in L$) if there is a sequence $\{X_n^*\}$ defined as in (1) and a rv X^* such that $F_{X^*} = F$ and $X_n^* \overset{d}{\to} X^*$. The class L contains the normal laws and all stable laws* and is a subclass of the class of infinitely divisible (inf. div.) laws*.

The term *self-decomposable* derives from the following characterization of class L distributions.

Theorem 1. The distribution of the rv X belongs to the class L if and only if for each $0 < c < 1$ there exists a rv X_c independent of X such that $X \overset{d}{=} cX + X_c$. In other words, if φ is the characteristic function* (ch.f.) of the d.f. F, then $F \in L$ if and only if for each $0 < c < 1$ there exists a ch.f. φ_c such that

$$\varphi(y) = \varphi(cy)\varphi_c(y).$$

From this characterization of the class L it is immediate that (a) if $F_X \in L$, then $F_{aX+b} \in L$ ($a, b \in \mathbb{R}$, $a \neq 0$); (b) if $F_X, F_Y \in L$ and X, Y are independent, then $F_{X+Y} \in L$; and (c) if $F_{X_n} \in L$, $n = 1, 2, \ldots$, and $X_n \overset{d}{\to} X$, then $F_X \in L$.

It is known that every self-decomposable distribution is absolutely continuous* [1, 3, 8] and unimodal [12]. See refs. 3 and 12 for the history of the development of these facts.

Since each distribution in L is inf. div., its ch.f. satisfies the Lévy–Khinchine formula*. On the other hand, the class L can be distinguished from among the inf. div. laws as

follows [7]. If λ is a finite measure on $\mathbb{R}$, set

$$\Phi_\lambda(x) = \int_{(e^x, \infty)} \frac{1 + y^2}{y^2} \, d\lambda(y),$$

$$\Psi_\lambda(x) = \int_{(-\infty, -e^x)} \frac{1 + y^2}{y^2} \, d\lambda(y).$$

Theorem 2. Let φ be the ch.f. of an inf. div. df F, and let λ be the Lévy spectral measure appearing in the Lévy–Khinchine formula for φ. Then $F \in L$ if and only if Φ_λ and Ψ_λ are convex functions on $\mathbb{R}$.

Another representation of the ch.f.'s of the self-decomposable laws was obtained by Urbanik [9].

Theorem 3. A function φ is the ch.f. of a self-decomposable distribution if and only if

$$\varphi(y) = \exp\left[ix_0 y \right.$$

$$+ \int_{-\infty}^{\infty} \left(\int_0^{xy} \frac{e^{iv} - 1}{v} \, dv - iy \tan^{-1} x \right)$$

$$\left. \times \frac{d\omega(x)}{\log(1 + x^2)} \right],$$

where $x_0 \in \mathbb{R}$, ω is a finite measure (or multiple of a df) on $\mathbb{R}$, and the integrand is defined by its limiting value $-y^2/4$ at $x = 0$.

Since the stable laws arise as the limits in distribution of normed sums (1) when the X_k are identically distributed, it is of interest in analyzing the statistical information inherent in a sequence $\{X_k\}$ of independent rvs to know what sort of stability conditions are satisfied by the limit distribution of a sequence of normed sums. To this end Urbanik [11] has used a sequence of stronger and stronger stability conditions to define subclasses $L_0 \supset L_1 \supset \cdots \supset L_\infty$ of the class L in such a way that $L_0 = L$, each L_m satisfies conditions (a) to (c), and L_∞ is the smallest class satisfying (a) to (c) and containing all the stable laws. Each of these classes L_m can be characterized by conditions similar to those appearing in Theorems 1 [11], 2 [5], and 3 [5, 11].

The definition of the class L, and Theorem 1 apply to rvs with values in $\mathbb{R}^k$, $k \geqslant 1$. Reference 10 contains an extension of Theorem 3 to this context. The extension of these ideas to Banach-space-valued rvs can be found in refs. 4 and 6.

References

[1] Fisz, M. and Varadarajan, V. S. (1963). *Zeit. Wahrscheinlichkeitsth.*, **1**, 335–339.

[2] Gnedenko, B. V. and Kolmogorov, A. N. (1954). *Limit Distributions for Sums of Independent Random Variables*, K. L. Chung, trans. Addison-Wesley, Reading, Mass., Chap. 6.

[3] Hsu, P. L. (1958). *Acta Sci. Nat. Univ. Pekinensis*, **4**, 257–270 (in Chinese). (Translation and comments by T. C. Sun in *Collected Works of Hsu Pao-Lu*, K. L. Chung, ed. Springer-Verlag, New York, 1982.)

[4] Kumar, A. and Schreiber, B. M. (1975). *Studia Math.*, **53**, 55–71.

[5] Kumar, A. and Schreiber, B. M. (1978). *Ann. Prob.*, **6**, 279–293.

[6] Kumar, A. and Schreiber, B. M. (1979). *J. Multivariate Anal.*, **9**, 288–303.

[7] Loève, M. (1963). *Probability Theory*, 3rd ed. D. Van Nostrand, Princeton, N.J., pp. 319–326.

[8] Tucker, H. G. (1962). *Pacific J. Math.*, **12**, 1125–1129.

[9] Urbanik, K. (1968). *Bull. Acad. Polon. Sci.*, **16**, 196–204.

[10] Urbanik, K. (1969). *Zastos. Mat. (Applicationes Math.)*, **10**, 91–97.

[11] Urbanik, K. (1973). In *Multivariate Analysis: Proceedings of the Third International Symposium*, P. R. Krishnaiah, ed. Academic Press, New York, pp. 225–237.

[12] Yamazato, M. (1978). *Ann. Prob.*, **6**, 523–531.

(CHARACTERISTIC FUNCTION
INFINITELY DIVISIBLE
LÉVY–KHINCHINE FORMULA
STABLE LAWS
UAN CONDITION)

B. M. SCHREIBER

LD50   *See* BIOASSAY, STATISTICAL METHODS IN

L₂ DESIGNS

Let there be $v = s^2$ treatments arranged in a $s \times s$ square. An L_2 design is an arrangement of the $v = s^2$ treatments in b blocks each of size k such that:

1. Every treatment occurs at most once in a block.
2. Every treatment occurs in r blocks.
3. Every pair of treatments, which occur in the same row or column of the $s \times s$ array, occur together in λ_1 blocks; while every other pair of treatments occur together in λ_2 blocks.

$v = s^2$, b, r, k, λ_1, and λ_2 are known as the parameters of the L_2 design and they satisfy the relations

$$vr = bk,$$

$$2(s - 1)\lambda_1 + (s - 1)^2\lambda_2 = r(k - 1).$$

As an example, let us consider $v = 9$ treatments arranged in a 3×3 row:

$$\begin{array}{ccc} 1 & 2 & 3 \\ 4 & 5 & 6 \\ 7 & 8 & 9 \end{array}$$

The nine blocks, given in the following table:

Block Number	Block Content
1	$(1, 6, 9, 2)$
2	$(6, 8, 2, 4)$
3	$(8, 1, 4, 9)$
4	$(9, 2, 5, 7)$
5	$(2, 4, 7, 3)$
6	$(4, 9, 3, 5)$
7	$(5, 7, 1, 6)$
8	$(7, 3, 6, 8)$
9	$(3, 5, 8, 1)$

can easily be verified to form an L_2 design

with parameters $v = 9$, $b = 9$, $r = 4$, $k = 4$, $\lambda_1 = 1$, and $\lambda_2 = 2$.

A two-associate-class association scheme is a relation of v treatments satisfying the following conditions:

1. Any two treatments are either first, or second associates, the relation of association being symmetrical.
2. Each treatment α has n_i ith associates $(i = 1, 2)$, the number n_i being independent of α.
3. If any two treatments α and β are ith associates, then the number of treatments that are ith associates of α, and kth associates of β, is p^i_{jk} and is independent of the pair of ith associates α and β.

Two-associate-class schemes are classified into five categories in ref. 3 and the association scheme is said to be an L_2 scheme if the number of treatments is $v = s^2$ arranged in a $s \times s$ square and two treatments are first associates if and only if they occur together in the same row or column of the $s \times s$ square array. Using L_2 association scheme, one can form a partially balanced incomplete block (PBIB) design* with two-associate classes as defined in ref. 2 and such designs are L_2 designs.

Ninety L_2 designs parameters and their solutions are tabulated by Clatworthy [5]. L_2 designs are generalized to L_i designs, which are two-associate-class PBIB designs; and extended L_2 designs, which are three-associate-class PBIB designs. Extended L_2 designs are introduced in ref. 9. L_i designs are PBIB designs with the L_i association scheme, which is obtained by superimposing $(i - 2)$ mutually orthogonal Latin squares on the $s \times s$ treatment array and defining two treatments to be first associates if and only if they occur together in the same row, or column, or with the same symbol of any of the Latin squares.

When s^2 treatments are tested in an experiment, the treatments will be written in a $s \times s$ square array and $2s$ blocks will be formed where s blocks correspond to the rows of the array and s blocks correspond to the columns of the array. Such a design is called a simple square lattice. p replications of a simple square lattice ($p \geqslant 1$) comprises an L_2 design.

Let $y_{ij(l)}$ be the observation on the jth plot of the ith block receiving lth treatment. Assuming the fixed-effects model

$$y_{ij(l)} = \mu + \beta_i + \tau_l + e_{ij(l)},$$

where μ is the general mean, β_i the ith block effect, τ_l the lth treatment effect, and $e_{ij(l)}$ are random errors which are independent, identically distributed (i.i.d.) $N(0, \sigma^2)$, the best linear unbiased estimator* of a treatment contrast $\sum_{l=1}^{v} a_l \tau_l$ is given by $\sum_{l=1}^{m} a_l \hat{\tau}_l$, where

$$\hat{\tau}_l = \frac{k - c_2}{\alpha} Q_l + \frac{c_1 - c_2}{\alpha} S_1(Q_l),$$

$$l = 1, 2, \ldots, v;$$

Q_l being the adjusted treatment total, $S_1(Q_l)$ being the total of $2(s - 1)$ adjusted treatment totals of the $2(s - 1)$ treatments which occur in the same row and column as the lth treatment in the $s \times s$ array of the treatments, $\alpha = r(k - 1)$,

$$k^2 \Delta = (\alpha + \lambda_1)(\alpha + \lambda_2) + (\lambda_1 - \lambda_2)$$
$$\times \{\alpha(3 - s) + (s - 1)\lambda_2$$
$$- 2(s - 2)\lambda_1\},$$

$$k \Delta c_1 = \lambda_1(\alpha + \lambda_2) + (\lambda_1 - \lambda_2)$$
$$\times \{(s - 1)\lambda_2 - 2(s - 2)\lambda_1\},$$

$$k \Delta c_2 = \lambda_2(\alpha + \lambda_1) + (\lambda_1 - \lambda_2)$$
$$\times \{(s - 1)\lambda_2 - 2(s - 2)\lambda_1\}.$$

The variance of the best linear unbiased estimate of $\tau_l - \tau_{l'}$ is given by

$$V(\hat{\tau}_l - \hat{\tau}_{l'})$$

$$= \begin{cases} \dfrac{k - c_1}{k - 1}\left(\dfrac{2\sigma^2}{r}\right), & \begin{array}{l}\text{if } l \text{ and } l' \text{ occur in the} \\ \text{same row or column of} \\ \text{the } s \times s \text{ treatment array} \end{array} \\[2em] \dfrac{k - c_2}{k - 1}\left(\dfrac{2\sigma^2}{r}\right), & \text{otherwise.} \end{cases}$$

The parameters c_1 and c_2 are also listed in tables.

Let $\mathbf{N} = (n_{ij})$ be the $v \times b$ incidence matrix of a L_2 design where $n_{ij} = 1$ or 0 according as the ith treatment occurs in the jth block or not. Then $\mathbf{NN'}$ has the eigen values $\theta_0 = rk$, $\theta_1 = r + (s-2)\lambda_1 - (s-1)\lambda_2$, and $\theta_2 = r - 2\lambda_1 + \lambda_2$ with respective multiplicities $\alpha_0 = 1$, $\alpha_1 = 2(s-1)$, and $\alpha_2 = (s-1)^2$. In ref. 6 it was shown that in a L_2 design if $\theta_1 = 0$, then k is divisible by s and every block of the design contains k/s treatments from each of the s rows of the treatment array. The number of common treatments (x) between any two blocks of a L_2 design satisfies the relation

$$\max\left[0, 2k - v, k - \theta_i\right]$$
$$\leqslant x \leqslant \min\left[k, \theta_i + \frac{2(rk - \theta_i)}{b} - k\right],$$

where $i = 1$ if $\lambda_1 > \lambda_2$ and $i = 2$ if $\lambda_1 < \lambda_2$ [1].

Using a complete set of mutually orthogonal Latin squares of order s, a L_2 design with parameters $v = s^2$, $b = s(s-1)$, $r = s - 1$, $k = s$, $\lambda_1 = 0$, $\lambda_2 = 1$ can always be constructed when s is a prime or a prime power. We illustrate this method with $s = 3$. Let the nine treatments $1, 2, \ldots, 9$ be arranged in a 3×3 square array as given earlier. Taking the two mutually orthogonal Latin squares*

A	B	C		α	γ	β
C	A	B		γ	β	α
B	C	A		β	α	γ

superimposing each of them on the treatment array and writing the blocks consisting of treatments occurring together with each of the Latin and Greek letters, we get the L_2 design

$$(1,5,9); \quad (2,6,7); \quad (3,4,8);$$
$$(1,6,8); \quad (3,5,7); \quad (2,4,9)$$

with parameters $v = 9$, $b = 6$, $r = 2$, $k = 3$, $\lambda_1 = 0$, and $\lambda_2 = 1$.

By using the treatment array, Clatworthy [4] constructed the following two series of L_2

designs:

$$v = b = s^2, \quad r = k = 2s - 1,$$
$$\lambda_1 = s, \quad \lambda_2 = 2,$$
$$v = s^2, \quad b = s(s-1), \quad r = 2(s-1),$$
$$k = 2s, \quad \lambda_1 = s, \quad \lambda_2 = 2.$$

The foregoing two series of L_2 designs exist for every s.

Raghavarao [7] described a method of constructing L_2 designs using the method of differences. A very interesting method of constructing L_2 designs with parameters

$$v = s^2, \quad b = 2s(s-1), \quad r = 2(s-1),$$
$$k = s, \quad \lambda_1 = 0, \quad \lambda_2 = 2$$

was recently given in ref. 8 when the solution of a balanced incomplete block design*

$$v^* = s = b^*, \quad r^* = s - 1 = k^*, \quad \lambda^* = s - 2$$

can be constructed with a certain property.

The following three sets of values are the unsolved parametric combinations of three L_2 designs:

$$v = 36, \quad b = 54, \quad r = 9, \quad k = 6,$$
$$\lambda_1 = 2, \quad \lambda_2 = 1$$
$$v = 36, \quad b = 40, \quad r = 10, \quad k = 9,$$
$$\lambda_1 = 3, \quad \lambda_2 = 2$$
$$v = 100, \quad b = 90, \quad r = 9, \quad k = 10,$$
$$\lambda_1 = 0, \quad \lambda_2 = 1$$

References

[1] Agrawal, H. L. (1964). *Calcutta Statist. Ass. Bull.*, **13**, 76–79. (Bounds on the number of common treatments between blocks of certain 2-associate-class PBIB designs are given.)

[2] Bose, R. C. and Nair, K. R. (1939). *Sankhyā*, **4**, 337–372. (Partially balanced incomplete block designs are introduced.)

[3] Bose, R. C. and Shimamoto, T. (1952). *J. Amer. Statist. Ass.*, **47**, 151–184. (2-associate-class schemes are classified.)

[4] Clatworthy, W. H. (1967). *Technometrics*, **9**, 229–243. (Construction of some families of L_2 designs.)

[5] Clatworthy, W. H. (1973). *Natl. Bur. Stand. Appl. Math. Ser. 63* (Washington, D.C.). (Tables listing parametric combinations of L_2 and other 2-associate-class PBIB designs.)

[6] Raghavarao, D. (1960). *Ann. Math. Statist.*, **31**, 787–791. (The distribution of the treatments in some L_2 designs is given.)

[7] Raghavarao, D. (1973). *Ann. Statist.*, **1**, 591–592. (Construction of some L_2 designs is given.)

[8] Sharma, H. C. (1978). *Calcutta Statist. Ass. Bull.*, **27**, 81–96. (Construction of some L_2 designs is given.)

[9] Singla, S. L. (1977). *Aust. J. Statist.*, **19**, 126–131. (Extended L_2 designs are introduced.)

(BALANCED INCOMPLETE BLOCK
   DESIGNS
CONFOUNDING
DESIGN OF EXPERIMENTS
GRAECO-LATIN SQUARES
GROUP DIVISIBLE DESIGNS
INTERACTION
LATIN SQUARES, LATIN CUBES,
   LATIN RECTANGLES, ETC.
PARTIALLY BALANCED INCOMPLETE
   BLOCKS DESIGNS)

Damaraju Raghavarao

L-DIVERGENCE   *See J*-divergences and related concepts

LEADING INDICATORS

In 1937, the U.S. Secretary of the Treasury, Henry Morgenthau, asked the National Bureau of Economic Research (a private Cambridge, Massachusetts firm specializing in business cycle research) for a numerical system that would signal an end to the 1937–1938 business decline. The National Bureau of Economic Research (NBER), under the direction of Wesley Mitchell and Arthur Burns, selected 21 economic time series which, based on past performance, seemed to be reliable indicators of economic trends. These series were characterized by Mitchell and Burns as *leading*, *coincident*, and *lagging* indicators of economic conditions. The initial list of indicators was published by NBER in May 1938 (see ref. 1).

Leading indicators are those economic time series whose changes in direction generally *precede* turns in measures of current aggregate economic activity. Coincident indicators are measures of *current* economic activity and lagging indicators are numerical measures whose turns *trail* the turns of the general economy.

In the United States, the current state of the economy is often measured by the gross national product (GNP: the total market value of all goods and services produced by the economy) or the Federal Reserve Board's Index of Industrial Production (IIP: an index of industrial output only). These coincident indicators contain broad swings of irregular duration known as business cycles. Business cycles arise as the result of the expansion and contraction of underlying economic processes. Leading indicators then are used to foretell turning points in the business cycles, which, in turn, are represented by broad swings in GNP, IIP, and other coincident indicators.

Indicators are initially selected by NBER on the basis of (1) the role they play in widely accepted theories of economic phenomena and (2) their empirical record. Indicators are subjectively rated using a 0 to 100 scale on each of six criterion variables:

1. Economic significance
2. Statistical adequacy
3. Consistency of timing at business cycle peaks and troughs
4. Conformity to business expansion and contractions
5. Smoothness
6. Promptness of publication

Time series with high scores become indicators.

At the time this article was written the most recent revision of the NBER indicator list occurred in 1977. There are presently 111 indicators which can be cross-classified according to timing (leading, coincident, lagging); economic process (employment and unemployment; production and income; consumption, trade, orders, and deliveries; fixed capital investment; inventories and inventory investment; prices, costs, and profits; money and credit); and performance at business cycle peaks and troughs (see ref. 3). The current and historical values of U.S. indicators, together with other important economic measures, are published monthly by the U.S. Department of Commerce in *Business Conditions Digest* (see ref. 3).

Since several indicators of the same type, say leading indicators, may provide different signals and therefore are difficult to interpret, groups of indicators are combined to form a broad view *index* of cyclical movement. Index components are chosen on the basis of their scores on the six criterion variables mentioned above. Composite indices tend to be more reliable and smoother indicators of business cycle peaks and troughs than their component series.

In 1981, the index of leading indicators (ILI) had 12 components. Among the leading indicators included in the index were: average workweek (production workers, manufacturing; hours); contracts and orders for plant and equipment ($1972); stock prices (500 common stocks; index: 1941–1943 = 10) and money supply (M2; $1972).

Before indicator data are published, the individual series are frequently adjusted. For example, seasonal variation is often removed by seasonal adjustment procedures. Series used in indices are each standardized so that their monthly swings are roughly the same order of magnitude. To construct the ILI, the standardized components are weighted by their (weighted) average scores on the six criterion variables previously discussed. The ILI is then "reverse trend adjusted" so that the long-term trend is the same as that of the index of coincident indicators—the latter index reaching its turning points at about the same time as the general economy.

How valuable are leading indicators as purveyors of short-term economic conditions? Do they move early enough to allow decision makers to capitalize on their messages? The reviews are mixed. Empirical studies suggest that leading indicators do turn before the turns in the general economy

(as officially determined by NBER), but lead times are not consistent and can be of shorter length than practically useful. Also, leading indicators often provide many false signals.

For example, using a one-month up-and-down criterion, Stekler and Schepsman compared turning points in the ILI with turning points in the IIP corresponding to NBER reference dates (see ref. 2). For the period 1948–1970, these authors found that the ILI indicated 24 peaks compared with 5 NBER identified peaks in the IIP (general economy). The average lead time of ILI turns preceding peaks in the economy was 1.6 months. In addition, the ILI indicated nine troughs over this period compared with five troughs identified in the IIP. The average lead time of ILI turns preceding troughs in the economy was 4 months.

Different sectors of the economy move at different times and this lack of consistency is reflected in the leading indicators. It is not uncommon for the ILI to increase (decrease) from one month to the next while several of its component series decrease (increase). Moreover, leading indicator values, although published on a monthly schedule, typically do not become public until at least a month after the period to which they apply and sometimes after the coincident indicators referring to the same date.

The lack of timing consistency, inherent month-to-month fluctuations, and delays in publication make interpreting leading indicators difficult. Economists confronting the same set of numbers frequently disagree on their meaning. Leading indicators must be evaluated with an awareness of the larger economic environment, determined, in part, by government policies, international events, and consumer behavior.

A good example of the efficacy of leading indicators is provided by the 1974–1975 recession. Figure 1 shows two coincident indicators of the economy—GNP ($1972; quarterly data) and the Federal Reserve Board's IIP (1967 = 100; monthly data)—for the period June 1973 through December 1975. Also shown, for the same period, are the ILI

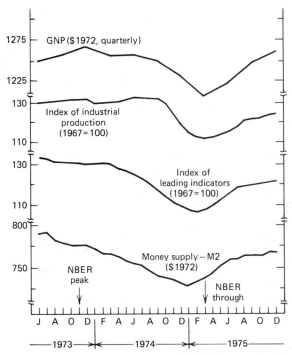

Figure 1   Leading indicator performance in the 1974–1975 recession. (From ref. 3.)

(1967 = 100; monthly data) and one of its component series, the money supply (M2, $1972; monthly data). The NBER designated cyclical peak and trough for the entire economy during this time period occurred in November 1973 and March 1975, respectively.

The 1974–1975 recession in the United States was the most severe since World War II. It is clear from Fig. 1 that the economy, as measured by GNP, began its downturn about the second quarter (June) of 1974, reaching a trough in the first quarter (March) of 1975. The monthly IIP figures, the other coincident indicator of economic health, began to decline after September 1974, reaching, again, a low point in March 1975. Both coincident indicators advanced steadily throughout the last three quarters of 1975, implying that the recession was ending and that economic recovery was well under way.

Examining the leading indicators in Fig. 1, we see that the ILI began its downward trend about March 1974—a few months before the decline in GNP and well before the

decline in IIP. The money supply (M2) began its descent even earlier (about December 1973). The leading indicators, in this case, appear to signal the imminent economic downturn clearly and well in advance of the actual demise. However, the ability of these leading indicators to signal an upturn is less satisfactory. For this example, the economy is well along toward recovery by the time enough leading indicator data are available to substantiate the upward trend.

In spite of their limitations (false signals, timing inconsistencies, publication delays), many private and public decision makers find leading indicators useful predictors of economic health. They are a primary source of information for forecasters of short-term economic trends.

References

[1] Mitchell, W. C. and Burns, A. F. (1938). Statistical Indicators of Cyclical Revivals. *Bull. No. 69*, National Bureau of Economic Research, New York. [See also Moore (1961) in the Bibliography below.]

[2] Stekler, H. O. and Schepsman, M. (1973). *J. Amer.*

Statist. Ass., **68**, 291–296. (Intermediate; empirical investigation of the ability of the ILI to signal changes in the U.S. economy.)

[3] U.S. Department of Commerce (1980, 1981). *Business Conditions Digest*. Washington, D.C. (Contains data used in Fig. 1; economic time series, including the NBER business cycle indicators.)

Bibliography

Granger, C. W. J. (1980). *Forecasting in Business and Economics*. Academic Press, New York. (Elementary; contains a nice summary of leading indicators.)

Moore, G. H., ed. (1961). *Business Cycle Indicators*, Vol. 1. Princeton University Press, Princeton, N.J. (Contains early contributions to the analysis of business conditions, including the Mitchell and Burns paper [1].)

Moore, G. H. and Shiskin, J. (1976). In *Statistics: A Guide to Business and Economics*, J. Tanur, ed. Holden-Day, San Francisco. (Elementary account of leading indicators by two people involved in their development.)

(ECONOMETRICS
INDEX NUMBERS
LAGGING INDICATORS
TIME SERIES)

DEAN WICHERN

LEAST FAVORABLE CONFIGURATION

In the theory of ranking and selection* we often deal with many parameters and using a least favorable configuration (LFC) of parameter values is one way of simplifying the problem. The LFC is that set of parameter values which minimizes the probability of a correct selection (or, in a wider context, the probability of a correct decision). Hence it is possible to put lower bounds on these probabilities (over certain parameter spaces) and to choose sample sizes to get acceptable lower bounds. The procedure is analogous to that of choosing sample sizes to get acceptable lower bounds on power for hypothesis tests.

In the following section we consider LFC in the context of a normal means selection problem, while in the section "Selection from Populations which are Continuous and

Stochastically Increasing" we generalize the selection problem to other continuous distributions and in the section "Least Favorable Configuration when Ties are Broken" to discrete distributions. In the final section we give a general definition of LFC.

NORMAL MEANS SELECTION PROBLEM

Consider k normal populations $N(\mu_i, \sigma^2)$, $i = 1, \ldots, k$, where the μ_i have some (unknown) ranking

$$\mu_{(1)} \leqslant \mu_{(2)} \leqslant \cdots \leqslant \mu_{(k-1)} < \mu_{(k)}, \quad (1)$$

the $\{(1), (2), \ldots, (k)\}$ being some fixed but unknown permutation of $\{1, 2, \ldots, k\}$.

The goal is, on the basis of N random observations from each population, to select that population with mean $\mu_{(k)}$.

Now, if our selection procedure consists of selecting that population with the largest sample mean, we see that the probability of correct selection (PCS) is given by

$$\Pr[\mathrm{CS} \mid \boldsymbol{\mu}, N] = \int \prod_{i \neq (k)} \Phi\big((x - \mu_i)\sqrt{N}/\sigma\big)$$

$$\times d\Phi\big((x - \mu_{(k)})\sqrt{N}/\sigma\big), \quad (2)$$

where $\boldsymbol{\mu} = (\mu_1, \mu_2, \ldots, \mu_k)$ and $\Phi(\cdot)$ is the standard normal cumulative distribution function.

Since $\Phi(\cdot)$ is an increasing function and $\mu_i < \mu_{(k)}$ for all $i \neq (k)$, we see that the expression given by (2) is greater than k^{-1}. Of course, we should like to have PCS considerably greater than k^{-1}, and a design problem is to choose N equal to N^*, say, so that

$$\Pr[\mathrm{CS} \mid \boldsymbol{\mu}, N^*] \geqslant P^*$$

where P^* is a predetermined constant with $k^{-1} < P^* < 1$.

However, N^* is obviously a complicated function of $\boldsymbol{\mu}$ which is unknown, so the *indifference zone* approach is to find N^* such that

$$\inf_{\boldsymbol{\mu} \in \bar{I}} \Pr[\mathrm{CS} \mid \boldsymbol{\mu}, N^*] \geqslant P^* \quad (3)$$

where $\bar{I}$ is called the preference zone and is

the complement (on the parameter space) of I, the indifference zone for μ. The usual indifference zone for this problem is given by

$$I = \{ \mu : \mu_{(k)} - \mu_{(k-1)} < \delta^* \} \quad (4)$$

for some predetermined δ^*.

In other words, the experimenter is "indifferent" about making a correct selection unless the "best" population has a mean at least δ^* greater than the next "best" population, but if it has, the experimenter wants the PCS to be at least P^*.

Let us denote by μ_{LF} a value of μ such that

$$\text{Pr}[\text{CS} \,|\, \mu_{LF}, N] = \inf_{\mu \in I} \text{Pr}[\text{CS} \,|\, \mu, N]. \quad (5)$$

The set of all such μ_{LF} is known as the *least favorable configuration* of μ values.

By examination of (2) we see that this set is given by

$$\mu_i = \mu_{(k)} - \delta^* \quad \text{for all} \quad i \neq (k), \quad (6)$$

for various values of $\mu_{(k)}$. In other words, we move the μ_i as close as possible to $\mu_{(k)}$ to get the LFC. The set of μ values satisfying $\mu_i = \mu_{(k-1)}$ for all $i \neq (k)$, $(k-1)$ is sometimes referred to as the *generalized least favorable configuration*.

We notice that

$$\text{Pr}[\text{CS} \,|\, \mu_{LF}, N]$$

$$= \int \left\{ \Phi\left(y + \delta^* \sqrt{N}/\sigma \right) \right\}^{k-1} d\Phi(y)$$

and this has been tabled (as a function of $\delta^* \sqrt{N}/\sigma$) by Bechhofer [2] in a pioneering paper on the subject of ranking and selection. This enables us to find N^* to satisfy (3).

We may easily generalize the problem above to that of selecting the $t(< k)$ "best" populations out of k. That is, we wish to select (not necessarily in order) the set of populations whose means are $\mu_{(k)}, \dots, \mu_{(k-t+1)}$. With an indifference zone given by

$$\mu_{(k-t+1)} - \mu_{(k-t)} < \delta^*$$

we find that the LFC is given by

$$\mu_{(1)} = \mu_{(2)} = \cdots = \mu_{(k-t)} = \mu_{(k-t+1)} - \delta^*$$

$$= \mu_{(k-t+2)} - \delta^* = \cdots = \mu_{(k)} - \delta^*.$$

As before, we move the "better" means $\mu_{(k-t+1)}, \dots, \mu_{(k)}$ and the "worse" means $\mu_{(1)}, \dots, \mu_{(k-t)}$ as close as possible to each other to get the LFC.

SELECTION FROM POPULATIONS THAT ARE CONTINUOUS AND STOCHASTICALLY INCREASING

Let $Y_1, \dots, Y_k$ be independent continuous random variables with distribution functions $F(y; \theta_i)$, $i = 1, \dots, k$. Suppose that each $F(y; \theta_i)$ is nonincreasing in θ_i for all y. (We say that the distribution is stochastically increasing in θ_i.) For the goal of selecting those populations with the t largest values of θ_i a common procedure is to select the populations with the t largest observed values of the Y_i.

The preceding section may be considered as a special case where the Y_i are sample means, the distributions are normal, and the θ_i are location parameters.

Corresponding to (2), with $\theta_{(1)}, \theta_{(2)}, \dots, \theta_{(k)}$ being the ordered θ_i we find the probability of correct selection to be

$$\text{Pr}[\text{CS} \,|\, \boldsymbol{\theta}, F(\cdot)]$$

$$= \int \prod_{i \leqslant k-t} F(y; \theta_{(i)}) d$$

$$\times \prod_{j > k-t} F(y; \theta_{(j)}) \quad (7)$$

$$= 1 - \int \prod_{j > k-t} F(y; \theta_{(j)}) d$$

$$\times \prod_{i \leqslant k-t} F(y; \theta_{(i)}). \quad (8)$$

Since $F(y; \theta_i)$ is nonincreasing in θ_i, we see, using (7), that the PCS is nonincreasing in $\theta_{(i)}$ for $i \leqslant k - t$ and, using (8), that it is nondecreasing in $\theta_{(j)}$ for $j > k - t$. Hence with a preference zone given by

$$\bar{I} = \{ \boldsymbol{\theta} : \theta_{(k-t)} \leqslant \theta_L^* < \theta_M^* \leqslant \theta_{(k-t+1)} \} \quad (9)$$

where θ_L^* and θ_M^* are specified, the LFC is given by

$$\theta_{(1)} = \theta_{(2)} = \cdots = \theta_{(k-t)} = \theta_L^* \quad \text{and}$$

$$\theta_{(k-t+1)} = \cdots = \theta_{(k)} = \theta_M^* . \quad (10)$$

If θ_L^* and θ_M^* are not specified but it is given that

$$\psi(\theta_L^*) \leqslant \theta_M^*$$

where $\psi(\theta)$ is a function satisfying $\theta < \psi(\theta)$, the LFC is given by (10) with $\theta_M^* = \psi(\theta_L^*)$ and θ_L^* chosen to give the infimum of (7) [subject to (10)].

When θ is a location parameter (as in the preceding section) it is usual to take $\psi(\theta) = \theta + \delta^*$ with $\delta^* > 0$. The probability [the infimum of (7)] which has to be tabled is then a function of δ^* and $F(\cdot)$ only, and not of θ_L^*.

For the same reason, when θ is a scale parameter, a useful form is $\psi(\theta) = \theta/\rho^*$ with $0 < \rho^* < 1$.

The results above may be found in Barr and Rizvi [1] and tables for the implementation of the procedure for various distributions are collected in Gibbons et al. [6].

LEAST FAVORABLE CONFIGURATION WHEN TIES ARE BROKEN

If the Y_i considered in the preceding section are discrete variables, we may find the selection procedure complicated by the presence of ties*. Suppose that (where y_i indicate observed Y_i values)

$$y_{i_1} \leqslant y_{i_2} \leqslant \cdots \leqslant y_{i_{k-t-p+1}} = y_{i_{k-t-p+2}}$$

$$= \cdots = y_{i_{k-t+q}} < y_{i_{k-t+q+1}} \leqslant \cdots \leqslant y_{i_k}$$

(with $q \geqslant 0$ and $p \geqslant 1$). Then a reasonable procedure is to select the populations corresponding to

$$y_{i_{k-t+q+1}}, \ldots, y_{i_k}$$

and to select at random q of the $(p + q)$

populations corresponding to

$$y_{i_{k-t-p+1}}, \ldots, y_{i_{k-t+q}} .$$

The PCS is then a much more complicated expression than that given by (7) but we may still derive the same results on LFC. The details are given in Bofinger [3].

GENERAL DEFINITION

Let us consider some decision procedure R which may be a selection procedure (as in the last three sections) or may be a ranking procedure or a procedure for partitioning populations into those which are better and those which are worse than a control. For various examples of such procedures, see Gibbons et al. [6], which gives the methodology in detail and gives references to the theory. Alternatively, a more theoretical treatment will be found in Gupta and Panchapakesan [8]. A short introduction is given by Gibbons et al. [6].

We say that a correct decision is made if the appropriate populations are selected, or the populations are ranked correctly or partitioned correctly and we consider the probability of correct decision denoted by

$$\Pr[\text{CD} \mid \phi, R], \quad (11)$$

depending on some parameter $\phi \in \Omega$. Suppose that we have an indifference zone I and preference zone $\bar{I} = \Omega - I$ for the parameter ϕ.

Definition. The LFC of ϕ values is the set of values ϕ_{LF} such that

$$\Pr[\text{CD} \mid \phi_{\text{LF}}, R] = \inf_{\phi \in \bar{I}} \Pr[\text{CD} \mid \phi, R]. \quad (12)$$

Unless expression (11) has appropriate monotonic properties in ϕ for $\phi \in \bar{I}$, it may be extremely difficult to find the values ϕ_{LF}.

In the previous sections it seems intuitively reasonable to require the distribution of Y_i to be stochastically increasing in θ_i, since we take a large Y_i to indicate a large value of θ_i, and the results on LFC are not really surprising. However, it is not always obvious how nuisance parameters may be specified for the LFC.

For example, consider the selection procedure of the section "Selection from Populations Which are Continuous and Stochastically Increasing," where the Y_i have a normal distribution with mean θ_i and variance η_i^2 ($i = 1, \ldots, k$) and we are given an upper bound, say η^2, on the η_i^2. The parameter ϕ now contains nuisance parameters η_i^2 as well as "selection" parameters θ_i. We may take the preference zone to be

$$\bar{I} = \left\{ \phi : \theta_{(k-t)} \leqslant \theta_L^* < \theta_M^* \leqslant \theta_{(k-t+1)}, \right.$$

$$\left. \eta_i^2 \leqslant \eta^2 \quad i = 1, \ldots, k \right\}.$$

It might be thought that the LFC would be given by (10) and $\eta_i^2 = \eta^2$. The fact that in general this is not so has been given some attention in the literature, particularly because of its connection with the two-stage normal means selection problem. (See Dudewicz and Dalal [5]). On the other hand, if

$$\eta^2 \leqslant 0.8488(\theta_M^* - \theta_L^*)^2$$

$$\times \min\left((t-1)^{-1}, (k-t-1)^{-1}\right) \quad (13)$$

it may be seen [4] that the LFC *is* given by (10) and $\eta_i^2 = \eta^2$.

A conclusion must be that although there is no difficulty in generalizing the idea of LFC as in (12), each case will have to be examined carefully since it is not likely that the PCS will be monotonic in the nuisance parameters for all values of the selection parameters.

References

[1] Barr, D. R. and Rizvi, M. H. (1966). *J. Amer. Statist. Ass.*, **61**, 640–646.

[2] Bechhofer, R. E. (1954). *Ann. Math. Statist.*, **25**, 16–39.

[3] Bofinger, E. (1976). *J. Amer. Statist. Ass.*, **71**, 423–424.

[4] Bofinger, E. (1979). *Aust. J. Statist.*, **21**, 149–156.

[5] Dudewicz, E. J. and Dalal, S. R. (1975). *Sankhyā B*, **37**, 28–78.

[6] Gibbons, J. D., Olkin, I., and Sobel, M. (1977). *Selecting and Ordering Populations: A New Statistical Methodology*. Wiley, New York.

[7] Gibbons, J. D., Olkin, I., and Sobel, M. (1979). *Amer. Statist.*, **33**, 185–195.

[8] Gupta, S. S. and Panchapakesan, S. (1979). *Multiple Decision Procedures: Theory and Methodology of Selecting and Ranking Populations*. Wiley, New York.

(INDIFFERENCE ZONE RANKING AND SELECTION)

Eve Bofinger

LEAST FAVORABLE DISTRIBUTIONS

In his paper [8] Lehmann used least favorable distributions (LFDs) to signify almost the same as in robust testing* now. (This meaning of the term is to be considered here.) However, in the setup based on ref. 8 —which uses parametric and less convenient neighborhoods than those of robust testing —LFDs are in general difficult to construct. Huber [4] avoided this drawback in terms of the suitable neighborhoods **1**(a) and **1**(b), discussed below, and succeeded in constructing least favorable pairs of distributions (LFPs). Besides ref. 4, Strassen's paper [13] is pioneering in this field. He deserves the credit for probabilizing interval arithmetic by total-alternating capacities (in the sense of Choquet) and also for proving the first Neyman–Pearson lemma for capacities. The breakthrough in this context, i.e., concerning the existence of LFPs, was achieved in ref. 7, where the importance of two-alternating capacities was demonstrated. More systematic work on the construction of LFPs has been done since ref. 4, e.g., in refs. 9–11. A treatment of LFPs from the point of view of *comparison of statistical experiments** is given in ref. 1. Although not identical, our approach will be intimately related.

Like its classical version, the Neyman–Pearson lemma* for capacities and the topic of LFPs seem to be more of philosophical than of practical impact. However, serious work has been done to make the theory applicable to practical problems: in robust estimation,* particularly in Huber's finite sample minimax theory for location parameters [5] and Rieder's asymptotic counterpart and generalization to arbitrary parameters [12].

MOTIVATION AND DEFINITION OF LFPS

Suppose that a cheat has a loaded coin. But what a pity: he forgot whether heads and tails come according to the probability distribution $P = (\theta, 1 - \theta)$ or according to $Q = (1 - \theta, \theta)$. As a matter of fact, even he does not know θ exactly. The coinmaker could only guarantee $|\theta - 0.3| \leqslant 0.1$. So actually the cheat has to decide between the two composite hypotheses $\mathscr{P}_0 = \{P : |\theta - 0.3| \leqslant 0.1\}$ and $\mathscr{P}_1 = \{Q : |\theta - 0.3| \leqslant 0.1\}$ linked by the same parameter θ (cf. Fig. 1). But how can he discover this except by tossing the coin?

In order to control both error probabilities at—say 0.05—he has to fix the number n of coin-tosses according to the worst possible case: i.e., the one for which the testing problem is hardest or in other words the one for which the two probability distributions are as close as possible. Roughly speaking, a pair $(Q_0, Q_1) \in \mathscr{P}_0 \times \mathscr{P}_1$ of probability distributions having these properties is called a *least favorable pair of distributions*. In the cheat's example this is obviously given by $Q_0 = (0.4, 0.6)$ and $Q_1 = (0.6, 0.4)$.

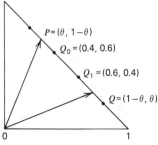

Figure 1

To get a notion of *robust testing*, let us start more generally with two elements P_0 and P_1 from the set $\mathscr{W}$ of all probability distributions on the real line $\mathbb{R}$ (associated with the system $\mathscr{B}$ of Borel sets), and let us assume that we have to decide between them on the basis of independent observations. Furthermore, let us suppose that we have for this purpose instead of the observations $X_1, \ldots, X_n$ with distributions $\mathscr{L}(X_i) = P_j$, $1 \leqslant i \leqslant n$ and $j = 0$ or $j = 1$, respectively, only erroneous observations $Y_1, \ldots, Y_n$, i.e., observations with distributions $\mathscr{L}(Y_i) = P_{i0}$ from a certain neighborhood $\mathscr{P}_0 \subset \mathscr{W}$ of P_0 [respectively, $\mathscr{L}(Y_i) = P_{i1}$] from a certain neighborhood $\mathscr{P}_1 \subset \mathscr{W}$ of P_1. (The trivial case $\mathscr{P}_0 \cap \mathscr{P}_1 \neq \varnothing$ will always be excluded!) What will happen? Clearly, a likelihood ratio* test $\prod_{i=1}^{n}\{p_1(y_i)/p_0(y_i)\} > t$ may fail to give a reasonable decision. A single factor $p_1(y_i)/p_0(y_i)$ caused by an erroneous value (outliers*) y_i could be equal or almost equal to 0 or ∞ and hence determine the outcome of the test. In the setting above, a likelihood ratio test of P_0^n versus P_1^n is not robust. The way out of this dilemma is its *robustized version* $\prod_{i=1}^{n}\{q_1(y_i)/q_0(y_i)\} > \tilde{t}$ defined in terms of the likelihood ratios $\pi(y) = q_1(y)/q_0(y)$ with respect to a LFP (Q_0, Q_1) owing to the following

Definition. A pair $(Q_0, Q_1) \in \mathscr{P}_0 \times \mathscr{P}_1$ is called *least favorable* if all likelihood ratio tests of Q_0 vs. Q_1 have the property that no matter which pair $(P, Q) \in \mathscr{P}_0 \times \mathscr{P}_1$ actually underlies, their error probabilities are bounded by those given in terms of (Q_0, Q_1), or formally if

$$\left. \begin{array}{l} \sup\{P[\pi > t], P \in \mathscr{P}_0\} = Q_0[\pi > t] \\ \sup\{Q[\pi \leqslant t], Q \in \mathscr{P}_1\} = Q_1[\pi \leqslant t] \end{array} \right\}$$

$$\text{for all} \quad t \geqslant 0.$$

If one of the neighborhoods shrinks to only one probability distribution—say $\mathscr{P}_0 = \{P_0\}$—then the other element Q_1 of the LPF $(Q_0, Q_1) = (P_0, Q_1)$ is called the *least favorable distribution*.

ERRONEOUS OBSERVATIONS AND NEIGHBORHOODS

Classical parametric statistics derives results under the assumption that these parametric models are exactly true. However, apart from some simple discrete models perhaps, such models are never exactly true. We may try to distinguish three main reasons for the deviations: (a) the occurrence of *gross errors*, such as blunders in measuring, wrong decimal points, errors in copying, inadvertent measurement of a member of a different population, or just "something went wrong"; (b) rounding and grouping and other *local inaccuracies*; (c) the model may have been conceived as an *approximation* anyway, e.g., by virtue of the central limit theorem*. [2, p. 88].

Neighborhoods $\mathscr{P}_v = N(P, \epsilon) = \{ Q \in \mathscr{W} : Q(B) \leqslant v(B) \text{ for all } B \in \mathscr{B} \}$ defined by means of certain monotone set functions v, satisfying $v(\varnothing) = 0$, help to formalize these three phenomena. In what follows we give the corresponding neighborhoods together with their vague interpretations. These rest on [14]. Neighborhoods 1(a) and 1(b) correspond to case (i): gross errors; Neighborhood 2 corresponds to case (ii): local inaccuracies; and Neighborhood 3 to case (iii): approximation. For further reading, see ref. 6 [Chap. 2].

1 (a) **Contamination Neighborhood.** $v(B) = \min\{(1 - \epsilon)P(B) + \epsilon, 1\}$, $(B \neq \varnothing)$. The random variable X with model distribution P is observed only with probability $1 - \epsilon$. With probability ϵ some jamming variable Z with unknown distribution R is observed.

   (b) **Total-Variation Neighborhood.** $v(B) = \min\{P(B) + \epsilon, 1\}$. Instead of X a random variable Y is observed which differs from X at least with probability ϵ.

2 **Local-Variation Neighborhood.** $v(B) = \min\{P(B^\epsilon), 1\}$, where $B^\epsilon = \{ y \in \mathbb{R} : |x - y| \leqslant \epsilon, x \in B \}$ is the closed ϵ-neighborhood of B. A random variable Y is observed which differs from X more than ϵ only with probability 0.

3 **Prokhorov Neighborhood.** $v(B) = \min\{P(B^\epsilon) + \epsilon, 1\}$ [a synthesis of neighborhoods 1(b) and 2]. A random variable Y is observed which differs from X more than ϵ at most with probability ϵ.

EXISTENCE OF LFPs

The common feature of the monotone set functions v defining the neighborhoods above is the validity of the weakening

$$v(A \cup B) + v(A \cap B) \leqslant v(A) + v(B)$$
$$\text{for all } A, B \in \mathscr{B} \quad (1)$$

of the well-known corresponding identity for probability distributions. This feature, which makes v a *two-alternating capacity*, is crucial for the existence of LFPs because of its equivalent:

Let $\pi(x)$ be any likelihood ratio.

Then there exists a probability distribution $Q_\pi \in \mathscr{P}_v$ which alone can satisfy $Q_\pi[\pi > t]$
$$= v[\pi > t] \text{ for all } t \geqslant 0. \quad (2)$$

The concept of *risk set* is an appropriate tool in the construction of LFPs. Let P and Q be two probability distributions with densities p and q, respectively. Then, owing to the Neyman–Pearson lemma, the sets $B_t = [q/p > t]$ determine the family $\{ B_t : t \geqslant 0 \}$ of most powerful (nonrandomized) tests of P vs. Q. The convex hull of the corresponding set $\{(P(B_t), Q(B_t^c)): t \geqslant 0\}$ of pairs of error probabilities, more precisely

$$R(P, Q) = \text{co}(\{(P(B_t), Q(B_t^c)): t \geqslant 0\}$$
$$\cup \{(1, 0), (0, 1)\})$$

is called the *risk set of the testing problem* (P, Q). It provides a measure of deviation of P and Q. More heuristically: The smaller the risk set, the harder the testing problem.

Example 1. For $P = (\theta, 1 - \theta)$ vs. $Q = (1 - \theta, \theta)$ in the cheat's example, note that

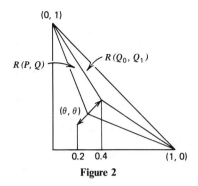

Figure 2

there are only two values of the likelihood ratio q/p, namely $(1 - \theta)/\theta$ and $\theta/(1 - \theta)$. Hence $B_t = \{\text{head}, \text{tail}\}$ for $t < \theta/(1 - \theta)$; $B_t = \{\text{head}\}$ for $\theta/(1 - \theta) \leqslant t < (1 - \theta)/\theta$ and $B_t = \varnothing$ for $t \geqslant (1 - \theta)/\theta$. The corresponding set is shown in Fig. 2.

Example 2. Let us consider a testing problem built of two Normal distributions: $P = N(-\mu, 1)$ vs. $Q = N(+\mu, 1)$ with $\mu > 0$. Then $B_t = \{x : \exp(2\mu x) > t\} = (\ln t/2\mu, +\infty)$ and $P(B_t) = 1 - \Phi(\ln t/2\mu + \mu) = \alpha$; $Q(B_t^c) = \Phi(\ln t/2\mu - \mu) = r_{(P,Q)}(\alpha)$ gives the parametric representation (parameter t) of the function $r_{(P,Q)}(\alpha) = \Phi(\Phi^{-1}(1 - \alpha) - 2\mu)$ defining the *lower boundary of the risk set $R(P, Q)$*.

For the graph of $r_{(P_0, P_1)}(\alpha)$ of another testing

problem, see Fig. 3. For further reading, see ref. 2 [pp. 198–204].

CONSTRUCTION OF LFDs

For the cheat's example the risk sets of the testing problems (P, Q) vary with $0.2 \leqslant \theta \leqslant 0.4$, whereby the LFP is associated with the smallest possible risk set (cf. Fig. 2). This statement is always true and quantifies the vague introductory formulations such as "the pair for which the two probability distributions are as close as possible."

In order to explain the basic idea of the construction we will restrict ourselves to a simple $(\{P_0\})$ vs. composite $(\mathscr{P}_1)$ testing problem with $\mathscr{P}_1 = \{Q \in \mathscr{W} : Q(B) \leqslant P_1(B) + \epsilon \text{ for all } B \in \mathscr{B}\}$. Like neighborhood **1**(a), this is a simple neighborhood to treat.

The starting point of the construction is the likelihood ratio p_1/p_0 of the testing problem (P_0, P_1) of the model. If one fixes the probability of an error of the first kind with $P_0[p_1/p_0 > t] = \alpha$, the probabilities $Q[p_1/p_0 \leqslant t]$ of an error of the second kind are bounded by $P_1[p_1/p_0 \leqslant t] + \epsilon = r_{(P_0, P_1)}(\alpha) + \epsilon$ according to the specified neighborhood. Geometrically, the latter means an upward shift of the lower bound-

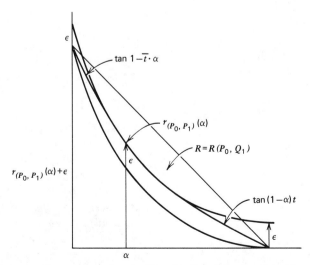

Figure 3 Construction of a candidate for the smallest possible risk set.

ary of the risk set of the original testing problem by ϵ. Now let us draw tangents onto the function $r_{(P_0, P_1)}(\alpha) + \epsilon$ through the points $(1, 0)$ and $(0, 1)$, and let us denote the absolute values of their slopes by $\underline{t}$ and $\bar{t}$, respectively. (Note that $\{P_0\} \cap \mathscr{P}_1 = \varnothing$ implies that $\underline{t} < \bar{t}$.) By construction,

$$R = \mathrm{co}\Big(\big\{\big((P_0[p_1/p_0 > t], P_1[p_1/p_0 \leqslant t] + \epsilon\big)$$

$$: \underline{t} \leqslant t \leqslant \bar{t}\big\} \cup \{(1, 0), (0, 1)\}\Big)$$

is a candidate for the smallest possible risk set. For an illustration, see Fig. 3. And indeed:

$$q_1(x) = \begin{cases} \underline{t} \cdot p_0(x) & \text{if} & p_1(x)/p_0(x) \leqslant \underline{t} \\ p_1(x) & \text{if} & \underline{t} < p_1(x)/p_0(x) \leqslant \bar{t} \\ \bar{t} \cdot p_0(x) & \text{if} & \bar{t} < p_1(x)/p_0(x) \end{cases}$$

defines a probability distribution $Q_1 \in \mathscr{P}_1$ which forms together with P_0 a testing problem satisfying $R(P_0, Q_1) = R$. The equality $[q_1/p_0 > t] = [p_1/p_0 > t]$ for all $\underline{t} \leqslant t < \bar{t}$ finally proves that Q_1 is least favorable.

The technique just sketched can be extended to composite vs. composite testing problems. It may, in the form which is also applicable to neighborhoods **2** and **3**, be called "perimeter method," since the appropriate objective for finding a LFP is the perimeter of the risk set (see ref. 10).

Returning to our special case, note that owing to $q_1(x)/q_0(x) = \max(\underline{t}, \min(\bar{t}, p_1(x)/p_0(x)))$, robustizing results in censoring. In general this extends to composite versus composite testing problems involving only the neighborhoods **1** and **2**.

References

[1] Bednarski, T. (1982). *Ann. Statist.*, **10**, 226–232.

[2] Ferguson, T. S. (1967). *Mathematical Statistics: A Decision Theoretic Approach.* Academic Press, New York. (A good general reference stressing the geometric point of view.)

[3] Hampel, F. R. (1973). *Zeit. Wahrscheinlichkeitsth. verwand. Geb.*, **27**, 87–104. (The first survey on robust estimation.)

[4] Huber, P. J. (1965). *Ann. Math. Statist.*, **36**, 1753–1758. (Pioneering paper on the construction of LFPs.)

[5] Huber, P. J. (1968). *Zeit. Wahrscheinlichkeitsth. verwand. Geb.*, **10**, 269–278. (A finite sample treatise on estimates derived from robust tests.)

[6] Huber, P. J. (1981). *Robust Statistics*. Wiley, New York. ("This first systematic, book-length exposition reorganizes, summarizes and extends a wealth of material, providing solid foundation in robustness.")

[7] Huber, P. J. and Strassen, V. (1973). *Ann. Statist.*, **1**, 251–263; *ibid.*, **2**, 223–224 (1974). [The classic paper on the existence of LFPs (technical).]

[8] Lehmann, E. L. (1952). *Ann. Math. Statist.*, **23**, 408–416.

[9] Österreicher, F. (1978). *Zeit. Wahrscheinlichkeitsth. verwand. Geb.*, **43**, 49–55.

[10] Österreicher, F. (1979). *Arbeitsber. Math. Inst. Univ. Salzburg*, **1**(4), 1–16.

[11] Rieder, H. (1977). *Ann. Statist.*, **5**, 909–921. (As refs. 1 and 9, a systematic technical treatment of the construction of LFPs.)

[12] Rieder, H. (1980). *Ann. Statist.*, **8**, 106–115. (Extension and asymptotic counterpart of ref. 5.)

[13] Strassen, V. (1964). *Zeit. Wahrscheinlichkeitsth. verwand. Geb.*, **2**, 273–305. (Well-motivated technical pioneering work.)

[14] Strassen, V. (1965). *Ann. Math. Statist.*, **36**, 423–439. (Important technical paper providing an interpretation of various neighborhoods.)

(COMBINATION OF DATA
HYPOTHESIS TESTING
OUTLIERS
ROBUST ESTIMATION
ROBUST TESTING)

F. ÖSTERREICHER

LEAST SIGNIFICANT DIFFERENCE

See LEAST SIGNIFICANT RANGE; MULTIPLE DECISION PROBLEMS; TUKEY'S SIMULTANEOUS COMPARISON PROCEDURE

LEAST SIGNIFICANT RANGE

Given k random samples (usually of equal size n) from normal populations with common variance σ^2, for testing the hypothesis of equality of the population means, the range of any subset of $p \geqslant 2$ sample means must exceed a certain value before we consider any of the p population means to be different. This value is called the least significant range for the p means.

(*k*-RATIO *t* TESTS,
MULTIPLE COMPARISONS)

LEAST SIGNIFICANT STUDENTIZED RANGE

The least significant studentized range is the least significant range* divided by $\sqrt{s^2/n}$, where s^2 is the sample estimate of the common variance σ^2 of the underlying normal populations and n is the common sample size.

(*k*-RATIO *t* TESTS
MULTIPLE COMPARISONS)

LEAST SQUARES

The method of least squares is a mathematical method which is used to find the linear or nonlinear regression* equation, expressing the relation between a dependent variable and one or more independent variables, for which the sum of squares of the residuals (deviations from regression) is a minimum. Because of the Gauss–Markov theorem, the method of least squares is also used to obtain best linear unbiased estimates (BLUE*) of the parameters of a statistical population from the ordered observations in a random sample. This article deals primarily with the regression problem (the following section) and its solution by the method of least squares (the section "Method of Least Squares"). A brief account is given of the origin and history of least squares; more details are presented in a six-part article by the author [18]. Several alternative methods are mentioned in the section "Alternative Methods." In the final section, brief comments are made on the relation between least-squares regression and BLUE. There is also an extensive bibliography; a still more extensive one is given in the article referenced above.

REGRESSION AND RELATED PROBLEMS

A very important problem in mathematical statistics is that of finding the best linear or nonlinear regression equation to express the relation between a dependent variable and one or more independent variables. Given are observations, each subject to random error, greater in number than the parameters in the regression equation, on the dependent variable and the corresponding values of the independent variable(s), which may be known exactly or may also be subject to random error. Related problems are those of choosing the best measures of central tendency and dispersion of the observations. The best solutions of all three problems depend on the distribution of the random errors. If one assumes that the values of the independent variable(s) are known exactly and that the errors in the observations on the dependent variable are normally distributed and homoscedastic, it is well known that the arithmetic mean* is the best measure of central tendency, the standard deviation is the best measure of dispersion, and the method of least squares is the best method of fitting a regression equation. Other assumptions lead to different choices. Most practitioners have tended to make the assumption of normality and not to worry about the consequences when it is not justified. Another problem arises when the data are contaminated by spurious observations (outliers) which come from populations with different means and or larger standard deviations. Many methods have been proposed for rejection of outliers* or modifying them (or their weights). Recently, much attention has been devoted to robust procedures* (including adaptive methods* that vary according to the value of some criterion computed from the data) which minimize the effect of the presence of outliers or departures from distributional assumptions.

METHOD OF LEAST SQUARES

This method derives its name from the fact that it minimizes a sum of squares (or other quadratic form) in the observations and the quantities (e.g., regression coefficients) being estimated. In the degenerate case the quantity being estimated is the true value α of a physical quantity on which we have several measurements y_i ($i = 1, 2, \ldots, n$), each subject to error, the error in the ith measurement being denoted by ϵ_i. We have the observational equations

$$y_i = \alpha + \epsilon_i = a + e_i \quad (i = 1, 2, \ldots, n),$$

$$(1)$$

where a and e_i are estimates of α and ϵ_i, respectively. Solving for e_i, squaring, summing over all n values of i, and dropping the subscripts, we obtain

$$\sum e^2 = \sum (a - y)^2. \quad (2)$$

To find the value of a that minimizes $\sum e^2$, we take the partial derivative with respect to a and set $\partial \sum e^2 / \partial a = 0$, which yields

$$a = \sum y/n = \bar{y}, \quad (3)$$

where $\bar{y}$ is the arithmetic mean of the y_i's. Thus the least-squares estimate of the true value α is the arithmetic mean of the observations, since the sum of squares of deviations from the arithmetic mean is less than from any other value. The corresponding estimate of the dispersion (due to error) is the standard deviation or root-mean-square deviation from the arithmetic mean.

The simplest nondegenerate regression problem is that of estimating the linear relation $y = \alpha + \beta x$, given a set of points (x_i, y_i), $i = 1, 2, \ldots, n$, where the values x_i are exact and the values y_i are subject to error, the error in the ith measurement again being denoted by ϵ_i. We have the observational equations

$$y_i = \alpha + \beta x_i + \epsilon_i = a + bx_i + e_i$$

$$(i = 1, 2, \ldots, n), \quad (4)$$

where a, b, and e_i are estimates of α, β, and ϵ_i, respectively. Solving for e_i, squaring, summing over all n values of i and dropping the subscripts, we obtain

$$\sum e^2 = \sum (a + bx - y)^2. \quad (5)$$

To find the values of the regression coefficients a and b that minimize $\sum e^2$, we take partial derivatives with respect to a and b and set $\partial \sum e^2 / \partial a = \partial \sum e^2 / \partial b = 0$, which yields

$$na + b\sum x = \sum y,$$

$$a\sum x + b\sum x^2 = \sum xy. \quad (6)$$

These equations are called the *normal equations*. Simultaneous solution of the normal equations for a and b yields

$$b = \frac{\left[n\sum xy - \sum x \sum y \right]}{\left[n\sum x^2 - \left(\sum x \right)^2 \right]},$$

$$a = \frac{\left[\sum x^2 \sum y - \sum x \sum xy \right]}{\left[n\sum x^2 - \left(\sum x \right)^2 \right]}$$

$$(7)$$

$$= \bar{y} - b\bar{x}.$$

The values of the coefficients b and a given by (7) are, of course, the slope and the y-intercept, respectively, of the least-squares regression line.

The method of the least squares can also be used to estimate the coefficients of linear regression equations involving two independent variables (planes) or three or more independent variables (hyperplanes), as well as coefficients of nonlinear regression equations involving one independent variable (curves), two independent variables (surfaces), or three or more independent variables (hypersurfaces). In a regression equation, the functions whose coefficients are to be determined need not be algebraic; they may be transcendental (e.g., exponential, logarithmic, or trigonometric). In each case, the normal equations are obtained by taking partial derivatives of $\sum e^2$ with respect to each of the unknown coefficients to be estimated, and thus the number of normal equations to be solved simultaneously is equal to the number of unknown coefficients. The whole procedure is conceptually quite simple, and the computational labor is now much less onerous than it was before the development of electronic computers.

The method of least squares was developed independently by Gauss in Germany, Legendre in France, and Adrain in America. Legendre [27], while not the first to use the method of least squares, was the first to publish it. He started with the linear form $E = a + bx + cy + \cdots$ and derived the normal equations without the explicit use of calculus by multiplying the linear form in the unknowns by the coefficient of each of the unknowns and summing over all the observations, then setting the sums equal to zero. Legendre offered no mathematical proof of the method of least squares, but nevertheless made a strong claim for its superiority. Gauss* [11] claimed priority in the use (although not in the publication) of the method of least squares, asserting that he had been using it already for 12 years and promising to publish his results later [12]. Adrain [1], apparently unaware of the work of Legendre [27] and of the (as yet unpublished) work of Gauss, independently developed the method of least squares and used it to solve several problems. Gauss [12] deduced the normal (Gaussian) law of error from the postulate that when any number of equally good direct observations of an unknown quantity x are given, the most probable value is their arithmetic mean. He then showed, in what has come to be known as his first proof, that the method of least squares, used by him since 1795, but named by Legendre [27], follows as a consequence of the Gaussian law of error.

Laplace* [23] gave a rudimentary form of the central limit theorem*, and used it in two later papers [24, 25] to show that, while the arithmetic mean is "most advantageous" only when the error law is normal (Gaussian), it is advantageous when the number of observations is large or when one is taking the average of results each based on a large number of observations, and hence in such cases one may use the method of least squares. Laplace [26], in his monumental work on the analytic theory of probabilities, advocated use of the method of least squares, although not to the exclusion of earlier methods [minimax* and least absolute values* methods (see the following section)].

Gauss [13] compared his 1809 formulation (first proof) of the method of least squares with that of Laplace [26], and concluded that neither is entirely satisfactory. He offered another exposition (second proof) based on minimizing a loss function proportional to the sum of the squares of the errors, and on this basis justified the use of the method of least squares, whatever the number of observations and whatever the distribution of their errors. Gauss's second exposition seems to the present writer to be no more satisfactory than his first. Nevertheless, his arguments apparently convinced most of his contemporaries, since the literature of the next few decades includes many writings on least squares but only a few on rival methods.

Dirichlet [7] quoted from Laplace [26, p. 348], to the effect that the best method of correcting one or more elements, on the basis of linear observational equations greater in number than the elements, depends, for a small number of observations, on the underlying law of error, but that this dependence disappears when the number of observations is very large, in which case the method of least squares is to be preferred to all others. He pointed out the Laplace's conclusion concerning the superiority of the method of least squares for very large numbers of observations is not only unwarranted by the evidence presented by Laplace, but is actually incorrect. In the case of a single element, for example, Laplace's conclusion does not follow from the fact that the arithmetic mean converges in probability to the true value as the number of observations increases, since the same is true of other measures of central tendency (e.g., the median). Whatever the sample size, Dirichlet pointed out, the question of whether or not the arithmetic mean is superior to the median depends on the ratio of two constants (their standard errors), one of which (the standard error of the arithmetic mean) depends on an integral over the whole range of the error curve, while the other (the standard error of the median) depends only on the maximum ordinate of the curve.

Merriman [31] compiled an annotated bibliography of writings on least squares up to that date. Several excellent books on the theory of least squares and its applications were published in the late nineteenth and early twentieth centuries. Particularly noteworthy are those of Merriman [32] on theory and Helmert [19] on applications to geodesy.

Eisenhart [8] has given an excellent review of the history, philosophy, and current status of the method of least squares. Plackett [37] has examined in detail, with the aid of correspondence, the circumstances in which the discovery of the method of least squares took place and the source of the ensuing controversy concerning priority between Gauss and Legendre. Modern computational procedures have been discussed by numerous authors, including Marquardt [30], Bauer [3], Businger and Golub [4], Golub [14], Golub and Wilkinson [15], Hanson and Lawson [16], Hudson [20], A. Jones [21], R. H. Jones [22], Furnival [9], Osborne [35], and Furnival and Wilson [10]. Various libraries of statistical computer programs (e.g., BMDP and IMSL) contain programs for computation, by the method of least squares, of simple linear regression, multiple linear regression* (including stepwise and all possible subsets regressions), and nonlinear regression.

Although the method of least squares has been much criticized, and although many alternative methods (the following section) have been proposed, the method of least squares, due to its mathematical tractability and its computational simplicity, has remained for 175 years the predominant method, and it is still given in almost every textbook of statistics, sometimes to the exclusion of all others. If all error distributions were mesokurtic (medium-tailed) and if spurious observations did not occur, that would be ideal. However, the method of least squares is not at all robust to the presence of outliers resulting from spurious observations or from leptokurtic (long-tailed) error distributions. Therefore, when outliers do occur, either they should be rejected before use of the method of least squares, or a more robust method should be used (see the next section). Results of using the method of least squares in the case of platykurtic (short-tailed) error distributions are not so disastrous, but some improvement can be made by use of a more appropriate method (see the following section).

ALTERNATIVE METHODS

Several methods of determining regression coefficients from inconsistent data (method of averages*, minimax method*, method of least absolute values*, and method of maximum likelihood*) were proposed before the method of least squares. Other methods (method of group averages*, most approximative method*, Cauchy's method of interpolation*, method of least pth powers*, Brown–Mood method*, and various adaptive methods*, including Harter's adaptive robust method*) have been proposed more recently. The method of least absolute values and the minimax method yield maximum likelihood estimates of the regression coefficients for double exponential and uniform error distributions, respectively, and give good results for error distributions that are long-tailed or short-tailed, respectively. Harter's adaptive method, which chooses among these two methods and the method of least squares, is robust to either long-tailed or short-tailed error distributions.

RELATIONS BETWEEN LEAST-SQUARES REGRESSION AND BLUE

Because of the Gauss–Markov theorem, the best linear unbiased estimates of the parameters of a statistical population are found by the method of least squares. If the values of the independent variable x are considered to be fixed (chosen in advance and not subject to error), then the least-squares estimates a and b of the regression coefficients α and β in the equation of the straight line $y = \alpha + \beta x$ are linear in the dependent variable y, and are, in fact, the BLUE of α and β [33, Theorem 13.2].

The Gauss–Markov theorem states, in effect, that, under general conditions which will not be given here, an unbiased estimator of a population parameter, based on a linear combination of sample observations, is "best" (has minimum variance) when the estimator is obtained by least squares. This theorem was first stated by Gauss [13] and restated, with clarification of the underlying assumptions, by Markoff [29]. It was extended by David and Neyman [5]. In a historical note, Plackett [36] summarized, in matrix notation, the work of these authors and that of Aitken [2]. Ogawa [34] pointed out that, by application of the Gauss–Markov theorem, the method of least squares can be used to obtain best linear unbiased estimators. Lloyd [28] used the method of least squares to determine the coefficients of the order statistics of complete samples in the best linear unbiased estimators of the location and scale parameters of symmetric populations, especially the rectangular (uniform) and normal populations. During the next 20 years, authors too numerous to mention here gave similar results for complete and censored samples from a great variety of populations, usually for small sample sizes, since determination of the coefficients in the BLUE requires tabulation of the expected values, variances, and covariances of the order statistics. Other authors proposed modifications when the means of the order statistics are known, but their variances and covariances are unknown, and when even the means are unknown. All of this work has been summarized by Harter [17] and by David [6]. Rao [38–40] has given unified theories of linear estimation and of least squares.

References

[1] Adrain, R. (1808). *Analyst*, **1**, 93–109.

[2] Aitken, A. O. (1935). *Proc. R. Soc. Edinb. A*, **55**, 42–48.

[3] Bauer, F. L. (1965). *Numer. Math.*, **7**, 338–352.

[4] Businger, P. and Golub, G. H. (1965). *Numer. Math.*, **7**, 269–276.

[5] David, F. N. and Neyman, J. (1938). *Statist. Res. Mem.*, **2**, 105–116.

[6] David, H. A. (1981). *Order Statistics*, 2nd ed. Wiley, New York.

[7] Dirichlet, P. G. L. (1836). *Abhand. Akad. Wiss. Berlin*, **33**, 67–68.

[8] Eisenhart, C. (1964). *J. Wash. Acad. Sci.*, **54**, 24–33.

[9] Furnival, G. M. (1971). *Technometrics*, **13**, 403–408.

[10] Furnival, G. M. and Wilson, R. W. (1974). *Technometrics*, **16**, 499–511.

[11] Gauss, C. F. (1806). *Monatl. Corresp. Beförd. Erd-Himmelskd.*, **14**, 181–186.

[12] Gauss, C. F. (1809). *Theoria Motus Corporum Coelestium in Sectionibus Conicis Solem Ambientium*. Frid. Perthes et I. H. Besser, Hamburg.

[13] Gauss, C. F. (1823). *Comment. Soc. Gotting.*, **5**, 23–90.

[14] Golub, G. H. (1965). *Numer. Math.*, **7**, 206–216.

[15] Golub, G. H. and Wilkinson, J. H. (1966). *Numer. Math.*, **9**, 139–148.

[16] Hanson, R. J. and Lawson, C. L. (1969). *Math. Comp.*, **23**, 787–812.

[17] Harter, H. L. (1971). In *Optimizing Methods in Statistics*, J. S. Rustagi, ed. Academic Press, New York, pp. 33–62.

[18] Harter, H. L. (1974). *Int. Statist. Rev.*, **42**, 147–174, 235–264, 282; *ibid.*, **43**, 1–44, 125–190, 269–278 (1975); *ibid.*, **44**, 113–159 (1976).

[19] Helmert, F. R. (1907). *Die Ausgleichsrechnung nach der kleinsten Quadrate mit Anwendumg auf die Geodäsie und die Theorie der Messinstrumente*, 2nd ed. B. G. Teubner, Leipzig.

[20] Hudson, D. J. (1969). *J. R. Statist. Soc. B*, **31**, 113–118.

[21] Jones, A. (1970). *Computer J.*, **13**, 301–308.

[22] Jones, R. H. (1970). *Ann. Math. Statist.*, **41**, 688–691.

[23] Laplace, P. S. (1810). *Mém. Cl. Sci. Math. Phys. Inst. Fr.*, Année 1809, 353–415; supplement, 559–565.

[24] Laplace, P. S. (1811). *Mém. Cl. Sci. Math. Phys. Inst. Fr.*, Année 1810, 279–347.

[25] Laplace, P. S. (1811). *Connaissance des Tems*, Année 1813, 213–223.

[26] Laplace, P. S. (1812). *Théorie Analytique des Probabilitiés*. Courcier, Paris.

[27] Legendre, A. M. (1805). *Nouvelles Méthodes pour la Détermination des Orbites des Comètes*. Courcier, Paris.

[28] Lloyd, E. H. (1952). *Biometrika*, **36**, 88–95.

[29] Markoff, A. A. (1912). *Wahrscheinlichkeitsrechnung*. B. G. Teubner, Leipzig. (translation of second Russian edition, 1908).

[30] Marquardt, D. W. (1963). *J. Soc. Ind. Appl. Math.*, **11**, 431–441.

[31] Merriman, M. (1877). *Trans. Conn. Acad. Arts Sci.*, **4**(1), 151–232.

[32] Merriman, M. (1884). *A Text-Book on the Method of Least Squares*. Wiley, New York.

[33] Mood, A. M. and Graybill, F. A. (1963). *Introduction to the Theory of Statistics*, 2nd ed. McGraw-Hill, New York.

[34] Ogawa, J. (1951). *Osaka Math. J.*, **3**, 175–213.

[35] Osborne, M. R. (1972). *Aust. Computer J.*, **4**, 164–169.

[36] Plackett, R. L. (1949). *Biometrika*, **36**, 458–460.

[37] Plackett, R. L. (1972). *Biometrika*, **59**, 239–251.

[38] Rao, C. R. (1971). *Sankhyā A*, **33**, 371–394.

[39] Rao, C. R. (1973). *Commun. Statist.*, **1**, 1–8.

[40] Rao, C. R. (1978). *Commun. Statist. A*, **7**, 409–411.

(ADAPTIVE METHODS
ARITHMETIC MEAN
BEST LINEAR UNBIASED ESTIMATES
  (BLUE)
BROWN–MOOD METHOD
GAUSS–MARKOV THEOREM
HARTER'S ADAPTIVE ROBUST METHOD
INTERPOLATION, CAUCHY'S METHOD
KURTOSIS
LAWS OF ERROR, I, II, AND III
LINEAR REGRESSION
MAXIMUM LIKELIHOOD
METHOD OF AVERAGES
METHOD OF GROUP AVERAGES
METHOD OF LEAST ABSOLUTE VALUES
METHOD OF LEAST pTH POWERS
METHOD OF MAXIMUM LIKELIHOOD
MINIMAX METHOD
NONLINEAR REGRESSION
ORDER STATISTICS)

H. Leon Harter

LE CAM LEMMAS *See* contiguity

LEGENDRE SYMBOL

The Legendre symbol is used in some combinatorial problems related to experimental design*. It is defined for a prime p as follows:

$$(a/p) = 1 \qquad \text{if } a \text{ is a quadratic residue of } p$$
$$= -1 \qquad \text{if } a \text{ is a nonquadratic residue of } p.$$

LEHMANN ALTERNATIVES

Lehmann [8] pointed out that comparison of different rank tests* for a nonparametric hypothesis H is hard because under most alternatives the test is no longer distribution-free. For instance, in the two-sample problem where a sample $X_1, \ldots, X_m$ is taken from a distribution function (df) F, $Y_1, \ldots, Y_n$ from G and $H : F = G$ is to be tested, the distribution of the ranks of the X's and Y's in the combined sample does not depend on F if H is true, but under shift alternatives* $G(x) = F(x - \Delta)$ the distribution depends not only on Δ but also on F. (Assumed throughout is that all df's are continuous.) To circumvent this difficulty, Lehmann proposed to consider alternatives of the form $G = f(F)$, with the function f specified. Then the distribution of ranks under the alternative does not depend on F, only on f. Thus different level α rank tests can be compared on the basis of their power at f, for various choices of f. One of the simplest choices is $G = F^\lambda$, $0 < \lambda \neq 1$. As an example, Lehmann computes the power of various rank tests at $G = F^2$ and at $G = F^3$. One can also ask for which Lehmann alternative (if any) a given rank test is most powerful (MP). As an example it is shown in ref. 8 that the one-sided Wilcoxon–Mann–Whitney test* is locally MP against $G = (1 - p)F + pF^2$ $(0 < p \leqslant 1)$ as $p \rightarrow 0$.

Lehmann considers the more general situation of $Z_1, \ldots, Z_n$ real-valued independent with Z_i having df $F_i = f_i(F)$, $i = 1, \ldots, N$, in which the f_i are specified but F

is not. Then the distribution of the ranks of the Z's depends on $(f_1, \ldots, f_N)$ but not on F. This reduces to the two-sample problem by taking $f_1 = \cdots = f_m =$ identity function and $f_{m+1} = \cdots = f_N = f$. As another example of the same approach to choice of alternatives, consider a bivariate population with df $H(x, y)$ and the hypothesis of independence $H(x, y) = F(x)G(y)$ with F, G unspecified. Lehmann proposes alternatives of the form $H(x, y) = h(F(x), G(y))$. Then if a sample $(X_1, Y_1), \ldots, (X_n, Y_n)$ is taken, the distribution of ranks (of each X_i among the X's, and similarly the Y_i) depends on h but is free of (F, G). As an application it is shown in ref. 8 that the rank correlation test* is locally MP against $h(F, G) = (1 - p) FG + p(FG)^2$, as $p \to 0$.

From the preceding example it is seen that one can speak of a Lehmann alternative (LA) to a certain hypothesis if both hypothesis and alternative are expressible as known functions of a set of unknown df's. More formally [8, Sec. 8], any specific LA is an orbit* in the parameter space under the action of a semigroup with identity whose orbits in the sample space are the rank orders of the observations.

The number of papers in which LAs have been used is very large and only a few will be indicated here. The bulk of applications has been in two-sample testing problems. Savage [10] studies probability of rank orders under LA $G = F^\lambda$, $\lambda > 1$, and finds a locally MP rank test which is even UMP* (i.e., for all $\lambda > 1$) under special circumstances. Uhlmann [17] derives a locally MP test vs. the two-sided LA $G = (1 - p)F + pF^2$ or with G and F interchanged. Shorack [14] (with references to earlier work using LAs) derives distributions of various statistics under $G = F^\lambda$, $1 - G = (1 - F)^\lambda$, and other LAs. Chikkagoudar and Shuster [3] study failure rates, taking $1 - G$ as various functions of $1 - F$. Steck [15] derives the distribution of the Kolmogorov–Smirnov statistic under $G = F^\lambda$. Halperin and Ware [5] take LA $\log G = (1 + r(m + n)^{-1/2})\log F$ and let $m, n \to \infty$. Saleh and Dionne [9]

study a test based on the frequency of X's and Y's between sample quantiles and show it to be locally MP at $1 - G = (1 - F)^\lambda$. Sequential two-sample tests based on LAs, mostly of the type $G = F^\lambda$, were studied by Bradley [2] in collaboration with Wilcoxon, Rhodes, and others (this expository paper has references to the original sources); by Berk and Savage [1]; by Savage and Sethuraman [12]; and by Sethuraman [13]. For the k-sample problem Savage [11] considered tests vs. LA $F_i = F^{\lambda_i}$, $\lambda_1 \leqslant \cdots \leqslant \lambda_k$; Karlin and Truax [7] considered the case where one population has slipped according to $G = (1 - \lambda)F + \lambda F^2$, and $G = F^{1+\lambda}$; Hoel [6] treated the problem of selecting the stochastically largest population among $F_1, \ldots, F_k$, where $F_i = F^{\lambda_i}$, $\lambda_1 \leqslant \cdots \leqslant \lambda_k$. The problem of independence was treated by Choi [4] with LA $H(x, y) = \Sigma p_i (F(x))^{a_i}(G(y))^{b_i}$. Confidence intervals derived from two-sample rank tests against various LAs were obtained by van der Laan [18] for a difference in location, and by Stedl and Fox [16] for $P(Y < X)$.

References

[1] Berk, R. H. and Savage, I. R. (1968). *Ann. Math. Statist.*, **39**, 1661–1674.

[2] Bradley, R. A. (1967). *Proc. 5th Berkeley Symp. Math. Statist. Prob.*, Vol. 1. University of California Press, Berkeley, Calif, pp. 593–607.

[3] Chikkagoudar, M. S. and Shuster, J. J. (1974). *J. Amer. Statist. Ass.*, **69**, 411–413.

[4] Choi, S. C. (1973). *Technometrics*, **15**, 625–629.

[5] Halperin, M. and Ware, J. (1974). *J. Amer. Statist. Ass.*, **69**, 414–422.

[6] Hoel, D. G. (1971). *Ann. Math. Statist.*, **42**, 630–642.

[7] Karlin, S. and Truax, D. (1966). *Ann. Math. Statist.*, **31**, 296–324.

[8] Lehmann, E. L. (1953). *Ann. Math. Statist.*, **24**, 23–43.

[9] Saleh, A. K. M. E. and Dionne, J.-P. (1977). *Commun. Statist. A*, **6**, 1213–1221.

[10] Savage, I. R. (1956). *Ann. Math. Statist.*, **27**, 590–615.

[11] Savage, I. R. (1957). *Ann. Math. Statist.*, **28**, 968–977.

[12] Savage, I. R. and Sethuraman, J. (1966). *Ann. Math. Statist.*, **37**, 1154–1160; *ibid.*, **38**, 1309, (1967).

[13] Sethuraman, J. (1970). *Ann. Math. Statist.*, **41**, 1322–1333.

[14] Shorack, R. A. (1968). *J. Amer. Statist. Ass.*, **63**, 353–366.

[15] Steck, G. P. (1969). *Ann. Math. Statist.*, **40**, 1449–1466.

[16] Stedl, J. and Fox, K. (1978). *Commun. Statist. B*, **7**, 151–161.

[17] Uhlmann, W. (1959). *Metrika*, **2**, 169–185.

[18] van der Laan, P. (1970). *Philips Res. Rep. Suppl. No. 5*, pp. 1–158.

PARTIAL LIST OF USERS OF LEHMANN ALTERNATIVES (1953–1979)

Alam, K. and Thompson, J. R. (1971). *Ann. Inst. Statist. Math. Tokyo*, **23**, 253–262.

Berk, R. H. and Savage, I. R. (1968). *Ann. Math. Statist.*, **39**, 1661–1674.

Bofinger, V. J. (1965). *Aust. J. Statist.*, **7**, 20–31.

Bradley, R. A. (1965). *Psychometrika*, **30**, 315–318.

Bradley, R. A. (1967). *Proc. 5th Berkeley Symp. Math. Statist. Prob.*, Vol. 1. University of California Press, Berkeley, Calif., pp. 593–607.

Brooks, R. J. (1974). *Biometrika*, **61**, 501–507.

Burr, P. and Young, D. H. (1977). *J. R. Statist. Soc. B*, **39**, 79–85.

Chikkagoudar, M. S. and Shuster, J. J. (1974). *J. Amer. Statist. Ass.*, **69**, 411–413.

Chio, S. C. (1973). *Technometrics*, **15**, 625–629.

Cox, D. R. (1972). *J. R. Statist. Soc. B*, **34**, 187–220.

David, H. A. (1973). *J. Amer. Statist. Ass.*, **68**, 743–745.

Davidson, R. R. (1969). *Biometrics*, **25**, 597–599.

Davies, R. B. (1971). *J. Amer. Statist. Ass.*, **66**, 879–883.

Elteren, P. v. (1960). *Bull. Int. Statist. Inst.*, **37**, 351–361.

Fu, K. S. and Chien, Y. T. (1967). *IEEE Trans. Inf. Theory*, **IT-13**, 484–492.

Godambe, V. P. (1961). *Ann. Math. Statist.*, **32**, 1091–1107.

Govindarajulu, Z. and Haller, H. S. (1972). *J. Indian Statist. Ass.*, **10**, 17–35.

Hall, W. J., Wijsman, R. A., and Ghosh, J. K. (1965). *Ann. Math. Statist.*, **36**, 575–614 (Sec. 18).

Halperin, M. and Ware, J. (1974). *J. Amer. Statist. Ass.*, **69**, 414–422.

Hoel, D. G. (1971). *Ann. Math. Statist.*, **42**, 630–642.

Karlin, S. and Traux, D. (1966). *Ann. Math. Statist.*, **31**, 296–324 (Sec. 10).

Krauth, J. (1971). *Ann. Math. Statist.*, **42**, 1949–1956.

Lehmann, E. L. (1953). *Ann. Math. Statist.*, **24**, 23–43.

Lehmann, E. L. (1959). *Testing Statistical Hypotheses.* Wiley, New York, Chap. 6, Sec. 12, problems 23–29.

Moses, L. E. (1964). *J. Amer. Statist. Ass.*, **59**, 645–651.

Narayana, T. V., Savage, I. R., and Saxena, K. M. L. (1978). *Canad. J. Statist.*, **6**, 41–47.

Odeh, R. E. (1972). *Biometrika*, **59**, 467–471.

Peto, R. (1972). *Biometrika*, **59**, 472–475.

Peto, R. and Peto, J. (1972). *J. R. Statist. Soc. A*, **135**, 185–198.

Pfanzagl, J. (1960). *Metrika*, **3**, 143–150.

Puri, M. L. (1965). *Rev. Int. Statist. Inst.*, **33**, 229–241.

Rao, U. V. R., Savage, I. R., and Sobel, M. (1960). *Ann. Math. Statist.*, **31**, 415–426 (Cor. 3.4 and 3.5).

Rosenblatt, J. R. (1960). In *Contributions to Probability and Statistics: Essays in Honor of Harold Hotelling*, I. Olkin et al., eds. Stanford University Press, Stanford, Calif., pp. 358–370.

Ruymgaart, F. H., Shorack, G. R., and van Zwet, W. R. (1972). *Ann. Math. Statist.*, **43**, 1122–1135.

Saleh, A. K. M. E. and Dionne, J.-P. (1977). *Commun. Statist. A*, **6**, 1213–1221.

Savage, I. R. (1956). *Ann. Math. Statist.*, **27**, 590–615.

Savage, I. R. (1957). *Ann. Math. Statist.*, **28**, 968–977.

Savage, I. R. and Sethuraman, J. (1966). *Ann. Math. Statist.*, **37**, 1154–1160; *ibid.*, **38**, 1309 (1966).

Sethuraman, J. (1970). *Ann. Math. Statist.*, **41**, 1322–1333.

Shorack, R. A. (1967). *Technometrics*, **9**, 154–158, 666–677.

Shorack, R. A. (1968). *J. Amer. Statist. Ass.*, **63**, 353–366.

Steck, G. P. (1969). *Ann. Math. Statist.*, **40**, 1449–1466.

Steck, G. P. (1974). *Ann. Prob.*, **2**, 155–160.

Stedl, J. and Fox, K. (1978). *Commun. Statist. B*, **7**, 151–161.

Stoker, D. J. (1954). *Proc. Kon. Ned. Akad. Wet.*, **57**, 599–614.

Sukhatme, B. V. and Deshpande, M. V. (1966). *J. Indian Soc. Agric. Statist.*, **18**, 57–69.

Uhlmann, W. (1959). *Metrika*, **2**, 169–185.

van der Laan, P. (1970). *Philips Res. Rep. Suppl. No. 5* (Sec. 4.2.2).

Weed, H. D. and Bradley, R. A. (1971). *J. Amer. Statist. Ass.*, **66**, 321–326.

Wegner, L. H. (1956). *Ann. Math. Statist.*, **27**, 1006–1016 (Sec. 8).

Wijsman, R. A. (1979). *Dev. Statist.*, **2**, 235–314 (Secs. 4.10 and 6.10).

Wilcoxon, F., Rhodes, L. J., and Bradley, R. A. (1963). *Biometrics*, **19**, 58–84.

Young, D. H. (1973). *Biometrika*, **60**, 543–549.

(COX'S REGRESSION MODEL
DISTRIBUTION-FREE METHODS
HYPOTHESIS TESTING
ORBIT
RANK TESTS
SURVIVAL ANALYSIS)

R. A. WIJSMAN

LEHMANN CONTRAST EXTIMATORS

E. L. Lehmann has been one of the giants in the postwar rapid growth of nonparametric statistics, one important area of which has been the development of estimators based on linear rank tests.

The foundation paper in rank estimation [3] introduced a class of estimators which includes the well-known Hodges–Lehmann estimators for one- and two-sample problems (see HODGES–LEHMANN ESTIMATORS).

Following that initial paper, Lehmann extended the techniques to estimation of contrasts in linear models [4, 5]. Since the mid-1960s many others have investigated rank estimators in various settings, much of the work deriving from the initial ideas of Lehmann.

If $\tau_1, \ldots, \tau_k$ represent the "main" effects of a k-level, fixed factor, then any linear model involving that factor will have the form

$$X_{u\ldots} = \tau_u + \text{(other main effects}$$

$$\text{and interactions)}$$

$$+ e_u \ldots \text{(error)}, \qquad u = 1, \ldots, k.$$

One problem of interest with such models is to estimate the difference in effects between two distinct groups of levels of a given factor. A contrast is any such difference that can be written as

$$\theta = \sum_{u=1}^{k} a_u \tau_u \qquad \text{with} \qquad \sum_{u=1}^{k} a_u = 0.$$

Thus $\theta = \tau_1 - (\tau_2 + \tau_3)/2$ is a contrast* between level 1 and a combination of levels 2 and 3, with $a_1 = 1$, $a_2 = a_3 = -\frac{1}{2}$.

Another useful way to write contrasts is as

$$\theta = \sum_{u=1}^{k} \sum_{v=1}^{k} d_{uv} \Delta_{uv}$$

with the definition $\Delta_{uv} = \tau_u - \tau_v$. For a given contrast the coefficients d_{uv} are not unique, but one version is $d_{uv} = a_u/k$.

ONE-WAY LAYOUT (K INDEPENDENT SAMPLES)

For $u = 1, \ldots, k$, let $\{X_{ui} : i = 1, \ldots, n_u\}$ be an independent, identically distributed (i.i.d.) sample with cumulative distribution function (CDF) $F(t - \tau_u)$, for some unknown F. The linear model is then simply $X_{ui} = \tau_u + e_{ui}$.

Define $\hat{\Delta}_{uv}$ to be the Hodges–Lehmann two-sample estimator of Δ_{uv}:

$$\hat{\Delta}_{uv} = \text{median}\{X_{ui} - X_{vj} : i = 1, \ldots, n_u ;$$

$$j = 1, \ldots, n_v\}.$$

Since $\hat{\Delta}_{vu} = -\hat{\Delta}_{uv}$, calculations are necessary only for $1 \leqslant v < u \leqslant k$. Lehmann [4] points out that although $\Delta_{uv} = \Delta_{uh} + \Delta_{hv}$, in general $\hat{\Delta}_{uv} \neq \hat{\Delta}_{uh} + \hat{\Delta}_{hv}$, and therefore calls these estimators "incompatible."

Instead, he proposes constructing

$$\hat{\Delta}_{u\cdot} = \sum_{v=1}^{k} \hat{\Delta}_{uv}/k \qquad (\hat{\Delta}_{uu} \equiv 0).$$

and "adjusted" estimators

$$\hat{\hat{\Delta}}_{uv} = \hat{\Delta}_{u\cdot} - \hat{\Delta}_{v\cdot}.$$

While these estimators are "compatible" they do possess the disadvantage that $\hat{\hat{\Delta}}_{uv}$ is influenced by observations from all other samples.

The final contrast estimator is then

$$\hat{\theta} = \sum_{u=1}^{k} a_u \hat{\Delta}_{u\cdot} = \sum_{u=1}^{k} \sum_{v=1}^{k} d_{uv} \hat{\Delta}_{uv}.$$

For unequal sample sizes, Spjøtvoll [10] provided the following improvement:

$$\Delta_{u\cdot}^* = \sum_{v=1}^{k} n_v \hat{\Delta}_{uv} \bigg/ \sum_{v=1}^{k} n_v.$$

Then $\hat{\tilde{\Delta}}_{uv}$ and $\hat{\theta}$ are defined as before, but with $\hat{\Delta}_{u\cdot}$ replaced by $\Delta_{u\cdot}^*$.

Provided that $\min\{n_u/N : u = 1, \dots, k\}$ tends to λ with $0 < \lambda < 1$, as $N \to \infty$, the Spjøtvoll versions of $\hat{\Delta}$ and $\hat{\theta}$ are consistent and their asymptotic relative efficiencies compared to the least-squares estimators are identical to those of the Hodges–Lehmann estimators for one- and two-sample problems.

A somewhat different approach, also considered by Lehmann [4], applies when the distribution function F can be assumed symmetric. Then let $\tilde{\tau}_u$ be the one-sample Hodges–Lehmann estimator of τ_u: $\tilde{\tau}_u = \frac{1}{2} \times \text{median}\{X_{ui} - X_{uj} : 1 \leqslant i \leqslant j \leqslant n_u\}$. Finally, define

$$\tilde{\Delta}_{uv} = \tilde{\tau}_u - \tilde{\tau}_v,$$

and

$$\tilde{\theta} = \sum_{u=1}^{k} a_u \tilde{\tau}_u = \sum_{u=1}^{k} \sum_{v=1}^{k} d_{uv} \tilde{\Delta}_{uv}.$$

The estimates $\tilde{\Delta}_{uv}$ are clearly compatible and $\tilde{\Delta}_{uv}$ depends only on samples u and v. Furthermore, $\tilde{\theta}$ and $\tilde{\Delta}_{uv}$ share the same consistency and efficiency properties as the Lehmann–Spjøtvoll estimators. The main disadvantage is the symmetry assumption mentioned above, which somewhat restricts applicability.

TWO-WAY LAYOUT (RANDOMIZED BLOCKS)

Let $\{X_{ui} : i = 1, \dots, b; \ u = 1, \dots, k\}$ be mutually independent and assume the CDF for X_{ui} is $F(t - \beta_i - \tau_u)$ for some unknown CDF F. The linear model in this case is $X_{ui} = \tau_u + \beta_i + e_{ui}$.

This corresponds to a fixed-effects two-way ANOVA model with one observation per cell, or a complete randomized blocks model. For purposes of estimating contrasts in $\{\tau_1, \dots, \tau_k\}$, the β_i are nuisance or block effects. It is assumed that no interaction exists.

Let $\hat{\Delta}_{uv}$ be the one-sample Hodges–Lehmann estimator applied to the differences $\{X_{ui} - X_{vi} : i = 1, \dots, b\}$. Thus

$$\hat{\Delta}_{uv} = \frac{1}{2} \text{median}\{X_{ui} - X_{vi} + X_{uj} - X_{vj} :$$
$$1 \leqslant i \leqslant j \leqslant b\}.$$

Given this different definition of $\hat{\Delta}_{uv}$, the development of $\hat{\tilde{\Delta}}_{uv}$ and $\hat{\theta}$ is identical to that in the one-way layout with the exception that the Spjøtvoll modification is not pertinent.

In this case $\hat{\theta}$ is consistent ($b \to \infty$) and the asymptotic relative efficiency compared to the least-squares estimate is as before.

The Doksum [2] contrast estimator for the two-way layout is similar to Lehmann's, the only difference being in the definition of $\hat{\Delta}_{uv}$, where

$$\hat{\Delta}_{uv} = \text{median}\{X_{ui} - X_{vi} : i = 1, \dots, b\}.$$

Adjusted estimators $\hat{\tilde{\Delta}}_{uv}$ and $\hat{\theta}$ are developed in the usual way from the current $\hat{\Delta}_{uv}$.

The Doksum version of $\hat{\Delta}_{uv}$ is the estimator related to the sign test for paired samples (*see* THOMPSON–SAVUR ESTIMATOR), so it is not surprising that Doksum's contrast estimators generally have lower asymptotic efficiencies than Lehmann's except with very heavy tailed distributions.

EXTENSIONS

Lehmann contrast estimators can readily be extended to linear models with three or more factors.

In a three-way layout with one observation per cell, if there are no interactions between the factor to be estimated and the other two, then the model can be represented by X_{ujl} having CDF $F(t - \alpha_j - \beta_l - (\alpha\beta)_{jl} - \tau_u)$ for $l = 1, \dots, b; \ j = 1, \dots, c; \ u = 1, \dots, k$. The linear model then is $X_{ujl} = \tau_u + \alpha_j + \beta_l + (\alpha\beta)_{jl} + e_{ujl}$, where α_j, β_l represent main effects of the other two factors, with interaction $(\alpha\beta)_{jl}$.

For purposes of estimating $\{\tau_u : u = 1, \dots, k\}$ we may combine the other factors into a single factor by setting $\beta_i = \alpha_l + \gamma_j + (\alpha\gamma)_{jl}$, using any convenient one-to-one relationship for $i \leftrightarrow (j, l)$. The estimation procedures and properties for a two-way layout apply directly. This generalizes trivially to four or more factors.

For extensions to within-cell replications in the two-way layout, let the CDF of X_{uih} be $F(t - \beta_i - \tau_u)$, $i = 1, \ldots, b$; $u = 1, \ldots, k$; $h = 1, \ldots, n_{ui}$. Of the several possible extensions, perhaps the most natural is to define

$$\hat{\Delta}_{uv} = \text{median}\{ X_{uih} - X_{vig} : h = 1, \ldots, n_{ui} ;$$

$$g = 1, \ldots, n_{vi} ;$$

$$i = 1, \ldots, b\},$$

and proceed to $\hat{\Delta}_{uv}$ and $\hat{\theta}$ in the usual way. The assumption of no interaction is still operative.

CONFIDENCE INTERVALS FOR CONTRASTS

Confidence intervals are a necessary adjunct to any estimation problem. Confidence intervals based on Lehmann's rank estimation methods, including for both *a priori* and *a posteriori* contrasts, have been investigated in several papers, including refs. 1, 6, 8, and 9.

Example. The techniques used in all these contrast estimators are very similar so that a single example should suffice. Consider a three-treatment, one-way layout, and estimation of the excess of treatment one over the average of the other treatments:

$$\theta = \tau_1 - (\tau_2 + \tau_3)/2$$

with $a_1 = 1$, $a_2 = a_3 = -\frac{1}{2}$. The hypothetical data are

$X_{1i} = 8.6, 7.8, 8.2 \quad (n_1 = 3)$

$X_{2i} = 7.9, 6.5, 6.9 \quad (n_2 = 3)$

$X_{3i} = 5.8, 6.6 \quad (n_3 = 2)$.

$\hat{\Delta}_{13} = \text{median}\{8.6 - 5.8, 8.6 - 6.6, 7.8 - 5.8,$

$\qquad\qquad 7.8 - 6.6, 8.2 - 5.8, 8.2 - 6.6\}$

$\quad = 2.0 = -\hat{\Delta}_{31}$,

$\hat{\Delta}_{12} = 1.3 = -\hat{\Delta}_{21}$, $\hat{\Delta}_{23} = 0.85 = -\hat{\Delta}_{32}$,

$\Delta_1^* = [3(1.3) + 2(2.0)]/8 = 0.9875$,

$\Delta_2^* = -0.2750$, $\Delta_3^* = -1.06875$.

Thus the Lehmann–Spjøtvoll estimator of θ is

$$\hat{\theta} = 0.9875 - (-0.2750 - 1.06875)/2 = 1.66.$$

References

[1] Crouse, C. F. (1969). *S. Afr. Statist. J.*, **3**, 35–48.

[2] Doksum, K. (1967). *Ann. Math. Statist.*, **38**, 878–883.

[3] Hodges, J. L., Jr. and Lehmann, E. L. (1963). *Ann. Math. Statist.*, **34**, 598–611.

[4] Lehmann, E. L. (1963). *Ann. Math. Statist.*, **34**, 957–966.

[5] Lehmann, E. L. (1963). *Ann. Math. Statist.*, **34**, 1494–1506.

[6] Lehmann, E. L. (1963). *Ann. Math. Statist.*, **34**, 1507–1512.

[7] Lehmann, E. L. (1964). *Ann. Math. Statist.*, **35**, 726–734.

[8] Marascuilo, L. A. (1966). *Psychol. Bull.*, **64**, 280–290.

[9] Sen, P. K. (1966). *Ann. Inst. Statist. Math.*, **18**, 319–336.

[10] Spjøtvoll, E. (1968). *Ann. Math. Statist.*, **39**, 1486–1492.

Bibliography

Hollander, M. and Wolfe, D. A. (1973). *Nonparametric Statistical Methods*. Wiley, New York, Chaps. 6 and 7. (A clear, applications-oriented treatment.)

Lehmann, E. L. (1975). *Nonparametrics: Statistical Methods Based on Ranks*. Holden-Day, San Francisco, Chaps. 5 and 6. (A very readable introduction to theory and methods.)

Sen, P. K. and Puri, M. L. (1971). *Nonparametric Methods in Multivariate Analysis*. Wiley, New York, Chaps. 6 and 7. (Advanced mathematical theory. All three contain excellent references.)

(ANALYSIS OF VARIANCE
DISTRIBUTION-FREE METHODS
RANK TESTS)

WALTER R. PIRIE

LEHMANN–SCHEFFÉ THEOREM

The Lehmann–Scheffé theorem [1] is an extension of the Rao–Blackwell inequality* and deals with properties of complete* sufficient* statistics.

If T is a sufficient statistic for parameter θ and the distribution of T is complete [i.e., $E[g(T)] = 0$ implies that $g(T) = 0$ for any function g], then the minimum variance unbiased (MVU) estimator* is given by $E(\hat{\theta} \mid T)$, where $\hat{\theta}$ is *any* unbiased estimator of θ. A related result also attributed to Lehmann and Scheffé states that if $\hat{\theta}$ is an unbiased estimator of θ and is uncorrelated with every statistic Z such that $E(Z) = 0$, then $\hat{\theta}$ is a MVU estimator of θ.

It follows from the Lehmann–Scheffé theorem that the MVU estimator of the reliability* $e^{-\lambda}$ in the case of an exponential distribution with parameter λ based on a sample $X_1, \ldots, X_n$ of size n is

$$\left[(n-1)/n \right]^T, \quad \text{where} \quad T = \sum_{i=1}^{n} X_i$$

rather than the maximum likelihood estimator $e^{-T/n}$.

Reference

[1] Lehmann, E. L. and Scheffé, H. (1950). *Sankhyā*, **10**, 305–340.

(MINIMUM VARIANCE
UNBIASED ESTIMATOR)

LEHMANN TESTS

One of the basic nonparametric problems is to test for the identity of two distributions. Let $X_1, \ldots, X_m$ be m independent random variables (RV) from a distribution F, defined on the real line $R = (-\infty, \infty)$. Also, let $Y_1, \ldots, Y_n$ be an independent set of n independent RVs from a distribution G, defined on R. Both F and G are of unspecified form, but may be assumed to be continuous. Based on the $N = m + n$ observations in the two samples, it is desired to test for

$$H_0 : F \equiv G \quad \text{against} \quad H_1 : F \neq G, \quad (1)$$

where H_1 includes a general class of alternatives, without necessarily being confined to location and/or scale differences or to stochastic ordering of X and Y.

Classical tests for (1) are based on the Kolmogorov–Smirnov (KS) and Cramér–von Mises (CvM) test statistics

$$K_N = (mn/N)^{1/2} \sup_{x \in R} |F_m(x) - G_n(x)| \quad (2)$$

and

$$V_n = (mn/N) \int_R \{F_m(x) - G_n(x)\}^2 \, dH_N(x), \quad (3)$$

where $F_m(x) (= m^{-1} \sum_{i=1}^{m} I(X_i \leq x))$ and $G_n(x) (= n^{-1} \sum_{i=1}^{n} I(Y_i \leq x))$ are respectively the *sample (empirical) distributions* for the X and Y observations, $H_N(x)$ $[= N^{-1}\{m F_m(x) + n G_n(x)\}]$ is the combined sample distribution and $I(A)$ stands for the indicator function of the set A.

Lehmann [3] has proposed another test for (1), based on a statistic closely resembling V_N in (3), where the theory of Hoeffding's [2] U-statistics plays the basic role. As a measure of distance between F and G, consider the functional

$$\Delta(F, G) = \int_R \{F(x) - G(x)\}^2$$
$$\times d\left\{ \tfrac{1}{2}(F(x) + G(x)) \right\}, \quad (4)$$

where $\Delta(F, G)$ is nonnegative for all F, G and is equal to 0 only when $F \equiv G$. Lehmann succeeded in providing an unbiased estimator of $\Delta(F, G)$ and using the same as a test statistic for testing H_0 in (1). Consider the statistic $\phi(X_1, X_2; Y_1, Y_2)$, defined by

$$\phi(a,b;c,d) = \begin{cases} 1, & \max(a,b) < \min(c,d) \\ & \text{or } \min(a,b) > \max(c,d); \\ 0, & \text{otherwise,} \end{cases}$$
$$(5)$$

so that

$$E\phi(X_1, X_2; Y_1, Y_2)$$
$$= \int F^2(x)[1 - G(x)] \, dG(x)$$
$$+ \int G^2(x)[1 - F(x)] \, dF(x)$$
$$= \tfrac{1}{3} + 2\Delta(F, G). \quad (6)$$

Thus $\phi(X_1, X_2; Y_1, Y_2)$ unbiasedly estimates $\tfrac{1}{3} + 2\Delta(F, G) = \theta(F, G)$, say, where $\theta(F, G)$

$\geqslant \frac{1}{3}$ and the equality sign holds only when $F \equiv G$. $\phi(X_1, X_2; Y_1, Y_2)$ may be termed a *kernel* of degree $(2, 2)$ (see Hoeffding [2]) and the (*generalized*) *U-statistic* corresponding to the kernel ϕ in (5) is

$$U_N = \binom{m}{2}^{-1}\binom{n}{2}^{-1} \sum_{1 \leqslant i < j \leqslant m} \sum_{1 \leqslant r < s \leqslant n}$$
$$\times \phi(X_i, X_j, Y_r, Y_s) \qquad (7)$$

and U_N unbiasedly estimates $\frac{1}{3} + 2\Delta(F, G)$, for all continuous F and G. Lehmann's test for (1) is based on the test statistic

$$L_N = N^{-1}mn\left(U_N - \tfrac{1}{3}\right). \qquad (8)$$

Note that if $s_1, \ldots, s_n$ stand for the ranks of the ordered $Y_1, \ldots, Y_n$ in the combined sample of size N, then by (5) and (7),

$$\binom{m}{2}\binom{n}{2}U_N = \sum_{k=1}^{n}\left\{(n-k)\binom{s_k - k}{2}\right.$$
$$\left. + (k-1)\binom{m + k - s_k}{2}\right\},$$
$$(9)$$

so that U_N (and hence, L_N) is a rank statistic, too; and remains invariant under any monotone transformation on the observations. Thus L_N is a genuinely distribution-free statistic (under H_0) and the null hypothesis* distribution of L_N is generated by the $N!$ equally likely realizations of the vector of ranks of the combined sample observations (over the set of permutations of $1, \ldots, N$). Several expressions for L_N in (8) have been suggested by Rényi [4], Sundrum [7], and Zajta [9], among others. Of these, the following statistic, due to Rényi [4],

$$W_N = \sum_{i=1}^{m}\binom{r_i - i}{2}\bigg/\left\{m\binom{n}{2}\right\}$$
$$+ \sum_{k=1}^{n}\binom{s_k - k}{2}\bigg/\left\{n\binom{m}{2}\right\} \qquad (10)$$

is the best suited for computation; here r_i $(1 \leqslant i \leqslant m)$ and s_k $(1 \leqslant k \leqslant n)$ denote the ranks of the ordered sample elements of the first and second samples, respectively, in the pooled ordered sample. By (4) and (6), the *critical region** of the test for (1) based on

L_N in (8) is specified by $L_N > L_{N,\alpha}$, where, for a given level of significance α $(0 < \alpha < 1)$,

$$P\{L_N > L_{N,\alpha} \mid H_0\}$$
$$\leqslant \alpha \leqslant P\{L_N \geqslant L_{N,\alpha} \mid H_0\}, \quad (11)$$

and a randomized test procedure may be adapted to achieve exact significance level α. Now $L_{N,\alpha}$ does not depend on F (continuous) and has to be determined by reference to the exact null hypothesis distribution of L_N, generated by the $N!$ equally likely realizations of the vector of combined sample rankings. This task becomes prohibitively laborious as N increases, and hence the large-sample distribution theory of L_N needs to be studied for suitable approximations to $L_{N,\alpha}$. For $m = n$ [$= 4(1), 16, 18, 20, 22, 23$], some of these critical values may be obtained from Zajta and Pandikow [10].

For numerical illustration, consider the following data pertaining to the time (in seconds) taken to complete a specific job individually by seven boys and seven girls (selected at random from a vocational training school):

Boys	52.8, 56.4, 54.6, 60.6, 59.1, 59.5, 61.3
Girls	55.2, 57.6, 53.7, 56.7, 52.2, 57.0, 58.8

Here the r_i are 2, 4, 5, 11, 12, 13, and 14, while the s_k are 1, 3, 6, 7, 8, 9, and 10; so that by (10), $W = 101/147$. Using the table of Zajta and Pandikow [10], we obtain that the critical value for W at $\alpha = 0.05$ is $113/147$, so that the null hypothesis that the boys and the girls have the same distribution is tenable.

Note that by (5),

$$\phi_{10}(x) = E\{\phi(X_1, X_2; Y_1, Y_2) \mid X_1 = x, H_0\}$$
$$= E\{\phi(X_1, X_2; Y_1, Y_2) \mid Y_1 = x, H_0\}$$
$$= \phi_{01}(x) = \tfrac{1}{3} \quad \text{for every} \quad x \in R.$$
$$(12)$$

Thus, in the terminology of Hoeffding [2], $\theta(F, G)$ is *stationary of order* 1 for (F, G)

$\in \mathcal{F}^* = \{(F, G): F \equiv G, \text{ continuous}\}$, and this explains why L_N does not have asymptotically a normal distribution under H_0, though for $F \neq G$, the asymptotic normality of the standardized form of U_N would follow easily. It follows from Sundrum [7] that under H_0 in (1),

$$\text{var}(U_N) = \binom{m}{2}^{-1}\binom{n}{2}^{-1}\left[\binom{N}{2} - 1\right]\frac{2}{45},$$

(13)

but this formula is of little help in specifying the asymptotic distribution of L_N in (8). However, Wegner [8] has shown that under H_0 in (1), $\frac{1}{2}L_N$ has the same limiting distribution as the classical Cramér–von Mises statistic (and the latter distribution has been tabulated by Anderson and Darling [1]). This is not surprising as the Lehmann statistic and the Cramér–von Mises statistic are asymptotically equivalent, and for equal sample sizes they are identical.

Modifications of the test statistic (and procedure) when F and G are not necessarily continuous (everywhere) and/or multivariate distributions have also been considered by Lehmann [3]. They all share the common problem of a complicated asymptotic null distribution. Lehmann [3] has also considered a two-sample test of $H_0: F \equiv G$, against scale alternatives, based on the test statistic

$$U_N^* = \frac{\displaystyle\sum_{1 \leqslant i < j \leqslant m}\sum_{1 \leqslant r < s \leqslant n} \phi^*(X_i, X_j; Y_r, Y_s)}{\binom{m}{2}\binom{n}{2}}$$

(14)

where

$$\phi^*(a, b; c, d) = \begin{cases} 1, & \text{if } |a - b| > |c - d|, \\ 0, & \text{otherwise,} \end{cases}$$

(15)

Note that $E(U_N^* H_0) = \frac{1}{2}$. Unfortunately, U_N^* is neither a rank statistic nor is genuinely distribution-free under H_0. A studentized version of U_N^* (using a jackknife-type estimator of its variance, which depends on F) is

due to Sen [5]. Let

$$V_{i0} = \frac{\displaystyle\sum_{j=1(\neq i)}^{n}\sum_{1 \leqslant r < s \leqslant n} \phi^*(X_i, X_j; Y_r, Y_s)}{\binom{m-1}{1}\binom{n}{2}}$$

(16)

for $i = 1, \ldots, m$, and

$$V_{or} = \frac{\displaystyle\sum_{1 \leqslant i \leqslant j \leqslant m}\sum_{s=1(\neq r)}^{n} \phi^*(X_i, X_j; Y_r, Y_s)}{\binom{m}{2}\binom{n-1}{1}},$$

(17)

for $r = 1, \ldots, n$. Now let

$$S_{10}^2 = \frac{1}{m-1}\sum_{i=1}^{m}[V_{i0} - U_N^*]^2, \quad (18)$$

$$S_{01}^2 = \frac{1}{n-1}\sum_{r=1}^{n}[V_{or} - U_N^*]^2. \quad (19)$$

Then, under H_0, $\frac{1}{4}(U_N^* - \frac{1}{2})(m^{-1}S_{10}^2 + n^{-1}S_{01}^2)^{-1/2}$ has asymptotically a normal distribution (when m/n does not converge to 0 or to $+\infty$ as $N \to \infty$), and this provides an asymptotically distribution-free test for $H_0: F \equiv G$, against scale alternatives. It is not precisely known whether U_N^* is locally most powerful against any specific (underlying) density which fits both F and G (with possibly different scale parameters) (see Sen [6]). This test has also not met the light of popularity in actual applications.

References

[1] Anderson, T. W. and Darling, D. A. (1952). *Ann. Math. Statist.*, **23**, 193–213.

[2] Hoeffding, W. (1948). *Ann. Math. Statist.*, **19**, 293–325.

[3] Lehmann, E. L. (1951). *Ann. Math. Statist.*, **22**, 165–179.

[4] Rényi, A. (1953). *Magy. Tud. Akad. Alkalm. Mat. Int. Közl.*, **2**, 243–265.

[5] Sen, P. K. (1960). *Calcutta Statist. Ass. Bull.*, **10**, 1–18.

[6] Sen, P. K. (1963). *Ann. Inst. Statist. Math. Tokyo*, **15**, 117–135.

[7] Sundrum, R. M. (1954). *Ann. Math. Statist.*, **25**, 139–145.

[8] Wegner, L. H. (1956). *Ann. Math. Statist.*, **27**, 1006–1016.

[9] Zajta, A. (1960). *Magy. Tud. Akad. Alkalm. Mat. Int. Közl.*, **5**, 447–459.

[10] Zajta, A. and Pandikow, W. (1977). *Biometrika*, **64**, 167–169.

(DISTRIBUTION-FREE METHODS
HYPOTHESIS TESTING
JACKKNIFE METHODS)

P. K. Sen

LEIPNIK DISTRIBUTION

A first-order approximation to the distribution of the circular serial correlation* $\tilde{R}_1$ of lag 1 (under the bivariate normal assumption on the distribution of the sample) given by

$$p_{\tilde{R}_1}(r) = \left[B\left(\tfrac{1}{2}, \tfrac{1}{2}(n+1)\right) \right]^{-1} (1 - r^2)^{(1/2)(n-1)}$$

$$\times (1 + \rho^2 - 2\rho r)^{-(1/2)n} \quad (-1 \leqslant r \leqslant 1),$$

where n is the sample size and ρ is the population correlation coefficient of the underlying bivariate normal distribution*. For $\rho = 0$ the distribution reduces to the distribution on an "ordinary" correlation coefficient* (for a bivariate normal distribution with $\rho = 0$ and a sample size of $n + 3$).

Bibliography

Johnson, N. L. and Kotz, S. (1970). *Distributions in Statistics: Continuous Univariate Distributions*, Vol. 2. Wiley, New York, Chap. 32.

Leipnik, R. B. (1947). *Ann. Math. Statist.*, **18**, 80–87.

Leipnik, R. B. (1948). *Biometrika*, **35**, 559–562.

(SERIAL CORRELATION)

LENGTH-BIASED SAMPLING

The length-biased sampling effect arises when a segment of N consecutive points of a renewal process* is contaminated by another point located at random on the length of the segment. See Littlejohn [1] for a theoretical discussion, and Shiavi and Negin [2] for an application in neurophysiological problems.

References

[1] Littlejohn, R. P. (1981). *Aust. J. Statist.*, **23**, 91–94.

[2] Shiavi, R. and Negin, M. (1973). *IEEE Trans. Biomed. Eng.*, **20**, 374–378.

(RENEWAL THEORY)

LEPTOKURTIC CURVE

A frequency curve with a positive coefficient of kurtosis*:

$$\gamma_2 = \left(\mu_4 / \mu_2^2 \right) - 3 > 0.$$

Such a curve is usually taller and slimmer than the normal curve in the neighborhood of the mode.

The Student t distribution with less than 9 degrees of freedom is markedly leptokurtic; in general, all Pearson's type VII frequency curves, given by

$$y = y_0 (1 + x^2 / a^2)^{-n}$$

(where the range of x is unlimited), are leptokurtic.

(KURTOSIS
MESOKURTIC CURVE
PLATYKURTIC CURVE)

L_1 **ESTIMATION** *See* METHOD OF LEAST ABSOLUTE VALUES

LETHAL DOSE FIFTY *See* METHODS IN BIOASSAY, STATISTICAL

LEVEL OF A TEST

The notion of the level (or *significance level*) of a test is an extension, for the case of composite hypotheses, of the concept of the size* of a test.

Let the null hypothesis be

$$H_0 : \theta \in \Theta_0$$

and the alternative

$$H_1 : \theta \in \Theta_1$$

where Θ_0 and Θ_1 are two disjoint sets of values of the parameter θ. A test of hypothesis H_0 is said to be of level α if for each $\theta \in \Theta_0$,

$$P_\theta \{ \text{reject } H_0 \} \leqslant \alpha$$

and for each $\delta < \alpha$ there exists at least one $\theta \in \Theta_0$ such that

$$P_\theta \{ \text{reject } H_0 \} > \delta,$$

where P_θ is the probability distribution determined by θ. That is,

$$\alpha = \sup_{\theta \in \Theta_0} P_\theta \{ \text{reject } H_0 \}.$$

When H_0 is a simple hypothesis, a level α test coincides with a size α test.

(HYPOTHESIS TESTING)

LEVEL OF SIGNIFICANCE

The customary ("classical") meaning of this term is a property of a statistical test procedure—namely, the probability of rejecting the hypothesis tested when it is, in fact, valid (*see* HYPOTHESIS TESTING). Most standard tests are constructed to control the level of significance at a prespecified value, commonly 0.01 or 0.05 (1% or 5%).

More recently (see, e.g., Gibbons and Pratt [1]), the term "level of significance" of a test has been applied to *data*, being the least numerical level of significance at which the hypothesis tested would be rejected by the test if applied to the data. This is also called the "*P*-value" for the data.

Note that the lower the level of significance, in this sense, the more "highly significant" of departure from the hypothesis—in conventional phraseology—is the result of the test.

The distinction between *significance*—which related to evidence of departure

(some sort)—and *importance*—which relates to size and nature of departure—is widely emphasized but often forgotten.

Reference

[1] Gibbons, J. D. and Pratt, J. W. (1975). *Amer. Statist.*, **29**, 20–25.

(HYPOTHESIS TESTING
INFERENCE, STATISTICAL)

LEVENE'S ROBUST TEST OF HOMOGENEITY OF VARIANCES

Many statistical techniques require for their validity the assumption that each of several population variances is the same. The standard test of homogeneity of variances, the Bartlett* test, is an effective tool, however, only if the underlying populations are approximately normally distributed. When the assumption of normality is violated the actual size can be many times larger than the nominal significance level. A procedure that is relatively insensitive to departures from normality is Levene's test [7]. It is robust in the sense that the actual size nearly coincides with the nominal significance level for a large variety of underlying distributions.

DESCRIPTION OF LEVENE'S TEST

Suppose that $k \geqslant 2$ independent samples, $X_{i1}, \ldots, X_{in_i}$, $i = 1, \ldots, k$, are taken. Sample i is assumed to be a collection of n_i independent, identically distributed random variables with distribution G_i, mean μ_i, and variance σ_i^2. Neither G_i nor μ_i nor σ_i^2 is assumed known. The null hypothesis of equal variances, $H_0 : \sigma_1^2 = \cdots = \sigma_k^2$, is to be tested at significance level α against the alternative that not all the variances are the same, $H_a : \sigma_i^2 \neq \sigma_j^2$ for some $i \neq j$. Denote absolute deviations from sample means by $Z_{ij} = |X_{ij} - \bar{X}_{i\cdot}|$, $j = 1, \ldots, n_i$, $i = 1$,

$\ldots, k$. Define the statistic,

$$W = \frac{n-k}{k-1} \frac{\sum\limits_{i=1}^{k} n_i (\bar{Z}_{i.} - \bar{Z}_{..})^2}{\sum\limits_{i=1}^{k} \sum\limits_{j=1}^{n_i} (Z_{ij} - \bar{Z}_{i.})^2},$$

where $n = \sum_{i=1}^{k} n_i$. Levene's test [7] consists of rejecting H_0 in favor of H_a if W exceeds $F_{k-1, n-k, 1-\alpha}$, the $(1-\alpha)$-quantile of the F distribution with $k-1$ and $n-k$ degrees of freedom. The test is thus a one-way analysis of variance* of the absolute deviations Z_{ij}. The use of Z_{ij} rather than, say, Z_{ij}^2 makes the test criterion less sensitive to heavy-tailed distributions G_i. Since the Z_{ij} are in general neither normally distributed nor independent (Z_{ij} and Z_{il} have a correlation of the order n_i^{-2} [4]), the null distribution of W is not Snedecor's F. Nonetheless, for a variety of distributions G_i, e.g., normal distributions and symmetric heavy-tailed distributions such as the double exponential and the Student's t with 4 degrees of freedom, the usual significance levels, $\alpha = 0.01$, 0.05, or 0.10, and sample sizes of at least $n_i = 10$, the Levene criterion is robust in size [2, 7]; i.e., the $(1-\alpha)$-quantile of the null distribution of W, as estimated by Monte Carlo methods, is approximately equal to $F_{k-1, n-k, 1-\alpha}$.

MODIFICATIONS OF LEVENE'S TEST

For skewed distributions G_i, such as the χ^2 with 4 degrees of freedom, and extremely heavy-tailed distributions, such as the Cauchy, the Levene criterion tends to have too many significant results: the actual size appreciably exceeds the nominal significance level [2]. For these settings, improved Levene-type procedures have been proposed by Brown and Forsythe [2] which modify the test statistic W by replacing the central location estimators $\bar{X}_i$ with more robust versions. (The critical value $F_{k-1, n-k, 1-\alpha}$ is not changed.) Specifically, consider W' based on $Z_{ij}' = |X_{ij} - \tilde{X}_i|$, where $\tilde{X}_i$ is the median of the ith sample, and W'' based on $Z_{ij}'' = |X_{ij} - \bar{X}_{i.}'|$, where $\bar{X}_{i.}'$ is the 10% trimmed

mean* of the ith sample (the mean of the subsample remaining after deleting the 10% largest and 10% smallest values in the ith sample). Monte Carlo studies [2] show that in addition to being robust in size whenever W is, both W' and W'' are robust for the very heavy tailed Cauchy distribution*. (In this case W'' is to be preferred for its superior power against alternatives in H_a.) For the skewed χ_4^2 distribution, W' is robust, whereas W'' rejects too often. In cases where W, W', and W'' are all robust in size, the W test typically has greatest power.

COMPARISONS

The performance of any homogeneity of variances test depends on the underlying distributions G_i. If the G_i can be assumed to have a known parametric representation, a test exploiting this parametric character will be preferred (e.g., the Bartlett test in the normal setting). If, on the other hand, specific distributional assumptions on the G_i cannot be made, a robust Levene or Levene-type test can be used successfully. The choice of test statistic W, W', or W'' will be determined by the general type of G_i anticipated. A simulation study of the performance of more than 50 proposed homogeneity of variances tests has been undertaken by Conover et al. [3]. Included in the study are the W and W' tests, a χ^2 test by Layard [6] involving kurtosis, a jackknife* procedure by Miller [8], and a random grouping test by Box [1]. Three tests are found to be relatively superior in terms of power while achieving robustness in size: the W' test, and two modifications of a proposed procedure by Fligner and Killeen [5], whereby the $|X_{ij} - \tilde{X}_i|$ are ranked, assigned normal scores, and then subjected to either a χ^2 test* or an analysis of variance* F test*.

References

[1] Box, G. E. P. (1953). *Biometrika*, **40**, 318–335.

[2] Brown, M. B. and Forsythe, A. B. (1974). *J. Amer. Statist. Ass.*, **69**, 364–367.

[3] Conover, W. J., Johnson, M. E., and Johnson, M. M. (1981). *Technometrics*, **23**, 351–361. (Includes an excellent bibliography of proposed tests of homogeneity of variances.)

[4] Fisher, R. A. (1920). *Monthly Notices R. Astron. Soc.*, **80**, 758–770.

[5] Fligner, M. A. and Killeen, T. J. (1976). *J. Amer. Statist. Ass.*, **71**, 210–213.

[6] Layard, M. W. J. (1973). *J. Amer. Statist. Ass.*, **68**, 195–198. (A Monte Carlo study computing size and power for the robust techniques of Layard, Miller, and Box. The Layard procedure has test statistic

$$S = \sum_{i=1}^{k} \nu_i \left[\ln S_i^2 - \frac{1}{n-k} \sum_{r=1}^{k} \nu_r \ln S_r^2 \right]^2$$

$$\times \left[2 + (1 - k/n)\hat{\gamma} \right]^{-1},$$

where by definition $\nu_i = n_i - 1$ and

$$S_i^2 = \frac{1}{\nu_i} \sum_{j=1}^{n_i} \left(X_{ij} - \bar{X}_{i \cdot} \right)^2, \qquad i = 1, \ldots, k,$$

and

$$\hat{\gamma} = n \sum_{i=1}^{k} \sum_{j=1}^{n_i} \left(X_{ij} - \bar{X}_{i \cdot} \right)^4 \bigg/ \left(\sum_{i=1}^{k} \nu_i S_i^2 \right)^2 - 3$$

is a consistent estimator of the common kurtosis when each G_i satisfies

$$G_i(x) = G\left(\frac{x_i - \mu_i}{\sigma_i} \right)$$

and H_0 is true. In this case, the asymptotic distribution of S is χ^2 with $k - 1$ degrees of freedom. The Layard test, motivated by these considerations, consists of rejecting H_0 if S exceeds $\chi^2_{k-1, 1-\alpha}$, the $(1 - \alpha)$-quantile of this distribution.)

[7] Levene, H. (1960). In *Contributions to Probability and Statistics: Essays in Honor of Harold Hotelling*, I. Olkin et al., eds. Stanford University Press, Stanford, Calif., pp. 278–292. (Includes a Monte Carlo study of power functions.)

[8] Miller, R. G., Jr. (1968). *Ann. Math. Statist.*, **39**, 567–582.

[9] Snedecor, G. W. and Cochran, W. G. (1980). *Statistical Methods*, 7th ed. Iowa State University Press, Ames, Iowa. [Includes (pp. 253–254) an example of using Levene's test.]

(ANALYSIS OF VARIANCE
BARTLETT'S TEST OF HOMOGENEITY OF VARIANCES
HOMOGENEITY
ROBUSTNESS
TESTS OF HOMOGENEITY)

R. E. GLASER

LEVERAGE

A leverage point is an observation whose fitted value is largely determined by the corresponding response value. More formally, if the magnitude of the derivative of the ith fitted value with respect to the ith response value is large (or large relative to the derivatives at other observations), then the ith observation is a leverage point.

The term *leverage* is used mainly in connection with linear regression*. Let $\mathbf{H} = \mathbf{X}(\mathbf{X}^T\mathbf{X})^{-1}\mathbf{X}^T$ and $h_i = x_i(\mathbf{X}^T\mathbf{X})^{-1}x_i^T$, where x_i is a row of the $n \times p$ matrix $\mathbf{X}$. Then

$$\hat{y}_i = (1 - h_i)x_i b(i) + h_i y_i \tag{1}$$

where $b(i)$ is the least-squares estimate obtained without the ith observation. It follows that the derivative of $\hat{y}_i$ with respect to y_i is just h_i. Therefore, the diagonal elements of the "hat" matrix* $\mathbf{H}$ are usually associated with leverage.

There is no general agreement on when h_i is "large." Hoaglin and Welsch [2] argue that since $\sum_{i=1}^{n} h_i$ is equal to the member of parameters being fitted (when $\mathbf{X}$ is of full rank), an individual h_i should not be too far from a balanced design (all $h_i = p/n$), and state that when h_i is greater than $2p/n$ the ith observation is a leverage point (provided that $n > 2p$). Belsley et al. [1] show that when the x_i are independent, identically distributed (i.i.d.) multivariate Gaussian, the distribution of $h_i/(1 - h_i)$ can be related to an F-statistic and this leads to cutoff values for h_i in the range of $3p/n$. Note that these leverage criteria depend on p and n.

Huber [3] uses (1) and notes that $\hat{y}_i$ is a convex combination of $x_i b(i)$, the ith predicted value, and y_i. He then suggests that when h_i is greater than 0.2, too much weight is being given to y_i and hence the ith observation is a leverage point. This cutoff is not dependent on p or n.

A useful compromise between these two general approaches is to consider h_i, $i = 1, \ldots, n$, as a batch of data to be analyzed by exploratory data analysis [5, 6]. Observations with outlying values of h_i would then be considered leverage points.

Since $nh_i/(1 - h_i)$ is essentially the Mahalanobis distance* from x_i to the mean of the remaining observations, leverage points (in linear regression*) may be thought of as observations which are far away (using the covariance metric) in factor space from the rest of the data.

Belsley et al. [1] and Velleman and Welsch [7] define partial leverage as the marginal contribution of the jth carrier (or explanatory variable) to h_i. Partial leverage is closely related to adjusted residuals (see Mosteller and Tukey [4]).

References

[1] Belsley, D. A., Kuh, E., and Welsch, R. E. (1980). *Regression Diagnostics*. Wiley, New York.

[2] Hoaglin, D. C. and Welsch, R. E. (1978). *Amer. Statist.*, **32**, 17–22; Corrigenda, **32**, 146.

[3] Huber, P. J. (1981). *Robust Statistics*. Wiley, New York.

[4] Mosteller, F. and Tukey, J. W. (1977). *Data Analysis and Regression: A Second Course in Statistics*, Addison-Wesley, Reading, Mass.

[5] Tukey, J. W. (1977). *Exploratory Data Analysis*. Addison-Wesley, Reading, Mass.

[6] Velleman, P. F. and Hoaglin, D. C. (1981). *Applications, Basics, and Computing of Exploratory Data Analysis*. Duxbury Press, Boston.

[7] Velleman, P. F. and Welsch, R. E. (1981). *Amer. Statist.*, **35**, 234–242.

Further Reading

See the following works, as well as the references just given, for more information on the topic of leverage.

Behnken, D. W. and Draper, N. R. (1972). *Technometrics*, **14**, 101–111.

Box, G. E. P. and Draper, N. R. (1975). *Biometrika*, **62**, 347–351.

Davies, R. B. and Hutton, B. (1975). *Biometrika*, **62**, 383–391.

Hampel, F. R. (1978). *1978 Proc. Statist. Computing Sec., Amer. Statist. Ass.*, Washington, D.C., pp. 59–64.

(INFLUENCE FUNCTION
INFLUENTIAL DATA
OUTLIERS)

ROY E. WELSCH

LEVIN'S SUMMATION ALGORITHM

The solution of a differential equation, the expansion of an integral, iterative algorithms, and other procedures may involve slowly convergent, or divergent sequences of approximants. For example, the series $1 - 1/3 + 1/5 - \cdots$ for $\pi/4$ has partial sums which converge slowly; the factorial series $1/n - 1!/n^2 + 2!/n^3 - \cdots$ for the expansion of the exponential integral $(\exp n) E_1(n)$ diverges.

Historically, slowly convergent series have been studied extensively but wildly divergent series, occurring in practice rarely, have only received intermittent attention up to about half a century ago when the computer revolution arrived. Series in mechanics, theoretical physics, and astronomy painfully and laboriously developed for a few terms are now extended through computers to as many as 20 to 60 or 80 terms (see, e.g., Van Dyke's [10] remarks on perturbation series in fluid mechanics, Baker's [1] account of Padé methods in theoretical physics).

Computer-extended series in statistical applications, such as moments of the sample standard deviation, Student's t, skewness and kurtosis measures, all turn out to exhibit divergent characteristics. The Levin algorithm has frequently proved useful.

LEVIN'S NONLINEAR TRANSFORMATION

Summation of a series $g(n) \sim e_0 + e_1/n + \cdots$ first of all directs attention to the partial sums

$$A_j = \sum_{s=0}^{j-1} a_s, \quad j = 1, 2, \ldots; \quad a_s = e_s/n^s,$$

$$(1)$$

where we assume that $e_s \neq 0$, $s = 0, 1, \ldots$. We also assume that the series has a unique source (definite integral, for example) which defines a value to $g(n)$ for n in a certain domain. It should be noted that inside its circle of convergence the series $e_0 + e_1 x + \cdots$ defines $g(1/x)$, but outside a summation technique may be needed to define the function; in this connection $x - x^2/2 + x^3/3 - \cdots$ defines $\ln(1 + x)$ for $|x| < 1$ and $x = 1$ but $10 - 10^2/2 + 10^3/3 - \cdots$ relates to $\ln 11$ only when subject to a suitable transformation such as the conversion of the partial sums of $g(1/x)$ to rational fractions.

Returning to (1), specifically in connection with divergent series such as the factorial for which $f(x) \sim 1 - 1!\,x + 2!\,x^2 - \cdots$. Here the sum of $2s + 1$ terms is clearly dominated by $(2s)!\,x^{2s}$ as also—$(2s - 1)!\,x^{2s-1}$ dominates the sum of a $2s$ terms. Crudely $A_{j+1} \sim c a_j$ where c is a constant, and this approximation captures the dominant part of the series—we are looking for the smaller contribution which escapes. An improvement is

$$A_{j+1} = \alpha_r + a_j\left(b_0^{(r)} + \frac{b_1^{(r)}}{j} + \cdots + \frac{b_{r-1}^{(r)}}{j^{r-1}}\right),$$

$$j = 0, 1, \ldots, r; \quad r = 0, 1, \ldots, \quad (2)$$

for which system there is the simple solution for the limit sequence

$$\alpha_r = \frac{\displaystyle\sum_{j=0}^{r}(-1)^j\binom{r}{j}\left(\frac{j+1}{r+1}\right)^{r-1}\frac{A_{j+1}}{a_j}}{\displaystyle\sum_{j=0}^{r}(-1)^j\binom{r}{j}\left(\frac{j+1}{r+1}\right)^{r-1}\frac{1}{a_j}},$$

$$(3)$$

the values $b_0^{(r)}, b_1^{(r)}, \ldots$, being of secondary interest. This set of approximants may converge, or may diverge but still provide a "best" approximant in some sense. Of course, one may truncate the series or its inverse and derive several sets of similar sequences, relabeling to conform to (3). This is a version of the t-algorithm of Levin [4]; a U-version for not so rapidly divergent cases replaces a_j in (2) by $j a_j$, which in turn replaces $((j + 1)/(r + 1))^{r-1}$ in (3) by $((j + 1)/(r + 1))^{r-2}$.

Note the simplicity of (3) in numerical application but its intractability in theory.

RELATION TO THE SHANKS' ALGORITHM

In problems such as radioactive decay, series with alternating decreasing coefficients, a model (corresponding to the "regression" concept) involving a linear function of exponentials (transients) seems appropriate. This is the essence of Shanks' [6] approach; he sets

$$A_j = \alpha + \sum_{t=0}^{r} p_t \exp(j q_t) \qquad (4)$$

so that depending on the sign of the real part of a typical parameter q_r, A_j may tend to α, may oscillate, or may tend to infinity (the usual characteristics of transients, interpreted in the broad sense). Shanks considered a number of series which had been of interest to mathematicians in the nineteenth century and earlier such as that for $\pi/4$, the factorial series, the series $\frac{1}{2}\pi(3\ln 3 - 5\ln 5 + \cdots)$ and others. They involved a certain amount of puzzlement because of a lack of clear definition relating to sums.

Note that (4) with $r = 0$ and $j = s$, $s + 1$, $s + 2$ leads to the well-known sequence accelerator

$$\alpha_s = \left(A_s A_{s+2} - A_{s+1}^2\right)/\left(A_s + A_{s+2} - 2A_{s+1}\right).$$

LEVIN'S ALGORITHM AND DIVERGENT SERIES

Theorems in this case are rare but there are some relating to convergent alternating series, and oscillatory and monotone sequences (see Sidi [9]). Our experience suggests that for series with alternating terms (a few initial anomolous signs being permissible) and diverging about as fast as the factorial, but not faster than the double factorial $(1 - x 2! + x^2 4! - \cdots)$ the algorithm diverges but goes through a "best" stopping point in the sequence; the location is discussed in the next paragraph.

In particular, for the factorial series $1 - 1! + 2! - \cdots$ Levin's t-algorithm gave 14 correct digits using 29 terms but faded to only 7 digits using 42 terms. There is a progressive loss of accuracy to be aware of through the intrusion of alternating signs in the numerator of (3).

For the double-factorial series $1 - 2! + 4! - \cdots$, 28 terms gave 6 correct digits, whereas 31 only gave 3 correct. There is further deterioration for the triple factorial series, but series as explosive as this are rare in current literature.

ILLUSTRATION: STANDARD DEVIATION IN SAMPLING FROM PEARSON TYPE III

Craig [2] developed algebraically the first few terms in the first four moments of the standard deviation $\sqrt{m_2}$ [m_2 the second central sample moment with expectation $(1 - 1/n)\mu_2$] in sampling from the type III density

$$f(x) = x^{\rho - 1}e^{-x}/\Gamma(\rho). \tag{5}$$

He also considered the sample skewness $\sqrt{b_1} = m_3/m_2^{3/2}$. E. S. Pearson [5] followed up the study by numerical work on the standardized standard deviation $s(n) = \sqrt{m_2/\mu_2}$ with a few remarks on the skewness. The labor and patience involved in these studies seems remarkable viewed from modern computerized approaches. One gets the impression that Craig pursued the moments of the skewness to the point of exhaustion.

In a recent letter [3] Craig says that he had the impression the series were divergent; Pearson summed the series using sufficiently large sample sizes n to give the appearance of fairly rapid convergence. By a computer approach [7, 8] we have developed series for the four moments of $s(n)$ to order n^{-24}, and $\sqrt{b_1}$ to order n^{-12} for the Pearson study populations, i.e., $\rho = 20, 8, 4$, and $8/3$ (see his Table 1). Illustrations of $E\sqrt{m_2/\mu_2}$ are given in Table 1. As far as the computations go, there is complete agreement with the Pearson values.

In the series, notice that the magnitude of e_{24} decreases as ρ increases (less skewness) and lies between $24! \sim 6.2\text{E}20$ and $48! \sim 1.2\text{E}61$; for $\rho = 1$, not reported here, $e_{24} \sim 8.6\text{E}60$. The sign pattern is alternating, strictly so for $\rho = 4$, $\frac{8}{3}$, and 1; anomalies creep in initially for ρ increasing and when normality is reached a pattern $- - + +$ (period 4) is established from e_2 onward. Thus the Levin algorithm is expected to be successful for $\rho \geqslant 1$. Sequences for the first

four moments for $n = 2, 3$ and $\rho = 4$ are given in Table 2. It will be seen that the sequence for μ_1' runs through a minimum, and very likely retreats from the true value thereafter. However, we can evaluate the moments fairly accurately for small n (Table 3).

(For a sample of 2, (x, y) say, we merely evaluate the rth noncentral moment $E|x - y|^r/2^r$ by quadrature. For integer ρ as a check with a similar formula for $Em_2^{3/2}$. For $n = 3$ the three-dimensional integrals are reduced by transforming to polar coordinates and evaluated by quadrature (program NAGFLIB:DO1DAF from the Harwell subroutine library).

The best stopping points $r = 12, 17, 13$, and 16 are flagged from Table 2 for $n = 2$; a check of consistency is found from comparisons for $n = 3$. It seems plausible to accept these stopping points for $n > 3$, and in addition here there is less demand on the algorithm because of the damping effect of larger n; for example, for $n = 10$ the sequence α_r, $r = 9$ through 25 for $\mu_1'(s(n))$ is stable at 0.91022.

Some comparisons for the moments, for three sample sizes, are given in Table 4. Internal checks arise by noting that $\mu_2(s(n)) = 1 - 1/n - \{E(s(n))\}^2$ with a similar formula for $\mu_4(s(n))$ in terms of lower-order moments. Thus at $n = 5$ these formulas give $\mu_2 = 0.1230$, and $\mu_4 = 0.06733$, whereas at $n = 20$ the values are $\mu_2 = 0.039092$ and $\mu_4 = 0.0055350$, in satisfactory agreement.

The skewness $\sqrt{b_1}$ is a more complicated function and the moment series alternate in sign, but increase in magnitude almost as fast as the double-factorial series. The Levin algorithm fades somewhat especially for small n, but this can be offset by going to n near to 20 for low-order moments and 50 or so for third and fourth moments. For example, with $\rho = 4$ and $n = 20$ the 10th through 13th t-approximants round off to 0.7009 for $E\sqrt{b_1/\beta_1}$ with simulation (100,000 runs) comparison 0.701; similarly, this sequence of approximants for $\text{var}\sqrt{b_1/\beta_1}$ yields values between 0.2822 and 0.2826, with simulation result 0.283. The third and fourth moments are not stable at $n = 20$.

Table 1. $E\sqrt{m_2/\mu_2} \sim e_0 + e_1/n + \cdots$ from Type III Populations[a]

	ρ = 20		ρ = 4		ρ = 8/3	
σ	e_s	τ_s	e_s	τ_s	e_s	τ_s
1	−7.875000000000000D − 01	0	−9.375000000000000D − 01	1	−1.031249996484375D 00	1
2	−5.742187499999999D − 02	13	7.519531250000000D − 01	11	1.391113255358887D 00	13
3	−7.281298828125000D − 01	7	−8.153375976562500D 00	19	−1.765763810949841D 01	26
4	4.783566146850586D 00	13	1.569145107269287D 02	33	4.584073532781980D 02	44
5	−6.109329180850983D 01	19	−5.211049717068671D 03	49	−2.038853466308149D 04	66
6	1.139841907074662D 03	25	2.536497460335679D 05	66	1.338138455220871D 06	89
7	−2.803292167558395D 04	31	−1.6739654470580102D 07	85	−1.197494978956125D 08	116
8	8.705141578232930D 05	38	1.424017396079829D 09	106	1.388569900486344D 10	145
9	−3.301137092509394D 07	45	−1.507704731102456D 11	128	−2.013846319933139D 12	177
10	1.490907092191074D 09	53	1.936423369141064D 13	153	3.559462355169827D 14	211
11	−7.868924823233592D 10	61	2.958364998751791D 15	179	−7.516529509507154D 16	248
12	4.781869348471500D 12	69	5.295251075487383D 17	207	1.866771945476233D 19	288
13	−3.305895100861481D 14	78	−1.095511020861536D 20	237	−5.383130554986099D 21	331
14	2.574569895186978D 16	87	2.594935963811900D 22	269	1.783187738689187D 24	377
15	−2.240045195251104D 18	97	−6.973039889731325D 24	303	−6.723821211594726D 26	426
16	2.162194346129203D 20	106	2.109500526643329D 27	338	2.863263656040230D 29	478
17	−2.301385253972923D 22	117	−7.136578445394941D 29	376	−1.367477937712182D 32	532
18	2.686873004026409D 24	127	2.683962467731926D 32	416	7.279783435838170D 34	590
19	−3.424884752123857D 26	139	−1.116177449969253D 35	458	−4.295948976131484D 37	651
20	4.746631081067046D 28	150	5.108358046321021D 37	501	2.796268326329912D 40	715
21	−7.126053540022001D 30	162	−2.561722667282540D 40	547	−1.998539833996209D 43	782
22	1.154978742048399D 33	174	1.402043913632886D 43	595	1.561951516152406D 46	851
23	−2.014780164974830D 35	187	−8.344270215322666D 45	645	−1.329838928074173D 49	924
24	3.772150396418360D 37		5.382152115711575D 48		1.229121132510263D 52	

[a] $e_0 = 1$ in each case; $\tau_s = |e_s/e_{s-1}|$.

Table 2. Levin's _t_ Algorithm for Moments of the Standard Deviation $s(n)$ in Sampling from $(x^3/6)\exp(-x)$

v^a	μ_1'	μ_2	μ_3	μ_4
		$n = 2$		
10	0.55593321	0.216979408	0.109127022	0.338290082
11	0.55146819	0.217735347	0.111842293	0.287604236
12	*0.54795510*	0.217559359	0.117050760	0.254133566
13	0.54527123	0.216927409	*0.118085591*	0.239257051
14	0.54337883	0.215054441	0.126930549	0.226671387
15	0.54227258	0.212195038	0.129926713	0.222375697
16	0.54184300	0.209325711	0.130391534	*0.221031467*
17	0.54226120	*0.202968694*	0.161554301	0.189106924
18	0.54347337	0.197941195	0.129696734	0.218904756
19	0.54498671	0.196032213	0.133182138	0.212733957
20	0.54800501	0.174952443	0.318695824	0.016216595
True Value	0.546875	0.200928	0.122032	0.220559
		$n = 3$		
11	0.70496471	0.173802068	0.073902068	0.154875741
12	*0.70420168	0.173746778	0.074916379	0.147406875
13	0.70367920	0.173619314	*0.074996013*	0.144721069
14	0.70335355	0.173288574	0.076346068	0.132694417
15	0.70319127	0.172853687	0.076630348	0.142193112
16	0.70314407	0.172484965	0.076606114	*0.142090521*
17	0.70322010	*0.171688912*	0.080111793	0.138387855
18	0.70338657	0.171175720	0.076554816	0.141853737
19	0.70356133	0.171053676	0.077062602	0.141099615
20	0.70388889	0.169019834	0.092344016	0.124569263
True Value	0.703901	0.171190	0.076434	0.141481

$^a\tau$ = number of terms used in the algorithm.

Table 3. Moments of $s(n)$, $\rho = 4$

	μ_1'	μ_2	μ_3	μ_4
$n = 2$	0.546875	0.200928	0.122032	0.220559
$n = 3$	0.703901	0.171190	0.076434	0.141481

For a sample of 2, (x, y) say, we merely evaluate the _r_th noncentral moment $E|x - y|^r/2^r$ by quadrature. For integer ρ as a check

$$E\sqrt{m_2} = \frac{1}{\Gamma^2(\rho)} \sum_{s=1}^{\rho} K_s,$$

$$K_s = \frac{s(\rho - 1)^{(s-1)}(2\rho - s - 1)!}{2^{2\rho - s}}$$

with a similar formula for $Em_2^{3/2}$. For $n = 3$ the three-dimensional integrals are reduced by transforming to polar coordinates and evaluated by quadrature (program NAGFLIB : DO1DAF from the Harwell subroutine library).

Table 4.   Comparisons of Approximants for Moments of $s(n)$ in Sampling from $(x^3/6)\exp(-x)^a$

Type of algorithm		μ_1'	μ_2	μ_3	μ_4
	L	0.8228(12)	0.1230(17)	0.03850(13)	0.06691(16)
$n = 5$	P	0.8223(22)	0.1233(22)	0.03930(20)	0.06796(20)
		0.8230(23)	0.1233(23)	0.03878(21)	0.06701(21)
	M.C.	0.822	0.123	0.0390	0.0674
	L	0.91023	0.071492	0.013567	0.020284
$n = 10$	P	0.91022	0.071497	0.013604	0.020309
		0.91023	0.071498	0.013576	0.020284
	M.C.	0.909	0.708	0.0133	0.0198
	L	0.954415	0.0390919	0.0042549	0.0055351
$n = 20$	P	0.954415	0.0390920	0.0042549	0.0055354
		0.954415	0.0390920	0.0042551	0.0055351
	M.C.	0.954	0.0387	0.00402	0.00535

[a]L = Levin, P = Padé, M.C. = simulations using 100,000 samples. The parenthetic entries for Levin show the number of terms of the series used; those for Padé refer to the convergent used in the Stieltjes continued fraction form $(n/n+)(p_1/1+)(q_1/n+)(p_2/1+)\cdots$ for the mean, and $(q_0/n+)(p_1'/1+)(q_1'/n+)\cdots$ for μ_2, and similarly for μ_3 and μ_4.

CONCLUSION

We have studied over a hundred applications including (a) $\sqrt{m_2}$ from half-normal*, logistic*, and rectangular* densities; (b) Student's t from a selection of Pearson type 1 populations; and (c) the coefficient of variation and moment estimator for the shape parameter for 30 or so Weibull* two-parameter densities.

The Levin algorithm in general works well for functions of sample moments involving mean and variance only; there is some deterioration for more complicated structures such as the skewness and kurtosis. From a computer point of view, complications can also arise with implicit moment functions arising in estimation problems.

There is always more difficulty in summing series for higher moments (third and higher) than for the mean and variance. The difficulty seems centered in anomolous patterns of sign and magnitude in the initial terms. In this connection we must remember that summation algorithms are in general nonlinear so that the response to changes in the initial terms are unpredictable, and indeed potentially catastrophic.

Convergency questions are supplanted by error analysis and validity studies, the approach being mainly through a discipline of experimental numerical science—mathematical error bounding is generally hopeless. Alternative summation techniques are essential (*see* PADÉ APPROXIMATION).

For the computer work on the factorial series (section 3) a variable precision package, effectively using 50 significant digits, was used on IBM 3033. For the moment series coefficients double-double precision arithmetic was used, and for most of the remaining work double precision only on the same computer.

References

[1] Baker, G. A., Jr. (1975). *Essentials of Padé Approximations*. Academic Press, New York.

[2] Craig, C. C. (1929). *Biometrika*, **21**, 287–293.

[3] Craig, C. C. (1981). Personal communication.

[4] Levin, D. (1973). *Intern. J. Comp. Math.*, **B3**, 371–388.

[5] Pearson, E. S. (1929). *Biometrika*, **21**, 294–302.

[6] Shanks, D. (1955). *J. Math. Phys.*, **34**, 1–42.

[7] Shenton, L. R., Bowman, K. O., and Sheehan, D. (1971). *J. R. Statist. Soc. B*, **33**, 444–457.

[8] Shenton, L. R., Bowman, K. O., and Lam, H. K. (1979). *Proc. Statist. Computing Sect. Amer. Statist. Ass.*, pp. 20–29.

[9] Sidi, A. (1979). *Math. Comp.*, **33**, 145, 315–326.

[10] Van Dyke, M. (1975). *SIAM J. Appl. Math.*, **28**, 720–734.

K. O. BOWMAN
L. R. SHENTON

LÉVY, PAUL-PIERRE

Born: September 1, 1886, in Paris, France.

Died: December 15, 1971, in Paris, France.

Contributed to: functional analysis, calculus of probability, study of stochastic processes.

Paul-Pierre Lévy was born September 1, 1886, in Paris and died December 15, 1971, in Paris. He was educated in the Lycée Louis le Grand and Lycée Saint-Louis Paris and then at the Ecole Polytechnique (1904–1906), where he had Henri Poincaré as a professor. He left this school as a mine engineer and then converted to research and academic teaching. He was a professor in the Mine School of Saint Etienne, Mine School of Paris, and Ecole Polytechnique until 1959. He gave lectures at the Collège de France and at the Sorbonne. He was a member of the Institut de France (Académie des Sciences). His scientific work started with functional analysis (he is the author of *Problèmes Concrets d'Analyse Fonctionnelle*, Gauthier-Villars, Paris, 1951). His doctorate thesis was titled: "Sur les équations intégro-différentielles définissant des fonctions de lignes," but soon after diverted much of his attention to the calculus of probability. This orientation was suggested by the material of his lecturers in Ecole Polytechnique. At the same time he gave many papers on the philosophy of mathematics and in one of them had the idea of theorems being possibly true but whose demonstration would never be achieved, needing an infinity of mental steps. That was about 10 years before Gödel's work. In the axiom of choice debate he was a supporter of the axiom. He believed in the "existence" of a set apart from the mathematician who deals with and defines it.

In the calculus of probability he discovered many important tools; the first of them is the characteristic function* (c.f.), which is the Fourier transform of the distribution function ($\int \exp itx \, dF(x)$). He substituted this characteristic function to that of Poincaré, which was the Laplace transform ($\int \exp tx \, dF(x)$) existing only for certain classes of distribution. He classified the characteristic functions, opening the vast field of indefinitely divisible laws of probability, developed the notions of stable, semistable, and quasistable laws, and could write expressions of these characteristic functions. He gave a very useful definition of the distance between two probability laws, and achieved many fundamental results in random topology, chiefly in the field of random series. He introduced the concentration function of a distribution and used it for studying many convergence problems. It was in working about random series that he had the intuition of the truth of the theorem which is known as Lévy–Cramér's or Cramér–Lévy's theorem, whose proof was given by Cramér in 1937: "If the sum of two independent random variables is a normal variable, each one of the two variables is normal."

All that was the starting point of the arithmetic of probability laws, to which he devoted a book (*Théorie de l'Addition des Variables Aléatoires*, 2nd ed., Gauthier-Villars, Paris, 1954). The infinitely divisible* laws led him to the study of stochastic processes* (he is the author of *Processus Stochastiques et Mouvement Brownien*, 2nd ed., Gauthier-Villars, Paris, 1965), on which he wrote many papers. He was, with Wiener, the initiator of what is now called the Wiener–Lévy process [i.e., a Gaussian random function $X(t)$ with $E(X(t)) = 0$ and $\operatorname{cov} X(t_1), X(t_2) = \min(t_1, t_2)$], that is, the typical Brownian motion process. He discovered a probability law (quite analogous to von Mises–Smirnov law) which gives the probability law of the area swept by a random vector whose end point has two coordinates which are two independent Wiener–Lévy processes. This result can be used for testing the discrepancy between a theoretical and an empirical distribution, and so is a typical statistical result. Paul Lévy discovered the so-called arc sine law*, which is the probability law of the period during which a Wiener–Lévy process is positive. An extension of that result to the case of a Wiener–Lévy's bridge is known as Gnedenko's theorem and leads again to a comparison of empirical and statistical distribution. The extension of this theorem to multidimensional distribution is still open. He also dealt with Markov chains, Markov processes, martingales*, and theory of games. He worked very hard until his last days. A very important part of his scientific production was written during his last 10 years.

One year before his death he wrote his memoirs under the title "Quelques aspects de la pensée d'un mathématicien" (Ed. Alfred Blanchard, Paris, 1971).

All his papers and memoirs are published by Gauthier-Villars, Bordas, and Dunod (*Oeuvres Complètes de Paul Lévy*, 6 vols.). The entire output is about 250 titles.

D. DUGUÉ

LÉVY CONCENTRATION FUNCTION

The Lévy concentration function $Q(X; \lambda)$ of a random variable X is a function of the positive variable λ defined by

$$Q(X; \lambda) = \sup_{-\infty < x < \infty} \Pr(x \leqslant X \leqslant x + \lambda)$$

The concentration function is a useful instrument in investigations of sums of inde-

pendent random variables. For instance, one can show that a sum S_n of independent random variables, $S_n = \sum_{i=1}^{n} X_i$, converges essentially almost everywhere as $n \to \infty$ if and only if $\lim_{n \to \infty} Q(S_n; \lambda) > 0$ for every $\lambda > 0$. The main interest in investigations of the concentration function has therefore been to find bounds for the concentration function of $Q(S_n; \lambda)$.

More generally, let X be a random vector in a vector space V and let E be a set in V. Then the concentration function is defined by $Q(X; E) = \sup_{x \in V} \Pr(X \in x + E)$. In the following, let $X_1, X_2, \ldots, X_n$ be independent random vectors in the d-dimensional space R^d or a Hilbert space H and let E be the sphere $\Sigma_\lambda = \{x; |x| \leqslant \lambda\}$. x is a vector in R^d or H, (x, x) the inner product and $|x| = (x, x)^{1/2}$. The distribution function of the random vector X is denoted by F (with or without an index) and we set $X^s = X - X'$, where X' is a random vector independent of X and with the same distribution. The distribution function of X^s is denoted by F^s. Note that with the notation above $Q(X; \lambda) = Q(X; \Sigma_{(1/2)\lambda})$ in the one-dimensional case. Define the censored variance of X (with respect to Σ_λ) by

$$D^2(X; \Sigma_\lambda) = \inf_a \left\{ \lambda^{-2} \int_{|x-a| \leqslant \lambda} |x - a|^2 \, dF(x) + \int_{|x-a| > \lambda} dF(x) \right\}$$

It is not difficult to show that

(a) $D^2(X; \Sigma_\lambda) \geqslant 1 - Q(X; \Sigma_\lambda)$,

(b) $(\tau/\lambda)^2 D^2(X; \Sigma_\tau) \leqslant D^2(X; \Sigma_\lambda)$
$$\leqslant D^2(X; \Sigma_\tau),$$

(c) $D^2(X + Y; \Sigma_\lambda)$
$$\geqslant \max\left(D^2(X; \Sigma_\lambda), D^2(Y; \Sigma_\lambda) \right),$$

if X and Y are independent random vectors,

(d) $\dfrac{1}{4} \left\{ \lambda^{-2} \int_{|x| \leqslant \lambda} |x|^2 \, dF^s(x) + \int_{|x| > \lambda} dF^s(x) \right\}$
$$\leqslant D^2(X; \Sigma_\lambda) \leqslant D^2(X^2; \Sigma_\lambda).$$

With the notations above the main inequality is stated in:

Theorem 1. If $X_1, X_2, \ldots, X_n$ are independent random vectors in the d-dimensional space R^d or in a Hilbert space, $S_n = \sum_{i=1}^n X_i$, then

$$Q(S_n; \Sigma_\lambda) \leqslant C \left\{ \sum_{i=1}^n D^2(X_i; \Sigma_\lambda) \right\}^{-1/2}$$

Here C is a universal constant.

Using formulas (a) and (b) one can infer:

Corollary. *If $\lambda_1, \lambda_2, \ldots, \lambda_n \leqslant \lambda$, then*

$$Q(S_n; \Sigma_\lambda)$$

$$\leqslant C\lambda \left\{ \sum_{i=1}^n \lambda_i^2 (1 - Q(X_i; \Sigma_{\lambda_i})) \right\}^{-1/2}.$$

The inequalities can be generalized to general symmetric and convex sets $E \in R^d$ [with appropriate definition of $D^2(X; E)$], but then the constant C depends on the dimension d.

If $X_1, X_2, \ldots, X_n$ are identically distributed, then obviously

$$Q(S_n; \Sigma_\lambda) \leqslant Cn^{-1/2} D^{-1}(X_1, \Sigma_\lambda)$$

An inequality of the type above was first given by Kolmogorov. Later Esseen, using characteristic functions, was able to improve and generalize earlier results. Kesten and Halász have given the inequalities in the next theorem, which here is stated only in the case $X_1, X_2, \ldots, X_n$ are identically distributed. Suppose that $X_1 \in R^d$ and define $D(\lambda) = \inf_{|e|=1} \{\lambda^{-2} \int \min[(x,e)^2, \lambda^2] dF_1^s(x)\}$. $D(\lambda)$ measures to what extent the distribution of X_1 is d-dimensional.

Theorem 2. If $X_1, X_2, \ldots, X_n$ are independent, identically distributed random vectors with values in R^d, then

$$Q(S_n; \Sigma_\lambda) \leqslant C(d) Q(X_1^s; \Sigma_\lambda) n^{-d/2}$$

$$\times \{D(\lambda)\}^{-d/2}$$

$C(d)$ is a constant depending on the dimension d.

It is also possible to give general lower bounds according to:

Theorem 3. If $X_1, X_2, \ldots, X_n$ are independent random vectors with values in R^d then $Q(S_n; \Sigma_\lambda) \geqslant C(d,r)\lambda^d \exp\{-4\int_{|x|>\lambda_i} dF_i^s\} \cdot \{\lambda^d + T^d + \mu(r)^d\}^{-1}$, where $T = \max(T_1, T_2, \ldots, T_n)$, T_i the 75% percentile of $|X_i^s|$, $\mu(r = \{\sum_{i=1}^n \int_{|x| \leqslant \lambda_i} |x|^r dF_i^s\}^{1/r}$ and $\lambda_1, \lambda_2, \ldots, \lambda_n$ arbitrary numbers > 0, r an arbitrary number such that $0 < r \leqslant 2$, and $C(d, r)$ a constant depending on the dimension d and on r.

From this theorem easily follows

Corollary. *If $X_1, X_2, \ldots, X_n$ are identically distributed and $E[|X_1|^r] < \infty$, then $Q(S_n; \Sigma_\lambda) \geqslant C(d,r)n^{-d/r}\lambda^d \{\lambda^d + (E|X_1^s|^r)^{d/r}\}^{-1}$.*

Bibliography

Esseen, C. G. (1968). *Zeit. Wahrscheinlichkeitsth. verwand. Geb.*, **9**, 290–308.

Halász, G. (1977). *Periodica Math. Hung.*, **8**(3), 197–211.

Hengarner, W. and Thoedorescu, R. (1973). *Concentration Functions.* Academic Press, New York. (This book gives the basic properties of the Lévy concentration function, upper and lower bounds, and applications mainly to convergence problems. It also includes a long reference list.)

JAN ENGER

LÉVY CONTINUITY THEOREM *See* CHARACTERISTIC FUNCTIONS

LÉVY DISTANCE (METRIC)

Given arbitrary n-dimensional distribution functions* Φ_1 and Φ_2, define

$$\epsilon_{ij} = \inf\{\epsilon \mid \Phi_j(C) < \Phi_i(C^\epsilon)$$
$$\text{for every closed set } C \subset R^n\}.$$

(Here C^ϵ denotes an ϵ-neighborhood of the set C.) The metric defined by

$$\rho(\Phi_1, \Phi_2) = \max(\epsilon_{12}, \epsilon_{21})$$

is called the Lévy distance. It was introduced by Lévy [1] in the one-dimensional case and by Prohorov [2] for the case of metric spaces. Given a sequence of distributions $\Phi_1, \Phi_2, \ldots, \Phi_n, \ldots$ the sequence converges weakly* to a distribution Φ if and only if $\lim_{k\to\infty}\rho(\Phi_k,\Phi)=0$.

References

[1] Lévy, P. (1937). *Théorie de l'Addition des Variables Aléatoires*. Gauthier-Villars, Paris.

[2] Prohorov, Ju. V. (1961). *Proc. 4th Berkeley Symp. Math. Statist. Prob.*, Vol. 2. University of California Press, Berkeley, Calif., pp. 403–419.

(CONVERGENCE OF SEQUENCES OF RANDOM VARIABLES)

LÉVY–KHINCHINE FORMULA

In the 1930s P. Lévy [4] and A. I. Khinchine [3] developed a formula for the representation of the characteristic function (ch. f.) of any infinitely divisible* (inf. div.) random vector (rv). A rv X is called *infinitely divisible* if for each $n = 1, 2, \ldots$ there are independent identically distributed rvs $X_{n1}, \ldots, X_{nn}$ such that X and $X_{n1} + \cdots + X_{nn}$ have the same distribution. Equivalently, if μ is the distribution (law) of X on $\mathbb{R}^k$ and φ is its ch. f., then X and μ are called infinitely divisible if for each n there is a probability measure ν_n with ch. f. ψ_n such that

$$\mu = \nu_n * \cdots * \nu_n \quad (n \text{ times})$$

($*$ denotes convolution*), meaning that $\varphi = \psi_n^n$.

LÉVY–KHINCHINE FORMULA I

The function φ on $\mathbb{R}^k$ is the ch. f. of an inf. div. rv (distribution) if and only if φ can be expressed in the form

$$\varphi(y) = \exp\left[i\langle x_0, y\rangle - \tfrac{1}{2}\langle Dy, y\rangle + \int_{\mathbb{R}^k} K(x, y)\, d\lambda(x)\right], \quad (1)$$

where $x_0 \in \mathbb{R}^k$, D is a nonnegative-definite $k \times k$ matrix, λ is a measure on $\mathbb{R}^k$ which is finite on the complement of every neighborhood of 0 in $\mathbb{R}^k$ and satisfies $\lambda(\{0\}) = 0$ and

$$\int_{|x|\leqslant 1} |x|^2\, d\lambda(x) < \infty,$$

and

$$K(x, y) = e^{i\langle x,y\rangle} - 1 - \frac{i\langle x, y\rangle}{1 + |x|^2}. \quad (2)$$

[Here $\langle x, y\rangle$ is the usual inner (dot) product on $\mathbb{R}^k$.] The parameters x_0, D, and λ in (1) are uniquely determined by φ.

It is sometimes helpful to have a representation such as (1) where the measure λ is totally finite. This can be done by modifying the kernel K. The proper version of the representation in this form is as follows.

LÉVY–KHINCHINE FORMULA II

The function φ on $\mathbb{R}^k$ is the ch. f. of an inf. div. rv (distribution) on $\mathbb{R}^k$ if and only if φ can be expressed in the form (1), where x_0 and D are as above, λ is a finite measure (or multiple of a probability distribution function) on $\mathbb{R}^k$ such that $\lambda(\{0\}) = 0$, and the function K in (2) is replaced by

$$K'(x, y) = K(x, y)\frac{1 + |x|^2}{|x|^2}. \quad (3)$$

The triple x_0, D, λ is unique given φ.

When this version of the formula is used, the measure λ is called the *Lévy spectral measure* of φ or of a corresponding rv X and its distribution μ. Note that the first two terms in the exponent of (1) correspond to the Gaussian* factor of φ or of μ having mean x_0 and covariance matrix D. The integral term represents a factor of "Poisson type."

The formulas of Lévy and Khinchine are often used to determine whether a sequence $\{X_n\}$ of inf. div. rvs converges in distribution to the inf. div. rv X_0. Denote by $F(x_0, D, \lambda)$ the distribution function of a rv whose ch. f. is given by (1) for a certain $x_0 \in \mathbb{R}^k$, matrix D, and measure λ, say as a Formula I.

Theorem 1. Let X_n be an inf. div. RV such that $F_{X_n} = F(x_n, D_n, \lambda_n)$, $n = 0, 1, 2, \ldots$. Then X_n converges in distribution to X_0 if and only if

1. $$\lim_{n\to\infty} x_n = x_0.$$

2. The restriction of λ_n to the complement of each closed neighborhood of 0 in $\mathbb{R}^k$ converges weakly* as $n \to \infty$ to the corresponding restriction of λ_0.

3. $$\lim_{\epsilon \to 0} \limsup_{n \to \infty} \left[\int_{|x| \leqslant \epsilon} \langle x, y \rangle^2 \, d\lambda_n(x) \right.$$

$$\left. + \langle D_n y, y \rangle \right] = \langle D_0 y, y \rangle,$$

$$y \in \mathbb{R}^k.$$

There are a number of important subclasses of the class of infinitely divisible laws. These include the stable* laws and the class L^* laws (self-decomposable). Each of these classes can be characterized by easily verified conditions on the Lévy spectral measures of its members.

Versions of the Lévy–Khinchine formulas valid for random elements of more abstract space than $\mathbb{R}^k$ can be found in refs. 1, 2, and 5.

References

[1] de Acosta, A., Araujo, A., and Giné, E. (1978). In *Probability on Banach Spaces*, J. Kuelbs, ed. Marcel Dekker, New York, pp. 1–68. (The most up-to-date treatment of the Lévy–Khinchine formula for Banach-space-valued RVs.)

[2] Heyer, H. (1977). *Probability Measures on Locally Compact Groups*. Springer-Verlag, New York, Chap. 4. (Treats inf. div. random variables with values in a locally compact group.)

[3] Khinchine, A. (1937). *Bull. Math. Univ. Mosc.*, **1**, 6–17.

[4] Lévy, P. (1937). *Théorie de l'Addition des Variables Aléatoires*. Gauthier-Villars, Paris (2nd ed., 1954).

[5] Parthasarathy, K. R. (1967). *Probability Measures on Metric Spaces*. Academic Press, New York. (A fine treatment of the analogs of Formula I for random variables with values in either a locally compact Abelian group or Hilbert space.)

Further Reading

Gnedenko, B. V. and Kolmogorov, A. N. (1954). *Limit Distributions for Sums of Independent Random Variables*, K. L. Chung, trans. Addison-Wesley, Reading, Mass., Chap. 3. (The reference most often cited for a proof of Formulas I and II for the case $k = 1$; somewhat cumbersome mathematically.)

Tortrat, A. (1971). *Calcul des Probabilités et Introduction aux Processus Aléatoires*. Masson, Paris, Chap. 13. (Develops Formula II.)

Tucker, H. G. (1967). *A Graduate Course in Probability*. Academic Press, New York, Chap. 6. (A readable, modern treatment of Formula I for the case $k = 1$.)

(CHARACTERISTIC FUNCTION
CONVOLUTION
INFINITE DIVISIBILITY
L CLASS LAWS
SPECTRAL MEASURE
STABLE LAWS
WEAK CONVERGENCE)

B. M. Schreiber

LÉVY PROCESS

A real-valued stochastic process* $\{X_t\}$, $0 \leqslant t < \infty$, with independent increments* is called a Lévy process if it is continuous in probability and if almost all sample functions* of this process are right continuous and possess left-hand limits for every $t \in [0, \infty]$. Wiener*, Poisson*, compound Poisson, and stable* processes are all particular cases of Lévy process.

Further Reading

Doob, J. L. (1953). *Stochastic Processes*. Wiley, New York.

Lévy, P. (1948). *Processus Stochastiques et Mouvement Brownien*. Gauthier-Villars, Paris.

LÉVY'S INEQUALITIES

These inequalities are related to sums of independent random variables centered at suitable medians* (compare with Kolmogorov's inequalities*).

Let $X_1, X_2, \ldots, X_n$ be a sequence of independent random variables, and $S_k = \sum_{j=1}^{k} X_j$; then for any real a,

$$\Pr\left[\max_{1 \leqslant k \leqslant n} \{ S_k + m(S_n - S_k) \} > a \right]$$

$$\leqslant 2 \Pr[S_n > a]$$

and

$$\Pr\{\max|S_k + m(S_n - S_k)| > a\}$$
$$\leqslant 2 \Pr[|S_n| > a]$$

where $m \equiv m(X)$ is a median* of the random variable X satisfying

$$\Pr[X \geqslant m(X)] \geqslant \tfrac{1}{2}, \quad \Pr[X \leqslant m(X)] \geqslant \tfrac{1}{2}.$$

(CONVERGENCE OF SEQUENCES OF
  RANDOM VARIABLES
KOLMOGOROV INEQUALITIES
LAWS OF ITERATED LOGARITHMS
LAWS OF LARGE NUMBERS)

LÉVY SPECTRAL MEASURE   *See* LÉVY–
KHINCHINE FORMULA

LEXIAN DISTRIBUTION

This is the convolution* of k binomial* (m, p_j) distributions $(j = 1, \ldots, k)$ (distribution of $X = X_1 + X_2 + \cdots + X_k$, where X_j is distributed binomial (m, p_j) and the X_j's are mutually independent). The expected value and variance of X are

$$E(X) = mk\bar{p}, \quad \text{where} \quad \bar{p} = k^{-1}\sum_{j=1}^{k} p_j$$

$$\text{var}(X) = mk\bar{p}(1 - \bar{p}) - mk\sigma_p^2,$$

$$\text{where} \quad \sigma_p^2 = k^{-1}\sum_{i=1}^{k}(p_i - \bar{p})^2.$$

The Lexian variance is less than that of a binomial $(mk, \bar{p})$.

Note that if X_j is distributed as binomial (m_j, p), then X is binomial $(\sum_{j=1}^{k} m_j, p)$ with expected value $\bar{m}kp$, and variance $\bar{m}kp(1 - p)$, where $\bar{m} = k^{-1}\sum_{j=1}^{k} m_j$.

The name "Lexian distribution" is sometimes given to that arising from a sequence of n repeated independent Bernoulli trials with the same probability p, where p is a random variable, equally likely to take any one of a set of k values, $p_1, \ldots, p_k$. The number (X) of successes in these n trials has a compound binomial distribution, studied by Bienaymé*. The expected value and the

variance of X are, respectively,

$$E(X) = n\bar{p}$$

and

$$V(X) = n\bar{p}(1 - \bar{p}) + n(n - 1)k^{-1}$$
$$\times \sum_{i=1}^{k}(p_i - \bar{p})^2,$$

where $\bar{p} = k^{-1}\sum_{i=1}^{k} p_i$. The Bienaymé variance is greater than the variance of a binomial random variable with parameters n and $\bar{p}$. [1, Chap. 3].

Reference

[1] Heyde, C. C. and Seneta, E. (1977). *I. J. Bienaymé: Statistical Theory Anticipated*. Springer-Verlag, New York.

(BIENAYMÉ, I. J.
BINOMIAL DISTRIBUTION
COMPOUND DISTRIBUTION
LEXIS, WILHELM)

LEXICOSTATISTICS

Natural languages occur in "families." For example, Spanish and French are related through their common ancestor, Latin, and all the languages descended from Latin are related to those descended from Greek and Sanskrit via an Indo-European protolanguage. Lexicostatistics is the statistical exploration of such relationships based on data about the vocabularies of the languages. (*See also* LINGUISTICS, STATISTICAL.)

The raw data consist of a list of *meanings* and the corresponding most commonly used words in each of the languages being studied. For every meaning the linguist decides whether or not each pair of words are *cognate*, that is whether they developed from a common ancestor word by a process of regular phonetic change. A measure of similarity for each pair of languages is the proportion of meanings whose words are judged cognate. With these similarity values the methods of numerical taxonomy* can be used to produce a family tree representing the historical evolution of the languages or to depict their relationships in various other ways.

In 1952, the linguist Morris Swadesh proposed that, by analogy with radioactive decay, word changes could be modeled by a simple Poisson process* so that if two languages had separated over a time t, then for any meaning

Pr[they have cognate words at time t]
$$= \exp(-rt).$$

For various languages whose separation times were known he calculated the cognation proportions based on a standard list of meanings and in each case found that about 80% of words were retained per millenium. Examples quoted by Lees [12] include: Old English of A.D. 900–1000 and modern English, 160 cognates in a list of 209 meanings giving $e^{-r} = 0.776$ for $t = 1.0$; Koine Greek and modern Cypriote, 143 cognates from 211 meanings giving $e^{-r} = 0.829$ since $t = 2.07$; and Ancient Classical Chinese of A.D. 950 and modern Mandarin; 167 cognates from 210 meanings giving $e^{-r} = 0.795$ since $t = 1.0$. This suggested that the replacement rate r might be a "universal constant" independent of meaning, language, or historical time. Thus an unknown separation time between two contemporary languages could be estimated from their proportion p of cognate words using $\hat{t} = -\log p/2r$, where t represents the evolution time of either language from their most recent common ancestor language. An extensive highly readable review of this method, termed *glottochronology*, was given by Hymes [10].

There were several major criticisms of Swadesh's work. One was based on failure to understand that the relationship between p and t was stochastic, not deterministic [6]. Another took the form of counterexamples demonstrating that replacement rates r differed much more widely than Swadesh's examples had suggested (e.g., Bergsland and Vogt [3]). Further criticisms concerned difficulties with the use of a standard list of meanings.

Soon empirical evidence was accumulated suggesting that replacement rates depended strongly on meaning, so Dyen et al. [9] proposed the model

$$p_{ij} = \exp(-t_i r_j),$$

where p_{ij} is the probability that language pair i (with separation time t_i) have cognate words for meaning j (with replacement rate r_j). This model was used in an extensive study by Kruskal et al. [11], who calculated maximum likelihood* estimates of the parameters t_i and r_j and found that replacement rates for any given meaning were remarkably similar in different language families (e.g., Indo-European and Cushitic). Other major modifications to Swadesh's model included treating r as a random variable [7, 14] and relaxing the need for one-to-one comparisons of words for the calculation of cognation proportions [13].

Languages do not acquire words only by inheritance or by the invention of new words; they also borrow from other languages. When peoples speaking different languages have frequent contact with one another, borrowing can be the major agent for vocabulary changes and lexical similarities may reflect geographical rather than historical affinities. In such cases multidimensional scaling* has proved to be a useful statistical tool for investigating spatial relationships between languages from lexical data [4].

In the last decade the mathematical literature has contained several probabilistic models for the evolution of languages. Extensions of the simple Poisson model have been made to accommodate borrowing [8], to allow different replacement rates for different languages and the substitution of noncognate words by cognates [5]. Arapov and Hertz [1, 2] have developed a promising theory of lexical change based on the idea, which they attribute to Zipf, that the probability of a word being replaced is inversely proportional to its frequency of usage.

Although the topic of lexicostatistics has received recent attention from mathematicians, it currently appears to be of little interest to linguists. Nevertheless, anthropologists, who need to summarize their field data and make inferences about the peoples they are studying, find these statistical tools useful.

References

[1] Arapov, M. V. and Hertz, M. N. (1972). *Inf. vopr. semiotiki, lingvistiki avtom. perevoda* (Moscow), **3**, 3–85 (in Russian).

[2] Arapov, M. V. and Hertz, M. N. (1974). *Mathematical Methods in Genetic Linguistics.* Moscow (in Russian).

[3] Bergsland, K. and Vogt, H. (1972). *Curr. Anthropl.*, **3**, 115–153.

[4] Black, P. (1976). *Cahiers Inst. Linguistique Louvain*, **3**, 43–92.

[5] Brainerd, B. (1970). *J. Appl. Prob.*, **7**, 69–78.

[6] Chrétien, C. D. (1962). *Language*, **38**, 11–37. (Countered by A. J. Dobson, J. B. Kruskal, D. Sankoff, and L. J. Savage, *Anthropol. Linguistics*, 205–212, 1972.)

[7] Dobson, A. J. (1978). *J. Amer. Statist. Ass.*, **73**, 58–64.

[8] Dobson, A. J. (1978). *J. Appl. Prob.*, **15**, 38–45.

[9] Dyen, I., James, A. T., and Cole, J. W. L. (1967). *Language*, **43**, 150–171.

[10] Hymes, D. H. (1960). *Curr. Anthropol.*, **1**, 3–43.

[11] Kruskal, J. B., Dyen, I., and Black, P. (1971). In *Mathematics in the Archeological and Historical Sciences*, Hodson, Kendall, and Tatu, eds. Edinburgh University Press, Edinburgh.

[12] Lees, R. B. (1953). *Language*, **29**, 113–127.

[13] Sankoff, D. (1970). *Language*, **46**, 564–569.

[14] Sankoff, D. (1973). In *Current Trends in Linguistics*, Vol. 11, T. Sebeck, H. Hoenigswald, and R. Longacre, eds. Mouton, The Hague.

(CLASSIFICATION
CLUSTER ANALYSIS
MULTIDIMENSIONAL SCALING
SIMILARITY MEASURES)

ANNETTE J. DOBSON

LEXIS, WILHELM

> **Born:** July 17, 1837, in Eschweiler, Germany.
>
> **Died:** August 24, 1914, in Göttingen, Germany.
>
> **Contributed to:** theoretical statistics, economics, population studies, sociology.

German statistician and economist. Graduated from the University of Bonn in 1859, Lexis's initial training was in mathematics and physics, but after a brief period in Bunsen's laboratory, he left for Paris and the study of the social sciences. Lexis's ensuing academic career is indicative of the range and diversity of his interests: appointed to the chair of economics at Strassburg, 1872; professor of geography, ethnology, and statistics at Dorpat, 1874; professor of economics at Freiburg, 1876, and Breslau, 1884; and professor of political science at Gottingen, 1887 (where he remained until his death).

Lexis's most important statistical contributions came during the brief period from 1876 to 1879. Dissatisfied with the uncritical and usually unsupported assumption of statistical homogeneity in sampling, often made by Quetelet and his followers, Lexis devised a statistic Q, now called the *Lexis ratio*, to test this assumption and demonstrate its frequent invalidity. If $X_{ij}(1 \leqslant i \leqslant m, 1 \leqslant j \leqslant n)$ are independent Bernoulli trials with success probabilities p_{ij}, where for each i the X_{ij} represent a sample from subpopulation i, then

$$Q = \frac{\sum_{i=1}^{m}(\hat{p}_i - \bar{p})^2/m}{\bar{p}(1 - \bar{p})/n},$$

where $N = mn$, $\bar{p} = \sum_{ij}p_{ij}/N$, and $\hat{p}_i = \sum_j X_{ij}/n$. It can be shown that the *theoretical* dispersion coefficient $D = \sqrt{E(Q)}$ has values $D < 1$ (subnormal dispersion), $D = 1$ (normal dispersion), or $D > 1$ (supernormal dispersion) whenever $p_{ij} \equiv p_j$ (Poissonian sampling), $p_{ij} \equiv p$ (Bernoulli sampling), or $p_{ij} \equiv p_i$ (Lexian sampling), respectively. In actual practice Q was replaced by

$$Q^* = \frac{\sum_{i=1}^{m}(\hat{p}_i - \hat{p})^2/m}{\hat{p}(1 - \hat{p})/n},$$

where $\hat{p} = \sum_{ij}X_{ij}/N$.

Lexis's results were first extended by von Bortkiewicz* and then placed on a sound mathematical basis by Chuprov and Markov, who defined a modified Lexis ratio

$$L = \frac{(N - 1)}{n(m - 1)} Q^*$$

and proved rigorously that $E(L) = 1$ [1],

var(L) $\leqslant 2/(m - 1)$ for $m \leqslant 5$, and that (in effect) the asymptotic distribution of $(m - 1)$ L is $\chi^2(m - 1)$ [8, 9].

The Lexis ratio is, up to multiplicative constant, the variance test for homogeneity of binomial distribution* (see Snedecor and Cochran [13, Sec. 11.7]), and, as such, is an early instance of the analysis of variance*. It is a sad commentary on the gulf between continental and English statistics that it was not until 1924 that it was pointed out, by Fisher, that mQ^* is exactly the same as Pearson's chi-squared* statistic; at the same time, Fisher was unaware of Markov's work which four years earlier, had settled the degrees of freedom* controversy between Fisher and Pearson before it had even begun. In fact, with the exception of Edgeworth and Keynes, the work of Lexis and his school was largely ignored in England.

Literature

For details of Lexis's life and work, see the obituaries by von Bortkiewicz [15] and Klein [7], and the articles by Heiss [4] and Oldenburg [10].

The history of the Lexis ratio is ably discussed by C. C. Heyde and E. Seneta [5, Sec. 3.4]; mathematical details, including some of the results of Chuprov and Markov, are given in J. V. Uspensky's *Introduction to Mathematical Probability* [14, pp. 212–230]. *The Correspondence Between A. A. Markov and A. A. Chuprov on the Theory of Probability and Mathematical Statistics* [11] is an invaluable source of information about the development of the ideas of Chuprov and Markov, and their assessment of the work of Lexis and von Bortkiewicz. Keynes's discussion of Lexis in his *Treatise on Probability* [6, Chap. 32] is also of interest. Other useful accounts of the Lexis ratio include those of J. L. Coolidge [2], H. L. Rietz [12, Chap. 6], and F. N. David, [3, pp. 152–159].

References

[1] Chuprov, A. A. (1916). *Iz. Akad. Nauk. Pgd. Ser. VI*, **10**, 1789–1798.

[2] Coolidge, J. L. (1925). *An Introduction to Mathematical Probability*. Clarendon Press, Oxford.

[3] David, F. N. (1949). *Probability Theory for Statistical Methods*. Cambridge University Press, Cambridge.

[4] Heiss, (1978). In *International Encyclopedia of Statistics*, W. Kruskal and J. Tanur, eds. Free Press, New York.

[5] Heyde, C. C. and Seneta, E. (1977). *I. J. Bienaymé: Statistical Theory Anticipated*. Springer-Verlag, New York.

[6] Keynes, (1921). *Treatise on Probability*. Macmillan, New York.

[7] Klein, (1914). *Jahresber. Dtsch. Math.-Ver.*, **23**, 314–317.

[8] Markov, A. A. (1916). *Iz. Akad. Nauk. Pgd. Ser. VI*, **10**, 709–718.

[9] Markov, A. A. (1920). *Iz. Akad. Nauk. Pgd. Ser. VI*, **14**, 1191–1198.

[10] Oldenburg, (1933). In *Encyclopedia of the Social Sciences*, Vol. 9. Macmillan, New York.

[11] Ondar, K. O., ed. (1981). *The Correspondence between A. A. Markov and A. A. Chuprov on the Theory of Probability and Mathematical Statistics*. Springer-Verlag, New York.

[12] Rietz, H. L. (1927). *Mathemtical Statistics*. Mathematical Association of America, Washington, D. C.

[13] Snedecor, G. W. and Cochran, W. G. (1980). *Statistical Methods*, 7th ed. Iowa State University Press, Ames, Iowa.

[14] Uspensky, J. V. (1937). *Introduction to Mathematical Probability*. McGraw-Hill, New York.

[15] von Bortkiewicz, L. (1915). *Bull. Int. Statist. Inst.*, **20**, 328–332.

(LEXIAN DISTRIBUTION)

SANDY L. ZABELL

LEXIS QUOTIENT   *See* DISPERSION THEORY

LEXIS RATIO   *See* DISPERSION THEORY; LEXIS, WILHELM

LIAPUNOV, ALEXANDER MIKHAILOVICH

Born: June 7 (N.S.), 1857, in Yaroslavl, Russia.

Died: November 3, 1918, in Odessa, USSR.

Contributed to: equilibrium-stability theory, probability theory.

Liapunov entered high school in 1870, having received his elementary education at home, initially from his father, who had been until 1855 an astronomer at Kazan University. He entered Petersburg University in 1876 at the flowering of the Petersburg mathematical "school" founded by Chebyshev*, who exerted a profound influence on the direction of Liapunov's academic development. This period saw the beginning of his association with A. A. Markov*. In 1885 he took up an academic appointment of Kharkov University, where he was heavily engaged in the teaching of mechanics, and was active in the affairs of the Kharkov Mathematical Society, as were his students V. A. Steklov and N. N. Saltykov. He left Kharkov for St. Petersburg in 1902 on his election to its Academy of Sciences to work within its framework, but counted his years in Kharkov as the happiest of his life. Honors bestowed on him included foreign membership of the Academy of the Lincei in Rome. His life ended in suicide, following the death of his wife from tuberculosis in Odessa, to where they had traveled for her health, and where he had a brother.

Liapunov's interest in probability was stimulated by his attendance at Chebyshev's lectures in 1879–1880 on his not completely rigorous proof of the central limit theorem* for independent (but not identically distributed) summands. Chebyshev's proof was modified in 1898 by Markov, and this modification, together with the fact that in his last years in Kharkov Liapunov taught probability theory, seems to be responsible for his papers on the central limit problem in 1900–1901. Although his few published writings on probabilistic topics, completely described in ref. 1, are confined to this problem, and to a single year, they culminate in that very general version of the central limit theorem known as Liapunov's theorem* [4], which has had a profound effect on probability theory. The first rigorous proof of a central limit theorem* by extensive use of characteristic functions (as is the case with Liapunov's proof) should be ascribed to Cauchy* and I. V. Sleshinsky (see ref. 2), but Liapunov's approach and conclusion

still have a number of remarkably novel features.

Let X_i, $i \geqslant 1$, be a sequence of independent random variables such that $\mu_i \equiv EX_i$ and $E|X_i - \mu_i|^{2+\delta}$ (for some $\delta > 0$ independent of i) are both well defined for $i \geqslant 1$. Then, writing $B_n^2 = \operatorname{var} S_n$, where $S_n = \sum_{i=1}^n X_i$, Liapunov's theorem states that

$$L_n^2 \equiv \left(\sum_{i=1}^n E|X_i - \mu_i|^{2+\delta} \right) / B_n^{2+\delta} \to 0$$

$$\Rightarrow F_n(x) \to \Phi(x)$$

uniformly for all x, as $n \to \infty$, where $F_n(x) = P[(S_n - ES_n)/B_n \leqslant x]$, $\Phi(x) = (2\pi)^{-1/2} \int_{-\infty}^x \exp(-y^2/2) dy$. The proof, in terms of discrete random variables, culminates in showing that $\Omega_n = \sup_x |F_n(x) - \Phi(x)|$ is bounded by a quantity that approaches zero as $n \to \infty$. In the case $\delta = 1$ he obtains $\Omega_n \leqslant \operatorname{const}.|L_n \log L_n|$; it follows that for identically distributed X_i, $\Omega_n \leqslant \operatorname{const}. \log n / \sqrt{n}$. Thus he not only satisfied Chebyshev's criterion (of rigorous proof by estimation of the error of approximation), but also initiated the topic of *rate of convergence* to a limit distribution. Liapunov's bounds have subsequently been extended and improved by many workers. In the course of the proof he obtains the inequality for $l > m > n \geqslant 0$ for the absolute moments of a random variable X:

$$(E|X|^m)^{l-n} \leqslant (E|X|^n)^{l-m} (E|X|^l)^{m-n},$$

which in the case $n = 0$ is usually known as Liapunov's inequality*, even though it was known to Bienaymé* [2].

Finally worth mentioning is the manuscript [7], pertaining to the classical linear model* of full column rank r where the residuals ϵ_i, $i = 1, \ldots, N$, are independently distributed with $\operatorname{var} \epsilon_i = k_i \mu^2$ (k_i assumed known), in which it is shown that $\hat{\mu}^2 = \sum_i (\hat{\epsilon}_i^2 / k_i)/(N - r)$ is consistent for μ^2 as $N \to \infty$, $\hat{\epsilon}_i$, $i = 1, \ldots, N$, being estimated residuals from weighted least-squares* fit, provided that $E(\epsilon_i^4)/(E(\epsilon_i^2))^2$ is uniformly bounded for all $i \geqslant 1$. This "ratio" condition resembles in nature his conditions in his central limit work; and he uses Markov's inequality* in the deduction.

References

[1] Gnedenko, B. V. (1959). *Istor. Mat. Issled.*, **12**, 135–160 (in Russian).

[2] Heyde, C. C. and Seneta, E. (1977). *I. J. Bienaymé: Statistical Theory Anticipated.* Springer-Verlag, New York.

[3] Krylov, A. N. (1919). *Ross. Akad. Nauk*, **13**, 6th Ser., 389–394 (in Russian). (Obituary.)

[4] Liapunov, A. M. (1901). Nouvelle forme du théorème sur la limite des probabilités. *Mém. Acad. Imp. Sci. St. Pétersbourg*, **12**(5), 1–24. [Russian translation in refs. 5 (pp. 219–250) and 6, (pp. 157–176).]

[5] Liapunov, A. M. (1948). *Izbrannie Trudy*. ANSSSR, Moscow. (Selected works, in Russian, with extensive commentaries and bibliographical material. Contains a biographical sketch and survey of Liapunov's nonprobabilistic work, based on refs. 3 and 8, by Smirnov, with two photographs.)

[6] Liapunov, A. M. (1954). *Sobranie Sochineniy*, Vol. 1. ANSSSR, Moscow. (The first of four volumes of collected works in Russian, published 1954–1965; also contains Liapunov's three earlier memoirs on the central limit problem.)

[7] Liapunov, A. M. (1975). On the formula of Gauss for estimating the precision of observations. *Istor. Mat. Issled.*, **20**, 319–322 (in Russian). (Commentary on this undated and unpublished manuscript discovered by O. B. Sheynin is given by him on pp. 323–328.)

[8] Steklov, V. A. (1919). *Izv. Ross Akad. Nauk.*, **13**, 6th Ser., 367–388 (in Russian). (Obituary.)

Further Reading

See the following work, as well as the references just given, for more information on the life and work of Liapunov.

Adams, W. J. (1974). *The Life and Times of the Central Limit Theorem*. Kaedmon, New York. (Pages 84–98 contain a biographical sketch and a survey of the approach and results of Liapunov's four memoirs on the central limit problem, following ref. 1. There are also two photographs of Liapunov.)

(CONSISTENCY
LIAPUNOV'S INEQUALITY
LIAPUNOV'S THEOREM
LIMIT THEOREM, CENTRAL
LINEAR MODEL
MARKOV'S INEQUALITY
RATES OF CONVERGENCE)

E. SENETA

LIAPUNOV'S INEQUALITY

A particular case of Jensen's inequality*. If $0 < s < t$, then for any random variable X (with $E|X| < \infty$),

$$\left(E|X|^s\right)^{1/s} \leqslant \left(E|X|^t\right)^{1/t}.$$

This inequality implies the following chain of inequalities for absolute moments for a random variable X:

$$E|X| \leqslant \left(E|X|^2\right)^{1/2} \leqslant \cdots \leqslant \left(E|X|^n\right)^{1/n}.$$

An elegant proof was given by Good [1]. For additional information, see Uspensky [2].

References

[1] Good, I. J. (1950). *Proc. Camb. Philos. Soc. Math.-Phys. Sci.*, **96**, 353.

[2] Uspensky, J. V. (1937). *Introduction to Mathematical Probability*. McGraw-Hill, New York.

(HOLDER'S INEQUALITY
JENSEN'S INEQUALITY
LIAPUNOV, A. M.
LINDEBERG–FELLER THEOREM
MINKOWSKI INEQUALITY)

LIAPUNOV'S THEOREM *See* LIMIT THEOREM, CENTRAL

LIEBERMAN–ROSS PROCEDURE

A procedure for obtaining upper confidence bounds on the reliability of *series systems* developed by Lieberman and Ross [1]. More precisely, let statistically independent components be connected in series where each component type ($i = 1, \ldots, N$) has an exponential distribution of failures with survival (reliability) function

$$R_i(T_0) = \Pr(T > T_0) = e^{-\lambda_i T_0},$$

$$i = 1, \ldots, N.$$

The system reliability is then given by

$$R(T_0) = e^{-\lambda T_0}, \qquad \lambda = \sum_{i=1}^{N} \lambda_i.$$

To obtain an upper bound on λ, Lieberman and Ross [1] propose to observe individual failure times T_{ij} of k_i items of each type of component, $i = 1, 2, \ldots, N$ and $j = 1, 2, \ldots, k_i$. The cumulative time U at which one would have first exhausted all components of one type had they been tested *sequentially* is given by

$$U = \min_i \left\{ \sum_{j=1}^{k_i} T_{ij} \right\}.$$

Define for each type of component a number n_i as

$$n_i = \text{largest} \left\{ j \leqslant k_i \mid \sum_{l=1}^{j} T_{il} \leqslant U \right\}$$

and let $K = \sum n_i$. It was shown by Lieberman and Ross that $2\lambda U$ possesses a chi-squared distribution with $2K$ degrees of freedom. This result allows one to estimate upper bounds on λ at any given confidence interval. (See Schoenstadt [3] for a detailed discussion of this method and its comparison with an alternative procedure for obtaining confidence bounds on λ devised by Mann and Grubbs [2].)

References

[1] Lieberman, J. and Ross, S. (1971). *J. Amer. Statist. Ass.*, **66**, 837–840.

[2] Mann, N. R. and Grubbs, F. E. (1974). *Technometrics*, **16**, 335–347.

[3] Schoenstadt, A. L. (1980). *J. Amer. Statist. Ass.*, **75**, 212–216.

(COHERENT STRUCTURES
EXPONENTIAL DISTRIBUTIONS
MANN–GRUBBS PROCEDURE
RELIABILITY)

LIFE INSURANCE

Essentially, an undertaking to pay an agreed amount on death of a specified individual, in return for payment of a single premium or for payments of premiums at specified times (usually equally spaced in time). Calculation of appropriate premiums is based, *inter alia*, on estimated mortality* rates, which are, in turn, predictions based on statistical data. Statistics on financial conditions affecting rates of interest are also relevant, although prediction here is usually much less accurate than it is for mortality.

(ACTUARIAL STATISTICS, LIFE)

LIFE TABLES

John Graunt, generally regarded as the founder of demography* and indeed of statistics itself, also developed the concept of a steady diminution by death, at successive ages, in the number surviving out of a group of infants all born at the same moment of time. To represent this trend, he wrote down from general observation a series of figures declining from 100 at age 0 to 64 at age 7 to 25 at age 27, to 6 at age 57, and so on. During the eighteenth century, various attempts were made to construct such a "life table" more precisely from actual experience. Most of these were based only on records of the numbers of deaths at each age, a method that can give accurate results only when the population is constant in size. Later, the correct procedure was found with the aid of counts of the population; by early in the nineteenth century it had been used to produce results for some national populations and for certain select groups of people. Table 1 shows specimen entries based on a work published in 1849, in which the data relate to women in England and Wales around the year 1841; besides the numbers living it shows the numbers attaining exact age x (l_x), the numbers dying between exact age x and exact age $x + 1$ (d_x), and the probability of death (or "mortality rate") (q_x), which is the result of dividing d_x by l_x. The table is constructed on the basis of an arbitrary number (4873) at the lowest age (0). The probability of death at that age, having been assessed from actual population and death statistics, was entered into the

Table 1. Extract from a Life Table

Age, x	Number Living, l_x	Number Dying before $(x + 1)$, d_x	Probability of Death, q_x
0	4873	646	0.133
1	4227	255	0.060
2	3972	134	0.034
3	3838	90	0.023
4	3748	66	0.018
10	3505	21	0.006
30	2986	30	0.010
50	2324	36	0.015
70	1271	77	0.061
90	66	18	0.273

right-hand column and then multiplied by 4873 to obtain the number of life-table deaths (i.e., 646), shown under the heading d_x. Deduction of this 646 from 4873 gave the number reaching age 1, namely 4227. A similar procedure was followed at each subsequent age. From such figures it is easy to compute the chances of survival from one age to another, and also the average length of time lived from any age onward; this average is called the "expectation of life."

Life tables reflect the characteristics of mortality* and so are different for men and women, for different races, nationalities, occupations, and periods of time. Their appropriateness to any group depends on the correct calculation of the probabilities of death, a process that requires strict attention to detail. In broad principle, these probabilities can be assessed in one of two ways:

1. By counting the numbers of deaths at each age in a given period and dividing them by the numbers of people alive. This can be estimated from a census taken at the middle of the period, provided that the numbers enumerated (usually at age on the last birthday) are suitably adjusted; the denominator should represent the numbers with an average age of x precisely; or

2. By keeping a record of all the people in a group over a period and from this ascertaining at each age their numbers at the beginning and end, the numbers entering and leaving the group in that period, and the numbers dying.

GENERATION AND SECULAR TABLES

In theory, the life history of a group of babies all born at one time could be portrayed in a table. In practice, this is not possible because of the difficulty of keeping in touch with all the people concerned; even if such an exercise could be carried out, it could not be completed in less than 100 years, at the end of which the results would, because of social and economic developments affecting mortality, be of historical interest alone. Most users require life tables which show the experience of a hypothetical group of people deemed to experience mortality at present-day rates throughout the whole of their lives; or similar tables based on the mortality rates expected to occur in the future. Nevertheless, by the use of census and registration records, which in many countries today, extend back over a century, medical and other statisticians have constructed for research purposes good approximations to historical life tables, and these are usually called "generation tables" (sometimes called "cohort tables"), in contrast to "secular" ones based on the mortality of a shorter period of time.

Both sorts of life tables have been used to test theories about possible mathematical relationships between the age x and the probability of death at age x. The best known formula for this relationship is that of Benjamin Gompertz, who in 1825 suggested that the "inability to withstand death" grows in geometrical progression. More complex formulas have subsequently been proposed, but none is of universal application. Formulas and other methods are, however, used in order to "graduate" mortality rates, that is, to eliminate casual irregularities in the run of the probabilities of death as the age increases; smoothness is required for life assurance and other purposes.

Life tables that depend mainly on age but also partly on the duration of time are sometimes constructed. For instance, applicants for life assurance are accepted only if they can be shown to be healthy. Those insured are thus less likely to die than the average person, but as time passes after the health test, their mortality tends to fall back toward the average; in other words, the degree of selection applied by the test wears off. Tables reflecting this concept are called "select" tables and they show the numbers living and dying classified by age and also by duration of assurance (the durations separated out usually extend no longer than five years).

EXTENSIONS OF THE LIFE-TABLE PRINCIPLE

The field of application of life tables is not restricted to human mortality and survival. Similar statements can be prepared, where the necessary data exist, for animals, for fish, or even for insects. Histories of inanimate objects, such as motor cars or railway engines, can be exhibited to show the numbers remaining in use and ceasing to be serviceable. Claims for fire or accidental damage insurance can be classified into those still outstanding and those settled at successive periods. The time scale in non-human applications will clearly vary according to the subject. It is not even essential

that the scale should be one of time; for example, married couples can be tabulated to show those with $0, 1, 2, \ldots$ children, the numbers having an nth child forming the decrement column. Some groups can have more than one decrement; thus bachelors can leave their category of unmarried men either by death or marriage, and employed persons can cease work through unemployment, ill health, or final retirement. All such decrements can be combined into one column, corresponding with the deaths in a conventional life table; but they may also be shown in separate columns for each of their component parts, thus forming what is called a "multiple decrement" table*—a device much used in the field of finances of social security and pension schemes.

The numbers living, shown in a life table, can be regarded as a population classified by age. It is indeed what a real population could be always if there were a constant number of births each year and if the mortality rate at each age remained at the specified value (conditions that have never applied in practice). It is sometimes called a "stationary population, and it represents a special case of a "stable" population (for a definition, *see* DEMOGRAPHY) whose growth rate is zero. Conversely, a stable population with a nonzero growth rate represents a development of the life table in which the intake at the lowest age is not fixed but rises (or falls) steadily. A further adaptation is to introduce increments into the life table at ages other than the minimum. This may be useful in the study of hierarchical structures such as occur in the armed forces, the numbers of persons occupying a particular rank being augmented at various ages by promotions from below as well as being depleted by promotions to higher ranks.

PRACTICAL USES OF LIFE TABLES

Life tables are the basis for a good deal of actuarial mathematics, concerned as it is with the payment of sums of money to those dying, at whatever future age, or of annuities to those who live on. Until the eighteenth century, without life tables, insurance could be only very limited in scope; moreover, the British government issued annuities without realizing that they were creating a large loss. Thereafter, however, it became possible to correct this loss and for insurers to issue, for example, whole-life assurance by level annual premiums. This type of provision was first available in Britain, and its worth was officially recognized in 1798 when that part of a person's income that was paid as a premium was exempted from income tax. Toward the end of the nineteenth century, the award of pensions on retirement began to be extended, from payments to public servants out of state revenues, to the field of private employment. To secure private pensioners' rights it became necessary to set up trust funds, the financial adequacy of which for their purposes is assessed by qualified actuaries. Subsequently, similar provision was secured through life assurance companies, already advised by actuaries. Multiple-decrement tables play a large part in their calculations, because the form and amount of benefit depend not only on the time at which it is claimed but also the cause of the claim—whether change of employment, retirement of one type or other, or death (on the occasion of which a widow's pension may become payable). A similar diversity of benefits is an important feature of national social security systems; although these differ from life assurance and pension funds in important ways, the life-table principle plays an important part in planning ahead, by means of forecasts of the likely numbers of contributors and beneficiaries.

The basis of forward estimates for social security schemes is usually a "projection" of the population (*see* DEMOGRAPHY *and* POPULATION PROJECTIONS). In the 1920s Raymond Pearl used algebraic formulas for this purpose, but a more useful method was developed by Arthur Bowley, who employed a life-table approach based on age. Subsequently, length of time married has sometimes been found valuable as a criterion for the estimation of future births, by the application of fertility rates to the numbers of couples still together after 1, 2, 3, . . . years

of marriage, arrived at by multiple-decrement techniques.

A recent application relates to the long-term deployment of personnel in a public corporation or large firm. The number of vacancies likely to occur through wastage are estimated, and then it is calculated to what extent these can be filled by promotion within the organization coupled with recruitment at the lowest level. It may then transpire that there is a need for special staff training; otherwise, recruitment of skilled personnel is likely to be required at some point in the future.

There are still many areas of the world for which it is not possible to construct valid life tables because of inadequacy or inaccuracy of death registration, or because a recent census was not successful. To mitigate this disadvantage, studies have been made on an international scale of the range of variation of the characteristics of all the reliable tables. From these, sets of "model" life tables have been developed, and from such evidence as is available for any area it is possible to assign one of the set as being the most relevant to that area. In Fig. 1, the shape of

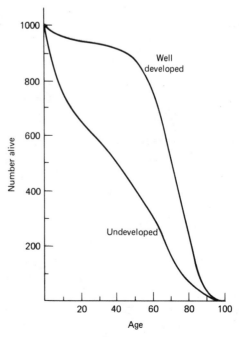

Figure 1

the "number living" column is contrasted for (1) a modern, economically well developed country, and (2) a relatively poor area that is just beginning to modernize.

Bibliography

Cox, P. R. (1975). *Life Tables: the Measure of Mortality. Life Tables: (2) Wider Applications.* In *Population Trends*, Vols. 1 and 2. Her Majesty's Stationary Office, London. (An account of the origin, nature, and use of life tables.)

Farr, W. (1841). *Fifth Report of the Registrar General.* London. (This contains a history of early life tables and an account of the construction of one of the first national tables.)

Graunt, J. (1666). *Natural and Political Observations . . . upon the Bills of Mortality.* London. (The original work, containing the first life table; reprinted in *J. Inst. Actuaries*, **90**, 1–61.)

Spiegelman, M. (1968). *Introduction to Demography.* Harvard University Press, Cambridge, Mass. (Like most demographic textbooks, this contains a chapter on life tables, with some useful comments on their construction.)

United Nations (1955). *Age and Sex Patterns of Mortality. Model Life Tables for Underdeveloped Countries.* United Nations, New York. (A valuable tool for those concerned with the countries of the Third World.)

(ACTUARIAL STATISTICS, LIFE
DEMOGRAPHY
GRADUATION
KEYFITZ METHOD OF LIFE-TABLE
    CONSTRUCTION
MANPOWER PLANNING, STATISTICAL
    TECHNIQUES
MATHEMATICAL THEORY OF
    POPULATION
POPULATION PROJECTIONS
STOCHASTIC DEMOGRAPHY
SURVIVAL ANALYSIS)

P. R. Cox

LIFE TESTING

The term "life testing" is used to describe experiments made to collect life-length data. Such data are used to estimate certain parameters, to make predictions, or to make decisions as to whether to accept or to reject a lot (batch) of items. Examples of the parameters of interest are the "mean time to failure," the "failure rate," the "survival function," and the "reliable life." Life testing is routinely undertaken in industrial environments such as the automobile, telecommunications, and electronics industries, in military and defense-related industries, and in biological- and health-related activities such as drug screening and bioassay experiments. In many instances, life testing is also undertaken to satisfy contractual and regulatory requirements.

There is an enormous amount of literature pertaining to this area, among the best sources of which are Proschan and Serfling [28], Mann et al. [22], Barlow and Proschan [3], Barlow et al. [4], Gross and Clark [17], Tsokos and Shimi [40], Bain [2], Elandt-Johnson and Johnson [13], Nelson [27], and Martz and Waller [23].

Since the purpose of a life test is to obtain some form of life-length data, a random sample of items is submitted to test, and the failure information of interest, together with time of withdrawal from test of any unfailed items, is recorded. In a typical situation the items are tested in an environment as nearly as possible identical to the environment in which they are designed to operate, i.e., an environment with *nominal* stress. Such tests are *ordinary life tests*. If the items in question have large times to failure, then an ordinary life test will involve an inordinate amount of test time. Thus it is a common practice to subject the items to a test environment with greater than nominal stress. Such tests are called *accelerated life tests* or *overstress life tests*; in some "military standards" documents, they are also referred to as *environmental tests*. Because of the fact that modern technology has been so successful in creating items having long life, accelerated life tests are now undertaken more often than are ordinary life tests. There are several strategies for performing accelerated life tests, and there are some special difficulties in making meaningful inferences from them. These will be described later.

With both ordinary or accelerated life

tests, it is common that not all the items under test will be observed until failure. That is, some of the items will be withdrawn or removed from the life test. When this happens, the life test is said to be a *censored-sample life test*. A life test without any censoring is said to be a *complete-sample life test*. In industrial life testing, censoring is often undertaken to save on test time or to save on the number of items that are tested until failure. In biological life testing, especially that involving human subjects, censoring is often due to causes that are beyond the control of the experimenter.

A life test (ordinary or accelerated), in which the number of items to be tested is fixed in advance, is called a *fixed-sample life test*. A fixed-sample test could be either a complete-sample test or a censored-sample test, depending on whether there are any removals or withdrawals during the test. Fixed-sample tests are typically used when the goal of life testing is the estimation of unknown parameters. In contrast to fixed-sample tests are the *sequential tests*, for which the number of items to be tested is a random variable. Sequential tests are used when the goal of life testing is to make a decision as to whether a batch of items satisfies or fails to satisfy a specified life requirement, such as the mean time to failure being equal to or greater than a specified value. Sequential tests have the advantage that the expected number of items that are tested is smaller than that required in fixed-sample tests having the same performance characteristics.

CENSORING PATTERNS

As stated earlier, censoring (*see also* CENSORED DATA) is said to occur when unfailed items are withdrawn from the life test. In some instances, the experimenter has control over the amount and the times of censoring; in others, he or she does not. Conceptually, the simplest form of censoring is that under the control of the experimenter, which is known as *type I censoring*.

Type I Censoring

In type I censoring, the life test is begun by observing the times to failure of the failed items among a predetermined number of items, say n, under an ordinary or an accelerated life test. The life test is terminated at a *predetermined* time, say t_0, or at the time of failure of the last of the n items, whichever occurs first. The number of items that fails by time t_0, say s, is a random variable, where $s \leqslant n$, and the lifetime data consist of the s observable times to failure $X_{(1)} \leqslant X_{(2)} \leqslant \cdots \leqslant X_{(s)}$, and the knowledge that $(n - s)$ items survived until time t_0. The random variable $X_{(i)}$ denotes the ith observed time to failure, $i = 1, \ldots, s, s \leqslant n$.

The main advantage of type I censoring is that the test procedure must terminate by time t_0, so that the experimenter knows in advance the time at which the experiment will be completed. A disadvantage of type I censoring is the possibility of observing no failures by time t_0. This poses some problems in obtaining the maximum likelihood estimators of the mean time to failure. There are, however, no difficulties in obtaining a Bayes estimator [22, pp. 399–404]. Thus, in choosing t_0, it is helpful if the experimenter has some prior knowledge about the failure behavior of the items.

Even though type I censoring is conceptually simple, a sample-theory (non-Bayesian) analysis of lifetime data from type I censored life tests poses several difficulties. These have been dealt with by Bartholomew [5] and by Yang and Sirvanci [41]. See also Proschan et al. [31]. A Bayesian analysis is straightforward provided that the user can specify meaningful priors for the parameters of interest.

Proschan et al. [31] have provided a computer program for calculating from type I censored data an upper confidence bound on failure rate λ which is exact for exponential data and conservative for data with decreasing failure rate. This is based on Bartholomew's [5] result applying to exponential data. Mann et al. [33] have used a direct approach to generate tables for determining such confidence bounds as functions of significance level, α, sample size, n, and a quantity that is equal to the maximum likelihood estimate of the exponential mean, $\theta = 1/\lambda$ divided by t_0.

Proschan et al. [31] give the following example of type I censored data. Suppose that $n = 10$ items are put on test and that one can assume nonincreasing failure rate. Failures are observed at times 4, 9, 11, 18, 27, and 38 preceding a fixed time $t_0 = 50$ of censoring. One can compute the maximum likelihood estimate of $\hat{\theta}$ as

$$\hat{\theta} = \tfrac{1}{6}\big[4 + 9 + 11 + 18 + 27 + 38 + 4(50)\big]$$

$$= 51.166.$$

Using the actual failure times, plus $n = 10$, $r = 6$, $t_0 = 50$, and $\alpha = 0.05$ as input, one obtains from the computer program 0.03510 as a conservative upper 95% confidence bound for λ (which is exact for exponential data). For the tables of Schafer et al. [33] one uses as input $\hat{\theta}/t_0 = 1.023$, $n = 10$, and $\alpha = 0.05$ to obtain 0.5714 and 28.5714 as lower 95% confidence bounds on θ/t_0 and θ, respectively, and hence 0.03510 as a 95% upper confidence bound for λ. The latter procedure requires interpolation.

Type II Censoring

In type II censoring, the life test is begun by observing the times to failure of the items failing among a predetermined number of items, say n, under an ordinary or an accelerated life test. The life test is terminated at the time of failure of the rth item, where $1 \leqslant r \leqslant n$, is also *predetermined*. Thus, with type II censoring, the time at which the life test is terminated is a random variable. The lifetime data from this test consist of observable times to failure $X_{(1)} \leqslant X_{(2)} \leqslant \cdots \leqslant X_{(r)}$, together with the knowledge that $(n - r)$ items have survived until time $X_{(r)}$. Since $r \leqslant n$, $X_{(r)} \leqslant X_{(n)}$. A disadvantage of type II censoring is that the experimenter does not know in advance how long it will take to complete the test. Methods for analysis of data from such tests were originally

proposed by Epstein and Sobel [15] and are summarized in Mann et al. [22, pp. 163–174].

Proschan et al. [31] use the data of the preceding section to illustrate the calculation of an upper confidence bound on failure rate λ from type II censored data. Here it is assumed that the life test is terminated at the time of the sixth (rth) failure. A conservative 95% upper confidence bound for λ at time 38 is given by

$$\frac{\chi^2_{0.95}(12)}{2\big[4 + 9 + 11 + 18 + 27 + 38 + 4(38)\big]}$$

$$= \frac{21.026}{518} = 0.0406.$$

where $\chi^2_{0.95}(12)$ is the 95% point of a chi-square distribution with $2r = 12$ degrees of freedom.

Combination of Type I and Type II Censoring

Another form of censoring combines the features of type I and type II censoring. Here the test is terminated either by a *predetermined* time, t_0, or at $X_{(r)}$, the time to failure of the rth item, whichever occurs first; r is also *predetermined*. Techniques for analyzing lifetime data from life tests which have both type I and type II censoring, and when the lifetimes have an exponential distribution are given by Epstein [14]. Fertig and Mann [16] give sampling plans in which the tables allow for type II censoring, but a maximum time on test is specified.

Progressive Censoring

Progressive censoring, which is also known as "multiple censoring" or "hypercensoring," involves elimination from further observation, some (although not all) of the surviving items, at various stages of a life test. Those items that are remaining after each stage of censoring are continued to be observed until failure, or a subsequent stage of censoring.

Suppose that the life test is begun by observing the times to failure of a predetermined number of items, say *n*, under an ordinary or an accelerated life test. Let *r* be the number of items which fail, and let censoring occur progressively in *k* stages at times $T_1 < T_2 < \cdots < T_k$. At the *i*th stage of censoring, r_i items are selected randomly from the survivors at time T_i, and are removed from further observation; thus $n = r + \sum_{i=1}^{k} r_i$. Progressive censoring may be type I, type II, or a mixture of both.

Techniques for analyzing data from progressively censored life tests are given by Cohen [9] and Mann [19, 20].

Random Censoring

This type of censoring usually occurs when a life test is performed on biological subjects, particularly human beings. Here the amounts and the times of censoring are not under the control of the investigator. *Random censoring* refers to the elimination from further observation of the surviving item or items, at *random* points in time, τ_i, $i = 1, \ldots, k$, where $k \leqslant n$, and *n* is the number of items under observation at the start of the life test.

The data from a randomly censored life test consist of the random variables Y_1, $Y_2, \ldots, Y_n$, where

$$Y_i = \min(\tau_i, X_i), \qquad i = 1, \ldots, n,$$

and $X_1, \ldots, X_n$ are the failure times of the *n* items if they are allowed to operate freely without censoring. Associated with each Y_i is an indicator variable δ_i, where $\delta_i = 0$, if the *i*th item is censored, that is, $Y_i = \tau_i$; and $\delta_i = 1$, otherwise, for $i = 1, \ldots, n$. It is common to assume that the τ_i's, $i = 1, \ldots, k$, are a random sample drawn independently of the X_i's, and have an unknown distribution which is different from the distribution of the X_i's.

The model for random censorship was proposed by Efron [12]; methods for the analysis of data from randomly censored life tests are given by Breslow and Crowley [7].

The competing-risk model is a special case of random censoring in which the censoring results because of a mode of failure that

"competes" with the mode of interest. The competing mode might be, for example, various types of accidents that cause the death of cancer patients. The extensive body of literature on competing risks is summarized by Birnbaum [6], David and Moeschberger [10], and Elandt-Johnson and Johnson [13, Part 3].

ACCELERATED LIFE TESTING

Accelerated life testing is performed to shorten the time period of a life test. This is achieved by inducing early failures in the items under test, by allowing them to operate in an environment that is more severe (accelerated) than the normal (*use conditions* or *nominal*) environment. A more severe environment can be created by increasing levels of one or more stresses that constitute the environment, to values which are greater than their usual values. In accelerated life testing, any one of the censoring patterns discussed in the section "Censoring Patterns" can be used. The main problem with inference from accelerated life tests is that statements about the failure behavior of the items at use conditions environment have to be made using life-length data from the more severe environments. One way of approaching this problem is to assume that the life distributions under the various environmental conditions come from the same family of distributions that is specified in advance, but that the parameters of the distributions change with the stresses according to a specified relationship with unknown parameters. Such relationships are known as *time transformation functions*, of which the "power law" is prominent; see Mann et al. [22, p. 425], Nelson [25, 27], Singpurwalla and Al-Khayaal [39], and Mann [21] for some recent contributions. Methods for estimating the parameters of the time transformation function when the unknown family of life distributions cannot be specified but the time transformation function can be specified are given by Shaked et al. [37]; a method for testing hypotheses on the unknown distributions is given by Sethuraman and Singpurwalla [34]. The techniques of the foregoing two papers have been combined by Shaked and Singpurwalla [35] to give a unified semiparametric approach to accelerated life testing. A nonparametric approach for inference from accelerated life tests when neither the time transformation function nor the family of failure distributions can be specified is considered by Proschan and Singpurwalla [29, 30].

There are several means by which accelerated environments can be obtained. All of these involve increasing the values of one or more of the stress levels which constitute the environment from their nominal values to more severe values. In choosing the higher values of the stress levels, care should be taken to ensure that the environment is not so severe as to introduce completely different failure modes. Also, testing under very severe environments will not give much information about the life behavior of the items under the severe environments, because all the items under test will fail instantly, implying a degenerate distribution (possibly at zero) for the failure times.

Fixed-Stress Accelerated Test

The simplest method, both physically and from the point of view of analysis and inference, of obtaining an accelerated environment for conducting a life test is to increase the value of one or more of the stresses and hold them *fixed* over time. Such tests are known as *fixed-stress* tests. The accelerated life test experiment involves choosing several such increased values of the stress, and then performing either a complete-sample or a censored-sample life test at each of the increased values. For example, if the environment consists of a single stress, described by a variable, say V, and if V_0 denotes the use conditions stress, then it is common to consider k values of V, say $V_1 < V_2 < \cdots < V_k$, for $k > 1$, and perform a life test at each of the k values. Note that $V_0 < V_1$. If the environment consists of two stresses, say

V and H, with V_0 and H_0 being the use conditions stresses, it is common to choose k values of V, $V_1 < V_2 < \cdots < V_k$, and l values of H, $H_1 < H_2 < \cdots < H_l$, and perform a total of m life tests, one at each combination of values of V and H, say (V_i, H_j), $0 \leq i \leq k$, $0 \leq j \leq l$, and excluding, if necessary, the value (V_0, H_0); note that $m \leq (k + 1)(l + 1)$. The number of choices for performing accelerated life tests increases as the number of stresses that constitute the environment increases. All the references described above pertain to fixed-stress accelerated tests.

Step-Stress Accelerated Tests

Instead of choosing a high value of the stress and holding it fixed over all time, in *step-stress testing* the higher values of the stress are introduced in stages, so that they form a step function over time. Specifically, consider the single stress environment of the preceding section with $V_0 < V_1 < \cdots < V_k$. Then in step-stress testing, the life test is started off by choosing any value of V, say V_i, $0 \leq i < k$, and observing the lifetimes of say n items under stress V_i, for τ_i units of time, where τ_i is *preselected*. At time τ_i, the stress level is increased to V_{i+j}, $1 \leq j \leq k - 1$, and the lifetimes of the remaining items observed for an additional preselected τ_{i+j} units of time. At time $\tau_i + \tau_{i+j}$, the stress level is again increased to V_{i+j+p}, $1 \leq p \leq k - i - j$, and so on.

Even though step-stress testing is commonly undertaken in practice, there is a limited amount of literature dealing with the analysis of lifetime data from such tests. The papers by DeGroot and Groel [11], Nelson [26], and Shaked and Singpurwalla [36] address the various issues that arise when dealing with data from step-stress tests.

Accelerated Tests with Continuously Increasing Stresses

In step-stress testing, the various higher values of the stresses are introduced in stages, holding the values of the stresses fixed over time within each stage. In tests with *continuously increasing stresses*, the stress increases over time according to some specified functional relationship. A simple version of such a test with a single stress environment may be to allow the stress to increase linearly over time, and observe the life lengths of the failed items. Even though such tests have been discussed in the literature (see Allen [1]) there do not appear to be any methods for analyzing data from such tests. In Figs. 1, 2, and 3, we illustrate the differences between the types of accelerated tests described above.

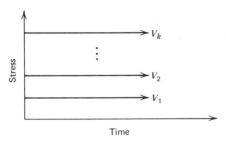

Figure 1   Fixed stress accelerated test.

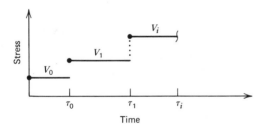

Figure 2   Step-stress accelerated test.

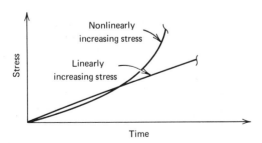

Figure 3   Accelerated test with continuously increasing stresses.

SEQUENTIAL LIFE TESTS

Sequential life tests for the exponential distribution* are so frequently used by agencies of the U.S. government that the underlying procedures have been incorporated (as MIL-STD-781C) into a series of documents called the Military Standards. Military Standards are issued by the Department of Defense, and they specify operating procedures for quality control, life testing, methods of environmental testing, procedures for accepting or rejecting a batch (lot) of items, etc., which must be used by all the agencies of the Department of Defense. Sequential life testing is performed mainly for the purpose of making a decision as to whether to accept or to reject a lot of items on the basis of their life-length characteristics. As a by-product, data from a sequential test can be used for estimation purposes.

In *sequential life testing*, instead of starting the life test by subjecting all the sample items to the test environment, the experimenter subjects only one item to test initially and following this, only one item at a time upon the failure of the preceding item. Two hypotheses concerning the exponential mean θ are specified, H_0: $\theta = \theta_0$ and H_1: $\theta = \theta_1$. At any given time, total test time is T and the total number having been subjected to test is n. Then, two boundaries, with intercepts and common slope all functions of θ_0 and θ_1, determine regions for acceptance of H_0 or of H_1 in the (n, T)-plane. As long as the value of the data point (n, T) stays within these boundaries, the test continues (until a prespecified truncation time).

The key feature of sequential testing is to save on the number of items that are tested by making an early decision on acceptance or rejection if the lot is very good or very bad, a prolonged amount of testing being required only if the lot quality is neither one of these two extremes.

Sequential life testing for the exponential life distribution was first introduced by Epstein and Sobel [15]. Even though MIL-STD-781C clearly specifies that the prescribed sequential procedures apply only when the underlying life distribution is exponential, there does remain the possibility that these could be misused. In a recent paper, Harter and Moore [18] point out the consequences of using the MIL-STD-781C procedures when the underlying life distribution is Weibull. Montagne and Singpurwalla [24] generalize Harter and Moore's results to the case in which the underlying life distribution has a monotone failure rate. Sequential life tests for situations in which there is some prior knowledge about the scale parameter of the exponential distribution have been considered by Schafer and Singpurwalla [32]. Sequential life-test procedures for other life distributions, such as the Weibull or the gamma, are not available.

Once a sequential life test is terminated, with a decision being made as to whether to accept or to reject the batch, the observed life lengths can be used to estimate the parameters of the life distribution in question. Methods for doing this have been discussed by Bryant and Schmee [8] and by Siegmund [38].

References

[1] Allen, W. R. (1959). *J. Operat. Res. Soc. Amer.*, 303–312.

[2] Bain, L. J. (1978). *Statistical Analysis of Reliability and Life Testing: Theory and Methods*, Dekker, New York.

[3] Barlow, R. E. and Singpurwalla, N. D. (1975). *Statistical Theory of Reliability and Life Testing*. Holt, Rinehart and Winston, New York.

[4] Barlow, R. E., Fussell, J. B., and Singpurwalla, N. D. (1975). *Reliability and Fault-Tree Analysis*. Ser. Appl. Math. SIAM, Philadelphia.

[5] Bartholomew, D. J. (1963). *Technometrics*, **5**, 361–374.

[6] Birnbaum, Z. W. (1979). *On the Mathematics of Competing Risks*. DHEW Publ. No. (PHS) 70-1351. Superintendent of Documents, U.S. Government Printing Office, Washington, D.C.

[7] Breslow, N. and Crowley, J. (1974). *Ann. Statist.*, **2**, 437–453.

[8] Bryant, C. and Schmee, J. (1979). *Technometrics*, **21**, 33–42.

[9] Cohen, A. C. (1963). *Technometrics*, **5**, 327–339.

[10] David, H. A. and Moeschberger, M. L. (1978). *The Theory of Competing Risks*. Griffin's Statist. Monogr. Courses No. 30. Charles Griffin, London.

[11] DeGroot, M. H. and Goel, P. K. (1979). *Naval Res. Logist. Quart.*, **26**, 223–235.

[12] Efron, B. (1967). *Proc. 5th Berkeley Symp. Math. Statist. Prob.*, Vol. 4. University of California Press, Berkeley, Calif., pp. 831–853.

[13] Elandt-Johnson, R. C. and Johnson, N. L. (1980). *Survival Models and Data Analysis*. Wiley, New York.

[14] Epstein, B. (1954). *Ann. Math. Statist.*, **25**, 555–564.

[15] Epstein, B. and Sobel, M. (1955). *Ann. Math. Statist.*, **26**, 82–93.

[16] Fertig, K. W. and Mann, N. R. (1980). *Technometrics*, **22**, 165–177.

[17] Gross, A. J. and Clark, V. A. (1975). *Survival Distributions: Reliability Applications in the Biomedical Sciences*. Wiley, New York.

[18] Harter, H. L. and Moore, A. H. (1976). *IEEE Trans. Rel.*, **R-25**, 100–104.

[19] Mann, N. R. (1969). *J. Amer. Statist. Ass.*, **64**, 306–315.

[20] Mann, N. R. (1971). *Technometrics*, **13**, 521–533.

[21] Mann, N. R. (1978). *Commun. Statist. A*, **7**, 107–111.

[22] Mann, N. R., Schafer, R. E., and Singpurwalla, N. D. (1974). *Methods for Statistical Analysis of Reliability and Life Data*. Wiley, New York.

[23] Martz, H. D. and Waller, R. A. (1982). *Bayesian Reliability Analysis*. Wiley, New York.

[24] Montagne, E. and Singpurwalla, N. D. (1982). On the Robustness of Exponential Sequential Life Testing Procedures. *Tech. Paper*, GWU/IRRA/TR-82/2/. The George Washington University, Washington, D.C.

[25] Nelson, W. B. (1975). *IEEE Trans. Rel.*, **R-24**, 103–107.

[26] Nelson, W. B. (1980). *IEEE Trans. Rel.*, **R-29**, 103–108.

[27] Nelson, W. B. (1981). *Life Data Analysis*. Wiley, New York.

[28] Proschan, F. and Serfling, R. J. (1974). *Reliability and Biometry*. Ser. Appl. Math. SIAM, Philadelphia.

[29] Proschan, F. and Singpurwalla, N. D. (1979). In *Optimization in Statistics*, J. Rustagi, ed. Academic Press, New York, pp. 385–401.

[30] Proschan, F. and Singpurwalla, N. D. (1980). *IEEE Trans. Rel.*, **R-29**, 98–102.

[31] Proschan, F., Barlow, R. E., Madansky, A., and Scheuer, E. M. (1968). *Technometrics*, **10**, 51–62.

[32] Schafer, R. E. and Singpurwalla, N. D. (1970). *Naval Res. Logist. Quart.*, **17**, 55–67.

[33] Mann, N. R., Schafer, R. E., and Harr, M. C. In *Survival Analysis*, Crowley, J. and Johnson, R. A., Eds. IMS Lecture Notes—Monograph Series.

[34] Sethuraman, J. and Singpurwalla, N. D. (1982). *J. Amer. Statist. Ass.*, **77**, 204–208.

[35] Shaked, M. and Singpurwalla, N. D. (1982). *IEEE Trans. Rel.*, **R-31**, 69–74.

[36] Shaked, M. and Singpurwalla, N. D. (1983). Accelerated Life Testing—Nonparametric Models for Step-Stress Testing. *J. Statist. Plan. Inf.* (to appear).

[37] Shaked, M., Zimmer, W. J., and Ball, C. A. (1979). *J. Amer. Statist. Ass.*, **74**, 694–699.

[38] Siegmund, D. (1979). *Naval Res. Logist. Quart.*, **26**, 57–67.

[39] Singpurwalla, N. D. and Al-Khayal, F. A. (1977). In *The Theory and Applications of Reliability*, Vol. 2, C. P. Tsokos and I. N. Shimi, eds. Academic Press, New York, pp. 381–399.

[40] Tsokos, C. P. and Shimi, I. N., eds. (1977). *The Theory and Applications of Reliability with Emphasis on Bayesian and Nonparametric Methods*, 2 vols. Academic Press, New York.

[41] Yang, G. and Sirvanci, M. (1977). *J. Amer. Statist. Ass.*, **72**, 444–447.

Acknowledgment

N.R.M.'s contribution was supported by the Office of Naval Research, Contract N00014-76-C-0723, Project NR 047-204. N.D.S.'s work was supported by the Office of Naval Research, Contract N00014-77-C-0263, Project NR 042-372; and the Army Research Office, Grant DAAG 29-80-C-0067, with George Washington University.

(ACCEPTANCE SAMPLING
CENSORED DATA
COMPETING RISKS
ESTIMATION
PROGRESSIVE CENSORING: DESIGN
PROGRESSIVE CENSORING: ANALYSIS
RELIABILITY
SEQUENTIAL ANALYSIS
SURVIVAL ANALYSIS
WEIBULL DISTRIBUTION)

NANCY R. MANN
NOZER D. SINGPURWALLA

LIKELIHOOD

The technical use of the term *likelihood* is due to R. A. Fisher, who introduced it [2] in connection with the estimation of a correlation coefficient ρ from a sample value r, with these words:

> Bayes (1763 [1]) attempted to find, by observing a sample, the actual *probability* that the population value lay in any given range. . . . Such a problem is indeterminate without knowing the statistical mechanism under which different values of ρ come into existence. . . . What we can find from a sample is the *likelihood* of any particular value of ρ, if we define the likelihood as a quantity proportional to the probability that, from a population having that particular value of ρ, a sample having the observed value r, should be obtained. So defined, probability and likelihood are quantities of an entirely different nature. . . .

Since likelihood is used in connection with statistical inference from real data, and since the set S over which any real data x can vary is necessarily finite, it is enough to define likelihood as follows: If the range of an observable x is a finite set S, and the range of an unknown parameter θ is Ω and if there is a known function $f: S x \Omega \to R$ such that when the true value of the parameter is θ_0, the probability of observing $x = x_0$ is $f(x_0, \theta_0)$, then the likelihood function of θ, given $x = x_0$, is $L(\theta; x_0) = f(x_0, \theta)$. Only ratios of likelihoods have meaning; $L(\theta; x_0)/L(\theta'; x_0)$ measures the weight of evidence for θ as against θ' provided by the observation $x = x_0$. Thus $L(\theta; x_0)$ can be multiplied by any positive bounded function of x_0 without changing its meaning, and two likelihood functions whose ratio is constant on Ω are regarded as equivalent. Since $f(x, \theta)$ is a probability, it is bounded, so that, for every x_0, $\sup_\theta L(\theta; x_0) = L(\hat{\theta}; x_0)$ exists and is bounded, and $L(\theta; x_0)/L(\hat{\theta}; x_0)$ is called the relative likelihood function of θ, given x_0. $\hat{\theta}$ is a function of x whose value may not belong to Ω, in which case a suitable extension of the range of $f(\cdot, \cdot)$ is required; when $\hat{\theta}(x_0)$ does belong to Ω and is unique it is called the maximum likelihood estimate of θ, given x_0.

Representation of observations by continuous real variables involves an approximate limiting process. Proper attention to this limiting process is necessary if the associated probability density function is to replace the function $f(x,\theta)$ of the preceding paragraph. For example, if x is taken to be normally distributed with mean θ_1 and standard deviation θ_2, with $\theta = (\theta_1, \theta_2)$ ranging over the closed upper half-plane, simple use of the probability density will suggest that the likelihood

$$L(\theta; x) = \left(\theta_2\sqrt{2\pi}\right)^{-1} \exp\left[-\tfrac{1}{2}\{(\theta_1 - x)/\theta_2\}^2\right]$$

has a complicated singularity at $(x, 0)$. This disappears when it is borne in mind that the distribution of x must in reality be discrete.

The meaning to be attached to a statement such as "$L(\theta; x_0)/L(\theta'; x_0) = 4$" can be explained in various ways. The most direct interpretation involves a "reference experiment" such as the following. An urn contains three similar balls, either (a) 2 black, 1 white, or (b) 2 white, 1 black. Two drawings are made with replacement, and both give a white ball. The evidence from the data x_0 in favor of θ as against θ' is exactly as weighty as the evidence from the two drawings in favor of condition (b) as against condition (a). From the definition of likelihood it follows that if x, y, and z are independent observables, with probability functions $f(x,\theta)$, $g(y,\theta)$, $h(z,\theta)$, and the separate likelihood functions, given $x = x_0$, $y = y_0$, $z = z_0$, are, respectively, $L(\theta; x_0)$, $L(\theta; y_0)$, $L(\theta; z_0)$, then since the joint probability function is $f(x,\theta)g(y,\theta)h(z,\theta)$, the joint likelihood function must be $L(\theta; x_0)$ $L(\theta; y_0)L(\theta; z_0)$. If $L(\theta; x_0)/L(\theta'; x_0) = 4$, while $L(\theta; y_0)/L(\theta'; y_0) = \tfrac{1}{2} = L(\theta; z_0)/L(\theta'; z_0)$, then on the combined data (x_0, y_0, z_0), the values θ and θ' would be equally credible. A likelihood ratio of $4:1$ is equivalent to two independent ratios each of $2:1$. Note here that likelihood measures the weight of evidence provided by the data; we may well have other grounds for preferring θ to θ', or vice versa.

Thus likelihoods on independent sets of data can be meaningfully multiplied. The sharp difference between likelihood and probability is brought out by the fact that likelihoods cannot be meaningfully added. If we wanted to find a likelihood for (θ or θ'), given $x = x_0$, we would need to find the probability of $x = x_0$ when the true value was either θ or θ'. Apart from the trivial cases when $\theta = \theta'$, we have no means of doing this. We might perhaps interpret (θ or θ') as meaning that we are given a definite random mechanism which picks θ with probability π, say, and θ' with probability $1 - \pi$. Given such a mechanism, and the associated special meaning for (θ or θ'), we could calculate the probability of $x = x_0$ as $\pi f(x_0, \theta) + (1 - \pi)f(x_0, \theta')$, giving the likelihood $\pi L(\theta; x_0) + (1 - \pi)L(\theta'; x_0)$—not the sum, but a weighted mean. But any such random mechanism, for any π, could be regarded as an interpretation of (θ or θ'), so that unless we had some specifiable reason for picking out one value of π, the likelihood, given $x = x_0$, of (θ or θ') would have to be regarded as indeterminate. As Fisher says, to speak of the likelihood of (θ or θ') is like speaking of the height of Peter or Paul— we do not know what it is until we specify which of the two is meant.

The fact that the likelihood of a disjunction of exclusive alternative values cannot be determined from the likelihoods of the individual values gives rise to the principal difficulty in using likelihood for inferences. For we often wish to make an inference concerning a function of θ, say $\xi(\theta)$, when nothing is known about the functional complement $\eta(\theta)$ [i.e., the function such that the transformation $\theta \to (\xi(\theta), \eta(\theta))$ is bijective]. If the likelihood $L(\theta; x_0)$ happens to factorize,

$$L(\theta; x_0) = L(\xi(\theta); x_0)L(\eta(\theta); x_0),$$

then the first factor can serve as a likelihood function for $\xi(\theta)$. Such cases are rare, although more commonly an approximate factorization is possible, giving an approximate likelihood function for $\xi(\theta)$. When the likelihood for θ does not factorize in this convenient way it is, strictly, not possible to make fully efficient inferences about ξ without reference to the value of η: further information that may come to hand concerning η may affect our inference about θ. It may happen, however, that we can find functions (y, z) of x such that $x \rightarrow (y(x), z(x))$ is a bijection and either (a) the marginal distribution of y depends only on ξ or (b) the conditional distribution of y, given the observed z, depends only on ξ. The likelihood function for ξ derived from the distribution of y is called in case (a) a marginal likelihood for ξ and in case (b) a conditional likelihood. It is reasonable to use such modified likelihoods provided that the information about ξ provided by z can safely be ignored.

An example arises in the case of the following 2×2 table:

Group	A	not-A	Total
I	a	b	m
II	c	d	n
Total	r	s	N

where the model is specified by $\theta = (\xi, \eta)$ and the probability of the data is taken to be

$$(m!/a!b!)\,p_1^a q_1^b \cdot (n!/c!d!)\,p_2^c q_2^d$$

and

$$\ln(p_1/q_1) = \eta + \tfrac{1}{2}\xi, \qquad \ln(p_2/q_2) = \eta - \tfrac{1}{2}\xi.$$

Unless m and n and r are exceedingly small, the distribution of $r = a + c$ depends mainly on η, while the conditional distribution of a, given r, depends strongly on ξ. It is reasonable, under these conditions, to use the conditional likelihood for ξ based on this conditional distribution, i.e.,

$$L(\xi; a, r) = \exp(a\xi/a!\,b!\,c!\,d!)$$

$$\Big/ \sum_u \exp[u\xi/u!\,(m-u)!\,(r-u)!$$

$$\times (n - r + u)!\,]$$

the summation in the denominator being taken over all values of u for which the factorial arguments are nonnegative.

In the theory of testing hypotheses as developed by Neyman and Pearson (*see* HYPOTHESIS TESTING), likelihood plays a key role. If hypothesis $H_0\colon \theta = \theta_0$ is tested against alternatives $H(\theta)$ using the critical function $C(x)$ ($= 1$ if the hypothesis is to be rejected, $= 0$ if not), then, as noted by Fisher, the (marginal) likelihood function for θ, given only that $C(x) = 1$, is the power function* of the test procedure. The Neyman–Pearson lemma* asserts that for maximum power at θ, subject to fixed "size" at θ_0, the function $C(x)$ must be a monotone function of the likelihood ratio $L(\theta; x)/L(\theta_0; x)$. And as noted by Pitman, in a sequence of hypothesis-testing situations, testing $\theta_i = \theta_{i0}$ against $\theta_i = \theta_{i1}$, $i = 1, 2, 3, \ldots$, if we wish to minimize the upper bound for the long-run frequency of errors of the second kind* subject to a fixed upper bound for the long-run frequency for errors of the first kind,

$$\bar{\alpha} = \lim_{n \to \infty} \Big(\sum_i \alpha_i\Big)/n,$$

then we should not make α_i independent of i, but rather make the critical likelihood ratio $L(\theta_i; x_i)/L(\theta_{0i}; x_i)$ independent of i.

The likelihood function statistic $\mathscr{L}(x)$ is defined as the function of the observable x whose value at x_0 is the equivalence class to which $L(\theta; x_0)$ belongs, equivalence being defined as above—i.e., two functions L and L' are regarded as equivalent if their ratio is constant on the parameter space Ω. Thus possible observations $x, x', x'', \ldots$ give the same value to $\mathscr{L}(x)$ if and only if, identically in θ,

$$L(\theta, x)/a(x) = L(\theta; x')/a(x')$$

$$= L(\theta; x'')/a(x'') = \cdots,$$

i.e., if and only if

$$f(x, \theta)/a(x) = f(x', \theta)/a(x')$$

$$= f(x'', \theta)/a(x'') = \cdots,$$

which implies that the conditional probability of the observation x, given the value of the likelihood function statistic,

$$f(x,\theta) \Big/ \sum_{x'} f(x',\theta) = a(x) \Big/ \sum_{x'} a(x'),$$

does not involve θ. Here the summation is over all x' giving the same value of $\mathscr{L}(x')$. Since the conditional distribution of x, given the value of $\mathscr{L}(x)$, does not depend on θ, it follows that $\mathscr{L}(x)$ is a sufficient statistic* for θ. The argument reverses, showing that $\mathscr{L}(x)$ must take the same value for any two observations x, x' giving the same value to any sufficient statistic. Thus $\mathscr{L}$ is minimal sufficient. Any statistic from which $\mathscr{L}$ (together with its distribution) can be determined will be sufficient.

For example, if the probability density function for a single observation x_i is

$$f(x_i,\theta) = (1 + x_i\theta)/2, \quad -1 < \theta, x_i < +1$$

and x denotes n independent observations $(x_i, i = 1, 2, \ldots, n)$, then

$$L(\theta; x) = \prod_i (1 + x_i\theta)/2$$

and $\mathscr{L}(x)$ is the class of polynomials $P(\theta)$ (with positive leading coefficient) whose roots are $\theta = -1/x_i$. Thus the full set of order statistics, $x_{(1)} \leqslant x_{(2)} \leqslant \cdots \leqslant x_{(n)}$, which determine and are determined by this set of polynomials, is a minimal sufficient statistic in this case.

It is a widely accepted principle in statistical inference that two possible observations x and x' from the same experiment which give the same value to a statistic which is sufficient for θ should give the same inference for θ. Thus two observations from the same experiment giving the same value for $\mathscr{L}$ should give the same inference. This is the *restricted likelihood principle*. The (general) likelihood principle (LP) asserts that the same holds even if x and x' are results of two distinct experiments. Birnbaum showed that a widely accepted conditionality principle (CP), together with a "mathematical equivalence" principle (MP) would imply LP. The mathematical equivalence principle asserts that if there are two experiments $(S,$ $\Omega, f(x,\theta))$ and $(S', \Omega, f'(x',\theta))$, with the same parameter space, and there is a bijection T from S to S' such that $f'(T(x),\theta)$ $= f(x,\theta)$, identically, then the inference from x about θ in the first experiment should be the same as the inference from $T(x)$ in the second experiment. Although CP is widely (although not universally) accepted, LP is not; criticisms of Birnbaum's derivation of LP have been directed mainly at MP. It has been pointed out, for example, that if S possesses an order structure which is not preserved by T, then there may be grounds for drawing a conclusion from x which would not follow from $T(x)$.

In Bayesian inference*, where a prior probability function $\pi(\theta)$ is assumed given for θ, the posterior probability function, given $x = x_0$ is

$$P(\theta \mid x_0) = K(x_0) f(x_0, \theta) \pi(\theta),$$

where $K(x_0)$ is a normalizer, determined so that $P(\theta \mid x_0)$ integrates with respect to θ to the value 1. Evidently, $K(x_0) f(x_0,\theta)$ belongs to the equivalence class $\mathscr{L}(x_0)$, so that $\mathscr{L}(x_0)$ determines $P(\theta \mid x_0)$ once π is given. In this sense the likelihood principle is an immediate consequence of the Bayesian approach, unless π is made to depend on the kind of experiment performed.

Up to this point this article has been concerned with the likelihood as a function of θ, taken as a whole. In the early development of the theory of likelihood, Fisher laid great stress on the value of θ which maximizes $L(\theta; x_0)$, the maximum likelihood* estimate $\hat{\theta}(x_0)$ of θ, which he initially called the "optimum" value for θ, given the observations x_0. It is obvious from its definition that, as compared with any other single value θ' for θ, $\hat{\theta}$ is better supported by the data x_0; but this tells us nothing about the frequency with which the true value of θ will be at or near $\hat{\theta}$. For example, if we observe $x = x_0$, with

$$x = \theta + e$$

and $\Pr(e = 0) = 0.001$, while $\Pr(e = r) = 0.000001$ for $r = 1, 2, \ldots, 999,000$, $\hat{\theta} = x_0$ will in many senses be the best choice, if a choice of a single value for θ has to be made; but the probability that this choice is incorrect is 0.999. For information about the frequency with which a collection of possible θ values will include the true value, we need to have properties of $\hat{\theta}$ in terms of probability rather than simply likelihood. To derive such properties we need to assume more than is specified in the most general specification of S, Ω, and f.

Results in this direction are concerned with frequency properties of the likelihood or score function

$$\text{Sc}(\theta; x_0) = \partial \log L(\theta; x_0)/\partial \theta,$$

and are subject to regularity conditions, such as differentiability of $\text{Sc}(\theta; x_0)$ for every x_0. The maximum likelihood estimate is usually obtained as a root $\hat{\theta}$ of the estimating equation $\text{Sc}(\theta; x_0) = 0$. This estimating equation is unbiased, in the sense that

$$E_\theta\big[\text{Sc}(\theta; x)\big] = 0,$$

exactly, for all sample sizes. Further, among unbiased estimating equations it is optimal in the sense that if $g(\theta; x) = 0$ is any unbiased estimating equation, standardized so that $E_\theta \, \partial g/\partial \theta = E_\theta \, \partial \text{Sc}/\partial \theta$, then var g $\geqslant$ var Sc. The variance var Sc is equal to $-E_\theta(\partial \text{Sc}/\partial \theta) = \mathscr{I}(\theta)$, called the Fisher information*. This will often be near the *observed information*, defined sometimes as $(\partial \text{Sc}/\partial \theta)$, with θ equal to its true value, and sometimes as $(\partial \text{Sc}/\partial \theta)_{\theta = \hat{\theta}}$.

For independent samples the central limit theorem* implies that Sc will have, for sufficiently large sample size n, an approximate normal distribution. If, in addition, it is, or can be made, approximately linear in θ, over the relevant range of nonnegligible probability, then use of Sc will lead approximately to shortest confidence intervals*.

For this reason, Sc is a convenient starting point from which other asymptotically equivalent estimating functions can be derived. One of these is

$$u(\theta, \hat{\theta}) = (\hat{\theta} - \theta)\sqrt{\mathscr{I}(\theta)},$$

where $\mathscr{I}(\theta)$ was defined above. The theorem of maximum likelihood as presented in textbooks proves that u is asymptotically

standard normal. Other asymptotically equivalent functions are

$$\text{Sc}(\theta; x)\Big/ \sqrt{\mathscr{I}(\hat{\theta})}\, , \qquad (\hat{\theta} - \theta)\sqrt{\mathscr{I}(\hat{\theta})} \qquad (1)$$

and

$$\text{Sc}(\theta; x)\Big/ \sqrt{I_\theta}\, , \qquad (\hat{\theta} - \theta)\sqrt{I_\theta}\, , \qquad (2)$$

in which the function of $\mathscr{I}(\theta)$ is estimated by $\mathscr{I}(\hat{\theta})$ or by the "observed" information $I_\theta = -\partial^2 \log L(\theta)/\partial \hat{\theta}^2$. Yet another such function uses the likelihood function itself, more particularly, $\pm \sqrt{-2 \log L(\theta)/L(\hat{\theta})}$.

The practical reasons for emphasizing the normal estimating functions is that in finite samples they are not equivalent and have different degrees of accuracy as standard normal approximations. If, however, a parameter $\phi = \phi(\theta)$ can be found that approximately linearizes the score function over its relevant range, or equivalently makes the likelihood $L(\theta)$ approximately normal, then (1) and (2) are all approximately equal and preferable to u and $\text{Sc}/\sqrt{\mathscr{I}(\theta)}$ as standard normal variates. This leads, for example, to a reasonably accurate summary of the data in the form of approximate confidence intervals $\phi = \hat{\phi} \pm K_\alpha/\sqrt{I_\phi}$, where K_α is the standard normal deviate at level α (e.g., 1.96 for a 95% confidence interval). These can, of course, be transformed back to equivalent confidence intervals for the original θ or for any other desired function $g(\theta)$.

Another approximation is concerned with the conditional distribution of $\hat{\theta} - \theta$. For a single parameter curved exponential family, the conditional distribution of $\hat{\theta} - \theta$ is approximately the normalized likelihood function, with a relative error proportional to $1/n$ if θ is chosen so that $\mathscr{I}(\theta) \equiv n$.

The preceding discussion of maximum likelihood was limited to a single scalar parameter. Some of the results can be extended to estimating a scalar parameter in the presence of "nuisance" parameters. Joint maximum likelihood estimation is also possible. An important result is the extension of the likelihood ratio to a vector parameter, whereby $-2 \log \Lambda$ has asymptotically the

$\chi^2_{(n-r)}$ distribution, where Λ is the likelihood ratio $\Lambda = \sup_\omega L(\theta; \theta_0)/\sup_\Omega L(\theta; x_0)$, Ω being n-dimensional and ω, r-dimensional.

Indications of the foregoing results and others are found in Fisher's works. Improvements and extensions continue to appear.

References

[1] Bayes, T. (1763). *Philos. Trans. R. Soc. Lond.*, **53**, 370–418.

[2] Fisher, R. A. (1921). *Metron*, **1**, 3–32.

Bibliography

See the following works, as well as the references just given, for more information on the topic of likelihood.

Barnard, G. A. (1958). *Biometrika*, **45**, 293–295.

Barnard, G. A. (1966). *Proc. 5th Berkeley Symp. Math. Statist. Prob.*, Vol. 1. University of California Press, Berkeley, Calif.

Barnard, G. A., Jenkins, G. M., and Winsten, C. B. (1962). *J. R. Statist. Soc. A*, **125**, 321–372 (with discussion).

Edwards, A. W. F. (1972). *Likelihood*. Cambridge University Press, Cambridge.

Fisher, R. A. (1922). *Philos. Trans. R. Soc. Lond. A*, **222**, 309–368.

Fisher, R. A. (1925). *Proc. Camb. Philos. Soc.*, **22**, 700–725.

Fisher, R. A. (1973). *Statistical Methods and Scientific Inference*, 3rd ed. Hafner Press, New York.

Hinkley, D. V. (1980). *Biometrika*, **67**, 287–292.

Hinkley, D. V. *Canad. J. Statist.*, **8**, 151–163.

Kalbfleisch, J. D. and Sprott, D. A. (1970). *J. R. Statist. Soc. B*, **32**, 175–208.

Sprott, D. A. (1975). *Sankhyā*, **37**, 259–270.

Sprott, D. A. (1980). *Biometrika*, **67**, 515–523.

(BASU THEOREMS
BAYESIAN INFERENCE
ESTIMATION
FIDUCIAL INFERENCE
HYPOTHESIS TESTING
INFERENCE, STATISTICAL
LIKELIHOOD PRINCIPLE
MAXIMUM LIKELIHOOD
SUFFICIENCY)

G. A. BARNARD
D. A. SPROTT

LIKELIHOOD PRINCIPLE

This principle has come into prominence particularly during the last 30 years because of the advocacy by some statisticians (see Barnard [2], Birnbaum [4], and Edwards [9]) of an approach to statistical inference based entirely on the likelihood principle, in supersession of both the frequentist or Neyman–Pearson–Wald* (N-P-W) and the Bayesian approaches. The principle is incompatible with the frequentist approach (see the section "Relation to Bayesian and Frequentist Approaches"). It is, however, consistent with the Bayesian approach and is in fact implied by the latter. It is therefore necessary to distinguish clearly between the "likelihood principle" and the "likelihood approach." All Bayesian statisticians accept the principle since it is an integral part of their methodology, but they do not accept the "likelihood approach," which involves rejection of the general validity of prior probabilities.

LIKELIHOOD PRINCIPLE

Let X denote the set of random variables under observation and x a set of realized values of X. The joint probability distribution of the random variables comprised in X is assumed to be specified by a function $f(x, \theta)$ in which $\theta = \{\theta_1, \theta_2, \ldots, \theta_k\}$ is an unknown parameter. The parameter space Ω is the set of all values that θ may assume. We make the simplifying assumption that all the random variables in X are either discrete, in which case $f(x, \theta)$ gives the probability that X assumes the value x, or that they are all continuous and have a joint probability density $f(x, \theta)$. In either case, the function $f(x, \theta)$ is called the probability density function (PDF). The likelihood function determined by any given outcome x is defined as the function on Ω equal to $cf(x, \theta)$, where c may be assigned any arbitrary positive value.

Two likelihood functions, whether arising from the same experiment or from different experiments E_1, E_2 with a common parameter space, are said to be equivalent if their ratio is positive and independent of Ω for all $\theta \in \Omega$, save at the points, if any, at which both functions vanish.

The likelihood principle (L) asserts that for a given experiment E, the evidential meaning of any outcome x, for inference regarding θ is contained entirely in the likelihood function determined by x. Hence all other features of the experiment, as, e.g., the sample space, are irrelevant.

It follows that if two experiments E_1 and E_2 have PDFs $f(x, \theta)$, $g(y, \theta)$ with a common parameter θ and common parameter space Ω, and if for some particular outcomes x_1 of E_1 and y_1 of E_2,

$$f(x_1, \theta) = h(x_1, y_1) g(y_1, \theta)$$
$$\text{for all} \quad \theta \in \Omega, \quad (1)$$

where $h(x_1, y_1) > 0$, then the outcomes x_1 and y_1 must result in an identical inference regarding θ. Some authors adopt the latter proposition as the formulation of the likelihood principle. This alternative formulation can be shown to be equivalent to the formulation (L).

Proposition (L) is, however, incomplete, as it does not say how the evidential meaning is to be determined from the likelihood function. There is a further part of the principle (Edwards [9] calls this the "law of likelihood") as follows:

Given an experiment E, the evidential support provided by an outcome x to different hypothetical values $\theta_1, \theta_2, \theta_3, \ldots$, of θ is proportional to the "likelihoods" of $\theta_1, \theta_2, \theta_3, \ldots$, i.e., to the values of the likelihood function at $\theta_1, \theta_2, \theta_3, \ldots$.

Proposition (L), together with the above-stated further part, really constitute the total "likelihood principle"; however, it has been customary to designate only part (L) as the likelihood principle.

RELATION TO BAYESIAN AND FREQUENTIST APPROACHES

Let $\pi(\theta)$ be the prior density of θ. Then given the outcome x, the posterior density of θ is, by Bayes' theorem,

$$f(\theta \mid x) = \frac{f(x, \theta) \pi(\theta)}{\int_\Omega f(x, \theta) \pi(\theta) \, d\theta}$$
$$= \frac{l(\theta) \pi(\theta)}{\int_\Omega l(\theta) \pi(\theta) \, d\theta},$$

where $l(\theta) = cf(x, \theta)$ is the likelihood function determined by x. Thus the posterior density, and hence the Bayesian inference, depend on the observed outcome only through the likelihood function. The likelihood principle is thus implied by the Bayesian approach.

On the other hand, the principle flatly contradicts the frequentist approach. For example, consider the criterion of unbiasedness. An estimator T is unbiased for θ if $E_\theta T = \theta$ for all $\theta \in \Omega$. But the computation of the "expectation" $E_\theta T$ involves the sampling distribution of T, which depends on the sample space S and the probability distribution $f(x, \theta)$ on it. And the same is true for every other optimality criterion considered in the frequency approach, as for example the variance of an estimator, type I and type II errors of a test procedure, the inclusion probability of a confidence interval, etc. According to the likelihood principle all such criteria are invalid as they depend on the sample space, which, according to that principle, is irrelevant for inference regarding θ. The frequentist approach is thus incompatible with the likelihood approach.

RELATION TO DECISION THEORY*

According to the statisticians who take this approach, the likelihood principle does not apply when the main object of the investigation is to take some decision as, e.g., in sampling inspections in industry; because in such investigations there arise other considerations, such as loss functions, loss in the long run, etc. According to this school, the likelihood principle applies only when the primary object of the investigation is inferential, i.e., to draw valid inferences from the observed data, as for example in scientific investigations. However, this distinction between inference and decision making is not acceptable to many statisticians, particularly those having the N-P-W approach.

GROUNDS IN FAVOR OF THE LIKELIHOOD APPROACH

The objections of the statisticians who advocate the likelihood approach against the frequentist and the Bayesian approaches are broadly the same as those which had been urged respectively by Jeffreys [13] and R. A. Fisher [10]. According to Jeffreys, for inference from given data, other outcomes that have not occurred cannot have any relevance. Hence for the purpose of inference the sample space is irrelevant, which thus rules out the frequentist approach. The Bayesian approach is rejected for the reasons advanced by Fisher.

Discarding both these approaches, therefore, it is argued that, in the discrete case, if an outcome x has a higher probability of occurrence under the hypothesis $H_1: \theta = \theta_1$ than under $H_2: \theta = \theta_2$, then when the outcome x occurs it seems reasonable to have greater belief in the proposition that H_1 is true than in the proposition that H_2 is true. This forms the basis for the principle in the discrete case. It is extended to the continuous case by treating the latter as the limit of discrete distributions.

WEAK LIKELIHOOD PRINCIPLE

There is a weaker version of the principle according to which two likelihood functions $f(x_1, \theta)$, $f(y_1, \theta)$ which are proportional to each other [i.e., satisfy (1)] are equivalent for inference regarding θ if, and only if, the outcomes x_1 and y_1 relate to the same experiment. But it is easily shown that there then exists a sufficient statistic t having a common value at x_1 and y_1 [7, p. 41]. Thus the weak principle implies nothing more than what is already contained in the sufficiency principle. Hence it is of lesser importance.

BIRNBAUM'S THEOREM

This theorem (cf. [4]), which asserts that the principles of conditionality and sufficiency together imply the likelihood principle (L), was considered to be of great importance, as many statisticians of the frequentist school accept the sufficiency principle and also some form of conditionality principle, and it was considered that as a result of the theorem the likelihood approach would gain general acceptance. But later, objections to the proof of the theorem were raised by Joshi [14] in the continuous case and on more general grounds by Durbin [8] and Kalbfleisch [16]. (See ANCILLARY STATISTICS regarding the last two references.) See also Birnbaum [5] for a related theorem of Birnbaum.

MARGINAL AND CONDITIONAL LIKELIHOODS

These are techniques of estimation (see Cox and Hinkley [7]) and do not involve the likelihood principle essentially.

COUNTEREXAMPLES

Some well-known counterexamples are one due to Stein [18], the stopping rule paradox [17, pp. 75–76], and one in survey sampling [12].

There is also a simple counterexample due to Stone [19] given originally against uniform priors but which, as pointed out by Fraser [11], is a strong counterexample against the likelihood approach. As shown by Joshi [15], Stone's example becomes sharper by considering a k-dimensional space instead of a 2-dimensional one. The example then becomes as follows. There is a discrete distribution such that each outcome x can arise from $2k$ values of θ, all having exactly the same likelihood. But there is a function $\delta(x)$ such that $P_\theta\{\delta(x) = \theta\} = 1 - \frac{1}{2}k$ for all θ. Thus although the $2k$ values are all equally credible according to the likelihood principle, the odds in favor of the particular one that coincides with $\delta(x)$ are $(2k - 1)$ to 1, e.g., more than 100 to 1 if $k > 50$.

References

[1] Barnard, G. A. (1962). *J. Amer. Statist. Ass.*, **57**, 308–309. (Comments on Birnbaum's paper [4].)

[2] Barnard, G. A. (1970). In *Foundations of Statistical Inference*, V. P. Godambe and D. A. Sprott, eds. Holt, Rinehart and Winston, Toronto, pp. 289–304.

[3] Barnard, G. A. (1975). *Biometrika*, **62**, 260–261. (Comments on Kalbfleisch's paper [15]. References 1–3 are representative of Barnard's views regarding the likelihood principle. His acceptance of the principle has always been subject to reservations regarding the domain of its applicability.)

[4] Birnbaum, A. (1962). *J. Amer. Statist. Ass.*, **57**, 269–305. (A foundational paper; essential reading for the study of the likelihood approach. However, Birnbaum's views had changed in later life; see ref. 6.)

[5] Birnbaum, A. (1972). *J. Amer. Statist. Ass.*, **67**, 858–861. (Contains proof of a related but less important theorem that the principle of sufficiency together with a principle of mathematical equivalence imply that of conditionality, thereby correcting a statement in ref. 4 that conditionality alone implies the principle.)

[6] Birnbaum, A. (1975). *Biometrika*, **62**, 262–264. (Comments on Kalbfleisch's paper [16].)

[7] Cox, D. R. and Hinkley, D. V. (1973). *Theoretical Statistics*. Chapman & Hall, London. (Advanced-level textbook on statistical inference.)

[8] Durbin, J. (1970). *J. Amer. Statist. Ass.*, **65**, 395–397.

[9] Edwards, A. W. F. (1972). *Likelihood*. Cambridge University Press, Cambridge. (A comprehensive exposition of the likelihood approach. Edwards accepts this approach in its entirety.)

[10] Fisher, R. A. (1930). *Proc. Camb. Philos. Soc.*, **26**, 528–535.

[11] Fraser, D. A. S. (1976). *J. Amer. Statist. Ass.*, **71**. (Comment during the discussion of Stone's paper [18].)

[12] Godambe, V. P. (1966). *J. R. Statist. Soc. B*, 310–319.

[13] Jeffreys, H. (1961). *Theory of Probability*, 3rd ed. Clarendon Press, Oxford. (A standard reference for a Bayesian development, which assumes universally valid prior distributions.)

[14] Joshi, V. M. (1976). *J. Amer. Statist. Ass.*, **71**, 345–346.

[15] Joshi, V. M. (1982). *Brit. J. Phil. Sci.*, **33**, 287–289.

[16] Kalbfleisch, J. P. (1975). *Biometrika*, **62**, 251–268.

[17] Savage, L. J. (1962). *The Foundations of Statistical Inference*, Methuen, London and Wiley, New York.

[18] Stein, C. (1962). *J. R. Statist. Soc. A*, **125**, 565–568.

[19] Stone, M. (1976). *J. Amer. Statist. Ass.*, **71**, 114–123.

Bibliography

See the following works, as well as the references just cited, for more information on the topic of the likelihood principle.

Birnbaum, A. (1970). *J. Amer. Statist. Ass.*, **65**, 402–403.

Edwards, A. W. F. (1969). *Nature (Lond.)*, **222**, 1233.

Hacking, I. (1965). *Logic of Statistical Inference*. Cambridge University Press, Cambridge. (Analyzes the notion of statistical evidence from a logician's viewpoint. Although Hacking's conclusions generally support the likelihood principle, he considers the validity of the principle an open question.)

Savage, L. J. (1976). *The Foundations of Statistical Inference*. Spottiswoode, Ballantyne, London. (Contains an exposition of Savage's Bayesian approach together with a critical discussion by others.)

Savage, L. J. (1976). *Ann. Statist.*, **4**, 441–474. (A scholarly article which, *inter alia*, discusses the extent to which Fisher had accepted the likelihood approach.)

(ANCILLARITY
BASU THEOREMS
BAYESIAN INFERENCE
DECISION THEORY
ESTIMATION
FIDUCIAL INFERENCE
HYPOTHESIS TESTING
INFERENCE, STATISTICAL
LIKELIHOOD RATIO TESTS
MAXIMUM LIKELIHOOD)

V. M. JOSHI

LIKELIHOOD RATIO TESTS

Let X denote the data available to the statistician where X has a distribution $f(X, \theta)$ for $\theta \in \Omega$ (which is absolutely continuous with respect to some vector μ). Here X and θ may be real or vector valued. It is desired to test the null hypothesis $H_0: \sigma \in \omega$ versus the alternative $H_1: \sigma \in \Omega - \omega$.

The likelihood ratio (LR) test of H_0 vs. H_1 is based on the statistic

$$\lambda(X) = \left(\sup_{\theta \in \omega} f(X, \theta) \right) \Big/ \left(\sup_{\theta \in \Omega} f(X, \theta) \right)$$

and has critical region (rejection region) $W_\alpha = \{X: \lambda(X) < C\}$, where C is chosen so that $\int_{W_\alpha} f(X, \theta) d\mu(X) \leqslant \alpha$, for all $\theta \in \omega$. Neyman and Pearson [8] proposed the method, which has a central place in hypothesis testing literature and practice.

If both H_0 and H_1 are simple (i.e., both ω and $\Omega - \omega$ consist of a single point), the likelihood ratio test is closely related to, although not necessarily identical to, the most powerful test given by the Neyman–Pearson lemma*. The two test functions $\lambda(X)$ and $f(X, \theta_0)/f(X, \theta_1)$ will order points for inclusion into the rejection region identically except possibly for those X's for which $\lambda(X) = 1$. In most practically interesting examples where α is small, the two tests will coincide since $\lambda(X) = 1$ implies that the likelihood under ω is at least as large as under $\Omega - \omega$; hence such X's will typically be in the acceptance region of both tests.

As an example of a likelihood ratio test, let $X = (X_1, \ldots, X_n)$ be a vector of independent, identically distributed normal random variables with mean μ and variance σ^2 (both unknown). It is desired to test $H_0: \mu = 0$ vs. $H_0: \mu \neq 0$. Setting $\theta = (\mu, \sigma^2)$, the ingredients of the LR test are as follows:

$$\Omega = \{(\mu, \sigma^2); -\infty < \mu < \infty, \sigma^2 > 0\}$$

$$\omega = \{(\mu, \sigma^2): \mu = 0, \sigma^2 > 0\}$$

$$f(X, \theta) = (2\pi\sigma^2)^{-n/2} \exp\left[-\frac{1}{2} \sum_{i=1}^{n} (X_i - \mu)^2 \Big/ \sigma^2 \right].$$

Denoting the maximum likelihood estimators* of θ under ω and Ω by $\hat{\theta}_\omega$ and $\hat{\theta}_\Omega$, respectively,

$$\hat{\theta}_\omega = \left(0, \frac{1}{n} \sum_{i=1}^{n} X_i^2 \right)$$

$$= \left(0, \frac{1}{n} \sum_{i=1}^{n} (X_i - \overline{X})^2 + \overline{X}^2 \right)$$

$$\hat{\theta}_\Omega = \left(\overline{X}, \frac{1}{n} \sum_{i=1}^{n} (X - \overline{X})^2 \right).$$

Hence

$$\sup_{\theta \in \omega} f(X, \theta)$$

$$= f(X, \hat{\theta}_\omega)$$

$$= \left[2\pi \left(\frac{1}{n} \sum (X_i - \overline{X})^2 + \overline{X}^2 \right) \right]^{-n/2} e^{-n/2}$$

and

$$\sup_{\theta \in \Omega} f(X, \hat{\theta})$$

$$= f(X, \hat{\theta}_\Omega) = \left[2\pi \left\{ \frac{1}{n} \sum (X_i - \overline{X})^2 \right\} \right]^{-n/2} e^{-n/2}.$$

Therefore,

$$\lambda(X) = \left[\frac{(1/n)\left\{ \sum (X_i - \overline{X})^2 + \overline{X}^2 \right\}}{(1/n) \sum (X_i - \overline{X})^2} \right]^{-n/2}$$

$$= \left[\frac{(1/n) \sum (X_i - \overline{X})^2}{(1/n)(X_i - \overline{X})^2 + \overline{X}^2} \right]^{n/2}.$$

The LR test is therefore to reject H_i if $\lambda(X) < C$ or equivalently to reject H_0 if

$$\frac{1}{n} \frac{\overline{X}^2}{\sum (X_i - \overline{X})^2} > C^{-2/n} - 1,$$

or finally if

$$|t| = \frac{\sqrt{n}\, |\overline{X}|}{\left\{ \frac{1}{n-1} \sum (X_i - \overline{X})^2 \right\}^{1/2}}$$

$$> \left[(n-1)(C^{-2/n} - 1) \right]^{1/2}.$$

This is the usual t-test and is of level α if $[(n-1)(C^{-2/n} - 1)]^{1/2}$ is chosen as the $100(1 - \alpha/2)$ percentile of the t-distribution* with $(n-1)$ degrees of freedom. This test is uniformly most powerful among unbiased tests* and also among invariant tests*.

More generally, consider the problem of testing $H_0: \beta_2 = 0$ vs. $H_1: \beta_2 \neq 0$ in a general linear model (*see* LINEAR MODELS) the form $Y = X\beta + \epsilon$, where Y and ϵ are n vectors, X is an $n \times p$ matrix of full rank, and β is a p vector with components β_1 and β_2 which are k and $p - k$ vectors, respectively. It is assumed that the components of ϵ are independent, identically distributed normal random variables with mean 0 and variance σ_Ω^2. Letting $S_\Omega = \inf_\Omega \| Y - X\beta \|^2$ and $S_\omega = \inf_\omega \| Y - X\beta \|^2$, where

$$\Omega = \{ (\beta, \sigma^2) : \beta \in R^k, \sigma^2 > 0 \}$$

and

$$\omega = \{ (\beta, \sigma^2) : \beta_2 = 0, \sigma^2 > 0 \},$$

$\lambda(X) = (S_\Omega / S_\omega)^{1/2}$ and the LR test is equivalent to the usual F test* of H_0; namely, reject if

$$F = \frac{(n-p)(S_\omega - S_\Omega)}{(p-k)S_\Omega} > F(\alpha),$$

where $F(\alpha)$ is the $100(1 - \alpha)$th percentile of an F distribution with $p - k$ and $n - k$ degrees of freedom. Here there is no uniformly most powerful unbiased test (unless $p - k = 1$), but the LR test is uniformly most powerful invariant.

It is true quite generally that when a testing problem is invariant, the LR test is invariant (see, e.g., Lehmann [6, Ex. 17, p. 252] or Eaton [3, Chap. 7]). It is not true, however, that the LR test is necessarily uniformly most powerful invariant (UMPI) even when such exists (see, e.g., Lehmann [6, Ex. 18, p. 252]). There are also certain problems (e.g., some GMANOVA settings) where there is no UMPI test, but where the LR test is dominated by another invariant test (see, e.g., Marden and Perlman [7]).

The LR test need not be similar or unbiased (see, e.g., Kendall and Stuart [5, p. 226]). Nonetheless, in a surprisingly large number of important examples, the LR test has good finite sample optimality properties.

In the large sample case there is a well-developed theory for the LR test. We first describe some of what is known in the so-called "regular" case, where derivatives and integrals may be interchanged at the whim of the printer, and the range of f does not depend on Θ. The conditions are essentially those that ensure asymptotic normality* and efficiency of maximum likelihood estimators.

Suppose that $\Omega \subset R^r$ and

$$\omega = \{ \theta \in \Omega : \theta_{k+1} = \theta_{k+1,0}, \theta_{k+1} = \theta_{k+2,0}, \\ \dots, \theta_r = \theta_{r,0} \}$$

for some k, where $\theta_{k+1,0}, \dots, \theta_{r0}$ are fixed

values. Hence the null hypothesis becomes $H_0 : \theta_{k+1} = \theta_{k+1,0}, \theta_{k+2} = \theta_{k+2,0}, \dots, \theta_r = \theta_{r,0}$. Wilks [13] showed that under H_0 the distribution of $-2 \log \lambda(X)$ is asymptotically central chi-square with $r - k$ degrees of freedom.

Wald [11, 12] showed for $\theta_1 \in \Omega$ that $-2 \log \lambda(X)$ has a noncentral chi-square distribution with $r - k$ degrees of freedom and noncentrality parameter

$$\sum_{i=k+1}^r \sum_{j=k+1}^r \alpha_{ij}(\theta_i - \theta_{i,1})(\theta_j - \theta_{j,1}),$$

where $\theta_1 = (\theta_{1,1}, \theta_{2,1}, \dots, \theta_{r,1})$ and for $i, j = 1, \dots, r$,

$$\alpha_{ij} = -E_{\theta_1} \left(\frac{\partial \log f(X,\theta)}{\partial \theta_i \, \partial \theta_j} \bigg|_{\theta = \theta_1} \right).$$

[Hence $A = (a_{ij})$ is the Fisher information matrix*.] These results allow determination of approximate critical values and power for large samples.

If, for example, X_i, $i = 1, \dots, r$, are independent normal variables with means θ_i, respectively, and variance equal to 1, the likelihood ratio for testing $H_0 : \theta_{k+1} = \theta_{k+2} = \dots = \theta_r = 0$ is given by

$$\lambda(X) = \exp\left(-\frac{1}{2} \sum_{i=k+1}^r X_i^2 \right).$$

In this example $-2 \log \lambda(X) = \sum_{i=k+1}^r X_i^2$ and the large-sample approximations above are exact.

In the general setting above, the LR test is consistent and asymptotically unbiased. It is also asymptotically efficient under a variety of definitions of efficiency, the most important being those of Pitman [9] and Bahadur [1]. (*See* BAHADUR EFFICIENCY *and* PITMAN EFFICIENCY.)

The remarks above hold essentially without change if $\omega = \{ \theta : \theta$ lies in some k-dimensional hyperspace$\}$. The nature of the results may change if either the regularity conditions fail to hold or if the null hypothesis is not of the foregoing kind.

For example, suppose that X_{ij}, $j = 1, \dots, n$, are a sample from a uniform distribution on $[0, \theta_i]$ for $i = 1, \dots, k$, and the null hypothesis is $H_0 : \theta_1 = \theta_2 = \dots = \theta_k$

= 1. The likelihood ratio is

$$\prod_{i=1}^{k} \left\{ \max_{1 \leq j \leq n} X_{ij} \right\}^{n}$$

if all X_{ij} are less than 1. Under H_0, the distribution of $-2\log\lambda$ is exactly distributed as a chi-square random variable with $2k$ degrees of freedom instead of k. See Hogg [4] for related results.

As an example of the sort of problem that happens in the regular case when the null hypothesis is not a hyperspace, consider the following due to Chernoff [2]. Let X be a single observation on a p-variate normal distribution with known covariance matrix and let the null hypothesis be that the mean vector lies on one side of a smooth $(k-1)$-dimensional surface. The distribution of $-2\log\lambda$ for θ lying on the surface is zero half the time and behaves like a chi-square with 1 degree of freedom half the time.

References

[1] Bahadur, R. R. (1960). *Sankhyā*, **22**, 229–252.

[2] Chernoff, H. (1954). *Ann. Math. Statist.*, **25**, 573–578.

[3] Eaton, M. L. (1972). *Multivariate Statistical Analysis*. Institute of Mathematical Statistics, University of Copenhagen, Denmark.

[4] Hogg, R. R. (1956). *Ann. Math. Statist.*, **27**, 529–532.

[5] Kendall, M. G. and Stewart, A. (1967). *The Advanced Theory of Statistics*, Vol. 2. Charles Griffin, London.

[6] Lehmann, E. L. (1959). *Testing Statistical Hypotheses*. Wiley, New York.

[7] Marden, J. and Perlman, M. (1980). *Ann. Statist.*, **8**, 25–63.

[8] Neyman, J. and Pearson, E. S. (1928). *Biometrika*, **20A**, Pt. I, 175–240; *ibid.*, Pt. II, 263–294.

[9] Pitman, E. J. G. (1949). *Lecture Notes on Nonparametric Statistical Inference*. Columbia University, New York.

[10] Rao, C. R. (1973). *Linear Statistical Inference and Its Applications*, 2nd ed. Wiley, New York.

[11] Wald, A. (1941). *Ann. Math. Statist.*, **12**, 1–19.

[12] Wald, A. (1941). *Ann. Math. Statist.*, **12**, 396–408.

[13] Wilks, S. S. (1962). *Mathematical Statistics*. Wiley, New York.

(ANALYSIS OF VARIANCE
BAHADUR EFFICIENCY

EXPONENTIAL FAMILIES
HYPOTHESIS TESTING
HYPOTHESIS TESTS
INFERENCE, STATISTICAL
INVARIANCE
INVARIANT TESTS
LIKELIHOOD PRINCIPLE
LINEAR MODELS
MAXIMUM LIKELIHOOD ESTIMATES
NEYMAN–PEARSON LEMMA
PITMAN EFFICIENCY
UNBIASED TESTS)

WILLIAM E. STRAWDERMAN

LILLIEFORS TEST

A modification of the Kolmogorov–Smirnov* test. Lilliefors [1] modified the corresponding Kolmogorov–Smirnov critical tables (see, e.g., Miller [2]) so that the test statistic W may be based on the sample mean $\overline{X}$ and the sample standard deviation S in place of μ and σ, respectively. To construct Lilliefors' test statistics, standardize the observations $X_1, \ldots, X_n$ by using

$$Z_i = \left(X_i - \overline{X}\right)/s, \qquad i = 1, \ldots, n.$$

Let $S(z)$ denote the sample cumulative distribution function (CDF) of the Z_i's. The test statistic is then

$$L = \sup_z |F_0(z) - S(z)|,$$

where $F_0(x)$ is the hypothesized continuous population CDF. Tables for critical values of the Lilliefors test statistic when $F_0(x)$ is normal are given in Lilliefors [1] and are reproduced in some modern statistical textbooks (e.g., Pfaffenberger and Patterson [3]).

References

[1] Lilliefors, H. W. (1967). *J. Amer. Statist. Ass.*, **62**, 399–402.

[2] Miller, L. H. (1956). *J. Amer. Statist. Ass.*, **51**, 111–121.

[3] Pfaffenberger, R. C. and Patterson, J. H. (1977). *Statistical Methods for Business and Economics*. Richard D. Irwin, Homewood, Ill.

(DEPARTURES FROM NORMALITY
KOLMOGOROV–SMIRNOV TESTS)

LIMIT THEOREM, CENTRAL

The *central limit theorem* (CLT) is generally regarded as a generic name applied to any theorem giving convergence in distribution to the normal law* for a sum of an increasing number of random variables. Results of this kind hold under far reaching circumstances and give the normal distribution its central place in the theory of probability and statistics.

Classical forms of the CLT deal with sums of independent random variables $X_1, X_2,$ .... . Suppose that $EX_k = 0$, $EX_k^2 = \sigma_k^2 < \infty$ for each k and write $S_n = \sum_{k=1}^n X_k$, $s_n^2 = \sum_{k=1}^n \sigma_k^2$. We shall use $\overset{d}{\to}$ to denote convergence in distribution and $N(\mu, \sigma^2)$ to denote the normal law with mean μ and variance σ^2, while $I(\cdot)$ is the indicator function*. The most basic central limit result is the following Lindeberg–Feller theorem*.

Theorem 1. We have $s_n^{-1} S_n \overset{d}{\to} N(0, 1)$ and $\max_{k \leqslant n} s_n^{-1} \sigma_k \to 0$ if and only if for every $\epsilon > 0$,

$$s_n^{-2} \sum_{k=1}^n E\big(X_k^2 I(|X_k| > \epsilon s_n)\big) \to 0.$$

The CLT draws much of its significance from its role as a rate of convergence result about the strong law of large numbers*. To see this role, take X_k, $k = 1, 2, \ldots$ as independent, identically distributed random variables with $E|X_1| < \infty$, $EX_1 = \mu$, and write $S_n = \sum_{k=1}^n X_k$. Then the strong law of large numbers* gives $n^{-1} S_n \to \mu$ almost surely as $n \to \infty$. If, in addition, $\operatorname{var} X_1 = \sigma^2 < \infty$, then Theorem 1 gives a concrete statement about the rate of this convergence, namely,

$$\sigma^{-1} n^{1/2}\big(n^{-1} S_n - \mu\big) \overset{d}{\to} N(0, 1)$$

as $n \to \infty$. This result is at the heart of statistical theory for it enables approximate confidence intervals* for μ to be constructed and hypotheses about μ tested (*see* HYPOTHESIS TESTING) using the sample mean $n^{-1} S_n$.

The central limit theorem was historically known as the law of errors through the work of Laplace* and Gauss* in the early nineteenth century on the theory of errors of observation. The result was first established for the case of Bernoulli trials* [X_k, $k = 1$, $2, \ldots$, independent and identically distributed (i.i.d.) with $\Pr(X_k = 1) = p$, $\Pr(X_k = 0) = 1 - p$, $0 < p < 1$]. The case $p = \frac{1}{2}$ was treated by de Moivre* in 1718 and the case of general p by Laplace in 1812. Effective methods for the rigorous proof of limit theorems for sums of arbitrarily distributed random variables were developed in the second half of the nineteenth century by Chebyshev*. His results of 1887 are based on the method of moments*. The first modern discussion was given by Liapunov* in 1900 and 1901 under considerably more general conditions and using characteristic functions*. These have subsequently continued to be used in most treatments, even for dependent variables. The sufficiency part of Theorem 1 is due to Lindeberg in 1922 and the necessity to Feller in 1935.

There are many generalizations of the Lindeberg–Feller result and many limit laws other than the normal can be obtained. These results are still loosely described as central limit results provided only that they involve convergence in distribution* to a proper law for the sum of an increasing number of random variables. We shall comment first on the case of independent variables. Note that $\{s_n^{-1} S_n, n \geqslant 1\}$ is a particular case of $\{\sum_{k=1}^n X_{nk}, n \geqslant 1\}$, the X_{nk}, $1 \leqslant k \leqslant n$, being independent for each fixed n. Note further that some restriction on the X_{nk} is necessary to obtain meaningful limit results because without any restriction we could let $\{Y_n, n \geqslant 1\}$ be an arbitrary sequence of random variables and set $X_{n1} = Y_n$ and $X_{nk} = 0$, $k > 1$ and every n. Any limit behavior could then be obtained for $\sum_{k=1}^n X_{nk} = Y_n$.

The usual restriction that is imposed is for the summands X_{nk} to be *uniformly asymptotically negligible* (UAN). That is, for every $\epsilon > 0$,

$$\max_{1 \leqslant k \leqslant n} \Pr(|X_{nk}| > \epsilon) \to 0,$$

or in other words, X_{nk} converges in probability to zero, uniformly in k. Under the UAN condition it is possible to provide detailed answers to the problems of what limit laws are possible for the $\sum_{k=1}^n X_{nk}$ and when they obtain. Comprehensive discussions of results of this kind are given in many texts, e.g., Feller [5], Gnedenko and Kolmogorov [6], Ibragimov and Linnik [12], Loève [13], and Petrov [17]. The following theorem is typical of the general results that emerge.

Theorem 2. Let $\{X_{nk}, 1 \leqslant k \leqslant n, n \geqslant 1\}$ be UAN summands while $\{a_n, n \geqslant 1\}$ is an arbitrary sequence of constants.

1. The family of possible limit laws for sequences $\{\sum_{k=1}^n X_{nk} - a_n, n \geqslant 1\}$ is the family of infinitely divisible* laws. [These laws are the ones for which the logarithm of their characteristic function is expressible in the form

$$iu\alpha + \int_{-\infty}^{\infty} \left(e^{iux} - 1 - \frac{iux}{1+x^2} \right) \frac{1+x^2}{x^2} \, d\Psi(x),$$

where α is a real constant and Ψ is a distribution function up to a multiplicative constant (see INFINITE DIVISIBILITY).]

2. In order that $\sum_{k=1}^n X_{nk} - a_n$ should converge in distribution to a proper law it is necessary and sufficient that

$$\Psi_n(x) = \sum_{k=1}^n \int_{-\infty}^x \frac{y^2}{1+y^2} \, d\Pr(X_{nk} \leqslant x + a_{nk})$$
$$\to \Psi(x)$$

at all points of continuity of Ψ, which is a distribution function up to a multiplicative constant, and where

$$a_{nk} = \int_{|x| \leqslant \tau} x \, d\Pr(X_{nk} \leqslant x)$$

for some $\tau > 0$. Furthermore, all admissible a_n are of the form $a_n = \alpha_n - \alpha + o(1)$ as $n \to \infty$, where α is an arbitrary real number and

$$\alpha_n = \sum_{k=1}^n \left\{ a_{nk} + \int_{-\infty}^{\infty} \frac{x^2}{1+x^2} \, d\Pr(X_{nk} \leqslant x + a_{nk}) \right\}.$$

Theorem 2 has various important applications, among which are the particular cases of convergence to the normal and Poisson* laws.

NORMAL CONVERGENCE. $\sum_{k=1}^n X_{nk} \xrightarrow{d} N(\mu, \sigma^2)$ and the X_{nk} are UAN if and only if for every $\epsilon > 0$ and a $\tau > 0$,

(a) $$\sum_{k=1}^n \Pr(|X_{nk}| > \epsilon) \to 0$$

(b) $$\sum_{k=1}^n \mathrm{var}(X_{nk} I(|X_{nk}| \leqslant \tau)) \to \sigma^2$$

$$\sum_{k=1}^n E(X_{nk} I(|X_{nk}| \leqslant \tau)) \to \mu$$

$$\text{as} \quad n \to \infty.$$

POISSON CONVERGENCE. If the X_{nk} are UAN, $\sum_{k=1}^n X_{nk}$ converges in distribution to the Poisson law with parameter λ if and only if for every ϵ, $0 < \epsilon < 1$, and a τ, $0 < \tau < 1$,

(a) $$\sum_{k=1}^n \Pr(|X_{nk}| > \epsilon, |X_{nk} - 1| > \epsilon) \to 0$$

$$\sum_{k=1}^n \Pr(|X_{nk} - 1| \leqslant \epsilon) \to \lambda$$

(b) $$\sum_{k=1}^n \mathrm{var}(X_{nk} I(|X_{nk}| \leqslant \tau)) \to 0$$

$$\sum_{k=1}^n E(X_{nk} I(|X_{nk}| \leqslant \tau)) \to 0$$

$$\text{as} \quad n \to \infty.$$

General results on convergence to normality in the absence of the UAN condition have been given by Zolotarev [21].

Perhaps the most important cases for applications are those where $X_{nk} = b_n^{-1} X_k - k^{-1} a_n$, $1 \leqslant k \leqslant n$, the X_k being (1) independent and (2) i.i.d., while the b_k are positive constants and the a_k are arbitrary constants. The classes of possible limit laws in these contexts are subsets of the infinitely divisible laws. In the case of condition (1) they are called the self-decomposable laws* and in the case of condition (2) the stable laws*. These laws, the circumstances under which they occur, and the sequences of norming constants required to produce them have been fully explored and details may be obtained from the texts cited above (*see also* INFINITE DIVISIBILITY). Corresponding but rather less detailed results have been obtained for the case of random vectors and random elements in locally compact Abelian groups. Theorems for random vectors in k-dimensional Euclidean space R^k can often be obtained from results involving only ordinary random variables in R^1 using the Cramér–Wold device*. Suppose that the random vectors $X_n = (X_{n1}, \ldots, X_{nk})$ and $X = (X_1, \ldots, X_k)$ satisfy

$$\sum_{j=1}^{k} t_j X_{nj} \xrightarrow{d} \sum_{j=1}^{k} t_j X_j$$

for each point $t = (t_1, \ldots, t_k)$ of R^k. Then it is easily seen from an examination of the characteristic functions that $X_n \xrightarrow{d} X$.

In addition to ordinary central limit results there are generalizations to what are termed *functional central limit theorems* or *invariance principles**. These provide a portmanteau from which many associated limit results can be read off. The simplest case is a result known as Donsker's theorem. Let X_1, $X_2, \ldots$ be a sequence of i.i.d. random variables with zero mean and unit variance defined on some basic probability space $(\Omega, \mathcal{F}, P)$ and write $S_n = \sum_{k=1}^{n} X_k$. For each integer n, and each sample point ω, the function $Z_n(t, \omega)$ is defined for $0 \leqslant t \leqslant 1$ by

$$Z_n(t, \omega) = n^{-1/2}(S_{k-1}(\omega)$$
$$+ n^{-1/2}[tn - (k-1)]X_k(\omega)$$
$$\text{if} \quad \frac{k-1}{n} \leqslant t \leqslant \frac{k}{n}.$$

For each ω, $Z_n(\cdot, \omega)$ is an element of the space C of continuous real-valued functions on $[0, 1]$ which is metrized using the uniform metric. Let P_n be the distribution of $Z_n(\cdot, \omega)$ in C. Then Donsker's theorem asserts that P_n converges weakly in C to W, the Wiener

measure* in C. This general result leads to many important conclusions. For example, suppose that h is a measurable map from C to R^1 and D_h denotes the set of discontinuities of h. If $\Pr(Z_n \in D_h) = 0$, then $h(Z_n) \xrightarrow{d} h(W)$. The classical central limit theorem for i.i.d. random variables follows by taking $h(x)(t) = x(1)$, $0 \leqslant t \leqslant 1$. One other important case is that of $h(x)(t) = \sup_{0 \leqslant s \leqslant t} x(s)$. A comprehensive discussion of this theory is provided in Billingsley [3].

Many of the results mentioned above have generalizations to dependent variables but necessary conditions are rarely available without independence. For example, Theorem 1 has the following generalization to the case of martingales*.

Theorem 3. Suppose that $\{S_n, \mathcal{F}_n, n \geqslant 1\}$ is a zero-mean, square-integrable martingale, $\{\mathcal{F}_n, n \geqslant 1\}$ being a sequence of σ-fields such that S_n is $\mathcal{F}_n$-measurable for each n. Write $S_n = \sum_{k=1}^{n} X_k$ as a sum of differences and put $V_n^2 = \sum_{k=1}^{n} E(X_k^2 | \mathcal{F}_{k-1})$, $s_n^2 = ES_n^2 = EV_n^2$. If

(a) $$s_n^{-2} V_n^2 \to \eta^2$$

in probability as $n \to \infty$ for some η^2 (> 0 almost surely), and

(b) $$s_n^{-2} \sum_{k=1}^{n} E\left(X_k^2 I(|X_k| > \epsilon s_n)\right) \to 0$$

as $n \to \infty$ for any $\epsilon > 0$, then $s_n^{-1} S_n \xrightarrow{d} N(0, 1)$.

This result, which is given in considerably more general form in Hall [8], reduces to the sufficiency part of the Lindeberg–Feller theorem* (Theorem 1) in the case of independent X_k for then condition (a) is automatically satisfied with $\eta^2 = 1$. Results on convergence to infinitely divisible* laws for the martingale context have been given by Brown and Eagleson [4]. Many other central limit results have been obtained for sequences of random variables satisfying various asymptotic independence conditions such as mixing conditions*. Useful sources are Ibragimov and Linnik [12], Philipp and Stout [19], and McLeish [15].

Convergence to normality also occurs in a very wide variety of statistical applications in which the classical CLT is indirectly applicable. Under appropriate conditions it holds, for example, for *U*-statistics* (e.g., Hoeffding [11]), rank statistics* (e.g., Hájek and Šidák [7, Chap. 5]) and various other robust methods such as *M*- or *L*-estimators*. It also holds in other related contexts such as for the posterior distribution* in a Bayesian setting (e.g., Heyde and Johnstone [10]).

There is a large literature on the rate of convergence to normality in the CLT and comprehensive discussions for the independence case are provided in Petrov [17] and Bhattacharya and Ranga Rao [1], the latter emphasizing the multivariate case. Many different convergence forms have been investigated including L_p metrics, $1 \leqslant p \leqslant \infty$, for the difference between the distribution function of the normalized sum and that of the standard normal law, and asymptotic expansions* for the distribution function of the normalized sum. The most useful of the results are those involving the uniform (L_∞) metric for the difference between distribution functions and corresponding nonuniform bounds. These are given in the next two theorems, the first being the celebrated Berry–Esseen theorem*, which dates from 1941–1942.

Theorem 4. Suppose that X_k, $k = 1, 2, \ldots$, are *i.i.d.* random variables with $EX_k = 0$, $EX_k^2 = \sigma^2$, $E|X_k|^3 < \infty$. Write $S_n = \sum_{k=1}^n X_k$ and let Φ denote the distribution function of the standard normal law. Then

$$\sup_{-\infty < x < \infty} |\Pr(S_n \leqslant x\sigma n^{1/2}) - \Phi(x)|$$
$$\leqslant CE|X_1|^3 \sigma^{-3} n^{-1/2},$$

where C is an absolute constant, $C \leqslant 0.82$.

Theorem 4 has a complete generalization to provide a nonuniform bound for the case of independent but nonidentically distributed random variables, due to Bikelis [2].

Theorem 5. Suppose that X_k, $k = 1, 2, \ldots$ are independent random variables with $EX_k = 0$, $E|X_k|^3 < \infty$, each k. Write $S_n = \sum_{k=1}^n X_k$ and $s_n^2 = \sum_{k=1}^n EX_k^2$. Then

$$|P(S_n \leqslant xs_n) - \Phi(x)|$$
$$\leqslant Cs_n^{-3}(1 + |x|)^{-3} \sum_{k=1}^n \int_0^{s_n(1+|x|)} dv$$
$$\times \int_{|u| > v} u^2 d\Pr(X_k \leqslant u),$$

C denoting a positive constant. In the case of identically distributed random variables with variance σ^2 this bound yields

$$|P(S_n \leqslant x\sigma n^{1/2}) - \Phi(x)|$$
$$\leqslant CE|X_1|^3 \sigma^{-3} n^{-1/2}(1 + |x|)^{-3}.$$

Many variants on these results appear in the literature; for recent contributions, see Maejima [14].

In the case of dependent variables, some similar but weaker results are available. For example, martingales are treated in Heyde and Brown [9], *m*-dependent sequences in Stein [20], ϕ-mixing processes in Philipp [18], and strong mixing processes in Oodaira and Yoshihara [16].

References

[1] Bhattacharya, R. N. and Ranga Rao, R. (1976). *Normal Approximations and Asymptotic Expansions*. Wiley, New York.

[2] Bikelis, A. (1966). *Litovsk. math. sb.*, **6**, 323–346 (in Russian).

[3] Billingsley, P. (1968). *Convergence of Probability Measures*. Wiley, New York.

[4] Brown, B. M. and Eagleson, G. K. (1971). *Trans. Amer. Math. Soc.*, **162**, 449–453.

[5] Feller, W. (1971). *An Introduction to Probability Theory and Its Applications*, Vol. 2. Wiley, New York.

[6] Gnedenko, B. V. and Kolmogorov, A. N. (1954). *Limit Distributions for Sums of Independent Random Variables*. Addison-Wesley, Reading, Mass.

[7] Hájek, J. and Šidák, Z. (1967). *Theory of Rank Tests*. Academic Press, New York.

[8] Hall, P. G. (1977). *Ann. Prob.*, **5**, 875–887.

[9] Heyde, C. C. and Brown, B. M. (1970). *Ann. Math. Statist.*, **41**, 2161–2165.

[10] Heyde, C. C. and Johnstone, I. M. (1979). *J. R. Statist. Soc. B*, **41**, 184–189.

[11] Hoeffding, W. (1948). *Ann. Math. Statist.*, **19**, 293–325.

[12] Ibragimov, I. A. and Linnik, Yu. V. (1971). *Independent and Stationary Sequences of Random Variables*. Wolters-Noordhoff, Groningen.

[13] Loève, M. (1977). *Probability Theory I*, 4th ed. Springer-Verlag, New York.

[14] Maejima, M. (1978). *Ann. Prob.*, **6**, 341–344.

[15] McLeish, D. L. (1977). *Ann. Prob.*, **5**, 616–621.

[16] Oodaira, H. and Yoshihara, K. (1971). *Kōdai Math. Sem. Rep.*, **23**, 311–334.

[17] Petrov, V. V. (1974). *Sums of Independent Random Variables*. Springer-Verlag, Berlin.

[18] Philipp, W. (1969). *Ann. Math. Statist.*, **40**, 601–609.

[19] Philipp, W. and Stout, W. F. (1975). Almost Sure Invariance Principles for Partial Sums of Weakly Dependent Random Variables. *Mem. Amer. Math. Soc. No. 161.*

[20] Stein, C. (1972). *Proc. 6th Berkeley Symp. Math. Statist. Prob.*, Vol. 2. University of California Press, Berkeley, Calif., pp. 583–602.

[21] Zolotarev, V. M. (1967). *Theory Prob. Appl.*, **12**, 608–618.

(ASYMPTOTIC NORMALITY
CHEBYSHEV
CONVERGENCE OF DISTRIBUTIONS
INFINITE DIVISIBILITY
LAW OF ITERATED LOGARITHM
LAWS OF ERROR, I, II, and III
LIAPUNOV, A. M.)

C. C. Heyde

LIMIT THEOREMS

Limit theorems in probability theory and statistics are generally regarded as theorems giving convergence of sequences of probability distributions or random variables (including stochastic processes). The kinds of convergence usually considered are convergence almost surely or with probability 1, in distribution, of densities, in L^p-norm, of moments, and in probability. Typically, theorems are the classical forms of the weak and strong law of large numbers*, the law of the iterated logarithm*, and the central limit theorem*. They all deal with the behavior for large n of a sum $S_n = X_1 + \cdots + X_n$ of independent, identically distributed ran-

dom variables $X_1, X_2, \ldots$. The asymptotic behavior of extreme values as $\max(X_1, \ldots, X_n)$ when $n \to \infty$ is also of interest. Important are theorems of the Cramér–Slutsky type, e.g., if $X_n \to 0$ in probability and $Y_n \to Y$ in distribution, then $X_n + Y_n \to Y$ in distribution. This kind of result often allows limit theorems to be used indirectly in a wide variety of statistical applications.

The classical results have been considerably generalized during the last 50 years. For example, a complete characterization of the possible limit distributions for sums of independent random variables under general "smallness" conditions was obtained in the 1930s (see the central limit theorem and Loève [10] and Petrov [11]). Various situations with dependent random variables have also been studied (see, e.g., the examples below).

The literature on limit theorems is very extensive. Such results are contained in almost every book in probability theory and mathematical statistics. Thorough mathematical treatments on the central limit theorem in particular, but also on other types of limit theorems, can be found in Loève [10] and Petrov [11], for example. Illuminating are the discussions in Feller [5, 6]. The detailed and basic book by Billingsley [1] on weak convergence is a standard reference for many results in connection with limit theorems for stochastic processes, also important for many problems in statistics. Rao [12] and Cox and Hinkley [2] contain many results for large samples in statistics. The classical treatise on mathematical statistics by Cramér [3] is still an important reference for limit theorems in statistics.

The classical occupancy problem gives an example of a sum of dependent random variables. The problem is to throw n balls at random into r boxes and find the distribution of Z_{rn} = the number of empty boxes. The problem has surprisingly many applications (for a detailed discussion, see Johnson and Kotz [8]). With $X_{jn} = 1$ if box j is empty, $X_{jn} = 0$ otherwise, one can write $Z_{rn} = X_{1n} + \cdots + X_{rn}$. The exact distribution (obtained by De Moivre* in the eighteenth century) is

$$\Pr[Z_{rn} = z] = r! \, S(n, r - z)/(z! \, r^n),$$

where $S(n, r - z)$ is a Stirling number of the second kind (see, e.g., Feller [5] or Johnson and Kotz [8]). Furthermore,

$$E[Z_{rn}] = r(1 - 1/r)^n,$$

$$\text{var}(Z_{rn}) = r(1 - 1/r)^n (1 - (1 - 1/r)^n)$$

$$+ 2\binom{r}{2}((1 - 2/r)^n$$

$$- (1 - 1/r)^{2n}).$$

For large n and r the exact distribution is not very useful for numerical purposes; therefore, approximations are of interest. As $X_{1n}, \ldots, X_{rn}$ are dependent random variables the conventional central limit theory is not directly applicable. Different cases occur depending on how $n, r \to \infty$. If $n, r \to \infty$ such that $n/r \to \alpha$, then

$$E[Z_{rn}] = r e^{-n/r} + O(1),$$

$$\text{var}(Z_{rn}) \sim r\left(e^{-\alpha}(1 - e^{-\alpha}) - \alpha e^{-2\alpha}\right).$$

Using the method of moments it was proved around 1950 that

$$(Z_{rn} - E[Z_{rn}])/(\text{var}(Z_{rn}))^{1/2} \xrightarrow{d} N(0, 1).$$

Note that the X's are not asymptotically independent because in that case one would have $\text{var}(Z_{rn}) \sim r e^{-\alpha}(1 - e^{-\alpha})$. Now suppose that $n, r \to \infty$ so

$$E[Z_{rn}] = r(1 - 1/r)^n \to \lambda > 0,$$

implying that $\text{var}(Z_{rn}) \to \lambda$. In this case Z_{rn} is a sum of many 0's and few 1's, a typical situation of convergence to the Poisson distribution. An elementary proof of this is given, e.g., in Feller [5]. Rényi [13] analyzed all different cases of $n, r \to \infty$ using characteristic functions. Many related problems have been studied by Rényi's technique, see Kolchin et al. [9].

The classical occupancy problem is a special case of the more general problem of finding the distribution of

$$Z_{rn} = \sum_{j=1}^{r} h_{jr}(Y_{jn}),$$

where $(Y_{1n}, \ldots, Y_{rn})$ has a symmetric multinomial distribution and $\{h_{jr}\}$ are given functions. The classical occupancy problem corresponds to letting $h_{jr}(x) = I \ (x = 0)$. In Holst [7] generalizations of such problems are studied.

Another example of a problem involving dependent random variables of a nontrivial kind occurs in connection with simple random sampling without replacement from a finite population. Suppose that to each element of a population of size r there is associated a real number, say a_{jr} for element j, $j = 1, 2, \ldots, r$. Draw without replacement n elements and add the corresponding a's. A basic problem is to find the probability distribution of the sum. This random variable can be written

$$Z_{rn} = \sum_{j=1}^{r} a_{jr} Y_{jn},$$

where $(Y_{1n}, \ldots, Y_{rn})$ is a random vector with 1's in n different randomly chosen places and 0's in the other $r - n$ places. Clearly, the Y's are dependent. No useful form of the exact distribution of Z_{rn} exists unless r is very small or the a's are very special, so approximations are needed. To formulate a limit theorem consider sequences of population vectors $\{(a_{1r}, \ldots, a_{rr})\}$ and random vectors $\{(Y_{1n}, \ldots, Y_{rn})\}$ and let $n, r \to \infty$. Under general conditions it was first proved by Erdös and Rényi [4] that

$$(Z_{rn} - E[Z_{rn}])/(\text{var}(Z_{rn}))^{1/2} \xrightarrow{d} N(0, 1)$$

(see also Holst [7]). The result is not surprising, but it is not a simple consequence of the classical central limit theorems for sums of independent random variables.

References

[1] Billingsley, P. (1968). *Convergence of Probability Measures*. Wiley, New York. (Advanced; mathematical treatment of weak convergence.)

[2] Cox, D. R. and Hinkley, D. V. (1974). *Theoretical Statistics*. Chapman & Hall, London. (Intermediate level; contains many different areas of statistics; many references.)

[3] Cramér, H. (1946). *Mathematical Methods of Statistics*. Princeton University Press, Princeton, N.J. (Intermediate level; classical treatise of mathematical statistics.)

[4] Erdös, P. and Rényi, A. (1959). *Publ. Math. Inst. Hung. Acad. Sci.*, **4**, 49–61.

[5] Feller, W. (1968). *An Introduction to Probability Theory and Its Applications*, Vol. 1, 3rd ed. Wiley, New York. (Elementary; classical treatise on probability theory.)

[6] Feller, W. (1971). *An Introduction to Probability Theory and Its Applications*, Vol. 2, 2nd ed. Wiley, New York. (Intermediate level; contains many different areas of probability theory.)

[7] Holst, L. (1979). *Ann. Statist.*, **7**, 551–557.

[8] Johnson, N. L. and Kotz, S. (1977). *Urn Models and Their Application: An Approach to Modern Discrete Probability Theory*. Wiley, New York. (Elementary; special treatment of urn models; many references.)

[9] Kolchin, V. F., Sevastyanov, B. A., and Chistyakov, V. P. (1978). *Random Allocations*. V. H. Winston, Washington, D.C. (Advanced; special treatment of limit theorems for urn models; references.)

[10] Loève, M. (1977). *Probability Theory I*, 4th ed. Springer-Verlag, New York. (Advanced; mathematical treatise on probability theory.)

[11] Petrov, V. (1974). *Sums of Independent Random Variables*. Springer-Verlag, New York. (Advanced; thorough treatment of limit theorems.)

[12] Rao, C. R. (1973). *Linear Statistical Inference and Its Applications*, 2nd ed. Wiley, New York. (Intermediate level; thorough treatment on mathematical statistics.)

[13] Rényi, A. (1962). *Publ. Math. Inst. Hung. Acad. Sci.*, **7**, 203–214.

(CONVERGENCE OF DISTRIBUTIONS
LAWS OF ERROR, I, II, and III
LIMIT THEOREM, CENTRAL)

LARS K. HOLST